Legal Aspects of Architecture, Engineering and the Construction Process

FIFTH EDITION

Justin Sweet
John H. Boalt Professor of Law
University of California (Berkeley)

West Publishing Company
St. Paul   New York   Los Angeles   San Francisco

WEST'S COMMITMENT TO THE ENVIRONMENT

In 1906, West Publishing Company began recycling materials left over from the production of books. This began a tradition of efficient and responsible use of resources. Today, up to 95 percent of our legal books and 70 percent of our college and school texts are printed on recycled, acid-free stock. West also recycles nearly 22 million pounds of scrap paper annually—the equivalent of 181,717 trees. Since the 1960s, West has devised ways to capture and recycle waste inks, solvents, oils, and vapors created in the printing process. We also recycle plastics of all kinds, wood, glass, corrugated cardboard, and batteries, and have eliminated the use of Styrofoam book packaging. We at West are proud of the longevity and the scope of our commitment to the environment.

 TEXT IS PRINTED ON 10% POST CONSUMER RECYCLED PAPER

Copyeditor: Deborah Cady
Artwork: Gloria Langer
Interior and Cover Design: Lois Stanfield, LightSource Images
Composition: Parkwood Composition
Production, Prepress, Printing and Binding by West Publishing Company

Library of Congress Cataloging-in-Publication Data

Sweet, Justin.
    Legal aspects of architecture, engineering and the construction
process / Justin Sweet.
        p.   cm.
    Includes index.
    ISBN 0-314-02706-8
    1. Construction contracts—United States.   2. Architectural
contracts—United States.   3. Engineering contracts—United States.
I. Title.
KF902.S93   1994
343.73′078624—dc20
[347.30378624]
                                                         93-34741
                                       CIP

To my wife Sheba

Table of Contents

CHAPTER FOUR

The Agency Relationship: A Legal Concept Essential to Contract Making 30

CHAPTER FIVE

Contracts and Their Formation: Connectors for Construction Participants 38

CHAPTER EIGHT

Introduction to the Construction Process: Ingredients for Disputes   104

CHAPTER FIFTEEN

Risk Management: Variety of Techniques 316

CHAPTER SIXTEEN

Intellectual Property: Ideas, Copyrights, Patents, and Trade Secrets   329

CHAPTER SEVENTEEN

Planning the Project: Compensation and Organization Variations   342

CHAPTER EIGHTEEN

Competitive Bidding: Theory, Realities, and Legal Pitfalls   379

CHAPTER NINETEEN

Sources of Construction Contract Rights and Duties: Contract Documents and Legal Rules   407

CHAPTER TWENTY-EIGHT

The Subcontracting Process: An "Achilles' Heel" 620

CHAPTER TWENTY-NINE

Design Professional as Judge: Tradition Under Attack 651

Preface to Fifth Edition

This edition continues the primary focus of the previous editions—to provide a bridge for architecture and engineering students between the academic world and the real world. My hope is to cushion the inevitable shock for students as they move from the world of teachers and books to the world of clients, contractors, developers, building inspectors, and loan officers. Cushioning requires that the reader understand how and why law and legal institutions operate. Such an understanding requires clear, concise, jargon-free text. But I will continue my policy of probing beneath the surface to achieve in-depth understanding. Finally, I will not talk down to my nonlawyer audience. Years of contact with architecture and engineering students has convinced me they can handle legal doctrines and policies.

Some legal rules cannot be understood without reading actual cases. Also, a feel for law and legal institutions requires reading actual cases. As a result, I have inserted carefully selected cases, edited as needed. Most are recent cases. About one-quarter are new for this edition. Also, I have continued my policy of supplementing the cases by summarizing instructional cases.

When the first edition of this treatise was published in 1970, there were no courses in Construction Law in American law schools. Currently, over a dozen law schools offer courses dealing with the legal problems incident to the construction process. To address this new audience, I have expanded coverage, presented authorities for textual statements that relate to Construction Law, noted supplementary sources for those who wish to pursue a topic in greater detail, and offered my views on selected legal issues.

Another audience has developed since this treatise was first published. Attorneys dealing with Construction Law, arbitrators seeking legal sources to make a proper award, and judges faced with resolving construction legal problems have found the treatise useful. The treatise has been cited in twenty court decisions—an additional reason to expand legal analysis and insert references to supplemental legal sources.

Multiple audiences inevitably generate a longer treatise. However, I have ruthlessly pruned where I could, have not added new cases without deleting old ones, and have reorganized to save space. Also, I have not forgotten that my primary audience consists of architecture and engineering students.

FORMAT AND OBJECTIVES

Structurally, this edition does not differ from earlier ones. Text remains my main communication tool. I shall continue to emphasize exploring reasons behind the law and contract provisions as well as neutrally present arguments for and against particular outcomes. I have continued my policy of minimizing textual footnotes. I have also added references to the vast literature in the engineering and management fields. These references are useful for nonlawyer readers and those who are legally trained. Problems (some simple and some complex) appear at the end of most chapters and in the middle of longer chapters. They are included to deepen understanding and aid class discussion. A new instructor's manual is available.

CHANGES IN THE FIFTH EDITION

There are a number of reasons for this new edition. At the federal level, Congress has enacted the Americans with Disabilities Act[1] and the Architec-

---

[1] Section 14.05(C).

tural Works Copyright Protection Act of 1990.[2] Also, federal agencies have sought to bring design professionals who perform site services under the Occupational Health and Safety Act (OSHA).[3]

State legislatures have been more active in regulating the payment process, principally in public but also in private construction contracts.[4] Recent legislation has been enacted to protect design professionals who act in emergencies (Good Samaritan statutes) from liability.[5] More important, some states have granted immunity to design professionals who perform site services from claims resulting from workplace accidents.[6]

I have added material dealing with

1. purchase orders[7]

2. letters of intent[8]

3. memoranda of understandings[9]

4. partnering[10]

5. teaming contracts[11]

6. build operate and transfer (BOT) contracts[12]

7. liquidating agreements[13]

I have amplified material on

1. environmental laws[14]

2. building codes[15]

3. contractual liability limitations in design service contracts[16]

4. joint checks[17]

5. the recoverability of extended home-office overhead[18]

6. the use of total cost and jury verdicts in contractor claims[19]

7. design-build[20]

8. construction management[21]

9. turnkey contracts[22]

Because of the increased interest in alternative dispute resolution, I have introduced material dealing with dispute review boards,[23] the claims master,[24] and expanded material dealing with other more recognized techniques, such as arbitration[25] and mediation.[26]

Since the fourth edition in 1992, the Engineers Joint Contract Documents Committee (EJCDC) has issued new standard contracts for design services (No. 1910-8). Earlier in 1990, it issued new standard contracts for construction services (Nos. 1910-8-a-1, 1910-8-A-2, and 1910-8). See Clark, *Commentary on Agreements for Engineering Services and Construction Related Documents*, EJCDC No. 1910-9 (1993).

All references to documents published by the American Institute of Architects (AIA), unless indicated otherwise, are to A101, A201, A401, A511, and B141, published in 1987. All except A511 are in the appendix.

LEGAL CITATION FORM

For the benefit of nonlawyer readers, let me outline the mechanics of citing case decisions. While legal citations seem complicated, they are, in reality, quite simple. There are four elements to a citation. A simple citation would be *Sniadach v. Family Finance Corp.*, 395 U.S. 337 (1969). First, the name of the case is given, usually with the plaintiff (the person starting the lawsuit) listed first, followed by the defendant. "U.S." is an abbreviation of the reporter system from which the case is taken, in this case the United States Supreme Court Reports. The number preceding the abbreviation of the reporter system (395) is the volume in which the case is lo-

---

[2]Section 16.04(D).
[3]Section 14.08(I).
[4]Sections 22.02(I) and 22.03.
[5]Section 14.09(E).
[6]Ibid.
[7]Section 11.04(C).
[8]Section 5.06(I).
[9]Section 5.06(H).
[10]Section 17.04(G).
[11]Section 17.04(H).
[12]Section 17.04(I).
[13]Section 28.08(B).
[14]Section 9.13.
[15]Section 19.02(C).
[16]Section 15.03(D).
[17]Section 22.02(J).
[18]Section 27.02(F).

---

[19]Ibid.
[20]Section 17.05(F).
[21]Section 17.05(D).
[22]Section 17.05(E).
[23]Section 30.18(E).
[24]Section 30.18(F).
[25]Sections 30.18(A) and (B).
[26]Section 30.18(C).

cated. The number following the abbreviation of the reporter system (337) states the page on which the case begins. The citation concludes with the year that the court announced the decision.

Many significant reported appellate opinions come from the federal circuit courts of appeal. *Moorehead Construction Co. v. City of Grand Forks*, 508 F.2d 1008 (8th Cir.1975) is illustrative. Following the name of the parties, an abbreviation of the series of reports in which the opinion is contained (F.2d) (Federal Reporter, Second Series) is given. Again, it is preceded by the volume (508) of the report. The reporter abbreviation is followed by the page on which the case begins (1008). The particular federal circuit court (8th) deciding the case follows. The citation is completed by the year (1975) of the decision.

A third type of federal court is the district court, the trial court of general jurisdiction. While state trial court opinions (except those from New York) are not collected and published, some trial court opinions of the Federal District Courts are published. An illustrative citation would be *Gevyn Construction Corp. v. United States*, 357 F.Supp. 18 (S.D.N.Y.1972). The opinion is collected in the Federal Supplement (F.Supp.). The name of the reporter series is preceded by the volume number of the report and followed by the page on which the opinion begins. This citation shows that the district court was the Southern District of New York (S.D.N.Y.). Larger states have different courts located in different cities. Again, the citation concludes with the date (1972) the opinion was issued.

To understand state court citations, it is essential to understand the role of the West Regional Reporter System. At one time, most state citations had two citations—one to the official reporter system selected by the court and the other to the unofficial reporter system, usually the West Regional Reporter System. West divided the states regionally and published sets of reporters based on these regional classifications. If a state still has an official report system, a typical citation would be *Anco Construction Co. v. City of Wichita*, 233 Kan. 132, 660 P.2d 560 (1983). This citation indicates that the case came from the Kansas Supreme Court (Kan.) and is also collected in the Pacific Regional Reporter (P.2d).

States are increasingly abandoning the official reporter system and cite only to the West Regional

Reporter System. An illustration of such a citation would be *Larry Smith v. George M. Gilmer*, 488 So.2d 1143 (La.App.1986). This citation, from the Southern Reports, indicates that this was a decision of the Louisiana Court of Appeals (its intermediate appellate court) and that there is no official reporter system in Louisiana.

Finally, two of the more populous states—California and New York—have an unofficial reporter published by West that collects all the reported opinions from those states. In California, it is the California Reporter; in New York, it is the New York Supplement. Decisions by the highest appellate court—in California, the Supreme Court, and in New York, the Court of Appeals—are found in the California Reporter and the New York Supplement as well in the official and regional reports. For example, the California Supreme Court case of *Pollard v. Saxe and Yolles Dev. Co.* is cited, 12 Cal.3d 374, 525 P.2d 88, 115 Cal.Rptr. 648 (1974). "Cal" is the official reporter. Cal.2d being the second series of that reporter. "P.2d" is the second series of the Pacific Regional Reports. "Cal.Rptr." collects all California appellate cases, Supreme Court decisions as well as those of the intermediate California Courts of Appeal. Similarly, a decision by the New York Court of Appeals would be found in its official reporter (N.Y. or N.Y.2d) and the West Regional Reports, the North Eastern Reporter (N.E. or N.E.2d) and the New York Supplement (N.Y.S. or N.Y.S.2d). The New York Supplement also contains decisions of the New York intermediate courts of appeal (the Appellate Division of the Supreme Court) and the Supreme Court (the New York trial court). Intermediate appellate court decisions from these states would have only two citations—one to the official reports and the other to the California Reporter and the New York Supplement. Intermediate appellate court decisions from these states are not included in the regional reports. (The Table of Abbreviation lists citation references.)

The number of reported appellate decisions has increased greatly. Most reporter systems have gone into second, third, or even fourth series of reports. This is indicated by the citation in the *Pollard* case which shows that the official reporter is in the third series, the unofficial regional reporter is in the second series, and the California Reporter was at that time still in its first series. (It now is in its second series.)

ACKNOWLEDGEMENTS

Persons too numerous to mention have helped me prepare this fifth edition. I would like to make special reference to Milton F. Lunch, Esq., who for many years was counsel to the National Society of Professional Engineers and is currently a consultant to the Victor Schinerer Company. Milt, who prepares Liability Update published by Schinerer, has been extraordinarily helpful to me. He has provided me with information as to new legislation, impending and new regulations, and important case decisions.

I would also like to mention Marc Schneier, Esq., the editor of Construction Litigation Reporter. Marc taught a course in Construction Law at the University of San Francisco Law School. He prepared a detailed commentary on the fourth edition along with recent cases that he felt should be included in a new edition. To him I owe a great debt.

The diagrams were prepared by Andrew Jacobson. Trained as an architect, Andrew came to law school and took my course in Construction Law.

I must also acknowledge the painstaking work by two student assistants who not only checked references but straightened out textual angularities. They are Ann B. Kim, a recent graduate of the University of California (Berkeley), and Matthew Truesdell, who completed his first year at Berkeley. Much of the credit further expanded Section 9.13 dealing with environmental law belongs to Ms. Kim.

As to the work of my secretary, Ms. Elizabeth Duke, who has participated with me in three prior editions, I can only say that she patiently and even cheerfully typed the innumerable drafts by an author who is a stickler on style. Without her invaluable work, this book could never have been completed.

I would also like to thank members of West Publishing Company's editorial staff: Stephanie Syata, Production Editor; Patricia Bryant, Developmental Editor; and Elizabeth Hannan, Editor.

Finally, many thanks are due to the users of the fourth edition and to the following panel of reviewers who provided invaluable suggestions for revising the text:

Chris Allaire
St. Martin's College

Thomas C. Ferrara
California State University, Chico

Edward Gibson
University of Texas at Austin

Georgia L. Holmes
Mankato State University

Lloyd S. Jones
Purdue University

Steve R. Sanders
Clemson University

Demetres A. Vlatas
Texas A & M University

Dr. Janet K. Yates
Polytechnic University

Table of Abbreviations

The Federal Supplement (F.Supp.) contains opinions of Federal (U.S.) District Courts. These courts are located throughout the United States. In some states they are located in one place. As an example, a reference in the F.Supp. to (D.Neb.) indicates an opinion of *the* Federal District Court for Nebraska. Others are located in different parts of the state. As an example, a reference to (E.D.Pa.) indicates the Eastern District of Pennsylvania; (M.D.) Middle District, (C.D.) Central District), (N.D.) Northern District, (S.D.) Southern District, and (W.D.) Western District.

A. Atlantic Reporter[3]

A.2d Atlantic Reporter, Second Series[3]

AAA American Arbitration Association

A.C. Appeal Cases (English)

A.D.2d Appellate Division, Second Series (New York)

ADR Alternative Dispute Resolution

A/E Legal Newsletter Architect/Engineer Legal Newsletter

AGC Associated General Contractors

AIA American Institute of Architects

AIA Doc. American Institute of Architects Document

A.L.R.2d American Law Reports, Second Series[7]

A.L.R.3d American Law Reports, Third Series[7]

A.L.R.4th American Law Reports, Fourth Series[7]

ASBCA Armed Services Board of Contract Appeals[10]

Ala. Alabama Reports[1]

Ala.App. Alabama Appellete Court Reports

Alberta L.Rev. Alberta Law Review[6]

Am.Jur.2d American Jurisprudence, Second Series[9]

Annot. Annotation

Ariz. Arizona Reports[1]

Ariz.App. Arizona Appeals Reports

Ariz.L.Rev. Arizona Law Review[6]

Ariz.Rev.Stat.Ann. Arizona Revised Statutes Annotated[8]

Ariz.Sess.Laws Arizona Session Laws[8]

Ark. Arkansas Reports[1]

Ark.App. Arkansas Appellate Reports

BCA Board of Contract Appeals (Federal)

B.C.Int'l & Comp.L.Rev. Boston College International and Comparative Law Review[6]

B.C.L.Rev. Boston College Law Review[6]

BNA Bureau of National Affairs (Publisher)

BOT Build Operate and Transfer

B.R. Bankruptcy Reporter

B.U.L.Rev. Boston University Law Review[6]

B.Y.U.L.Rev. Brigham Young University Law Review[6]

Baylor L.Rev. Baylor Law Review[6]

Brooklyn L.Rev. Brooklyn Law Review[6]

CCD Construction Change Directive

CCH Commerce Clearing House

CDOS California Daily Opinion Service

CFR Code of Federal Regulations

---

[1]Highest Court of State
[2]Highest Court of U.S.
[3]Regional Reporter
[4]All New York Cases
[5]All California Cases
[6]Periodical
[7]Annotation to important cases
[8]Statute
[9]Encyclopedia
[10]Federal Agency Appeals Board

CGL Commercial General Liability or Comprehensive General Liability

CI Rules American Arbitration Association, Construction Industry Arbitration Rules

CM Construction Manager

CMa Construction Manager: Adviser

CMAA Construction Management Association of America

CMc Construction Manager: Constructor

CO Change Order

CPM Critical Path Method

Cal. California Reports

Cal.2d California Reports, Second Series[1]

Cal.3d California Reports, Third Series[1]

Cal.4th California Reports, Fourth Series[1]

Cal.App.2d California Appellate Reports, Second Series

Cal.App.3d California Appellate Reports, Third Series

Cal.App.4th California Appellate Reports, Fourth Series

Cal.Exec.Order California Executive Order

Cal.Rptr. California Reporter[5]

Cal.Rptr.2d California Reporter, Second Series[5]

Cal.Stat. Statutes of California[8]

Cal.W.L.Rev. California Western Law Review[6]

Calif.L.Rev. California Law Review[6]

Campbell L.Rev. Campbell Law Review

Cl.Ct. Claims Court (Federal)

Clev.St.L.Rev. Cleveland State Law Review[6]

Colo. Colorado Reports[1]

Colo.App. Colorado Court of Appeals Reports

Colo.Rev.Stat. Colorado Revised Statutes[8]

Columb.L.Rev. Columbia Law Review[6]

Comp.Gen.Dec. Comptroller General Decisions

Conn. Connecticut Reports[1]

Conn.App. Connecticut Appellate Reports

Conn.Gen.Stat.Ann. Connecticut General Statutes Annotated[8]

Conn.Supp. Connecticut Supplement

Constr.Contractor Construction Contractor[6]

Constr.L.J. Construction Law Journal[6]

Constr.Lawyer Construction Lawyer[6]

Constr.Litig.Rep. Construction Litigation Reporter[6]

Constr.Mgmt. & Econ. Construction Management & Economics[6]

CornellL.Rev. Cornell Law Review[6]

Ct.Cl. Court of Claims (Federal)

D/B Design/Build

DOTCAB Department of Transportation, Civil Aeronautics Appeal Board[10]

DSC Differing Site Condition

Det.C.L.Rev. Detroit College Law Review[6]

Duke L.J. Duke Law Journal

Duq.L.Rev. Duquesne Law Review[6]

EEO Equal Employment Opportunity

EJCDC Engineers Joint Contract Documents Committee

EPA Environmental Protection Agency (Federal)

Emory L.J. Emory Law Journal[6]

Eng.B.C.A. Corps of Engineers, Board of Contract Appeals[10]

Eng.Rep. English Reports

Exec.Order Executive Order (Federal)

FAACAP Federal Aviation Agency, Contract Appeals Board[10]

FAR Federal Acquisition Regulations

FHA Federal Housing Authority

FIDIC Federal Internationale Des Ingenieurs-Conseils

F.2d Federal Reporter, Second Series

Fed.Cir. United States Circuit Court, Federal Circuit

Fed.Reg. Federal Register

Fed.Rules Federal Rules

Fla. Florida Supreme Court[1]

Fla.Dist.Ct.App. Florida District Court of Appeals

Forum Forum[6]

GAAP Generally Accepted Accounting Principles

GAO General Accounting Office (Federal)

GMC Guaranteed Maximum Cost

GMP Guaranteed Maximum Price

GSA General Services Administration (Federal)

---

[1]Highest Court of State
[2]Highest Court of U.S.
[3]Regional Reporter
[4]All New York Cases
[5]All California Cases
[6]Periodical
[7]Annotation to important cases
[8]Statute
[9]Encyclopedia
[10]Federal Agency Appeals Board

GSBCA   General Services Board of Contract of Contract Appeals[10]

Ga.   Georgia Reports[1]

Ga.App.   Georgia Appeals Reports

Ga.L.Rev.   Georgia Law Review[6]

Gonz.L.Rev.   Gonzaga Law Review[6]

H.R.Doc.   House of Representative Document (Federal)

HUD   Department of Housing and Urban Development (Federal)

Harv.Bus.Rev.   Harvard Business Review[6]

Harv.L.Rev.   Harvard Law Review[6]

Hastings L.J.   Hastings Law Journal[6]

Haw.Adm.Rules   Hawaii Administrative Rules

Hawaii   Hawaii Reports[1]

Hous.L.Rev.   Houston Law Review[6]

IBCA   Interior Board of Contract Appeals[10]

Idaho   Idaho Reports[1]

Ill.   Illinois Reports[1]

Ill.2d   Illinois Reports, Second Series[1]

Ill.Ann.Stat.   Illinois Annotated Statutes[8]

Ill.App.2d   Illinois Appellate Court Reports, Second Series

Ill.App.3d   Illinois Appellate Court Reports, Third Series

Ill.Rev.Stat.   Illinois Revised Statutes[8]

Ind.   Indiana Reports[1]

Ind.App.   Indiana Appellate Court Reports

Ind.L.Rev.   Indiana Law Review[6]

Ins.Couns.J.   Insurance Counsel Journal[6]

Int'l Constr.L.Rev.   International Construction Law Review[6]

Iowa   Iowa Reports[1]

Iowa Code Ann.   Iowa Code Annotated[8]

Iowa L.Rev.   Iowa Law Review[6]

J. of Constr.Eng'g & Mgmt   Journal of Construction Engineering and Management[6]

J.Mar.L.Rev.   John Marshall Law Review[6]

J.Pat.Off.Soc.   Journal of Patent Officials Society[6]

K.B.   King's Beach (England)

KC News   Kellogg Corp News

Kan.   Kansas Reports[1]

Kan.App.2d   Kansas Appellate Reports, Second Series

Kan.Stat.Ann.   Kansas Statutes Annotated[8]

Ky.   Kentucky Reports[1]

Ky.L.J.   Kentucky Law Journal[6]

Ky.Rev.Stat.   Kentucky Revised Statutes[8]

LDA   Local Development Authority

L.R.-H.L.   Law Reports, House of Lords (England)

L.W.   United States Law Week

La.   Louisiana Reports[1]

La.App.   Louisiana Court of Appeals

Law & Contemp.Probs.   Law and Contemporary Problems[6]

Loy.L.A.L.Rev.   Loyola (Los Angeles) Law Review[6]

Loy.L.Rev.   Loyola Law Review[6]

MBE   Minority Business Enterprise

MLDC   Model Land Development Code

Marq.L.Rev.   Marquette Law Review[6]

Mass.   Massachusetts Reports[1]

Mass.App.   Massachusetts Appeals Court Reports

Mass.Gen.Laws Ann.   Massachusetts General Laws Annotated[8]

McKinney's N.Y.Gen.Obligation Law   McKinney's General Obligations Law (New York)[8]

Md.   Maryland Reports[1]

Md.App   Maryland Appellate Reports

Md.L.Rev.   Maryland Law Review[6]

Me.   Maine Reports[1]

Mem.St.U.L.Rev.   Memphis State University Law Review[6]

Mich.   Michigan Reports[1]

Mich.App.   Michigan Appeals Reports

Mich.Comp.Laws Ann.   Michigan Compiled Laws Annotated[8]

Minn.   Minnesota Reports[1]

Minn.L.Rev.   Minnesota Law Review[6]

Minn.Stat.Ann.   Minnesota Statutes, Annotated[8]

Misc.   Miscellaneous New York Reports

Misc.2d   Miscellaneous New York Reports, Second Series

Miss.   Mississippi Reports[1]

---

[1]Highest Court of State
[2]Highest Court of U.S.
[3]Regional Reporter
[4]All New York Cases
[5]All California Cases
[6]Periodical
[7]Annotation to important cases
[8]Statute
[9]Encyclopedia
[10]Federal Agency Appeals Board

Miss.L.J. Mississippi Law Journal[6]

Mo.App. Missouri Appeal Reports

Mo.L.Rev. Missouri Law Review[6]

Mont. Montana Reports[1]

N.C. North Carolina Reports[1]

N.C.App. North Carolina Court of Appeals Reports

N.C.Gen.Stat. North Carolina General Statutes[8]

N.D. North Dakota Reports[1]

N.E. North Eastern Reporter[3]

N.E.2d North Eastern Reporter, Second Series[3]

N.H. New Hampshire Reports[1]

N.H.Rev.Stat.Ann. New Hampshire Revised Statutes Annotated[8]

N.J. New Jersey Reports[1]

N.J.L. New Jersey Law Reports

N.J.Stat.Ann. New Jersey Statutes Annotated[8]

N.J.Super. New Jersey Superior Court Reports

NLRB National Labor Relations Board (Federal)

N.M. New Mexico Reports[1]

N.M.Stat.Ann. New Mexico Statutes Annotated[8]

N.S.P.E. National Society of Professional Engineers

NTP Notice to Proceed

N.W. North Western Reporter[3]

N.W.2d North Western Reports, Second Series[3]

N.Y. New York Reports[1]

N.Y.2d New York Reports, Second Series[1]

N.Y.Gen.Obl. New York General Obligations Act[8]

N.Y.S. New York Supplement[4]

N.Y.S.2d New York Supplement, Second Series[4]

Neb. Nebraska Reports[1]

Neb.L.Rev. Nebraska Law Review[6]

Nev. Nevada Reports[1]

O.C.G.A. Official Code of Georgia Annotated

OFPP Office of Federal Procurement Policy

OSHA Occupational Safety and Health Act

OSHC Occupational Safety and Health Commission (Federal)

Ohio App.2d Ohio Appellate Reports, Second Series

Ohio App.3d Ohio Appellate Reports, Third Series

Ohio Misc. Ohio Miscellaneous Reports

Ohio N.U.L.Rev. Ohio Northern University Law Review[6]

Ohio Ops. Ohio Opinions

Ohio St. Ohio State Reports, Second Series[1]

Ohio St.3d Ohio State Reports, Third Series[1]

Okl.St.Ann. Oklahoma Statutes Annotated[8]

Op.Cal. Att'y Gen. Opinions of the California Attorney General

Op.Colo. Att'y Gen. Opinions of the Colorado Attorney General

Or. Oregon Reports[1]

Or.App. Oregon Court of Appeals Reports

Or.Rev.Stat. Oregon Revised Statutes[8]

P. Pacific Reporter[3]

P.L. Public Law (U.S.)[8]

P.2d Pacific Reporter, Second Series[3]

PSBCA Postal Services Board of Contract Appeals

Pa. Pennsylvania State Reports[1]

Pa.Commw. Pennsylvania Commonwealth Court Reports

Pa.Super. Pennsylvania Superior Court Reports

Pac.L.J. Pacific Law Journal[6]

Pub.Cont.L.J. Public Contract Law Journal[6]

Pub.L. Public Law (U.S.)[8]

R.I. Rhode Island Reports[1]

Real Est.L.J. Real Estate Law Journal[6]

S. Senate Bill (Federal)

S.C. South Carolina Reports[1]

S.C.L.Rev. South Carolina Law Review[6]

S.C.R. Supreme Court Reports (Canada)

S.Ct. Supreme Court Reporter (U.S.)[2]

S.D. South Dakota Reports[1]

S.E.2d South Eastern Reporter, Second Series[3]

S.W. South Western Reporter

S.W.2d South Western Reporter, Second Series[3]

Santa Clara Computer & High Tech.L.J. Santa Clara Computer and High Technology Law Journal[6]

Seton Hall Legisl.J. Seton Hall Journal of Legislation[6]

So. Southern Reporter[3]

So.2d Southern Reporter, Second Series[3]

---

[1]Highest Court of State
[2]Highest Court of U.S.
[3]Regional Reporter
[4]All New York Cases
[5]All California Cases
[6]Periodical
[7]Annotation to important cases
[8]Statute
[9]Encyclopedia
[10]Federal Agency Appeals Board

Stan.L.Rev. Stanford Law Review[6]

Stat. Statutes-at-Large (U.S.)[8]

Temp.L.Rev. Temple Law Review[6]

Tenn. Tennessee Reports[1]

Tenn.App. Tennessee Appeals

Tenn.Code Ann. Tennessee Code Annotated[8]

Tex. Texas Supreme Court[1]

Tex.Ct.App. Texas Court of Appeals

Tex.Civ.Stat.Ann. Texas Civil Statutes Annotated[8]

Tex.Prop.Code.Ann. Texas Property Code Annotated

Tex.Tech.L.Rev. Texas Tech. Law Review[6]

Thurgood Marshall L.Rev. Thurgood Marshall Law Review[6]

U.C.C. Uniform Commercial Code

UC Davis L.Rev. University of California at Davis Law Review[6]

UCLA L.Rev. University of California, Los Angeles Law Review[6]

UMKC L.Rev. University of Missouri-Kansas City Law Review[6]

U.S. United States Supreme Court Reports[2]

U.S.C.A. United States Code Annotated[8]

U.Bridgeport L.Rev. University of Bridgeport Law Review[6]

U.Dayton L.Rev. University of Dayton Law Review[6]

U.Kan.L.Rev. University of Kansas Law Review[6]

U.Puget Sound L.Rev. University of Puget Sound Law Review[6]

U.Rich.L.Rev. University of Richmond Law Review[6]

Univ.Cinn.L.Rev. University of Cincinnati Law Review[6]

Univ.Pa.L.Rev. University of Pennsylvania Law Review[6]

Utah 2d Utah Reports, Second Series[1]

Va. Virginia Reports[1]

Va.Code Ann. Virginia Code Annotated[8]

Vernon's Mo.Ann.Stat. Vernon's Missouri Statutes Annotated[8]

Vill.L.Rev. Villanova Law Review[6]

Vt. Vermont Reports[1]

Vt.L.Rev. Vermont Law Review[6]

WBE Women's Business Enterprise

W.Va. West Virginia Reports[1]

Wash. Washington Reports[1]

Wash.2d Washington Reports, Second Series[1]

Wash.App. Washington Appellate Reports

Wash.L.Rev. Washington Law Review[6]

Wash.& Lee L.Rev. Washington and Lee Law Review[6]

Wash.Rev.Code Ann. Washington Revised Code Annotated

Wash.U.J.Urb. & Contemp.L. Washington University Journal of Urban and Contemporary Law[6]

West Ann.Cal.Bus. & Prof.Code West Annotated California Business and Professions Code[8]

West Ann.Cal.Civ.Code West Annotated California Civil Code[8]

West Ann.Cal.Civ.Proc.Code West Annotated California Code of Civil Procedure[8]

West Ann.Cal.Com.Code West Annotated California Commercial Code[8]

West Ann.Cal.Govt.Code West Annotated California Government Code[8]

West Ann.Cal.Health and Safety Code West Annotated California Health and Safety Code[8]

West Ann.Cal.Lab. Code West California Labor Code[8]

West Ann.Cal.Penal Code West Annotated Penal Code[8]

West Ann.Cal.Pub.Cont.Code West Annotated California Public Contract Code[8]

West Ann.Cal.Pub.Res.Code West Annotated California Public Resources Code[8]

West Ann.Colo.Rev.Stat. West Annotated Colorado Revised Statute[8]

West Ann.Fla.Stat. West Annotated Florida Statutes[8]

West Ann.Ind.Code West Annotated Indiana Code[8]

West Ann.Minn.Stat. West Annotated Minnesota Statutes[8]

West Ann.Tex.Prop.Code West Annotated Texas Property Code[8]

West Ann.Wash.Rev.Code West Annotated Washington Revised Code[8]

West Okla.Stat.Ann. West Oklahoma Statutes Annotated[8]

---

[1]Highest Court of State
[2]Highest Court of U.S.
[3]Regional Reporter
[4]All New York Cases
[5]All California Cases
[6]Periodical
[7]Annotation to important cases
[8]Statute
[9]Encyclopedia
[10]Federal Agency Appeals Board

West Wis.Stat.Ann. West Wisconsin Statutes Annotated[8]

Willamette L.Rev. Willamette Law Review[6]

Wis. Wisconsin Reports[1]

Wis.2d Wisconsin Reports, Second Series[1]

Wis.L.Rev. Wisconsin Law Review[6]

Wm.Mitchell L.Rev. William Mitchell Law Review[6]

[1]Highest Court of State
[5]All California Cases
[6]Periodical
[8]Statute

Table of Cases

C H A P T E R   O N E

Sources of Law: Varied and Dynamic

SECTION 1.01 Relevance

Law consists of coercive rules created and enforced by the state to regulate the citizens of the state and provide for the general welfare of the state and its citizens. Law is an integral part of modern society and plays a major role in the Construction Process. Because this treatise examines the intersection between law and the Construction Process, it is important to be aware of the various sources of law and the characteristics and functions of the law.

Many illustrations can be provided, but suppose a man who owns property wishes to build a house to provide shelter for his family. Without assurance that stronger persons will not use force to seize materials with which he is building or throw him out of the house after it is built, it would take an adventurous or powerful person to invest time and materials to build the house. Similarly, workers would be reluctant to pound nails or pour concrete if they were fearful of being attacked by armed gangs. Here, criminal law protects both the property owner from those who might take away his property and the workers from those who might harm them.

Similarly, contractors would hesitate to invest their time or money to build houses if they did not believe they could use the civil courts to enforce their contracts and help them collect for their work if owners do not pay them. Workers would be less inclined to work on a house if they were not confident that they could use the civil courts to collect for work that they had done or to compensate them if they were injured on the job.

Finally, some would be unwilling to engage in construction activity if they were not confident that an impartial forum would be available if disputes arose over performance. Were the state not to provide such a forum, participants might settle their disputes by force.

Today various sources of law seek to provide these needs. These sources of law are spotlighted in this chapter. Chapter 2 focuses on the American judicial system.

SECTION 1.02 The Federal System

Very large countries, such as the United States, Canada, Australia, Nigeria, and India, employ a federal system of government. Even smaller countries, particularly those with distinct religious, linguistic, ethnic, or national communities (for example, Switzerland), may choose to live under a federal system. A federal system of government gives local entities limited autonomy to deal with cultural and other activities. The alternative would be domination by the majority or by small, often economically inefficient or weak nation-states.

In a federal system, power is shared; the exact division of power between the central government and constituent members varies. For example, Canada has a looser federal system than the United States, with Canadian provinces having more autonomy than American states.

A federal system may be created in a large country where persons living in one part of the country hold political, social, or economic views distinct from the rest of the country. For example, New

1

York might choose to provide greater social benefits for its citizens than Texas. Similarly, some states may wish to execute murderers, while others may not. In a federal system, diverse views may be accommodated within one country.

Also, depending on its characteristics, a federal system can bring citizens closer to those who govern them. A rancher in Wyoming may resent being controlled by a legislature in Washington but may be more willing to submit to laws that come from Cheyenne. When the entities of a federal system are themselves large, such as Canadian provinces or American states, citizens may see even the state capital as being too remote and prefer to be governed by those they elect in their cities or towns.

The federal system recognizes the need for a central government to deal with certain issues on behalf of *all* citizens. For example, the American federal government controls currency, foreign relations, and defense, to mention a few of the areas it controls exclusively. The constitutional framers did not want each state to have its own currency, foreign policy, or military forces.

Other functions can be shared by the federal government and state governments, such as the enactment of tax laws. Similarly, federal and state laws deal with crime, labor relations, and—to use an illustration more germane to construction—worksite safety. To avoid duplication of enforcement efforts and to relieve construction contractors from inconsistent regulations, the federal government, though dominant, can delegate workplace safety standards and their enforcement to the states so long as the states meet federal standards. Similar delegations are found in environmental protection laws.

Despite general federal supremacy over the states, the U.S. Constitution reserved some authority to the states. For example, except for contracts made by the federal government or those affected with a strong federal interest, state law determines which contracts will be enforced, the remedies granted for breach of contract, the conduct that gives rise to civil liability, and the laws that relate to the ownership of property—all core legal concepts in the Construction Process. As a result, most law that regulates construction is determined by the state in which the project is located or in which the activities in question are performed.

This can and sometimes does lead to variations in legal rules that relate to construction. However, the dominance of nationally created standard con-

tract forms, the willingness of courts in one state to look at and often follow the decisions from another state, and the unification of areas of private law, such as the sale of goods, have minimized the actual variation in state laws that relate to construction. Some laws vary greatly, such as mechanics' lien laws and licensing, both of which are regulated by state statutes.

SECTION 1.03 Constitutions

When organizations are created, whether political, economic, or social, they usually attract members because of their goals, the methods they choose to achieve the goals, and the procedures by which they operate. Often these factors are set forth in a basic set of principles or rules providing a constitutional framework that induces persons to join and regulates the operations of the organization. Although this basic set of rules may be referred to by different labels, it provides a constitutional framework intended to be durable and, although not immutable, amendable only with the consent of most of the members of the organization.

Most organizations need rules to govern day-to-day operations. These rules, though created with less formality than constitutional rules and more responsive to changing conditions, must stay within the basic framework of the constitutional rules. For example, a business corporation will usually have articles of incorporation that operate as a constitution of the corporation and are agreed on by those who create the corporation and available to those who wish to purchase shares. The corporation is also likely to have bylaws that regulate corporate activities and that are often more detailed and more easily changed. In addition, a large corporate organization may need standard operating procedures and hierarchical authority mechanisms to control the corporate operation.

For two centuries, the U.S. Constitution has regulated the power among the branches of the federal government, the federal government and the states, and all governments and their citizens. Despite universal admiration for this durable document, the wealth of federal laws that have been enacted attests to the importance of legislation as a method of dealing with changing circumstances and the varying allocation of political power.

On the whole, state constitutions are longer than the federal Constitution and are changed more frequently. While not identical, they share common characteristics with the U.S. Constitution, mainly the separation of powers among the legislative, executive, and judicial branches of government and the importance of protecting citizens from abuse of power by the states.

Although constitutional law does not play a significant role in the Construction Process, a few areas exist where the federal Constitution or state constitutions have been invoked. For example, constitutional law has been used to challenge laws requiring that minority contractors be given a certain portion of federal construction contracts.[1] It has, along with state constitutions, been invoked to challenge the validity of state legislation that relieves certain parties who engage in design and construction from liability after a designated number of years have elapsed following completion of a project.[2] Constitutional doctrines, with little success, have challenged public contract rules.[3]

SECTION 1.04 Legislation

Legislation, mainly at the state level, plays an increasingly larger role in the Construction Process. Those who wish to understand how law affects design and construction must pay increasingly careful attention to legislation.

Legislation is the most democratic lawmaking process. It expresses the will of the majority of the citizens of the state. (Perhaps the only source of law *more* democratic is the law contracting parties create to regulate their private rights and duties.)

Legislatures are political instrumentalities. Persons and organizations seek to influence lawmakers. Though often denigrated as an area where power and money control, legislation often reflects popular attitudes. Lawmakers who vote in ways not popular with their constituents are likely to be removed from office. Also, legislation can be enacted quickly to respond to what the legislature believes to be important social and economic needs of its citizens.

The legislative process functions at different governmental levels. Most are aware of the federal, state, county, and city legislative bodies. But special districts, such as school and sewerage districts, cannot be ignored. They often do not attract the attention of the voters and the media, and those that are elected are often unknown community figures. This can mean a lesser degree of voter responsiveness to their activities and, unfortunately, greater opportunities for corruption. But these special districts are often important to the Construction Process, as they commission construction work.

Legislators do not have absolute freedom to enact legislation. The checks and balances so central to the American political process bar legislators from absolute power even though the legislatures do represent the political will of the citizens. Laws must be constitutionally enacted, and the constitutionality of a law is determined by the courts. In their role of interpreting legislation, the courts can indirectly control the legislature and the political desires of the majority. Similarly, local units such as counties, cities, or special districts are limited by the legislation that has created them.

As stated, state legislation affects design and construction. Licensing and registration laws determine who can design and, increasingly, who can build. Other state statutes affect design and construction. Dispute resolution has been increasingly affected by legislative activities encouraging arbitration. Similarly, construction litigation has been affected by state laws that preclude certain types of indemnification provisions in construction contracts[4] and by legislation that cuts off liability after the expiration of a designated time.[5] Finally, local legislative bodies exercising their power to protect the public and regulate land use have had an increasingly important effect on construction. Local authorities must approve many types of projects. Through housing and building codes, they also control the quality of construction.

The Uniform Commercial Code (U.C.C.), particularly Article 2, deals with transactions in goods. The Code was developed by the American Law Institute and the Commissioners on Uniform State Laws, both private organizations devoted to unification of private law. At present, all states but Louisiana have adopted the Code. (Louisiana, influenced by its civil law tradition, has adopted only some parts of the Code.) Article 2 regulates the sale

---

[1]See *City of Richmond v. J. A. Croson Co.*, 488 U.S. 469 (1989).
[2]See Section 23.03(G).
[3]Sections 22.06(K) (withholds in public contracts) and 34.03(J) (termination in public contracts).

[4]Section 32.05(D).
[5]See Section 23.03(G).

of materials and supplies used in a project. While it does not regulate construction itself, inasmuch as it does not govern services, it can be influential in a construction dispute. Reference is made to it in this treatise.

SECTION 1.05 The Executive Branch

The separation of powers central to the American political system previously mentioned is designed to prohibit any of the three branches—executive, legislative, and judicial—from having dominant power. Although laws come out of the legislature, the executive and judicial branches participate in lawmaking. The executive branches—the president at the federal level, the governor at the state level, and the mayor in some municipal systems—are elected by all the citizens. Federal and state chief executives and sometimes local chief executives can veto legislation. To override a veto requires more than a simple majority of the legislature, usually two thirds.

In this treatise, the executive branch plays a limited role. The chief executive can issue executive orders to those under his control. For example, the first prohibition of discrimination in employment resulted from a presidential executive order.[6] Also, an executive order by the governor of California in 1978[7] changed the dispute resolution system for California state public contracts and led to legislative change.[8]

Perhaps most important, the chief executive, such as the president or the governor, has the power to appoint the heads of administrative agencies or administrative boards that play a significant role in construction. Examples of such regulatory instrumentalities are those that deal with workplace safety, registration of design professionals, and licensing of contractors, to name only some of the most important.

SECTION 1.06 Administrative Agencies

One great change in the American governmental structure has been the emergence of administrative regulatory agencies at every level of government, but particularly at federal and state levels. Such agencies developed for a variety of reasons. First, special activities and industries were thought best regulated by experts in those activities and industries. Second, legislatures were often unable or unwilling to involve themselves in the details of regulation. Third, regulation through agencies was thought to be a better alternative than no regulation (this changed in the 1980s, with some industries being deregulated) or government operation of these activities and industries.

The constitution for a regulatory agency is the legislation creating it. In that sense, the agency is a creation of the legislature. However, the chief executive usually appoints, often with the advice and consent of the legislature, key agency officials. Legislatures, particularly the Congress, monitor agency performance through exercising an oversight function. This is accomplished through committee hearings during which complaints are heard and agency officials are asked to give explanations.

Agencies operate through issuance of regulations and through disciplinary actions. Such activities are subject to judicial review. During the period when many of these agencies were created, courts extended considerable and almost total deference to such agencies because of presumed agency expertise. As the agencies became more active and as complaints about their aggressiveness and preoccupation with problems that some thought trivial became more vocal, agencies' powers and activities were looked at more carefully by the courts and legislatures.

Regulation through administrative agencies has generated intense controversy. Some agencies are attacked as being under the control of those they are supposed to regulate. This is asserted by pointing to agency employees being selected from the regulated industries or agency employees looking forward to being hired by members of the industry after they leave agency employment. On the other hand, agencies have sometimes been attacked for overzealous regulation, and such attacks had some success in the 1980s.

Administrative agencies play an important role in some aspects of the Construction Process. They are often given responsibility for implementation of laws enacted by the legislatures. For example, licensing and registration of professionals are largely in the hands of state regulatory agencies,

---

[6]The first was Exec. Order No. 8802, 6 Fed. Rules 3109 (1941). The history is collected in *Farmer v. Philadelphia Elec. Co.*, 329 F.2d 3 (3d Cir.1964).
[7]Cal. Exec. Order B50-78 (1978).
[8]West Ann.Cal.Civ. Code §1670. See Section 30.20(B).

which determine levels of enforcement, the nature of examinations, the requirements for licensing and registration, and other matters relating to such professions. Also, the preventive aspects of workplace safety and environmental protection are largely within the control of federal and state agencies. Social insurance for workplace injuries is operated by state workers' compensation agencies. At local levels, administrators play important roles in land-use control and construction quality. In areas affecting public safety, the most important source of law is regulations issued by administrative agencies.

SECTION 1.07 Courts: The Common Law

Courts are an important part of the lawmaking process. Exercising their power of judicial review, they determine whether legislation is constitutional and interpret the legislative enactments. They have the principal responsibility for granting remedies under legislative systems. Courts also pass upon and interpret administrative regulations and grant remedies for violations.

The sense in which courts are a source of law central to the construction process is principally based on the dual function of American appellate courts. On one hand, the appellate courts review trial court decisions by providing a forum for those litigants who are dissatisfied with judgments of the trial courts. More important, the process by which appellate courts judge the correctness of decisions made by the trial courts has had an immense impact on design and construction. American courts follow precedent to resolve disputes. An understanding of how courts make law and resolve disputes requires an understanding of the precedent system discussed in Section 2.14.

SECTION 1.08 Contracting Parties

Sources of law usually are public bodies. Yet contract law, with the broad autonomy granted the contracting parties to determine the terms of their exchange, grants lawmaking power to those who make contracts.

Even if autonomy is seen as an inherent liberty in a free society, the state still plays a role—enough of a role to create a "partnership" between contracting parties and the state. Despite broad autonomy given the parties, the state still determines who can contract, creates the formal requirements,

and provides remedies when contracts are not performed.

If the contracting parties can make their contract a source of law, this is the most *democratic* form of lawmaking. The contracting parties freely determine the terms of their exchange and voluntarily limit their freedom of action by agreeing to perform in the future, with the coercive arm of the state operating if they do not.

Chapter 5 is devoted to some aspects of contract law and Chapter 6 to contract remedies. At this point, it is sufficient to mention the emergence of the bipolar analysis of contract law, one that draws a sharp distinction between *negotiated* contracts and contracts of *adhesion*. The former emphasize autonomy or freedom of contract based on consent being freely given by the contracting parties. The latter involve contracts made with the terms, or at least most of them, largely dictated by *one* of the parties and merely adhered to by the other. This distinction is discussed in Section 5.04(C).

SECTION 1.09 Publishers of Standardized Documents

Adhesion contracts, represented by prepared printed contracts, are prepared by *one* party to the contract largely, if not exclusively, with its interests in mind. Yet many prepared printed forms are used in contracts for design and construction that should not be considered adhesion contracts in the sense that the term is used in Sections 1.08 and 5.04(C). Often contracts for design and construction services are based largely on documents published by professional associations such as the American Institute of Architects (AIA), the Associated General Contractors (AGC), and the Engineers Joint Contract Documents Committee (EJCDC), a consortium of professional engineering associations, to mention some of the most important.

These associations have no official status, and some would contend that they cannot be considered sources of law. Certainly contracting parties need not use the standardized documents or, if they do use them, are free to make drastic changes. But the frequency with which these documents are used largely unchanged justifies classifying these associations as sources of law. Even if in a *technical* sense they should not be placed along with courts, the realities of construction contracting make them even more important.

SECTION 1.10 Summary

Understanding how law bears on design and construction requires a recognition of the wide variety of law sources. Those involved in design and construction should be aware of the laws and legal institutions that have placed their mark on design and construction. Trends as to which sources of law predominate and the forms that legal intervention take must be appreciated. A half century ago, it could be confidently stated that the principal sources of law in design and construction were the contracting parties and the courts. Increasingly, other instrumentalities such as legislatures and regulatory agencies are leaving their imprint on design and construction.

CHAPTER TWO

The American Judicial System: A Forum for Dispute Resolution

SECTION 2.01 State Court Systems: Trial and Appellate Courts

Each state has its own judicial system. Courts are divided into the basic categories of trial and appellate courts. Within each category, there may be a subclassification based on the amount or type of relief sought or on the nature of the matter being litigated.

The basic trial court, frequently called the court of general jurisdiction, hears all types of cases. Depending on the state, it may be called a superior, district, or circuit court. The bulk of the work before such courts consists of criminal cases, personal injury cases, commercial disputes, domestic matters (divorce, custody, and adoption), and probate (transfer of property at death).

A court of general jurisdiction may review determinations of administrative agencies (zoning, employment injuries, licensing, etc.). The presiding official will be a judge, and in certain matters there may also be a jury. The division of functions between judge and jury is discussed in Sections 2.10 and 2.11. Courts of general jurisdiction are usually located at county seats.

Many states have established subordinate courts of limited jurisdiction. Municipal or city courts often have jurisdiction to hear disputes that involve less money than a minimum figure set for courts of general jurisdiction. There can also be a jury in such cases. As a rule, the dockets are less congested in these courts of limited jurisdiction than are the dockets in the courts of general jurisdiction. The procedures in municipal or city courts are essen-

tially the same as those in the courts of general jurisdiction.

Some state legislatures have established small claims courts that provide expeditious and inexpensive procedures for disputes involving small sums. Usually the party seeking a remedy from such a court will be able to start the legal proceeding by paying a small filing fee and filling out a form provided by the clerk of the small claims court. The procedures are usually informal. There are no juries. Although the judges are lawyers, attorneys are less important in such disputes, and in some state, attorneys are not permitted to represent the parties.

Some states still use justices of the peace, a vestige of a rural, dispersed society. These judges often have had little or no legal training. Although they are being phased out in favor of small claims, municipal, or city courts, they still exist in sparsely populated areas of some states. The future may witness their elimination.

At least one appeal is usually possible from a decision of a trial court. Generally, the appeal is to the next highest court. For example, a party losing a decision in a municipal court may have a right to appeal to the court of general jurisdiction or to an appellate division consisting of judges of the court of general jurisdiction. Appeals from courts of general jurisdiction are made to an appellate court. Beginning in the more populous states and increasingly in other states, appeals first must go to an intermediate appellate court. In states with such courts, the state supreme court generally is given the discretion to decide whether it will hear an appeal from an intermediate appellate court. In states without intermediate appellate courts, an appeal is made from the court of general jurisdiction to the supreme court of the state.

State court judges are either appointed by the governor or elected. Even states that elect judges have many judges who were initially appointed. Generally, judges are reelected and leave office only when they retire or die. Such vacancies are filled by interim appointments, and appointees are usually elected when they go before the voters.

Some states use an elective process in which all judges—or at least appellate judges—periodically submit their records to the electorate but do not run against other candidates.

SECTION 2.02 Federal Court System

The American judicial system includes two systems—federal and state—which to a degree exist side by side. Although each state has its own judicial system, the federal courts also operate in each state. The federal courts have jurisdiction to decide federal questions, that is, disputes involving the federal Constitution or federal statutes. They also have jurisdiction to hear civil disputes between citizens of different states, often called *diversity of citizenship* jurisdiction. In theory, the amount in controversy in each type of case must exceed $50,000 exclusive of interest and costs, but the amount in controversy is important principally in diversity of citizenship cases.

Many claims brought before federal courts can also be brought in state courts, creating concurrent jurisdiction. Some matters, however, can be brought only in the federal courts. The principal areas of exclusive federal jurisdiction are as follows:

1. admiralty
2. bankruptcy
3. patent and copyright
4. actions involving the United States
5. violations of federal criminal statutes

Where exclusive federal jurisdiction exists, there is no amount in controversy requirement.

The federal courts operate under more modern and less formal procedural rules than do most state courts. For substantive law (laws that establish legal rights and duties), the federal courts use the substantive law of the state in which they sit or some other applicable state law, unless the case involves federal law, such as a federal constitutional provision or a federal statute.

The basic trial court in the federal court system is the district court. Each state has at least one, and the populous states have a number of district courts located in the principal metropolitan centers.

The district court is presided over by a federal district court judge, and juries are used in certain cases. A party may appeal a decision of the district court to one of the eleven circuit courts that hear appeals from that district. Another circuit court hears appeals mainly from decisions of U.S. administrative agencies. Another hears appeals from specialized federal courts.

A party dissatisfied with the result of a decision by the circuit court of appeals may ask that the U.S. Supreme Court review the case. In general, the Supreme Court determines which cases it will review. If it decides not to hear the case, the matter is ended. If it decides to hear the case, briefs and oral arguments are presented to the Court. The Supreme Court usually rejects petitions for hearings unless the matter involves either a conflict between federal circuit courts or an important legal issue.

Federal court judges are appointed by the president of the United States and are confirmed by the Senate. In essence, the judges serve for life or until voluntary retirement.

Within the federal system, a few courts deal with specialized matters. For example, the United States Court of Federal Claims (formerly the Claims Court) hears claims against the U.S. government, appeals from which are taken to the U.S. Court of Appeals for the Federal Circuit. Usually these cases relate to tax disputes or to disputes between government contractors and government agencies that award government contracts. Another specialized court is the Tax Court, which hears disputes between taxpayers and the government.

SECTION 2.03 Statute of Limitation: Time to Bring the Lawsuit

Statutory provisions require that legal action be commenced within a specified period of time. Such statutes in the context of the Construction Process are discussed in Sections 14.09(C) and 23.03(G).

SECTION 2.04 Hiring an Attorney: Role and Compensation

Usually a person with problems that may involve the law or the legal system will consult an attorney. After an inquiry into the facts and a study of the law, the attorney advises a client of her legal rights and responsibilities. Although the attorney may also give an opinion on the desirability of instituting legal action or defending against any action that has been asserted against a client, the litigation choice is usually made by the client.

Unless the law sets the fee, which is rare, attorney and client can determine the fee. Sometimes the fee is a specified amount for the entire service to be performed, such as uncontested divorces or simple incorporations. The fee may be a designated percentage of what is at stake, such as in the probate of an estate. Often the fee is based on time spent by the attorney, typically computed on an hourly basis. If no specific agreement on the fee is reached, the client must pay a reasonable amount.

Hourly rates charged depend on the attorney's skill, the demands on the attorney's time, the amount at stake, the complexity of the case, what the client can afford, the locality in which the attorney practices, and the outcome in the event of litigation. Although hourly rates vary greatly, at present an experienced attorney located in a large city is likely to charge from $150 to $300 an hour. (The second edition of this treatise published in 1977 stated the hourly rate as $50 to $75.)

A potential client should ask a prospective attorney the likely charges for legal services. Such advance inquiry can avert possible later misunderstandings or disputes over the hourly rate. The client may wish to set a maximum figure for a particular legal service. Attorneys are reluctant to accept maximum figures, because it is often difficult to predict the amount of time needed to provide proper legal service.

Commonly in personal injury or death cases and occasionally in commercial disputes, attorneys use contingent fee contracts. The lawyers are not paid for legal services if they do not obtain a recovery for the client. Usually a client agrees to reimburse the attorney for out-of-pocket costs, such as deposition expenses, filing fees, and witness fees. For taking the risk of collecting nothing for time spent, the attorney will receive a specified percentage of any recovery. In personal injury cases, the percentage can range from 25% to 50%, depending on the locality, the difficulty of the case, and the reputation of the attorney. In some metropolitan areas, attorneys will charge 25% to 33% if they obtain a settlement without trial and 33% to 40% if they go to trial.

The contingent fee system is much criticized, especially where the attorney obtains an astronomical sum in a tragic injury or death case. Such a system gives the attorney an entrepreneurial stake in the claim. Some feel this is unprofessional and may influence the attorney's advice as well as raise questions when attorney and client disagree over settlement. Yet some defend the contingent fee as the only method by which poor clients can have their claims properly presented. States are beginning to regulate contingent fees.

A retainer is an amount paid by a client to an attorney either at the commencement of or periodically during the attorney-client relationship. Although the retainer can be an agreed-upon value for the legal services performed by the attorney or an amount to pay for having a call on the attorney's services unrelated to fees for services, the retainer payment is more commonly an advance payment on any fees that the client is obliged to pay the attorney. Sometimes it covers routine services but does not include extraordinary services, such as litigation. The attorney and the client should agree on the function and operation of any retainer.

Some attorneys, particularly in criminal or matrimonial matters, label the retainer as nonrefundable. If the attorney's services are not requested under the terms of such a retainer, no refund is available. Attorneys who use nonrefundable retainers assert that once having been retained, they are barred from being retained by the other party, since retention can generate conflict-of-interest problems later. Such attorneys assert that the client has an option to use the attorney whose services she purchases by paying the retainer.

Opponents of the nonrefundable retainer assert that such a payment limits the right of the client to choose her own attorney. They also contend that it is unconscionable and unethical for attorneys to receive compensation when they do not perform services.

A New York attorney was disciplined because he used a nonrefundable retainer. His two-year suspension was upheld by the intermediate appellate court of New York.[1]

In the past, bar associations published schedules of recommended or minimum fees. Such schedules violate the antitrust laws and are no longer used. For a discussion of architectural and engineering fee schedules, see Section 11.03.

The high cost of legal services and an increasing use of prepaid group health plans has led to the suggestion of prepaid group legal service plans.

The law recognizes the need for a client to be as candid in communicating with an attorney as with a member of the clergy, a doctor, or a spouse. For this reason, the attorney must keep confidential any communication made by the client, unless the communication was to plan or commit a crime or if the client challenges the competence of the attorney's performance.

SECTION 2.05 Jurisdiction of Courts

Jurisdiction—the power to grant the remedy sought—means power over the person being sued and over the subject matter in question. This section discusses state court jurisdiction.

The U.S. Constitution requires that state and federal governments ensure that a defendant receives due process, the opportunity of knowing what the suit is about, who is suing, and where and when the trial will take place. Usually due process is accomplished by having a process server hand to the defendant in the state where the trial is to take place a summons ordering the defendant to appear and a complaint stating the reasons for the lawsuit. The power of the state to compel the defendant to appear does not extend beyond the border of the state. If the defendant does not come within the state, traditionally, the state court in which the lawsuit has been brought would not have jurisdiction over the person or entity against whom the claim has been made. The plaintiff would have to sue in the state where the defendant actually could be handed the legal papers.

The requirement of physical presence in the state where the court sits can place great hardship on a plaintiff who may be forced to begin a lawsuit in a state that is far from the evidence and witnesses. For this reason, most states have enacted long-arm statutes that permit plaintiffs in certain cases to sue in their home state even though they cannot hand the legal papers to the defendant within the plaintiff's state.

Long-arm statutes are often used to sue defendant motorists who reside in a state other than the one where the accident occurred or where the injured party resides. Suppose an Iowa driver were involved in an accident in Illinois that injured an Illinois resident. The Iowa driver returns to Iowa after the accident. The injured Illinois resident can always sue in Iowa. Usually the injured party would also be given the opportunity to sue in Illinois by mailing a notice of the lawsuit to the other driver in Iowa or by filing legal papers with a designated state official in Iowa or Illinois.

Sometimes long-arm statutes are used to sue businesses that have their legal residence in a state other than that of the plaintiff. To illustrate, suppose a New York company sells, directly or indirectly, electric drills to Wyoming purchasers. A Wyoming buyer is likely to be able to sue the New York seller in a Wyoming court for injuries resulting from the purchase and use of a defective drill.

Jones Enterprises, Inc. v. Atlas Service Corp.[2] illustrates the effectiveness of such a statute in a construction context. It involved a claim brought in the federal court in Alaska by a prime contractor against an out-of-state supplier, an out-of-state sub-supplier to the supplier, and an out-of-state structural engineer retained by the supplier to design precast concrete work. The claim resulted when the structure constructed in Alaska by the prime contractor and made with materials designed and shipped to Alaska by the out-of-state defendants collapsed. The defendants were not served legal papers in Alaska, as they were not physically present there. But the plaintiff relied on an Alaska long-arm statute.

None of the out-of-state defendants were qualified to do business in Alaska, none had agents in Alaska, and all the work had been performed by the defendants outside Alaska. However, the court held that all the defendants were subject to the jurisdiction of the Alaska federal court in which the

---

[1] *In the Matter of Edward M. Cooperman*, 591 N.Y.S.2d 855 (App. Div. 1993).

[2] 442 F.2d 1136 (9th Cir. 1971).

action had been brought. Although the supplier and the subsupplier to the supplier had had some minimal contact with Alaska, the structural engineer had never even been in Alaska in connection with the project. But because he knew that the building was to be constructed in Alaska and he sent his design into the stream of commerce aimed at Alaska, the court concluded that he, along with all of the other out-of-state defendants, was subject to the Alaska court.

Jurisdiction may also involve the subject matter of the lawsuit. For example, if land is located in State A and there is a dispute as to the ownership of the land, only State A will have jurisdiction, even though persons who may claim an interest in the land may not be residents of State A and may not be served with the legal papers in State A.

States usually limit their jurisdiction to matters of concern to that state. A state may have jurisdiction because property is located in the state, the injury occurred in the state, the plaintiff or defendant is a resident of the state, or a contract was made or performed within the state.

Another aspect of jurisdiction relates to the type of remedy or decree sought. Early in English legal history, two sets of courts existed—law courts and equity courts. Juries were used in the former but not in the latter. In addition, equity court judges could award remedies not available in the law courts. Parties seeking equitable relief had to show that their remedy in the law courts was inadequate. As a result of this dual set of courts, two sets of procedural and substantive rules emerged. Generally, the procedures and remedies tended to be more flexible in the equity courts.

The division between law and equity was incorporated into the U.S. judicial system. However, a gradual merger of the two systems has occurred in most states. For all practical purposes, in most states only one set of courts exists, although some differences remain. The most important is that certain remedies may be given only by an equity court. The most important of these remedies are specific performance (under which one party is ordered by a court to perform as promised in a contract), injunctions (orders by the court that persons do or do not do certain things), and reformation (rewriting of a written document to make it accord with the parties' actual intention).

As Section 2.10 explains, there are no juries when an equitable remedy is requested.

Contractual provisions sometimes specify the judicial forum for resolving disputes under the contract. Such clauses are generally upheld if they result from negotiation between contracting parties who are aware of the nature of the provision and are able to protect themselves. In the absence of any forum selection clause, the courts would apply their jurisdictional rules.

A distinction must be drawn between clauses that seek to bar a forum from one that would be available under the jurisdictional rules of the state and those that seek to create jurisdiction where the state laws would not do so. The law will more likely allow a party to waive its right to use its courts if this was done deliberately and knowingly than it would allow contracting parties to create jurisdiction that would not already be present. The latter can place an unwelcome burden on the courts. Yet forum selection clauses can be the result of an abuse of power. Wisconsin and Virginia enacted statutes that limit the enforcement of forum selection clauses.

A Wisconsin law enacted in 1991 invalidates clauses in contracts for the improvement of land in Wisconsin that requires "that any litigation, arbitration or other dispute process on the contract occur in another state." Under such a statute, a contract for building a structure in Wisconsin could not contain a provision requiring that disputes be resolved in New York.[3]

Virginia law permits a Virginia contractor who makes a contract to perform work in Virginia to bring a claim in a Virginia court even though there is a contract provision to the contrary.[4] For example, suppose a contract between a Virginia subcontractor and a New York contractor contains a provision that all lawsuits must be brought in New York. Under the Virginia statute, the Virginia subcontractor could nevertheless bring an action for breach of contract in Virginia.

SECTION 2.06 Parties to the Litigation

Generally, the party commencing the action is called the plaintiff, and the party against whom the action is commenced is called the defendant. In

---

[3]1991 Wis. Laws 200, codified at West Wis. Stat. Ann. Act 200 § 779.035 (1m) (b) (1992).
[4]Va. Code Ann. § 8.01-262.1 (1992).

construction disputes, it is common for the defendant to assert a counterclaim against the plaintiff or to make claims against third parties arising from the same transaction. For example, suppose an employee of a subcontractor sues the prime contractor based on a claim that the prime contractor has not lived up to the legal standard of conduct and this caused a loss to the claimant. The prime contractor may, in addition to defending the claim made by the employee of the subcontractor, sue the architect and the owner, claiming that the former was negligent and the latter was responsible for failure of the architect to live up to the legal standard, or sue the subcontractor employer of the claimant based on indemnification.

The prime contractor in asserting these claims is called a cross-complainant in most state courts and a third-party plaintiff in the federal courts. Those against whom these claims are made would be called cross-defendants in most state courts and third-party defendants in the federal courts.

Cross claims by defendants were difficult to maintain in the law courts in England and under early American procedural rules. However, equity courts freely permitted a number of different parties to be involved in one lawsuit as long as the issues did not become too confusing. They sought to resolve disputes relating to the same transaction in one lawsuit. Ultimately, the rules developed in the equity courts prevailed, and it is generally possible to have multiparty lawsuits if jurisdiction can be obtained over all the parties and if litigation can proceed without undue confusion or difficulty.

SECTION 2.07 Prejudgment Remedies

Not uncommonly, the defendant against whom a judgment has been obtained has insufficient assets to pay the judgment. A judgment may be uncollectible if assets are hidden or heavily encumbered by prior rights of other creditors. Most states have statutes creating prejudgment or, as they are sometimes called, provisional remedies. Plaintiffs (or defendants who are asserting counterclaims) may be able, in advance of litigation, to seize or tie up specific assets of the opposing party to ensure that if they prevail, they can collect the judgment. Such assets may be attached if in the possession of the defendant or garnished if in the hands of third parties. It is not uncommon for a plaintiff to attempt to tie up assets such as bank accounts or other liquid assets in advance of the litigation.

The party whose assets are tied up can dissolve (have removed) the attachment or garnishment by posting a sufficient bond. Parties whose assets have been tied up may ultimately prevail in the litigation. If so, they may have unfairly suffered damage from the seizure or tying up of their assets. To protect against the risk of not finding assets out of which the prevailing party can be indemnified for such losses, the law frequently requires that the attachment or garnishment be accompanied by a bond.

Even if a party can avoid having assets tied up in advance of litigation by posting a bond, such a protective device is often not feasible where the party whose assets are being tied up or seized is poor. As a result, great hardship can occur to poor people and their families if the wage earner's wages are seized or if before the court hearing the sheriff repossesses a car, a television set, or furniture being purchased under a conditional sales contract. Under such arrangements, the buyer takes possession, but the seller retains ownership until all payments have been made. As a result of this, the U.S. Supreme Court has looked closely at prejudgment remedies to ensure that they give the defendant wage earner or debtor due process.[5]

Another prejudgment remedy is a preliminary injunction—an equitable decree ordering a defendant to cease doing something that the plaintiff claims is causing or will wrongfully cause injury to the plaintiff. Such an injunction can be issued before a full trial if the plaintiff can show that it is needed to preserve the status quo until the trial, that the plaintiff would suffer irreparable loss if it were not issued, and that there is a strong likelihood that the plaintiff will prevail at the trial.

Suppose an executive of a company that makes scientific instruments has threatened or begun to violate a promise made in her employment contract not to compete with her employer after leaving the company. If such a promise not to compete is reasonable, a court of equity will enforce it by ordering the executive to live up to the promise. The employer would very likely seek a preliminary injunction because irreparable harm can be caused the employer between the time the executive starts

---

[5]*Fuentes v. Shevin*, 407 U.S. 67 (1972).

to work for a competitor and the issuance of a court order prohibiting the executive from doing so after a full trial. If the employer can show a strong likelihood that the clause will be enforced, the court should award a preliminary injunction. The employer must post a bond to protect the executive from any damage caused by the issuance of the preliminary injunction if the executive should prevail after a full trial.

SECTION 2.08 Pleadings

The lawsuit is usually begun by handing to the defendant a summons and complaint, called service of process. The defendant has a specified time to answer. The defendant's answer may assert that the plaintiff has not stated the facts correctly or that defenses exist even if the allegations of the plaintiff's complaint are true. In some states, the legal sufficiency of the complaint may be tested by filing a demurrer—a statement that even if the facts as alleged by the plaintiff are true, the plaintiff has no valid claim.

In addition to answering the plaintiff's complaint, the defendant in some cases may assert a claim against a third party by filing a crosscomplaint or may file a counterclaim against the plaintiff asserting that the defendant has a claim against the plaintiff.

In some states, the plaintiff will be required to submit a reply to the answer. Most states require an answer to a counterclaim. In early American legal practice, there were a substantial number of other pleadings that might be filed. The tendency in American procedural law today is to reduce the number of pleadings.

The pleadings should inform each party of the other party's contentions and eliminate from the trial matters on which there is no disagreement. By an examination of the pleadings, an attorney or a judge should be able to determine the salient issues to be explored in the litigation. Pleadings should streamline the litigation and avoid the proof of unnecessary matters. Trial preparation should be more efficient and settlement expedited if each side knows the issues on which the other intends to present evidence at the trial.

The complaint should be a concise statement of facts on which a plaintiff is basing a claim and the specific remedy sought. Unfortunately, complaints are often prolix and contain factual statements and serious charges that the plaintiff's attorney may not be able to prove or that have not been checked out carefully. The inexperienced litigant should understand some reasons for this.

Frequently, the lawsuit is commenced before the plaintiff's attorney has had sufficient time to investigate thoroughly and use the discovery procedures described in Section 2.09. Unsure of the facts, an attorney may plead a number of different legal theories as a protective measure.

The attorney wants to be certain that the complaint will be legally sufficient, that is, that the judge will not uphold a claim made by the opposing attorney that no legal claim has been stated. The plaintiff's attorney may decide to use the identical language of pleadings that have been held to be sufficient in the past. Such language may not be tailor-made to the particular case in which it is used. If the language is taken verbatim from an old pleading form book, it may contain archaic and exaggerated language.

The plaintiff's attorney may believe that the only way to get the defendant or the defendant's attorney to consider seriously a client's claim is to start a lawsuit immediately, make strong accusations, and ask for a large amount of money. Unfortunately, exaggeration and overstatement are frequently used legal weapons. But when litigants are not made aware of the reasons for such language, its use may simply increase the already existing hostility between the litigants and make settlement more difficult.

If the defendant has been served with a summons and complaint, failure to answer or to receive an extension of time to answer within the time specified by law allows the plaintiff to take what is called a default judgment, and the case is lost by default.

The plaintiff who obtains a default judgment can collect on this judgment in the manner described in Section 2.13. Most states permit the judge to set aside the default judgment. To do so, the defendant must present cogent reasons why it would be unfair to enforce the judgment despite failure to respond to the pleadings in the designated time. Modern courts are *more* willing than courts a generation ago to set aside default judgments if the defendant can show that there is a valid defense, that failure to answer was the result of an excusable mistake, and that the amount of the judgment is substantial. Still, default judgments are rarely set aside. For this reason, a defendant who wishes to contest a claim made in a complaint should have an attorney answer the complaint within the required time.

SECTION 2.09 Pretrial Activities: Discovery

The parties and their attorneys wish to discover all facts that are material to the lawsuit. *Discovery* is the legal process to uncover information in the hands of the other party. One method of discovery is written interrogatories. One party's attorney sends a series of written questions to the other party, who has a specified period of time in which to respond.

The other method is to take a deposition, that is, to compel the other party or its agent to appear at a certain time and place and answer questions asked by the attorney seeking discovery. The attorney is also likely to demand that the person being questioned bring all relevant documents relating to the matter in dispute.

The person being questioned usually brings an attorney. Questions are asked under oath. Questions and answers are transcribed by a reporter, and a transcript is made available to both parties.

Generally, the transcript does not substitute for testimony at the trial. However, the transcript can be used to impeach (that is, to contradict) the witness if the testimony at the trial is inconsistent with the statements made under oath in the discovery process.

A proper use of discovery should avoid surprise at the trial, reduce trial time, and encourage settlement. However, the sweeping range of inquiry and the wide latitude given to demand that documents be produced has made the process time consuming and very costly.

Witnesses who will be out of jurisdiction at the time of trial or who are very sick and may not be alive when the case is tried can also be examined under oath and their testimony recorded and transcribed. This deposition process can substitute for testimony that would otherwise be unavailable at the time of trial.

In many states, the pretrial conference is another step in the litigation process. Usually such a conference is conducted in the judge's chambers in the presence of the judge, the attorneys, and sometimes the parties. The main purposes of the pretrial conference are, like those of the pleadings, to narrow the issues, avoid surprise, and encourage settlement.

SECTION 2.10 The Jury

A dispute not settled or abandoned before trial will be submitted to a judge and sometimes to a jury. Generally, the right to a jury trial is constitutionally guaranteed, except in claims where decrees by an equity court are sought. The parties can agree to try the case without a jury. The principal present use of juries is in criminal and personal injury trials.

The use of juries in commercial disputes, although available, is less common. In commercial disputes, the use of juries is criticized when the case is extraordinarily complex. This is a problem that comes up frequently in construction disputes, which often can involve jury trials of up to six months. Yet American courts have been unwilling to deprive a claimant of its right to a jury trial as guaranteed by the Constitution, despite the complexity of the case and the difficulty the jurors might have in deciding the case properly.[6]

One court held that it was improper to submit a complex claim to a jury. The case was estimated to last one year. The claim involved an alleged conspiracy by twenty-four Japanese electronic companies and one hundred co-conspirators to maintain an artificially low price for fifteen years. The claimant brought claims under the American antitrust laws. Pretrial activity went on for nine years and generated millions of documents and over 100,000 pages of depositions. The court based its denial of a jury trial on the likelihood that the defendant's right to receive due process could be violated by such a jury trial.[7]

Historically, juries have consisted of twelve lay persons selected from the community to pass upon guilt and sometimes sentence in criminal matters and to decide disputed factual questions in civil matters. Juries have been attacked as inefficient. Jury trials take longer, cost more, and have a higher

---

[6]*In re U.S. Financial Securities Litigation*, 609 F.2d 411 (9th Cir. 1979), cert. denied sub nom. *Gant v. Union Bank*, 446 U.S. 929 (1980).

[7]*In re Japanese Electronic Products Antitrust Litigation*, 723 F.2d 238 (3d Cir. 1983).

degree of reversals by appellate courts. Some states reduce jury size from twelve to six members. In criminal matters, the decision must be unanimous. Some states require only a five-sixths decision in civil disputes.

In many state courts, attorneys are allowed to question prospective jurors to determine their impartiality. An attorney dissatisfied with a particular juror can ask that the juror be excused for cause. The judge rules on whether there is proper cause to excuse the juror. Usually the attorneys can strike (that is, excuse) a designated number of jurors peremptorily (that is, for no reason except for one racially motivated). In the federal courts, jury examination is typically conducted by the judge.

Jurors need give no reason for their decision in a case. While the judge instructs the jury as to the law, jurors are, for all practical purposes, free to do as they choose. Some criticize this, while others feel that it is a useful safety valve, especially in criminal matters, to allow members of the community to excuse law violators.

Until recently, jurors did not make the details of their deliberations public. But of late, jurors, especially in controversial cases, openly describe the jury process to representatives of the media and anyone else who may be interested or willing to pay.

Some feel that juries are easily manipulated by clever and persuasive attorneys. Others feel that jurors are sensible and generally come to the right result. The jury system will always be controversial.

SECTION 2.11 Trials: The Adversary System

In some jurisdictions, the parties are required before the actual trial to submit their dispute to nonbinding arbitration. Usually the arbitrator is a volunteer local attorney, and usually the requirement exists only for claims within a certain dollar amount.

Generally, parties do not seek an actual trial after the results in the nonbinding arbitration. They are either satisfied with the outcome, are principally concerned in having a neutral look at their dispute, or are deterred by the cost of going to trial.

The trial is usually conducted in public and is begun with an opening statement by the plaintiff's attorney. Sometimes the defendant's attorney also makes an opening statement. In the opening statement, the attorney usually states what she intends to prove and also seeks to convince the jury that the client's case is meritorious. Opening statements are less common in trials without a jury.

After opening statements, the plaintiff's attorney may call witnesses who give testimony. Also, during this phase, physical exhibits and documents can be offered into evidence. In civil actions, the other party can be called as a witness.

An attorney cannot ask leading questions of the witnesses except to establish less important preliminary matters or where the witness is of low mentality or is very young. A leading question is one that tends to suggest the answer to the witness. Such questions can be and usually are asked witnesses called by the other party.

Witnesses are supposed to testify only to those matters that they have perceived through their own senses. Usually, witnesses cannot express opinions on technical questions unless they are qualified as experts. Expert testimony is discussed in greater detail in Section 14.06.

The hearsay rule precludes witnesses from testifying as to what they have been told if the purpose of the testimony is to prove the truth of the statement. The danger in hearsay testimony is that the party who made the statement is not in court and cannot be cross-examined as to the basis on which the statement was made. However, there are many exceptions to the hearsay rule.

After the attorney has finished questioning the witness, the other party's attorney will cross-examine and try to bring out additional facts favorable to her client or to discredit the testimony given on direct examination. Cross-examination can be an effective tool to catch a perjurer or show that a witness is mistaken. However, when used improperly, it can create sympathy for the witness or reinforce testimony of the witness.

Documents play a large role in litigation. Under modern rules of evidence, it is relatively easy to introduce into evidence documents to be considered by the court. Documents are hearsay testimony. They are usually writings made by persons who are not in court. However, there are exceptions to the hearsay rule that permit the admission of documents such as official records and business entries. Often the attorneys will stipulate to the admissibility of certain documents.

After the plaintiff's attorney has presented her client's case, the defendant's attorney will present the defendant's case. After this is done, the plaintiff is given the opportunity to present rebuttal evidence. After all the evidence has been presented, the judge submits most disputed matters to the jury (when one is used). The judge instructs the jury on the law using instructions sometimes difficult for a juror to understand. The jury meets in private, discusses the case, takes ballots, and decides who prevails and how much should be awarded to the winning litigant. (In controversial cases, this is often followed by a press conference. Refer to Section 2.10.)

The adversary system, though much criticized because of its expense, its often needless consumption of time, and the hostility it can engender, is central to the American judicial process. Each party, through its attorney, determines how its case is to be presented and can present its case vigorously and persuasively. Also, each party, mainly through cross-examination and oral argument, has considerable freedom to attack the other's case. Generally, the judge acts as an umpire to see that the procedural rules are followed.

One justification for the adversary system is that when the smoke has cleared, the truth will emerge. This assumes well-trained advocates of relatively equal skill and a judge who ensures that procedural rules are followed.

Another justification for the adversary system is that it gives the litigants the feeling that they are being honestly represented by someone of their choosing and are not simply being judged by an official of the state. The system can give the litigants a sense of personal involvement in the process rather than simply being passive recipients of decisions handed them by the state. But excessive partisanship can generate bitter wrangles and delays. Also, it can mean that a trial resembles military combat rather than a reasoned pursuit of the truth.

Even under the adversary system there are limits on advocacy. The law prescribes rules of trial decorum administered by the judge, who can punish those who violate these rules by citing violators for contempt. The person cited can be fined and, in some cases, imprisoned. In some court systems, an attorney can be sanctioned for certain types of conduct that does not rise to the level required to hold the attorney for contempt of court. Usually the sanction carries with it a fine.

The legal profession can discipline its members who violate professional rules of conduct. As obvious illustrations, an attorney must not bribe witnesses, encourage or permit perjured testimony, or mislead the judge on legal issues. An attorney is the champion for the client yet part of the system of the administration of justice and must not do anything that would subvert or dishonor that system.

Attorneys frequently make objections in hotly contested litigation. This is always irksome to participants, but it is often, though not always, necessary. If objections are not made at the proper time, the right to complain of errors may be lost. The judge should be given the chance to correct judicial mistakes on the spot.

The possibility of drama and tension in the litigation process can increase when a jury is used. Some attorneys employ dramatic tactics to influence the jury. Although some enjoy the combat of litigation, litigation is at best unpleasant and at worst traumatic for the litigant or witness who is not accustomed to the legal setting, the legal jargon, or the adversary process. This can be intensified if the trial attracts members of the public or is reported in the media.

A trial is an expensive way to settle disputes. In addition to attorneys' fees, witness fees, court costs, and stenographic expenses, there are less obvious expenses to the litigant. Much time must be spent preparing for and attending the trial. The litigant may have to disrupt business operations by searching through records. For these and other reasons, most lawsuits are settled out of court.

SECTION 2.12 Judgments

A judgment is an order by the court stating that one of the parties is entitled to a specified amount of money or to another type of remedy. Some judges rule immediately after the trial. Usually, the judge takes the matter under consideration. Often judges write opinions giving the reasons for their decisions.

SECTION 2.13 Enforcement of Judgments

If the plaintiff obtains a money award, the defendant should pay the amount of money specified in the judgment. However, if the defendant does not pay voluntarily, the plaintiff's attorney will deliver the judgment to a sheriff and ask that property of the defendant in the hands of the defendant or any third party be seized and sold to pay the judgment. It is often difficult to find property of a defendant. In some states, defendants can be compelled to answer questions about their assets.

Even if assets can be found, exemption laws may mean that certain assets may not be taken by the plaintiff to satisfy the judgment. Legislatures declare certain property exempt from seizure. Statutes usually contain a long list of items of property that cannot be seized by the sheriff to satisfy a judgment. Such items are considered necessary to basic existence and include automobiles, personal clothing, television sets, tools of trade, the family Bible, and other items that vary depending upon the state and the time in which the exemption laws were passed.

Homestead laws exempt the house in which the defendant lives from execution to satisfy judgments. Financial misfortunes notwithstanding, a debtor and those dependent on the debtor should have shelter. Sometimes homestead rights are limited to a certain amount. Under certain circumstances, a homestead can be ordered sold.

In many cases, judgments are not satisfied because the defendant has no property or the property in the defendant's possession is exempt. Collection of a judgment when the defendant does not pay voluntarily can be difficult, uncertain, and costly—another reason for settlement.

SECTION 2.14 Appeals: The Use of Precedent

The party who appeals the trial court's decision is called the appellant. The party seeking to uphold the trial court decision is called either the appellee or the respondent.

The attorney for each party will submit a printed brief to the appellate court. The court can permit a typewritten brief. The brief of the appellant seeks to persuade the appellate court that the trial judge has made errors, and the brief for the respondent seeks to persuade the appellate court to the contrary or that any errors that have been committed were not serious enough to warrant reversal.

Usually the attorneys make brief oral arguments before the appellate court. Sometimes these arguments consist of a summary of the briefs. Sometimes the attorneys are questioned by the judges on specific points that appear in the briefs or trouble the judge. Introduction of evidence is not permitted in appeals.

In the American system, most appellate judges write opinions stating their reasons for the way they have decided the case. Decisions vary in length. Usually, the legal basis for a decision in a trial court or appellate court is a statute, a regulation, or a prior case precedent. The latter requires amplification.

All decision makers seek guidance from the past. Nonjudicial decision makers, such as committees, boards of directors, and organizations of all types, often seek to determine what they have done in the past as a guide for what they should do in the present. Similarly, a refusal to make a particular decision may be based upon the fear that it will be looked upon as a precedent and the basis of a claim by others for similar treatment. However, in such decision making, it is not likely that the precedents of the past *must* be followed. English and American courts must follow precedent, subject to an exception described later in this section. Within a particular judicial system, judges must decide matters in accordance with earlier decisions of higher courts within the system.

For example, *all* judges in the state of California must follow the precedents set by the highest member of the system—the state supreme court. Trial judges and those who serve on the intermediate appellate courts must follow decisions of the intermediate appellate courts.

As stated, most appellate judges give reasons for their decisions in written opinions. All members of the system are expected to follow the reasoning as well as the results in earlier cases decided by a higher court. However, California judges are not compelled to follow decisions from other state courts, although they may look to decisions of other state courts for guidance. Nor are they compelled to follow the decisions of federal courts except the decisions of the U.S. Supreme Court.

Despite reverence for precedent, some European countries forbid following precedent, and some decision makers in this country, such as labor and commercial arbitrators, avoid it. This suggests that arguments can be made for and against following precedent.

A legal system should be reasonably predictable. To plan their activities, persons and organizations wish to know the law in advance. If they can feel confident that future judges will follow earlier decisions in similar cases, they should be able to predict and plan more efficiently. Knowing in advance what a judge or a court will do should encourage settlement of disputed matters. A fair judicial system should treat like cases alike.

Another argument for following precedent relates to conservation of judicial energy. If a matter has been thoroughly reviewed, analyzed, and reasoned in an earlier decision, there is no reason for a later court to replow the same ground. This reason for following precedent assumes great confidence in the decision makers of the past and a relatively static political, economic, and social order in which the system operates.

Those who attack the precedent system maintain that the sought-after certainty is illusory because courts can avoid precedent. They also argue that a precedent system tends to become rigid and unresponsive to changing needs. Finally, they contend that the precedent system can be an excuse for judges avoiding hard questions that should be examined.

Precedents must be followed only in similar cases. If the facts are different, precedents can be distinguished by judges and held not to control the current case. Even under a system of following precedents, there are times when earlier precedents become so outmoded that they are overruled. When American judges cannot accept the result that applying precedent compels, they create exceptions to the precedent. The greater degree of dissatisfaction, the greater the number of exceptions likely to be created. When the exceptions seem to swallow up the precedent, an activistic court may recognize the realities and overrule the precedent.

Good courts seek to avoid unthinking rigidity at one extreme and whimsical decision making at the other. The important feature of a precedent system is reasonable predictability that can accommodate change. The skillful attorney reads the precedents but also looks at the facts and any relevant political, social, and economic factors that may bear

upon the likelihood that the precedent will be distinguished or overruled.

Usually all the members of the appellate court agree. Sometimes a dissenting judge gives reasons for not agreeing with the majority. Sometimes the majority of the court that agrees on the disposition of the case cannot agree on the reasons for the decision, and one or more majority judges may write a concurring opinion.

SECTION 2.15   International Contracts

Increased worldwide competition exists for building and engineering contracts. As a result, those who design and construct may be engaged by a foreign national, often the sovereign itself, in a foreign country where the work is to be performed. These international contracts raise special problems discussed in Sections 8.09 and 30.21. One topic deserves mention here, however.

Legal systems vary not only as to substantive law but also as to the independence of their judiciary and dispute resolution processes. In some countries, the judiciary is simply an arm of the state that follows to a large degree the will of the head of state or the agency with whom the contract has been made.

A foreigner may not have confidence in the impartiality of such a dispute resolution process. This can even be a problem in the United States or within a particular state. The U.S. Constitution recognized this by allowing the removal of a case from a state court to the federal court if there is diversity of citizenship. This problem can be particularly difficult when the designer or contractor is a company from an industrialized country doing business in a less developed country.

Lack of confidence in the judiciary often necessitates a contract clause under which disputes will be resolved by some international arbitration process. Even where there is more confidence in the independence of the judiciary, such as in a transaction where disputes would normally be held before a court in the United States or a Western European country, unfamiliarity with the processes and the possibility of the application of unfamiliar law may lead to contractual provisions dealing with disputes. For example, European legal systems do not use the adversary system. They use a more inquisitorial system under which the judge is likely to be a professional civil servant who plays a more active role in resolving disputes. Similarly, laws in Western European countries are often based on brief yet comprehensive civil codes quite different from those an American lawyer may encounter in domestic practice. Under such conditions, it is common for the parties to provide by contract that disputes will be resolved by international arbitration with the applicable law of a neutral and respected legal system.

Forms of Association: Organizing to Accomplish Objectives

SECTION 3.01 Relevance

People engaged in design or construction should have a basic understanding of the ways in which individuals associate to accomplish particular objectives. The professional in private practice should know the basic elements of those forms of associations professionally available. Anyone who becomes an employee or executive of a large organization should understand the basic legal structure of that organization. The design professional who deals with a corporate contractor on a project should understand corporate organization. Those who are shareholders in or transact business with a corporation should understand the concept that insulates shareholders from almost all liabilities of the corporation.

The following are the most important organizational forms:

1. sole proprietorship
2. partnership
3. corporation
4. joint venture
5. unincorporated association
6. loose association

SECTION 3.02 Sole Proprietorship

The sole proprietorship, although not a form of association as such, is the logical business organization with which to begin the discussion. It is the simplest form and is the form used by many private practicing design professionals.

The creation and operation of the sole proprietorship is informal. By the nature of the business, the sole proprietor need not make arrangements with anyone else for the operation of the business. Generally, no state regulations apply except for those requiring registration of fictitious names or for having a license in certain businesses or professions. Sole proprietors need not maintain records relating to the operation of the business except those necessary for tax purposes. Sole proprietors have complete control over the operation of the business, taking the profits and absorbing the losses. They rent or buy space for the operation of the business, hire employees, and may buy or rent personal property used in the business operation.

The proprietorship continues until abandonment or death of the sole proprietor. Continuity of operation in the event of death sometimes can be achieved through a direction in the proprietor's will that an executor continue the business until it can be taken over by the person to whom it is sold or given by will. This ensures that the business is a continuing operation during the handling of the estate and that there is no costly hiatus in operation.

The proprietor may hold title to property in his own name or in any fictitious name. Interest in the business can be transferred, the only exception being that in some states the spouse of the proprietor may have an interest in certain types of property used in the enterprise.

Capital must be raised by the proprietor or by obtaining someone to guarantee the indebtedness. This differs from a corporation, which can issue shares as a method of raising capital.

SECTION 3.03 Partnership

A. Generally: Uniform Partnership Act

A partnership is an association of two or more persons to carry on a business for profit as co-owners. Unlike a corporation, it is not a legal entity. All states except Louisiana have adopted the Uniform Partnership Act, which sets forth the rights and duties between the partners themselves if the partners have not specified to the contrary in the partnership agreement. It also deals with claims of third parties.

B. Creation

Although the partnership agreement need not be evidenced by a sufficient written memorandum, it is advisable that it be. In addition to reasons for having any agreement in writing, other reasons exist for expressing partnership agreements in writing. The transfer of an interest in land must be evidenced by a written memorandum, as must agreements that by their terms cannot be performed within a year. If the creation of the partnership is accompanied by a transfer of land, that portion of the partnership agreement dealing with the land should be evidenced by a written memorandum.

If the duration of the partnership expressly exceeds a year, the partnership agreement should be evidenced by a written memorandum. Agreements to answer for the debts of another must, under certain circumstances, be evidenced by a written memorandum. Partnership agreements with provisions of this type should be evidenced by a written memorandum.

In stating that certain agreements must be evidenced by a written memorandum, a distinction is frequently misunderstood. Suppose a partnership is created orally for a period of more than a year. The rights and duties of the partners will still be governed by the oral agreement and by the Uniform Partnership Act. However, partners have no legally enforceable obligation to perform in the future, although they will have to perform those obligations that accrued before the decision to terminate the oral partnership.

Third parties will still have whatever rights the law would give them against the partners despite the requirement that the partnership agreement be evidenced by a written memorandum. The fact that the law required the partnership agreement to be so evidenced because it was to last over a year will have a limited impact on the partners and, in most cases, no impact on third parties.

C. Operation

Generally, the partners decide who is to exercise control. Most matters can be decided by majority vote. However, certain important matters, such as amendment of the partnership agreement, must be by unanimous vote. Sometimes partners wish to have something other than the majority rule apply. In a large partnership, a small group of partners may be given the authority to decide certain matters without requiring that a majority of the partners approve such decisions.

D. Fiduciary Duties

Partnership is an important illustration of the fiduciary relationship. As fiduciaries, each partner must be able to trust and have confidence in the other. Neither must take advantage of this trust for selfish reasons or in any way betray the partnership. For example, a partner must account to the partnership for any secret profits made. A partner must not harm the partnership because of any undisclosed conflict of interest. For example, where a partnership composed of design professionals represents the owner in dealing with a construction company, it would be improper for one of the partners to have a financial interest in the construction company unless it was disclosed in advance to the partners and to the owner-principal.

Because the fiduciary obligation is of great importance, it may be helpful in the partnership agreement to spell out the permissible scope of outside activities of individual partners. In the absence of an understanding among the partners as to permissible outside activities, any activity that raises a conflict of interest or competes in any way with the partnership would not be proper. For further discussion, see Section 11.04(B).

E. Performance Obligations, Profits, Losses, Withdrawal of Capital and Interest

In the absence of any agreement to the contrary, the partners are to devote full time to the operation of the business and share profits and losses equally even if one works more than the others. If a specified proportion of the profits is allocated by partnership contract to specified partners, that same percentage will apply to losses. If some arrangement other than this is desired, it must be expressed in the partnership agreement.

Normally, partners cannot withdraw capital during the life of the partnership unless permitted by the partnership agreement. A partner may collect interest for money or property lent to the partnership as well as for capital contributions advanced upon request of the partnership.

F. Authority of Partner

Section 9(1) of the Uniform Partnership Act states:

> Every partner is an agent of the partnership for the purpose of its business, and the act of every partner, including the execution in the partnership name of any instrument, for apparently carrying on in the usual way the business of the partnership of which he is a member binds the partnership, unless the partner so acting has in fact no authority to act for the partnership in the particular matter, and the person with whom he is dealing has knowledge of the fact that he has no authority.

Clearly, the authorized acts of the partner will bind the partnership. The more difficult problems arise when the partner's acts are not authorized.

Here the doctrine of apparent authority can charge the partnership with unauthorized acts by a partner. The partnership can create apparent authority by making it appear to third parties that a partner has authority he does not actually possess (see Section 4.06). Certain extraordinary acts, such as criminal or other illegal acts, are not charged to the partnership. However, because of the vast range of authority—both actual and apparent—given to the partners and the fiduciary obligation owed by each partner to the other, the character

and integrity of prospective partners are of great importance.

G. Liability of General Partnership and Individual Partners

The liability of the partnership and the partners for debts of the partnership and individual debts of the partners is complicated and confusing. Clearly, creditors of the partnership can look to specific partnership property and, if this is insufficient, to the property of the individual partners. Creditors of individual partners who obtain a court judgment may satisfy the judgment out of the partner's interest in the partnership property, but the partnership can avoid losing the property by paying the judgment.

Even more complicated is the liability of an incoming partner for obligations created before becoming a partner and of an outgoing partner for obligations incurred after leaving the partnership. Generally, the partner is not liable for obligations incurred before becoming a partner. Partners can eliminate liability for obligations that occur after leaving the partnership by informing those who have dealt with the partnership of their departure. But because of the uncertainty of the law in this area, it is best for those who enter existing partnerships to determine what obligations exist at the time they enter the partnership and those who leave partnerships to notify those who have dealt with the partnership of their departure.

Because of the ease in which partnerships can be dissolved and reconstituted, considerable confusion exists as to the right of creditors of predecessor partnerships to hold successor partnerships for the obligations of the predecessor. Sections 17 and 41 of the Uniform Partnership Act should be consulted.

A partner who incurs liability or pays more than a proper share of a debt should receive contribution from the other partners. Similarly, a partner who incurs liability or pays a partnership debt should be indemnified by the partnership. Usually these are expressly dealt with in the partnership agreement.

H. Transferability of Partnership Agreement

A partner has some interests that are transferable. Profits can be assigned, but not specific partnership property or management and control.

I. Dissolution and Winding Up

Partnerships do not have the legal stability of corporations. Section 31 of the Uniform Partnership Act lists a formidable number of events that will dissolve the partnership. Section 32 lists an equally formidable number of events that will give a court the power to dissolve a partnership on application of a partner or a purchaser of a partner's interest. Sections 37–42 are rules for winding up a partnership.

J. Limited Partnership

Most states permit the creation of a limited partnership made up of one or more general partners and one or more limited partners. Such an organization permits investors to receive profits but limits their liability to their investment without having to use the corporate form. Generally, the general partner manages or controls the corporation while the limited partners are investors.

Most states adopted the Uniform Limited Partnership Act of 1916. Most of those states adopted a new version, the Uniform Act of 1976, with modifications made in 1985. Currently all states but Louisiana have adopted one of the two versions of the Uniform Act, with many states modifying the act adopted. The details of such statutes are beyond the scope of this treatise.

SECTION 3.04 Profit Corporation

A. Use

The corporate form is used by most large and medium-sized businesses in the United States and has become the vehicle by which many small businesses are conducted. In addition, practicing design professionals are increasingly choosing the corporate form, where possible, for their business organization. The corporate form takes on even greater significance in light of the increasing number of design professionals who are employed by corporations. Because many complexities exist in corporation law that cannot be discussed in this treatise, the discussion here must be brief and simple.

B. General Attributes

Although the partnership is merely an aggregate of individuals who join together for a specific purpose, the corporation is itself a legal entity. It exists as a legal person. It can take hold and convey property and sue or be sued in its corporate name. As shall be seen, the other important corporation attributes are centralization of management in the board of directors, free transferability of interests, and perpetual duration. In addition, the corporation offers the advantage of limiting shareholder liability for the debts of the corporation to the extent of the obligation to pay for the corporate shares purchased. Although not all of these attributes are available to all types of corporations, as a general introductory statement, these attributes set the corporation apart from sole proprietorship and partnership.

C. Preincorporation Problems: Promoters

Often persons called promoters set into motion the creation of a corporation. Although they may continue in control, sometimes they merely organize the corporation and turn control over to others. Usually they are compensated by the corporation after the corporation has been organized. This compensation may be cash, shares of stock, stock options, or positions within the corporation.

During the promotion phase, promoters often make contracts with third parties, who should consider the possibility that the corporation will not be formed or will not adopt the contract made by the promoter. Consideration of such possibilities may induce the third party to insist that the promoter be held individually liable if the corporation is not formed or does not adopt the promoter's contracts. A third party who has doubts about the finances of the prospective corporation may wish to hold the promoter liable even if the contract is adopted by the corporation.

D. Share Ownership

The corporation is a separate legal entity owned by the shareholders, who do not own any part of specific corporation property. Shareholders have a right against the corporation that is governed by statutes, the articles of incorporation, and the wording of the shares. One of the great strengths of modern corporation law is the variety of types of shares that can be employed by the corporation—preferred and common, par and no-par, and voting and nonvoting.

Ordinarily, shares of stock in a corporation are freely transferable (one of the major advantages of incorporation). Usually restrictions exist on transferability of shares in closely held corporations whose shares are not sold to the general public. This enables the shareholders to keep control of the company by eliminating the possibility of outsiders becoming shareholders. Restrictions on transferability sometimes accompany shares purchased by executives or employees as part of a stock option plan.

E. Piercing the Corporate Veil

One of the principal functions of the corporation is to shield shareholders from corporate obligations. Ordinarily, the liability of shareholders is limited to paying for shares they have purchased. This limitation of liability can operate unfairly where a third party relies on what appears to be a solvent corporation. The corporation thought to be solvent may be merely a shell. The assets of the corporation may not be sufficient to pay the corporate obligations. Someone injured by the acts or failure to act of a corporation may find that the corporation is unable to pay for damages because the amount of capital paid into the corporation was very small.

The corporate form can be disregarded and, to use a picturesque phrase, the corporate veil pierced and shareholders held liable. A court may do so if unjust or undesirable consequences would result by interposing the corporation as an entity between the injured party (or a creditor) and the shareholders. This is more likely to be done where a person has suffered physical harm through corporate activities; where a creditor could not reasonably have been expected to check on the credit of the corporation; and where the one-person, family, or closely held corporation is used. The latter types of corporations have been popular because they combine control with limitation of liability. Ordinarily, such corporations are valid and protect the shareholders from liability. However, this is so only if the corporation is used for legitimate purposes and the business is conducted on a corporate basis. Also, the enterprise must be established on an adequate financial basis so that the corporation is able to respond to a substantial degree for its obligations. If not, circumstances such as those mentioned can result in piercing the corporate veil. Similar problems can arise when a corporation organizes a subsidiary corporation and holds all the shares of the subsidiary.

F. Activities, Management, and Control

The state law under which the corporation was created determines the outer limits of corporate activities and organization. Because state laws generally extend considerable latitude to the corporation on such matters, as a rule, activities and organization are governed by the articles of incorporation.

The articles of incorporation are the constitution of the corporation. They generally set forth the permissible activities of the corporation, the organization and management of the corporation, and the rights of the shareholders. Sometimes these matters are phrased in general terms in the articles of incorporation and articulated more specifically in the corporate bylaws. In large corporations, a set of corporate documents is likely to exist that delineates the chain of command and states the authority of corporate officers and employees to handle particular corporate matters. The ultimate power within the corporation lies with the shareholders, who delegate this power to an elected board of directors. In theory, the board of directors controls long-term corporate policies, while the day-to-day operations are handled by the corporate officers.

This model of corporate control may vary depending on the type of corporation. In a smaller corporation, the board of directors, or even large shareholders, may exert an influence on day-to-day operations. If the corporation is a large, publicly held corporation with thousands of shareholders, the actual power is largely with the management. Even though theoretically the shareholders can displace the board of directors, the diffusion of share ownership often makes it difficult for this to be done and gives the board of directors and the officers of the corporation effective power, with the shareholders having little control over policy or corporate acts. However, this does not make the corporation immune from takeover bids by other corporations. In the 1980s, there were many well-publicized takeovers by soliciting shares from shareholders at above-market prices. Increasingly, investors with large blocks of shares, such as pension funds, are playing a more active role in the management of the corporation.

Some corporations have cumulative voting of shareholders for directors. A shareholder asked to choose a slate of seven members of the board may cast all seven votes for one candidate. Cumulative voting is required in some states and has the effect of protecting minority interests in the corporation.

The directors choose the officers of the corporation. Usually they will select a president, vice president, secretary, and treasurer. Larger corporations might also, through the board of directors, designate persons to serve as general counsel or controller or as other corporate officers.

The officers are in charge of the day-to-day running of the corporation. The larger the corporation, the more likely it is that many of the details will be delegated to other employees of the corporation.

There has been a marked development in American law toward extending fiduciary obligations of fair dealing to directors and to officers. Directors and officers owe each other fiduciary duties of fair dealing and a fiduciary duty to the shareholders. It is not proper, for example, for a member of the board or for an officer to use inside information to purchase shares of corporate stock from shareholders who are not aware of the inside information known by the director or officer. Directors and officers are not permitted to take advantage of economic opportunities that should be made available to the corporation.

The articles of incorporation usually provide for annual shareholder meetings as well as for periodic meetings of the board of directors. Minutes must be kept of all board proceedings. The administrative burden of operating a corporation can be formidable.

G. Profits and Losses

Usually the board of directors determines the disposition or distribution of profits made by the corporation. Sometimes control is limited by the articles of incorporation, especially the rights of preferred shareholders. In addition, the corporation may have obligations to the creditors of the corporation based on contracts or on agreements made between the corporation and lenders such as shareholders or banks. Profits may be reinvested in the corporation and used for corporate purposes.

Subject to statutory regulations, the board determines when (and how large) a share of the profits is to be paid as dividends to the shareholders. The board may pay dividends only out of certain specified funds. Sometimes the board issues a stock dividend instead of a cash dividend. In such cases, the shareholders receive additional shares in the corporation instead of money. Statutory or other limitations may exist relating to the redemption of shares by the corporation and to the repurchase by the corporation of its own shares.

If the board unlawfully issues dividends, the board members are liable to the corporation and to those creditors of the corporation harmed by the unlawful declaration of dividends.

Normally, individual shareholders are not responsible for losses of the corporation because of the insulation from personal liability given corporate shareholders. A shareholder who paid for shares in accordance with the purchase agreement with the corporation is not liable for any obligations of the corporation.

H. Life of Corporation

One advantage of the corporation is its perpetual life. Sole proprietorships end with the death of the sole proprietor. Often partnerships end with the death of any of the partners. The corporation will continue despite the death of shareholders. A court can dissolve a corporation in the event of a deadlock among the board of directors or under other circumstances.

I. Dissolution

The dissolution of a corporation is complicated and is governed largely by statute. The bankruptcy laws play a large role in disposing of assets of the corporation upon dissolution. This treatise does not discuss dissolution.

SECTION 3.05　Nonprofit Corporation

Nonprofit corporations are similar in organization and operation to profit corporations. The major difference is that no profits can be distributed to shareholders by a nonprofit corporation. Examples of nonprofit corporations are hospitals, most educational institutions, and charities.

Capital for a nonprofit corporation is raised by donations, grants, and, occasionally, the sale of shares. Usually there are members instead of shareholders. The members elect the board of directors or trustees; the board selects officers to run the corporation. Nonprofits have articles of incorporation, bylaws, meetings, and other similarities to profit corporations. Generally, the shareholders are insulated from personal liability for the debts of the corporation. Nonprofit corporations are exempt from taxes because they have no profit. Nonprofit corporations engaging in certain types of political or profit-making activities can lose their tax exemptions.

SECTION 3.06 Professional Corporation

Although design professionals traditionally have practiced as sole proprietors or partnerships, increasingly, states have permitted them to practice through a professional corporation. The corporation form has been used mainly to take advantage of tax laws that allow employees of corporations to receive fringe benefits without having them included within taxable income.

Some states that have allowed professionals to incorporate have made it clear that they cannot use the corporate form to shield their individual assets, a principal reason for using business corporations.

Herkert v. Stauber[1] demonstrates the interrelationship of legal issues in construction disputes as well as the ambit of professional corporation statutes. The dispute involved a claim by an owner against a professional corporation that had agreed to design and build an apartment house for the elderly. The owner claimed that the contract required the professional corporation to obtain approval of a federal agency that would assist in funding the project. One of the issues involved the extent to which individual shareholders in the professional corporation were liable for damages caused by breach of this obligation by the corporation.

The applicable professional incorporation law did not allow the corporate form to limit liability for negligent performance of professional services. Did failure to obtain funding from a federal agency constitute negligent performance of *professional* services?

The court did not have to determine whether the breach was negligent because it concluded, citing an earlier edition of this treatise, that obtaining financing for construction is not a professional service. It limited professional architectural services to *design* and *supervision*. In addition to pointing to the absence of any professional training in this field, it pointed to the absence of standards to determine whether failure to obtain financing was negligent. As a result, the shareholders were insulated from liability.

SECTION 3.07 Joint Venture

Joint ventures can be created by two or more separate entities who associate together, usually to engage in one specific project or transaction. Such arrangements are contractual and can be expressed by written contracts or implied by acts. In large construction projects, two contractors may find that they cannot handle a particular project individually but can do so if they associate themselves in a joint venture. Another type of joint venture can be created by an organization that performs design and an organization that constructs. Sometimes joint ventures are needed to bid on "design-build" projects.[2]

Usually the agreement under which such a joint venture is created is complex and sets forth in detail the rights and duties of the joint venturers. Joint ventures created informally by acts have gaps as to specific rights and duties that are filled in by the law. When such gaps exist, it is likely that principles of partnership will be applied. Refer to Section 3.03.

Suppose there is a dispute as to the existence of a joint venture. Because joint ventures are considered very much like partnerships for one transaction, the relationship is often examined to determine whether partnershiplike attributes are present. For example, sometimes it is stated that there must be a community of interest, a common proprietory interest in the subject matter, the right of each to govern policy, and a sharing of profits and losses.

SECTION 3.08 Unincorporated Association

Individuals sometimes band together to accomplish a collective objective without using any of the forms of association described thus far, such as

[1]106 Wis.2d 545, 317 N.W.2d 834 (1982).

[2]See Section 17.04(F).

partnerships or corporations. Instead they may organize an unincorporated association, such as a fraternal lodge, a social club, a labor union, or a church. Design professionals may perform professional services for these associations or become involved in the associations themselves.

Generally, unincorporated associations are not legal entities. For this reason, early American law did not allow them to hold property in the association name, make contracts, sue in the name of the association, or be sued as a group. They were merely a group of individuals who banded together to accomplish a particular purpose.

In most states, statutes have removed many of the former procedural disabilities. Although they are still not entities, unincorporated associations are often permitted to contract, to hold property, and to sue or be sued in the name of the association.

Usually such groups have constitutions, bylaws, and other group-related rules that govern the rights and duties of the members. They elect officers who have specified authority, such as to hire employees and run the activities of the association. All the members generally are responsible for those contracts entered into by the officers who were authorized by the members. Liability rests on agency principles.[3]

Establishing the authority of the officers is often difficult because formalities are often disregarded in these organizations. It may be necessary to show evidence of which members voted for resolutions authorizing the officers to make a particular contract. Although the officers may have certain inherent authority by virtue of their positions, important projects, such as engaging a design professional or a contractor, usually are beyond the scope of their inherent authority.

Officers who make the contract may be individually liable if the contract was clearly made by them as individuals. It may be possible to hold the officers for having misrepresented their authority if the contract was not authorized. Because of the potential risk to officers and to members, many contracts contain provisions that limit liability to certain designated property held in trust for the association in those states where they are not permitted to own property in the association name. Where the association is permitted to hold property in its own name, liability is often limited to that property.

SECTION 3.09 Loose Association: Share-Office Arrangement

Sometimes design professionals use the term *association* to describe an arrangement under which they are independent sole proprietors but join together to share offices, equipment, and clerical help. They may also do work for each other. Sometimes such loose associations are known as share-office arrangements.

During the sharing arrangement, several legal problems can develop. First, suppose one associate performs services for another and the latter is not paid by his client. In the absence of any specific agreement or well-documented understanding dealing with this risk, the associate who performs services at the request of another should be paid.

Second, suppose the associate performing services at the other associate's request does not perform properly and causes a loss to the client. The client would have a claim against the associate who did the work and the associate with whom it dealt. If the latter settled the claim or paid a court judgment, he would have a valid claim for indemnification against the associate who did not perform properly.

Third, if those in the share-office arrangement create an impression to the outside world that they are a partnership, they will be treated as such. For example, if the letterhead, building directory, and telephone directory list them as Smith, Brown and Jones, such listings may create an apparent partnership. If so, each member of the association can create partnershiplike contracts and tort obligations.

When a loose association terminates or one participant withdraws, all associates in the first case and the withdrawing associate in the case of withdrawal can freely compete with former associates or those remaining. However, the associates in the first case and the withdrawing associate in the second cannot do any of the following:

1. Take association records that belong to another associate.
2. Represent that they still are associated with former associates if this is not the case.

---

[3]See Sections 4.05 and 4.06.

3. Wrongfully interfere with contractual or stable economic relationships existing at the time the association ended or one associate withdrew.

SECTION 3.10 Professional Associations

Professional associations of participants in the Construction Process, such as the American Institute of Architects (AIA), the National Society of Professional Engineers (NSPE), and the Associated General Contractors (AGC), have had a substantial impact on design and construction. Although they are not organizations like the others mentioned in this chapter created for the purposes of engaging in design and construction, some mention should be made of their activity and some of the legal restraints upon them.

These professional associations engage in many activities. They speak for their members and the professions or industries associated with them. In addition, they seek to educate their members in matters that relate to their activities. One of the most important activities is publishing standard forms for design and construction that used not only as the basis for contracts for design and construction but also to implement the construction administrative process. It must be kept in mind that membership in these professional associations is not required for one to design or build.

By and large, professional associations can choose their members. However, as these associations become more important, the law is beginning to set limits on their power to determine who will be admitted to membership[4] and what the legitimate grounds are for discipline. Restraints placed on professional associations have related to attempts by public authorities to encourage competition in the professional marketplace. This has resulted in attacks by public officials on what were called the rules of ethical conduct that determine whether an applicant would be admitted to the association and can be disciplined for violation of ethical rules of the association.

Often ethical rules limit the power of one professional to supplant another and bar or discipline members, based on "ungentlemanly conduct such as competing with a fellow member on the basis of price." Such disciplinary action has been found to violate agreements between an association and federal antitrust officials that the association would not limit competition.[5]

---

[4]*Pinsker v. Pac. Coast Soc'y of Orthodontists*, 12 Cal.3d 541, 526 P.2d 253, 116 Cal.Rptr. 245 (1974).
[5]*United States v. Am Soc'y of Civil Eng'rs*, 446 F.Supp. 803 (S.D.N.Y.1977).

PROBLEMS

1. A and B were partners in an architectural firm. B's brother C was a struggling young architect, hardly able to pay his bills. B was approached by a prospective client to build a 150,000-dollar residence at a 15,000-dollar commission. B suggested that the client go to C because C needed the business much more than A and B. A is unhappy with this. Does he have any legal recourse against B?

2. Smith and Jones were young architects who decided to share expenses and office space. Smith was wealthy, while Jones was barely able to make ends meet. Their offices were in the Atlas Building, and the door was painted with the inscription "Smith and Jones, Architects." This was also the listing in the telephone directory and in the office directory of the Atlas Building.

Actually, each had his own clients, and neither did any work for the other. Together they hired a secretary, who also acted as a bookkeeper. Each income item was allocated to the person who had performed the services, and expenses were allocated based on who used the services or supplies. Each had a separate checking account, and there was a third account in the name of Smith and Jones. The latter account was used to pay for the rent, the cost of the secretary, and other office expenses.

On occasion, each would consult the other with regard to professional design matters. It was not uncommon for one to call in the other while a client was present to get informal advice on a design question.

a. Jones ordered a computer that cost $20,000. The order was placed in the name of Smith and Jones. Jones did not pay for the machine, and the seller claims that he has a legal right to collect from

Smith as well as from Jones. Do you agree? Give your reasons. Would your answer depend in any way on whether the computer company knew of Smith's wealth and supplied the computer based on his good credit?

b. Jones designed an exclusive residence for a client. His negligence in design required that certain work be redone at an expense of $10,000 to the client. The client's lawyer ascertained that Jones had no assets or professional liability insurance and that any judgment against him would be uncollectible.

The client's lawyer now asserts that he has the right to recover from Smith because Jones and Smith appear to the world as a partnership. Is he correct? Would your conclusion be changed or reinforced if it could be shown that during a conference between Jones and his client, Smith was called in to offer some suggestions relating to design? (Assume that the suggestions did not relate to matters that ultimately were the basis for the claim and that Smith was not negligent in any way.)

C H A P T E R   F O U R

The Agency Relationship: A Legal Concept Essential to Contract Making

SECTION 4.01 Relevance

Agency rules and their application determine when the acts of one person bind another. In the typical agency problem, there is a principal, an agent, and a third party. The agent is the person whose acts are asserted by the third party to bind the principal. There can be legal problems relating to the rights and duties of the principal and agent as between themselves as well as disputes between the agent and the third party. But because the third party v. principal part of the agency triangle is most important—and most troublesome—these problems will be used to demonstrate the relevance of agency law.

The design professional, whether in private practice or working as an employee, may be in any of the three positions of the agency triangle.

For example, suppose the design professional is a principal (partner) of a large office. The office manager orders an expensive computer. In this illustration, the design professional, as a partner, is a principal who may be responsible for the acts of the agent office manager.

Suppose the design professional is retained by an owner to design a large structure. The design professional is also engaged to perform certain functions on behalf of the owner in the construction process itself. Suppose the design professional orders certain changes in the work that will increase the cost of the project. A dispute may arise between the owner and the contractor relating to the power of the design professional to bind the owner. Here the design professional falls into the agent category, and the issue is the extent of her authority.

Suppose X approaches a design professional in private practice regarding a commission to design a structure. X states that she is the vice president of T Corporation. The design professional and X come to an agreement. Does the design professional have a contract with T Corporation? Here the design professional is in the position of the third party in the agency triangle.

The agency concept is basic to understanding the different forms by which persons conduct their business affairs, such as partnerships and corporations.

SECTION 4.02 Policies Behind Agency Concept

A. Commercial Efficiency and Protection of Reasonable Expectations

As the commercial economy expanded beyond simple person-to-person dealings, commercial necessity required that persons be able to act through others. Principals needed to employ agents with whom third parties would deal. Third parties will deal with an agent if they feel assured that they can look to the principal. The agent may be a person of doubtful financial responsibility. The concept of agency filled the need for giving third persons some assurance that they can hold the principal.

Agency exposes the principal to risks. The principal may be liable for an unauthorized commitment made by the agent. Suppose the principal authorizes its agent to make purchases of up to $1,000 but the agent orders $5,000 worth of goods. From the third party's standpoint, this 5,000-dollar purchase may be reasonable in light of the position of the agent or what the latter had been ordering in the past. The law protects the principal from unauthorized commitments but not at the expense of reasonable expectations of the third person. That problems such as this can develop did not destroy the unquestioned usefulness of the agency concept. Such problems required the law to create rules and solutions to handle such questions that would be in accord with commercial necessity and common sense.

B. Relationships Between Principal and Agent

As a rule, the relationship of principal and agent is created by a contract expressed in words or manifested by acts. The agent may be a regular employee of the principal or an independent person hired for a specific purpose and not controlled as to the details of requested activities. Several aspects of the agency relationship make it different from the ordinary commercial, arm's-length relationship.

An arm's-length transaction is one whereby the parties are expected to protect themselves. In such a transaction, no general duty is imposed on one party to protect the other party, nor is any duty imposed to disclose essential facts to the other party. Although there are some exceptions, generally, commercial dealings are at arm's length. On the other hand, principal and agent have a fiduciary relationship—one of trust and loyalty.[1] In such a relationship, one person relies on the integrity and fidelity of the other. The latter must not take unfair advantage of the trust in her by benefiting at the expense of the other.

Often the arrangement under which an agent performs services is sketchy and does not specifically delineate all the rights and duties of principal and agent. In such a case, terms often must be implied by law.

SECTION 4.03 Other Related Legal Concepts

Confusion often occurs because of the overlapping nature of terms such as *principal-agent*, *master-servant*, and *employee-employer*. This chapter deals

only with the principal-agent relationship and the power of agency under which one person may bind another to a contractual obligation.

SECTION 4.04 Creation of Agency Relationship

Generally, agency relationships are created by a manifestation by the principal to the agent that the agent can act on the principal's behalf and by some manifestation by the agent to the principal that she will so act. However, these elements of consent are often informal. They need not be in writing, except in cases where statutes require that the agent's authority be expressed in writing if the transaction to be consummated by the agent is one that would also require a writing.

SECTION 4.05 Actual Authority

The agent is ordinarily authorized to do only what it is reasonable to believe the principal wants done. In determining this, the agent must look at the surrounding facts and circumstances. If, for example, the principal has authorized the agent to purchase raw materials to be used in a particular manufacturing process and the agent learns that the principal has decided not to proceed with the project, it is unreasonable for the agent to believe that the authority to buy the materials still exists.

The agent must consider the situation of the principal, the general usages of the business and trade, the object the principal wishes to accomplish, and any other surrounding facts and circumstances that would reasonably lead to a belief that the agent can or cannot do something. This is largely a matter of common sense. Sometimes the agent is given specific authority to do those incidental acts that are reasonably necessary to accomplish the primary act. For example, when an agent is given authority to purchase a car for the principal, it is likely that the agent also has authority to buy liability insurance for the principal. The agent, when possible, should seek authorization from the principal to perform those acts that are not expressly authorized.

Sometimes emergencies arise that make it difficult or impracticable for the agent to communicate with the principal. In such cases, the agent is given authority to do those necessary acts to prevent loss

---

[1]See Section 11.04(B).

to the principal with respect to the interests committed to the agent's charge.

SECTION 4.06 Apparent Authority: *Frank Sullivan Co. v. Midwest Sheet Metal Works*

Most difficult principal-agency cases arise under the doctrine of apparent authority, which exists when the principal's conduct reasonably leads a third party to believe that the principal consents to acts done on its behalf by the person purporting to act for it.

It may be useful to draw some initial distinctions essential to an understanding of apparent authority problems. First, suppose the asserted agent neither is an employee of the principal nor has been hired by the principal to perform any specific task. This can be illustrated by *Amritt v. Paragon Homes, Inc.*[2] The plaintiff homeowner sued for damages caused by faulty construction of a home built for her by Romero, a contractor. Clearly, Romero was liable but evidently could not pay a court judgment. The homeowner sued Paragon, the manufacturer of the prefabricated home erected by Romero, and Sewer, the manufacturer's local distributor.

The homeowner had arranged through Sewer to purchase the materials, plans, and drawings for such a home. Evidently, Sewer represented himself to the homeowner as the agent of Paragon and implied that Paragon would not be responsible unless the homeowner used Romero to construct the home. The agreement was on a form supplied by Paragon and filled in by Sewer. The homeowner borrowed money from and gave a mortgage to Paragon for the construction costs. The homeowner made payments by depositing funds in escrow with Paragon, and Paragon demanded a completion certificate before disbursing the funds. Sewer inspected the construction work for Paragon.

Paragon, Sewer, and Romero were all separate legal entities. Romero worked for himself, as did Sewer. The court held, however, that Sewer and Romero were agents for Paragon because both had "the outward trappings of apparent authority to be

Paragon's agents."[3] The court concluded that the homeowner was "reasonably led to believe they were Paragon agents."[4]

The court, quoting the Restatement of the Law of Agency,[5] pointed to Paragon's conduct—supplying all materials and forms through Sewer, requiring that the house be inspected by Sewer and built to its plans, and disbursing all construction payments—and stated that the combination of these activities could reasonably have led the homeowner to believe Paragon had authorized Sewer and Romero to be its agents. Although the court spoke in terms of apparent authority, it actually found an apparent agency.

Because of the many legal entities involved in a construction project, one party not infrequently seeks to bind an entity other than the one with whom it has dealt on the same theory as that used in *Amritt v. Paragon Homes, Inc.* This is especially likely when consumers find that the persons with whom they have dealt are unable to respond when those persons have not performed in accordance with their legal obligation. The law is more likely to permit a claimant, as in the *Amritt* decision, to recover from the only solvent defendant that can be connected to the transaction if that solvent party has organized the transaction, controlled it, and sought to eliminate responsibility for a risk legitimately a part of its enterprise by the use of a separate legal entity that lacks the financial capability to respond for product defects. The conclusion reached in the *Amritt* decision is also more justified if the party dealing with the contractor is a consumer of limited commercial experience.

A second problem can stem from a client's engaging a professional person. For example, the owner may engage a design professional to perform services relating to design and contract ad-

---

[2]474 F.2d 1251 (3d Cir.1973).

[3]Id. at 1252.
[4]Ibid.
[5]The Restatements of the Law are collections of rules, comments, and illustrations about a particular legal subject. They are published by the American Law Institute, a private organization made up of lawyers, judges, and legal scholars. The Institute's function is to collect case law from all the states and distill it into rules. Unless adopted by a legislature or followed by a court, it is *not* law. The principal Restatements for the purposes of this treatise are those of Torts, Contracts, and Agency.

ministration for a particular project. The latter is rarely an employee of the owner but has certain designated authority. This matter is treated more fully in Section 17.05(B).

The third and perhaps most difficult problem relates to whether an admitted employee of an employer has the authority to engage in certain acts and thereby to bind the employer. Here there is an agency relationship in that the agent has some authority to bind the principal. But the problems that develop relate to the extent of that authority.

The following case deals with apparent authority of an employee in the construction contract context.

FRANK SULLIVAN COMPANY v. MIDWEST SHEET METAL WORKS

United States Court of Appeals, Eighth Circuit, 1964. 335 F.2d 33.

BLACKMUN, Circuit Judge.

Midwest Sheet Metal Works, a Minnesota partnership, instituted this diversity suit against Frank Sullivan Company, a Boston contractor, to recover damages for breach of contract. The jury returned a verdict in favor of Midwest for $85,000.

. . . The controversy arises out of the project for the extension and remodeling of the United States Post Office and Customs House at Saint Paul. Minnesota law controls. Although Sullivan asserts twenty separate points on its appeal, these come down essentially to four primary issues:

. . . [first and second issues omitted]

3. The authority of Sullivan's agent to sign the agreement.

. . . [fourth issue omitted]

The prime contractor on the project was Electronic & Missile Facilities, Inc., of New York City (EMF). On December 4, 1961, EMF and Sullivan executed a lengthy and detailed subcontract whereby Sullivan undertook all plumbing, heating apparatus, air conditioning, ventilation work on the job for an agreed price of $1,650,000. This contract was executed on behalf of Sullivan by Francis J. Sullivan (Frank) as its president.

In the fall of 1961, before the formal execution of the agreement between EMF and Sullivan, Frank had contact with Michael J. Elnicky, the dominant partner of Midwest, about Midwest's taking on the sheet metal and air conditioning portion of Sullivan's subcontract. Sullivan had even invited a quotation from Midwest for this work. On November 1 Midwest quoted a figure in excess of a million dollars. Frank by telephone told Elnicky that this bid was about $200,000 too high. Elnicky indicated he might reduce his price somewhat but could not approach Sullivan's suggested figure. Frank testified that he then told Elnicky, "Well, look, we have got a fellow going out

there, and I will show you that these are not prices that we dreamed up, these are prices that we used in making our bid, and we got them confirmed by letters by reputable people."

Near the close of 1961 EMF told Sullivan that it was imperative that the work in Saint Paul be started. Sullivan promised EMF that it would get a superintendent, a foreman, and men and material on the job by the end of January. Sullivan sent John Sullivan (Jack) from Boston to Saint Paul in early January. On this trip he conferred with EMF's superintendent on the project. Jack was back in Minnesota later in the same month with Byers who was to be Sullivan's general superintendent on the job. Before he left Boston on this second trip Jack had been instructed by Frank to look over the labor situation in the area, to check in with prospective subcontractors, to see Elnicky and give him quotations which "will back up the reduction of his bid," and to "get the job started." Frank gave Jack the job estimates which had been prepared by Sullivan but he was not given and had not seen the prime contract.

Upon their arrival Byers called upon local union business agents, purchased material and tools, received other equipment from Boston, and placed four steamfitters on the job.

On Monday, January 22, Jack came to Midwest's office. This was the first time Elnicky saw him. Elnicky and some of his employees testified that Jack told him on this visit that he was "part of the [Sullivan] organization," and had "a piece of it." Jack denied that he made any such statement. Elnicky conceded that he made no attempt to check Jack's authority with the Sullivan home office and that he did not ask for written evidence of it.

Elnicky also testified that Jack early in this first meeting suggested that Midwest price off "the whole works"; in any event, Elnicky indicated that he was interested in taking over the entire Sullivan job. Jack did not object and said that he would try to get a copy of the prime contract for him. The two men met again on Tuesday when Jack permitted Elnicky and his people to review the bids Sullivan had received. By Wednesday Jack obtained a copy of the prime contract from EMF's office on the job. He gave it to Elnicky who kept it overnight. Meanwhile, Jack talked with other prospective subcontractors. Elnicky and Jack met further, and sometimes socially, during the same week. Jack told him that Sullivan would have to have a minimum of $100,000 if Elnicky took over. On Thursday Jack told Frank by telephone of the discussions he was having with Midwest. As to this conversation Frank testified that he told Jack that this could not be done, that EMF wanted Sullivan on the job, and that Jack should "pick up what you got and come home." Midwest finished its estimating on Saturday, January 27. That evening Jack was at Elnicky's home with Elnicky and two of the latter's men. They discussed costs and what Elnicky might offer to do the job but no conclusion was reached.

Early in the afternoon of the next day, Sunday, Elnicky came to Jack's hotel room. Jack was planning to return to Boston. Elnicky arrived with a fifth of Scotch. The men were together for three and one-half hours, discussing the job and drinking the entire fifth. Byers was present but left for a time to get the copy of the prime contract which had been left elsewhere. Elnicky made an offer of $1,550,000 to perform the work. This was discussed as was the question of what to do with the equipment and materials which Sullivan already had on the job. Jack then started to write something out. Elnicky dictated part of it. Several drafts were made. Later Jack dictated a draft to a hotel typist and he and Elnicky signed it. Elnicky then took Jack to the airport. The typed draft is Midwest's Exhibit 5* and is the document in controversy. It was admitted in evidence over Sullivan's objection.

The testimony as to the execution of the exhibit is in sharp conflict. Jack testified that he told Elnicky that this agreement was subject to approval by Frank and EMF, that there was no sense in working out other details until this was done, that it was his intention that Exhibit 5 be merely a proposal by Elnicky to Sullivan, and that he did not intend thereby to turn the Sullivan contract over to Midwest. Byers testified that Jack told Elnicky that it had to be approved by Frank and EMF; that Elnicky acknowledged this; and that he, Byers, had expressed a hope that Frank would approve it so he "could go home." Elnicky flatly denied that Jack had said Exhibit 5 was subject to approval by Frank and EMF.

The next day, Monday the 29th, Jack called Byers from Boston to see if Elnicky had someone on the job as the writing provided. Byers called Elnicky who told him he would have someone there on Tuesday. On Tuesday Jack and Elnicky conferred by telephone. Byers then sent Elnicky's men away from the job. On the same day, January 30th, Jack, signing on behalf of the Sullivan Company, wrote Midwest that "the agreement made on January 28th, 1962, between the Midwest Sheet Metal Company and the Frank Sullivan Company is hereby cancelled" and that they would try to arrange for Midwest to quote on the job's ventilating, air conditioning, and refrigeration. Jack testified that he wrote this letter without discussion with Frank. Midwest's receipt of that letter led to the present suit.

Jack Sullivan's status is obviously of vital importance. In January 1962 he was 29 years of age. He was a high school graduate and had had one year of "night college." He held a plumber's union card and a journeyman plumber's license. He had worked for Sullivan for ten years. He had started there as a plumber and, by 1959, was a job superintendent. About mid-1960 he became an "outside superintendent." In this capacity he traveled to various jobs, examined labor and material situations and reported back to Sullivan. He did no hiring or firing. He did not order materials. He did make recommendations. He was not an officer, director, or shareholder of Sullivan. He was not related to Frank. He had nothing to do with obtaining or negotiating subcontracts. He had had no experience in sheet metal or air conditioning. He had done no estimating. He had been given no

---

*   "Hotel Saint Paul
           St. Paul, Minnesota
           January 28, 1962
"The Midwest Sheet Metal Co.
340 Taft St.
Minneapolis, Minnesota
   "The Midwest Sheet Metal Company of 340 Taft St., Minneapolis, Minnesota, agrees to take over Frank Sullivan Company's contract with Electronic Missile Facilities, Inc., in the amount of One Million Five Hundred Fifty and 00/100 Dollars ($1,550,000).
   "The cost of the bond will be paid for by the Frank Sullivan Company.
   "The Midwest Sheet Metal Company will man the above mentioned project on January 29, 1962, to show good faith regarding this contract.

The above agreement is made between
                    Midwest Sheet Metal
                    Midwest Sheet Metal
                      M. J. Elnicky
                    Frank Sullivan Co.
                      John Sullivan"
The discrepancy between the words and the figures is readily apparent.

specific authority to sign any contract for Sullivan or to assign Sullivan's subcontract with EMF.

In January 1962 Elnicky was about 52. He had been in sheet metal and similar work for many years and had run his own business since 1947.

[The Court held] . . . that Exhibit 5 . . . was sufficiently clear and definite to be valid and to constitute an enforceable contract.

C. Jack's authority. Midwest does not contend that the record supports a finding that Jack possessed actual authority to act on behalf of Sullivan. The issue is one of apparent authority. Sullivan asserts that the evidence as a matter of law was insufficient to support a finding of apparent authority and that the court's submission of this issue to the jury and its rulings and instructions consistent therewith were prejudicially erroneous.

Jack certainly assumed the mantle and the posture of responsibility and authority. There is evidence to the effect that he professed an ownership interest in Sullivan, possessed and produced the bids and the cost estimates the company has assembled, permitted Elnicky to review them, obtained a copy of the EMF-Sullivan subcontract for Elnicky, mentioned to others than Elnicky his interest in contracting out the entire Sullivan portion of the job, demonstrated a permissive attitude toward Midwest's interest in taking on the full subcontract, was in contact with Elnicky's performance bond man and asked him to confirm his comments by letter, accepted Elnicky's entertainment favors, bargained continuously for a week, and even wrote the letter of cancellation.

But apparent authority must be founded on something more than the conduct and statements of the agent himself. Liability can be imposed upon a principal only "for that appearance of authority caused by himself." 2 Williston on Contracts (3d Ed.1960), § 277A, pp. 222–24; . . . Of course, a degree of reasonableness and of diligence is required of one who deals with the agent. . . .

We find in this record adequate support for the submission of the issue of apparent authority to the jury. Accepting the evidence, as we must, in the light most favorable to the prevailing plaintiff, we have, apart from and in addition to Jack's own acts and statements, all the following: (1) Sullivan sublet part of every job it had for there were certain types of work (air conditioning and sheet metal, for example) which it never performed; (2) Sullivan sent Jack to Saint Paul to get the job started; (3) Sullivan instructed Jack to get in touch with area people in the construction industry and to obtain subcontract offers; (4) Sullivan placed Jack in possession of the breakdown of costs it had prepared and of the bids it had received; (5) Frank told Elnicky that he had a man going out to

Minnesota; (6) Jack possessed the Sullivan name; (7) Jack was the highest person in authority in Sullivan's employ on the job; (8) so far as the Saint Paul job was concerned, Byers followed Jack's instructions; (9) Jack, while he was in Saint Paul the week of January 22, was in telephone communication with Boston, and, specifically, talked with Frank; and (10) Frank knew that Jack was discussing with Elnicky a complete takeover by Midwest and yet did nothing to disavow his status to Elnicky. And, for what it is worth, Jack was still with the Sullivan Company at the time of the trial.

Of course, there is an opposing factual argument, namely, that Sullivan had given Jack no instructions to turn over the entire job; that Jack was not supplied with a copy of the EMF-Sullivan subcontract and had to obtain it from the EMF man on the job; that the preliminary conversations between Sullivan and Elnicky had only to do with a limited area of work; that Elnicky knew who Frank was and was in communication with him; and that Elnicky did not inquire of Jack or of Frank as to Jack's authority. But this is just another argument for the trier of fact. The jury was not persuaded.

Sullivan places great emphasis on Elnicky's failure to make inquiry as to Jack's authority, and it urges the important nature of the contract as demonstrated by the amount involved and the time required for its performance. This argument, however, cuts both ways. If the job was so large and so important, a jury might properly infer that Sullivan's top man on the project was there with workable authority. We feel that the cases Sullivan cites in support of its argument are distinguishable on their facts. Hill v. James, supra, concerned a traveling salesman, a fact which the court stressed, and a questionably completed contract. The court readily recognized that a principal may clothe even such a salesman with apparent authority. Dispatch Printing Co. v. National Bank of Commerce, supra, concerned a Saint Paul newspaper's Minneapolis advertising solicitor-collector and his indorsing checks drawn in favor of the newspaper and depositing them in his own account. Mooney v. Jones . . . concerned a mortgagor's representation, claimed to be made on behalf of the mortgagee, to a contractor that the latter's lien waiver would not affect his lien rights for future work. Language in these opinions relative to an agent's actual authority and a third person's duty to inquire is accepted law but the Minnesota court has consistently recognized the established principles of apparent authority. Illustrative is the court's comment in Mooney itself . . . that "the record is insufficient to show that National Guardian did anything to hold Gustafson out as its agent". . . . And in Sauber v. Northland Ins. Co., supra, Chief Justice Knutson, in speaking for the court, said, "Apparent authority exists by virtue of conduct on the part of the principal which warrants a finding that a third party, acting in good faith, was justified in relying on the assumption that the agent had authority to act." [footnote omitted]

The situation here strikes us as one where, as the negotiations developed, Jack sensed the opportunity to bring his company out of the project at a convenient profit of $100,000, without additional cost or participation, and further sensed that this would be a feather in his cap if he could bring it about. Whether Sullivan sensed this, desired it, and encouraged it, we shall, of course, never know with positive assurance. But the record supports just such an inference by the jury. We are not at liberty to overturn that body's conclusion.

... Affirmed.

The *Sullivan* opinion reflects the differing roles of trial and appellate courts. It notes arguments and evidence for and against the conclusion that Elnicky could have reasonably believed that Jack Sullivan had authority to make the contract. A reasonable inference is that the appellate court or at least some judges of that court might have come to a different conclusion. But where the issue is factual—that is, what Elnicky might have reasonably believed—and where there is conflicting testimony and evidence at the trial, the appellate court is likely to affirm the judgment of the trial court.

As seen in the *Sullivan* decision, apparent authority can protect a party who has relied on appearances. However, a third party unsure of an agent's authority should check with the principal where feasible. The application of the apparent authority doctrine is uneven, and the doctrine's existence cannot justify carelessness by third parties when dealing with agents.

SECTION 4.07 Termination of Agency

If authority is conferred for a specified period, it will terminate at the end of that period. If no time is specified, authority continues for a reasonable time. If authority is limited to performing a specified act or to accomplishing a certain result, it terminates when the act or result is completed.

Sometimes the terms of the authorization specify that the authorization is to continue until a certain event occurs. If so, the occurrence of the event will terminate the agency unless the event would not come to the attention of the agent. Sometimes the loss or destruction of certain subject matter terminates the agency.

Sometimes the agency terminates when principal and agent consent to terminate the relationship. This can occur if either principal or agent manifests to the other that the relationship will no longer continue. In some cases, the agency can also be terminated by death of the principal, and almost always by death of the agent. However, agency relationships sometimes are created by contract for fixed terms. The principal may, despite a fixed-term contract, terminate the agency and terminate the power of the agent to bind it, but such a termination will be a breach of contract with the agent.

These events effect *actual* authority of the agent. If termination of authority has taken place but manifestations to third parties that the agent has authority still exist, the agent may be able to bind the principal by the doctrine of apparent authority, which will continue only so long as the third party should not realize that the agent no longer has authority to bind the principal.

SECTION 4.08 Disputes Between Principal and Third Party

Sometimes the principal *wishes* to be bound by a transaction between the principal's agent and the third party. The latter may refuse to deal with the principal, claiming it did not know it was dealing with an agent. The agent may not have informed the third party that she was an agent, and the third party may not have had reason to know that the agent was acting on behalf of someone else. The undisclosed principal may disclose its status and hold the third party, unless the facts would make it unjust to permit the principal to assert its status. (Once the principal is disclosed, the third party can hold the principal.)

Suppose the agent did not have actual authority to enter into a transaction but the principal discovers the transaction. In such a case, the principal may ratify the transaction within a reasonable time and bind the third party (and itself) by notifying the third party and affirming the agent's unauthorized acts.

SECTION 4.09 Disputes Between Agent and Third Party

Disputes between an agent and a third party are relatively rare. If the agent is acting on behalf of an undisclosed principal, the third party has the right to sue the agent individually. The third party may lose this right if it pursues its remedy against the principal, once the principal becomes disclosed. If the agent has misrepresented her authority and the third party is unable to hold the principal, the third party may have an action against the agent for misrepresentation of authority.[6]

Klepp Wood Flooring Corp. v. Butterfield[7] demonstrates the liability exposure of an architect who acts as an agent for a client. The owner commis-

---

[6]See Section 17.05(B).
[7]176 Conn. 528, 409 A.2d 1017 (1979).

sioned the defendant architect to design an expansion for a private school and play an undefined role in procuring contractors. The architect asked the plaintiff, a flooring contractor, to submit a bid for the floor installation work. The architect told the contractor his bid was accepted. No formal contract was prepared. The contractor performed and billed the *architect.*

The trial judge held the architect liable as a general contractor for an undisclosed principal. The flooring contractor relied on the reputation of the architect and expected payment from him. The appellate court found evidence to sustain the trial court's decision that the contractor did not realize he was dealing with an agent and had no knowledge of the identity of any principal. The court held that to avoid *personal* liability, the agent must disclose that he is acting as an agent and must disclose the identity of the principal, the third party having no duty to discover these facts.

PROBLEMS

1. A has authority to buy normal office supplies for an engineering firm. She enters into a written contract with T under which T will furnish an intercom system for the office at a cost of $5,000. Can T enforce this contract against the firm? What additional facts would be helpful in deciding this question, and why would these facts be helpful?

2. Here are the salient facts in a court case: On June 18, 1953, R. J. McDonald was the owner of a 1952 Hudson automobile. On that date, he procured an insurance policy on the car from defendant. Among other things, the policy covered damages caused by collision or upset. The policy ran for two years, and the premium was paid for that length of time.

On November 20, 1953, McDonald sold the car to his brother-in-law, John E. Sauber. The transfer was completed in a bank at Farmington. After transferring the title card to Sauber, McDonald handed him an envelope containing the insurance policy. Sauber then called Northland Insurance Company, defendant herein, on the telephone about the insurance. His testimony is that a woman answered the telephone. She inquired whether she could help him, and his testimony in that regard was as follows:

I was informed, naturally, it was the Northland Insurance Company; I didn't know her name or whether she said this was Northland Insurance Company, but she knew I was talking to the right place; the purpose was, I told her I had purchased the car and it was transferred to me and I was the new owner of the car and I had the insurance policy and I wanted to know if it was all right I would drive the car with this insurance and she said it is perfectly all right, go ahead and that is about the summary of the whole deal; I was the new owner of the car and it was insured by them people.

A Northland clerk corroborated the phone call but insisted she did not state the policy would be transferred to him. She stated she told him to bring in the policy and fill out some forms. Assume the car was driven, an injury resulted, and Northland denied coverage, claiming first that no assurance had been given and second, that any assurance was not authorized. Would the insurance company be liable if a jury believed Sauber's testimony? What facts are relevant to the issue of apparent authority?

Contracts and Their Formation: Connectors for Construction Participants

SECTION 5.01 Relevance

The private design professional will make contracts with, among others, clients, consultants, employees, landlords, and sellers of goods. In addition, contracts are made between owners and prime contractors, prime contractors and subcontractors, contractors and suppliers, employers and employees, buyers and sellers of land, and brokers and property owners.

Contracts for design and for construction are analyzed in other parts of the treatise. Sections 5.01 through 5.10 provide a framework for such analysis. Sections 5.11 and 5.12 present a brief treatment of employment contracts and contracts incident to the acquisition of land ownership, respectively, both relevant to the Construction Process.

SECTION 5.02 The Function of Enforcing Contracts: Freedom of Contract

The principal function of enforcing contracts in the commercial world is to encourage economic exchanges that lead to economic efficiency and greater productivity. This is accomplished by protecting the reasonable expectations of contracting parties that each will perform as promised. Although many and perhaps most contracts are and would be performed without resort to court enforcement, the availability of legal sanctions plays an important role in obtaining performance.

Generally, American law gives autonomy to contracting parties to choose the substantive content of their contracts. Because most contracts are economic exchanges, giving parties autonomy allows each to value the other's performance.

To a large degree, autonomy assumes and supports a marketplace where market participants are free to pick the parties with whom they will deal and the terms on which they will deal. In addition to the parties being in the best position to determine terms of exchange, the alternative of state-prescribed rules for economic exchanges not only would lead to rigidity but also would place a heavy burden on the state. Also, parties are more likely to perform in accordance with their promises if they have participated freely in making the exchange and determined its terms. Finally, such autonomy—often called freedom of contract—fits well in a free society that encourages individual enterprise. However, broad grants of autonomy assume contracting parties of relatively equal bargaining power, equal accessibility to information, and a relatively free marketplace. When such conditions exist, there is a check on overreaching. A party who believes the other party's terms to be unreasonable can deal with others. If the parties do arrive at an agreement under such conditions, the give-and-take of bargaining should ensure a contract that falls within the boundaries of reasonableness.

The development of mass-produced contracts and the emergence of large blocs of economic power often dealing with parties with limited or no bargaining power have made this earlier model of the negotiated contract the exception. If the state, through its courts, enforces adhesion contracts (contracts presented on a take-it-or-leave-it basis), the state is according almost sovereign power to those who have the economic power to dictate contract terms. For this reason, many inroads have been made on contractual freedom by federal and state legislation, regulations of administrative agencies, and courts through their power to interpret contracts and determine their validity.

SECTION 5.03 Preliminary Definitions

The promisor is a person who makes a promise, and the promisee is the person to whom the promise is made. In most two-party contracts, each party is a promisee and promisor, both making and receiving promises.

The offeror is a person who makes an offer, and the offeree is a person to whom the offer is made. All other definitions will be given as the particular term is discussed.

SECTION 5.04 Contract Classifications

A. Express and Implied

Sometimes contracts are classified according to the method by which they are created. Using this classification, there are express contracts and implied-in-fact contracts. In express contracts, the parties manifest their assent or agreement by oral or written words. In implied-in-fact contracts, assent is manifested by acts rather than by words.

U.S. v. Young Lumber Co. illustrates this in a construction project context.[1] The employee of a subcontractor brought equipment he had rented to the job site to replace equipment that a third party had supplied but had removed. The employer admitted that the employee's equipment was on the job site and was used but contended that there was no express agreement under which it would pay the employee rent for use of the equipment.

In ruling for the employee, the court stated:

> . . . even if there was no express oral agreement for equipment rental, an implied contract may be inferred from the conduct of the parties. Implied contracts arise . . . where there is circumstantial evidence showing that the parties intended to make a contract. Where one performs for another a useful service of a character that is usually charged for, and such service is rendered with the knowledge and approval of the recipient who either expresses no dissent or avails himself of the service rendered, the law raises an implied

promise on the part of the recipient to pay the reasonable value of such service.[2]

Unless the agreement is required to be in writing, the implied-in-fact agreement is as valid as an express contract.

B. Subject Matter

Sometimes contracts are classified by the transaction involved, such as sales of land, sales of goods, loans of money, leases, service contracts, professional service contracts, insurance and family contracts. Although traditionally, American contract law has been thought of as a unitary system, different transactions are treated differently. For example, the seller of *goods* is in many instances held to a *successful outcome* standard, usually referred to as implied warranty of fitness, while those who sell professional *services* are usually held to a less strict standard,[3] one that compares what was done with what *others* would have done. Subject matter variations are manifested by regulatory legislation over certain types of contracts. Judicial opinions often treat one transaction differently from another.

C. Bargain and Adhesion

Contracts are sometimes classified as negotiated or adhered to. The latter are referred to as contracts of adhesion.

A negotiated contract arises when two parties with reasonably equivalent bargaining power enter into negotiations, give and take, and *jointly* work out a mutually satisfactory agreement.

The adhesion contract has no or minimal bargaining. The dominant party hands the contract to the weaker party on a take-it-or-leave-it basis. At its extreme, the weaker party who wishes to enter into the transaction must accept all the terms of the stronger party.

Sometimes important terms can be negotiated, although the balance will be dictated by one party, with little opportunity for bargaining. For example, the purchaser of a new automobile may be able to bargain on price but will have to accept the standardized terms of the dealer as to all or almost all

---

[1]376 F.Supp. 1290 (D.S.C.1974).

[2]Id. at 1298. Where such an arrangement is considered nonconsensual, recovery can be based on unjust enrichment.

[3]See Section 14.05.

other aspects of the purchase. The rare buyer who reads the standardized terms and objects to them will have great difficulty in persuading the dealer to change the terms.

Sometimes the adhesion contract is accompanied by monopoly power. In a competitive economy, a weaker party who does not want to accept harsh terms may deal with others. However, in many transactions, the weaker party will find the same terms used by the competitors of the person whose terms were unpalatable or will find no competitor. Frequently, such adhesion contracts are printed, and the person with whom the weaker power is dealing (such as a salesperson or clerk) lacks authority to vary the printed terms of the contract. Modern courts recognize the difference between the negotiated contract and the contract of adhesion.

Interpretation of contracts between clients and their design professionals[4] and between contractors and owners[5] is discussed later in this treatise. It is useful, however, to note how courts have treated adhesion contracts. At the very least, ambiguous terms will be interpreted against the party who supplied the adhesion contract.[6] Even more important are legal rules that will seek to determine the reasonable expectation of the party who was presented a contract on a take-it-or-leave-it basis and interpret the contract in accordance with that interpretation. This may even be done if the party to whom such a contract has been presented was aware of the terms and the likely intention of the party who supplied them.[7] Finally, it is more likely that terms in an adhesion contract will be considered unconscionable and will not be enforced by a court before whom such a contract has been presented.[8]

SECTION 5.05 Capacity to Contract

Capacity usually relates to age and mental awareness. Since it rarely affects contracts for design or construction, it is not discussed in this treatise.

[4]Section 11.04.
[5]Section 20.02.
[6]Sections 11.04, 20.02.
[7]*Graham v. Scissor Tail, Inc.*, 28 Cal.3d 807, 623 P.2d 165, 171 Cal.Rptr. 604 (1981).
[8]Section 5.07(D).

SECTION 5.06 Mutual Assent

A. Objective Theory of Contracts: Manifestations of Mutual Assent

Early English contract law stated there had to be a meeting of the minds—actual agreement as to the existence of a contract and its terms—before there could be a valid contract.

As a rule, each party believes it has made a contract, though perhaps differing as to the exact nature of performance. Suppose one person harbors a secret intention not to be bound and yet manifests to the other party an intention to be bound. The party who relied on the objective manifestation should be protected.

The law protects the reasonable expectations of the innocent party through the objective theory. A party is bound by what it manifests to the other party. Secret intentions are not relevant. If one party, innocently or otherwise, misleads the other into thinking that it has serious contractual intention, a contract exists despite the lack of actual agreement of the parties. The same principle holds true if one party is only joking when entering into negotiations and making an agreement with the other party. Unless the other party should reasonably have realized that the negotiations were not serious, the party who is not serious will be held to the agreement.

B. Offer and Acceptance in the Assent Process: *Western Contracting v. Sooner Construction*

Typically the process by which contracting parties make an agreement involves communications, oral and written, that culminate with one party making an offer and the other party accepting it.

Sometimes before the parties can be said to be offering and accepting, they must be in a negotiating framework. This is illustrated by *Contempo Construction Co. v. Mountain States Telephone and Telegraph Co.*[9] The contractor had contracted with the City of Phoenix to repair a street. Before the contractor could perform, two utilities had to remove poles and other equipment. At a preconstruction conference attended by representatives of the contractor, the city, and the two utilities, the con-

[9]153 Ariz. 279, 736 P.2d 13 (App.1987).

tractor's representative asked those who represented the utilities when they would complete their work on the street. One said it would be completed by the following week; the other said that it would probably be completed within sixty days. The latter had stated, "We're going to do our best, but—we're going to shoot for that."[10]

Failure of the utilities to complete their work in accordance with the representations that had been made delayed the project. Ultimately, the contractor was terminated. It sued the utilities for breach of contract and for commission of a tort. But the court held that the utilities:

> . . . made no statements that could be construed as either an offer or an acceptance.[11]

The court noted that most of the questioning had been done by Mr. Sing, the City of Phoenix' representative. Even his statements, according to the court, did not request a promise or performance in return. Finally, the court characterized the conference as an information-gathering session, not negotiations to enter into a contract.

Where parties exchange communications at the same time and place and an agreement is reached, rarely is it necessary to determine whether there has been an offer and acceptance. But where parties communicate with each other at a distance, it is often necessary to determine whether the parties have arrived at an agreement, and this necessitates an examination of whether there has been an offer and acceptance.

Offers are differentiated from preliminary negotiations or proposals for offers. Although a number of abstract definitions of an offer exist, it is more useful to consider the *effect* of an offer. An offer creates a "power of acceptance in the offeree." The offeree can create a legally enforceable obligation without any further act of the offeror. How is it determined whether the offeree has a reasonable belief that he has a power of acceptance?

Some cases scrutinize the language of the offer, especially a written offer, looking for definite words of commitment on the part of the offeror, such as "I offer" or "I promise." However, parties only rarely express themselves in legal terms.

The entire written proposal is examined to see whether a reasonable person receiving it would think he can "close the deal." Factors include the certainty of the terms, any indication that the proposer will not have to take further action, the past dealings between the parties, and the person to whom the offer is made. For example, if nothing is stated on essential terms such as price, quantity, or quality, the bargaining is probably in a preliminary stage. The first proposal is intended merely to start the negotiating mechanism, and more negotiating will take place before agreement is reached.

If the proposer is negotiating with a number of other persons at the same time and the person to whom a particular proposal is directed knows of this, it is likely that the latter realizes, or should realize, that the proposer wishes to have the last word rather than risk being obligated to a number of persons. In such a case, the communication is not an offer.

Even if stated to be irrevocable, the common law concluded that offers are revocable. At any time before acceptance, an offeror can withdraw the offer, provided the offeror communicates the revocation directly or indirectly to the offeree. This rule of revocability has generated many exceptions, the most important being reliance on the offer and the firm offer under § 2–205 of the Uniform Commercial Code dealing with the sale of goods. The latter makes a firm written offer in a signed writing by a merchant irrevocable for the time stated or, if not stated, for a reasonable time not to exceed three months.

Suppose a written offer states that it will be open for a specified period, say ten days. At the expiration of this period, the power of acceptance terminates. If the offer is made by letter, it may not be clear when the time period begins. Does it begin on the date of the letter offer, at the time the letter is mailed, at the time when it would normally be received, or at the time when it is actually received? Usually it begins when it is actually received. It is better to use a specific terminal date, such as "You have 10 days from May 1, 1993." (See (C) later in this section.)

If the duration of the offer is not stated specifically, the offer remains open for a reasonable time. Facts considered are the state of the market for the goods or services in question, the need of the offeror to be able to deal with others in the event the

---

[10]736 P.2d at 15.
[11]Id. at 15–16.

offeree does not decide to accept, and custom and usage in the particular transaction. For example, the reasonable time to accept an offer for the sale of shares of stock in a fluctuating market will usually be shorter than the reasonable time to accept an offer for the sale of land. The time to accept the offer to sell perishable goods will be shorter than the time to accept an offer to sell durable goods. Such amount of time is judged from the viewpoint of the offeree, whose reasonable belief as to the duration of the offer governs.

An offer can be revoked by the offeror. In most states, revocation must be communicated to the offeree. A few states, notably California, make a revocation effective when *placed* in the means of communication, such as mail, telegram, or fax machine, similar to an acceptance. After a valid revocation, an offeree who is still interested must make another proposal to the original offeror.

Generally, an offer terminates if it is rejected by the offeree. Sometimes the rejection is explicit. For example, the offeree may communicate a lack of interest. In such a case, the offeror is free to deal with others unless the power of acceptance has been created by an option that has been pur-

chased.[12] Usually rejection is implied if the offeree makes a counteroffer. The normal expectation of the offeror in the event of a counteroffer is that the offeree is no longer interested in doing business on the basis of the original offer. This implication may be negated if the counteroffer makes clear that the offeree is still considering the original offer. A counteroffer both creates a power of acceptance in the original offeror and usually terminates the power of acceptance of the original offeree.

The acceptance must be sent or communicated while the power of acceptance still exists. Any acceptance after that time does not operate to create the contract without a further act by the offeror.

The acceptance must be unequivocal and must not propose new or different terms. The following federal trial court opinion illustrates this rule and provides an illustration of a negotiation over the supply of construction material. It also illustrates a subcontractor "quote" used by the prime and failure by prime and subcontractor to agree on terms, a topic explored in Section 28.02.

_____

[12]Restatement (Second) of Contracts § 37 (1981).

WESTERN CONTRACTING CORP. V. SOONER CONSTRUCTION CO.

United States District Court, Western District of Oklahoma, 1966. 256 F.Supp. 163.
[Ed. note: Footnote changed to *.]

DAUGHERTY, District Judge.

This is an action by the plaintiff, Western Contracting Corporation, against the defendant, Sooner Construction Company, for breach of an alleged subcontract between the parties on a runway project

_____

*

Received after Bidding
Haskell Lemon Construction Co.
Road Building and Paving
Phone Windsor 6-3357 P.O. Box 7118
OKLAHOMA CITY, OKLAHOMA
Western Contracting Corp.
March 25, 1963   Sioux City, Iowa

RECEIVED
Mar 28 1963

Western Contracting Corporation
400 Benson Building
Sioux City, Iowa
Gentlemen:
Congratulations on receiving the Tinker Field contract. I hope that it will prove to be a most successful job for you.

We would like very much to make a contract with you to do your asphalt paving work on this job. This letter is

to confirm the prices which we quoted you at Fort Worth.
Item 9. Prime 48,150 gallons @ .19 Furnish, deliver and apply .. no brooming or blotting.
Item 10. Tack coat, 260 gal. @ .37
Item 11. Tack coat, 318 gal. @ .37
Item 13. Hot mix surface 17,220 T. @ $8.32, less 50¢/ ton discount for payment by 10th of month
Item 14. Asphalt (85–100) 215,350 gal. @ .13
No quotation on item 12, 15, 16, 17 as they may be deleted.
We will be happy to help you in any way possible on this contract. We would appreciate your contacting us when you establish your job office here. We hope the job will prove to be both pleasant and profitable and that we will have the opportunity of working with you.
Yours very truly,
/s/ Haskell Lemon
Haskell Lemon
Haskell Lemon Construction Co.

ker Air Force Base, Oklahoma. There was no written subcontract signed by the parties.

. . . From the evidence the Court finds that Sooner orally quoted certain unit prices on asphalt paving to Western prior to Western submitting its bid on the project for the prime contract. Western was successful on its bid. Thereafter, Sooner confirmed its orally quoted prices to Western by a letter dated March 25, 1963.* The oral quotes and the letter quotes were the same. Regarding the item of hot mix surface the price quote of Sooner was 17,220 tons at $8.32 per ton less a 50¢ per ton discount if payment is made by the 10th of the month. Sooner asked for the subcontract during these activities but the request was denied. During the period from March 25, 1963, the date of the above mentioned letter, until July 15, 1963, Western opened an office at Tinker and the parties had various contacts, telephone calls, and discussions regarding the possibility of a subcontract. Western was obtaining asphalt quotes elsewhere. On July 15, 1963, at Tinker a meeting was had attended by a Mr. Hastie for Western, a Mr. Lemon for Sooner, a Mr. Pybas, a superintendent of Sooner, and a representative or representatives of the United States Corps of Engineers. At the meeting discussions were had regarding the specifications, equipment and rolling stock. Hastie testified that after this meeting he for Western and Lemon for Sooner reached an oral agreement on the subcontract, following which a form of subcontract, unsigned by Western, was forwarded by Hastie to Sooner for execution and return to Western for execution by Western at its home office in Iowa. On the hot mix surface this written subcontract submitted by Hastie contained a price of $7.82 per ton thereon but did not provide for payment by the 10th of the month. Rather, it provided for partial payments to Sooner, less a retained percentage of 10%, as Western was paid on estimates by the owner and final payment to Sooner upon complete performance of the subcontract within 45 days after final payment is received from the owner by Western. Lemon denied that an oral subcontract was agreed upon on July 15, 1963 with Hastie and denied that any discussions were even had whereby Sooner would agree to the $7.82 price with the retainage provision and final payment provision as above set out instead of payment for the hot mix surface by the 10th of the month. Pybas, who testified that he was with Lemon at all times going to, at and from the Tinker meeting on July 15, 1963, also denied any oral agreement on the subcontract or any discussions about the discounted price of $7.82 being agreeable without payment by the 10th of the month. Lemon testified that shortly after receiving the written subcontract from Hastie he called Hastie on the phone several times and objected to the lower price of $7.82 per ton without payment being provided for by the

10th of the month in accordance with his quoted terms. Hastie acknowledged several telephone conversations after July 15, 1963, with Lemon regarding the retainage and that Hastie suggested in one of these conversations a reduction of the retainage to only 50% of the work. Hastie further testified that Lemon never gave him an answer to this suggestion.

* * *

In late September, 1963, certain developments took place. Sooner sent a signed subcontract to Western to which it attached certain amendments, six in number, one of which called for full payment for each of the three phases of the work to be done by Sooner within 45 days of completion by Sooner of each phase in lieu of Sooner's requirement in its written confirmation of payment by the 10th of the month and Western's requirement in the written subcontract it prepared and submitted of the 10% retainage and the final payment in 45 days. Western wrote Sooner a letter advising Sooner that it was delinquent in the performance of its subcontract and that if Sooner did not correct this default in performance within five days Western would exercise its rights under the subcontract. Then on September 29 or 30, 1963, a meeting was held in Oklahoma City which brought a Mr. Shaller down from Iowa for Western. Shaller as manager of heavy construction for Western was over Hastie. At this meeting the six amendments were discussed one by one and Shaller disapproved the amendment about full payment in 45 days following each phase of completion as well as two other amendments and in his own hand wrote "out" opposite each of the three amendments so disapproved. Shaller approved the other three amendments. In the language of Shaller, finally at this meeting "things were terminated." Under date of October 2, 1963, Western made a subcontract with Metropolitan Paving Company for larger unit prices as to all items (the price for hot mix surface—the largest item—was $8.53 per ton) and sues herein for the difference amounting to $16,957.08 plus interest, overhead and profit and other expenses.

* * *

15 Oklahoma Statutes, Section 71 provides:
"An acceptance must be absolute and unqualified, or must include in itself an acceptance of that character, which the proposer can separate from the rest, and which will include the person accepting. A qualified acceptance is a new proposal."

Anderson v. Garrison . . . provides:
"In order that a counteroffer and acceptance thereof may result in a binding contract, the acceptance must be absolute, unconditional, and identical with the terms of the counter-offer."

* * *

Thus, the importance of what actually took place on July 15, 1963 between Hastie and Lemon—whether they reached an oral agreement or not—becomes apparent. Hastie says they reached an oral agreement. Lemon and Pybas say they did not. Hastie does not claim that the alleged oral agreement was reached in the presence of the representatives of the Corps of Engineers who attended the meeting, therefore, these representatives are not able to give any assistance to the problem. The Court is of the opinion and finds and concludes that Hastie and Lemon did not on July 15, 1963, reach an oral agreement on the subcontract or discuss and settle the price differential on the hot mix surface with reference to the two alternative prices quoted by Sooner and the effect of the method of payment on the same. It is believed that when Western prepared the subcontract shortly after July 15, 1963, it sought to take advantage of the lower quoted price without meeting the condition attached to the same regarding payment. To this Sooner promptly objected and Sooner did not sign and return the subcontract as requested by Western. Hastie admits that several telephone conversations immediately followed his mailing the subcontract he prepared, these telephone conversations coming from Lemon and that the subject matter of the calls had to do with the retainage provision of the submitted subcontract as it affected the price of the hot mix surface.

* * *

The Court, therefore, finds and concludes from the evidence that Sooner quoted a price of $7.82 a ton on hot mix surface provided payment was received for the same by the 10th of the month—otherwise the price would be $8.32 a ton; that the weight of the evidence indicates that this alternative quote and payment condition of Sooner was not changed or discussed on July 15, 1963; that Western in submitting a subcontract to Sooner set out the $7.82 per ton price but did not meet the payment condition attached to the same; that Sooner immediately and admittedly by several telephone conversations objected to this feature of the subcontract as submitted; . . . that while a period of several weeks lapsed before Sooner submitted its subcontract with amendments such lapse of time transpired in the face of and after objections were made by Sooner to the subcontract submitted by Western and particularly with reference to the use of the lower quoted price on hot mix surface without complying with the requested method of payment in the use of such price; that at the meeting on September 30, 1963, Western would not agree to the originally quoted alternative price of Sooner or its modification as later proposed by Sooner and terminated the matter . . . The Court is of the opinion that the parties never reached a meeting of the minds on the price and method of payment for the hot mix surface, the principle item involved. . . . In simple summary, Sooner quoted an alternative price on hot mix surface to Western, Western submitted a subcontract containing a price and method of payment different from each alternative, Sooner submitted an amended contract with still a different price and method of payment, this was not acceptable to Western and the matter was terminated by Western.

Plaintiff is, therefore, not entitled to the judgment it seeks.

As demonstrated in the preceding decision, when the offeree (the person to whom the offer is made) proposes different terms, it usually means no agreement as yet has been reached. The increased use of standardized forms and letterhead with printed provisions along the margins can create problems.

Suppose a letter of acceptance contains printed provisions along the margins that differ from the offer. A leading case held that such a letter was not an acceptance, because it proposed different terms.[13] (The printed terms on the letterhead probably were not intended to be part of the agreement.) The mirror image rule—that the acceptance must be a ''mirror'' of the offer—often frustrated the common intention of the parties.

To remedy this problem and to deal with transactions where buyer and seller in the sale of goods will sign only their own forms that rarely agree, the Uniform Commercial Code adopted § 2–207. Inclusion of additional or different terms does not necessarily preclude the communication from being an acceptance if it appears that the offeree is accepting the terms of the offer and does not condition his acceptance expressly on the offeror agreeing to additional or different terms. Those additional terms are considered proposals for additions to the contract that are sometimes binding without further communication of the offeror. Section 2–207 also provides that if the conduct of the

---

[13]*Poel v. Brunswick-Balke-Collender Co. of N.Y.*, 216 N.Y. 310, 110 N.E. 619 (1915).

parties indicates that the parties have made a contract although the writings of the parties do not match, a contract has been concluded on the terms that do match, with the supplemental terms being supplied by law.

As offerors usually wish to know whether their offers have been accepted, acceptance, as a rule, must be communicated to the offeror. The offeror can deprive himself of the right to receive actual notice of acceptance by stating, "If you accept, sign the letter and you need not communicate any further with me."

C. Contracts by Correspondence

Generally, to be valid, offers, revocations, and rejections must be communicated to the other party. Because contract law protects reasonable expectations, contracting parties must know where they stand. However, a special rule has been developed for acceptances.

When parties began to make contracts by correspondence, English courts adopted the mailbox, or dispatch, rule. If the offeree used the same means of communication as the offeror had employed, the acceptance would be effective when placed in the means of communication. It did not have to be actually received. For example, if the offeree mails an acceptance to the offeror in response to an offer received by mail, the address on the letter of acceptance is correct, and the proper postage is placed on the letter, the contract is formed at the time the letter is mailed. This places the risk of a delayed or lost letter on the offeror and protects the offeree's expectation that when the letter has been mailed, a contract has been formed.

Although early cases scrutinized the means of communication used by the offeree, generally the mailbox rule applies if a reasonable means of communication is employed.

One aspect of this rule is often ignored. The rule applies only if the offeror has not *specifically* required that the communication actually be *received*. For example, if the offeror makes an offer by mail but states that word *must* be received by a specified date as to whether the offeree accepts, the contract is not formed unless the offeror receives the offeree's acceptance by the stated date.

D. Reasonable Certainty of Terms

As mentioned in Section 5.06(B) one test for determining whether an offer has been made is the clar-

ity and completeness of the terms of the proposal. Even when an agreement on terms has been reached and no issue exists as to the existence of an offer, agreed terms that lack reasonable certainty of meaning preclude a valid contract.

Why require that even agreed terms be reasonably clear in meaning before a valid contract can exist? First, vagueness or incompleteness of terms may indicate that the parties are still in the bargaining stage. Second, a third party, such as a judge or an arbitrator, should be able to determine without undue difficulty whether the contract has been performed. But *reasonable* certainty does not require terms so clear that there can be *no* doubt as to their meaning.

Janzen v. Phillips[14] involved an action by a landscape architect to recover his fee. The architect and his client were somewhat hazy in their initial dealings. There had been no definite discussion of costs, and the defendant stated that he wanted a "first class job," and work went ahead on this basis.

Difficulties developed when the interim billings exceeded what the client thought he would have to pay. After discussions regarding costs, the architect sent a letter to the client in which he stated that he would "substantially landscape most of your property" for $6,500.00, or "nearly so."[15] The court was concerned with the terms *substantially* and *nearly so*. Without these terms, clearly, assent to this letter would have created a valid contract. The court stated that where it appears the parties have intended to make a contract, courts will tread lightly in concluding that their efforts were not successful because of the absence of reasonable certainty. These terms, according to the court, have a definite meaning and are frequently used to indicate that the cost would be *about* $6,500 and the work would be essentially completed for that price. Some latitude was expected, and the architect would have some leeway from the 6,500-dollar figure.

Another court was less willing to be flexible when an event occurred that neither party anticipated. In *Goebel v. National Exchangors, Inc.*,[16] the contract stated that the architect would be paid $500 per apartment for each apartment con-

---

[14]73 Wash.2d 174, 437 P.2d 189 (1968).
[15]437 P.2d at 191.
[16]88 Wis.2d 596, 277 N.W.2d 755 (1979).

structed. Because of costs climbing substantially beyond the budget of the owner, the project was abandoned, but not until the architect had performed substantial services. The court held the 500-dollar figure to be of no help in determining the compensation, because the testimony of the parties revealed that the number of apartments was "totally indefinite, ranging anywhere from 68 to 100 or more units." This, according to the court, precluded a valid contract from having been formed. However, the court also held that the architect should have been given the opportunity to establish his claim based on having made an implied contract allowing him to recover the reasonable value of his services.

Relevant factors in resolving "certainty" disputes are the importance of the term in question, whether performance has commenced, how far it has proceeded, and whether it appears that the party claiming that the contract is invalid appears to be looking for an excuse to avoid contractual obligations.

E. A More Formal Document

Suppose an informal writing is concluded but the parties expect that a more formal document will be drawn up. Have they intended to bind themselves, with the formal written document only a memorial of their actual agreement, or did they not intend to bind themselves until assent to the more formal writing? This depends on the intention of the parties.

An expression of their intention will control. But suppose specific statements of intentions do not exist. If the transaction is complicated and nonroutine and the basic agreement is sketchy, it is likely that the parties do not intend to bind themselves until the more formal writing is prepared and signed. Commencement of performance by both parties very likely means that a contract has been formed without a formal agreement. Performance by one party is more ambiguous.

Sometimes parties begin performance on the basis of a letter of intention or order to proceed. If no final agreement is made, the letter often states that compensation will be based on a specified formula or reasonable compensation. If the letter of intent does not deal with payment in the event no contract is concluded, the party receiving the other party's performance must restore the benefit where

possible or pay the value of the benefit conferred, based on the principle of unjust enrichment. See Section 5.06(I).

Modern courts tend to find a contract at the earliest possible stage if the court is convinced that the parties intended to be bound and only minor gaps need to be filled.

F. Agreements to Agree

Frequently, parties to a contract are unable to specifically define every aspect of their relationship in the contract, yet they may wish to create a legally enforceable agreement. To accomplish these two objectives, they may specify that certain terms will be agreed upon by the parties at a later date. For example, a long-term supply of goods contract may specify that the shipping arrangements or delivery dates will be agreed upon at a later date. The parties have *agreed to agree.*

Older cases refused to enforce contracts that contained such provisions. Judges saw themselves as simply enforcing agreements that the parties made rather than making agreements for the parties.

Modern courts are more willing to enforce contracts that contain agreements to agree as long as the parties are not still in the negotiation stage and the matters to be agreed upon are not essential. For example, § 2–204(3) of the Uniform Commercial Code states that a contract for sale of goods is valid even though certain terms are left open, "if the parties have intended to make a contract and there is a reasonably certain basis for giving an appropriate remedy."

Where courts have faced contracts that contain agreements to agree, as a rule, the parties are not disputing the to-be-agreed-to provisions. One party contends that it is not bound because there is a fatal defect that rendered the contract invalid. If performance has begun and the matter to be agreed to is not so essential that it indicated the parties were still in the bargaining stage, courts are likely to enforce the agreement. The agreed-to provision will be supplied by a standard of reasonableness that will often take into account trade usage.

G. Agreements to Negotiate in Good Faith

Sometimes an agreement to agree is interpreted to be an agreement to negotiate in good faith. The parties do not state that they will agree but state that they will negotiate in good faith on certain matters. Although some courts will not enforce such provisions, there is an increasing tendency to enforce them by requiring the parties to negotiate in good faith. If the parties in good faith seek to agree but cannot, further performance should not continue, and any performance that has been rendered should be dealt with by principles of restitution based on unjust enrichment. However, if one of the parties does not negotiate in good faith, the court may determine what good-faith negotiation would have been produced by way of agreement.[17]

H. Memorandum of Understanding

In the course of a complicated and often lengthy negotiation, the parties may draw up and agree to what they call a memorandum of understanding, which indicates that they have reached agreement on certain terms but are still negotiating over others. Generally, such memoranda are not binding. If the negotiations have not been concluded, all matters are open to negotiation.

The purpose of a memorandum of understanding is to give the parties some indication of those things that have come to the stage of tentative agreement to aid the parties in seeing what still remains to be negotiated. While one party may *assert* that the other party is not negotiating in good faith if it seeks to reopen terms to which there has been agreement, the agreement process should not be placed in this straightjacket.

I. Letter of Intent

During the course of a complex negotiation, the parties may wish to commence performance before there has been a final agreement. The party asked to commence performance may wish some assurance that if the negotiations are not successfully concluded, it will still be paid for what it has done. A letter of intent issued by the other party will be a directive to proceed and will provide a compensation formula for work performed that will be binding even if the parties do not reach a final agreement. If one party commences performance after the other party issues a letter of intent, the parties have agreed upon a contract that will govern their relationship until it is terminated, even if they do not successfully conclude the negotiations.

Despite this seeming simplicity, letters of intent have proved troublesome in operation, as demonstrated by *Quake Construction, Inc. v. American Airlines, Inc.*,[18] which involved the construction of an airport expansion. The prime contractor orally informed the subcontractor that it had been awarded the subcontract to build a restaurant, rest room, and locker facilities. Evidently, the prime contractor wanted the subcontractor to enter into sub-subcontracts to induce sub-subcontractors to reveal their license numbers. To do this, the prime sent a letter of intent to the subcontractor a few days before performance was to begin. The letter stated that the prime had "elected to award the contract" to the subcontractor. It also stated:

> A contract agreement outlining the detailed terms and conditions is being prepared and will be available for your signature shortly.[19]

The work was described, and there were provisions dealing with time deadlines, insurance and compliance with all MBE, WBE, and EEO goals. The letter of intent concluded by stating that the prime reserved "the right to cancel this letter of intent if the parties cannot agree on a fully executed sub-contract agreement."[20] The prime and the subcontractor discussed and agreed to certain changes in the written form contract that were made in the contract to reflect the parties' agreements. Although the prime advised the subcontractor that it would prepare and send the written agreement to the subcontractor for its signature, this was never done.

At a preconstruction meeting, the prime told all present that the subcontractor had been awarded the subcontract. Yet a day later, the owner informed the subcontractor that its involvement in the project was terminated.

The trial court dismissed the complaint made by the subcontractor against the prime based upon the cancellation clause. The intermediate appellate court ruled that the letter of intent was ambiguous and ordered that the case be sent back to the trial court to determine the intent of the parties as to

---

[17]*Purvis v. United States*, 344 F.2d 867 (9th Cir.1965).

[18]141 Ill.2d 281, 565 N.E.2d 990 (1990).
[19]565 N.E.2d at 992.
[20]Id. at 993.

whether the letter constituted a binding contract. In a twenty-page double-columned opinion, the Supreme Court of Illinois found the letter of intent ambiguous and ordered that the parties be allowed to introduce extrinsic evidence. When letters of intent are prepared or issued, thought should be given as to whether they are simply tentative agreements that will be valid if the final agreement is not concluded or whether they themselves are binding contracts.

J. Illustrative Case: *Sand Creek v. CSO*

The preceding subsections can be illustrated and the pressure to commence performance before the contract has been worked out is demonstrated by *Sand Creek Country Club, Ltd. v. CSO Architects, Inc.*[21] The architect had been engaged by the owner to design a proposed expansion of the country club. On July 25, 1988, the architect sent the owner a letter stating in part that he was sending architects from his office to start work on the project as soon as possible and that he intended to begin work but to delay billings until financing would be in place. The letter also discussed the proposed fee structure that was to be included in an AIA contract. He concluded by stating that the architect would prepare "an AIA Contract in accordance with the guidelines set forth in this letter, and it is hereby understood that nothing contained in this letter shall bind either until said AIA document is executed by both parties."[22] The letter was signed by the architect and accepted by the owner.

On September 9, 1988, the owner ordered the architect to begin work on the construction drawings. He also asked the architect to "get together an American Institute of Architects (AIA) Contract" and bill him for the work prepared to date.[23] The architect sent the bill but received no response. Eventually, the architect stopped work and filed suit against the owner, requesting payment for the work he had performed.

The owner argued that there was no binding contract, based upon language in the letter of July 25 stating the parties would not be bound until the

AIA document was executed. However, the trial court found that there had been an agreement that the architect would perform services and that the owner would pay for them. It rejected the argument by the owner that the parties had simply made an unenforceable agreement to agree.

The appellate court noted that the letter was not the only evidence of whether the parties had made an agreement. Evidence that the owner had given the architect the go-ahead and instructed the architect to bill supported the finding of the trial court that the parties had made a binding agreement. Even if the parties had not intended to be bound until they had signed the AIA contract, they intended that the architect begin work before that time and be paid.

In effect, the court concluded that the parties had made a letter of intent under which the architect would be compensated even if no final agreement resulted. There may have been contradictions between terms expressed in the July 25 letter and AIA provisions. But note that the letter states that the AIA contract would be prepared in accordance with the guidelines set forth in the letter. If there were anything in the AIA document that was not consistent with the letter, the letter would take precedence.

Another way of interpreting the July 25 letter would be to consider that the parties had made a binding contract and were incorporating an appropriate AIA document to the extent it was not inconsistent with the July 25 letter. This would use the AIA document to fill any gaps and provide administrative and legal provisions that would not normally be found in the letter. To do so, however, the court would have to conclude that the parties intended to bind themselves *immediately* and were using the AIA document as a set of general conditions that would regulate their relationship.

The difficulty with this interpretation is that the July 25 letter stated that *neither* party would be bound until the AIA document was executed by both parties. This statement would be evidence of an intention that neither party intended to be bound until both parties had reviewed and agreed to the AIA document. In such a case, the July 25 letter, as indicated earlier, simply operated as a letter of intention that would be effective if the parties were not successful in negotiating their contract.

---

[21]582 N.E.2d 872 (Ind.Ct.App. 1991).
[22]Id. at 874.
[23]Ibid.

SECTION 5.07 Defects in the Mutual Assent Process

A. Fraud and Misrepresentation: Duty to Disclose

The negotiation process frequently consists of promises and factual representations made by each party. If important factual representations or promises are made to deceive the other party, the deceived party may cancel the transaction or receive damages if it has reasonably relied upon the representations or promises.

Sometimes such representations are made not with the intention of deceiving but are negligent or even innocent. As the conduct becomes less morally reprehensible, the remedies to the deceived party will be fewer. For example, in fraudulent representation, the deceived party may receive punitive damages designed to punish the conduct and not to compensate the victim. Although the victim of negligent misrepresentation will be able to recover damages to compensate for the loss or cancel the transaction, the only relief likely to be accorded someone who relies upon an innocent misrepresentation is cancellation of the contract.

Suppose one party to a contract wishes to cancel the transaction because of a claim that the other party should have disclosed important facts that, had they been known, would have persuaded the former not to enter into the transaction. Although early cases rarely placed a duty to disclose on contracting parties, there is an increasing tendency to do so where the matter that is not disclosed is important and where the party knowing of the facts should have realized that the other party was not likely to ascertain them.

B. Economic Duress

A valid contract requires consent freely given. Consent cannot be obtained by duress or compulsion. Early common law duress was physical, such as obtaining a deed or a contract by threatening to kill or injure the other party.

Modern duress in the commercial world principally involves economic duress or (as it is sometimes called) business compulsion, which can exist if one party exerts *excessive* pressure beyond permissible bargaining and the other party consents because it has no real choice. Courts have found economic duress to exist in very limited cases. Their hesitance undoubtedly relates to the inherent pressures involved in the bargaining process and the fear that many contracts could be upset if the economic duress concept were used too frequently. Yet, as indicated by *Rich & Whillock v. Ashton Development* (reproduced in Section 27.13), the law seems more willing to set aside contractual adjustments or releases even in the hard bargaining commercial world.[24]

C. Mistake

The doctrine used most often in attacking the formation process is mistake. Here again, there are no fixed or absolute rules. There may be mistakes as to the terms of the contract. One party may not read the agreement, or because of mistake or fraud, the agreement may not reflect the earlier understanding of the parties. Also, parties are often mistaken as to the basic assumptions on which the contract was made. Everyone who makes a contract has certain underlying assumptions that, if untrue, make the contract undesirable.

A few generalizations can be made. Although there is always some carelessness in the making of a mistake, the greater the degree of carelessness, the less likely it is that the party making the mistake will be given relief. A comparison of the values exchanged is relevant. If one party is getting something for almost nothing, mistake is more likely to be employed to relieve the other party from the contract. The question of when the mistake is uncovered is important as well. If the mistake is discovered before there has been reliance by the other party or before there has been performance by the other party, there is a greater likelihood that the contract will not be enforced.

In extreme cases, courts will relieve a party from certain risks. However, the law also seeks to prevent parties from avoiding performance of their promises by dishonest claims of mistake when it proves economically advantageous to do so. Steering a course between these two policies as well as supporting the policy that persons should be able

---

[24]*Centric Corp. v. Morrison-Knudsen Co.*, 731 P.2d 411 (Okla.1986) (collecting many cases).

to rely on agreements has meant that there are few fixed legal rules that can be used to predict probable results in such cases.

A determination of whether relief will be granted requires a careful evaluation of the facts, with principal scrutiny paid to the relative degree of negligence, any assumption of risk, the disparity in the exchange values in the transaction, and the likelihood that the status quo can be restored.

D. Unconscionability

Equity courts refused to enforce contracts that were unconscionable—contracts that shocked the conscience of the judge. Borrowing this concept, the Uniform Commercial Code in § 2–302 gave the trial judge the power to strike out all or part of an unconscionable contract for the sale of goods. Gradually, the concept is being accepted in *all* contracts. But contracts or clauses that have been struck down (and few have) have tended to

1. Be consumer rather than commercial transactions.
2. Involve sharp dealing in the bargaining process rather than harsh terms.
3. Involve clauses that exculpate a party from its obligation or make vindication of the other party's legal rights very difficult rather than price terms.

As yet the doctrine has not made much of a direct impact on contracts for design and construction except for anti-indemnity statutes, discussed in Section 32.05(D). Yet it has been held that placing a duty on the owner to seek verification of a bid in the event the bid appears to have been a mistake is done to avoid an unconscionable contract.[25] The doctrine's development, as noted in Section 19.02(E), has, however, encouraged courts to examine all the circumstances that surround the making of a contract and to intervene more frequently in the bargain.

E. Formation Defects and Restitution

One of the important uses of restitution relates to the conferring of benefits when for a variety of rea-

sons a claim cannot be based on a valid contract. There may be no one with whom a valid contract can be made. For example, suppose a doctor passing a construction site sees a passerby struck by a falling piece of lumber, rendering the passerby unconscious. The doctor administers medical services in an attempt to save the victim. As the doctor is a person trained in such emergencies, he should be encouraged to act. His training, the emergency, and the fact that he is not likely to be rendering services gratuitously takes him out of the volunteer category and allows him to recover.[26]

Such cases are rare. More commonly, restitution claims are made where parties intended to make a contract but did not do so. In the context of design or construction services, two principal formation defects require a person who has performed services to look to restitution. First, the contract fails because the terms lack sufficient certainty (usually in contracts for design services), the parties could not agree on terms while work progressed,[27] or the parties did not get around to formalizing the contract by executing a formal writing where it was their intention to formalize it that way or there was a misunderstanding over the terms of the contract.[28] In all these instances, the parties believed they had or would have a valid contract.[29] Where benefit is conferred under these circumstances, the party who has conferred the benefit will very likely be able to recover in restitution.[30]

The measure of recovery may vary, from the *value* of the expenditures to the *benefit* actually conferred, depending on the equities. It may even, in rare cases, be based on expenditures that did not benefit the defendant.[31] In a close case as to

---

[25]*Sulzer Bingham Pumps, Inc. v. Lockheed Missiles & Space Co., Inc.,* 947 F.2d 1362 (9th Cir. 1991), reproduced in Section 18.04(E).

[26]*Cotnam v. Wisdom,* 83 Ark. 601, 104 S.W. 164 (1907). Admiralty law allowed a recovery to the owner of a ship that altered its course to help a stricken ship that lacked a medical person. *Peninsula & Oriental Steam Navigation Co. v. Overseas Oil Carriers, Inc.,* 553 F.2d 830 (2d Cir.1977). Reluctance to help is likely to be traceable to fear of a malpractice claim by the victim. Many states have enacted Good Samaritan statutes immunizing health care persons who help in emergencies.

[27]*Comm. v. Goodman,* 6 Ill.App.3d 847, 286 N.E.2d 758 (1972).

[28]*Anderco, Inc. v. Buildex Design, Inc.,* 538 F.Supp. 1139 (D.D.C.1982) (misunderstanding of meaning of terms).

[29]*Dyer Constr. Co. v. Ellas Constr. Co.,* 153 Ind.App. 304, 287 N.E.2d 262 (1972).

[30]Refer to notes 27, 28, and 29 supra, where all recovered.

[31]*Minsky's Follies of Florida, Inc. v. Sennes,* 206 F.2d 1 (5th Cir.1953) (failure to have a written memorandum).

whether a valid contract has been made, performance by both parties or even one party may tip the scales in favor of a conclusion that a valid contract did exist. In these illustrations, neither party had breached. As a result, recovery here is more likely to be limited to actual benefit conferred.[32]

The second defect requiring restitution that is relevant to design and construction services (more common in construction) relates to contracts made illegally, principally where public contracts require that awards be made competitively. Because of the varying judicial attitude toward the importance of rigorously enforcing those rules and the various levels of illegality (compare a bribed official awarding a contract[33] with a technical irregularity by well-meaning officials[34]), court decisions may not appear consistent. If there is *any* trend, it is toward granting recovery in favor of those who have performed in good faith whose contract was invalid because of technical irregularities.[35]

SECTION 5.08 Consideration as a Contract Requirement

A. Functions of Consideration

No legal system enforces all promises. Some promises are neither important enough to enforce nor intended to create enforceable rights. For these reasons, the law will not enforce a promise to attend a social event.

Enforcement of some promises can disrupt important institutions. Suppose an eight-year-old boy can obtain judicial relief to enforce a promise by his father to take him to the zoo. This can disrupt the order needed for a family unit.

Some promises are made hastily or impulsively, without appreciation for the consequences. Altruistic promises are often made without regard for the burden they can place on the promisor. Some

promises are best left to other sanctions. For example, the law generally does not use contract law to hold politicians to their campaign promises. Instead, the ballot box and public opinion are mechanisms to pressure public officials to live up to their promises. Moreover, society may be better off if public officials have flexibility to deal with public matters.

Some contracting parties have used their bargaining power to impose harsh obligations on weaker parties, promising little or nothing of value in return. In extreme cases, such overreaching may be restrained by denying enforcement of the weaker party's promise.

The common law has made the consideration doctrine the chief vehicle for determining which promises will be enforced by the courts. It operates as a limit on complete contract freedom.

B. Definitions: Emergence of Bargain Concept

Early law required benefit to the promisor (the one who had made the promise). Later decisions, however, enforced a promise where the promisor did not receive pecuniary benefit but the promisee suffered a detriment.

In this century, the concept of bargain developed. Rather than look for benefit to the person making the promise or detriment to the person to whom the promise had been made, the courts examined the transaction to see whether there had been a bargain. If so, usually there is consideration. Fairness of the bargain is not a requirement, except in certain consumer transactions and where equitable remedies are sought.

Most commercial contracts involve bargains, and consideration is rarely a problem. However, sometimes a guarantee or surety promise—that is, one to answer for the obligations of another—can raise consideration questions because it is often made after the contract that is being guaranteed. For example, in *Northern State Construction Co. v. Robbins*,[36] the contractor made a construction contract with a corporate owner of questionable financial responsibility on February 27, 1962. On March 1, 1962, several shareholders in the corporation executed a written guarantee of the corporate owner's performance under the construction contract.

The contractor's fears about the financial instability of the corporate owner proved well

---

[32]But in *Coastal Timber, Inc. v. Regard*, 483 So.2d 1110 (La.App.1986), the plaintiff recovered the reasonable value of labor and materials supplied, including profit. Louisiana law appears more favorable than other states to claimants.

[33]*Manning Eng'g, Inc. v. Hudson County Park Comm'n*, 74 N.J. 113, 376 A.2d 1194 (1977) (no recovery).

[34]*Layne Minnesota Co. v. Town of Stuntz*, 257 N.W.2d 295 (Minn.1977) (no recovery but based upon absence of benefit).

[35]*Blum v. City of Hillsboro*, 49 Wis.2d 667, 183 N.W.2d 47 (1971).

[36]76 Wash.2d 357, 457 P.2d 187 (1969).

grounded. Upon completion of the work, the corporate owner was unable to pay the outstanding balance of the contract price. The contractor brought an action against the guarantors.

The action was unsuccessful because the court concluded that the guarantee occurred *after* the construction contract had been made. This, according to the court, meant that the guarantors received nothing for their promise inasmuch as the owner was already committed to perform the contract. Had the construction contract and the guarantee been part of one transaction and the agreement by the prime contractor induced in part by the guarantee, the guarantee would have been enforceable. A different result would be reached in other states.[37]

C. Reliance

Section 90(1) of the Restatement (Second) of Contracts states

> A promise which the promisor should reasonably expect to induce action or forbearance on the part of the promisee or a third person and which does induce such action or forbearance is binding if injustice can be avoided only by enforcement of the promise. . . .

This doctrine, developed originally as a means of enforcing gift promises that were relied upon and promises made to charitable organizations, is now used in a variety of transactions and operates as a supplement to the bargain concept.

Reliance has extended into commercial areas and has been used to enforce promises of bonuses made by employers to employees and promises made by manufacturers to franchised dealers. In bonus cases, the reliance is the employee's remaining on the job to obtain the bonus. In the dealer franchise cases, reliance frequently consists of expansion of the dealer's facilities and investment of capital. Although it is possible to find a bargained-for exchange in these cases, some courts have avoided the often tortured reasoning necessary to find a bargain and instead have used the reliance concept to make the promise enforceable.

The reliance concept has entered the construction process. Section 28.02 describes its use as a means of enforcing bids made by subcontractors to prime contractors.[38]

D. Past Consideration

Although it is often stated that promises motivated by benefits having been received in the past are not enforceable, there are inroads in this rule. In some states, a promise based on past benefits not conferred as a gift may be enforced "to the extent necessary to prevent injustice."[39]

In the commercial world, there are exceptions to the traditional denial of nonenforceability. Promises to pay a debt that has been either discharged by bankruptcy[40] or barred by the Statute of Limitations are enforced.[41]

SECTION 5.09 Promises Under Seal

When most people were unable to write, a method was needed to accomplish conveyances and other legal acts. The method adopted in early English legal history was the seal, an impression made by a ring or other similar instrument in soft wax that had been placed on the document. The seal emphasized the seriousness of the act being performed. Consideration was not necessary. A sealed instrument was a powerful document.

In the United States, the seal began to lose its power when, instead of being a formal impression in wax, it was reduced to either putting the word *seal* on the document or using some abbreviation such as *LS*. This did not have the trappings of formality possessed by the old seal. As a result, in many states, the seal was denied any function in the preparation of the documents. In other states, a document under seal creates enforceability or creates a presumption of consideration.

---

[37]Restatement (Second) of Contracts § 88 (1981).

[38]Similarly, see *Bethlehem Fabricators, Inc. v. British Overseas Airways Corp.*, 434 F.2d 840 (2d Cir. 1970) (promise made by owner to subcontractor enforced because of subcontractor's reliance).

[39]Restatement (Second) of Contracts § 86 (1981).

[40]Id. at § 83. Formal requirements to pay debts barred by bankruptcy are contained in 11 U.S.C.A. § 524.

[41]Restatement (Second) of Contracts § 82 (1981). Many states require a written promise.

SECTION 5.10 Writing Requirement: Statute of Frauds

A. History

In 1677, the English Parliament passed the Statute of Frauds, which requires that a sufficient written memorandum be signed by the defendant before there can be judicial enforcement of certain specified transactions. One reason for the writing requirement is to protect litigants from dishonest claims and protect the courts from the burden of hearing claims of questionable merit. Another is that the requirement of a writing acts as a cautionary device. It warns persons that they are undertaking serious legal obligations when they assent to a written agreement. The Statute of Frauds has been adopted in all American jurisdictions in one form or another. The tendency has been to expand the classifications that require a writing.

B. Transactions Required To Be Evidenced by a Sufficient Memorandum

Transactions singled out in the original statute were promises by an executor or administrator to pay damages out of his own estate; promises to answer for the debt, default, or miscarriages of another; agreements made on consideration of marriage; contracts for the sale of land or an interest in land; agreements not to be performed within a year from their making; and contracts for the sale of goods over a specified value. Some states require contracts for the performance of real estate brokerage services and contracts to leave property by will be evidenced by a written memorandum. Finally, statutes regulating home improvements often require that there be a written contract.[42]

Classification of the transaction has been one device by which courts have cut down the effectiveness of this statute. For example, a contract to construct a twenty-story building will not require a writing if the court concludes that the contract *could* be performed within a year. Many other limitations and exceptions on the various transactions have been made by the courts.

C. Sufficiency of Memorandum

The memorandum must be signed by the party to be charged. It must contain the basic terms but need not express the entire agreement. It need not have been signed at the time the agreement was made but can be signed later. For example, a letter written and signed by a party after a dispute has arisen will provide the needed memorandum if the letter contains sufficient terms. In goods transactions, the statute is satisfied by an admission in the course of litigation.[43] As technology changes, the law will have to decide whether audio or visual tapes will satisfy the requirement.[44]

D. Avoiding the Writing Requirement

Sometimes oral agreements that generally require a memorandum are enforced despite the absence of a memorandum. Section 2–201 of the Uniform Commercial Code states that acceptance of part payment for the goods is sufficient to make enforceable an oral contract for *those* goods.

In cases involving the sale of land, the part performance doctrine developed as a substitute for a writing. For example, if the buyer took possession and made improvements, the court could order the seller to convey despite the absence of a writing. However, part payment by the purchaser was not sufficient part performance.

In service contracts, full performance by the parties would bar the use of the Statute of Frauds as a defense. However, this exception was not applied in a home improvement contract required to be in writing.[45]

Some jurisdictions enforce oral agreements if one party has reasonably relied on and changed its position based on the promised written agreement or if the other party has been unjustly enriched by the performance.[46]

The Statute of Frauds has become eroded in many states. If a claim seems genuine and if the

---

[42]West Ann.Cal.Bus. & Prof.Code § 7159.

[43]U.C.C. § 2–201(3b). But the contract is enforced only to the quantity admitted.
[44]*Ellis Canning Co. v. Bernstein,* 348 F.Supp. 1212 (D.Colo.1972) held that an audio tape satisfied the Statute of Frauds.
[45]*Caulkins v. Petrillo,* 200 Conn. 713, 513 A.2d 43 (1986).
[46]*Monarco v. Lo Greco,* 35 Cal.2d 621, 220 P.2d 737 (1950).

claimant has relied reasonably on the oral agreement, a strong likelihood exists that the claimant will be permitted to prove the agreement.

SECTION 5.11 Employment Contracts

Application of general legal doctrines to design and construction contracts are discussed later. However, two other types of contracts important in the Construction Process—those that relate to employment and those that relate to acquiring ownership of real property by purchase—have particular rules that were developed to meet specific needs and contractual practices. With this in mind, Sections 5.11 and 5.12 touch on some of the salient legal characteristics of those contracts.

A. Contracts for Indefinite Periods

The common law classified employment agreements for an indefinite period as agreements at will. Such contracts can be terminated at any time by either party. In essence, the contract governed the rights and duties of the parties so long as they were both willing to continue the relationship.

The at-will rule was based in part on the difficulty the employer would have enforcing any contract against the employee, because it was likely that the employee would not be financially able to respond to any court judgment. Courts would not *order* an employee to go back to work for an employer, and only rarely would they order an employee not to compete with the employer even if the employee promised not to compete. The at-will rule may be in accord with the intention of the parties.

Increasingly, dismissal of employees has led to successful claims by the dismissed employee despite the at-will rule. One court allowed such a claim because the employee had been discharged in bad faith because she would not date her foreman (an early precursor of sexual harassment claims).[47] Another court allowed the employee to recover based upon the employer's failure to follow its disciplinary review procedures as set forth in its employee handbook.[48] Another theory, de-

veloped in California, protected a long-term employee who had been discharged after thirty-two years of service by holding that there had been an implied contract to discharge only for good cause.[49] However, California refused to allow the employee in another case to assert a tort claim based upon the breach of the covenant of good faith and fair dealing that would have exposed the employer to expanded tort remedies of noneconomic losses and possible punitive damages.[50] Yet California allowed punitive damages when an employer dismissed its employee in violation of public policy.[51]

The preceding cases demonstrate that in some states, employees receive protection from wrongful discharge despite the at-will rule. Expanded employee protection has led to employers developing protective policies to avoid such claims, such as carefully articulated discharge review procedures, more candid job performance evaluations, and frank reference letters.

B. Modification

Suppose an employee receives an offer of a better job during the term of employment and informs the employer, and the employer promises to pay more money for the same work for the balance of the term. Generally, the employer's promise is not enforceable. The employer received nothing for the promise other than the performance that the contract entitled the employer, and the employee did no more than what was required under the employment contract. Denial of enforcement was based on what was called the "preexisting duty rule."

Perhaps the real basis for such decisions was the presence of actual or subtle duress on the employer. Yet in many situations, there was no duress or unfairness. As a result, exceptions developed to the normal rule in employment cases. For example, if the employee's duties are changed in the slightest, the promise to pay more money would be enforced. Also, if the old contract has been cancelled by the parties, the preexisting duty rule does not apply.

---

[47]*Monge v. Beebe Rubber Co.*, 114 N.H. 130, 316 A.2d 549 (1974).
[48]*Pine River State Bank v. Mettille*, 333 N.W.2d 622 (Minn.1983).

[49]*Pugh v. See's Candies, Inc.*, 116 Cal.App.3d 311, 171 Cal.Rptr. 917 (1981).
[50]*Foley v. Interactive Data Corp.*, 47 Cal.3d 654, 765 P.2d 373, 254 Cal.Rptr. 211 (1988).
[51]*Tameny v. Atlantic Richfield Co.*, 27 Cal.3d 167, 610 P.2d 1330, 164 Cal.Rptr. 839 (1980) (employee refused to participate in illegal price fixing).

In some states, statutes were passed stating that the modifications did not require additional consideration to be enforceable if the modifications were in writing. These exceptions reflect a judicial and statutory recognition that employees' rights to try to improve their positions are more important than being held to their contracts.

C. Writing Requirement

The Statute of Frauds requires that contracts that by their terms cannot be performed within a year from its making be evidenced by a written memorandum signed by the party from whom performance was legally demanded. A writing is not required if the work can be performed within a year. The employment of an engineer for a large project that could take two years need not be in writing if the contract *could* be performed within a year.

D. Public Employment

Even without civil service rules requiring a fair hearing, some public employees cannot be discharged without a hearing because of expanded application of federal constitutional obligations to states and public agencies.

Public employees are increasingly joining unions or employee associations, which has led to legislation permitting collective bargaining by some employees. When negotiations break down, strikes often result. Although strikes by public employees are many times forbidden, public officials are often unwilling to invoke legal sanctions against such strikes.

E. Government Regulation of Private Employment

Private law is of diminished importance in employment relations. Collective bargaining is, to a certain degree, controlled by federal and state legislation. Increasingly, state and federal governments regulate private employment, mainly in the area of wages, hours, working conditions, and employee selection. Laws limit the work hours of minors and require extra pay for work over a stated minimum number of hours per day or week. Minimum wage laws set a wage floor.

During the 1950s, many states enacted fair employment practices acts. The policy of nondiscrimination was given a substantial impetus by the enactment of the Federal Civil Rights Act of 1964. This act applies, with some exceptions, to employers engaged in interstate commerce. Given the interstate characteristics of most businesses, it realistically affects almost all employers who have eight or more employees. The Act makes it an unfair employment practice

> to fail or refuse to hire or to discharge any individual, or otherwise to discriminate against any individual with respect to his compensation, terms, conditions, or privileges of employment, because of such individual's race, color, religion, sex, or national origin.[52]

In 1967, the federal government enacted comparable legislation prohibiting discrimination on the basis of age unless age is a bona fide occupational qualification or unless there is a bona fide seniority or retirement plan.[53]

In 1990, the Americans with Disabilities Act[54] was signed into law. It prohibits discrimination against disabled workers in all aspects of employment and requires access for disabled persons to public transportation and public accommodations. While some states already bar discrimination against disabled persons, the enactment of this federal law will undoubtedly broaden job protection rights of the disabled.

F. Remedies

(See Section 6.09.)

SECTION 5.12 Contractual Acquisition of Interests in Land: Special Problems

A. Broker Contracts

Function. Although it is not necessary to hire a broker to assist a seller or buyer, typically a person wishing to buy or sell land will contact a real estate broker to assist in finding a buyer or seller. Ordinarily, the broker is hired to bring prospective buyers or sellers to the principal. The broker is not given the authority to actually sell the land.

Sometimes brokers try to find prospective buyers or sellers without any advance agreement as to

---

[52]42 U.S.C.A. § 2000e–2.
[53]29 U.S.C.A. § 623.
[54]Americans with Disabilities Act of 1990. Pub. L. No. 101–336, 104 Stat. 327.

commissions. More often, a listing or exclusive listing is made by owners of land with the real estate broker, or similar arrangements are made by prospective buyers. Because it is more common for the seller than the buyer to obtain a real estate broker, the discussion centers on the use of the broker by the prospective seller of property.

Listings. Generally, the owner of land promises to pay the broker a specified commission if the broker procures a buyer ready, willing, and able to purchase the property. The owner of the property is not obliged to sell to the person making an offer to purchase. Federal, state, or local open occupancy laws may limit the seller's right if the reason for refusing to sell relates to race, color, religion, sex, or national origin. To avoid payment if the sale is not consummated, owners can condition their obligation to pay the fee upon the consummation of a contract for purchase, the actual transfer of title, or the receipt of a specified payment.

Listings fall into three categories: general, exclusive agency, and exclusive right to sell. A general listing, usually for an unspecified time, permits owners the greatest freedom. Owners can authorize any number of agents, sell the land themselves, or withdraw before a purchaser is found. An exclusive agency gives the chance to one agent, but owners can sell the land themselves. The tightest listing is an exclusive right to sell, with the agent receiving a commission no matter who makes the sale.

Multiple listings allow a number of agents to find a buyer, with the fee being divided between the agent who obtains the listing and the one who finds the buyer.

Fees. Traditionally, brokers have received a specified percentage of the sale price as their compensation. This method of computing the fee, like a similar method traditionally used by design professionals, has been criticized. Often it does not relate to the amount of time spent by the broker. Some owners resent having to pay a full fee when it took little effort to sell the property at the asking price, not an uncommon phenomenon in a rising market. Also, such a method can induce the broker to seek a low asking price which, although producing a lower fee, can result in a quick sale with minimal effort.

As the trend continues toward alternative methods of providing professional services and continued effort by governmental officials to encourage competition, it is likely that alternative fee methods will develop. Some brokers provide a service that is not conditioned upon a sale but is compensated for by a fee based on the time spent or by a flat fee. Some states have enacted legislation requiring that brokers tell prospective clients that it is possible to negotiate a fee different from the one customarily used in the community. (Most indications are that such laws have very little effect on actual fees.)

Legal Problems. The problems that arise most frequently relate to the one-sidedness of the agreement. In the typical brokerage arrangement, the broker does not promise to do anything. The promise is made by the owner, and for that reason, the contract is typically classified as unilateral. If it is unilateral, earlier cases allowed the listing to be revoked by the owner at any time prior to full performance. However, doctrines have been developed to protect the broker from expending substantial amounts of effort and spending money in performance of the contract only to find that the agreement has been revoked before the broker's completion of performance.

Some brokers protect themselves today from revocation by language in the listing agreement that binds brokers to use their best efforts to secure a buyer ready, willing, an able to buy the property. This technique is intended to convert the contract into a two-sided contract (a bilateral contract) where *both* parties have obligations. Another technique employed by brokers is the inclusion of a provision allowing the owner to withdraw the property but to condition this right on payment of the broker's fee.

Formalities. In some states, agreements by which the owner of land promises to pay a commission if a broker procures a buyer or a renter must be evidenced by a written memorandum. In most categories of transactions required to be evidenced by a sufficient written memorandum, the courts have been astute in finding exceptions that would in effect enforce an oral transaction.[55] In contracts in-

---

[55]See Section 5.10(D).

volving brokerage services, courts have generally refused to employ exceptions.

Fiduciary Relationship. Hiring a broker creates an agency relationship between the client and the real estate broker, with mutual fiduciary obligations. The parties must disclose pertinent information and any conflict of interest to each other, not make profits at the other's expense, and behave fairly toward each other. There has been a tendency to hold brokers liable to the buyer for secret profits made by telling the buyer the seller will take only a price higher than the seller wishes, buying at the lower price from the seller and reselling at a higher price to the buyer.

B. Contracts to Purchase an Interest in Land

When Are Parties Bound? One problem that recurs in contracts for the purchase of land is whether the parties intend to wait until a more formal writing is signed before they bind themselves. Ordinarily, the history of a contractual transaction begins with preliminary negotiations. One party puts forth a proposal to which the other party may either agree or make a counter proposal. At some stage, there is general agreement as to the basic terms of the transaction. For example, the parties may agree as to the land that is to be sold, the price, the financing terms, and the date of transfer of possession.

Many other elements exist to this contract that are not likely to be considered by the parties at this stage. These elements will be agreed upon when the agreement is transformed into a more formal written contract. The type of deed, the allocation between the parties of the risk of loss between the time the contract is formed and the delivery of the deed, the furnishing of title insurance, and provisions for an escrow arrangement are not likely to be discussed at the time the preliminary agreement is reached.

Sometimes the parties may make a notation informally on a piece of paper as to the basic terms to which there *has* been agreement. They may, in addition, sign this informal writing. Have they intended to bind themselves by this informal writing so that neither party can withdraw at its own discretion? Is the informal writing only a tentative agreement, with the binding contract to be formed

when both parties have assented to a more formal writing that will contain details omitted in the informal writing?

Whether the parties are to be bound at the preliminary stage or only upon their assent to a more formal document depends on the intention of the parties. However, criteria exist to determine such intention. Objective criteria are useful if the matter is disputed and both parties are likely to state that they had contrary intentions. The law examines the complexity of the transaction, whether either or both parties had acted in reliance upon the earlier informal agreement, and whether the law provides sufficient guidelines to fill in the gaps not yet resolved.

Generally, the law finds that the parties in land transactions do *not* intend to bind themselves until there is assent to a more formal writing. A transaction involving the sale of an interest in real property usually involves much money and is not routinely made by one or both parties. It is better to delay binding the parties until they or their advisers have had an opportunity to examine provisions dealing with the problems not anticipated at the time the informal agreement is negotiated. However, parties *can* bind themselves at the time of the informal document, particularly if standard provisions accepted by the community can satisfactorily fill in the gaps.

Agreement to Agree. (See Section 5.06(F).)

Options and Contingencies. In these contracts, purchasers may not be willing to bind themselves before inquiring into such matters as financing, costs of construction, and the uses to which the property can be made. They may, instead of binding themselves, wish to obtain an option under which the seller will agree to sell, while the buyer retains the freedom of choosing to buy.

As noted in Section 5.06(B), most offers are revocable even if the offer states that it will not be revoked for a specified period of time. To make the offer irrevocable and to create a binding option, the person giving the option—in this case the seller—must receive something for the option. Usually this is accomplished by the buyer's paying a nominal amount, such as one dollar or ten dollars. If a written option is made for a reasonable period for a purchase at an adequate price, no inquiry will be made into the question of whether this nominal amount was bargained for or if this was a price actually paid for the option. Some options are bought for an amount of money that reflects the value of the option. Although the payment of the nominal amount does not always reflect the value of the option, payment of this amount usually will make the option irrevocable.[56]

The seller may not be willing to grant a one-sided option. If the buyer cannot obtain an option, the buyer may agree to buy the land subject to the ability to obtain financing of a specified amount and at a specified rate. Buyers may want to condition their obligations to purchase upon obtaining a variance from the zoning laws, upon selling their own homes, or upon other important events occurring before they are obligated to complete the purchase. Sometimes the buyer's promise is phrased in this way: "I will buy if I obtain financing of 80% of the purchase price for a loan not to exceed 20 years at a rate not to exceed 8%." (The first edition in 1970 stated $7^1/_2$%, but the fourth edition in 1989 stated 12%. This gives some idea of the fluctuation in long-term interest rates.) In such a situation, the buyer may not have made a promise at all but has given a conditional acceptance. Neither party is bound to perform. However, more likely, the buyer has made an absolute promise to perform but inserted a number of contingencies in the contract.

Sometimes the occurrence of contingencies that condition the obligation to perform is largely within the control of the purchaser. Suppose the contingency is the buyer selling his own home. Buyers may not try hard to sell their house if they have second thoughts about a house they may have contracted to buy. They may not make attempts to obtain financing if financing is a contingency spelled out in the contract to purchase the land.

Because occurrence of such conditions is to a substantial degree within the buyer's power, does this destroy the two-sidedness of the contract? It may appear that the seller is bound *absolutely* to perform, while the buyer can retain the power to determine whether or not to perform. For this reason, some courts have held the contract too one-sided for enforcement. Most, though, enforce the contract by implying promises of best efforts or of

---

[56]Restatement (Second) of Contracts § 87 (1981).

good faith on the part of the buyer to cause the conditions to occur. Contingencies are common, and generally the parties intend to arrive at a valid agreement.

Sometimes sellers wish to protect themselves by the use of contingencies. They may not wish to sell their house unless they are able to purchase another house or be admitted to a community for senior citizens. The same principles discussed in connection with protective contingencies for the buyer apply if the seller obtains this protection.

Formal Requirements.   (See Section 5.10.)

Destruction Before Conveyance or Transfer of Possession.   Often a substantial period of time must elapse before the title is actually conveyed or possession is transferred. Suppose the house is destroyed before transfer of either title or possession? Does the buyer still have to pay? Is a seller who cannot deliver the premises in the state promised by the contract guilty of breach of contract?

Unless covered in the contract, the result varies from state to state. Most states force the buyer to complete the purchase even if the house has been destroyed. Some relieve the buyers even if they have gone into possession. Some have adopted the Uniform Vendors and Purchasers Risk Act, which places the risk of loss on the buyer if *either* the title has been conveyed or the purchaser has gone into possession.

The parties should consider this matter at the time the contract is made. If the contract allocates the loss to one party, that party should protect himself against this loss by insurance.

Delay.   Usually the contract states a date for the payment of money and the issuance of the deed. Yet performance is often delayed, usually because of delay in obtaining the loan or clearing title.

Suppose no contract provision deals with delay. The injured party can accept delayed performance and receive damages for the delay. Although delay can terminate the other party's obligation to perform, the frequency of delay in such transactions is likely to preclude such a drastic result in actual practice. If a claim is made for the land or money *itself*, that is, for specific performance, in the absence of a contract provision to the contrary, time would not be "of the essence." Although delay is a breach, it does not terminate the other party's obligation to perform.

Contracts usually state that time is of the essence. Suppose the party guilty of delay has relied on promised performance by the other party, either by making other commitments or by paying money under the contract. In such a case, the law may not permit termination, particularly in the case of delay in making payments by the buyer if the latter tenders the amount due with interest. Decisions, though, will vary, making generalizations precarious. About the best that can be said is that if the delay is truly minimal and not caused by fault and if the innocent party is not damaged by the delay, termination will very likely be refused.

Even though late performance *may* not end the contract, the party guilty of unexcused delay must pay any damages caused by the delay. In the case of delayed performance by the buyer, interest is usually the measure of recovery; by the seller, it is usually loss of use or rental value.

Implied Warranty.   (See Section 24.10.)

Remedies.   (See Section 6.06.)

C. Escrow

In many transactions involving the purchase of real property, an escrow arrangement is used. The seller deposits the deed and any other documents that are ultimately to go to the buyer with a third party, commonly a title company or a lending institution. The instructions to the third party escrow holder may be to deliver these documents to the buyer when the buyer has delivered to the escrow holder the purchase price (and perhaps has met other conditions agreed to by the parties). The escrow holder acts as a clearinghouse to ensure that the buyer does not have to pay without getting the deed and that the seller does not have to deliver the deed without being paid. Usually the escrow holder is a stranger, although the attorney of one of the parties may be used as the escrow holder.

Escrows are used for a number of reasons. If both parties begin their performance by opening an escrow and making a deposit into the hands of the escrow holder, it is less likely that the deal will fall through. The advance execution of the deed makes it less likely that there will be delay and difficulty if the seller dies, as compared with those situations where the seller does not make out the deed until the purchaser has put up the money. The escrow assures each party that when it performs it takes

no risk that the other party may not perform. Escrows are also useful when there are a number of interested parties to the transaction. In many transactions involving the sale of land, there will be a prior lender who will want the debt paid off and a new lender who is advancing money to the purchaser. A substantial portion of the new loan will be used to pay off the prior lender. A number of mechanical details can be taken care of by escrow holders, especially when the escrow holders are professional escrow companies.

Sometimes problems develop because the escrow instructions on standard printed forms frequently vary from the contract of sale. The documents *as a whole* must be examined to find the common intention of the parties. The escrow instructions and the contract for the sale should be consistent.

D. Title Assurance

By Seller.   Purchasers of land want to be reasonably certain that they will acquire undisputed ownership of the land. They want the title, usually marketable title, promised by the seller. This is a title sufficiently free from dispute that it would be accepted by most buyers in the community. Buyers may be concerned with the possibility that someone will assert a claim against the land that they will either have to buy off or litigate. If the seller does not have the title promised, the buyer has an action against the seller for breaching the warranty to deliver a marketable title free of encumbrances other than those specified. However, the seller may be out of the jurisdiction or unable to pay any judgment awarded to the buyer.

By Attorney.   The buyer often wants better protection than a legal claim against the seller. For this reason, the buyer may require that the seller furnish a title abstract (a summary of all the recorded transactions that pertain to the land). The buyer asks his attorney to review the abstract to see whether the seller has the title claimed and to find any encumbrances on the land. This system is used in small communities in the United States.

The principal defect of such a system is that any recovery against the attorney must be based on the attorney's negligence, which may be difficult to show. Even if the attorney was negligent, it may be that the damages caused by the attorney's negligent performance are substantially greater than his ability to pay. The examining attorney does not assume responsibility for the correctness of the abstract that has been compiled by an abstract company. Any action for an improper copying would have to be brought against the abstract company. The abstract is not a reproduction of the original documents on record but is a condensed statement of the key facts in each transfer.

By Title Company.   For the reasons just discussed, much of the title assurance work in larger cities and in many states is handled by title insurance. The title insurance company issues a policy at the time of the transfer, insuring the title against certain defects that appear in the land records. The title insurance company checks through the recorded documents pertaining to the property. If it makes a mistake, the title insurance company is usually capable of paying damages.

Certain defects, or "clouds on title," cannot be discovered from a search of the land records. Such title defects usually are not covered by title insurance. However, most buyers are satisfied by the title policy, because such defects are rare.

Title insurance is not absolute insurance. The policy only warrants that an expert company has checked the records and can pay if it makes a mistake. A buyer who wishes to know the scope of the insurance should read the title insurance policy.

E. Security Interests

Purpose.   Land transactions frequently involve substantial sums. Usually the purchaser is unable to pay the entire purchase price and frequently borrows money from a lender to buy the property.

Lenders choose borrowers who are likely to repay the loan. However, lenders want a security interest in the property out of which they can be repaid if the borrower is unwilling or unable to pay back the debt. For example, a purchaser who wishes to buy property for $100,000 may have only $10,000 to put down on the property and may wish to borrow the balance from a bank. If the bank will lend $90,000, the buyer will usually sign a note for the $90,000, with an agreement to make specified payments at specified intervals. The buyer will also give a security interest to the lender. The security interest may be created by a mortgage or, in some states, by what is called a deed of trust.

Operation.   Generally the buyer (borrower) of the land takes title and the lender has a security inter-

est in the land. If the borrower does not pay in accordance with the loan obligation, the holder of the security interest has the right to foreclose on the security. In some states, this can be done without judicial sale.

In most cases, the lender will go to court and ask for a judicial sale of the property. The lender will be paid out of the proceeds of the property to the extent of the unpaid debt. If the sale does not net enough money to pay off the debt, a deficiency judgment is awarded by the court to the lender against the borrower. In some states, deficiency judgments are prohibited by law in certain transactions. In these states, the lender can recover only the amount obtained by the judicial sale.

Secured transactions can become very complicated. Many parties may demand the proceeds of the sale, which can result in difficult priority problems. Sometimes there are both a first and a second mortgage—sometimes even a third. There may be construction loans on the property, and mechanics' liens and tax liens may be asserted. In such cases, litigation is often necessary to unravel the competing claims.

Equity of Redemption.   Usually the mortgagor-debtor has an *equity of redemption*. The foreclosure sale will not be effective to transfer ownership permanently until a specified period set by statute has elapsed. During this statutory period, the mortgagor-debtor may pay the amount owing and redeem the property. During the statutory period, the purchaser at a foreclosure sale does not have good title. This can affect the amount the buyer is willing to bid at the foreclosure sale.

Land Contract.   A land contract is a security device under which the transfer of ownership or the delivery of the deed is withheld until all payments have been made. It may be used when the buyer does not have enough money to make a substantial down payment. Suppose the buyer pays for a long period and then is unable to pay. The buyer may have paid much more than the value received by occupying the property. The buyer may have also made improvements. It seems unfair to have to forfeit the interest in the land for failure to make some payments. For this reason, a few states treat the land contract as a security device with an equity of redemption. Most allow the seller to retake and keep all payments.

PROBLEMS

1. A is a president of Acme Corporation and has two years left on a written four-year contract at an annual salary of $100,000. He is approached by a competitor and offered $125,000 a year. A tells this to the board of directors of Acme who raise his salary to $125,000. A agrees. Is this modification of the original agreement legally enforceable? If not, how can it be made enforceable?

2. The following is a statement of the facts in a court case. After reading it carefully, answer questions a through f.

A is an established firm specializing in a form of foundation work known as grouting, which consists of pressurized injection of a cement-based mixture into the soil underlying a building for the purpose of arresting subsidence and in some cases actually raising foundation walls. In the summer of 1990, B, a small construction firm, entered into a contract with Brazilian Embassy to stabilize and partially reconstruct a building in the District of Columbia known as the Brazilian Annex. The Annex was a relatively old building constructed in large part on filled ground. The structure had sunk on all four sides with the result that the floors bowed in the middle. B's task, among other things, was to stabilize the structure to prevent further sinking and to raise certain parts of the foundation, particularly the northeast corner, in order to partially alleviate the unevenness of the floors. This lawsuit arises from B's decision to ask A to perform this aspect of the job using A's grouting technique in lieu of alternative methods available.

During early August 1990, Davis, a vice president of A, and Downey, B's president, discussed the project in a number of conferences and telephone calls. On August 14, 1990, Davis submitted a written proposal to which was annexed a standard set of conditions. The proposal, to the extent here material, made no guarantee that efforts either to stabilize or to lift the building would be successful, made no commitment as to time of completion of the job, and contemplated that the work would be billed at a per diem rate without any stated limitation on the total price of the job. After further conversations with Davis—the precise contents of which are hotly disputed—Downey sent Davis a telegram on August 25 indicating that A's proposal was accepted subject to "verbally agreed changes" and that a signed revision would follow. The next day Downey prepared and signed an edited version of A's written proposal to be mailed to Davis. The purport of the revision was to indicate that A was committed to stabilize the building and to lift the northeast corner by at least one and one-half inches and that the job would be undertaken in approximately ten days with a maximum payment of $20,000. A never received this document or inquired as to why it had not been received as promised.

Both parties proceeded on the assumption that they had come to some type of agreement, and work on the site began August 28. Although A was eventually able to stabilize the perimeter of the building, it was unable, despite protracted effort, to achieve the desired rise in the northeast corner of the building. As the work proceeded, A reported in writing daily to B, and B therefore had full knowledge that A was proceeding without marked success. At the end of eleven days of work, A billed B at the per diem rate in A's proposal, and the work was paid for in the total amount of $9,936.75. B at no time indicated to A that it should stop working. Downey constantly reiterated, however, that B had only $20,000 to pay for the work. A never consented to the $20,000 cap and urged B to seek an adjustment in the contract price from the Brazilian Embassy which B consistently refused to do. After approximately twenty-five days of continuous work, A concluded that it would not be possible to lift the building one and one-half inches and, requiring the equipment for another job, informed B that it was terminating work. It did no more work.

a. What would A argue to support its claim?

b. What would B argue to support its claim?

c. Was there a contract? If so, what were its terms?

d. Suppose there were no contract. Can A recover from B for the work A performed? Can B recover the payments it made? (Please study Section 5.07(E).)

e. How could this misunderstanding have been avoided?

f. How does this case differ from *Western Contracting Corp. v. Sooner Construction Co.* set forth in Section 5.06(B)?

CHAPTER SIX

Remedies for Contract Breach: Emphasis on Flexibility

SECTION 6.01 An Overview

Determining the remedies available for breach of contract is much more difficult than determining whether a valid contract has been made. It is exasperatingly difficult to predict *before* trial what the law will require that a breaching party must pay. Even after all the evidence has been produced at the trial, it can be very difficult to determine the precise remedy that the party entitled to relief should receive. This sometimes leads to a trial procedure under which remedies are considered *before* going into the issues of whether a valid contract has been formed and whether it has been breached. If very little can be accomplished remedially, it makes little sense to have a long, protracted lawsuit to establish the claim.

Although the function of awarding a remedy for contract breach is to compensate the injured party (the exceptions to this are discussed in Section 6.04), measuring the losses incurred and gains prevented can be very difficult. There may be agreement on general objectives, yet the implementation of these objectives can be difficult because of the great variety of fact situations, the different times at which a breach can occur in the history of a contract, the variety of causes that may generate the breach, and the different judicial attitudes toward breach of contract itself.

The common law has developed conventional formulas that are applied in particular cases designed to implement the basic compensation objective. These formulas may not appear to achieve a just result when they are applied. Borderline cases make it difficult to determine which formula should be applied. The range of remedies, both as to type

and amount, can give discretion to juries, trial judges, and appellate courts. For example, if one party seems to have taken unfair advantage of the other, doubts may be resolved in favor of the latter. The reason for the breach, though perhaps not exculpating the breaching party, can be influential in measuring the award. The relative abilities of plaintiff and defendant to bear the loss can be an influential factor, although often an unstated one.

Generally, when juries are used, they are given considerable latitude to determine the award. The jury award is likely to be upheld if it is based on substantial evidence or reasonable inferences from the evidence, unless the award seems to result from passion or prejudice. Although compromise can play a strong part in jury determinations, the failure to require a jury to be specific as to the reasons for its award can give the jury a mechanism for achieving what it believes to be a just result even if it would differ from what the law appears to require.

SECTION 6.02 Relationship to Other Chapters

This chapter explores *generally* the judicial remedies awarded for breach of contract. Succeeding chapters apply these basic principles to Construction Process disputes.

Usually claims by a design professional are for services rendered for which payment has not been made. Claims by the owner against the design professional usually relate to a delay caused by the design professional, defective design, or a failure to monitor the contractor's performance. These topics are considered in greater detail in Sections 12.14 and 14.10.

Claims by owners against contractors usually involve losses asserted because of unexcused contractor delay (covered in Chapter 26) or failure by the contractor to build the project as required (covered in Chapter 24). Claims by the contractor against the owner are usually based on payment for work performed that has not been received or on increased cost of performance attributable to the owner (covered in Chapters 22, 23, and 25). Measuring the value of the claim is discussed in Chapter 27.

This chapter, in addition to providing a broad sketch of contract remedies generally, looks briefly in Sections 6.09 and 6.10 at special remedial problems relating to employment contracts and contracts to purchase an interest in real property (discussed in Sections 5.11 and 5.12, respectively).

SECTION 6.03 Money Awards and Specific Decrees: Damages and Specific Performance

Judicial remedies can be divided into judgments that simply state that the defendant owes the plaintiff a designated amount of money (sometimes called the money award) and judgments that specifically order the defendant to do something (specific performance) or to stop doing something (injunction). It is important to recognize the essential differences between money awards and specific decrees.

The ordinary court judgment (the money award) is not a specific order to the defendant to pay this amount to the plaintiff. In the absence of voluntary compliance by the defendant with the court decree, the plaintiff must take the initiative to ask law enforcement officials to seize property of the defendant, now a judgment debtor, that the law does not exempt. If this is done—often a costly and frustrating process—the property is sold to satisfy the judgment. Any amount remaining after payment of the judgment and costs of sale is paid to the defendant. Enforcement of court judgments in this fashion is often costly where successful and often is unsuccessful.

The specific decree is a much more effective remedy. Failure by the defendant to comply with the court order can be the basis for citing the de-

fendant for contempt of court. The defendant is brought before the judge to explain why she has not complied with the decree. If the explanation is not satisfactory, the judge can punish the defendant by fine or imprisonment or coerce the defendant into performing by stating that the defendant must pay a designated amount or stay in jail until she performs.

With some exceptions,[1] the claims central to this treatise will, if successful, result in money awards. It is the principal remedy sought when claims are made that someone has not performed a contract for design or construction services. Courts have been very reluctant to issue orders (specific performance) requiring design professionals or contractors to perform in accordance with their contractual obligations. By the time the dispute reaches the courts, continued contact between the disputing parties is likely to generate additional disputes that can place a substantial burden on the judge. (A specific decree is a personal order by the judge.)

SECTION 6.04 Compensation and Punishment: Emergence of Punitive Damages

As stated in Section 6.01, the basic purpose in awarding damages for breach of contract is to compensate a party who has suffered losses or who has been prevented from making gains. The common law did not look upon contract breach as an immoral act that might justify punishment. This may have been based on the importance in a market-oriented society of persons engaging in economic exchanges.

Excessive sanctions may discourage persons from making contracts. In the past few years, courts have been willing to award punitive damages for certain types of contract breach, justifying this by concluding that a particular breach was tortious. Wrongful dismissals claims discussed in Section 5.11(B) have begun to generate punitive damages, which are awarded to the claimant to prevent the conduct from being repeated by the defendant or to make an example of the defendant to deter others.

---

[1]See Sections 11.02(B), 28.07(D) (mechanics' liens); Sections 6.09(B), 6.10 (injunctions and specific performance); Section 30.12 (specific performance of arbitration award).

Punitive damage awards have been quite rare in construction disputes. However, awarding punitive damages in the construction contract context, particularly against contractors, has been increasing. Specific illustrations are given in Section 27.10.

SECTION 6.05 Protected Interests

Increasingly, law looks upon the varying ways in which the party who makes a contract can be protected from breach of the other party. This protection is frequently described as encompassing restitution, reliance, or expectation.

The most protected of the three interests, *restitution,* seeks to restore the status quo that existed before the contract was made by awarding the plaintiff any benefit the plaintiff has conferred on the defendant. It is most protected because it is relatively easy to establish and it involves both a loss to the plaintiff and a gain to the defendant. Its importance in measuring claims by design professionals and contractors for services performed justify its inclusion in Sections 12.14(C) and 27.02(E).

A plaintiff protects its *reliance interest* by obtaining reimbursement of expenses from the defendant that have been incurred either in reliance upon the contract or expenses incurred before the contract was made that have become valueless because of the breach. Such a remedy looks backward to a point in time even before the contract was made (something similar to restitution) but also looks forward. If the defendant can establish that the expenditure sought would never have been reimbursed in the venture engaged in by the plaintiff, the plaintiff cannot recover from the defendant.

Suppose a contractor failed to build a building in which the owner intended to manufacture Nehru jackets, a sartorial meteor of the 1960s. Suppose further that the owner invested a large amount of money to promote these jackets. If the contractor can establish that market saturation or changing fashions meant not *one* jacket would have been sold, the owner could not recover the marketing costs, as they would not have been reimbursed from sale proceeds.

The third protected interest, *expectation,* looks forward and seeks to place the plaintiff in the position it would have found itself had the defendant performed. One way of accomplishing this objective is to order the defendant to perform as prom-ised, a remedy generally unavailable in contracts to perform design or construction services. The same objectives can be achieved by a money award, however.

Suppose a design professional or contractor refuses to perform its contract. Awarding an amount that would enable the owner to hire someone to perform identical design or construction services would protect the owner's expectation interest. As this is a formula used very frequently in claims by owners, it is discussed in greater detail in Chapter 27.

SECTION 6.06 Limits on Recovery

Using rules described in this section, the law limits recovery by a claimant who has suffered losses that can be connected to the other contracting party's failure to perform in accordance with the contract. These rules often provide insurmountable obstacles to the claimant, a phenomenon that has prompted some to contend that the law does not give adequate protection to those who suffer losses because of contract breach. Undoubtedly, these obstacles can make recovery difficult. Yet they can also be looked upon as rules that encourage parties to make contracts without inordinate fear that their nonperformance will expose them to unpredictable or devastating damage claims.

These doctrines *can* limit recovery but do not *invariably* do so. To the extent that any generalization can be made, modern law seems *more* willing to grant greater compensatory damages than earlier periods in American legal history.

A. Causation

The claimant must show that the defendant's breach has caused the loss. Losses may be caused by more than one actor, and on occasion, causal factors can include other events and conditions. For example, the contractor's performance may be delayed by the owner, by strikes, by material shortages, *and* by the contractor's poor planning. It may be difficult if not impossible to establish the amount of the loss caused by contributing causes or conditions.

The defendant will be responsible for the loss if its breach was a substantial factor in bringing about the loss. It need not be the sole cause of the loss. Generally, this question is determined by the finder of fact, sometimes the jury, and sometimes the trial judge. Any judicial finding regarding causation not based upon guesswork will be upheld.

Causation problems can be particularly difficult when claims are made for breach of a contract to design or to construct, mainly because of the number of different entities that participate in the process and also because of the variety of conditions which may affect causation.

One common fact pattern involves tracing responsibility for a defect both to defective design and improper workmanship. Similarly, delayed performance by the contractor may be traceable to poor management by the contractor and excessive design changes made by the design professional. These difficult problems are treated in Sections 24.06 and 26.06, respectively.

B. Certainty

A claimant must prove the extent of losses with *reasonable* certainty. One court described the certainty rule in these terms:

> Courts have modified the "certainty" rule into a more flexible one of "reasonable certainty." In such instances, recovery may often be based on opinion evidence, in the legal sense of that term, from which liberal inferences may be drawn. Generally, proof of actual or even estimated costs is all that is required with certainty.
>
> Some of the modifications which have been aimed at avoiding the harsh requirements of the "certainty" rule include: (a) if the fact of damage is proven with certainty, the extent or the amount thereof may be left to reasonable inference; (b) where a defendant's wrong has caused the difficulty of proving damage, he cannot complain of the resulting uncertainty; (c) mere difficulty in ascertaining the amount of damage is not fatal; (d) mathematical precision in fixing the exact amount is not required; (e) it is sufficient if the best evidence of the damage which is available is produced; and (f) the plaintiff is entitled to recover the value of his contract as measured by the value of his profits.[2]

Certainty in the context of construction claims is examined in Section 27.04.

C. Foreseeability: Freak Events and Disproportionate Losses

A series of improbable events can combine to lead to large losses. In the contract context, suppose a contractor is hired to build a high-rise office building. Shortly after the work is completed and accepted, the electrical system fails because of improper workmanship, causing the elevators to be out of service for two hours. During those two hours, the building owner has scheduled an interview with a prospective tenant who is thinking of leasing three floors of office space at a rental very attractive to the owner. Because the elevators are out of service, the prospective tenant decides to rent elsewhere. Should the contractor be liable to the building owner for the extraordinary lease profits that the owner lost?

Clearly, the contractor's breach, perhaps a minor one, set off a chain of events that culminated in the loss of extraordinary rental profits. The improper installation caused an electrical failure, causing the elevators to malfunction. The prospective tenant was in the building at that very time. The prospective tenant's decision not to rent may have been caused by the tenant's irascibility or ignorance. This series of possible but improbable events has caused the building owner to suffer a loss. Is it fair to transfer this loss to the contractor?

The loss was an extraordinary loss and one that could not be reasonably foreseen at the time the contract was made. Contrast this with an electrical malfunction that causes a fire. Such a loss falls more easily into the type of loss that can be reasonably foreseen at the time a contract is made. Also, such a risk is commonly covered by property insurance taken out by the owner and public liability insurance maintained by the contractor.

The leading common law case dealing with what are sometimes called consequential damages is the English case, *Hadley v. Baxendale*,[3] which gave rise to the requirement that losses be reasonably foreseeable as a *probable* result of the breach. A shipper claimed against a carrier when the latter's delay in returning a shaft that had been sent to the factory for repair caused the shipper's plant to be

[2]*M & R Contractors & Builders, Inc. v. Michael*, 215 Md. 340, 138 A.2d 350, 355 (1958).

[3]156 Eng.Rep. 145 (1854). See Restatement (Second) of Contracts § 351 (1981).

shut down. The amount paid for shipping the shaft was disproportionately small compared with the cost of an entire plant shutdown. The court concluded that the carrier is liable for losses *naturally* resulting from the breach and other losses, the possibility of which is brought to the carrier's attention at the time a contract is made.

Advance awareness of the risk gives the carrier the chance to adjust its rates for performance or decide to forgo the transaction. Similarly, is it fair to place the responsibility for a building shutdown on an electrician who is called to make a minor repair in a high-rise building but who fails to perform in accordance with the contract obligations? This may be too onerous in light of the small profit earned. Care must be taken, however, to distinguish risks that are insurable and the premiums paid to transfer the risk to the insurer as part of a performing party's overhead.

D. Avoidable Consequences (The Concept of Mitigation)

The rule of avoidable consequences is another limitation. A claimant cannot recover those damages that the claimant could have reasonably avoided. Sometimes this rule is expressed as one that requires the victim of a contract breach to do what is reasonable to mitigate or reduce the damages. This limiting rule relates to the requirement that the breaching party is responsible only for those losses that its breach has caused. If the loss could have been reasonably avoided or reduced by the claimant, the claimant cannot transfer that loss to the breaching party.

This limitation has not been a favored one. The law not only has placed the burden of establishing that the loss could have been avoided by the claimant on the breaching party but also has been hesitant to give the rule much scope. For illustrations in the construction context, see Section 27.07. (The avoidable consequences doctrine raises special problems in employment contracts and is discussed in Section 6.09.)

E. Lost Profits

Claims for lost profits, extraordinary or ordinary, have caused particular difficulty. They involve the preceding limitations on contract recovery, causation, certainty, foreseeability, and avoidable consequences.

Profits on the *very* contract in question, such as claims by a contractor for lost profits on the construction contract, are part of the routine measurement used in contractor claims (discussed in Sections 27.02, 27.04, and 27.06).

This subsection comments on claims by contractors for profits on *other* contracts it asserts it would have obtained. (Depriving the owner of lost profits, though similar to those made by the contractor for profits on other contracts, is covered in Section 27.06(B).)

Suppose a contractor is unjustifiably terminated, either before starting performance or during performance. Suppose further that the contractor claims that completion of this contract would have earned additional profits through the award of other contracts. Recovery of such profits requires that a formidable number of hurdles be overcome.

First, even if the contractor can establish the loss of other contracts, were the contracts lost because of the owner's refusal to allow the contractor to complete this contract or for other reasons? Second, how certain is it that the contractor would have received not only the other contracts but also the amount of profit that might have been earned? Third, was it reasonably foreseeable at the time the contract was made that unjustified termination of the contract would have caused the contractor to lose other contracts? Fourth, could the contractor by reasonable effort have avoided losing any contracts that were lost?

Some states deny recovery of lost profits by new businesses, proof being too uncertain that it would have earned profits. The modern tendency has been to treat this issue as any other factual issue. Even a new business can *try* to prove it would have earned additional profits from other contracts.

F. Collateral Source Rule

Suppose the defendant has breached but the plaintiff has been compensated by a third party. Can the defendant show that the plaintiff has not suffered a loss?

If the loss is caused by a tortious defendant— that is, one who has not lived up to the legal standard of care—the defendant cannot reduce or eliminate its liability by pointing to the loss having been compensated by a third party. The third party in such cases is called a *collateral source*. Any acts of that source are collateral and cannot be taken into account. Typically, tort cases involve harm to persons, and the collateral source is an insurance company, an employer who provides medical benefits, or the state that provides replacement for lost earnings or medical losses. Because the tortious defendant is considered a wrongdoer, the collateral source rule deprives the defendant of a windfall created by contract protection purchased by the injured party or benefits awarded by the state.

Whether the collateral source rule deprives parties who have breached contracts of the right to point to the loss having been compensated by a third party is not clear. The rule and its application to construction contracts are discussed in Section 27.08.

There is a trend toward treating tort and contract claims similarly and basing the application of the rule on other criteria.

G. Contractual Control: A Look at the U.C.C.

The preceding discussion has assumed an absence of any controlling contract clauses regulating the remedy. Yet it is becoming more common for contracts to regulate the remedy. Some contracts specify or liquidate the damages when establishing actual damages would be difficult, if not impossible, to prove. Some contracts contain language insisted upon by one of the contracting parties to either exculpate itself from responsibility or limit its exposure.

When remedies are specified, it is common to state in the contract that they are not exclusive. For example, the American Institute of Architects (AIA) provides remedies for termination. Yet A201, ¶ 13.4.1, states that remedies specified are not exclusive. Similarly, evidence must be clear that a specified remedy is to be exclusive before it is given that effect.

Contractual control of remedies is treated in other parts of this treatise.[4] The Uniform Commercial Code, which governs transactions in goods, has been having increasing impact on construction contract disputes. Section 2–719 allows the parties to provide for remedies by contract but states

> Where circumstances cause an exclusive or limited remedy to fail of its essential purpose, remedy may be had as provided in this Act.

Frequently, sellers of goods seek to limit their obligation for breach to repair and replacement of defective goods. In *Coastal Modular Corp. v. Laminators, Inc.,*[5] a contractor was allowed to recover damages from a supplier of defective panels for a construction project despite a contract providing that the remedy was limited to repair and replacement. The court held that the remedy had failed of its "essential purpose" under § 2–719. The contractual remedy would apply only if the defect had been discovered while the work was in progress. Here the defect had been discovered after completion. Repair and replacement would not have been a viable remedy.

The limited remedy was applied in *Price Brothers Co. v. Charles J. Rogers Construction Co.*[6] in a slightly different context. The pipe subcontract stated that the supplier-installer would pay only for damages relating to *above-ground* repair and replacement. The pipe failed, and the contractor incurred large expenses removing and replacing the defective pipe *below* ground. The contractor sought to recover these expenses from the supplier-installer.

The court held that the contractual remedy did not fail under § 2–719. The event, though rare, was foreseeable. Also, the contractor received the benefit of a lower price. The stated remedy in the contract *would* have failed had the replacement pipe *also* been defective.

H. Noneconomic Losses

To this point, with the exception of the discussion relating to punitive damages, this chapter has focused on compensatory economic losses—losses that can be established precisely or roughly in the marketplace. Increasingly, claims for breach of contract, sometimes tied with tort claims, are made for noneconomic losses, such as emotional distress, caused by a breach of contract.

Generally, contract law does not grant recovery for such losses. Denial has usually been based on the lack of foreseeability. However, a more acceptable rationale is that contracting parties should not be liable for potentially open-ended and freak losses that are extremely difficult to measure in economic terms. A homeowner who cannot take possession of a new residence when promised can certainly suffer emotional distress. But the likelihood and gravity of such distress generally depends on the emotional and psychological makeup of the homeowner.

Some decisions that have classified the contract breach as tortious have allowed recovery for emotional distress. *McCune v. Grimaldi Buick-Opel, Inc.*[7] allowed recovery where an employer had failed to perform its contractual obligation to provide medical insurance for an employee, exposing the employee to emotional distress caused by bill-collection tactics of the hospital. Such a contract was found to be not simply a commercial one but also one geared to provide mental solicitude to one of the parties. Increasingly, though modestly, noneconomic losses are beginning to be awarded in contracts or conduct related to the Construction Process. For cases in a construction context, see Section 27.09.

SECTION 6.07   Cost of Dispute Resolution: Attorneys' Fees

This treatise has suggested that contract remedies should compensate the injured party. Yet full compensation is rarely achieved. The rules that determine whether and how much damages are avoided sometimes make it difficult to achieve full compensation. Even more important, the winner cannot recover its cost of litigation, an item that has risen to staggering proportions. American common law (judge-made) does not transfer the prevailing party's attorneys' fees to the other party unless the claim is based on a contract that contains a provision providing for a recovery or recovery is based on a statute granting attorneys' fees. Under extraordinary circumstances, such as commission of an *intentional* tort or the losing party's claim having been vexatious or frivolous, such costs can be recovered.

Denial of attorneys' fees is based on the reluctance of American law to discourage citizens from using the legal system. Although an unsuccessful claimant may have to bear the cost of its own fees, the law absolves the claimant of the responsibility of bearing the other party's costs.

This rule has led to increased legislation authorizing attorneys' fees as part of costs in consumer

and civil rights cases. It has led to reciprocal attorneys' fees legislation in some states. These statutes provide that if one party can recover attorneys' fees under a contract, the other party can do so even though the latter is not specifically granted this in the contract. Some statutes grant attorneys' fees to the prevailing party where a mechanics' lien has been sought. Some give attorneys' fees when claims are made on surety bonds. A few statutes allow attorneys' fees *generally* for contract actions.

Documents published by the AIA do not, except for indemnification, provide for attorneys' fees, each party bearing its litigation costs.

AIA documents require arbitration under the Construction Industry Arbitration Rules administered by the American Arbitration Association. These rules do not specifically provide for attorneys' fees. They deal mainly with other costs of arbitration, including the arbitrator's fee. Although the arbitrator would very likely have freedom to award attorneys' fees, arbitrators customarily do not award these fees.[8]

Local law, currently in a state of ferment, must be consulted by those who draft contracts or those who must counsel as to the availability of attorneys' fees to the prevailing party. Although discussion has centered on attorneys' fees, the other formidable costs of construction litigation—particularly complex litigation that involves defects and impact claims—cannot be ignored. These claims generate immense costs to reproduce, classify, analyze, and store documents as well as the staggering costs involved in conducting pretrial discovery. To these costs must be added the costs of preparing exhibits and retaining expert witnesses and the nonproductive costs incurred by personnel in preparing for the lawsuit. Those who draft contracts may wish to deal with these costs. Such costs should remind the parties they must use every effort to avoid litigation.

SECTION 6.08   Interest

Many cases have involved recoverability of prejudgment interest. Claimants often seek interest from a time earlier than entry of court judgment. Perhaps this is due to the ease with which a claim

---

[7]45 Mich.App. 472, 206 N.W.2d 742 (1973).

[8]But see *Harris v. Dyer,* 292 Or. 233, 637 P.2d 918 (1981), discussed in Section 30.13.

for interest can be tacked onto a claim for work performed or defective work. In any event, states vary considerably in their treatment of claims for prejudgment interest.

The varying legal rules dealing with prejudgment interest are caused by the differing judicial attitudes toward the desirability of *complete* compensation on one hand and on the other hand punishing a defendant when the latter exercised a good-faith judgment not to pay that turned out to be incorrect.

Although case holdings and statutes have many slight variations, they follow three principal rules. One limits prejudgment interest to liquidated (specific) amounts or unliquidated amounts that are easily determinable by computation with reference to a fixed standard contained in the contract without reliance on opinion or discretion.[9] A party against whom a claim is made should be able to avoid prejudgment interest by tendering the amount due. Unless the amount is known or easily determined, tender cannot be made. This ignores the likely possibility that payment is not made because of a dispute over the validity of the claim, not the amount due were the claim held to be valid.

Some courts and statutes give the judge or jury discretion to determine whether interest should be awarded.[10] Often discretion takes into account whether the refusal by the defendant to pay was vexatious, what the inflation rate is, and how important money use is to the party entitled to the payment.

Some jurisdictions, either by court decision or increasingly by statutes, give prejudgment interest unless special circumstances would make it unjust to award it.[11] Undoubtedly, this reflects increased legislative recognition that delay in payment causes a serious loss that should be compensated through interest. Again, local law must be consulted.

Frequently, special rules deal with claims against public authorities. Until recently, interest could not be recovered against the federal government. This changed in 1978.[12]

With costs of financing now more recognizable as an important element in construction costs, contracts should specify that payments will bear interest from the time they are due. As to claims that are difficult to evaluate, the issue is not so clear. Suppose each party has a good-faith belief in the merit of its position and for that reason refuses to settle. Ultimate determination that refusal to pay was unjustified permitting recovery of prejudgment interest can place a heavy burden on the party who has asserted an honest reason for not paying. On the other hand, undoubtedly the loss should have been paid, and not awarding prejudgment interest denies full compensation to the party whose position was ultimately vindicated. Balancing compensation against punishment makes resolution of this issue difficult.

Unless the contract or an applicable statute states otherwise, in most states, the percentage will be the amount specified by law to be paid on legal judgments. (See Section 22.02(L).)

SECTION 6.09 Special Problems of Employment Contracts

Section 5.11 discussed employment contracts. This section discusses remedies for breach of those contracts.

A. Money Awards

The application of two rules discussed in Section 6.06 have particular significance in employment contracts. The first is the rule of avoidable consequences discussed in (D), which noted that the courts have not been solicitous toward claims by a breaching party that the loss could have been avoided or reduced by efforts of the claimant.

The law has been even less solicitous toward employers who have wrongfully dismissed an employee and who assert the employee could have reduced or avoided the loss. The employee recov-

---

[9]*E.C. Ernst, Inc. v. Koppers Co.*, 626 F.2d 324 (3d Cir.1980). See Annot., 60 A.L.R.3d 487 (1974).

[10]West's Ann.Cal.Civ.Code § 3287(b) (grants judge discretion to award interest from date no earlier than date legal action commenced).

[11]See *Hedla v. McCool*, 476 F.2d 1223 (9th Cir.1973).

---

[12]41 U.S.C.A. § 611 allows interest from the date the claim is received until payment. The rate is set by the Secretary of the Treasury. From Jan. 1, 1993, through June 30, 1993, the rate is $6^1/_2$%. See 53 Fed.Reg. 62418 (1992). (In 1984, it was $12^1/_8$%.)

ers the full amount of salary that would have been earned for the period of service less the amount the *employer affirmatively proves* the employee has earned or with reasonable effort could have earned from other employment. However, before projected earnings and other employment opportunities not sought or accepted by the employee can be applied to reduce the damage award, the employer must show that the other employment was comparable or substantially similar to the service the employee would have performed under the contract. The employee may reject or fail to seek available employment of a different or inferior status. In addition, as was noted in Section 5.11(B), punitive damages may be awarded for wrongful dismissal in extreme cases.

The other legal doctrine that has particular importance in employment contracts is the collateral source rule, discussed in Section 6.06(F). Employees who are wrongfully terminated from their employment often receive social benefits, such as unemployment insurance payments under either state or private plans. Although no clear answer has emerged in the cases, it is likely that these payments will be considered a collateral source and not applied to reduce the amount of the claim.

B. Equitable Remedies

Direct court orders that an employee work for an employer or that an employer take back an employee are not issued for breach of an employment contract. (Reinstatement *is* common under collective-bargaining agreements.) However, based upon a famous English decision, *Lumley v. Wagner*,[13] in some states, *negative* injunctions—that is, orders barring a breaching employee from working for the plaintiff's *competitor*—are issued. Usually such orders involve employment of a highly unique type where a replacement cannot be easily obtained, such as in contracts for performance of services by professional entertainers or athletes. However, such negative injunctions are sometimes limited by statute.[14] The modern tendency is to avoid them, as they can do by indirection something not directly allowed.[15]

In another area, equitable injunctions play a significant role in remedies for breach of an employment contract. Employees sometimes develop or learn technical or commercial data that an employer wishes to keep secret. Employees, particularly those who work in high-technology industries, commonly sign employment agreements under which they promise that if the employment relationship is terminated, they will not compete with the employer or work for a competitor for a designated period of time. Such contracts also provide that employees will not divulge technical or commercially valuable information within a specified period after the relationship terminates. Even without express promises, the law may apply a fiduciary obligation not to divulge trade secrets.

Usually the employer who wishes to enforce such a contract or implied obligation seeks a court order prohibiting an employee from competing or disclosing trade secrets. In deciding whether such an injunction should be granted, the law must balance the legitimate interests of an employer in protecting valuable industrial and commercial information with the often unreasonable burden such restraints place on employees whose only source of livelihood may be using what they have learned while working for their employer.

The law has sought to minimize restraints on the use of all information—commercial and technical. A reconciliation of such policies has resulted in a rule under which covenants not to compete or to disclose industrial or commercial secrets are enforceable by injunction if reasonable. To decide whether a restraint is reasonable, the duration of the restraint, the extent of the restraint geographically and competitively, and the legitimate interests of the employer, the employee, and the public are examined.

SECTION 6.10   Sales of Land

Because land is regarded as unique, courts will usually grant a request for "specific performance" if the seller, without legal excuse, refuses to go through with the transaction. (As a corollary to this, the seller may be able to obtain a judgment for the purchase price if the buyer is unjustified in refusing to perform.) The court order will order that the seller give a deed to the buyer. Failure to do so will put the seller in contempt of court. Sell-

---

[13]42 Eng.Rep. 687 (1852).

[14]See West Ann.Cal.Civ.Code § 3423 (barring negative injunctions except in special cases).

[15]See Restatement (Second) of Contracts § 367 (1981).

ers can be jailed until they purge themselves of contempt by executing the deed.

In some states, the court can execute the deed where the seller refuses to perform. When asked to award specific performance, the law looks at the fairness of the transaction—a factor that is not the subject of inquiry when the court is asked to award a money judgment for damages. In addition to other requirements, specific performance will not be awarded if innocent third parties have purchased the property.

If the seller refuses to perform and specific performance is either not requested or not granted, the buyer should be put in the position it would have been in had the seller performed. Usually this means that the buyer is entitled to the benefit of the bargain—the difference between the contract price at the time conveyance should have been made and the fair market value of the property. If the contract price had been $100,000 and the buyer can show that the fair market value is $125,000, the buyer receives $25,000. The buyer may be able to recover additional losses caused by the seller's breach if the losses are proved with reasonable certainty and if the seller could foresee that such losses probably would result.

The same measure of damages will apply if the buyer refuses to perform. The seller recovers the benefit of the bargain—the difference between the contract price and the market price and any other foreseeable expenses or losses resulting from the breach.

Some jurisdictions use the English rule, dividing breaches into ordinary breaches and bad-faith breaches. Ordinary breaches are usually caused by sellers' being unable to transfer the title that they thought they had when they agreed to sell. There is no moral blameworthiness attached to a seller's nonperformance in such cases. A bad-faith breach is a seller's refusal for no *good* reason, for example, if the seller can sell to someone else for more money. In such bad-faith breaches, the buyer can recover benefit-of-the-bargain damages. A buyer who cannot show that the breach was in bad faith can recover only out-of-pocket reliance expenses incurred. In 1983, California, formerly a leading English rule jurisdiction, eliminated the bad-faith requirement for recovery of the benefit of the bargain.[16] This may presage a gradual elimination of the English rule.

---

[16] West Ann.Cal.Civ.Code § 3306.

PROBLEMS

1. Professor Best, an assistant professor at a state university, made a contract with Engineering Books, Incorporated, a publisher of engineering texts. The contract provided that Best would supply a manuscript dealing with a research project she had conducted. The project dealt with an analysis of disputes that resulted in the course of building facilities for waste-water treatment. Best was to receive ten complimentary copies and 15% of all the revenues received by the publisher attributable to the book.

Best finished the manuscript and submitted it to the publisher. The publisher decided that the manuscript, while of a quality that could be considered publishable, would not be published, because it did not believe that the cost could be recovered.

When Professor Best was notified by the publisher, she was enraged, because failure of the manuscript to be published would have a significantly adverse effect on her chances for tenure in her de-

partment. She had also hoped that publication of the manuscript would result in her receiving consultation fees and invitations to conferences and thought that the book would earn substantial royalties. She had not published any other books.

Professor Best contacted some other potential publishers, who, when they heard that the book had been rejected despite the contract to publish it, were cool about publishing the book. She may be able to get the book published and marketed if she pays the publisher for these services.

Professor Best does not believe that she can establish with any certainty that her progress as a professor will be impeded, but she is reasonably certain that her career has suffered a setback. She had asked a number of colleagues if they would review the book. If the book is not published, she feels that she will suffer a diminution of respect in their eyes. She spent approximately one hundred hours lining up potential reviewers—time she could have used in developing her consulting business.

What remedies might she seek? What remedy or remedies do you believe a court would award her?

2. Alice Sullivan was about to take her first job as an architect. On November 28, 1993, she had accepted an invitation to be associated with the firm of Rauch and Burns (R&B), whose offices are located in Denver, Colorado. One reason Alice wanted to work for R&B was its undoubted expertise in the design of prisons and correctional institutions. R&B was nationally known for its work in this area. Alice had recently seen a study that indicated a big boom ahead in the construction of such projects.

The firm gave Alice an employment contract on November 30, 1993. All terms were agreeable to Alice. But she was puzzled by paragraph 17, which provided the following:

> If the employment relation is terminated for any reason, employee agrees she will not join any other architectural firm as a principal or associate for one year. This provision is agreed upon as necessary to protect R&B proprietary data, including design concepts and prospective clients. If this covenant is breached, R&B can choose either to seek past damages agreed to be $500.00 a day and an injunction *or* agreed damages for each day of violation of $300.00 a day but not to exceed $50,000.

Would this clause be enforceable if Alice were to sign the contract? Should Alice suggest modifications? (Her bargaining position in this regard is not very good.)

Losses, Conduct, and the Tort System: Principles and Trends

SECTION 7.01 Relevance to the Construction Process

During the Construction Process, events can occur that might harm persons, property, or economic interests. Workers or others who enter a construction site may be injured or killed. The owner or adjacent landowners can suffer damage to land or improvements. Participants in the process may incur damage to or destruction of their equipment or machinery. The owner or other participants in the project may incur expenses greater than anticipated. Investors in the project or those who execute bonds on participants may also suffer financial losses.

After completion of a project, persons who enter or live in the project might be injured or killed because of defective design, poor workmanship, or improper materials. Those who invest in the project may find investment value reduced for similar reasons.

The construction project is a complex undertaking involving many participants. It is one that has a high risk of physical harm to those actively engaged in it. Sometimes such harm is caused by failure of participants to live up to the conduct required by law. Losses sometimes occur by human error that does not constitute wrongful conduct. Losses sometimes occur because of unpredictable and unavoidable events for which no one can be held accountable. Because of the varying causes of losses, the many participants in the project, and the complex network of laws, regulations, and contracts, placing responsibility is a difficult undertaking.

SECTION 7.02 Tort Law: Some Background

A. Definition

A tort has been defined as a civil wrong, other than a breach of contract, for which the law will grant a remedy, typically a money award. This definition, though not very helpful, mirrors the difficulty of making broad generalizations about tort law in the United States. One reason is the incremental or piecemeal development of tort law necessitated by new activities causing harm. For this reason, much of American tort law consists of a collection of wrongs called by particular terms, which were given legal recognition in order to deal with particular problems.

Some basic distinctions are essential. Although tort law and criminal law have features in common—each regulating human conduct—they operate independently. Crimes are offenses against the public for which the state brings legal action in the form of criminal prosecution. Prosecution is designed to protect the public by punishing wrongdoers through fines or imprisonment and to deter criminal conduct.

The tort system is essentially private. Only individual victims can use the system. Any sanctions imposed against those who do not live up to the standard of tort conduct are for the benefit of the victim. For example, the automobile driver who violates the criminal law may be fined or imprisoned. Those who are injured because of such criminal conduct are likely to institute a civil action to transfer their losses. This civil action is part of the tort system.

B. Function

Tort law has different functions. The particular function most emphasized at any given time depends on social and economic conditions in which the system operates.

The principal functions are to compensate accident victims, to deter unsafe or uneconomic behavior (encourage investment of up to but not beyond the point at which incremental safety costs equal incremental injury costs), to punish wrongful conduct, or to protect social norms from a sense of outrage generated by perceived injustice through providing a dispute resolution mechanism or a combination of any of these goals.[1]

In seeking to implement these functions, often the law must choose between goals and interests of individuals and groups. One person's desires may come at the expense of another. One person may wish to drive a car at a high rate of speed, which exposes others on the road to risks of danger. Property owners may wish the freedom to maintain their property as they wish. But this freedom may come at the expense of those who enter the land and are injured.[2] A manufacturer may wish complete freedom to design a product that will earn the highest profit. But this freedom may come at the expense of buyers of the product who suffer harm from using it.[3] Adjusting these conflicts can reflect conscious decisions to select or favor one competing interest or goal over another—a form of social engineering.

Tort rules may severely limit a property owner's freedom or the freedom of a manufacturer in order to give greater protection to those who are injured on unsafe property or by defective goods. The historical description in (D) suggests trends in social engineering.

C. Some Threefold Classifications

Two important threshold concepts are threefold. The first concept describes the interests considered

sufficiently important to merit tort protection and often described as follows:

1. *Personal*, sometimes defined to include psychic or emotional interests.
2. *Property*, tangible and intangible.
3. *Economic*, unconnected to harm to a person or damage to property.

The second concept classifies the conduct of the person causing the loss:

1. *Intentional*, including not only the desire to cause the harm but also the realization that the conduct will almost certainly cause the harm.
2. *Negligent*, usually defined as failure to live up to the standard prescribed by law.
3. *Nonculpable*, though in a sense wrongful, in which the actor neither intends harm nor is negligent.

These threefold classifications play important roles in determining which victims will receive reparation from those causing the loss. On the whole, harm to a person is *most* deserving of protection, with harm to property being considered second in importance. At the conduct end, intentional conduct that causes harm is *least* worthy of protection, followed by negligent conduct. These classifications are gross, and many subtle distinctions must be made.

D. Some Historical Patterns

Earliest English private law developed during the feudal period in which land dominated society. As a result, property law developed before any significant developments in tort or contract law. The feudal period was dominated by agriculture and a largely illiterate population, with little need for a developed contract or tort law system.

Although early English legal history saw the development of laws that dealt with finance, banking, and maritime matters, what is known today as tort law—as well as much of what is known today as contract law—did not develop until the Industrial Revolution in the late eighteenth and early nineteenth centuries. The Industrial Revolution moved manufacturing out of cottages and into factories, necessitating a transportation system and migration of workers from the farms and villages to the towns and cities.

The changes brought significant developments in tort law. The preindustrial agrarian society, with

---

[1]Smith, *The Critics and the "Crisis": A Reassessment of Current Conceptions of Tort Law*, 72 Cornell L.Rev. 765 (1987).
[2]See Section 7.08.
[3]See Section 7.09.

its emphasis on property, was most concerned with property rights and keeping the peace. The important torts were those that were intentional, as they could invite retribution, and those that invaded property interests. As a result, although persons may not have always acted at their peril in the sense that they would have to account for any damages their activities caused, much liability was "strict." Trespass, an invasion of a property owner's right to exclusive possession of its property, did not require any showing of fault. Any trespass, whether innocent or deliberate, was wrongful. Both the importance of property rights and the inability to deal with subtle concepts such as negligence and fault contributed to the rather simple and often harsh structure of early tort law.

Keeping the peace required protection against serious intentional torts, such as trespass, assault (apprehension of harm), battery (harmful or offensive touching), or false imprisonment (deprivation of freedom of movement). These serious matters could lead to breaches of the peace, and such conduct had to be eliminated or at least minimized through making the actor pay for the harm caused.

Unintentional or negligent conduct did not rate very high on the interest-protection scale, both for reasons mentioned and because more important matters required attention. Matters such as harsh words, offensive conduct, or careless jostling were not sufficiently important in such a society to receive protection.

The Industrial Revolution generated factory and transportation accidents as well as migration to population centers. Now the law had to deal with conduct that became serious in crowded towns and cities. Many matters of an earlier day that were too trivial to be dealt with now required attention.

Even more important, difficult choices had to be made when commercial activity caused harm. The law chose to protect new and useful commercial and industrial activities from potentially crushing liability by not making liability as "strict" as it had been in preindustrial times. With some important exceptions,[4] a person who suffered a loss could not transfer the loss to the person causing it unless the injured person was free of negligence and could establish that the person causing the harm did not live up to the negligence standard of conduct. Transferring the loss *only* upon a showing of negligence and the development of other legal doctrines were designed to free useful activities from responsibility. These rules were less concerned with compensating victims of these activities.

Liberalization, as the term was then used, was designed to free economic activity from the shackles of heavy state mercantilistic controls. Rules that protected industrial and commercial activity may also have been generated by the belief that such activities brought long-run social and economic advantages. The nineteenth century emphasized the moral aspects of individual responsibility, and shifting a loss required wrongdoing. But as the toll in human misery and economic deprivation rose because of industrial, commercial, and transportation activities and as the automobile replaced the horse, changes were inevitable.

At the beginning of the twentieth century, many industrial countries sought to remove industrial accidents from the tort system (briefly treated in Sections 7.04(C) and 31.02). Workers' compensation brought a more humanitarian approach to industrial accidents. In the 1920s, manufacturers began to lose protection, and the consumer revolution of the 1960s has for all practical purposes made manufacturers the insurers of losses caused by their defective products.[5] Similarly, some earlier protection given property owners began to be diluted in the late 1960s,[6] although responsibility was not as extensive as that of manufacturers.

People who furnished services, such as architects and engineers, were in the backwash of these changes. (The effect on professional liability is better postponed until Chapter 14.) The modern era has sought to give security from certain risks to all members of society, and the tort system has reflected this, along with the other goals noted in Section 7.02(A). Compensation of victims rather than unshackling enterprises became predominant. The shift is sometimes described as enterprise liability. Under it an enterprise can and should bear the normal risks of its activity. These risks can be predicted, computed, and insured. The social costs of the activity can through pricing be spread to all

---

[4]See Section 7.04.

[5]This is a deliberate but only slight exaggeration of the current state of the law. (See Section 7.09.)
[6]See Section 7.08.

who benefit from the enterprise by using its products.

What of the post-industrial tort law? Some have advocated its abolition. They would replace it with social insurance, under which all victims of accidents would be compensated by the state or by private insurers.

Some have advocated, and many states have adopted, no-fault handling of road accidents. The victim recovers up to a threshold amount of medical expenses (which is quite low) *without* showing any fault. Above this threshold, the victim can use tort law. (Low thresholds have meant inflated medical expenses and very little reduction in lawsuits.)

Some have felt that enterprise liability has placed too heavy a burden on manufacturers and professionals. Much has been written about the "malpractice crisis," and legislative activity has moderated the harsh treatment given professionals and manufacturers by the courts. In addition, some have questioned the expansion of liability without fault and have suggested a return to a more negligence oriented tort standard. Such calls for a brake on expanded liability are based upon assertions that either liability cannot be insured against or that insurance premiums and claims expenses make useful services or products unprofitable. This has been seen in decisions by drug companies to pull particular drugs off the market and increased reluctance by medical professionals to pursue certain specialties, such as obstetrics and gynecology.

As noted in Section 7.02(A), deterring unsafe and inefficient conduct is one goal of tort law. With the advent of liability insurance, some believe that the deterrent function is no longer accomplished by tort law. They would prefer *direct* control by legislation, similar to the Occupational Safety and Health Act (OSHA) enacted by Congress. Though much criticized mainly for obsessive interest in detailed rules, the Act is a direct attempt to make industrial activities safer than would the indirect method of the tort system.

Although predictions are dangerous, it is likely that the tort system in some form will continue to serve, if not the primary, at least an ancillary role in compensating victims and regulating activity.

E. Some General Factors in Determining Tort Liability

Some have examined the unruly and disparate thousands of tort cases and have attempted to ar-

ticulate factors that affect tort liability. One scholar listed the following items as important:[7]

1. The moral aspect of the defendant's conduct.
2. The burden of recognizing a legal right upon the judicial system.
3. The capacity of each party to bear or spread the loss.
4. The extent to which liability will prevent future harm.

F. Coverage of Chapter

Tort law is simply too diverse and immense to cover in this treatise. By and large, intentional torts[8] such as trespass,[9] assault, battery, false imprisonment, intentional infliction of emotional distress,[10] defamation,[11] invasion of privacy, and

---

[7]See W. PROSSER, *TORTS*, 16–23 (4th ed. 1971). A subsequent edition added "a recognized need for compensation and historical development." W. PROSSER and P. KEETON, *TORTS*, 20–26 (5th ed. 1984).

[8]Comprehensive treatment of intentional torts can be found in W. PROSSER and P. KEETON, supra note 7 at pp. 33–159.

[9]*In re Catalano,* 29 Cal.3d 1, 623 P.2d 228, 171 Cal.Rptr. 667 (1981), held that a union official did *not* violate the *criminal* trespass statute when he refused to leave the site when ordered to do so by the owner. The official was conducting a safety inspection, a power given the union under its contract.

[10]See Section 6.06(H).

[11]The tort of defamation, usually subdivided into libel and slander, sometimes arises in the construction context. For example, *Diplomat Elec. Inc. v. Westinghouse Elec. Supply Co.,* 378 F.2d 377 (5th Cir.1967), involved a communication by a supplier of a subcontractor made to the prime contractor and owner that stated that the supplier had not been paid by the subcontractor. The court held that if the plaintiff subcontractor could prove that the communication was false, he would be able to recover his losses from the defendant supplier based upon the tort of defamation.

A defamation case involving a design professional was *Priestley v. Hastings & Sons Publishing Co. of Lynn,* 360 Mass. 118, 271 N.E.2d 628 (1971). Here the defendant newspaper published defamatory statements made by a town official regarding the plaintiff architect who had designed and administered construction of a public high school. Earlier decisions of the U.S. Supreme Court had granted privileges to newspapers based on the First Amendment's protection to the press. The plaintiff was required to prove under certain circumstances that the defamatory material was published maliciously. The Massachusetts court concluded that free press protection required malice in this case because the architect, though not a public official, became involved in a matter of public or general concern. This case and earlier U.S. Supreme Court cases upon which the holding was based con-

interference with contract or prospective advantage[12] are not discussed.[13] Emphasis is placed on negligence and strict liability, as those concepts relate to the Construction Process, starting with design and culminating with the finished project.

SECTION 7.03 Negligence: The "Fault" Concept

A. Emergence of Negligence Concept

The Industrial Revolution was the precipitating factor for tort law moving from principal emphasis on intentional torts and the need to keep the peace to a system that would deal with increased accidents brought about by industrialization. This movement culminated with the recognition of negligence as the principal basis for liability. To legitimately transfer the plaintiff's loss to the defendant, the plaintiff was required to establish that the defendant did not perform in accordance with the legal standard of conduct which, though somewhat inaccurate, was called the "fault" system.

Today the negligence concept largely governs road accident losses, losses caused by the possessor of land failing to keep the land reasonably safe, losses that occur in the home, losses caused by the activities of professional persons, and, for the most part, losses caused by participants in the Construction Process.

B. Elements of Negligence

To justify a conclusion that the defendant was negligent, the plaintiff must establish the following:

1. The defendant owed a duty to the plaintiff to conform to a certain standard of conduct in order to protect the plaintiff against unreasonable risk of harm.[14]
2. The defendant did not conform to the standard required.[15]
3. A reasonably close causal connection existed between the conduct of the defendant and the injury to the plaintiff.[16]
4. The defendant invaded a legally protected interest of the plaintiff.[17]

C. Standard of Conduct: The Reasonable Person

Objective Standard and Some Exceptions. Nineteenth-century English and American courts rejected a standard based on the subjective ability of the defendant. It was not sufficient for the defendant to show that it did the best it could. To protect the community, its members are held to a standard that can exceed what they are able to do. In this sense, negligence is not synonymous with fault. The community standard requires that the defendant do what the reasonable person of ordinary prudence would have done. Such a standard can hold the defendant liable despite its having done the best it could. Negligence, then, or much of it, is not congruent with morality. The standard holds persons who live in the community to an average community standard. For example, an inexperienced driver is expected to drive as well as the average driver.

Exceptions do exist, and a person can be held to a lower or higher standard than that of the community. Usually, such exceptions are created by designating special subcommunities smaller than the general community and then applying an objective subcommunity standard. For example, children generally are not held to the adult standard but are held only to the standard of children of

---

cluded that private persons can become public figures if they inject themselves into public matters. A subsequent decision of the U.S. Supreme Court seemed to narrow the protection given the press by concluding that private persons who through no desire of their own find themselves thrust into public matters may be able to recover for defamation without any showing that the defamatory communication was made maliciously. *Gertz v. Robert Welch, Inc.*, 418 U.S. 323 (1974).

[12]This intentional tort is noted briefly in Section 14.08(F). See *Custom Roofing Co. v. Alling*, 146 Ariz. 388, 706 P.2d 400 (App.1985), which upheld a punitive damage award against a supplier in a claim by a contractor. The award was based on the supplier's wanton conduct and indifference to the rights of others. The supplier had failed to supply material, knowing it would cause the contractor to lose its contract.

[13]Most law relating to torts is called *common law* in that the principal sources of law are reported appellate decisions. Legislative bodies increasingly bar certain conduct, such as improper reasons for refusing to sell, rent, or hire.

[14]See Section 7.03(E).
[15]See Section 7.03(C).
[16]See Section 7.03(D).
[17]See Section 7.03(F).

similar age and experience. But children who engage in *adult* activities, such as driving an automobile, are held to an adult standard. Similarly, persons with physical disabilities, such as blindness, are not expected to conduct themselves in the same way as the average community member who does not suffer from such a disability. Blind persons would be expected to conduct themselves as would average members of the blind community. But persons with mental or emotional disabilities are generally held to the community standard.

Because the application of these standards is generally performed by a jury, some tolerance of human weakness and some exceptions to the objective standard other than those mentioned may find their way into jury decisions.

Another exception applies to those persons who, because of special training or innate skill, are expected to do better than the average person. Physicians are held to a higher standard in dealing with medical matters than are ordinary members of the community. Architects and engineers are held, as a rule, to the standards of their subcommunity (discussed in greater detail in Chapter 14). Professional truck drivers are expected to drive better than ordinary drivers. The combination of objective and subjective standards in such cases is reflected by statements that defendants are judged by what they knew or should have known or by what they did or should have done.

Unreasonable Risk of Harm: Some Formulas. Courts and commentators seek to refine vague community or reasonable-person standards by articulating factors that should sharpen the inquiry into whether the standard of conduct has been met. One approach is to evaluate the magnitude of the risk, the utility of the conduct, and the burden of eliminating risk.

To determine the magnitude of risk, the relevant factors are the gravity, the frequency, and the imminence of the risk. Clearly the likelihood and the severity of harm that can result are important factors in determining the type of conduct that should be expected to avoid these risks. Driving at an excessive speed on the highway clearly creates a high risk because accidents often result from excessive speed and their consequences are usually serious. Railroad crossings are dangerous because, though the likelihood of a train's striking an automobile may be small, the consequences of such an occur-

rence are serious. Conversely, although throwing a soft rubber ball into a crowd may not cause *serious* harm to *anyone*, it is likely to cause *some* harm to *someone*.

The other factors recognize that imposing liability on actors restricts human freedom. The social utility of the conduct being regulated is an important criterion in determining whether the legal standard has been met. Suppose a bank robber carelessly jostles a bank patron in the course of robbing the bank. If the law holds the bank robber responsible for the harm caused the patron, one factor is likely to be that bank robbing is not considered a useful activity and can be regulated by tort law as well as by criminal law. Conversely, the same carelessness by a bank security guard in the performance of a socially useful activity, such as organizing the patrons of the bank so that they can make an orderly retreat in the face of a fire, would not expose the bank or guard to the same liability as the bank robber.

The burden of eliminating the risk recognizes that almost all risks can be eliminated or minimized if sufficient resources are mobilized. Undoubtedly, the impact of road accidents would be minimized if guardrails were installed on all public highways. Yet doing so would involve an immense expenditure that might not be commensurate with the gain that could be realized. Likewise, the burden of eliminating the risk in this manner would take into account not only the cost but also the aesthetic deprivation caused by universal installation of guardrails.

Where serious harm can be avoided by minimal effort or expenditure, failure to do so will very likely be negligent. For example, if serious burns can be avoided by installing a five-dollar mixer valve in the bathroom fixtures, failure to do so would very likely be negligent.[18]

Common Practice: Custom. Suppose the defendant conformed to or deviated from common practice, or what is sometimes called the custom in the community. Frequently, defendants attempt to exculpate themselves by showing that they performed as others do. This arises most frequently in claims against a manufacturer. Often the manufac-

---

[18]*Schipper v. Levitt & Sons, Inc.*, 44 N.J. 70, 207 A.2d 314 (1965) (builder-vendor strictly liable without need to show negligence).

turer establishes that it performs in accordance with industry practices.

Although compliance with customary practices is *evidence* of compliance with the legal standard of care, it is not conclusive. The customary standard itself may be careless and create unreasonable risk of harm. For example, suppose that most pedestrians jaywalk or that all workers refuse to wear hard hats in a hard-hat area. Similarly, failure to conform to customary practices is not *conclusive*, and a defendant may be exonerated despite deviation from customary practices if good reasons existed for deviation and the defendant conducted its activities with reasonable care.

One important exception to this relates to the standard of conduct expected of professional persons. As this is more appropriately dealt with in examining liability of design professionals, major discussion is postponed until Chapter 14. It is sufficient to state at this point that when defendants are judged by the subcommunity of their profession, customary practices of the profession become the standard.

Violations of or Compliance with Statutes. In the exercise of its responsibility to protect all citizens, government frequently prohibits certain conduct and attaches civil or criminal sanctions for violations. The Construction Process is governed by a multitude of laws dealing with land use, design, construction methods, and worker safety. What effect do violations of those statutes have on *civil* liability?

It is possible, though uncommon, for the statute to expressly declare that violations of the statute *determine* civil liability. More commonly, the statute expressly imposes criminal sanctions only. In such cases, courts can and do look at the statute as a legislative declaration of proper community conduct. To have *any* relevance, however, some preliminary questions must be addressed.

First, the person suffering the harm must be in the class of persons that the legislature intended the statute to protect. Many statutes are intended to protect members of the community at large by achieving public peace and order rather than to protect any particular group or individual. For example, statutes sometimes prohibit certain businesses from operating on Sunday. Such statutes are not designed to protect those who suffer physical harm while a business operates in violation of the

Sunday closing laws. Similarly, although statutes require that automobiles be registered, the purpose of such laws is to raise revenue and not to impose liability on the driver of an unregistered car even though driving properly.

Sometimes the legislation is designed to protect an extremely limited class of persons that may *not* include the injured party. For example, legislation requiring that dangerous machinery be shielded may be designed for the benefit of employees and not of those who enter the plant for other purposes. Similarly, as seen in Section 31.04(F), some states hold that safety regulations imposed on a contractor-employer are not designed to protect employees of *other* employers on the site.

Second, did the statute deal with the particular risk that caused the injury? For example, suppose a statute limits the time a train may obstruct a street crossing. This is designed to deal with traffic delays and not with the risk of personal harm caused by the delaying train. On the whole, the tendency has been to broadly define the particular risk.

The violation of law can be excused in most instances by showing extraordinary circumstances that made compliance more dangerous than violation. For example, a statute may require that drivers always drive on the right unless they are passing another vehicle or making a left turn. Suppose the driver veers to the left lane to avoid hitting a child. This technical violation will be excused and have no bearing on negligence.

Where the policy expressed by the statute is particularly strong, the statute may expressly eliminate any possibility of a violation being excused. Violations of such statutes are conclusive on the question of negligence. Such statutes often impose liability despite assumption of risk or contributory negligence by the injured party which, as shall be seen in Section 7.03(G) often is a defense. This form of strict liability may result if the statute was intended to protect someone from his own immaturity or carelessness. Illustrations are those prohibiting child labor or requiring safety measures in construction work.

Much can depend on the particular law violated. Some laws seem anachronistic and continue to exist only because the legislature lacks the energy to modernize rules. Violation of such laws may have very little impact.

Suppose it is concluded that the preliminary requirements have been met. The plaintiff is in the class of persons to be protected, the statute was intended to cover the risk in question, and the violation was not excused. Most courts hold the violation to be negligence *per se* and conclusive on the question of negligence. The trier of fact, whether judge or jury, need not decide whether there had been negligent conduct.

With the exception of special protective statutes of the type described earlier, a *per se* violation may or may not preclude a defense such as assumption of risk or contributory negligence. It does not preclude the defendant from showing that the violation of the statute did not cause the harm. For example, in *Hazelwood v. Gordon*,[19] an employee fell down a flight of stairs, which were too narrow at the bottom. The narrowness of the stairs, together with an inadequate handrail, violated a city ordinance. However, the court held that the injured party could not recover from the property owner, because her injury was not *caused* by a violation of the ordinance but was caused by her negligently placing her foot on the top step of the staircase knowing the stairs were dangerous.

Some jurisdictions, however, find that a statutory violation is simply *evidence* of negligence to be given to the jury and weighed along with other evidence to determine whether the defendant lived up to the legal standard of care. Some states that employ negligence *per se* hold violations of local ordinances, traffic regulations, or administrative regulations to be only *evidence* of negligence. For example, *Bostic v. East Construction Co.*[20] dealt with administrative regulations for fire safety. The court held that administrative regulations can be the basis for negligence *per se* but are less likely to be. The court also held that the regulations must be understandable, and the regulations involved in this case were not sufficiently clear. (In any event, the court seemed to believe that the failure to comply with the fire regulations did not cause the injury.)

Compliance with the statutory standard does not necessarily preclude a finding of negligence. The statutory standard is a minimum. Additional precautions can be required. For example, it may not be a statutory violation to park a car on the shoulder of a highway so long as a taillight functions. But under certain circumstances, such as on an extremely foggy night, this conduct may be below the legal standard and be negligent. Such instances are rare.

Res Ipsa Loquitur. Proof of negligence can be by direct testimony of witnesses who testify based on their own observations of the defendant's conduct. However, sometimes this evidence is not available. Absence of direct evidence does not preclude the plaintiff from establishing indirectly (that is by circumstantial evidence) that the defendant was negligent. This can be accomplished by showing facts relating to the accident that tend to show, in the absence of an explanation by the defendant, that the accident was probably caused by the defendant's negligence. For example, suppose a tool falls from a scaffold and injures a passerby. Once the facts are established and it is known that the contractor's workers were working on the scaffold, it is more likely than not that the accident was caused by the contractor's negligence.

The unfortunate Latin term *res ipsa loquitur* used to describe this process of indirect proof of negligence was employed in an English case in which a passerby was struck by a flour barrel falling from a warehouse window.[21]

Much controversy has developed regarding the *res ipsa* concept, mainly centered around judicial and scholarly statements of *res ipsa* requirements. In reality, the supposed requirements are not truly requirements, because the doctrine is sometimes applied despite absence of some of them. However, the stated requirements may provide some assistance in understanding when the concept will be used.

It is usually stated that *res ipsa* requires the following:

1. An event is one that ordinarily does not occur in the absence of someone's negligence.
2. The event must be caused by an agency or instrumentality within the exclusive control of the defendant.
3. The accident must not be due to any voluntary action or contribution by the plaintiff.

[19]253 Cal.App.2d 179, 61 Cal.Rptr. 115 (1967).
[20]497 F.2d 712 (6th Cir.1974).
[21]*Byrne v. Boadle*, 159 Eng.Rep. 299 (1863).

Some courts have suggested that the evidence must be more readily accessible to the defendant than to the plaintiff. But as stated, these requirements cannot be taken as conclusive, as they are not always found in cases where the doctrine is applied.

The use of circumstantial evidence does not, as a rule, shift the burden of proof from plaintiff to defendant. But if the facts indicate that it was likely that the defendant's negligence caused the accident, the matter will be submitted to the jury. The application of the doctrine does not preclude defendants from introducing evidence that they were not negligent. For example, in the illustration given earlier, the contractor can show that the tool fell because of a strong gust of wind for which it was not responsible. But the doctrine helps plaintiffs get their cases before the jury.

Sometimes the inference of negligence on the part of the defendant is so strong that it may persuade the jury that the plaintiff has met his burden of proof. Similarly, a very strong inference of negligence may justify the judge directing a verdict for the plaintiff.

D. Legal Cause: Cause in Fact and Proximate Cause

A reasonably close connection must exist between conduct of the defendant and the harm to the plaintiff. Legal cause is divided into two separate though related questions:

1. Has the defendant's conduct caused the harm to the plaintiff? (This is usually referred to as "cause in fact.")
2. Has the defendant's conduct been the "proximate cause" of the harm to the plaintiff?

Cause in Fact. The first, considered less complicated, is a factual question decided by the finder of fact, usually the jury. Even this supposedly simple question of causation can raise difficult issues. First, the defendant's conduct need not be the sole cause of the loss. Many acts and conditions join together to produce a particular event. Suppose an employee of a subcontractor suffers a fatal fall while working on a scaffold high above the ground. Any one of the following events could be considered a cause of the death in the sense that without any of the events the fatal fall would not have occurred:

1. The worker's need to pay medical bills causing the worker to take this risky job.
2. Defective scaffolding supplied by a scaffolding supplier.
3. Failure by the subcontractor or prime contractor to remove the scaffolding when complaints were made about its unsafe condition.
4. Weather conditions that made the scaffold particularly slippery on the day of the accident.
5. A low-flying plane that momentarily distracted the worker.
6. The worker's refusal to wear a safety belt.
7. The subcontractor's or prime contractor's failure to enforce safety belt rules.

Although the list can be amplified, remove *any* link in the causation chain and the worker would not have been killed. Yet it would be unfair to relieve any actor whose failure to live up to the legal standard played a significant role in the injury simply because other actors or conditions also played a part in causing the fall.

Liability in such a case would depend on a conclusion that any of the defendants *substantially* caused the injury. Suppose a claim had been made against the scaffold supplier and the prime contractor. Each could have been a substantial cause of the injury. In an indivisible injury, each would be liable for the entire loss, and whether the party paying the claim would be entitled to recovery from the other party would depend on whether a right to contribution or indemnity existed.[22]

Cause in fact requires that the harm would not have occurred *without* the defendant's failure to live up to the legal standard. Put another way, the defendant will usually be exonerated if the injury would have happened even if the defendant had lived up to the legal standard of conduct. For example, suppose the prime contractor did not supply a safety belt to the worker as required by law but the worker would have been killed because he would have refused to wear it. Under these conditions, it is likely that the prime contractor would not be held liable because the contractor's negligence did not *cause* the harm.

---

[22]A discussion of concurrent liability for indivisible loss is found in Sections 24.06 and 27.14. A claim by one concurrent wrongdoer against the other is discussed in Sections 32.03 and 32.05.

The "but for" defense has one important exception. Suppose two builders are constructing houses on adjacent lots. Each is simultaneously negligent, causing fires to begin on each building site. The fires join together, roar down the street, and burn a number of homes. Either fire would have been sufficient to burn the houses. But neither builder will be able to point to the "but for" rule as a defense. Each will be liable for the entire harm.[23]

Proximate Cause. Proximate cause, though related to cause in fact, serves a different function. Cause-in-fact judgments are factual and best made by commonsense decisions of juries. Proximate cause, on the other hand, involves a legal policy that draws liability lines to relieve those whose failure to live up to the legal standard of conduct causes harm. Proximate cause serves a similar function as the requirement that there be a duty on the defendant to act to protect the plaintiff (discussed in Section 7.03(E)). Both duty and proximate cause can minimize crushing liability burdens on those engaged in useful activities and prevent liability from going "too far."

Proximate cause can involve the following:

1. Harm of a different type than reasonably anticipated.
2. Harm caused to an unforeseeable person.
3. Harm caused by the operation of intervening forces.

As an example of the first, suppose a contractor installs a sheltered walkway around a project. It can foresee that defective planking could cause a sprained ankle, but will it be liable if a pedestrian pushing a baby stroller falls on a defective plank, causing the baby to tumble from the stroller and fracture its skull? Suppose defective wiring on a high-rise building causes a power failure and shuts down all the elevators. A person who intends to submit a bid on a public project on the top floor cannot reach the awarding authority's office in time to submit the bid. Can the person recover the lost profits on the contract from the supplier of the electric wire or the owner of the building?

These freak accidents, though they occur infrequently, are dramatic enough to excite scholarly interest when they reach appellate courts. The most famous illustration of the second type of case was *Palsgraf v. Long Island Railroad Co.*[24] It involved a passenger running to catch one of the defendant's trains. Some employees of the defendant sought to help the passenger board the train but did so carelessly, dislodging a package from the passenger's arms, which fell on the rail. The package contained fireworks that exploded with some violence. The concussion overturned some scales many feet down the platform. This was unfortunate for the plaintiff, who was struck by a scale, but fortunate for legal scholars and generations of law students who dissected the subsequent appellate court decision. In a four-to-three opinion, the court held that the plaintiff was outside the zone of risk. As an unforeseeable plaintiff, the railroad company did not owe any duty toward her despite its employees being careless toward someone else.

A case illustrating the third type was *Petition of Kinsman Transit Co.*, which involved two claims and two appeals. A negligently moored ship in the Buffalo River was set adrift by floating ice, picking up another ship on the way. Bridge attendants were warned but inexplicably failed to lift a bridge. Both ships collided with the bridge, and the bridge collapsed. The ships and ice blocked the river channel by creating a dam. Water and ice backed up, damaging factories on the bank as far up the river as the original mooring. One claim involved flood damage to the factories, and the other involved pecuniary loss due to the necessity of transporting ship cargos around the blockage. The court granted a recovery to the factory owners[25] but denied recovery for those ships incurring additional expenses to go around the blockage.[26]

This defense was attempted in a construction context in *Diamond Springs Lime Co. v. American River Constructors.*[27] Riverfront land flooded when a dam collapsed because of heavy rains. The contractor did not perform in accordance with the specifications but contended it should be relieved because the heavy rains were a superseding cause of the loss. The court held that the contractor would not be relieved, because the heavy rains that

---

[23]See Section 24.06.

[24]248 N.Y. 339, 162 N.E. 99 (1928).

[25]*Petition of Kinsman Transit Co.*, 338 F.2d 708 (2d Cir.1964).

[26]*Kinsman Transit Co. v. City of Buffalo*, 388 F.2d 821 (2d Cir.1968).

[27]16 Cal.App.3d 581, 94 Cal.Rptr. 200 (1971).

caused the flooding were a reasonably foreseeable peril.

It would serve no useful function to explore the many formulas used in these cases, such as direct cause, foreseeability, hindsight, and superseding causes. It is sufficient to indicate that freak accidents cause unusual harm, and lines must be drawn. One treatise, after cataloging the various formulas, suggested that those proposing formulas are groping for something that it is difficult if not impossible to put into words. It suggested the need for a method

> . . . of limiting liability to those consequences which have some reasonably close connection with the defendant's conduct and the harm which it originally threatened, and are in themselves not so remarkable and unusual as to lead one to stop short of them.[28]

Even though proximate cause analytically is considered an issue of law because it deals with a policy of limiting liability, usually juries are given a vague instruction relating to proximate cause similar to the instruction they are given relating to the standard of conduct. It is hoped that the jury will use common sense in deciding such freak cases.[29]

E. Duty

Duty is related to proximate cause. The first duty cases were decided in the mid-nineteenth century in a legal climate favorable to new industries developing after the beginning of the Industrial Revolution. Like proximate cause, it was an attempt to draw a line beyond which recovery would not be granted. Unlike proximate cause, it tended to emphasize the relationship between individuals that imposes upon one a legal obligation to watch out for the other. Sometimes in a freak accident of the type described in Section 7.03(D), courts conclude that the defendant did not owe *any* duty to the plaintiff. Even if the defendant did not live up to the standard required by law and even if that failure caused harm to the plaintiff, the plaintiff cannot recover.

The duty concept has found its way into the Construction Process. For example, professional persons such as architects, engineers, and surveyors often supply information as part of their professional services. As seen in Section 7.07(D), persons other than those who have requested and paid for the information may suffer losses if the information furnished is incorrect. Whether those third persons can recover from the person supplying information is sometimes phrased in terms of duty. For example, did the surveyor owe a duty to someone who relied on an inaccurate survey?[30]

Similarly, suppose a manufacturer's defective products harm persons with whom the manufacturer has no contractual relationship, such as those who have purchased the products from retailers, nonbuyers who use the products, or others injured by the products? As seen in Section 7.09, early American decisions denied recovery against a manufacturer because of the absence of any contractual relationship between the person harmed and the manufacturer. Such decisions sometimes spoke of an absence of duty owed by the manufacturer to the injured party.

The two preceding illustrations involve one corollary of the duty concept frequently described as the "privity rule." Nineteenth-century English law would grant a third party recovery for damages caused by another's breach of contract only if there was privity between the third party and the contract breaker.[31]

The privity requirement arose in nineteenth-century England and insulated even a negligent party who was said not to be in privity with the party injured. Privity usually referred to the contractual relationship between an injured party and the party against whom the claim had been brought. It could also mean the existence of a relationship under which one party owed a duty to another party to look out for the other's physical safety. Illustrations of such relationships at common law were passenger-carrier, innkeeper-guest,

---

[28]W. PROSSER and P. KEETON, *TORTS*, 300 (5th ed. 1984).

[29]Perhaps a layperson's view based on a gut reaction of what is going too far is the best solution to these vexatious problems. In commenting favorably on a hindsight test, one commentator suggested a line be drawn: short of the remarkable, the preposterous, the highly unlikely, in the language of the street, the cock-eyed and far-fetched, even when we look at the event, as we must, after it has occurred. (Id. at 299.)

[30]See *Rozny v. Marnul*, discussed in Section 7.07(G).
[31]*Winterbottom v. Wright*, 152 Eng.Rep. 402 (1842). This is no longer the law. See Section 7.09.

shipper-seaman, employer-employee, shopkeeper-visitor, host-social guest, jailer-prisoner, and school-pupil. The privity doctrine protected manufacturing and commercial activity.

F. Protected Interests and Emotional Distress

As Section 7.02(C) indicates, the particular loss suffered often controls liability. Even if all the requirements for negligence are met, the particular harm that has resulted may not receive judicial protection. Courts speak of whether particular interests are protectable in the process of deciding whether the defendant must respond for certain losses caused the plaintiff.

Harm to the person is most worthy of protection. Death not only ends one's life but also can have a severe financial and emotional impact on the deceased's survivors. Physical injury often means medical expenses and diminished earnings as well as pain and suffering. Often those who suffer physical harm are low-income persons who may not procure insurance to protect themselves and their dependents. As a result, for these as well as humanitarian reasons, it may be important to extend protection to those who suffer physical harm.

Harm to property such as damage or destruction is also considered worthy of protection because of the importance placed on property in modern society. But as the harm moves away from personal and property losses, the interest receives less protection. Economic harm such as diminished commercial contractual expectations, lost profits, or additional expenses to perform contractual obligations, while often protected, is considered less worthy of protection than harm to person or to property. Even less protection is accorded emotional distress and psychic harm, sometimes called noneconomic losses.

A number of reasons for caution exist when claims are made for economic loss, and even more when claims are made for emotional distress. Economic losses result from many causes other than the conduct of the defendant and are difficult to prove. Liability can place crushing burdens on the defendant. Although courts have cautiously extended protection for economic harm, they have been even less willing to extend protection for emotional distress. Part of this is the undoubted difficulty of establishing the genuineness of the claim. Another is the difficulty of placing an eco-

nomic value on it. Because of these difficulties, the law has been unsettled in this area, and decisions vary from jurisdiction to jurisdiction.

With some exceptions,[32] the plaintiff cannot recover for mental disturbance caused by the defendant's negligence in the absence of accompanying physical injury or consequences. Clearly, however, there can be recovery for mental disturbance if the negligence of the defendant has inflicted an immediate physical injury and the mental disturbance such as pain and suffering was caused by the physical injury. Suppose physical harm *follows* fright or shock of the plaintiff, such as a miscarriage suffered after negligent conduct by the defendant had caused a mental disturbance. Many American cases have required the plaintiff to show that there had been some physical impact on the plaintiff caused by the defendant's negligent acts. Others have not. The recent cases have eliminated the requirement for impact, and the impact rule is likely to disappear.

Suppose physical injury resulting from fright occurs when the injured person feared for his own safety. Here most states allow recovery. But fear for someone else's safety has divided the courts. The classic example has involved a mother, though not in danger, seeing her child seriously harmed. Some courts allow recovery. Others do not.[33]

Emotional distress and mental disturbance can arise in the context of construction work. Suppose a worker is on a roof. The roof partially collapses, but the worker is not injured. The worker may have suffered emotional distress from the fear that he was about to fall or that he might have fallen. Also, suppose a worker experiences traumatic shock when he sees someone else fall from the roof of a building.

Emotional distress in a construction context occurred in *Olivas v. United States*.[34] The injured employee was working on a project for the United States. He was ordered to enter a blast valve for cleaning. While in the valve, an Air Force officer negligently ordered the valve activated. Olivas

---

[32] Refer to Sections 6.06(H) and 27.09 for claims for emotional distress based on contract breach. See also W. PROSSER and P. KEETON, *TORTS*, 361–362 (5th ed. 1984).
[33] W. PROSSER and P. KEETON, supra note 32 at 365–366.
[34] 506 F.2d 1158 (9th Cir.1974).

knew that the valve door would be closed in 40 seconds, that it could not be stopped, and that he would be crushed to death unless he could be extricated. Fortunately, he was dragged out by a foreman and suffered only minor injuries. However, the trial court concluded that imminent danger of serious injury caused him to sustain an anxiety neurosis of a traumatic type with psychoneurotic reactions to stress. This neurosis caused him permanent total disability.

One reason this judgment was affirmed on appeal was the presence of some physical injury. Where physical harm has been suffered, often emotional distress can be the basis for additional compensation. However, the court also seemed willing to recognize the genuineness of the claim and its pecuniary loss as justification for granting recovery.

G. Defenses

Assumption of Risk. Assumption of risk completely bars recovery even if the defendant has been negligent. Advance consent by the plaintiff relieves the defendant of any obligation toward the consenting party. The plaintiff has chosen to take a chance. Suppose the plaintiff voluntarily entered into a relationship with the defendant with knowledge that the defendant would not protect the plaintiff from the risk. In such cases, the plaintiff impliedly assumed the risk. Sometimes the plaintiff is aware of a risk created by the negligence of the defendant but proceeds voluntarily to encounter it.

Express agreements to assume the risk are often given effect if knowingly and freely made by parties of relatively equal bargaining power.[35] But sometimes the relationship is one regulated by law and this freedom is denied. For example, workers are frequently prohibited from signing agreements to assume the risk of physical harm.

Most cases involve implied assumption of risk. Did the plaintiff know and understand the risk? Was the choice free and voluntary? Voluntariness has generated considerable controversy where

workers take risks under the threat that they will be discharged if they do not continue working.[36] Under such circumstances, some American courts conclude that the worker even under such pressure has assumed the risk of performing dangerous work.[37] Although statutes often preclude assumption of risk in employment relationships, the concept can be applied in third-party actions brought by workers against persons *other* than their employer.[38]

The assumption-of-risk defense is not favored. As a result, the defendant must plead and prove that the plaintiff assumed the risk.

Contributory Negligence. Until quite recently, most states would bar the plaintiff if the defendant established that the plaintiff's negligence played a significant role in causing the injury. Juries were thought to use techniques to mitigate the harshness of this defense. Where the plaintiff's negligence was slight and much less than that of the defendant, juries sometimes decided that the plaintiff was not contributorily negligent or that the contributory negligence did not cause the harm. Where both parties were negligent but the defendant much more so, the jury may have compromised by finding that there has not been contributory negligence but diminishing the amount of the plaintiff's award to take the plaintiff's negligence roughly into account.

The contributory negligence rule has been criticized because slight negligence bars what would otherwise appear to be a just claim. A number of states, notably Wisconsin and Louisiana, enacted legislation early in the century that modified this rule by requiring that the negligence of the plaintiff and the defendant be compared rather than automatically barring the plaintiff from recovery if the plaintiff has been negligent. Within the past twenty years, most states have enacted statutes creating comparative negligence, and a few courts have created comparative negligence without the benefit of legislation.

The details of comparative negligence laws vary considerably. The essential feature is that the plaintiff's negligence does not automatically bar recov-

---

[35]*Delta Air Lines, Inc. v. Douglas Aircraft Co.*, 238 Cal.App.2d 95, 47 Cal.Rptr. 518 (1965). However, a release signed by a person seeking admission to a university hospital was not upheld in *Tunkl v. Regents of the Univ. of California*, 60 Cal.2d 92, 383 P.2d 441, 32 Cal.Rptr. 33 (1963).

[36]See Section 31.08(A).
[37]Ibid.
[38]Third-party actions are discussed in Section 31.02.

ery but only diminishes recovery. A plaintiff whose negligence reaches a specific percentage, such as 50%, is barred in some states. In other states, the comparison is "pure." An 80% negligent plaintiff would be entitled to recover 20% of the loss from a 20% negligent defendant. Comparative negligence has been used in admiralty law and in claims by employees of certain common carriers subject to the Federal Employers Liability Act.

Independent Contractor Rule.   Unlike the two preceding defenses, the defense of the independent contractor rule does not involve conduct or risk taking by the party who suffered the loss. But because of its importance in construction legal problems, it merits brief mention at this point.

Suppose a homeowner hires a contractor to do remodeling work and the contractor negligently drops tools from the roof, injuring a passerby. Unless one of the many exceptions to the independent contractor rule applies, the homeowner will have a defense if sued by the passerby based on the negligence of the *contractor*. Sometimes one person is vicariously liable for the negligence of another.[39] But no vicarious liability exists if the party against whom the action is brought can establish that the negligent party was an *independent contractor*. An independent contractor is asked to achieve a result and is not controlled as to means by which this is accomplished. Vicarious liability requires the right to control the details of the work.

Because many independent contractors are one-person or small operations unable to respond for losses caused by their activities, many exceptions to the independent contractor rule exist.[40] Another reason for exceptions is that the independent contractor rule can enable financially solvent employers to insulate themselves from liability for activities that are essentially part of their business by hiring an independent contractor.

SECTION 7.04 Nonintentional Nonnegligent Wrongs: Strict Liability

A. Abnormally Dangerous Things and Activities

As indicated earlier,[41] preindustrial law was often strict in the sense that liability did not require neg-

ligence. Despite the emergence of negligence as the dominant basis for liability that developed in the nineteenth century, some pockets of law found liability "strictly."

One—keeping of animals—is of little importance to construction. (But watch strict liability for keeping a vicious dog on the site to deter vandals.) The other—strict liability for abnormally dangerous things and activities—does find application to modern construction problems and merits some comment.[42] *Rylands v. Fletcher*,[43] an 1868 English decision, held that a mill owner *not* shown to have been negligent was liable when he built a reservoir whose waters broke through to an abandoned coal mine and flooded parts of the plaintiff's adjoining mine.

Ultimately, most American decisions in some form or another found strict liability where the defendant owned dangerous things or engaged in dangerous activities. The prototype abnormally dangerous activity is often stated to be blasting. The activity may not be negligent, because the social utility and the high cost of avoiding the harm makes the conduct reasonable despite the high risk of serious harm. The modern rationale for strict liability for such activities is that such enterprises should pay their way and are better risk bearers than the persons who have been harmed by the activity.

As modern tort law emphasizes victim compensation and loss spreading through insurance and pricing goods and services, strict liability (liability without a need to establish negligence) has become more important in construction-related activities. Strict liability for defective products plays an increasingly important role in construction-related litigation (discussed in Section 7.09). (Also, the professional standard to which design professionals are held is increasingly under attack, the assertion being that they should be held to a "stricter" standard.[44]) This trend is reflected in federal laws that make those who generate, transport, or dispose of hazardous waste strictly liable for violation of statutes and regulations that regulate these activities.[45]

An illustration in a modern context can be seen in *Doundoulakis v. Town of Hempstead*, a New York

---

[39]See Section 31.05(A).
[40]See Section 31.05(C).
[41]See Section 7.02.

[42]Sometimes the type of activity can determine whether a person who hires an independent contractor is liable for that person's negligence. See Section 31.05(C).
[43]L.R.-3 H.L. 330 (1868).
[44]See Section 14.07.
[45]See also Section 9.13(A).

case.[46] The claim in this case centered around a proposal by a local entity to lay out a public park. The park was to be on a significantly lower elevation than the filled land on which the claimants' homes had been constructed. To build this park, 1.5 million cubic yards of sand fill had to be deposited on 146 acres. The public entity contracted with an engineer to prepare the plans and specifications and to operate as a supervising engineer and with a dredging company to execute the land-filling contract.

The land required the contractor to pump a dredge mix of 85% water and 15% land under pressure onto the park site. The mixture was brought from an ocean outlet. To impound the waters, dikes were constructed around most of the landfill site. Settlement of the sand and level of water were controlled by a system of exit weirs designed to discharge up to 40,000 gallons of water per minute. On the eleventh day of the operation, damage was observed to the property of the claimants—landowners adjacent to the proposed park. The claimants asserted that the subterranean percolation or seepage of the water from the landfill raised the underground water table, increasing the pressure and loosening the bulkhead anchorages designed to protect their land. The defendants—the town, the contractor, and the design engineer—asserted that the collapse of the bulkheads was due to their dilapidation as well as to an increase in pressure caused by heavy rainfalls.

The principal issue before the Court of Appeals was whether hydraulic dredging and landfilling, as done in this case, were abnormally dangerous activities giving rise to strict liability and not requiring a showing that the defendants had been negligent. The court held that the hydraulic dredging and landfilling were not normal activities and that a new trial would be needed to determine whether the activities in this case were abnormally dangerous. The court also held that if a claim on strict liability was sustained, those liable would include not only the owner of the land (the public entity) but also the design engineer and the contractor. In discussing strict liability, the court stated:

> Imposing strict liability upon landowners who undertake abnormally dangerous activities is not uncommon [citation omitted]. The policy consideration may be simply put: those who engage in activity of sufficiently high risk of harm to others, especially where there are reasonable even if more costly alternatives, should bear the cost of harm caused the innocent [citation omitted]. Determining whether an activity is abnormally dangerous involves multiple factors. Analysis of no one factor is determinative. Moreover, even an activity abnormally dangerous under one set of circumstances is not necessarily abnormally dangerous for all occasions (see Restatement, Torts 2d, § 520, Comment f).

> Guidelines, however, are identifiable. The many cases and authorities suggest the numerous factors to be weighed. Particularly useful are the six criteria listed in Restatement of Torts Second (§ 520): "(a) existence of a high degree of risk of some harm to the person, land or chattels of others; (b) likelihood that the harm that results from it will be great; (c) inability to eliminate the risk by the exercise of reasonable care; (d) extent to which the activity is not a matter of common usage; (e) inappropriateness of the activity to the place where it is carried on; and (f) extent to which its value to the community is outweighed by its dangerous attributes."[47]

B. Vicarious Liability

Sometimes one person is responsible for the negligence of another. The most common illustration is the employment relationship. Despite the absence of negligence by the employer, the employer is liable for harm caused by the negligent conduct of the employee as long as that conduct is within the scope of the employment. Sometimes the employer can be held liable for intentional torts committed by the employee. For example, a subcontractor was held liable for injuries sustained by two employees of the general contractor as a result of an assault committed on them by employees of the subcontractor.[48] The ruling was made despite the assault having occurred after the assaulting employees had completed their work shift.

One reason sometimes given for vicarious liability, or as it is sometimes called, *respondeat superior*, is the incentive it gives to an employer to choose employees carefully. The real basis for vi-

---

[46]42 N.Y.2d 440, 368 N.E.2d 24, 398 N.Y.S.2d 401 (1977).

[47]368 N.E.2d at 27.
[48]*Rodgers v. Kemper Constr. Co.*, 50 Cal.App.3d 608, 124 Cal.Rptr. 143 (1975).

carious liability is that enterprises should pay for losses they cause. Also, vicarious liability is premised on the ability to pay, or the deep pocket. The pocket of the employer is likely to be deeper than that of the employee.

The hirer of an independent contractor is not liable for the latter's negligence. Sometimes it is difficult to determine whether the person who has been engaged is an employee or an independent contractor. For reasons of enterprise liability and deep-pocket notions (as shall be seen in Section 31.05), vicarious liability has been expanding and the independent contractor rule has been diminishing in effectiveness.

C. Employment Accidents and Workers' Compensation

Although the principal thrust of this chapter is the role of tort law in distributing losses and responsibilities, one exception is the workers' compensation law, which was to supplant tort law in workplace accidents. The employer's liability under workers' compensation is strict. Recovery against the employer does not require the latter's negligence. For these reasons, it may be useful to briefly describe certain doctrines that deal with employment injuries.

The injured worker did not fare well under English or American common law. Expansion of the Industrial Revolution generated employment injuries but only limited legal protection to those injured. The employer's duties were limited: to furnish the worker a safe place to work and safe appliances, tools, and equipment and to warn the worker of any known danger connected with the work. The injured worker faced, in addition to limited employer duties, obstacles to compensation. First, the worker had to prove that the employer was negligent. Many accidents occurred because of the nature of the work and not the negligence of the employer.

Even if negligence were established, the worker faced the "unholy trinity" defenses of contributory negligence,[49] assumption of risk,[50] and the fellow servant rule. The first barred a worker whose injury was substantially caused by the worker's neg-

ligence unless the employer's conduct had been willful or wanton. The second barred the worker from recovery if taking the job was an assumption of the risk of injuries normally incident to the employment. A worker who stayed on a job under protest after the worker knew or appreciated the danger assumed the risk. The third defense meant that with some exceptions, a worker could not recover for injuries caused by the negligence of a fellow servant.

These formidable barriers often meant hardship to injured workers and their families. Following German social insurance law, American states began in the early twentieth century to enact what were then called workmen's compensation laws designed to replace tort law with social insurance for industrial accidents. Though some early statutes were held unconstitutional, currently all states have workers' compensation laws.

Although there are considerable variations among the states, certain common issues have arisen. Are employer and worker covered under the workers' compensation law? Many statutes exclude agricultural workers and domestics as well as employers who have only a few employees. The injury must arise out of the employment. Do injuries that occur on company picnics or while parking the worker's car in the company lot arise out of the employment? Doubts generally are resolved in favor of the employee.

Some states cover occupational diseases. Others do not. The worker need not show negligence by the employer. Nor are workers precluded from recovering by their own negligence, by their having assumed the risk, or by the injury having been caused by their fellow workers. Some states deny recovery if the worker was guilty of willful misconduct or intoxication and such misconduct caused the injury.

Most states require that the employer obtain compensation insurance. A few states have set up state funds to pay compensation awards. In some states, an employer can be a self-insurer if it makes adequate proof of financial responsibility. All these requirements are designed to ensure that there will be a financially responsible entity.

The particular problems of the construction industry have been recognized. Many states enacted "subcontractor under" or "statutory employer" statutes under which a prime contractor is the em-

---

[49]See Section 7.03(G).
[50]Ibid.

ployer of subcontractor employees under certain circumstances, such as a showing that the subcontractor did not procure the required insurance. This allows recovery against the prime contractor's compensation insurer. Special rules for the construction industry recognize that many subcontractors are not financially sound and may not obtain the requisite insurance. Compensation coverage may be diluted if a prime contractor subcontracted out work that would normally be performed by the prime contractor. This may be done to reduce the number of employees for whom the prime contractor would have to obtain insurance coverage or be exempt as having too few employees.

Typically, the award consists of a proportionate amount of the employee's wages as well as reimbursement for medical expenses incurred. Sometimes the employee is also able to receive a specific monetary award for designated injuries. More intangible, noneconomic losses—such as amounts to compensate for the emotional distress caused by disfigurement or for pain and suffering—are generally not recoverable. Workers' compensation awards provide *part* (estimated at from one third to three quarters of economic loss) and not *full* compensation. The recovery is intended to ensure that the injured worker does not become a burden on others. Compensation recoveries are less, and in many states much less, than can be recovered in a tort action. This has led to demands for federal minimum standards for compensation awards.

Workers' compensation remedies generally supplant whatever tort remedy the worker may have had against the employer. This is accomplished by statutory immunization of the employer from tort claims by the workers. Immunity gave the employer something in exchange for giving up existing legal protection in employment accidents.

Elimination of all tort suits would create a total compensation system and keep disputes within the administrative agencies charged with the responsibility for handling such claims. This has not been accomplished. Most states permit injured workers or compensation insurers who have paid them to institute tort actions against *third parties* whose negligence caused the injury.[51] A factory worker may

be able to recover in tort against the manufacturer of a defective product. A worker on a construction project may recover in tort against one of the many entities involved in the project other than his own employer. Because of the limited recovery available under workers' compensation, as noted in Section 31.02, it is becoming increasingly common for injured construction workers to institute third-party actions against anyone they can connect with their injury except their employer.

Workers' compensation claims are handled by an administrative agency rather than a court. Hearings are informal and usually conducted by a hearing officer or examiner. The employee can represent himself or be represented by a layperson or lawyer. Fees for representation are usually regulated by law. Although awards by the agency can be appealed to a court, judicial review is extremely limited, and very few awards are overturned. Third-party actions, on the other hand, because they involve tort claims, are brought to court, and incredibly complicated lawsuits often result.

Workers' compensation took on political and economic dimensions in the 1990s. In many states, insurance rates jumped drastically because of the expansion of claims falling within the workers' compensation system to claims involving, for example, job-connected stress. Employers in some states where the insurance rates were high complained that they could not compete with employers in states with lower rates and often threatened to move their operations to states that did not generate such expensive claims. Others complained that many claims were fraudulent, that too much of the ultimate payout went into the pockets of attorneys and health care providers, and that insurers charged exorbitant administration overhead. These factors have led to attempts to legislate changes designed to cure some of the problems, for better or worse. However, the efforts of those who would be adversely affected by change often thwart such attempts.

D. Product Liability

Another and perhaps more spectacular illustration of strict liability is imposed on manufacturers for defective products (discussed in greater detail in Section 7.09).

---

[51] A few states do not permit third-party actions against those in a common employment or design professionals. See Section 31.02.

SECTION 7.05 Claims by Third Parties

A. Lost Consortium

Suppose a spouse is seriously injured. In addition to causing economic loss, the injury can cause losses of an intangible nature to the other spouse or a child of the injured person. Such losses are sometimes called *consortium,* a term that encompasses the services and society lost and, in the case of the spouse, sexual relations.

Although the husband could recover for loss of consortium when his wife was injured, the law generally did not give corresponding rights to his wife. Slowly the courts have equalized consortium rights of husband and wife. However, a child cannot recover for lost consortium of either parent.[52]

B. Survival and Wrongful Death Statutes

Before the mid-nineteenth century, the death of the wrongdoer precluded any claim being made against the wrongdoer's estate. The wrong died with the wrongdoer. Similarly, death of the party harmed also barred any claim he would have had. However, statutes generally changed this common law rule. The easiest were those that allowed claims to be made against the estate of the wrongdoer.

More problems developed because of the patchwork of statutes affecting the right of the claimant. Those statutes are divided into survival statutes and wrongful death statutes. The former gave rights—often limited ones—to the estate of the deceased for harm suffered between the wrongful act and the death. The second provided compensation to those dependent upon the deceased for their economic losses. Local statutes must be consulted.

The claim is usually measured by any loss the deceased has suffered prior to death resulting from the defendant's negligence and, more important, any loss to the estate or survivors. Most statutes limit recovery to pecuniary losses, although some permit a limited amount of noneconomic losses to be recovered. Recovery for economic losses to the survivors is based on potential earnings of the deceased during his working life that would have been available to the survivors. The amount recov-

ered for the death of a person with high earnings or high earning capacity is often large. A few states limit the amount of recovery for wrongful death.

SECTION 7.06 Immunity

A. Charitable Organizations

Initially, American law granted immunity from tort liability to charitable organizations. Immunity was originally based on the charitable and nonprofit characteristics of the organization. The availability of public liability insurance, the recognition that even charitable institutions take on some of the characteristics of commercial enterprises, and the injured party's needs are factors that led to virtual abolition of this immunity.

B. Employers and Workers' Compensation

The worker's sole remedy against the employer is under workers' compensation. Employers have immunity, subject to a few exceptions, from tort claims made by their workers. That this has complicated .construction accident claims is shown in Chapter 31.

C. Public Officials

Judges and high public officials are usually granted absolute or qualified immunity from claims against them.

D. Sovereign Immunity

For various reasons—some metaphysical and some practical—English law immunized the sovereign from being sued in royal courts. This doctrine was adopted early in the nineteenth century by the U.S. federal courts, and it soon became established that the federal government could not be sued without its consent.

In 1946, the Federal Tort Claims Act was adopted. This legislation gave individuals the right to sue the United States for certain wrongs it committed. The Act had two important exceptions: (1) the federal government could not be sued for certain intentional torts, and (2) certain discretionary functions or duties performed by government officials could not give rise to tort liability. *Dalehite v. United States* clarified this by denying liability when the conduct was a policy or planning decision, also holding that the government could not

---

[52]A few states do allow recovery. See Note, 68 Marq.L.Rev. 174 (1984).

be held liable unless it was shown to have been negligent.[53] *Feres v. United States* barred a claim by a service person against the United States.[54] This has been extended to a manufacturer who followed government specifications.[55] Strict liability cannot be the basis for any claim against the federal government. Negligence under applicable state law must be established.

States generally adopted the English rule of sovereign immunity, although many states have given consent to be sued for specific claims.[56]

Municipal corporations, such as counties and cities, receive immunity for governmental acts but not for those they have performed in a private or proprietary capacity. The cases employing this distinction are often confused and contradictory.

Immunity has frequently been criticized, and beginning with the mid-twentieth century, perhaps slightly more than one third of the states have abolished it. States that abolished sovereign immunity frequently enacted comprehensive statutes modeled to some degree on the Federal Tort Claims Act. Sovereign immunity as it bears on the Construction Process is discussed in Section 31.08(B).

Modern justifications are usually based on either relieving already burdened public entities of serious financial responsibilities or precluding judicial intrusion into the running of government. Those who oppose immunity contend that it is better to spread the loss among all the taxpayers in the public entity than to concentrate it on the person who has suffered it. As to the fear of judicial intrusion, opponents of immunity contend that the current trend is toward increased accountability rather than relieving persons from their negligent acts. Yet expanded liability of public entities has generated a drastic increase of the cost of the insurance they procure and in some instances has made insurance unavailable. While it is not likely that the law will reverse its field and restore sovereign immunity wholesale, these concerns are likely to limit expansion of governmental liability or, in special cases, restore it.

---

[53]346 U.S. 15 (1953).
[54]340 U.S. 135 (1950).
[55]*Boyle v. United Technologies Corp.*, 487 U.S. 500, cert. denied, 488 U.S. 994 (1988).
[56]For canvassing of state law, see Comment, 58 Wash.L.Rev. 537 (1983).

Even when immunity has been eliminated, claims against governmental units require particular attention. Often such claims must be made within a shorter period of time than claims against private persons. In addition, such claims must be presented, as a rule, to legislative bodies of the governmental unit for their review before court action can be begun.

SECTION 7.07 Misrepresentation

A. Scope of Discussion

Although misrepresentation problems can occur in the contract formation process, this section treats the liability of persons whose business it is to make representations. A surveyor makes representations as to boundaries, a geotechnical engineer as to soil conditions, a design professional as to costs and the amount of payment due a contractor.

B. Representation or Opinion

Representations should be distinguished from opinions. For example, an architect may give his best considered judgment on what a particular project will cost. The prediction, however, may not be intended by him or understood by the client to be a factual representation that will give the client a legal claim in the event the prediction turns out to be inaccurate. If the statement is merely an opinion and not a representation of fact, it is reasonably clear that the person making the representation will not be liable simply because he is wrong.

C. Conduct Classified

The person making the misrepresentation may have had a fraudulent intent. He may have made the representation knowing that it was false, with the intention of deceiving the person to whom the representation was made. The representation may not have been made with the intention to deceive but may have been made negligently. Finally, the representation may have been made with due care but turned out to be wrong. This is sometimes referred to as an innocent misrepresentation.

D. Person Suffering the Loss

Another classification relates to the person who was harmed, the person to whom the representation was made, or a third party. For example, the geotechnical engineer may make a representation of soil conditions to a client. If the representation is incorrect, the harm may be suffered by the client or, in some cases, by third parties, such as a contractor or a subsequent purchaser or occupant.

E. Type of Loss

Cases can also be classified by the harm that resulted from the misrepresentation. A misrepresentation of soil conditions might result in a cave-in that kills or injures workers. It might also cause damage to property or economic loss unrelated to personal harm or damage to the client's property. The client may have to pay for damage caused to an adjacent landowner's property. A subcontractor may incur additional costs during the excavation because of the misrepresentation.

F. Reliance

In addition to the representation having to be material or serious, it must have been relied upon reasonably by the person suffering the loss. If there is no reliance or if the reliance is not reasonable, there is no liability for misrepresentation. A principal application of this doctrine is discussed in Chapter 25.

Often the owner makes representations as to soil conditions to the contractor and then attempts to disclaim responsibility for the accuracy of the representation. The disclaimer is an attempt to transfer the risk of loss for any inaccurate representations to the contractor. It is intended to negate the element of reliance, a basic requirement of misrepresentation. Generally, but not invariably, as seen in Section 25.05, such disclaimers are successful in placing the risk of loss on the contractor. However, they cannot relieve the person making the representation from liability for fraud, and they may not be effective if the representations were negligently made.

G. Some Generalizations

Generally, the more wrongful the conduct by the person making the representation, the greater the likelihood of recovery against the party. For example, a fraudulent misrepresentation will always create liability to the party to whom it was made or to third parties. A negligent one, in addition to providing the basis for a claim by the party who has paid for the representation, may be the basis for a claim by third parties. An innocent misrepresentation, being least culpable, is the most diffi-

cult on which to base a claim. Third parties are rarely able to recover, and even the other party to the recovery will be able to recover damages only if there is a warranty of accuracy.

Liability to third parties often depends on the type of harm suffered. If personal harm such as death or injury results, the absence of a contractual relationship between the person suffering the harm and the person making the misrepresentation is not likely to constitute a defense. In a Louisiana case, a structure collapsed during construction, resulting in a large number of personal injury claims. One party against whom a claim had been made was the architect who, in applying for a building permit, certified that he would inspect the work and verify that certain structural requirements were met. He did neither and was held liable for having made a negligent misrepresentation.[57]

Where harm is economic, lack of privity (contractual relationship) between the claimant and the person who made the misrepresentation causes the greatest difficulty. One reason is the wide range of individuals affected by the representations of persons in the business of making them. Such persons can be exposed to enormous liability, often disproportionately high to the remuneration paid for the services. For example, a certified public accountant may make an audit report that causes thousands of investors to buy shares in a particular company. Were the accountant accountable to *all* these investors, he would face enormous risk exposure.

In the construction context, the misrepresentation cases typically involve claims against surveyors and those who provide geotechnical information (but see Section 14.08(D) for expanded use of misrepresentation against architects and engineers). Those who provide services often are liable to third parties for economic losses when the representations made were found to be negligent, provided the professional making the representation can reasonably foresee the type of harm that is likely to occur and the persons who may suffer losses. This trend is reflected in *Rozny v. Marnul*,[58] in which a surveyor was held liable to a homeowner who had built a house and garage relying

---

[57]*Stewart v. Schmieder*, 376 So.2d 1046 (La.App.1979). See also 386 So.2d 1351 (La.1980), holding the city liable for negligently approving the permit application. See Section 14.08(D).
[58]43 Ill.2d 54, 250 N.E.2d 656 (1969).

on a survey the surveyor had prepared for a developer. The developer sold the lot to the homeowner and evidently the survey along with it. The house and garage encroached upon a neighbor's lot. After reviewing the legal history of such claims against professionals, the court stated that the factors to be considered were as follows:

1. The express, unrestricted, and wholly voluntary "absolute guarantee for accuracy" appearing on the face of the inaccurate plat.[59]
2. Defendant's knowledge that this plat would be used and relied on by others than the person ordering it, including plaintiffs.[60]
3. The fact that potential liability in this case was restricted to a comparatively small group and that ordinarily only one member of that group would have suffered loss.
4. The absence of proof that copies of the corrected plat were delivered to anyone.
5. The undesirability of requiring an innocent reliant party to carry the burden of a surveyor's professional mistakes.
6. The fact that recovery here by a reliant user whose ultimate use was foreseeable would have promoted cautionary techniques among surveyors.

SECTION 7.08 Duty of the Possessor of Land

A. Relevance

Tort law determines the duty owed by the possessor of land—that is, the one with operative control—to those persons who pass by the land or enter upon it. Before, during, or after completion of a construction project, members of the public will pass by the land or, with or without permission, enter upon the land with the potential of being injured or killed by a condition on the land or by an activity engaged in by the person "in con-

trol" of the land. Workers also may suffer injury or death because of the condition of the land or activities on it. Persons who live in or enter a completed project may suffer injury or death because of something related to the land or the Construction Process. Liability for such harm depends on the particular nature of the obligation owed to the plaintiff by the possessor of land near which or on which the physical harm was suffered. It can depend on the injured party's permission to be near or on the land and the purpose for being there.

The term *possessor of land* is used in this section without exploring the troublesome question of whether the owner, the prime contractor, or the subcontractors fall into this category during the Construction Process (discussed in Section 31.03).

B. To Passersby

Passersby can expect that the conditions of the land and activities on it will not expose them to unreasonable risk of harm. Whether the possessor has measured up to the legal standard will depend on factors discussed in Section 7.03(C).

C. To Trespassing Adults

The trespasser enters the land of another without permission. In so doing, the trespasser is invading the owner's exclusive right to possess the land. Veneration for landowner rights led to a very limited protection for trespassers by English and American law. The possessor was not liable if trespassers were injured by the possessor's failure to keep the land reasonably safe or by the possessor's activities on the land.

Exceptions developed as human rights took precedence over property rights. For example, possessors who know that trespassers use limited areas of their land must conduct their activities in such a way as to discover and protect trespassers from unreasonable risk of harm. Railroads were required to be aware of persons who crossed the tracks at particular places. Another exception was applied frequently to railroads for dangerous activities conducted on the land.

Discovered trespassers are entitled to protection. The landowner must avoid exposing such trespassers to unreasonable risk of harm.

Despite their unfavored position, trespassers are not outlaws. Possessors cannot shoot them or inflict physical harm on them to protect their property.

---

[59]A subsequent California case that applied the rule in the *Rozny* decision held that the absence of any guarantee would not affect liability for negligent misrepresentation. *Kent v. Bartlett*, 49 Cal.App.3d 724, 122 Cal.Rptr. 615 (1975).

[60]The *Rozny* decision was distinguished (involved facts that made it different from those in earlier precedent) in *Bushnell v. Sillitoe*, 550 P.2d 1284 (Utah 1976), in which the claimant was a person for whose guidance the survey had not been prepared.

What steps can possessors take to protect their land from trespassers? Suppose a contractor puts up a barbed wire fence. Suppose a contractor keeps savage dogs on a fenced-in site at night to protect the site and construction work from vandals.

A case that excited controversy was *Katko v. Briney*.[61] The defendant owned an uninhabited farmhouse containing some antiques and old jars. The house had been broken into and the contents removed several times. After requests to law enforcement authorities were unproductive, the defendant installed a spring gun aimed at the legs of anyone who entered the house and sought to enter a particular room. The plaintiff, thinking the house uninhabited and looking for old fruit jars, entered the house and was severely injured by the spring gun's discharge. Charged with a felony, the plaintiff pleaded guilty to a misdemeanor and received a sixty-day suspended jail term.

The plaintiff then sued the defendant for having caused the injury. A jury award against the landowner for compensatory and punitive damages was upheld because the privilege to protect property did not extend to the infliction of serious bodily harm. (For an interesting trespasser case, see (G).)

D. To Trespassing Children

Special rules have developed for trespassing children. Frequently, children do not realize that they are entering the land of another. Sometimes they are not aware of the dangerous characteristics of natural and artificial conditions on the land that they enter. Possessors must conduct their activities in such a way as to avoid unreasonable risk of harm to trespassing children. However, controversy frequently develops regarding the extent to which limited protection given trespassers as to artificial conditions on the land should be applied to trespassing children.

The law's strong protection for landowners and their rights has, for the most part, been qualified by humanitarian concerns for children of tender years injured or killed when they confront or deal with dangerous conditions on the land of another. The Restatement (Second) of Torts reflects this diminution of landowner protection and states that the possessor of land is liable for injuries to trespassing children caused by artificial conditions on the land if

(a) the place where the condition exists is one upon which the possessor knows or has reason to know that children are likely to trespass, and

(b) the condition is one which the possessor knows or has reason to know and which he realizes or should realize will involve an unreasonable risk of death or serious bodily harm to such children, and

(c) the children because of their youth do not discover the condition or realize the risk involved in intermeddling with it or in coming within the area made dangerous by it, and

(d) the utility to the possessor of maintaining the condition and the burden of eliminating the danger are slight as compared with the risk of children involved, and

(e) the possessor fails to exercise reasonable care to eliminate the danger or otherwise protect the children.[62]

Children often trespass on construction sites. In the process of doing so, they may engage in an activity that can result in injury or death. This section has presented an overview, although it may be useful to look at this problem more specifically. In so doing, it is not important to focus sharply on the legal theories, such as attractive nuisance and the playground doctrine, that are sometimes applied by courts to determine the nature of the duty owed trespassing children. What is important is to examine the following:

1. Recurrent fact patterns.
2. The clash of important policies, such as humanitarian protection for children and freedom of landowners, that has divided many appellate courts.

As to the first, cases have involved the following:

1. A child who was injured when concrete blocks on which he was climbing collapsed.[63]
2. A child who fell when a scaffold collapsed.[64]

---

[61]183 N.W.2d 657 (Iowa 1971).

[62]Section 339 (1965).
[63]*Goben v. Sidney Winer Co.*, 342 S.W.2d 706 (Ky.1961) (jury question).
[64]*Bloodworth v. Stuart*, 221 Tenn. 567, 428 S.W.2d 786 (1968) (jury question).

3. A child who was injured when another child threw a clod of dirt found at the construction site.[65]

4. A child who was injured by an exploding cartridge that he had found on a construction site and had taken home.[66]

5. A child who suffocated in a cave-in of an exposed excavation.[67]

6. A child who drowned while playing in an excavation that had become filled with rainwater.[68]

Although generally courts have upheld trial courts that ruled for contractors, the appellate court decisions are often made by divided courts in which the judges have exchanged sharp views on the policies involved in making such decisions. Some judges show solicitude toward children and their propensity to play where they should not. One judge stated:

> It is the instinct of children of the age of appellee to play. Building material, stacked as this was, is peculiarly attractive to them. This is a fact known of everyone. In a populous community this instinct is more than likely to find vent in availing itself of such temptation. Warnings are not enough to make the premises reasonably safe. The material should be stacked so, with the knowledge that the premises will be probably so used in spite of warnings and precautions of the lot owner, that the children playing thereabout will not be subjected to the hazards of falling timbers and material insecurely put up.[69]

In concluding that a particular dispute should have gone to the jury another judge stated:

> Nor does it make a difference that no children actually were present at the very moment of [a visit by the contractor's representative] (which may have been during school hours). In a closely built-up residential neighborhood children are as much a part of the natural scene as grasshoppers. Their intrusive appearance upon and around the unenclosed premises of such an area is to be expected.[70]

The dissenting judge in the same case took a different approach, stating:

> The majority opinion takes the view that the builder of the structure should have employed certain security measures in order to protect children who might intrude and play on the structure as young Goben did. To require such a practice would make building costs, which are mounting skyward by the hour, well nigh prohibitive for the average person.[71]

The economic burden that can be placed on contractors was emphasized in another case in which the court, after pointing to the contractor's awareness that children played on the site, noted that there were no fences and no signs. The court also pointed to the absence of a security guard, who would have cost $609 a week in a 75,000-dollar job.[72]

Judges less sympathetic to trespassing children also emphasize the responsibility of even young children to know what is dangerous and of parents to keep their children away from construction sites. In many cases, part of the responsibility for the child's injury must fall upon parents who do not supervise the child properly.

E. To Licensees

A licensee has a privilege of entering or remaining on the land of another because of the latter's consent. Licensees come for their own purposes rather than for the interest or purposes of the possessor of the land. Examples of licensees are persons who take shortcuts over property with permission, persons who come into a building to avoid inclement weather or to look for their children, door-to-door salespeople, and social guests.

Some anomalous exceptions exist, such as the firefighter who enters a building at night to put out a fire or the police officer who enters to apprehend a burglar. Logically, such persons should be considered as benefiting the possessor of land, but many cases hold that they are simply licensees.

Early cases held that the only limitation on the possessor's activity was to refrain from intention-

---

[65]*Kirven v. Askins*, 253 S.C. 110, 169 S.E.2d 139 (1969) (affirmed trial court judgment for contractor despite jury verdict for child).

[66]*Concrete Constr., Inc. of Lake Worth v. Petterson*, 216 So.2d 221 (Fla.1968) (contractor owed duty to child but injury too remote).

[67]*Gagnier v. Curran Constr. Co.*, 151 Mont. 468, 443 P.2d 894 (1968) (should not have been submitted to jury).

[68]*Martinez v. C. R. Davis Contracting Co.*, 73 N.M. 474, 389 P.2d 597 (1964) (jury question).

[69]*Louisville Ry. Co. v. Esselman*, 93 S.W. 50, 52 (Ky.1906).

[70]*Goben v. Sidney Winer Co.*, supra note 63 at 711.

[71]Id. at 713.

[72]*Bloodworth v. Stuart*, supra note 64.

ally or recklessly injuring a licensee. However, most courts today require that the possessor of land conduct activities in such a way as to avoid unreasonable risk of harm.

The possessor has a duty to repair known defects or dangerous conditions or to warn licensees of nonobvious dangerous conditions. Licensees cannot demand that the land be made reasonably safe for them. The possessor need not inspect the premises, discover dangers unknown to the possessor, or warn the licensee about conditions that are known or should have been known to the licensee.

F. To Invitees

An invitee receives the greatest protection. The possessor must protect the invitee not only against dangers of which the possessor is aware but also against those that could have been discovered with reasonable care. Although not an insurer of the safety of invitees, the possessor is under an affirmative duty to inspect and take reasonable care to see that the premises are safe. Sometimes the possessor can satisfy the obligation by warning the invitees of nonobvious dangers.

Who qualifies for such protection? Some cases have limited invitees to those persons who furnish an economic benefit to the possessor. More jurisdictions, however, seek to determine whether the facts imply an invitation to the entrant. However, the invitation concept does not include those invited as social guests.

The line between licensee and invitee is difficult to draw and sometimes seems arbitrary. Although most courts consider police officers and firefighters licensees, courts hold building inspectors to be invitees.[73] The tendency is to place increased responsibility on possessors engaged in industrial or commercial activities, based on charging enterprises with the normal harm their activities cause.

G. Movement Toward General Standard of Care

Undoubtedly, the various categories that determine the standard of care are difficult to administer. Exceptions develop within the categories, and the ap-

plication of the categories is often uneven. For this reason, there is some movement toward a rule that would require the possessor of land to avoid unreasonable risk of harm to *all* who enter upon the land. The Supreme Court of California stated:

> Without attempting to labor all of the rules relating to the possessor's liability, it is apparent that the classifications of trespasser, licensee, and invitee, the immunities from liability predicated upon those classifications, and the exceptions to those immunities, often do not reflect the major factors which should determine whether immunity should be conferred upon the possessor of land. Some of those factors, including the closeness of the connection between the injury and the defendant's conduct, the moral blame attached to the defendant's conduct, the policy of preventing future harm, and the prevalence and availability of insurance, bear little, if any, relationship to the classifications of trespasser, licensee and invitee and the existing rules conferring immunity. . . .

> We decline to follow and perpetuate such rigid classifications. The proper test to be applied to the liability of the possessor of land in accordance with section 1714 of the Civil Code is whether in the management of his property he has acted as a reasonable man in view of the probability of injury to others, and, although the plaintiff's status as a trespasser, licensee, or invitee may in the light of the facts giving rise to such status have some bearing on the question of liability, the status is not determinative.[74]

The California decision has been influential in causing some jurisdictions to abolish the classifications.

California's elimination of the threefold common law classification led to the enactment of California Civil Code § 846, designed to encourage those who own property suitable for recreational use to allow the public to enter by limiting the owner's tort liability. This statute was invoked by an owner who was building two homes near a scenic beach. Some persons on their way to a beach picnic decided to explore the houses under construction. The roofs of the two homes were connected by two loose boards. The plaintiff's sister panicked after climbing on one of the roofs. The plaintiff rescued her by guiding her across the boards to the other roof. The sister crossed safely,

---

[73]W. PROSSER and P. KEETON, *TORTS*, 428–432 (5th ed. 1984).

[74]*Rowland v. Christian*, 69 Cal.2d 108, 443 P.2d 561, 567–68, 70 Cal.Rptr. 97, 103–4 (1968).

but the plaintiff was injured when he fell to the ground.

The court held that the statute could not be invoked by the owner. Even though the homes were located near scenic beaches suitable for recreational use, the homes themselves were not suitable for recreational use. The case was to be decided based on general negligence principles.[75] (Testimony at the trial suggested it was negligent to use loose boards to connect the roofs and not to fence the site.)

H. Nondelegability of Legal Responsibility

Suppose the possessor hires someone to make the land reasonably safe and that person does not do so. Clearly, the possessor who fails to use reasonable care to select a contractor will be liable. Similarly, the possessor who does not remove an incompetent contractor or does not inspect the work properly will be liable.

Suppose the possessor has lived up to the legal standard of care. Subject to many exceptions, the employer of an independent contractor is not liable for the latter's negligence. However, as seen in Section 31.05(C), one common exception involves using an independent contractor to comply with an important responsibility, such as making the land reasonably safe. Such a responsibility is nondelegable. It is too important to be shifted to an independent contractor.

SECTION 7.09 Product Liability

A. Relevance

The historical development of legal rules relating to the liability of a manufacturer for harm caused by its products manifests a shift from protection of commercial ventures toward compensating victims and making enterprises bear the normal enterprise risks. Manufacturer's liability has become important in the Construction Process, as harm can be caused by defective equipment or materials.

B. Some History: From Near Immunity to Strict Liability

In 1842, the English case of *Winterbottom v. Wright*[76] held that an injured party could not recover from the maker of a defective product in the absence of privity between the injured party and maker. As noted in Section 7.03(E), privity was usually a contractual relationship, but it could also be based upon a status that placed a duty upon one party to watch out for another.

The privity requirement protected an infant manufacturing industry developing during the Industrial Revolution. It allowed contracting parties to know their liability exposure and deal with inordinate risks by contract. Privity permitted the manufacturer to be secure in the belief that it would not be held liable to persons other than those with whom it dealt. It could relieve itself from inordinate risks with those it dealt with by contract disclaimers. It could not do so with the many third parties who might be injured by its products or activities.

Protection was no more tolerable than the sad plight of the uncompensated industrial accident victim who ultimately received protection through workers' compensation. By the early twentieth century, the privity rule was no longer acceptable. Injured parties needed a solvent defendant from whom they could recover. Ultimate responsibility should be on the manufacturer. The latter can insure against predictable losses and pass the cost to those who benefited from the enterprise, such as owners or users of the enterprise's activities.

The major turning point occurred in New York in 1916. Before 1916, New York had held that a negligent manufacturer could be liable despite the absence of privity if there was a latent defect in the goods sold or if the goods sold were inherently dangerous. In 1916, the New York Court of Appeals held in *MacPherson v. Buick Motor Co.*[77] that this exception included goods that were dangerous if made defectively. The *MacPherson* case, one involving an automobile, led to abolition of the privity rule in the United States.

Still the plaintiff had the difficult task of establishing that goods were negligently made by the manufacturer. But the *res ipsa loquitur* doctrine proved of great assistance. It permitted the plaintiff to present its case to the jury even though the plaintiff introduced no direct evidence of negligent conduct by the manufacturer.

---

[75]*Potts v. Halsted Financial Corp.*, 142 Cal.App.3d 727, 191 Cal.Rptr. 160 (1983).
[76]Supra note 31.

[77]217 N.Y. 382, 111 N.E. 1050 (1916).

Res ipsa did not, as a rule, deprive the defendant of the opportunity of introducing evidence that it had not been negligent. This was typically done by seeking to establish that the defendant had followed common industry practices and had used a system designed to ensure that its products were safe. Yet juries typically ruled for victims of adulterated food or beverages or defective products.

A more efficient way of placing this risk on the manufacturer was needed. The concept that first accomplished this purpose was implied warranty. This doctrine, borrowed largely from commercial law, held sellers liable when their goods were not merchantable or, under certain circumstances, not fit for the purposes for which buyers bought them. It eliminated the need to establish negligence. Implied warranty would under certain circumstances extend protection to third parties.

Warranty first was used in food and drug cases. It then began to be employed when harm was caused by manufactured goods. This concept, at least for a time, put normal enterprise risks on the enterprise.

Early in the use of the implied warranty, some courts recognized that though useful, it was essentially a commercial doctrine that was inappropriate in determining who should bear the risk of physical harm caused by defective products. In addition, warranty carried with it technical rules more appropriate to commercial transactions. As a result, a few courts began to treat claims by injured parties against manufacturers as involving strict liability in tort.[78] Soon other courts fell into line, and the strict liability concept gradually supplanted implied warranty as a risk distribution device.

C. Section 402A of the Restatement of Torts

One factor that led to the replacement of implied warranty by strict liability in tort was the decision by the American Law Institute, a private group of judges, scholars, and lawyers, to restate the law of torts by publishing Section 402A in 1965. This section states:

1. One who sells any product in a defective condition unreasonably dangerous to the user or consumer or to his property is subject to liability for physical harm thereby caused to the ultimate user or consumer, or to his property, if

   a. the seller is engaged in the business of selling such a product, and

   b. it is expected to and does reach the user or consumer without substantial change in the condition in which it is sold.

2. The rule stated in Subsection (1) applies although

   a. the seller has exercised all possible care in the preparation and sale of his product, and

   b. the user or consumer has not bought the product from or entered into any contractual relation with the seller.

Section 402A has had a significant impact. Increasingly, manufacturers of defective products are held liable without any privity between the injured party and the manufacturer and without the injured party having to establish that the manufacturer had been negligent.

Yet these standards, those of defective conditions unreasonably dangerous, mask a number of difficult questions. What is defective? Does the *unreasonably dangerous* requirement in effect put the burden of establishing negligence back on the plaintiff? Can responsibility for marketing a high-risk product be satisfied by warning the user? What if products are dangerous no matter how much care is taken in their manufacture, such as cigarettes, whiskey, and drugs? Additional complications developed when governments, particularly the federal government, required that cigarettes contain a label warning the user of smoking's health hazards. Did these laws preempt (that is, replace) state product liability tort law? The U.S. Supreme Court held that such labeling laws did not. Such laws did supersede state laws dealing with the same topic, however, such as failure to warn and fraudulent misrepresentation claims, but not those based upon express warranty or conspiracy.[79] And what of those who sell blood for transfusions, a necessary activity but one that cannot eliminate the risk of infected blood? (Legislatures often protect those engaged in this activity.)

---

[78]The leading case is *Greenman v. Yuba Power Products, Inc.*, 59 Cal.2d 57, 377 P.2d 897, 27 Cal.Rptr. 697 (1963).

[79]*Cipollone v. Liggett Group, Inc.*, _____ U.S. _____ , 112 S.Ct. 2608 (1992).

D. Product Use

Generally, a manufacturer's responsibility extends only to reasonably foreseeable use of its products. For example, a manufacturer of casements to be used as window frames is not liable when workers use them as ladders.[80] However, if product misuse is reasonably foreseeable, a jury can find that the manufacturer had an obligation at least to warn the user or even to design the product with this in mind.[81]

E. Parties

Many entities play significant roles in the manufacturing and distribution of products. Manufacturers buy component parts and materials from other suppliers. They may obtain independent design and testing services. The product itself may be sold or installed by independent retailers or installers. Sometimes products are distributed through wholesalers who sell to retailers.

It is generally assumed that the manufacturer is best able to spread the loss and avoid harm. But some retailers have this capacity. For example, many products are distributed through large national retail chains that design or set performance standards for products in contracts with smaller manufacturers.

The purchaser of the product is not the only one who may be injured by a defective product. Members of the purchaser's family may use the product. Social guests may be injured if a television set explodes in the living room. A defectively designed car can injure drivers, passengers, other vehicles, or pedestrians.

Section 402A of the Restatement of Torts took no position as to whether persons other than users or consumers can use strict liability to recover against the manufacturer. Nor did it take a position as to whether the seller of a component part would be liable. Generally, strict liability protection is being given to all those who the defendant could have reasonably anticipated would be injured by the product. Similarly, the trend seems toward holding responsible all those who play significant roles in product manufacture.

F. Defenses

Contributory negligence, where it still exists, is generally not a defense available to a manufacturer. But if the latter establishes that the injured party voluntarily assumed the risk, the injured party cannot recover. Where comparative negligence applies, it has been applied in strict liability claims.[82]

Sometimes product manufacturers expect or require that those to whom they sell will take steps to ensure that the product will be used safely. For example, suppose user instructions make it clear that guards are to be used around dangerous machinery. Similarly, suppose a manufacturer of automobiles requires that the car dealer prepare the car in a designated way for the customer. Some courts have held that the manufacturer's duty to make a product that is reasonably safe cannot be delegated to others.[83] However, negligent conduct by the purchaser under some circumstances can be a superseding cause that may relieve the manufacturer. This superseding cause may consist of the purchaser making changes that would affect the way the machine was used.[84]

G. Economic Losses

Is the manufacturer liable for economic losses such as delay damages, lost profits, or injury to the product itself? Clearly, a manufacturer is responsible if there is an express warranty. After some early uncertainty, the trend is to bar recovery for economic losses in claims based on strict liability.[85]

H. Disclaimers

In the commercial world, sellers frequently seek to limit their risk by disclaiming responsibility for certain losses or by limiting the remedy. One reason for shifting from implied warranty to strict liability was the desire to avoid disclaimers frequently part of the commercial transaction where the party injured was not a real participant in the transaction.

---

[80]*McCready v. United Iron & Steel Co.*, 272 F.2d 700 (10th Cir.1959) (negligence action).
[81]*Ford Motor Co. v. Matthews*, 291 So.2d 169 (Miss.1974).

[82]*Daly v. General Motors Corp.*, 20 Cal.3d 725, 575 P.2d 1162, 144 Cal.Rptr. 380 (1978).
[83]*Vandermark v. Ford Motor Co.*, 61 Cal.2d 256, 391 P.2d 168, 37 Cal.Rptr. 896 (1964); *Bexiga v. Havir Mfg. Corp.*, 60 N.J. 402, 290 A.2d 281 (1972).
[84]*Schreffler v. Birdsboro Corp.*, 490 F.2d 1148 (3d Cir.1974).
[85]*Jones & Laughlin Steel Corp. v. Johns-Manville Sales Corp.*, 626 F.2d 280 (3d Cir.1980). See Section 14.08(E).

Disclaimers are likely to be given effect between the parties if the parties are business entities of relatively equal bargaining strength. However, disclaimers are not likely to be given effect in consumer purchases.

I. Design Defects

Early product liability cases involved manufacturing defects. To determine whether a product is defective, the product causing the injury is compared to either the design plans or other products made from the same design. If there is a deviation, the product is defective.

Even in the absence of negligence, some products will have unintended manufacturing defects. It is not economically feasible to eliminate all risks of randomly defective products. It is better to predict the likelihood of defects, calculate the liability exposure, insure or self-insure against these risks, and include the cost in the product price. In addition to the exasperating question of whether there was a manufacturing or design defect (was it a bad weld or cheap material that made welding difficult), judging the design has no neat test as does manufacturing.

The law has struggled painfully with design defect definitions. California had earlier rejected the Restatement definition of "unreasonably dangerous" as reinstituting the discarded negligence test.[86] It then held that two tests must be applied. First, did the product meet the expectations of an ordinary consumer as to product safety? Second, even if it did, the manufacturer is liable if it cannot establish (using *negligencelike* criteria) that on balance, benefits of the design outweighed the risks. The court allowed a hindsight look at the design and reserved judgment on whether the state of the art being followed would be a defense. The court did not decide whether products are defective because they lack adequate warnings or directions.[87]

New Jersey expressly permitted hindsight to help the plaintiff prove a defective design. Failure to provide an appropriate warning will make the product defective. The state of the art is not a defense to claims based upon design defects. The

New Jersey Supreme Court held that the manufacturer must reduce risk to the greatest extent possible consistent with the product's utility.[88] Pennsylvania finds liability if a *judge* determines the product was defectively designed and leaves criteria for making this determination uncertain.[89]

Even within these jurisdictions there will be confusion. It will be even harder for uncommitted courts to decide which standard to adopt. How can a designer determine whether a proposed design is defective? Even worse, how does an attorney advise his client after a claim has been made? It is no wonder attempts are being made legislatively to deal with this problem. (See (L).)

J. Government-Furnished Design

Suppose someone suffers physical harm because of a defectively designed product procured by the federal government, the design compelled by the United States in its contract with the manufacturer. The federal government is immune from tort liability except to the extent that it has deprived itself of that immunity in the Federal Tort Claims Act. The Act requires that the government be negligent, and negligence is *not* required to establish the liability of the manufacturer for a defective product. Inasmuch as the federal government is likely to be immune, those who have suffered personal harm or their survivors frequently sue the manufacturer of the product.

The manufacturer is likely to assert two defenses: (1) the *government specification* defense if it followed nonobviously defective government specifications[90] and (2) the *government contract* defense, which gives the manufacturer the same immunity that would be given to the government.

These defenses were relatively uncontroversial until the explosion of the environmental movement in the 1970s. The realization that acts committed many years ago can create seriously harmful risks

[86]*Cronin v. J.B.E. Olson Corp.*, 8 Cal.3d 121, 501 P.2d 1153, 104 Cal.Rptr. 433 (1972).
[87]*Barker v. Lull Eng'g Co.*, 20 Cal.3d 413, 573 P.2d 443, 143 Cal.Rptr. 225 (1978).

[88]*Beshada v. Johns-Mansville Products Corp.*, 90 N.J. 191, 447 A.2d 539 (1982).
[89]*Azzarello v. Black Bros. Co.*, 480 Pa. 547, 391 A.2d 1020 (1978). For an evaluation of design defect cases, see Diamond, *Eliminating the "Defect" in Design Strict Products Liability Theory*, 34 Hastings L.J. 529 (1983).
[90]The U.S. Supreme Court held (5–4) that the contractor who drew attention to the problem was not liable. *Boyle v. United Technologies Corp.*, 487 U.S. 500, *cert. denied*, 488 U.S. 994 (1988).

led to many claims based on exposure to unsafe chemicals and hazardous wastes. The most controversial have been those that have related to Agent Orange, a defoliant used by the U.S. Armed Forces in Vietnam manufactured in accordance with government specifications. A federal court held that the manufacturer will be given a defense if it can show that it followed government specifications in the manufacture of the defoliant and that the United States knew as much as or more than the manufacturer of the hazards.[91]

K. Beyond Products: Sellers of Services

Attempts to hold those who perform services, such as architects and engineers, strictly liable or liable based on implied warranties have not been successful.[92] This creates anomalies and different standards. Those who mass-produce homes[93] or lots[94] have been held strictly liable in some states. Those who manufacture products are held strictly liable. But suppose a claimant sues an independent designer of a defectively designed product? Or suppose the manufacturer sues the independent designer? In either case, the designer would not be held to a standard of strict liability or implied warranty. These anomalies may lead some courts to hold those who are in the business of designing to enlarged, more strict liability.[95]

L. Future Developments

Considerable dissatisfaction has been expressed with the evolution of product liability law. Manufacturers have complained that the options available to them are all unsatisfactory. One alternative is simply not to insure when the cost of premiums makes the price uncompetitive or is beyond the financial capacity of the manufacturer. Another is to overdesign a product that will pass even hindsight judicial or jury review evaluation. The product line can be dropped, which can mean diminished competition, fewer consumer choices, and higher prices.

Additionally, the cost of defense is staggering, including attorneys' fees, costs of testing, and experts. As if this were not enough, courts are beginning to award punitive damages when a jury decides that design choices did not take safety or public needs into account.[96]

In the 1990s, attacks were made on expanded product liability law by those who contended that such restraints on American industry made American products uncompetitive internationally. Attackers pointed to the costs incurred by manufacturers, such as increased cost of testing, ballooning insurance premiums, and the need to set aside reserves for uninsured losses and claims overhead, and compared American manufacturers to international competitors who did not face these same costs. They also pointed to manufacturers' decisions to withdraw products from the market because the costs associated with liability made the products unprofitable.

Defenders of product liability law asserted that competing manufacturers in other countries were highly regulated by government agencies. When the costs to American manufacturers are compared to those with whom they compete, the defenders claimed, the differentials are not great. If the cost of production in the United States is truly higher than elsewhere, defenders said, prices in the United States should be higher and American manufacturers should be less profitable. Yet, they claimed, this is not the case.

Defenders also pointed to the policy rationale for product liability law: that the need for safe products acts as a spur to manufacturers to develop safe products that would be attractive to consumers and as a disincentive to produce unsafe products. Defenders of American product liability law further claimed that no evidence supports the argument that such laws stifle innovation.

Clearly, emphasis upon competitiveness points to product liability law as a form of government

---

[91]*In re "Agent Orange" Product Liability Litigation*, 818 F.2d 179 (2d Cir.1987), *cert. denied* sub nom. *Krupkin v. Dow Chem. Co.*, 487 U.S. 1234 (1988).

[92]*LaRossa v. Scientific Design Co.*, 402 F.2d 937 (3d Cir.1968). See also Section 14.05.

[93]*Schipper v. Levitt & Sons, Inc.*, 44 N.J. 70, 207 A.2d 314 (1965); *Kriegler v. Eichler Homes, Inc.*, 269 Cal.App.2d 224, 74 Cal.Rptr. 749 (1969). But see *Wright v. Creative Corp.*, 30 Colo.App. 575, 498 P.2d 1179 (1972). For a thorough discussion, see Comment, 33 Emory L.J. 175 (1984). See Section 24.10.

[94]*Avner v. Longridge Estates*, 272 Cal.App.2d 607, 77 Cal.Rptr. 633 (1969).

[95]See Comment, 55 Calif.L.Rev. 1361 (1967); Comment, 15 Cal.W.L.Rev. 305 (1979).

[96]*Grimshaw v. Ford Motor Co.*, 119 Cal.App.3d 757, 174 Cal.Rptr. 348 (1981).

regulation. When that is revealed, it can be seen more clearly which groups are struggling over this issue. Plaintiffs' lawyers and consumer organizations argue for retaining the expanded liability of manufacturers, while manufacturers and many economists stress the need for a free market.

This debate has reached the attention of the courts. While the law may not be rolled back by the courts, it is not likely to be expanded.

Many legislatures have enacted statutes that have limited common law liability. Recommendations have been made for a uniform federal statute dealing with manufacturer's liability for defective products.

SECTION 7.10 Remedies

A. Compensation

The principal function of awarding tort damages is to compensate the plaintiff for the loss. In the ordinary injury case, the plaintiff is entitled to recover economic losses and certain noneconomic losses.[97] Economic losses are, for example, lost earnings and medical expenses. The principal noneconomic loss is pain and suffering.

Recovery for emotional distress has always been given hesitantly. However, where there is physical injury, there has been no difficulty in allowing recovery for pain and suffering. Often the plaintiff's attorney will seek to obtain a large award for pain and suffering by asking the jury to use a per diem or even per hour method to compute the pain and suffering award. Breaking down the period of pain and suffering into small units can generate a large award.

Many have suggested that pain and suffering not be recoverable or that limits be placed on recovery for pain and suffering. This has been especially attractive to those seeking to minimize malpractice liability of doctors. In the 1970s, many states reduced the liability of health care providers. One method was to cap noneconomic losses.

Recovery of often open-ended pain and suffering damages has been justified by the large amount of the damage award that usually goes to pay the victim's attorney. Plaintiff advocates emphasize that pain and suffering are real and that placing an economic value on them, though difficult, can give victims a sense that the legal system has taken adequate account of the harm they have suffered.

B. Collateral Source Rule

Often an accident victim's hospitalization costs are paid by an employer, a health insurer, or the government. Many victims recover lost wages through disability insurance. Although the many types of benefits create varied results, on the whole, most sources of compensation are considered collateral and are not taken into account when determining the victim's loss.[98]

Considerable criticism has been made of what is known as the collateral source rule because it can overcompensate. However, in tort cases, three principal justifications have been made for the rule. First, the defendant is a wrongdoer who should not receive any credit for benefits provided by third parties. Second, accident victims frequently must use a large share of their award to pay their attorneys. Third, accident victims often receive benefits because they have planned for them or because they are part of payment for their services.

C. Punitive Damages

Tortious conduct that is intentional and deliberate—close to bordering on the criminal—can be punished by awarding punitive damages. Such damages are designed not to compensate the victim but to punish and make an example of the wrongdoer to deter others from committing similar wrongs. In some areas of tort law, such as defamation, punitive damages play an important role because compensatory damages are often difficult to measure. There has been a tendency to award punitive damages for wrongful refusal to settle claims by insurance companies with their own insureds to ensure that there is fair dealing in the claim settlement process. A few courts have awarded punitive damages in claims against manufacturers of defective products where the manufacturer seemed unwilling to place a high value on human life.[99]

---

[97]See Section 27.09 for this in the context of a *contract* breach.

[98]Refer to this rule in the contract claim context in Section 6.06(F).
[99]See note 96, supra.

One criticism of punitive damages is that these damages are awarded to a claimant over and above his actual losses. Since the purpose of punitive damages is essentially a public one—that is, to deter wrongful conduct by making an example of the defendant—some have suggested that the damages go into the public coffers and not into the pocket of the claimant. To deal with this, a number of states have enacted legislation that requires that a designated percentage—usually between 30% and 75%—be paid to the state.

D. Attorneys' Fees: Cost of Litigation

Generally, accident victims are not able to recover their attorneys' fees or other costs of litigation from the wrongdoer. This has led to expanded damages through pain and suffering and the collateral source rule as well as the contingency fee contract under which attorneys risk their time if they do not obtain a recovery.[100]

E. Interest

Because tort damages are rarely liquidatable, it is difficult to receive interest from any period of time before award of judgment. However, as indicated earlier,[101] statutes vary considerably, and in some states, the trial judge has discretion to award interest from the date legal action was commenced.

---

[100]See Section 2.04.
[101]See Section 6.08.

P R O B L E M S

1. C was constructing a new school in a neighborhood where there was considerable vandalism. To protect materials on the site and to avoid liability for possible injuries, C put up a cyclone fence around the site and a series of locked gates. Inside the fence he kept a fierce watchdog. He placed a sign at various positions along the fence stating that trespassers should beware of the vicious dog.

a. One night the employee charged with the responsibility of locking the gate did not do so. As a result, the dog left the site and attacked a ten-year-old child who was walking by the site. Will the contractor be held liable? If so, what would be the recovery?

b. Suppose a ten-year-old child climbed the fence, entered the site, and was mauled by the dog. Would the contractor be liable?

2. B purchased a six-step stepladder at a local hardware store. Several days later she decided to use the ladder to trim a hedge. She started up the ladder carrying a connected electric hedge trimmer. The fourth step collapsed, and B fell to the ground. Her arm was lacerated by the hedge trimmer, and in falling she struck her eye on a sharp thistle on the hedge. What would B have to show to recover from the hardware store? What if B weighed 350 pounds? What if the fall caused B to be electrocuted?

Introduction to the Construction Process: Ingredients for Disputes

SECTION 8.01 Relationship to Balance of Treatise

The rest of the treatise deals with the impact of law on the Construction Process. To provide a deeper understanding of such interaction, it is important at the outset to have an overall picture of certain salient characteristics of the construction industry, with special reference to the main participants in the Construction Process. This chapter provides such an introduction.

SECTION 8.02 The Main Actors: Eternal Triangle

A. The Owner

As a rule, the owner is the entity that provides the site, the design, the organizational process, and the money for the project. With some exceptions, such as real estate developers, owners tend to be "one-shot" players, not the repetitive players found in the contractor and design professional segments of the industry. Because of the broad nature of those who commission construction, some differentiation between them is essential.

Most important is the differentiation between public and private entities. A private owner can select its design professional (by competition, competitive bid, or negotiation), its contractor (by competitive bid or by negotiation), and its contracting system (single contract or multiple prime, separate contracts) in *any manner* it chooses. Public agencies, on the other hand, are limited by statute or regulation. As a rule, they must hire their designers principally on the basis of design skill and design

reputation rather than on the basis of fee. Construction services generally must be awarded to the lowest responsible bidder through competitive bidding. Often a public entity must use separate or multiple prime contracts because of successful efforts in the legislatures by specialty trade contractors.

Other important differentiations exist between public and private owners. Public contracts have traditionally been used to accomplish goals that go beyond simply getting the best project built at the best price in the optimal period of time. Contracts to build public projects have often been influenced by the desire to improve the status of disadvantaged citizens, to remedy past discrimination, to give preferences to small businesses, to place a floor on labor wage rates, and to improve economic conditions in depressed geographical areas. Considerations exist that are not likely to play a significant role in the award of private contracts.

A public entity is more likely than are its private counterparts to be required to deal fairly with those from whom they procure design and construction services. Yet public entities, having the responsibility for public monies, usually impose tight controls on how that money is to be spent. As a result, such transactions have often generated intense monitoring by public officials and by the press to avoid the possibility that public contracts will be awarded for corrupt motives or favoritism. In addition, public projects are more controversial. To whom the project is awarded, the nature of the project, and the project's location often excite fierce public debate and occasional treks to the courthouse. Public owners are not expected to allow those from whom they procure goods and services to make profits on unperformed work or excessive profits.

Public owners often seek to control dispute resolution by contract or by law. Some public owners now require arbitration. Many experienced public owners, such as federal contracting agencies and similar agencies in large states, have developed a specialized dispute resolution mechanism, often using specialized regulatory, arbitral, or judicial forums.

Another differentiation is between experienced and inexperienced owners. An experienced owner engages in construction, if not on a routine basis, at least on a repeated one. This owner is familiar with common legal problems that arise, construction legal and technical terminology, and standard construction documents. It may also have a skilled internal infrastructure of attorneys, engineers, risk managers, and accountants.

An inexperienced owner, though often experienced in its business, is likely to find the construction world strange and often bewildering. Such an owner lacks the internal infrastructure and may, even if it has resources to hire such skill, not even know whether it should do so and, if so, how it can be done. The prototype, of course, is an owner building a residence for her own use.

Many other owners are "inexperienced." Although as a general rule, public owners are more experienced than private ones, care must be taken to differentiate between, for instance, the U.S. Corps of Engineers and a small local school district. The former has experienced contracting officers, contract administrators, and legal counsel and operates through comprehensive agency regulations, standard contracts, and an internal dispute resolution mechanism. The small local school district, on the other hand, may be governed by a school board composed of volunteer citizens, be run by a modest administrative staff, and have a part-time legal counsel who may be unfamiliar with construction or the complicated legislation that regulates the school district.

Similar comparisons can be made between private owners. Compare General Motors building a new plant with a group of doctors building a medical clinic or a limited partnership composed of professional persons seeking tax shelters through building or renting out commercial space. The

clinic or limited partnership at least can buy the skill needed to pilot through the shoals of the Construction Process. But a greater dichotomy can be seen if a comparison is made between General Motors and a private individual building a residence she intends to occupy. The latter as a rule does not obtain technical assistance because of costs.

Looking next to awarding construction contracts, it is likely that inexperienced private owners will prefer competitive bidding, not because they must do so but because they will not know the construction market well enough to sit across the negotiating table from a contractor. On the other hand, an experienced owner may be able to review the contractor's proposal and be aware of the market and other factors necessary to negotiate. (Ironically, often the public agency, which is in the best position to negotiate, *must* use the competitive process, a reflection of the distrust in public officials.)

An inexperienced owner will need to engage an architect or engineer to design and administer the construction contract much more than will an experienced owner. The latter may have sufficient skill within its own internal organization and not need an outside adviser experienced in construction.

An inexperienced owner will more likely use standard construction contracts such as those published by the American Institute of Architects. It will do so because it does not wish to spend the money for an individualized contract, it does not have an attorney who can draft such a contract, or it wishes to acquiesce to the suggestions of its architect.

The phenomenon of the inexperienced owner dealing with complex standard contract terms can generate many legal problems. The "form" may not "fit," leading to interpretation problems. In addition, if the persons operating under the contract are not familiar with or do not understand the contract, provisions are likely to be disregarded, leading to claims that those provisions have been waived.

Performance may also be affected in other ways by this differentiation. Inexperienced owners may make many design changes that mount the cost of construction and increase the likelihood for disputes over additional charges. They may also refuse legitimate contractor requests for additional compensation because they are unaware of those provisions of a contract that may provide the basis for additional compensation. Inexperienced owners may not keep the careful records so crucial when disputes arise.

Where the language must be interpreted, doubts will very likely be resolved in favor of the inexperienced owner. An experienced owner could have made the contract language clear. A claim made by a contractor that it be excused from default or be given additional compensation because of the occurrence of unforeseen events during performance is *less* likely to be successful against an inexperienced owner.

B. The Contractor

The contracting industry is highly decentralized, with a large number of small and medium-sized firms. Despite concentration in other industries, construction is largely local, with most contractors serving a single metropolitan area. Few construction companies are even regional, let alone national or international.

Half a million companies engage in construction. The average company is family-owned with an average of five to ten permanent employees. Most workers are hired for a particular job through unions or otherwise. Contractors obtain their work by competitive bidding. Profit margins are usually low and bankruptcies high. Because construction requires outside sources of funds, it is often at the mercy of the changing monetary and fiscal policies.

Two out of three contractors are specialty contractors, and one out of two workers is in a specialized trade, such as plumbing, electrical work, masonry, carpentry, plastering, and excavation. This means that in many construction projects, the contractor acts principally as a coordinator rather than as a builder. Its principal function is to select a group of specialty contractors who will do the job, schedule the work, police specialty trades for compliance with the schedule and quality requirements, and act as a conduit for the money flow. Additionally, in a fixed-price contract, the contractors provide security to the owner by giving a fixed price.

The volatility of the construction industry adds to the high probability of construction project disputes. Because the fixed-price or lump-sum contract is so common, a few bad bids can mean financial disaster. Contractors are often underfinanced. They may not have adequate financial capability or equipment when they enter into a project. They spread their money over a number of projects. They expect to construct a project with finances furnished by the owner through progress payments and with loans obtained from lending institutions.

Labor problems, especially jurisdictional disputes, are common. Many of the trade unions have restrictive labor practices that can control construction methods. Some contractors are union; others are nonunion; still others are double-breasted, having different entities for union and nonunion jobs.

Some contractors do not have the technological skill necessary for a successful construction project. Often the technological skill, if there is any, rests with a few key employees or officers. The skill is often spread thinly over a number of projects and can be effectively diminished by the departure of key employees or officers for better paying jobs.

The construction industry has attracted a few contractors of questionable integrity and honesty. These contractors will try to avoid their contractual obligations and conceal inefficient or defective performance. Such contractors are skillful at diverting funds intended for one project to a different project.

C. The Design Professional

Design professionals, the third element of the construction triangle, find themselves in the often uncomfortable position of working for the owner yet being expected to make impartial decisions during construction. They, like the contractors, have financial problems because there is usually not enough work to go around.

D. The Industry

All of the main actors—owners, contractors, and design professionals—suffer because of a chronically sick building industry particularly affected by rapid movements of the economy, changes in public spending policies, and swings of monetary policy, all of which can affect the interest rates, a formidable factor in most construction.

SECTION 8.03 The Supporting Cast on a Crowded Stage

A. The Owner Chain: Spotlight on Lender

Since owners rarely have funds to provide for major (or even minor) construction work, an important actor along the owner chain is the entity providing lending for the project. Availability of funds is essential to the participants who provide services and materials. Often the owner's inability to obtain funds is the reason a project does not proceed, despite the design professional's having spent time on the design. Similarly, available funding may not be adequate to deal with such common events in construction as design adjustments, escalation of prices, changed conditions, and claims. Inadequate funding can create a breakdown in the relationship among the project's participants which in turn leads to claims.

With regard to funding, it is important again to differentiate between public and private projects. Public projects are usually funded by appropriations made by a legislature or administrative agency. In controversial projects, challenges may be made to the project itself, to the method by which the contract was awarded, or to compliance with other legal requirements. If challenges are made to the project, funds may never become available or may generate delay in payment.

Another problem generated in public projects relates to the project's being funded by a mixture of funds from various public agencies or even a combination of public agencies and private entities. For example, the interstate road system was funded 90% by the federal government and 10% by the states, while wastewater projects were funded 75% by the federal government, $12^1/_2\%$ by the state, and $12^1/_2\%$ by local entities. Mixed funding can generate problems because one funding entity requires certain types of contract clauses that may be different from those required by other contract clauses.

In private projects, funds are usually provided by construction loans made by lenders for short periods, usually only for the time necessary to complete the project. The interest rates are short-term, usually floating, and generally higher than those charged by permanent lenders (most commonly insurance companies and occasionally ordinary lending sources).

The private construction lender wants the loan to be repaid and usually takes a security interest in the project that can be used to obtain repayment if the loan is not repaid. The lender makes an economic evaluation to determine whether a commercial borrower will derive enough revenue to repay the loan. Also, such decisions are made by permanent lenders.

In addition, the lender will wish to be certain that the borrower has put up enough of its own money so that if the project runs into problems, the borrower will not walk away from the project and avoid the loan obligation by going bankrupt—leaving the lender with only the security interest and the unpleasant prospect of taking over a defaulted project.

Similarly, the lender wants to be certain that advanced funds go into the project and enhance the value of the project by the amount of money advanced. It does not wish to see the borrower diverting funds for other purposes.

The lender also wishes to be assured that there are no delays, which can often be caused by faulty design or poor workmanship. Costly design changes, claims for extras by the contractor, and additional expenses due to unforeseen conditions are events the lender wishes to avoid. Similarly, the lender wishes to be assured that those who provide services, such as contractors and subcontractors and those who provide materials and equipment, such as suppliers, are paid and do not assert liens against the project. If liens are asserted, it may be difficult for the construction lender to obtain a permanent "take-out" lender.

Finally, the lender usually wants to know how the project is proceeding and may, in addition to approving the construction contracts made at the outset, insist on receiving copies of contract addenda, modifications, requests for changes, change orders, claims, and even routine correspondence.

The seller often retains a security interest in the land. The owner may be constructing a commercial structure in which space has been leased in advance to tenants. The owner's creditors may have an interest in the construction project. They may hope to collect their debts from profits made by the project or by having the land seized or sold to pay the owner's debts.

B. Contractor Chain

The contractor chain in the single contract system involves, in addition to the prime contractor, a large number of subcontractors and possibly sub-subcontractors. Each one of these contractors, as well as the prime contractor itself, purchases supplies and rents equipment. The material and equipment suppliers also have a substantial stake in the construction project.

The use of surety bonds for prime contractors, and often for subcontractors, brings a number of surety bond companies into the picture. The creditors of the contractors, other than suppliers, are often involved. There may be taxing authorities to whom contractors owe taxes as well as persons who have lent money to the contractors. The contract chain would not be complete without reference to the trade unions, which have a substantial stake in the construction project.

C. Design Professional Chain

In addition to the owner and contractor chains, there is a somewhat shorter chain, beginning with the design professional, who may hire consultants. Some consultants, such as geotechnical engineers, may have been hired by the owner and be in the owner's chain. In larger projects, an employee of the design professional might be on the site daily.

Consultants also have a stake in the construction project. They wish to be paid, and their negligence may cause damage to any number of persons who are also affected by or involved in the Construction Process.

D. Insurers

Insurers are also important actors on this crowded stage. All the major participants, such as design professionals, owners, and contractors, are likely to have various types of liability insurance. In addition, either the owner or the prime contractor will carry some form of property insurance on the work as it proceeds and on materials and equipment not yet incorporated into the project.

Figure 8.1 illustrates the participants in a complex building project, excluding sureties, lenders, and insurers.

SECTION 8.04 The Construction Contract

Building a construction project is a complicated undertaking. It is hoped that the construction documents, particularly plans and specifications, will be clear and complete. At their best, they should give a good indication of the contractor's duties. Unfortunately, even the best design professionals cannot do a perfect job of drafting the construction documents that encompass the complete construction obligation.

The contract documents must be interpreted often in American construction by the person who designed the project and who was selected and is paid by one party. The inevitable interpretation issues can induce corner-cutting contractors to bid low and submit a large bill for extras. The variety of contract documents creates additional dispute possibilities, inconsistencies, and ambiguities.

The competitive bidding process so often used to select a contractor and the frequent use of the fixed-price contract play a significant role in dispute generation. The former emphasizes price rather than quality, and the latter places tensions on the relationship by placing the risk of many unknown and abruptly shifting factors on the contractor.

SECTION 8.05 The Delivery Systems

Although there have always been variations in construction delivery systems, construction has been dominated by what is sometimes called the traditional construction method, also known as "design-award-build" or "design-bid-build." Under this system, the owner acquires the right to improve land, commissions an architect or engineer to prepare the design, and engages the contractor to agree to execute the completed design, either through competitive bidding or through negotiation, under a fixed price or lump-sum contract. Commonly, the American construction delivery system employs the design professional to monitor construction as it proceeds and to play a central role in interpreting the contract, resolving disputes, and issuing certificates of payment and completion. Depending upon the type of project, some parts of the design are prepared by consultant design professionals, and much of the actual construction is performed by subcontractors.

Although variations on the system have always existed, increasing emphasis on the variations began after World War II, particularly in the 1970s. Principal variations include awarding the contracts for construction to separate contractors or multiple prime contractors, commencing construction before

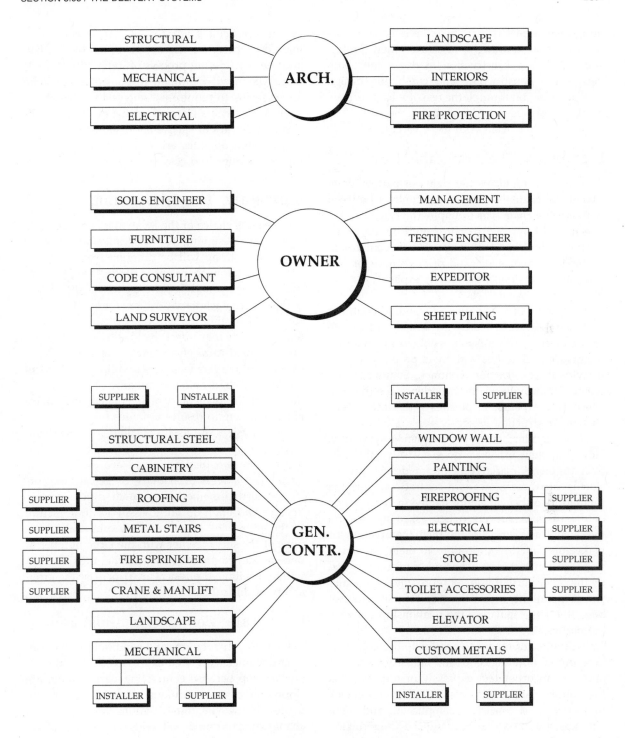

PROJECT ORGANIZATION

design is completed (known as "phased construction" or "fast tracking"), inserting a new professional called a construction manager into both design and construction, and combining design and construction through the use of what has been called a turn-key or design-build. Section 17.04 deals in greater detail with these variations.

SECTION 8.06 The Applicable Law

The law that regulates the rights of parties along any of the chains should be the contract between those on the chain. For example, the basic law between contractor and owner should be the prime contract. Similarly, the contracts between design professional and owner and prime contractor and subcontractor should be found, explicitly or implicitly, in their contracts.

There is an increasing use of standard contracts such as those published by the American Institute of Architects (AIA). It is impossible to anticipate all the problems and deal with them properly in each individual construction contract. Standard contracts rely heavily on the experiences of the past and on the expertise of persons with wide experience in construction projects.

Good standard contracts are planned carefully. However, their existence does not solve the contract problems for lawyers or for design professionals. First, some standard contracts acquire the reputation for being heavily slanted in favor of one of the parties. Lawyers, when asked to pass on these contracts, may reject them completely or make substantial modifications to them. Second, a standard contract is often unread or misunderstood by the other party if it is not represented by a lawyer. A possibility exists that the contract will not be *all* the law regulating the relationship between the parties.

A number of other laws regulate the construction project. Building codes and industry standards are often incorporated, expressly or impliedly, in the construction documents. Building codes lack uniformity and consist of complicated and often cumbersome rules that regulate the Construction Process. There are zoning laws and subdivision laws. There are tort doctrines, such as those relating to nuisance and soil support, that affect the use of land. Title and security problems are often difficult. Because of history and the archaic language of surety bonds, interpretation of a surety's obli-

gation is difficult. The rights of injured persons or injured property owners are governed by tort law and the bewildering process of indemnification. If more were needed, ultimate responsibility for some losses seems to require a "slug fest" between insurance carriers, all armed with unreadable policies with hordes of special endorsements. Is it any wonder that disputes are common and litigation time consuming and costly?

SECTION 8.07 The Construction Site

The owner's choice of the location of the project is important. Designing for a locality where craft unions are strong will be different from localities where nonunion labor is used. If the site is in a city with environmentally sensitive citizens, the project will have characteristics different from those designed for localities where this scrutiny is less intense. The site may affect the likelihood that bribes must be paid to receive contracts or permits. The site may also affect weather conditions, which can influence design and performance. The site may determine the availability of an efficient and honest judicial system to resolve disputes.

The physical site itself contributes to the likelihood of disputes and Construction Process difficulties. No two pieces of land are exactly alike, generating a high probability of subsurface surprises. Testing methods for soil conditions are expensive and often do not give an accurate picture of the entire site. In addition, the physical limitations of the site, together with the large number of persons and contracting parties who must perform within this limited physical area, increase the probability of difficulties.

SECTION 8.08 Contract Administration

Even if the general terms and conditions of the contract documents are well expressed, difficulties often develop because contracting parties often are sloppy in contract administration. Decisions are made on the site, modifications are agreed to, changes are ordered—all without the formal requirements frequently expressed in the general terms and conditions of the written contract documents. Telephone conversations are often used to resolve difficulties and continue the work, but a dispute may arise at a later date as to what was said during the conversations.

In the process of the dispute, one party will often point to the contract clauses requiring that certain directions be given in writing or that certain modifications be expressed in writing. The other party will then state that throughout the entire course of administration these formal requirements were disregarded. These are the seeds from which disputes and lawsuits develop.

SECTION 8.09 International Transactions

Crucial differences exist between foreign and domestic owners. Contracts made with foreign owners are most likely to involve the sovereign or one of its agencies. Under such circumstances, the persons contracting to provide design or construction services must be aware of the sovereign's power to regulate foreign exchange rates, import or export of goods or money, local labor conditions, the necessity for bribes, the lack of an independent judiciary, and the risk of expropriation.

Even contracting with a private owner in a foreign country involves risks that are not significant when contracting with domestic owners. Problems may still exist of unfamiliar laws, different subcontracting practices, different laws and customs regulating the labor market, and different legal solutions.

SECTION 8.10 Unresolved Disputes and Litigation

Disputes between parties may be resolved without litigation when the desire to maintain goodwill in the parties' future dealings is present. This element—the necessity of future relations—may be missing in many construction projects. A dispute may involve a number of parties, such as insurance companies and sureties, that must consent to any settlement. The uncertainty of both the law and the facts and variety of legal issues discourages settlement.

Unless the parties value the goodwill of the owner, not common in much construction, the need to compromise is often absent, another factor discouraging settlement and leading to the courtroom. If there is an arbitration provision, one or both of the parties may not trust the arbitration process. Even if the arbitrator makes an award, the party against whom the award is made may not perform. Such awards must be confirmed by a court.

In summary, a dispute-prone process such as construction will have the propensity to call on the legal system to enforce contracts or obtain compensation for losses. Participants in the process must be aware of this. They must do all they can to avoid disputes, to seek to settle those that do develop, and to be aware of the role law plays in the process.

C H A P T E R   N I N E

Limits on Ownership: Land Use Controls

A variety of controls are examined in this chapter. Sections 9.01 through 9.05 look at *private* land use controls, created by contract or tort law and enforced by courts. Sections 9.06 through 9.15 deal with *public* land use controls enacted mainly by local entities through powers given them by the state. Administration is handled largely through local regulatory commissions or legislative bodies. Courts, though less important, play mainly a passive, oversight role, determining whether local laws or administrations meet the state legislative and constitutional requirements.

SECTION 9.01 Nuisance: Unreasonable Land Use

The law protects the rights of landowners' (those protected can also include family members or tenants or employees) to enjoy their land free of unreasonable interference caused by the activities of landowners in the same vicinity. This law regulates the use of land by prohibiting uses that unreasonably interfere with the use of land by others. The unfortunate term *nuisance* has developed as the means of describing this land use control.

A landowner (landowner as used here can include a tenant) must make reasonable use of the land and not deprive landowners in the vicinity of the reasonable use and enjoyment of their land. To a large degree, the granting of legal protection through nuisance depends on the activity or use complained of and the effect such activity or use has on other landowners.

Looking first from the perspective of the complaining landowner, the interference can take many forms. At one extreme, there can be actual physical damage to the land. For example, blasting activities by a neighbor or the discharge of solid, liquid, or gaseous matter on the land may change the physical characteristics and shape of the complaining landowner's property and seriously affect the land's use.

The offensive activity may consist of disturbing the comfort and convenience of the complaining landowner. For example, activities on the offending land can create loud noise or the emission of noxious odors that may disturb the occupants of adjacent or nearby land.

Moving to even more intangible interference, the neighbor's complaint may be based on simply knowing the nature of the activities on the adjacent or nearby land. For example, a neighbor's mental tranquility may be disturbed by the knowledge that an adjacent landowner maintains a house of prostitution or a meeting place for people whose activities are bizarre or unconventional.

Moving to the outer extreme of intangible interference, one neighbor may be extremely disturbed by the appearance of a neighbor's house.

An objective standard is used. Would the acts or activities in question have disturbed a reasonable person under such circumstances? Could the interference have been avoided had the complaining landowner taken reasonable measures? For example, would closing the windows have reduced or eliminated excess noise caused by the neighbor? If so, would such protective measures unduly interfere with the use and enjoyment of the land?

Generally, the more the activity in question causes an actual physical result, such as changing the characteristics or contours of the land or making a physical impact on the land, the more likely the law will give relief. As the interference moves toward more intangible matters, such as peace of mind and aesthetic judgments, the less likely the law will accord protection. Although such interests may be protected, the law would require that there be a significant interference with the use and enjoyment of the complaining landowner's property.

The likelihood of legal protection may depend on the extent and duration of the interference. Temporary interference or interference that occurs only at long intervals is less likely to justify relief.

Legal protection does not require that the complaining party show that the acts or activities in question were *intended* to interfere with the use and enjoyment of adjacent or nearby landowners. The conduct need not be negligent. However, the *more* intentional and *more* negligent the conduct, the more likely the activity in question is a nuisance.

The social utility of the activity is important. Certain commercial and industrial activities are necessary and are encouraged by society. The necessity for these activities may overcome minor inconveniences to those in the vicinity. If the activity has little social value, slight interference with the use and enjoyment of the adjacent landowner's land can constitute a nuisance.

Often the crucial issue is the remedy to be given the complaining owner when the defendant has made an unreasonable use of the land. The choice is between giving a money award for damages to the complaining party and ordering the offending party to cease the activities—an equitable injunction.

The money damage award must be inadequate before an injunction can be awarded. However, it is often difficult to establish the precise value of the damages caused in such cases. This is especially true if the offensive activity consists of noise or emission of noxious fumes or if the activity has caused losses other than physical harm. As a result, it has become relatively easy to obtain injunctive relief in nuisance cases, especially if the nuisance is a continuing one.

Suppose an industrial activity is being conducted on a large scale and has adversely affected the occupancy and use of nearby land. Some early court decisions affirmed orders that required a cessation of the offensive activities. Such an order can close industrial plants on which the economy of the town in which the plant is located may be dependent. The modern tendency is to award what are called *permanent damages* based on the diminution in the value of the land affected by the industrial activity rather than ordering that the offending activity cease.[1]

Class actions are increasingly being used, and some landowners maintain the action for the benefit of all landowners similarly situated. Class actions through the aggregation of small claims against the offending landowner can justify the often large expenses needed to mount a lawsuit against the industrial activity adversely affecting a large number of landowners in the vicinity.

SECTION 9.02 Soil Support

The owner of unimproved land has the right to lateral support by adjacent land. Because this is an *absolute* property right, the owner of unimproved land can receive legal relief if lateral soil support was withdrawn by excavation without the owner having to show that the excavating landowner did not use proper methods.

This rule of property—an absolute rule as contrasted to a flexible one used in nuisance—does not apply if the affected land has been improved, adding weight to the soil. In such a case, the excavator is liable only if the adjacent landowner can show that the excavator did not comply with the legal standard of care. If the adjacent landowner can show that the soil support would have been withdrawn by the neighboring excavation regardless of the added weight caused by the improvement, the adjacent landowner can recover without showing that the excavator did not comply with the legal standard.

Frequently, the result in excavation cases is controlled by state legislation. For example, in Illinois,[2] the excavator must notify the adjacent landowners of its intent to excavate, the depth of excavation, and when the excavation will begin. Legal rights depend on the depth of the excavation.

---

[1] *Boomer v. Atlantic Cement Co.*, 26 N.Y.2d 219, 257 N.E.2d 870, 309 N.Y.S.2d 312 (1970).
[2] Ill.Ann.Stat. ch. $111^1/_2$, § 3301.

The excavator should comply with any public controls such as the Illinois statute mentioned, give notice of the excavation to the other party, seek to work out an advance agreement on protection and responsibility, and use reasonable care in excavating.

SECTION 9.03 Drainage and Surface Waters

Construction frequently affects and is affected by drainage at the site and adjacent land. Drainage changes can cause troublesome collection of surface waters. Three surface-water rules have emerged in the United States.[3] The first is the "common enemy" rule. Each landowner can treat surface water as a common enemy with the right to deal with it as it pleases without regard to neighbors. The second is the "civil law" rule. A landowner who interferes with the natural flow of water is strictly liable for any damage caused the neighbors.

The preceding rules are absolute. One gave great freedom to deal with surface water. Another created absolute liability. A third rule, that of "reasonable use," is gradually supplanting the absolute rules. Each case in jurisdictions with such a rule is decided on its own facts. The landowner affecting the water flow must act reasonably. In making this determination, the law considers the following:

1. Was there a reasonable necessity for such drainage?
2. Was reasonable care taken to avoid unnecessary damage to the land receiving the water?
3. Did the benefit accruing to the land drained reasonably outweigh the resulting harm?
4. When practicable, was the diversion accomplished by reasonably improving the normal and natural system of drainage, or if such a procedure was not practicable, was a reasonable and feasible artificial drainage system installed?

SECTION 9.04 Easements for Light, Air, and View

Light, air, and view are important considerations in the construction of buildings and residences. *Fontainebleau Hotel Corp. v. Forty-Five Twenty-Five, Inc.*[4] involved an addition built by one luxury hotel in such a way as to cut off sun, light, and view to an adjacent hotel. The hotel on which an addition was being built—the Fontainebleau—was constructed in 1955; its adjacent competitor, the Eden Roc, was constructed in 1956. The addition was to be a fourteen-story tower. During the winter months from two o'clock in the afternoon and for the remainder of the day, the tower would cast a shadow over the cabana, swimming pool, and sunbathing areas of the Eden Roc. The Eden Roc sought to obtain an order restraining the Fontainebleau from proceeding with the construction of the addition at a time when the addition was roughly eight stories high.

The trial court granted the requested order, but the appellate court reversed. The court stated that no American decision, in the absence of some contractual or statutory obligation, had held that a landowner "has a legal right to the free flow of light and air across the adjoining land of his neighbor."[5] The court concluded that the construction of a structure that serves a useful and beneficial purpose could not be restrained by the court even though it cut off light and air and interfered with the view of a neighbor. Despite the partially spiteful motivation for the structure, relief was denied the Eden Roc.

Wisconsin refused to follow the absolute property rule. It applied the flexible nuisance doctrine to preclude an owner from building in such a way as to reduce the effectiveness of a neighbor's solar heating panels.[6]

An owner who constructs to take advantage of view should consider either obtaining easements from adjacent landowners or buying adjacent land.

---

[3]*Butler v. Bruno*, 115 R.I. 264, 341 A.2d 735 (1975) presents a complete discussion of the history of the three American water law rules.

[4]114 So.2d 357 (Fla.Dist.Ct.App.1959).
[5]Id. at 359.
[6]*Prah v. Maretti*, 108 Wis.2d 223, 321 N.W.2d 182 (1982). This case is noted in 21 Duq.L.Rev. 1159 (1983) and 48 Mo.L.Rev. 769 (1983).

Some California communities have adopted tree ordinances. An owner whose view is impeded by a tree on a neighbor's land can request the tree be removed or trimmed. If done, the costs are divided. If the tree owner is unwilling, a local "tree commission" decides the dispute.

Suppose a neighbor puts up a "spite fence"—a high, unsightly fence that interferes with a neighbor's view and the entrance of light to the neighbor's land—and creates an eyesore. States have not been uniform in their treatment of spite fences. Some emphasize the right of a landowner to use its land as it wishes and permit spite fences. Other states hold that if the dominant motive in erecting the fence is malicious and if the fence serves no useful purpose, the neighbor can recover damages and be granted a court decree ordering that the fence be removed. In some states, statutes limit the height of any fence erected maliciously or for the purpose of annoying a neighbor. The increasing tendency is to restrict the use of spite fences.

SECTION 9.05 Restrictive Covenants

Generally, an individual grantor (one who conveys) of land has no interest in how the land is used after the deed has transferred ownership. But suppose the grantor lives or will live in the vicinity of the land transferred? Or, more important, suppose the grantor is a residential developer who is selling lots or houses to a large number of buyers? In either case, the grantor may wish to control land use after the title is transferred by deed. The grantor who is not a developer may wish to control the land use to enhance or maintain the use or value of land it owns in the vicinity of the land transferred. The developer-grantor may wish to assure buyers in the development that the land will retain its residential character to protect the enjoyment of and investment in the land. Although restrictive covenants can be created in agricultural or commercial property, this section emphasizes their use in the development of residential property.

To accomplish the land-planning objectives, the developer-grantor can obtain an express promise relating to land use. There are no particular legal problems between the developer and the original buyer in accomplishing this result by contract, with the exception of racial or religious restrictive cov-

enants (mentioned later in the section). The contract in such cases can and usually does contain promises that limit the buyer's right to use the land. In land law, such promises are called restrictive covenants.

Land ownership in a housing development is likely to be transferred. Because use affects value, the developer wishes to assure *buyers* that restrictions on use will bind future buyers and to assure *buyers or their successors* that they will have enforcement rights if any buyers violate the restrictive covenants. As a marketing device, tightly drawn enforceable restrictive covenants can persuade buyers to buy because of the protection of restrictive covenants. To accomplish this, a developer will include restrictions in the deeds that give buyers of lots or residences in the development at the time the covenants are violated the right to enforce these restrictions.

To be effective over a long period of time, such restrictions must be "tied to" or "run with" the land. All buyers, present and future, must be bound to the restrictive covenants. All owners, present and future, must be able to obtain judicial enforcement of these restrictive covenants. Early English land law that sought to preserve property values by limiting land use through such restrictive covenants became unsatisfactory. A number of technical requirements often frustrated attempts to enforce these covenants.

The ineffectiveness of the system led to the nineteenth-century development of equitable servitudes in England and subsequently in the United States. In the famous case of *Tulk v. Moxhay,*[7] the English court held that a purchaser of land who *knew* of the restrictions on the use of the land would be ordered by the Equity Court not to violate these restrictions even though there was nothing in the deed that restricted the land use. Equitable servitudes avoided technical requirements of covenants "running with the land." They created a viable private system of land use controls that had great impact on the development of cities and that still exist despite the proliferation of public land use controls.

---

[7]41 Eng.Rep. 1143 (Ch.1848).

In the United States, an equitable servitude requires an intent to benefit the adjoining land and notice of the servitude or restriction to buyers of the burdened land, that is, the land that is limited in use. Intent to benefit is usually shown by a uniform, common scheme of restrictions. Generally, this scheme is shown by a legal document recorded in the land records, although it can be shown in some states by a uniform pattern of use throughout the development.

Courts have generally given considerable freedom to developers and buyers in the creation of restrictive covenants. However, contractual freedom is not absolute. In *Shelley v. Kraemer*, the U.S. Supreme Court held that state enforcement of racial restrictive covenants violated the Equal Protection Clause of the federal Constitution.[8]

Should the constitutional tests that determine the validity of *public* land use controls (as seen in Section 9.15) be applied to *private* restrictive covenants? Some contend that buyers do not have adequate notice of the restrictions, the agreements are contracts of adhesion and are not consensual, and *in effect*, broad-scale enforcement of private restrictive covenants is tantamount to private zoning. Private restrictions, especially those that limit construction to single-family residences, can frustrate plans to have integrated housing in urban and suburban areas.

One writer urged that constitutional requirements for zoning ordinances be applied to decisions of those who administer such systems, such as homeowner associations and architectural committees, as they often wield power analogous to local authorities.[9] However, courts generally regard restrictive covenants as a useful private planning device and have not as yet been willing to apply public land use standards to reviewing restrictive covenants and actions of homeowner associations.

Private restrictive covenants can be rigid. Restrictions must be effective for a sufficiently long time to protect buyer investments. Yet the future is difficult to predict. To deal with the need for flexibility, many planned developments create homeowner associations composed of owners in the development. These associations can modify exist-

ing restrictions under certain circumstances and pass on requests to deviate from the restrictive covenants. They are often given the power to seek injunctions for violations even though they do not own land in the affected area.

Restrictive covenants have begun to deal with aesthetic and architectural aspects of a development. In *Hanson v. Salishan Properties, Inc.*,[10] leases for beachfront lots contained provisions dealing with design aspects that affected the view of others in the development. In addition to being required to comply with specific covenants set forth in the lease relating to these matters, tenants were referred to an architectural checklist drafted by the architectural committee that set forth general concepts relating to aesthetics and view for the development. The lease required that plans be approved by the architectural committee.

A tenant submitted preliminary plans for a home that were approved by the architectural committee. Some upland tenants sought an injunction against construction of the home that they claimed violated the restrictive covenants. The Oregon court concluded that the committee acted properly and that its decision would not be disturbed. The court did not have to articulate standards for judicial review of decisions of architectural committees because the court concluded that the architectural committee's decision was correct.

The association or committee generally must act reasonably and in good faith. It must act in accordance with predetermined standards. If there are no predetermined standards, its decisions must be consistent with the standards set forth in the original declaration of restrictions and with existing neighborhood conditions.

Sometimes private restrictive covenants prohibit owners from modifying or changing the exterior in such a way as to detract from architectural harmony or historic tradition. For example, *Gaskin v. Harris*[11] involved a restrictive covenant that required all exteriors to be Old Santa Fe or Pueblo-Spanish architecture. One homeowner wished to build a modern Oriental swimming pool, and other homeowners objected. New Mexico upheld the restrictive covenant despite the pool builder's claim that the pool had already been constructed and that

---

[8]334 U.S. 1 (1948).
[9]Comment 21 U.C.L.A.L.Rev. 1655 (1974).

[10]267 Or. 199, 515 P.2d 1325 (1973).
[11]82 N.M. 336, 481 P.2d 698 (1971).

the restrictive covenant was too vague to be enforceable.

Other techniques, though used less often, can soften some of the rigidity of restrictive covenants. A landowner burdened by a restrictive covenant can seek relief or may resist an injunction sought by other landowners by showing that conditions have so changed that the servitude should be removed or modified. The court will not order an injunction if the conditions in the neighborhood have so changed that enforcement of the restrictive covenant would be unreasonable. Denial of enforcement on the basis of changed conditions requires a showing that enforcement of the covenant would harm the burdened land much more than nonenforcement would harm the other land in the development. Reasons for denial can include a significant increase in noise and traffic, the unsuitability of the burdened land for the restricted purpose, and rezoning of the land for a more permissive use.

Servitudes can be eliminated by agreement of all the affected landowners or occasionally by benefited landowners failing to seek judicial relief for past violations. The servitude can also be eliminated if a public agency takes the land for a use inconsistent with the restriction. This is accomplished by condemnation, and those property owners that would have benefited by continued enforcement generally receive compensation from the agency exercising the power of eminent domain.

Despite the proliferation of public land use controls, private restrictive covenants are still important. They are often the principal land use controls in many existing developments and are used in new developments to some degree as protection from the risks of zoning law changes.

SECTION 9.06 Development of Land: Expanded Public Role

In the individualistic system preceding restrictive covenants and after the development and enforcement of restrictive covenants, the law had a minimal and mainly passive role. Individual judgment and the free market predominated. The present dominant system of land use control has expanded the law's role. Today the state itself largely determines the permissible use of land.

It is essential to compare the passive approval of private restrictive covenants and the more active public developmental control to understand why the latter has dominated modern land use controls. The essentially political nature of public control is discussed in Sections 9.07 and 9.08.

Other differentiations exist between private restrictive covenants and public land use controls. The principal objective of restrictive covenants is to preserve property values and maintain the characteristics of the neighborhood. Although public land use controls take these factors into account, local authorities must also concern themselves with fiscal matters. As property taxes have been the principal revenue-raising measure available to local authorities—at least until the taxpayer revolt against property taxes in the late 1970s—permitting industrial, commercial, or expensive residential use can increase the tax base and raise revenue.

Local authorities must furnish fire and police protection, educational facilities, and parks as well as water and sewerage facilities. Local authorities are also responsible for the overall quality of life within and around the community. They must consider communitywide problems such as housing and crime. Public land use planning is more complicated and difficult than private controls.

Flexibility is another differentiating aspect of public land use controls. Although some flexibility is needed in private restrictive covenants, the private system emphasizes stability. On the other hand, public controls must take into account and devise methods for dealing with changes that occur at an astounding pace in urban life.

The developer thinks mainly of profits and creates restrictive covenants to sell lots or homes. The public planning process should marshal the best thinking of many disciplines to accomplish legitimate planning of objectives. One judge described the public planning process as bringing to bear upon planning decisions

> the insights and the learning of the philosopher, the city planner, the economist, the sociologist, the public health expert and all the other professions concerned with urban problems.[12]

---

[12]*Udall v. Haas*, 21 N.Y.2d 463, 235 N.E.2d 897, 900, 288 N.Y.S.2d 888, 893 (1968).

The passive role did not seem adequate for these and other reasons. As a result, the government took primary responsibility for urban land development, which then became a political process.

Legislatures, both state and local, consist of persons elected by and responsible to voters. The public land use control system that evolved in the twentieth century not only was representatively democratic but also often became directly democratic through decisions made by the voters themselves.

SECTION 9.07 Local Control of Land Development

Although there have been and will be continued attempts to bring the federal government into land use planning, the power over public land use controls essentially belongs to each state. But exercise of state powers is limited because property rights have received special protection in the United States. Owners cannot be deprived of their property without due process of law. If an owner's right to use the land is *too* limited, such a limitation can constitute a "taking" of the property, and the state must pay the owner.

The U.S. Supreme Court has held that a landowner's property could be "taken" by a temporary land use regulation or one later invalidated by the courts.[13] In this respect, temporary takings are no different from permanent takings. However, the court stated that the holding does not apply to normal delays that are part of the building process (e.g., waiting for building permits, planning approval, or zoning hearings).

In 1992, the U.S Supreme Court again examined the issue of whether land use control or environmental rules have so reduced the value of the land to the owner that they constitute a taking under the Fifth and Fourteenth Amendments to the U.S. Constitution entitling the landowner to just compensation. *Lucas v. South Carolina Coastal Council*[14] involved lots on the Isle of Palms, east of Charleston, South Carolina. While the beachfront had been a critical area since 1977 and subject to regulation, the lots were landward of the restricted area when

Lucas bought them and were zoned for development as residential home sites.

In 1988, pursuant to a new beachfront statute, construction of improvements, with some exceptions, was prohibited in such a way as to bar Lucas from developing his property. The statute was designed to protect life and property by creating a storm barrier that dissipates wave energy and contributes to shoreline stability, both to protect the tourist industry and to provide a habitat for plants and animals. Lucas conceded that the ban was necessary to protect life and property against serious harm but asserted that the new law had completely extinguished his property's value, entitling him to compensation. The trial court found that the ban had made Lucas's lots valueless, but the South Carolina Supreme Court reversed the determination.

The U.S. Supreme Court by a five-member majority held that compensation would be required unless the regulatory provision violated "restrictions that background principles of the State's law of property and nuisance already place[d] upon land ownership."[15] The court remanded the case to the South Carolina Court to determine what existing law permitted.

The case was criticized for its failure to articulate rules that would determine whether there had been a taking. To some degree, the implementation of its holding was clarified in *Reahard v. Lee County*,[16] which involved the creation of open space (also discussed in Section 9.10(H)). As that section will show, public entities sometimes seek to preserve open space by barring any development of particular space. In *Reahard*, Reahard's parents had purchased 540 acres in 1944. During the mid-1970s, the family had subdivided, developed, and sold tracts of the parcel, retaining approximately forty acres. Reahard inherited the site in 1984 and sought to continue development for single-family homes. However, in December 1984, Lee County (Florida) classified the forty acres as part of a resource protection area, which limited the forty acres to a single residence or for uses of a "recreational, open space, or conservation nature."[17]

Reahard filed a complaint, contending that this had been a taking, entitling him to compensation.

---

[13]*First English Evangelical Lutheran Church of Glendale v. County of Los Angeles*, 482 U.S. 304 (1987), 793 (1990).
[14]_____ U.S. _____ , 112 S.Ct. 2886 (1992).

[15]112 S.Ct. at 2888.
[16]968 F.2d 1131 (11th Cir.1992).
[17]Id. at 1133.

Among the trial court's findings was one that simply stated that as a result of the adoption of the plan, there was "a substantial deprivation of the value" of Reahard's property, resulting in a taking.[18] The federal appellate for the Eleventh Circuit initially noted the difficulty of determining whether a regulation "goes too far."[19] The court must determine whether the regulation has denied an owner economically viable use of his own property. To do this, the court must determine the economic impact of the regulation on the property owner and the extent to which the regulation has interfered with investment-backed expectations. The appellate court held that the trial court neither analyzed the factors nor set forth findings necessary for such analysis. A proper taking analysis according to the appellate court would address certain questions:

In this case, those questions are: (1) the history of the property—when was it purchased? How much land was purchased? Where was the land located? What was the nature of title? What was the composition of the land and how was it initially used?; (2) the history of development—what was built on the property and by whom? How was it subdivided and to whom was it sold? What plats were filed? What roads were dedicated?; (3) the history of zoning and regulation—how and when was the land classified? How was use proscribed? What changes in classifications occurred?; (4) how did development change when title passed?; (5) what is the present nature and extent of the property?; (6) what were the reasonable expectations of the landowner under state common law?; (7) what were the reasonable expectations of the neighboring landowners under state common law?; and (8) perhaps most importantly, what was the diminution in the investment-backed expectations of the landowner, if any, after passage of the regulation?[20]

The court then remanded the case to the trial court to address these issues and determine whether there had been a taking.

Regulators must consider whether the regulations proposed will constitute a taking. If so, enacting the regulations will be costly. Yet these

standards still leave a considerable amount of room for regulators to enact land use and environmental controls.

To avoid land use control being a taking, most public control over land use is predicated on the police powers of the state to protect its citizens and the state's concern for the general health, safety, and welfare of the community. To bring legislated land use controls within the police powers of the state, the state legislation stressed objectives such as reduced street congestion, better fire and police protection, and promotion of health and general welfare as the reasons for restricting land use.

Land use control was accomplished by state Enabling Acts, which expressed goals in general terms and allowed local authorities to make specific rules and administer them. Generally, local authorities exercised this power. They enacted and administered ordinances to regulate land development. The traditional method for accomplishing this was to create districts in which only certain activities are permitted.

The original Enabling Acts passed in the 1920s recognized that much of the work needed to create and administer public land use controls could not be done by the local legislative governing bodies. These bodies generally are composed of unpaid or modestly paid citizens elected to local legislative bodies. The elected officials would not have the time or expertise to actually draft and administer a development system.

The original Enabling Acts allowed two agencies to be created to deal with these matters. A planning commission was authorized to draft a master plan and detailed ordinances. A board of adjustment was authorized to deal with appeals from decisions by the local administrator or public officials. The structures, personnel, and names of these agencies varied considerably from locality to locality.

Members were generally given tenure for a specific number of years to insulate them from improper pressures. Beyond tenure requirement, localities varied in the attributes of such agencies. Compensation, qualifications, staff, and support would largely depend on the size of the locality. Even in larger cities, however, members of planning commissions were frequently interested citizens who served without pay and often with minimal staff. The final authority locally in the en-

---

[18]Id. at 1134.
[19]Id. at 1135.
[20]Id. at 1136.

actment of ordinances and appeals from decisions was usually vested in the local governing body such as the city council.

At this point and in other parts in this chapter, reference is made to the Model Land Development Code (MLDC),[21] a Model Enabling Act approved in 1975 by the American Law Institute, a group of scholars, lawyers, and judges. The MLDC gives greater responsibility to the Local Development Authority (LDA), the agency implementing and operating the development program. The Code removed the local governing body, such as the city council, from decisions relating to specific development plan proposals and day-to-day operation of the system given to the LDA.

Development control is given to local authorities because land use is largely a local matter. Local decision makers are expected to be knowledgeable about the community and its needs and represent the persons who will be most affected by the developmental rules.

Should an increasingly interdependent urban society give such controls to often small political units? Insularity and unwillingness to take metropolitan social needs into account have led to increased demands for regional or statewide controls. The MLDC has recognized this, and certain provisions are designed to encourage increased activity by regional and statewide authorities.

On the other hand, criticism has been made of control by governing bodies of large cities that can be insensitive to the needs of local neighborhoods. Although local neighborhoods were not given decision-making powers in the MLDC, neighborhood organizations were recognized and given greater stature.

The Copley Place project in Boston, a half-billion-dollar private project, demonstrated the increased activity of neighborhood groups in the 1970s. Before the state authorities would grant needed air rights over a freeway, they required the developer to negotiate agreements with surrounding neighborhoods. From 1977 to 1980, fifty public hearings were held with neighborhood groups representing a wide stratum of interests. Issues thrashed out related to the scale of the project and its effect on neighborhoods, traffic, parking, shadows, wind generation, and jobs. The process is one repeated in many cities where large-scale development is proposed.

Land use legislation and administration engender controversy. Much of land value depends on use, and land can involve large economic stakes. Such decisions often polarize the community. Land use debates frequently feature clashes between those who favor growth and those who oppose it and between those who advocate the rights of landowners and those who champion social control through the political process. Land use controls can exclude certain persons and activities (discussed in greater detail in Section 9.15). From a business vantage point, controls can restrict competition.

Udall v. Haas[22] illustrates some of these features. The municipality had enacted a building zoning ordinance that reclassified a landowner's property from business to residential. The new classification would have permitted only public and religious buildings and residences with a designated minimum size. The event that led to the rezoning appears to have been the submission to local authorities of a preliminary sketch for the development of a bowling alley combined with either a supermarket or a discount house to be built on a vacant lot that had been zoned for business use. The powerful economic effect of zoning on property value can be demonstrated by the observation made by the New York Court of Appeals that more than 60% of the value of the land, or over $260,000, was wiped out by the rezoning. Describing the process by which the rezoning was accomplished as a "race to the statute books," the court invalidated the rezoning as not being in compliance with the Master Plan.

Another factor that has exacerbated abuse potential has been judicial reluctance to carefully review decisions of local legislative instrumentalities.

SECTION 9.08 Requirements for Validity

Generally, local land use control requires an Enabling Act followed by the enactment of a master or local land development plan by the local gov-

---

[21]This will also be referred to as "the Code." This code is not law. It can be adopted in whole or in part by state legislatures. Sometimes such model codes have great influence; sometimes they are ignored.

[22]Supra note 12.

erning body. In addition to serving planning functions to justify a police power interference with ownership, the master plan is designed to avoid *spot zoning*, different land use for small zones. Spot zoning cannot be justified as a legitimate overall planning device and is susceptible to abuse.

The adoption of local land use controls is generally preceded by notice to property owners and the public hearing at which all interested citizens may express their views. Public hearings in these matters frequently generate lengthy and heated debate.

Although the basic power to regulate land use resides in local legislative bodies, citizens are given greater direct power over such decisions, a power used increasingly to limit growth. Sometimes local voters can use the initiative process. This is direct enactment of law by the voters. The process requires the filing of a designated number of signatures—usually 5% to 25% of the voters—on petitions to place a proposed local ordinance on the ballot. Voters then determine whether the proposed ordinance will be adopted.

Sometimes direct voter power can be exercised *after* a local ordinance is passed by the local legislative body by a referendum that submits the ordinance to the voters for their review. The process can begin with the presentation of a petition with a designated number of voter signatures within a certain number of days after the passage of the challenged ordinance or permit decision. Typically, the petition must be filed before the effective date of the ordinance or permit. The petition compels the local governing body to either repeal the ordinance or change its decision on the permit or place the issue on the ballot for voter decision.

Exercise of either initiative or referendum can be a costly process and one not likely to be undertaken unless the stakes are high or citizens are greatly distressed with decisions made by their city council or board of supervisors.

Increasingly, attacks are made on the land use controls as unconstitutional. On the whole, these attacks have not been successful.

SECTION 9.09 Original Enabling Acts and Euclidean Zones: The MLDC

Most zoning enabling statutes passed in the 1920s dealt mainly with the physical characteristics of de-

velopment. They tended to divide the territory of the municipality into districts of different contemplated uses. Each district was to have uniform regulations. This system was sometimes called Euclidean Zoning after the U.S. Supreme Court case that first validated a zoning plan.[23] Euclidean zones required homogeneous use. Only particular uses were permitted in each district. For example, residential districts permitted only residences, commercial districts only commercial activities, and industrial districts only industrial activity. Major categories were further divided. For example, industry would be divided into heavy industry and light industry. Commercial use was divided into different categories. Residential use was divided into single-family homes, two-family homes, and multiple-family uses. Homogeneous districts were based on the assumption that differing uses within a district would be injurious to property values.

Some cities allow multiuse districts, or what is sometimes called cumulative zoning. For example, residences only might be permitted in the residential zones, but commercial zones would also permit residences and industrial zones would permit residential and commercial uses as well.

In addition to creating districts of permitted use, the Enabling Acts of the 1920s regulated matters such as height, bulk, and setback lines. Rigid patterns of land and development were assumed. Single homes were to be placed on gridlike lots.

The assumption of most early Enabling Acts was that landowners could develop their land as they wished provided they did not contravene specific restrictions expressed in the ordinances. The statutes authorized but did not compel the local authority to control development decisions. There was no exhortation or incentive to owners to undertake *desirable* development. The statutes expressed prohibitions designed to avoid *undesirable* development.

The original Enabling Acts were relatively indifferent to the effect of no or poor local planning on areas *outside* the municipality. Local public interest dominated.

Much has changed since the 1920s, and the development of the MLDC was a response to such changes. The Code seeks to go beyond simply pro-

---

[23]*Village of Euclid v. Ambler Realty Co.*, 272 U.S. 365 (1926).

hibiting poor planning and establish guidelines for desirable planning. The Code recognizes the need for regional and statewide planning. It seeks to develop flexibility and new planning methods (discussed in Section 9.10) to replace the rigid district techniques of the original enabling statutes. The Code more openly recognized aesthetics, environmental problems, and the preservation of historical sites as proper planning and developmental factors (discussed in Sections 9.10 and 9.11).

SECTION 9.10 Flexibility: Old Tools and New Ones

As stated in the preceding section, the controls of early zoning rarely went beyond height, bulk, and setback regulations. But the ugly, inefficient, and expensive explosion of urban sprawl just before and after World War II compelled planners to find methods to improve the quality of urban life. Planners were asked to enact land use controls that would create and preserve a healthy, aesthetic community with proper regard for the finite quality of resources and the historical patrimony of the community. This and the following two sections deal with traditional and new methods to accomplish these more ambitious objectives.

A. Variances and Special Use Permits

A variance permits the parcel of land to be used differently than prescribed in the zoning ordinance. Issuance requires the landowner to meet requirements that will appear in the cases discussed later in this subsection. Usually the landowner asserts that the use permitted by the ordinance was not economically feasible and this would cause great hardship.

Zoning ordinances often allow permits to be issued for a special use. These permits can be issued by the local land development authority only if the special uses are set forth in the ordinance. Sometimes the zoning ordinances spell out specific uses that can justify the issuance of a special use permit. Sometimes the standards for issuance of these permits are described in general terms.

Ordinances often distinguish use and area variances. To avoid an unconstitutional confiscatory ordinance, use variances are given if the land as zoned cannot yield a reasonable return. But because a use variance is a more drastic disruption of planning, a stronger burden is placed upon the

landowner who seeks a use variance than where an area variance is sought.

The line between use and area variances is often difficult to draw. The *Broadway, Laguna* case involved an application submitted by a developer of the luxury apartment building for a variance of the floor area ratio imposed by San Francisco's zoning law. The ratio of lot area to rentable floor space was not to exceed 1:4.8. The developer's proposal had a floor area ratio of 1:5.51. To obtain a variance, the developer contended that it had recently discovered unusual subsurface conditions that would require excavation costs two and one-half times more than anticipated and that its development would possess attractive architectural features beyond Code requirements. The ordinance stated:

> The Zoning Administrator shall grant the requested variance in whole or in part if, from the facts presented in connection with the application, or at the public hearing, or determined by investigation, it appears and the Zoning Administrator specifies in his findings the facts which establish: (1) that there are exceptional or extraordinary circumstances or conditions applying to the property involved or to the intended use of the property that do not apply generally to other property or uses in the same class of district; (2) that owing to such exceptional or extraordinary circumstances the literal enforcement of specified provisions of the Code would result in practical difficulty or unnecessary hardship; (3) that the variance is necessary for the preservation of a substantial property right of the petitioner possessed by other property in the same class of district; (4) that the granting of the variance will not be materially detrimental to the public welfare or materially injurious to the property or improvements in the vicinity; and (5) that the granting of such variance will be in harmony with the general purpose and intent of this Code and will not adversely affect the Master Plan.

The zoning administrator concluded that none of the five requirements had been met, but his decision was reversed by the Board of Permit Appeals. The trial court would not disturb the Board's decision, and the Neighborhood Association that had brought the lawsuit appealed.

The California Supreme Court[24] noted that the floor area ratio was an important density control

---

[24]*Broadway, Laguna, Vallejo Ass'n v. Bd. of Permit App.*, 66 Cal.2d 767, 427 P.2d 810, 59 Cal.Rptr. 146 (1967).

measure. The court was not persuaded that discovery of the subsurface conditions or the assertion that attractive features that went beyond the Code would satisfy any of the first three criteria.

The court noted the distinction between

those circumstances which prevent a builder from profitably developing a lot within the strictures of the planning code and those conditions which simply render a complying structure *less profitable than anticipated*. If conditions which merely reduce profit margin were deemed sufficiently "exceptional" to warrant relief from the zoning laws, then all but the least imaginative developers could obtain a variety of variances, and the "public interest in the enforcement of a comprehensive zoning plan" [citation] would inevitably yield to the private interest in the maximization of profits.[25]

In addition to demonstrating the increasing willingness of some modern courts to review local grants of variance, *Broadway, Laguna* also demonstrates the rigidity of criteria such as the floor area ratio. Perhaps the issuance of a permit conditioned upon certain things being done by the developer would have been a better solution than mechanically applying the floor area ratio.

The purpose for the special exception permit sought may influence whether it should be issued. For example, in *New York Institute of Technology, Inc. v. LeBoutillier*,[26] a college wished to obtain a special exception permit to use a large mansion it had recently acquired for educational activities. The court recognized that it would take less of a showing to grant a permit to educational institutions but concluded that the denial of the permit in this case was justified. The principal reason for not issuing the permit was that the school could have offered the instruction on its original grounds and offering instruction at the recently acquired site would have substantially increased the traffic in the residential part of the town.

For a variance, § 2–204 of the MLDC requires that the land not be reasonably capable of economic use under the general development provision and that the development not significantly interfere with the enjoyment of other land in the vicinity. In commenting on this section, the Code drafters stated that they had rejected the phrase

"unnecessary hardship" and used "economic use" so that the applicant would be required to prove in dollar-and-cents evidence that it cannot realize any reasonable monetary gain from the land as regulated.

B. Nonconforming Uses: Billboards

Though not considered a technique for avoiding the overrigidity of the Euclidean grid, the nonconforming use has had that effect. Most zoning enactments created carefully segregated districts, but the cities on which these grids were to be imposed did not follow such a neat arrangement. Junkyards existed in residential or shopping areas. Stores were found in residential areas and factories next to retail sales stores. To ensure similarity of use within districts, early planners sought to eliminate nonconforming uses. Although planners felt that police powers of the state could justify elimination of existing uses as well as prohibition of future uses, the political process dictated that the two be treated differently. To avoid entire zoning plans being turned down by local governing bodies, a system for protecting existing lawful uses was developed.

At first rules were adopted based on the hope that time would eliminate nonconforming uses. These rules precluded nonconforming uses from being changed. Nonconforming structures could not be altered, repaired, or restored, and a use could not be reinstituted after it had been abandoned. Implementation of these rules was difficult. Defining a nonconforming use was not easy. Things became more complex when courts sought to differentiate a nonconforming use in a conforming building from a nonconforming use in a nonconforming building.

In any event, nonconforming uses did not wither away. Permission for a nonconforming use created a monopoly, such as a small grocery store in a residential neighborhood. The next attack was to enact amortization ordinances ordering nonconforming uses to cease after a period of time during which the nonconforming user could recover its investment. Amortization periods ranged from one year for billboards to twenty-five years for gas stations to from fifty to sixty years for substantial buildings. Although courts generally enforced amortization ordinances, such ordinances were not used extensively. A 1971 survey by the American

---

[25] 427 P.2d at 815, 59 Cal.Rptr. at 151.
[26] 33 N.Y.2d 125, 305 N.E.2d 754, 350 N.Y.S.2d 623 (1973).

Society of Planning Officials showed that less than one-third of the responding municipalities adopted such ordinances and only 27 of 159 that had adopted them had actually employed them, mainly against billboards.

Many of these amortization periods for billboards expired in the early 1990s. This generated claims by billboard owners that such ordinances were unconstitutional takings (see Section 9.07), entitling them to full compensation under the federal and state constitutions. Billboard owners also attacked these ordinances as being censorship of free speech under the First Amendment.

As noted in Section 9.07, total destruction of the value of the property would be a taking, but the law is not clear as to whether the property owner is entitled to compensation for a partial taking when the value of the property is substantially reduced but not destroyed.

The communities that have enacted amortization ordinances argue that the amortization periods give the billboard owners the opportunity to continue reaping profits for a significant period of time and that the property value has not been completely destroyed.

Though courts generally enforced amortization ordinances, judicial opinions expressed skepticism about the fairness of such ordinances. Research revealed a patchwork quilt created by variances and special use permits rather than homogeneity of uses. Some expressed doubt regarding the desirability of homogeneous zones. As a result of these factors, elimination of nonconforming uses by amortization looked less attractive. Put another way, can there really be nonconforming uses in an urban environment that essentially was nonconformist?

C. Rezoning

The only alternatives for a landowner who cannot meet the often stringent requirements for a variance and who does not wish or is not able to obtain a conditional use permit is to seek rezoning. Rezoning can cause a hardship to the neighboring property owners, and under rigid Euclidean theory, an all-or-nothing choice will have to be made between the landowner and the neighboring property owners.

D. Contract and Conditional Zoning

Some have proposed increased use of contract or conditional zoning. Suppose the developer wishes to build a shopping center. The governing body or administrative agency can rezone the land or issue a permit. Contract zoning allows the governing body to obtain in exchange for its action the applicant's promise to do certain things. In conditional zoning or issuance of a special permit, the effectiveness of rezoning or the permit is conditioned upon the applicant's doing certain things required by local authorities. The promised acts or conditions can deal with noise abatement, traffic control, setback lines, erection of fences, or any other devices to enable the center to blend into the surrounding neighborhood. As an alternative to making promises to do these things, the developer can promise to pay the local authority to have these designated things done by the latter.

Currently, conditions for approval of a large-scale project such as a mixed-use development or a high-rise office building may include protection of historic views, employment preferences to local citizens and minorities, inclusion of low-income housing units or subsidy for them elsewhere, or a promise of child-care facilities at the proposed project or elsewhere.

Although contract or conditional zoning gives flexibility, courts have divided as to its validity. Contract zoning is especially vulnerable to attack, because it appears that the lawmaking power is being traded away. As a result, conditional zoning is more likely to be the method selected.

As to development conditions, the U.S. Supreme Court requires that the condition be directly related to a detrimental aspect of the project and that it must substantially advance a legitimate state interest.[27] This ruling has raised concern among city planners that cities could be liable for monetary damages to developers if they impose conditions that the court subsequently finds to have no substantial connection to the burdens caused by the development project.

Section 2–103 of the MLDC allows the imposition of conditions for the issuance of a special de-

---

[27]*Nollan v. California Coastal Commission*, 483 U.S. 825 (1987).

velopment permit. These conditions can include the requirement that the developer set up mechanisms for future maintenance of the property, such as a homeowner association.

E. Floating Zones

Floating zones provide flexibility within fixed ones. The boundaries of such districts are not determined by ordinance but are fixed by approval of a petition by a property owner to develop a specific tract for a specifically designated use.

F. Bonus Zoning

The *Broadway, Laguna* case discussed earlier involved a claim by the developer that its apartment had special features that justified an area variance. In some cities, a developer can receive dispensation from normal development requirements by providing bonus features. For example, in 1971, the Bankers Trust Building in New York City was given certain planning dispensations in tower height and floor area in exchange for including a large, elevated open plaza with desirable architectural features and a two-level covered arcade of shops.

The bonus features were attractive to the developer not only because it could get dispensation mainly in floor area but also because those features provided amenities that could make the project more attractive to tenants and obtain a higher rental.

G. Planned Unit Development

The gridlike Euclidean zoning, with its emphasis on individual lots, is inappropriate for planning large developments on unimproved land. Section 2–211 of the MLDC authorizes the designation of specialized planned areas in which development will be permitted only in accordance with a plan of development for the entire area. The drafters of the Code contemplated relatively undeveloped land where local planners anticipate some development in the future but wish to discourage small scattered uncontrolled developments. The designation can also be applied to urban areas anticipating major development.

If a specially planned area has been designated, no development can take place until the Land Development Agency adopts a precise plan for the area, which may include street locations, utilities, dimensions and grading of parcels, and siting of structures as well as the location and characteristics of permissible types of development. Developers owning land in the area can obtain a development permit if their development is consistent with the plan specified in the planned area ordinance.

H. Open Space

Various techniques have evolved to preserve open space. Greenbelts, developed in England, are buffer zones of open space between developed areas. Cluster zoning allows more than concentrated density, usually in the form of high-rise multiple dwellings, in exchange for commonly used open space. The *overall* density does not exceed density limits.

A Seattle greenbelt ordinance constituted a taking of private property without just compensation in violation of the Washington Constitution and the Fifth Amendment of the U.S. Constitution. The ordinance deprived landowners of all profitable use of a substantial portion of their land. Under the ordinance, 50% to 70% of greenbelt zone lots had to be preserved in or returned to a natural state. The owner of a lot located within the greenbelt zone could not make any profitable use of that portion of land required to be preserved under the ordinance.[28]

For a decision involving an attempt to create open space, refer to *Reahard v. Lee County*,[29] discussed in Section 9.07. *Reahard* noted the desire on the part of local authorities to limit land development and to preserve open space, often at the expense of property owners. As in many of these taking cases, the issue is whether the loss in value should be borne by the property owner or whether the city should have to exercise its power of eminent domain and pay just compensation. What cannot be ignored, however, is the potential impact that a requirement of just compensation would place on environmental protection.

Localities sometimes seek to preserve open space by creating an open space district. Palo Alto, Cali-

---

[28]*Allingham v. City of Seattle,* 109 Wash.2d 947, 749 P.2d 160 (1988).
[29]968 F.2d 1131 (11th Cir.1992).

fornia, created such a district to protect its undeveloped foothills in 1972. The districts were limited to open space uses, such as public recreation, enjoyment of scenic beauty, and conservation of natural resources. Single-family homes could be built but only on ten-acre minimum lots with no more than 3.5% of impervious area and building coverage. Previously, the land had one-acre minimum lot requirements. The ordinance was challenged in California and federal courts. The ordinance was upheld in *Eldridge v. City of Palo Alto.*[30]

SECTION 9.11 Aesthetics and Control

Early in the history of public land use control, doubts existed as to the constitutionality of limits on property rights designed to create a more aesthetic and pleasing environment. Police powers could be used to restrict development property rights to protect public health, safety, and welfare. But could property rights be limited to accomplish aesthetic objectives?

Berman v. Parker, a 1954 decision of the U.S. Supreme Court that upheld an urban redevelopment program, ended any doubt over the validity of regulation designed to create a more aesthetically pleasing environment. Although aesthetics were not directly involved, the opinion included language that was later used to justify aesthetics as a land use control objective. It stated:

> It is within the power of the legislature to determine that the community should be beautiful as well as healthy, spacious as well as clean, well-balanced as well as carefully patrolled.[31]

In residential developments, regulations have been enacted to promote uniformity by specifying certain design requirements, such as attached garages, two-level houses, or styles of exteriors. Sometimes ordinances sought to avoid subdivision monotony by prohibiting a house from looking too much like the surrounding houses.

An ordinance passed by the Village of Olympia Fields, Illinois, combined both uniformity and non-uniformity by requiring that a permit would not be issued in the case of a design that was excessively

similar, dissimilar, or inappropriate in relation to nearby property.[32]

Sometimes architectural compliance with standards is determined by an Architectural Commission created by the local governing body. The commission passes on designs, and its approval is required before building permits can be issued.

State ex rel. *Stoyanoff v. Berkeley*[33] involved the denial of a building permit because the architectural board created by the City of Ladue, Missouri, had concluded that the design submitted did not conform to the architectural standards required by the local ordinance. The ordinance required the design to

> conform to certain minimum architectural standards of appearance and conformity with surrounding structures, and that unsightly, grotesque and unsuitable structures, detrimental to the stability of value and the welfare of surrounding property, structures and residents, and to the general welfare and happiness of the community, be avoided, and that appropriate standards of beauty and conformity be fostered and encouraged.[34]

The design was for a single-family home in the shape of a truncated pyramid with triangular windows. The neighborhood included expensive homes in Colonial, French Provincial, and English styles. The ordinance was sustained and the Review Board's decision upheld.

An increasing number of cities have sought to protect views from developers' projects. Illustrations of views protected have been Seattle's nearby bay and mountains, the state capitol in Austin, Texas, the rolling hills outside Cincinnati, the view of the Capitol in Washington, D.C.,[35] and the view of the Rocky Mountains from downtown Denver.[36] But Scottsdale, Arizona's, attempt to ban building

---

[30]57 Cal.App.3d 613, 129 Cal.Rptr. 575 (1976).
[31]348 U.S. 26, 33 (1954).

[32]The ordinance was found to be invalid as an improper delegation of decision making to the architectural advisory committee without providing adequate standards. See *Pacesetter Homes, Inc. v. Village of Olympia Fields*, 104 Ill.App.2d 218, 244 N.E.2d 369 (1968).
[33]458 S.W.2d 305 (Mo.1970).
[34]Id. at 306–307.
[35]*Wall St. J.*, July 22, 1987, at 23.
[36]Upheld in *Landmark Land Co. Inc. v. City and County of Denver*, 728 P.2d 1281 (Colo.1986), appeal dismissed sub nom. *Harsh Inv. Corp. v. City and County of Denver*, 483 U.S. 1001 (1987).

from a nearby mountain partly within the city was successfully challenged.[37]

SECTION 9.12 Historic and Landmark Preservation

National concern has been increasing over the destruction of historically significant buildings. A survey of historic buildings made in 1933 included 12,000 buildings. By 1970, over one half had been razed. The pressure to demolish or substantially alter historic structures is especially strong in fast-changing urban areas. The response has been activity at every governmental level, but especially at the local level of government.

The principal steps taken have been to survey historic landmarks, create historic preservation districts, acquire ownership of historically significant structures or easements over their facades, and grant tax concessions to owners who will preserve the historic character of their structures. Illustrations of historic district designations are Vieux Carré by the City of New Orleans,[38] a four-block area surrounding the Lincoln house in Springfield, Illinois,[39] and the Old Town District of San Diego, California.[40]

In addition to creating historic districts, some local entities have designated specific buildings as historic landmarks despite the possibility that the designation could be considered spot zoning. Illustrations of buildings that could justify such a designation are New York City's Grand Central Station,[41] Lincoln's home in Springfield, Illinois, and Jefferson's home in Monticello, Virginia.

Designation usually means that the landowner cannot demolish or alter the existing structure. To alleviate the possible hardship that such a limitation can cause, such ordinances can provide that the owner is expected to realize at least a designated return on the property. If it proves to be an economic hardship because of a lesser return, a commission is given discretion to ease the hardship by effectuating a real estate tax rebate or the commission is afforded the additional right of producing a buyer or lessee who can profitably utilize the premises without the sought-for alteration or demolition. If these remedies prove unrealistic or unobtainable, the city is given the power to condemn the property.

The most serious constitutional attack on ordinances creating historical districts or designating landmark buildings has been the claim by the landowner that the use limitation created by the ordinance or designation violates the constitutional prohibition against taking private property without paying just compensation. If the action by local authorities is considered a taking, the owner must be compensated. As compensation can be costly, especially at a time when local government faces continuing fiscal problems, ordinances and designations are usually based on the police powers. Courts have been hesitant to consider anything short of actual appropriation to be a taking as long as the law can be justified as an exercise of police powers.

Yet the Pennsylvania Supreme Court, in *United Artists Theatre Circuit, Inc. v. City of Philadelphia*,[42] showed some dissatisfaction with the flexibility given public entities to designate a particular building as historic. The Boyd Theater had been designated historic because it was an example of art deco architecture, was done by an important architectural firm, and "represent[ed] a significant phase in American cultural history and in the history of Philadelphia."[43] The majority of the court, consisting of four judges, asked whether the cost of designation—that is, the reduction in the property owner's value—should be borne by all taxpayers or by the owner. This particular ordinance, according to the court, would prohibit any change except painting or papering—including even the moving of a mirror from one wall to another—without a permit. The majority held that this constituted an unconstitutional taking under the Pennsylvania Constitution. Three concurring judges would have focused attention upon the radical nature of the ordinance and would have limited the

---

[37]*Corrigan v. City of Scottsdale*, 149 Ariz. 538, 720 P.2d 513 (1986) (owner entitled to damages as well as invalidation of ordinance), cert. denied, 479 U.S. 986 (1986).

[38]Upheld in *Maher v. City of New Orleans*, 516 F.2d 1051 (5th Cir.1975).

[39]Upheld in *Rebman v. City of Springfield*, 111 Ill.App.2d 430, 250 N.E.2d 282 (1969).

[40]Upheld in *Bohannan v. City of San Diego*, 30 Cal.App.3d 416, 106 Cal.Rptr. 333 (1973).

[41]Upheld in *Penn Central Transp. Co. v. New York City*, 438 U.S. 104 (1978).

[42]528 Pa. 12, 595 A.2d 6 (1991).

[43]595 A.2d at 8.

ordinance to requiring a permit only for changes to the exterior and to the interior only if they affected the exterior.

Undoubtedly, the historic importance of this theater did not approach the historic value of illustrations presented earlier in this section. The extreme character of the ordinance must also have had a significant impact on the court's decision not to uphold the ordinance. Yet it is likely that this decision will remain a rather isolated one, as it appears to be out of the mainstream of court decisions in the taking area. Nevertheless, it does reflect concern that the power to declare property as historic can be abused.

A method of relieving against the hardship of placing the entire burden on the owner of the landmark building has been the development in New York City of Transferable Development Rights as a means of preserving historic buildings without incurring the high cost of condemnation. Under this plan, a landmark owner who is prevented from using its property to its full permitted use, such as the maximum floor area ratio permitted by law, is compensated by being given a transferable developmental right for a specific transfer district. The developmental right equals the excess potential the landowner was precluded from using because its building had been designated a historical landmark. It can use this excess potential in the transfer district building even if it exceeds normal density limits for the building. An owner who does not wish to use the development right can transfer it.

SECTION 9.13 The Environmental Movement and Owner Liability

During the late 1960s, the American public began to demand that government make strong efforts to protect the natural environment. The goals of the environmental movement are evident in the preamble of the National Environmental Policy Act (NEPA),[44] enacted by Congress in 1969:

The Congress, recognizing the profound impact of man's activity on the interrelations of all components of the natural environment, particularly the profound influences of population growth, high-density urbani-

zation, industrial expansion, resource exploitation, and new and expanding technological advances and recognizing further the critical importance of restoring and maintaining environmental quality to the overall welfare and development of man, declares that it is the continuing policy of the Federal Government, in cooperation with state and local governments, and other concerned public and private organizations, to use all practicable means and measures, including financial and technical assistance, in a manner calculated to foster and promote the general welfare, to create and maintain conditions under which man and nature can exist in productive harmony, and fulfill the social, economic, and other requirements of present and future generations of Americans.[45]

The strength of this movement has led Congress to enact several key pieces of environmental legislation, including the Clean Air Act,[46] Clean Water Act,[47] Resource Conservation and Recovery Act (RCRA),[48] and the Comprehensive Environmental Response, Compensation, and Liability Act (CERCLA).[49] While these laws provide much needed protection of our natural resources, they also create significant obligations and liabilities for landowners and developers.

Given their complexity, it would be impractical—if not impossible—to provide a comprehensive discussion of environmental laws and regulations. The following statutes provide insight into the kinds of obligations and liabilities that environmental laws impose on owners and developers.

A. National Environmental Policy Act (NEPA)

NEPA is a broad-based environmental statute that imposes continuing responsibility on the federal government "to improve and coordinate Federal plans, functions, programs, and resources" in order to protect the natural and human environment.[50] As a means to that end, Section 102 of NEPA requires all federal agencies to

---

[44]42 U.S.C.A. §§ 4321–4370c.

[45]Id. § 4331(a).
[46]42 U.S.C.A. §§ 7401–7671q.
[47]33 U.S.C.A. §§ 1251–1387 (also known as the Federal Water Pollution Control Act).
[48]42 U.S.C.A. §§ 6901–6992k, amending the Solid Waste Disposal Act.
[49]42 U.S.C.A. §§ 9601–9675.
[50]42 U.S.C.A. § 4331(b).

include in every recommendation or report on proposals for legislation and other major Federal actions significantly affecting the quality of the human environment, a detailed statement by the responsible official on—

    (i) the environmental impact of the proposed action;

    (ii) any adverse environmental effects which cannot be avoided should the proposal be implemented;

    (iii) alternatives to the proposed action;

    (iv) the relationship between local short-term uses of man's environment and the maintenance and enhancement of long-term productivity; and

    (v) any irreversible and irretrievable commitments of resources which would be involved in the proposed action should it be implemented.[51]

This detailed statement—commonly known as an Environmental Impact Statement (EIS)—must be made available to the public, the president, and the Council on Environmental Quality and must accompany the proposal through the agency review process.[52]

An owner or developer making a proposal that requires approval by a federal agency faces a number of potential hurdles under NEPA. First, to determine whether the action will "significantly" affect the environment, the appropriate federal agency typically prepares an abbreviated EIS—commonly known as an Environmental Assessment (EA). If the agency determines that the proposed action will *not* significantly affect the environment, the project may proceed without a full-scale EIS. Citizens groups or other parties may, however, challenge the agency's determination, thereby delaying the project.

Second, if the agency determines that the proposed action *will* significantly affect the environment, the owner or developer must await completion of a comprehensive EIS. The EIS process can be extremely costly and time-consuming. Moreover, once the EIS is completed, the federal agency may decide to cancel or significantly alter the proposed action. In addition, if the agency decides to proceed with the action as proposed, citizens groups or other parties may once again challenge the decision in court.

Since its enactment in 1969, NEPA has created considerable litigation. From the mid-1970s to the early 1980s, an average of 140 lawsuits were filed annually under NEPA; in the mid- to late 1980s, that figure fell to 80.[53] Despite these figures, very few projects are enjoined under NEPA. When agency determinations are challenged, courts usually extend great deference to agencies based on the latter's expertise and experience. Therefore, the most significant burden of NEPA appears to be the time and expense of preparing the EA and EIS.

B. Comprehensive Environmental Response, Compensation, and Liability Act (CERCLA)

CERCLA, commonly known as Superfund, governs the cleanup of hazardous waste sites. Sections 104 and 106 of CERCLA give the federal government broad power to clean up contaminated sites. Under Section 104, the president is authorized to take removal and remediation action whenever

(A) any hazardous substance is released or there is a substantial threat of such a release into the environment, or

(B) there is a release or substantial threat of release into the environment of any pollutant or contaminant which may present an imminent and substantial danger to the public health or welfare.[54]

Similarly, under Section 106, the president is authorized to "require the Attorney General of the United States to secure such relief as may be necessary to abate such danger or threat" and to "take other action . . . as may be necessary to protect public health and welfare and the environment."[55]

Section 107 of CERCLA imposes liability on four categories of "potentially responsible parties" (also known as PRPs):

(1) the owner and operator of a vessel or a facility,

(2) any person who at the time of disposal of any hazardous substance owned or operated any facility at which such hazardous substances were disposed of,

(3) any person who by contract, agreement, or otherwise arranged for disposal or treatment . . . of hazardous substances . . . and

---

[51]Id. § 4332(C).
[52]Id.

[53]Council on Environmental Quality, *Environmental Quality: Twentieth Annual Report,* Appendix B.
[54]42 U.S.C.A. § 9604(a).
[55]Id. § 9606(a).

(4) any person who accepts or accepted any hazardous substances for transport to disposal or treatment facilities, incineration vessels or sites selected by such person, from which there is a release, or a threatened release which causes the incurrence of response costs, of a hazardous substance.[56]

Courts have interpreted these categories to make up a broad range of parties. For example, the first category of PRPs—owner and operator—has been held to include "current owners and operators of a facility . . . bankruptcy estates, absent landowners/lessors, lessees, foreclosing banks, corporate officers, parents, and successors."[57] As potentially liable parties, owners and developers should be aware of any hazardous substances that may exist on their property.

The costs of liability under CERCLA can be extremely high. Potentially responsible parties are liable for the removal or remedial costs incurred by the federal government, other necessary response costs, damages for injury to or destruction or loss of natural resources, and the costs of certain health assessments.[58] In addition, parties who fail to provide for proper removal or remedial action may be liable for punitive damages of up to three times the amount of costs incurred by the federal government as a result of such failure.[59] Moreover, parties may be subject to costly civil and criminal penalties for violating CERCLA's notification, recordkeeping, or financial responsibility requirements and for violating or refusing to obey settlement agreements, administrative orders, consent decrees, or interagency agreements.[60]

CERCLA provides few defenses to liability. The statutory defenses—an act of God, an act of war, and certain acts or omissions of a third party other than an employee or agent[61]—are very narrowly interpreted. Liability is both joint and several, and traditional notions of causation carry little weight in CERCLA case law. Consequently, even *de minimis* (very little) contributors may be held liable for

the entire cost of cleaning up a hazardous waste site. Moreover, since liability is strict and retroactive, "liability attaches regardless of the time when the material was deposited."[62]

To make matters worse for owners and developers, standard comprehensive general liability (CGL) insurance policies may not cover response costs under CERCLA. One commentator has written:

CGL policies generally require insurers to "pay on behalf of the Insured all sums which the Insured shall become legally obligated to pay *as damages* because of . . . property damage." The courts, particularly federal courts, have split over the meaning to be given the term "as damages." Some courts have found that "as damages" is an unambiguous term of art in the insurance context that obligates insurers to pay only legal damages, thereby precluding payment of CERCLA response costs, which are equitable.

However, a greater number of (particularly state) courts have held that the term is open to interpretation. . . .[63]

In response to this uncertainty, insurance companies have begun to charge what one writer called "skyrocketing" rates for environmental liability coverage.[64] Moreover, many companies now refuse to provide such coverage altogether. Currently, standard CGL policies provide for CERCLA exclusions, but the insurance market is very fluid.

C. Resource Conservation and Recovery Act (RCRA)

Enacted as amendments to the Solid Waste Disposal Act of 1965, RCRA regulates solid waste—primarily hazardous solid waste—from generation to disposal (or, as it is often said, "from cradle to grave"). RCRA provides specific, albeit confusing, definitions of solid waste[65] and hazardous waste.[66] RCRA directs the EPA to promulgate regulations for these hazardous wastes but provides exemp-

---

[56]Id. § 9607(a).
[57]McSlarrow et al., *A Decade of Superfund Litigation: CERCLA Case Law From 1981–1991,* 21 Environmental Law Reporter 10367, 10390 (1991).
[58]42 U.S.C.A. § 9607(a).
[59]Id. § 9607(c)(3).
[60]Id. § 9609.
[61]Id. § 9607(b).

[62]Nash, *An Economic Approach to the Availability of Hazardous Waste Insurance,* Annual Survey of American Law 455, 463 (1991).
[63]McSlarrow et al., supra note 57, at 10407.
[64]Nash, supra note 62, at 469.
[65]42 U.S.C.A. § 6903(27).
[66]Id. § 6903(5).

tions for domestic sewage, legal point source discharges, irrigation return flows, and other potentially hazardous materials.

RCRA regulates three categories of hazardous waste handlers: (1) generators, (2) transporters, and (3) owners and operators of treatment, storage, and disposal (TSD) facilities. Hazardous waste generators are required to keep accurate records, store the waste in particular containers, label the waste in a specific manner, provide information about the waste and its characteristics, and utilize a manifest system to track the hazardous waste until it is delivered to a TSD facility. Similarly, transporters are required to ensure that the manifest system is accurate, that spills are minimized and properly handled, and that hazardous wastes are not mixed or stored in violation of RCRA.

Owners and operators of TSD facilities are the most closely regulated parties under RCRA. In addition to recordkeeping, storage, labeling, information, and manifest requirements, TSD owners and operators must meet financial responsibility standards and provide records of past regulatory compliance. These additional criteria are intended to ensure that the owners will not become insolvent, leaving behind "orphan" sites for the government to clean up.

Like CERCLA, RCRA presents owners and developers with potentially significant costs and penalties. RCRA authorizes the EPA to issue administrative orders to comply with the law and to assess civil penalties of up to $25,000 per day for violations of the statute. The exact penalty is determined based on a variety of factors, including the potential for harm and the extent of deviation. The scope and severity of liability often make it difficult for businesses to obtain liability insurance. Moreover, federal prosecutors have become increasingly willing to pursue criminal convictions for violations of RCRA and other environmental laws. In 1991 alone, the EPA tallied 125 indictments, 75 convictions, $14.1 million in criminal fines, and 550 months of prison time for convicted environmental defendants.[67] Although large companies tend to pay the most in criminal fines, most of the defendants sentenced to time in jail were owners and operators of much smaller companies.[68]

D. State Law

The environmental movement has also led state legislatures to enact environmental laws often based on, if not more protective than, federal environmental laws. For example, roughly three fourths of the states have enacted NEPA type of laws. In California, the law includes detailed requirements for agency procedures, substantive requirements for environmental impact reports (EIRs) and agency review, and judicial review standards.[69] Although NEPA does not require federal agencies to modify projects in response to EIRs, the California statute does require state agencies, before approving a project, to mitigate or avoid significant environmental effects identified in the EIR.[70] Similarly, many states have enacted stringent hazardous waste laws, modelled after CERCLA and RCRA, which may create additional regulatory burdens for owners and developers. For example, under Massachusetts law, parties liable for hazardous waste cleanup are required to pay the Commonwealth for costs of assessment, containment, and removal of hazardous wastes and for damages for injury to and destruction of natural resources. Unlike CERCLA, however, Massachusetts law also holds such parties liable to any third party for damages to its real or personal property.[71]

SECTION 9.14 Judicial Review

Although action or inaction of a local governing body can be challenged politically, such as by initiative, referendum, or the election process, the opponents of an ordinance or those who wish to contest a permit decision often go to court. A preliminary and often serious obstacle is whether the challenger has "standing" to challenge the ordinance judicially.

Judicial remedies generally are limited to persons who have suffered or are likely to suffer in-

---

[67]*Few Big Firms Get Jail Time for Polluting*, Wall St. J., December 9, 1991, at B1.

[68]Id.
[69]California Environmental Quality Act, West Cal. Pub. Res. Code § 21000 et seq.
[70]Id. § 21081.6.
[71]Mass. Gen. Laws Ann. ch. 21E, § 5.

jury to themselves. Persons cannot seek legal relief when the injuries they suffer are no different from those suffered by the general public. Nor are judicial remedies generally given to those who wish to assert the rights of third parties.

Rules that look at whether the plaintiff has standing to sue have a number of objectives. One is to avoid collusive lawsuits. Such suits are not between true adversaries but are between apparent adversaries who attempt to use the judicial process to establish a precedent both desire.

Other more important reasons exist for limiting judicial relief. Limitations of this type are designed to sharpen the issues. Sharper issues are more likely to result when the plaintiff can show that it suffered specific injuries than if it broadly claims the defendants have acted illegally.

It is easier for a court to determine whether the denial of a permit for a particular project to be built in a particular place violates the law than to determine whether in general an overall plan that limits development or growth is illegal. The remedy to be awarded will be more difficult to frame and more drastic when there is a broad judicial attack rather than a claim of *specific* injury to the plaintiff. In the background is a desire to conserve judicial energy and avoid the court's deciding abstract matters of great public importance that should be addressed to other arms of the government.

In land use litigation, a person whose property has been directly affected can always challenge the ordinance or official decision. Increasingly, standing status has been accorded nearby property owners and neighborhood associations.

In *Warth v. Seldin*,[72] the U.S. Supreme Court faced the question of whether certain individuals and groups could question the validity of ordinances claimed to be exclusionary (see Section 9.15) and designed to keep out low- and moderate-income housing from Penfield, New York, a suburb of Rochester. A divided court held that the various groups from Rochester representing low-income persons, Rochester taxpayers, and residential construction firms did not have standing to judicially challenge the Penfield ordinance. Nor did the court grant standing to individuals who claimed they had been denied low-income housing in Penfield.

Individuals who could show restrictions to particular proposed projects in which they would have resided or which *they* would have built would have standing. The court concluded that none of the plaintiffs had shown that they would have benefited personally from court intervention.

The *Warth* decision undoubtedly will make it more difficult to use the federal courts to challenge local developmental controls. Persons who wish to challenge local ordinances will have to go farther and seek permits to show in some way that they would have been able to live in the area had it not been for the ordinances.

If there is standing, the principal grounds for legal challenge have been the following:

1. The Enabling Act did not authorize the act in question.
2. The ordinance or permit decision was unconstitutional.
3. The making of decisions by nonelected officials, such as an architectural review commission, was invalid because the local governing body delegated power without proper standards.

Successful judicial challenges to decisions of local authorities have been rare.[73]

In *Village of Belle Terre v. Boraas*,[74] Belle Terre, New York, a village of less than one square mile consisting of 700 people and 220 houses, restricted the entire village to single-family dwellings. The village prohibited three or more unrelated persons from living together. The ordinance was challenged by six unrelated students from a nearby university who wished to live together.

In upholding the challenged ordinance, the Supreme Court stated:

> A quiet place where yards are wide, people few, and motor vehicles restricted are legitimate guidelines in a land-use project addressed to family needs. . . . The police power is not confined to elimination of filth, stench, and unhealthy places. It is ample to lay out zones where family values, youth values, and the blessings of quiet seclusion and clean air make the area a sanctuary for people.[75]

---

[72]422 U.S. 490 (1975).

[73]The most successful attacks have been in New Jersey and Pennsylvania. See Section 9.15.
[74]416 U.S. 1 (1974).
[75]416 U.S. at 9.

The court concluded that the ordinance was a rational means of achieving these objectives. This decision will make constitutional attacks on local development control even more difficult than in the past.

SECTION 9.15 Housing and Land Use Controls

A. Residential Zones

When zoning took center stage in the 1920s, one justification for local control was the need to develop and maintain residential neighborhoods that would keep their value and enable persons to live among their own kind. In the early days of zoning, this was accomplished principally by limiting residences in certain districts to single-family detached homes and severely curtailing the multifamily living unit.

Social aspects of residential zoning were revealed in the first opinion of the U.S. Supreme Court that passed on the validity of zoning laws. In *Village of Euclid v. Ambler Realty Co.*,[76] the U.S. Supreme Court, after justifying the limitation of property rights by the police power to provide a safe and more pleasant community, expressed the American deification of the single-family home and disdain for apartment living.

Homeownership has always been a prized goal for upward mobile classes and a status symbol of achievement. Before the advent of modern *public* land use controls, private mechanisms existed to develop and maintain quality residential neighborhoods of individually owned homes. Affluent Americans wanted and were able to buy homes in districts exclusively devoted to attractive, well-built homes on spacious lots. Such home buyers generally sought to live with persons of their socioeconomic background. They built or bought in neighborhoods where their children could play and go to school with children of their cultural background and social class. The free market encouraged the development and maintenance of such neighborhoods. Only the affluent could afford fine residential homes because, at least in normal times, the demand for such homes was large and the supply limited. As a result, high prices kept out persons of low or moderate income.

Buyers wanted assurance that there would be no change in the residential environment. To induce them to buy in such neighborhoods, developers used restrictive covenants. Buyers would be willing to pay higher prices to live in districts that were carefully restricted to quality single-family homes. Buyers wanted assurance that property values would be maintained, that schools would be kept high quality, and that neighbors shared their interests and values.

Political power of local communities to control land use enabled communities to resist market forces that might otherwise have impinged on the model of the single-family detached home. Unlike restrictive covenants, public land use controls clash with free-market concepts. This is demonstrated by the frequent alliance of developers and low-income groups who unite to attack zoning laws.

From its inception, residential zoning excluded certain types of people from designated districts. This was justified as a proper planning control designed to encourage investment and the development of good residential neighborhoods. It was not until the 1960s that strong attacks were made on zoning laws as *excluding* low- and moderate-income families from districts reserved to the more affluent (discussed in greater detail in (C)).

B. Subdivision Controls

Many communities grew when developers subdivided raw land for homes. Subdividing increased population and placed added burdens on the community to provide streets, fire and police protection, schools, parks, and other municipal services. Although many of these costs could have been provided by special property assessments, often many local communities exacted conditions for approval of subdivisions. These conditions could include dedication of part of the raw land for public use for streets, schools, and recreation. Sometimes in-lieu payments were conditions for subdivision approval. These payments—sometimes called exactions— were used to reimburse the community for providing facilities and services in the subdivision or services that the community had to furnish to residents of the subdivision.

Some states have gone further and have permitted local communities to require land or a

---

[76]Supra note 23.

money exaction for parks and recreation facilities "to bear a reasonable relationship to the use of the park and recreational facilities by the future inhabitants of the subdivision." Such an exaction was upheld with the suggestion that a city could exact a fee to purchase park land some distance from the subdivision but that could be used by subdivision residents.[77]

Subdivision exactions place upon new subdivision residents the cost for additional municipal services they will require. But if exactions substantially increase construction costs and prices of subdivision houses, they can, in addition, be exclusionary (developed in greater detail in (C)).

C. Exclusionary and Inclusionary Zoning

Attacks on zoning laws began with changing demographic patterns during the mid-1950s. In *Oakwood at Madison, Inc., v. Township of Madison*, the court recognized this and stated:

> Madison Township, among other municipalities, is encouraging new industry. Industry is moving into the county and region from the central cities. Population continues to expand rapidly. New housing is in short supply. Congestion is worsening under deplorable living conditions in the central cities, both of the county and nearby. The ghetto population to an increasing extent is trapped, unable to find or afford adequate housing in the suburbs because of restrictive zoning.[78]

Focusing on zoning laws as a cause for social injustice took the form of describing white suburbs surrounding central cities as "tight little islands" that refused to do their fair share of housing the poor.[79]

The decline of central cities and the flight to the suburbs caused fiscal problems for central cities. (By the 1980s, some cities saw a reversal of this trend as the middle classes moved back and gentrified decaying inner-city neighborhoods.) Cities were left with the increased costs of social services for the poor and a declining tax base because the middle classes, along with many businesses,

moved to the suburbs. Central cities looked upon the suburbs as parasite communities composed of persons who earned their livings in central cities and used the central cities' cultural and recreational facilities. These parasite communities, however, did not bear their proper share of the urban costs that were increasing because of the need to care for poor people.

Suburbs sought to attract light industry that paid high property taxes but did not require many employees of low or moderate income. The latter were to be avoided. They lived in modest homes or apartments that brought in less revenue than the expenditures—mainly in social services and education—required to take care of these people. Persons of middle income and above wanted to live with those who shared their social and cultural values. Some techniques used to accomplish these purposes were as follows:

1. large lot requirements
2. minimum house size requirements
3. exclusion of multiple dwellings
4. exclusion of mobile homes
5. unnecessarily high subdivision requirements or in-lieu exactions[80]

Judicial attacks have been made on many ordinances by those who wish to build or those who wish to be able to live in suburbs. Usually these attacks are based on claims that such ordinances violated federal and state constitutions.[81]

---

[77] *Associated Home Builders v. City of Walnut Creek*, 4 Cal.3d 633, 484 P.2d 606, 94 Cal.Rptr. 630 (1971).

[78] 117 N.J.Super. 11, 17, 283 A.2d 353, 356 (1971).

[79] Sager, *Tight Little Islands: Exclusionary Zoning, Equal Protection, and the Indigent*, 21 Stan.L.Rev. 767, 791–792 (1969).

---

[80] *Building the American City*, Report of National Commission on Urban Problems, H.R.Doc. No. 91–34, 91st Cong., 1st Sess. 211–216.

[81] The first challenge to an exclusionary technique, a minimum dwelling size restriction, failed in *Lionshead Lake, Inc. v. Wayne Township*, 10 N.J. 165, 89 A.2d 693 (1952), appeal dismissed 344 U.S. 919 (1953). New Jersey subsequently invalidated many exclusionary techniques in *Southern Burlington County N.A.A.C.P. v. Township of Mt. Laurel*, 67 N.J. 151, 336 A.2d 713 (1975), a lengthy but interesting opinion. For a later decision, see 92 N.J. 158, 456 A.2d 390 (1983), called Mt. Laurel II, which blasted the inaction of Mt. Laurel. Repeating its earlier decision, the court expanded a "builder's remedy" it created in an earlier case, allowing the builder greater density in exchange for mixing low- and moderate-income housing with expensive housing. As to reinvestment in decaying areas displacing residents, see Salsich, *Displacement and Urban Reinvestment: A Mt. Laurel Perspective*, 53 U. cin.L.Rev. 333 (1984).

In 1965, the Pennsylvania Supreme Court in *National Land and Investment Co. v. Kohn*,[82] invalidated a four-acre minimum lot requirement. Noting that evaluating the constitutionality of a local zoning ordinance was not easy and declining to be "a super board of adjustment" or "a planning commission of the last resort," the court nevertheless saw itself as a judicial overseer "drawing the limits beyond which local regulation may not go."[83]

The court stated that controlling density is a legitimate exercise of police power and that each ordinance must be examined individually in light of the surrounding circumstances and the needs of the community. Rejecting the justification for a four-acre minimum, the Pennsylvania Supreme Court focused on the exclusionary aspects of the ordinance and in oft-quoted language stated:

> The briefs submitted by each appellant in this case are revealing in that they point up the two factors which appear to lie at the heart of their fight for four acre zoning.
>
> The township's brief raises (but, unfortunately, does not attempt to answer) the interesting issue of the township's responsibility to those who do not yet live in the township but who are part, or may become part, of the population expansion of the suburbs. Four acre zoning represents Easttown's position that it does not desire to accommodate those who are pressing for admittance to the township unless such admittance will not create any additional burdens upon governmental functions and services. The question posed is whether the township can stand in the way of the natural forces which send our growing population into hitherto undeveloped areas in search of a comfortable place to live. We have concluded not. A zoning ordinance whose primary purpose is to prevent the entrance of newcomers in order to avoid future burdens, economic and otherwise, upon the administration of public services and facilities can not be held valid. Of course, we do not mean to imply that a governmental body may not utilize its zoning power in order to insure that the municipal services which the community requires are provided in an orderly and rational manner.

> The brief of the appellant [homeowners] creates less of a problem but points up the factors which sometime lurk behind the espoused motives for zoning. What basically appears to bother [homeowners] is that a small number of lovely old homes will have to start keeping company with a growing number of smaller, less expensive, more densely located houses. It is clear, however, that the general welfare is not fostered or promoted by a zoning ordinance designed to be exclusive and exclusionary. But this does not mean that individual action is foreclosed. "An owner of land may constitutionally make his property as large and as private or secluded or exclusive as he desires and his purse can afford. He may, for example, singly or with his neighbors, purchase sufficient neighboring land to protect and preserve by restrictions in deeds or by covenants inter se, the privacy, a minimum acreage, the quiet, peaceful atmosphere and the tone and character of the community which existed when he or they moved there." (quoting an earlier opinion)[84]

Pennsylvania and New Jersey have gone far toward what some have called socioeconomic public land use controls. Courts in these states have taken an active role in attempting to compel local communities to confront and deal with housing problems. This activistic approach has been criticized as an undue interference in local affairs and an unwarranted attempt by courts to venture into complicated social and fiscal problems more appropriate for legislative treatment. There has been some question regarding the right of courts and legislative bodies to frustrate the desires of persons to choose the types of people near whom they wish to live.

Other courts have not yet been as willing as Pennsylvania and New Jersey courts to step into local land use decisions. This is demonstrated by cases discussed in (D) that permit local communities to limit growth through the land use control process.[85]

The preceding discussion examined zoning laws that make it legally or economically impossible to build housing for low- and moderate-income persons. Some cases have dealt with attempts by

---

[82]419 Pa. 504, 215 A.2d 597 (1965). For a more recent case, see Appeal of Elocin, Inc., 501 Pa. 348, 461 A.2d 771 (1983).
[83]215 A.2d at 607.

[84]215 A.2d at 612–13 (the court's footnotes omitted).
[85]Also see *Village of Belle Terre v. Boraas*, discussed in Section 9.14.

local communities to impede the building of such housing by not extending reasonable cooperation needed to get such projects built.[86]

The introduction of the term *exclusionary* has been accompanied by increasing use of what is called inclusionary zoning. New Jersey compelled each community to do its fair share to house persons of low or moderate income.[87] Failure by local communities in New Jersey to do so can result in the communities' zoning laws being invalidated. This, in a sense, is similar to affirmative action employment programs. The court is asking local communities not simply to refrain from excluding certain people but to affirmatively use efforts to *include* them. In 1969, Massachusetts attempted to require municipalities to provide for such housing by an "Anti-Snob" Zoning Law.[88]

Efforts have been made at local levels of government to include those often excluded. Typically, such ordinances require that developers of more than a certain number of units include a designated percentage for low- and moderate-income tenants. Often they inhibit development.[89]

D. Phased Growth

The once accepted goal of growth as the proven road to prosperity and social mobility began to be questioned seriously in the 1960s. Some advocated goals that would look less at material measurements such as the gross national product and more at measurements of the quality of life. Different measurement processes and the realization of the finite nature of resources were factors that led to the development of the environmental movement discussed in Section 9.13.

From the standpoint of urban planning, growth became synonymous with urban sprawl—the often uncontrolled development of land at the outskirts of major American cities. What often resulted was an unplanned, market-oriented development of

isolated and scattered parcels of land on the fringes of suburbia followed by the gradual urbanization of the intervening undeveloped areas. Urban sprawl has been criticized as unaesthetic, wasteful of valuable land resources, and unduly increasing the cost of providing municipal services.

Rethinking of goals and priorities reflected themselves in public land use planning. Communities sought to avoid the disruptive effect of an influx of people into their communities.

Plans attacked as exclusionary and defended as phased growth often have common characteristics. However, a crude comparison can be made between these two classifications. Suburbs whose plans are exclusionary seek to be tight little islands. They wish to keep out low- and moderate-income families for social and fiscal reasons. However, phased growth plans such as those discussed in this subsection are designed to keep out *all* people, not simply undesirable ones.

Another preliminary caution is that all land use controls tend to exclude because density control is a legitimate and frequent objective. For this reason, terms such as *exclusionary* and *phased growth* must be looked at in light of their value-laden characteristics. Exclusionary zoning has been the label critics of suburbs have developed when they attack homogeneous suburbs as tight little islands. Yet phased growth, though frequently a label used to justify an exclusionary plan, carries with it socially desirable goals of protecting nature and wildlife and preserving the values and attributes of small-town life.

The first significant case to pass on a phased growth plan involved a system devised by Ramapo, New York, a suburb of New York City some twenty-five miles from downtown Manhattan. The master plan adopted in 1966 contemplated an eighteen-year period during which public facilities would be built and after which the town would be fully developed. In 1969, an ordinance was passed to control residential development on any vacant lots or parcels. Applications for permits were to be made to the town board, which was to take into account the availability of major public improvements and services such as sewers or approved substitutes, drainage facilities, parks, or recreational facilities including public school sites, improved roads, and firehouses. The degree of availability of each facility to the site was to be

[86]*Village of Arlington Heights v. Metro. Housing Dev. Corp.*, 429 U.S. 252 (1977) (violation requires racially discriminatory intent or purpose).
[87]*Southern Burlington County N.A.A.C.P. v. Township of Mt. Laurel*, supra note 81.
[88]Mass.Ann.Gen.Laws ch. 40B, §§ 20 et seq.
[89]*Board of Supervisors v. DeGroff Enterprises, Inc.*, 214 Va. 235, 198 S.E.2d 600 (1973). For a comprehensive discussion, see Kleven, *Inclusionary Ordinances*, 21 U.C.L.A.L.Rev. 1432 (1974).

measured and scored on a scale of 0 to 5, and no permit could be issued unless a minimum of 15 points were obtained. A developer could advance the date for development by providing facilities itself or by obtaining a variance. Defenders of the Ramapo approach pointed to the advantage of each parcel owner's knowing the *eventual* use and density classification of every parcel. The plan only postponed development. Backers of the Ramapo approach pointed to specific criteria governing planning board decision making rather than the fuzziness of most permit-issuing criteria.

The Ramapo approach was upheld by the New York Court of Appeals.[90] The court held that the plan was within the powers given the town by the state enabling laws, with the effective end to be served by the time controls within the state's police powers and not so unreasonable as to constitute a taking without compensation. (By the 1980s, the economic slowdown forced Ramapo to substantially modify the plan to *encourage* development.)

A second case involved an attempt by a suburban community to limit growth by controlling the number of units that could be built. The plan was upheld,[91] but again by the 1980s, the plan was *too* effective and changes had to be made to encourage growth.

A third case citing the fragile ecology of a small rural town upheld an ordinance drastically increasing the minimum lot size to block a large development of second homes for urban residents.[92]

---

[90]*Golden v. Planning Bd. of Town of Ramapo*, 30 N.Y.2d 359, 285 N.E.2d 291, 334 N.Y.S.2d 138 (1972), appeal dismissed 409 U.S. 1003 (1972). Both majority and dissent called for regional land use planning.

[91]*Constr. Indus. Ass'n v. City of Petaluma*, 522 F.2d 897 (9th Cir.1975).

[92]*Steel Hill Dev., Inc. v. Town of Sanbornton*, 469 F.2d 956 (1st Cir.1972).

C H A P T E R   T E N

Professional Registration and Contract or Licensing: Evidence of Competence or Needless Entry Barrier?

SECTION 10.01 Overview

Legal requirements for professional practice is the logical place to begin a study of how law intersects with design and the design professions. This chapter deals with professional registration laws. It also examines state occupational licensing that can and often does include contractors, a latecomer into the field of occupational regulation.

SECTION 10.02 Public Regulation: A Controversial Policy

Public regulation of professional activity can take many forms. One form of indirect public regulation—professional liability—is discussed in Chapter 14. This chapter discusses the *direct* legal requirements that determine whether a particular person or business entity may use a particular title or perform design or construction services.

Statutes enacted by legislative bodies and regulations promulgated by state administrative agencies set criteria for professional practice and administer systems that determine who may legally perform these services.

In the United States, the states control registration. Generally, the factors considered by a state before granting the contractor a license are similar to those considered by the federal government when it selects contractors to perform services. But state licensing laws will not be applied to those contractors who work exclusively for the federal government.[1] The federal government must oper-

ate free of interference by the states to give effect to the supremacy clause of the U.S. Constitution.

There is no federal regulation system for registration and licensing of design professionals. Local regulation, where it exists, is designed principally to raise revenue and not to regulate the professions. A few states have given local authorities the power to impose competence requirements on contractors.

A. Justification for Regulation

The police and public welfare powers granted to states by their constitutions permit states to regulate who may practice professions and occupations. States exercise this power by setting requirements for those who wish to practice professions or occupations. Such requirements are usually expressed in licensing or registration laws.[2]

Some members of the public who seek to have design or construction work performed are unable to judge whether those who offer to perform design or construction services have the necessary competence and integrity. One purpose of licensing laws, whether they apply to architects, engineers, surveyors, or contractors, is to give members of the public some assurance that those with whom they deal will have at least minimal competence and integrity. The public generally may suffer property damage or personal harm because of poor design or construction. One court, in denying recovery to an architect licensed where he practiced but not

[1]*Gartrell Constr. Inc. v. Aubry*, 940 F.2d 437 (9th Cir.1991) following *Leslie Miller, Inc. v. Arkansas*, 352 U.S. 187 (1956).

[2]The term *registration* is commonly used for the design professions and is used in this chapter interchangeably with the term *licensing*, except in the sections dealing with contractors.

where the project was located, refused to allow an exception for an isolated transaction and stated:

> One instance of untrained, unqualified, or unautho-
> rized practice of architecture or professional engineer-
> ing—be it an isolated transaction or one act in a
> continuing series of transactions—may be devastating
> to life, health or property.[3]

In passing on a contractor licensing law, one court stated that the law was designed to

> prevent unscrupulous or financially irresponsible con-
> tractors from deceiving and taking advantage of those
> who engage them to build. . . . It often happens that
> fly-by-night organizations begin a job and, standing in
> danger of losing money, leave it unfinished to the own-
> er's detriment. Or they may do unsatisfactory work,
> failing to comply with the terms of their agreement. The
> licensing requirement is designed to curb these evils;
> the license itself is some evidence to the owner that
> he is dealing with an honest and qualified builder.[4]

These undeniably desirable objectives are sought to be accomplished by requiring that those who wish to perform certain services have had specified education and experience and are able to demonstrate competence by passing examinations. Persons in business occupations may also have to establish financial responsibility. In addition, *after* entry into the profession, professional misconduct or gross incompetence may justify suspending or revoking a license.

Another objective was revealed when states went into comprehensive occupational licensing. When this occurred, licenses were needed not only by doctors, lawyers, architects, and engineers but also by barbers, beauticians, auto mechanics, and shoe repairers, to name but a few. Such laws were designed to bring status and dignity to those in useful occupations.

B. Criticism of Licensing Laws

Occupational licensing laws can deny some their only means of livelihood. As a result, imposition or tightening of licensing laws is often accompanied by "grandfathering" those who already are in

those occupations. This can relieve the hardship caused by denying persons the right to perform their best and often only skill. It can also dilute standards.

More important, even in the early history of occupational licensing, critics attacked the state's role of determining who and how many persons would be allowed to enter professions or occupations. Occupational licensing laws have been described as "fence-me-in" legislation designed to limit competition and protect those already in the profession or occupation. This criticism became even more strident when the states went, as noted in Section 10.02(A) into "wholesale" licensing. Licensing as a method of "turf protection" surfaced again in 1987 as the architectural profession sought to stop interior designers from being licensed. Any group with sufficient organization and political strength to demand that standards be set for its group can generate a new licensing law.[5]

The competition-inhibiting aspect of licensing laws was demonstrated in 1973 when 2,149 general contractors took and failed the Florida Construction Industry Licensing Board examination. Either all Florida applicants were incompetent or the Board sought to limit competition by barring new entrants to the field. After indignant protest from builders who had failed, the Board abruptly reversed itself and curved the grades so that 88% were given passing marks and contractors licenses.[6]

Florida again moved into the spotlight in 1987. After many complaints about shoddy workmanship performed by contractors, it overhauled its licensing examinations, upgraded tests, and added new sections in accounting, business administration, and insurance. Some 93% failed the new examination.[7]

Critics also noted that the agencies that administer licensing laws are often dominated by members of the profession being regulated. This can generate practices that keep the number of practitioners low to improve the economic status of those already in their profession.

---

[3]*Food Management, Inc. v. Blue Ribbon Beef Pack, Inc.*, 413 F.2d 716, 723–24 (8th Cir.1969).
[4]*Sobel v. Jones*, 96 Ariz. 297, 394 P.2d 415, 417 (1964).

[5]See Friedman, *Freedom of Contract and Occupational Licensing, 1890–1910: A Legal and Social Study*, 53 Calif.L.Rev. 487 (1965).
[6]See *Wall St. J.*, January 8, 1975, at 1.
[7]11 Constr. Contractor, ¶ 333 (1987).

The Florida experiences also demonstrate the difficulty of determining the proper level of competence necessary to receive state permission to engage in a particular occupation. This issue has triggered protest and lawsuits from minority groups who contend that their low representation in the professions results, among other things, from examination questions and grading that either are designed to limit their entry or have that effect. The long experience requirements for designers have been criticized as either inadequate as proper training or solely designed to provide registered designers with cheap labor.

New criticism of licensing and registration laws surfaced during the galloping inflation of the mid-1970s. Many pointed to professional licensing laws, among other protections given to the professions, as causing the high cost of professional services. Educational and experience preexamination requirements mean many unproductive years at great expense. This can lead to fees designed to recoup these losses after entry. Similarly, education, practice, and testing requirements limit the number of persons who can perform these professional services. This reduces supply and increases fees. Another undesirable by-product is that high professional fees tend to limit design services to those clients who can afford to pay these fees and to high-cost projects.

Will the regulatory process, though criticized for the reasons given, continue to proliferate? In 1974, the Federal Trade Commission conducted a study of television repair in three cities, one with a traditional occupational licensing law, a second with a modified licensing system, and the third with no law. That study, though limited, tended to show that the cost of repairing television sets in the District of Columbia, a *then* unregulated area, was substantially lower than the cost of repairing television sets in the others, with no proven diminution in competence.[8] Yet shortly after this, the District of Columbia enacted a *stringent* licensing law regulating auto mechanics and television repairers.[9]

Some believe that the political process and self-interest of the professions and occupations will lead to *more,* not less, regulation. Yet recent developments in Arizona show the perils of political predictions.

The winds of deregulation blew fiercely, if fitfully, in Arizona. In 1981, Arizona concluded that licensing laws were not needed in commercial and industrial construction where the market could provide a cheaper and more effective regulatory mechanism. The legislature stated:

> The legislature finds that regulation of the commercial and industrial construction business, including public works, . . . is not necessary for the protection of the public health, safety and welfare, and that it is in the public interest to deregulate such business. It is the purpose and intent of the legislature to continue the registrar of contractors agency in order to protect the public health, safety and welfare by providing for the continuing licensing, bonding and regulation of contractors engaged in residential construction. . . .[10]

Yet in 1986, effective July 1, 1987, Arizona restored much of the regulation that had been eliminated in 1981.[11] The Arizona experience illustrates not only the controversial nature of occupational licensing, but also the apparently inexorable movement toward greater state control.

Finally, wholesale occupational licensing laws can be a method by which the state can exercise pervasive control over its economy and its citizens. Enterprises or individuals thought to be operating against the public interest but not reachable by the penal laws can be put out of business by license suspension or revocation.

C. Importance of Attitude Toward the Regulatory Process

The attitude of lawmakers toward the regulation process clearly is influential. Legislators who take a beneficent view are likely to enact more licensing laws. Similarly, such legislators may seek to bar judicial doctrines—such as substantial compliance, discussed in Section 10.10(C)—that reduce the effectiveness of such laws. Of course, the attitude of courts that must often pass on these laws will influence decisions. For example, courts are often called on to determine the constitutionality and

[8]*Economic Report: Regulation of the Television Repair Industry in Louisiana and California* (Federal Trade Commission, November 1974).
[9]*Washington Post,* July 23, 1975, at D-1.

[10]1981 Ariz.Sess.Laws, ch. 221 § 1 (repealed by Laws 1986, ch. 318, § 21).
[11]Ariz.Rev.Stat.Ann § 32-1101.

meaning of the legislation. They also decide whether particular conduct has violated the statute, whether substantial compliance with licensing laws is adequate to excuse a violation, and whether a party who has performed work, though unlicensed, will be able to recover compensation.

D. Judicial Attitudes Toward Registration Laws

Different states express variant attitudes. For example, one court passing on a particular licensing practicing system noted that it expressed "grave policy."[12] Yet in that same year, another granted recovery to a contractor who through technical default had allowed his license to lapse, stating:

> It performed in all other respects competently and without injury to any person. . . . We are not involved in aiding an incompetent or dishonest artisan. . . . The defendant received full value under the terms of the contract. The licensing law should not be used as a shield for the avoidance of a just obligation.[13]

Even a particular state over time can change its attitude toward licensing laws. For example, in 1957, California precluded an unlicensed subcontractor from collecting from the prime contractor who knew it was unlicensed despite the latter's having been paid by the owner.[14] Yet cases decided by the California courts in 1966, 1973, and 1985 were more tolerant toward technical contractor noncompliance.[15] (As seen in Section 10.10(C), in 1989, the California legislature reacted negatively toward this line of decisions.) On the whole, licensing laws still are considered to express important and desirable policy. Criticism that has been made, however, has begun to be reflected in judicial decisions.

SECTION 10.03 Administration of Licensing Laws

Modern legislatures articulate rules of conduct by statute and create administrative agencies to administer and implement the laws.

Agencies created to regulate the professions can make rules and regulations to fill deliberate gaps left by the statutes and particularize the general concepts articulated by the legislature. For example, the licensing laws state that there must be examinations to determine competence. The details of the examinations, such as the type, duration, and frequency, are determined by the agency.

In addition to having quasi-legislative powers, the agencies have quasi-judicial power. They may, subject to judicial review, decide disputed questions, such as whether a particular school's degree will qualify an applicant to take the examination or whether certain conduct merits disciplinary sanction. They also have power to seek court orders requiring that persons cease violating the licensing laws.

These agencies, called boards or commissions, have great power over the professions they regulate. They are often controlled by the members of the profession they are supposed to regulate.

Increasingly, consumer movements are suggesting or demanding that lay persons be given significant policy-making roles. The principal control on their quasi-judicial decisions has been the scrutiny given to these decisions by courts when judicial review is sought (discussed in greater detail in Section 10.04(B)).

SECTION 10.04 The Licensing Process

A. Admission to Practice

Requirements imposed by states or territories vary considerably. Some states and territories require citizenship and residency. Most have minimum age requirements, usually ranging from 21 to 25. All require a designated number of years of practical experience that can be substantially reduced if the applicant has received professional training in recognized professional schools. All states require at least one examination, and some require two. Most inquire into character and honesty. Some require interviews. As to interstate practice, see Section 10.06(G). Section 10.08(A) covers contractors.

Most states permit practice through a corporate form,[16] but a substantial number do not.

---

[12]*Hedla v. McCool*, 476 F.2d 1223, 1228 (9th Cir.1973).

[13]*Vitek, Inc. v. Alvarado Ice Palace, Inc.*, 34 Cal.App.3d 586, 110 Cal.Rptr. 86, 92 (1973).

[14]*Lewis & Queen v. N. M. Ball Sons*, 48 Cal.2d 141, 308 P.2d 713 (1957).

[15]*Latipac, Inc. v. Superior Court*, 64 Cal.2d 278, 411 P.2d 564, 49 Cal.Rptr. 676 (1966); *Vitek, Inc. v. Alvarado Ice Palace, Inc.*, supra note 13; *Asdourian v. Araj*, 38 Cal.3d 276, 696 P.2d 95, 211 Cal.Rptr. 703 (1985).

---

[16]See Section 3.06.

B. Post-Admission Discipline: *Duncan v. Missouri Board for Architects, Professional Engineers and Land Surveyors*

Although the regulatory emphasis has been on carefully screening those who seek to enter the professions, all states can discipline persons who have been admitted. Discipline can be a reprimand or suspension or revocation of the license.

Grounds for disciplinary action vary considerably from state to state, but they are generally based on wrongful conduct in the admissions process, such as submitting inaccurate or misleading information, or conduct after admission that can be classified as unprofessional or grossly incompetent.

Florida's attempt to institute disciplinary proceedings against Markel, a Florida architect, is instructive. Markel was charged with placing his name and seal on drawings that he neither prepared nor directly supervised. He was also charged with having held himself out to the public as being a member of an architectural partnership when his partner was not a registered architect or engineer.

The hearing before the state licensing board's hearing officer was hotly contested. The hearing officer concluded that Markel had violated the Florida law and recommended revocation of Markel's license, a recommendation followed by the State Board of Architecture.

Markel appealed to the Florida intermediate court[17] and later to the Florida Supreme Court.[18]

The state supreme court held that the basis for severe disciplinary action was not clear-cut. As a result, it refused to revoke, noting that revocation is an extreme disciplinary measure. Revocation should be invoked only when the architect's conduct has been wholly inconsistent with approved professional standards. Upon remand, the intermediate appellate court suspended Markel for sixty days.[19]

Perhaps Markel was out of favor with the registration authorities who were seizing on a close question to justify a severe sanction. But the long history of the disciplinary action also demonstrates that the professional against whom discipline is sought can avoid this by wearing down the licensing agency by exhausting every legal right.

Finally, the Markel history shows that courts are more reluctant to deprive a person of her livelihood than the licensing authorities.

Yet the disciplinary action in the aftermath of the Hyatt Regency tragedy of 1981 in which many persons were killed or injured shows that in spectacular accidents, courts will back up strong sanctions imposed by a licensing authority. The decision of the intermediate Missouri court follows.

---

[19]*Markel v. Florida State Bd. of Architecture*, 274 So.2d 12 (Fla.Dist.Ct.App.1972).

---

[17]*Markel v. Florida State Bd. of Architecture*, 253 So.2d 914 (Fla.Dist.Ct.App.1971).
[18]*Markel v. Florida State Bd. of Architecture*, 268 So.2d 374 (Fla.1972).

DUNCAN v. MISSOURI BOARD for ARCHITECTS, PROFESSIONAL ENGINEERS AND LAND SURVEYORS

Missouri Court, Eastern District, Division Three, 1988. 744 S.W.2d 524.
[Ed. note: Footnotes renumbered and some omitted.]

SMITH, Judge.

On July 17, 1981, the second and fourth floor walkways of the Hyatt Regency Hotel in Kansas City collapsed and fell to the floor of the main lobby. Approximately 1500 to 2000 people were in the lobby. The walkways together weighed 142,000 pounds. One

hundred and fourteen people died and at least 186 were injured. In terms of loss of life and injuries, the National Bureau of Standards concluded this was the most devastating structural collapse ever to take place in this country. That Bureau conducted an investigation of the tragedy and made its report in May 1982.

In February 1984, the Missouri Board for Architects, Professional Engineers and Land Surveyors filed its complaint seeking a determination that the engineering certificates of registration of Daniel Duncan and Jack Gillum and the engineering certificate of authority of G.C.E. International were subject to discipline pursuant to Sec. 327.441 RSMo 1978. The Commission, after hearing, found that such certificates were subject to suspension or revocation. Upon remand for assessment of appropriate disciplinary action, the Board ordered all three certificates revoked.[20] Upon appeal the trial court affirmed. We do likewise.

G.C.E. is a Missouri corporation holding a certificate of authority to perform professional engineering services in Missouri. Gillum is a practicing structural engineer holding a license to practice professional engineering in Missouri. He is president of G.C.E. Duncan is a practicing structural engineer holding a license to practice professional engineering in Missouri and is an employee of G.C.E.

Gillum-Colaco, Inc., a Texas corporation, contracted with the architects of the Hyatt construction to perform structural engineering services in connection with the erection of that building. By subcontract the responsibility for performing all of such engineering services was assumed by G.C.E. The structural engineer, G.C.E., was part of the "Design Team" which also included the architect, and mechanical and electrical engineers. Gillum was identified pursuant to Sec. 327.401.2(2) RSMo 1978, as the individual personally in charge of and supervisory (sic) of professional engineering activities of G.C.E. in Missouri. His professional seal was utilized on structural engineering plans for the Hyatt. Duncan was the project engineer for the Hyatt construction in direct charge of the actual structural engineering work on the project. He was under the direct supervision of Gillum.

* * *

We will not attempt to set forth in detail the extensive evidence before the Commission. Some review of that evidence is, however, required. The atrium of the Hyatt was located between the 40 story tower section of the hotel and the function block. Connecting the tower and function block were three walkways, suspended from the atrium ceiling above the atrium lobby. The fourth floor walkway was positioned directly above the second floor walkway. The third floor walkway was to the east of the other two walkways. As originally designed the fourth and second floor walkways were to be supported by what is referred to as a "one rod" design. This consisted of six one

and one quarter inch steel rods, three on each side, connected to the atrium roof and running down through the two walkways. Under this design each walkway would receive its support from the steel rods and the second floor walkway would not be supported by the fourth floor walkway. At each junction of the rods and the walkways was a box beam-hanger rod connection. These were steel to steel connections and the design of such connections is an engineering function, because the design includes the performance of engineering calculations to determine the adequacy of the connection to carry the loads for which it is designed.[21]

Connections are basically of three kinds, simple, complex, and special. All connections are the responsibility of the structural engineer. Simple connections are those which have no unusual loads or forces. They may be designed by looking up the design in the American Institute of Steel Construction (AISC) Manual of Steel Construction and following directions found therein. This can be done by a steel fabricator utilizing non-engineering personnel. Complex connections are those where extreme or unusual loads are exerted upon the connection or where the loads are transferred to the connection from several directions. These connections cannot be designed from the AISC manual and require engineering expertise to design.

Special connections are a hybrid having characteristics of each of the other two. A simple connection becomes special where concentrated loads are placed thereon and the AISC manual no longer provides all the information necessary to properly design the connection. Such connections may also become special where the connections are "non-redundant." A "redundant" connection is one where failure of the connection will not cause failure of the entire system because the loads will be carried by other connections. A "non-redundant" connection which fails will cause collapse of the structure. The box beam-hanger rod connections were "non-redundant." The Commission found the box beam-hanger rod connections to be special connections.

The steel fabricator on the Hyatt project, Havens Steel Company, had engineers capable of designing simple, complex or special connections. The structural engineer on a project may, as a matter of custom, elect to have connections designed by the fabricator. To do this, he communicates this information to the fabricator by the manner in which he portrays the connection on his structural drawings. The adequacy of the connection design remains the responsibility of the

---

[20][Ed. note: See Note, 55 UMKC L.Rev. 108 (1986).]

[21][Ed. note: See Figure 10.1.]

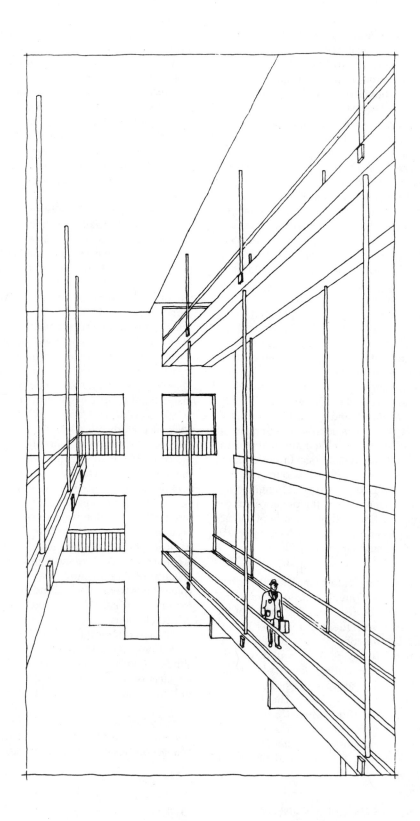

structural engineer. The Commission found that the structural drawings (S405.1 Secs. 10 and 11) did not communicate to the fabricator that it was to design the box beam-hanger rod connection, and did communicate to the fabricator that those connections had been designed by the engineer. Duncan testified that he intended for the fabricator to design the connections. Havens prepared its shop drawings on the basis that the connections shown on the design drawings had been designed by the structural engineer. Certain information concerning loads and other aspects of the box beam-hanger rod connections which appeared on Duncan's preliminary sketches was not included on the final structural drawings sent to the fabricator. The Commission also found that Duncan's structural drawings did not reflect the need for a special weld, did not reflect the need for stiffeners and bearing plates, and reflected that the hanger rods should be of regular strength steel rather than high strength. These factual findings are not contested. The hanger rods and the box beam-hanger rod connections shown on the structural drawings did not meet the design specifications of the Kansas City Building Code. That finding of fact by the Commission is also not contested.

Because of certain fabricating problems Havens proposed to Duncan the use of a "double rod" system to suspend the second and fourth floor walkways.[22] Under this system the original six rods would be connected only to the fourth floor walkway. A second set of rods would then connect the second floor walkway to the fourth floor walkway. The effect of this change was to double the load on the fourth floor walkway and the box beam-hanger rod connections on that walkway. There was evidence that one of the architects contacted Duncan to verify that the double rod arrangement was structurally sound and was advised by Duncan that it was. Appellants dispute that the architect's testimony clearly establishes such an inquiry and contend that the conversation dealt rather with the aesthetic nature of the change. Our review of the record causes us to conclude that the architect did testify to receiving assurances that the new design was structurally safe. It is difficult to understand why the architect would consult the structural engineer if his only concern was the aesthetics of the new design. The Commission further found that the records of G.C.E. failed to contain a record of a web shear calculation which Duncan testified he made and which would normally be a part of the G.C.E. records. Duncan's testimony reflected the need for such a calcula-

tion before approval of the double rod arrangement. The Commission also found certain additional necessary tests or calculations were not made. Appellants do not challenge that finding. It is a reasonable inference from the evidence that Duncan did not make the engineering calculations and tests necessary to determine the structural soundness of the double rod design.

Havens prepared the shop drawings of the structural steel fabrication. These drawings were returned to Duncan for review and approval. They contained the fabrication of the box beam-hanger rod connections based upon the structural drawings previously submitted by Duncan and bearing the seal of Gillum. The Commission found, and appellants do not dispute, that its own internal procedures called for a detailed check of all special connections. The primary reason for such a procedure is to provide assurance for the owner that the fabricator is conforming to the contract and that any engineering work conforms to acceptable standards. A technician employed by G.C.E. checked the sizes and materials of the structural members for compliance with design drawings. He called to Duncan's attention questions concerning the strength of the rods and the change from one rod to two. Duncan stated to the technician that the change to two rods was "basically the same as the one rod concept." Duncan did not "review" the fourth floor box beam connection shown on the Havens shop drawings nor did he, in accord with usual engineering practice, assemble its components to determine what the connection looked like in detail. The Commission found, again not disputed, that appellants did not review the shop drawings for compliance of the box beam-hanger rod connection with design specifications of the Kansas City Building Code; did not review the shop drawings for conformance with the design concept as required by the contract of G.C.E. and the specifications on the Hyatt project nor for compliance with the information given in the contract documents. Duncan and Gillum approved the shop drawings.

While construction of the Hyatt was in progress the atrium roof collapsed. Investigation into that collapse established that the cause was poor construction workmanship. During the course of their investigation of the atrium roof collapse, appellants discovered that they had made certain errors in their design drawings and that they had failed to find discrepancies in their review of shop drawings involving the atrium. These errors and discrepancies were in areas other than the walkway design. The owner and architect directed G.C.E., for an additional fee, to check the design of the entire atrium. G.C.E. undertook that review. Gillum assured the owner's representative

---

[22][Ed. note: See Figure 10.2.]

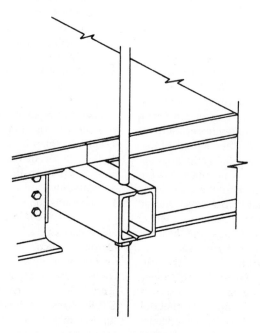

Original detail

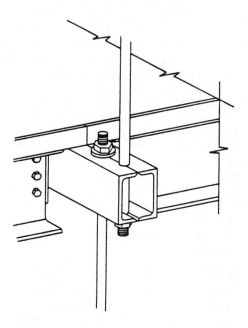

As built

that "he would personally look at every connection in the hotel." Appellants were also specifically requested by the construction manager to inspect the steel in the bridges including the connections. Duncan subsequently advised him that had been done. In their report to the architects, appellants advised "we then checked the suspended bridges and found them to be satisfactory." This report was a culmination of a design check of the "structural steel framing in the atrium as per the request of Crown Center [the owner]." Appellants did not do a complete check of the design of all steel in the atrium nor a complete check of the suspended bridges. Gillum reviewed the report prepared by Duncan and took no exception. At a meeting with the owner and architect, Gillum stated that his company had "run a detailed, thorough re-analysis of all of the structure. And to determine if there was any other areas that were critical or had any kind of a design deficiency or detail deficiency." Duncan reported at that meeting: "We went back, myself and another engineer, and checked all the atrium steel. . . . Everything in the atrium checked out very well [with one non-relevant exception]." Appellants checked only the atrium roof steel.

Approximately a year after completion of the Hyatt Regency the second and fourth floor walkways collapsed. The cause of the walkway collapse was the failure of the fourth floor box beam-hanger rod connections.

[Ed. note: The engineers challenged the constitutionality of the licensing laws. They contended that the "gross negligence" standard was so vague as to deny them due process under the U.S. Constitution. The court rejected this contention, concluding that the "phrase provides a guideline sufficient to preclude arbitrary and discriminatory application." See Section 10.06(A).]

The Commission rejected the definition utilized in the first category of cases (difference in degree) and utilized a definition recognizing that gross negligence is different in kind from ordinary negligence. Appellants do not disagree with this selection of category. The Commission defined the phrase in the licensing context as "an act or course of conduct which demonstrates a conscious indifference to a professional duty." This definition, the Commission found, requires at least some inferred mental state, which inference may arise from the conduct of the licensee in light of all surrounding circumstances. Appellants have posited a definition purportedly different that would define the phrase as "reckless conduct done with knowledge that there is a strong probability of harm, and indifference as to that likely harm." We are not persuaded that the two definitions are in fact different. An act which demonstrates a conscious indif-

ference to a professional duty would appear to be a reckless act or more seriously a willful and wanton abrogation of professional responsibility. The very nature of the obligations and responsibility of a professional engineer should appear to make evident to him the probability of harm from his conscious indifference to professional duty and conscious indifference includes indifference to the harm as well as to the duty. The structural engineer's duty is to determine that the structural plans which he designs or approves will provide structural safety because if they do not a strong probability of harm exists. Indifference to the duty is indifference to the harm. We find no error in the definition utilized by the Commission. It imposes discipline for more than mere inadvertence and requires a finding that the conduct is so egregious as to warrant an inference of a mental state unacceptable in a professional engineer.

The appellants also challenge the Commission's findings that each of several different acts or omissions constituted gross negligence justifying discipline. These findings were apparently made by the Commission in an abundance of caution for in a footnote it recognized that "it is only after a complete analysis of their overall performance within the system that any judgment of their conduct can be made under the terms of the licensing statute." This is clearly true. It is the combination of a series of acts and omissions which created the structurally unsound walkways. Any one of those acts or omissions alone might well not have compromised the structural integrity of the walkways if the series of acts and omissions had not existed in combination. For the Commission to require a finding that each of these acts or omissions had to be grossly negligent to support discipline (if in fact the Commission so required) would place upon the Board a greater burden than was required. It is apparent, however, that the Commission found the overall conduct of appellants grossly negligent and if that finding is supported by competent and substantial evidence we are bound by it.

* * *

Appellants in this connection challenge one of the Commission's findings on the basis that the collapse of the walkways was not caused by a certain specific failure. This is raised in connection with the finding that the failure to delineate special strength steel for the rods was grossly negligent. The rods themselves did not fail. In making this assertion appellants rely on the elements of a common law cause of action for negligence, i.e., duty, breach, proximate causation, and injury or damage. They assert that proximate causation is not present. In the first place we are not dealing with a civil cause of action for negligence. We are dealing instead with a determination of whether appellants negligently breached their duty in the design of the walkways. That breach occurred at the latest when their design was incorporated into the building with their approval and they were subject to discipline whether or not any collapse subsequently occurred. It is the appellants' conscious disregard of their duty for which discipline is being imposed not the result of that breach. . . . Damage or injury is not an element of this disciplinary proceeding and proximate cause is the legal concept that authorizes civil recovery for damage resulting from negligence. It is not in and of itself an aspect of "negligence," only an aspect of a civil cause of action for negligence. Related to that concept is the fact that indeed there was damage caused by the breach. By statute and under the contract the owner of the building was entitled to a building structurally safe and sound. . . . The owner did not receive such a building because of appellants' breach of their professional responsibility. The owner received a defective building. Whether the walkways collapsed or not, the owner was damaged because it received less than it was entitled to and that damage was proximately caused by appellants' acts and omissions. Further we have previously stated that gross negligence is not required in each act of appellants; it is their overall conduct in regard to the Hyatt construction which justifies discipline.

* * *

The statutory provisions make clear that Missouri has established a stringent set of requirements for professional engineers practicing in the state. The thrust of those requirements is professional accountability by a specific individual certified engineer. These requirements establish the public policy of the state for the protection of the public. They require that plans for construction of structures in this state which require engineering expertise be prepared by or under the direct supervision of a specified certified engineer and that that engineer bear personal and professional responsibility for those plans. The affixing of his seal on the plans makes him responsible for the entire engineering project and all documents connected therewith unless he specifically disclaims responsibility for some document relating to or intended to be used for *any* part of the engineering project. It would be difficult to imagine statutory language more clearly evidencing the total responsibility imposed upon the engineer, and accepted by him when he contracts to provide his services. The statutory statement that the right to engage in the profession is a personal right based upon the individual's qualifications in no way impacts upon the responsibilities imposed upon an engineer. Rather the assessment of the individual

"qualifications" of the engineer include his willingness and ability to accept the responsibilities imposed on him by the statutes.

[Ed. note: The court rejected the engineers' contention that the custom of relying on the fabricators to design the connections precluded any finding they were grossly negligent. The court noted that the employees of the fabricator are exempt from licensing requirements.]

* * *

The public policy of this state as it pertains to the responsibility of engineers has been established by the General Assembly in Chapter 327. That Chapter imposes upon the engineer a non-delegable duty of responsibility for projects to which he affixes his seal. . . .

The purpose of disciplinary action against licensed professionals is not the infliction of punishment but rather the protection of the public. . . . Chapter 327 has established the responsibility a certified engineer bears when he undertakes a contract in his professional capacity. Sec. 327.191 authorizes noncertificated engineers to perform engineering work "under the direction and continuing supervision of and is checked by" a certificated engineer. It is a misdemeanor for a certified engineer to affix his seal to plans which have not been prepared "by him or under his immediate personal supervision." Sec. 327.201. A corporation may engage in engineering activities if it has assigned *responsibility* for proper conduct of its professional engineering to a registered professional engineer. Sec. 327.401. Gillum was the engineer designated by G.C.E. as having that responsibility. An engineer affixing his seal to plans is personally and professionally responsible therefor. Sec. 327.401. Affixing his seal to plans imposes upon the engineer responsibility for the *whole* engineering project unless he, under seal, disclaims such responsibility. Gillum made no such disclaimer here. The entire thrust of Chapter 327 is to place individual personal and professional responsibility upon a known and identified certificated engineer. This is the responsibility the engineer assumes in exchange for the right to practice his profession. It is the assumption of this responsibility for which he is compensated. The statutory framework is established to protect the public and to hold responsible licensed engineers who fail to afford that protection. It is clear that the statute expresses the intent to impose disciplinary sanctions on the engineer responsible for the project whether the improper conduct is that of himself or attributable to the employees or others upon whom he relies. This case differs, therefore, from the cases relied upon by Gillum and G.C.E. where the statute did not impose such non-delegable responsibility. The Commission did not err in finding that Gillum and G.C.E. were subject to discipline for the acts or omissions of Duncan.

* * *

We now turn to the sufficiency of the evidence to support the Commission's findings that discipline was warranted. In so doing we note again the concession of appellants that except for five findings the findings of fact of the Commission are supported by evidence. We have previously stated our analysis of the five disputed findings. We review the sufficiency within the legal principles and framework heretofore explicated.

We look first to Duncan. He was the project engineer for the Hyatt and as such had primary responsibility within his company for designing and approving those aspects of the Hyatt which required structural engineering expertise. The design of the connections in the walkways and the design of the walkways themselves were included in that responsibility. The walkways were intended to carry pedestrian traffic. They were suspended above the main lobby of the hotel, recognized to be the main point of congregation within the hotel. The walkways each weighed approximately 35 tons and were comprised of heavy and largely non-malleable materials such as steel, concrete, glass and wood. The connections in the walkway were nonredundant so that if any one within a single walkway failed they all would fail and the walkway would collapse. Duncan had never designed a system similar to the Hyatt walkways. It is self-evident that the walkways offered a potential of great danger to human life if defectively designed. The Commission could properly consider the potential of danger in determining the question of gross negligence. That which might constitute inadvertence where no danger exists may well rise to conscious indifference where the potential danger to human life is great. This is simply to say that the level of care required of a professional engineer is directly proportional to the potential for harm arising from his design and as we have previously stated indifference to harm and indifference to duty are closely related if not identical.

The structural drawings of Duncan furnished to the fabricator contained several serious errors. Under standard engineering practice Duncan could either design the box beam-hanger rod connections or cause the drawings to reflect his intention that they be designed by the fabricator. These drawings did neither. They appeared to be connections fully designed by the engineer and were reasonably so interpreted by the fabricator. Duncan testified that he intended the fabricator to design the connections. The drawings did not contain information indicating that the connec-

tions were to be designed by the fabricator and omitted important engineering load calculations necessary to enable the fabricator to design the connections. The drawings failed to properly identify the type of weld required, the need for bearing plates and/or stiffeners, and erroneously identified the hanger rods as standard rather than high-strength steel. The box beam-hanger rod connections and the hanger rods themselves on all three walkways, as shown by the structural drawings, did not meet the design specifications of the Kansas City Building Code. That Code is intended to provide a required level of safety for buildings within the City. It is difficult to conclude that gross failure to comply with that Code can constitute other than conscious indifference to duty by a structural engineer.

Because of certain difficulties in fabrication Havens requested a change to the double rod configuration. This request was transmitted to Duncan who approved it and verified its structural soundness and safety to the architect. He did so without having conducted all necessary engineering tests and calculations to determine the soundness and safety of the double rod arrangement. His concern was with its architectural acceptability not its structural acceptability. The result of this change was to double the load on the fourth floor walkway and impose a similar increase on the connections which were already substantially below Code requirements.

Havens supplied Duncan with its shop drawings. Under the contract, and under the statute, review and approval of the shop drawings is an engineering function. Appellants' normal in-house procedures called for detailed check of all special connections during shop drawing review. Duncan was aware of the change to the two-rod system but did not review the box beam-hanger rod connection on the fourth floor walkway. Duncan did not, as is standard practice, look for an assembled detail of the connection and did not assemble the components, either in his mind or on a sketch, to determine what the connection looked like in detail. The shop drawings did not reflect the use of stiffeners or bearing plates necessary to bring the connections within Code requirements. No review was made nor calculations performed to determine whether the box beam-hanger rod connection shown on the shop drawings met Code requirements. Shop drawing review by the engineer is contractually required, universally accepted and always done as part of the design engineer's responsibility. The box beam-hanger rod connections and the hanger rod shown on the shop drawings did not meet design specifications of the Code.

Following the atrium roof collapse appellants were requested by the architect and owner to recheck all the steel in the atrium. They reported that they had done so and included in that report was the statement "we then checked the suspended bridges and found them to be satisfactory." In fact appellants did not do a complete check of the design of all steel in the atrium and did not do a complete check of the suspended "bridges," i.e., walkways. As finally built, the hanger rods and the box beam-hanger rod connections did not meet the requirements of the Code. The walkway collapse was the result of the failure of the fourth floor box rod connections. The third floor walkway, which did not collapse, had a "high probability" of failure during the life of the building.

The determination of conscious indifference to a professional duty, i.e., gross negligence, is a determination of fact. The conduct of Duncan from initial design through shop drawing review and through the subsequent requested connection review following the atrium roof collapse fully supports the Commissions' finding of conscious indifference to professional duty. The responsibility for the structural integrity and safety of the walkway connections was Duncan's and that responsibility was non-delegable....

He breached that duty in continuing fashion. His reliance upon others to perform that duty serves as no justification for his indifference to his obligations and responsibility. The findings of the Commission as to Duncan's gross negligence are fully supported by the record.

The Commission also found Duncan subject to discipline for misconduct in misrepresenting to the architects the engineering acceptability of the double rod configuration when he performed no engineering calculations or other engineering activities to support his representation. The Commission found such representation to have been made either knowing of its falsity or without knowledge of the truth or falsity. The Commission found Duncan's misrepresentation to be the willful doing of an act with wrongful intention which it had defined as misconduct. We find no error in either the factual findings or legal conclusion of the Commission. Duncan's representation to the architect concerning a material fact, without a basis for knowledge of its truth or falsity, could properly be viewed as either misconduct as an engineer or gross negligence. In either event it subjected Duncan to disciplinary action.

The Commission found Gillum subject to discipline for gross negligence under the vicarious liability theory and also personally grossly negligent in failing to assure that the Hyatt engineering designs and drawings were structurally sound from an engineering standpoint prior to impressing thereon his seal and in failing to assure adequate shop drawing review. It further found Gillum to be subject to discipline for unprofessional conduct and gross negligence in his refusal to accept his responsibility as mandated by Chapter 327 and his denial that such responsibility existed. All of these findings arise from the same basic attitude of Gillum that the responsibility imposed by Chapter 327 is not in keeping with usual and customary engineering practices and that that responsibility did not mandate his personal involvement in the design of the Hyatt. In essence he placed the responsibility for the improper design of the connections on Havens and took the position that the structural engineer was entitled to rely on Havens expertise. What we have heretofore said in regard to the requirements of Chapter 327 and the responsibility imposed upon an engineer thereby sufficiently deals with Gillum's contentions.

His argument here that utilization of his seal without disclaimer could not impose responsibility upon him for the shop drawings of another entity prepared after impression of the seal is clearly rejected by the language of the statute. By section 327.411.2 the owner of the seal is responsible for the "whole ... engineering project" when he places his seal on "any plans" unless he expressly disclaims responsibility and specifies the documents which he disclaims. The shop drawings were part of the documents comprising the engineering project and were "intended to be used for any part or parts of the ... engineering project...." Gillum was by statute responsible for those drawings and he accepted such responsibility when he entered into the contract and utilized his seal. His refusal to accept a responsibility so clearly imposed by the statute manifests both the gross negligence and unprofessional conduct found by the Commission. These findings are further bolstered by the evidence of Gillum's participation in the misrepresentations concerning, and nonperformance of, a review of the atrium design upon direct request of the architect and owner. Although we have found that a specific finding of misconduct and discipline therefor cannot be based upon the atrium design review because not charged in the complaint, the evidence is relevant and persuasive on Gillum's overall mental approach to his responsibilities as an engineer and the cavalier attitude he adopted concerning the Hyatt project.

Appellant G.C.E. is, for reasons heretofore stated, subject to discipline for the conduct of its employees and particularly for the conduct of the engineer assigned the responsibility for the "proper conduct of all its ... professional engineering ... in this state...."

The findings of misconduct against Gillum arising from the "atrium design review" is reversed.[23] In all

---

[23][Ed. note: This ground for discipline was not alleged in the complaint.]

Other cases have also involved sufficiently flagrant misconduct or incompetence to justify a drastic agency decision. For example, a court affirmed an agency decision revoking the license of a professional engineer who had performed welding without being certified as required by the state administrative code.[24] In the same case, however, the court would not affirm the agency's revocation where the professional engineer designed and supervised the construction of a garage that collapsed. The court held that the incompetence did not have to consist of continued and repeated acts. However, incompetence must refer to some demonstrated lack of ability to perform professional functions. Although recognizing that there was an admitted error in the design of the roof supports for the garage, the court noted that the error was not obvious and that this was the first failure the engineer had experienced in eleven years of practice.

A revocation of an architect's license was affirmed based on serious design errors leading to the failure of the basement wall.[25] The architect also caused construction delay, failed to obtain a building permit, misplaced the building in reference to the lot line and secured the owner's endorsement of payment without informing him of the facts. Another court affirmed a six-month suspension based on a deficient ventilation plan, a superficial inspection of the premises, a superficial scanning of the architectural plan, and reliance on the judgment of two relatively inexperienced employees.[26] In addition, the engineer, after finding

other respects the order of the Commission and the discipline imposed by the Board is affirmed.

KAROHL, P. J., and KELLY, J., concur.

---

that the original certifications were in error and serious defects existed, did not disclose this to appropriate city officials.

Many members of the design professions operate under economically unstable conditions. Swings in the economic cycle can prove devastating. As a result, individual bankruptcies of architects and engineers are not uncommon when the economy turns for the worse. Those in the contracting business face similar risks.

The California experience illustrates how difficult it can be to reconcile the state's right to protect the public through its licensing laws with the interest of the federal government in effectuating the bankruptcy laws designed in part to give the bankrupt a fresh start. Section 7113.5 of the California Business and Professions Code had stated that grounds for disciplinary action existed if the contractor had settled a lawful obligation for less than the full amount of the obligation by having all or part of the debt discharged in bankruptcy. Yet *Grimes v. Hoschler*[27] held that this would frustrate the fresh-start purpose of the federal bankruptcy law. Risking one's license could discourage a contractor from discharging its business debts in bankruptcy. The contractor would very likely pay its discharged business debts to avoid losing its license. The licensing law would not only discourage contractors from availing themselves of the provisions of the Bankruptcy Act but also deny a fresh start to those licensees who are adjudicated bankrupts.

Shortly after this decision, the California legislature bowed to the federal law and the court decision by modifying Section 7113.5, barring disciplinary action being taken against a contractor who settled his debts for less than the full amount by bankruptcy.[28]

---

[24]*Vivian v. Examining Bd. of Architects, etc.*, 61 Wis.2d 627, 213 N.W.2d 359 (1974).
[25]*Kuehnel v. Wisconsin Registration Bd. of Architects & Professional Engr's*, 243 Wis. 188, 9 N.W.2d 630 (1943).
[26]*Shapiro v. Bd. of Regents*, 29 A.D.2d 801, 286 N.Y.S.2d 1001 (1968). Cases involving architect license suspensions are collected in 58 A.L.R.3d 543 (1974) and those involving engineer licenses in 64 A.L.R.3d 509 (1975).

---

[27]12 Cal.3d 305, 525 P.2d 65, 115 Cal.Rptr. 625 (1974).
[28]1975 Cal.Stat., ch. 818, § 2.

The issue surfaced again in 1986. In *Parker v. Contractors State License Board,*[29] the licensing board sought to suspend a bankrupt contractor who had deducted benefit contributions from his employees' wages but did not forward them to the union. However, the union trust fund claim against him had been discharged by bankruptcy. The court held that the board could not discipline the contractor for going bankrupt, as that would prevent him from obtaining a fresh start. But disciplinary action could be based on fraud or a violation of the California Labor Code. Factors other than the bankruptcy could indicate the contractor was unsuitable to hold a license. The court sent the case back to the trial court to determine whether the activity of the contractor was fraudulent or simply failure to pay for lack of money.

The effectiveness of post-admission disciplinary powers has been limited. Attempts to suspend or revoke are almost always challenged by the design professional or contractor whose means of livelihood are being taken away. Challenges often mean costly appeals. Often administrative agencies charged with responsibility for regulating the profession or occupation are underfunded and understaffed. Suspension and revocation are unpleasant tasks. Even when action is taken, courts closely scrutinize decisions of administrative agencies. They *should* extend considerable deference to the agency decision. In matters as important as these, however, courts seem to make a redetermination of what is proper. The combination of agency lethargy and overextensive judicial scrutiny when they do act may be one reason, among many, for increased professional liability. The latter can supplement or even replace the regulatory licensing process as a means by which incompetent practitioners are eliminated.

SECTION 10.05 Types of Licensing Laws

It must be emphasized again that licensing rules are determined by each state. As a result, considerable variety exists in the regulatory controls in both prohibited conduct and sanctions for violations. Also, legislation is frequently changed. Case decisions cannot always be relied on because sub-

sequent legislation may have been enacted to change the result in a particular case. Despite these dangers, some broad patterns have emerged as to the types of controls that have been enacted.

A. Licensing of Architects and Engineers: Holding Out and Practice Statutes

Licensing laws fall into two main categories. Some statutes regulate the professional title and are called "holding out" statutes. For example, in 1965 Ohio enacted such a statute stating that

> [n]o person shall use the title "landscape architect" . . . unless he is registered . . . or holds a permit.

A court passing on the validity of this legislation stated:

> A practitioner, upon qualifying, becomes entitled to use the label, "landscape architect." Only the title, not the practice or profession, is restricted to licensees. Unregistered members may continue to practice, but without employing the title. Thus, appellant's allegation that the law prohibits him from practicing his profession is not well taken; only his use of the title "landscape architect" is proscribed.[30]

Often a holding out statute is the first step toward the second category of licensing laws—the practice statute, one that bars professional practice. For example, Iowa originally enacted a holding out statute. A court decision in 1962 granted recovery to an out-of-state architect because the statute did not preclude practice but only holding oneself out as licensed.[31] Very likely, reaction to that decision generated legislation in 1965 that changed Iowa's registration laws to a practice statute.[32]

Projects that affect the public generally are more likely to require the services of a licensed design professional. A licensed design professional is more likely to be required where the project is one for human habitation or one in which large numbers of persons will gather.

Increasingly, states have enacted legislation requiring public entities to hire only *licensed* archi-

---

[29]187 Cal.App.3d 205, 231 Cal.Rptr. 577 (1986).

[30]*Garono v. State Bd. of Landscape Architect Examiners,* 35 Ohio St.2d 44, 298 N.E.2d 565, 567 (1973).
[31]*Davis, Brody, Wisniewski v. Barrett,* 253 Iowa 1178, 115 N.W.2d 839 (1962).
[32]*Food Management, Inc. v. Blue Ribbon Beef Pack, Inc.,* supra note 3.

tects or engineers for construction projects for human occupations that cost more than a designated amount.[33] Some states exclude less complex structures, such as single-family residences, agricultural buildings, or storefronts. Improvements that cost less than a specified amount—typically $10,000 to $30,000—may be exempt. Local laws must be checked.

B. Contractor Licensing

The proliferation of occupational licensing has led to an increased number of state licensing statutes for contractors. Until recently, only one half of the states had such statutes, although the number seems to be increasing steadily. The statutes raise different problems and are discussed in Sections 10.08 through 10.11.

C. Variations on the Traditional Contracting System

The traditional system for delivering construction services, which was described in Section 8.05, will be discussed again in Section 17.03(A). Modern variations on this system are covered in Section 17.04. Some of these variations, such as the use of construction management (Section 17.04(D)) and combining designing and building (Section 17.04(F)), have raised difficult licensing and registration problems, which are discussed in those sections.

SECTION 10.06 Statutory Violations

A. Preliminary Issue: Constitutionality

Because of the controversial nature of professional regulation, constitutional attacks are common. By and large, such legislation is upheld based on the police powers granted by state constitutions and constitutional provisions permitting the state to legislate in the interest of public welfare.[34] Some

successful attacks have been made that have usually involved language or procedural problems and not the power of the state to regulate the professions and occupations.[35]

B. Holding Out

Rodgers v. Kelley[36] involved an action by the plaintiff for design services for a project that was abandoned because of excessive costs. The clients claimed the plaintiff violated the Vermont holding out statute. The plaintiff contended that he had never signed his name or in any way represented that he was an architect. He testified that he had twelve years of experience in the architectural field but contended that so long as he did not label himself, his plans, or his business with the title "architect" he had not violated the statute. However, the court stated:

> But "holding oneself out as" an architect does not limit itself to avoiding the use of the label. The evidence is clear that, in the community of Stowe, this plaintiff was known as a proficient practitioner of all of the architectural arts with respect to homebuilding, at least. It was a business operation from which he received fees. He presented himself to the public as one who does the work of an architect. This constitutes holding oneself out as an architect, and is part of the very activity sought to be regulated through registration.

C. Practicing

Which activities fall within architectural practice? Design services can range from simply sketching a

---

[33]Tex.Civ.Stat. Art. 249a, § 16 (Supp. 1993) (over $100,000 for new construction and $50,000 for alterations or additions to an existing building).

[34]*Richmond v. Florida State Bd. of Architecture*, 163 So.2d 262 (Fla.1964); *State v. Beck*, 156 Me. 403, 165 A.2d 433 (1960); *State v. Knutson*, 178 Neb. 375, 133 N.W.2d 577 (1965); *Pine v. Leavitt*, 84 Nev. 507, 445 P.2d 942 (1968); *Chapdelaine v. Tennessee State Bd.*, 541 S.W.2d 786 (Tenn.1976) (state required surveyors but not engineers to register).

[35]*H&V Eng'g, Inc. v. Idaho State Bd. of Professional Eng'r and Land Surveyors*, 747 P.2d 55 (Idaho 1987) ("gross negligence" standard for disciplinary action too vague). But see *Duncan v. Missouri Bd. for Architects, Professional Eng'r and Land Surveyors* reproduced in Section 10.04(B) (upheld such a standard against constitutional challenge); *New Jersey Builders Ass'n v. Mayor*, 60 N.J. 222, 287 A.2d 725 (1972) (definitions too vague for enforcement); *Jenkins v. Manry*, 216 Ga. 538, 118 S.E.2d 91 (1961) (improper discrimination between self-employed and employed persons and an exemption for public employees).

[36]128 Vt. 146, 259 A.2d 784, 785 (1969). See also *State ex rel. Love v. Howell*, 281 S.C. 463, 316 S.E.2d 381 (1984) (architectural designer held to be holding out he was an architect). In a later opinion, the court found Howell guilty of civil contempt. It sent the case back to the trial court to determine the sanction, which could include a fine, imprisonment, or shutting down Howell's business operation. 285 S.C. 53, 328 S.E.2d 77 (1985).

floor plan or planning to place a residence on a designated site all the way to construction documents with sufficient detail to obtain a bid from the contractor. In addition to the varying design activities, a differentiation can be made between those services that are part of the design process and those services performed by a design professional during construction.

Kansas Quality Construction, Inc. v. Chiasson[37] involved services during the design phase. The plaintiff was engaged in the business of building apartments. It represented that it had the necessary "knowhow" related to preliminary assistance and advice to assist owners in getting apartments "off the ground." The defendants wished to build an apartment complex and made an agreement with the plaintiff under which the latter would develop a plat layout showing buildings, recreational area, and parking. The plaintiff would also *cause* plans to be drawn for loan submission and assist the owner in obtaining financing. The plaintiff would "[e]rect apartment buildings on owner's ground in accordance with approved plans and specifications" on a turnkey basis. The contract also provided that if the owner contracted with another builder, the plaintiff would receive a fee of $4,000 for services performed. Plaintiff performed the preliminary services, but the defendant selected another builder.

One of the defenses to the plaintiff's claim to the 4,000-dollar fee was that the plaintiff was performing architectural services without being licensed. The Illinois statute defined architecture as including the offering or furnishing of professional services in connection with the construction or erection of any building, structure, or project. The court stated:

> We are of the opinion that offering to develop a plat layout . . . to show placement, total buildings, etc., is not so connected with "construction" that it partakes of doing what architects do. If this is so, then performance likewise is outside of the Act. To be sure, there is a nexus between the layout and construction in the sense that placement must precede construction—but then so must the idea itself. "Construction" at the very least, means getting off the ground by going either up

or down, not just thinking about it, and plaintiff's plat layout here, "to show building placement, total buildings, recreation area, parking, etc." is certainly not construction nor connected with it to the extent that it falls within the statutory definition. The layout was developed first to please defendants and upon their approval, to convince the municipal authorities of the project's desirability and to gain their approval. To be sure, a plat layout can at times be a very complicated business calling for the collaboration of a battery of experts—surveyors, engineers, planners, landscapers, even lawyers—but in bringing their varied expertise to bear, none are thereby transmogrified into architects—and vice versa. As is said in 5 Am.Jur.2d 665, § 3, Architects, "[T]he making of a survey of existing conditions, with recommendations and preliminary sketches and layouts, with regard to work needed on a hotel building, constituted nonarchitectural services."

> Likewise . . . causing plans of the proposed building to be drawn for loan submission is not a statement by plaintiff that it will perform architectural services, rather, it is an agreement that they will cause such plans to be drawn for such purpose—loan submission—which may or may not be eventually connected with the projected construction and which may or may not be drawn by an architect—it all depends on how much detail the banker wants.[38]

Noting that the plaintiff had used a non-Illinois architect to prepare these plans, the court indicated that the plaintiff was not "playing architect." Similarly, the court concluded that obtaining financing "has even less to do with architecture—if anything at all."[39]

Some of the language in the opinion may be traceable to the court's belief that the defendant had put forth a number of nonmeritorious defenses to a legitimate obligation. There was little consideration for the public purpose protection so often stated to be the justification for registration laws.

The differentiation between those services performed during the designing phase and those performed during construction can be difficult. It has

[37]112 Ill.App.2d 277, 250 N.E.2d 785 (1969).

[38]250 N.E.2d at 787. Similarly, a California court held creation of preliminary sketches not to be architectural services. See *Walter M. Ballard Corp. v. Dougherty*, 106 Cal.App.2d 35, 234 P.2d 745 (1951).
[39]As to financing, see *Herkert v. Stauber*, 106 Wis.2d 545, 317 N.W.2d 834 (1982) (earlier edition of treatise cited and followed), discussed in Section 3.06.

been stated that services performed during the design phase are more likely to be considered within the licensing laws than those performed during the construction phase.[40] This is undoubtedly based on the specialized training and professional skill possessed by a design professional being manifested mainly in the design phase.

The design professional does bring skill to the Construction Process by interpreting documents, advising on changes, and judging performance. Some of these skills are intimately connected to other design process skills. The generalization may be true if emphasis is placed on construction methods and organization, more properly the province of the contractor. Yet the language in the *Kansas Quality* case described earlier that seems to emphasize professional service "in connection with the construction or erection of any building" can be taken to mean that those professional skills exercised during construction may also fall within the licensing laws.

This differentiation may be of little practical importance. Design professionals rarely perform services connected only with construction. They perform services connected either with design and construction or with design alone. However, the problem may come to a head when the law will be asked to determine whether construction managers must have a design professional license, a contractor's license, or neither.[41]

Sometimes the person performs design services for registered design professionals and does not deal directly with the public. For example, a Mississippi case involved an attempt by the state registration board to restrain Rogers from performing engineering services or holding himself out as a mechanical designer.[42] Rogers was not a registered engineer. He operated a small office and did all his work for other architects and engineers and never in any way dealt with members of the public. Mississippi exempted persons who performed design services, such as drafters, under the direct control of a registered design professional. In concluding that Rogers was not violating the registration laws, the court stated:

Inasmuch as Rogers' work was supervised and controlled by the architect or engineer for whom he was doing the work, obviously his status was not that of an independent contractor or practitioner of engineering, but rather a helper or employee.[43]

Sometimes the design services are incidental to the selling of equipment or mechanical systems. For example, in *Dick Weatherston's Associated Mechanical Services, Inc. v. Minnesota Mutual Life Insurance Co.*,[44] an insurance company was dissatisfied with the advice it had received from its architects regarding the air-conditioning system. It approached Weatherston, a mechanical contractor with a degree in engineering, for his advice. Weatherston made some design changes but made it clear that he was not a registered engineer. These changes were ultimately accepted by architects and engineers retained by the insurance company.

Weatherston claimed and proved that he had a valid contract with the insurance company to supply air-conditioning systems. He brought legal action, claiming the insurance company had repudiated the contract. The insurance company claimed Weatherston had been practicing engineering without a license. However, the court concluded after noting that the company knew he was not licensed that his action was brought on the contract to supply the air-conditioning equipment and not a contract for engineering services. The court seemed to assume that often those who sell or install mechanical equipment provide some engineering advice to buyers. When it appears that the principal basis for the claim is the sale of the equipment, the absence of a license under these circumstances should have no effect.

Does testifying as an expert constitute professional practice? For example, in one case a party challenged a lower court verdict because an unlicensed architect had been permitted to testify.[45] The claim was made that testifying was practicing and violated the licensing laws.

The possession of a license generally does not, however, determine whether a witness will be permitted to testify as an expert. The court permitted

[40]Annot., 82 A.L.R.2d 1013 (1962).
[41]See Section 17.04(D).
[42]*State Bd. of Registration v. Rogers*, 239 Miss. 35, 120 So.2d 772 (1960).

[43]120 So.2d at 776.
[44]257 Minn. 184, 100 N.W.2d 819 (1960).
[45]*W. W. White Co. v. LeClaire*, 25 Mich.App. 562, 181 N.W.2d 790 (1970).

the testimony by concluding that testifying was not practicing. However, registration or absence of it goes to the *weight* of the testimony and not to its admissibility.[46]

D. Architecture and Engineering Compared

States generally regulate architecture and engineering separately. Each profession and the agencies regulating them sometimes differ over where one profession begins and the other ends. These conflicts demonstrate the economic importance of registration as well as the secondary role sometimes played by the public interest.

The two professions can be differentiated by project types and their use. One court stated:

One prominent architect, in explaining the difference between architecture and engineering, said in effect that the entire structure and all of its component parts is architecture, if such structure is to be utilized by human beings as a place of work or assembly. He pointed out that, if the authorities were going to erect a courthouse as the building in which the [case] was being tried, they would obtain the service of an architect; but, if it was proposed to construct a power plant . . . they should employ an engineering firm. . . .

All of the architects and those who were registered as both architect and engineer agreed that the overall plan of a building and its contents and accessories is that of the architect and that he has full responsibility therefor. As one witness answered it, he is the commander in chief.[47]

In *State v. Beck*,[48] the Maine Supreme Court stated aesthetics to be the principal difference between engineering and architecture. To that court an architect was "basically an engineer with training in art." The court also stated:

While categorically an engineer, the architect—without disparagement toward the professional engineer—is required to demonstrate that he possesses and utilizes a particular talent in his engineering, to wit, art or aesthetics, not only theoretically but practically, also, in coordination with basic engineering.[49]

The court then cited a Louisiana case that stated that an engineer "designs and supervises the con-

struction of bridges and great buildings, tunnels, dams, reservoirs and aqueducts."[50]

The Maine court then described architectural projects:

Architects are commonly engaged to project and supervise the erection of costly residences, schools, hospitals, factories, office and industrial buildings and to plan and contain urban and suburban development. Health, safety, utility, efficiency, stabilization of property values, sociology and psychology are only some of the integrants involved intimately. Banking quarters, commercial office suites, building lobbies, store merchandising salons and display atmospheres, motels, restaurants and hotels eloquently and universally attest to the decisive importance in competitive business of architectural science, skill and taste. A synthesis of the utilitarian, the efficient, the economical, the healthful, the alluring and the blandished is often the difference between employment and unemployment, thriving commerce and a low standard of existence. Basic engineering no longer suffices to satisfy many demands of American health, wealth or prosperity.[51]

Statutes sometimes use definitions that recognize some differences between architecture and engineering. For example, in Texas the practice of architecture is defined as services that apply "the art and science of developing design concepts, planning for functional relationships and intended uses, and establishing the form, appearance, aesthetics, and construction details, for any building or buildings, or environs. . . ."[52] The statute dealing with engineering services in Texas speaks of the "application of special knowledge of the mathematical, physical or engineering sciences. . . ."[53]

Some statutes permit engineers to perform architectural services incident to engineering work. One case[54] involved an attempt by the state licensing authorities to restrain two officers of a construction company who were licensed engineers from

---

[46]Expert testimony is discussed in Section 14.06.
[47]*State Bd. of Registration v. Rogers,* supra note 42 at 774.
[48]Supra note 34.
[49]165 A.2d at 435.

[50]*State v. Beck* quoted from *Rabinowitz v. Hurwitz-Mintz Furniture Co.,* 19 La.App.811, 133 So. 498, 499 (1931).
[51]165 A.2d at 437.
[52]Tex.Civ.Stat. Art. 249a, § 10(a).
[53]Id. at Art. 3271a, § 2(4).
[54]*Dahlem Constr. Co. v. State Bd. of Examiners,* 459 S.W.2d 169 (Ky.1970). The result was undoubtedly assisted by a special Kentucky statute requiring that architects design nursing homes.

designing a seventy-eight-bed nursing home. The engineers contended that the design work they were performing was incident to their engineering services to build the structure. The court recognized that some design services are required for any structure and that this is the proper function of the exemption for architectural services incident to engineering work. However, the building of a nursing home with both the aesthetic considerations of the human beings who would live there and the aesthetic aspects of positioning the project was architectural, and the defendants could not perform these services.

Turf fights between architects and engineers seem to be endemic to the design professions. As a rule, these disputes are resolved by the legislatures, with some intervention by the courts and regulatory agencies. For example, the Michigan constitution required that a majority of the members of any regulatory board be members of the profession being regulated. Subsequently, Michigan passed a statute that required only three members of the seven-member board regulating architects to be architects. To deal with the prior constitutional requirement, the legislature simply declared architects and engineers to be members of the same profession.

Such a legislative declaration was rejected by a Michigan court that stated that the legislature could not declare something that is not so.[55] The court noted that architects and engineers have different educational requirements and serve different functions.

Turf fights again exploded in 1992. The attorney general of Texas issued an opinion that examined a Texas statute requiring registered architects to prepare architectural plans and specifications for new public construction with a value over $100,000. He ruled that the statute did not bar a licensed professional engineer from preparing plans and specifications that require the application of engineering principles for a new building that is to be constructed and owned by a public entity if the building will be used for education, assembly, or office occupancy and if the construction costs exceed $100,000.[56] In essence, the attorney general

concluded from other Texas legislation regulating engineers that an overlap exists between the work of an architect and an engineer in building design and that legislation prior to 1989 had allowed licensed engineers to prepare building designs and specifications. He ruled that the 1989 statute requiring that a licensed architect do so under certain circumstances did not bar engineers.

That these turf fights can be resolved harmoniously between the two professions was demonstrated in New Mexico in 1992 by an agreement made between the agencies regulating engineers, surveyors, and architects. The agreement recognized the overlap between architecture and engineering and that there are some projects where either an architect or an engineer will be able to meet the needs of the owner and provide a safe, serviceable building. The agreement defined the incidental practice of architecture and engineering "as those services performed on buildings which have a construction valuation of no more than $250,000 or an occupant load of no more than 50." According to the agreement, a single professional seal would meet the requirement on building plans submitted for permits within these limits. The agreement resulted from an order by a court in 1985 that an architect-engineer joint practice committee be established to resolve these turf disputes.

For several years, the committee worked on an agreement, which resulted in the agreement made in 1992 and adopted by the regulatory agencies on March 2, 1992. The agreement was also designed to permit the regulatory boards to focus their efforts on unlicensed practice rather than interdisciplinary disputes.[57]

E. Statutory Exemptions

A substantial number of states do not require projects that do not cost more than a designated amount of money be designed by a licensed architect. Obviously, in a rapidly inflating period, fixed limits of this type soon become outmoded.

Such statutes can generate interpretation problems because the cost of a project evolves through various stages. The owner has a budget. The design professional gives cost predictions. The contract price can be higher. Not uncommonly the ultimate

[55]*Nemer v. Mich. State Bd. for Architects*, 20 Mich.App. 429, 174 N.W.2d 293 (1969).
[56]Opinion No. DM-161, Office of the Attorney General of Texas (August 27, 1992).
[57]Press release, Office of the Governor, Santa Fe, New Mexico (May 13, 1992).

payment for the project exceeds the contract price because of extras or delay claims.

The statute itself may determine which of these figures determines whether the project is exempt. In one case, the unregistered design professional gave a good-faith cost prediction that the project could be built for less than the statutory limit. The fact that the ultimate cost for various reasons was in excess of the statutory limit would not bar the person performing design services from recovery.[58] The person performing design services should be able to know at the time she performs them whether she is violating the law.

A similar standard should apply if the exemption relates to square footage of the structure. However, this exemption should not be abused by the person performing design services. Clearly, if that person knows the ultimate cost will go beyond the exempt amount or the ultimate space will exceed the amount allowed, the exemption should not protect the person performing those services.

The California Business & Professions Code § 5536(a) (4)(b) exempts "single-family dwellings of woodframe construction not more than two stories and basements in height," multi-family dwellings with no more than four units of a similar size and construction unless the units form apartment and condominium complexes of over four units, garages and structures appurtenant to exempt dwellings, and agricultural or ranch buildings unless public officials deem "that an undue risk to the public health, safety or welfare is involved."

F. Possessor of License

Design professionals traditionally performed as sole proprietors or partners. Increasingly, for tax reasons, design professionals are being permitted to perform design services through the corporate structure, a concept discussed in Section 3.06. Whether the business entity chosen is a partnership or a corporation, it will be necessary to determine who must be licensed. One court held that a part-nership itself must be licensed even though the partners were individually licensed.[59] Another court held that a corporation need not be licensed if the managing agent was licensed.[60]

The determination of who actually must hold a license when partnership or corporation forms are used requires an evaluation of the licensing statutes, administrative regulations, and case decisions. For this reason, any design professional who operates through a partnership or corporation should seek legal advice.

G. Out-of-State Practice

The practice of architecture and engineering as well as the performance of construction services increasingly transcends state lines. As a result, it is common for a person licensed in one state to perform services in another.

Sometimes design professionals who perform services on a multistate basis obtain licenses in each state in which they perform services or in which projects for which they perform services are located. This has become easier because of standardized examinations and increased reciprocity.

Some registration laws do not require professional persons licensed in a foreign state to become registered if they perform work only for an isolated transaction or perform work not to exceed a designated number of days. Often such exemption requires an easily obtained temporary license.

As interstate practice has become more common, an increasing number of cases have dealt with attempts by design professionals licensed in one state to recover for services related to a project in another. For example in *Johnson v. Delane*,[61] an engineer licensed in the state of Washington obtained a commission to prepare plans and specifications for a project to be built in Idaho. The engineer was not to perform any supervisory function. He obtained the commission while visiting in Idaho. He performed the requisite design services in Washington and delivered the plans and specifications to the client in Idaho. The court held that he was not practicing architecture in Idaho, and he recovered for his services.

---

[58]*State v. Spann,* 270 Ala. 396, 118 So.2d 740 (1959). But see *Sample v. Morgan,* 66 N.C.App. 338, 311 S.E.2d 47 (1984), petition allowed as to additional issues, 310 N.C. 626, 315 S.E.2d 692 (1984), which limited the contractor's recovery to the amount his license allowed him to build when the ultimate cost exceeded this limit because of owner-directed changes.

[59]*Nickels v. Walker,* 74 N.M. 545, 395 P.2d 679 (1964).
[60]*Hattis Assoc., Inc. v. Metro Sports, Inc.,* 34 Ill.App.3d 125, 339 N.E.2d 270 (1975).
[61]77 Idaho 172, 290 P.2d 213 (1955).

Is the purpose of Idaho licensing laws to protect Idaho clients from retaining unqualified architects and to protect Idaho citizens from being exposed to risk of harm due to structures that may not comply with local building laws and codes? If so, it would seem that the project being constructed in Idaho for Idaho clients should necessitate the licensing of the architect under Idaho laws.

The result in *Johnson* is supportable if the architectural registration requirements for Washington and Idaho are similar and if there are no peculiarities of Idaho building codes that might make them substantially different from those in effect in Washington. As to Idaho's interest in ensuring that structures meet Idaho building code standards, the defendant client would have had a defense to any action by the architect, whether licensed in Idaho or not, that there had been a failure to comply with local building codes.

The decision can be supported on another ground. It is likely that the architect could have registered or been allowed a temporary license to perform the services for one project. If so, granting him recovery despite his not being registered in Idaho would avoid his being uncompensated and prevent unjust enrichment of the client by, in effect, excusing his failure to use this method of complying with Idaho law.

Some state laws do not require a license for performance of services by out-of-state design professionals if they are licensed in the state where they practice principally and as long as there is a licensed design professional in overall charge of the project. However, the licensed local design professional must not be a figurehead but must actually perform the usual design professional functions. Sometimes out-of-state architects or engineers "associate" a local architect as a means of ensuring that they will be able to collect their fee.

In *Food Management, Inc. v. Blue Ribbon Beef Pack, Inc.*,[62] plaintiff was an Ohio engineering corporation that had entered into a written turnkey contract with the defendant for the design and construction of a meat-packing plant to be built in Iowa. The plaintiff was not licensed in Iowa but entered into a written contract with an Iowa architect-engineer. This agreement designated the Ohio corporation as the Principal Consultant and referred to the local architect-engineer as the Associate Engineer. The contract stated in part:

> C. Consideration of the Associate Engineer's Work: The Principal Consultant shall give thorough consideration to all reports, sketches, drawings, specifications, proposals and other documents presented by the Associate Engineer, and shall inform the Associate Engineer of his decision within a reasonable time so as not to delay the work of the Associate Engineer.
>
> D. Standards: The Principal Consultant shall furnish the Associate Engineer with a copy of any design and construction standards he shall require the Associate Engineer to follow in the preparation of drawings and specifications for This Part of the Project.

The project was ultimately abandoned because of excessive cost. When the plaintiff sued for the services performed, the client claimed the plaintiff was not licensed to practice in Iowa. One of the contentions made by the plaintiff was that the engineering or architecture performed in Iowa had actually been performed by the Iowa architects and engineers. The court held that the Iowa design professionals were not "in charge" and denied recovery to the Ohio corporation.

In *Hedla v. McCool*,[63] a Washington architect performed design services for a project to be built in Alaska and had the plans approved by an Alaskan engineer. The court emphasized the importance of the policy expressed in Alaska's licensing law. The court distinguished *Johnson v. Delane*[64] (discussed earlier) by noting that the Alaska statute was broader, that this was not an isolated transaction, and that conditions in Alaska were different from those in the state of Washington. The court was more persuaded by the holding in the *Food Management* case, which had not looked kindly on out-of-state architects associating a local architect when the out-of-state architects were principally responsible for design decisions.

One important consideration permeating these cases is whether the unlicensed person seems to have performed properly. If so, denial of recovery, despite the importance of license compliance, may

---

[62]413 F.2d 716 (8th Cir.1969).

[63]476 F.2d 1223 (9th Cir.1973). Similarly, a local association failed in *O'Kon and Co. Inc. v. Riedel*, 588 So.2d 1025 (Fla.Dist.Ct.App. 1991).
[64]Supra note 61.

seem unjust. For example, in *Delane,* the work appears to have been performed correctly. But in *Hedla v. McCool,* where recovery was denied, the costs overran considerably and the design was never used—factors that make recovery less attractive.

Perhaps the expansion of reciprocity and the ease with which out-of-state design professionals who are registered in their home states can be allowed to practice in other states will make it more difficult for design professionals to recover if they do *not* use these techniques for legitimating their projects in states where they are not licensed. In any event, any design professionals considering performing design services either in another state or for a project that will be built in another state should receive legal advice on the proper process for ensuring that they are not violating the laws of the state where the services are being performed or the project is located.

This subsection has dealt with whether there *has been* a violation by the out-of-state professional. Section 10.07(B) discusses two cases that involved attempts to recover for services *illegally* performed by design professionals registered in another state.

H. Subterfuge

Sometimes artificial arrangements are made to bypass registration laws. Two cases reflect differing judicial attitudes toward such subterfuge. In *Snodgrass v. Immler,*[65] the owner had approached an unlicensed architect and asked him to design a house. The architect told the owner that he was not licensed and would be violating the licensing statute if he did. But the owner entered into a contract with a licensed architect under which the latter would employ the unlicensed architect to draw plans "under the supervision" of the licensed architect. A contract was also made between the licensed and unlicensed architects. The fee was to be 5% of the construction cost, of which the unlicensed architect would receive 4% and the licensed architect 1%.

Problems developed when costs overran and the design submitted by the unlicensed architect was abandoned. The unlicensed architect sought recovery for his services claiming that he was an intended beneficiary of the contract between the licensed architect and the client. In denying recovery, the court noted that the licensed architect was merely a front and characterized the two contracts as shams. It did not appear that the licensed architect played any significant part in the relationship between the unlicensed architect and the client. (He "sold" his license.)

Scott-Daniels Properties, Inc. v. Dresser[66] involved roughly similar facts. The client wished to build a motel in Minnesota. He had heard of Dresser's reputation and contacted him. After being satisfied that Dresser was an outstanding architect and a former pupil of Frank Lloyd Wright, the client retained Dresser to design and supervise the motel. Evidently the client knew Dresser was not licensed in Minnesota.

The actual identity of the design professional was somewhat cloudy. Dresser had formed a corporation, and the corporation had a licensed architect as an employee. The client contended that his contract was with Dresser *personally,* while Dresser contended that the contract was with the corporation, with Dresser, an employee, to do the work. The trial court concluded that the client contracted with the corporation, and this was affirmed on appeal. However, it appears that the registered architect was not in responsible charge of the work (a requirement of the Minnesota law) and that the corporation was simply a sham through which Dresser, an unlicensed architect, could practice architecture.

The two cases reflect an observation made in the preceding subsection. In *Scott-Daniels,* it appears that the court was convinced that the architect had performed properly and the motel was actually built. On the other hand, in *Snodgrass,* the court wondered why the client had not defended on the grounds of the unlicensed architect's design having substantially exceeded the estimated budget. The client may have thought it would be easier to defeat the unlicensed architect's claim by showing failure to comply with the registration laws than to establish that the budget had been substantially exceeded through no fault of the client.

Other devices are sometimes used in an attempt to bypass licensing laws. For example, suppose the

[65]232 Md. 416, 194 A.2d 103 (1963).

[66]281 Minn. 179, 160 N.W.2d 675 (1968).

unlicensed architect assigns her right to recover under the contract to a third party. Normally, rights under a contract can be transferred by assignment.

A court that believed circumvention would frustrate the purpose of the registration laws barred an assignee (the person to whom the assignment had been made) from recovery.[67] Similarly, an attempt to exempt a project based on a statute exempting projects under a designated dollar amount by dividing one contract into two was not successful.[68] An arbitration award in favor of an unlicensed contractor was not confirmed by a court.[69]

Another attempt to avoid the licensing requirement for contractors was demonstrated by *Hydrotech Systems, Ltd. v. Oasis Waterpark*.[70] In this case, the claimant was a subcontractor who had furnished goods and services to a prime contractor for a water-oriented amusement park. The claimant contracted to design and construct a surfing pool using wave equipment.

The claimant alleged that it was concerned about its not having a license and originally wished only to sell and deliver its equipment and avoid involvement in design or construction of the pool. The owner insisted, however, that the claimant's unique experience was essential. To induce the claimant to contract for these services, the owner promised that the prime contractor would arrange for a California contractor to work with the claimant on any construction activities that required a license. The claimant asserted that the defendants—the owner and the prime contractor—never intended to honor these promises and that such actions constituted fraud.

While earlier California cases had appeared to allow claims by the claimant (the unlicensed contractor) for fraud, the Supreme Court distinguished those cases in *Hydrotech*. It pointed to the strong public policy expressed in the licensing statute. The court would not allow an unlicensed contractor to

"circumvent the clear provisions and purposes of Section 7031 simply by alleging that when the illegal contract was made, the other party had no intention of performing.[71] The court concluded that the strong public policy evidenced by the licensing laws applied whether the promise by the other party to pay for the work was honest or deceitful.

Circumvention through subterfuge raises ethical and moral questions. But success will depend on the troublesome and complex questions of whether *denial* of recovery for work performed will create unjust enrichment.

I. Substantial Compliance

Increasingly, attempts by unlicensed contractors to recover have been based on asserted "technical" noncompliance with the law. This is not, strictly speaking, a contention that the law has not been violated. As a result, this doctrine is discussed in Section 10.07(B) and, inasmuch as contractors have been the principal users of this approach, in Section 10.10(C).

SECTION 10.07 Sanctions for Licensing Law Violations

A. Criminal and Quasi-Criminal Sanctions

Licensing laws usually carry criminal sanctions. Violations can be punished by fine or imprisonment. However, use of the criminal sanction is relatively rare. Perhaps occasional instances have occurred where fines have been imposed, although it is quite unlikely, though possible, that imprisonment will be ordered. Where a penal violation is found, it is likely that sentence will be suspended or probation granted.[72]

A quasi-criminal sanction can be invoked. If the licensing authorities obtain a judgment from the court ordering that the unlicensed design professional cease practice and the order is disobeyed, failure to comply will be contempt of court and punishable by a fine or imprisonment.

---

[67]*Walker v. Nitzberg*, 13 Cal.App.3d 359, 91 Cal.Rptr. 526 (1970).
[68]*Cochran v. Ozark Country Club, Inc.*, 339 So.2d 1023 (Ala.1976).
[69]*Loving & Evans v. Blick*, 33 Cal.2d 603, 204 P.2d 23 (1949). But see *Parking Unlimited, Inc. v. Monsour Medical Found.*, 299 Pa.Super. 289, 445 A.2d 758 (1982) (award upheld where it appeared that the arbitrator held recovery justified).
[70]52 Cal.3d 988, 803 P.2d 370, 227 Cal.Rptr. 517 (1991).

[71]803 P.2d at 375, 227 Cal.Rptr. at 523.
[72]Restitution as a condition to probation was permitted in *State of Washington v. Bedker*, 35 Wash.App. 490, 667 P.2d 1113 (1983). Usually this is not done, because the defendant does not have protections available in civil actions such as pleading requirements, discovery, and a jury trial. The court ordered restitution to correct only *dangerous* conditions.

Just as law enforcement officials rarely have the staff or resolve to seek criminal sanctions, the regulatory agencies rarely seek a judicial order that the violator is in contempt.

In 1985, the California Code of Civil Procedure § 1029.8 added another quasi-criminal sanction. It *trebles* any actual damages (additional damages cannot exceed $10,000) and allows the court to award attorneys' fees if an unlicensed person negligently "causes injury or damage to another." This sanction does not apply to persons who believe in good faith that they are licensed and to persons who have failed to renew their license when renewal would have been automatic.

Although the statute appears to be directed to personal injury claims, it can be read broadly to include damage to property and even economic losses. (Section 253c of the California Insurance Code was added at the same time, stating that coverage of such a treble damage award cannot be insured against.) Such a statute will encourage smaller claims to be taken to court.

B. Recovery for Work Performed

The infrequent use of criminal or quasi-criminal sanctions means that the principal sanction for unlicensed practice of architecture, engineering, or construction work has been to deny recovery for work performed. Attempts by unlicensed persons to recover for their work has been the main legal battleground.

To determine whether persons not authorized to perform particular services can recover for work performed, recourse first must be made to the statutes that regulate the conduct. The statute may deny the use of the courts to recover for work performed where there has been a violation of the licensing laws. For example, § 7031 of the California Business and Professions Code bars a contractor from bringing an action in any court for recovery under a contract unless it establishes that it was a duly licensed contractor "at all times during the performance of such act or contract."

Explicit statutory treatment of collection rights is rare. As a result, the court must deal with the effect of making a contract to perform illegal services. This requires an examination of the objectives of the statute, the strength of the public policy expressed in the statute, the intended beneficiary of the statute, and whether denial of recovery for services performed under such a contract will create unjust enrichment.

Generally, neither party can enforce an illegal contract. The law provides no help if the parties are equally guilty. If a citizen bribes a public official, the citizen cannot sue to have the promised performance made nor can the public official sue to recover the promised bribe.

Parties protected by the statute are given enforcement rights. For example, suppose an unlicensed architect contracted with the client. The client can enforce the contract against the unlicensed architect and maintain any action for faulty performance.[73] But suppose the unlicensed person has conferred a benefit on the other party. The party performing may have performed in accordance with the contract and seeks payment, not on the illegal contract but based on restitution. Leaving the parties in status quo could cause unjust enrichment.

Clearly, the client is enriched. But is enrichment unjust? For example, if the client *knew* the architect was unlicensed, any enrichment may not be unjust. However, if this factor were sufficient to grant recovery based on unjust enrichment, the legislative purpose expressed through the licensing laws may be frustrated. The unlicensed architect might continue to violate the law, hoping that she either would be paid or would be able to recover through restitution based on unjust enrichment.

Tennessee Code § 62-6-103 recognizes this dilemma. It allows an unlicensed contractor to recover if it can show clear and convincing proof of actual documented expenses.[74] Profits cannot be recovered. Very likely overhead would not be recoverable.

Whether the law should not only refuse to enforce such contracts but also deny recovery for work performed is particularly difficult when illegality results from noncompliance with laws licensing contractors. Suppose there is a contract between an unlicensed subcontractor and a prime contractor who knew the subcontractor was unli-

---

[73]*Hedla v. McCool,* supra note 63; *Domach v. Spencer,* 101 Cal.App.3d 308, 161 Cal.Rptr. 459 (1980) (unlicensed contractor); *Cohen v. Mayflower Corp.,* 196 Va. 1153, 86 S.E.2d 860 (1955).
[74]Interpreted in *Wiltcher v. Bradley,* 708 S.W.2d 407 (Tenn.App.1986).

censed. If the purpose of the law is to protect homeowners who may not be able to determine the competence of contractors, the unlicensed subcontractor should recover. Yet, as shall be seen, there are different judicial attitudes when the contract is made between contractors rather than between a contractor and an owner.[75]

Despite some recent criticism of licensing laws noted in Section 10.02, the laws generally are still received favorably by the courts. Even in the absence of an express statutory provision dealing with the right to collect, unlicensed persons generally are not able to recover either on the contract or on unjust enrichment.[76] (Yet, as seen in (C), one influential state allowed recovery, undoubtedly a reflection both of expanded notions of restitution based on unjust enrichment and of greater skepticism toward licensing laws.)

Two cases that involve design professionals practicing outside the state in which they were licensed—an increasing phenomenon—demonstrate the turmoil in resolving these issues.

In *Markus & Nocka v. Julian Goodrich Architects, Inc.,*[77] the plaintiff was an architect licensed in Massachusetts who specialized in hospital design. The defendant was a Vermont architect retained to perform design services for hospital facilities in Vermont. The hospital directed the Vermont architect to hire a specialist in hospital architecture, and the Vermont architect hired the plaintiff.

The plaintiff made a study of the design needs, consulted with the Vermont hospital staff, and prepared revisions of preliminary sketches and specifications. The plaintiff's staff made numerous trips to Vermont in performing these services. Ultimately, the design recommendations of the plaintiff were not accepted by the hospital staff, and the

project was put out to bid based on the design work of the defendant.

The court held that the Massachusetts architects had violated the Vermont registration laws. Because the architectural contract between plaintiff and defendant violated Vermont law, the contract was illegal and the provision for the payment of compensation was unenforceable. The court noted that the construction was to be undertaken in Vermont and that many visits had been made to the Vermont site, along with consultations with Vermont hospital personnel.

As the consulting contract was illegal, could the plaintiff consulting architects collect for their services? The court held that they could not. In denying recovery the court stated:

> The underlying policy is one of protecting the citizens of the state from untrained, unqualified and unauthorized practitioners.[78]

The court would not apply an exception used by some states that permit consulting architects to recover if their performance was one single, isolated act. The court noted that such an interpretation would weaken the registration laws. The court pointed out that the Vermont registration laws were phrased in broad, positive terms and provided no such exception.

The court also noted that the registration laws for medicine and engineering *specifically* authorized consulting services in Vermont by those properly licensed out of state but that there was no corresponding provision relating to out-of-state architectural consultants. The court noted that Vermont law provided that architects licensed in another state whose standards were not below those of Vermont could be admitted in Vermont without examination.

The court made it clear that not all consultations across state lines on Vermont projects would necessarily violate Vermont registration laws. But the court concluded:

> . . . when the nonresident architect presumes to consult, advise and service, in some direct measure, a Vermont client relative to Vermont construction, he is putting himself within the scope of the Vermont architectural registration law.[79]

---

[75]Compare *Enlow & Son, Inc. v. Higgerson,* 201 Va. 780, 113 S.E.2d 855 (1960); *Triple B Corp. v. Brown & Root, Inc.,* 106 N.M. 99, 739 P.2d 968 (1987) (granted recovery) with *Lewis & Queen v. N. M. Ball Sons,* 48 Cal.2d 141, 308 P.2d 713 (1957) and *Frank v. Fischer,* 108 Wash.2d 468, 739 P.2d 1145 (1987) (denied recovery). See Section 10.10(A).

[76]*Food Management, Inc. v. Blue Ribbon Beef Pack, Inc.,* supra note 62. The decision cites many cases from a number of jurisdictions denying recovery. See also *Southern Metal Treating Co. v. Goodner,* 271 Ala. 510, 125 So.2d 268 (1960). Earlier edition of this treatise cited in *Gerry Potter's Store Fixtures, Inc. v. Cohen,* 46 Md.App. 131, 416 A.2d 283, 285 (1980).

[77]127 Vt. 204, 250 A.2d 739 (1969).

---

[78]250 A.2d at 741.

[79]Id. at 742.

Costello v. Schmidlin[80] involved an action for consultation fees by Costello, who was licensed in New York but not licensed in New Jersey, the state in which the project was being built. Costello was an engineer who specialized in the construction of swimming pools. In addition to being licensed in New York (his principal place of business), he was licensed in Maryland, Illinois, and New Mexico. He agreed to perform consulting services for a New Jersey municipal swimming pool complex through a consulting contract with the New Jersey architect.

Costello's design was used in the bid invitations. However, the initial bids exceeded the cost budget. Costello redesigned, and the project was constructed. There was no assertion that Costello's work did not conform to his contract obligations.

When Costello was not paid, he asserted a legal claim in federal court. The trial court judge, applying New Jersey law, held that the services were performed under an illegal contract because Costello was not licensed in New Jersey, and Costello could not recover.

The appeals court reviewed some decisions in which design professionals had been denied recovery and noted that in most cases the design professional was not licensed in any state. More important, the court emphasized earlier decisions that had noted the distinction between unlicensed persons dealing with members of the public and dealings with persons *within* a profession, such as dealings between design professionals or contractors. Here there was a contract between an unlicensed engineer and a licensed architect, the latter *knowing* that the engineer was not licensed. The court allowed recovery despite the admittedly illegal contract, pointing to the lack of *express* denial of compensation in the licensing laws. (Many cases deny recovery *without* a statute.)

Were recovery based on unjust enrichment, the court should have determined the reasonable value of Costello's services. However, Costello was allowed to recover an amount based on the admittedly illegal contract.

The courts in the *Markus* and *Costello* cases reflect genuine differences of opinion regarding licensing laws. The *Markus* opinion emphasized the need to ensure that all persons who perform design services for Vermont clients on Vermont projects meet Vermont standards of competence. The federal court in the *Costello* case seemed more concerned with doing justice between the two parties than with arriving at a result that might frustrate the operation of the licensing laws.

Perhaps the variant holdings can in some way be traceable to an important factual distinction in the two cases. In the *Markus* case, the unlicensed professional's work was not used. But in the *Costello* case, the work was used and there was no indication that there had been any deviation from Costello's obligations. Just as in the subterfuge cases discussed in Section 10.06(H), courts will be favorably inclined toward the unlicensed person if it appears that that person has performed properly and the other party seems to be using failure to comply with the licensing laws as an excuse for not paying. This observation does not necessarily mean that when these facts occur courts will *always* allow recovery. Obviously, much may depend on the language of the statute and the court's attitude toward the importance of licensing laws. But where it appears that the work was proper and that it was used, many courts will make a serious attempt to allow recovery.

The differing attitudes expressed in the *Markus* and *Costello* decisions and some trends described in (C) do not change the undeniable fact that as stated,[81] the general rule is that unlicensed persons cannot use the courts to recover for work they have performed. Yet a number of issues lie beneath the surface and must be explored.

If the contract is not absolutely void, an owner may be permitted to sue for damages for breach of such a contract.[82] More important, the desire to protect the status quo in licensing law violations was generally thought to bar the innocent party (the owner) from recovering any payments that had been made to the party performing services without a proper registration or license.[83] Payment was thought to mean that the work had been performed properly. Also, it would take egregious illegal conduct to justify recovery of payments that had been made.[84] While it was generally thought

---

[80]404 F.2d 87 (3d Cir.1968).

[81]See note 76 supra.
[82]See note 73 supra.
[83]*Medak v. Cox*, 12 Cal.App.3d 70, 90 Cal.Rptr. 452 (1970).
[84]*Albany Supply & Equip. Co. v. City of Cohoes*, 18 N.Y.2d 968, 224 N.E.2d 716 (1966).

that licensing violations did not fall into that category, some recent cases have allowed the owner to recover payments that had been made.[85] This may signal increased importance of registration and licensing laws and a tougher attitude toward those who perform under such illegal contracts.

Suppose the owner sends a check to the unlicensed party but stops payment on it. Had *payment* been received, it could not have been recovered. A court held that the unlicensed contractor cannot recover, because the payment had not been received.[86]

Some courts give partial relief to the unlicensed person, permitting her to set off compensation for work that has been performed against any claim made against the unlicensed person. For example, suppose an unlicensed architect had furnished services for which she was still owed $10,000. Suppose that the client on an unrelated matter had $15,000 coming from the unlicensed architect. In such a case, if the client brought a claim for the $15,000, some courts would permit the setoff of $10,000 for work performed under the illegal contract to reduce the claim against the unlicensed architect to $5,000.[87]

A claim usually seeks a money award. But other uses can be made of the claim. For example, suppose the unlicensed person files a mechanics' lien based on a statute giving security to the person who has improved another's property. Although it can be asserted that this remedy is given to avoid unjust enrichment, some courts, pointing to the need for a valid contract, have denied the right to impose a lien.[88] Yet one court allowed a lien despite the illegality of the contract when the obligation was restitutionary based on unjust enrichment.[89]

Suppose the unlicensed person and the other party make a settlement. One court enforced the settlement, even though it would not have enforced the *original* illegal contract.[90] The court concluded that the settlement had been made in good faith, analogous to payment.

Suppose the parties to the contract arbitrate the dispute and the arbitrator grants an award to the unlicensed person. Will a court confirm it? One court, stressing the importance of the licensing statute (the case was decided in 1949) refused to confirm the award.[91] Another case, decided in 1982, stressing that the arbitrator could resolve the legal issues, confirmed the arbitration award.[92]

Suppose the claimant presents its claim to the Bankruptcy Court when the other party to the contract has been adjudged a bankrupt. An opinion allowed the claim, for whatever it will be worth, based on the broad discretion given to the Bankruptcy Court.[93] The court was influenced by the bankrupt having been enriched and having acquiesced in continued performance by the claimant *after* the bankrupt knew the performing party was not licensed. (This claim will come at the expense of the other unsecured creditors, *not* of the bankrupt.)

The special problems of contractor licensing laws are treated in Sections 10.08–10.11. At this point, the substantial compliance doctrine, developed in claims by contractors whose violations were technical (such as failure to renew because of an office manager's mental breakdown[94]), merits mention. The case of *Wilson v. Kealakekua Ranch, Limited,*[95] though not *expressly* discussing substantial compliance, dealt with failure to renew, a staple in claims of substantial compliance.

Wilson sued the ranch for architectural services performed. Though Wilson testified he was licensed, cross-examination revealed he had failed to pay the renewal fee of $15. On that basis, the trial

---

[85]*Ransburg v. Haase,* 224 Ill.App.3d 681, 586 N.E.2d 1295 (1992) (to protect owners and discourage architects); *Mascarenas v. Jaramillo,* 111 N.M. 410, 806 P.2d 59 (1991) (consumer knew contractor was not licensed).
[86]*Vedder v. Spellman,* 78 Wash.2d 834, 480 P.2d 207 (1971).
[87]*Sumner Dev. Corp. v. Shivers,* 517 P.2d 757 (Alaska 1974); *S & Q Constr. Co. v. Palma Ceia Dev. Org.'n,* 179 Cal.App.2d 364, 3 Cal.Rptr. 690 (1960).
[88]*Sumner Dev.t Corp. v. Shivers,* supra note 87; *Chickering v. George R. Ogonowski Constr. Co., Inc.,* 18 Ariz.App. 324, 501 P.2d 952 (1972).
[89]*Expert Drywall, Inc. v. Brain,* 17 Wash.App. 529, 564 P.2d 803 (1977). Cf. *Bastian v. Gafford,* 98 Idaho 324, 563 P.2d 48 (1977).

[90]*Shelton v. Grubbs,* 116 Ariz. 230, 568 P.2d 1128 (1977).
[91]*Loving & Evans v. Blick,* supra note 69.
[92]*Parking Unlimited, Inc. v. Monsour Medical Found.,* supra note 69.
[93]*In re Spanish Trails Lanes, Inc.,* 16 B.R. 304 (Bkrtcy.Ariz.1981).
[94]*Latipac, Inc. v. Superior Court,* 64 Cal.2d 278, 411 P.2d 564, 49 Cal.Rptr. 676 (1966).
[95]57 Hawaii 124, 551 P.2d 525 (1976).

court granted the ranch's motion to dismiss Wilson's claim.

First, as is common when the court is inclined to grant recovery, the appellate court pointed to the absence of any *specific* legislative language precluding recovery. Next, it distinguished statutes to raise revenue from those "for protection of the public agent against incompetence and fraud." Recovery is usually denied in the latter cases.

According to the court, however, the true issue is one of legislative intent. The court concluded that the legislature must not have intended to create a forfeiture out of proportion to the offense that would benefit the other party. Here the penal sanction was a fine not to exceed $500 or imprisonment for one year. Denying recovery would mean a forfeiture of $34,000 by Wilson gained by the ranch.

The court pointed to Wilson's having been licensed and no indication that had he reapplied there would have been reinvestigation or reexamination; this according to the court, shows that *renewal* is simply to raise revenue. Renewal is automatic unless a charge has been filed against the architect. Even if this were not true, the court returned to denial of fees as disproportionate and punitive, noting that Wilson is still exposed to the penal sanctions.

True, the *Wilson* decision is not a frontal attack on the basic rule *denying* liability. But its determination that the statutory requirement for renewal is to raise revenue and its highlighting the absence of any direct command from the legislature to *deny* fees must be taken as a sign that claims for recovery by unregistered persons are to be given respectful judicial consideration.

C. A Checkerboard Future

The influential Supreme Judicial Court of Massachusetts seems to be moving in the direction of allowing recovery to unlicensed persons. *Town Planning & Engineering Associates, Inc. v. Amesbury Specialty Co., Inc.* involved a contract for design services made by an unlicensed designer who had been approached by a client. The unlicensed designer retained two registered professional engineers as well as a registered land surveyor to perform services related to the project. However, much of the work was done by the designer and his six drafters. The project was ultimately abandoned when the bids were too high. At first the client told the unlicensed designer to renegotiate

the bids or get lower ones, but shortly thereafter the client terminated the contract, claiming it was illegal. Subsequently, the client negotiated directly with a contractor, obtained a reduced bid for a somewhat scaled-down project, and to some degree used the unlicensed designer's plans.

The court noted that the plaintiff's work had been properly performed. In holding that the plaintiff could recover, the court stated:

> If there was a violation here, it was punishable as a misdemeanor under the statute. Violation of the statute, aimed in part at least at enhancing public safety, should not be condoned. But we have to ask whether a consequence, beyond the one prescribed by statute, should attach, inhibiting recovery of compensation, and we agree with the judge in his negative answer to the question in the present case. To find a proper answer, all the circumstances are to be considered and evaluated: what was the nature of the subject matter of the contract; what was the extent of the illegal behavior; was that behavior a material or only an incidental part of the performance of the contract (were "the characteristics which gave the plaintiff's act its value to the defendant . . . the same as those which made it a violation . . . of law"); what was the strength of the public policy underlying the prohibition; how far would effectuation of the policy be defeated by denial of an added sanction; how serious or deserved would be the forfeiture suffered by the plaintiff, how gross or undeserved the defendant's windfall. The vector of considerations here points in the plaintiff's favor.

Our cases warn against the sentimental fallacy of piling on sanctions unthinkingly once an illegality is found. As was said in *Nussenbaum v. Chambers & Chambers, Inc.*, 322 Mass. 419, 422, 77 N.E.2d 780, 782 (1948): "Courts do not go out of their way to discover some illegal element in a contract or to impose hardship upon the parties beyond that which is necessary to uphold the policy of the law." Again the court said in *Buccella v. Schuster*, 340 Mass. 323, 326, 164 N.E.2d 141, 143 (1960), where the plaintiff was allowed to recover his compensation for blasting ledge on the defendant's property although he had not given bond or secured a blasting permit as required by law: "We do not reach the conclusion that blasting without complying with the requirements . . . is so repugnant to public policy that the defendant should receive a gift of the plaintiff's services." Professor Corbin adds: "The statute may be clearly for protection against fraud and incompetence; but in very many cases the statute breaker is neither fraudulent nor incompetent. He may have rendered excellent service or delivered goods of the highest quality, his non-compliance with the statute seems nearly harmless, and the real defrauder seems to be the defendant who is enriching himself at the plaintiff's expense. Although many courts yearn for a mechanically applicable rule, they have not made one in the present instance. Justice requires that the penalty should fit the crime; and justice and sound policy do not always require the enforcement of licensing statutes by large forfeitures going not to the state but to repudiating defendants." 6A A. Corbin, Contracts § 1512, at 713 (1962) (footnote omitted).[96]

While Massachusetts courts may be more willing to allow recovery to unregistered design professionals and unlicensed contractors under certain circumstances, courts in Illinois and New Mexico, as noted in Section 10.07(B), thought so highly of registration laws that they went beyond denying recovery to allow recovery of payments made to the unregistered design professional.[97]

Another factor makes spotting the trend hazardous. A constant tension exists between legislatures which desire strict enforcement of licensing laws and courts that tend to be more tolerant of human frailties and less willing to allow someone to receive services without paying for them. As seen in Section 10.10(C), the California legislature made clear in 1989 that it no longer wished courts to apply the substantial compliance doctrine after a number of court decisions over a period of more than twenty years had done so.

While Massachusetts indicates that inroads have been made on the general rule denying recovery, it is still common to see courts in different states come to different conclusions when claims are brought by unregistered designers[98] and unlicensed contractors.[99]

D. Summary

Although it must still be generally stated that unlicensed persons are not able to recover based on either the contract they have made or unjust enrichment, some factors may motivate a court to determine that the statute was not violated, a subterfuge will be given effect, substantial compliance is sufficient, or recovery will be had despite a licensing violation. Such factors are as follows:

1. The work for which recovery is sought conforms to the contract requirements.
2. The party seeking a defense based on a violation of the licensing law is in the same business or profession as the unlicensed person.
3. The unlicensed person apparently had the qualifications to receive the license.

SECTION 10.08 Should Contractors Be Licensed?

A. Purpose of Licensing Laws

To protect the public, the state regulates professions, occupations, and businesses. Those who wish to build must retain a contractor and often a designer. Licensing laws should provide a representation to the public that the holder meets a minimal level of competence, honesty, and financial capacity. The state wishes to protect the general public from being harmed by poor construction work and seeks to accomplish this by allowing only those who meet state requirements to design and to build. Yet as seen in this chapter, these efforts are not costless. Educational and experience requirements and the administration of these regulatory systems increase the cost of providing these services. They also reduce the pool of contractors.

Professional designers usually must meet certain educational and training requirements before they are allowed to take the examination that will determine whether they will be registered. Although these requirements do raise the cost of entering the profession, it can be contended that they improve the quality of designers.

As a rule, contractors need meet only experience requirements and not educational ones. (The California application fee cannot exceed $150.[100]) In

---

[96]369 Mass. 737, 342 N.E.2d 706, 711–712 (1976).

[97]See *Ransburg v. Haase,* supra note 85; *Mascarenas v. Jaramillo,* supra note 85. Similarly, see *Kansas City Community Center v. Heritage Industries, Inc.,* 972 F.2d 185 (8th Cir.1992) (no benefit to client and law designed to protect client).

[98]See *Tucker v. Whitehead,* 155 Ga.App. 104, 270 S.E.2d 317 (1980) (dictum denied right to recovery). But see *West Baton Rouge Parish School Bd. v. T. R. Ray, Inc.,* 367 So.2d 332 (La.1979) (cannot arbitrate but can recover based on the unjust enrichment if contract made in good faith; corporate architect's licensed employee quit after contract made).

[99]See *Revis Sand & Stone, Inc. v. King,* 49 N.C.App. 168, 270 S.E.2d 580 (1980) (denial). But see *C. B. Jackson & Sons*

Constr. Co. v. Davis, 365 So.2d 207 (Fla.Dist.Ct.App. 1978) (recovery based on unjust enrichment).

[100]West Ann.Cal.Bus. & Prof.Code § 7137.

California, for example, they must have four years of experience in the trade they wish to enter. Under certain circumstances, particular education courses taken can fulfill up to three years of this four-year experience requirement. Applicants in California take a one-day written examination, half of which is devoted to law and the other half to trade practices. As seen in the Florida experiences described in Section 10.02, the "pass" line can be drawn arbitrarily. Can examiners determine what is a reasonable level of classroom performance? Can the classroom simulate actual construction? Even these minimal examination requirements can be waived under certain circumstances under the California Business and Professions Code § 7065.1.

How will the examiners determine the integrity of the person who applies for a license? This is undoubtedly accomplished by letters of reference (of dubious value) and perhaps other investigations. Yet will investigations uncover much that will reflect the integrity of the contractor? How will the regulatory agency determine financial responsibility? In California, financial solvency requires a working capital that exceeds $2,500.[101] Until 1980, an applicant had to post a bond for $2,500 in California. Effective in 1980, the applicant must post a bond of $5,000 ($10,000 if a swimming pool contractor).[102] Is this bond, which covers more than complaints by homeowners, adequate to establish financial capacity?

California did augment these relatively modest bond requirements. The California Business and Professions Code requires a contractor to post a bond that equals any unsatisfied final judgment against the contractor that is substantially related to the qualifications, functions, or duties of the license applied for.[103] Again, respecting the supremacy of the federal bankruptcy law,[104] if there has been a discharge of the unsatisfied judgment in a bankruptcy proceeding, the bond need not be furnished.[105] Finally, under Section 7071.8, a contractor who has been disciplined by having had its license suspended or revoked can be forced to file an additional bond of not less than $15,000 or ten times the ordinary bond if it seeks to have its license restored.[106]

What must a contractor do to keep a license in effect? In California, the license must be renewed every two years.[107] The cost of renewal for an active contractor shall not exceed $200, and renewal requires posting of a bond.[108] An inactive contractor can renew for $75.[109] There is no provision for periodic examination or continuing education.

Are the undeniable costs that such a system imposes more than balanced by helping *those who need help* to determine who is competent? This raises additional questions. Who needs the protection, and who gets it? Clearly, such legislation was intended to protect those who are inexperienced in the world of construction or who cannot without great cost obtain sufficient information to make judgments. Owners that fall into either category are more likely to make judgments based on whether a contractor is licensed and bonded.

Perhaps the inexperienced owner building a residence, a duplex, or a four-unit apartment or the naive owner about to engage in home improvement can be helped by the existence of licensing laws. But does that justify an across-the-board licensing requirement and an expensive licensing apparatus? Particular abuses can be dealt with by already abundant consumer legislation.

In addition to these questions, does this system deter unlicensed persons from building? As seen in Section 10.10, often an unlicensed contractor *can* use the courts to recover for services performed.

B. Harmful Effects

Suppose contractor licensing laws neither screen out the incompetent nor only protect those who need protection. Do they do any harm?

Contractor licensing laws can artificially reduce the pool of contractors, as demonstrated by the Florida experience described in Section 10.02.

Contractor licensing statutes usually draw lines between general contractors and the specialized trades. In addition to the difficulty of drawing these lines, the segregation of specialty trades by licensing tends to reduce the likelihood of better

---

[101]West Ann.Cal.Bus. & Prof.Code § 7067.5.
[102]Id. at § 7071.6.
[103]Ibid.
[104]See § 10.04(B).
[105]West Ann.Cal.Bus. & Prof.Code § 7071.6.

[106]Id. at § 7071.8.
[107]Id. at 7140.
[108]Id. at 7137.
[109]Ibid.

organized, more efficient organizational structures, something so needed in construction.

Other dangers exist. It is not uncommon for licenses to be "bought" and "sold" as any other valuable commodity. An owner who wishes to build a residence or small commercial structure and avoid the cost of a full-time prime contractor sometimes engages an individual with a license, gives that person very little overall control, and limits her activity to periodic visits. This is done to avoid the owner's being liable for violating the licensing laws and gives a licensed contractor a saleable commodity for a relatively small investment.

The issuance of a license can be a false representation by the state that may deceive the unwary consumer, that is, state-sponsored "consumer fraud." Consumers may believe they are dealing with a competent, honest, and financially reliable builder because the builder is licensed. To obtain business from a gullible public, many builders advertise that they are licensed and bonded. Elimination of contractor licensing laws might induce consumers to protect themselves.

SECTION 10.09 Contractor Licensing Laws

Along with the general proliferation in occupational licensing, there has been an increase in contractor licensing laws. At present about one half of the states have such laws. Perhaps the deregulation spirit seen briefly in Arizona[110] may signal a slowdown or even rollback. But ideological movements such as deregulation often lack the staying power of organized groups, such as consumer groups and contractors, both prime movers in occupational licensing.

Yet inevitably, licensing laws demonstrate the trade-offs inherent in the political process. Even though in 1987 Arizona went back to a traditional contractor licensing law, it specified fifteen exemptions. For example, owners who improve their own property by themselves or jointly with a licensed contractor are exempt if the structure is not intended for sale or rental. Similarly, employees of

owners of condominiums, townhouses, cooperatives, or apartments of four units or fewer need not have a contractor's license. Also, a surety company that undertakes to complete a contract under the terms of a bond needs no license, provided the work is performed by licensed contractors.[111]

The sanctions imposed on the unlicensed contractor are discussed in detail in Section 10.10. However, many states have statutes that specifically bar an unlicensed contractor from recovering for work performed, something usually not found in architectural and engineering registration laws.

Yet, some legislatures have sought to induce contractors to become licensed by granting benefits only to licensed contractors. For example, California permits only a *licensed* contractor to hold a supplier of goods to any orally communicated price proposal made under certain circumstances.[112]

Contractor licensing laws have also regulated the formation and performance of construction contracts. For example, in California, a contractor must notify the owner of lien law provisions and the person to whom complaints may be made.[113] Home improvement contracts must contain a notice that the owner or tenant has a right to require the contractor to post a performance and payment bond.[114] The contractor must notify the owner that the contractor's failure to begin work under a home improvement contract within twenty days of the agreed date without excuse violates the state licensing laws.[115]

California licensing laws contain grounds for disciplinary action against the contractor that appear to be no more than simple breach of contract. For example, the following can be grounds for disciplinary action:

1. abandonment of project[116]
2. willful departure from plans[117]

---

[110]Refer to Section 10.02(B). Until 1987, a license had been needed *only* for residential and small apartment construction. Ariz.Rev.Stat.Ann. § 32-1101.

[111]Ariz.Rev.Stat.Ann. § 32-1121.
[112]West Ann.Cal.Comm.Code § 2205.
[113]See West Ann.Cal.Bus. & Prof.Code § 7018.5. Failure to comply with a predecessor statute did not bar the contractor from recovery. *Gonzales v. Concord Gardens Mobile Home Park, Ltd.*, 90 Cal.App.3d 871, 153 Cal.Rptr. 559 (1979).
[114]West Ann.Cal.Bus. & Prof.Code § 7159(f).
[115]Id. at § 7159(j).
[116]Id. at § 7107.
[117]Id. at § 7109.

3. material failure to complete project for price stated[118]
4. failure to prosecute work diligently[119]

These provisions cause public intervention into relations between owner and contractor (discussed in greater detail in Section 10.11).

SECTION 10.10 The Unlicensed Contractor: Civil Sanctions

A. Recovery for Work Performed

The chief sanction for violating licensing laws has been to bar the use of the courts to collect for services.[120] Yet for reasons to be noted in (E), denials of recovery, while still common, are less likely when a contractor seeks judicial relief than when such relief is sought by a design professional. This occurs *despite* the specific legislative direction in many contractor licensing laws to bar the state courts to unlicensed contractors.

These specific legislative directives, such as found in California or Washington, do *not* mean contractors *never* recover. Such a harsh sanction has led these states to create a substantial compliance doctrine to avoid unjust enrichment (discussed in (C)). Courts have divided on the use of such laws as a defense by a knowledgeable participant in the construction industry.[121] The recent cases show the usual variant results when unlicensed contractors seek recovery.[122]

Unwillingness by the courts in some states to impose the harsh sanction of barring recovery for work performed may have led Tennessee to enact

a statute that permits an unlicensed contractor to recover if it can show clear and convincing proof of actual documented expenses.[123]

The construction industry, particularly the portion that consists of contractors and subcontractors, witnesses chronic financial difficulties and frequent bankruptcies. The tension between protecting consumers from financially insecure or irresponsible contractors and the policy in the federal bankruptcy law to give debtors a fresh start has led to disputes over whether bankruptcy is grounds for taking away the contractor's license. As noted in Section 10.04(B), this conflict has been resolved in favor of the bankruptcy policy's permitting a fresh start.

B. Exceptions

Barring the unlicensed contractor from using the courts to recover for services it has performed does not mean that the unlicensed contractor cannot use other avenues. Section 10.07 noted that at least in some states an unlicensed person or entity can make a settlement and have it enforced, obtain an arbitration award and have it enforced, or present a claim in a bankruptcy proceeding. Most of those cases involve unlicensed *contractors.*

In addition, an unlicensed subcontractor has been allowed to sue a prime contractor for delay damages in states that bar the unlicensed contractor from using the courts to recover under a contract.[124] The reasoning of these cases—that recovery is *not* sought for *performance*—should allow an unlicensed *prime* contractor to recover from an owner for delay damages.

Other cases demonstrate the ambivalence of courts when asked to deny recovery to unlicensed contractors who have done their work properly. For example, courts have divided as to the effect of noncompliance when the dispute does not involve an ordinary member of the public but involves those knowledgeable about construction.[125]

Another evidence of the greater solicitude shown unlicensed contractors is reflected by *Moore*

---

[118]Id. at § 7113. Failure need not be willful to constitute grounds for disciplinary action. *Mickelson Concrete Co. v. Contractors' State License Bd.*, 95 Cal.App.3d 631, 157 Cal.Rptr. 96 (1979).

[119]West Ann.Cal.Bus. & Prof.Code § 7119.

[120]Another California sanction is trebling the award against an unlicensed contractor and giving the court discretion to award attorneys' fees. See Section 10.07(A).

[121]See Section 10.07(B), note 75.

[122]Compare *C. B. Jackson & Sons Constr. Co. v. Davis*, 365 So.2d 20 (Fla.Dist.Ct.App.1978) (recovery), and *Lignell v. Berg*, infra note 128 (sophisticated owner), with *Cochran v. Ozark Country Club, Inc.*, 339 So.2d 1023 (Ala.1976) (had been paid $35,000, sought $28,000 more: denied); *United Stage Equipment, Inc. v. Charles Carter & Co.*, 342 So.2d 1153 (La.App.1977) (paid all but $1,300: denied); *Revis Sand & Stone, Inc. v. King*, 49 N.C.App. 168, 270 S.E.2d 580 (1980) (need to protect public).

[123]Tenn.Code Ann. § 62-6-103.

[124]*American Sheet Metal, Inc. v. Em-Kay Eng'g Co.*, 478 F.Supp. 809 (E.D.Cal.1979); *Gaines v. Eastern Pac.* 136 Cal.App.3d 679, 186 Cal.Rptr. 421 (1982).

[125]See Section 10.07(B); supra note 75.

v. Breeden.[126] The applicable law required that the owner give notice to the contractor that a license is required before the owner can defend any claim made against it based on the contractor's having been unlicensed. It was held that this notice requirement applies even if the contractor knew it was required to be licensed. This interpretation allowed the court to grant a recovery of almost $300,000 to an unlicensed contractor.

A similar judgment in favor of the claimant was based on the claimant's not being a contractor at all but simply a foreman who had been engaged by the owner to perform construction work.[127]

These cases should not convey the impression that courts simply disregard contractor licensing laws. They do show the difficulty *some* courts have had in reconciling the legislative requirement for a license with the merits of a particular claim, the denial of which would appear to create unjust enrichment.

C. Substantial Compliance

A number of states recognize the substantial compliance doctrine. In these states, excusable errors by the contractor relating to obtaining or renewing a license will not preclude a recovery.[128] The California experience is instructive.

In *Latipac, Inc. v. Superior Court,*[129] failure to renew a license was due to the inadvertence of an office manager who had had a mental breakdown. California has struggled with this doctrine but on the whole has used it to allow recovery where it appears that the work was properly performed and failure to obtain a license did not relate to competence but to bureaucratic error.[130]

Finally, the California Supreme Court decided *Asdourian v. Araj*[131] in 1985. The contractor had made a contract in his own name instead of the business under which he had obtained his license. Nor had he made a written contract for home improvements as required by California law. Yet the court allowed him to recover by a liberal use of the substantial compliance doctrine and a narrow reading of the statutory requirements for a written contract. The court, over one dissenting opinion, showed little respect for the public policy supposedly behind contractor licensing laws. It also scolded intermediate appellate courts, which had not made broad use of the substantial compliance doctrine and which were more inclined to follow the literal meaning of the statutes.

The court noted that the substantial compliance doctrine had been followed for almost five decades without any response by the legislature. The legislature took up the challenge in 1989, amending Section 7031 of the California Business and Professions Code to make clear that the doctrine of substantial compliance does not permit an unlicensed contractor to recover for its services.[132]

Those jurisdictions that are quick to employ the substantial compliance doctrine are inclined to see contractor licensing laws as traps for unwary contractors who perform proper work and then face what appears to be a technicality that may bar them from recovering for their work. Yet those same jurisdictions are often emphatic in their desire to protect consumers from those who might prey on them. Again, as noted in Section 10.07(B), the absence of a license may not bar a contractor from recovering where it appears that it did not mislead the owner and did its work properly.

D. Interstate Transactions

One difficulty generated by the sporadic adoption of contractor licensing laws—not nearly as universal as professional registration laws—is illustrated by *Conderback, Inc. v. Standard Oil Co. of California,*[133] a case that involved the extraterritoriality of the California licensing statute. The contractor was not licensed in California, the state in which its principal place of business was located. It sought to use the California courts to recover for services per-

---

[126]209 Va. 111, 161 S.E.2d 729 (1968).

[127]*Sobel v. Jones,* 96 Ariz. 297, 394 P.2d 415 (1964).

[128]*Jones v. Short,* 696 P.2d 665 (Alaska 1985); *Lignell v. Berg,* 593 P.2d 800 (Utah 1979); *Murphy v. Campbell Investment Co.,* 79 Wash.2d 417, 486 P.2d 1080 (1971); *Expert Drywall, Inc. v. Brain,* 17 Wash.App. 529, 564 P.2d 803 (1977). For Washington law, see Comment, 14 Gonz.L.Rev. 647, 659–661 (1979). For a case vehemently rejecting the substantial compliance doctrine, see *Brady v. Fulghum,* 309 N.C. 580, 308 S.E.2d 327 (1983).

[129]64 Cal.2d 278, 411 P.2d 564, 49 Cal.Rptr. 676 (1966).

[130]See *Vitek, Inc. v. Alvarado Ice Palace, Inc.,* 34 Cal.App.3d 586, 110 Cal.Rptr. 86 (1973).

[131]38 Cal.3d 276, 696 P.2d 95, 211 Cal.Rptr. 703 (1985).

[132]Cannistraci, *An Unlicensed Contractor Cannot Sue in California to Recover the Value of Construction Services,* 11 Constr. Lawyer No. 3, August 1991, p. 21.

[133]239 Cal.App.2d 664, 48 Cal.Rptr. 901 (1966).

formed in Washington, which at that time did not have a licensing law. California requires that a contractor be licensed before maintaining an action for the performance of construction services. But California would not bar its courts to the unlicensed contractor, as that would be applying California law to a Washington transaction. Since Washington—the state with the greater interest—had no licensing law, California would not impose its policy in this transaction.

Two cases noted in Section 10.07(B) came to contrary results when dealing with the collectibility for design services where a professional was licensed in one state but performed work in another where he was not licensed. Two other cases involving a similar problem for interstate contractors denied recovery, one over a strong dissent.[134]

E. Observations

The contractor licensing laws and cases that interpret them demonstrate different attitudes from those that involve architects, engineers, and surveyors. More appears to be expected of design professionals than of contractors. Contractors are often small businesses that may not be aware of the existence of licensing laws and may not have the staff to ensure that they are always in compliance. The recent enactment of many contractor licensing laws and the difficulty of formulating standards for determining competence may play a part in some judicial reluctance to bar a contractor. Even in cases where contractors *are* precluded from recovery, generally all that has been lost has been part payment, with contractors often having collected most of their compensation.

SECTION 10.11 Indirect Effect: Forum for Consumer Complaints

Some make a case for contractor licensing laws by pointing to the power they give the consumer-homeowner when a contractor is not responsive to complaints that work has been done poorly. A

criminal sanction, though rarely imposed, can be the basis for restitution, at least in aggravated cases.[135] More important, a threat to go to the licensing authorities may provide an incentive for the contractor to take the complaint seriously. If the dispute cannot be resolved amicably, there is likely to be an informal hearing before a representative of the state licensing agency. In effect, these agencies have become small claims courts.

The consumer protection aspect of contractor licensing would be more effective if licensing agencies made available a record of complaints against particular contractors.

Although perhaps not typical, California's experience is instructive. In 1981, California permitted the Registrar of Contractors to make public information related to complaints on file against a licensed contractor but not those complaints that were resolved in favor of the contractor. The legislature, recognizing the controversial nature of such a law, used a "sunset" approach by stating that the law would expire on July 1, 1985, unless reenacted by the legislature.[136] The legislature did not reenact the law in 1985 but did so in 1986.

This indirect function has its dangers. It can provide too powerful a tool to an owner who uses it in an abusive way to obtain more than promised under the contract or required by law. The system can be costly to operate, particularly where a high level of consumer dissatisfaction exists. This induced California to require arbitration of such disputes.[137] This was enacted in 1979[138] and strengthened in 1987.[139]

Currently, the Registrar of Contractors can, with the agreement of both parties, refer the alleged violation in any dispute based on it to arbitration under certain circumstances. To do so, the licensee must not have had a history of repeated or similar violations or any outstanding disciplinary actions filed against it, and the claim must be between $500 and $25,000.

Again, reflecting the heavy use of the complaint system since 1987, California permits (funds avail-

---

[134]*Meridian Corp. v. McGlynn/Garmaker Co.*, 567 P.2d 1110 (Utah 1977) (3 to 2); *Jary v. Emmett*, 234 So.2d 530 (La.App.1970), writ refused 256 La. 374, 236 So.2d 502 (1970). But see *Johnson v. Delane*, 77 Idaho 172, 290 P.2d 213 (1955) (engineer licensed in Washington recovered for design of Idaho project).

[135]*State of Washington v. Bedker*, 35 Wash.App. 490, 667 P.2d 1113 (1983) (probation conditioned on payment of amount to remedy defects affecting safety). Refer to Section 10.07.
[136]West Ann.Cal.Bus. & Prof.Code § 7124.6.
[137]Id. at § 7085.
[138]1979 Cal.Stat., ch. 1013, § 16.
[139]1987 Cal.Stat., ch. 1811, § 3.

able) the licensing board to contract with licensed professionals—defined as including engineers, architects, landscape architects, and geologists—to conduct a site investigation of consumer complaints. In enacting this statute, the legislature declared that the licensing board had failed to serve the needs of consumers who made complaints against licensed contractors. One reason was the failure by the board to communicate with consumers and manage its workload effectively. The legislature also noted a tremendous backlog of uninvestigated consumer complaints (increasing on a daily basis). Lack of accessibility and complaint backlog, according to the legislature, precluded consumers from recovering their losses and correcting damages to their homes or businesses. This legislative declaration of policy seems to ignore the availability of courts.

SECTION 10.12 The Trained but Unregistered Design Professional: Moonlighting

A. Unlicensed Persons: A Differentiation

This chapter has spoken of persons who violate the registration for licensing laws as if all of those who do so can be placed in the same category. Yet differentiation can be made between those who violate these laws who have been educated and trained as design professionals and others, such as contractors, developers, design-builders, or self-styled handypersons.

Ordinarily, a long period elapses between the beginning of architecture or engineering training (architecture will be used as an illustration) and registration. As a result, there are many persons with substantial education and training as design professionals but who for various reasons do not work under the supervision of a registered architect yet engage in the full spectrum of design services. To be sure, the law does not differentiate these persons from those without design education and training. Yet that these persons often do perform these services in violation of the law demonstrates not only the financial burden involved in education and training before registration but also that a market exists for such services.

This section is directed toward some of the legal problems faced by what is referred to as the moonlighter, who is likely to be a student, a recent grad-

uate, or a teacher. Some of the discussion anticipates material to be discussed in greater detail in the balance of the treatise, such as professional liability, exculpation, and indemnification. However, a focus on the range of legal problems that such individuals face is also useful.

B. Ethical and Legal Questions

A threshhold issue before the discussion of collection for services performed by moonlighters (C) and liability of moonlighters (D) is whether such work should be performed. A differentiation will be drawn between work by a moonlighter that does not violate the law because the project is exempt and the requirement that a registered designer perform services that do violate the registration laws. The former course raises no ethical or legal questions; the latter does.

The discussion in this section should not be taken as a suggestion that moonlighters should violate the law by engaging in design services prohibited by law. However, two factors necessitate discussion of moonlighting. First, arrangements between clients and moonlighters demonstrate that each group feels it will benefit from such an arrangement or such arrangements would not be made. That there is a good deal of moonlighting demonstrates either that the registration laws only make illegal conduct that should not be prohibited or great disrespect for the registration laws.

Second, the many cases that involve attempts by unlicensed persons to recover for services that they have performed and the periodic success of their claims indicate some dissatisfaction with the registration laws even by those charged with the responsibility of enforcing them indirectly by denying recovery for services performed in violation of them.

Anyone tempted to moonlight and violate registration laws must take into account—in addition to the risk of criminal and quasi-criminal sanctions—the risks discussed in the balance of this section—going unpaid for services rendered or being wiped out financially because of claims by the client or third parties.

C. Recovery for Services Performed

Clearly, the safest path is to design only those projects exempt from the registration laws. Undertaking nonexempt projects runs substantial risks. This subsection looks at these risks and ways of minimizing the likelihood that the moonlighter will not be able to recover for services performed.

Because payments made generally cannot be recovered by the client,[140] it is best to specifically provide that interim fee payments will be made and to make certain that they are paid. If payment has not been made in full, it is possible at least in some jurisdictions to recover for services in a restitutionary claim based on unjust enrichment. Such a claim is more likely to be successful if the client is aware of the moonlighter's nonregistered status, if the moonlighter has had substantial education and training in design, if the work appears to have been done properly, and if it appears that the client has benefited by a lower fee. The stated purpose of registration laws is to deter unqualified persons from performing design services and to protect the public from hiring unqualified professionals. A client who is aware of the status of the designer engaged has given up protection accorded by law, and to allow the client to avoid payment where the services have been properly rendered can create unjust enrichment.

If the project will expose the general public to the risk of physical harm, it is more difficult to justify either entering into such an arrangement or allowing recovery for services performed under one. But as a practical matter, this is not the type of project that moonlighters will design. The uncertainty of recovery and the high cost of litigation even if there is a recovery should deter moonlighters as a whole from doing nonexempt work. If the deterrent function will be accomplished, assuming that is one of the desirable results of these laws, there appears to be no reason to deny recovery where services have been performed by someone with the requisite education and training.

D. Liability Problems

This discussion assumes that the moonlighter will not be carrying professional liability insurance. As a result, any liability that moonlighters incur will be taken out of their pockets if there is anything in their pockets to take.

As shall be seen in Section 14.05, the professional is generally expected to perform as others in the profession would have performed. Should this standard be applied when a moonlighter is knowingly engaged by a client?

Differentiation must be made between an express agreement dealing with a standard of performance and one that is implied by law. Suppose the parties agree that the standard of performance will be greater or lesser than that usually required of a *registered* design professional. Would the court enforce any agreement under which the standard would be less than that required by a registered professional? This depends on a balancing of laws favoring autonomy, which would enforce such an agreement, and the strength of the policy expressed in the registration laws, which would deny enforcement of such an agreement. Although the resolution of such a question would not be easy, it is likely that the law would not give effect to an express agreement under which someone performing services in violation of law would be judged by some standard less than would be used in the event that the person doing the design were registered.

If no express agreement is made regarding the standard of performing, a case can be made for judging the performance by what the parties are likely to have intended. If the client knowingly engaged a moonlighter who received a lower fee, should the client get what she pays for—a lower level of performance for a lower price? However, again it is unlikely that the court in determining the implied terms of such an illegal agreement would prefer autonomy to a result that might appear to frustrate the registration laws.

Another approach, related but somewhat different from the one discussed in the preceding paragraphs, would be for the moonlighter to seek to persuade the client to exculpate the moonlighter from the responsibility from any performance that would violate the contract or to limit the moonlighter's exposure to a designated portion for the entire fee. Because by definition the moonlighter is not insured, the moonlighter may be able to persuade the client that she would not be in a position to pay for any losses that she caused, and this is taken into account in setting the fee.

Suppose the client will agree to exculpation or liability limitation. The moonlighter must take particular care to express these provisions very clearly,

---

[140]As a general rule, payments made under an illegal contract cannot be recovered by the payor; however, several recent cases have allowed the payor to recover payments that had been made. See, for example, *Ransburg v. Haase*, 224 Ill.App.3d 681, 586 N.E.2d 1295 (1992); *Mascarenas v. Jaramillo*, 111 N.M. 410, 806 P.2d 59 (1991).

as courts will at the very least construe them against the moonlighter. Even more, there is a risk that a court that feels strongly about the registration laws will not enforce such a clause if it would frustrate those laws. Keep in mind, however, that because the moonlighter is not likely to have either insurance or assets, a claim by the client is likely to be rare.

Third-party liability raises different problems, inasmuch as any arrangement for exculpation or a liability limitation between the moonlighter and the client will not affect the rights of third parties. But again, the moonlighter without insurance or assets is unlikely to be sued by a third party. A moonlighter concerned about the liability to third parties either because of moral considerations or because the moonlighter does have assets should attempt to seek indemnification from the owner. But as shall be seen, such clauses are also narrowly interpreted and in some states are unenforceable.[141]

From the standpoint of liability either to the client or to third parties, it does not appear that there is a great risk if moonlighters do not have assets of their own. However, if moonlighters do have assets or if they are concerned about causing loss to someone for which they ought to pay, the liability problems of moonlighters are likely to mean that the moonlighters will be judged by the standard of registered professionals.

---

[141]See Sections 32.05(D) and (E).

PROBLEMS

1. A was registered as a licensed architect in the state in which she had her offices. In that state, licenses must be renewed every five years. Renewals are usually granted unless a large number of complaints have been made with the licensing board.

Three months before A's license was required to be renewed, A entered into a written contract for the performance of design services for C. A was paid approximately 60% of her fee through interim payments. When the time came for final billing after the services were completed, A realized that she had forgotten to renew her registration. She called the licensing board, and it informed her that she would have been renewed had she applied at the proper time; A then took steps to renew her registration.

However, C discovered that A had not renewed her registration and now refuses to pay the balance of the fee. She also demands that A return to her the interim fee payments made.

Should A be entitled to recover the balance of the fee? If your conclusion is that she should not, should she be obligated to repay any interim fee payments made? Would you like to know any additional facts? If so, what and why?

2. You are a graduate architecture student. You have been asked to design a single-family residence by a developer for whom you wish to work after you become registered. What would be the factors you should take into account? Why would they be relevant?

Contracting for Design Services: Pitfalls and Advice

SECTION 11.01 Authority Problems

A. Private Owners

Negotiations to provide design services are always made with individuals. It is important to determine whether the individual with whom a design professional negotiates has authority to make binding representations and to enter into a contract. The general agency doctrines that regulate these issues were discussed in Chapter 4. Reference should be made to that chapter, particularly the material in Section 4.06 that deals with apparent authority.

Design professionals will be motivated by many considerations when they decide whether to enter into a contract to perform design services. One factor will be representations made by the prospective client. For example, a design professional may be influenced by representations that relate to the likelihood that the project will be approved by appropriate authorities, that adequate financing can be secured, and that the design professional will work out the design with a particular representative of the client or to other facts that will influence the design professional to contract to perform design services for the client.

In the best of all worlds, the persons at the top of the client's organization would be contacted to determine the authority of the person with whom the design professional is directly dealing. However, it may not be easy to determine who is at the top of the client's organization, and it may not always be politic to take this approach. The realities of negotiation mean that the design professional will have to rely largely on appearances and common sense to determine whether the person with whom he is dealing can make such representations. This may depend upon the position of the person

in the corporate structure, the importance of the representation, and the size of the corporation. The more important the representation and the more important the contract, the more likely it is that only a person high up in the organizational structure will have authority to make representations that will bind the corporation. If the corporation is particularly large, a person in a relatively lower corporate position may have authority to make representations. For example, in a large national corporation, the head of the purchasing department may be authorized to make representations that would require a vice president's approval in a smaller organization.

One useful technique is to request that important representations be incorporated in the final agreement. If the person with whom the design professional is dealing is unwilling to do so, that person may not have authority to make the representations that have been made.

This approach has an additional advantage. In the event of a dispute that goes to court, the parol evidence rule (discussed in Section 11.04(E)) may make it difficult to introduce evidence of representations that are not included in the final agreement. Even if admitted into evidence, failure to include it in the written agreement may make it difficult to persuade judge or jury that the representations *were* made if denied by the client.

Next it is important to examine the authority to *contract* for design services. This problem will be approached by looking at the principal forms of private organizations that are likely to commission design services.

Sole Proprietor. The sole proprietor clearly has authority to enter into the contract. Dealings with an agent of the sole proprietor are controlled by concepts of agency and scope of actual or apparent authority. Generally, sole proprietorships are small business operations. For a sole proprietorship, a contract for design services or construction is probably a serious and important transaction. For that reason, agents of a sole proprietor are not likely to have such authority.

Partnership. Usually partnerships are not large businesses. Generally, only the partners have the authority to enter into contracts for design services or construction. The partners may have designated certain partners to enter into contracts. Unless this comes to the attention of the design professional, it is unlikely that such a division of authority between the partners will affect him. Although partners have authority to enter into most contracts, it is advisable to get all the partners to sign the contract.

If the partnership is a large organization, agents may have authority to enter into a contract for performance of design services and construction. It would probably be best, however, to have some written authorization from a partner that the agent has this authority.

Corporation. The articles of incorporation or the bylaws generally specify who has the authority to make designated contracts and how such authorization is to be manifested. The more unusual the contract or the more money involved, the higher the authority needed. Frequently contracts that deal with land, loans, or sales or purchases not in the normal course of business must be authorized by the board of directors. Lesser contracts may require authorization by higher officials, while for contracts involving smaller amounts of money or those in the usual course of business, subordinate officials may be authorized to contract.

Contracts needing board approval are passed by resolution and entered into the minutes of the board meeting. In addition, bylaws or any chain-of-authority directive will frequently state which corporate officials must actually sign the contract. Again, the importance of the contract will usually determine at what echelon the contract must be signed and how many officials must sign it. The corporate bylaws or state statutes may require that the corporate seal be affixed to certain contracts.

For maximum protection, the design professional should check the articles of incorporation, the bylaws, and any chain-of-authority directive of the corporation. The design professional should see who has the authority to authorize the contract and whether the proper mechanism, such as the appropriate resolution and entry of the resolution in the minutes, was used. Then the design professional should determine whether the person who wants to sign or has already signed for the corporation has authority to do so.

In very important contracts, it may be wise to take all of these steps. It is not unreasonable for the design professional to request that the corporation attach to the contract a copy of its articles and bylaws and a copy of the resolution authorizing the particular project in question. The design professional can request that the contract itself be signed by the appropriate officers of the corporation, such as the president and secretary of a smaller corporation or the vice president and secretary of a larger corporation.

Sometimes such precautions need not be taken. The project may not seem important enough to warrant this extra caution. The design professional may have dealt with this corporation before and is reasonably assured that there will be no difficulty over authority to contract. However, laziness or fear of antagonizing the client is not a justifiable excuse.

For the risks involved in dealing with the promoter of a corporation not yet formed, refer to Section 3.04(C).

Unincorporated Association. Dealing with an unincorporated association seems simple but may involve many legal traps. To hold the members of the association, it is necessary to show that the persons with whom the contract was made were authorized to make the contract. In such cases, it is vital to examine the constitution or bylaws of the unincorporated association and to attach a copy of the resolution of the governing board authorizing that the contract be made. The persons signing the contract should be the authorized officers of the association.

It may be wise to obtain legal advice when dealing with an unincorporated association, unless the unincorporated association is a client for whom the design professional has worked in the past and in whom the design professional has confidence.

Spouses or Unmarried Cohabitants. A design professional may deal with a married or unmarried couple living together. Even with spouses, no presumption exists that one is agent of the other.[1] But one can bind the other if the former acts to further a common purpose or if unjust enrichment would result without a finding of agency.[2] It is advisable to have both spouses or members of an unmarried couple sign as parties to the contract.

B. Public Owners

Reference should be made to Section 8.02(A), which deals with public owners and describes some of the special rules for dealing with public entities. Authority issues arise with some frequency because of the need to protect public funds and preclude improper activity by public officials.

The law has generally provided little protection to those who deal with public entities, rarely invoking the apparent authority doctrine. Similarly, application of estoppel concepts, which would bar the public entity from establishing that representations were unauthorized or that contracts were not made by proper persons or in accordance with law, is also rare.[3] Yet there has been some relaxation of the protections accorded public entities. Restitution claims based on unjust enrichment are beginning to have some success,[4] as are assertions that a public entity cannot rely on provisions of a contract where it has misled the contractor.[5]

---

[1]*Oldham & Worth, Inc. v. Bratton,* 263 N.C. 307, 139 S.E.2d 653 (1965).
[2]*Capital Plumbing & Heating Supply Co. v. Snyder,* 2 Ill.App.3d 660, 275 N.E.2d 663 (1971) (wife agent for lien purposes).
[3]*School Dist. of Phila. v. Framlau Corp.,* 15 Pa.Comwlth. 621, 328 A.2d 866 (1974) (school board president); *County of Stephenson v. Bradley & Bradley, Inc.,* 2 Ill.App.3d 421, 275 N.E.2d 675 (1971) (chairman of board of supervisors). See also *M.A.T.H., Inc. v. Housing Auth. of East St. Louis,* 34 Ill.App.3d 884, 341 N.E.2d 51 (1976), which held a promise to renegotiate after completion made by the president of the housing authority to be beyond the latter's authority and unenforceable.
[4]*Coffin v. Dist. of Columbia,* 320 A.2d 301 (D.C.App.1974) (dictum: recovery limited to the amount of money that the contracting officer had authority to commit); *Saul Bass & Assoc. v. United States,* 205 Ct.Cl. 214, 505 F.2d 1386 (1974) (recovery based on an informal agreement despite failure to execute a formal contract).
[5]*Emeco Indus., Inc. v. United States,* 202 Ct.Cl. 1006, 485 F.2d 652 (1973); *Manloading & Management Assoc., Inc. v. United States,* 198 Ct.Cl. 628, 461 F.2d 1299 (1972). See also *Hueber Hares Glavin Partnership v. State,* 75 A.D.2d 464, 429 N.Y.S.2d 956 (1980) (dictum).

This should not encourage carelessness in dealing with public entities. There are still substantial risks in relying on representations by government officials and entering into contracts with public entities.

SECTION 11.02 Financial Capacity

A. Importance

A crucial factor in determining whether to undertake design work is the client's financial capacity. The relationship between the design professional and client can deteriorate when the latter cannot or will not pay interim fee payments as required, is unable or unwilling to obtain funds for the project, or does not possess the financial capacity to absorb strains on the budget caused by design changes or market conditions that increase the cost of labor and materials. If the client will not have the financial resources to pay any court judgment that may be awarded, the contract is risky and should be avoided.

Despite constant warnings to design professionals, many enter into contracts and perform work and find that they are unable to collect for their services. Professionals do not like to confront financial responsibility openly. They prefer to assume that clients are honorable people who will meet their obligations. Usually this is the case. However, there are too many instances of uncollected fees to justify cavalier disregard of this problem. Design professionals must seriously confront the problem of financial capacity.

B. Private Clients

Retainers and Interim Fees. Most standard contracts published by the professional associations provide for initial retainers to be paid by the client at the time the contract is signed and for interim fee payments to be made during performance. A major purpose of such provisions is to give the design professional working capital.

Another purpose is to limit the scope of the financial risk taken by the design professional. If services and efforts do not run very far beyond the money paid, risk of nonpayment is substantially reduced. The difficulty is that design professionals frequently do not insist that the client comply with these contract terms. This is *absolutely* essential. If the matter is explained properly to the client, there should be no difficulty. It should alert the design professional to possible danger when the client seems to be insulted when businesslike requests are made for advance retainers and interim fee payments when due. Clients who react adversely to such requests are often clients who either do not have the money or will not pay even if they do have the money.

Client Resources. AIA Doc. B141, ¶ 4.3, permits the architect to request that the client furnish evidence that financial arrangements have been made to fulfill the owner's obligations. This power can be exercised at any time during the contract period. This is both a promise and a condition to the architect's continued performance. Failure by the client to comply would permit the architect to suspend performance. If suspension continued or if the client indicated an inability or unwillingness to furnish this information, the architect could treat it as a material breach terminating the obligation to perform and giving remedies for breach of contract. (See Sections 27.02 and 34.04.) The design professional can and should ask for this information *before* deciding whether to perform services. It may be useful to run a credit check if the client is new or if the information furnished does not appear trustworthy.

One-Person and Closely Held Corporations: Individual Liability of Officers. Many small businesses are incorporated, and the shares of the business are held entirely by the proprietor of the business. The proprietor is permitted to do this by law. One purpose is to insulate personal assets from the liabilities of the corporation. Some corporations are small, closely held corporations, with the shares owned by a family or by the persons actually running the business. As noted in Section 3.04(E), the law can pierce the corporate veil and treat the inadequately capitalized one-person or closely held corporation as a sole proprietorship or partnership. If this is done, the design professional can recover from the shareholder or shareholders.

Piercing a corporate veil is rarely done, however. If a credit check reveals that the corporation is merely a shell with very few assets, it may be necessary to demand that the shareholders assume personal liability. To do so, the sole shareholder or shareholders should sign both as representatives of the corporation and as individuals. If those signing as individuals are solvent, this is a reasonably se-

cure method of assuring payment if the corporation is unable to pay the contractual obligations. If the individuals signing are not solvent, individual liability will be of little value. A refusal to sign as an individual may be a warning that the corporation is in serious financial trouble.

A Surety or Guarantor. Chapter 33 discusses the role of the surety in the construction phase of the project. If the design professional believes the person with whom he is dealing is not financially sound, the design professional may wish to obtain a financially sound person to guarantee the obligation of the corporation or the individual *before* entering into the contract. Subject to some exceptions, promises to pay the debt, default, or miscarriages of another must be in writing. The design professional who requests and obtains a third person to act as surety for the client should make certain that the surety signs the contract or a separate surety contract. Legal advice should be obtained where a surety or guarantor is involved.

Real or Personal Property Security. Another method of securing the design professional against the risk that the client will not pay is to obtain a security interest in real or personal property. This can be done by obtaining a mortgage or a deed of trust on the land on which the project is to be constructed or on other assets owned by the client. This may seem like a drastic measure. If that much insecurity is involved, it may be advisable not to deal with the client at all. The law regarding creating and perfecting security interests is beyond the scope of this treatise. Legal advice should be obtained if such protection is needed.

Client Identity. Some clients move quickly in and out of various different but related legal forms. What appears to be a partnership turns out to be a corporation of limited resources. The design professional may find that the corporation with whom he dealt is insolvent while other solvent related corporations are controlled by the client. It can be difficult for the design professional to know for whom he is working or who has legal responsibility. This can be complicated by the different entities described in the preliminary correspondence, the contracts, and the communications during performance. Design professionals should know the exact identity and legal status of the client.

Spouses or Unmarried Cohabitants. As mentioned in Section 11.01(A), dealing with spouses or cohabiting persons can raise agency questions. Design professionals may discover that the person with whom they dealt has no assets while that person's spouse or other cohabitant owns all the assets. It is important to have both spouses or cohabitants sign the contract.

Mechanics' Liens. Mechanics' liens—more important to contractors and suppliers—are treated in Section 28.07(D). Their occasional utility to professional designers can tempt those who are about to perform design services to avoid some of the techniques previously mentioned for fear of losing a client. For that reason, they are discussed here briefly.

State statutes often give persons a security interest in property that they have improved to the extent of any debt owed to the improver by the owner of the property or someone who has authority to bind the owner. The lien holder can demand a judicial foreclosure of the property and satisfy the obligation out of the proceeds.

Design professionals face two problems when they assert a mechanics' lien. Both problems relate to improvement of the owner's property as the basis for awarding a specific remedy of foreclosure against the land itself rather than simply giving the design professional a money award as described in Sections 2.12 and 2.13.

The first problem—a problem not faced by contractors—involves whether or not design services have improved the land. This problem generally has been surmounted by mechanics' lien legislation that specifically names design professionals as lien recipients. But not all states have done this.

The second issue recognizes the unstable world of design. It is not uncommon for the design professional to perform design services for a project that is abandoned before site work is initiated. (The beginning of the site work warns those who might advance credit to the owner that there may be a lien that will take precedence against any subsequent security interest.) As noted, improvement of the land is the basis for awarding mechanics' liens.

Courts have not been uniform when faced with the issue of whether a lien can be asserted despite the project's having been abandoned. Most courts have ruled that no lien has been created.[6] But some cases go in the other direction.[7]

In a number of states, the mechanics' lien laws have been amended to grant liens even though the project is never begun.[8] Such laws are usually supported by professional associations of architects and engineers who assert that design professionals perform services and should be paid. Clearly, design professionals would have a right to recover an ordinary court judgment unless they have assumed the risk that they would not be paid if the project did not go forward.[9] However, the professional associations seek to go further and create a specific remedy, that is, a lien against the property for which the design was commissioned. Often ignored is the fact that this specific remedy places design professionals ahead of those ordinary unsecured creditors who may have also performed services or supplied goods and have not been paid. Participants in the construction industry have sought to enlarge their lien rights, but it should not be ignored that this often comes at the expense of others.

Lien rights cannot be asserted against public improvements. Many technical requirements exist to create a valid lien. Notices have to be given, filings have to be made, and foreclosure actions must be taken within specified times. Without strict compliance, there is no lien. Other persons may have equal or prior security rights in the land. The land value may not be enough to pay the lien claims in their entirety.

Mechanics' lien statutes vary considerably from state to state and are frequently changed by legislatures. Design professionals should seek legal advice to see whether they are within the class of persons accorded liens and to ascertain the steps needed to perfect a lien. It is unwise to undertake work for a client who may not be able to pay with the hope that in the event of nonpayment, there

---

[6]*Mark Twain Kansas City Bank v. Kroh Bros. Dev. Co., et al.,* 19 Kan.App.2d 714, 798 P.2d 511 (1990); *Brownstein v. Rhomberg-Haglin & Assoc.,* 824 S.W.2d 13 (Mo. 1992); *An-*

thony and Associates v. Muller, 598 A.2d 1378 (R.I. 1991). *Goebel v. National Exchangors, Inc.,* 88 Wis.2d 596, 277 N.W.2d 755 (1979).
[7]*Design Associates, Inc. v. Powers,* 86 N.C.App. 216, 356 S.E.2d 819 (1989).
[8]Fla.Stat.Ann. § 713.03 (1988) (a design professional who has a direct contract to perform services relating to a specific parcel of real property has a lien for the amount owing to him regardless of whether such real property is actually improved). Similarly see West Cal.Civ. CodeAnn. § 3081.2 enacted in 1990. A permit must have been issued. Under § 3081.10, there are no liens against single-family residences that cost less than $100,000. See also West Tex.Prop. CodeAnn. § 53.021(c).
[9]See Section 12.13(C).

will be a right to a mechanics' lien. The possibility of being able to assert a mechanics' lien is never a substitute for a careful consideration of the financial responsibility of the client and collection of interim fee payments.

C. Public Owners

Those who deal with public entities run the risk that money will not be appropriated or bonds not issued to pay for the project or design services. An Illinois decision held that an architect could not recover for his services where the evidence showed he was risking compensation on the passage of a bond issue.[10] The court pointed to a statement made by the architect during negotiations that he was gambling with the county, which convinced the court that he took this risk.

The best protection against risks is to check carefully on the availability of funds and make certain that work does not begin until it is relatively certain that funds will be available to compensate the design professional.

SECTION 11.03 Competing for the Commission: Ethical and Legal Considerations

Associations of design professionals historically have sought to discipline their members for ungentlemanly competition,[11] both in obtaining a commission for design services and in replacing a fellow member who has been performing design services for a client.

As an illustration, AIA Document J330 in effect in 1958 listed among its mandatory standards of ethics:

9. An Architect shall not attempt to supplant another Architect after definite steps have been taken by a client toward the latter's employment.
10. An Architect shall not undertake a commission for which he knows another Architect has been employed until he has notified such other Architect of the fact in writing and has conclusively determined that the original employment has been terminated.

In addition, J330 stated that the architect would not compete on the basis of professional charges.

These restraints were considered part of ethics. Undoubtedly they are traceable to the desire to preserve these professions *as professions* and to avoid certain practices that would be accepted in the commercial marketplace. Although the associations must have recognized that there will be competition for work, it was hoped that competition would be conducted in a gentlemanly fashion and would emphasize professional skill rather than price.

Beginning in the 1970s, attacks were made on these ethical standards by public officials charged with the responsibility of preserving competition. One by one, association activities that were thought to impede competition came under attack, particularly those that dealt with fees. Fee schedules, whether required or suggested, were found to be illegal.[12] Disciplinary actions for competitive bidding were also forbidden.[13]

Standard 9 in J330 attempted to inhibit competition *before* any valid contract had been formed.[14] Standard 10, though less susceptible to the charge that it is anticompetitive, could have had the effect of limiting the client's power to replace one architect with another.

It is clear that design professionals do not like to compete on the basis of price. Although their associations cannot use their disciplinary power to attack this practice, current federal legislation precludes head-to-head fee competition.

In 1972, Congress enacted the Brooks Act.[15] The Act determines how contracts for design services can be awarded by federal agencies. It declares that the policy of the federal government is to negotiate on the basis of "demonstrated competence and qualification for the type of professional services required and at fair and reasonable prices."

Those who perform design services submit annual statements of qualifications and performance data. The agency evaluates the statements, together with those submitted by other firms requiring a

[10]*County of Stephenson v. Bradley & Bradley, Inc.,* supra note 3.
[11]For discussion of the boom in official competitions, see *N.Y. Times,* May 5, 1988, p. 81, col. 2.

[12]*Goldfarb v. Virginia State Bar,* 421 U.S. 773 (1975).
[13]*Nat'l Soc'y. of Professional Eng'rs v. United States,* 435 U.S. 679 (1978).
[14]*United States v. Am. Soc'y. of Civil Eng'rs,* 446 F.Supp. 803 (S.D.N.Y. 1977) (barred society from disciplining members who supplanted another member).
[15]40 U.S.C.A. §§ 541–544.

proposed product, and discusses with no fewer than three firms "anticipated concepts and the relative utility of alternative methods." The agency then ranks the three most qualified to perform the services. It will attempt to negotiate a contract with the highest qualified firm and takes into account "the estimated value of the services to be rendered, the scope, complexity, and professional nature thereof."

If a satisfactory contract cannot be negotiated with the most qualified, the agency will undertake negotiations with the second most qualified, and if that fails, with the third. The U.S. Justice Department has proposed repeal, arguing that the statute restricts competition. Many states have had comparable legislation regulating the award of contracts for design services by state agencies.

Despite the Brooks Act and various state statutes adopting the same approach, some procurement agencies have sought to incorporate price at the first level—that of determining competence and qualification. A well reasoned and researched opinion of the Colorado attorney general concluded that a public entity cannot require cost proposals in connection with selecting the most qualified architects, engineers, or land surveyors for a project.[16] However, an Ohio intermediate appellate court upheld a decision by the governor of Ohio that there can be two most qualified firms and that a tie in such a case may be broken at the first stage by using competitive price proposals. The low bidder then becomes the most qualified firm.[17]

One commentator, while obviously despairing at the prospect, made the following suggestions to design professionals forced to compete on price:

1. Make bids only on projects where there is a clearly defined scope of work.
2. Select your clients carefully.
3. Bid only on projects where your experience would favor selection in the absence of price competition.

4. Evaluate your competition.
5. Plan to use the most competent and qualified people you have.
6. Define your work with great precision.
7. Make detailed cost estimates.

The commentator also suggested that it is crucial to know when to walk away from a prospective commission and to be prepared to do so.[18]

One professional liability insurer refused to defend a claim against its insured related to a competitive design bid. The insured's commercial general liability carrier successfully defended the claim. The latter recovered a pro rata portion of its defense costs from the professional liability carrier.[19]

One troubling question that arose as a result of the increased use of Construction Management in the 1970s related to whether contracts for the performance of these service were more like those for design, which did not require competitive bidding, or more like contracting, which did. This is discussed in Section 17.04(D).

Tort law protects contracts or advantageous relationships from improper interference by third parties. The famous English decision *Lumley v. Gye*[20] allowed the plaintiff impresario to recover damages from the defendant who enticed an opera singer to breach her contract with the plaintiff and sing for the defendant. Whether there has been an improper interference with a contract requires balancing the protection given a contract with the freedom of parties to seek to persuade others that they can perform services better than a competitor who has been retained.

States that wish to give less protection to the contract or advantageous relationship require malicious intent or bad faith.[21] But the trend has been toward requiring merely an intent to do the act that will have the effect of interfering with a legally protected interest of the plaintiff.[22]

Williams v. Chittenden Trust Co.[23] illustrates this in the context of a client-design professional rela-

---

[16]Op. Colo. Att'y Gen., Sept. 2, 1992.
[17]*Ohio Ass'n of Consulting Eng'rs v. Voinovich.*, 83 Ohio App.3d 601. 615 N.E.2d 635 (1992). The state and national professional societies filed an appeal to the Supreme Court of Ohio based upon the Ohio mini-Brooks Act, which they asserted bars a procedure under which price is a factor in the selection of the most qualified firm. This appeal was denied. 66 Ohio St.3d 1459, 610 N.E.2d 423 (1993).

[18]Lekamp, Professional Liability Perspective, vol. 10, no. 11 (November 1991).
[19]*American Motorists Ins. v. Republic Ins.*, 830 P.2d 785 (Alaska 1992).
[20]2 El.&Bl. 216, 118 Eng.Rep. 749 (1853).
[21]*Dehnert v. Arrow Sprinklers*, 705 P.2d 846 (Wyo.1985).
[22]*Blivas Page, Inc. v. Klein*, 5 Ill.App.3d 280, 282 N.E.2d 210 (1972).
[23]146 Vt. 76, 484 A.2d 911 (1984).

tionship. The plaintiff architect had contracted with the owner to design a condominium project. After approximately one third of the units were completed, the owner encountered financial difficulty. He contacted the defendant—another architectural firm—about the possibility of reducing construction costs. The defendant analyzed the design prepared by the plaintiff, recommended some changes, and incorporated those changes into plans and specifications it prepared at the owner's request. The redesigned condominiums were very similar to those designed by the plaintiff but were smaller in size and less expensive to build. In the process of redesign, the defendant architect used the original plans and drawings prepared by the plaintiff.[24] The owner then informed the plaintiff that the project had been abandoned and canceled its contract. It then entered into a contract with the defendant for the remainder of the project.

The court held that one may not intentionally interfere with another's contract or with "reasonable expectancies of profit," even though the contract is terminable at will.[25] The plaintiff had shown sufficient evidence that the defendant had intentionally and improperly caused the owner not to perform its contract with the plaintiff. The court noted that at the first meeting between the owner and the defendant, the defendant had been told that the plaintiff was the designer of the project and had been given copies of the drawings before the defendant prepared his own. The evidence indicated that the defendant knew or had reason to know of the existence of the contract between the plaintiff and the owner. A jury could have concluded that the defendant realized he was interfering with the plaintiff's contract. The court concluded by noting that even though the owner and not the defendant had initiated the first contact, the owner had not yet decided at that time to terminate its contract with the plaintiff. The defendant then offered to perform architectural services at a price substantially below that charged by the plaintiff and at a time when the plaintiff's project was suffering from serious financial difficulties.

Wrongful interference with a contract is sometimes the basis for a claim against a design profes-

sional who suggests to the owner that a contractor be terminated (discussed in Section 14.08(F)).

SECTION 11.04 Professional Service Contracts: Some Remarks

A. Profits and Risk

Ordinarily, professional advisers do not have potentially high profit returns. They serve their client and expect to be paid for their services. This does not mean that advisers are always paid for their work. Nonpayment because the client does not have the financial resources is a risk taken by all who perform services.

This section is directed toward defenses by the client to any claim for compensation for services rendered based on an assertion that no money is due. Such defenses are usually predicated on an asserted understanding that if the project did not go forward—usually because public approval or funds could not be obtained—the professional designer would not be paid for the work.

Although autonomy gives contracting parties the power to make an arrangement under which the professional designer may risk the fee, any conclusion that this risk has been taken should be arrived at only if weighty evidence supports this conclusion. This is based not only on customary practices but also on the conclusion that professional designers do not as a rule recover profits but recover only payment for services they render.

B. Good Faith, Fair Dealing, and Fiduciary Relationships

The law increasingly finds that all contracting parties owe each other the responsibility of good faith and fair dealing.[26] This is particularly so when objectives sought by the design professional and the client require close cooperation. Each party should help the other achieve its goals under the contract.

The retention of a professional designer by a client—like the relationship between employer and employee and between partners—creates a fiduciary relationship. The core of a fiduciary relationship is trust and confidence. To achieve these, the per-

---

[24]See Section 12.11.
[25]484 A.2d at 915.

[26]Restatement (Second) of Contracts § 205 (1981): Uniform Commercial Code § 1-203. See Section 19.02(D).

son to be protected by the relationship must believe in the undivided loyalty of its fiduciary.

The language in AIA Document A111 forms the basis of a construction contract where payment is based on the cost of the work plus a fee. Article 3 states in part:

> The Contractor accepts the relationship of trust and confidence . . . and covenants with the Owner to cooperate with the Architect and utilize the Contractor's best skill, efforts and judgment in furthering the interests of the Owner . . . and to perform the Work in the best way and most expeditious and economical manner consistent with the interests of the Owner . . .

The client should be able to trust its professional adviser and receive honest professional advice. Both parties—design professional and client—should be candid and open in their discussions, and each should feel confident that the other will not divulge confidential information.

The fiduciary relationship can be contrasted with the arm's-length relationship. In a commercial setting, parties generally deal at arm's length: each party must look out for itself. On the other hand, a fiduciary relationship is a close one, and the parties must be able to trust each other.

Some aspects of the fiduciary relationship are obvious. Design professionals should not take kickbacks or bribes. They should not profit from professional services other than by receiving compensation from the client.

Funds held by the design professional that belong to the client should be kept separate. Commingling will be a breach of the fiduciary obligation. In such a case, any doubts about to whom the money belongs or for whom profitable investments were made will be resolved in favor of the client.

Financial opportunities that come to the attention of the design professional as a result of the services that he is performing for the client should be disclosed to the client if the services would be an opportunity falling within the client's business.

Although the design professional should not disclose confidential information, there may be extreme cases where the design professional has a duty to disclose information even in the face of an *express* contract prohibiting such disclosure. For example, in *Lachman v. Sperry-Sun Well Surveying Co.*,[27]

a client brought an action against its engineer who, in violation of an express covenant not to disclose information, notified adjacent landowners of his client's slant drilling. The court held that such a covenant would not be enforced.

One of the most troubling concepts relates to conflict of interest. One cannot serve two masters. The client should be able to trust its design professional to make judgments based solely on the best interests of the client. Advice or decisions by the design professional should be untainted by any real or apparent conflict of interest. The client must believe the design professional serves it and it alone.

For these reasons, design professionals should not have a financial interest in anyone bidding on a project for which they are furnishing professional advice. Likewise, they should not have a financial interest in any contractor or subcontractor who is engaged in a project for which they have been engaged. Design professionals should not have any significant financial interest in manufacturers, suppliers, or distributors whose products might be specified by them. Products should not be endorsed that could affect specification writing, nor should designated products be specified because manufacturers or distributors of those products have furnished free engineering. The purpose of these restrictions is to avoid conflict of interest. Design professionals cannot serve their clients loyally if they might personally profit by their advice.

Generally, a client who is fully aware of a potential conflict of interest can nevertheless choose to continue to use the design professional. Consent to a conflict of interest should be binding only if it is clear that the client knows all the facts and has sufficient understanding to make a choice. A design professional who intends to rely on consent by the client must be certain that such requirements are met.

Some conflicts of interest are clear; others are murky. For example, suppose an architect retained by an airport authority is also retained by airline tenants who plan to rent space in the airport. The city attorney of San Francisco concluded that there was no conflict of interest, principally because the airport authority knew that the architects were working or might work for airline tenants, that such arrangements were common, and that the ar-

---

[27]457 F.2d 850 (10th Cir. 1972).

chitects were not "vested with authority to exercise any discretion on behalf of the City."[28]

If after full disclosure the client freely chose to proceed, the opinion may be correct. But the *client's* having the power to make decisions—a point emphasized in the opinion—does not eliminate the architect's advisory function. Can the architect advise impartially if he hopes to obtain commissions from airport tenants? The city attorney's conclusion that such dual retentions are efficient and common is questionable.

The client clearly can avoid responsibility for any act by the design professional that is tainted by a breach of the fiduciary obligation. For example, if a contract is awarded to a bidder with whom the architect or engineer colluded, the contract can be set aside by the client. If the architect or engineer issued a certificate for payment dishonestly or in violation of his fiduciary obligation, the certificate can be set aside if the client so desires.

The breach of a fiduciary obligation can give the client grounds to dismiss the design professional. Any bribes or gifts taken by the design professional can be recovered by the client. Any profit that the design professional has made as a result of breaching the fiduciary obligation must be given to the client even if the profit was generated principally by the skill of the design professional. Obviously, the design professional must take his fiduciary obligation seriously and make every effort to avoid its breach or the *appearance* of such a breach.

Undoubtedly, some of the freedom that courts have felt to develop the fiduciary doctrine occurred because of the frequently brief and incomplete agreements made between clients and design professionals.

Today there is an increasing use of detailed written agreements that deal with almost every aspect of the design professional-client relationship. If there is a detailed written agreement, is there less need for courts to employ the fiduciary concept?

Many aspects of the fiduciary relationship relating to trust and confidence are not likely to be found in standard contracts, because the assumption is that they are so obvious that they need not be expressed. In any event, the fiduciary concept

has been expanding, and the tendency to have longer and more detailed standard form contracts is not likely to affect such expansion.

C. Variety of Types: Purchase Orders

Some professional relationships are not accompanied by careful planning in a contractual sense. Such relationships sometimes are created by a handshake without any exploration of important attributes of the relationship. If such attributes are discussed and resolved, the resolution is often not expressed in tangible form.

At the other extreme, the relationship often is cemented by assent to a pre–prepared standardized form supplied frequently by architect or engineer and occasionally by the client. Often the client will not understand these standardized agreements. On occasion the actual agreement of client and design professional may be different from provisions of the standardized form. Under any of the circumstances mentioned, difficulties can exist if disagreements arise between client and design professional over the services to be performed by the latter.

Another contracting system that can create problems, particularly for the design professional, involves the use of purchase orders as a method for commissioning professional services by architects and engineers. A client will send a purchase order to the design professional and ask that a copy be signed and returned to the client. Purchase orders ordinarily contain a set of standard terms and conditions that relate more to delivery of "off-the-shelf" goods than to professional services. Their use may cause the design professional and his client to fail to achieve a clear mutual understanding as to the services to be performed, the compensation to be received, and responsibility if things go wrong.

From a substantive standpoint, the use of a purchase order more geared to the delivery of goods may impose upon the design professional stricter warranties as to the outcome of the services than are required normally of design professionals.[29] Such warranties and guaranties may be excluded from professional liability insurance coverage.

[28]Letter from Thomas M. O'Connor to Mr. Robert J. Dolan, Clerk, Board of Supervisors, September 7, 1972.

[29]See Sections 14.05 and 14.07.

Probably the best approach for a design professional who receives a standard purchase order is to send the form back unsigned and begin discussions with the client over scope and responsibilities before commencing performance.

D. Interpretation

Contracts between owners and contractors and between contractors and subcontractors generate more interpretation disputes than those between design professionals and clients. For this reason, the bulk of the discussion in this treatise relating to contract interpretation is found in Chapters 19 and 20. However, a few interpretation guides relating to contracts between design professionals and clients may be useful here.

One attribute of the design professional-client relationship that sets it apart from ordinary commercial contracts is the relative inexperience of many clients in design and design services. Another is the way such agreements are made. Not uncommonly, such relationships are created by a vague, informal agreement sometimes followed by assent to a pre-prepared contract form supplied by the design professional. These two attributes frequently generate honest misunderstandings between design professionals and clients.

The most important lodestar—the common intention of the parties at the time the contract was made—is determined by an examination of any discussions the parties may have had before entering into the agreement, the language in any written contract to which both parties have assented, the facts and circumstances surrounding the making of the agreement, the conduct of the parties after the relationship has begun, and any custom or usage of the trade of which both parties were aware or should have been aware.

Although discussion between the parties—the admissibility of which is determined by the parol evidence rule described in (E)—and surrounding facts and circumstances are important, emphasis here is upon the language of the contract. Although courts still invoke the "plain meaning" rule—the search for evidence outside the writing requiring a preliminary determination that the meaning is not "plain" on its face—by and large courts today will look at any relevant evidence to determine the meaning of language selected by the parties.[30]

As a rule, the most important evidence of the intention of the parties is any written agreement to which the parties have assented. This is true whether the agreement is a carefully negotiated contract between client and design professional of relatively equal bargaining power and experience or an agreement between an unsophisticated client and an experienced design professional where the latter has supplied a pre-prepared form contract.

In the former, there is likely to be a neutral reading of the language. In the latter, the interpretation guide that contract terms are generally interpreted against the party who has prepared the agreement will be used (contra proferentum).[31] This interpretation guide can be justified either because the drafter carelessly or deliberately caused the language ambiguity or because the law assumes that the party who prepares the agreement is in a position to force unfair or onerous terms on the other party. This doctrine may be an implicit recognition that the party who does not prepare the language may not take the time or have the ability to examine carefully the language selected by the other party.

This discussion assumes that either both or at least one party had an intention as to specific language. When lengthy standard contracts published by a third party such as the AIA are used, it is possible that neither party had any idea of the purpose of a particular clause, often an important datum invoked to interpret the clause. As this "absence of intention" can arise in construction contracts, discussion of this topic is postponed until Section 20.02.

The interpretation guide that resolves the dispute against the party who supplied the ambiguous terms will also be applied to contracts published by the professional association of the design professional.[32] Ambiguous language is likely to be interpreted to be consistent with the reasonable expectations of the client rather than the literal interpretation that the language might otherwise bear.

Suppose a design professional is retained by a large institutional client such as a public agency or large private corporation. In such a case, the client often prepares the contract for professional services

---

[30]See Section 20.02.

[31]*Williams Eng'g, Inc. v. Goodyear*, 496 So.2d 1012 (La.1986).
[32]*Malo v. Gilman*, 177 Ind.App. 365, 379 N.E.2d 554 (1978); *Durand Assoc., Inc. v. Guardian Inv. Co.*, 186 Neb. 349, 183 N.W.2d 246 (1971).

and presents it to the design professional on a take-it-or-leave-it basis. Ambiguous language in such a case should be interpreted in favor of the design professional.

As to other interpretation guides, handwritten portions of a contract are preferred to typewritten portions or printed portions, and typewritten portions are given more weight than printed provisions in a form contract. Where parties choose language in connection with a particular transaction, the language is more likely to reflect their common intention.

Specific provisions are given more weight than general provisions. For example, suppose a contract between architect and client states that the architect will perform services in accordance with normal architectural professional standards. Suppose further that the same agreement provides that the architect will visit the site at least once a day. If normal professional standards would not require that the architect visit the site this often, a conflict exists between the two provisions. In such a case, the provision specifically relating to the number of site visits will control the general provision requiring that the architect perform in accordance with accepted professional standards.

Words are generally used in their normal meanings unless both parties know or should know of trade usages that have grown up around the use of certain terms. This rule is important in the design professional's relationship to the client. Often a word used in an agreement will have a definite meaning to the design professional but have a different meaning, or at least an indefinable meaning, to the client. For example, suppose that the agreement between design professional and client states that the design professional will make periodic visits to the site during construction. Suppose further that the design professional shows evidence of a custom that visits customarily occur at weekly intervals during a certain stage of the project. If the client does not know or have reason to know of this custom, the custom cannot be used to interpret the phrase "periodic visits." Although parties can be said to contract with reference to established usages in a trade, if one party is not or should not have been aware of the usage, it would be unfair to permit evidence of usage to be used to interpret the language in question.

Another interpretation guide—that of practical interpretation—looks at the *practices* of the parties.

Such practices may give good evidence of what the parties intended. For example, suppose the language to be interpreted is "periodic visits." Suppose the architect visited the site weekly and this was known by and acquiesced in by the client. A court would consider this persuasive evidence that the parties agreed that the visits would be weekly.

Orput-Orput & Associates, Inc. v. McCarthy[33] illustrates several interpretation guides. It involved a disputed oral agreement between architect and client over the existence of a cost limitation (covered in Section 12.03). One issue related to the amount of compensation due the architect. The original construction contract price was approximately $1.2 million. After the work was completed, the client and contractor entered into negotiations that resulted in a final total payout of slightly over one million dollars. The contract between architect and client stated that the fee would be computed on the basis of "the construction cost of the Project." The client contended that the actual expenditure was to govern the fee, while the architect contended that the original contract price controlled.

The appellate court noted that the contract did not assist the court in determining the intention of the parties. The court held that the language should be given a "fair and reasonable interpretation." The trial court had listened to seven witnesses on the question of whether the amount charged was fair and reasonable. The appellate court then noted that though contracts are generally construed against the party who prepares them, the architect's interpretation was fair and reasonable and the trial court's judgment in favor of the architect on this point would be sustained.

Another issue in the *Orput* case involved the question of whether certain services were additional or fell within the basic fee. In addition to looking at the testimony of the architect, the appellate court noted that the architect had billed the client in accordance with the interpretation asserted by the architect and that these billings had been received without protest by the client until commencement of the lawsuit.

Although the court was openly skeptical of contentions made after litigation commenced, the holding of the court can also be taken as a recog-

---

[33]12 Ill.App.3d 88, 298 N.E.2d 225 (1973).

nition of the doctrine of practical (that is, determined by the *practices* of the parties) interpretation. The billings by the architect unobjected to by the client were evidence of what the parties intended when they made the agreement.

E. The Parol Evidence Rule and Contract Completeness

One aspect of the parol evidence rule relates to the provability of asserted prior oral agreements when the parties have assented to a written agreement. A client may contend that the design professional agreed to perform certain services that were not specified in the written agreement. The latter's attorney may contend that the writing expressed the entire agreement and that parol evidence or oral evidence is not admissible to "add to, vary or contradict a written document." Sometimes such oral agreements are provable. Sometimes they are not.

Courts generally do not consider that written agreements between clients and design professionals are complete expressions of the entire agreement.[34] Undoubtedly, one of the reasons that clients are usually allowed to testify as to promises not found in the written contract is the close relationship between the design professional and the client. Under such circumstances, the client may have been reasonable in not insisting that oral promises be included in the writing.[35]

This emphasis upon the close relationship between the contracting parties can be demonstrated by a Mississippi case in which the principal issue was whether a contractor should have read an addendum that contained a provision that the contractor had been told by the architect would be deleted.[36] In excusing the contractor's failure to read the addendum and his relying solely upon the statement of the architect, the court noted that the contractor had been assured the provision had been removed by the senior architect in the firm, "a man in whom he reposed considerable confidence following over a decade of business

dealings."[37] The contractor, according to the court, expected to have been told the truth and not to have been misled. The court concluded stating that this was "a case of dealings between gentlemen, two honorable men."[38] This approach clearly can be used when the issue is whether the client can testify as to conversations about future conduct that were not included in a written contract.

Sometimes a court will refuse to listen to *any* evidence outside the writing if it is convinced that the writing is so clear in meaning that it needs no outside assistance. Most courts will permit either party, especially the client, to show prior oral agreements not included in the writing.

Although attorneys for design professionals frequently rely on the parol evidence rule, such reliance is often misplaced. This is especially so if the agreement is sketchy and does not spell out the details adequately and if the client was not represented by an attorney during the negotiations.

Reducing an arrangement to writing does not necessarily protect the design professional from assertions of additional oral agreements. However, the more detail included in the agreement and the greater the likelihood that the client understood the terms or had legal counsel, the greater the probability that the client will *not* be allowed to prove the claimed oral agreement. The parol evidence rule does not apply to agreements made after the written agreement has been signed by the parties.

The parol evidence rule relates only to the provability of antecedent or contemporaneous agreements. If such agreements are admitted into evidence, the trial court or the jury—depending on who makes the determination of fact—must decide whether the evidence shows that such an agreement was made. Very often the attorney for the design professional places heavy reliance on the parol evidence rule and does not adequately prepare for the more important question of whether the asserted agreement took place.

Most parol evidence cases require a determination of whether the writing was intended to be the complete and final repository of the entire agreement of the parties. To preclude the court from determining that a particular writing was not intended to be complete, contracts prepared by at-

---

[34]*Spitz v. Brickhouse*, 3 Ill.App.2d 536, 123 N.E.2d 117 (1954); *Malo v. Gilman*, supra note 32.

[35]*Lee v. Joseph E. Seagram & Sons, Inc.*, 552 F.2d 447 (2d Cir.1977).

[36]*Godfrey, Bassett & Kuykendall, Architects, Ltd. v. Huntington Lumber & Supply Co.*, 584 So.2d 1254 (Miss.1991).

---

[37]Id. at 1257.

[38]Ibid.

torneys or by professional associations often contain integration or merger clauses.

Before 1966, the AIA Standard Documents did not contain an integration clause, reflecting the then emphasis on the close professional relationship between architect and client. Beginning in 1966, the AIA Documents tended to become more concerned with liability and litigation, and this undoubtedly was the reason for inclusion of B141 ¶ 9.6, which currently states:

> This Agreement represents the entire and integrated agreement between the Owner and the Architect and supersedes all prior negotiations, representations or agreements, either written or oral. . . .

In commercial contracts, such clauses are generally successful in accomplishing the drafter's objective unless the party attacking the clause asserts that the other party fraudulently induced the agreement.

The effect of such a clause in the context of the design professional-client relationship is likely to vary due to both the way in which such relationships are originated and the general disrespect some courts have for the parol evidence rule.

Pipe Welding Supply Co. v. Haskell, Conner & Frost,[39] involved a dispute between architect and client over a cost overrun. Without objection or discussion of the parol evidence rule, the court permitted the client to testify that he had been assured that there would be no more than 10% or 15% de-

viation from the cost estimates despite a provision stating that the architect does not warrant that bids will not vary from estimates. Also, the contract had been made on a form published by the American Institute of Architects, which contained an integration clause.

Yet three years later the Alaska Supreme Court decided *Lower Kuskokwim School District v. Alaska Diversified Contractors, Inc.*[40] This case involved a construction contract that was expressed in a thirty-six page printed contract along with various supplemental conditions. The contractor sought to introduce testimony of what had transpired at a pre-bid conference relating to the time for completion. The testimony the contractor sought to introduce was directly contradictory to the written agreement. The court made short shrift of the contractor's contention that this testimony should have been admitted by concluding that the transaction had been integrated in a final writing despite the absence of an integration clause. (In denying the motion for reconsideration, the court found a provision in the invitation to bidders that it considered the functional equivalent of an integration clause.)

Although actual outcomes may vary, these two cases suggest that an integration clause is not likely to have much effect in a dispute between the design professional and the client but is likely to be determinative in a dispute between an owner and a contractor.

---

[39]96 A.D.2d 29, 469 N.Y.S.2d 221 (1983) affirmed 61 N.Y.2d 884, 462 N.E.2d 1190, 474 N.Y.S.2d 472 (1984).

[40]734 P.2d 62 (Alaska 1987) denied motion for reconsideration, 778 P.2d 581 (Alaska 1989) cert. denied 493 U.S. 1022 (1990).

PROBLEMS

1. X comes to see you at your architectural office. You have been in practice for three months and have not designed any projects. X gives you details regarding the construction of a five-story office building that he wishes you to design. He asks you to begin work immediately. What steps would you take toward protecting the collectibility of your fee?

2. Y approaches you and asks you to perform certain design services. He tells you that he is the vice president of Comac Corporation, a company that makes electronic equipment. He gives you his calling card, which identifies him. He tells you that it will take some time to get the formal contracts

signed by the appropriate officers of the corporation and that you should begin work immediately. Y also tells you that a written, authorized contract will be issued shortly. You note that he is driving a new car with "Comac Corporation" neatly embossed upon each door.

You start work. Three weeks later you bill Comac for $5,000 for schematic design work. Comac refuses to pay because it says Y was not authorized to bind the company. Would Comac have a legal defense? Give your reasons. What other facts would you like to have?

C H A P T E R   T W E L V E

Professional Design Services: The Sensitive Issues

SECTION 12.01 Range of Possible Professional Services: Fees and Insurance

The range of *potential* professional services, already substantial because of the complexity of design and construction and the centrality of the design professional's position, has been increasing because of the increased public control over design and construction and the greater participation by the design professional in economic and financial aspects of the project. A glance at any pre-prepared standard contract for professional services, especially those labeled as "additional," demonstrates this.

Compensation is discussed in Chapter 13, but one aspect of compensation relates to the potential list of professional services. Although the percentage of construction cost method of compensating the design professional is beginning to lose its domination, it is still common for the fee to be computed by this formula. However, and not without occasional surprise to the client, this amount is called the *basic* fee. In addition, services are frequently designated as "additional services," which entitle the design professional to compensation *in addition to* the basic fee. In many cases, the design professional will be willing to perform any services that the law allows and within her professional skill, provided there is compensation for performance of these services.[1] But disputes may arise

over whether the services requested fall under the basic fee or entitle the design professional to additional compensation.

Before 1958, Standard Documents published by the American Institute of Architects (AIA) that dealt with architectural services contained no description of additional services. From 1958 to 1963, nine additional services were listed. Steady increases culminated in 1977 with twenty-two and in 1987 with nine contingent and twenty optional additional services.

Such a formidable list can be the breeding ground for honest disputes. Although the extensive list may reflect instances where architects were unjustifiably denied compensation for services beyond normal services, its use can cause misunderstanding.

An argument can be made for an extensive list of additional services. If the bulk of the additional services is rarely used, including it within the basic design services can increase the basic design fee when those services are not needed and not performed. Separating the rarely performed services, it is argued, benefits the client by requiring that it pay only for those services used.

The formidable list of additional services includes some that seem performed with relative frequency. If so, and if the client is not made aware of this, the client may complain that it is paying *more* than it expected.

Another factor—professional liability insurance (discussed in Section 15.05)—must be taken into account in planning services to be performed. Design professionals should be aware of which services can be covered by professional liability insurance and which are generally excluded.

---

[1] But suppose the design professional is precluded by law or does not wish to perform requested services? A literal reading of AIA Doc. B141, ¶ 3.4.20, would seem to give the client the power to insist the architect do almost anything that relates to the project.

As a general rule, insurers wish to insure only those services routinely performed and connected to the professional training and experience of the insured design professional. Such services include analyzing the client's needs, preparing a design solution, and performing site services to see that the design is executed properly. Services that go beyond these may very well be excluded unless there is a *special* endorsement.

Similarly, the client should be aware of which services are included. Although it is possible to agree that services be rendered that are *excluded* from insurance coverage, this should be a deliberate choice made by the contracting parties and not an inadvertently made choice.

At the outset of the professional relationship or as soon as possible thereafter, the client and design professional should determine in advance which of the possible professional services are to be performed by the design professional and whether those services fall within the basic fee. After this is done, the written agreement should reflect the actual agreement, and performance should be in accordance with the contract requirements or any subsequent modifications.

SECTION 12.02 Design Services: Roles of Design Professional

Although the *basic* elements of design preparation have not engendered legal problems, except to the extent the law has had to determine the standard of performance (see Section 14.05), some mention should be made of *basic* design services.

Usually the client comes to the designer with a problem that it hopes the designer can solve. The client describes its needs and, as a rule, what it *wishes* to spend. After consultation with the client, the designer develops a schematic design and might revise the client's budget.

Next the designer studies the design and prepares drawings and possible models that illustrate the plan, site development, features of construction equipment, and appearance. The designer is also likely to prepare outline specifications and again possibly revise the predicted costs. In small projects, schematic design and design development are still designated as *preliminary* studies.

After the client approves the design development, the designer prepares working drawings and specifications that cover in detail the general construction, structure, mechanical systems, materials, workmanship, site development, and responsibility of the parties. Often the designer will supply or draft general conditions and bidding information (discussed in Section 12.07 dealing with services of a legal nature).

The final preconstruction phase is generally called the bidding or negotiation phase. The designer helps the client obtain a construction contractor through bidding or negotiation. These phases are discussed again in Section 13.02, as phases often determine timing of interim fee payments.

During construction, the designer interprets the contract documents, checks on the progress of the work to issue payment certificates, participates in the change order process, and resolves disputes. Most, but not all, of those services are part of *basic* design services to see the project through to *proper* completion.

The different types of services performed by an architect or an engineer raise serious and often complex questions regarding the design professional's status. Also, the relevance of the design professional's status arises in many different contexts. As a result, a conclusion that has been reached in one context may not necessarily be controlling in another.

As a general rule, during the design phase the design professional is considered an independent contractor concerning the preparation of plans and specifications. From the standpoint of risk allocation, the owner who retains the design professional is not vicariously liable for the negligence of the design professional in planning the design. This results from the design professional's particular professional qualifications and skill and her independence as to details from the owner who commissions her. Similarly, the design professional would not be an agent of the owner authorized to make contracts on behalf of the owner. Even if she were considered an agent, her authority would be very limited. She would certainly not have actual or even apparent authority to enter into contracts on the owner's behalf.[2]

---

[2]*Blue Cross and Blue Shield of South Carolina v. W. R. Grace & Co.*, 781 F.Supp. 420 (D.S.C.1991) (collecting authorities) The case was settled after the opinion of the federal district court.

If all the design professional did was design, the problem of status would be relatively simple. However, once the design professional is involved in procuring a contractor or performing certain services during construction, the problems become more complex.

Post-design services are dealt with in greater detail in Section 12.08(B). At this point, however, the multifaceted role of design professionals and the complexity and occasional dissatisfaction this engenders are demonstrated by the New Engineering Contract (NEC) issued by the English Institution of Civil Engineers (ICE) in 1993. The ICE notes that the consulting engineer traditionally has performed four principal functions:

1. designing
2. managing the project
3. supervising construction
4. adjudicating disputes

Noting dissatisfaction with one person serving all these functions and the tendency for employers (owners) to separate the roles and give them to different individuals and organizations, the NEC defines the roles separately so that they can be allocated to different individuals or organizations as the preferences of the employer for a particular project dictate.

SECTION 12.03 Cost Predictions

A. Inaccurate Cost Prediction: A Source of Misunderstanding

The relative accuracy of cost predictions is crucial to both public and private clients. Clients may be limited to bond issues, appropriations, grants, loans, or other available capital. Unfortunately, many clients think that cost estimating is a scientific process by which accurate estimates can be ground out mechanically by the design professional. For this reason, design professionals should start out with the assumption that cost predictions are vital to the client and that the client does not realize the difficulty in accurately predicting costs.

The close relationship between design professional and client often deteriorates when the low bid substantially exceeds any cost figures discussed at the beginning of the relationship or even the last cost prediction made by the design professional. Not uncommonly, the project is abandoned, and the client often feels that it should not have to pay

the design professional. The design professional contends that cost predictions are educated guesses and their accuracy depends in large part on events beyond the control of the design professional. Clients frequently request cost predictions before many details of the project have been worked out. Clients frequently change the design specifications without realizing the impact such changes can have on earlier cost predictions. Design professionals point to unstable labor and material costs. The amount a contractor is willing to bid often depends on supply and demand factors that cannot be predicted far in advance. Design professionals contend that they are willing to redesign to try to bring costs down but that unless it can be shown that they have not exercised the professional skill that can be expected of persons situated as they were, they should be paid for their work.

An indication that this is a sensitive area can be demonstrated by the terminology used relating to cost predictions. The most commonly used expression, though used less frequently in contracts made by professional associations, is *cost estimates*. The term *estimate* is itself troublesome. In some contexts it means a firm proposal intended to be binding, and in others it is only an educated guess. Probably many clients believe that cost estimates will be in the ballpark, while design professionals may look on such estimates as educated guesses.

Professional associations seek to minimize the likelihood that their members will go unpaid if a project is abandoned because of excessive costs (explored in greater detail in (C)). Part of this effort is reflected in the terminology chosen. For example, in its B141 published in 1977, the AIA sought to differentiate between "statements of probable construction costs, detailed estimates of construction costs, and fixed limit of construction costs." Only the latter, according to the AIA, created a risk assumption by the architects that they will go unpaid if the project is abandoned because of excessive costs. In 1987, the AIA changed to *preliminary cost estimates*. The Engineers Joint Contracts Document Committee (EJCDC) refers only to "opinions of probable construction costs." The potential for misunderstanding is demonstrated in the two reported appellate decisions reproduced in (C).

B. Two Models of Cost Predictions

The many reported appellate cases demonstrate not only the frequency of misunderstanding but also the difficulties many design professionals have predicting costs. Recognition of this, along with other factors explored in Chapter 17, has led the sophisticated client to seek more refined methods of controlling and predicting costs. It is important at the outset to distinguish between what can be called the traditional method and these more refined methods.

The traditional method usually involves the design professional's using rough rules of thumb based on projected square or cubic footage, modulated to some degree by a skillful design professional's sense of the types of design choices that will be made by a particular client. Through the development of the design, cost predictions are almost likely to be given, but they are not going to be based on much more than these rough formulas, refined somewhat as the design proceeds toward completion.

The cost predictions should be more accurate as time for obtaining bids or negotiating with a contractor draws near. In this model, there is a great deal of suspense when bids are opened or when negotiations become serious. Under this system, a greater likelihood exists of substantial if not catastrophic differences between the costs expected by the client and the likely costs of construction as reflected through bids or negotiations. It is this model that gives rise to the bulk of litigation.

The other model—that of more efficient techniques—is likely to be used by sophisticated clients who are aware of the difficulties design professionals have using the model just described. Clients are likely to engage someone who will be able to give a more accurate cost prediction as the design evolves. Sometimes this is done by hiring a skilled cost estimator—something close to the quantity surveyor used in England—as a separate consultant. Sometimes it is done by engaging a construction manager (CM) who is supposed to have a better understanding of the labor market, the materials and equipment market, and the construction industry, as well as the Construction Process itself. Using a CM is intended not only to free the designer from major responsibility for cost predictions but also to keep an accurate, ongoing cost prediction. Sometimes the CM agrees to give a guaranteed maximum price (GMP) that may vary as the design evolves. To do this, the CM may obtain firm price commitments from the specialty contractors.

This second model—a fine-tuned model—is designed to avoid the devastating surprises that are common under the traditional model. To determine whether the design professional bears the risk of losing the fee, a differentiation between the two models is vital, the risk being greater in the traditional model.

C. Creation of a Cost Condition: *Griswold & Rauma v. Aesculapius* and *Malo v. Gilman*

When clients assert that they have no obligation to pay the design professional, they are asserting a cost condition. They are claiming that their obligation to pay was *conditioned* on the accuracy of a design professional's cost prediction. A cost condition is a gamble by the design professional that the cost prediction will be reasonably accurate. If it is not, the fee is lost unless the client prevented the condition from occurring, such as making excessive design changes, or is willing to dispense with the cost condition. In such a case, the client need make no showing that the design professional has not lived up to the professional standard of care.

Clients sometimes contend that the design professional not only has gambled the fee but also has *promised* that the project will be brought in within a designated price. Failure to perform in accordance with this promise makes the design professional responsible for any damage to the client that was reasonably foreseeable at the time the contract was made. Subsection (C) emphasizes the question of whether a cost condition has been created, while Subsection (F) discusses the question of damages for breach.

The Parol Evidence Rule and Contract Completeness. The issue that has arisen most frequently in cost cases relates to the parol evidence rule (discussed in Section 11.04(E)). This rule determines whether a writing assented to by contracting parties is the sole and final repository of the parties' agreement. If so, testimony relating to agreements made before or at the time of the written contract is not admitted into evidence by the judge. In the context of a cost condition, the issue is whether the client will be permitted to testify that it had made an earlier oral agreement or had an understanding that it could abandon the project and not pay for design services if the low construction bid substantially exceeded the cost prediction of the design professional.

The client generally will be permitted to testify that such an agreement had been made if the agreement between the design professional and the client is oral, if the agreement is written but nothing is stated as to the effect of accurate cost predictions, or in some cases even if this problem is dealt with in the agreement. (See *Malo v. Gilman* later in this section). Permitting such testimony is based on the conclusion that such agreements are not, as a rule, the final and complete repository of the entire agreement between design professional and client.[3]

Standardized contracts prepared by the AIA or EJCDC include language that seeks to protect the design professional from assertions of the existence of a cost condition. Although such language has been useful to design professionals, its presence is not ironclad protection against clients being permitted to testify to their understanding that they could abandon the project if the costs were excessive and not pay the design professional for services. The cases reproduced later in this section involve clauses that sought to make the writing complete. Yet, as shall be seen, testimony seems to have been freely admitted. (The *current* AIA language is discussed later in this section.)

Keep in mind that permitting the testimony does not end the matter. Issues are still likely to exist as to whether the agreement took place, the nature of the agreement, and whether the condition has occurred or been excused.

Some Preliminary Issues. Before discussing cost conditions—or what the AIA calls a fixed limit of construction cost—other issues should be addressed. Are cost and cost control essential elements of professional service? What is the legal effect of giving a cost estimate? Do other obligations exist relating to cost and cost control?

Two cases, each involving identical AIA documents that required the architect to give preliminary cost estimates *when requested*, came to different conclusions. Texas held that the fiduciary relationship did not require the architect to take cost into account and advise the owner as to estimated cost.[4] Michigan held such service inherent

in the client-design professional relationship.[5] The unusual facts in the Texas case, the increased responsibility placed on fiduciaries, and the usual expectations of the client make it likely that the Michigan result will be followed.[6]

Although a cost estimate usually is simply an opinion and not a guarantee,[7] it can be a factual representation by the design professional. (This may be the reason the AIA in 1987 changed from a statement of probable construction cost to a preliminary cost estimate.) If it is sufficiently certain to be relied on reasonably by the client, an inaccurate estimate is a misrepresentation by the design professional. New York held that a cost prediction could constitute an intentional misrepresentation if it was guaranteed or if the client relied on the architect's opinion as an expert.[8]

An innocent misrepresentation that induces the making of the contract permits rescission of the contract. If the contract is made, the design professional cannot recover for services and must repay any amounts received. If her design was used by the client, the design professional can recover in restitution based on unjust enrichment. If the representation was negligently made, the client would also have a claim for damages (discussed in Section 12.03(F)).

Even if an estimate is not a guarantee or even if a cost condition is not created, there can be other cost-related obligations. For example, the Louisiana Supreme Court held that the engineer who designed a water slide was liable for *damages* despite a contractual provision stating that the engineer was simply giving opinions to be used only as a guide.[9] The court concluded that the engineer had breached the contract "not by giving an inaccurate initial estimate, but by failing to employ a professional estimator, failing to look at other water slides, failing to advise the owners about other contractual possibilities and failing to provide revised cost estimates."[10]

---

[3]*Stevens v. Fanning,* 59 Ill.App.2d 285, 207 N.E.2d 136 (1965). Many cases are collected in 20 A.L.R.3d 778 (1968).
[4]*Baylor Univ. v. Carlander,* 316 S.W.2d 277 (Tex.Ct.App.1958).

[5]*Zannoth v. Booth Radio Stations,* 333 Mich. 233, 52 N.W.2d 678 (1952).
[6]See *Williams Eng'g, Inc. v. Goodyear, Inc.,* 496 So.2d 1012, 1017 (La.1986), which followed the Michigan case.
[7]AIA Doc. B141, ¶ 5.2.1.
[8]*Pickard & Anderson v. Young Men's Christian Ass'n,* 119 A.D.2d 976, 500 N.Y.S.2d 874 (1986).
[9]*Williams Eng'g v. Goodyear, Inc.,* supra note 6. This case is discussed in Section 12.03(F).
[10]Id. at 1017.

(AIA Doc. B141, ¶ 3.4.10, states that detailed cost estimates are optional additional services.)

Existence of Cost Condition. Although cost condition cases generally involve the client claiming an *express* agreement based on a cost agreed to or set forth in the contract under which it could abandon the project and not pay a fee, a cost condition can be created by implication without any *specific* agreement as to the effect of inaccurate costs.[11]

Certainly design professionals do not operate in the dark. If they know what funds are available and are aware of the remoteness of obtaining additional funds, any cost specified may be "hard." In such a case, a cost condition may be created despite the absence of an *express* agreement under which this risk is taken.

Evidence that bears on the softness or hardness of any projected costs discussed by the design professional and the client or expressed in their agreement is crucial. This evidence, such as labeling the amounts as merely an estimate or using a cost range, may indicate that the amount or range specified is what is hoped for rather than a fixed-cost limitation. Where the amount is "soft," design professionals are being exhorted to use their professional skill to bring the project in for the amount specified. Where it is "hard," the client may be informing the design professional that the latter is risking the fee on ability to accomplish this objective.

This distillation of appellate cases appeared in a legal journal dealing with architectural cost predictions:[12]

> Courts have admitted evidence of custom in the profession. Architects have been permitted to introduce evidence that customarily architects do not assume the risk of the accuracy of their cost predictions. Also, courts have been more favorably disposed toward holding for the architect if the project in question has

involved remodeling rather than new construction, because estimating costs in remodeling is extremely difficult. The same result should follow if the type of construction involves experimental techniques or materials.

Courts sometimes distinguish between cases and justify varying results on the basis of the amount of detail given to the architect by the client in advance. Generally, the greater the detail, the easier it should be for the architect to predict accurately. However, it is much more difficult for the architect to fulfill the desires of the client within a specified cost figure if the client retains a great deal of control over details, especially if these controls are exercised throughout the architect's performance. For this reason, some courts have held that a cost condition is not created where the architect is not given much flexibility in designs or materials.

Some courts have looked at the stage of the architect's performance in which the cost condition was created. If it is created at an early stage, it is more difficult for the architect to be accurate in his cost predictions. Generally, the later the cost limit is imposed in good faith, the more likely it is to be a cost condition. But courts should recognize that if it is imposed later, creation—or, more realistically, imposition—may be an unfair attempt by the client to deprive the architect of his fee.

Occasionally the courts have applied the rule that an ambiguous contract should be interpreted against the person who drew it up and thus created the ambiguity. If the client is a private party, the contract is usually drafted or supplied by the architect. Courts have looked at the building and business experience of the client. If the client is experienced, he should be more aware of the difficulty of making accurate cost estimates. If he has building experience, the client is more likely to be aware of the custom that architects usually do not risk their fee upon the accuracy of their cost estimates.

Courts have sometimes cited provisions for interim payments as an indication that the architect is not assuming the risk of losing his fees on the accuracy of his cost estimates. However, standard printed clauses buried in a contract are not always an accurate reflection of the understanding of the party not familiar with the customs or the forms. If payments have actually been made during the architect's performance, this is a clearer indication that the client is not laboring under the belief that he will not have to pay any fee if the low

---

[11]*Stanley Consultants, Inc. v. H. Kalicak Constr. Co.*, 383 F.Supp. 315 (E.D.Mo.1974) (dictum). In *George Wagschal Assoc., Inc. v. West*, 362 Mich. 676, 107 N.W.2d 874 (1961), a cost condition was found in a consultant contract because the consultant knew of the client's budget limit. But if the contract expressly negates a cost condition, it will not be implied. *Kurz v. Quincy Post Number 37, Am. Legion*, 5 Ill.App.3d 412, 283 N.E.2d 8 (1972).

[12]Sweet & Sweet, *Architectural Cost Predictions: A Legal and Institutional Analysis*, 56 Calif.L.Rev. 996, 1006–1007 (1968).

bid substantially exceeds the final cost estimate. A few cases have looked for good faith on the part of the client. For example, if the client has offered some payment to the architect for his services, this may impress a court as a show of fairness and good faith. [footnotes omitted]

Standard Contracts and Disclaimers: A Look at AIA Standard Contracts. Professional associations have dealt with cost problems by inclusion of language in their standard contracts designed to protect users from losing fees when cost problems develop. The protective language has ranged from the brief statement that cost estimates cannot be guaranteed to the currently elaborate contract language contained in AIA Doc. B141 starting with ¶ 5.2 set forth in Appendix A. The length and complexity of the contract language demonstrates the seriousness of the problem.

Paragraph 5.2 requires the client to include in its budget amounts that take into account the possibility that the bids may substantially exceed the budget. Paragraph 5.2.1 requires the architect to use her best judgment to evaluate any client-supplied project budget or architect-created cost estimates but states that the architect does not warrant accuracy.

In addition to ¶ 9.6 stating that the writing is complete, ¶ 5.2.2 states that a project budget is not a fixed-cost limit and requires that any such limit be in writing and signed by the parties. Although not clear, it is likely that the ordinary signature to the contract as a whole will not be sufficient and that the drafters intended that such a limit be signed or at least initialed separately. If such a fixed-cost limit (not an ordinary budget) is established, the architect is given the right to determine "materials, equipment, component systems, and types of construction" to be included and to make reasonable adjustments in project scope. The architect can include pricing contingencies and alternate bids.

Paragraph 5.2.3 requires an adjustment if bid or negotiation is delayed more than ninety days after submission of construction documents. Finally, ¶ 5.2.4 is a contractual expression of the obligation of good faith and fair dealing. It states that if the lowest bona fide bid exceeds any fixed-cost limit, the owner must do one of the following:

1. Approve an increase in the fixed limit.
2. Authorize rebidding or renegotiating.
3. Terminate and pay for work performed and termination expenses if the project is abandoned.
4. Cooperate in project revision to reduce cost.

If (4) is chosen, under ¶ 5.2.5, the architect must bear the cost of redesign, specified as the limit of the architect's responsibility, and she is to be paid for any other service rendered.

Taken as a whole, AIA treatment of cost predictions uses a multifaceted approach. First, there is the attempt to preclude any oral testimony of a fixed limit. Second, there is the requirement that the client give up certain design choices if a fixed limit is employed. Third, additional protection is given the architect if the low bid substantially exceeds the cost limit.

That such clauses can help the architect is shown by *Griswold & Rauma*, reproduced later in this section.[13] Earlier cases, such as *Stevens v. Fanning*,[14] and some recent cases, such as *Malo v. Gilman*, also reproduced in this section, reflect a different attitude toward such protective clauses. Much will depend on the facts surrounding the transaction, especially the degree of variance, the extent of client changes in the design, and the way each party behaved before and after the problem surfaced.

Two Illustrative Cases. Two cases, both probably using the same standard form but coming to different conclusions, are reproduced here.

---

[13]See also *Torres v. Jarmon*, 501 S.W.2d 369 (Tex.Ct.App.1973).
[14]Supra note 3.

GRISWOLD AND RAUMA, ARCHITECTS, INC. v. AESCULAPIUS CORP.

Supreme Court of Minnesota, 1974. 301 Minn. 121, 221 N.W.2d 556.
[Ed. note: Footnotes omitted.]

PETERSON, Justice.

Plaintiff brought this action to enforce and foreclose a lien for $19,438.65 for architectural services provided defendant. Defendant answered by denying that it owed plaintiff anything and filed a counterclaim to recover $17,436.04 already paid plaintiff, defendant's theory being that plaintiff was entitled to no fee because it breached its contract by grossly underestimating the probable cost of construction of the as yet unbuilt building. The trial court found for defendant. We reverse, for reasons requiring an extended recital of the factual setting out of which the litigation arose.

Plaintiff is a corporation engaged in providing architectural services. Defendant is also a corporation, the principal stockholders of which are Drs. James Ponterio, P. J. Adams, and A. A. Spagnolo. Defendant owns a medical building in Shakopee which it rents to the Shakopee Clinic, which in turn is operated by Drs. Ponterio, Adams, and Spagnolo.

In early 1970 defendant decided to expand the Shakopee Clinic to allow for a larger staff of doctors. As a result the members of the corporation and their business manager, Frank Schneider, contacted various architectural firms and selected plaintiff.

At his first meeting with the doctors in February or March 1970 David Griswold, one of plaintiff's senior architects, was shown a rough draft of the proposed addition and given a very general idea of what the doctors wanted. Although the evidence is conflicting, it appears that at this meeting the doctors talked in general terms of a budget of about $300,000 to $325,000.

After a number of subsequent conferences with the doctors and Mr. Schneider, plaintiff prepared and delivered to the doctors on May 8, 1970, a document entitled "Program of Requirements." This document, which outlined and discussed the requirements of the project as then contemplated, contained the following final section:

"BUDGET:

"The design to evolve from this program will indicate a certain construction volume that can be projected to a project cost by the application of unit (per square foot and per cubic foot) costs; and eventually, as the design is developed in detail, by an actual materials take-off. Inevitably the projected cost must be compatible with a budget determined by available funds. It is obvious that adjustment of either the program or the budget may be necessary and that possibility must be recognized.

"The project budget established, as currently understood, is $300,000. It has not been stated if this is intended to include non-building costs such as furnishings, equipment and fees—which may be approximately 25% of the total expenditure—as well as construction costs. Advice in this respect will eventually be necessary.

"The essential principle to be considered in the design development is as previously stated in the paragraphs of the section titled PROJECT OBJECTIVES.

"The construction shall be as economical as possible within the limitations imposed by the desire to build well and provide all of the facility required for a medical service."

In spite of the suggestion at the end of the second paragraph quoted above, neither party at any time thereafter sought to define more particularly what the budget was intended to include. Mr. Schneider testified, however, that he believed the original budget figure included the cost of construction, architects' fees, and the remodeling of the old building. In contrast, Dr. Ponterio testified that it was his belief that the original budget figure did not include a communications system valued at $12,755, architects' fees, or the $20,000 remodeling of the existing building.

On or about May 22, 1970, plaintiff submitted to defendant two alternate preliminary plans, designated SK-1 and SK-2. Plan SK-1 projected the programmed services to be housed partly in the existing building and partly in the new building. Plan SK-2 projected the programmed services as being housed entirely in the new building. Plan SK-2, as specifically shown on the plans, involved a larger plan in terms of area than SK-1. Defendant indicated its preference for plan SK-2, the larger and more elaborate of the two.

On June 1, 1970, plaintiff provided defendant with a "Cost Analysis" of the plan chosen by defendant. This cost analysis showed the dimensions of the project in square feet as then contemplated, and computed the cost of the project at two different rates per square foot. At the higher rate per square foot, the cost came to $322,140, plus an estimated $20,000 for remodeling the old building, totaling $342,140. At the lesser rate per square foot, the cost came to $284,575, plus $15,000 for the remodeling of the old building, totaling $299,575. The cost analysis memorandum also noted that "the best procedure for projecting costs is by a materials take-off" which was to be done "when sufficient information is available."

It is undisputed that subsequent to the June 1, 1970, cost estimate, no further cost estimates were ever conveyed to defendant. What is disputed is whether in the ensuing months there was any discussion as to whether the project was coming within the budget. According to the testimony of Mr. Schneider, defendant was assured at all times during the preparation of the building plans and in all discussions with Griswold that the construction would come within the budget. Dr. Ponterio also emphasized that Griswold constantly mentioned the budget figure of $300,000 at their meetings. Griswold, however, denied that he had ever assured defendant that the project was coming within the budget.

Although the architectural services began in March and the first billing was May 6, 1970, no written contract was forwarded until June 23, 1970. At that time a standard American Institute of Architects (AIA) contract was forwarded, calling for payment at plaintiff's standard hourly rate and for reimbursement of expenses and recognizing that a lump sum fee for the construction phase would be negotiated prior to its commencement. The following provisions of the contract have relevance to this case:

"SCHEMATIC DESIGN PHASE
"1.1.3 The Architect shall submit to the Owner a Statement of Probable Construction Cost based on current area, volume or other unit costs."
"DESIGN DEVELOPMENT PHASE
"1.1.5 The Architect shall submit to the Owner a further Statement of Probable Construction Cost."
"CONSTRUCTION DOCUMENTS PHASE
"1.1.7 The Architect shall advise the Owner of any adjustments to previous Statements of Probable Construction Cost indicated by changes in requirements or general market conditions."
"THE OWNER'S RESPONSIBILITIES
"2.8 If the Owner observes or otherwise becomes aware of any fault or defect in the Project, or nonconformance with the Contract Documents, he shall give prompt written notice thereof to the Architect."
"CONSTRUCTION COST
"3.4 . . . Accordingly, the Architect cannot and does not guarantee that bids will not vary from any Statement of Probable Construction Cost or other cost estimate prepared by him.
"3.5 When a fixed limit of Construction Cost is established as a condition of this Agreement, it shall include a bidding contingency of ten per cent unless another amount is agreed upon in writing. . . .
"3.5.1 If the lowest bona fide bid . . . exceeds such fixed limit of Construction Cost (including the bidding contingency) established as a condition of this Agreement, the Owner shall (1) give written approval of an increase in such fixed limit, (2) authorize rebidding the Project within a reasonable time, or (3) cooperate in revising the Project scope and quality as required to reduce the Probable Construction Cost. In the case of (3) the Architect, without additional charge, shall modify the Drawings and Specifications as necessary to bring the Construction Cost within the fixed limit. The providing of this service shall be the limit of the Architect's responsibility in this regard, and having done so, the Architect shall be entitled to his fees in accordance with this Agreement."
"PAYMENTS TO THE ARCHITECT
"6.3 If the Project is suspended for more than three months or abandoned in whole or in part, the Architect shall be paid his compensation for services performed prior to receipt of written notice from the Owner of such suspension or abandonment, together with Reimbursable Expenses then due and all terminal expenses resulting from such suspension or abandonment."

Between June 1, 1970, and November 25, 1970, when the bids were opened, plaintiff worked actively with defendant both in the design development and construction documents phases of the project, plaintiff continuing to bill defendant on an hourly basis without objection by defendant. Although the project was substantially increased in size and scope during this period, plaintiff did not furnish and defendant did not request any up-to-date cost projections. Significantly important changes in the project made during this period include:

1. Replacement of offices with examining rooms necessitating additional plumbing;
2. More extensive X-ray space;
3. A doubling of the size of the laboratory;
4. Addition of a sophisticated communications system;
5. Addition of 2,100 feet of finished space in the basement (to provide facilities originally projected for the remodeled old building);
6. Enlargement of the structure as follows:

waiting area   1710' to 1724'
first floor   5880' to 6650'
basement   6260' to 6814'

All of the changes were discussed, approved, and understood by defendant. Defendant alleges, however, that it was under the impression that all such changes would be included in the original budget figure.

Bids were opened on November 25, 1970. The low construction bid was Kratochvil Construction Company at $423,380. Deductive alternates agreed to by defendant would bring the total low bid cost, including carpeting, down to $413,037.

Subsequent to the opening of the bids, the doctors called a meeting with Griswold at which the doctors informed Griswold that they could not complete the building according to the cost evidenced by the bids. Thereafter, Griswold met with the doctors and offered suggestions as to how the low bid figure could be reduced. Approximately $42,000 of reductions were projected, so that the final bid as reduced totaled $370,897. This figure included construction, carpeting, and remodeling of the old building and would meet the program of requirements without reducing size in any way. Griswold also pointed out to the doctors that the project cost could be further reduced by eliminating "bays" (series of examination rooms) from the building at a saving of approximately $35,000 per bay. Defendant was willing to accept the $42,000 reduction but was not in favor of eliminating any bays from the proposed project.

From November 25, 1970, when the bids were opened, until October 12, 1971, plaintiff and defendant met and corresponded many times in connection with various possible revisions of the project. During this time defendant never indicated that the project was abandoned and in fact, in January 1971 made a payment of $12,000 on its bill. During this time defendant never asked to be excused from the balance of the bill, and defendant offered to help plaintiff by paying interest on the open account if plaintiff required bank financing by reason of nonpayment. During this time defendant advised plaintiff that ground breaking would be deferred for 6 months, and that request by plaintiff for a further payment was "well taken" but that no further payment would be recommended until construction began.

The building, in fact, was never constructed. Plaintiff filed its lien on April 30, 1971, and commenced this action to enforce it in October 1971.

In analyzing the facts of this case we have considered Minnesota decisions as well as decisions from other jurisdictions. From these decisions we have extracted a number of factors which we believe are relevant to a determination of what the effect on compensation of an architect or building contractor should be when the actual or, as here, probable cost of construction exceeds an agreed maximum cost figure.

One very significant factor is whether the agreed maximum cost figure was expressed in terms of an approximation or estimate rather than a guarantee. Where the figure was merely an approximation or estimate and not a guarantee, courts generally permit the architect to recover compensation provided the actual or probable cost of construction does not substantially exceed the agreed figure.

Another significant factor is whether the excess of the actual or probable cost resulted from orders by the client to change the plans. Where the client ordered changes which increased the actual or probable construction costs, courts are more likely to permit the architect to recover compensation notwithstanding a cost overrun.

A third factor is whether the client has waived his right to object either by accepting the architect's performance without objecting or by failing to make a timely objection to that performance.

A fourth factor, applicable in a case such as this where the planned building was never constructed, is whether the architect, after receiving excessive bids, suggested reasonable revisions in plans which would reduce the probable cost. Courts have held that if the architect made such suggestions and the proposed revisions would not materially alter the agreed general design, then the architect is entitled to his fee, again provided that the then probable cost does not substantially exceed the agreed maximum cost figure.

Considering this case in light of these factors, we conclude that the trial court erred in denying plaintiff's motion for amended findings of fact, conclusions of law, and order for judgment.

First, it does not appear that plaintiff guaranteed the maximum cost figure. The trial court did not expressly state whether the agreed cost figure of $300,000 to $325,000 was an established estimate or a guarantee. However, the intention of the parties, as evidenced by the record, especially by contract provision 3.4, quoted earlier, more reasonably supports an established cost estimate than a guaranteed cost figure.

Secondly, the probable cost of the project in our view did not substantially exceed the cost figure. At trial architect Griswold testified and Mr. Schneider agreed, that the project could be completed for approximately $360,000 to $370,000 without reducing the square footage of the project. A probable construction cost of $370,000 exceeds the agreed cost estimate maximum found by the trial court, $325,000, by only 13 percent. It seems difficult to classify such a degree of cost excess as substantial. A review of cases cited in Annotation , 20 A.L.R.3d 778, 804 to 805, suggests that most courts would not consider such a degree of cost excess substantial.

Thirdly, we think it relevant that defendant approved substantial changes beyond the original plans. Defendant not only adopted and acknowledged these changes but was often active in advocating them (especially in expanding the X-ray facilities and changing the doctors' offices into extra exam rooms). It is true that defendant contends that it was under the impression at all times that such changes were within the original cost estimate. However, this impression seems unreasonable and unjustified in view of the scope of the changes.

Finally, we think it relevant that plaintiff showed defendant how they could reduce the cost of the lowest bid below $370,000. For example, defendant would have reduced the cost drastically by simply agreeing to eliminate one of the "bays" or a fraction of a bay from the project. Provision 3.5.1 of the contract required the parties to revise the project scope and quality if necessary to reduce the probable construction cost. While it might be against public policy to allow an architect, under such a provision, to reduce substantially the area of a proposed project, it seems that, barring a specifically guaranteed area, a reasonable reduction in the size of the project should be allowed when necessary to meet the construction cost.

Reversed and remanded.

SHERAN, C.J., took no part in the consideration or decision of this case.

MALO
v. GILMAN

Court of Appeals of Indiana, 1978. 177 Ind.App. 365, 379 N.E.2d 554.
[Ed. note: Footnotes renumbered and some omitted.]

STATON, Judge.

Edward Malo, an architect, completed plans and specifications for Arnold Gilman, who sought to construct an office building. The construction bids received totaled $105,000, 50% more than the preliminary estimated cost of $70,000, which appeared in the contract between the men. Gilman was unable to secure financing and the building was never built. Malo brought an action to recover his fee as architect. Gilman counterclaimed for the $500 he had paid Malo. The trial court found for defendant Gilman and granted his counterclaim. We affirm the judgment.

Malo agreed to provide architectural services to Gilman in the design and construction of an office building. Between May, 1967, and November, 1968, Malo expended considerable time and effort on the project. A verbal agreement was reached on July 28, 1967. Gilman was assured that costs of construction could be kept below $20.00 per square foot. After talking to prospective tenants, Gilman decided he required 3,500 square feet in the building. A standard American Institute of Architects (A.I.A.) form contract was signed on May 14, 1968.[15] Among the terms was the following:

> "It is recognized that this written contract ratifies the similar verbal contract entered into July 28, 1967.
>
> "The preliminary estimated cost of this project is Seventy Thousand Dollars, ($70,000.00)."

The contract also contained the following standard clause:

> "3.4 Statements of Probable Construction Cost and Detailed Cost Estimates prepared by the Architect represent his best judgment as a design professional familiar with the construction industry. It is recognized, however, that neither the Architect nor the Owner has any control over the cost of labor, materials or equipment, over the contractors' methods of determining bid prices, or over competitive bidding or market conditions. Accordingly, the Architect cannot and does not guarantee that bids will not vary from any Statement of Probable Construction Cost or other cost estimate prepared by him."

Malo completed the plans and specifications for the project in September, 1968. In October, 1968, bids were solicited. The lowest total of bids received was approximately $128,000, which was negotiated down to $105,000. The bids were never accepted. Gilman indicated that the bids were unacceptable to him and that he was unable to secure financing. In mid-December, Gilman sold the land on which the building was to have been erected.

Malo demanded payment of his fee for architectural services in the sum of $9,132.60. Gilman refused to pay. Malo brought an action to collect his fee. Gilman counterclaimed for the $500 he previously had paid Malo. The trial court denied Malo's claim, but granted Gilman's counterclaim.

We hold that the judgment of the trial court can be affirmed on either of two alternate theories: (1) that parol evidence was properly admitted to show a maximum cost limitation of approximately $70,000.00, which was exceeded unreasonably by Malo's plans for construction; or (2) that the estimated cost figure appearing in the contract placed a reasonable limit on the actual cost of the project, which limit was exceeded unreasonably. In either event, architect Malo breached the contract and is not entitled to compensation under the contract.

I. PAROL EVIDENCE TO SHOW A MAXIMUM COST LIMITATION

On appeal, Malo argues that no fixed price agreement appeared in the "fully integrated contract" for architectural services. The only figure appearing in the contract, $70,000.00, was merely a preliminary estimated cost figure, which was not binding on the architect.[16] Further, even if the trial court properly allowed evidence of a $20 per square foot cost limitation, Malo claims his final design plans contained 5,400 square feet,[17] 50% more space than originally projected. In

---

[15]Since the contract was prepared by Malo, it must be strictly construed against him and in favor of Gilman. *Oxford Development Corp. v. Rausauer Builders, Inc.* (1973), 158 Ind.App. 622, 304 N.E.2d 211.

[16]For a discussion of the legal issues raised by cost estimates, see Sweet & Sweet, *Architectural Cost Predictions: A Legal and Institutional Analysis*, 56 Calif.L.Rev. 996 (1968).

[17]Of this space, less than 3,900 square feet was on the main floor, with the remaining square footage comprising a "partially finished" basement.

that case, the bids totaling $105,000 were in the right price range for a building costing $20 per square foot.

Gilman contends that evidence showing the existence of a $20 per square foot cost limitation (or $70,000 to $78,000 total for the project) was properly admitted, since the contract failed to contain a maximum cost limitation. Further, no significant changes in the project occurred to increase its size or cost.

Normally parol evidence may not be considered if it contradicts or supersedes matters intended to be covered by the written agreement. However, parol evidence may be admitted to supply an omission in the terms of the contract. . . . Many contracts for architectural services, as here, fail to include specific requirements such as the size, style, and character of the building, the number of rooms, the quality of the materials to be used, and, finally, the maximum cost. Yet, according to section 1.1.1 of the A.I.A. form contract,

"The Architect shall consult with the Owner to ascertain the requirements of the Project and shall confirm such requirements to the Owner."

hus, depending on the specific needs of the owner, these requirements may be integral parts of the contract for architectural services. . . . A contract that fails to set out the details agreed upon, then, is not a complete and integrated statement of the agreement. . . . Parol evidence may be considered to determine the agreement with respect to these matters. . . .

Ordinarily, the maximum cost of a project is agreed upon prior to commencement of design. The owner who plans to construct a building has in mind a figure for the maximum cost of construction, particularly where, as here, he must secure outside financing. The architect must design the project, keeping in mind this maximum cost limitation. Evidence of the maximum cost limitation should be admissible where the contract fails to show that figure.[18] As noted in an

annotation to the *Spitz* case, 49 A.L.R.2d 679, 680 (1956):

"In the great majority of the cases where the question has been raised the evidence has been held admissible, usually on the ground that the written contract failed to disclose the parties' intention as to the cost of the structure contemplated, and that such contemplated cost was an element which must have entered into the negotiations."

Indiana has not yet decided a case on this point. However, the following cases allowed the introduction of parol evidence to show a maximum cost limitation when the contract failed to contain one: [Citing cases].

We agree that parol evidence of a maximum cost limitation may be introduced where the contract fails to contain such a limitation. The question of fact, whether architect Malo agreed to design a building, the cost of which could not exceed $20 per square foot (or $78,000 for 3,900 square feet), was resolved by the trial court in favor of Gilman. On examining the record, we cannot say that the finding of fact was incorrect as a matter of law.

Gilman testified that, from the beginning, he received repeated assurances that Malo would have "no problem" designing a building costing less than $20 per square foot. The amount of usable space Gilman required was approximately 3,500 square feet. By multiplying the figures, Gilman and Malo arrived at $70,000 as a "firm figure" for maximum cost. Gilman sought to reduce costs, accepting the use of a cost-saving "Uni-Roof" design, and suggesting a cost-cutting relocation of the basement.

Malo testified that he was aware that Gilman was interested in "getting the best price on the market." Yet he claimed that there was no ceiling on the cost of the project, that "it could be as much as a million dollars." Other evidence presented concerning the cost of the project included two bids received on the basis of Malo's preliminary drawings; one was $80,000; the other (which would incorporate the Uni-Roof) was $62,000 to $70,000. Finally, one of plaintiff's witnesses, who was present at a meeting where the bids were tabulated, testified that no specific "cost talk" was discussed, that he did not recall the figure of $70,000, but that $80,000 kept "ringing a bell."

The evidence fails to show that Gilman would have paid *any* sum of money to construct the proposed building. On the contrary, the evidence clearly shows that he wished to construct the building as cheaply as possible. The trial court correctly resolved the factual question of a maximum cost limitation in favor of Gilman. Under the terms of the agreement, then, Malo lost his right to recover compensation when he designed a building impossible of construction within the maximum cost limitation.

---

[18]A line of cases, among them *Wick v. Murphy* (1952), 237 Minn. 447, 54 N.W.2d 805, has admitted parol evidence to show the cost of the project agreed upon by the parties, based upon a different line of reasoning. As the courts interpret the contract, the fee for architectural services depends on the actual construction cost. Where no contract has ever been let, the terms of the contract become ambiguous; that is, the fee cannot be determined based on actual construction cost. Parol evidence is admissible to resolve the ambiguity.

We note that the contract contains a diagram, with percentage plotted against cost of the project, to determine the architect's fees. The contract utilizes the line marked "Group B." We note that the 7.75%, which is marked on the graph, corresponds to a cost figure of $70,000. This is clearly the figure intended to be used to calculate Malo's fee.

II. ESTIMATED COST FIGURE EXCEEDED UNREASONABLY

The contract for architectural services contained an estimated cost figure of $70,000. After negotiation, the lowest construction bids totaled $105,000, a figure 50% higher. Appellee Gillman argues that so great a discrepancy should bar Malo from receiving his fee for architectural services.

The court in *Caldwell v. United Presbyterian Church* [20 Ohio Ops.2d 364, 180 N.E.2d 638 (1961)] supported such a theory. In that case, the building would have cost at least $57,800 to build, or $12,800 (almost 30%) in excess of the $45,000 cost limitation on the project. The court concluded,

> "that plaintiff has not substantially complied with the terms of his written contract and, therefore, is not entitled to recover in this action. . . ."

Id. 180 N.E.2d at 642.

Citing many cases, a federal Court of Appeals decision declared as a general rule

> "That there may be no recovery for engineering or architectural services where the actual cost of the structure substantially or unreasonably exceeds the estimated cost limitation, unless the cost excess is attributable to the owner's action. . . ."

Food Management, Inc. v. Blue Ribbon Beef Pack, Inc., supra, 413 F.2d at 726.

More recently, *Durand Associates, Inc. v. Guardian Investment Co.,* (1971), 186 Neb. 349, 183 N.W.2d 246, involved construction bids which exceeded the estimate by 55%. The court interpreted the section of the contract in which the architect explicitly refused to guarantee the cost estimate, declaring that the figure represented a "reasonable approximation of the cost of the project." *Id.* at 250. This did *not* mean that an architect would never be bound by his estimate. Such a situation would be "contrary to public policy because it would mean that no matter how large the bid for doing the work, defendants would be obligated to pay an architectural fee based on that amount. . . ." *Id.* at 250. The court held,

> "that an architect or engineer may breach his contract for architectural services by underestimating the construction costs of a proposed structure. The rule to be applied is that the cost of construction must reasonably approach that stated in the estimate unless the owner orders changes which increase the cost of construction. . . ."

Id. at 251.

III. CONCLUSION

Under either theory, the trial court could have found that Malo breached his contract for architectural services and denied Malo compensation under the contract, and found that Gilman was entitled to recover the $500 he paid Malo under the contract. The judgment of the trial court is affirmed.

BUCHANAN, C.J. (by designation), concurs.

HOFFMAN, J., concurs in Part II only.

Negligent Cost Predictions. If a cost condition is created, it is not necessary to determine whether the design professional met the appropriate standards of performance. However, in many cases where the cost estimate is wide of the mark, it is likely that a design professional did not live up to the legal standard of performance.

In *Malo v. Gilman* (reproduced earlier in this section), one reason for barring the architect's claim for compensation was that the cost estimate had been unreasonably exceeded. Yet New York held that negligence will not be inferred when there is a discrepancy of 33% to 45% where 10% to 25% discrepancy is common.[19] That court held that a

judge should not instruct the jury that a cost estimate that is very far out of line proves that the architect did not do her work properly. Negligence could not be inferred simply because the discrepancy was greater than usual.

If the client does show that the design professional has been negligent in preparing the cost estimates, it is not necessary to establish that a cost condition has been created. It is implied that the design professional will perform in accordance with the standards of her profession. If she does not do so and if the project is abandoned, the client can recover any fees it has paid and need not pay any additional fees. However, if the client *uses* the design prepared by the design professional, the client must pay the net benefit of the architect's work. This is computed by subtracting any damages caused by the breach from the amount the client has been benefited by use of the design.

---

[19]*Pipe Welding Supply Co. v. Haskell, Connor & Frost,* 96 A.D.2d 29, 409 N.Y.S.2d 221 (1983) affirmed 61 N.Y.2d 884, 462 N.E.2d 1190, 474 N.Y.S.2d 472 (1984).

Stanley Consultants, Inc. v. H. Kalicak Construction Co.[20] illustrates a clear case of a failure to estimate properly. The project was a sixty-one-unit housing project to be built in Zaire, Africa. The cost estimate was $8 million, and the only bidder submitted a bid of $16 million. The design professional had prepared cost estimates without data from Zaire when such data were available. The driveways contained impassable grades. The sewer lines were twenty feet above the surface with eight-foot supports, and the sewer lines were to be fifty feet above a river without any supports being designated. Structures were located outside the property lines, and the design professional failed to take into account an easement to a religious shrine and created an encroachment. The court found ample evidence that the design professional had not lived up to the reasonable standard of professional skill. Such a finding could be the basis not only for denying the design professional compensation for services rendered but also for holding the design professional responsible for damages. See (F).

D. Interpretation of Cost Condition

Is the design professional allowed some tolerance in determining whether the cost condition has been fulfilled? One court required a reasonable approximation.[21] Roughly 10% seems to be accepted, although this tolerance figure may be reduced if the project is large and the fee justified continual detailed pricing takeoffs. Before 1977, ¶ 3.4.1 of AIA Doc. B141 included a bidding contingency of 10% unless another amount was fixed in writing. Now ¶ 4.2 requires a bidding contingency in the budget.[22] The degree of tolerance permitted may also depend on the language used to create the cost limitation. The more specific the amount, the more likely a small tolerance figure will be applied.

E. Dispensing with the Cost Condition

A cost condition generally is created for the benefit of the client. If it so chooses, the client can dispense with this protection. Courts that conclude that the client has dispensed with the cost condition usually state that the condition has been waived. Where this occurs, the condition is excused and the design professional is entitled to be paid even if the cost condition has not been fulfilled.

Excusing a condition can occur in a number of ways. A condition is excused if its occurrence had been prevented or unreasonably hindered by the client. For example, if the client does not permit bidding by contractors or limits bidding to an unrepresentative group of bidders, the condition is excused. The most common basis for excusing the condition has been excessive changes made by the client during the design phase.[23]

The client proceeding with the project despite the awareness of a marked disparity between cost estimates and the construction contract price can excuse the condition. By proceeding, the client may be indicating that it is willing to dispense with the originally created cost condition.[24] However, proceeding with the project should not *automatically* excuse the condition. It may be economically disadvantageous to abandon the project. Proceeding in such cases may not indicate a willingness to dispense with the condition. In such cases, any recovery of the design professional should be based on restitution measured by any benefit conferred.

F. Nonperformance as a Breach: Recovery of Damages

The client occasionally seeks damages based on the breach of a promise of accuracy or negligence in making the cost prediction. Refer to (C). An understanding of the basis of such claims requires that *promises* be differentiated from *conditions*.

The creation of a cost condition does not necessarily mean that the design professional promises to fulfill it. Design professionals can risk their fees on the accuracy of the cost prediction, though they may not wish to be responsible for losses caused by nonperformance.

Courts seem to *assume*, however, that a fixed-cost limit constitutes a promise by the design pro-

---

[20]383 F.Supp. 315 (E.D.Mo.1974).

[21]*Durand Assoc., Inc. v. Guardian Inv. Co.*, 186 Neb. 349, 183 N.W.2d 246 (1971).

[22]Despite a 10% tolerance figure, a court allowed the architect to recover where there was a 13% difference in *Griswold and Rauma, Architects, Inc. v. Aesculapius Corp.* (reproduced in Section 12.03(C)). But this factor was taken *with others* in ruling for the architect.

[23]*Koerber v. Middlesex College*, 128 Vt. 11, 258 A.2d 572 (1969). See also *Griswold and Rauma, Architects, Inc. v. Aesculapius Corp.* (reproduced in Section 12.03(C)).

[24]One factor considered in *Kurz v. Quincy Post Number 37, Am. Legion*, supra note 11.

fessional that the project would cost no more than the designated amount. Under this assumption, if costs substantially exceed predicted costs, the design professional has breached even though she has lived up to the professional standard in making cost predictions. The breach entitles the nonbreaching party to recover foreseeable losses caused by the breach that could not have been reasonably avoided by the nonbreaching party.

Suppose the project is abandoned. The client can recover any interim fee payments based on restitution.[25] In spite of this, restitution of fees paid does not occur often. For example, in *Stanley Consultants, Inc. v. Kalicak Const. Co.*[26] (discussed in (C)), the architect had received $18,000 in interim fee payments and the court considered this adequate compensation for the architect's services. The architect's breach of promise to perform in accordance with the professional standard—something clearly found by the trial court—entitled the client to restitution of any interim fee payments made. Yet the client did not seek restitution of the fees paid. Even where restitution *is* sought, clients often do not press these claims.[27] Denial of a design professional's claim does not necessarily mean services have been performed without any remuneration.

Abandonment of the project may cause other client losses, such as wasted expenditures in reliance on the design professional's promise to bring the project in within a designated cost.[28]

Redesign followed by construction very likely causes delay. Delay damages are recoverable if they can be proved with reasonable certainty and were reasonably foreseeable at the time the agreement was made.[29]

Suppose the project is constructed and the client seeks to recover the difference between the cost prediction and the actual cost. The property may be worth its cost. If so, the client has suffered no loss. Proceeding with the project knowing the costs would substantially exceed the predictions *may* show that the loss could have been reasonably avoided by abandoning the project.

Kellogg v. Pizza Oven, Inc.,[30] involved a client who rented space for a restaurant. The lease provided that the landlord would pay up to $60,000 for the cost of an improvement to the landlord's building. The architect had negligently estimated costs at $62,000, but the project cost $92,000. The tenant-client had to pay the balance of approximately $30,000. The tenant recovered this amount less a 10% tolerance for errors from the design professional because of the excess cost.[31]

Where the project is built, the architect should not be held for the excess of actual costs over predicted costs. This is not based on the client's having proceeded despite its knowledge that the costs will be more than anticipated. The client should not be required to give up the project to reduce damages for the design professional. However, the client benefits by ownership of property presumably of a value equal to what it has paid for the improvement. Damages should not be awarded unless the client can prove that the economic utility of the project was reduced in some ascertainable manner because of the excessive costs.

Williams Engineering, Inc. v. Goodyear, Inc.[32] also involved a claim for damages in a commercial context similar to *Kellogg*, noted earlier in this section. Williams, the engineer, had been retained to design a recreational water slide. Time was important, as the client hoped to open for business in the summer of 1979. Williams gave a preliminary estimate of some $409,000 but suggested that the client proceed on a "fast track" basis to expedite comple-

[25]*Durand Assoc., Inc. v. Guardian Inv. Co.*, supra note 21 (by implication).
[26]Supra note 20.
[27]But see *Malo v. Gilman* (reproduced in Section 12.03(C)).
[28]The client's claim in *Durand Assoc., Inc. v. Guardian Inv. Co.*, supra note 21, included excavation costs and losses suffered on a steel prepurchase made to avoid an anticipated price rise. These losses should have been recovered if they were reasonably foreseeable and not avoidable. Because they were incurred by an affiliated company of the client and not by the client, they were denied.
[29]*Impastato v. Senner*, 190 So.2d 111 (La.App.1966), denied recovery for delay, but *Hedla v. McCool*, 476 F.2d 1223 (9th Cir.1973), allowed recovery. AIA Doc. B141, ¶ 5.2.5, states redesign to be the limit of the architect's responsibility. This should not relieve the architect from responsibility for delay damages caused by negligence.

[30]157 Colo. 295, 402 P.2d 633 (1965). Similarly, see *Kaufman v. Leard*, 356 Mass. 163, 248 N.E.2d 480 (1969).
[31]The landlord ended up with improvements probably worth considerably more than the $60,000 he spent. Perhaps the design professional should have claimed against the landlord based on unjust enrichment. The unjust enrichment argument was also rejected in *Kaufman v. Leard*, supra note 30.
[32]Supra note 6.

tion.[33] Also, he suggested the contractor be hired on a cost-plus basis.

The design phase was completed on April 18, 1979. During design and construction, the engineer submitted written invoices for his fee based on a percentage of the estimated cost of $409,000. On August 1, 1979, three days before the slide opened, the engineer's bill still showed the cost as $409,000. But twenty days after the water slide was opened for business, the client was billed for construction costs of almost $888,000. The project was still only 82% complete. Even worse, the engineer submitted a new bill based on the projected cost of almost $1,000,000.

The client paid $824,000 for the water slide. The water slide turned out to be a financial failure; the client lost its lease and had to pay to have the slide removed. Although the basic reason for the failure was lack of customers, the delayed opening and some design features contributed to the financial disaster.

In addition to resisting the engineer's claim for additional fees, the client claimed and testimony supported that the engineer breached by failing to employ a professional estimator, failing to look at other water slides, failing to advise the client about the other contract possibilities, and failing to provide revised cost estimates. The client contended that it would have modified the design or given up the venture had it been apprised of ultimate cost.

The client initially sought damages of $634,000. As the trial proceeded, it reduced its claim to $409,000, by chance the amount it thought it was going to have to pay for the completed water slide. After an eight-day trial, the jury awarded the engineer additional professional fees of $25,000, attorneys' fees of $23,000, $2,800 in litigation costs, and expenses of expert witnesses of $6,000. It also decided that the engineer had breached the contract and that the client should have damages of $125,000 plus expert witness fees of $3,000. This created a net award of $71,200 in favor of the client.

The intermediate court of appeals reversed the jury's award granting the engineer additional fees and increased the damages to the client to $205,000.[34] The Louisiana Supreme Court, though noting the engineer did not operate in bad faith, held that he had committed a misrepresentation by his failure to re-

veal "even an approximation of the true cost of the project until after the construction was completed."[35] The court could not use the frequently used damage measure of the difference between the market value of the project and the excess cost, as the water slide had become worthless and had to be removed. The court affirmed the decision of the intermediate court, denying any additional engineering fees to the engineer and awarding the client over $205,000 in damages.

The intermediate appellate court's decision induced representatives of the design professions to submit briefs as "friends of the court." The briefs contended that the engineer did not guarantee the cost and could not be responsible for damages for the overrun. In response, the Louisiana Supreme Court stated damages were awarded not because of any guarantee of estimates but because the engineer was *negligent*, as shown by the testimony at the trial.

The case illustrates not only the difference between a cost condition and a promise but also the risk involved in designing for commercial ventures. Also, the protective language in the contract, much the same as the court commonly found in contracts prepared by design profession associations or by design professionals with good bargaining power, did not protect the design professional when he did not watch costs.

G. Relationship Between Principal Design Professional and Consultant

Section 12.10(B) discusses the use of consultants by the prime design professional. One problem that can arise is whether the consultant bears the risk that the prime design professional will not be paid by the client. While nonpayment can result from many causes, often this occurs when a project is abandoned because of excessive costs. *George Wagschal Associates, Inc. v. West*[36] involved a legal action by the consulting engineer against the principal design professional (the architect) for services he had performed in the design of a school that was never built to these plans because of excessive cost. The cost overrun apparently was due to engineering overdesign.

The architect evidently had not been paid by the school district and contended that the engineer

---

[33]See Section 17.04(B).
[34]480 So.2d 772 (La.App.1985).

[35]496 So.2d at 1018.
[36]Supra note 11.

should share this loss with him. The court upheld this contention and seemed to hold that the principal design professional and consulting engineer were jointly engaged in the project. The architect risked his fee by his contract with the school district, and the engineer risked his fee because he knew of the budgetary constraints.[37]

The AIA, representing mainly prime design professionals, and the Engineers Joint Contract Documents Committee (EJCDC), representing professional engineers, have differed as to the right of a consulting engineer to be paid if the architect has not been paid. The EJCDC argues that consultants generally do not take this risk. It notes that the prime design professional selects the client and has the best opportunity to evaluate its capacity to pay. It also points to AIA Doc. B141, ¶ 4.3, which allows the architect to request that the client provide a statement of funds available for the project and of the source of funds.

The AIA contends that frequently a long-term relationship exists between prime design professionals and consultants that resembles a partnership even if not cast in legal terms. It believes that under such conditions, the risk of nonpayment should be shared.

Obviously the standardized documents published by these associations reflect their positions, as does their unwillingness to endorse the documents of the other.

In 1987, the AIA issued a new C141. Paragraph 10.3.6 states that the payment to the consultant "will be made promptly after the Architect is paid by the Owner." In most jurisdictions, this language would not be specific enough to create a payment condition and make the consultant take the risk that she would not be paid if the prime design professional is not paid.[38] But ¶ 10.3.6, after noting that the architect must make reasonable and diligent efforts to collect from the owner, states that regardless of whether the owner pays in full, "the Architect will pay the Consultant in proportion to the amount received. . . ." If the architect gets nothing, she pays nothing, but if the architect recovers half of her fee, she must pay the consultant half of the latter's fee.

Also in 1987, ¶ 11.4.3 was added. It is a blank that includes a direction to users to insert any "provisions as to conditions, contingencies . . . and other particulars concerning payments. . . ." This is a laudable attempt to individualize the sensitive payment conditions problem, one that should be worked out in advance by the prime design professional and the consultant.

H. Advice to Design Professionals

Many design professional-client relationships deteriorate because of excessive costs. Design professionals should try to make the cost prediction process more accurate. The chief method chosen by the professional associations has been to use the contract to protect their members from losing their fees where their cost predictions are inaccurate. The professional associations have included provisions in their contracts that are supposed to ensure that fees will not be lost when cost predictions are inaccurate and that fees will be lost only where the cost predictions are made negligently.

Protective language should be explained to the client. Design professionals who give a reasonable explanation to a client are not likely to incur difficulty over this problem. They should inform the client how cost predictions are made and how difficult it is to achieve accuracy when balancing uncontrollable factors. They should state that best efforts will be made but that for various specific reasons, the low bids from the contractors may be substantially in excess of the statement of probable construction costs. The suggestion should be made that under such circumstances, the design professional and the client should join to work toward a design solution that will satisfy the needs of the client. In helping the client to be realistic about desires and funds, the design professional should request that the client be as specific as possible as to expectations about the project.

If these steps are taken, some clients may be lost. It may be better to lose them at the outset than to spend many hours and either not be paid or be forced to go to court to try to collect. Without an honest discussion with the client at the outset, the design professional takes risks.

---

[37]If the lost fee resulted from engineering overdesign, should the architect be able to transfer his lost fee to the engineer?

[38]Since the problem of the payment condition occurs most commonly in disputes between prime and subcontractor, it is dealt with in Section 28.06.

Design professionals should also consider greater flexibility in fee arrangements. If the stated percentage of construction costs is used, it may be advisable to reduce or eliminate any fee based on construction costs that exceed cost predictions.

During performance, the design professional should state what effect any changes made by the client will have on any existing cost predictions. It is hoped that not every change will require an increase in the cost predictions. If the client approves any design work, the request for approval should state whether any change has occurred in cost predictions.

PROBLEM

A, an architect, had been retained by C to design a small commercial building. The contract was AIA Doc. B141 (as set forth in Appendix A). C requested that the building be five stories and contain a specified amount of square feet. She also requested that the building have enough luxury features so that she could attract high-class tenants who would pay high rentals. As she hoped to attract law firms as tenants, she stated that sufficient space should be segregated for a law library that could be used by all of the lawyer tenants.

In the course of her design performance, A had informed C that she planned to use a very luxurious type of wood paneling in the larger offices and that there would be murals painted in the entrance hall. She also stated that there would be a sauna on each floor. She planned to install piped-in music and a number of other luxury features. All of this was agreeable to C.

The price was not to exceed one million dollars. The construction documents were submitted to five bidders, and the lowest bid was $1.3 million. A then stated that she would replace the luxurious wood panels with a less expensive type of wall construction. She suggested that the murals and saunas be eliminated. She suggested elimination of the music system. Plumbing features would be of cheaper quality. In addition, A wanted to eliminate the library and thus increase the rental area as well as cut costs. She also stated that the size of the windows would be reduced and that a number of other features would be changed to cut the costs.

C was unhappy about all these changes. She stated that she could not charge the projected rent unless the luxury features were retained. Does A have the right to make these changes? Are there any limits to A's rights to make deletions or substitutions? Give illustrations of changes that would be permissible and those that would not.

SECTION 12.04 Assistance in Obtaining Financing

Frequently, a building project requires lender financing. To persuade a lender that a loan should be granted, the client generally submits schematic designs or even design development, economic feasibility studies, cost estimates, and sometimes the contract documents and proposed contractor. This submission includes materials prepared by the design professional.

Must the design professional do more than permit the use of design work for such a purpose? Does the basic fee cover such services as appearing before prospective lenders, advising the client as to who might be willing to lend the client money, or assisting the client in preparing any information that the lender may require? Must this be done only if requested as *additional* service and paid for accordingly?

The professional education and training of a design professional does not include techniques for obtaining financing for a project. Nor is it likely that the design professional will be examined on this activity when seeking to become registered. It should not be considered part of *basic* design services.

Clearly this is true if the design professional has been retained by a large institutional client with personnel experienced in financial matters. The result does not change, however, even if the client does not have the skill within its own organization. Rarely does the client engage a professional designer because the client expects the designer to have and use skills relating to obtaining funds for the project.

Most firms give financing advice to some clients, many furnish financial contacts, and almost all arrange financing occasionally. Yet extra charges for such services are rare. Perhaps the activity is not burdensome, or it merely reflects economic realities of practice.

To understand how the AIA deals with this issue in its B141, it is important to note the function of dividing basic from additional services. Although this division will be discussed later,[39] one function of the long list of additional services is to exclude them by implication from basic services.

Does Article 3—the long list of additional services—specifically include services relating to obtaining financing for the project? Although ¶ 3.4.2 lists financial feasibility as an additional service, this does not include obtaining money for the project. More likely it means the economic viability of the project.[40] As a result, Article 3 does not by implication exclude such services from basic services. But Article 2, dealing with basic services, does not include these services among basic services. This omission, taken together with the reasons advanced earlier in this section, supports the conclusion that the architect does not perform these services as basic services under Article 2.

But *must* the architect perform such services as an additional service if requested by the owner? Paragraph 3.4.20—the catchall provision dealing with additional services—is broad enough to empower the owner to demand the architect perform these services. If ordered and performed, such services are additional and entitle the architect to compensation beyond any basic fee.

This reveals a deficiency in ¶ 3.4.20. It can compel the architect to perform services for which she may lack professional skill. Also, performance of these services may expose her to liability if she does not do a good job or if the owner can establish that the architect *promised* to obtain financing but did not do so.[41]

If the architect performs these services without authorization, she cannot recover additional compensation. Her performance has indicated she considered it part of basic services.

Services outside basic design services can involve the risk of increased liability if the services are performed but the client asserts they were not done properly. (See Section 12.05.)

SECTION 12.05 Economic Feasibility of Project

As noted in Section 12.04, AIA Doc. B141, ¶ 3.4.2, states that financial feasibility studies are additional services. They should be interpreted as studies that deal with the profitability of the project, not simply the availability of funds. The determination by a lender will be affected by the general economic feasibility of the venture. But if the architect should hesitate before making representations as to the availability of funds, she should certainly avoid venturing into financial feasibility. Although ¶ 3.4.2 makes clear that these are not part of basic services, the architect must perform such studies if requested to do so by the owner. If such studies are requested and made, the design professional will be exposed to claims by the client if the venture is unsuccessful. Such predictions are treacherous, ones for which design professionals are rarely trained by education and experience.

In *Martin Bloom Associates, Inc. v. Manzie*, the plaintiff architect sued the defendant clients for design services he rendered to the defendants before the project was abandoned by the clients because they could not obtain financing. The clients owned land in Las Vegas, Nevada, which they wished to develop for investment purposes. The architect produced a written contract under which he was to provide design services and contract administration for a designated compensation fee. Here the issue was not the architect's right to additional compensation but his responsibility for the accuracy of his representations as to financing and profitability.

The clients claimed that the architect had represented that the clients would have no difficulty in obtaining financing and that the project would be a profitable one. The architect's version differed. He contended that the entire agreement was in the letter and that no representation had been made as to profitability.

First, the trial judge concluded that the written agreement was not the complete contract between the parties. The judge then stated:

> Throughout the two days of trial, the Court had the opportunity to observe the manner of the witnesses while they testified, the consistency of their versions, both internally and compared to those of other witnesses, the probability or improbability of their versions and their interest in the outcome of the lawsuit. Based

---

[39]See Section 13.01(G).
[40]See Section 12.05.
[41]See *Herkert v. Stauber*, 106 Wis.2d 545, 317 N.W.2d 834 (1982) (citing and following earlier edition of treatise), discussed in Section 3.06.

on these factors, the Court credits the testimony of Manzie [the client] that representations of Bloom [the architect], both express and implied, induced Manzie to enter into the agreement and also constituted part of the agreement itself.[42]

The case illustrates the danger of exceeding one's professional capabilities. If the architect had, as determined by the trial judge, made representations relating to financial feasibility and profitability, he may very well have ventured into an unpredictable area beyond his professional skill, services excluded from coverage under his professional liability policy. As seen in Section 12.03, architects and engineers often lose fees when their cost predictions are inaccurate. But at least cost predictions, though certainly difficult, should be within the professional competence of a design professional. Economic feasibility goes beyond this and should be avoided.

SECTION 12.06 Securing Approval of Public Authorities

Greater governmental control and participation in all forms of economic activity have meant that the design professional increasingly deals with federal, state, and local agencies. Must the design professional assist the client or its attorney in preparing a presentation to be made for the planning commission, zoning board, or city council? Must she appear at such a hearing and act as a witness if *requested*? Is the design professional who does these things entitled to compensation in *addition* to the basic fee?

Reasonable cooperation by the design professional in matters relating to public land use control can be expected by the client. Design professionals are expected to have expertise in matters that are often at issue in these public hearings. It may be within the client's reasonable expectations that the architect or engineer will render reasonable assistance and advise the client on these matters. Each contracting party should do all that is reasonably necessary to help obtain the objectives of the other.

The increased legal controls and their effect on design professionals' services are reflected in AIA Doc. B141. Documents drafted in the 1970s were not models of clarity or completeness in dealing with the design professional's role in obtaining approvals. But in 1987, the AIA made clear that services related to this activity did not fall under basic services but, if authorized, gave the architect compensation in addition to the basic fee. For example, ¶ 3.4.4 states that "special surveys, environmental studies and submissions required for approvals of governmental authorities or others having jurisdiction" are optional additional services. Similarly, ¶ 3.3.8 states that providing "services in connection with a public hearing" are contingent additional services. The differences between optional and contingent additional services are discussed in Section 13.01(G).

Reflecting the different nature of engineering projects and their funding sources, EJCDC No. 1910-8 states the following are additional services requiring advance authorization:

> 3.1.1. Preparation of applications and supporting documents (in addition to those furnished under Basic Services) for private or governmental grants, loans or advances in connection with the Project; preparation or review of environmental assessments and impact statements; review and evaluation of the effect on the design requirements of the Project of any such statements and documents prepared by others; and assistance in obtaining approvals of authorities having jurisdiction over the anticipated environmental impact of the Project.

> 3.1.17. Preparing to serve or serving as a consultant or witness for OWNER in any litigation, arbitration or other legal or administrative proceeding involving the Project (except for assistance in consultations which is included as part of Basic Services).

Again, it may be useful or desirable to perform such services without requesting additional compensation. This section deals solely with the questions of whether design professionals are obligated to perform the services and whether they are legally entitled to be paid additionally for doing so.

SECTION 12.07 Services of a Legal Nature

Some design professionals volunteer or are asked to perform services for which legal education, training, and licensure may be required. This may be traceable to the high cost of legal services, and

---

[42]389 F.Supp. 848, 852 (D.Nev.1975).

the uncertainty as to which services can be performed only by a lawyer.

Undoubtedly, one of the most troublesome activities is drafting or providing the construction contract. It may not be *wise* for lay persons to draw contracts for themselves, although the law allows professional designers to supply or draft contracts for their *own* services. Suppose the designer drafts or selects the *construction* contract for a client? Lawyers often contend that design professionals are practicing law when they perform such activities. To meet such complaints, the AIA includes language on its standard forms of agreement stating that the document has important legal consequences and that "consultation with an attorney is encouraged with respect to its completion or modification."

AIA Doc. B141, ¶ 2.4.2, requires that the architect assist the owner in the preparation of "bidding information, bidding forms, the Conditions of the Contract, and the form of Agreement between the Owner and the Contractor." In addition, ¶ 4.8 requires the owner to furnish legal services as necessary. The AIA clearly wants architects not to perform legal services.

Some provisions in construction contacts can be considered "architectural" or "engineering" in that architectural or engineering training and experience are essential to reviewing or even drafting such provisions. This may tempt professional designers to play an aggressive role rather than realize that their *advice* will certainly be useful to the client or its lawyer.

A blurred border exists between legal and nonlegal services in the land use area. Although land use matters can and do involve legal skills, the architect may be knowledgeable about the politics and procedures in land use matters, particularly if she has had experience in dealing with land use agencies.

Another illustration of the overlap in matters relating to land use can be demonstrated by the hearings required before public agencies charged with the responsibility of issuing permits. On the whole, such hearings tend to be political rather than legal. The orchestration of such a hearing may be best in the hands of someone with political skill and experience. This is the responsibility of the owner, although the architect or engineer can be of great value in advising the owner or even in orchestrating the hearing and the strategy for it.

The line between legal and nonlegal services can also be demonstrated by the not uncommon phenomenon of the design professional being asked to advise the client as to whether a surety bond should be required. Because of their experience, design professionals may be put in the position of being able to determine whether a particular contractor should be bonded. Their experience may also lead them to have strong opinions on the general desirability of surety bonds. Repetitive involvement with such matters may give design professionals confidence that they can handle them.

Here the balance tips strongly in favor of considering these legal services. Sometimes bonds are required for public projects. The use of surety bonds may to a large degree depend upon other legal remedies given subcontractors and suppliers, such as the right to assert a mechanics' lien or to stop payments. The design professional should simply answer specific questions of a nonlegal nature rather than advise generally on such matters. If *asked* to advise on nonlegal issues, the services a design professional provides in response to such a request are additional (such services are excluded from professional liability insurance coverage).

Perhaps the most educational "horror" story involved a project that was abandoned because of excess costs and the unavailability of funds, but not until after a contract had been made with the contractor with no specific language allowing the owner to cancel the construction contract if funds could not be obtained.

The contractor demanded lost profits. When the owner refused, the contractor demanded arbitration. The contractor pointed to AIA Doc. A201 (incorporated by reference in AIA Doc. A101, which was the only document the owner had signed or seen). Nevertheless, an arbitration was held and the arbitrator awarded lost profits to the contractor. As if this were not bad enough, the architect who had furnished the contract documents sued for services he had performed. The owner contended:

1. When the architect furnished the contract he was acting as a lawyer.
2. A competent lawyer would have included protective language in the contract with the contractor and would certainly have pointed out the arbitration clause to the owner.
3. The architect must pay the arbitration award and be denied any fees for his services—legal

services being performed in violation of law and the design services having been performed negligently because of the cost overrun.

The case was settled, according to some sources, on terms close to the amount demanded by the owner.

SECTION 12.08 Site Services

A. Relation to Chapter 14

Section 12.08 concentrates on the reasonable expectations of the client as to the design professional's role on the site while construction proceeds. It focuses on the belief that the client *may* have that the design professional has been paid to *ensure* that the client receives all it is entitled to receive under the construction contract. To accomplish this, the client may expect the design professional to watch over the job, see to it that the contractor performs properly, and be responsible if the contractor does not. As seen in (B), a design professional may take a different view of her function. The purpose of Section 12.08 is to explore this problem with a view toward seeing what the law has done when faced with resolving the often different expectations of client and design professional.

Site services are but a part of the total professional services performed by the design professional. Disputes over design performance may involve determining whether the design professional has performed in accordance with her professional or contractual obligation.

The architect's site services and the frequently used disclaimers of responsibility for the work of the contractor or for safety often play a significant part in the architect's liability to clients or third parties who have suffered losses that they can trace to the failure of the design professional to perform site services properly. Liability will be discussed in Chapter 14.

B. Supervision to Observation: *Watson, Watson, Rutland/Architects v. Montgomery County Bd. of Educ.*

While contracts between design professionals and their clients include, as noted in Section 12.08(A), a lengthy list of services, this section treats the *general* role of the design professional during construction with special reference to site visits. The details of site visits are often dealt with in the contract between the design professional and her client. Provisions dealing with this topic can include requiring the design professional to take an active role in the process, performing such tasks as solving execution problems, directing how work is to be done, searching carefully to determine failure to follow the design, and seeing that safety regulations are followed. They can also limit the design professional to a passive role under which she "walks the site" periodically to check to see how things are going, perhaps measures progress for payment purposes, and reports to the client about work in progress. If the design professional takes an active role, she will be diluting the authority and responsibility of the contractor and expose herself and her client to liability. Yet some clients may prefer this route because they see it as a means of ensuring a more successful project.

The professional associations representing design professionals have favored placing the authority and responsibility for executing the design solely upon the contractor and limiting the architect or engineer to a more passive role. To implement this, the standard documents published by professional associations include language that clearly gives sole responsibility for executing the design to the contractor. Such language disclaims any responsibility on the part of the design professionals for how the work is to be done or for the contractor's failure to comply with the contract documents. The effect of such disclaimers, as shall be seen later, can generate concern both by clients (What am I getting for my money?) and by courts (Is the design professional taking any responsibility?).

As noted, the AIA has, as has the EJCDC, selected a passive mode for architects and engineers. This can be demonstrated by a brief look at the history of AIA documents.

Before 1961, the architect had general supervision of the work. In response to the specter of the expanded liability to third parties, the AIA dropped the phrase "general supervision" in favor of language under which the architect observed rather than supervised, made visits "at intervals appropriate to the stage of construction or as otherwise agreed"[43] rather than conducted exhaustive on-site inspections, and did her best. She was *not*

---

[43]AIA Doc. B141, ¶ 2.6.5.

responsible, however, for the contractor's failure to perform in accordance with the contract documents or for methods of executing the design.

The AIA justified this language change by suggesting that the architect was no longer the masterbuilder exercising almost total domination of the Construction Process from beginning to end.[44] Instead, said the AIA, the architect has turned over the responsibility for executing the design to the contractor who presumably has the necessary skill to accomplish this properly and safely. Even more, in sophisticated construction, the architect may simply be part of the management team that may consist of a project manager, a construction manager, and a field representative of owner or lender. This shift in role and function, though principally in response to increased liability of third parties, was also designed to recognize the shift in organization for most construction, the architect no longer being masterbuilder but being *called in* to interpret or decide disputes and making periodic observations to check on the progress of the work, mainly to issue certificates for progress payments.

The design professional associations contend that such terminological changes simply reflect actual practices and were not intended to change design professional site responsibilities. Yet it seems clear that such changes were directed toward the courts with the hope that they would provide defenses to what the professional associations thought were unmeritorious claims against design professionals. The AIA's approach is explored in the *Watson* case (reproduced in this section).

The following issues can arise as to the specific obligation to visit the site:

1. When must the design professional be on the site (continuous presence versus periodic visits)?
2. What is the intensity level of checking for compliance (intense inspection to ferret out deficiencies versus casual observation)?
3. If deficiencies or noncompliance is discovered, what must the design professional do (direct work be corrected versus report to owner)?

The following issues can arise as to the contractual disclaimers:

---

[44]See Sweet, *The Architectural Profession Responds to Construction Management and Design-Build: The Spotlight on AIA Documents*, 46 Law & Contemp.Probs. 69 (1983).

1. Do they totally exculpate the design professional from any responsibility for failure by the contractor to comply with the design documents? (Are they a basis for summary judgment resolving the issue of responsibility without a full trial?)
2. Do they exculpate the design professional even if she discovers the noncompliance by the contractor?
3. Do they exculpate the design professional if she has failed to comply with her contractual obligation? (Is she responsible for what she would have detected had she complied with her contractual commitment?)

The exact nature of the design professional's site visits depends, of course, upon the contract, both its express and implied terms. The contract may specify continuous versus noncontinuous presence, the frequency and timing of visits, the intensity of inspection, and action to be taken upon discovery of deficiencies.

Kleb v. Wendling[45] is instructive. It held the architect to a supervision standard of the contract despite the use of the less comprehensive term *administration*. An architect who has agreed to supervise or oversee the construction of a building must "prevent gross carelessness or imperfect construction." Merely detecting defective workmanship does not relieve the architect of a duty to prevent it. This appears to require that the architect provide continuous monitoring of the contract or service. In this case, the architect's representative had visited the site daily, while the architect had visited the site at least twice a month. (The legal effect of permanent on-site observation is discussed in (D).) This strict standard may have been traceable to the owner's absence during construction and a representation by the architect that he would keep a "closer tab on the general contractor than he normally would because he doubted the general contractor's competence."

As to compliance with the standard applicable, *Kleb* pointed to the reputation of the contractor for corner cutting and the absence of the owner as indications that careful monitoring was expected.

Other factors likely to be examined to determine whether the duration, frequency, and timing of site visits are adequate are as follows:

---

[45]67 Ill.App.3d 1016, 385 N.E.2d 346 (1979).

1. Size of the project.[46]
2. Distance between the site and the design professional's home office.[47]
3. When crucial steps are undertaken, such as pouring concrete or covering work.[48]
4. Type of construction contract. (Cost contracts require more monitoring.)
5. Experimental design or unusual materials specified.
6. Extent to which owner has a technical staff that will take over some of these responsibilities.

7. Observation of contractor's performance during visits.[49]
8. Contractor's record of performance on the project.

The *Watson* case (reproduced here) notes the important cases dealing with the responsibility of the design professional for failure of the contractor to comply with the contract documents, with special reference to the effect of disclaimers. *Watson* also demonstrates the complexity of a defect claim with multiple parties and multiple claims.

---

[46]*First Nat'l Bank of Akron v. Cann*, 503 F.Supp. 419, 437 (N.D.Ohio 1980) (citing earlier edition of treatise) affirmed 669 F.2d 415 (6th Cir.1982).
[47]*Warde v. Davis*, 494 F.2d 655 (10th Cir.1974).
[48]*Tectonics*, 89-3 BCA (CCH) ¶ 22,119 (PSBCA 1990).

[49]*Chiaverini v. Vail*, 61 R.I. 117, 200 A. 462 (1938) (visits when contractor *not* on site inadequate).

WATSON, WATSON, RUTLAND/ARCHITECTS, INC. v. MONTGOMERY COUNTY BOARD OF EDUCATION, et al.

559 So.2d 168 [Ed. note: footnote renumbered.] Supreme Court of Alabama, 1990.

MADDOX, Justice.*

This case arises out of property damage incurred when the roof of Brewbaker Junior High School in Montgomery leaked. The Montgomery County Board of Education (hereinafter "the School Board") filed this action against Bear Brothers, Inc., the general contractor; United States Mineral Products Company (hereinafter "U.S. Mineral"), the manufacturer of the roofing membrane; and W. Murray Watson, W. Michael Watson, and J. Michael Rutland, the architects for the school project. Bear Brothers joined Dixie Roof Decks, Inc. (hereinafter "Dixie"), the roofing subcontractor, as a third-party defendant. The trial court substituted the corporate entity Watson, Watson, Rutland/Architects, Inc. (hereinafter "the Architect"), for the individually named architects.

The School Board alleged negligence and breach of contract against the Architect and breach of contract and breach of guaranty against the Bear Brothers and U.S. Mineral. The Architect filed a cross-claim against U.S. Mineral and Dixie for indemnity in case the Architect were held liable to the Board. The trial court entered summary judgment for U.S. Mineral and

Dixie on that cross-claim, holding that a one-year statute of limitations applied to that cross-claim and that the statute barred the claim.

The School Board settled with Bear Brothers and U.S. Mineral for a total of $100,000. At trial against the Architect, the court granted the Architect's motion for a directed verdict as to the School Board's negligence claim, on the ground that the statute of limitations had expired. The breach of contract claim was submitted to the jury, and the jury returned a verdict of $24,813.08 against the Architect. The court entered a judgment based on that verdict.

The Architect appealed from that portion of the judgment based on the verdict in favor of the School Board and from the dismissal of its cross-claim against U.S. Mineral and Dixie. The School Board then cross-appealed from that portion of the trial court's judgment based on the Architect's directed verdict on the negligence claim.

At the center of this dispute is the architectural agreement between the Architect and the School Board; it included the following language:

"ARTICLE 8. Administration of the Construction Contract. The Architect will endeavor to require the Contractor to strictly adhere to the plans and specifications, to guard the Owner against defects and deficiencies in the work of Contractors, and shall promptly notify the Owner in writing

---

*Justice Maddox did not attend oral arguments, but he has listened to the tape of the arguments and has carefully examined the record and the briefs.

of any significant departure in the quality of materials or workmanship from the requirements of the plans and specifications, but he does not guarantee the performance of the contracts.

"* * * *

"The Architect shall make periodic visits to the site and as hereinafter defined to familiarize himself generally with the progress and quality of the Work and to determine in general if the Work is proceeding in accordance with the Contract Documents. On the basis of his on-site observations as an Architect, he shall endeavor to guard the Owner against defect and deficiencies in the work of the Contractor. The Architect shall not be required to make continuous on-site inspections to check the quality of the Work. Architect shall not be responsible for construction means, methods, techniques, sequences or procedures, or for safety precautions and programs in connection with the Work, unless spelled out in the Contract Documents, and he shall not be liable for results of Contractor's failure to carry out the work in accordance with the Contract Documents.

"* * * *

"The Architect shall not be responsible for the acts or omissions of the Contractor, or any Subcontractors, or any of the Contractor's or Subcontractor's agents or employees, or any other persons performing any of the Work."

The following issues respectively have been argued orally and in the parties' briefs:

On the Architect's appeal:

(1) Does the exculpatory language in the architectural agreement absolve the Architect from liability for damages arising from the failure of the contractor to follow the plans and specifications?

(2) If the answer to Issue 1 is no, then is the Architect entitled to cross-claim for indemnity against the roofing subcontractor and the manufacturer of the roofing membrane?

On the School Board's cross-appeal:

(3) When did the statutory period of limitations applicable to the school board's negligence claim begin to run?

I

The Architect argues that the exculpatory language in Article 8 of the architectural agreement absolves the Architect from liability to the School Board because all roof leaks involved were attributable to the faulty workmanship of the contractor. The Architect points out that Article 8 provides for two types of inspection services by the Architect; that is, the owner could elect to receive only *general* site inspection by the Architect, or the owner could elect to pay an additional fee for continuous on-site inspections (known as the "clerk

of the works" alternative); the School Board elected not to pay for the second option.[50]

This Court has construed language virtually identical to that at issue here, in *Sheetz, Aiken & Aiken, Inc. v. Spann, Hall, Ritchie, Inc.*, 512 So.2d 99 (Ala.1987). In that case, the contract in question contained the following language:

"The ARCHITECT shall not be responsible for the acts or omission of the contractor, or any subcontractors, or any other contractor or subcontractors, agents or employees, or any person performing any work."

The Court, in *Sheetz* stated: "The contract expressly states that Spann is not responsible in any fashion for the acts or omission of the contractor, subcontractors, agents, or employees performing the work." 512 So.2d at 102.

A case analogous to this one is *Moundsview Indep. School Dist. No. 621 v. Buetow & Associates, Inc.*, 253 N.W.2d 836 (Minn.1977), where the contract language regarding inspection duties was virtually identical to that used here; the trial court entered a summary judgment for the architect, and the Minnesota Supreme Court affirmed. The supreme court found it significant that the school district had elected not to obtain continuous supervisory services from the architect through the "clerk of the works" clause. The court held that the contractual provisions "absolved [the architect] from any liability, as a matter of law, for a contractor's failure to fasten the roof to the building with washers and nuts." 253 N.W.2d at 839.

Other courts have also held that similar language absolved the architect from liability as a matter of law.

[Citing cases.]

The Architect contends that imposition of liability upon it here would be nothing short of a holding that the Architect was the guarantor of the contractor's work.

The School Board responds to the Architect's argument by saying that it has never contended that the Architect should be a *guarantor* of the contractor's work, but the School Board strongly contends that where a contract is for work and services, there is an implied duty to perform with an ordinary and reasonable degree of skill and care, citing, *C. P. Robbins & Associates v. Stevens*, 53 Ala.App. 432, 301 So.2d 196 (1974). The School Board also argues that its claim was covered under the terms of the agreement because its claim was limited to those deviations from the plans and specifications that should have been obvious to one skilled in the construction industry. According to

---

[50]Article 8 is modeled after language found in American Institute of Architects (AIA) contracts used throughout the country.

the School Board, if the Architect's construction of the contract language is accepted, the Architect would not be accountable to anyone for its failure to make a reasonably adequate inspection so long as it made an "inspection," no matter how cursory, on a weekly basis.

The School Board points out that the contract also included the following language:

> "On the basis of his on-site observations as an Architect, he shall endeavor to guard the Owner against defects and deficiencies in the work of the Contractor."
>
> "The Architect shall have authority to reject Work which does not conform to the Contract Documents."
>
> "The administration of the contract by the Architect is not normally to be construed as meaning the furnishing of continuous inspection which may be obtained by the employment of a Clerk of the Works. However, the administration shall be consistent with the size and nature of the work and must include, at least, one inspection each week, a final inspection, and an inspection at the end of the one year guarantee period shall be required on all projects."

It is apparent that our focus must be drawn to the exculpatory language contained in Article 8 of the agreement. Most courts in other jurisdictions that have considered similar exculpatory clauses have recognized that while such clauses do not absolve an architect from all liability, an architect is under no duty to perform continuous inspections that could be obtained by the employment of a "clerk of the works."

The critical question in most of the cases we have reviewed from other jurisdictions seems to focus on the extent of the obligation owed when an architect agrees to perform the type of inspection that the architect agreed to perform in this case. Some of the courts hold, as a matter of law, that the agreement does not cover particular factual situations, and at least one court makes a distinction based on whether a failure of a contractor to follow plans and specifications is *known to the architect during the course of the construction.*

A fair reading of Article 8 obviously operates to impose certain inspection responsibilities upon the architect to view the ongoing construction progress. The frequency and number of such inspections is made somewhat specific by the contractual requirement that these visits to the site be conducted at least once each week. The difficulty arises in construing the particular terminology used, such as *"endeavor to require," "familiarize himself generally* with the progress and quality of the work," and *"endeavor* to guard the owner against defect and deficiencies." When coupled with the contract language stating that "[t]he administration of the contract by the Architect is not normally

to be construed as meaning the furnishing of continuous inspection which may be obtained by the employment of a Clerk of the Works," these phrases make it obvious that the Architect's duty to inspect is somewhat limited, but we cannot agree with the argument on appeal that there could never be an imposition of liability under an agreement similar to Article 8 no matter how serious the deviation of the contractor from the plans and specifications.

We begin our interpretation of Article 8 by stating the general rule that one must look at the contract as a whole in order to determine the intent of the parties. . . . In this case, the School Board presented some evidence by persons knowledgeable of construction practices, and they testified concerning the cause of the leaks and gave general testimony concerning the conditions at the construction site, which, they allege, should have put the Architect on notice that the plans and specifications were not being followed. However, the only testimony in the record we find concerning the obligation of the Architect to conduct an inspection pursuant to the agreement was to the effect that, according to one architect/witness who had read the agreement, it did not call for an inspection to discover the defect alleged in this case.

Some courts have determined, as a matter of law, that certain factual situations are not covered under an agreement to inspect like that agreement involved here, and some courts have emphasized that the contract is one with a professional and have appeared to apply a rule requiring the presentation of expert testimony in order to prove the contract meaning. For example, the *Moundsview* court said, "An architect, as a professional, is required to perform his services with reasonable care and competence and will be liablein damages for any failure to do so." 253 N.W.2d at 839. In [*Mayor v. City Council of Columbus v. Clark-Dietz & Associates-Engineers, Inc.,* 550 F.Supp. 610 (N.D.Miss.1982)] the court held that the architect's liability was limited by the contract language, but also held that the architect still had a duty to supervise construction by observing the general progress of the work. The court did not impose liability on the architect, because it explicitly found that the architect had not been negligent in its limited duties.

The same AIA contract language was involved in *First Nat'l Bank of Akron v. Cann,* 503 F.Supp. 419 (N.D.Ohio 1980), aff'd, 669 F.2d 415 (6th Cir.1982), where the court wrote:

> "That exhaustive, continuous on-site inspections were not required, however, does not allow the architect to close his eyes on the construction site, refrain from engaging in any inspection procedure whatsoever, and then disclaim

liability for construction defects that even the most perfunctory monitoring would have prevented."

503 F.Supp. at 436.

Although the contract here clearly made the Architect's inspection duty a limited one, we cannot hold that it absolved the Architect from all possible liability or relieved it of the duty to perform reasonably the limited contractual duties that it agreed to undertake. Otherwise, as the School Board argues, the owner would have bought nothing from the Architect. While the agreement may have absolved the Architect of liability for any negligent acts or omissions of the contractor and subcontractors, it did not absolve the Architect of liability arising out of its own failure to inspect reasonably. Nor could the Architect close its eyes on the construction site and not engage in any inspection procedure, and then disclaim liability for construction defects that even the most perfunctory monitoring would have prevented, or fail to advise the owner of a *known* failure of the contractor to follow the plans and specifications.

The issue here is whether the Architect can be held liable for its failure *to inspect and to discover* the acts or omissions of the contractor or subcontractors in failing to follow the plans and specifications. Under the terms of the contract, the Architect had at least a duty to perform reasonable inspections, and the School Board had a right to a remedy for any failure to perform that duty. There is no question that the Architect performed inspections; the thrust of the School Board's argument is that these inspections were not as thorough as the Architect agreed they would be.

As we have already stated, the only evidence concerning the Architect's duty under the agreement was from an architect who stated that under his interpretation of the contract the Architect was not under a duty to inspect for the specific defect that was alleged in this case.

As was pointed out in *Moundsview,* an architect is a "professional," and we are of the opinion that expert testimony was needed in order to show whether the defects here should have been obvious to the Architect during the weekly inspections. Just as in cases dealing with an alleged breach of a duty by an attorney, a doctor, or any other professional, unless the breach is so obvious that any reasonable person would see it, then expert testimony is necessary in order to establish the alleged breach. The nature and extent of the duty of an architect who agrees to conduct the inspection called for by the subject agreement are not matters of common knowledge. The rule of law in Alabama concerning the use of expert testimony is as follows: " '[E]xpert opinion testimony

should not be admitted unless it is clear that the jurors themselves are not capable, from want of experience or knowledge of the subject, to draw correct conclusions from the facts. The opinion of the expert is inadmissible upon matters of common knowledge.' " *Wal-Mart Stores, Inc. v. White,* 476 So.2d 614, 617 (Ala.1985) (quoting C. Gamble, *McElroy's Alabama Evidence* § 127.01(5) (3d ed.1977)).

The breach alleged in this case involved architectural matters that would not be within the common knowledge of the jurors, yet the School Board presented no expert testimony regarding the Architect's inspections and any deficiencies in those inspections. The School Board presented no expert testimony regarding the standard of care imposed within the architectural profession by the weekly inspection provision contained in this contract, and there was no expert evidence that that standard was breached by the Architect. In fact, the only expert that testified concerning the "weep holes" (which were the source of the leaks) was an architect who stated that it was not within the standard of care under the weekly inspection provision to keep track of each and every weep hole.

In a case startlingly similar to this one, a New York court, while holding that an architect could be found liable under the AIA contract for a failure to notify the board of education about leaks in a school's roof, did so only after finding that there was evidence that the architect had knowledge during the course of the construction that the contractor was not following the plans and specifications. In *Board of Educ. of Hudson City School Dist. v. Sargent, Webster, Crenshaw & Folley,* 146 A.D.2d 190, 539 N.Y.S.2d 814, 817–18 (1989), the court stated:

> "While this very clause in the standard AIA architect/owner contract [the clause absolving the architect from responsibility for the contractor's failures] has been given exculpatory effect in this State and other jurisdictions (*see, Jewish Bd. of Guardians v. Grumman Allied Inds.,* 96 A.D.2d 465, 467, 464 N.Y.S.2d 778, *aff'd,* 62 N.Y.2d 684, 476 N.Y.S.2d 535, 465 N.E.2d 42; *Shepard v. City of Palatka,* 414 So.2d 1077 [Fla.Dist.Ct.App. 1981]; *Moundsview Ind. School Dist. No. 621 v. Buetow & Assoc.,* 253 N.W.2d 836 [Minn.1977]), *none of the cases involved defects known by the architect during the course of construction,* which he failed to apprise the owner of under the contractual duty to 'keep the [School District] informed of the progress of the work'. We decline to extend the application of the clause in question to an instance such as this, *where the trier of facts could find that the architect was aware of the defect and failed to notify the owner of it.*" [Emphasis added.]

The key distinction between that case and this one is that in *Sargent, Webster* there was evidence that the architect *knew* of the defects during construction and failed to notify the owner. Clearly, the architect there would have breached his contractual duty if he failed to notify the owner of a *known* defect. It is undisputed in this case that during construction the Architect did not know of the defects with the weep holes. In fact, the School Board does not suggest that the Architect did know, only that under the provisions of the contract the trier of fact could have found that the Architect should have conducted an inspection that would have discovered the defect.

We conclude that, although the Architect had a duty under the contract to inspect, exhaustive, continuous on-site inspections were not required. We also hold, however, that an architect has a legal duty, under such an agreement, to notify the owner of a *known* defect. Furthermore, an architect cannot close his eyes on the construction site and refuse to engage in any inspection procedure whatsoever and then disclaim liability for construction defects that even the most perfunctory monitoring would have been prevented. In this case, we hold, as a matter of law, that the School Board failed to prove that the Architect breached the agreement.

In making this judgment, we would point out the value, and in some cases, the necessity, for expert testimony to aid the court and the trier of fact in resolving conflicts that might arise.

II

We now address the School Board's argument on its cross-appeal that the negligence claim should have been submitted to the jury. The School Board stated at oral argument that its negligence claim was asserted as an alternate theory of recovery in case its contract claim was not allowed. The trial court ruled that this negligence claim was barred by the statute of limitations.

* * * *

The trial court was correct, therefore, in directing a verdict for the Architect on the School Board's negligence claim.

III

Because we hold for the Architect on the School Board's contract and negligence claims, we determine that the issue of whether the Architect should be allowed to assert its cross-claim against the roofing subcontractor and the manufacturer of the roofing membrane is moot.

IV

For the above-stated reasons, that portion of the judgment based on the jury's verdict against the Architect on the School Board's contract claim is reversed, and a judgment is rendered for the Architect on that claim. That portion of the judgment based on the directed verdict in the Architect's favor on the School Board's negligence claim is affirmed.

88-220 REVERSED AND JUDGMENT RENDERED.
88-273 AFFIRMED.
HORNSBY, C. J., and JONES, SHORES, HOUSTON and STEAGALL, JJ., concur.

Some observations on the *Watson* decision and some discussion of other important decisions (some of which were noted in *Watson*) may be useful in understanding the legal aspects of the general obligation of the design professional to perform site services.

After concluding that the exculpatory language did not provide a complete shield to the architect, the court faced the question of whether the inspections were as thorough as the architect agreed they would be. First, it should be noted that the AIA documents do not specifically require the architect to make inspections until the contractor claims that the work has been substantially completed.[51]

Watson spoke of the duty to *inspect* and discover the acts or omissions of the contractor in failing to follow the plans and specifications. Yet the court was not able to determine whether the architect had conducted a proper inspection in the absence of expert testimony showing whether the defects should have been obvious to the architect during his weekly inspections. It analogized the claim against the architect here to malpractice cases against professionals that require expert testimony unless the subject matter is a matter of common knowledge. However, *Board of Education of the Hudson School District v. Sargent, Webster, Crenshaw and Folley,*[52] noted in the *Watson* decision, rejected a contention by the architect that the claim could not be sustained because of the absence of expert testimony that the architect's performance fell short

---

[51] AIA Doc. A201, ¶¶ 4.2.2, 4.2.9.

[52] 146 A.D.2d 190, 539 N.Y.S.2d 814, appeal denied 75 N.Y.2d 702, 551 N.E.2d 107, 551 N.Y.S.2d 906 (1989).

of accepted professional standards of architectural practice. The court recognized that a person bringing a claim against a professional often can choose whether to base the claim on commission of a tort or on breach of contract. But it noted that a tort claim against the architect for failure to notify would still be based upon the breach of a specific contractual undertaking, which does not require proof of failure to meet the professional standard of care.

Whether the claim is approached as one based upon breach of contract or professional malpractice, evidence will have to be introduced that bears upon whether the design professional did what she promised or what the law requires. However, the issue of whether the design professional conducted a proper "inspection" does not generate the technical issues for which expert testimony is required.[53]

The second issue relates to the troublesome question of the interrelation of claims based upon breach of contract and those based upon the tort of negligence. In *Watson,* the school board alleged negligence and breach of contract against the architect. It lost its claim based upon negligence because the claim had not been filed within the time required by the Statute of Limitations. Typically, the claimant has a longer period to file a claim based upon breach of contract than one based upon negligence. Also, the contract claim may be easier to maintain, since it does not require expert testimony. But a claim based upon tort may provide a larger judgment than one based upon breach of contract. While this distinction is relatively unimportant in defect cases, it would not be if the claimant sought recovery for emotional distress or if the claim was argued under a legal theory that might provide for punitive damages.[54]

Third, *Watson* seeks to impose an objective standard in determining whether there has been a breach of the obligation to inspect and discover. The court's call for expert testimony is designed to determine what other architects would have done under similar circumstances. The court noted that the language could not exculpate the architect "no matter how serious the deviation of the contractor from the plans and specifications."

The AIA sought to employ a subjective standard to measure performance even if the disclaimers do not exculpate the architect. Yet the court was frustrated by the AIA's attempt to provide a loose standard as exemplified by phrases such as "endeavor to require," "familiarize himself generally with the progress and quality of the work," and "endeavor to guard the owner." Frustration with the imprecision of these terms undoubtedly motivated the court to employ an objective standard.

Fourth, the contract in *Watson,* a modified AIA Doc. B141, specified that the "inspection" would be made at least once a week. While the B141 in effect at the time this contract was made used somewhat different language, it is useful to note that in 1987, ¶ 1.5.4 required the architect to visit the site "at intervals appropriate to the stage of construction or as otherwise agreed." Parties commonly reject this formulation and include a requirement that visits be made at specific times. However, requiring weekly visits can generate excessive costs because of visits being made when they are not necessary.

At the time of the *Watson* decision, other courts were dealing with these issues. *Hunt v. Ellisor & Tanner, Inc.,*[55] cited in many subsequent decisions, held that the disclaimer is simply designed to ensure that the architect is not held to be a guarantor for the mistakes of the contractor. It rejected the *Moundsview* decision and its conclusion that the disclaimers, as a matter of law, relieved the architect. *Moundsview* held that it was proper to grant the architect a summary judgment (a resolution of the case without a full trial). In doing this, *Hunt* pointed to the responsibility of the architect "to visit, to familiarize, to determine, to inform and to endeavor to guard" and found these obligations to be nonconstruction responsibilities of the architect to provide information and not to perform work. It then concluded that the exculpatory paragraph emphasized the architect's nonconstruction responsibility and that the architect as a provider of information does not insure or guarantee the contractor's work.

While it is likely that the AIA intended to provide a contractual method by which the architect would obtain summary judgment solely by virtue of the exculpatory language, the method of arriving at the result in *Hunt* is an imaginative but strained interpretation of the contract provisions. Undoubtedly, reactions such as those shown in

---

[53]See Section 14.06(B).
[54]See Sections 27.09 and 27.10.

[55]739 S.W.2d 933 (Tex.Ct.App.1987).

Watson and *Hunt,* as well as others that will be seen shortly, reflect a negative reaction against this attempt by the AIA to shield the architect from responsibility for the work of the contractor.

Two New York cases demonstrate an attitude toward disclaimers similar to that in *Hunt.* The *Sargent* case[56] (noted earlier in this section) concluded that the exculpatory language was designed to avoid the architect's being held as a guarantor for the work of the contractor. It conceded that cases had concluded that the exculpatory language could be the basis for granting a summary judgment to the architect, but it distinguished them by noting that they did not involve cases where the defects had been known to the architect during construction and where the architect failed to keep the owner informed of the progress of the work.

It should be noted that *Sargent* did not go as far as did *Watson.* It did not hold the architect responsible where she has not been paid for the type of close supervision that would reveal most if not all defects in the contractor's performance. It noted that the owner bargained and paid for the architect to make periodic inspections and to convey information relating to those defects that the architect discovered as a result of her visits that would enable the owner to ameliorate the problems being created.

A New York trial court decision, the *Diocese of Rochester v. R-Monde Contractors, Inc.,* is also instructive.[57] In that case, the architect contended that its contract did not impose a duty to inspect during these periodic visits. This, as noted earlier, is certainly the intention of the AIA. But the court stated that even though the architect was not obligated to supervise or make exhaustive or continuous on-site inspections, it was obligated to visit the site to check on the progress and quality of the work, to inform the client of its findings, and to guard the client against defects in the work. Similarly, the court noted that the architect was required to issue payment certificates that required it to be familiar not only with the quantity of the work done but also with the work's quality. These duties apparently convinced the court that there was some obligation to inspect during periodic vis-

its and that the exculpatory provision as decided in *Hunt* was designed simply to make clear that the architect was not the insurer or guarantor of the contractor's work. The court also noted that exculpatory provisions are disfavored, that ambiguities are to be resolved against the maker, and that a contrary holding would leave owners little recourse against architects "who fail to fulfill their contractual duties to make timely and proper inspections."[58]

The court recognized that it was going beyond the holding in *Sargent* and concluded that *Sargent* would not immunize the architect from liability if it failed "to learn of a defect due to a breach of its own contractual duty to conduct adequate periodic inspections."[59] It also noted that the architect issued certificates of payment that included work that the architect had not inspected. The court pointed to provisions in the architect's contract that stated that issuance of a certificate constituted a representation that to the best of the architect's knowledge, the quality of the work was in accordance with the contract documents. It stated that this raised factual questions as to whether the architect "failed to fulfill its obligation to inspect and, if so, whether such omission was a proximate cause of the failure to discover the allegedly defective manner in which the insulation was installed."[60]

These decisions indicate a trend toward making the issue of the architect or engineer's performance a factual one that must be resolved by evidence. This trend appears to reverse the tendency of some earlier decisions which had used exculpatory language to justify awarding the architect a summary judgment making a trial unnecessary.

The cases reproduced and summarized provide some generalizations regarding the site role of design professionals who perform services under contracts such as those published by the AIA. First, the design professional breaches her contract if she does not do what she has promised to do as to site visits, their frequency, and their intensity. Second, if her breach is a substantial cause of harm to the client, she must compensate the client for its losses. (Third-party claims are discussed in Section 14.08.) Third, if she exceeds her contractual commitment,

[56]Supra note 52.
[57]148 Misc.2d 926, 562 N.Y.S.2d 593 (Sup.Ct.1989), affirmed 166 A.D.2d 891, 561 N.Y.S.2d 659 (1990).

[58]562 N.Y.S.2d at 596.
[59]Id. at 597.
[60]Ibid.

such as visiting more frequently, making more intensive examination of the contractor's work, directing the contractor as to how work should be performed, or acting positively as to safety, she has led the client to believe she would continue to perform those services even if not contractually obligated to do so. She must compensate the client if she is negligent in the performance of these services or if the client relies to its detriment by not having others perform those services. Fourth, the exculpatory provisions will not protect the design professional if she became aware of the contractor's failure to perform properly—performance including both compliance with the contract and laws such as building or safety laws—and did not take reasonable action to eliminate or minimize the likelihood that the finished project would not comply with the contractual requirements, that third persons on or about the site might suffer physical harm, or that others nearby would suffer damage to their property.

Fifth, language protecting the design professional in the contract between the design professional and her client will not be effective in a claim by a contractor against an architect if the specifications in the construction contract provided that the architect would supervise the work. This has been interpreted to include giving instructions when problems develop and inspecting the work as it is being performed.[61]

The cases discussed here demonstrate the inherent difficulties of standard form contracts. It is difficult to quarrel with the basic allocation expressed in AIA documents of responsibility for executing the design. That the architect checks the work periodically and issues certificates of payment should not disturb the division of authority and responsibility under which the architect designs and the contractor builds. However, general language often cannot take into account unforeseen circumstances. If the architect has not breached her obligation to observe (or inspect) and has not learned of defects, the exculpatory language should be sufficient to grant her a summary judgment. But where she has learned of defects and not reported them or where she has not complied with her contractual obligations, the exculpatory language should not provide a shield.

While it must be recognized that contracts drafted by the professional associations will inevitably be designed to favor members of those associations, such standard contracts cannot anticipate every problem.

Also, it is clear that the AIA's attempt to employ a subjective standard and to avoid the requirement of inspections until substantial completion has not been successful in court. This may be due to the recognition that such contracts as those published by the AIA seek to give marginal advantages to the architect or to the feeling that the owner pays for a service and should receive it. But it also should be noted that the courts recognize that the owner should not receive more than the services for which it has paid. If the owner wishes a continuous presence on site, it will have to pay for it.

C. Submittals

During the Construction Process, the contractor is usually required to submit information generally known as *submittals* to the design professional. The best-known submittals are shop drawings. (The term comes from the fabrication process: shop drawings instruct the fabrication "shop" how to fabricate component parts.) AIA Doc. A201, ¶ 3.12.1, defines shop drawings as data especially prepared for the work that are "to illustrate some portion of the Work." Submittals also include product data, defined in ¶ 3.12.1 as including "illustrations, standard schedules, performance charts, instructions, brochures, diagrams and other information" that illustrate the materials or equipment the contractor proposes to use. Finally, ¶ 3.12.3 defines samples as physical examples that illustrate "materials, equipment or workmanship and establish standards by which the Work will be judged."

An owner who is aware of the submittal system sees the process as a double and even triple checking system that should reduce the likelihood of defects or accidents. Often submittals are prepared by the subcontractor and reviewed by the contractor and the design professional or her consultant. Although submittals should reduce potential loss, they also generate additional expense.

The prime design professional is even more concerned about submittals. Submittal review exposes the design professional to claims and even liability when persons are injured, often the result of construction methods that are not within her expertise

---

[61]*Colbert v. B. F. Carvin Constr. Co.*, 600 So.2d 719 (La.App.1992).

and not her responsibility in many if not most construction contracts. In addition, project failure often can be traced to submittal review.[62]

The prime design professional must rely heavily on the expertise of her consultants and will be held accountable to her client or to the third parties if the consultant does not perform properly.[63]

Finally, the submittal process can expose the prime design professional to claims by the client and the contractor that improper delay in reviewing submittals delayed the project or disrupted the contractor's planned sequence of performance.[64]

The consultant who reviews submittals, such as a structural engineer, sees the submittal process as one that exposes her to the risk of license revocation or other disciplinary action. The *Duncan* case reproduced in Section 10.04(B) dealt with disciplinary action against a licensed engineer based on shop drawing review, among other things.

In addition, the consultant may, like the prime design professional, be exposed to liability to the owner or to third parties. She may also have to indemnify the prime design professional against whom a claim has been made.

The contractor often relies on the subcontractor to prepare submittals. Just as the prime design professional will be held responsible for the acts of her consultant, the contractor will be held responsible for the acts of the subcontractors.[65] More important, the contractor may believe that approval of its submittals transfers certain risks and responsibilities to the owner, a position often contrary to contract language.

Subcontractors and prime contractors see submittals as forcing them to fill in gaps in the design created by the design professional as well as exposing them to liability.

Before noting the varied purposes of the submittal process, imagine a building project *without* a submittal process. The contractor would build in accordance with its beliefs as to its contractual obligations. It would not have to state *in advance* what it proposes to do. It would simply be judged by what it does. If it fulfills its contract commitment,

it earns the contract price. If it does not, the owner can exercise its legal remedies. But submittals, like the site monitoring performed for the owner by the design professional, should reduce the *likelihood* that the project will not be built as required.

Although submittals have ancillary purposes, such as to persuade a vendor to process a purchase order[66] or allow different manufacturers to show their products and how such products join and relate to other products,[67] the principal purpose of a submittal is to obtain a representation from the contractor as to how it plans to execute the aspects of the design for which submittals are required.

In traditional construction projects,[68] such as those for which AIA documents are used, the contractor determines how the design is to be executed. The architect does not control the contractor's methods of executing the design. The architect monitors performance by site visits mainly to determine progress in order to issue certificates for payments. Until substantial completion, however, the architect makes visits to determine whether she sees anything to indicate that the *completed project* will not comply with contract requirements. Similarly, as to submittals, AIA Doc. A201, ¶ 4.2.7, describes the architect's review to be only for "the limited purpose of checking for conformance with information given and the design concept." The architect does not check the methods by which the contractor intends to execute the design, in accordance with ¶ 3.3.1, which states that the contractor is solely responsible for and controls the "construction means, methods, techniques, sequences and procedures."

The submittals do not, however, lend themselves to a neat breakdown between design and its execution. As a result, information may be submitted (or omitted) that relates to execution of the completed project or the temporary work needed to accomplish the completed project. *Waggoner v. W. & W. Steel Co.*[69] demonstrates this. An accident resulted from inadequate temporary connections while the structural steel was being assembled and erected, a choice made by the contractor. The court

---

[62]Rubin and Ressler, *"To Build a Better Mousetrap"—The Search to Define Responsibility for Shop Drawing Review,* 5 Constr. Lawyer No. 4, April 1985, p. 1.
[63]See Section 12.10(B).
[64]See Section 26.10.
[65]See Section 28.05(C).

[66]*Day v. National U.S. Radiator Corp.,* 241 La. 288, 128 So.2d 660 (1961).
[67]*Alabama Soc. for Crippled Children & Adults, Inc. v. Still Constr. Co., Inc.,* 54 Ala.App. 390, 309 So.2d 102, 104 (1975).
[68]See Section 17.03.
[69]657 P.2d 147 (Okla.1982).

commenced its opinion by stating the issue to be whether the architect was responsible for ensuring "that the contractor employ safe methods and procedures in performing his work."[70] Under the AIA contract used in that project, this was the responsibility of the contractor. The site visits by the architect did not change this.

The plaintiffs contended that the shop drawings that were submitted and approved by the architect "should have included specifications for temporary bracing and connections."[71] They contended that the architect was negligent when he approved shop drawings that did not provide for proper temporary connections on the expansion joints. They asserted that the architect or his structural engineer reviewing the submittal should have refused to approve it until the reviewer determined that the contractor would use proper temporary connections.

The court pointed to contract language stating that the *contractor* was solely responsible for construction methods and that it was not the responsibility of the architect to require this information be furnished.

The court stated that it was the duty of the contractor and not the architect to see that the shop drawings included how the temporary connections were to be made. It would be more correct to state that there was no requirement that the contractor *communicate* its plan as to temporary connection to the architect through submittals. The architect could assume that a failure to include provisions in the shop drawings for temporary connections did not indicate that the contractor would use unsafe connections but indicated only that this was not information that was required by the submittal process. (Had the architect known that unsafe methods were to be employed, he should have taken some affirmative action to avoid structural collapse and harm to workers.)

Another lesser known function of the submittal process is to make more definite the contract document requirements for which submittals are required. Language and even graphic depiction have communication limitations. (Try to describe or draw even a simple house!) As the contractor plans its *actual* performance rather than review the de-

sign prior to making its bid, its submittal indicates its interpretation of the contract, particularly in sensitive areas such as steel detailing. Submittal review can be the process by which some potential interpretation disputes are exposed.

In addition, submittals as to product data and samples may be the method by which the contractor proposes substitutions if they are permitted by the contract. For example, AIA Doc. A201, ¶ 3.12.8, states that the contractor will be relieved of responsibility for deviations if it specifically informs the architect of the deviation at the time of the submittal and the architect approves the specific deviation.

Finally, although not entitled to it by the contract, the submittal process can be the method by which the contractor seeks information from the prime design professional or her consultants, particularly as to engineering techniques.

Much of what has been suggested regarding the submittal process and its use in construction is reflected in AIA documents. For example, B141, ¶ 2.6.12, contained in the AIA document for design services, states that the architect's review and approval are only for the "limited purpose of checking for conformance with information given and the design concept." It also states that the review "shall not constitute approval of safety precautions or, unless otherwise specifically stated by the Architect, of construction means, methods, techniques, sequences or procedures." Similar language appears in A201, ¶ 4.2.7.

B141 and A201 reflect the AIA's desire to avoid exposing the architect to liability. The principal liability exposure generated by the submittal process is the architect being responsible for defects or accidents that can be traced to her approval of submittals, as in *Waggoner*.

Another method the AIA uses to avoid liability exposure is A201, ¶ 3.12.4, which states that submittals are not contract documents but merely demonstrate how the contractor proposes to conform to the design concept expressed in the documents. Yet ¶ 3.12.8 states that the contractor is relieved from responsibility for deviations if it informs the architect in writing of these deviations and the architect approves them. If after receiving the architect's approval the contractor performs as it has proposed, the owner should not be able to claim a breach by pointing to ¶ 3.12.4 and its declaration that submittals are not contract documents.

---

[70] 657 P.2d at 148.
[71] 657 P.2d at 151.

(The troublesome question of the legal effect of compliance with approved submittals is discussed later in this section.)

Finally, A201, ¶ 3.12.10, recognizes that some submittals (though not defining them) are not intended to be reviewed and approved but are simply for information.

B141 and A201 also focus on other liability exposure. The AIA seeks to minimize the likelihood of successful contractor claims that it has suffered losses because the architect unreasonably delayed passing on submittals. B141, ¶ 2.6.12, states that the architect shall act with reasonable promptness and should be given "sufficient time" in her professional judgment "to permit adequate review." Similar language is found in A201, ¶¶ 4.2.7 and 3.12.5. More important, A201, ¶ 3.10.2, requires that the contractor and architect agree to a schedule of submittals that allows the architect reasonable time for review.

The effect of review and approval of submittals depends on the purpose of the submittal. In making a submittal, the contractor can be asking one or a number of different questions. It can be asking whether what it proposes to do "is okay." If it performs as it says it will, has it performed its contractual obligation?

As mentioned earlier, this can create an interpretation issue. The submittal process can flush out problems and make the contractual obligations of the contractor more concrete or specific. Relieving the contractor if it complies can be based on ¶ 4.2.11, which gives the architect the power to interpret the requirements of the contract documents on request of either party. To be sure, ¶ 4.2.11 looks to interpretation issues that arise on the site and not in the submittal process. Also, interpretations during the submittal process may bypass the process in ¶ 4.3 for resolving disputes. Yet the submittal process can also serve a concretization function. The architect has no power to *change* the contract, except under those contracts such as A201 that give her the power to make minor changes that do not affect time or price.[72] If concretization is accomplished and the contractor performs in accordance with the proposal it makes in the submittal, it has complied with the contract documents.

Yet the preceding conclusion—that the contractor who complies with an approved submittal has performed in accordance with its contract commitments in A201—is troublesome. Paragraph 3.12.8 states that the contractor is not relieved from responsibility for deviations in the contract documents by the architect's approval of submittals unless the architect has been given specific notice of the proposed deviation and gives written approval to it. It also states that the contractor shall not be relieved of responsibility for errors or omissions in the submittals even if approved by the architect.

Other provisions of A201 are relevant, however. Paragraph 3.2.3. states that the contractor will perform the work in accordance with the contract documents and *approved submittals*. Yet, ¶ 3.3.3 states that the contractor will not be relieved from its obligation to perform in accordance with the contract documents by approvals required, presumably approved submittals. Paragraph 3.12.6 states that the contractor will not perform any work that requires submittal and review until the architect has given approval.

These provisions place the contractor on the horns of a dilemma. The contractor cannot perform certain work until submittals have been approved, since it must perform in accordance with approved submittals. Yet it appears that its noncompliance with the contract documents will be a breach even if its submittals have been approved unless it can point to the deviation under ¶ 3.12.8 and the architect has given written approval to the specific deviation. Other than that, its work can be challenged even if its submittals indicate how it proposes to do the work and those submittals have been approved.

The specific exception of ¶ 3.12.8 undoubtedly recognizes that review of submittals can often be perfunctory. Yet it does not seem fair to deny *any* legal effect to approval and allow the owner to challenge the work as not complying with the contract documents only if the limited exception of ¶ 3.12.8 applies.

This analysis assumes that the information submitted is data that the contractor must furnish and that the architect must review. (The architect should, though, inform the owner of any approval that is in effect an interpretation ruling of which the owner would want to be aware.)

A submittal may be a request by the contractor for technical advice from the design professional.

---

[72] AIA Doc. A201, ¶ 7.4.1.

The contractor may be stating that this is what it proposes to do and asking the design professional, "What do you think?" or "Can you help me out?" If this request relates to construction means, methods, or techniques, the design professional need not respond. But if the design professional knows that the contractor proposes to proceed in a way that will lead to project failure or accidents, the design professional must draw attention to this danger. This duty to act is similar to her obligation when she visits the site. During site visits, the design professional need not search for potential trouble in operations under the contractor's control, but she cannot close her eyes to obviously dangerous conditions or to information that clearly indicates likely project failure.[73] Similarly, she should not be able to close her eyes to such information revealed during the submittal process even though she has no obligation to review that information. For example, had the submittal in the *Waggoner* case discussed earlier in this section been a clear signal to the architect that unsafe temporary connections were to be used or if industry custom required this information be shown in the submittals and it was not, the architect should have directed attention to this problem.

Whether such a duty to speak out exists depends on the balance of expertise. For example, if the subject of the submittals relates to steel detailing largely within the expertise of the structural engineer, there is a clear duty to take some action.[74] However, if the submittal relates to an activity in which the balance of expertise lies on the contractor's side, such as vertical transportation, approval by the design professional should not relieve the contractor from responsibility. Similarly, the design professional need not answer any questions on such matters that the submittals may ask.

Finally, some information contained in submittals is simply to inform the design professional of what the contractor intends to do but is not a request for approval or even a request for information or advice. This is what is contemplated by A201, ¶ 3.12.10, which speaks of information submittals. In such a case, only a clear indication of the almost certain likelihood of project failure or harm to third parties would place a duty to take action on the design professional.

D. Use of Project Representative

The standard contracts published by the professional associations do not require the design professional to have a continuous presence on the site. If this is required, parties can agree to have a permanent representative such as project representative or a resident engineer on the site continually. Although the principal problems involved in the use of a project representative relate to the representative's authority,[75] such use can bear upon the responsibility of the design professional.

At the very least, the presence of a full-time project representative means that the design professional is not expected to be on the site continuously. It should also mean that the frequency of visits may be diminished. However, this should not mean that the design professional need not visit the site at times when it would otherwise be appropriate.[76] Also, the design professional must select competent representatives and monitor their work.[77]

E. Use of Construction Manager

Although newer construction organization methods are discussed in Chapter 17, it may be useful at this point to note the development of Construction Management. The role of the design professional has been substantially changed by standard contracts, particularly in the area of predicting costs, scheduling, safety, visiting the site, and passing on submittals. This has left a vacuum that in some projects has been filled by a Construction Manager (CM).

A CM should provide construction experience and skill, particularly at the design stage but also during design execution. There are two models of cost estimating: one uses rough formulas, and the other does a continuous evaluation of those elements that construction costs comprise. Similarly, during construction the CM should bring greater skill in monitoring contract compliance and sched-

---

[73]See Sections 14.08(G) and 14.11(B).
[74]*Duncan v. Missouri Bd. for Architects, Professional Eng'rs and Land Surveyors* (reproduced in Section 10.04(B)).

[75]See Section 17.05(B).
[76]*Central School Dist. Number 2 v. Flintkote Co.*, 56 A.D.2d 642, 391 N.Y.S.2d 887 (1977).
[77]*Town of Winnsboro v. Barnard & Burk, Inc.*, 294 So.2d 867 (La.App.1974), cert. denied 295 So.2d 445 (1974).

ules. Is work being done in accordance with safety regulations? Will or does the end product meet the contract document requirements? It is likely that the future will see increased use of specialized professionals unconnected with the creation of the design to perform many services mentioned in this chapter.

F. Statutory and Administrative Regulations

Increasingly, state statutes and regulations have dealt with design professional services, mainly those related to site services. Florida requires that a structural inspection plan be submitted to public authorities prior to issuing a building permit for any building greater than three stories or 50 feet in height or that has an assembly occupancy classification that exceeds 5,000 square feet in area and an occupant content of greater than 5,000 persons.[78] The plan must be prepared by the engineer or architect of record. Its purpose is

> to provide specific inspection procedures and schedules so that the building can be adequately inspected for compliance with the permitted documents.[79]

The public agency must require that a state-certified inspector perform structural inspections of the shoring and reshoring for conformance with plans submitted to public authorities. The architect or engineer of record can perform these special inspections if she is on the list of those qualified. The costs of the special inspector are borne by the owner.

California enacted statutes in 1971 for engineers[80] and in 1985 for architects[81] that deal with site services. The former currently requires that where supervision by a licensed engineer is required, the engineer must perform periodic observations "to determine general compliance." Supervision does not include responsibility "for the superintendence of construction processes, site conditions, operations, equipment, personnel, or the maintenance of a safe place to work or any safety in, on, or about the site." Periodic observa-

tion by an engineer is defined by the statute as a visit by the engineer to the site of the work. The statute regulating architects states that those who sign plans are not obligated to observe the construction of the work but the architect and client are not precluded from making a contract dealing with these services. The statute defines construction observation services as

> periodic observation of completed work to determine general compliance with the plans, specifications, reports, or other contract documents.[82]

The statute states that these services do not encompass construction processes, site conditions, and activities similar to those set forth in the statute dealing with engineers.

Unlike the *statutory* approach used in California, Hawaii, by *administrative* regulation, requires an application made for a building or construction permit "involving the public safety or health" to state that all plans and specifications shall include a warranty by the certificate holder stating that construction would be under her supervision. If she cannot make such a warranty, she must notify the registration board within fifteen days and include the name of another licensed professional who will continue with construction supervision.[83]

State control over every aspect of the Construction Process is increasing.[84] Those who engage in that process should obtain legal advice on the requirements for performing particular activities. In addition, the legislative definitions such as those found in the California statutes show an attempt by the design professions to obtain statutory language that seeks to limit the obligations of design professionals. However, those statutory definitions relate only to the public law aspects of construction. Those statutes do harmonize with language in standard contracts published by associations of design professionals. (Very likely the statutes were "pushed" by those associations.) But those statutes would not bar design professionals and their clients from making contracts that place a greater responsibility on the design professional than found in the statutes.

[78]West Ann.Fla.Stat. § 553.71(7).
[79]Id. at § 553.79(5)(a).
[80]West Ann.Cal.Bus. & Prof.Code § 6703.1.
[81]Id. at § 5536.25(b).

[82]Id. at § 5536.25(c).
[83]Hawaii Adm. Rules § 16-82-8(b)(c).
[84]For discussion of building codes, see Section 14.05(C).

PROBLEM

Suppose you are the architect who has been retained to design and supervise the construction of a luxury residence that is to cost $950,000. The project is to take about four months, and you are to receive a fee of 10% of the cost. When and how often would you be expected to visit the site? On what things would the duration of your visit depend? What would you do while you were at the site? Would the frequency of your visits in any way depend on whether the contractor has been found to be deliberately skimping on the job or has been late in the scheduled performance?

SECTION 12.09 Hazardous Materials

The environmental movement noted in Section 9.13 has had a significant effect on the Construction Process. A proposed site might be found to have been contaminated by hazardous or even toxic materials stored there years ago or leaking from an adjacent site. Similarly, during a renovation project, asbestos or other hazardous materials might be encountered that could prove dangerous to workers or others in the building. Even in new projects, greater emphasis on sensitivity to environmental factors could generate concern over materials and equipment to be incorporated in the project.

As the public became more concerned over environmental matters, regulatory efforts were directed toward this problem. Although this is not the place to discuss these regulations in detail, we can point to some of the factors involved in such regulation.

Regulation often included requirements that reports be submitted before certain work involving hazardous materials could be performed and that only persons certified by state or local authorities could perform dangerous work, such as asbestos removal. Also, the uncertain liability connected with environmental problems led insurers to exclude many hazardous material activities from liability coverage. Much of the difficulties encountered by insurers related to the uncertainty of what could be considered safe exposure levels and the fact that the claims may come many years after work has been performed.

These factors led those actively involved in the Construction Process, such as design professionals

and contractors, to seek methods to exculpate themselves from responsibility for these environmental concerns with the hope that liability would fall on the owner.

Illustrations can be found in the AIA Doc. B141 published in 1987. For example, ¶ 4.6 requires the owner on request of the architect to furnish geotechnical information that includes "evaluations of hazardous materials" with "reports and appropriate professional recommendations." In addition, ¶ 4.7 requires the owner to furnish "water pollution tests, tests for hazardous materials, and other laboratory and environmental tests, inspections and reports required by law or the Contract Documents." Finally, buried in the Miscellaneous Article, ¶ 9.8 states that the architect and her consultants

> shall have no responsibility for the discovery, presence, handling, removal or disposal of or exposure of persons to hazardous materials in any form at the Project site, including but not limited to asbestos, asbestos products, polychlorinated biphenyl (PCB) or other toxic substances.

The uncertainties created by the uncovering of environmental risks has led participants in the Construction Process to use contract language to avoid responsibility. The hope is that these clauses will be effective, at least to contracted connected parties, even though liability exposure of other sorts, such as claims by third parties or public officials, may not be affected. Because the owner has the upside and downside risks of ownership, it is not unreasonable to place ultimate responsibility on the owner and let the owner deal with transferring or distributing this risk through insurance.[85]

SECTION 12.10 Who Actually Performs Services: Use of and Responsibility for Consultants

A. Within Design Professional's Organization

The design professional may operate through a corporation or a partnership or be a sole proprietor. The actual performance of the work may be done by the sole proprietor, a principal[86] with whom the

---

[85]AIA Doc. A201, ¶ 10.1.4.
[86]"Principal" is used to define a partner in a partnership or a person with equivalent training, experience, and managerial control in a corporation.

client discussed the project, another principal, employees of the contracting party (whether the contracting party is a sole proprietor, a partnership, or a corporation), or (as discussed in Section 12.10(B)), a consultant hired by the design professional. Are the client's obligations conditioned on the performance being rendered by any particular person? Can certain portions of the performance be rendered by persons other than the design professional without affecting the obligation of the client to pay the fee?

Services of design professionals, especially those relating to design, are generally considered personal. A client who retains a design professional usually does so because it is impressed with the professional skill of the person with whom it is dealing or the firm that person represents. The client is likely to realize that licensing or registration laws may require that certain work be done by or approved by persons possessing designated licenses.[87] Yet the client is also likely to realize that some parts of the performance will be delegated to other principals in the firm with whom the client is dealing, employees of that firm, or consultants retained by the design professional organization.

As to design, unless indicated otherwise in the negotiations or in the contract, the client probably expects the design professional with whom it has dealt to assemble and maintain a design "team" and to control and be responsible for the design. The fleshing out of basic design concepts, such as the construction drawings and specifications, is likely to be actually executed by other employees of the design professional's organization. Although contract administration is probably less personal than design, it is still likely that the client will expect that the principal contract administration decisions will be made, if not by the person with whom it has dealt, at least by another principal in the design professional organization. However, just as in design, the client is likely to realize that the person who has overall responsibility will not actually perform every aspect of contract administration. (As to interpretations and disputes, see Section 29.06.)

These conclusions can depend on the client's knowledge of the size and nature of the design professional's organization. They may also depend on whether the client has ever dealt with design professionals before and whether it knew of the division of labor among principals, employees, and consultants.

B. Outside Design Professional's Organization: Consultants

Because of the complexity of modern construction, the high degree of specialization, and the proliferation of licensing laws, design professionals frequently retain consultants to perform certain portions of the services they have agreed to provide. Although the principal legal issues have been whether consultant fees are additional services not covered by the basic fee[88] and whether the design professionals are responsible for the consultants they hire—a point to be mentioned later in this subsection—inquiry should be directed initially to whether the design professional can use a consultant to fulfill the obligations owed the client.

Clients generally prefer that highly specialized work be performed by highly qualified specialists, often outside the design professional's organization. Suppose, though, the client insists that all design services be performed *within* the design professional's organization. The client may believe that it would not be able to hold the design professional accountable if the consultant did not perform properly. In such a case, the client might feel at a serious disadvantage if forced to deal with or institute legal action against a consultant it has not selected and with whom it has no direct contractual relationship. Clearly, the client can insist on this in the negotiations, and the contract could so provide. But suppose nothing is discussed or stated specifically in the agreement.

It is unlikely that any client who is or should be aware of the customary use of consultants for certain types of work can insist on consultants not being used. Standard contracts published by professional associations contain language relating to professional consultants. Such language usually does not directly relate to the question of whether consultants may be used but deals indirectly by providing that certain consulting services are covered under the basic fee and others are additional services. Inclusion of such language should indi-

---

[87]For discussion of the effect of licensing laws on who is permitted to do the work, refer to Section 10.05.

[88]See Sections 12.01 and 13.01(G).

cate to the client the likelihood that some consultants will be used and that possibly others will be used if the client consents.

One person can be held for the wrongdoing of another. The most common example is the liability of an employer for the negligent acts of its employee committed in the scope of employment. Although many reasons exist for vicarious liability, one reason sometimes given is that the employer controls or has the right to control the details of the employee's activities.

If the negligent actor is not controlled or subject to the control of the person who has hired the actor, the actor is an independent contractor and not an employee. Subject to many exceptions, the employer of an independent contractor is generally not liable for the negligence of the latter. For example, the businessperson who hires an independent garage to service its fleet of trucks generally will not be held liable for the negligence of the garage. An opposite result would follow if the businessperson had a repair service as part of its organization.

Generally, the consultant is an independent contractor. As a rule, the consultant is asked to accomplish a certain result but can control the details of how it is to be accomplished. If an architect retains a structural engineer as a consultant, the latter's negligent conduct that causes injuries to third persons will generally not be chargeable to the architect.

The client who retained the architect is not a third party in the same sense. Permitting the architect to use a consulting engineer as a substitute to perform certain portions of the work should not relieve the architect of responsibility to the client for proper performance of that work. The architect's obligation to the owner is based on their contract.[89]

A trilogy of consultant cases decided by the Supreme Court of Oregon are instructive. The first, *Scott & Payne v. Potomac Insurance Co.*,[90] involved a claim by an owner against an architect based on failure in the heating system that the owner claimed was caused by defective design. The architect's insurer refused to defend, claiming that the negligent act occurred in a period not covered by the policy. The architect settled the claim and brought an action against his insurance company when it refused to reimburse him. To recover, the architect found himself in the strange position of claiming he was liable. One argument made by the insurer was that the architect was not negligent because he relied on the advice of a heating engineer. In rejecting this argument, the court stated that the architect should possess skill in all aspects of the building process and cannot shift responsibility to a consultant.

The second case in the trilogy was *Johnson v. Salem Title Co.*,[91] in which an injured third party sued an architect when a wall defectively designed by the consulting engineer collapsed. The court concluded that the engineer was an independent contractor which would normally give the architect a defense. However, the court held the architect liable because of an exception to the independent contractor rule for nondelegable duties created by safety statutes. In this case, the architect was required to comply with building codes, and this duty could not be delegated to the engineer. By nondelegation the court did not mean that the architect could not use the engineer to fulfill code requirements but meant that the architect could not divest himself of ultimate responsibility if the code were not followed.

The action was brought not by the client against the architect but by an injured party. Had the action been brought by the client for its losses, the independent contractor rule would not have been relevant.

The third case in the Oregon trilogy, *Owings v. Rosé*,[92] involved the owner's suffering a loss when the floor cracked because of defective design by the consulting engineer. The architect paid the owner's claim and then sought and was given indemnity

---

[89]*Harold A. Newman Co. v. Nero,* 31 Cal.App.3d 490, 107 Cal.Rptr. 464 (1973). Similarly, see *South Dakota Bldg. Auth. v. Geiger-Berger Assoc.,* 414 N.W.2d 15 (S.D.1987) (architect denied indemnification against consulting engineer because it fixed its seal to the engineer's drawings and it failed to comply with its contractual obligations to prepare architecturally sound plans), and *Brooks v. Hayes,* 133 Wis.2d 228, 395 N.W.2d 167 (1986) (prime liable to owner for negligence of subcontractor even though latter was independent contractor).

[90]217 Or. 323, 341 P.2d 1083 (1959).
[91]246 Or. 409, 425 P.2d 519 (1967).
[92]262 Or. 247, 497 P.2d 1183 (1972).

from the consulting engineer. In passing on the indemnity claim, the court stated that the architect was liable to the owner on the basis of *Scott & Payne v. Potomac Insurance Co.*[93]

Design professionals can exculpate themselves from liability to clients for errors of their consultants. One method—a novation—is a tripartite contract under which the consultant is substituted for the design professional for that part of the work. Another is to seek and obtain from the client exculpation from any responsibility for the conduct of the consultants. This is more likely to be obtainable if the client designates that particular consultants be used. It does not seem unreasonable for the principal design professional in such a case to seek exculpation from the client.

Following this approach to its logical extreme, the principal design professional can suggest or insist that the client contract directly with consultants. Clearly this would relieve the principal design professional unless she had information regarding the consultant that should have been communicated to the owner. Although this approach is often used in the retention of geotechnical engineers, principal design professionals frequently prefer to keep overall professional control and are not anxious for clients to contract separately with consultants.

It is likely that the principal design professional will not be relieved from liability to the client for the acts of consultants. Probably the best the principal design professional can do is to ensure that the consultant is obligated to perform in an identical manner to the principal design professional's obligation to the client. The principal design professional must consider the financial responsibility of consultants. Having a good claim against a consultant may be meaningless if the consultant is not able to pay the claim. The principal design professional should undertake an investigation of the financial capacity of consultants employed. If the consultant is a small corporation, the principal design professional should bind the individual shareholders to the contract so that they are personally liable. The principal design professional probably should require that the consultant carry and maintain adequate professional liability insurance.

---

[93]Supra note 90.

Does the consulting engineer assume the risk of not being paid if the principal design professional—the architect—is not paid by the client? Refer to Section 12.03(G).

PROBLEM

A has been hired by C to design a project at a cost of about $300,000. A is the senior partner of a five-partner architectural firm. Later C asks what she, A, will do personally. A states that she will not draft anything but will delegate all the design work to draftspersons in her office. A states that she will look at and approve all work that will be done and will offer various suggestions to the draftspersons. As to supervision, she will send out one of her partners from time to time who has good engineering training and knows construction better than A does. C is unhappy and tells A that she wants A to participate more in the work. What are C's legal rights in this regard?

SECTION 12.11 Ownership of Drawings and Specifications

Who owns, more properly, who has use rights of the tangible manifestations, usually written but increasingly computer generated, created by the design professional necessary to build the project?

Clients sometimes contend, and many agencies of state and local government insist, that the party who pays for the production of drawings and specifications should have exclusive right to their use.

On the other hand, design professionals contend that they are selling their ideas and not the tangible manifestations of these ideas as reflected in drawings and specifications. This, along with the desire to avoid implied warranties to which sellers of goods are held, is the basis for calling the tangible manifestations "instruments of service."

Design professionals contend that the subsequent use of their drawings and specifications may expose them to liability claims. If another design professional completes the project, the original designer may be denied the opportunity to correct design errors as they surface during construction. Design professionals contend that most projects are one of a kind and that in reality design is a trial-and-error process. Similarly, liability exposure can result if the design is used for an addition to the project or for a new project for which it may not be suitable. Even if the design professional is absolved, absolution may not come until after a lengthy and costly trial.[94] (As discussed later in this section, this can be dealt with by indemnification.) In addition, reuse without adaption may compromise the aesthetics or structural integrity of the original design.

Another reason a design professional may wish to retain ownership of the design is to receive credit as an author and to protect her professional reputation whenever her creative works are displayed to the public. In addition, employees of the architect want recognition for their work. Often draftspersons, project designers, and others wish to make copies of their work to show prospective employers and clients.[95]

Perhaps most important, design professionals contend that the extent of use by the owner is an important factor in determining the value of design professionals' services and, correspondingly, their fee, a factor recognized by the frequent use of the percentage of construction cost to determine compensation. (See Section 13.01(B).) Had the design professional known that drawings and specifications would be used again, a larger fee would have been justified.[96]

In the absence of any specific provision in the contract dealing with reuse, the client who has paid for the services has the exclusive right to use the tangible manifestations of the design services performed by the design professional.[97] Although cases are rare, this result stems from the analogy to a sale of goods. In ownership terms, the client "owns" the drawings and specifications.

Suppose a design professional establishes a custom that drawings and specifications belong to the person creating them and that the client is allowed

their use only for the particular project. The client would not be bound by it unless it knew or should have known of such a custom.[98]

It has been held that an AIA design services document supports the establishment of a custom that in the architectural profession the architect retains ownership of the plans unless there is an express agreement to the contrary.[99]

In examining the question of whether AIA documents, particularly AIA Doc. B141, can support a custom that the architect typically retains ownership of the contract documents she prepares, one court quoted an expert witness who had incorrectly stated that B141 "is jointly sponsored by the associations of general contractors, realtors and architects".[100] However, B141, unlike A201, is solely the product of the American Institute of Architects and is not endorsed by any other association.

If the design professional intends to claim ownership of the plans and specifications, she should include a contractual provision in the agreement with the client. Such provisions are included in standard contracts published by design professional associations. For example, AIA Doc. B141, ¶ 6.1, states that the drawings, specifications, and "other documents" prepared by the architect for the project are instruments of the architect's service and the architect is deemed their author. The owner can retain copies for information and reference in connection with its use and occupancy. Use "by the Owner or others on other projects, for additions to this Project or for completion of this Project by others, unless the Architect is adjudged to be in default under this Agreement," is prohibited without written permission by the architect and appropriate compensation paid to her.

Similarly AIA Doc. A201, ¶ 1.3.1, protects the architect from unauthorized use of protected documents by the contractor or any subcontractor or supplier without specific written consent of the owner and the architect. Contractors and suppliers

---

[94]See *Karna v. Byron Reed Syndicate*, 374 F.Supp. 687 (D.Neb.1974) (designer absolved when project's use unforeseeably changed). See also West Ann.Cal.Bus. & Prof.Code § 5536.25(a) which gives the architect a defense if her design is changed without her approval.
[95]Ellickson, *Ownership of Documents: Does It Matter Who Owns the Drawing or the Design?* AIA Documents Supplement Service, July 1991, at 1-4.
[96]*Garcia v. Cosicher*, 504 So.2d 462 (Fla.Dist.Ct.App.1987) (contract provided for repeat fees in event of owner reuse).
[97]5 Am.Jur.2d Architects § 11 (1962).

[98]*Meltzer v. Zoller*, 520 F.Supp. 847 (D.N.J.1981).
[99]*Aitken, Hazen, Hoffman, Miller, P.C. v. Empire Constr. Co.*, 542 F.Supp. 252, 261 (D.Neb.1982).
[100]*Kunycia v. Melville Realty Co.*, 755 F.Supp. 566, 572 (S.D.N.Y.1990) This misconception has also found its way into the scholarly literature. See Winnick, *Copyright Protection for Architecture After the Architectural Works Copyright Act of 1990*, 41 Duke L.Rev. 1598, 1643 (1992).

can reproduce a limited number of applicable portions of the protected documents, but any such copies must bear the statutory copyright shown on the protected documents.

Under the AIA system, expanded in 1987, the protected documents do not include those prepared by consultants. In B141, the owner promises that it will respect the architect's rights in the documents but also that others will do so. This exposes the owner to a claim for improper use or reproduction by anyone in the Construction Process. Similarly A201, ¶ 1.3.1, makes the contractor a guarantor that subcontractors and suppliers will respect the architect's rights in the documents. Although the architect's rights are usually conditioned on her not being in default under her contract with the owner, B141, ¶ 6.1, narrows this condition to require that the architect have been "adjudged to be in default." Taken literally, this means the architect's rights would be conditioned not simply on her having *been* in default but on her having been *adjudged,* presumably by a court, to have been in default.

Legitimate reasons exist for giving the design professional exclusive right to reuse the drawings and specifications, mainly on the basis of use determining value of the professional services. Yet, the prohibition against the client's using the materials for additions to or completion of the project can be looked on as a device to discourage the client from retaining a new architect or at least to make the client pay compensation if it replaces the original architect. It is as if an implied term of the original retention agreement gave the design professional an option to perform any additional design services required by an addition to the original project. Hiding such "options" in the paragraph dealing with ownership of drawings and specifications can make courts suspicious of the fairness of such standardized contracts.

The Engineers Joint Contracts Documents Committee (EJCDC), a consortium made up of a division of the National Society of Professional Engineers, the American Consulting Engineers Council, and the American Society of Civil Engineers, published a new edition of its standard form of agreement between owner and engineer for professional services in 1992. The document—No. 1910-1, ¶ 8.2, captioned "Reuse of Documents"—though generally similar to the approach of the AIA, includes language not found in comparable AIA documents. For example, the EJCDC provision would appear to give the engineer and her consultants the discretion to reuse the documents whether or not the project is completed. Though the owner may make and retain copies for information and reference in connection with the use and occupancy of the project, the EJCDC provision states that the documents are not intended to be suitable for reuse on extensions of the project or on any other project. The owner who uses the documents without written verification or adoption by the engineer or her consultants does so at its own risk and will indemnify and hold harmless the engineer or her consultants from claims relating to this use. Any verification or adoption entitles the engineer to "further compensation at rates to be agreed upon by OWNER and ENGINEER."

Even if the design professional obtains ownership of the design documents, the owner may have a statutory right to use them. For example, in 1992, the California legislature enacted legislation that provides that in the event of damage that is covered by insurance to a single-family home as a result of declared national disasters, the architect must release a copy of the plans to the homeowner's insurer or to the homeowner upon request and verification that the plans will be used solely for the purpose of verifying the fact and amount of damage for insurance purposes. Under this legislation, the homeowner cannot use the plans to rebuild without the prior written consent of the architect. If the architect does not consent, the architect who has drafted the original plans and released them is not liable if the plans are subsequently used by the owner to rebuild all or a part of the residence.[101]

Are there any limitations on the *design professional's* right to reuse the drawings and specifications? Suppose the design professional plans to use the documents to build an identical residence near the completed residence that would diminish the exclusivity of the original residence. While the owner could obtain exclusivity by contractual protection, as done in the Trump Plaza contract noted later in this subsection, this is rarely anticipated in the negotiations. It is likely that the law would im-

---

[101]West Ann.Cal.Bus. & Prof.Code § 5536.3.

ply a promise by the design professional not to re-use the drawings and specifications in any way that would significantly diminish the value of the original residence.

The AIA reports that Donald Trump obtained an out-of-court settlement against his architect barring the architect and a competing developer from utilizing the design of Trump Plaza in the development of a site one block south of the plaza. According to the AIA, the written agreement stated that the drawings were not to be used for any other buildings, although it allowed the architect to retain ownership of the drawings. The settlement required changes in the facade of the second building to distinguish it from Trump Plaza.[102]

The hotly contested issue of drawing ownership may justify a contractual compromise that avoids an all-or-nothing solution. One possibility would be *joint* ownership with specific delineation of reuse rights. For example, the owner could be given the right to use the drawings to construct, maintain, repair, or modify the project. The design professional, in the case of owner reuse for another project, could be given an agreed additional fee if the original design and designer were used, and indemnification plus a small fee if another design professional were selected. The design professional could be given reuse rights only if she did not produce a project with similar, distinctive features that would diminish the uniqueness of the original project.

This section has dealt with contractual provisions designed to protect the design professional's exclusive right to use the documents she has prepared. Claims by design professionals may have other substantive bases. As shall be seen in Sections 16.03 and 16.04, a design professional, as author, can be given federal copyright protection. In some jurisdictions, unauthorized use by a third party may be a tortious conversion, which can be the basis for a claim.[103]

Remedially, generous protection is provided by the federal copyright law, as seen in Section 16.04(D). If the claim is based on a breach of contract, in addition to a possible injunction (also permitted under the federal copyright law), the design professional would be able to recover the market value of her services that have been used improperly by the defendant—very likely the cost of producing the documents. A claim based on the tort of conversion would be more expansive than one based on breach of contract. The remedy could include not only the cost to prepare the documents but also, in the case of an intentional conversion, any enrichment or benefit the defendant has received as a result of the conversion and, under extreme circumstances, punitive damages.

SECTION 12.12 Time

If there is no specific provision dealing with time for performance, the parties must perform within a reasonable time. However, the cooperative nature of design, the client providing a program, and the design professional creating a design subject to client approval make it difficult to determine who is responsible for delay. Delay may be caused by failure of public officials to move the administrative process along or by lenders deciding whether to make a loan. Even if a schedule is created, it is likely to require frequent adjustment. For these reasons, claims for a delay during design are difficult to sustain.

For delays in contract administration, claims are more likely to be made by contractors. This is dealt with in Chapter 26.

In 1977, the AIA added ¶ 1.8 (currently ¶ 1.1.2) to B141 requiring the architect to perform services as ''expeditiously as is consistent with the professional skill and care.'' On the owner's request, the architect must submit a schedule adjusted as required as the project proceeds. The language has sufficient escape hatches to make it quite unlikely that the architect will be held strictly to any schedule that may be submitted and approved. In 1987, the owner was made responsible for delay it caused.

Some architects have expressed reservations about the current ¶ 1.1.2. They fear the client may justify a refusal to make interim payments or reduce payments by an allegation that the architect has not completed the work according to the schedule. On the other hand, such a schedule, loose as it may be, may spur a slow-performing architect to develop the design.

---

[102]Supra note 95.
[103]*Williams v. Chittenden Trust Co.*, 484 A.2d 911 (Vt.1984).

SECTION 12.13 Cessation of Services: Special Problems of the Client-Design Professional Relationship

A. Coverage

Contracts to perform design or construction services are service contracts, and many of the same legal rules apply. Principles discussed in Chapter 34 dealing with construction contracts also apply to contracts for design services. However, some aspects of the client-design professional relationship have generated special legal rules discussed in this section.

B. Specific Contract Provision as to Term

Often contracts specifically define the duration of the contract relationships. However, this rarely exists in contracts to perform design services because of the high likelihood of delay traceable to the large number of participants in the Construction Process, including not only owners, designers, and contractors but also lenders and public authorities. (Refer to Section 12.12 dealing with time and schedules.)

At the outset, a number of issues must be differentiated. First, when do the contractual obligations between client and design professional end? Second, when does delay in the performance of design services entitle the design professional to additional compensation? Third, when has the design professional's dispute resolution power ended? The first is discussed in this section, the second in Section 13.01(G), and the third in Section 29.05.

Usually the design professional's services are divided into phases. Since the construction phase is last, it is the focal point for determining when the client-design professional relationship has terminated. Also, the construction phase is *itself* divided into segments, principally to enable the contractor to be paid as it works. Phases are important for other purposes as well.

Generally, the owner would like the design professional to perform until the project is "completed." At this point, the owner takes possession of the work. After that point, except for post-completion services such as furnishing as-built drawings, the owner no longer has a need for the design professional's services.

In most construction contracts executed on standard contracts published by the professional associations, completion has two stages: *substantial* and *final.* AIA Doc. B141, ¶ 2.6.1, states that the construction phase terminates

> at the earlier of the issuance to the Owner of the final Certificate for Payment or sixty days after the date of Substantial Completion of the Work . . .

Reference must also be made to A201, the General Conditions for Construction, ¶ 4.2.1, which provides that the architect will perform until final payment is due and "with the Owner's concurrence, from time to time during the correction period described in Paragraph 12.2." Paragraph 4.2.1 is not identical to B141, ¶ 2.6.1. The latter states that termination of the construction phase must occur no later than sixty days after substantial completion. The architect will continue to perform services beyond the sixty-day period, but such services will be considered additional services under B141, ¶ 3.4.18, or if the project is extended beyond the time specified in ¶ 11.5.1.

The final certificate benchmark recognizes that the function of the design professional has been completed when the design professional has certified that final payment is due. The alternative benchmark—sixty days after substantial completion—recognizes that issuance of a final certificate may be delayed because of the contractor's unwillingness to correct punchlist items. But that services performed after the sixty-day period are "additional" may surprise the client.

C. Conditions

A condition is an event that must occur or be excused before a party is obligated to begin or continue performance. Often contracting parties do not wish to begin or continue performance unless certain events occur or do not occur. For example, an owner may not wish to start construction until it has obtained a loan or permit. Design professionals may wish to condition their obligations on their ability to rent additional space or hire an adequate staff. Yet each party may wish to make a binding contract in the sense that neither can withdraw at its own discretion.

A condition that is within the *sole* power of one of the parties may prevent a valid contract from being formed. Such a conclusion is avoided by implying an obligation to use good faith to seek occurrence of the condition. However, the creation of a condition does not affect the validity of the contract as long as the condition is described with reasonable certainty.

In contracts for design services, conditions are frequently asserted by the client, the nonoccurrence giving the client the power to terminate the relationship and, depending on the language of the condition, preclude the design professional from being compensated for services rendered before termination.

Sometimes the client asserts that a condition was created though not expressed in the written contract (discussed in Section 12.03 dealing with costs). Generally, the client can testify as to an oral condition so long as the condition does not directly contradict the written contract or unless the written contract is clearly the final and complete repository of the entire agreement. Even if the condition is expressed in the written agreement, problems of reconciling it with other contractual provisions may exist. *Parsons v. Bristol Development Co.*[104] is instructive in this regard. The contract provided that a condition precedent to payment by the owner was the owner's obtaining a loan to finance the project. The architect commenced work and was paid part of his fee. He then began to draft final plans and specifications for the building. However, the owner was unable to obtain a construction loan and abandoned the project.

The owner pointed to the condition in the contract as a basis for its refusal to pay for any unpaid services that the architect had rendered. The architect pointed to language in the contract stating that if any work designed by the architect is abandoned, the architect "is to be paid forthwith to the extent that his services have been rendered." Such language frequently appears in documents published by the AIA and is known as a "savings clause." (Currently it is found in B141, ¶¶ 8.3 and 8.6.) However, the court interpreted the two apparently inconsistent provisions in favor of the more specific one dealing with the condition of obtaining a construction loan.

The case demonstrates the tendency of courts to construe language against the parties that supplied the language and to protect clients from having to pay for design services when the project is abandoned. Such client protection is based on the understandable reluctance to force a client to pay for services that it ultimately does not use. But professionals generally expect to be paid for their work. Fee risks *can* be taken, but any conclusion that the risk was taken should be supported by strong evidence that the design professional assumed the risk.

Courts facing claims of finance conditions often come to different results depending on the language, the surrounding circumstances, and judicial attitude toward outcomes that either deny any payment for work performed or force a party to pay for services it cannot use.[105]

D. Suspension

Suspension of performance in contracts for design and construction is common. The owner may wish to suspend performance of either design professional or contractor if it is having financial problems or it wishes to rethink the wisdom of the project. Those who perform services in exchange for money, such as the design professional and the contractor, may wish to have the power to suspend performance if they are not paid. Suspending performance can provide a powerful weapon to obtain payment for work that has been performed. Also, suspending performance reduces the risk that further performance will go uncompensated.

The common law did not develop clear rules that allowed suspension. The performing party either must continue performing or under proper circumstances could discharge its obligation to perform.

Recognition of an interim remedy—suspension—was not well developed until the adoption of the Uniform Commercial Code by states beginning in the early 1950s. Section 2-609 of the Code allowed a party to a contract involving goods to demand assurance under certain circumstances and withhold its performance until reasonable assurance was provided. This doctrine has carried over to contracts that do not fall within the jurisdiction of the Uniform Commercial Code, such as those that involve design or construction.[106]

[104]62 Cal.2d 861, 402 P.2d 839, 44 Cal.Rptr. 767 (1965).

[105]Compare *Campisano v. Phillips*, 26 Ariz.App. 174, 547 P.2d 26 (1976) (architect assumed risk), with *Vrla v. Western Mortgage Co.*, 263 Or. 421, 502 P.2d 593 (1972) (architect recovered).
[106]Restatement (Second) of Contracts § 237, illustration 1; §§ 251, 252 (1981).

Contracts for design services frequently include provisions granting the client the power to suspend the performance of the design professional. Usually, such clauses give the design professional an immediate right to compensation for past services and a provision stating that suspension will become a termination if it continues beyond a certain period. If the design professional has sufficient bargaining power, she will seek to include a provision stating that she can suspend if she is not paid and that if suspension continues for a designated period, she can terminate her performance under the contract. Because of the uncertainty of the common law as to suspension, it is desirable to include specific provisions granting this power to either party.

In 1987, the AIA revised B141 dealing with suspension. Currently, ¶ 8.2 appears to give the owner the power to suspend at its own convenience. If suspension continues for more than thirty consecutive days, the architect is compensated for services prior to notice of suspension. If the project is resumed, the architect receives an equitable adjustment in her compensation. There is no specific provision dealing with protracted suspension becoming grounds for termination. However, as seen in Section 12.13(E) dealing with abandonment, language in B141 can be invoked to deal with protracted suspension.

B141, ¶ 8.5 gives the architect the power to suspend if the owner fails to make payment, with the suspension's being effective after seven days' written notice to the owner. The seven-day period allows for cure by the owner and continuation of performance.

E. Abandonment

Because of the uncertainties inherent in the Construction Process, project abandonment is not a rare occurrence. When it occurs, some clients refuse to pay for services that have been performed by their design professional. Sometimes, as seen in the *Bristol* case described in Section 12.13(C), the client points to language in the contract or an oral agreement under which the design professional forfeits her compensation in case of project abandonment.

In the absence of any contract provision giving the owner the power to abandon or any common law power to abandon based on changed circumstances or frustration of purpose, abandonment by the client is a breach of contract. The design pro-

fessional is entitled to recover for the work she has performed and any profit on unperformed work if that can be established. For that reason, contracts usually deal with abandonment, the client usually wishing to reduce its exposure if it does abandon and the design professional wishing to protect her right to be paid for work performed if there is an abandonment.

B141, ¶ 8.3, gives the owner the right to terminate by giving not less than seven days' written notice if the project is permanently abandoned. If this is done, the architect, under ¶¶ 8.6 and 8.7, recovers payment for work performed, reimbursement of certain expenses incurred, and termination expenses as provided in ¶ 8.7.

Paragraph 8.3 also gives the architect the power to terminate if the owner has abandoned the project for more than ninety consecutive days.

The latter provision reflects the common problem of the client's refusing to admit that a project has been abandoned when all the evidence indicates that it will never be resumed. If this uncertain status continues for ninety consecutive days, the architect can terminate the contract and recover under ¶¶ 8.6 and 8.7.

Is client abandonment a breach of contract? Does the owner impliedly promise that the project will not be stalled for more than ninety consecutive days? Is the ninety-day consecutive period a protracted suspension that would give the architect a power to terminate and give her the rights granted by ¶¶ 8.6 and 8.7 or those granted by the common law? More likely the power to abandon granted by the contract is a privileged act invoking only the contract remedy and not remedies granted by the common law for breach.[107]

F. Termination Clauses

Contracts frequently contain provisions under which one or both parties can terminate their contractual obligations to perform. Termination does not necessarily—nor does it usually—extinguish any claim either party may have for the other's failure to perform.

Termination clauses vary. Some provide that one party can terminate if the other commits a *serious*

---

[107]*Furst v. Bd. of Educ.*, 20 Ill.App.2d 205, 155 N.E.2d 654 (1959).

breach of the contract. Some allow termination powers for *any* breach, a method intended to foreclose any inquiry into the seriousness of the breach. However, such a power can be abused. For that reason, careful judicial inquiry is likely to be made into the way in which the contract was made. If such a clause would operate unfairly, at the very least it will be interpreted against the stronger party to the contract and may be found unenforceable.

Some contracts provide that either party can terminate by giving a specified notice without any need to show that the other party has breached the contract. This codifies what the common law called "contracts at will," under which the contract regulates rights and duties of the parties only as long as both wish to continue. For a close, confidential relationship, such as one created when a design professional is retained by a client, such a provision may be reasonable.

However, AIA Doc. B141, ¶ 8.1, is a "default" termination clause requiring a substantial failure to perform through *no* fault of the terminating party. Paragraph 8.4 states that the owner's failure to make payments is considered substantial nonperformance and cause for termination. (Nonpayment under ¶ 8.5 grants the architect a power to suspend performance.) As noted, the AIA allows the owner to abandon the contract by giving a seven-day written notice, something resembling construction contract provisions that allow the owner to terminate for its own convenience. AIA Doc. B141, ¶¶ 8.6 and 8.7, provide a remedy: the architect is entitled to payment for services performed before termination along with reimbursable expenses due and termination expenses.

Termination of a contract can result from a material breach without any specific power to terminate. Although the same result may not be reached in every jurisdiction, it is likely that a contractual termination clause will *not* be exclusive and will not limit any common law right to terminate.[108]

Default termination clauses often specify that the breaching party be allowed time to cure any default. As this opportunity to cure arises more fre-

quently in construction contracts, it is discussed in Section 34.03(F).

Often termination clauses require that a written notice of termination be given. For example, AIA Doc. B141 ¶ 8.1 allows termination "upon not less than seven days' written notice." Although it is clear that termination does not actually become effective until expiration of the notice period, it is not always clear what the rights and duties of the contracting parties are during the period between receipt of notice and the effective date of termination. This may depend on the purpose of the notice.

The notice period can serve as a cooling-off device. Termination is a serious step for both parties. If it is ultimately determined that there were insufficient grounds for termination, the terminating party has committed a serious and costly breach. A short notice period can enable the party who has terminated to obtain legal advice and to rethink its position.

If the notice period provides a cooling-off period, performance should continue during the notice period but cease when the notice period expires. Only if the termination is retracted *during* the notice period should the parties continue performance after the effective date of termination.

If the right to terminate requires a contract breach, notice can have an additional function. It may be designed to give the breaching party time to cure past defaults and provide assurances that there will be no future defaults. If cure is the function of the notice period, actual termination should occur only if the defaults are not cured by the expiration of the notice period. During the notice period, the parties should continue performance. If it appears that there is no reasonable likelihood that past defaults *can* be cured and reasonable assurances given, performance by the defaulting party should continue only at the option of the party terminating the contract. The latter should not be forced to receive and perhaps pay for substandard performance.

Probably the principal purpose of a notice is to wind down the work to allow the parties to plan new arrangements made necessary by the termination. A short continuation period can avoid a costly shutdown of the project or the unavoidable expenses that can result if the design professional must stop performance immediately. The notice period can enable the client to obtain a successor

---

[108]*North Harris County Junior College Dist. v. Fleetwood Constr. Co.*, 604 S.W.2d 247 (Tex.Ct.App.1980) (dictum). Cf. *Glantz Contracting Co. v. General Elec.*, 379 So.2d 912 (Miss.1980).

design professional while retaining the original professional for a short period. It can also enable the design professional to make work-force adjustments, to get employees back to home base, to cancel arrangements made with third parties, and to allow time to line up work for employees.

If making adjustments is the principal reason for the notice period, each party should be able to continue performing during the notice period. However, if relations have so deteriorated that continued performance would likely mean deliberately poor performance by either or both parties during the notice period, neither should be compelled to perform during the notice period. Certainly work that cannot be finished before the effective date of termination should not be begun.

Contracting parties should decide in advance what function the notice period is to serve and what will be the rights and duties of the parties during this period. Once this determination is made, the contract should reflect the common understanding of the parties, and any standardized contracts should be modified accordingly.

G. Material Breach

Commission of a material (that is, serious) breach empowers the other party to terminate the contract. Whether a breach is material will depend on a number of factors discussed in Section 34.04(A). Generally, a material breach by the client is an unexcused and persistent failure to pay compensation or cooperate in creating the design. A material breach by the design professional is likely to be negligent performance or excessive delays. As in construction contracts, termination of contracts to perform design services—the principal problem involving client abandonment of the project—is relatively rare.

H. Subsequent Events

Contract law normally places the risk of performance being more difficult or expensive than planned on the party promising performance. But sometimes events occur after the contract is made that go far beyond the assumptions of the parties at the time of contract making. If so, the law will relieve a party who is affected by these events unless the contract clearly allocates this risk to the performing party. This is more commonly a prob-

lem in construction contracts and is dealt with in Section 23.05A. However, the highly personal nature of the performance of design services makes it important to discuss one problem that rarely surfaces in the performance of *construction* contracts.

Suppose a key person is no longer available to perform professional services. Is that person's continued availability so important that her inability to perform—because of either disability, death, or an employment change—will terminate the contract? The issue usually arises if the client wishes to terminate its obligation because a key design person is no longer available. That key person can be the sole proprietor, a partner, or an important employee of a partnership or professional corporation.

Contract obligations generally continue despite the death, disability, or unavailability of persons who are expected to perform. Only in clear cases of highly personal services will performance be excused.

Unavailability of key design persons can frustrate contract expectations. For example, in the absence of a contrary contractual provision, the death of a design professional who is a party to the contract, such as a sole proprietor or partner, will terminate the obligation of each party. The personal performance of that particular design professional was very likely a fundamental assumption on which the contract was made. A successor to the design professional can, of course, offer to continue performance, and this may be acceptable to the client. However, continuation depends on the consent of both successor and owner. Without agreement, each party is relieved from further performance obligations.

Suppose the person expected to actually perform design services is an employee of a large partnership or professional corporation. That person's unavailability may still release each party, but it would take stronger showing that the unavailable design professional was *crucial* to the project and that her continued performance was a *fundamental assumption* on which the contract was made.

The parties should consider the effect of the unavailability of key design personnel and include a provision that states clearly whether the contract continues if that person dies, becomes disabled, or for any other reason becomes unavailable.

AIA Doc. B141, ¶ 9.5, binds owner and architect and their *successors* to the contract. Under this obscure language it appears that the parties contemplate successors stepping in if for some reason a contracting party, such as the architect, can no longer perform. This appears to require that the client continue dealing with the partnership if the partner with whom the client had originally dealt dies, becomes disabled, or leaves the partnership.

Continuity may be desirable. However, the close relationship required between design professional and client may mean that the client does not wish to continue using the partnership if the person in whom it had confidence and with whom it dealt is no longer available. Similarly, a successor may not *want* to work for the client. Specific language should be included dealing with this issue.

I. The Lender's Perspective

In the course of constructing the project, the owner may have difficulty paying its construction loan. If so, two events can occur. First, the owner defaults and the lender may decide to take over the project to salvage something from the financial disaster. Second, cutting off the money flow may give the design professional, usually the architect, the power to terminate her obligation to perform further under the contract.

It may be important to the lender to keep the architect on the project. The architect may no longer wish to continue performing or may believe she is entitled to additional compensation. Even if the lender can persuade the architect to continue and even if the renegotiation is successfully completed, costly delay is likely to result.

The lender would prefer as part of the takeover to receive an assignment of the owner's rights to the architect's services without the need for renegotiation. However, either the law or a contractual provision barring assignment without the architect's consent may make this course of action unavailable.

To avoid this impediment to the assignment, the lender may insist as part of the loan commitment package that it receive a promise by the architect to consent to any assignment the owner may make to the lender in the event of default. Also, the lender may demand that the architect agree to transfer the right to use the drawings and specifications to the lender in the event of lender takeover. If the lender is successful in its demands, the agreement between the architect and the owner should reflect these undertakings.

PROBLEMS

1. Client and Architect have executed AIA Doc. B141 (found in Appendix A). After Architect worked on the design for two months, Client told her that the project was suspended. Architect would prefer to be paid for what she has done and not be called upon to resume performance. She has found Client difficult to deal with and would rather work elsewhere. Architect claims that the project has been abandoned and not simply suspended. How would it be determined whether the project has been suspended or abandoned, and what would be the result of either under B141?

2. A and C entered into a written contract by which A agreed to perform designated design services for C. The contract contained the following provision: "Either party may terminate the contract by giving the other party seven days' notice in writing." What are the rights and duties of A and C if either gives the other written notice of termination?

SECTION 12.14 Judicial Remedy for Breach: Special Problems of Client-Design Professional Relationship

A. Coverage

Basic judicial remedies for contract breach were discussed in Chapter 6. Chapter 27 discusses claims in the context of the construction contract. This section applies basic legal doctrines to special problems found in the relationship between client and design professional.

B. Client Claims

The principal claims that clients make against professionals relate to defective design. A breach of contract by the design professional entitles the client to protect its restitution, reliance, and expectation interests. (Refer to Section 6.05.)

Although clients occasionally seek to protect their restitutionary interest by demanding return of any payments made, the principal problem relates to the client's expectation interest. If the project is designed defectively, the client is entitled to be put in the same position it would have been had the design professional prepared a proper design. The first issue that can arise is whether the client can measure its expectation loss by proving the cost of correcting the defective work or whether it is limited to the difference between the project as it should have been designed and the project as it was designed. Claims against contractors that involve this issue are discussed in Sections 22.06(B) and 27.03(D).

This issue can be demonstrated by looking at *Bayuk v. Edson*.[109] In this case, the owner complained about faulty design consisting of, among other things, an improperly designed floor, closets too small, outside doors constructed for a milder climate than where the house was built and of an unusual type that could not be constructed by artisans in the area, unaesthetic kitchen tile, sliding doors that did not fit properly in their tracks, and a fireplace that became permanently cracked.

A number of witnesses testified that it would not have made economic sense to repair the defects. One witness testified that tearing out and repairing would cost more than the cost of rebuilding the house in its entirety. The plaintiff produced an expert real estate appraiser who fixed the value of the house without the defects at $50,000 to $60,000 and with the defects at $27,500 to $31,500. The trial court awarded a judgment of $18,500, the least of the possible remainders. This was affirmed by the appellate court.

Suppose, however, that it would not have been economically wasteful to correct the defective work. This would entitle the owner to the cost of correction. In claims against a design professional for improper design, application of this standard involves the betterment rule, based on the cost of correction sometimes unjustly enriching the owner. For example, in *St. Joseph Hospital v. Corbetta Construction Co., Inc.*,[110] the hospital sued its architect, the contractor, and the supplier of wall paneling that had been installed when it was disclosed that the wall paneling had a flame spread rating some seventeen times the maximum permitted under the Chicago Building Code.

After the hospital had been substantially completed, it was advised that it could not receive a license because of the improper wall paneling. The city threatened criminal action against the hospital for operating without a license. The hospital removed the paneling and installed paneling that met Code standards. The jury awarded $300,000 for removal of the original paneling and its replacement by Code-complying paneling and an additional $20,000 for architectural services performed in connection with removal and replacement.

In reviewing the jury award of $320,000, the appellate court noted that had the architect complied with his obligation to specify wall paneling that would have met Code standards, the construction contract price for both paneling and cost of installation would have been substantially higher. The court stated that the hospital should not receive a windfall of the more expensive paneling for a contract price that assumed less expensive paneling. The paneling that *should* have been specified together with installation would have cost $186,000, while the paneling specified with installation cost $91,000. This, according to the court, should have reduced the judgment by $95,000. The court reduced the award an additional $21,000 for items that were installed when the panels were replaced that were not called for under the original contract. As a result, the judgment was reduced some $116,000.

That awarding the full cost of correction would put the client in a better position than it would have been had there been full performance is an affirmative defense that must be established by the design professional. The design professional must show that the owner would have proceeded with the project at the same site and under the same conditions regardless of the increased cost.

A variation of the betterment rule is one sometimes called the extended life rule or added value. For example, if a roof has to be replaced after the repairs turn out to be ineffective, the client has received a new roof instead of a roof that would have used up some of its useful life value. If the design professional can prove that the client now has an asset with an extended life value, this can be deducted from the damages.

Under either the betterment or the extended life rule, the client is entitled to any additional cost incurred because the corrected work has taken place after the initial work has been performed if the price for the cost of correction has increased.

While the betterment and extended life rules have a logical attractiveness, they are not always

---

[109]236 Cal.App.2d 309, 46 Cal.Rptr. 49 (1965).
[110]21 Ill.App.3d 925, 316 N.E.2d 51 (1974). See also Karge, *Architect-Engineer Damages: The Added First Benefit Theory,* 9 Constr.Lawyer No. 4, November 1989, p. 1.

applied, particularly if the party who has to pay for the cost of correction is a consumer or a client of limited financial capacity. In addition, any close questions as to whether the work would have been done and cost incurred originally or as to the extended life value may be resolved in favor of such a client. In addition, the law sometimes offsets the cost to the client of having to correct a defective condition many years before it would have had to correct this condition had there not originally been a design defect.

Another claim sometimes made by clients is unexcused delay in preparing the design or performing administrative work during construction. Delay can harm the contractor, and most delay disputes are between an owner and a contractor. Claims for delay during the design phase are difficult to establish because of the likelihood of multiple causes. But suppose the project is completed late because of negligence by the architect in passing on submittals of the contractor. There are two main damage items. First, the owner will lose the use of its project for the period of unexcused delay. Second, the contractor may make a claim against the owner or design professional for any loss suffered because of delay wrongfully caused by the design professional. (The second claim is discussed in Section 26.10.)

The first damage item, that of loss of use caused by the delay, was before the court in *Miami Heart Institute v. Heery Architects & Engineers, Inc.*[111] This case demonstrates the liability exposure of design professionals when they have delayed completion of a project without justification. In the case, the architect had agreed to design plans for the building of a new hospital structure that would house patients. During construction, the patients were housed in older buildings on the premises. The architect's failure to comply with code requirements caused a ten-month delay in the issuance of a certificate of occupancy.[112]

The court rejected the contention by the architect that damages for loss of use can be recovered only

when there is a delay by the contractor. It also rejected the contention that lost-use value can be recovered only if the project itself was to be rented to others. It concluded that the proper measure of recovery should be measured by the reasonable rental value of the structure during the period of delay.

The court was faced with the fact that the patients who would have been in the new structure were housed in the old structure for the period of the delay. With this in mind, the court concluded that the proper measure of recovery was the difference between the reasonable rental value of the new structure and that of the old structure for the period of delay. Undoubtedly, the patients were inconvenienced in still being housed in the old structure and the quality of the medical services may have been poorer than if they had been in the new structure. Yet to establish the actual economic value of such losses would be almost impossible. Of course, some of this may be factored into the reasonable rental value amounts. Clearly, this is one case where it would have been helpful to have had liquidated damages (discussed in Section 26.09(B) dealing with unexcused delay by the contractor).[113]

There were other claimed damages. The court held that the betterment rule would prevent the owner from recovering for additional expenses incurred to make the building meet code requirements. It noted that the plans called for two feet of electrical wire, but proper plans would have required twelve feet of wire. But the owner's expense in purchasing the additional ten feet cannot be chargeable to the architect. The owner would have incurred this expense regardless of the architect's defective plans.

Finally, the court concluded that the owner could also recover any delay damage claims it paid

---

[111]765 F.Supp. 1083 (S.D.Fla.1991).

[112]The court's opinion in this case was in response to a preliminary dispute relating to the measure of recovery. The opinion made the assumption that the owner can establish that the architect was liable. This had not yet been established.

[113]See *E.C. Ernst, Inc. v. Manhattan Constr. Co. of Texas,* 387 F.Supp. 1001 (S.D.Ala.1974) affirmed in part vacated in part 551 F.2d 1026, reh'g granted in part and opinion modified, 559 F.2d 268 (5th Cir.1977), cert. denied sub nom. *Providence Hosp. v. Manhattan Constr. Co. of Texas,* 434 U.S. 1067 (1978) (court measured damages for delay by the liquidated damages clause in the contract between the owner and the contractor, the court noting that the architect had participated in selecting the liquidated damages amount).

to the contractors and subcontractors as a result of the architect's breach.

Delay can also cause less direct, or what are sometimes called consequential, damages. For example, delay in completion of an industrial plant may generate losses caused by the inability of the plant owner to fill the orders of its customers, exposing the plant owner to claims by the customers and causing it to lose profits on the transactions. (Such losses were discussed in Section 6.06.)

As noted in Section 15.03(C), design professionals faced with such losses and having sufficient bargaining power may seek to deal with this problem by limiting the liability of the design professional in such a way as to preclude consequential damages from being recoverable. Clients with strong bargaining power sometimes insist that contracts contain provisions stating that the design professional *will* be liable for consequential damages. In the absence of such a provision, the law would determine whether consequential damages can be recovered. The inclusion of such a provision will create greater liability exposure for the design professional.

The law does place some limits on recovery of such damages. Such a provision might strip the design professional of the protections granted by the law. While such liability would be covered under professional liability insurance policies, currently such policies have large deductibles that would require the design professional to pick up some of the loss. Certainly design professionals should avoid such clauses, and courts should think carefully before enforcing them. While parties are generally allowed to determine the terms under which they will contract, placing these risks upon a design professional who may earn compensation that bears no relationship to the risk may be unconscionable.

Suppose the client claims its design professional exceeded her authority in ordering changes in the work or accepting defective work without authority of the client. As this usually involves claims by the contractor as well, this topic and the measure of recovery for a valid claim are discussed in Section 21.04(D).

C. Design Professional Claims

The principal claims made by a design professional against the client relate to the latter's failure to pay for services performed. These claims have not raised difficult valuation questions. Design professionals commonly seek to protect their restitution interests and recover the reasonable value of their services. Occasionally, clients have resisted this claim by contending that they did not use the plans and specifications drafted and thereby have not been enriched. Such defenses have been generally unsuccessful.[114] (The more difficult problem—the problem of a contractor who seeks to protect its restitution interest when the owner has breached—is discussed in Section 27.02(E).)

Suppose the design professional seeks to protect her expectation interest. To do this, she must be put in the position she would have been had the client performed as promised. This could be computed by what she could have earned had she fully performed less the expense that she has saved or her expenditures in part performance plus her profit, less what she has already been paid.

AIA Doc. B141, ¶¶ 8.6 and 8.7, provide a contractual method of measuring compensation in the event of termination. This is not the exclusive compensation formula. Such clauses recognize that in the normal case, the architect is satisfied if she has paid for what she has performed and granted termination expenses. But common law damages are not excluded.

---

[114]*Barnes v. Lozoff*, 20 Wis.2d 644, 123 N.W.2d 543 (1963) (measured by rate of pay in community).

PROBLEM

A has agreed to perform design services for C. A is to be paid 6% of the construction cost. After the professional had been working on schematic designs, the design professional repudiated the contract and stated that she would perform no further. Which of the following items of damages can the client recover:

1. A payment made by the client to the design professional when the contract was signed.

2. The additional cost that would be incurred in hiring a new design professional.

3. Delay damages such as increased cost of obtaining a construction loan, loss of rentals from pro-spective tenants, and increase in construction contract costs.

Compensation and Other Owner Obligations

SECTION 13.01 Contractual Fee Arrangements

A. Limited Role of Law

A number of different fee arrangements are available to design professionals and owners. The choice among these arrangements is principally guided by criteria that are professional and not legal. Put another way, design professionals are generally in a better position to determine the type of fee structure than are their attorneys. Nevertheless, the law does play a limited role.

Principally, the law interprets any contractual terms that bear on fee computation when the contracting parties disagree. If no fee arrangement is specified in the contract and the parties cannot determine an agreed-upon fee subsequent to performance, the law may be called on to make this decision.

Despite this limited role, certain legal principles must be taken into account in choosing a fee arrangement or in predicting the legal result if a dispute arises. For example, faced with the question of whether certain services come within the basic design fee, the law may choose to protect the reasonable expectation of the *client* if the design professional selected the language. Likewise, any fee arrangement that measures compensation by a stated percentage of construction cost may be interpreted to favor the client. This can result from the belief that such a fee formula can be unfair to the client, given the design professional's incentive to run up costs. In rare cases, the contractual method selected will be disregarded because su-

pervening events occur that neither party contemplated.

B. Stated Percentage of Construction Costs: Square Footage

Though no longer universal (fixed-fee and cost methods of compensation being used increasingly), the stated percentage of construction costs is still a common method of fee computation. In such a method, the fee is determined by multiplying the construction costs by a designated percentage set forth in the contract. (A closely related variant to this method used in real estate residential developments is a stated dollar amount per square foot of the residence.) Because the stated percentage of construction cost typically covers only basic design services, this amount is often augmented by payments for additional services and reimbursable expenses. (But see (I) on fee limits.)

This method has been criticized. It can be a disincentive to cut costs and may reward the design professional who is less cost conscious. It can be too rigid, as projects and time spent can vary considerably. It may not reflect time spent. It also tends to subsidize the inefficient client at the expense of the efficient one.

Despite constant criticism, the fee method is still used. Clients seem accustomed to it. The method can avoid bargaining over fee. In *normal* projects it may be an accurate reflection of the work performed. Although it may undercompensate on some projects and overcompensate on others, some design professionals feel that the fees average out.

The fee method can avoid extensive record keeping. It also is much less likely to generate a client demand to examine the design professional's records, a common feature of a cost type of fee formula.

Percentages vary, with the figure selected in major part reflecting the amount of work the design professional must perform and, increasingly these days, the risk of failure. The latter is addressed in greater detail in Chapter 14. The former depends on a number of factors.

If the design professional has worked for the client before, past experience may be relevant in determining the stated percentage. A client who is inexperienced and inefficient may require more work than an efficient client who has dealt with construction before. Sometimes the percentage is based on whether the project is residential or commercial, whether the construction contracts are single or separate, and whether the construction contract price is fixed or a cost type. Smaller projects may have a minimum fee.

To avoid criticism that the fee method encourages high costs and discourages cost reduction, some design professionals use a flexible percentage. One method is a percentage that *declines* as the costs increase. The actual percentage for the entire project is determined from a schedule that has variable percentages depending on the ultimate construction cost. For example, a fee schedule may provide for a 5% fee if the costs do not exceed one million dollars, a fee of 4% if the ultimate cost is between one million dollars and $1.5 million, and 3% if the cost is over $1.5 million.

Another method to encourage cost consciousness is to employ a sliding scale under which the highest percentage is applied to a cost up to a specified amount and then the percentage reduces on succeeding amounts. For example, the fee can be 8% on the first million dollars of cost, 7% on the next $4 million, and 6% on all amounts over $5 million.

As to the construction cost multiplied by the percentage, the AIA states:

5.1.1 The Construction Cost shall be the total cost or estimated cost to the Owner of all elements of the Project designed or specified by the Architect.

5.1.2 The Construction Cost shall include the cost at current market rates of labor and materials furnished by the Owner and equipment designed, specified, selected or specially provided for by the Architect, plus a reasonable allowance for the Contractor's overhead and profit. In addition, a reasonable allowance for contingencies shall be included for market conditions at the time of bidding and for changes in the Work during construction.

5.1.3 Construction Cost does not include the compensation of the Architect and Architect's consultants, the costs of the land, rights-of-way, financing or other costs which are the responsibility of the Owner as provided in Article 4.[1]

(For equitable adjustments, see (J).)

Courts have interpreted fee provisions, but because of the different provisions that can be or are employed, generalizations are perilous. In close cases, courts are likely to favor the position of the client if the design professional selected the contract language. But the client does not always succeed. For example, in *Orput-Orput & Associates, Inc. v. McCarthy,*[2] the court held the construction cost to be the construction contract price and not the amount the client ultimately paid the contractor after a renegotiation. *Simonson v. "U" District Office Building Corp.*[3] held that work done for tenants after completion and after final settlement was made was included in construction cost. Suppose the actual owner payout to the contractor exceeds the original contract price because of defective design for which the design professional was responsible. Although this is likely to mean increased design service, this increase should not enlarge the design professional's fee.

Suppose responsibility is *shared* by the design professional and the contractor based on the contractor's *not* having directed attention to obvious design defects. An apportionment of responsibility may be appropriate, with the design professional's fee increased by that portion of the corrected work cost chargeable to the contractor. However, the difficulty of making such allocation and primary responsibility for design being that of the design professional will likely preclude such an apportionment. Yet the major issue is likely to be who bears responsibility for the cost of correction and other losses. If an apportionment has to be made

---

[1]AIA Doc. B141.
[2]12 Ill.App.3d 88, 298 N.E.2d 225 (1973).
[3]70 Wash.2d 35, 422 P.2d 1 (1966).

for this purpose, there seems to be no reason why that apportionment cannot be used to determine the design professional's fee.

Techniques for imparting flexibility to what otherwise can be a rigid fee method are mentioned in (J).

C. Multiple of Direct Personnel Expense: Daily or Hourly Rates

Personnel multipliers determine the fee for basic and additional services by multiplying direct personnel expense by a designated multiple ranging from two to four, the average being 2.5 to 2.7. The multiple gives the design professional administrative overhead and profits. AIA Doc. B141, ¶ 10.1.1, defines direct personnel expense as

> the direct salaries of the Architect's personnel engaged on the Project and the portion of the cost of their mandatory and customary contributions and benefits related thereto, such as employment taxes and other statutory employee benefits, insurance, sick leave, holidays, vacations, pensions, and similar contributions and benefits.

Once fringe benefits were truly on the "fringe." Today, they can constitute as much as 25% to 40% of the total employee cost. The contract must *clearly* specify personnel compensation cost beyond the actual salaries or wages.

There are obvious disadvantages to daily or hourly rates. Because a day is a more imprecise measurement than an hour, if either of these methods is used, it is likely to be the hourly rate. Such a method requires detailed cost records that set forth the following:

1. The exact amount of time spent.
2. The precise project on which the work was performed.
3. The exact nature of the work.
4. Who did the work.

Differential hourly rates may be used for work by personnel of differential skills.

When compensation is based on cost incurred by the design professional, the client often prescribes the records that must be kept, how long they must be kept, and that they be made available to it. The B141 that had been published in 1977 stated that the architect would keep records based on GAAP (generally accepted accounting principles).

Currently, ¶ 10.6.1 simply provides that records pertaining to additional services will be available to the owner "at mutually convenient times." Records need not be kept on the basis of GAAP. This change was made to protect architects who are likely to keep records on a cash basis. Although cash-basis records are acceptable for tax purposes, they are not acceptable under GAAP.

Clients will likely change ¶ 10.6.1 to specify with greater particularity the standards under which the records will be kept, the types of records that will be kept, the power to inspect and make copies, and the length of time records must be kept.

D. Professional Fee Plus Expenses

This form of fee arrangement is analogous to the cost type of contracts discussed in greater detail in Section 17.02(B). One advantage of a cost type of contract for *design* services is that the compensation is not tied to actual construction costs and there should be incentive to reduce *construction* costs. It can be a disincentive, however, to reduce the cost of *design* services.

The cost type of contract, or what the AIA calls "professional fee plus expenses," necessitates careful definition of recoverable costs. Costs, direct and indirect, can be an accounting nightmare. Disputes can arise over whether certain costs were excessive or necessary. In cost contracts, advance client approval can be required on the size of the design professional staff, salaries, and other important cost factors. Cost contracts require detailed record keeping. (As seen in (I), a ceiling can be placed on costs.)

Suppose the design professional estimates what the costs are likely to be in such a contract. Although the client may wish to know approximately how much the design services are likely to cost, an estimate can easily become a cost ceiling.[4] If an estimate is given—*not* intended as a ceiling—it should be accompanied by language that indicates the assumptions on which the estimate is based and that it is not a fixed ceiling or a promise that design costs will not exceed a designated amount.

---

[4]*Ballinger v. Howell Mfg. Co.*, 407 Pa. 319, 180 A.2d 555 (1962).

E. Fixed Fee

Design professional and client can agree that compensation will be a fixed fee determined in advance and incorporated in the contract. Before such a method is employed, a design professional should have a clear idea of direct cost, overhead, and profit as well as appreciate the possibility that contingencies may arise that will affect performance costs. A fixed fee should be used only where the scope of design services is clearly defined and the construction project well planned. It works best in repetitive work for the same client.

Does the fixed fee cover only *basic* design services? Does it include additional services and reimbursables? Standard contracts published by professional associations usually limit the fixed fee to *basic* design services. A design professional who intends to limit fixed fees to basic services should make this clear to the client. See (I) for discussion of fee provisions that place an absolute ceiling on compensation.

F. Reasonable Value of Services or a Fee to Be Agreed Upon

The fee will be the reasonable value of the services where the parties do not agree on a compensation method. If there is no agreed valuation method for *additional* services, compensation is the reasonable value of the services. The reasonable value of a design professional's services will take into account the nature of the work, the degree of risk to the design professional, the novelty of the work, the hours performed, the experience and training of the design professional, and any other factors that bear on the value of these services, including overhead and a reasonable profit. Proving the reasonable value of services requires detailed cost records. Leaving the fee open is generally inadvisable.

Where this issue does arise, each party usually introduces evidence of customary charges made by other design professionals in the locality as well as evidence that bears on factors outlined in the preceding paragraph. In many cases, a great variation exists between the testimony of the expert witnesses for each party. It is not unusual for the court or jury to make a determination that falls somewhere in between.

The parties can agree to jointly determine the fee at the completion of performance. Where the proj-

ect has gone well, where the parties wish to work with each other again, and where adequate records have been kept, agreement on fees may be reached easily. However, when such fortunate events have not occurred, an agreement on fees may prove difficult.

At one time, such agreements were considered unenforceable as simply "agreements to agree." However, if the work has been performed, it is likely that the parties must negotiate in good faith to determine a fee or, more likely, to determine what the parties would have agreed to had they bargained in good faith and made an agreement on compensation. The same result can follow if the parties cannot agree and the matter is submitted to arbitration.

G. Additional Services

Whether certain services fall within the basic design fee or are additional services has been discussed earlier,[5] with emphasis on whether particular services were part of *basic* design services or were *additional*. This section treats the following:

1. methods of authorizing additional services
2. compensation for such services

The first depends on the agreement between design professional and client. In the *absence* of any specific method of authorization designated in the contract, the client must request that these services be performed and the design professional must perform or agree to perform them or have them performed by a consultant.

Prior to the AIA's revision of its documents in 1987, the mechanism for ordering additional work was relatively simple. For example, in 1977, AIA Doc. B141, ¶ 1.7 required that services be performed "if authorized or confirmed in writing by the Owner." Under this language, the work was to be performed and additional compensation paid if either authorized in advance in writing by the owner or authorized orally and confirmed in writing by the owner. In either case, advanced authorization was required.

As the AIA prepared to issue a new B141, it became concerned that many architects often could not bill for additional services performed because

---

[5]See Sections 12.01 and 12.04.

they did not comply with paperwork requirements. It sought to remedy this problem by creating in 1987 a system expressed in ¶ 3.3.1 that would entitle the architect to additional compensation for certain services *without* advance owner authorization. It did this by dividing additional services into nine "contingent" (¶ 3.3) and twenty optional (¶ 3.4) additional services.

Contingent additional services are those "required due to circumstances beyond the Architect's control." The architect must "notify the Owner prior to commencing such services." If the owner does not wish to have these services performed, "the Owner shall give prompt written notice to the Architect." The services listed in ¶ 3.3 are performed *contingent* on not receiving *written* notice by the owner not to perform them. (Early in the drafting process, these services were classified as *mandatory* additional services.) B141 assumes that services listed in ¶ 3.3 generally are needed and desired by the owner. The architect will perform them after waiting a reasonable time during which the owner has the opportunity of ordering that they not be performed.

The notice given by the architect that he intends to perform these services need *not* be in writing, but the owner's decision not to have them performed *must* be given by "prompt written notice," a reflection of the paperwork problem.

Optional additional services as specified in ¶ 3.4 employ procedures similar to those used in 1977—advance written authorization or confirmation.

There has been criticism of both the mechanism employed and the lengthy lists of additional services.[6] Some clients may believe that some of the contingent (¶ 3.3) and optional (¶ 3.4) services are included in basic services. Also, the lists assume a great rarity in construction—the perfect trouble-free project. Finally, these extras can appear to be a hidden method to increase the architect's fee beyond that expected by the client. For a defense of the formidable list, refer to Section 12.01.

In any event, disputes can be minimized if the design professional and client agree on which services are basic and which earn additional compensation.

---

[6]J. SWEET, *SWEET ON CONSTRUCTION INDUSTRY CONTRACTS* § 6.6 (2d ed. 1992).

In the absence of any contractual formula for compensating additional services, compensation will be based on the reasonable value of the services. However, it is common to deal contractually with compensation for additional services.

For example, AIA Doc. B141, ¶ 11.3.2, provides a blank for compensation for additional services. It suggests that individuals who will perform services be identified and a formula for compensating them be specified. Paragraph 11.5.3 provides an automatic adjustment of the rates and multiples set forth for additional services "in accordance with normal salary review practices of the Architect."

Additional services of consultants are dealt with in ¶ 11.3.3. A blank is provided for the insertion of a multiple times the amount "billed to the Architect." Although practices vary, it is common for that multiple to take into account additional overhead, profit, and risk. The add-on will range from 10% to 20% of the amount billed. The obligation of good faith and fair dealing requires that the architect check the legitimacy of any bills submitted by consultants to make sure they are in line with professional compensation practices.

H. Reimbursables

The following are illustrations of reimbursables:

1. Transportation and living expenses in connection with out-of-town travel.
2. Long-distance communications.
3. Fees paid to secure approval of authorities having jurisdiction.
4. Reproductions, postage, and handling of construction documents.
5. Data processing.
6. Overtime work.
7. Renderings, models, and mockups when requested by owner.
8. Computer-aided design.

For the current AIA list of reimbursables, see AIA Doc. B141, ¶ 10.2, reproduced in Appendix A.

Some reimbursables are modest items that do not require advanced approval or authorization by the client. However, AIA Doc. B141 does not require advanced client authorization for expenses of computer-aided design and drafting equipment time. The reimbursable list should be studied carefully by the owner to maintain cost control.

Incurring obligations for the client and paying them can impose an administrative burden on the design professional. Sometimes design professionals charge the client a markup for handling reimbursables. For example, suppose the design professional incurred expenses of $1,000 for traveling in connection with the project and long-distance calls. Under a markup system, the design professional might bill the client $1,000 plus an additional 10% or $100, making a total of $1,100. The markup percentage can depend on the number of reimbursables and the administrative overhead incurred in handling them. A design professional who wishes to add an overhead markup should explain this to the client in advance and obtain client approval.

I. Fee Ceilings

Owners are often concerned about the total fee, particularly if the fee is cost based. But even in a compensation plan under which the design professional is paid a fixed fee or a percentage of compensation, the client may be concerned about additional services, reimbursables, or increases based on an unusual jump in the *construction* cost. Public owners with a specified appropriation for design services may seek to limit the fee to a specified amount. It may be useful to look at two cases that have dealt with fee limits.

In *Hueber Hares Glavin Partnership v. State,*[7] the contract limited the fee. It also *excluded* recovery for work to correct design errors. The Appellate Division held the language unambiguous and the fee limit not an estimate. The fee limit could not be exceeded by costs attributed to design errors. By dictum, the court stated that the city would have been precluded from asserting the fee limit had it ordered extra services knowing the fee limit had been reached.

Harris County v. Howard[8] involved an AIA document. The upset price was also held to unambiguously include additional services and reimbursables. The public owner inserted a detailed recital on the fee limit and what it included. The court *rejected* the architect's contention that his having been paid

$20,000 *over* the limit showed the limit did *not* include additional services or reimbursables.

Substantial changes in project scope should eliminate any fee ceiling that may have been established.[9]

J. Adjustment of Fee

Generally, the law places the risk that performance will cost more than planned on the party who has promised to perform. Careful planners with strong bargaining power build a contingency into their contract price that takes this risk into account. Suppose a client directs significant and frequent changes in the design. Suppose for *any* reason the design services must be performed over a substantially longer period than planned. It is unlikely that the law will give the design professional a price adjustment under a fixed-price contract. However, the AIA has sought to protect architects from these risks in AIA Doc. B141.[10]

K. Deductions from the Fee: Deductive Changes

Acts of the design professional may cause the client to incur expense or liability, and the client may wish to deduct expenses incurred or likely to be incurred from the fee to be paid to the design professional. Suppose a design professional commits design errors that cause a claim to be made by an adjacent landowner or by the contractor against the client. Suppose the client settles any claims or wishes to deduct an amount to reimburse itself in the event it must pay the claims.

The right to take deductions or offsets in such cases can be created either by the contract or by law. An illustration of the first is the frequent inclusion of provisions in construction contracts that give the owner the right to make deductions and offsets against the contractor. This is usually *not* found in contracts between design professionals and clients. However, even in the absence of such provisions, the client may be able to take deduc-

---

[7]75 A.D.2d 464, 429 N.Y.S.2d 956 (1980).
[8]494 S.W.2d 250 (Tex.Ct.App.1973).

[9]*Herbert Shaffer Assoc., Inc. v. First Bank of Oak Park,* 30 Ill.App.3d 647, 332 N.E.2d 703 (1975).
[10]AIA Doc. B141, ¶ 3.3.2, makes significant scope changes additional services. Under ¶ 3.4.18, services performed sixty days after substantial completion are also additional services, as are any services performed after the termination date set forth in ¶ 11.5.1.

tions for expenses incurred or likely to be incurred by the client as a result of any contractual breach by the design professional. The amount deducted must not be disproportionate to the actual or potential liability of the client.

It is likely that the construction contract price will be increased or decreased during actual performance through issuance of change orders. AIA Doc. B141, ¶ 10.5.1, states that no payments will be withheld from the architect "on account of the cost of changes" unless the architect is "found to be liable." Presumably, this would encompass only *judicially* determined negligent design. Deductive changes, though they may reduce the construction contract price, do not affect the architect's fee even if the fee is based on a stated percentage of construction costs. Deductive change orders not only are *not* likely to reduce the extent of the architect's services but also may *increase* it.

In addition, ¶ 10.5.1 states that there will be no reduction from the architect's compensation if the total contractor payout is reduced because of deductions for liquidated damages or other sums withheld from the contractor, such as deductions for damages suffered because of the contractor's breach. Such deductions do not reduce the contract price; they only reimburse the client for its losses when it does not receive performance it has been promised by the contractor.

L. Project Risks

In determining an appropriate compensation, the design professional should take into account the projected risk the commission creates. While risk management is discussed in greater detail in Chapter 15, it is important to note in this chapter that the scope of risk should play an influential if not dominant role in determining whether a project should be undertaken and the appropriate compensation for it. Risk means the likelihood that claims will result and the cost of dealing with such claims.

One factor that should be taken into account, particularly in a fee structure not based on cost, is the likelihood that the client will be cooperative and efficient in performing its part of the contract for design services. Another relates to the technical problems that may be encountered relating to both design and administration. Design professionals should be aware of the Architecture and Engineer-

ing Performance Information Center headquartered at the University of Maryland. The center collects, analyzes, and disseminates information about the performance of buildings, civil structures, and other constructed facilities. It uses its data to analyze trends over time relating to insurance claims. It also notes the likelihood particular clients will press claims and the organizational structures (the traditional method, cost projects, design-build) that seem prone to increased claims. It can also give information regarding the likelihood there will be particular types of problems in particular projects.

M. The Fee as a Limitation of Liability

Looking ahead to professional liability, the fee can serve another function. Some design professionals seek to limit or actually limit their liability exposure to their client to the amount of their fee.[11]

SECTION 13.02 Time for Payment

A. Service Contracts and the Right to Be Paid as One Performs

In service contracts, the promises exchanged are the payment of money for the performance of services. Unless such contracts specifically deal with this question, the performance of all services must precede the payment of any money. Put another way, the promise to pay compensation is conditioned on the services being performed.

Such a rule operates harshly to the person performing services. First, if the performance of services spans a lengthy time period, the party performing these services may need a source of financing to perform. Second, the greater the performance without being paid, the greater the risk of being unpaid. For these reasons, the law protects manufacturers and sellers of goods by giving them the right to payment as installments are delivered. However, this protection was not accorded persons performing services. If the hardships and risks described are to be avoided, contracts for professional services must contain provisions giving design professionals the right to be paid as they perform.

---

[11]See Section 15.03(D).

State statutes generally provide that employees are to be paid at designated periodic intervals. However, such statutes do not protect those that perform design services who are not employees of the owner.

B. Interim Fee Payments

Design professionals commonly include contract clauses giving them the right to interim fee payments. This avoids the problems described in the preceding paragraphs. From the client's standpoint, interim fee payments can create an incentive for the design professional to begin and continue working on the project.

Usually, interim payments in design professional contracts become due as certain defined portions of the work are completed. Although the current AIA Doc. B141 leaves blanks for interim fee payments, in 1977, B141a (Instruction Sheet) suggested interim fee payments as follows:

- Schematic Design Phase   15%
- Design Development Phase   35%
- Construction Documents Phase   75%
- Bidding or Negotiation Phase   80%
- Construction Phase   100%

(These percentages do not appear in the current B141a.) Any schedule used should depend on the breakdown of professional services and the predicted work involved in each phase.

Dividing the design services and allocating a designated percentage of the fee to each service can make it appear that the contract is *divisible*. A divisible contract matches specified phases of the work to specified compensation or a specified portion of the total compensation. In a truly divisible contract, the amount designated is earned at the completion of each phase and the value of the work for each completed phase cannot be revalued.

As an illustration, suppose the contract were considered divisible and the design professional unjustifiably discharged after completion of the construction documents phase. In such a case, the design professional will recover 75% of the fee if this were the amount specified even if the reasonable value of services exceeded this amount. Conversely, the client would not be permitted to show that the reasonable value of the services was less than 75% of the fee. Interim fee payments provisions should *not* make these contracts divisible. The amounts chosen are usually rough approximations and not agreed final valuations for each phase. This is especially true if the standard form contract specifies the phases and allocates a percentage of the fee for each phase. Such payment should be considered only provisional.[12]

C. Monthly Billings

In many projects, months may elapse before a particular phase is completed. To avoid overly long periods between payments, contracts for such projects should provide for monthly billings within the designated phases.[13]

D. Late Payments

Financing costs are an increasingly important part of performing design professional services. Late payments and reduced cash flow can compel design professionals to borrow to meet payrolls and pay expenses. The contract should provide that a specified rate tied to the actual cost of money be paid on delayed payments.[14]

In the absence of a contractually specified rate or formula, the interest is the "legal rate." In some states, this is the amount of interest payable on court judgments. During the inflationary period of the 1970s, the actual cost of borrowing was substantially in excess of the legal rate. As a result, many states increased the legal rate to make it more reflective of the actual costs of borrowing money. But in the 1990s, inflation dropped drastically, and as a result, the legal rate currently may be substantially greater than the actual cost of borrowing money. (This is not likely to be the case for short-term commercial loans.)

Among the welter of consumer protection legislation that has spewed forth from the Congress is the federal Truth in Lending Act. The Act requires those who lend money to disclose the details of the

---

[12]*Herbert Shaffer Assoc., Inc. v. First Bank of Oak Park,* supra note 9. But see *May v. Morganelli-Heumann & Assoc.,* 618 F.2d 1363 (9th Cir.1980) (architect contract divisible).

[13]See AIA Doc. B141, ¶¶ 10.3.2, 10.4.1.

[14]AIA Doc. B141, ¶ 11.5.2, provides blank spaces to be filled in for the rate of interest and when payable. If the blanks are not filled, late payments invoke the legal rate at the architect's principal place of business. Before 1977, ¶ 6.5 specified the "legal rate" payable sixty days from date of billing.

cost of the loan. The AIA has expressed concern over this law and has pointed to the law's possible application in a note to AIA Document B141, ¶ 11.5.2, which provides for late payments. The note states that certain consumer protection laws may affect the validity of the late payment provision and suggests that specific legal advice be obtained if changes are made in the clause or if disclosure must be made.

It is possible to conceive of the architect or the engineer—while generally not thought of as being a lender—as extending credit by allowing payments to be made after services are performed. If the amounts specified are considered a finance charge rather than a charge for late payment, disclosure requirements must be made in accordance with the Truth in Lending Act.

This is illustrated by *Porter v. Hill*,[15] a case decided by the Oregon Supreme Court in 1992. An attorney had included at the end of a billing statement a provision stating that a late payment charge of one and one-half percent per month would be added to the balance due if the amounts are more than thirty days overdue. The client challenged this provision, claiming it was a financing charge and required that there be disclosures in accordance with the Truth in Lending Act.

The court held that two factors are examined to determine whether the charge is a late payment charge to which the disclosure requirements do not apply: (1) whether the terms of account require the consumer to pay the balance in full each month and (2) whether the creditor acquiesces in the extension of credit by allowing the consumer to pay the account over time without demanding payment in full.

The court concluded here that under these standards, this was a late payment charge and not a financing charge. It pointed to the fact that the amount was described as a late payment charge and not a finance charge. It also noted that the written agreement required the client to pay the balance in full upon billing and that the attorney had not given his client the option of paying over time, subject to the one and one-half percent payment charge. Instead, the court noted, the attorney demanded full payment each month and, when he

was not paid, brought an action to collect the full amount due.

Design professionals should ensure that their bills state that the client is required to pay the balance in full each month. Also, if the design professional allows clients to pay the accounts over time without demanding payment in full and no effort is made to collect the full amount due, any late charge may be considered a finance charge invoking the onerous disclosure requirements of the Truth in Lending Act. Legal advice should be sought on this issue.

E. Suggestions Regarding Interim Fee Payments

When clients delay payments, the design professional should make a polite and sometimes strong suggestion that payments should be made when due. If the design professional, as noted in Section 13.02(D), acquiesces to a pattern of delayed payments, he may find that he has extended credit and is subject to the Truth in Lending Act. Also, a pattern of delayed payments should make the design professional seriously consider exercising any power to suspend further performance until payments are made.[16] If the suspension continues for a substantial time period, the design professional should consider terminating the contract.

In cases of suspension or termination of performance, it is desirable to notify the client of an intention to either suspend or terminate unless payment is received within a specified period of time. This gives the client an opportunity to make the payment. It also shows the client that failure to make interim fee payments as promised will not be tolerated.

SECTION 13.03 To Whom Payments Are to Be Made

Can the design professional assign the right to receive compensation for services rendered for the client? Suppose the design professional wishes to transfer his right to receive compensation to a

---

[15]314 Or. 86, 838 P.2d 45 (1992).

[16]Restatement (Second) of Contracts § 237, illustration 1; §§ 251, 252 (1981) (power to suspend until adequate assurance given). Refer to Section 12.13(D).

lender as security for a loan or to placate an impatient creditor.

Although early contract law set up restrictions that hampered the assignment of rights to receive money, modern law permits and even encourages such transfers. Permitting assignment enables parties to cash in contract-created rights. The person obligated to pay the money (the obligor) is not unduly burdened if it must pay to the person to whom the right has been transferred (the assignee) rather than the other party to the contract who has transferred the right (the assignor).

Unless special statutes compel a different result, the assignability of contract rights can be prohibited by contract provisions. Contracts commonly contain provisions stating that rights under the contract are not assignable without consent of the party who would have to perform the obligation. The law interprets such nonassignability clauses narrowly. Unless it is quite clear that the obligation to pay money cannot be transferred, general nonassignability clauses prohibit substituted performance but not the payment of money. A carefully drafted clause precludes assignability of money payment rights except where the assignment is made as security for a debt.

Most standard agreements used by design professionals include provisions requiring that consent be given to any assignments. Such provisions are not enforceable where the right assigned is the payment of money and the right is assigned as security for a loan.[17]

SECTION 13.04 Payment Despite Nonperformance

Denying a contracting party recovery for services performed unless it has performed *all* the obligations under the contract can create forfeiture (loss of contractual payment rights, substantially exceeding harm caused by breach) or unjust enrichment (the other party retaining and using the performance without paying for it). Legal doctrines have developed that minimize the likelihood of forfeiture and unjust enrichment.

Where it is not clear from the contract whether *exact* performance is a promise or a condition, the law is likely to classify the nonperformance as a breach creating a right to recover damages rather than a failure of a condition that bars recovery for work performed under the contract. The party who has breached can recover if it has *substantially* performed. Failure to *fully* perform does not bar recovery if caused by the other party's prevention or hindrance of performance or if the latter has failed to extend reasonable cooperation necessary to performance.[18] Sometimes the party to whom performance is due may have waived its right to performance by indicating that it was satisfied with less than exact performance.

Repudiation of the contract, denying its validity, or communicating an unwillingness or inability to perform excuses full performance. It would make little sense for the party to complete performance when the other party has indicated it does not wish performance and will not accept it.

Increasingly, the law is recognizing the right of a party in default to recover despite failure to perform under the contract.

These problems arise more commonly in construction contracts, and the legal doctrines have been developed with those contracts in mind. For this reason, a detailed discussion of these doctrines is postponed until Section 22.06.

SECTION 13.05 Other Client Obligations

AIA Doc. B141, Art. 4, lists other client obligations, including furnishing specialized services, information, surveys, and reports. A client engaging in a construction project for the first time may be surprised to discover that under ¶ 4.6 it must on the architect's request furnish services of a geotechnical engineer and other subsurface information and under ¶ 4.7 tests, inspections, and reports required by law or the contract documents. As noted in Section 12.09, it must furnish information relating to hazardous materials and pollution under ¶¶ 4.6 and 4.7. The client may be even more surprised to

---

[17]U.C.C. § 9-318(4). See *Mississippi Bank v. Nickles & Wells Constr. Co.*, 421 So.2d 1056 (Miss.1982) (AIA Doc.); *Aetna Cas. & Sur. Co. v. Bedford-Stuyvesant Restoration Constr. Corp.*, 90 A.D.2d 474, 455 N.Y.S.2d 265 (1982).

[18]*Carroll Fiscal Court v. McClorey*, 455 S.W.2d 547 (Ky.1970) (architect recovered despite abandonment of project because public entity did not make good-faith effort to obtain matching funds).

find that under ¶ 4.9 it warrants accuracy of information it has furnished the design professional, a standard that the client will not ordinarily be able to demand from the party from whom it has purchased the information. Finally, ¶ 4.3 requires the client to comply with any request by the architect for "evidence that financial arrangements have been made to fulfill the Owner's obligations" under the contract. It is advisable to explain such provisions to an inexperienced client in advance. Strong owners often change these provisions.

In addition to express provisions, obligations can be implied into contracts. The client impliedly promises not to interfere with the design professional's performance and to cooperate. For example, the client should not refuse the design professional access to information that is necessary for the performance of the work. Refusal to permit the design professional to inspect the site would be prevention and a breach of the implied obligation owed by the client to the design professional.

Positive duties are owed by the client. The client should exercise good faith and expedition in passing on the work of the design professional and in approving work at the various stages of the latter's performance. It should request bids from a reasonable number of contractors and should use best efforts to obtain a competent bidder who will agree to do the work at the best possible price. If conditions exist that will require acts of the client, such as obtaining a variance or obtaining financing, the client impliedly promises to use best efforts to cause the condition to occur.[19]

Although the legal issue was defamation and not implication of terms, *Sharratt v. Housing Innovations, Inc.*,[20] raised an interesting problem. A brochure advertising a housing project incorrectly listed the *associate* architect as the *principal* architect. The principal architect brought legal action against the publisher, claiming that he had been defamed because he had informed others in the design profession and construction community that *he* had been named as principal architect. The court held that the architect had been defamed and could recover damages from the publisher of the brochure, inasmuch as the publication, though not on its face defamatory, injured the architect when surrounding facts and circumstances were taken into account.

Suppose the client made an incorrect announcement similar to the brochure in the *Sharratt* case. The law implies the client will make a correct announcement of the engagement of the architect. (AIA Doc. B141, ¶ 9.9, requires the owner to "provide professional credit for the Architect on the construction sign and in the promotional materials for the Project.") If an inaccurate announcement was made, damages can be mitigated by a retraction and correction. If the mistake cannot be or is not corrected, the architect can recover the damages to his reputation caused by such a breach and any lost commissions he can show he has suffered. If the breach was tortious, such as one made maliciously or in bad faith, the architect can recover punitive damages.

---

[20]365 Mass. 141, 310 N.E.2d 343 (1974).

---

[19]Ibid.

PROBLEMS

1. A contracted to perform design services for the construction of an office building that was to have a contract price of one million dollars. A was to be paid 6% of the construction cost of the project. The construction contract contained a provision for liquidated damages under which the contractor was charged $500 for each day of unexcused delay. After performance was completed, it was determined that there were fifty days of unexcused delay. The owner deducted $25,000 from the final balance owed the contractor. Use AIA Doc. B141 in Appendix A to determine the amount of A's fee.

2. You are about to receive your first commission. Your client has asked you about methods of compensation and the method you would prefer. What would be your answer? Why? What additional facts would be useful?

Professional Liability: Process or Product?

SECTION 14.01 Claims Against Design Professionals: On the Increase

The possibility that a design professional will find herself in court has increased dramatically. One out of every three practicing architects is likely to find herself in litigation. Sometimes litigation is necessitated by the client's failure to pay fees for professional services. This chapter concentrates on claims made by clients and others that they have suffered losses that should be transferred to the design professional. This section explores the many reasons for increased claims against design professionals.

A. Changes in Substantive Law

Some defenses that had proved useful when claims were made by parties other than the client (collectively referred to as third parties) have proved of diminished value or have largely disappeared. For example, the requirement of privity between claimant and design professional and acceptance of the project terminating liability of the design professional—both valuable in avoiding third-party claims—have proved *much* less effective. The requirement that expert testimony be introduced to establish that a professional has not lived up to the standards of her profession has loosened. These substantive changes have resulted from increasing emphasis on ensuring that victims receive compensation and deterring wrongful conduct rather than on protecting socially useful activities.

B. Procedural Changes

With little difficulty and minimal costs, a claimant can bring legal action against a number of defen-dants in the same lawsuit. Similarly, those against whom claims have been brought can bring other defendants into that lawsuit with relative ease and without much expense. What has resulted is a complicated lawsuit with a host of parties defending and asserting claims.

Also, statutes of limitations designed to protect defendants from stale claims based on activities that took place many years before the claims have provided increasingly less protection.

C. Ability of Design Professionals to Pay Court Judgments

Claimants usually do not assert legal action against persons who they believe will be unable to pay for a court judgment or are not insured. Although expansion of liability and increased cost of insurance premiums have begun to reduce the percentage of design professionals who carry insurance, many will have professional liability insurance or sufficient resources to respond to court judgments. As a result, claims increase.

D. Access to Legal System

A person who seeks relief through the legal system usually engages a lawyer. Easier access to legal services will mean more claims and litigation. Increased accessibility to lawyers began by giving those charged with major crimes a lawyer even if they cannot afford to hire one. Programs were later developed to give legal representation to the poor when they sought to use the legal system. Along with this was increased emphasis on informing persons of their legal rights. For example, construction trade unions routinely inform their members of their legal rights and encourage them to use the legal process. Increasingly, members are provided legal services through prepaid legal insurance plans.

Another reason for easier access to the legal system is the much-maligned contingent fee contract (discussed in Section 2.04). Under such a contract, the client pays the lawyer for her time only out of any recovery obtained, and the client's investment is generally limited to expenses (in injury cases often paid by the attorney).

Another reason for more claims is that the prevailing party does *not* recover *its* costs of defense, including attorneys' fees, unless it can point to a contract providing for such recovery or to a statute granting attorneys' fees to the prevailing party. This can encourage legal claims by assuring the claimant that it does not run the risk of having to pay the *other* party's legal expenses if the claim fails.

E. Societal Changes

Americans today are less willing to accept their grievances silently. They are more inclined to use the legal system if they feel they have a grievance. The high ratio of lawyers to population in the United States, perhaps the highest in the world, demonstrates this.

American society has become increasingly urban and impersonal. There is much less likelihood that disputants are part of a cohesive social unit. Those units usually provide informal mechanisms for adjustment of rights and discourage resort to outside processes.

A sense of alienation in a large impersonal society makes people feel powerless with no one to protect them or help them. This was undoubtedly one element in the rise of consumerism in the 1960s. Aggressive use of the legal system responded to this phenomenon.

F. Enterprise Liability: Consumerism

Compensating victims rather than protecting enterprises has been the dominant modern tort motif. This has led to liability rules based on the belief that it is better that victims recover from the *enterprise* that is in the best position to avoid or spread the losses to those who benefit from the enterprise. Much of this drew unconsciously from the emphasis on security, which became a dominant objective

after World War II. Although much was accomplished through social welfare legislation, such as unemployment insurance, public housing, public welfare, and job security, emphasizing compensating victims through expansion of tort rights was a useful adjunct.

Liability expansion is also traceable to the consumer movement of the 1960s. Those who felt consumers were being supplied shoddy goods and services advocated increased liability as a means of bringing home to the business sector the importance of dealing fairly with consumers.

In the 1990s, a backlash resulted from the belief by some that expanded liability and increased litigation placed too heavy a burden on commercial enterprises and insurers. As noted in Section 7.08(L), much of this backlash was cloaked in the garb of competitiveness as American enterprises increasingly competed with products and services performed by enterprises in foreign countries. This backlash in the product liability field was demonstrated by the increased tendency of state legislatures to enact legislation that would limit the exposure of those who manufacture products. In the field of professional services, some states enacted legislation designed to limit the liability of health care providers. Still the strong tendency toward compensating victims continues to play a significant (perhaps even dominant) role in American tort law.

G. Design Centrality

Increased awareness exists of the centrality of design as a regulator of human conduct and allocator of societal resources. For example, California sought to reduce water consumption by legislation limiting the flushing capacity of toilets.[1] Public officials have stated that much of the responsibility for avoiding fires or minimizing fire losses falls on those who design structures. A fire chief stated that "good fire protection in high-rise buildings begins on the architect's drawing board." He also stated that it was the builder's responsibility to design "a safe building, not a firetrap."[2]

Law enforcement officials have stated that assaults can be reduced if those who design business

---

[1]West Ann.Cal.Health & Safety Code § 17921.3.
[2]*New York Times*, August 4, 1974, p. 11.

and residential areas plan properly. A director of the National Institute of Law Enforcement and Criminal Justice stated:

> Better environmental design can do much more [to reduce crime]. New housing projects, schools, shopping centers and other areas can be designed, for example, with more windows looking out on streets, fewer hidden corridors, and other crime discouraging features.[3]

In the context of design professionals, this is demonstrated by recent reported cases that involved claims based on negligent prison design. *State v. Gathman-Matotan*[4] was a wrongful death action against a prison in the aftermath of a prison riot. The prison asserted a cross-complaint against the architect who had designed the prison renovation. The prison contended that the architect had impliedly warranted that it would design a control area secure enough to prevent a prisoner takeover.

Similarly, the centrality of design was shown by three negligence claims against architects by estates of inmates who had committed suicide based on failure to design in such a way as to prevent suicides.[5]

H. Codes

Liability has also expanded because of a proliferation of detailed building and housing codes. Violation of these codes, although not conclusive on the question of negligence, makes it relatively easy to establish that the design professional is liable not only to the client but also to third parties who suffer foreseeable harm because of the violation.

Although most building codes are enacted by local authorities, the enactment of the Federal Americans with Disabilities Act will also be used

as the basis for predicating liability on design professionals. See Section 14.05(C).

I. Expansion of Professional Services

Design professionals are expected to either provide or volunteer to provide services that go beyond core design services. These services can relate to availability of funds for the project, the likely profitability of the project, and the likelihood of approval by public officials and agencies who control building and construction. This can increase claims when clients suffer disappointments or losses.

J. Site Services

A design professional's varied site services discussed in Chapter 12 and in the balance of the treatise make the design professional more vulnerable to a claim traceable not only to design but also to the way in which the design was executed, both as to compliance with the design and the methods of accomplishing it.

SECTION 14.02 Overview of Chapters 14 and 15

The preceding section outlined the reasons that professional liability has expanded and the likelihood that more claims will be made against design professionals. Chapter 14 deals with professional liability claims mainly by clients but also by third parties. Chapter 15 discusses techniques to avoid or reduce liability risks through risk management.

A. Applicable Law

Professional liability is usually determined by state law. Depending on various factors, professional liability rules will be those of the state in which the design professional has its principal place of business, where the actual design is created, or where the project is located. Because both legal rules and actual outcomes may vary depending on the applicable state law, discussion of liability must be general and not directed to a particular state except to the extent that illustrative cases may do so.

B. Types of Harm

It is useful to divide claims into those that involve personal harm, those that involve harm to property, and those that involve economic loss not connected to personal harm or property damage. As a

---

[3]Ibid. See also O. NEWMAN, *DEFENSIBLE SPACE: CRIME PREVENTION THROUGH URBAN DESIGN* (Collier Books, 1973).

[4]98 N.M. 740, 653 P.2d 166 (1982) (court refused to imply a warranty into a professional service contract).

[5]*Easterday v. Masiello*, 518 So.2d 260 (Fla.1988) (claim barred by prison's acceptance of project with patent defects); *La Bombarbe v. Phillips Swager Assoc.*, 130 Ill.App.3d 896, 474 N.E.2d 942 (1985) (architect relieved as duty to design was for ordinary use, not for suicide); *Tittle v. Giattina, Fisher & Co., Architects, Inc.*, 597 So.2d 679 (Ala.1992) (design of jail facility not proximate cause of death). As to acceptance of the project relieving the design professional from responsibility, see Section 14.09(B).

general rule, the law is more likely to sustain claims based on personal harm than claims based on harm to property or economic loss and to sustain claims related to property harm than those involving economic loss. Traditionally, though less so today, tort law dealt with harm to person or property.

This hierarchy of protection can help predict the outcome of any lawsuit, particularly where a court is asked to veer from established rules and recognize new legal rights. This is more likely to be done when a claim is based on personal harm and the need for compensation more urgent. Once this inroad has been made, some courts stop at that point. But more often, they decide that there is no particular reason to limit it to cases involving personal harm. Over time, what was once an established rule may disappear through this process. Disappearance is usually gradual, however. This preserves the appearance of stability while the need for change is accommodated.

SECTION 14.03 Claims Against Design Professionals: Some Illustrations

Another way to demonstrate the liability explosion described in Section 14.01 is to note the types of claims that have been made against design professionals. Although *most* of the claims to be noted were successful, some cautionary remarks are essential.

Any conclusion that particular conduct gave rise to a valid claim was made in the context of a particular contract and particular facts. Conduct *held* to be below the standard required in a particular case does not mean that this conduct will always fall below the legal standard.

Many appellate opinions simply conclude that the determination made by the finder of the facts—either the trial judge or jury—was within its discretion. An appellate court reviewing a lower court decision may not agree with the decision but will respect the differentiation between the role of trial and appellate courts, the former resolving factual disputes and the latter, as a rule, deciding questions of law.

As to negligence in the design phase, one writer stated:

Architects might fail to use due care in various ways. The architect may inadequately consider the nature of the soil under the building; he may design an inade-

quate foundation; he may design a roof too weak to support the weight it will foreseeably have to bear; he may insulate or soundproof the building inadequately. The architect may negligently design a sewer so that waste is carried toward rather than away from the house, he may design windows too small or too large, he may fail to put a handrail on a stairway, or he may specify that nails rather than bolts be used to secure a sundeck. . . . In addition the architect may negligently fail to notice a defect in the work of a consultant he has hired to help prepare the plans and specifications. The architect would also probably be liable for damage caused by his failure to hire a consultant where a reasonable architect would have done so.

Negligence in design can be based on negligently incomplete specifications as well as upon complete but erroneous ones. The plans and specifications must be complete and unambiguous. For example, specifications are negligently prepared if they are so indefinite that a contractor can bid as if he were going to use first class materials and then build using inferior materials. If measurement of a material is involved, the specification must distinguish between dry and liquid states, or loose or tight packing, where there is any chance of ambiguity.[6]

Since that analysis there have been other illustrations provided in judicial opinions. For example, the following cases involved claims, many of which were successful:

1. Misrepresenting existing topography.[7]
2. Relying on an out-of-date map and building on land not owned by the owner.[8]
3. Specifying material that did not comply with building codes.[9]

---

[6]Comment, 55 Calif.L.Rev. 1361, 1370–71 (1967). The author's footnotes to cases are omitted. Cases involving surveyors' mistakes are collected in Annot., 35 A.L.R.3d 504 (1971).

[7]*Mississippi Meadows, Inc. v. Hodson*, 13 Ill.App.3d 24, 299 N.E.2d 359 (1973) (dictum).

[8]*Jacka v. Ouachita Parish School Bd.*, 249 La. 223, 186 So.2d 571 (1966). Here the architect was relieved because the client was obligated to and did furnish the out-of-date map.

[9]*St. Joseph Hosp. v. Corbetta Constr. Co.*, 21 Ill.App.3d 925, 316 N.E.2d 51 (1974); *Johnson v. Salem Title Co.*, 246 Or. 409, 425 P.2d 519 (1967); *Edward J. Seibert, A.I.A., Architect and Planner, P.A. v. Bayport Beach and Tennis Club Assoc., Inc.*, 573 So.2d 889 (Fla.Dist.Ct.App.1990), review denied, 583 So.2d 1034 (1991).

4. Positioning the building so as to violate setback requirements.[10]
5. Failing to inform client of potential risks of using certain materials.[11]
6. Failing to advise client as to potential problems with new product.[12]
7. Failing to inform the client that it could not make a reliable judgment when materials were known to the architect and failing to collect information about such materials.[13]
8. Failing to inform the client that the design professional had acquired information that water temperature was too low for the operation of a pump.[14]
9. Drafting ambiguous sketches causing extra work.[15]
10. Designing a house that could not be accomplished by tradespeople in the community where the project was to be built.[16]
11. Designing closets not large enough for the clothing to be contained in them.[17]
12. Designing a project that greatly exceeded the client's budget.[18]
13. Specifying untested material solely because of seller's representations.[19]
14. Designing inadequate solar heating system.[20]
15. Failing to consider energy costs.[21]

16. Failing to disclose an underground high-voltage live wire.[22]
17. Failing to include owner as named insured and omitting indemnity clause.[23]
18. Failing to advise a need for use permit.[24]
19. Failing to design a prison that would make it "takeover" proof.[25]
20. Failing to design a prison that would avoid prisoner suicide.[26]
21. Failing to know of local building codes and safety laws.[27]
22. Specifying a competition diving board for a grade school and failing to warn users of dangers.[28]

Claims, again mostly successful, relating to the Construction Process phase were as follows:

1. Allowing non-code approved material to be installed.[29]
2. Ordering excess fill to be placed without consulting a soil tester.[30]
3. Failing to make changes needed to comply with codes.[31]
4. Failing to condemn defective work.[32]
5. Performing scheduling and coordination incompetently.[33]

[10]*Armstrong Constr. Co. v. Thomson*, 64 Wash.2d 191, 390 P.2d 976 (1964).
[11]*Banner v. Town of Dayton*, 474 P.2d 300 (Wyo.1970).
[12]*White Budd Van Ness Partnership v. Major-Gladys Drive Joint Venture*, 798 S.W.2d 805 (Tex.Ct.App.1990), cert. denied 112 S.Ct. 180 (1991).
[13]*Richard Roberts Holdings Ltd. v. Douglas Smith Stimson Partnership and Others* (decision by official referee in the United Kingdom and summarized in 5 Constr.L.J. 223 (1989)).
[14]*Green Island Assoc. v. Lawler, Matusky & Skelly Eng'rs*, 170 A.D.2d 854, 566 N.Y.S.2d 715 (1991).
[15]*General Trading Corp. v. Burnup & Sims*, 523 F.2d 98 (3d Cir.1975).
[16]*Bayuk v. Edson*, 236 Cal.App.2d 309, 46 Cal.Rptr. 49 (1965).
[17]*Ibid.*
[18]*Stanley Consultants, Inc. v. H. Kalicak Constr. Co.*, 383 F.Supp. 315 (E.D.Mo.1974).
[19]*New Orleans Unity Soc'y v. Standard Roofing Co.*, 224 So.2d 60 (La.App.1969) (dictum).
[20]*Keel v. Titan Constr. Corp.*, 639 P.2d 1228 (Okl.1981).
[21]*Bd. of Educ. v. Hueber*, 90 A.D.2d 685, 456 N.Y.S.2d 283 (1982) (unsuccessful).

[22]*Mallow v. Tucker, Sadler & Bennett, Architects & Eng'rs, Inc.*, 245 Cal.App.2d 700, 54 Cal.Rptr. 174 (1966).
[23]*Transit Cas. Co. v. Spink*, 94 Cal.App.3d 124, 156 Cal.Rptr. 360 (1979) (trial court finding for client not challenged on appeal), disapproved on other issues in *Commercial Union Assurance Co. v. Safeway Stores*, 26 Cal.3d 912, 610 P.2d 1038, 164 Cal.Rptr. 709 (1980).
[24]*Chaplis v. County of Monterey*, 97 Cal.App.3d 249, 158 Cal.Rptr. 395 (1979).
[25]*State v. Gathman-Matotan*, supra note 4.
[26]*Easterday v. Masiello*, supra note 5; *La Bombarbe v. Phillips Swager Assoc.*, supra note 5; *Tittle v. Giattina, Fisher & Co. Architects, Inc.*, supra note 5.
[27]*Ins. Co. of North Am. v. G.M.R., Ltd.*, 499 A.2d 878 (D.C.App.1985).
[28]*Francisco v. Manson, Jackson & Kane, Inc.*, 145 Mich.App. 255, 377 N.W.2d 313 (1985).
[29]*St. Joseph Hosp. v. Corbetta Constr. Co.*, supra note 9.
[30]*First Ins. Co. of Hawaii v. Continental Casualty Co.*, 466 F.2d 807 (9th Cir.1972).
[31]*Mississippi Meadows, Inc. v. Hodson*, supra note 7.
[32]*Skidmore, Owings & Merrill v. Connecticut General Life Ins. Co.*, 25 Conn.Sup. 76, 197 A.2d 83 (1963).
[33]*Peter Kiewit Sons' Co. v. Iowa Southern Utilities Co.*, 355 F.Supp. 376 (S.D.Iowa 1973).

6. Failing to exercise supervisory powers properly.[34]

7. Failing to warn an experienced contractor of general precautions not known in the industry.[35]

8. Failing to engage and check with a consultant.[36]

9. Failing to stop work after discovering contractor using unsafe methods.[37]

10. Issuing payments or certificates negligently.[38]

11. Failing to warn of bankruptcy when paying for materials in contractor's possession.[39]

12. Failing to observe design deviation when checking shop drawings.[40]

13. Failing to give instructions, issue change orders, and conduct inspections.[41]

SECTION 14.04 Specific Contract Standard

A. Likelihood of Specific Standard

Contracting parties can by agreement determine the standard of performance. Because the client-design professional relationship is created by agreement, a primary source of any agreed-upon standard is the contract itself. Most disputes between the client and the design professional do not involve a *specific* contractually designated standard. Rather, they involve a *general* standard *not* set forth in the contract, such as the professional standard to be described in Section 14.05, or an outcome-oriented standard, such as implied warranty described in Section 14.07.

Why do most contracts fail to *specifically* state *how* the design professional is to perform? First,

many relationships are created without any written agreement—by handshake arrangements. Second, many are made by casual letter agreements drafted by the design professional that are not likely to describe specific standards. Third, those relationships created by assent to standard contracts published by professional associations such as the American Institute of Architects (AIA) do not, despite their completeness as to services and what the design professional is *not* responsible for, specifically describe *how* the work will be done. However, in 1992 the Engineers Joint Contracts Documents Committee (EJCDC) inserted ¶ 1.1 in its No. 1910-1 (owner-engineer contract) stating that the standard of care will be that ordinarily used by members of the engineer's profession under similar conditions at the same locality but that the engineer makes no "warranties, express or implied, under this Agreement or otherwise."

Even if the standard of performance is discussed in advance—something that is rare—it might not be included in the written contract: the design professional will not want it included, or the client may think it unimportant to do so. Both may believe that any assurances as to outcome are simply nonbinding opinions or expectations.

It is common for large organizations of design professionals to include language in their standard agreements specifying the professional standard described in Section 14.05. This is done for two reasons. First, the design professional may wish to refine elements of the professional standard that may operate to her advantage. For example, she may include a provision that states that she is to be compared to other design professionals in her community to avoid any contention that she is to be compared to design professionals with whom she competes. Second, and more commonly, the specification of the professional standard is accompanied by language stating that the architect will not be held to any express or implied warranty standard, a standard often more rigorous than the professional standard.[42] Clients who are strong and experienced, such as public entities, often resist such clauses, seeking to preserve the possibility of later claiming the design professional should be held to a stricter standard.

---

[34]*Aetna Ins. Co. v. Hellmuth, Obata & Kassabaum, Inc.*, 392 F.2d 472 (8th Cir.1968).

[35]*Vonasek v. Hirsch & Stevens, Inc.*, 65 Wis.2d 1, 221 N.W.2d 815 (1974) (dictum).

[36]*Cutlip v. Lucky Stores, Inc.*, 22 Md.App. 673, 325 A.2d 432 (1974).

[37]*Associated Eng'rs, Inc. v. Job*, 370 F.2d 633 (8th Cir.1966). See Annot., 59 A.L.R.3d 869 (1974).

[38]*Aetna Ins. Co. v. Hellmuth, Obata & Kassabaum, Inc.*, supra note 34.

[39]*Travelers Indem. Co. v. Ewing, Cole, Erdman & Eubank*, 711 F.2d 14 (3d Cir.1983) (unsuccessful).

[40]*Jaeger v. Henningson, Durham & Richardson, Inc.*, 714 F.2d 773 (8th Cir.1983).

[41]*Colbert v. B.F. Carvin Constr. Co.*, 600 So.2d 719 (La.App.1992).

---

[42]See Section 14.05.

If challenged, which is more common if clients are strong and experienced, such as public entities, the design professional will justify such a clause by stating that its professional liability insurance coverage will not include contractual risks that deviate from the professional standard. If the matter cannot be resolved one way or the other, the language may be omitted, leaving the standard to that applied by law. Yet sometimes contractually specified standards of performance are created. Looking at some of them can provide a useful backdrop to the nonspecific standards described in Sections 14.05 through 14.07.

B. Client Satisfaction

Sometimes the client's obligation to pay arises only if the client is satisfied with the work of the design professional. Such a contract may be interpreted to be a promise by the design professional to satisfy the client. Though relatively one-sided and one that design professionals generally seek to avoid, if clear evidence exists of such an agreement, the agreement will be enforced.

If satisfaction is a condition to the client's obligation to proceed and to pay, the client need not pay unless it is satisfied or waives this performance measurement. Any legal obligation that may arise must be based on unjust enrichment created by the client using the work of the design professional.

If satisfaction has been *promised*, failure to perform requires that the design professional compensate the client for any losses the latter may have suffered because of the breach. For example, if a breach caused the project to be delayed or abandoned, the design professional is accountable for any foreseeable losses that can be established with reasonable certainty and that could not have been reasonably avoided.

Although the design professional may, however unwisely, risk the fee, strong evidence of such a risk assumption should be produced before she must respond for losses caused by her failure to satisfy the client.

Two standards of satisfaction exist.[43] If performance can be measured *objectively*, the standard

is reasonable satisfaction. Would a reasonable person have been satisfied? Objective standards are more likely to be applied where performance can be measured mechanically. For example, if an engineer agreed with a manufacturer that the manufacturer would pay if satisfied with the performance of a particular machine designed by the engineer, the obligation to pay would require that a reasonable manufacturer be satisfied.

More personal performance invokes a *subjective* standard. Suppose an artist agrees to paint a portrait that will satisfy the person commissioning the portrait. The latter must exercise a good-faith judgment and must be genuinely dissatisfied before she is relieved of the obligation to pay. If, for example, the person refused to view the portrait or give it sufficient light to judge its quality, any judgment was not exercised in good faith.

In practice, the two standards may not operate differently. In the preceding example, if judge or jury thought the performance satisfactory, it would take a strong showing on the part of the person commissioning the portrait that there has been genuine dissatisfaction.

As a rule, the *subjective* satisfaction standard arises in the design phase, particularly in aesthetic matters. However, in standard commercial projects, an objective standard may be invoked.

What about a design professional's performance during construction? If the performance in question related to delicate matters, such as how the design professional handled the contractor or public officials or how she dealt with site conflicts, a subjective standard may be applied. Roughly speaking, though, design is more likely to be measured subjectively, while contract administration is more likely to be measured objectively—another illustration of the important difference between design and nondesign site services.[44]

Suppose the client is justified in refusing to pay. This may create a forfeiture. The loss in such a case to the design professional may substantially exceed the loss to the client that would occur if the client, though dissatisfied, were required to accept the design professional's work. Various legal doctrines, among them waiver and estoppel, can be em-

---

[43]*First Nat'l Realty Corp. v. Warren-Ehret Co.*, 247 Md. 652, 233 A.2d 811 (1967) (collecting authorities). For a careful analysis, see *Morin Building Products Co. v. Baystone Constr. Co.*, 717 F.2d 413 (7th Cir.1983).

---

[44]*Jaeger v. Henningson, Durham & Richardson, Inc.*, 714 F.2d 773 (8th Cir.1983) (expert testimony not needed for claim based on site services).

ployed to avoid forfeiture. But if the language makes clear that the risk of forfeiture was clearly assumed by one party to the contract, the clause will be enforced even if this would create a forfeiture. (Exceptions to this exist in some jurisdictions relating to the purchase of land.)

C. Fitness Standard

A more specific performance standard than satisfaction can be contractually created. The parties may agree that the completed project will be suitable or fit for those purposes for which the client entered into the project.

The client who plans a luxury residence usually wants a house suitable for a person of her means and taste. In addition to wanting the normal requirements for any residence, such as structural stability, shelter from the elements, and compliance with safety and sanitation standards, the client may want a house that is admired by those who enter it or a residence that can facilitate closing business deals or making business contracts. The client may hope that the opulence of the residence will make social events successful.

The client who plans a commercial office building wishes to make a profit from the rental of space. To accomplish this, suitable tenants at an economically adequate rent must be found. Such a client assumes that the planned use of the structure will be permitted under zoning laws and that the structure will comply with the applicable building codes and zoning regulations relating to materials, safety, density, setback regulations, and other land use controls. In addition, the client who builds an unusually designed office building may hope that the structure will be the subject of national architectural interest.

The client who wishes to build an industrial plant generally assumes that the plant when completed will be adequate to perform anticipated plant activities. The building is expected to comply with applicable laws relating to public health and safety.

Proper design requires that the design professional consider client objectives such as those discussed in the preceding paragraphs. Some items mentioned in those paragraphs will be discussed and included in the client's program. Some of the matters discussed in the preceding paragraphs would be assumed and probably not discussed.

One would not expect client and design professional to discuss the necessity of complying with building codes or regulations dealing with health and safety. Yet beyond these basic objectives, discussions may have taken place of economic and social goals less directly connected with basic design objectives.

One *possible* standard to measure the performance of the design professional is whether the project accomplishes the objectives of the client. Put another way, is the structure suitable for the client's anticipated needs, or is it fit for the purpose for which it was built? Is the building an architectural success? Has the client been able to attract good tenants? Has plant production increased? Are the social events successful?

To determine whether suitability or fitness performance standards will be used to measure the design professional's obligation, it is important to look at any antecedent negotiations, discussions, or understandings that may have preceded the client-design professional contract or may have occurred during the course of the design professional's activities.

Suppose there were discussions of client objectives during precontract negotiations or during the design professional's design performance. Were any assurances given by the design professional that related to the fitness or suitability of her design to accomplish particular client objectives *promises* that the design would accomplish the objectives or just statements of *opinion* that these objectives would be achieved?

Suppose the design professional made statements relating to such matters. The design professional might have stated that a particular luxury residence would create an artistic stir within the client's social circle. The design professional might have expressed a belief that suitable tenants could be found or that someday a particularly unusual office building would be considered an architectural landmark. She might have assured the client that the client could conduct certain activities on the premises of the structure being built.

To determine whether a statement is a promise or an expression of opinion, the law looks at the definiteness with which the statement was made ("I am certain your cost per unit will decrease" versus "It's my considered opinion that you will improve productivity"), the degree to which the design professional's performance can bring about that objective ("People will like the exterior design" versus "Your parties will be great successes"), and the degree and reasonableness of any reliance by the client on the statement (using certain types of machinery in a plant versus redecorating the interior of a house at great expense for the new social season). The more definite the statement, the more within the control or professional expertise of the design professional is the outcome, and the more likely there has been justifiable reliance, the more likely it is the law will find there has been a promise.

If the statement was made before the formation of a contract, a design professional may contend that such statements cannot be proved because of the parol evidence rule. Although results are not always consistent, by and large, the client will be permitted to testify as to these statements.

The design professional should avoid assuring the client that particular objectives will be obtained unless she is willing to risk the possibility of being held accountable if the objectives are not achieved. Assurances of certain matters should be given, such as the design meeting public land use controls, such as zoning laws and building codes. But the design professional should avoid venturing into areas that are beyond her expertise and require difficult predictions of the future.

D. Quantitative or Qualitative Performance Standards

Sometimes the contract between design professional and client contains a specific performance standard. For example, an engineer may make a contract with a manufacturer under which it was specifically agreed that the machine designed by the engineer would produce a designated number of units of a particular quality within a designated period of time.

Suppose the performing party finds that it is extremely difficult to meet the performance standard or that the performance standard will require an amount of time and money not anticipated by the performing party. In some extreme cases, the performance standard may be impossible to meet.

In such cases, two legal issues may arise. First, is the design professional entitled to be paid for the effort made in trying to accomplish the performance specifications? It may be an onerous contract, but if this is the risk assumed, there will be no recovery. Like satisfaction contracts, interpretation doubts are resolved in favor of the performing party to avoid forfeiture. But if this cannot be done, the performing party will be uncompensated unless she can show that efforts expended, although not fulfilling the performance standards, have contributed a benefit to the other party. Although it may be beneficial to the other party to be shown that the performance standards were not possible, in most cases, there will be no unjust enrichment and no recovery for the performing party.

Second, has the performing party breached by not accomplishing the objective? Again this is a question of whether the accomplishment of the objective is not merely a condition to the client's obligation to pay but also a promise on the part of the performing party.

Denial of recovery generally is a sufficient burden for the performing party. For this reason, the performing party should not be held to a promise to accomplish what turns out to be either an extremely difficult or an impossible performance. But in *Gurney Industries, Inc. v. St. Paul Fire & Marine Insurance Co.*,[45] a designer-builder was required to pay damages for failure to meet performance standards. There was no indication that the standards were beyond the state of the art or involved inordinate expense to meet. The contract doctrines that relieve contracting parties where their performance becomes impossible or impracticable may grant relief from any promise to fulfill performance specifications that are beyond the state of the art or disproportionately expensive.[46]

It is obvious that performance standards place a heavy risk on the performing party. Yet they are attractive to clients because they objectively measure whether the client is getting what it was promised. Such a standard can be an effective sales device for the designer.

E. Indemnification

The frequency of indemnification in construction contracts necessitates more complete treatment in Chapter 32. For the purposes of this section, it should be noted that clients increasingly demand that the design professional indemnify *them* against

---

[45]467 F.2d 588 (4th Cir.1972).
[46]See Section 23.03(D).

claims that the client believes to be the responsibility of the design professional. Any design professional indemnification creates another specific contractual obligation, one potentially broader than the professional standard.

F. Cost Overruns Due to Design

An Ohio public entity had made a proposal requiring design professionals to pay the cost of construction change orders if the change orders were caused by the design professional. The professional societies and professional liability insurers objected to this policy, stating that the policy erroneously assumed that change orders are necessarily caused by negligence of the design professional and that the professional liability insurance policies would not cover such a form of strict liability. After these objections were made, the Ohio agency withdrew the proposal. However, it is unlikely that this is the end of such attempts to charge cost overruns to the design professional.

Attempts have been made, mainly by clients but also by design professionals, to recognize the likelihood of cost overruns due to design and to deal with them in various ways in the contract. One public contract in Virginia specified that the design professional would be responsible for cost overruns related to design if the overruns exceeded three percent of the contract price, without any need to show that the design professional had not performed in accordance with the professional standard. Even more, as noted in the preceding paragraph, some clients insist that all overruns due to design be the responsibility of the design professional whether there has been failure to perform in accordance with the professional standard or not. This form of strict liability would almost certainly not be insurable under normal professional liability policies.

It would be unfair to the contractor if the contractor submits its bid without knowing that the architect is strictly responsible for all design overruns. If the contractor knows of such strict liability, it could infer that all questions of interpretation that arise during performance would be resolved against it. This would affect the bid.

Looking at this problem from the standpoint of a design professional, those interested in limiting design professional liability have suggested inclusion of provisions stating that the design will in-

evitably contain errors, omissions, conflicts, and ambiguity that will require clarification and correction during construction. They suggest the client be advised that the production of perfect documents is an impossibility and that some design decisions are more efficiently deferred for the benefit of the client until there is construction and actual field conditions. Finally, such persons advise that a contract contain a provision relieving the design professional from liability for all cost overruns up to a stated percentage of the construction cost.

G. Contractual Diminution of Legal Standard

Specific contractual standards are usually higher than the professional standard. Suppose, however, that the agreement between the client and the design professional specifies a lower standard. Since this more directly involves the extent to which the standard set by law can be varied, it is discussed in Section 14.05(C).

SECTION 14.05 The Professional Standard: What Would Others Have Done?

A. Defined and Justified: *City of Mounds View v. Walijarvi*

The Minnesota Supreme Court in *City of Eveleth v. Ruble*[47] stated:

> (1) In an action against a design engineer for negligence, the applicable legal principles are held to be:
>
> (a) One who undertakes to render professional services is under a duty to the person for whom the service is to be performed to exercise such care, skill, and diligence as men in that profession ordinarily exercise under like circumstances.
>
> (b) The circumstances to be considered in determining the standard of care, skill, and diligence to be required include the terms of the employment agreement, the nature of the problem which the supplier of the service represented himself as being competent to solve, and the effect reasonably to be anticipated from

---

[47]302 Minn. 249, 225 N.W.2d 521 (1974). See also *Chrischilles v. Griswold*, 260 Iowa 453, 150 N.W.2d 94 (1967); *Milton J. Womack, Inc. v. State House of Representatives*, 509 So.2d 62 (La.App.1987) cert. denied 513 So.2d 1208, (La.1987); *State v. Gathman-Matotan*, supra note 4.

the proposed remedies upon the balance of the [water] system.

(c) Ordinarily, a determination that the care, skill, and diligence exercised by a professional engaged in furnishing skilled services for compensation was less than that normally possessed and exercised by members of that profession in good standing and that the damage sustained resulted from the variance requires expert testimony to establish the prevailing standard and the consequences of departure from it in the case under consideration.[48]

---

[48]225 N.W.2d at 522.

CITY OF MOUNDS VIEW v. WALIJARVI

Supreme Court of Minnesota, 1978. 263 N.W.2d 420.
[Ed. note: footnotes omitted.]

TODD, Justice.

[Ed. note: The city became apprehensive because of dampness in the basement of an addition that was being added to a city building. The architect wrote to the city that its design would, if executed properly, generate a "water-tight and damp-free" basement. But problems grew worse, and corrective work was needed.

The city sued the architect, based on claims of negligence, express warranty, and implied warranty.

The trial court held that the language in the letter asserted to constitute a warranty to be merely an expression of opinion and that Minnesota did not recognize implied warranty of a perfect plan or an entirely satisfactory result in an architectural service contract. The trial court granted the architect's motion for summary judgment (no trial needed) on the warranty claims. The Minnesota Supreme Court held that the express warranty claim failed because the contract required that modifications be written and signed by both parties and no evidence had been introduced of any written agreement by the city. The opinion then dealt with implied warranty.]

3. As an alternative basis for recovering damages from the architects, the city urges that we adopt a rule of implied warranty of fitness when architectural services are provided. Under this rule, as articulated in the city's brief, an architect who contracts to design a building of any sort is deemed to impliedly warrant that the structure which is completed in accordance with his plans will be fit for its intended purpose.

As the city candidly observes, the theory of liability which it proposes is clearly contrary to the prevailing rule in a solid majority of jurisdictions. The majority position limits the liability of architects and others rendering "professional" services to those situations in which the professional is negligent in the provision of his or her services. With respect to architects, the rule was stated as early as 1896 by the Supreme Court of Maine (*Coombs v. Beede*, 89 Me.187, 188, 36 A.104 [1896]):

"In an examination of the merits of the controversy between these parties, we must bear in mind that the [architect] was not a contractor who had entered into an agreement to construct a house for the [owner], but was merely an agent of the [owner] to assist him in building one. The responsibility resting on an architect is essentially the same as that which rests upon the lawyer to his client, or upon the physician to his patient, or which rests upon anyone to another where such person pretends to possess some skill and ability in some special employment, and offers his services to the public on account of his fitness to act in the line of business for which he may be employed. The undertaking of an architect implies that he possesses skill and ability, including taste, sufficient to enable him to perform the required services at least ordinarily and reasonably well; and that he will exercise and apply in the given case his skill and ability, his judgment and taste, reasonably and without neglect. But the undertaking does not imply or warrant a satisfactory result."

The reasoning underlying the general rule as it applies both to architects and other vendors of professional services is relatively straightforward. Architects, doctors, engineers, attorneys, and others deal in somewhat inexact sciences and are continually called upon to exercise their skilled judgment in order to anticipate and provide for random factors which are incapable of precise measurement. The indeterminate nature of these factors makes it impossible for professional service people to gauge them with complete accuracy in every instance. Thus, doctors cannot promise that every operation will be successful; a lawyer can never be certain that a contract he drafts is without latent ambiguity; and an architect cannot be certain that a structural design will interact with natural forces as anticipated. Because of the inescapable possibility of error which inheres in these services, the law has traditionally required, not perfect results, but rather the exercise of that skill and judgment which can be reasonably expected from similarly situated professionals. As we stated in *City of Eveleth v. Ruble*, 302 Minn. 249, 253, 225 N.W.2d 521, 524 (1974):

> "One who undertakes to render professional services is under a duty to the person for whom the service is to be performed to exercise such care, skill, and diligence as men in that profession ordinarily exercise under like circumstances."

See, also, *Kostohryz v. McGuire*, 298 Minn. 513, 212 N.W.2d 850 (1973).

We have reexamined our case law on the subject of professional services and are not persuaded that the time has yet arrived for the abrogation of the traditional rule. Adoption of the city's implied warranty theory would in effect impose strict liability on architects for latent defects in the structures they design. That is, once a court or jury has made the threshold finding that a structure was somehow unfit for its intended purpose, liability would be imposed on the responsible architect in spite of his diligent application of state-of-the-art design techniques. If every facet of structural design consisted of little more than the mechanical application of immutable physical principles, we could accept the rule of strict liability which the city proposes. But even in the present state of relative technological enlightenment, the keenest engineering minds can err in their most searching assessment of the natural factors which determine whether structural components will adequately serve their intended purpose. Until the random element is eliminated in the application of architectural sciences, we think it fairer than (sic) the purchaser of the architect's services bear the risk of such unforeseeable difficulties.

The city suggests that many of the design-related tasks performed by modern architects are routine and carry no risk of error if they are performed with professional due care. It is argued that with respect to such tasks, the premise on which the traditional rule rests is inoperative, making the adoption of the implied warranty theory fully proper. We note, however, that architectural errors in relatively simple matters are quite easily handled under the existing cause of action for professional negligence.

Moreover, if implied warranties are held to accompany only uncomplicated architectural endeavors, the finder of fact will be forced in every case to determine, as a preliminary matter, whether the alleged architectural error was made in the performance of a sufficiently simplistic task. Defects which are found to be more esoteric would presumably continue to be tried under the traditional rule. It seems apparent, however, that the making of any such threshold determination would require the taking of expert testimony and necessitate an inquiry strikingly similar to that which is presently made under the prevailing negligence standard. We think the net effect would be the interjection of substantive ambiguity into the law of professional malpractice without a favorable tradeoff in procedural expedience.

In addition, we observe that the ills which spurred the creation and expansion of the implied warranty/strict liability doctrine are not really present in this case or in the architect-client relationship generally. The implied warranty of fitness originated primarily as a means of facilitating the legitimate interests of the consuming public and bringing common-law remedies into step with the practicalities of modern industrialism. The outmoded requirement of contractual privity, coupled with manufacturers' sweeping disclaimers of liability, frequently operated to deny effective remedies to those who purchased commercial products at the bottom of a multi-tiered production and distribution network. See, generally, Prosser, Torts (4 ed.) §§ 97, 98. The introduction of the implied warranty doctrine created an effective remedy by allowing plaintiffs to proceed directly against the offending party without reliance on express contractual warranties.

The relationship between architect and client is markedly different. For a client, architectural services are hardly produced by a faceless business entity, insulated by a network of distributors, wholesalers, and retailers. Architects and clients normally enjoy a one-to-one relationship and communicate fairly extensively during the course of the relationship. When a legal dispute arises, the client has no trouble locating the source of his problem, and a remedial device like the implied warranty is largely unnecessary.

Finally, while it is undoubtedly fair to impose strict liability on manufacturers who have ample opportunity to test their products for defects before marketing them, the same cannot be said of architects. Normally, an architect has but a single chance to create a design for a client which will produce a defect-free structure. Accordingly, we do not think it just that architects should be forced to bear the same burden of liability for their products as that which has been imposed on manufacturers generally.

For these reasons, we decline to extend the implied warranty/strict liability doctrine to cover vendors of professional services. Our conclusion does not, of course, preclude the city from pursuing its standard malpractice action against the architects and proving that the basement area of the new addition was negligently designed. That issue remains for the trier of fact in the district court.

Affirmed.

OTIS, J., took no part in the consideration or decision of this case.

That the court felt compelled to justify at some length its decision made four years earlier in *City of Eveleth v. Ruble* demonstrates some dissatisfaction with the professional standard. This is reflected in Section 14.06(A), dealing with exceptions to the requirement of expert testimony; in Section 14.07, comparing the professional standard with that of implied warranty; and in Section 14.11(A), analyzing some of the current controversies relating to professional liability.

Some refinements should be made in the professional standard rule. First, against whom is the conduct of the professional measured? Usually it is assumed that it is measured against the conduct of others in the professional's locality. For example, if the professional practices in a small town, it is usually assumed that she should not be compared with professionals who practice in a large city. To some degree, this is based on the reasonable expectations of the client, and to some degree, on the likelihood that advances in the profession come first to large urban areas.

It is more appropriate, however, to measure the conduct against those professionals with whom the defendant competes. Some design professionals in smaller cities compete against design professional firms in large metropolitan areas. Similarly, the choice made by the client may be one based on a comparison of those design professionals who are established in a particular specialty, and those specialty firms may be scattered in small towns, cities, and large urban areas. A case has held that an architect would not be measured solely against those in the locality in which she practices.[49] This may set a trend for a more refined application of the professional standard.

Similarly, differentiations are not always made regarding the precise nature of the standard to which the design professional will be held. Two authors developed a classification system for determining whether engineers are liable for design errors. They divided the state of the art into the cutting edge, the open literature, acceptance by the profession, and what they refer to as the undergraduate horizon. The authors concluded that in the ordinary case, the challenged engineer should be measured against the "informed" engineer, who, they assert, practices in accordance with the state of the art accepted by the profession.[50]

The burden of establishing that there has been a failure to comply with the professional standard is generally on the claimant who seeks to have her loss transferred to the professional.[51]

B. Expert Testimony

As noted in the preceding section, expert testimony is *usually* required to support a finding that the professional standard has *not* been met.[52] Since this has been a central issue in professional liability claims, it is discussed in Section 14.06.

---

[49]*Hill Constr. Co. v. Bragg,* 291 Ark. 382, 725 S.W.2d 538 (1987). Opinion after remand on different issue, 297 Ark. 537, 764 S.W.2d 41 (1989). South Carolina refused to use the local standard in a claim against a doctor, *King v. Williams,* 276 S.C. 478, 279 S.E.2d 618 (1981), and against an accountant, *Folkens v. Hunt,* 290 S.C. 194, 348 S.E.2d 839 (Ct.App.1986). But a federal court applying South Carolina law by dictum stated that the local standard would be applied in a claim against the geotechnical engineer. *Georgetown Steel Corp. v. Union Carbide,* 806 F.Supp.74 (D.S.C.1992).

[50]Peck & Hoch, *Liability of Engineers for Structural Design Errors: State of the Art Considerations in Defining the Standard of Care,* 30 Vill.L.Rev. 403 (1985).

[51]*Coulson & C.A.E., Inc. v. Lake L.B.J. Mun. Util. Dist.,* 734 S.W.2d 649 (Tex.1987). For subsequent opinions on different issues, see 771 S.W.2d 145 (Tex.Ct.App.1988) and 781 S.W.2d 594 (Tex.1990).

[52]See Farrug, *The Necessity of Expert Testimony in Establishing the Standard of Care for Design Professionals,* 38 DePaul L.Rev. 873 (1989) (attacking Illinois law, which does not require expert testimony in cases of professional malpractice).

C. Building and Housing Codes: Americans with Disabilities Act

Design professionals are generally expected to consider and comply with building and housing codes.[53] Failure to design in accordance with such codes is almost always considered a violation of the obligation the design professional owes her client. Code compliance is considered to be one aspect of meeting the professional standard imposed by law on design professionals. In addition, the code violation may create private rights to those who are damaged by the violation. This occurs if the statute specifically designates that a violation gives private rights to individuals damaged by the violation or if the statute is so interpreted.

Building codes can determine design, materials, and construction methods. Design specifications, performance standards, or a combination of the two is used. Codes provide *minimum* standards to protect against structural failures, fire, and unsanitary conditions.

Housing codes are of relatively recent vintage. They have gone beyond structure, fire, and basic sanitation to include light, air, modern sanitation facilities, maintenance standards, and occupancy density rules. Sometimes they are used to measure the implied warranty of habitability owed tenants by landlords.

While there is some movement toward statewide and even federal building codes, local codes predominate. In addition to reflecting the extreme local variations in subsurface conditions and climate, local control reflects strong democratic roots in the local community. (Yet, as shall be seen in this subsection, the federal government recently enacted the Americans with Disabilities Act, part of which deals with building requirements.)

Codes generally fall into one of four major types developed by various private associations, such as fire underwriters and local building officials. Choice within these types is often regional, with one part of the country choosing a particular type. Yet frequent local modification and variant local interpretation have made for variations. These, along with unrealistic and unneeded standards, often dictated by local special interest groups, have

undoubtedly played a part in high construction costs.

As shall be seen in some of the cases noted in this section, codes are often difficult to understand and are subject to uneven if not whimsical local interpretation. Often, unrealistically high code standards exclude low- and moderate-income persons from certain communities. Moreover, the codes are not always enforced and can lead to bribery.

Some of the strengths and weaknesses of local building codes can be demonstrated by two cases. *Greenhaven Corp. v. Hutchcraft & Assocs.*, an Indiana case,[54] involved a claim by an architect against a client for fees for providing plans for remodeling a building. The preliminary plans called for two exits from the top floor, but the owner's representative, a contractor, requested that the plans be altered to provide for only one exit. The architect complied with this request. This would have violated the building code, but the project was abandoned.

The court noted that the architect impliedly promises to draw plans and specifications that comply with local codes. But it held that parties can contract for nonconforming plans, that contracting parties can make any agreement, as long as it is not illegal or contrary to public policy.

The court stated that neither party contended that an agreement which permits nonconforming plans was contrary to public policy. Also, the court stated that the public was not harmed by such an agreement. Before the building could have been occupied, the Fire Marshall would have had to approve by issuing an occupancy permit. If it had not done so, the parties might have been able to obtain a variance or change the plans to conform to the requirements. In either case, the court found the public was protected regardless of the agreement between the parties. It also noted that nonconforming plans are encouraged by the availability of variance procedures.

The court showed little respect for the building codes. But its language also recognized the flexibility achieved through the variance process. While a variance process provides for flexibility—something that can be very useful if codes become too

---

[53]See Sections 14.03, 14.05(C).

[54]463 N.E.2d 283 (Ind.App.1984).

rigid and out of touch with reality—such a system can also lead to corruption.

The second case, *Edward J. Seibert, A.I.A., Architect and Planner, P.A. v. Bayport Beach and Tennis Club Association, Inc.*[55] arose in Florida. The issue was whether the architect had designed in accordance with the local building code. The dispute also involved a fire exit. The architect had designed each second-floor dwelling unit with one independent unenclosed set of stairs leading directly from the front door of each unit to the ground. The completed plans were submitted to the city's chief building inspector, who was also the chief code enforcement officer. The inspector's interpretation of the building code led him to conclude that because of the size of the units and because the exit was unenclosed, the single unenclosed exit design complied with the code. After the fire department approved the plans, the building inspector issued a building permit, and the units were ultimately built according to these plans.

Two years later, the individual condominium owners assumed control of the association and filed a claim against a number of parties, including the architect, for various defects. While the jury exculpated the architect for many of what it found to be defects, it found him liable for the improperly designed fire exit.

The court extensively examined the testimony of experts called by the parties to determine whether there had been a violation. The association presented an expert witness—a structural engineer who testified that under his interpretation, the exit did not comply with the code.

The architect presented two experts who testified to the contrary. One was a professional engineer who testified he had been employed by the Southern Building Code Congress, which had promulgated the code, and that during his employment he was responsible for building code changes, hearings, plans, reviews, and code interpretation. The engineer testified that the architect's design complied with the code and was consistent with safe design as intended by the code. Yet the jury, as noted, found the architect liable for the cost of adding a second exit.

The appellate court noted that the jury needed to know what the code required and that such information be presented to the jury by means of expert testimony. The court stated that expert testimony could be presented if the jury needed to understand the evidence better or to determine a key issue in the case, but experts could not testify as to how the code should be interpreted. This was a question of law. The court held that the trial court should have interpreted the meaning of the code and instructed the jury concerning that meaning.

Finally, the appellate court held that the trial judge should have concluded that the architect had complied with the code because the chief building inspector and chief code enforcement officer stated that the design did comply. This was relied upon by the architect, and the building was constructed in accordance with the approved design. The court noted that even though a code could be interpreted in more than one way, the law will follow the interpretation of the agency with the authority to implement the code. The court concluded by stating that a judgment should have been entered in favor of the architect.

It is difficult to quarrel with the conclusion that the architect can rely on the interpretation of the building inspector. However, the fact that codes can be interpreted in various ways can make compliance difficult and encourage corruption.

The recent enactment of the Americans with Disabilities Act[56] has generated building standards as part of an overall policy to protect disabled persons, a group that has been subjected to long-term discrimination in employment, housing, education, transportation, public accommodations, recreation, and health services. The Act comprehensively defines the treatment of disabled persons, imposes significant new obligations on employers, and (most importantly for design professionals) mandates that commercial and public accommodations be accessible to people with disabilities.

Title 3 of the Act states that existing buildings, new constructions, and alterations are governed by the requirement that individuals with disabilities have access to almost all businesses and public places. The rules exempt private clubs, religious in-

---

[55]Supra note 9.

[56]Pub.L. No. 101-336, 104 Stat.327 (1990).

stitutions, residential facilities covered by fair-housing laws, and certain owner-occupied inns.

Many of the details will be expressed in regulations (Americans with Disabilities Accessibility Guidelines) provided by the U.S. Architectural and Transportation Barriers Compliance Board. The guidelines are based largely on existing American National Standards Institute (ANSI) standards. Construction completed after January 26, 1993, must be readily accessible to and useable by individuals with disabilities. Even more important, the Act requires removal of architectural barriers in existing buildings if the structure is considered to be a place of public accommodation. There are some exceptions, such as removal not being readily achievable. Achievability is defined by the guidelines as "easily accomplishable and able to be carried out without much difficulty or expense." The regulations take into account the nature and cost of removal, the financial resources of the party involved, and the impact of removal on the operation of the site and upon profitability. These factors permit a case-by-case approach.

Alterations to a place of public accommodation built after January 26, 1992, must ensure to the maximum extent feasible that the altered portions are readily accessible to and useable by disabled persons, including those who use wheelchairs. The Act defines certain areas of a building as having a "primary function." In those areas there are additional requirements. Not only must the design conform to Title 3 requirements, but the path of travel to the altered area to the maximum extent feasible must be accessible to and useable by the disabled. But if alterations to the path of travel cost more than 20% of the cost of the alterations, this is considered disproportionate and not required. The barriers are not limited to mobility impairments but can include those that impair vision, hearing, and reading.

If there is a violation of Title 3, the court can issue an order requiring the owner to modify physical facilities. The court can award monetary but not punitive damages, civil penalties, and attorneys' fees. In considering civil penalties up to $100,000 under Title 3, the court can take into account any good-faith efforts to comply.

Many state and local laws deal with the same topic. A state or locality can request that the federal attorney general certify that its code complies with the Act. If the certification is given, compliance with local or state building codes will also comply with the federal act. But if local and state requirements are less stringent, the federal act will supersede them.

In effect, the Act has created a form of national building code, but no entity is obligated under the Act to review and approve drawings and specifications for compliance. This means that the parties seeking to comply may not know whether they have discharged their obligation until a complaint is filed.

This legislation should provide a significant market for the services of design professionals. Some clients may seek to shift the *ultimate* risk of compliance to the design professional by providing that the design professional must certify that the design complies with the law. This certification may be uninsurable. While liability resulting from negligent failure to produce a design that complies with codes is clearly insurable, contract provisions that turn the duty of compliance into a guaranty of compliance can create the possibility of exclusion from coverage.

D. Contractual Diminution of Standard: Informed Consent

As noted in Sections 14.04 and 14.05, the professional standard is *residual,* applying only where the parties have not contractually agreed to a different standard. Usually, any specific standard is more strict; that the design professional did what others would have done will not in itself relieve the design professional. Suppose the design professional can point to a lower specific standard in the contract. Although rare, exploration of this possibility uncovers difficult problems.

Suppose the design professional would have selected X, a material that would have been designated by other professional designers because of a combination of its durability, low maintenance, cost, and appearance. But the client being aware of the trade-offs orders that Y be used simply because it is less expensive. Suppose the material selected needed replacement earlier than the client expected. The client should not be able to recover any correction costs against the designer. The client and the designer agreed to a standard different from the professional standard—in this case, a lower standard. Assuming that this choice does not expose others to unreasonable risk of harm, the law will let the contracting parties decide whether the professional standard or something less will satisfy the designer's contractual obligations.

But suppose an applicable building code requires X, and Y is thought to be unsafe. If the design professional makes the client aware of the code but the client insists on Y, what are the consequences? Several issues are presented:

1. Will the design professional be denied recovery for services that relate to selection of this material?
2. Will the design professional be given a defense if the client sues her for the cost of corrective work?
3. Will this selection expose the design professional to negligence claims by any third parties who suffer losses because of this choice?
4. Will the client and the design professional (and any contractor who knowingly violates the code) who violated the building laws be exposed to criminal prosecution?
5. Will this selection be grounds for disciplining the design professional under the registration laws?

The answers to these questions should be yes. Because of the importance of design, the transaction cannot, like the preceding one, be simply regarded as a private one between client and design professional. In matters of safety, public protection takes precedence.

Bowman v. Coursey[57] illustrates the difficulty in determining whether a particular agreement required performance less than that called for by the professional standard and whether such an agreement is simply a private arrangement or one that involved public safety.

Coursey hired Bowman to design a warehouse. Bowman's original plans included pilings beneath the walls in the floor slab. As Coursey wanted to reduce the cost, the plans were revised. Coursey wanted to eliminate the pilings beneath the floor area. Bowman told Coursey that the revisions would very likely mean that the floor would be subject to settlement. Coursey asked how long it would take for the expected settlement to occur

and, when it occurred, how the settlement could best be handled. After consulting with the soils engineer and an engineer for the lender, Bowman revised the design so that the floor slab would be reinforced to minimize the effect of any settlement. The revised plans were submitted to and approved by the local regulatory agency.

After construction began, problems developed with the quality of the construction work. Experts criticized the wall construction and the foundation design. A dispute arose over whether the architect Bowman had been negligent in determining the capacity of the piles beneath the wall area. Work on the project ceased. Coursey brought legal action against Bowman. A number of experts testified at the trial, most of whom testified that they would not have designed as did Bowman. They agreed that some settlement was inevitable, but none would say that the design was necessarily wrong or that the building was inadequate. One testified that the owner should be made aware of the possibility of settlement. An expert testifying for Bowman stated that although this was not the best design possible, it was adequate. He indicated that as long as Coursey knew what he was getting, allowing the floor to float was not a bad condition. Settlement, according to that witness, would occur in a controlled fashion because of the stiffness of the walls. This might be acceptable to the owner.

The court concluded that if the warehouse were built according to the revised plans, it would be less than perfect but the imperfections were not due to a breach by the architect Bowman but by the client Coursey's preferring the imperfections to further financial outlays. When warned that there would be settlement, Coursey's response was that he could live with it. The court noted that additional pilings would have been necessary to prevent settlement, but none of the witnesses concluded that the design was clearly wrong or engineeringly unsound. The design did not violate the building code, nor would it present a danger or hazard to persons working within it. The court concluded that the problems that Coursey would face were a trade-off for reduction in building expenses.

Although the court concluded that Bowman had not been negligent, its conclusion was based principally on Bowman's having pointed out all the problems to Coursey and Coursey's deciding to accept certain risks in exchange for a lower construction cost. But the court emphasized that the

[57]433 So.2d 251 (La.App.1983), cert. denied 440 So.2d 151 (1983).

controlled settlement that was likely to occur would not present a danger or hazard to persons within the building and that the design did not violate the building codes. Yet the expert testimony indicated that the experts, though testifying in careful and guarded terms, indicated that they would not have designed the warehouse in the way chosen by Bowman with Coursey's concurrence. In essence, the parties agreed to a design below the standard other professionals would have used.

This problem also exposes another issue—an increasingly debated one. Must any client's consent be informed? Taking a leaf from the law regulating the relationship between physician and patient, one writer advocated that architects be required to inform their clients of the costs and benefits of design choices.[58] This can, under certain circumstances, be related to the requirement of good faith and fair dealing which augments the express obligations required of contracting parties.

The uncertain dimensions of such a duty and the increasing concern for expanded liability even under the generally more protective professional standard may be reasons why informed consent has not as yet received overt approval.

E. Tort and Contract

Often the claimant prefers tort law, with its more limited foreseeability defense and its more potent remedies. For example, it is very difficult to recover for emotional distress and almost impossible to recover punitive damages if the claim is based upon breach of contract. However, a tort claim may justify recovery of damages for emotional distress and, if a breach of contract is also considered a tort, recovery of punitive damages.

Yet there are advantages to bringing the claim for breach of contract if that alternative exists. The time limit for commencing a lawsuit is longer—often very much longer—when the claim is based upon a breach of contract rather than upon a tort theory. (However, if the period to start a tort claim does not begin until the claimant discovers it has a claim against the defendant, the time period may, in states with a short contract period, be longer than for a breach of contract.)

Unlike claims by patients against their doctors, claims by clients against their design professionals

are more likely to involve specific provisions of a contract. If the client is permitted to bring its claim in tort, it might be able to bypass certain contract defenses, such as the parol evidence rule, the statute of frauds, or exculpatory provisions found in the contract between the client and its design professional.

In addition, claims based upon the commission of a tort sometimes run afoul of the economic loss rule. While this is more commonly a problem when claims are made by third parties, as discussed in Section 14.08(E), it can also be a problem when the claim is by one party to a contract against the other. Usually there is no difficulty when patients sue their doctors, since there will be a claim for physical harm to which claims for economic losses can be attached. Claims by owners or contractors against design professionals usually do not involve physical harm but are more likely to involve purely economic losses, those based upon disappointed expectations. In such cases, a tort claim for economic losses may not be sustained.

Also, claims against design professionals are generally based upon the failure of the design professional to perform in accordance with the professional standard discussed in Section 14.05(A). While there is some tendency toward incorporating into the contract language that deals with how the services are to be performed, claims are most commonly based upon a term implied by law—the professional standard. In such cases, it has been relatively easy for courts to focus upon the professional relationship, assert that the duty is created by that relationship, and conclude that the contract is simply a method of implementing the relationship. Although a tort claim requires that negligence be established while a breach of contract claim is, at least in theory, "strict," claims by clients against their design professionals, as mentioned earlier, will employ the same standard—the professional standard—whether brought based upon tort or based upon contract. Only if the contract contains a refinement of the professional standard or imposes some form of strict liability will liability be based upon a standard other than the professional standard.

While some courts seem to look for the gravamen, or essence, of the claim,[59] it is likely that the

---

[58]Note, 30 Hastings L.J. 729 (1979).

[59]*Lesmeister v. Dilly*, 330 N.W.2d 95 (Minn.1983); *Mac-Fab Products, Inc. v. Bi-State Dev. Agency*, 726 S.W.2d 815 (Mo.App.1987). See generally, W. PROSSER & P. KEETON, TORTS, 664–667 (5th ed. 1984).

client will be able to assert a claim based upon the commission of a tort or the breach of a contract, whichever is more advantageous to the claimant. Very likely the client will be able to assert alternative claims. Only at the stage where the judge determines that the client must make an election of remedies will the client be forced to choose the theory upon which it is basing its claim.

Yet, as seen in Section 12.08, many contracts between clients and design professionals, especially those published by the AIA, clearly express the rights and obligations of both parties and also provide contractual exculpation for the design professional. Where such exculpations appear unfair, it is likely that the client will be given the option of maintaining the claim under tort law.

Looking at a few cases may help understand the difficult issue of tort versus contract in a professional relationship.

Collins v. Reynard[60] involved a claim by a client against her attorney. The client claimed that in drafting certain documents, her attorney failed to protect her financial interest and caused her to suffer a loss. The complaint against the lawyer was brought both on a theory of breach of contract and on a theory of commission of a tort. The issue of whether such a claim could be brought in tort wended its way through the Illinois judicial system until it reached the supreme court. In its initial decision, the court held that a complaint for lawyer malpractice could be brought only in contract and not in tort. However, the court granted the plaintiff's petition for a rehearing, and the case was reargued.

After reargument, the majority of the court concluded that the client could bring her claim in either contract or tort. (It appears that the choice to bring the claim in tort as well as contract was done to obtain a more expansive remedy.)

The majority of the court reversed its original decision because of a long line of Illinois cases permitting recovery in tort for lawyer malpractice. The court pointed to the importance of respecting precedent as a means of providing certainty in the law. It appears that the majority of the court felt powerless to change long-established practice and cus-

tom. It stated, "Logic may be a face card but custom is a trump."[61]

The court noted that tort law is based upon a duty that exists wholly apart from a contractual undertaking and follows its own rules. It also noted, however, that some breaches of contract have crossed the line and have become "cognizable in tort." The court took pains to make clear that its ruling in this case was limited solely to lawyer malpractice claims, as it did not wish to disturb its earlier decisions stating that certain product liability claims could not be maintained in tort.[62]

Three judges concurred but sought to distinguish the earlier cases. They stated that those cases have no application in claims by clients against their attorneys. For example, they noted that tort law could not be applied to claims brought against architects and builders because of the economic loss rule.[63] But they singled out claims by clients against their attorneys for special treatment. They stated that tort law provides a flexible standard by which to assess performance and that the duty exists without regard to any terms of contract that might be entered into between the lawyer and the client. For these reasons, the concurring judges thought it inappropriate to apply the economic loss rule (increasingly called the commercial loss rule) to attorney malpractice claims.

This issue came before a Wisconsin court in the context of a design professional in *Milwaukee Partners v. Collins Engineers, Inc.*[64] An engineer had been retained to inspect a building to determine the building's structural soundness prior to the owner buying it. When the owner decided to sell the building some seven years later, a subsequent study by another engineer indicated that the building had structural deficiencies and, according to the second engineer, had them at the time it was inspected by the original engineer.

As is so often in these cases, the issue was whether the claim was barred by the passage of time. The engineer pointed to the six-year period of limitations for contract claims, but the court concluded that the claim was actually one brought in

[60]154 Ill.2d 48, 607 N.E.2d 1185 (1992).

[61]607 N.E.2d at 1186.
[62]607 N.E.2d at 1186–87.
[63]See Section 14.08(E).
[64]165 Wis.2d 355, 485 N.W.2d 274 (App.1992).

tort even though it flowed from a breach of contract.

The court stated that design professionals have a common law duty to exercise care similar to that of other professionals. The duty is not *created* by the contract, but the contract implements fulfillment of the duty. The court cited Section 552(1) of the Restatement (Second) of Torts, which states that professionals who supply false information for the guidance of others are liable in tort for pecuniary loss if the information is justifiably relied on and if the professional did not exercise reasonable care.

Having classified the claim as one that could be maintained in tort, the court directed attention to the statute of limitations for tort claims. In Wisconsin, the period of limitations is the same for breaches of contract and for commissions of certain torts. But the tort statute applies a discovery rule, one that does not start the period until the injury is discovered or should have been discovered. This is more favorable to the claimant. The case was remanded to determine when the injury was or could have been discovered.

As noted earlier in this section, the law will give the claimant the widest possible berth when bringing a claim against a professional whom it has retained, thus allowing the claimant to maintain an action either in contract or in tort, depending upon which theory is most advantageous. While this is accomplished by stating that the duty exists independent of the contract and the contract is only a means of executing the duty—a doubtful fiction at best—it seems clear that the underlying basis for claims by a client against the design professional is their contract. As indicated in Section 14.04(A), the contract may be explicit as to specified duties and how they are to be performed or it may be explicit as to duties but silent as to the standard. Many retention arrangements are silent as to specific duties and how they are to be performed.

A contract that covers *in detail* the duties of each party and allocates the risks of losses likely to occur is a *plan* under which the parties agree to exchange their performance and apportion risks in a specific manner. To disregard that plan by allowing tort claims exposes design professionals to risks they did not plan to undertake and for which they were not paid. Clients should not have an *election* to bring their actions in contract or tort. In planned transactions, they should be required to base their claims on breach of the contract.

This suggestion is compatible with a desire to encourage persons to plan their transactions by freeing them from the risk of open-ended tort exposure. Contracting parties who have suffered emotional distress can be protected by classifying their claims (though based on contract) as torts. Similarly, if there is a need to punish or deter, the claim can be based on tort law. But other claims based on failure to perform under the design professional-client relationship should be based *solely* upon the contract.

SECTION 14.06 Expert Testimony

One important component of the professional standard is the need, at least as a general rule, for expert testimony. (A) looks at the exceptions to that requirement, (B) examines expert testimony generally and particularly in the context of claims against design professionals, (C) looks at criticism of the current system, and (D) approaches the problem from the vantage point of the expert witness.

A. Purpose and Exceptions to General Rule

In the judicial system, judges and juries make decisions. To do so, they may have to hear, evaluate, and judge testimony and exhibits that relate to technical matters unfamiliar to them. To assist them, the law permits evidence of opinion testimony as an exception to the rules of evidence that *generally* bar opinion testimony. But before such opinion testimony can be admitted, the issue must be one that is too difficult for a judge or jury to decide *without* technical assistance. Also, the person permitted to give opinions must possess the necessary education and experience to be an "expert" on the issue for which the judge or jury needs help. Yet the expert is generally not allowed to give opinions that could determine the outcome of the case. This power is still retained by the judge and (when used) the jury. The importance of limiting expert testimony can be seen by the *Seibert* case, discussed in Section 14.05(C), where the court held that the expert would not be allowed to testify as to the proper interpretation of a building code.

The professional standard—generally—requires expert testimony to support any conclusion that the design professional has not performed in accordance with ordinary professional standards. There are important exceptions to this general rule. For example, in *City of Eveleth v. Ruble,* discussed in Section 14.05(A), the plaintiff city had retained defendant engineer to design a new water treatment plant. After completion, two difficulties developed. First, the intake system that took water from a lake and processed it for distribution into the city's transmission lines and storage reservoir proved inadequate. Second, pressure in the cast-iron distribution lines leading from the water treatment plant to users and storage facilities caused some of the leaded joints in the line to give way. Without any expert testimony, the trial court awarded the city damages against the engineer.

As to the diminished intake capacity, the court noted that the engineer knew that the city decided to build a new water plant to increase the intake capacity to a designated number of gallons per minute. The intake capacity was inadequate because there was a failure to anticipate changes in the lake level. The existing intake line that was laid on the bottom of the lake in approximately 40 feet of water was not the 18-inch line expected by the engineer but in some places was 16 inches and in other places only 12 inches in diameter.

The Minnesota Supreme Court concluded that the trial judge was able to assess the validity of these excuses "without the aid of expert testimony." In drawing this conclusion the court stated:

> In our judgment, no expert opinion is needed to demonstrate that a design engineer charged with the responsibility of analyzing the piping and other structural characteristics of an existing plant should be as certain of the dimensions of the intake line as circumstances would possibly permit before recommending a plan the function of which depended upon this critical measurement. It would seem clear that the examination of a photograph of the line would be of little value. Incomplete drawings made available to the Engineer at its request indicate that the intake line was 18 inches in diameter at the point of terminus with the old plant, but we believe that common knowledge would reject this as adequate basis for careful analysis. We find the explanation given for the failure of the Engineer's employees who entered the lake for the purpose of measuring the intake line unsatisfactory when the record shows that others employed for this same purpose by the City were able to obtain the true dimensions of the intake line without difficulties disproportionate to the importance of the task.

It seems to us that when a professional designs a water plant, in particular a plant intended to deal with increased community needs, that professional should, as a matter of reasonable care, be certain of the size of the piping which provides the plant with raw water and should be equally certain that, ultimately, there will be an adequate supply of water, both in the sense of supply to the plant and in the sense of supply upon which the plant may draw, to meet the operational expectations of the design; or, in avoidance, should more convincingly demonstrate why, under the circumstances, it was good professional practice not to do so.[65]

As to the excess pressures in the distribution lines causing the joints to give way, the trial court found that the high service pumps that distributed the clear water to the storage tanks caused sudden surges "with consequent water hammer" resulting in blowing out the lead-sealed joints. Evidently, the cast-iron distribution lines were buried in the ground and had been in use for approximately sixty years. The court concluded that it would not be fair to hold the engineer to any "express or implied commitment that the City's transmission lines would be trouble-free following the installation of the new facility."[66] The court stated that "it would not be reasonable to expect the Engineer to guarantee the performance of these lines under these circumstances." The court relieved the engineer by concluding that expert testimony was required because determining what type of pressure system should be used with pipes of this age was a technical question. In employing the professional standard, the court concluded:

> Would a design engineer, in the exercise of that degree of care, skill, and diligence to be expected from this profession, knowing the line pressures which would be created by the operation of the high service pumps and knowing the age of the line and the character of its construction, have reasonably anticipated that it would fail in use? If uncertain, would he have employed tests or techniques of inspection which could have been employed and which would have been employed by a design engineer applying the requisite standards of skill and care which should have revealed the deficiencies in the line? If not, would such an engineer, being unable to ascertain the facts, have recommended the installation of devices such as those now recommended as a precaution against possible

---

[65]302 Minn. 249, 225 N.W. 2d 521, 527–28 (1974).
[66]225 N.W.2d at 529.

but unpredictable ruptures? We do not think that common knowledge affords answers to these questions.[67]

The wide range of services often performed by the design professional in both design and construction phases will inevitably raise questions as to which services are sufficiently technical to *require* expert testimony and which are sufficiently nontechnical that the lay judge or jury needs no assistance. In many cases, the plaintiff is not able or chooses not to offer expert testimony and claims it is not needed, while the defendant contends that the lack of expert testimony bars any judgment of professional malpractice.

Claims based on design services as a rule will require expert testimony. Choices that involve excavation, design,[68] foundation sufficiency, structural stability, equipment and components,[69] protection against the elements,[70] energy efficiency,[71] and surface water disposal[72] are all matters that require expert testimony. Often they are technical areas for which the prime design professional will retain consultants. Whether such services are performed in-house or by outside consultants, they require decisions that should be made by professionals with specialized education and experience. Often these services must be performed only by persons registered by the state, though this is not determinative.

Seaman Unified School District v. Casson Construction Co.[73] involved water damage to a gymnasium following heavy rains. The owner sued the prime contractor, architect, and consulting engineer. The issue was the location of the sidewalk. Disregard-

ing testimony of three experts supporting the design choice made by the architect, the trial court noted that it was common knowledge that water runs downhill. Concluding that the facts of the drainage are too complicated to be resolved by mere application of this commonly known fact, the appellate court stated:

> The evidence reveals that the plan provided the sidewalk leading to the northwest stairwell was to have a slope of one inch per ten feet to the south and one and one half inches per ten feet to the east. The plan also specified a six tenths of 1 percent overlawn slope to the north for the area immediately east of the walk. This water was not scheduled to run off in a direct line downhill but was to be shunted diagonally by the pitch of the sidewalk and then diverted to the north by the land contours and the plan-specified grade east of the walk. Considering these facts and the size of the surface area to be drained, we do not think it is within the common knowledge of laymen to decide whether such specifications were proper. This conclusion, together with the testimony of three architects that the plan was in accordance with local architectural standards and the absence of any testimony by an *architect* to the contrary, convinces us that the trial court erred in finding directly opposite to these experts' opinion testimony.[74]

Yet even where the services seem to be technical in nature, such as a defective soils report, one case excused the requirement that there be expert testimony stating it would not be needed:

> . . . when the conduct of [the professional] is so unprofessional, so clearly improper, and so manifestly below reasonable standards dictated by ordinary intelligence, as to constitute a prima facie case of either a lack of the degree of skill and care exercised by others in the same general vicinity or failure to reasonably exercise such skill and care.[75]

Laypersons can infer negligence by applying common sense.

As to site services unconnected to design, the issue becomes more cloudy. Does it take expert testimony to assist a judge or jury in deciding whether the design professional should have certified a par-

[67]225 N.W.2d at 530.

[68]*Nauman v. Harold K. Beecher & Assoc.*, 24 Utah 2d 172, 467 P.2d 610 (1970). But a claim based on a failure to discover a structural barrier in a renovation project did not require expert testimony. See *Milton J. Womack, Inc. v. State House of Representatives*, supra note 47.

[69]*Dresco Mechanical Contractors, Inc. v. Todd-CEA, Inc.*, 531 F.2d 1292 (5th Cir.1976), reproduced in part in Section 14.06(B); *John Grace & Co. v. State Univ. Constr. Fund*, 64 N.Y.2d 709, 475 N.E.2d 105, 485 N.Y.S.2d 734 (1984).

[70]*South Burlington School Dist. v. Calcagni-Frazier-Zajchowski Architects, Inc.*, 138 Vt. 33, 410 A.2d 1359 (1980).

[71]*Bd. of Education v. Hueber*, 90 A.D.2d 685, 456 N.Y.S.2d 283 (1982).

[72]*National Cash Register Co. v. Haak*, 233 Pa.Super. 562, 335 A.2d 407 (1975), reproduced in part in Section 14.06(B). For a collection of cases, see 3 A.L.R.4th 1023 (1981).

[73]3 Kan.App.2d 289, 594 P.2d 241 (1979).

[74]594 P.2d at 245.

[75]*Nicholson & Loup, Inc. v. Carl E. Woodward, Inc., et al.*, 596 So.2d 374, 381 (La.App.1991).

ticular amount to be paid, detected a design deviation in a shop drawing or submittal, noted a deviation from an approved schedule, verified a changed condition, coordinated the work of separate contractors, to name some of the tasks often given to the design professional?

Two cases held that no expert testimony was needed to support claims based on construction-phase services, which, according to the courts, did not require specialized technical skill possessed only by those with particular education and training.[76] These courts lumped all construction-phase services into the uninformative "supervision" category, failing to differentiate those activities that require professional skill, such as preparing change orders, interpreting contract documents, or resolving disputes.

At the other extreme Virginia held that expert testimony would be required to support a claim by the owner against its architect based on the architect's asserted failure to detect defects in a complex project and the latter's alleged delay in processing proposed change orders.[77]

The increasing number of claims against design professionals based on their site services should develop rules that separate those services that require professional education and training from those that can be and often are performed by those without such training, such as inspectors and even construction managers. The former should require *expert* testimony. The latter should not. Care must be taken to distinguish those activities that are part of the design professional's "judging" role. In attacking a decision made by a design professional, expert testimony may not be needed because the issue is not whether the design professional did as others would have but the finality of her decision.[78] Unfortunately this distinction is not always drawn.

Claims by those who engage one entity to both design and build or who buy from a builder-vendor raise special problems. At the outset, a differentiation must be made between a claim based on professional malpractice and one based on im-

plied warranty, a standard that usually measures the obligation of the builder-vendor or developer. In the latter, negligence is not the standard.[79] As a result, no expert testimony is required.

B. Admissibility of Testimony: *Nat. Cash Register Co. v. Haak; Dresco Mechanical Contractors, Inc. v. Todd-CEA*

Before proceeding to the principal focus of this section—the issue of who can testify and the type of testimony needed—a few preliminary remarks relating to admissibility generally must be made.

First, a party intending to call an expert witness must notify the other party in advance of trial of the identity and qualifications of the expert and the issues about which expert testimony will be elicited. This enables its opponent to investigate or use the deposition process to check on the qualifications of the experts that the other side plans to use.

Second, the expert must give her opinion based on firsthand knowledge (such as an expert medical witness who has conducted her own examination), on facts admitted into evidence, or on a combination of firsthand knowledge and evidence. If the expert has no firsthand knowledge, the opinion of the expert in many states is based on a hypothetical set of facts that are included in the question and that are based on evidence that has been admitted or evidence that the party calling the expert plans to introduce. Sometimes the expert can observe the testimony of witnesses and be asked to assume the truth of previous testimony as a basis for her opinion. This can simplify the often bewilderingly complex hypothetical question.

Third, a witness who is not registered in accordance with the registration laws of the state very likely will be permitted to testify.[80] Local law must be consulted.

The qualifications of an expert to give an opinion—an increasingly difficult issue because of overlapping specialization in professions—is discussed in the following case.

---

[76]*Jaeger v. Henningson, Durham & Richardson, Inc.,* 714 F.2d 773 (8th Cir.1983) (approval of shop drawings); *Bartak v. Bell-Galyardt & Wells, Inc.,* 629 F.2d 523 (8th Cir.1980) (checking on contractor compliance).
[77]*Nelson v. Commonwealth,* 235 Va. 228, 368 S.E.2d 239 (1988).
[78]See Section 29.09.

[79]See Section 24.10.
[80]*South Burlington School Dist. v. Calcagni-Frazier-Zajchowski Architects, Inc.,* supra note 70.

NATIONAL CASH REGISTER CO. v. HAAK

Superior Court of Pennsylvania, 1975. 233 Pa.Super. 562, 335 A.2d 407.
[Ed. note: footnotes omitted.]

SPAETH, Judge.

[Ed. note: After completion of construction of a manufacturing plant, sinkholes adjacent to dry wells developed, threatening the integrity of the building. The problem was diagnosed by Gannett-Fleming, an engineering firm. Corrective work was accomplished. The owner brought a claim against the architect for negligent design.] The controversy on this appeal centers around the adequacy of the testimony of the four expert witnesses called by appellant [owner].

Charles W. Pickering was admitted as an expert in "civil engineering in hydraulics" (the court's characterization) or as a "civil engineer familiar with hydraulics" (defense counsel's characterization). He was employed by Gannett-Fleming and was the one who had examined the site for that firm and had recommended the removal of the dry wells and the installation of the new system. His opinion on the cause of the sinkhole activity was unequivocal:

It is my opinion that the Surface Water System as was installed on the NCR site has accelerated the formation of sinkhole activity on the site.

[I]f the present system were to be continued in use, . . . the formation of sinkholes would continue and with the ultimate possibility or ultimate meaning at sometime, some point in time, because it is a natural phenominal [sic] it cannot be predicted, that at sometime there may be serious damage caused to the major facilities on the site.
Q. Would that include the building?
A. Yes.

Timothy Saylor was admitted as an expert in geology. He was also employed by Gannett-Fleming, and his testimony, where relevant to the issues to be considered here, corroborated Mr. Pickering's.

Professor Jacob Freedman of Franklin & Marshall College also testified as an expert in geology. His qualifications indicated extensive experience in his field covering over 25 years, including being frequently called in as a consultant on the geology of Lancaster County, especially with regard to "water problems, foundations problems, [and] studies of quarries." His testimony corroborated Mr. Pickering's and Mr. Saylor's with regard to the causal relationship between the dry wells and the sinkholes. He then added the following at the end of re-direct examination:

I probably ought to say and I probably haven't said this in my discussion so far along these lines, that I have and I know no geologist who has ever recommended dry wells for construction in an area like this; that, in fact, we urge people not to use this kind of system; and if they have French drains—in fact, as I say, we frequently get calls at school; and anytime anybody mentions a French drain we tell them they are in for trouble, and French drains and dry wells are very similar in their properties. So these are systems that should not have been installed because they lead to trouble.

This brief statement (which was not given in direct response to any question, but which also was not objected to) was immediately explored on recross-examination:

Q. This is your opinion, Professor, is that correct?
A. Let's say it is not only opinion. It is an observation.
Q. And I understand you to say that no geologist that you know of recommends this system for this type of area?
A. I know all my colleagues invade [sic; "inveigh"?] against them.
Q. These are all of your colleagues at Franklin & Marshall?
A. Everybody I have talked to.
Q. At Franklin & Marshall?
A. No, at other places too.
Q. Pardon me?
A. Other places, other colleagues. We discuss these at meetings and I have been a consultant on a situation where I have seen the result of one of these things.

That was the complete recross-examination of this witness.

Appellant's final witness was F. James Knight, a registered engineer, who was qualified as an "engineering geologist." He was employed by Gannett-Fleming and had supervised the repairs of the sinkholes on appellant's site. He testified as to the extent of the damage and the extent of the repairs necessary, and expressed his professional opinion that the dry wells had "contributed significantly to the accelerated formation of sinkholes . . . on that tract."

At the close of appellant's case an oral motion for compulsory nonsuit was made on the ground that appellant had failed to present sufficient expert testimony to establish the standard of care required of an architect in that locality with respect to the design of surface water disposal systems. The court granted the motion, ruling that "while [appellant] has presented testimony of experts in the field of geology and engineering, they have not presented any testimony in the field of architecture tending to prove that [appellees'] professional services departed from accepted practice in this profession or that [appellees] failed to meet the standards of their professional duties." A motion to take off the nonsuit was denied by the court *en banc* with an opinion by Judge BUCHER, who was also the trial judge. In that opinion the court held that "testimony in the field of architecture" was necessary. It also suggested a second reason for the nonsuit, by asking, "[D]id [appellant] prove any negligence on the part of [appellees'] that justified submitting the case to the jury?" Both of these issues are before us on this appeal.

I.

The failure to present an architect as an expert witness was not fatal to appellant's case.

The court below states the general rule as being that "expert testimony is necessary to establish negligent practice in any profession." Wohlert v. Seibert, 23 Pa.Super. 213 (1903), is cited for this proposition (as it has been in other opinions), but in fact its test is more analytical:

> The crucial test of the competency of a witness offered as an expert to give testimony as such is the resolution of the question as to whether or not the jury or persons in general who are inexperienced in or unacquainted with the particular subject of inquiry would without the assistance of one who possesses a knowledge be capable of forming a correct judgment upon it.
> *Id.* at 216.

The opinion then goes on to restate the rules for the standard of care to which a physician is held; it does not discuss any other profession or professions in general. The same observation may be made of the section of Wigmore most cited on professional experts:

> On any and every topic, only a qualified witness can be received; and where the topic requires special experience, only a person of that special experience will be received [cross-reference omitted]. If therefore a topic requiring such special experience happens to form a main issue in the case, the evidence on that issue must contain expert testimony or it will not suffice. Wigmore on Evidence (3d ed. 1940) § 2090(a) at 453.

We have no doubt that expert testimony was required in this case. The "subject of inquiry" was the standard to be applied to one who holds himself out as competent to design and supervise the construction of a surface water disposal system in Lancaster County. This is certainly a subject that "requires special experience." However, there is nothing inherent in the nature of that experience that makes it unique to architects. What the jury needed was not "the assistance of one who possesses a knowledge" of architecture but of surface water disposal systems. Whether the person offering that assistance happened to be an architect, or engineer, or geologist, or something else, was unimportant; what was important was what he knew.

The error committed by the court below was that it literally analogized the instant case to one of medical malpractice. The court reasoned that in medical malpractice cases there must be expert testimony from physicians as to the appropriate standards of medical practice. From this the court reasoned that in a suit against architects, only architects are competent to testify as to the appropriate architectural standards. However, in medical malpractice cases the expert generally must be a physician because only a physician is trained to perform the medical functions that are the subject matter in controversy. In the instant case the subject matter in controversy (the design and installation of a surface water disposal system on the site in question) is not within the exclusive realm of one profession. To the contrary, it is within the realm of at least three professions: architects, engineers, and geologists. Therefore, a member of any of these professions (if otherwise qualified) was competent to state what the appropriate design and installation standards were.

A case similar to this one is Bloomsburg Mills Inc. v. Sordoni Construction Co., 401 Pa. 358, 164 A.2d 201 (1960). There the plaintiff hired some architects to design and supervise the construction of a weaving mill for nylon and rayon. This type of manufacturing requires a constant temperature and humidity. It was thus necessary to build the roof with a "vapor seal" to prevent condensation and leakage of moisture. The plaintiff alleged that the roof was defective in that the vapor seal did not function properly, and that the roof was otherwise inadequately sealed. The defendant appealed a verdict in favor of the plaintiff, claiming insufficient evidence of negligence. As part of this claim the defendant attacked the competency of the plaintiff's expert witness. The expert had had extensive experience with a large roofing manufacturing company and was at the time of the trial a professional roofing consultant. (He had, in fact, been requested by the defendants to submit a bid on the project in question but had declined.) He was not, however, an architect. Nevertheless the Supreme Court held him qualified to testify against the defendant architects. For other decisions in accord, *see* Abbott v. Steel City Piping Co., 437 Pa. 412, 263 A.2d 881 (1970) (witness with extensive experience in masonry competent to testify as to how a certain wall should be built despite lack of formal engineering degree); Willner v. Woodward, 201 Va. 104, 109 S.E.2d 132 (1959) (heating engineer competent to testify against architect regarding a heating and air conditioning duct); Cuttino v. Mimms, 98 Ga.App. 198, 105 S.E.2d 343 (1958) (engineer and contractor both competent to testify against architect in case based on faulty construction); Covil v. Robert & Co. Associates, 112 Ga.App. 163, 144 S.E.2d 450 (1955) (engineer competent to testify against architect who had drawn plans for water works where issue was whether a certain pipe joint was properly secured).

The court below dismissed *Bloomsburg Mills*, saying that the improper design of a roof is a "common problem" that "would hardly require expert testimony." As suggested by our preceding statement, however, in fact the problem was quite complex and involved a special type of structure. This is further apparent from the Supreme Court's quite lengthy and detailed discussion of the construction problems presented, and of the expert witness's qualifications, none of which would have been appropriate had the case presented only a "common problem," requiring no expert testimony at all. We thus find *Bloomsburg Mills* persuasive authority, and conclude that the failure of appellant to present an architect as an expert witness was not fatal to its case.

II.

Although, as noted above, the opinion of the court below asks the question, "[D]id [appellant] prove any negligence . . . that justified submitting the case to the jury?", in fact that question is not there addressed. Instead, the entire opinion deals only with whether the trial judge was correct in holding that in an action against an architect acting in his professional capacity, the plaintiff must produce testimony by another architect, which is the issue that we have just disposed of. In these circumstances we have made our own ex-

amination of the record, and have concluded that appellant did prove sufficient evidence of negligence to send the case to the jury.

Messrs. Pickering, Saylor, and Knight all testified unequivocally that in their respective professional opinions the sinkholes were aggravated by the dry well system, and that considerable damage to appellant's property resulted. However, none of them testified as to the standards of skill required of one who undertakes to design and supervise the installation of a surface water disposal system in the particular area in question, and, as observed in the preceding section of this opinion, it was essential to appellant's case that there be some testimony by a qualified expert on that point.

Appellees have contended that there was no such testimony. (They do not challenge the testimony of Messrs. Pickering, Saylor, and Knight.) This contention, however, overlooks the testimony of Professor Freedman, which we have already quoted in relevant part, *ante* at 409. It is indeed true that Professor Freedman's testimony was not developed in an orderly manner, and in fact it appears to have emerged almost by chance. It is, nevertheless, in the record, and the professor was cross-examined with respect to it. Summarized, the testimony was that neither the professor, nor any other geologist, nor any one consulted about surface water disposal, "has ever recommended dry wells for construction in an area like this . . . [W]e frequently get calls . . . and . . . we tell them they are in for trouble . . . [T]hese are systems that should not have been installed because they lead to trouble." When we bear in mind that we must give appellant every reasonable benefit from the evidence, *Shirley v. Clark, supra*, this testimony may fairly be read as a statement of opinion, by a qualified expert, that appellants violated the professional standards of care to which they were obliged to conform.

Order reversed.

Other recent cases have also been liberal in admitting expert testimony.[81]

The setting in which expert testimony is taken is best understood by reference to judicial opinions that look carefully at the testimony and determine whether it meets the legal requirements. A portion of such an opinion is reproduced here.

---

[81]*Tomberlin Assoc., Architects, Inc. v. Free*, 174 Ga.App. 167, 329 S.E.2d 296 (1985) (civil engineer allowed to testify in claim against architect based on soil erosion); *Keel v. Titan Constr. Corp.*, 721 P.2d 828 (Okla.1986) (physics professor could testify as to adequacy of design of solar system); *Wessel v. Erickson Landscaping Co.*, 711 P.2d 250 (Utah 1985) (structural engineer allowed to testify in claim against landscape architect as to design of retaining wall); *Perlmutter v. Flickinger*, 520 P.2d 596

(Colo.App.1974) (engineer and contractor permitted to testify about skylight design); *White Budd Van Ness Partnership v. Major-Gladys Drive Joint Venture*, supra note 12 (noting overlap between engineering and architecture permitted engineer to testify as expert in a claim against architect).

DRESCO MECHANICAL CONTRACTORS, INC. v. TODD-CEA

United States Court of Appeals, Fifth Circuit, 1976. 531 F.2d 1292.

CLARK, Circuit Judge.

[Ed. note: A boiler exploded, causing property damage. The owner brought legal action against Todd, who had designed and manufactured the boiler and combustion controls, based on Todd's negligence. Todd claimed the explosion resulted from design negligence by Austin, a consulting engineer retained by the owner who had specified that a dual timer system be used. The owner prevailed against Todd, and Todd asserted a claim against Austin. The jury verdict against Austin was set aside by the trial judge and an appeal taken.]

Todd's action against Austin rested on a theory of negligent design. To establish a cause of action on such a theory under Georgia law, Todd was required to establish that Austin's acts or omissions breached its duty to observe the standard of care it owed to those with whom it dealt and that this breach was a proximate cause of the occurrence from which damage was suffered. As an engineering firm, Austin's duty of care was that of a responsible *professional* person or organization. It is

> . . . the obligation to exercise a reasonable degree of care, skill, and ability, which generally is taken and considered to be such a degree of care and skill as, under similar conditions and like surrounding circumstances, is ordinarily employed by their respective professions.

* * *

It was Todd's burden to establish the failure to observe this standard by the introduction of expert opinion evidence. *See, e.g., Shea v. Phillips,* 213 Ga. 269, 271(2), 98 S.E.2d 552 (1957).

Todd presented the testimony of several witnesses to show that Austin had established the original requirement for a "completely redundant . . . dual flame safeguard system" and had refused to approve a change to a single timer system when Todd's engineers described it to them as "unsafe." Witness Horton Rucker, an employee of an engineering firm not involved in the suit, explained why he believed the explosion to have been caused by the dual timer system. He stated the dual timer presented a "bad situation" because of the possibility that the timers would get out of synchronization and send improper signals to the system. He said he had no personal experience with such a system in his twenty-five years of experience in the field of boiler and burner-control engineering.

James Warren of Honeywell, Inc., the manufacturer of the timers, testified that he "[didn't] believe [he] would try using these particular programmers in parallel" because "as long as everything [was] going well they would probably work all right. But if things started going wrong, they would get false messages and get confused."

In short, Todd's proof presents a convincing theory of the *cause* of the explosion and for the proposition that Austin participated in—or perhaps even dictated—the decision to install the system whose malfunction lay at the heart of that theory. What is missing is any sort of unequivocal statement by any witness that Austin's specification of the dual timer system or its manner of reviewing Todd's detailed plans failed to conform to the ordinary accepted practices of the engineering profession. An examination of Rucker's testimony demonstrates considerable lack of communication between counsel and witness, and neither the questions nor the answers establish whether Rucker was describing a "bad situation" inherent in the design or a "bad situation" that obviously resulted in this instance. Rucker stated that he was "totally unfamiliar" with dual timer systems prior to his inspection of the system involved in the case *after* the explosion. However, he did not intimate that this lack of familiarity was based upon the fact that such systems were not safe to use or not customarily put to use by others. Thus, he could have no basis for a professional judgment that a dual timer design was inherently unsafe or failed to conform to engineering practice. Likewise, Warren's testimony does not refer to the professional standard. It is true that he said he "didn't think" he would use such a system, but neither his position nor any reasons for it are free from ambiguity.

In marked contrast to these statements is the testimony of Austin's independent expert witness, Alderman. He stated unequivocally that the original Austin specifications were "excellent" and "complied generally and extrapolated on" the generally accepted practice of mechanical engineers in Georgia during the relevant time period. He further stated that the specifications for the dual flame safeguard system in specific met the accepted standard and that it was the standard practice for consulting engineers to review builders' wiring diagrams only for general compliance with the specified concept. He concluded his direct testimony as follows:

Q. Mr. Alderman, will you state whether or not the use of dual timers if they are properly wired is a practice which can be carried out safely in a flame safeguard system, if they are properly wired?

A. There's no doubt in my mind that an expert wiring control designer could successfully wire dual flame scanners, relays and timers.

Q. Is the practice of not checking the wiring, details of shop drawings for workability, is that the general accepted practice among consulting mechanical engineers in this State and was the practice in 1969, 1970 and '71?

A. It is the common practice.

In reviewing the correctness of the trial court's decision to take a case from the jury by directed verdict or judgment notwithstanding the verdict,

the Court should consider all of the evidence—not just that evidence which supports the non-mover's case—but in light and with all reasonable inference most favorable to the party opposed to the motion. If the facts and inferences point so strongly and overwhelmingly in favor of one party that the Court believes that reasonable men could not arrive at a contrary verdict, granting of the motions is proper. On the other hand, if there is substantial evidence opposed to the motions, that is, evidence of such quality and weight that reasonable and fair-minded men in the exercise of impartial judgment might reach different conclusions, the motions should be denied, and the case submitted to the jury.

. . . [I]t is the function of the jury as the traditional finder of the facts, and not the Court, to weigh conflicting evidence and inferences, and determine the credibility of witnesses.

Boeing Company v. Shipman, 411 F.2d 365, 374–75 (5th Cir.1969) (en banc). When this test is applied here it requires the conclusion that there was no substantial evidence opposed to the motion on the element of Austin's violation of the standard of care imposed on it. Therefore, the trial court properly refused to allow the jury's verdict to stand.

C. Critique of System

The expert testimony system has been severely criticized. The complex "hypothetical" question has led to overtechnical appellate review and confusion of jurors. In addition, complexity has made errors more likely because of the increasing specialization of professions, a point demonstrated by *Haak*. When administered too strictly, no experts may be found, a particular difficulty when expert testimony is required. The pool of experts being limited to *local* experts not only has made it more difficult to obtain experts but also has not taken into account the increasingly statewide or national standards of practice.

Even more criticisms have been made of the entire system itself, based largely on the adversary system discussed in Section 2.11. Each party looks not for the best qualified experts but for the experts who will best support its case. This, coupled with the high compensation paid experts, has led to skepticism as to the professional honesty of many experts.

Not many years ago there was intense criticism of what was called the "conspiracy of silence," the unwillingness of professionals to testify against one another. This was particularly difficult where local standards were employed. This led to the development of professional expert witnesses (called forensic design professionals because of their ability to persuade judges and juries of the soundness of their professional conclusions). Although this has certainly helped overcome the "conspiracy of silence," it has had some unfortunate results.

Such experts have been looked on as "hired guns"—too quick to find fault. (To those challenging design professionals, these experts are looked on as paladins.) Of course, other experts can be produced by the design professional charged with malpractice. But in the end, judge and jury are often confused. How can experts with such outstanding credentials differ so sharply as to the cause of the harm and the standards of professional practice?

The junk science debate which surfaced in 1991 relates to the power of trial court judges to admit scientific testimony by witnesses whose work has not been subject to peer review or published in professional journals but rather has been generated solely for use in litigation mainly involving claims against drug companies. The Federal Rules of Evidence state that expert testimony not accepted by the majority of a scientific community can be allowed if this testimony will assist the trier of fact. But a trial court refused to hear testimony based upon work that had not been subject to peer review or published in professional journals, based upon what is called the generally accepted rule. This was affirmed by the Circuit Court of Appeals.[82] The U.S. Supreme Court reversed, rejecting the more stringent requirements and leaving the issue of admissability largely to the discretion for the trial judge.[83]

This dispute and the high cost of producing expert witnesses have again revived calls for greater use of expert panels from which trial court judges can select expert witnesses rather than rely upon the testimony of experts called by the parties. Yet a report indicates that only 20% of the federal court judges have appointed their own experts. A system of court-appointed experts also raises problems. Not only does it operate to interfere with the adversary system, but questions are raised as to how judges will select the experts, whether the parties will be able to cross-examine the experts or offer their own expert testimony, and who will pay the fees of the experts.

D. Advice to Expert Witnesses

Space does not permit a detailed discussion of all the problems seen from the perspective of a design professional asked to be an expert witness. Some brief comments can be made, however.

A clear, written understanding should precede any services being performed. Such a writing should include the following:

1. Specific language making clear that the expert will give her best professional opinion.
2. Language that covers all aspects of compensation for time to prepare to testify, travel time, and actual time testifying before a court, board, or commission. (Many experts use an hourly rate for preparation time and a daily rate for travel and testimony time.)
3. Specifying expenses to be reimbursed, using a clear and administratively convenient formula for reimbursing costs of accommodations, meals, and transportation.
4. A minimum fee if the expert is not asked to testify. Some attorneys retain the best experts, use the experts whose opinions best suit their case, and, by having retained the others, preclude them from testifying for the *other* parties.

Details as to appearance, description of qualifications, methods of answering questions, explanations for opinions, and defending opinions on cross-examination are usually provided by the attorney calling the expert. It is important to recognize who is being addressed and the reason for seeking expert opinions. The expert should assist the judge or jury—persons often inexpert in evaluating technical material. For that reason, opinions and explanations must be understandable by persons who must evaluate them. The expert should *never* speak *down* to judge and jurors.

SECTION 14.07 Implied Warranty: An Outcome Standard

As noted in Section 14.05, the law generally looks at the process by which the services were to be performed by a design professional—the professional standard. Where a professional is retained to provide *design* services, American law has concluded (not without criticism, as explored in Section 14.11(A)) that the *likely* understanding between the client and the professional designer is not that a *successful outcome* will be achieved when professional services are purchased but that the professional will perform as would other professionals.[84]

A minority view has developed under which the design professional must design to achieve the communicated or understood expectations of the client. The issue in such cases is whether the design professional has met the client's purposes of which she was aware or of which she should have been aware. Liability—like the liability for any breach of contract or any manufactured product—is strict under this doctrine. There is no need to compare what the challenged design professional did to what other design professionals similarly situated would have done and no need to introduce expert testimony.

While early cases using the implied warranty concept involved contracts to design and build,[85] others have involved consumer-oriented transac-

[82]*Daubert v. Merrell Dow Pharmaceuticals, Inc.*, 951 F.2d 1128 (9th Cir.1991), reversed, 61 L.W. 4805 (1993).
[83]Ibid.

[84]Quoted and followed in *Kemper Architects v. McFall, Konkel & Kimball Consulting Eng'rs*, 843 P.2d 1178, 1186 (Wyo.1992).
[85]*J. Ray McDermott & Co. v. Vessel Morning Star*, 431 F.2d 714 (5th Cir.1970) (Admiralty); *Hill v. Polar Pantries*, 219 S.C. 263, 64 S.E.2d 885 (1951).

tions directly or by analogy.[86] There have also been a few cases involving a design professional retained to perform ordinary design services.[87]

Until recently the standardized contracts prepared by the design professional associations did not specify how the design professional's performance would be measured. But as noted in Section 14.04(A), in 1992, the Engineers Joint Contracts Documents Committee (EJCDC) included language stating that the professional standard applied. This was done to exclude any outcome-oriented express or implied warranties that would measure the engineer's performance.[88]

Some clients, both public and private, have sought to include language under which the design professional would perform to some successful outcome standard, such as provisions stating that the design professional would perform in accordance with the "highest standards of professional service." This could be taken to create an express warranty of a successful outcome or at least require that it be determined whether the challenged design professional had done better than others would have. Usually in such cases the design professional who is aware of the risk will seek to persuade the client not to include such language by stating that it is a contractually assumed liability and would not be covered under the terms of the professional liability insurance policy. But often the realities of the market mean that the design professional would have to accept this language if the client insists upon it.

The problem becomes more complicated when courts sometimes use the concept of implied terms to impose an obligation of a successful outcome. For example, one case stated that it was implied that the architect would specify "reasonably good materials" and that it would "perform its work in a reasonably workmanlike manner" and "in such a way as reasonably to satisfy such requirements as it had notice the work was required to meet."[89]

In addition, there are cases that are simply confusing, mixing professional standard language and implied warranty language.[90] As if this were not confusing enough, one case stated that the implied warranty standard did not require a *favorable* result but simply required a *reasonable* result.[91] This may not differ much from the professional standard. Finally, some claims for implied warranty are rejected when the implied warranty sought was one over which the design professional had no control.[92]

But is there much of a difference in outcome between the two standards? Although this is discussed in greater detail in Section 14.11(A), it should be noted here that the design professionals, their professional associations, and their insurers think so. Much of their polemical work is devoted to protecting the professional standard. But in many cases, failure by the design professional to design in accordance with the known or foreseeable purposes of the client will be a breach of the professional standard. Yet the struggle between the process-oriented professional standard and the outcome-oriented implied warranty is likely to continue.[93]

---

[86]*White Budd Van Ness Partnership v. Major-Gladys Drive Joint Venture,* supra note 12 (application of Texas Deceptive Trade Practices Act with its strict liability to licensed professionals); *Beachwalk Villas Condominium Ass'n Inc. v. Martin,* 305 S.C. 144, 406 S.E.2d 372 (1991).

[87]*Broyles v. Brown Eng'g Co.,* 275 Ala. 35, 151 So.2d 767 (1963); *Board of Educ. v. Del Biano & Assocs.,* 57 Ill.App.3d 302, 372 N.E.2d 953 (1978) (using implied term); *Tamarac Dev. Co. v. Delamater Freund & Assocs. P.A.,* 234 Kan. 618, 675 P.2d 361 (1984).

[88]EJCDC No. 1910-1, ¶ 1.8.

[89]*Board of Educ. v. Del Biano & Assocs.,* supra note 87, 372 N.E.2d at 958.

[90]*Bloomsburg Mills, Inc. v. Sordoni Constr. Co.,* 401 Pa.2d 358, 164 A.2d 201 (1960).

[91]*E. C.Ernst, Inc. v. Manhattan Const. Co. of Texas,* 551 F.2d 1021, rehearing denied in part, granted in part, 559 F.2d 268 (5th Cir.1977), cert. denied sub nom *Providence Hosp. v. Manhattan Constr. Co. of Texas,* 434 U.S. 1067 (1975).

[92]*Allied Properties v. John A. Blume & Assocs.,* 25 Cal.App.3d 848, 102 Cal.Rptr. 259 (1972). In this case, the client contended that the naval architect impliedly warranted that he would design a pier reasonably suitable for use by small craft. The evidence indicated hotel patrons did not use the pier for boats because of the rough weather, the length of the pier, the depth of the water, the availability of ample moorings nearby, and the fact that the formal atmosphere of the hotel did not appeal to boating people.

[93]See Jones, *Economic Loss Caused by Construction Deficiencies: The Competing Regimes of Tort and Contract,* 59 U. of Cinn.L.Rev. 1051, 1070, 1073 (argues for resulted-oriented standard if probability of success is high); 20 Mem.St.U.L.Rev. 611 (1990) (argues against implied warranty and strict liability).

SECTION 14.08 Third-Party Claims: Special Problems

Third-party claims raise problems that do not arise when the claimant is connected to the design professional by contract. The proliferation of third-party claims has generated more litigation, varying state rules, and judicial opinions of divided courts than have claims by clients against design professionals. This reflects rules in transition with inevitable strains and contradictions.

A. Potential Third Parties

The centrality of the design professional's position in construction—both designing and monitoring performance—generates a wide range of potential third-party claimants. At the inner core are the other major direct participants in the process who work on the site itself, such as contractors and construction workers. Around the core are those who supply money, materials, or equipment, such as lenders and suppliers. Next are those who "backstop" direct participants, such as sureties and insurers. Yet farther from the core are claimants who will ultimately take possession of the project, such as subsequent owners, tenants, and their employees. Farthest from the core are those who may enter or pass by the project during construction or after completion, such as members of the public or patrons.

The wide variety of potential claimants, the varying distance from core participants, the type of harm, the difference between those who have other sources of compensation, such as workers—all combine to ensure complexity.

B. Contracts for Benefit of Third Parties

This section emphasizes the increasing use of tort law by third parties who suffer losses related to the Construction Process, but third parties make claims that are sometimes based on the assertion that they are intended beneficiaries of contracts to which they are not parties. Until the mid-nineteenth century, American law generally did not permit persons not a party to a contract to maintain legal action for the contract's breach even though they may have suffered losses from the breach. But American law has steadily expanded the rights of third parties to recover for contract breach.[94]

The use of this doctrine has found its way into construction claims. In the context of claims against the design professional, the doctrine is sometimes employed by contractors and subcontractors who assert that they are intended beneficiaries of the contract made between the owner and the design professional.

Such claims have had spotty success,[95] with most cases concluding that contracting parties usually intend to benefit themselves. Yet the possibility of such a theory being invoked successfully has led contracting parties to seek to use the contract itself to bar claims by third parties as intended beneficiaries.[96]

Even if a claimant can establish that it is an intended beneficiary, its claim may fail because of provisions in the contract.[97] To complicate matters further, an architect sued by a contractor was allowed to use a provision in the contract between the contractor and the owner as a defense.[98] (Many of these problems could be avoided if owner, contractor, and design professional were all parties to one contract.)

C. Tort Law: Privity and Duty

Sections 14.08(C), (D), and (E) all relate to attempts by third parties, usually contractors or subcontractors, to assert tort claims against design professionals. Section 14.08(C) approaches the problem from the perspective of privity, the absence of which has been the principal justification for denying such claims historically and, to some degree, today. Sec-

---

[94]Restatement (Second) of Contracts, § 302 (1979).

[95]*John E. Green Plumbing & Heating Co. v. Turner Constr. Co.*, 742 F.2d 965 (6th Cir.1984) (contractor v. construction manager). But see *A. R. Moyer v. Graham*, 285 So.2d 397 (Fla.1973) (contractor could sue architect in tort but not as intended contract beneficiary). The cases, including those involving construction contracts, are canvassed in Prince, *Perfecting the Third Party Beneficiary Standing Rule Under Sec. 302 of the Restatement (Second) of Contracts*, 25 B.C.L.Rev. 919 (1984). See also Burch, *Third-Party Beneficiaries to the Construction Contract Documents*, 8 Constr. Lawyer No. 2 (April 1988), p. 1.
[96]See AIA Docs. B141, ¶ 9.7, and A201, ¶ 1.1.2. But the latter gives third-party rights to the architect.
[97]*Prichard Bros., Inc. v. Grady Co.*, 407 N.W.2d 423 (Minn.App.1987) reversed 428 N.W.2d 391 (Minn.1988) (contractor can bring tort claim against architect and is not limited to third-party beneficiary claim).
[98]*Bates & Rogers Constr. Corp. v. Greeley & Hanson*, 109 Ill.2d 225, 486 N.E.2d 902 (1985) (architect sued by contractors could invoke "no damage" clause in prime contract as defense).

tion 14.08(D) examines negligent misrepresentation, one avenue by which the requirement of privity has been avoided. Section 14.08(E) discusses the economic loss rule, a traditional principle of tort law that limited tort law to the protection of person and property. While this can also be applied in disputes between contract-connected parties, as seen in Section 14.05(D), in most disputes today, the economic loss rule is asserted by those against whom claims have been made as a reason to bar the claimant from basing its claim upon tort law. Because of the interrelationship of these doctrines, some overlap will occur in the three sections.

These sections all document the conflicting worlds of contract law, with its emphasis on planning and certainty, and tort law, with its emphasis on preventing harm and compensating victims.

A claimant employing tort law as a basis of transferring its loss to another must show that the latter owed a duty to protect the claimant from the unreasonable risk of harm. Often, particularly where the wrongful conduct was a breach of contract, the duty concept was phrased as one requiring privity between claimant and the person against whom the claim was made. Usually, though not exclusively, privity was based on a contract between the claimant and the person against whom the claim was made. The defense of absence of duty or lack of privity seeks to avoid unlimited liability and to free persons from the fear of being held accountable for any harm they might cause. As noted in Section 7.09(B), the requirement of privity had been an insurmountable obstacle in the nineteenth century and early twentieth century to even a claimant who suffered personal harm because of another's negligence. However, that section noted that the privity defense was eliminated in personal injury claims early in the twentieth century. The early cases doing so were claims against manufacturers by those injured from negligently manufactured products.

By the mid-twentieth century, participants in the Construction Process, such as design professionals, could no longer invoke the privity doctrine as a defense to claims based on personal harm.[99]

Tort law is not limited to claims for personal harm. In the Construction Process, claimants often suffer economic losses unconnected with personal harm or damage to property. Often they seek to recover these losses from participants with whom they do not have a contract. For example, contractors and sureties assert claims against design professionals, sometimes (as noted in Section 14.08(B)) as intended beneficiaries of a contract that the design professional has breached and increasingly by invoking tort law. States that permit third-party claims saw no reason to distinguish pure economic losses from harm to person or property. They were willing to compensate those who suffered economic losses because of the negligence of those against whom claims were made.[100] The leading case pointed to the life-or-death power the architect has over the contractor.[101] Requiring that the harm and the person suffering it be reasonably foreseeable was the method such courts felt would protect the person against whom a claim would be made from unlimited liability.

Other states still require privity in claims for purely economic losses.[102] They point to tort law's historic function of protecting personal and property rights and the need to draw some lines beyond which the law should not compel a person to pay

---

[99]*Miller v. DeWitt*, 37 Ill.2d 273, 226 N.E.2d 630 (1967).

[100]*E. C. Ernst, Inc. v. Manhattan Constr. Co. of Texas*, Supra note 91 (subcontractor v. architect); *Donnelly Constr. Co. v. Oberg/Hunt/Gilleland*, 139 Ariz. 184, 677 P.2d 1292 (1984) (contractor v. architect); *Cooper v. Jevne*, 56 Cal.App.3d 860, 128 Cal.Rptr. 724 (1976) (condominium purchasers v. architect); *Conforti & Eisele, Inc. v. John C. Morris Assoc.*, 175 N.J.Super. 341, 418 A.2d 1290 (1980) (contractor v. design professional); *Schoffner Ind., Inc. v. W. B. Lloyd Constr. Co.*, 42 N.C.App. 259, 257 S.E.2d 50 (1979), cert denied 298 N.C. 296, 259 S.E.2d 301 (1979); *Forte Bros., Inc. v. National Amusements, Inc.*, 525 A.2d 1301 (R.I.1987) (contractor v. architect). Many cases followed the influential opinion of the federal trial court judge in *United States v. Rogers & Rogers*, 161 F.Supp. 132 (S.D.Cal.1958). For surety claims, see Section 22.07. See also Annot., 65 A.L.R.3d 249 (1975).

[101]*United States v. Rogers & Rogers*, supra note 100.

[102]*Georgetown Steel Corp. v. Union Carbide*, 806 F.Supp. 74 (D.S.C.1992) (applying South Carolina law, the court noted that the fact that the parties were commercial sophisticates who dealt at arm's length and had their own experts was reason to refuse to allow lessee of owner to sue owner's engineer); *Bagwell Coating, Inc. v. Middle South Energy, Inc.*, 797 F.2d 1298 (5th Cir.1986) (Mississippi law) (contractor v. construction manager); *Peyronnin Constr. Co. v. Weiss*, 137 Ind.App. 417, 208 N.E.2d 489 (1965) (contractor v. engineer); *Delta Constr. Co. v. Jackson*, 198 So.2d 592 (Miss.1967) (contractor v. engineer); *Bernard Johnson, Inc. v. Continental Constructors, Inc.*, 630 S.W.2d 365 (Tex.Civ.App.1982) (contractor v. architect).

for another's loss. Foreseeability was considered insufficient protection. Also, decisions barring such claims frequently pointed to the claimant's often having a contractual right against the party with whom it contracted.

One method of avoiding the privity rule has been to invoke the doctrine of negligent misrepresentation, which can apply to those who furnish information that is relied upon by third parties (discussed in Section 14.08(D)).

D. Negligent Misrepresentation: *John Martin Co. v. Morse/Diesel* and *Ossining Union Free School Dist. v. Anderson, La Rocca Anderson*

The design professional's performance consists in part of supplying information. Sometimes representations are made to the client, such as those involving cost estimates,[103] the availability of funds for the project,[104] and the likelihood that the project will meet the client's needs.[105] This section

---

[103]See Section 12.03.
[104]See Section 12.04.
[105]See Section 12.05.

emphasizes claims by third parties, such as contractors and subcontractors, that they relied on negligent misrepresentations made by the design professional.[106]

The following case examines the problem of negligent misrepresentation in the construction context. It notes a division in the American cases as to the rights of a third party to bring a tort claim based upon negligent misrepresentation. The case was considered sufficiently important by participants in the Construction Process that an amicus curiae (friend of the court) brief was submitted on behalf of the construction manager by the American Institute of Architects, the National Society of Professional Engineers, the American Consulting Engineers Council, and state chapters of those organizations. A similar brief was submitted on behalf of the contractor by two Tennessee contractor associations.

---

[106]As to surety claims, see Section 22.07. For contractor claims against the owner for implied warranties created by the design, see Section 23.05(E).

JOHN MARTIN COMPANY, INC. v. MORSE/DIESEL, INC., . .

Supreme Court of Tennessee, at Knoxville, 1991. 819 S.W.2d 428.
OPINION

GARY R. WADE, Special Judge.
[Ed. note: footnotes renumbered.]
. . . . The issue is whether a subcontractor who has been fully paid by the owner for the performance of his contractual duties may make a separate claim in tort against a construction manager for economic loss caused by negligent misrepresentations.

We hold that a subcontractor, despite a lack of privity, may make such a claim against the construction manager based upon negligent misrepresentation, whether the negligence is in the form of direction or supervision. The judgment of the Court of Appeals, vacating the grant of summary judgment in favor of the defendant and remanding the cause for trial, is affirmed. Costs are adjudged against the defendant, Morse/Diesel, Inc.

Provident Insurance Company ("Provident") initially contracted with the defendant, Morse/Diesel, Inc., as a consultant to assist in the planning of an addition to its offices in Hamilton County. The defen-

dant reviewed the proposed architectural designs, suggested modifications, and prepared cost estimates for construction.

When the decision was made to proceed with the project, Provident engaged the defendant as construction manager. By the terms of their construction management contract, the defendant was to act in behalf of the owner in the employment of the necessary subcontractors, the coordination of their schedule, and the supervision of their work. The defendant had the specific responsibility to review, consider, and approve the plans and specifications used by the subcontractors in the performance of their tasks. Leslie Littlefield was the defendant's on-site construction superintendent.

The plan of construction was in two phases: first, the substructure, which included excavation of the site and construction of the foundation and basement; and second, the superstructure, which included all construction above street level. The entire project was to be constructed by the "fast track" method. That is, construction began on each phase of the building addition as component architectural plans and specifications were completed and approved.

Provident, through the defendant, entered into a number of contracts with various contractors. On August 19, 1981, the defendant, acting as agent for Provident, executed a subcontract employing the plaintiff, John Martin and Company, Inc., to provide concrete and rough carpentry for the superstructure. Although not a party to the agreement, the defendant was to supervise and direct the work. As such, the defendant was authorized to act on behalf of Provident in determining the means, techniques, sequences, and procedures for construction. The defendant had the responsibility to review and approve all shop drawings, the plaintiff's subcontractors and suppliers, and plaintiff's daily work progress.

The concrete floors to the superstructure were required to reach specific elevations provided for by the architectural plans. The defendant measured the floor elevations. As the plaintiff poured the concrete, the elevations were consistently low. More concrete was required. As a result, the plaintiff experienced additional costs and delays. The plaintiff claimed the extra labor and material were due to faulty plans. The defendant responded that the need for more concrete was due to the natural sag of the steel which, by the terms of the subcontract, was the responsibility of the plaintiff; its position was that the plaintiff had miscalculated the amount of concrete necessary. In consequence, the defendant refused to approve change orders for the additional concrete. The plaintiff did not finish its work until over a year beyond its contracted completion date.

The plaintiff, seeking compensation for its losses, filed suit against Provident and the defendant on both contractual and tort theories. Eventually, the plaintiff settled with Provident; it received the payments authorized by either the contract or approved change orders. The plaintiff and Provident signed a covenant not to sue. . . . The grant of the defendant's application for permission to appeal is based upon our view that this specific issue is one of first impression in this state.

The defendant's principal argument is that the plaintiff cannot recover on a theory of negligence because any losses were purely commercial in nature. It contends that principles of negligence should not be applied to contract-based interests. The defendant submits that there are sound public policy reasons, not considered by the Court of Appeals, which should preclude recovery for economic losses in the absence of privity. The *amicus curiae* brief [filed by various architectural and engineering associations] in support of the defendant's position argues in favor of the "economic loss doctrine": a general principle that prohibits the recovery of purely economic damages for negligence when the plaintiff lacks privity of contract with the defendant. . . . The *amicus* contends that it is a time-honored rule. The only exception, it submits, is in connection with an action for negligent misrepresentation, limited to those instances where the plaintiff has justifiably relied upon misinformation, usually provided by a professional, in determining whether to enter into or participate in a business transaction. . . . *See* Restatement (2d) of Torts § 552. Because the negligent misrepresentation alleged here is not in regard to "the transaction" but relates to the day-to-day performance of an executed contract, the *amicus* asserts that the doctrine's exception would not apply. To permit this action, it contends, would altogether do away with the economic loss doctrine and would subject architects, engineers, and construction managers to an unlimited number of potential claimants not remotely involved in the original transaction.

The plaintiff submits that the defendant, as construction manager with exclusive control over the project, had the continuing duty to properly supervise and direct. It argues that it had no choice but to rely upon the defendant's negligent misrepresentations in relation to the design and negligent supervision in regard to the construction. An *amicus curiae* [various contractor associations] who filed its brief in support of the plaintiff, maintains that the holding of the Court of Appeals is in accordance with the established law of this state and should be sustained based upon public policy reasons. Because this state's lawyers, title examiners, accountants, and surveyors, despite the lack of privity, have all been held responsible for economic losses suffered by their negligence, the plaintiff asserts that no special consideration should be afforded to architects, engineers, or professional construction managers. . . .

I.

This is not a products liability case. In this instance, the theory of recovery is that the defendant negligently supplied information intended for the guidance of others; the plaintiff relied upon the misrepresentation in the performance of his contracted service and experienced business losses as a result. Many of those cases relied upon by the defendant involve claims of product liability in respect to design defects; the losses suffered were caused by defective products, not misguidance or misdirection in the performance of services. . . . The question here is not one of harm to person or property, but of economic loss.

Whether the negligence alleged is based upon misrepresentation or supervision, the applicable law in Tennessee, absent privity, is found in the Restatement (2d) of Torts § 552 (1977):

1. One who, in the course of his business, profession or employment, or in any other transaction in which he has a pecuniary interest, supplies false information for the guidance of others in their business transactions, is subject to liability for pecuniary loss caused to them by their justifiable reliance upon the information, if he fails to exercise reasonable care or competence in the obtaining or communicating of the information.

2. Except as stated in Subsection (3), the liability stated in Subsection (1), is limited to loss suffered (a) by the person or one of a limited group of persons for whose benefit and guidance he intends to supply the information or knows that the recipient intends to supply it; and
(b) through reliance upon it in a transaction that he intends the information to influence or knows that the recipient so intends or in a substantially similar transaction.

By the use of this standard, liability in tort would result when, despite lack of contractual privity between the plaintiff and the defendant,

1. the defendant is acting in the course of his business, profession, or employment, or in a transaction in which he has a pecuniary (as opposed to gratuitous) interest; and
2. the defendant supplies faulty information meant to guide others in their business transaction; and
3. the defendant fails to exercise reasonable care in obtaining or communicating the information; and
4. the plaintiff justifiably relies upon the information.

The defendant is liable only to those, whether in contractual privity or not, for whose benefit and guidance the information is supplied. The information may be either direct or indirect. In that regard, the foreseeability of use is critical to liability.

Because the misinformation is negligently rather than intentionally supplied, courts have been careful to limit liability to only those whose use of the information is reasonably foreseeable:

By limiting the liability for negligence of a supplier of information to be used in commercial transactions to cases in which he manifests an intent to supply the information for the sort of use in which the plaintiff's loss occurs, the law promotes the important social policy of encouraging the flow of commercial information upon which the operation of the economy rests.

Restatement (2d) of Torts § 552 Comments (1977).

Contributory negligence is, of course, a defense to any action based in negligent misrepresentation.*

II.

Although there is a split of authority among the states, Tennessee appears to have embraced § 552 of the Restatement (2d) as the guiding principle. In 1962, this Court, citing a number of Tennessee cases, acknowledged the erosion of the privity doctrine in this type of case:

It is true the old rule was that there was no duty of care upon a defendant to a plaintiff not in privity. . . . But it can hardly be said that such a general rule any longer exists.
. . .

Such a duty has been imposed and defendant held liable to a plaintiff not in privity in a number of classes of cases, such as that of a supplier of goods or services which, if negligently made or rendered, are "reasonably certain to place life and limb in peril."

Such a duty, and liability for its breach, has also been imposed upon defendants in favor of a plaintiff not in privity in cases of reasonably foreseeable risk of damage to tangible property. . . .

[S]uch a duty and consequent liability have been imposed on a defendant in favor of plaintiff not in privity where the risk of harm from negligent performance of a contract was to an intangible interest of such plaintiff. *Glanzer v. Shepard*, 233 N.Y. 236, 135 N.E. 275, 23 A.L.R. 1425.

Howell v. Betts, 211 Tenn. 134, 362 S.W.2d 924, 925–926 (1962) (some citations omitted). . . .

Despite the recognition of a cause of action absent privity, the holding in *Howell* was in favor of the defendant. A surveyor whose erroneous plat had been used in a sale some 24 years after its preparation was not held liable for the losses to the purchaser of the property:

On principle and authority, we think the rule of liability cannot be extended to a case like that before us. If these surveyors could be held liable to such an unforeseeable and remote purchaser 24 years after the survey, they might, with equal reason, be held liable to any and all purchasers to the end of time. We think no duty so broad and no liability so limited should be imposed.

Id., 362 S.W.2d at 926; *see also Ultramares Corp. v. Touche*, 255 N.Y. 170, 174 N.E. 441 (1931).

In 1970, this Court permitted an action in tort based upon negligent misrepresentation when there

---

*[Ed note: In many states, negligence by the plaintiff would only *reduce* the award.]

was no privity between the parties. *Tartera v. Palumbo*, 224 Tenn. 262, 453 S.W.2d 780 (1970). . . .

Because this Court has previously dispensed with privity as a prerequisite for actions in tort based upon negligent misrepresentation against title examiners, surveyors, and attorneys, the rule must extend to other professions whose business is to supply technical information for the guidance of others:

> The standard of care applicable to the conduct of audits by public accountants is the same as that applied to doctors, lawyers, architects, engineers, and others furnishing skilled services for compensation and that standard requires reasonable care and competence therein.

Vineyard v. Timmons, 486 S.W.2d 914, 920 (Tenn.App.1972) (privity existed as plaintiff was client of defendant accountant); . . .

The Restatement makes no distinction based upon the nature of the profession. Neither do we.

III.

The record in this case is voluminous. Briefing has been extensive both by the parties and by their supportive professional organizations. We are grateful, of course, for the assistance provided by the *amicus curiae* briefs. Much is made, however, about whether Tennessee falls among the majority or the minority on the view that a cause of action exists in tort even in the absence of privity between the parties. Persuasive arguments, both to limit and extend the bounds of liability, are made on grounds of public policy. . . .

We have examined the competing views. Our conclusion is that there is no clear majority; instead, we find a split of authority among the states. A number of cases from other states have held that an action may be maintained against an engineer, architect, con-

tractor or other design professional despite the lack of contractual privity between the parties.[107] A recent example is in *Kennedy v. Columbia Lumber & Mfg. Co., Inc.*, 299 S.C. 335, 384 S.E.2d 730 (1989). In that case, where a home buyer was permitted to sue the builder despite a lack of privity, the Supreme Court of South Carolina indicated its disapproval of an intermediate court's decision one year earlier which had recognized the economic loss doctrine as a bar to the suit.[108] Other states have adhered to the traditional view that purely economic losses cannot be recovered in tort absent privity.[109]

The position this state has adopted, as recognized by the Court of Appeals decision, is based in great measure upon cases authored years ago by former Justice Benjamin Cardozo. . . . [The Court discussed the leading New York cases, *Glanzer v. Shepard* and *Ultramares v. Touche*, described in the *Ossining* case reproduced in this subsection.] . . .

In summary, whether the action for economic loss is based upon negligent supervision or negligent misrepresentation, § 552 of the Restatement (2d) of Torts is the applicable standard in this state. Privity is not a prerequisite. The Court of Appeals is affirmed. The summary judgment in favor of the defendant is vacated and the cause is remanded for trial.

REID, C. J., and DROWOTA, O'BRIEN and DAUGHTREY, J. J., concur.

---

[107][Court cited and summarized 23 cases.]
[108]*Carolina Winds Owners' Ass'n., Inc. v. Joe Harden Builder, Inc.*, 297 S.C. 74, 374 S.E.2d 897 (S.C.App.1988) (overruling recognized in *Beachwalk Villas Condominium Ass'n., Inc. v. Martin*, 406 S.E.2d 372 (S.C.1991).
[109][Court cited four cases.]

It is important to note that misrepresentations under this case are not limited to professional opinions such as surveyor reports or those made by geotechnical engineers. That they should be so limited was the position taken by the professional associations for design professionals but rejected by the Supreme Court of Tennessee. The court summarized its conclusion by stating that "the action for economic loss is based upon negligent supervision or negligent misrepresentation." Earlier in its opinion, the court spoke of a claim being "based upon negligent misrepresentation, whether the negligence is in the form of direction or supervision." The construction manager in this case re-

viewed the proposed architectural designs. Also, the subcontractor argued that it had no choice but to "rely upon the defendant's negligent misrepresentations in relation to the design and negligent supervision in regard to the construction." This decision would appear to allow the negligent misrepresentation exception to the privity requirement to be used in cases involving almost anything that a design professional does in the course of her professional services.

A person supplying information may be inhibited from doing so if she would be exposed to indeterminate liability to a large number of third-party claimants. Historically, the principal professionals who have sought legal protection have been accountants and others who supply financial information on which many may rely. As a result, liability for negligent misrepresentation has been more limited than claims based on ordinary negligence. Most important, the defense of foreseeability (i.e., the party supplying the information could not reasonably foresee that it would be relied on by particular persons) is more likely to be available if a claim is based on negligent misrepresentation than if it is based on ordinary negligence.

Some of the uncertainties created by the negligent misrepresentation exception to the privity rule

are demonstrated in New York. New York was thought to be a state that required privity for a claim based upon negligent conduct that causes purely economic losses. The following case held that New York does not limit the negligent misrepresentation exception to accountants. It also used a narrower standard for determining whether a third-party action can be brought for negligent misrepresentation than did Section 552 of the Second Restatement of Torts, which was cited and followed in the *John Martin* case reproduced earlier in this section.

OSSINING UNION FREE SCHOOL DISTRICT v. ANDERSON LaROCCA ANDERSON

Court of Appeals of New York, 1989. 73 N.Y.2d 417, 539 N.E.2d 91, 541 N.Y.S.2d 335.
OPINION

KAYE, Judge.

At issue is a question that has long been a subject of litigation: in negligent misrepresentation cases, which produce only economic injury, is privity of contract required in order for plaintiff to state a cause of action? Whether defendants are accountants (as in several recent cases) or not (as here), our answer continues to be that such a cause of action requires that the underlying relationship between the parties be one of contract or the bond between them so close as to be the functional equivalent of contractual privity. Such a bond having been alleged in the present action against engineers, we reverse the Appellate Division order and deny defendants' motion to dismiss the complaint.

Viewing the facts presented in a light most favorable to plaintiff, as we must at this stage of the proceeding, plaintiff school district alleges that in 1984, it began a general study and structural evaluation of its buildings. To that end, it entered into a written agreement with an architectural firm, codefendant Anderson LaRocca Anderson, whereby Anderson was hired to provide an evaluation and feasibility study of plaintiff's buildings; the contract authorized Anderson's retention of consultants. Anderson retained the defendants, Thune Associates Consulting Engineers and Geiger Associates, P.C., as engineering consultants to assist in various aspects of the work it had undertaken for the school district. Although the school board authorized the retention of Thune and Geiger, neither defendant had a contract with the school district.

This litigation arises from certain reports made by defendants following tests done on school district premises in order to determine the structural soundness of the high school annex. Specifically, defendant Thune and thereafter, at the school district's request to Anderson, a second engineering firm—defendant Geiger—tested the concrete at various locations throughout the building. Both reported that there were serious weaknesses in the building, particularly the concrete slabs that formed the building's superstructure, and Anderson informed the school district of those findings.

It is alleged that defendants were aware that plaintiff would rely on their findings and that the intended purpose of defendant's reports was in fact to enable the school district to determine what measures should be taken to deal with structural problems in its buildings. For safety reasons, the school district closed the annex and, purportedly at substantial expense, obtained other facilities for the dislodged activities. The school district, however, later retained a third independent expert to check the results, and that expert advised plaintiff that the annex had been constructed with a lightweight concrete known as "Gritcrete" rather than the 2,500 pound per square inch cement defendants had assumed and reported. According to plaintiff, this information was available to defendants in the original building design drawings and specifications which had been furnished to them. Had defendants read these materials rather than acting on their mistaken assumption as to the type of concrete used, they would not have made the reports and rendered the advice that eventuated in the unnecessary and expensive closing of the annex.

The school district then began this lawsuit against Anderson and both engineering consultants. Claims of negligence and malpractice were asserted against all three. A claim for breach of contract was also asserted against Anderson, the only party with which plaintiff school district had a contract. [Court gave the procedural history and summarized the decision of the intermediate appellate court. The decisions dealt with the tort claim by the school district against the consulting engineers.]

Courts have long struggled to define the ambit of duty or limits of liability for negligence, which in theory could be endless. While much of this struggle has been couched in the rhetoric of foreseeability of harm, under some circumstances foreseeability has appeared particularly inadequate for defining the scope of potential liability. In negligent misrepresentation cases especially, what is objectively foreseeable injury may be vast and unbounded, wholly disproportionate to a defendant's undertaking or wrongdoing. . . . In reaching the policy judgment called "duty", courts have therefore invoked a concept of privity of contract as a means of fixing fair, manageable bounds of liability in such cases.

[Court gave a history of the privity rule in cases involving personal harm, starting with *Winterbottom v. Wright* and highlighting *MacPherson v. Buick Motor Co.*, 217 N.Y. 382, 111 N.E. 1050 (1916). Refer to Section 7.09(B).]

In theory, there appeared to be no reason why the privity bar should be dispensed with in cases such as *MacPherson* but retained in certain other types of negligence cases, and in *Glanzer v. Shepard*, 233 N.Y. 236, 135 N.E. 275 we said as much. The defendants in *Glanzer* were public weighers hired by the sellers of beans to provide plaintiff with a certificate stating the weight of the beans. When defendants negligently misstated the weight, the buyers sued for the overpayment they had made in reliance on defendants' inaccurate statement. Rejecting defendants' claim that *MacPherson* applied only to products posing a risk of physical danger, Judge Cardozo wrote: "We do not need to state the duty in terms of contract or of privity. Growing out of a contract, it has none the less an origin not exclusively contractual. Given the contract and the relation, the duty is imposed by law". (*Id.*, at 239, 135 N.E. 275.)

In *Glanzer*, the particular relationship that was held to warrant imposition of a legal duty of care was found in the fact that "[t]he plaintiffs' use of the certificates was not an indirect or collateral consequence of the action of the weighers. It was a consequence which, to the weighers' knowledge, was the end and aim of the transaction" and that a copy of the certificate was sent to plaintiffs "for the very purpose of inducing action." (*Id.*, at 238–239, 135 N.E. 275.) While noting that the plaintiffs might be analogized to third-party beneficiaries of the contract between the bean sellers and the defendants, the opinion stressed that the result was reached more simply by analyzing the duty owed by the defendants under the circumstances: "The defendants, acting, not casually nor as mere servants, but in the pursuit of an independent calling, weighed and certified at the order of one with the very end and aim of shaping the conduct of another. Diligence was owing, not only to him who ordered, but to him also who relied." (*Id.*, at 242, 135 N.E. 275.)

The ambit of duty was not, however, determined simply by the class of persons who relied on the negligent misrepresentations. As Chief Judge Cardozo wrote in *Ultramares Corp. v. Touche*, 255 N.Y. 170, 174 N.E. 441, what was determinative in *Glanzer* was that there was a bond between plaintiff and defendant that was "so close as to approach that of privity, if not completely one with it." (*Id.*, at 182–183, 174 N.E. 441.) In *Ultramares*, by contrast, the range of potential plaintiffs was "as indefinite and wide as the possibilities of the business that was mirrored in the summary." (*Id.*, at 174, 174 N.E. 441.) That distinction led Chief Judge Cardozo to observe: "If liability for negligence exists, a thoughtless slip or blunder, the failure to detect a theft or forgery beneath the cover of deceptive entries, may expose accountants to a liability in an indeterminate amount for an indeterminate time to an indeterminate class. The hazards of a business conducted on these terms are so extreme as to enkindle doubt whether a flaw may not exist in the implication of a duty that exposes to these consequences." (*Id.*, at 179–180, 174 N.E. 441.)

While plaintiffs' reliance might have been objectively foreseeable both in *Glanzer* and in *Ultramares*, the court chose to circumscribe defendants' liability for negligent misstatements by privity of contract or its equivalent, because of concern for the indeterminate nature of the risk. That very concern, which has echoed in the law since *Winterbottom*, was also at the root of our recent decisions. . . .

In none of these cases, however, has the court erected a citadel of privity for negligent misrepresentation suits. Thus, the rule is not, as erroneously stated by the Appellate Division, that "recovery will not be granted to a third person for pecuniary loss arising from the negligent representations of a professional with whom he or she has had no contractual relationship". (135 A.D.2d at 520, 521 N.Y.S.2d 747.) The long-standing rule is that recovery may be had for pecuniary loss arising from negligent representations where there is actual privity of contract between the parties or a relationship so close as to approach that of privity.

Nor does the rule apply only to accountants. We have never drawn that categorical distinction, and see no basis for establishing such an arbitrary limitation now. It is true that in many of the cases involving claims for negligent misrepresentation, the defendants are accountants. Indeed, in attempting to fashion a rule that does not expose accountants to crippling liability, we have noted the central role played by that profession in the world of commercial credit. But while the rule has been developed in the context of cases involving accountants, it reflects our concern for fixing an appropriate ambit of duty, and there is no reason for excepting from it defendants other than accountants who fall within the narrow circumstances we have delineated. Notably, *Glanzer* itself did not involve a suit against accountants.

The remaining question, then, is whether under *Glanzer* and our subsequent cases, defendants owed a duty of care to plaintiff that was breached by their alleged negligent performance. We have defined this duty narrowly, more narrowly than other jurisdictions (*see, e.g., Rosenblum, Inc. v. Adler*, 93 N.J. 324, 461 A.2d 138). We have declined to adopt a rule permitting recovery by any "foreseeable" plaintiff who relied on the negligently prepared report, and have rejected even a somewhat narrower rule that would permit recovery where the reliant party or class of parties was actually known or foreseen by the defendants (*Credit Alliance Corp. v. Andersen & Co.*, 65 N.Y.2d at 553, n. 11, 493 N.Y.S.2d 435, 483 N.E.2d 110). It is our belief that imposition of such broad liability is unwise as a matter of policy (*see*, Siliciano, *Negligent Accounting and the Limits of Instrumental Tort Reform*, 86 Mich.L.Rev.1929) or, at the least, a matter for legislative rather than judicial reform.

Instead, we have required something more, and we have articulated the requirement in various ways. In *Glanzer*, it was described as reliance by the plaintiff that was "the end and aim of the transaction." (233 N.Y.2d, at 238–239, 135 N.E. 275; *see also, White v. Guarente*, 43 N.Y.2d 356, 362, 401 N.Y.S.2d 474, 372 N.E.2d 315 ["one of the ends and aims of the transaction."].) In *Ultramares*, we spoke of a bond "so close as to approach that of privity". (255 N.Y. at 182–182,174 N.E. 441.) Most recently, in *Credit Alliance*, we spelled out the following criteria for liability: (1) awareness that the reports were to be used for a particular purpose or purposes; (2) reliance by a known party or parties in furtherance of that purpose; and (3) some conduct by the defendants linking them to the party or parties and evincing defendant's understanding of their reliance (*Credit Alliance Corp. v. Andersen & Co.*, 65 N.Y.2d at 551, 493 N.Y.S.2d 435, 483 N.E.2d 110).

For present purposes, the facts asserted in plaintiff's submissions satisfy these prerequisites. Plaintiff alleges that through direct contact with defendants, information transmitted by Anderson, and the nature of the work, defendants were aware—indeed, could not possibly have failed to be aware—that the substance of the reports they furnished would be transmitted to and relied upon by the school district. Plaintiff asserts that that was the very purpose of defendants' engagement.

Though under contract to Anderson, defendants allegedly undertook their work in the knowledge that it was for the school district alone, and that their findings would be reported to and relied on by the school district in ongoing project—the evaluation of the structural soundness of the school buildings. Defendants were retained to visit plaintiff's premises, examine its buildings, and prepare reports of their findings upon which action would be taken. The engagement of consultants was provided for in the contract between the school district and Anderson; the retention of defendants specifically was authorized by the school board, and they were so informed; in seeking compensation, Geiger itself wrote "we were hired by [the school district]", and it sent a bill directly to the school district.* Plaintiff further alleges that defendants had various types of contact directly with the school district. That, as well as the contents of some of Anderson's communications with defendants, constitutes conduct linking defendants to plaintiff and evidencing their understanding of plaintiff's reliance.

Not unlike the bean weighers in *Glanzer*, defendants allegedly rendered their reports with the objective of thereby shaping this plaintiff's conduct, and thus they owed a duty of diligence established in our law at least since *Glanzer* not only to Anderson who ordered but also to the school district who relied.

Accordingly, the order of the Appellate Division should be reversed, with costs, the motions to dismiss the complaint as against defendants Thune Associates Consulting Engineers and Geiger Associates, P.C. denied, and the certified question answered in the negative.

SIMONS, ALEXANDER, TITONE, HANCOCK and BELLACOSA, J. J., concur.

WACHTLER, C. J., taking no part.

Order reversed, etc.

---

*Given plaintiff's factual allegations of known reliance on the findings of both engineering consultants, Geiger's argument that it was not engaged until after the school district had decided to close the annex, and thus could not be liable for plaintiff's losses, does not at this juncture entitle it to dismissal of the complaint.

While both the *John Martin* and the *Ossining* decisions claim to follow the leading New York decisions, there is a clear distinction between them. The *John Martin* decision followed the Restatement of Torts (Second) Section 552 requirements: first, that liability be restricted to the limited number of persons for whose benefit and guidance the defendant "intends to supply the information," and second, that supplier "intends the information to influence or knows that the recipient so intends." Yet, the opinion itself appears to base protection principally upon foreseeability of use, a relatively broad standard.

In the *Ossining* decision, the court took pains to reject both a rule permitting recovery by any foreseeable plaintiff and "a somewhat narrower rule that would permit recovery where the reliant party or class of parties was actually known or foreseen by the defendants." It thought these two standards would impose overly broad liability. Instead, the court required that the information be used for a particular purpose, that the reliance be by a known party in furtherance of that purpose, and that there be some conduct by the defendants linking them to the party or parties.

The court then found this necessary bonding in the direct contact between the engineers and the school district, the undoubted awareness by the engineers that their reports would be submitted and relied upon, that they were retained to visit the premises, examine the buildings, and repair reports, and finally that the retention by the architect of the consulting engineers was authorized by the owner. One of the consulting engineers himself wrote that he was hired by the school district, had sent a bill directly to the district, and had had various other types of direct contact with the district.

Whether the differentiation noted between the *John Martin* and *Ossining* decisions will amount to much in the actual processing of such claims remains to be seen. In any event, it seems clear that design professionals who furnish information and exercise the usual functions in planning the design and monitoring its execution will have a difficult time in many jurisdictions asserting that they are immune from claims by contractors or subcontractors because of the absence of privity.

E. Economic Loss Rule: Losses Unconnected to Personal Harm or Damage to Property

As seen in Section 14.08(C), in some states the requirement of privity can bar a third party's claim for economic loss. However, the economic loss rule not only bars third-party claims but also bars the use of tort law where claimant and defendant are contractually connected, such as a claim by a client against its design professional. But for convenience, this doctrine is treated in this section dealing with third-party claims. Illustrations of economic losses relating to defective products (or buildings) are

1. Damage to the defective product itself.
2. Diminution in the value of the product.
3. Natural deterioration of the product.
4. Cost of repair or replacement of product.
5. Loss of profits caused by use of a defective product.[110]

While the economic loss rule has long been a part of English and American tort law (as noted in Section 7.09), its modern development stems from the explosion of product liability in the 1960s. During that period, the courts had to face claims by purchasers of defective products against manufacturers when the product itself did not function properly and caused economic losses to the purchaser. In the leading case of *Seely v. White Motor Co.*, the court barred recovery of lost profits suffered by a purchaser of a defective truck.[111] In rejecting this claim, Chief Justice Traynor stated that the manufacturer can protect itself from liability for physical harm by meeting the tort standard. But it should not be charged for the performance level in the purchaser's business without showing a breach of warranty. The user cannot charge the success of her venture to the manufacturer.

In an admiralty case, the U.S. Supreme Court held that a charterer of a supertanker could not recover in tort from the turbine manufacturer who had designed and manufactured the turbines installed in the charterer's vessels for damage to the turbines themselves.[112] Speaking for the Court, Justice Blackmun stated that three views had developed in the state courts. The first would not allow tort to be used in such a claim, preserving a proper role for the law of warranty if a defective product causes purely monetary harm. At the other end of the spectrum, a minority of courts would allow recovery where the product itself was in-

---

[110]See 8 Constr.Litig.Rep. 110 (1987).

[111]*Seely v. White Motor Co.*, 63 Cal.2d 9, 403 P.2d 145, 45 Cal.Rptr. 17 (1965).

[112]*East River Steamship Corp. v. Transamerical Delaval, Inc.*, 476 U.S. 858 (1986).

jured, those courts seeing no distinction between harm to person and property and purely economic loss. Those jurisdictions are not concerned about unlimited liability, because a manufacturer can predict and ensure against product failure.

Some courts sought a compromise solution, allowing recovery if the defective product creates a situation potentially dangerous to persons or property.

The Supreme Court followed the majority rule, concluding that the intermediate position was too indeterminate to allow manufacturers to structure their business behavior. When harm to the product itself occurs, the resulting loss due to repair costs, decreased value, and lost profit is the failure of the purchaser to receive the benefit of its bargain, a reason to reject the minority rule allowing such a claim to be brought in tort. Also, the minority view did not adequately keep tort and contract law in their proper compartments and failed to maintain a realistic limitation on damages.

Justice Blackmun stated that tort law was concerned with safety but not with injury to the product itself. In such a case, the commercial user

... stands to lose the value of the product, risks the displeasure of its customers who find that the product does not meet their needs, or, as in this case, experiences increased costs in performing a service. Losses like these can be insured .... Society need not presume a customer needs special protection.[113]

In these transactions, as had been suggested by Chief Justice Traynor, contract law and the law of warranty function well.

The Supreme Court was not persuaded that the limitations set forth in the minority of cases granting recovery for economic losses—that of foreseeability—would be an adequate brake on unlimited liability. The Supreme Court stated:

Permitting recovery for all foreseeable claims for purely economic loss could make a manufacturer liable for vast sums. It would be difficult for a manufacturer to take into account the expectations of persons downstream who may encounter its product.[114]

The modern use of the economic loss rule has been in product liability cases. But it has begun to play an important role in construction disputes that involve claims by third parties—usually prime and subcontractors—against other participants in the Construction Process, mainly design professionals and construction managers. When these third-party claims began to be made in the 1960s, the issue principally was whether there had to be privity between the claimant and the party against whom the claim has been made, as discussed in Section 14.08(D). Then, as seen there, the issue in these third-party claims was often phrased as one involving the right of a third party to sue for negligent misrepresentation. This subsection deals with the interjection of the economic loss rule as a defense against third-party tort claims.

The Supreme Court of Ohio found itself facing this issue in *Floor Craft Floor Covering, Inc. v. Parma Community General Hospital Association.*[115] The case involved a claim by a separate contractor (multiple prime) against the owner and architect after the contractor was forced to redo certain flooring. The claim was based upon the architect's having negligently specified flooring and sealant incompatible with the construction methods required in the project. The court in a 4–3 decision held that the economic loss rule barred the third-party tort claim.

The majority examined cases from other jurisdictions, many, if not most, allowing such claims. But the majority followed cases from Virginia that had not allowed such tort claims to be brought against an architectural firm. In holding that tort law should not be applied, the majority was influenced by Virginia cases holding that no common law duty requires an architect to protect a contractor from purely economic loss. The majority quoted from Virginia cases stating that tort law is not designed to compensate for losses suffered as a result of a breach of a duty assumed only by agreement. Such claims had to be based upon the law of contracts.

The majority was also influenced by the possibility that allowing third-party tort claims could deny the party against whom the claim has been made protection of contract provisions in its contract with the owner, such as those limiting the forum to a particular court or a no-damage-for-delay clause. The majority concluded by stating that eco-

---

[113]476 U.S. at 871–872.
[114]Id. at 874.

[115]54 Ohio St.3d 1, 560 N.E.2d 206 (1990).

nomic losses are a subject for contractual negotiation and the absence of privity bars tort claims for such losses against design professionals involved in drafting plans and specifications. Nor did the court find any nexus similar to that found in the *Ossining* case (reproduced in Section 14.08(D)) to substitute for contractual privity.

Three judges dissented, an indication of the extent of controversy over the use of the economic loss rule as a bar to tort claims. The vigorous dissenting opinion cited many cases, holding that tort claims can be made in such cases. It chastised the majority for following an unsound and moribund rule and for extending deference to a federal diversity case (one involving citizens of different states in which state law controls) that had, for all practical purposes, been found to have been an incorrect prediction of what the New York court would rule if faced with such an issue.[116] The dissent saw no reason to distinguish economic losses, since it believed that the law of torts treats all injuries by use of an economic process, transferring money from the tortfeasor to the victim. It stated that tort law is directed toward compensating individuals who suffer losses because of another's unreasonable conduct. It rejected the majority's emphasis on the need to hold parties to their contracts by stating that granting design professionals immunity for misperformance will not encourage the performance of contractual duties. It suggested that design professionals who wish to protect themselves from third-party liability can contract directly with potential claimants such as builders and others to limit malpractice liability exposure or can assert that contract provisions in contracts between builders and owners were for the benefit of design professionals.

Clearly, the economic loss rule has added an additional complication to third-party claims. Yet when the three issues are before the courts—the privity requirement, the liability for negligent misrepresentation, and the economic loss rule—in the majority of American jurisdictions, contractors and subcontractors are likely to be able to bring tort claims against design professionals.

F. Interference with Contract or Prospective Advantage—The Adviser's Privilege

Intentional interference with a contract or prospective advantage is increasingly invoked in claims against design professionals. This intentional tort was discussed in Section 11.03 in the context of retaining a design professional. In this section, reference will be made to its use when the construction contract is awarded or during its performance.

In the award stage, the design professional can play a pivotal role. She may advise the owner as to which bid should be accepted when a contract is to be awarded by competitive bidding. She may also give advice when a contract is awarded by negotiation. During performance, the design professional plays a significant role both in advising the owner and in making decisions. The immunity sometimes granted to the design professional when she makes decisions is discussed in Section 14.09(D). The design professional's role generally in the decision-making process is discussed in Chapter 29.

During performance, the design professional may be asked to advise the owner on subcontractor selection or removal. She may be asked to advise the owner during a negotiation with the contractor over compensation for changes. She may also be required to certify, as in AIA Doc. A201, ¶ 14.2.2, that there is sufficient cause to terminate the contractor. Other illustrations can be given, but these are sufficient to show the design professional in her advising role and the capacity her advice may have to harm others. But much of the complexity in claims against the design professional stems from the varied roles she plays—agent, quasi-arbitrator, and individual architect.[117]

Because of the open-ended exposure the tort can create, the law has tended to protect only against intentional interference. However, the important case of *J'Aire Corp. v. Gregory* permitted the tenant in an airport to maintain an action for negligent interference with its restaurant business against a contractor whose negligence delayed the installation of an air-conditioning system.[118]

---

[116]560 N.E.2d at 215–16. The case is *Widett v. U.S. Fid. & Guar. Co.*, 815 F.2d 885 (2d Cir.1987).

[117]*Lundgren v. Freeman*, 307 F.2d 104 (9th Cir.1962) (a leading case).
[118]24 Cal.3d 799, 598 P.2d 60, 157 Cal.Rptr. 407 (1979).

As noted in Section 11.03, the substantive basis for an intentional interference claim can range from actual malice or bad faith to simply intending to perform the act that interferes with the contract or prospective advantage.

As illustrations, a subcontractor sued the architect who advised the owner not to approve that particular subcontractor.[119] Similarly, a rejected bidder sued the architect.[120] Another case involved a claim by a contractor against an architect who did not confirm a preliminary approved product substitution by issuing a change order. When the architect ordered the contractor to remove the substituted product, the contractor refused, and on the architect's advice, the owner terminated the prime contractor.[121] Another case involved a claim by a mechanical subcontractor and its supplier against an engineer for arbitrarily requiring equipment specifications the supplier could not have met.[122] Another involved a claim by the contractor that the engineer communicated directly with the contractor's suppliers, encouraged the suppliers to threaten to terminate or actually terminate their delivery of supplies, and encouraged the owner to treat nondelivery by the suppliers as grounds to terminate the contract.[123] Finally, an investigating engineer was sued by a soils engineer when he claimed that the report given by the investigating engineer to the owner was false and that it interfered with anticipated future relations with the owner.[124]

By and large these claims have not been successful. Some defendants have been able to invoke the Restatement (Second) of Torts § 772,[125] which states that interference has not been improper if the person against whom the claim has been made has in good faith given honest advice when requested to do so. Another defense granted has been the privilege given to a person who is asked to prepare for litigation.[126]

Despite the relative lack of success of such claims, the claims still expose design professionals to liability for an intentional tort with the possibility of punitive damages.[127]

G. Safety and the Design Professional: *Krieger v. Greiner*

The responsibility of design professionals for safety has become a contentious issue. Does the design professional engaged to perform the normal site services have any responsibility for harm to persons or to property when the principal cause of the harm has been negligence by the contractor or a subcontractor? This problem has surfaced principally in two forums. The first is the judicial system, where the problem is triggered by a claim by or on behalf of an injured worker, a member of the public, or an adjacent landowner that seeks to transfer losses that it has suffered to the design professional based upon a claim of negligence. This negligence usually consists of not taking reasonable steps to prevent contractors from performing work in a

[119]*Kecko Piping Co., Inc. v. Town of Monroe,* 172 Conn. 197, 374 A.2d 179 (1977) (architect given defense of adviser's privilege).
[120]*Commercial Indus. Constr., Inc. v. Anderson,* 683 P.2d 378 (Colo.Ct.App.1984) (owner privileged to reject bid). See also *Riblet Tramway Co. Inc. v. Ericksen Assoc.,* 665 F.Supp.81 (D.N.H.1987) (engineer advised public entity that bidder was not qualified: defense of adviser's privilege granted).
[121]*Dehnert & Arrow Sprinklers, Inc.,* 705 P.2d 846 (Wyo.1985) (claim unsuccessful, as bad faith not shown). See also *Victor M. Solis Underground Utility & Paving Co., Inc. v. City of Loredo,* 751 S.W.2d 532 (Tex.Ct.App.1988) (defense of engineer's superior contract right and duty to protect city granted).
[122]*Waldinger Corp. v. CRS Group Eng'rs, Inc.,* 775 F.2d 781 (7th Cir.1985) (engineer entitled to privilege unless it could be shown that it intended to harm the plaintiffs or further engineer's personal goals).
[123]*Santucci Constr. Co. v. Baxter & Woodman, Inc.,* 151 Ill.App.3d 547, 502 N.E.2d 1134 (1986).

[124]*Western Technologies, Inc. v. Sverdrup & Parcel, Inc.,* 154 Ariz. 1, 739 P.2d 1318 (App.1986) (report absolutely privileged as made by potential witness in anticipation of litigation).
[125]*Kecko Piping Co., Inc. v. Town of Monroe,* supra note 119; *Williams v. Chittenden,* 145 Vt. 76, 484 A.2d 911 (1984) (adviser's privilege not a defense, as conduct went beyond advice).
[126]*Western Technologies v. Sverdrup & Parcel, Inc.,* supra note 124.
[127]*Custom Roofing Co., Inc. v. Alling,* 146 Ariz. 388, 706 P.2d 400 (App.1985) (punitive damages justified by wanton conduct and indifference to the rights of others). The cases are catalogued and discussed by Schneier, *Tortious Interference with Contract Claims Against Architects and Engineers,* 10 Constr. Lawyer, No. 2, May 1990, p. 3.

way that unreasonably exposes the claimant to personal harm or property damage.

The other forum in which this issue arises is workplace safety laws, such as the Occupational Health and Safety Act (OSHA), or equivalent state regulatory agencies. Agencies charged with administering such laws increasingly charge design professionals with responsibility for unsafe workplaces.

Claims for compensation by those who have suffered losses often involve the effect of contract language that seeks to exculpate the design professional from any responsibility for safety. This is similar to the exculpations discussed in Section 12.08 that seek to exculpate the design professional from responsibility for the contractor not performing properly under its contract.

Tort claims brought through the judicial process on behalf of persons who seek to transfer their losses to the design professional are examined first.

A tort claim requires that the defendant owe a duty to the claimant to act in a way that avoids exposing the claimant to unreasonable risk of harm. When the privity requirement was dropped in the late 1950s, design professionals switched to another defense. They asserted that they owed no duty to third-party claimants, because their project monitoring was directed toward a project's fulfilling the contract obligations promised to the *owner*.

(Of course, it was more complicated than that, something that is demonstrated in the cases reproduced ahead and the analysis in Section 14.11(B).)

Design professionals complained that courts were unjustifiably placing responsibility on them if anything went wrong in the Construction Process. Courts did this by focusing on the construction contract, with its many powers given the design professional, such as to reject work, stop the work, and provide general supervision. These powers, they contended, existed only to implement their "monitoring" function. The design professional, they asserted, had no *right* or *duty* to tell the contractor *how* the work was to be done. Nor was she paid, in the ordinary project, to be a "safety" engineer.

As a result of adverse court decisions, standard agreements made by the professional associations were changed in the 1960s to seek to make clear that the contractor, not the design professional, decided how the work was to be done and that the design professional did not provide continuous on-site observation and certainly did not supervise the work. These changes, designers claimed, simply reflected the true allocation of responsibility for work on the site.

The following case sets out the issue in such cases.

KRIEGER v.
J. E. GREINER CO.

Court of Appeals of Maryland, 1978. 282 Md. 50, 382 A.2d 1069.
[Ed. note: Footnotes omitted.]

SMITH, Judge.

[Ed. note: Krieger was a construction worker injured when a steel column collapsed. He claimed that the erection subcontractor did not support the 780-pound reinforcing bars and the steel column. He asserted that the work was performed under the supervision of defendant Greiner, the prime engineer, and defendant Zollman, a consulting engineer. He asserted that each knew or should have known that the subcontractor was performing in a defective and dangerous manner and that the defendants had stopped work on other occasions when the work did not conform or was performed dangerously. Krieger and his wife brought legal action against Greiner and Zollman. The court noted the split of authority, summa-

rized the leading cases, and stated that the "weight of authority is on the side of nonliability."]

2. This Case

The Kriegers in their attempt to recover here have proceeded upon three theories, (1) that under the contracts between the Commission and Greiner and Zollman these engineers are responsible for supervision of the methods of construction, and hence are responsible for safety; (2) that the engineers under their contracts with the Commission are specifically responsible for safety; and (3) that, aside from the contracts, the engineers have assumed responsibility for safety.

a. The contracts with the engineers

The principles relative to construction of contracts are well known and have been enunciated by this Court numerous times: the clear and unambiguous language of an agreement will not give way to what the parties thought the agreement meant or intended it to mean; where a contract is plain and unambiguous, there is no room for construction, and it must be presumed that the parties meant what they expressed; and when the language of a contract is clear, the true test of what is meant is not what the parties to the contract intended it to mean, but what a reasonable person in the position of the parties would have thought it meant. *Board of Trustees v. Sherman*, 280 Md. 373, 380, 373 A.2d 626 (1977); *Billmyre v. Sacred Heart Hosp.*, 273 Md. 638, 642, 331 A.2d 313 (1975), and cases there cited. It will be recalled that in the declaration the Kriegers alleged that Greiner and Zollman each "expressly agreed to see that the method of the construction work conformed to all Federal, State, and local laws and ordinances." The provision of the Greiner contract from which this is derived states:

> **"V. General Conditions**
>
> "1. General Compliance with Laws
>
>   The Consultant will observe and comply with all Federal, State, and Local Laws or Ordinances that affect those employed or engaged by him on the Project, or the materials or equipment used, or the conduct of the work, and will procure all necessary licenses, permits and insurance."

There is also an allegation in the declaration that Greiner and Zollman "expressly agreed, by their respective contracts to be responsible for all damage to life and property due to their activities or those of their agents or employees." The relevant portion of the Greiner contract on this subject appears in the next succeeding paragraph to that which we have just quoted. It states:

> "2. Responsibility for Claims and Liability
>
>   The Consultant will be responsible for all damage to life and property due to his activities or those of his agents or employees, in connection with the services required under this Agreement and will be responsible for all parts of his work, both temporary and permanent, until the services under this Agreement are declared accepted by the Commission, it being expressly understood that the Consultant will indemnify and save harmless the Commission, its members, officers, agents, and employees of, from and against all claims, suits, judgments, expense, actions, damages, and costs of every name and description, arising out of or resulting from the services of the Consultant under this Agreement."

The contract then contains a covenant against contingent fees, and goes on to provide for termination, ownership of documents, and the like. The contract with Zollman contained identical provisions under "General Conditions." To us it is obvious that this language does not attempt in any manner to place upon the engineers a duty to see that the contractors obey all laws in connection with the work. It is simply an agreement on the part of the engineers that insofar as the work *they* perform on behalf of the Commission is concerned they will comply with such laws. Likewise, this quoted language does not purport to hold the engineers responsible for life and property generally in connection with the construction of the bridge. It is an agreement on the part of the engineers that as to the work *they* perform they will be responsible for damage to life and property.

Aside from the two paragraphs of the contract which we have just quoted, obviously taken out of context in the declaration, the recitals of contract provisions in the narr. [the abbreviation of *narratio*, the Maryland name for a portion of the legal pleading in which the plaintiff states the basis of its claim] are substantially correct.

We have carefully examined each of the contracts in question. We find no provisions in these contracts imposing any duty on the engineers to supervise the *methods* of construction. Some mathematics instructors have been heard to observe that there is more than one solution to a given problem and thus they are unable to say that any given method of solving a problem is the only correct solution, being able only to determine that the correct answer is produced. The same reasoning would apply to methods of construction. One skilled contractor may prefer one method for performing a given task while another such contractor may choose what seems to him a simpler, less expensive way of reaching the same end result, either of which procedures would be a proper method. It could well be, however, that one method might not have occurred to an engineer or another contractor.

We likewise find nothing in the contracts imposing any duty on the engineers to supervise safety in connection with construction.

The duty of the engineers under their contracts is to assure a certain end result, a completed bridge which complies with the plans and specifications previously prepared by Greiner. It will be observed that many of the cases which have held architects and engineers responsible for safety have done so upon the basis of the construction by the courts of the contracts existing between the engineer or architect and the owner. We hold that a fair interpretation of the contracts between the Commission and Greiner and Zollman is that the duties of those engineers do not include supervision of construction methods or supervision of work for compliance with safety laws and regulations. Hence, the Kriegers may not recover from the engineers under the contracts between the owner and its engineers.

b. Assumed responsibilities

Our determination relative to the contractual provisions does not end the matter, however, because in their declaration the Kriegers alleged that Greiner and Zollman had "inspectors and engineers [on the job who] had exercised their right to stop the work on numerous other occasions when said work ... was being performed in a negligent and dangerous manner which was unsafe for the workmen employed on the project" and that the Commission "required [Greiner and Zollman] to supervise the work as aforesaid in order to insure the safety of ... the workmen on the job." These alleged facts, if true, might provide a

basis for recovery. The declaration, however, refers only to the two contracts with the Commission discussed above, which are treated here under Rule 326 "as if incorporated in the pleading." As we have shown, those contracts negate the allegation that the Commission required the engineers to supervise the work for safety. However, the Kriegers should have an opportunity to amend their declaration to allege—if they have a proper foundation for such an allegation—some other possible hypothesis for recovery, such as an amendment of the contract between the engineers and the Commission, or a separate, supplemental agreement between them.

JUDGMENT REVERSED AND CASE REMANDED FOR FURTHER PROCEEDINGS CONSISTENT WITH THIS OPINION; ...

LEVINE, J., concurs in the result with opinion in which ELDRIDGE, J., joins [Ed. note: Concurring opinion omitted.]

Two other cases are instructive. In *Caldwell v. Bechtel, Inc.,*[128] Caldwell, an employee of the contractor, contracted silicosis while mucking in a tunnel in part of a metropolitan subway system. His claim was against Bechtel, the consultant who had been engaged by the public entity to provide safety engineering services. Those services included overseeing the enforcement of safety provisions and relevant safety codes and inspecting job sites for violations. The worker contended that the consultant was obligated to provide overall direction and supervision of safety measures and that it was aware or should have been aware of the danger posed by high levels of silica dust and inadequate ventilation in the tunnels and failed to take the necessary steps to rectify the situation.

The court approached the duty issue from a tort law perspective and concluded that the consultant's contractual authority with regard to job safety "created a special relationship between Bechtel and Caldwell under which Bechtel owed a duty to Caldwell to take reasonable steps to protect him from the foreseeable risk to his health posed by the dust-laden Metro tunels."[129]

The court concluded that Bechtel's attempt to emphasize the contract would return the law to the days when privity of contract was required for tort

liability. The duties undertaken by Bechtel are important, according to the court, because under them, Bechtel assumed a contractual duty to the public entity but also placed itself in the position of assuming a duty to the worker. The court pointed to particular circumstances in this case, including Bechtel's contract, its superior skills, and its ability to foresee the harm that might be expected to befall the worker. The court saw these as creating a duty in Bechtel to take reasonable steps to prevent harm to the worker. The court was not willing to impose a duty on Bechtel solely because of the relationship between a professional and others on the site, as suggested by the author of this treatise, but concluded that the contract provided the initiating source of duty. Bechtel was on the site and in charge of safety engineering. This transcended its contractual purposes and under these circumstances, Bechtel owed a duty to Caldwell to protect him from the unreasonable risk of harm.

The second case that delved into the responsibility of the design professional for worker safety was *Waggoner v. W&W Steel Co.*[130] The court saw the issue as whether the architect was responsible for ensuring that the contractor employ safe methods and procedures in performing preliminary work. The death of two workers and the injury of

---

[128]631 F.2d 989 (D.C.Cir.1980).
[129]631 F.2d at 993.

[130]657 P.2d 147 (Okla.1982).

another resulted from the contractor's failure to provide interior columns and cross beams that would have provided lateral bracing for the outside columns. While the steel was being erected, a gust of wind hit the unsecured and unbraced steel, causing it to collapse. The claimants asserted that the architect owed a duty to workers, since he undertook to supervise the project.

The court held that while privity was not required before a tort claim could be brought in claims for physical injuries to third parties, the actual liability depends upon whether a duty has been created, and this depends upon the nature of the architect's undertaking and his conduct.

The court then quoted extensively from general conditions of the contract for construction, very likely those of the American Institute of Architects (AIA). The contract provided that the responsibility for executing the contract and reasonable precautions for safety belonged to the contractor and not to the architect. The fact that the architect periodically visited the site did not remove the responsibility from the shoulders of the contractor. The court also rejected the contention by the plaintiffs that the architect's approval of shop drawings that did not provide for temporary connections on the expansion joints constituted negligence and made the architect responsible. The court pointed to language in the contract stating that the contractor represented that he had checked all shop drawings for the requirements of the contract and also noted that the contract makes the contractor solely responsible for "construction means, methods, techniques, sequences and procedures."[131] But the contract specified that the architect's approval was only for conformance with the design concept and the information given in the contract documents.

Most important from the perspective of design professionals, based mainly upon the exculpatory provisions in the contract, the court found that there was no question of fact for the jury to decide. The design professional was able to receive a summary judgment and avoid the cost of a full trial and the uncertain outcome of a jury verdict.

As mentioned earlier, another forum where this question has become controversial is the Occupational Safety and Health Review Commission, which passes upon violations of OSHA. In 1977, the Commission held that an architect who does not perform construction work is exempt from OSHA job-site regulations.[132] This exemption was extended to a consulting structural engineer in 1992.[133] The Commission required that the design professional actually perform construction work or substantially supervise it.

Efforts by public regulatory agencies to place more responsibility for workplace safety upon design professionals has not been successful where the design professional performs site services in the traditional manner, such as that outlined in AIA documents. But it is likely that we will see an increasing number of attempts to place responsibility on design professionals.

The problem of workplace safety and the role of design professionals is not limited to the United States. The ministers of the European Community (EC) have enacted a directive that would place on architects and engineers in Europe the burden of controlling safety on construction sites. The directive orders EC member nations to pass legislation implementing the mandate as a directive. Opposition from the architectural associations in Europe stated that European architects are not educated to recognize unsafe conditions nor do they contract to provide safety programs. Those representing architects insist that the responsibility for health and safety on sites must remain with the contractors.[134] Whether their attempt to overturn the directive will succeed remains to be seen.

Undoubtedly, the increasing emphasis on design professional responsibility for safety in the EC and the United States results from the feeling that worker safety and unsafe site conditions are matters of great importance and that the more people

---

[131]Id. at 151.

[132]Sec. of Labor v. Skidmore, Owings & Merrill, 5 O.S.H.Cas. (BNA) 1762 (O.S.H.Rev.Comm'n.1977).

[133]Sec. of Labor v. Simpson, Gumpertz & Heger, Inc., 15 O.S.H.Cas. (BNA) 1851 (O.S.H.Rev.Comm'n. 1992). (The Department of Labor appealed this case in 1993 to the First Federal Circuit Court of Appeals, claiming the design professional was liable even if he did no work and was not on the site when the accident happened, because his acts led the contractor to believe a proposed procedure was safe. See also Sec. of Labor v. Kulka Constr. Management Corp., 15 O.S.H.Cas. (BNA) 1870 (O.S.H.Rev.Comm'n. 1992).

[134]Press Release, Architects' Counsel of Europe (Conseil des Architectes d'Europe, 19 June 1992).

who are responsible, the better. The difficulty with this position is that it would place responsibility on persons who are not trained for the job.

That the issue continues to be hotly debated is demonstrated not only by both the *Krieger* and *Caldwell* decisions' having generated scholarly comment[135] but also in the varied conclusions in cases since they were decided. But as emphasized elsewhere in this treatise, *factual* differences, such as different contract terms, different actual practices, and whether *obviously* unsafe practices were occurring, may account for different conclusions. Yet even when these *are* taken into account, judicial opinions can still reflect different emphases, such as *Krieger* on the contract language and disclaimers of responsibility and *Caldwell* on tort concepts of compensation, foreseeability, and avoiding harm to persons.

Recent cases have tended to exonerate the design professional when the contractor, not the design professional, is responsible for means, method, and sequences by which the design is accomplished or for safety.[136] This can reconcile the

apparently contradictory holdings in the *Krieger* and *Caldwell* decisions. In *Krieger*, these responsibilities were given to the contractor, while in *Caldwell*, they were shared by the contractor and the engineer, the latter not only having the power to inspect for safety violations and to stop the work but also possessing the special skills "of one engaged in the profession of safety engineering."

As a result, design professionals whose site activities are of a more passive nature, such as those outlined in AIA documents, will not be held responsible for personal harm occurring to workers. If nothing further is alleged, the design professional is entitled to a summary judgment without there being a trial on the issue of negligence. But if the claimant alleges that (1) the contract obligated the design professional to take an active role in overseeing construction safety, (2) she undertook these functions even though her contract did not require her to do so, or (3) she was aware of obviously unsafe practices, the dispute must proceed to trial. If those allegations are established, the trier of fact—jury or judge—must determine whether the design professional has been negligent.[137]

---

[135]See Goldberg, *Liability of Architects and Engineers for Construction Site Accidents in Maryland—Krieger v. J. E. Greiner Co.; Background and Unanswered Questions,* 39 Md.L.Rev. 475 (1980); Comment, U.Kan.L.Rev. 429 (1982) (*Caldwell* criticized as too broad). The most recent cases are presented in Precella, *Architect Liability: Should an Architect's Status Create a Duty to Protect Construction Workers from Job-Site Hazards?* 11 Constr.Lawyer, No. 3, August 1991, at p. 11. The author suggests the best theories for claimants are assumption of duty and knowledge of the dangerous condition. She does not approve of finding a duty simply on the basis of status.

[136]For cases relieving the design professional based on her having a duty only to the owner, see

1. *Ramey Constr. Co., Inc. v. Apache Tribe of Mescalero Reservation,* 673 F.2d 315 (10th Cir.1982) (no duty to contractor).

2. *Swartz v. Ford, Bacon and Davis Constr. Corp.,* 469 So.2d 232 (Fla.Dist.Ct.App.1985) (no duty to worker).

3. *Yow v. Hussey, Gay, Bell & Deyoung Int'l, Inc.,* 201 Ga.App. 857, 412 S.E.2d 565 (1991).

4. *Hanna v. Huer, Johns, Neel, Rivers and Webb,* 233 Kan. 206, 662 P.2d 243 (1983).

5. *Welch v. Grant Dev. Co.,* 120 Misc.2d 493, 466 N.Y.S.2d 112 (1983) followed in *Kerr v. Rochester Gas and Elec. Co.,* 113 A.D.2d 412, 496 N.Y.S.2d 880 (1985); and *Jaroszewicz v. Facilities Dev. Co.,* 115 A.D.2d 159, 495 N.Y.S.2d 498 (1985).

6. *Kemp v. Bechtel Constr. Co.,* 720 P.2d 270 (Mont.1986).

7. *Marshall v. Port Auth. of Allegheny County,* 106 Pa.Commw. 131, 525 A.2d 857 (1987) affirmed 524 Pa.1, 568 A.2d 931 (1990).

8. *Porter v. Stevens, Thompson & Runyan, Inc.,* 24 Wash.App. 624, 602 P.2d 1192 (1979).

9. *Hortman v. Becker Constr. Co.,* 92 Wis.2d 210, 284 N.W.2d 621 (1979).

[137]For cases suggesting or holding that under certain circumstances the design professional would owe a duty to the worker, see

1. *Swartz v. Ford, Bacon and Davis Constr. Co.,* supra note 136 (dicta).

2. *Phillips v. United Eng'rs and Constructors, Inc.,* 500 N.E.2d 1265 (Ind.App.1986) (CM assumed responsibility by conducting safety meetings, touring the site, and noting safety violations and unsafe practices.)

3. *Balagna v. Shawnee County,* 233 Kan. 1068, 668 P.2d 157 (1983).

4. *Duncan v. Pennington County Housing Auth.,* 283 N.W.2d 546 (S.D.1979) (architect knew of OSHA violation and was expected to deal with safety).

5. *Riggins v. Bechtel Power Corp.,* 44 Wash.App. 244, 722 P.2d 819 (1986), petition for review denied, 107 Wash.2d 1003 (1986) (CM who had been hired as safety consultant to develop the site safety program owed duty to worker: jury issue as to the extent of the duty). For a discussion of CM liability for site safety, see Comment, 32 Loy.L.Rev. 447 (1986). In this Comment, the author is concerned that CM liability for site safety has gone beyond assumed undertakings and is based solely upon the relationship.

H. Action Taken on Site

Pastorelli v. Associated Engineers, Inc.[138] involved an injury to an employee of a racetrack caused by a falling heating duct. The accident occurred after the work had been completed and accepted by the injured employee's employer.[139] The injured employee sued the prime contractor, the sheet metal subcontractor who installed the duct, and engineers who prepared the plans and who agreed to "supervise the contractor's work throughout the job."

The duct was 20 feet long and weighed 500 pounds. It had not been attached directly to the roof itself or to the joists of the clubhouse but had been suspended from the ceiling of the clubhouse by the attachment of semi-rigid strips of metal called hangers, which were then attached to the ceiling. The ceiling was of sheathing and was nailed to the joists, leaving a considerable airspace between sheathing and roof. The specification required that sheet metal work be erected "in a first class and workmanlike manner" and that "the ducts be securely supported from the building construction in an approved manner."[140]

The trial judge concluded that the duct had not been properly installed. The engineer prepared and submitted periodic inspection reports while the work was in progress. The trial judge stated that the engineer's employee who prepared the reports

> . . . testified that his employer assigned to him the task of supervising the installation of said systems, and that in pursuance of his duties he visited the job site on one, two or three occasions each week to inspect the work of the contractor as it was being done. He also

testified, however, that he never observed any of the ducts being hung from the ceiling in said clubhouse, stating that whenever he visited the clubhouse the ducts were either on the floor or already installed. He also admitted that he never climbed a ladder to determine whether the hangers by which they were suspended were attached by nails or lag screws and never tested any of the hangers to see how securely they were attached.[141]

After holding prime and subcontractor negligent, the judge noted that the engineer's employee knew that the safety of persons in the clubhouse required that the ducts be attached to the joists and that he made no attempt to ascertain whether they were so installed. The judge also noted that the employee made no visits at a time when he *could* determine how they were being installed. Holding the engineer negligent, the judge stated:

> In other words, he failed to see that they were properly installed and took no steps after their installation to ascertain how and by what means they were secured. In my opinion he failed to use due care in carrying out his undertaking of general supervision.[142]

Suppose the engineer's employee had determined that the ducts were not properly secured. What should he have done? Should he have directed the employees of the sheet metal subcontractor to correct the work? Should he have gone to the superintendent of the prime contractor with a similar request? Should he have ordered the work terminated until proper corrective measures were taken? Should he have gone to a building inspector with a suggestion that the latter order that the work be corrected? These questions show the difficult position in which the design professional can find herself when she does determine that work is not being properly performed, especially when people could be injured or killed as a result.

If the design professional simply observes, one would think that her principal responsibility would be to call the attention of defective work to the contractor and have the contractor transmit this to the subcontractor in a case such as *Pastorelli*. Yet, if it is likely that correction will not come quickly enough and if it is likely that persons will be in-

---

Yet care should be taken to distinguish claims made by injured workers under the Illinois Structural Work Act, discussed in Section 31.04(F). See *Coyne v. Robert H. Anderson & Assocs., Inc.*, 215 Ill.App.3d 104, 574 N.E.2d 863 (1991) (architect with duty to monitor for compliance with safety regulations was "in charge"); *Sasser v. Alfred Benesch & Co.*, 216 Ill.App.3d 445, 576 N.E.2d 303 (1991) (duty to inspect, administer, and stop certain work raised factual issue of whether architect was "in charge."
[138]176 F.Supp. 159 (D.R.I.1959).
[139]Earlier American law immunized architect and contractor from most accidents that occurred after the project was completed and accepted by the owner. In *Pastorelli v. Associated Eng's, Inc.*, supra note 138, the court rejected this rule and held that completion and acceptance did not furnish a defense to the architect. This is discussed in Sections 14.09(B) and 31.07.
[140]176 F.Supp. at 162.

[141]Id. at 162–63.
[142]Id. at 167.

jured, registering a complaint indirectly would not be sufficient. The owner should be given the information, along with advice as to what should be done. The owner can then determine the course to be followed. If the danger was imminent and the situation urgent, perhaps the building inspector should be called, or if the design professional has the power, she should stop the work.[143]

A second injury case, *Cutlip v. Lucky Stores, Inc.*,[144] involved an action by the wife of a subcontractor's employee who had been killed during construction. Although the case had many legal issues, one part related to a site visit by the architect. The superintendent of the prime contractor called the architect and told him that either certain steel beams were not the proper length or the concrete piers on which the beams were to be erected were not in the correct location. Two solutions were proposed. The first was to send the steel back to be refabricated and the second to pour a new concrete base and pillar. The testimony was contradictory, but evidently the superintendent, claiming he had been authorized by the architect, poured the new base to avoid delay. Two days later, the worker was killed when the steel superstructure collapsed.

Although the architect retained a structural engineer to make structural drawings as well as to be available for consultation, the architect reserved to *himself* supervising the structural work of the building. An architect who testified as an expert for the plaintiff stated that in his opinion, architects such as the architect in the case were not qualified to perform structural engineering inspections. That expert witness also expressed the opinion that when the problem developed, the architect should have contacted the structural engineer and authorized him to redesign. In addition to noting this, the court pointed to the fact that the architect could have stopped the work at any time. The appellate court concluded that there had been sufficient evidence to submit the matter of negligence to the jury. (The jury had ruled against the architect.) The

court seemed persuaded that it was negligent to agree to perform services for which the architect was not qualified. Note the observation by the expert witness that whether it was proper or not for the architect to have agreed to inspect structural work of this sort, at the very least, he should have contacted the consulting structural engineer before making any decision relating to structure.

I. Safety Legislation

Often legislation is enacted after a tragic construction site accident. The L'Ambiance Plaza in Hartford, Connecticut collapsed during construction in 1987. Twenty-eight workers died and sixteen were injured. As a result, Connecticut enacted § 29-276b of its general statutes. After setting forth criteria for significant structures, it requires that for those structures local building officials hire an independent engineering consultant to review the plans and specifications for code compliance. The prime contractor and major subcontractors are required to keep a daily log and keep it available to both local building officials and engineers or architects involved with the building. All design professionals and general contractors must give signed approval that completed construction substantially complies with the plans. If a building has more than three stories, § 29-276c requires that the design professional review the implementation of her design and observe the construction.

In 1988, partly as a result of the L'Ambiance Plaza collapse, legislation was introduced into the U.S. Congress that would have amended the Occupational Safety and Health Act (OSHA) to provide stricter controls over construction and certify substantial compliance with the design before a certificate of occupancy can be issued. One bill introduced in 1988[145] would have required the owner to retain a state registered engineer to provide oversight to ensure worker safety and would have required safety plans and detailed reporting requirements. It was opposed by the design professionals because of (1) liability implications, (2) lack of training of engineers in safety, and (3) noncoverage of insurance. The bill died in committee. In 1989, a new bill was introduced.[146] It would have

---

[143]In 1970, the AIA General Conditions of the Contract A201 eliminated the power of the architect to stop the work and reposed this power solely in the owner. The clause doing this, currently ¶ 2.3.1, reflects the tendency in AIA documents to give the architect a more passive role in the hope of eliminating or reducing liability. Refer to Section 12.08(B).

[144]22 Md.App. 673, 325 A.2d 432 (1974).

[145]S.2518, 100th Cong., 2d Sess. (1988).

[146]S.930, H.R.2254, 101st Cong., 1st Sess. (1989).

deleted the requirement that the owner retain an engineer for general oversight but would have required the contractor to employ a certified safety specialist (CSS) for the safety function with authority to stop work where there is imminent danger to workers. The CSS would need to take a training course or receive on-the-job training and would be certified by a federal agency. Also, the contractor would have to obtain the advice of a qualified design professional if the contractor determined that a particular phase of the work involved special safety hazards.

These bills have not passed, the opposition typically coming from professional and trade associations representing active participants in the process. It seems likely that the future will see more legislation regulating workplace safety with increased design professional involvement.[147]

J. Summary

Increasingly, third parties, including public safety officials, have been asserting claims against design professionals. Claims based on the assertion that the claimant is an intended beneficiary of the contract between the design professional and the owner, though occasionally successful, have not proved as successful as tort claims. But while third-party claims in tort have met obstacles (as noted in this section), the trend except in site accident claims is clearly toward increased liability exposure for design professionals. Finally, safety legislation is becoming more common, and this poses new legal problems for design professionals.

SECTION 14.09 Special Legal Defenses

The principal issue in claims against design professionals relates to the standard of performance and whether there has been compliance. This section briefly outlines a number of defenses that have been used by the design professional when claims are made by the client or third parties.

A. Approval by Client

Suppose the design professional asserts that the design has been approved by the client and this re-lieves her of any liability, even for negligent design. Ordinarily, such a defense is not successful.[148]

The client retains a design professional because of the latter's skill in design, a skill not *usually* possessed by the client. The very purpose of engaging an expert would be defeated if the approval by the nonexpert client relieved the expert design professional from her negligence. Approval *does* authorize the design professional to proceed with the next phase of services, and client-directed changes *after* approval may justify the design professional who is compensated by a method unrelated to costs receiving additional compensation.

There may be an unusual circumstance that could justify approval relieving the design professional. Suppose the design professional points out a design dilemma to the client. The designer may inform the client that particular material may prove unsatisfactory for specified reasons but asks the client's approval of that material because it is less costly. If the client is apprised of all the risks and authorizes that particular material to be used, the client assumes the risk. Even if the design professional has not done what others would have done, approval by the client with *full* knowledge of the risks and with the ability to evaluate them should relieve the design professional.[149]

Claims by third parties raise other problems. Even if approval by the client would bar its claim, this would not bar a third-party claim against the design professional. If approval by the client were *itself* negligent, any third party who suffered harm could recover against the client as well and may give the design professional an indemnification claim against the client.

As a general rule, client approval of the design does not relieve the design professional from liability for negligent design.

B. Acceptance of the Project

Acceptance of the project usually involves the owner's taking possession of the completed project. It

---

[147]Refer to Section 12.08(F).

[148]*Eichler Homes, Inc. v. County of Marin,* 208 Cal.App.2d 653, 25 Cal.Rptr. 394 (1962); *Simpson Bros. Corp. v. Merrimac Chemical Co.,* 248 Mass. 346, 142 N.E. 922 (1924); *Bloomsburg Mills, Inc. v. Sordoni Constr. Co.,* supra note 90.
[149]*Bowman v. Coursey,* supra note 57, discussed in Section 14.05(C).

can—although under standard contracts it usually does not[150]—imply that the owner is satisfied with the work and bar any claim for existing defects or defects that may be discovered in the future.

For purposes of this chapter, acceptance can affect any claim third parties have against those who have participated in design and construction. Although some early cases[151]—and occasionally recent ones[152]—have barred some such claims after acceptance, the modern tendency is to hold that acceptance does not bar third-party claims against design professionals.[153] If acceptance bars claims, it is because the intervening act of the owner—that of acceptance—is a *superseding cause* relieving even negligent participants of liability. This rationale is particularly weak when the claim is asserted against the design professional—the person who often decides whether the project has been completed and should be accepted. Even if the owner decides to accept, acceptance may not be negligent. Although acceptance by the owner can be looked on as an intervening cause, it is in most cases a nonnegligent one that should not immunize a design professional from her negligence.

Suppose, though, that acceptance by the owner precluded the design professional from correcting design errors. Should the design professional be immunized from third-party claims? If the owner *knew* that there were defects and *barred* their correction, the owner's intentional acts operate as an intervening cause to immunize the design professional from a third-party claim. The owner's acceptance should give the design professional a claim for contribution or indemnity against the owner if the latter's acceptance precluded correction of defects.

Immunization because of the acts of others—even negligent acts—is not favored today. Acceptance rarely bars third-party claims against a design professional. As to accident claims by third parties, see Section 31.07.

C. Passage of Time: Statutes of Limitations

Sometimes the design professional can defend by establishing that legal action was not started within the time required by law. This is usually accomplished by invoking the statute of limitations as a bar to the claim. As this bar can apply to all claims—not only to those against design professionals—and since most construction-related claims are brought against other participants such as the owner or contractors, full discussion of this defense is postponed until Section 23.03(G). A few observations should be made in this section dealing with claims against design professionals.

Statutes of limitations usually prescribe a designated period of time within which certain claims must be brought. One of the troublesome areas in construction claims relates to the point at which that period *begins*. This can be a formidable problem in all construction-related claims, inasmuch as the defect may be discovered long after the design is created and executed. As a result, even in claims that do not involve the design professional, the law has had difficulty selecting from a number of base points, such as the wrongful act, the occurrence of damage, discovery of a substantial defect, or discovery of any defect.

This troublesome question is exacerbated when claims are made against the design professional. The latter usually develops the design, obtains approval by the client, gives it to the contractor to execute, and often issues a certificate that there has been substantial or final completion. In addition, the changes process often means that the design is changed during construction. All of this means that there are additional base points that can be used to determine when the period commences.

Essentially, design is a trial-and-error process. Although some states hold that the period commences when the negligence occurs, the trial-and-error aspects of design could, along with the continuation of the professional relationship between client and design professional, lead to a

---

[150]AIA Doc. A201 ¶ 4.3.5 (final payment does not bar most owner claims).

[151]*Sherman v. Miller Constr. Co.*, 90 Ind.App. 462, 158 N.E. 255 (1927).

[152]*Phifer v. T. L. James & Co., Inc.*, 513 F.2d 323 (5th Cir.1975); *Easterday v. Masiello*, 518 So.2d 260 (Fla.1988) (defense as to patent defects); *Shetter v. Davis Bros., Inc.*, 163 Ga.App. 230, 293 S.E.2d 397 (1982) (many exceptions); *Fauerso v. Maronick Constr. Co.*, 661 P.2d 20 (Mont.1983).

[153]*Pastorelli v. Associated Engr's, Inc.*, 176 F.Supp. 159 (D.R.I.1959); *Green Springs, Inc. v. Calvera*, 239 So.2d 264 (Fla.1970); *Theis v. Heuer*, 264 Ind. 1, 280 N.E.2d 300 (1972); *McDonough v. Whalen*, 365 Mass. 506, 313 N.E.2d 435 (1974); *Totten v. Gruzen*, 52 N.J. 202, 245 A.2d 1 (1968); *Strakos v. Gehring*, 360 S.W.2d 787 (Tex.1962); *Andrews v. Del Guzzi*, 56 Wash.2d 381, 353 P.2d 422 (1960).

determination that the period commences on completion of the project.[154]

New York refused to apply this rule in a claim by a third party who brought a tort action based on professional negligence against a designer, with the court holding that the period began when the wrongful act was discovered (usually the time of injury).[155] The court held that the completion base point is proper in dealing with claims between client and design professional but not those brought by third parties.

In addition to the complexity resulting from a differentiation between claimants drawn in New York, additional confusion can result because the period for beginning action is usually longer for claims based on breach of contract than for those based on other wrongful conduct, such as negligence. This becomes a serious issue because claimants against design professionals are often given the option of bringing actions in tort or in contract.[156]

Generally, a defense based on the passage of time—though still of value to the design professional—has provided limited protection. This has led to the enactment of completion statutes that seek to cut off liability after a designated period of time following substantial completion of the project. (The statutes are discussed in Section 23.03(G).)

The dubious utility and confusion engendered by application of the statutes determining when an action must be commenced have led to contractual attempts to create a private statute of limitations. If reasonable, these provisions can regulate the relationship between client and design professional. Contractual provisions dealing with this problem can, as do AIA Documents B141, ¶ 9.3, and A201, ¶ 13.7, seek to control the commencement of the statutory period. In addition, some contract clauses stipulate the period during which the claim must be asserted as well as the commencement of the period. If the clause is reasonable, it will be enforced.[157]

D. Decisions and Immunity

The design professional is frequently given the power to interpret the contract documents, resolve disputes, and monitor performance. When claims are brought against a design professional by the client or others for activity that can be said to resemble judicial dispute resolution (such as deciding disputes or issuing certificates),[158] design professionals sometimes assert that they should receive quasi-judicial immunity.

This is based on the analogy sometimes drawn between judges and the design professionals performing judgelike functions. For example, the judge is given absolute immunity from civil action, even for fraudulent or corrupt decisions. The corrupt judge may be removed from office or subject to criminal sanctions. But a disappointed party cannot institute civil action against a judge. Immunity protects judges from being harassed by vexatious litigants and encourages them to decide cases without fear of civil action being brought against them.

Quasi-judicial immunity for design professionals has had a troubled history both in England and in the United States. The English House of Lords reversed an earlier decision and held that the architect can be sued by the owner for a negligently issued certificate.[159]

American decisions have not been consistent. Some have granted quasi-judicial immunity, while others have not.[160] Where immunity is given, the design professional cannot be sued for decisions made unless they were made corruptly, dishonestly, or fraudulently. However, immunity

---

[154]*Sosnow v. Paul*, 43 A.D.2d 978, 352 N.Y.S.2d 502 (1974), affirmed, 36 N.Y.2d 780, 330 N.E.2d 643, 369 N.Y.S.2d 693 (1975).
[155]*Cubito v. Kreisberg*, 69 A.D.2d 738, 419 N.Y.S.2d 578 (1979), affirmed, 51 N.Y.2d 900, 415 N.E.2d 979, 434 N.Y.S.2d 991 (1980).
[156]Refer to Section 14.05(E).

[157]*Therma-Coustics Mfg., Inc. v. Borden, Inc.*, 167 Cal.App.3d 282, 213 Cal.Rptr. 611 (1985). See also U.C.C. § 2-725(1), which allows the parties to reduce the four-year period to "not less than one year." The parties cannot extend the period.
[158]See Section 22.07.
[159]*Sutcliffe v. Thackrah*, 1974 A.C. 727.
[160]Authorities are collected in *City of Durham v. Reidsville Eng'g Co.*, 255 N.C. 98, 120 S.E.2d 564 (1961). See also *Blecick v. School Dist.*, No. 18, 2 Ariz.App. 115, 406 P.2d 750 (1965); Annot., 43 A.L.R.2d 1227 (1955).

does not protect against negligent delay in making a decision.[161]

Again, the peculiar position of the design professional—independent contractor as designer, agent of the owner, decider of disputes, and an individual participating in construction projects—has caused difficulty. If immunity is granted, it should be based on both parties to the contract—owner and contractor—agreeing to give certain judging functions to the design professional. Yet the cases that have been quickest to deny immunity have been actions instituted against a design professional by the client—the party who has selected and paid the design professional.[162] Courts that have denied immunity in such cases seem more impressed with the client's selection of and payment to the design professional as indicating the design professional's principal responsibility being to protect the owner. This, in addition to the increasing tendency to hold professional persons accountable, may not mean that immunity will never be available when the claim is made by the client but reduces the likelihood that such a defense will be successful.

One would think that claims by the contractor and certainly third parties should be at least as successful as those of the client. Third parties have not selected the design professional and have been forced in most cases to accept the design professional as judge. Yet a recent case granting immunity was one involving a claim brought by a contractor against the architect.[163]

There has been criticism of immunity even where limited to good-faith decisions. Can the design professional truly be neutral in rendering a decision when she has been selected and paid by the client? Also, as indicated earlier, the contractor rarely has much choice in these matters. The English House of Lords in *Sutcliffe v. Thackrah*[164] dealt with a claim by the client that the architect had negligently overcertified. Lord Reid drew a distinction between a dispute resolver who is a judge or arbitrator and an architect. According to Lord Reid, a true dispute resolver is a passive recipient of information and arguments submitted by the parties. An architect, on the other hand, is a professional engaged to act at her client's instructions and to give her own opinions. The judge or true arbitrator does not investigate but simply decides matters submitted to her. Lord Reid concluded that deciding whether work is defective is not judicial. He noted here there had been no dispute, that the architect was not jointly engaged by the parties, that the parties did not submit evidence to the architect, and that the architect made his own investigation and came to his own decisions.

The reasoning by Lord Reid may not be applicable in the United States because of different, though perhaps marginal, American practices. First, in the United States, the architect is "approved" by both parties although *engaged* by the owner. Although the owner retains the design professional, the contractor knows who the design professional will be when it enters its bid or negotiates to perform the work. Under AIA Doc. A201, the contractor has a limited power of veto over any successor architect.[165] Also, A201 *seems* to contemplate the parties submitting evidence to the architect.[166]

The desirability of immunity was also passed on by an American court in *E. C. Ernst, Inc. v. Manhattan Construction Co. of Texas*.[167] The case involved a claim against the architect based upon his rejection of certain equipment proposed by the contractor. The court stated:

> The arbitrator's "quasi-judicial" immunity arises from his resemblance to a judge. [The court here is speaking of the architect as arbitrator.] The scope of his immunity should be no broader than this resemblance. The arbitrator serves as a private vehicle for the ordering of economic relationships. He is a creature of contract, paid by the parties to perform a duty, and his decision binds the parties because they make a specific, private decision to be bound. His decision is not socially momentous except to those who pay him to

---

[161]*E. C. Ernst, Inc. v. Manhattan Constr. Co. of Texas*, 551 F.2d 1026 (5th Cir.1977) rehearing denied in part and granted in part, 559 F.2d 268 (5th Cir.1977). cert. denied sub nom *Providence Hosp. v. Manhattan Const. Co. of Texas*, 434 U.S. 1067 (1975).

[162]*Newton Inv. Co. v. Barnard & Burk, Inc.*, 220 So.2d 822 (Miss.1969).

[163]*E. C. Ernst, Inc. v. Manhattan Constr. Co. of Texas*, supra note 161.

[164]Supra note 159.

[165]AIA Doc. A201, ¶ 4.1.3.

[166]Id. at ¶¶ 4.3.2, 4.4.4.

[167]Supra note 161.

decide. The judge, however, is an official governmental instrumentality for resolving societal disputes. The parties submit their disputes to him through the structure of the judicial system, at mostly public expense. His decisions may be glossed with public policy considerations and fraught with the consequences of stare decisis [a Latin phrase for precedent]. When in discharging his function the arbitrator resembles a judge, we protect the integrity of his decisionmaking by guarding his fear of being mulcted in damages. . . . But he should be immune from liability only to the extent that his action is functionally judge-like. Otherwise we become mesmerized by words.[168]

The court then concluded that such immunity as possessed by the architect as arbitrator did not extend to unexcused delay or *failure* to decide, with immunity limited to "judging."

Determining whether to grant immunity must also take into account the finality of the design professional's decision (taken up in greater detail in Section 29.09). Under many contracts, the initial decision by the design professional can be taken to arbitration. If so, any wrong or even negligent decision can be corrected by the arbitrators. However, if arbitration is not used and the dispute goes to court, the decision by the design professional is likely to have a certain degree of finality. Giving the decision substantial finality and the decision maker immunity may be granting too much power to the design professional—power that can be abused. In Section 29.10, the suggestion is made that the process under which the design professional interprets the contract and decides disputes can be justified principally by expediency—the need to move construction along. This justification will not be adversely affected even if the decision were given very little finality and even if the design professional is stripped of any immunity.

It is often difficult to determine whether the design professional is acting as *agent* or *judge*. Perhaps it is simply better to jettison immunity as the English have done. Such immunity as exists protects only against the negligent decision and not against one made in bad faith. Also, as shall be seen in Section 15.03(E), one way of dealing with this problem is to provide immunity by contract.

E. Recent Legislative Activities

Legislatures have been more receptive to pleas by associations representing design professionals that expansion of liability has had an undesirable effect not only on the professions but also upon the public. Much of the legislative effort came in the backwash of similar legislative protection given health care providers in the 1970s and 1980s. Some, such as the requirement that the claimant must post bond if a malpractice claim is made against a design professional under certain circumstances,[169] may not extend much protection.

A number of states have enacted legislation that may be of some value in protecting design professionals from claims. For example, in 1979, California enacted Section 411.35 of its Code of Civil Procedure. Section 411.35 requires the attorney for a claimant who files a claim for damages or indemnity arising out of the professional negligence of a licensed architect, a licensed engineer, or a licensed land surveyor to file a certificate stating that the attorney has reviewed the facts of the case, that she has consulted certain designated licensed design professionals who she reasonably believes are knowledgeable in the relevant issues, and that she has concluded "that there is reasonable and meritorious cause for the filing of such action." Provisions seek to protect the identity of the design professional who has been consulted. This particular statute expires on January 1, 1997.

Georgia has gone farther, enacting legislation requiring that a complaint charging professional malpractice be accompanied by an affidavit of an expert setting forth at least one negligent act or omission.[170]

A significant number of states have enacted "Good Samaritan" laws to protect design professionals from potential liability when they provide expert services in emergencies. The statutes vary as to the type of emergency that will invoke the pro-

---

[168]Id. at 1033.

[169]West Cal.Ann.Civ.Proc.Code § 1029.5.

[170]The Georgia statute was interpreted in *Kneip v. Southern Eng'g Co.*, 260 Ga. 409, 395 S.E.2d 809 (1990). See also *Housing Authority of Savannah v. Greene*, 259 Ga. 435, 383 S.E.2d 867 (1989) (statute not limited to medical malpractice actions although originally part of Medical Malpractice Act). See also *Jordan, Jones & Goulding, Inc. v. Wilson*, 197 Ga.App. 354, 398 S.E.2d 385 (1990) (affidavit required by nonclient against engineer).

tective legislation and who can benefit from such legislation. As an illustration, under California law, an architect or engineer who, at the request of a public official but without compensation or expectation of compensation, voluntarily provides structural inspection services at the scene of a declared emergency "caused by a major earthquake" is not liable for any harm to person or property caused by her good faith but negligent inspection of the structure. Protection does not extend to gross negligence or willful misconduct. The inspection must occur within thirty days of the earthquake.[171] Some states have extended this protection to those who provide inspection or other services following any natural disaster.[172]

Most significantly, beginning in the mid-1980s, associations of design professionals were successful in persuading a number of states to grant immunity to design professionals from third-party claims by injured workers who are covered by workers' compensation for site services unless the design professional has contracted to oversee safety or undertook to do so.[173] The injured worker has rights under workers' compensation laws. Limitation of third-party actions—that is, tort claims that can be made by the injured worker against those who are not immunized by workers' compensation laws—may be of considerable importance in reducing liability.

SECTION 14.10 Remedies

A. Against Design Professionals

The remedies available when the design professional does not perform in accordance with the contract or tort obligations were set forth in Sections 7.10 and 12.14.

B. Against Co-Wrongdoers

In construction, losses are commonly caused by a number of participants and conditions. As to defective work, see Section 24.06, and as to construction accidents, see Chapters 31 and 32. However, a few generalizations can be made in this section.

First, suppose the client asserts a claim against the design professional, but the loss has been caused in part by the negligence of the client. If the client is allowed to pursue a tort claim, generally the negligence of the client and that of the design professional will be compared, and the amount recoverable by the client will be reduced by the amount attributable to its own negligence, as noted in Section 7.03(G). For example, if the loss was $100,000 and a determination was made that 30% of the loss should be chargeable to the client, the client would recover $70,000. A few states apply the contributory negligence rule to bar the entire claim.

Suppose the claim by the client is based on breach of contract. Some states would still make an apportionment, while other states would not.[174] Those that do not apportion place the loss on the party whose breach substantially caused the loss.

Apportionment becomes more complicated when responsibility falls on a number of defendants and, in some instances, on the claimant itself. If a particular loss can be connected principally to a particular defendant, that defendant will pay the loss that it has caused. However, if the loss is indivisible and cannot be apportioned among the wrongdoers, it is likely that a tort claimant will be able to recover the entire loss from any of the defendants, with the ultimate responsibility determined by contribution laws and indemnification.[175]

---

[171]West Cal.Ann.Bus. & Prof.Code §§ 5536.27, 6706.

[172]Tenn.Code Ann. § 62-2-109; Vernon's Mo.Ann.Stat. § 44.023. The latter also provides that volunteers can receive their incidental expenses for up to three days' performance of voluntary services.

[173]Conn.Gen.Stat.Ann. § 31-293(b); West Fla.Stat.Ann. § 440.09(6); Kan.Stat.Ann. § 44-501(f); West Okla.Stat.Ann. tit. 85, § 12; West Wash.Rev.Code Ann. § 51.24.035 (immunity not granted if responsibility is specifically assumed by contract terms that were mutually negotiated or the design professional actually exercised control over the portion of the premises where the worker was injured). None of these statutes grant immunity for negligent design.

[174]For cases allowing an apportionment, see *S. J. Groves and Sons Co. v. Warner Co.*, 576 F.2d 524 (3rd Cir.1978); *Grow Constr. Co. v. State*, 56 A.D.2d 95, 391 N.Y.S.2d 726 (1977); and *Environmental Growth Chambers, Inc.*, ASBCA 25845, 83-2 BCA ¶ 16,609. See also *Shepard v. City of Palatka*, 414 So.2d 1077 (Fla.Dist.Ct.App.1981) (dictum). For cases refusing to employ apportionment, see *Broce-O'Dell Concrete Products, Inc. v. Mel Jarvis Constr. Co.*, 6 Kan.App.2d 757, 634 P.2d 1142 (1981); *Hunt v. Ellisor & Tanner, Inc.*, 739 S.W.2d 933 (Tex.Ct.App.1987).

[175]*Northern Petrochemical Co. v. Thorsen & Thorshov, Inc.*, 297 Minn. 118, 211 N.W.2d 159 (1973). But see *Shepard v. City of Palatka*, 414 So.2d 1077 (Fla.Dist.Ct.App.1981) (dictum loss would be apportioned between architect and client).

SECTION 14.11 Current Controversies: Some Observations

It would be impossible to comment at length on the issues related to professional liability addressed in this chapter. But this section highlights issues that have generated heated controversies.

A. The Professional Standard: Should Professionals Be Treated Differently?

The professional standard in essence permits local professional practice to be the legal standard. It has been subjected to intense criticism. Manufacturers of products and those who build and sell homes[176] are held to strict liability. Owners warrant to the contractor that the design they have supplied will be sufficient.[177] Should professionals be given special dispensation when they have caused harm?

This specialized treatment may have had some justification, according to its attackers, when professionals were practicing in one-person offices or in small firms and in times when the professionals largely learned from those practicing in their localities. Today, professionals such as doctors, lawyers, and many of the design professionals practice in large organizations, make large profits, compete nationally, and receive education and training that use state or national, not local, practices. Those who attack the professional standard point to this as a reason why professionals today do not deserve special treatment. They argue that design professionals must accomplish the objective for which they are retained. Any increased liability that *may* result can be handled by insurance and the cost spread to those who use the professional services. Even if insurers generally exclude contractual risks that go beyond negligence, those who insure are part of a competitive industry that will respond to market pressures. In any event, critics argue that insurance practices should no more determine the appropriate legal standard than professional practices. Attackers also stress the following:

1. Antitrust laws are increasingly applied to the professions. The law views their members as businesspersons rather than professionals.

2. The relationship between design professional and client has taken on a more commercial character. This should permit design professionals to reduce what they perceive to be open-ended liability by allowing them to contract for a standard lower than the professional standard, by limiting their liability, and by exculpating themselves from certain liability. See Section 15.03(D).

3. The current standard greatly increases the cost of litigation by requiring a parade of well-paid experts.

4. The line between design, to which the professional standard applies, and site monitoring, where it may not, is so blurred that it adds an added dimension of uncertainty.

5. To obtain a commission, the design professional often stresses it is *better* than others with whom it competes yet asserts it should be measured by the professional standard that looks to *average* local practices.

6. The professional standard is often unrefined, with insufficient attention devoted to clear analysis that takes into account the differences among the cutting edge of the profession, the material available in the literature, what becomes accepted in the profession as part of the knowledge possessed by an informed professional, and what is taught at the undergraduate[178] level.

Indictment of the professional standard has had an effect on the law, such as the loosened requirements for expert witnesses[179] and the beginning of a tendency to hold that some design professionals are engaged in ultrahazardous activities for which they are strictly liable.[180]

Counter arguments must exist that have persuaded courts to hold to the professional standard in the face of more strict standards being applied to others. Even modern cases in an atmosphere in

---

[176]Refer to Section 7.09 and See Section 24.10.
[177]See Section 23.05(E).

[178]Peck & Hoch, *Liability of Engineers for Structural Design Errors: State of the Art Considerations in Defining the Standard of Care*, 30 Vill.L.Rev. 403 (1985). Prof. Jones, whose paper is cited at note 93 supra, argues for a result-oriented liability if the probability of success is high. He contends that most services performed by a design professional fall into this category. Most design professionals would dispute this.
[179]Refer to Section 14.06.
[180]Refer to Section 7.04(A).

which compensation is emphasized have still applied the professional standard. Why? Perhaps most important is the belief that professionals operate amidst great uncertainty. Will the professional be judged harshly in the event of an unsuccessful outcome when even the *best* professional services would not have been able to create a satisfactory outcome? Does any *outcome* standard run the risk of holding the professional to professional performance that was more appropriate at the time of trial than at the time the professional services were performed?

Defenders contend that the client expects good *professional* service rather than insurance. Those who defend the professional standard fear that any warranty standard will expose design professionals to *unreasonable* expectations of the client that will be resolved in favor of the client by a jury, particularly if the client is unsophisticated in the world of design and construction. Also, what will be the *exact* nature of the warranty if an outcome standard is used? Might it not extend far beyond the function of design services such as seen in *Allied Properties v. John A. Blume & Associates.*[181]

Defenders of the professional standard argue that an outcome standard will generate overdesign at unneeded costs to reduce or avoid the risk of liability. Increased liability also means higher cost of service, resulting in fewer practitioners, higher prices, and more uninsured professionals.

Those who defend the design professions state that the high compensation that *some* professionals receive is rare in the design professions. Only "strong" design professionals will be able to contract out of any outcome standard. The ordinary professional designer lacks the bargaining power to obtain a more limited standard or believes that it is inappropriate to begin a professional relationship by demanding or requesting one.

So it stands. The reason for announcements of the professional standard in clear terms by courts means that it is unlikely that the standard will be abolished in the near future. Dissatisfaction at the privileged position accorded professionals is likely to result in chipping away at its protections as well as sophisticated clients demanding contractual protection beyond the standard.

---

[181]Supra note 92.

B. The Design Professional's Duty to Workers

The cases cited in Section 14.08(G) demonstrate the contrary perspectives from which this hotly contested issue can be viewed. It is important to recognize that this is a multilevel problem.

The *Krieger* decision reflects what appears to be the emerging rule, that the issue in the ordinary retention should be viewed almost exclusively as a contractual one. Design professionals are retained and paid to use their best efforts to see that the project is built in accordance with the contract documents. The contract makes clear that the contractor, not the design professional, is responsible for how the design is executed. This approach requires an examination of the contracts, mainly the ones for design services but also for construction. If they do not reveal that the design professional was accepting responsibility for construction methods (Bechtel, the design professional in the *Caldwell* case, was a safety engineer as well), the design professional has *no* duty to workers, and there need be no inquiry into her conduct.

The other perspective is to *start* with tort law, with its function of compensating victims, deterring wrongful conduct, and avoiding harm. That body of law jettisoned the privity rule to ensure that claimants can find a person from whom compensation can be recovered and to avoid technical defenses that bar scrutiny into the conduct of those persons who very likely have caused the harm. Why reinstitute this barrier by looking mainly at the purpose for engaging a design professional and focusing almost exclusively on the contract?

To be sure, the contract for design services is important. Without it, vague or detailed, the design professional would have no business on the site and not be in a position to look out for danger to workers or anyone else. The contract plays a significant role in determining what the design professional should have seen. The design professional cannot be expected to *look* for problems. That is not her function. If she sees unsafe practices or would have seen them had she done what the contract obligated her to do, her conduct will be judged. Did she act reasonably? She may satisfy this obligation by complaining to the contractor's superintendent, by bringing this matter to the client's attention, or even by inviting public officials who deal with safety matters to deal with the question.

But if viewed as a tort problem, she has a responsibility to act reasonably.[182]

While most of the material dealing with this troublesome problem is found in case decisions and legal journals, there is other relevant material. For example, a California attorney general opinion addresses the question of the duty of a registered engineer retained to investigate the integrity of a building who determines there are structural deficiencies but who is advised by the owner that no disclosure or corrective action is intended and that the professional information is to remain confidential. The question was whether the professional engineer has a duty to warn occupants or notify local building officials that she has uncovered evidence of structural deficiencies that create an imminent risk of serious injury to the occupants.

The attorney general could not find any duty to warn based upon the registration laws but concluded that there was such a duty because the common law has imposed a duty on the part of professionals to protect third parties whose lives may be endangered when the professional becomes aware of this risk.[183]

To be sure, as indicated in Section 14.08(G), the tendency of courts has been to refuse to place a duty upon design professionals who perform the traditional services on the site and are not in charge of or directly involved in safety to protect workers. Yet this attorney general's opinion reflects those cases noted in Section 14.08(G) that do place a duty upon the design professional to take reasonable steps when she knows that the workers are being exposed to an unreasonable risk of harm.

A second level looks at judicial procedures. Those who support the "no duty to workers" rule seek a defense that will bar a full trial and have the issue decided by looking solely at the pleading or the pleadings and affidavits submitted to support a motion for summary judgment. They wish to escape the high cost of trial even if they believe that the design professional will ultimately prevail if her conduct is judged.

Those who support the tort orientation see no reason why the design professional should escape being judged. They contend that if the facts clearly show that she could not have been expected to know of the unsafe practices or that she did what was clearly adequate, she will be able to avoid a full-scale trial. But they object to a rule of law that relieves her. They prefer that this question be treated as a factual question like any other.

A third level relates to workers' compensation law. As noted, one of the principal functions of modern tort law is to see that victims are compensated. Most workers will be compensated under workers' compensation law—a social insurance system that pays a certain portion of the economic losses through an administrative process that does not focus on wrongful conduct. Workers cannot sue their employers in tort, with workers' compensation being their exclusive remedy. Those who feel that the compensation system is inadequate seek to encourage the development of third-party claims that the injured worker can make to fully recover for the harm suffered. This is usually more than the often inadequate amounts recovered under workers' compensation law.

On the other hand, workers' compensation is sometimes used as an argument by those who oppose the design professional's having a duty to workers.[184] Those opposing such a duty argue that the worker will not be uncompensated, and they see no reason why a worker in a construction-related accident should be placed in a better position than a worker in an industrial accident who may not have as many third parties from whom she can seek tort recovery. They contend that the broadening of third-party claims has hopelessly overcomplicated often simple accident cases and has generated lawsuits with horrendous costs to all participants.

At a fourth level, operations on a construction site are stressed. Those who support the duty state that safety is everyone's business, and the more people concerned, the less likely injuries will occur.

Those who oppose a "duty to workers" rule claim that it will induce design professionals to venture into areas where they do not have the expertise and which can only cause blurred lines of responsibility as well as expose the design professional and the client to claims that intervention into

---

[182]A fuller expression of the author's views can be found in Sweet, *Site Architects and Construction Workers: Brothers and Keepers or Strangers?* 28 Emory L.J. 291 (1979).
[183]68 Op.Cal.Att'y Gen. 250 (1985).

---

[184]*Balagna v. Shawnee County,* 233 Kan. 1068, 668 P.2d 157, 170–172 (1983) (dissent).

these matters is a breach of the contract between the owner and the contractor.

Probably the best solution in an ideal world would be to have a total, enclosed social insurance system under which all workers receive adequate compensation without the necessity of going to court but without having rights against third parties. (Currently workers' compensation laws are extraordinarily controversial in light of the high workers' compensation premiums, the large percentage of awards and costs that go to attorneys and doctors, the large overhead and profit of some insurers, and the relatively low amounts that end up in the pockets of injured workers or dependents of those who have been killed.) Until this Utopia is achieved (it may be a long way off), it is hard to support a result that would revive the dead privity doctrine and put professional designers in a favored position under which their principal function precludes their conduct being judged.

C. Injection of Tort Law into the Commercial World: A Wild Card

The commercial world—the world of business dealings between merchants where much of the world's work is accomplished—certainly does not escape from tort law intervention. For example, under certain limited circumstances, the tort law will enter when a person has wrongfully induced another to breach a contract or impede a prospective economic advantage. On the whole, tort law interferes here only when there has been intentional wrongdoing or where there has been harm to persons or property.[185] Tort law, at least in some states (refer to Sections 14.08(C) and (E)), has been reluctant to shift economic losses caused by negligent conduct and has left that largely to contract and commercial law.

In the construction world, as in others, tort law has begun to play an increasingly important role in allocating purely economic losses. For example, as seen in Section 14.08(D), negligent representation has been the basis of claims made by persons who have relied upon the representations of those

in the business of furnishing information, such as surveyors or geotechnical engineers and even architects, engineers, and construction managers.[186] Even more, there has been the tendency to expose the design professional, among other participants, to claims by other participants, particularly contractors, sureties, and even prospective occupiers of projects, such as buyers or tenants.

Perhaps these tendencies cannot be rolled back in a legal world where tort law and accountability have become dominant factors in private law even in commercial disputes. Those courts faced with these decisions or asked to expand existing rules should consider the effect of injecting this wild card.

As an illustration, some courts allow a contractor to maintain a tort action against a geotechnical engineer for misrepresentation.[187] This was done in the context of a transaction in which it was clear that the construction contract placed the entire risk of unforeseen subsurface conditions on the contractor.[188] Allowing the tort action will induce the geotechnical engineer to request indemnity from the client, increase her contract price to take this risk into account, or price her work to encompass performance designed to ensure that her representations are accurate even if "excessive" caution would not be justified. Either way, the system of allocating risks is frustrated, with the client perhaps paying twice for the same risk by increased contractor bids (tort recovery is too uncertain to permit the prudent contractor to reduce the bid because of potential tort recovery) and the higher compensation to the geotechnical engineer.

Perhaps even worse, one court allowed a tenant who lost full use of its premises because of delays caused by the contractor to establish negligence and use that as the basis for a claim against the contractor.[189] What effect does this holding have

---

[185]A subsequent tenant was allowed to bring a tort claim against the architect when the floor settled, walls became damaged, and premises became untenantable. See *A. E. Inv. Corp. v. Link Builders, Inc.,* 62 Wis.2d 479, 214 N.W.2d 764 (1974).

[186]*Rozny v. Marnul,* 43 Ill.2d 54, 250 N.E.2d 656 (1969).
[187]*M. Miller Co. v. Central Contra Costa Sanitary Dist.,* 198 Cal.App.2d 305, 18 Cal.Rptr. 13 (1961). But see *Texas Tunneling Co. v. City of Chattanooga,* 329 F.2d 402 (6th Cir.1964).
[188]See Section 25.05.
[189]*J'Aire Corp. v. Gregory,* supra note 118. This case sought to put some limits on recovery by barring recovery for losses that involve ordinary business risks; a subsequent decision did not draw this line. *Chameleon Eng'g Corp. v. Air Dynamics, Inc.,* 101 Cal.App.3d 418, 161 Cal.Rptr. 463 (1980) (supplier of subcontractor sued by prime for not supplying components.

on the potential risk of construction contract participants?

Delay is dealt with at length in the construction contract with its time extensions, damage liquidations, or no-damage clauses. Will these clauses apply to tort claims? How can contractors or design professionals faced with this risk deal with it? They can hope the law will protect them. They may decide to build a contingency into their contract price to deal with this risk or demand indemnification if they have the bargaining power. They may, if they can identify potential claimants, seek exculpation from them or demand that the owner do so. All of these approaches add transaction costs and *must* increase contract prices. It must also be kept in mind that the party who has suffered economic losses usually can transfer such losses to the party with whom it has contracted—the tenant to the landlord or the contractor to the owner.

The tort wild card has caused unnecessary chaos in the construction world. It is bad enough to allow major participants, such as the design professional, owner, and contractors, to sue each other in tort. It is much worse to allow more *remote* participants, such as suppliers, potential buyers, tenants, lenders, or even sureties, to use tort law as a means of shifting their *economic* losses to the major participants in the Construction Process.

The law should encourage persons to enter into commercial transactions. One way is to limit the exposure for consequential damages suffered by the other *party*. But this protection is diminished in claims by third parties in *tort*. That tort usually in this area requires negligence does not compensate for this added exposure, particularly when claims against design professionals usually use a tortlike standard anyway.[190]

D. The Effect of Expanded Professional Liability

This section has outlined some of the arguments for and against expanded professional liability. But

the focus of this subsection is to point to the effect this expansion has had.

Expanded liability can be looked on as a method of eliminating incompetent practitioners from the professions to supplement the unarguably ineffective registration laws or, at least in this area, the inefficient marketplace. But does it have this effect?

Does expanded liability drive out the practitioner who should be removed from the profession? Are incompetents sued more often than those who are competent? Are they likely to be forced out by increased insurance rates, decisions by insurers not to insure them, or unwillingness of prospective clients to engage them if they cannot be insured? At best, these are unprovable. Very likely the answer to all three is no.

Has expanded liability improved professional practice? Undoubtedly, design professionals are more careful, perhaps too careful. The result can be overdesign, an unwillingness to take design risks, and mediocre design. Again it will be difficult to assemble anything beyond anecdotal evidence and polemics to uncover the truth.

Of course, expanded liability can be and has been justified as a process for allocating responsibility to the persons who are responsible and who can best spread the loss. The Construction Process, with its wealth of participants, its overlapping functions, and unclear lines of responsibility, makes it unlikely that expansion of professional liability will place responsibility on the party on whom it ought to be placed. All that can be certain is that expanded liability has led to hopelessly complicated and unpredictable lawsuits with the inevitable rise in the overhead of performing professional services.

What is also certain is that expanded liability has generated overprotective contract language that may simply drive prospective clients to others, such as those who design and build or manage construction. It, along with other causes of high operational overhead, such as taxes and other regulations, has slowly reduced the ranks of sole practitioners and small partnerships and has led to increased specialization. Expanded liability has led to an emphasis upon risk management (explored in Chapter 15).

[190]Refer to Sections 14.05–14.07. Prof. Jones, op. cit. supra at 93 at 1070–73, 1092–93 also suggests tort not be available to these claimants.

PROBLEMS

1. A was the architect for a large office building. When she selected material to be used for wall construction, she did not realize that the fire insurance rates were dependent on the type of wall construction material. She specified wall material based on cost, safety factors, and aesthetics. Had she specified different materials of approximately the same cost, there would have been a reduction in the fire insurance rate. When the owner found this out, she claimed that the additional insurance premium should be borne by the architect because the architect should have known that insurance premiums are based on materials used in construction. Is the owner correct in her statement? What additional facts would be helpful in resolving this problem?

2. A was the architect for a lavish theater to be built in New York. The theater owner wanted a dramatic staircase to be designed that would run from the lobby to the first and second balconies. Two types of floor coverings for the stairway were available. Type A was more pleasing aesthetically than Type B and cost 20% less. The principal advantage of Type B was that persons walking on the stair were less likely to slip. The architect chose Type A because it cost less and looked better, even though it was less safe than Type B. Would the architect be liable if a theater patron slipped and suffered injury when ascending the stair from the lobby to the first balcony? What facts would be necessary to answer this question?

3. Suppose an architect designs a building without knowing the federal government recently enacted a statute that gives a tax rebate to those who use particular types of insulation to conserve energy. The insulation that would qualify for the rebate was not specified. Would the architect be responsible to the client for any lost tax rebate? Suppose the contractor knew of this and said nothing? Suppose it was likely that the client would be aware of new legislation of this type?

Risk Management: Variety of Techniques

Chapter 14 chronicled the expansion of professional liability. An often ignored technique to avoid claims is cultivation of a good client relationship. Honesty in approach, respect for the client's intelligence, appreciation of the proper role of a professional adviser, and common courtesy (answering phone calls and letters) are perhaps the best techniques to avoid claims. These are nonlegal considerations. This chapter suggests legal and planning approaches to deal with this phenomenon.

SECTION 15.01 Sound Economic Basis: Bargaining Power

This treatise is not an appropriate place to discuss in detail the economics of the design professions. To implement some if not most of the approaches suggested in this chapter, the design professional must be able to choose which commissions to accept and to request or even demand that particular contract language be excluded or included. Doing this requires sufficient economic strength to pick and choose among projects and contracts. Although this chapter cannot deal with methods to achieve this power, the approaches suggested will be of no value unless economic strength can be attained and mobilized.

SECTION 15.02 Evaluating the Commission: Participants and Project

As a rule, there are more design professionals than commissions. Usually this means that most design professionals will take whatever work they can get.

But there are design professionals who can pick among projects. This section is directed to them.

It is most important to evaluate the client and its financial resources. A client with limited resources, particularly one with extravagant expectations, may not be able to withstand the shocks of added costs such as those generated by design changes, delays, claims, or other circumstances that will increase the ultimate contract payout. Such a client will be more inclined to abandon a project before construction. If construction does begin, it may be quicker to point at the design professional and other participants if the project does not proceed as planned. The AIA gives architects the power to request information on the financial resources of the owner at any time during their performance.[1] Yet it is likely that most architects will not exercise this power, particularly at the start of the professional relationship. Although it is understandable that an architect may wish to make these inquiries at the *beginning* of performance, such inquiries may be crucial *during* performance if financial troubles appear imminent.

Another important client criterion is experience in design and construction. An inexperienced client may be more likely to make claims because it does not realize the uncertainties inherent in construction and the likelihood that adjustments will have to be made. Such a client may also be mesmerized by a fixed-price contract and be unduly rigid as to

---

[1]AIA Doc. B141, ¶ 4.3.

price adjustments. Although construction contract pricing disputes principally affect the relationship between owner and contractor, when that relationship sours, there will be more administrative burdens placed on the design professional and a greater likelihood of claims.

The project, too, must be evaluated. One that involves new materials, untested equipment, and novel construction techniques must be viewed as creating special risks. If disappointments develop, claims (including those against the design professional) are more likely to be made.[2]

It is important to evaluate the other key participants, such as the prime contractor, the principal subcontractors, and consultants, for technical skills, financial capacity, and integrity. The construction contract is another factor that must be taken into account. A tight fixed-price contract, a rigid time schedule with a stiff liquidation of damage clause for delay, a multiple prime contractor arrangement, and a fast-track sequence all contain the seeds for controversy and, possibly, liability exposure.

It may be helpful to develop a point system for evaluating these factors. If the points reach a certain level, the project should not be undertaken without careful executive review. Beyond the next numerical benchmark, the commission should be refused unless changes are made. At a point beyond even that, the commission should be refused.

SECTION 15.03 Contractual Risk Control

Section 15.02 noted the importance of a general appraisal of the contract. This section looks at specific contract clauses that can be useful in risk management.

A. Scope of Services

The contract should make clear exactly what the design professional is expected to do. Reference was made in Sections 12.01 and 13.01(G) to the difference between basic and additional services. Here the emphasis is on services the client may expect the design professional to perform that the design professional does not feel are part of his undertak-

ing. Perhaps most important is the design professional's role in determining how the work is being performed and the responsibility of the design professional for the contractor not complying with the contract documents.

B. Standard of Performance

Usually the design professional wishes to be held to the professional standard discussed in Section 14.05. When this is the case, the contract language as well as any other communications should not use words such as *assure, insure, guarantee, achieve, accomplish, fitness,* or *suitability* or any language that appears to promise a specific result or achievement of the client's objectives. It is even better to include language that specifically incorporates the professional standard and, where possible, language that justifies it.

While Section 15.03(D) deals with liability limitations, one suggestion that relates to the standard of performance should be mentioned here. Many cost overruns result from design changes. Some have suggested that there be a toleration figure— an amount usually based upon some percentage of the construction contract for which the design professional would not be responsible. This is based on the assumption that design is essentially a trial-and-error process and some things truly cannot be discovered until the design is actually being executed. The design professional would be responsible over any such toleration figure if performance does not measure up to the obligation imposed upon him by the contract or by law.[3] (Usually the amount specified is approximately 3%.)

C. Exclusion of Consequential Damages

The law does not charge a breaching party with *all* the losses caused by its breach. Usually the breaching party is not chargeable with what are sometimes called consequential or less direct damages. This limits a contracting party's responsibility.

Yet the foreseeability requirement—the standard used most frequently in determining whether consequential damages can be recovered—has been applied by modern courts in such a way as

---

[2]The resources of the Architectural and Engineering Performance Information Center at the University of Maryland might be useful in evaluating problems that may arise on a project.

[3]See EJCDC No. 1910-1 Exhibit H, ¶ H8.7.4.4, entitled "Agreement Not to Claim for Cost of Certain Change Orders."

to diminish protection given by earlier courts. For that reason, the design professional should seek to exclude his liability for consequential damages in the contract. This can be done expressly or by limiting the responsibility of the design professional to correction of work caused by defective design or, when that is not economically feasible, to the diminished value of the project.[4]

D. Limiting Liability to Client

Some have advocated that the design professional include a clause in the contract under which the designer's liability to the *client* is limited to a designated amount or a specific portion of the fee, whichever is greater or, in some contracts, whichever is less, or the professional liability insurance coverage. Some clauses allow the client to buy out of the limitation by paying a specified amount. Very likely this is done not to encourage a buyout but to help enforce the clause by giving the client an alternative.[5] Although limits do not affect third-party claims, client claims still constitute the bulk of liability exposure.

EJCDC No. 1910-1 includes Exhibit H as a possible addition to the contract. Among the various "Allocation of Risks" included is ¶ H8.7.4, prototype language limiting the engineer's liability to either his compensation, the amount of insurance proceeds, or a specific amount.

Although the principal debate has centered on the enforceability of such a clause, the bargaining realities make the *main* problem obtaining client consent. This is difficult to do when dealing with private owners and almost impossible when dealing with public owners, where accountability is sensitive.

The issue of enforceability has stirred much debate. Some claim that such clauses are invalid, based on the undesirability of professional persons using a contract to reduce their professional responsibility and the assumption of modern courts that those who obtain *any* exculpation *must* be in an oppressive/dominant bargaining position. Those who suggest that such contract clauses would not be enforced also point to the possibility that enforcement would provide an incentive for professionals to perform carelessly. Those who suggest that such clauses *are* enforceable point to the absence of any strong bargaining power on the part of most design professionals. Also, they note that the clauses are not totally exculpatory clauses but simply limit liability to a particular amount, often more than trifling. They assert that the law is no longer treating professionals differently than commercial enterprises. They point to attacks by law enforcement officials on the professions through enforcement of antitrust laws. They also point to court decisions that have permitted exculpation between commercial parties who presumably know what is in their best interests. A clearly drafted liability limitation clause, particularly if the client is given an opportunity to purchase the full protection provided by law for an extra fee, is likely to be enforced.[6]

Contracting parties in this context have sufficient bargaining strength and intelligence to bargain freely over risks. They should be given the freedom to apportion risks as they see fit. Any client who agrees to such a limitation generally has legitimate reasons for doing so. The client may decide to go along with such a limitation because

---

[4]Upheld in *Wood River Pipeline Co. v. Willbros Energy Services Co.*, 241 Kan. 580, 738 P.2d 866 (1987).

[5]See *Cregg v. Ministor Ventures*, 148 Cal.App.3d 1107, 196 Cal.Rptr. 724 (1983) (upheld in consumer rental storage space contract where consumer given opportunity to insure with storage company at a higher rate or self-insure at a lower rate).

---

[6]*Georgetown Steel Corp. v. Union Carbide*, 806 F.Supp. 74 (D.S.C.1992) (arm's-length transaction between commercial sophisticates, with a chance for the client to buy out); *Long Island Lighting Co. v. IMO Deleval, Inc.*, 668 F.Supp. 237 (S.D.N.Y.1987) (insurance limits upheld, parties sophisticated corporate entities with experience in construction and negotiation); *Burns & Roe, Inc. v. Central Maine Power Co.*, 659 F.Supp. 141 (D.Me.1987); *Federal Reserve Bank of Richmond v. Wright*, 392 F.Supp. 1126 (E.D.Va.1975); *Markborough Cal., Inc. v. Superior Court*, 227 Cal.App.3d 705, 277 Cal.Rptr. 919 (1991) (enforced, based on statute permitting such a clause if opportunity to negotiate over it). *W. William Graham, Inc. v. City of Cave City*, 289 Ark. 105, 709 S.W.2d 94 (1986) (dictum that clause would be upheld but strictly construed); In an unpublished opinion, a federal trial court upheld such a clause but stated it would be carefully scrutinized and narrowly interpreted. See *Maine Power Co. v. Foster Wheeler Corp.*, no. 83-0056-P (D.Me. April 15, 1986). The *Markborough* decision is the centerpiece of a thorough review of liability limitations in Ashcraft, *Limitation of Liability—The View After Markborough*, 11 Constr.Lawyer No. 3, August 1991, p. 3.

those are the only terms under which a particular design professional will perform. It may also receive a reduced fee generated by lower professional liability insurance premiums. It may be willing to accept such a provision because it feels that it is fair and inclusion of it in a contract will get the professional relationship off on a positive basis. For *whatever* reason, if the parties have chosen freely to deal on that basis, there is no reason for the law to interfere.

A number of legislatures have enacted anti-indemnity statutes that invalidate indemnity clauses that relieve one party from ultimate responsibility for its sole negligence. Some also bar attempts by design professionals to indemnify themselves either generally or for their design negligence.[7]

Such legislation should not affect liability limitations. They are not exculpations. The amount selected usually bears a rational connection to the fee or the project costs. One anti-indemnity statute allows liability limitation if bargained for.[8]

Such clauses, where used, should make clear which expenditures come within the liability limitation, particularly if the design professional has agreed to indemnify the client. The clause should make clear that the liability limitation does not *automatically* entitle the client to recover the specified amount. Damages must be established, with the liability limitation being a ceiling on what can be recovered.

E. Immunity: Decision Making

Some American courts grant design professionals quasi-judicial immunity when they decide disputes under the terms of the construction contract. Many standard agreements published by the professional associations incorporate language that relieves the design professional from any responsibility if decisions are made in good faith.[9] They attempt to

incorporate into the contract limited quasi-judicial immunity. If such clauses are to be effective, they must be incorporated in the contracts both for design services and for construction services.

F. Contractual Statute of Limitations

As noted earlier, judicial claims can be lost simply by the passage of time, accomplished by statutes of limitations. Parties to a contract can include a private statute of limitations as long as it is reasonable and will not deprive one party of any viable judicial remedy. Some design professionals incorporate language in their contracts that states when the period begins and specifies the period itself.[10] Such statutes of limitations can be useful as risk-management tools.

G. Third-Party Claims

Third-party claims are sometimes based on the assertion that the claimant is an intended beneficiary of the contract. This contention can be negated by appropriate contract language.[11] Such language may even have some effect on tort claims.[12]

H. Dispute Resolution

Some believe that the most important risk-management tool is to control the process by which disputes will be resolved. Many American standardized construction contracts give first instance dispute resolution to the design professional and frequently provide for an appeal to arbitration. Because arbitration has become such an important feature of construction contract dispute resolution, it is covered in detail in Chapter 30.

I. The Residue

Although a wish list can be created,[13] the realities of contract bargaining necessitate that emphasis

---

[7]West Ann.Cal.Civ. Code § 2782. See Section 32.05(D).
[8]Id. at § 2782.5. See *Markborough Cal. Co. v. Superior Court,* supra note 6.
[9]AIA Doc. B141, ¶ 2.6.16, grants the architect immunity for his interpretations and decisions made in good faith, an immunity echoed in A201, ¶ 4.2.12. However, neither B141 nor A201 grants immunity for resolving claims and disputes under B141, ¶ 2.6.19, and A201, ¶ 4.3. Presumably, the AIA believes that ¶ 4.2.11 and making the decision subject to arbitration are sufficient protection. However, some parties delete arbitration from such contracts.

[10]AIA Doc. B141, ¶ 9.3, and A201, ¶ 13.7, seek to control when the statutory period of limitations commences.
[11]AIA Doc. B141, ¶ 9.7, and A201, ¶ 1.1.2, seek to bar persons who are not a party to the contract from asserting rights as intended beneficiaries. The latter provides an exception for the architect.
[12]*Harbor Mechanical, Inc. v. Arizona Elect. Power Coop., Inc.,* 496 F.Supp. 681 (D.Ariz.1980).
[13]Illustrations are an exculpation for consultants, a favorable choice of applicable law, waiver of a jury trial, a power to suspend work for nonpayment, insurance premiums as a reimbursable, and a stiff late payment formula, to mention a few.

be placed on those that are most useful in risk management.

J. Some Suggestions

The clauses noted in this section are more likely to be enforced if they are drafted clearly and express specific reasons for their inclusion, if the client's attention is directed to them, and if suggestions are made to the client to seek legal advice if it has doubts or questions about them.

SECTION 15.04 Indemnity: Risk Shifting or Sharing

This important risk-management device usually involves all key participants in the Construction Process. From the vantage point of the design professional, it seeks to shift any or a part of any loss he suffers related to claims by third parties such as workers, members of the public, or adjacent landowners to another participant, usually the contractor, but possibly the client. This process is discussed in Chapter 32.

The indemnification in AIA Doc. A201 that will be discussed in greater detail in Section 32.05(G) stemmed from attempts by the owner and the architect to receive indemnification from the contractor in the contract for construction. Currently, AIA Doc. B141—the standard form of agreement for design services—does not contain any indemnification provisions. Yet owners increasingly demand indemnification from their design professionals, and design professionals, where they have the bargaining power to do so, increasingly seek indemnification from the owner in the contract for design services.

In 1992, the EJCDC included an indemnification clause in its Document No. 1910-1 at ¶¶ 8.7.1 and 8.7.2. The former requires the engineer to indemnify the owner, and the latter, the owner to indemnify the engineer and his consultants. In either case, indemnification is limited to claims "caused solely by the negligent acts or omission" of the engineer and owner, respectively. In its Guide Sheet, the EJCDC does not recommend that the engineer agree to indemnify the owner, since this would extend the engineer's liability "beyond what it would be at common law." However, the EJCDC recognizes that many owners expect some form of indemnity and strongly urges that indemnification not go beyond that specified in its documents.

Over the past several years, bills have been introduced in the U.S. Congress to seek in some way to indemnify design professionals who engage in high-risk services for the federal government. In 1990, H.R. 4925 was introduced and referred to committee. The bill responded to complaints by engineering firms doing hazardous waste cleanup that the proposal by the Environmental Protection Agency for indemnification was too limited in dollars and time to provide sufficient protection to those firms against the possibility of catastrophic losses because of the nature of hazardous waste services.

Under the bill, firms engaged in certain activities would have been indemnified up to the greater of ten billion dollars or ten times the contract value. Other provisions dealt with the provisions of time in which claims against the engineers would have to be brought. The bill included an exclusion for gross negligence, intentional misconduct, or bad faith. The bill was not enacted.

Similarly, in 1991, the same congressman who had introduced the earlier bill introduced another bill, H.R. 1474, again intended to protect design professionals from excess liability when engaged in certain federal contracts. The bill would have applied to all federal design professional contracts, not just those with an extremely high risk of claims. Under the bill, the federal government would have indemnified the design professionals for any losses in excess of commercial insurance or self-insurance coverage. The bill would have required the design professional to carry professional liability insurance of the greater of 1% of the contract value or one million dollars. The bill was not enacted.

It is evident that the associations representing design professionals have been active in seeking to obtain indemnification for design professionals who are engaged in federal work that carries with it enormous liability risks. Those in favor of these bills principally look upon them as methods to assure that the federal government can obtain essential services from qualified design professionals.

To understand the reasons for demands for indemnification, it is important to recognize the manner in which the law would deal with shifting or sharing of risk *without* an express indemnification provision. Suppose a claim is made against the owner for harm caused principally by the design professional. In such a case, the owner might assert a claim against the design professional in the original lawsuit or a subsequent lawsuit where this can be done. A number of possible theories exist for a claim against the design professional. The owner might assert that the design professional has breached his contract, that the design professional was guilty of a tort, or that indemnification for such a claim was implied in the contract or justified by restitution based on unjust enrichment (equitable indemnification). In addition, the owner might have a claim under the laws of some states for contribution between wrongdoers.

All of these bases for recovery, particularly the ones relating to indemnification or contribution, can raise complicated legal issues. In addition, it is not clear that the owner could recover its costs in defending the claim. These uncertainties are the reason contracting parties frequently seek express indemnification in their contracts.

SECTION 15.05 Professional Liability Insurance: Risk Spreading

A. Requirement of Professional Liability Insurance

The law does not require design professionals to carry professional liability insurance.[14] However, clients increasingly require that design professionals have and maintain professional liability insurance. The AIA does not specifically require insurance but does state that any excess premium above that usually paid is a reimbursable.[15] Clients may have a claim against the design professional for losses relating to the project or because they have satisfied claims of third parties that are directly traceable to the design professional's failure to perform in accordance with the legal standards. To make any claim collectable, they may require that the design professional carry professional liability insurance. If the design professional has adequate professional liability insurance, third parties injured as a result of his conduct may choose to bring legal action against the design professional directly rather than against the owner.

Even if not required to, many design professionals carry such insurance. One reason is to protect their nonexempt assets from being seized if a judgment is obtained against them. Another is that many design professionals do not wish to see persons go uncompensated who suffer losses because of the design professional's failure to live up to the legal standard.

B. Volatility of Insurance Market

This section describes certain characteristics of professional liability insurance policies. Yet these characteristics, particularly coverage, exclusions, and premiums, are not static. At times, many desire to enter the insurance market. At these times, entrants are tempted by the investment of front-end premiums that can earn large returns. Payout to such entrants can be postponed to far into the future. In such a market, premiums are cut to obtain the business.

Yet when the investment portfolio does not generate large returns and claims begin to come in, the insurance business looks much less attractive. When this occurs, some insurers drop out of the market, or those that remain limit coverage, broaden exclusions, and raise premium rates. In addition, these insurers will make a careful selection of those for whom they will issue insurance and defend claims more vigorously.

Yet the cyclical nature of the insurance market means that after such a period of retrenchment and caution, entrants will again be tempted to get back into the insurance market. For this reason, it is difficult to describe with any certainty the types of policies available over an extended period. Those who wish to insure must obtain expert advice on the particular state of the market, coverage, and premiums. Similarly, those who impose contractual requirements for insurance must seek to determine whether any such insurance is available.

C. Regulation

All states regulate insurance, with varying details. State laws usually provide requirements of capital and financial capacity to insure the solvency of insurers, and they increasingly regulate claims settlement practices. State insurance regulatory agencies determine which insurers will be permitted to do business in the state. In some states, regulators can determine coverage, exclusions, and premiums.

Courts have also regulated insurance, mainly through interpreting insurance policies when dis-

---

[14]This term will be used instead of the more commonly used errors and omissions insurance.
[15]AIA Doc. B141, ¶ 10.2.1.5.

putes arise. The law interprets ambiguities against the insurer unless the insurance policy is one negotiated between a strong insured, such as a group of hospitals, and an insurer. Some courts seek to determine the reasonable expectations of the insured or some average insured. They will protect that expectation, some courts even disregarding insurance policy language in order to protect that expectation. Finally, there are many technical contractual requirements before the insurer's obligation to defend and indemnify matures, such as complying with warranties or representations as to activities, giving notice of claims, furnishing proof of loss, or cooperating with the insured in defending claims, to name some. Technically, a failure to comply with any of these requirements can result in loss of coverage. But law often, though not always, protects the insured by requiring the insurer to show that the failure to comply prejudiced the insurer.

D. Premiums

Premiums have become an increasingly important overhead cost for design professionals. For example, a recent study stated that architects who carry insurance pay approximately 3% of their gross billings for insurance premiums and structural engineers currently pay some 6% of their gross billings for insurance premiums. The amount of premiums for any individual insured is determined by a number of factors, such as the type of services performed, the experience of the insured, the locality in which work or projects are located, the gross billings of the insured, the contracts under which services are performed (one insurer states it reduces the premium if the contractual limitations of liability discussed in Section 15.03(D) are incorporated in a certain percentage of the insured's contracts), and the claims record of the insured.

Premiums can rise because of underwriting predictions (selection of risks) by insurers, a decline in the value of the insurer's portfolio, a downturn in the economy that can induce insureds to cancel coverage, high claims payouts, and increased cost of defending claims.

E. Policy Types: Occurrence or Claims Made

In the Construction Process, a long time lag can exist between the act or omission claimed to be the basis for liability and the making of the claim. Cov-

erage in cases depends on whether the policy is a "claims made"[16] or an "occurrence" policy. Claims made policies cover only claims made during the policy period regardless of when the act giving rise to the claim occurred. An occurrence policy, on the other hand, gives coverage if the act or omission occurs during the policy period.

Professional liability insurance is currently written on a claims made basis. This avoids the "tail" at the end of any occurrence policy. Proper insurance underwriting and rate making require that the insurer predict payouts in a designated period. Occurrence policies, particularly in states that do not commence the statutory period for bringing claims until discovery that there is a claim and against whom it can be made, can result in coverage many years after the premium has been fixed. This led the professional liability insurers to use claims made policies. It is easier to predict the value of claims that will be made during the period of the policy than to look far into the future, a prediction needed for setting premiums under an occurrence policy.

At any given time, a variety of claims made policies can be available. Most are considered hybrid. Such policies specifically protect against claims made during the policy period and require that the act giving rise to the claim occur during the policy period. California rejected a challenge to such a policy, noting that the policy terms were conspicuous and that the insured could expect only the coverage provided.[17]

Insurers usually offer retroactive, or prior acts, coverage, giving coverage regardless of when the act occurs. Retroactive coverage usually requires the insured to represent that he is unaware of any

---

[16]Sometimes a claims made policy is described as a "discovery" policy. To avoid the confusion that can result from statutes of limitations sometimes being described as "discovery" statutes, the *claims made* term will be used.

[17]*Merrill & Seeley, Inc. v. Admiral Ins. Co.*, 225 Cal.App.3d 624, 275 Cal.Rptr. 280 (1990). In that regard, the California Insurance Code Section 11580.01 requires that an application or proposal for a claims made policy must "recite prominently and conspicuously at the heading thereof that it is an application or proposal for a claims-made policy." Also, each policy must contain a prominent and conspicuous statement to that effect, along with language suggesting that the insured carefully study the policy and discuss coverage with his insurance agent or broker.

facts that could give rise to a claim or that he was insured when the act occurred and that he has carried insurance throughout his career.

International Insurance Co. v. Peabody International Corp. involved a claim by an insured with retroactive coverage. The insured argued that it did not have notice of the pending claim, because it considered the demands of its client, the claimant, to remedy contractual shortcomings as nothing more than a negotiating ploy rather than a claim. However, the court responded that the definition in the policy was unequivocal and covered a demand for money or services. The application to the policy provided that "[t]he Insured has no knowledge of such act, error or omission on the effective date of this policy". The court concluded that the insured knew of the acts that ultimately erupted as the basis of his client's claim.[18]

Other claims made policies cover all claims made during the policy period regardless of when the act gave rise to the claim.

Although most courts have sustained claims made policies,[19] one case struck down a claims made clause because its retroactive coverage applied only if the insured had carried a policy for an earlier period with the company that was insuring the insured at the time the claim was made.[20]

F. Coverage and Exclusions

In the absence of a special endorsement, professional liability insurance policies generally cover liability for performing normal professional services.[21] Tasks commonly undertaken by design professionals are known to the insurer. The insurer is aware of the standard to which the design professional will be held. These known and predictable elements are needed for loss predictions and intelligent rate making. For this reason, services undertaken by the insured as part of a joint venture are typically excluded from coverage. A joint venture, being a type of partnership, means that the negligence of one joint venturer will be charged to the other and create liability exposure to the insurer. Often, a special endorsement can be obtained if the insurer has a chance to investigate the arrangements under the joint venture.

The professional activities covered are core design services, such as preparing drawings and specifications. Coverage can also encompass site services, such as monitoring the work as it proceeds for issuing payments and completion certificates.

American Motorists Insurance Co. v. Republic Insurance Co.[22] involved the question of whether the architect's preparation and submission of a competitive design/build bid is a professional service and covered under the architect's professional liability policy. The case involved a competitive design system, with the insured being the successful bidder. After the award of the bid, an unsuccessful bidder sued the insured for misrepresentations and other tortious conduct. The insured's professional liability insurer refused to defend, but the action was defended successfully by the insured's comprehensive general liability insurer. The latter then sought to recover a pro rata share of its defense costs from the professional liability insurer.

The professional liability insurer argued that bid preparation and submission are merely preparation to render professional service and not the actual rendering of such services. However, the Alaska Supreme Court did not accept this contention, noting that only an architect using his specialized knowledge, labor, or skills could have prepared the bid. The bid consisted of two booklets approximating 160 pages in length and considerable detail as to the design. The court noted that there was no definition of professional services in the policy and in such cases, the term was considered ambiguous, requiring a construction that favored coverage.

Though coming under an exclusion for professional services in a comprehensive general liability (CGL) policy, the same issue came before the court in *Camp Dresser & McKee, Inc. v. Home Insurance Co.*[23] The original claim was made by an injured

---

[18]747 F.Supp. 477 (N.D.Ill.1990).
[19]*Zuckerman v. Nat. Union Fire Ins.*, 100 N.J. 304, 495 A.2d 395 (1985) (describing history and analysis of claims made policies); *Stine v. Continental Cas. Co.*, 419 Mich. 89, 349 N.W.2d 127 (1984). See also Parker, *The Untimely Demise of the "Claims Made" Insurance Form: A Critique of Stine v. Continental Casualty Co.*, 1983 Det.C.L.Rev. 25 (1983) (commenting on the intermediate appellate decision in the *Stine* case) and Comment, 22 U.C.L.A.L.Rev. 925 (1975).
[20]*Jones v. Continental Cas. Co.*, 123 N.J.Super. 353, 303 A.2d 91 (1973).
[21]See Annot., 83 A.L.R.3d 539 (1978).

---

[22]830 P.2d 785 (Alaska 1992).
[23]30 Mass.App. 318, 568 N.E.2d 631 (1991).

worker at a plant who attempted to throw ash onto a head pulley of a conveyor with a missing safety guard, exposing the worker's hand and arm to injury. Camp Dresser, the insured, acted as a project manager to coordinate the work and perform services similar to that of a construction manager. Camp Dresser's professional liability insurer defended the case and paid for the settlement in excess of the $150,000 deductible under the professional liability policy. But Camp Dresser also notified its CGL insurer because that policy had no deductible.

The court held in favor of Camp Dresser, noting that the fact that the services performed by Camp Dresser are usually performed by engineers or other professionals does not compel the conclusion that the contracts are those for professional services. The court held that professional services require specialized knowledge and mental rather than physical skills. Claims of ordinary negligence or negligent management and control were not expressly precluded, and the word *supervisory* in the exclusion is reasonably susceptible to ambiguous interpretation. The word can be construed narrowly as describing supervision of purely professional activities or broadly as describing management or control of the aspects of a project involving professional and nonprofessional activities. Since the policy did not make clear in what sense the term was used, Camp Dresser received coverage.

These cases demonstrate that the borderline between professional and nonprofessional services can mean that there is coverage in both types of policies. Yet it can also mean that under certain circumstances, neither policy will cover. An insured who can anticipate this and can be concerned that it will have to pay the formidable costs of defense might think of procuring business legal expense insurance.

More important are the long lists of exclusions incorporated in policies. Coverage excludes contractually assumed risks. One policy excluded work not customarily performed by an architect as well as activities relating to boundary surveys, subsurface conditions, ground testing, tunnels, bridges, and dams. Also excluded were failure to advise on or require insurance or surety bonds and failure to complete construction documents or to act on submittals in the time promised unless those losses were due to improper design. In addition, the policy excluded liability for guarantees, esti-

mates of probable construction costs (currently, one leading professional liability insurer does not exclude this, thereby demonstrating the changing nature of the insurance market and liability coverage), and copyright, trademark, and patent infringements.

Coverage for activity that relates to the handling of hazardous materials is fluid. Initially, it was almost impossible to obtain such coverage. Currently, insurers are increasingly willing to cover pollution claims relating to alleged design negligence of a design professional. Again, consultation with a competent insurance counsellor is vital in such a market.

The insured should review his contractual commitments to determine whether they are covered by his professional liability insurance policy. Some insureds submit unusual contracts to the insurer for its approval or to discuss coverage.

The usual professional liability insurance is not likely to cover many services performed by a construction manager or by those who both design and build. Special endorsements or specially tailored policies for these activities will be needed.

A complicated coverage problem can develop when several causes contribute to the loss. In some states, if one of the causes is covered and the other excluded, the insured receives coverage. This is demonstrated by *Comstock Insurance Co. v. Thomas A. Hanson & Associates, Inc.*,[24] in which the architect had been held liable to the owner because of negligent design, which was covered, and negligent cost estimates, which was excluded under that policy. (As noted earlier, currently, some professional liability insurers do not exclude this.) Under the Illinois law, which governed this transaction, if one cost is covered and the other excluded, there is coverage.

G. Deductible Policies

Insurance companies increasingly seek to reduce their risk by excluding from coverage claims, settlements, or court awards below a specified amount. Policies that exclude smaller claims are called deductible policies. Generally, the higher the deductible, the lower the premium cost. A high deductible means that substantial risks are borne by

[24]77 Md.App. 431, 550 A.2d 731 (1988). Some states use a proximate or dominant cause analysis.

the insured. The recent tendency to raise deductible amounts makes the insured increasingly a self-insurer for small claims. Some policies include in the deductible the cost of defense. For example, if the deductible is $5,000 and a claim of $3,000 is paid to a claimant, any cost of defense up to the deductible amount is borne by the insured.

Under some policies, the insured must pay all claim expenses until the deductible is reached. Other policies do not require payment until the claim is resolved.

Deductible policies can create a conflict of interest between insurer and insured. When small claims are made, the insurer may prefer to settle the claim rather than incur the cost of litigation. The insured may oppose such a settlement because of a belief that it is an admission of negligence and payment will come out of the insured's pocket. One insurer gives the insured the right to veto a settlement recommended by the insurer but provides that if the insured's ultimate liability exceeds that settlement proposal, the insurer's liability will not exceed the amount of the proposed settlement and any expense costs incurred *prior* to the settlement. In effect, such a provision gives the insurer the right to determine settlement.

Usually the deductible amount applies to each occurrence. But where the policy was not clear in this regard, one court applied the deductible of $10,000 to *each* claim, and in an accident with eight claimants, the total deductible amounted to $80,000.[25]

H. Policy Limits

American policies generally limit insurance liability.[26] Suppose a claim for $50,000 is made and the policy limits are $100,000. The claimant offers to settle for $40,000. The insurance company exercises its right to veto settlement and refuses to settle. The claim is litigated, and the claimant recovers $125,000. The insured may contend that had the insurer settled, it would not have had to pay amounts in excess of the $100,000 policy limit. Generally, insurance companies are liable for amounts over the policy limit if their refusal to settle was unreasonable in light of all the circumstances.[27]

Policies also contain an aggregate limit that applies to all claims made during the policy period. The maximum amount available to settle all claims arising out of one negligent act is usually the limit of liability per claim, but the maximum amount for all claims is the policy's aggregate limit of liability.

Some policies apply cost of defense to the aggregate limit. With the cost of defense so high, an insured who wishes to continue reasonable coverage might find it useful to increase its aggregate limit if defense costs are charged off against the aggregate limit.

I. Notice of Claim: Cooperation

Insurance policies usually state that the insured must notify the insurance company when an accident has occurred or when a claim has been made. The notice is to enable the insurer to evaluate the claim and gather evidence for a possible lawsuit.

In addition to requiring that the insured notify the insurer, policies usually require that the insured cooperate with the insurer in the handling of the claim. The insured must give honest statements (a false swearing clause may make any dishonest statement the basis for denying coverage). He must also make reasonable efforts to identify and locate witnesses and supply the insurer with material that may be important in defending the claim. The insured must also comply with any notice to attend hearings that have to do with the claim, such as depositions and trials.

While insurance policies often state that coverage will be denied if there is a failure to cooperate (as stated in subsection (C)), courts are not always willing to make failure of cooperation a sufficient basis to deny coverage unless the insurer can prove that it was prejudiced by the lack of cooperation. It should be noted, however, that the insured design professional has a disincentive for failure to cooperate. The design professional may be liable to the extent of the deductible and in certain unusual cases may be liable for any amount that exceeds the policy limits.

---

[25]*Lamberton v. Travelers Indem. Co.*, 325 A.2d 104 (Del.Super.1974).

[26]For discussion of whether the surety can be liable for more than the bond limits, see Section 33.10(E).

[27]*Comunale v. Traders & General Inv. Co.*, 50 Cal.2d 654, 328 P.2d 198 (1958); *Crisci v. Security Ins. Co.*, 66 Cal.2d 425, 426 P.2d 173, 58 Cal.Rptr. 13 (1967) (insured also recovered for her emotional distress).

J. Duty to Defend

In most liability insurance policies, the insurer promises to defend and indemnify. One of the most difficult areas of law surrounds the question of when the insured has a duty to defend. The duty to defend is broader than the duty to indemnify, as policies usually state that the insurer will defend claims that are groundless, false, or fraudulent. Whether there is a duty to defend first depends upon the complaint made against the insured. If coverage appears from the complaint, the insurer must defend. However, the complaint does not limit the duty to defend. The insured will have to defend if facts are brought to its attention that indicate that there is a possibility of coverage.

Suppose a claim is made or liability determined that is less than the deductible specified in the policy. Some policies provide for the cost of defense in such a case, but if the amount paid on the claim is *less* than the deductible amount, defense costs are considered part of the deductible up to the deductible amount.

Suppose the policy limit is $100,000 and a claim is made for $200,000. Any recovery over $100,000 must be paid by the insured. If the insurer is willing to pay the policy limits, will the insurer be obligated to defend the claim?

Professional liability policies generally require the insurer to defend even though the latter is willing to pay the policy limit. The professional liability policy is designed to furnish the *dual* protection of paying the claim and defending the claim, subject to policy limits and deductibles.

Policies may differ as to claim expenses. Some provide "first dollar defense" coverage. Under such provisions, the company will pay all claim expenses. In some of these policies, the amount of claim expenses reduces the aggregate limit of liability coverage. (As noted in (G), some policies require the insured to pay for the claims defense costs up to the amount of any deductible.) Some policies provide that the insurer will pay 80% of claim expenses and the design professional 20%. Finally, it has been held that the cost of defense can be prorated between acts that are covered under the policy and acts that are not covered.[28] The in-

sured must evaluate any options available as to the often formidable cost of defense.

In addition to the often formidable attorneys' fees involved in defending claims, there are other expenses. Exhibits must be prepared, and expert witness fees must be paid. Sometimes transcripts must be made of testimony taken in advance of trial or at the trial. Bonds sometimes may have to be provided at stages of the legal action. The insured should determine whether the insurer is obligated to pay for these expenses.

Claim defense can raise conflict of interest between the insurer and the insured. As noted in (H), claims that can be settled within the policy limits can create a conflict of interest. The insured may wish to settle to avoid the possibility of a judgment in excess of his insurance coverage. Similarly, as noted in (G), a claim within any deductible can also raise conflict of interest as to settlement.

In addition, the insurer need not indemnify if the trial reveals liability based on acts excluded from coverage. For example, suppose the claimant's loss could be attributable either to defective plans and specifications, which would be covered, or to an express warranty of a successful outcome, which would not. As the defense is within the control of the insurer, the insurer's attorney could use that control to obtain a judicial conclusion that placed liability on conduct not covered. (Also, the attorney is ordinarily more interested in preserving his relationship with the insurer than his concern for the interests of the insured.) This led a California court to conclude that if conflict of interest arises, the insurer must pay for an independent attorney to be provided the insured.[29] Even if the insurer need not pay for the cost of an independent attorney for the insured, it may be advisable for the insured to retain independent counsel at his expense.

K. Settlement

As mentioned in (I), the typical professional liability policy, although granting the power to settle to the insured, places sharp restraints on that power

---

[28]*Insurance Co. of North America v. Forty-Eight Insulations, Inc.*, 633 F.2d 1212 (6th Cir.1980); *National Steel Constr. Co. v. National U. Fire Ins. Co.*, 14 Wash.App. 573, 543 P.2d 642 (1975).

[29]*San Diego Navy Federal Credit Union v. Cumis Ins. Soc'y, Inc.*, 162 Cal.App.3d 358, 208 Cal.Rptr. 494 (1984). Shortly after this decision, the California legislature enacted Civil Code Section 2860, which codifies the right to independent counsel where there is certain conflict of interest but also regulates who can be appointed as independent counsel and his fees.

by making the insured take certain risks if he refuses to settle when suggested to do so by the insurer.

Settlement provisions typically state that the insurer will not settle without consent of the insured. If the insured fails to consent to a recommended settlement and elects to contest the claim and continue legal proceedings, the insurer's liability for the claim does not exceed the amount for which the claim would have been settled, plus claims expense incurred up to the date of such refusal. In other words, the insured will take the risk of the settlement having been a good one.

L. Multiparty Policies

In a transaction as complex as construction, with its host of participants, inefficiencies can develop if each party carries its own liability insurance. This has led some owners to require "wrap up" policies for those engaged in construction and projectwide professional liability insurance for all the professionals involved in the project. Use of such insurance is relatively rare.

M. Termination

Most insurance policies permit the insurer to terminate by giving a designated notice, often as short as thirty or forty-five days. Increasingly, legislatures and courts deny the insurer an *absolute* right to terminate.

SECTION 15.06 Preparing to Face Claims

Design professionals should anticipate the likelihood that claims will be made against them. With this in mind, the design professional must be able to document in the clearest and most objective way that a proper job was done. For example, expanded liability should not deter design professionals from using new designs, materials, or products. Design professionals must, however, prepare for the possibility that if things go wrong, they will be asked to explain their choice.

Using new materials as an example, the design professional should accumulate information directed toward predicting the performance of any contemplated new materials. Information should be obtained from unbiased persons who have used the materials on comparable projects. A list of such

persons can be requested from manufacturers, whose representatives should be questioned about instances where bad results were obtained. The manufacturer can be notified as to intended use of the project along with a request for technical data that include limitations of the materials. Sometimes it is possible to have a manufacturer's representative present when new material is being installed to verify installation procedures. Any representations or warranties obtained should be kept readily accessible.

Design professionals should be able to reconstruct the past quickly and efficiently. A system for efficient making, storing, and retrieving of memoranda, letters, and contracts is essential. Legal advice should determine the proper time to preserve records. If major design decisions have to be made, the design professional should indicate the advantages and disadvantages and obtain a final written approval from the client. Records should show when all communications are received and responses made. If work is to be rejected, the design professional should support his decision by communications to client and contractor. Similarly, if any previous approvals are to be withdrawn, written notice should be given to all interested parties. Records should be kept of all conferences, telephone calls, and discussions that may later have to be reconstructed in the event of a dispute.

In this regard, the instability of many design professional relationships can be troublesome. Design professionals dissolve partnerships frequently. Where dissolution occurs, records that should be kept are often lost or destroyed. When rearrangements occur, those involved should separate records and see that those who may need them have them.

Design professionals need competent legal services at prices they can afford at all stages of their practice. Certainly, legal advice only *after* disputes have arisen is insufficient.

Younger groups of design professionals should consider negotiating with those who provide legal services for prepaid legal services plans.

The operations of the design professional, including contracts used, records kept, and compliance with laws regulating employers, should be evaluated periodically.

PROBLEMS

1. A contract between an architect and his client stated that all claims are barred unless they are made within six months after final payment to the contractor is due. Would such a clause be enforced? What added facts would be relevant? If you conclude that it is not enforceable, could it have been modified in such a way as to make it enforceable?

2. An engineer entered into a written contract with a large manufacturer to design a warehouse. The fee was to be 5% of the cost, *estimated* to be one million dollars. The engineer demanded that his liability be limited to 50% of his fee or $10,000, whichever is greater. The client agreed. Would such a clause be enforced? Would it be enforced if it were the full fee *paid* or $10,000, whichever is *less?*

CHAPTER SIXTEEN

Intellectual Property: Ideas, Copyrights, Patents, and Trade Secrets

SECTION 16.01 Relevance to Design Professional

Design professionals use their training, intellect, and experience to solve design, construction, and manufacturing problems of their clients and employers. Usually design professionals reduce the proposed design solution to tangible form. These forms, whether sketches, renderings, diagrams, drawings, specifications, computer software, or models, communicate the design solution to the client or employer and others concerned. Often the design solution is followed by the completion of the product being designed, such as the construction project, the industrial process, or the machine. Additionally, some aspects of the design solution, such as the floor plans, sketches, diagrams, or pictures, may be used to advertise the project or the product. To sum up, the three steps are as follows:

1. The intellectual effort by which the solution is conceived.
2. Communication of the solution.
3. Development of the end product.

If the client owns the tangible manifestation of the design solution or the end product itself it may wish that it not be copied or used without its permission. To preserve uniqueness of the project, whether a residence or a building, the client may not want another project of an identical design constructed. The client might feel wronged if the construction documents are used by others without payment to it if the client paid for and received exclusive ownership rights.

Similarly, the manufacturer who invests funds to develop a product or process may not want others to copy it without permission. The manufacturer may wish to recoup the money invested to develop the process or product either directly through royalties or by retaining a competitive advantage the research investment has given.

The design professional may wish to obtain similar protection. Dealings with the client were discussed in Section 12.11. Likewise, the design professional may want protection against third parties copying the construction documents, diagrams, or drawings to be used in developing a product or the end products themselves.

The protection accorded those who create or hire others to create tangible manifestations of intellectual effort is the subject of this chapter.

SECTION 16.02 An Overview

A. Specificity of Discussion

Certain legal concepts relating to intellectual ideas will be explored in greater detail than others. For example, patents, though of great importance to engineers, will be discussed only briefly. Patent law is a highly technical area. Inventors who wish to obtain or enforce a patent require a patent lawyer. For this reason, only the basic principles and certain salient features of patent law will be mentioned.

Obtaining copyright protection, on the other hand, is a relatively simple process. Persons who wish to *acquire* copyright protection, in contrast to legal *enforcement* of copyright remedies, can generally do so without the assistance of an attorney. For this reason, more detail will be given to copyrights than to patents.

B. Purpose of Protection

Copyrights and patents are given authors and inventors for their writings and discoveries. The primary purpose of granting them is to foster social and industrial development for the public good. This development is accomplished by granting individuals monopoly rights to reward them for their contributions, monopolies that would otherwise be antithetical in a competitive system. One judge stated:

> The economic philosophy behind the clause empowering Congress to grant patents and copyrights is the conviction that encouragement of individual effort by personal gain is the best way to advance public welfare through the talents of authors and inventors. . . . Sacrificial days devoted to such creative activities deserve rewards commensurate with the services rendered.[1]

On the other hand, society can suffer from excessive protection. Much intellectual and industrial progress depends on free interchange of ideas and free use of the work of others. Commercial and industrial ventures can be frustrated or impeded if entrepreneurs are compelled to pay tribute to persons who claim that their ideas, designs, or inventions have been used in some way by the entrepreneur. The law attempts to reward truly creative and inventive work without unduly limiting the free flow of ideas and use of industrial and scientific technology. Patent law, for example, gives a seventeen-year monopoly to the inventor of a novel, original, and nonobvious invention in exchange for disclosure to the public. The period was chosen as a compromise that adequately rewards an inventor but does not unduly perpetuate the stagnation that can accompany monopoly.

C. Exclusions from Coverage: Trademarks and Shop Rights

The creation of an effective and universally recognized trademark or trade name is an intellectual act. However, design professionals are less concerned with trademarks and trade names. For this reason there will be no discussion of common law or statutory trademarks or trade names.[2]

Shop rights, a doctrine under which an employer under certain circumstances has limited rights in the inventions of employees, will not be discussed. For all practical purposes it has been preempted by near universal use of standard form employment contracts.

SECTION 16.03 Copyright Law of 1976[3]

A. Common Law Copyright Abolished

Section 301 of the federal Copyright Act has preempted *state* common law copyright laws that gave the author the power to determine when and if the work would be made available to the public. (See Section 16.04(C).) Preemption was intended to promote uniformity both by replacing state law with federal law and by eliminating the frequently difficult question of when the work had become dedicated to the public, an act that deprived the author of a common law copyright.

Common law copyright remains only for works of authorship not fixed in a tangible medium of expression that can nevertheless be copyrighted.

Such works include choreography that has never been filmed or notated, extemporaneous speech, original works of authorship communicated solely through conversations or live broadcasts, and a dramatic sketch or musical composition improvised or developed from memory and without being recorded or written down.

B. Statutory Copyright

Classification of Copyrightable Works. The classification of works that can be copyrighted was changed from the close-ended thirteen to an open-

---

[1]*Mazer v. Stein*, 347 U.S. 201, 219 (1954).

[2]The Lanham Act, 15 U.S.C.A. § 1051 et seq., a federal statute, permits registration of trade names, trademarks, and service marks as well provides for remedies. State law also deals with trademarks.

[3]In 1976 Congress enacted Pub.L. No. 94-553, 90 Stat. 2541, which replaced the 1909 Copyright Act and went into effect in 1978. References are to sections in the current Act. The statute is also found in 17 U.S.C.A. § 101 et seq.

ended seven. Section 102 permits copyright of the following categories:

1. literary works
2. musical works, including any accompanying words
3. dramatic works, including any accompanying music
4. pantomimes and choreographic works
5. pictorial, graphic, and sculptural works
6. motion pictures and other audiovisual works
7. sound recordings

Drawings or plans would fall under (5), while specifications fall under (1). In that regard, literary works need not be "literary" so long as they express concepts in words, numbers, or other symbols of expression.[4]

Copyright Duration: More Protection. Under the Copyright Act of 1909, the copyright holder had protection for twenty-eight years, with the right to renew for an additional twenty-eight years. Much criticism had been made of copyright duration, and longer life expectancy made it inadequate. To bring American law in line with that of most foreign countries, § 302 gives copyright protection for the life of the author plus fifty years thereafter. However, if a work has been one made for hire (discussed in greater detail in Section 16.04(D)), the duration of the copyright is seventy-five years after the year of its first publication or one hundred years from the year of its creation, whichever expires first.

Codification of Fair Use Doctrine. Section 107 expressly recognizes fair use. It permits reproduction "for purposes such as criticism, comment, news reporting, teaching (including multiple copies for classroom use), scholarship or research." Factors described as bearing on whether the use is a fair one include the following:

1. The purpose and character of the use, including whether such use is of a commercial nature or is for nonprofit educational purposes.

2. The nature of the copyrighted work.
3. The amount and substantiality of the portion used in relation to the copyrighted work as a whole.
4. The effect of the use on the potential market for or value of the copyrighted work.

Obtaining a Copyright. Although the form of copyright notice was not changed, § 405(a) gives some relief where the notice has been omitted in certain circumstances,[5] and § 406 gives relief if the notice contains an error in name or date.

The law does not require registration with the Copyright Office before commencement of an infringement action. But § 412 precludes recovery of statutory damages or attorneys' fees if the copyrighted work is not registered within three months after first publication of the work.

Section 407 stiffens the requirement that copyrighted works be deposited with the Library of Congress. The Act requires that two complete copies of the best edition be deposited, although failure to deposit will not affect the validity of the copyright. A person who fails to deposit within three months *after* a demand is subject to a fine of not more than $250 for each work and a fine of $2,500 if refusal is willful or persistent.

Remedies for Infringement: Increase in Statutory Damages. In addition to permitting an injunction to prevent or restrain infringement, § 503 allows a court to order an impounding or destruction of infringing copies and articles by which infringement has been accomplished. Section 504 allows recovery of actual damages and profits made by the infringer attributable to the infringement.[6] In establishing the infringer's profits, the copyright owner is required to present proof only of the gross revenue, and the infringer must prove deductible expenses and elements of profit attributable to factors other than the copyrighted work. Statutory

---

[4]*Apple Computer, Inc. v. Franklin Computer Corp.*, 714 F.2d 1240 (3d Cir.1983) held certain computer software copyrightable. An appeal to the U.S. Supreme Court was taken. The case was settled.

[5]Relief from the requirement of attaching a copyright notice was denied because § 405(a) grants relief only if the notice is omitted from a relatively small number of copies. *Donald Frederick Evans & Assoc., Inc. v. Continental Homes, Inc.*, 785 F.2d 897 (11th Cir.1986).

[6]An application of the statute will be found in *Aitken, Hazen, Hoffman, Miller, P.C. v. Empire Constr. Co.*, 542 F.Supp. 252 (D.Neb.1982), discussed in Section 16.04(D).

damages can now be awarded up to $10,000, with $50,000 for *willful* infringement. The statutory damages for an *innocent* infringer can be reduced to $100.

Works Commissioned by U.S. Government. Some had advocated that there be no copyright in works commissioned by the U.S. government. However, Congress rejected this position and gave procuring agencies discretion to determine whether to give the design professional copyright ownership. Copyright protection will be denied only if the copyrighted work is authored by an employee of the government.

SECTION 16.04 Special Copyright Problems of Design Professionals

A. Attitude of Design Professionals Toward Copyright Protection

Design professionals vary in their attitude toward the importance of legal protection for their work. Some design professionals want their work imitated. Imitation may manifest professional respect and approval of work. When credit is given to the originator, imitation may also enhance the professional reputation of the person whose work is copied. Some design professionals are messianic about their design ideas and would be distressed if their work were not copied. Many design professionals believe that free exchange and use of architectural and engineering technology are essential.

Even design professionals who want imitation or who do not object to it draw some lines. Some design success is predicated on exclusivity. Copying the interior features and layout of a luxury residence or putting up an identical structure in the same neighborhood is not likely to please the architect or client. The same design professional who would want her ideas to become known and used might resent someone going to a public agency and without authorization copying construction documents required to be filed there. This same design professional is likely to be equally distressed if a contractor were to copy plans made available for the limited purpose of making a bid. Much depends on what is copied, who does the copying, and whether appropriate credit is given to the originator.

B. What Might Be Copied?

Design professionals may wish protection for ideas, sketches, schematic and design drawings, computer software, two-dimensional renderings, three-dimensional models, construction documents sufficiently detailed to enable contractors to bid and build, and the completed project itself. Ideas themselves cannot receive legal protection, and legal protection for the executed project as noted in (D) only recently has received limited protection. The principal problems relate to tangible manifestations of design solutions that are a step toward the project. These tangible manifestations vary considerably in the amount of time taken to create them and the amount of time and money saved by the infringer who copies them.

C. Common Law Copyright and Publication

Traditionally, design professionals sought protection through common law copyright. Perhaps this was traceable to a lack of understanding of statutory copyright. Some design professionals believe that statutory copyright protection against infringement requires the expensive registration of often formidable construction documents after they are created.[7] Whatever the reason, the bulk of the cases involving design works involve claims of common law copyright. Common law copyright was abolished by the 1976 Federal Copyright Act. For historical reasons, a brief comment on common law copyright is adequate.

The principal difficulty in perfecting a common law copyright had been the frequent claim made by the alleged infringer that the work copied had already been published. If dissemination of and the facts and circumstances surrounding the work indicated to a reasonable person that the creator had dedicated the work to the public, common law copyright was lost.

Common law copyright did not provide much protection. Design professionals should welcome its abolition. Now they are limited to statutory copyright which, though it too has its weaknesses, is substantially better.

---

[7]The prior law did *not* require registration until an infringement action had been commenced. See 17 U.S.C.A. 411.

D. Statutory Copyright

Looking first at the beginning (creation of ideas) and end (execution of completed projects) of a design professional's services, it is clear that the former is not subject to copyright protection. Before the 1976 Act, the copyright holder of technical drawings had no exclusive right to complete the project. However, drawings or models for an unusual structure that could be classified as a work of art, such as the Washington Monument or the Eiffel Tower, gave the copyright holder the exclusive right "to complete, execute and finish it." But the Copyright Office took the position that works of design professionals were copyrightable only as technical drawings and not as works of art. On the other hand, if a structure incorporated features such as artistic sculpture, carving, or pictorial representation that could be identified separately and were capable of existing independently as works of art, such features were eligible for registration. Thus, limited aspects of a building received copyright protection as a work of art. Under some circumstances, certain features of design could be sufficiently novel to justify a design patent.

Until the enactment of the Architectural Works Copyright Protection Act of 1990,[8] architects were not given copyright protection of the actual three-dimensional design, that is, the completed project itself. Change resulted from the desire of the Congress to bring American law into line with the requirements of the Berne Convention, which provides the most comprehensive protection for copyright. Signatories to the convention agree to protect the copyright of citizens of other signatories to the convention. But to bring America into line with the Berne Convention, it was necessary that American law protect works of architecture and recognize the moral rights of artists, as noted in Section 16.05.

Under the new act, an architectural work is the abstract three-dimensional design for a building.

---

[8]Title VII of Pub.L. No. 101-650, 104 Stat.5089, codified at 17 U.S.C. §§ 101, 102, 106, 120, and 301. See J. DRATLER, JR., INTELLECTUAL PROPERTY LAW: COMMERCIAL, CREATIVE AND INDUSTRIAL PROPERTY § 5.02(4) (1991). For more detailed discussion of the act, see Note, 41 Duke L.Rev. 1598 (1992); Comment, 70 Neb.L.Rev. 873 (1991), and Pink, *The Expanding Scope of Copyright Protection for Architectural Works*, 12 Constr. Lawyer, No. 4, November 1992, p. 1.

The law provides copyright protection for original design elements of such buildings. In addition, the new act makes clear that injunctive relief can be granted to enforce a copyright even if it means destruction of buildings.

Protection does not extend to all work designed by design professionals. It covers designs for building that are habitable structures, such as houses and office buildings. It also includes structures that are not inhabited by human beings but are used by them, such as churches, pergolas, gazebos, and garden pavilions. It does not cover structures designed by engineers such as interstate highway bridges, cloverleafs, canals, dams, and pedestrian walkways. The Berne Convention does not require such extended protection, and Congress determined that protection is not necessary to stimulate creativity in those fields. Extensive monuments or commemorative structures, such as the Washington Monument and Statue of Liberty, are protected if they permit entry and temporary use by people.

The protections accorded by the new act have limitations. The act does not cover features of architectural works required by utilitarian function. An architectural work does not include "individual standard features." This would preclude protection for features such as common windows, doors, and other staple buildings components. Protection does not extend to forbidding the use of pictures of buildings if the building is located in or ordinarily visible from a public place. As a result, people can make use of and reproduce photographs, posters, and other pictorial representations of architectural works. The copyright owner cannot prevent the owner of a building embodying the work from altering or destroying the building or authorizing others to do so. Similarly, the new act ensures that local entities can exercise their police powers to enforce laws regarding landmarks, historical preservation, and zoning or building codes.

The new act applies to all architectural works created after December 1, 1990. It also applies to architectural works created before that date that were unconstructed and embodied in unpublished plans or drawings on that date. However, such protection expires unless the building is constructed before December 31, 2002.

Before the 1976 Act, it was assumed that the person commissioning copyrightable works was entitled to the copyright in the absence of an agreement to the contrary. This was one of the reasons for frequent inclusion of clauses in contracts between design professionals and their clients giving the former ownership rights.

Section 201 gives copyright protection to the author of the work. However, if the work is made for hire, the employer or other person for whom the work was prepared is considered the author unless the parties have agreed otherwise in a signed written agreement. Section 101 defines a "work for hire" as prepared by an employee or a work specially ordered or commissioned. The legislative history did not include a client commissioning a design professional as work for hire.

Yet the law that has emerged since the 1976 Act has made it clear that the design professional who operates independently and is retained by a client to prepare the design owns the copyright in the absence of an agreement to the contrary. The standard that applies is the common law of agency. Was the work prepared by an employee or by an independent contractor?[9] If the design professional is an independent contractor she owns the copyright. A number of factors are used in making this determination. The ones that seem to have emerged as most significant in the cases involving independent design professionals are whether the skills involved are beyond the capacity of a layperson, whether the client paid employee benefits such as health, unemployment, or life insurance benefits, and whether the client paid employment taxes and withheld federal and state income taxes.[10]

A federal trial court opinion dealt with the remedy granted an architect under federal copyright law for the infringement of architectural drawings by his developer client.[11] The developer had purchased a tract of land and decided to build an apartment complex on one of the four parcels he had acquired. He contemplated building another apartment complex on one of the other parcels in the future. The architect prepared the drawings and specifications, and the first apartment complex was built.

Without permission of the architect, the developer copied and revised the plans to produce another set of plans from which he built another apartment complex adjacent to the first.

After the architect discovered that his plans had been copied, he billed his client $36,000. He then commenced an action for copyright infringement and sought his actual damages and the profits made by the developer and others who had participated in building the second apartment complex.

As to actual damages, the architect was entitled to the fair market value of his architectural plans. But the architect had not established a market value for architectural plans for apartment complexes. As a result, the judge looked to what the developer would have paid the architect and what the architect would have expected to receive as the fair market value for those plans.

The judge noted that the developer had paid some $13,000 for the architect's services in preparing the architectural plans for the first apartment complex. This had been based on an hourly rate. The architect's bill for the second apartment complex of $36,000 was based on the estimated construction cost and a percentage fee of $7\frac{1}{2}\%$. However, the judge rejected this method. The $7\frac{1}{2}\%$ was based not only on the preparation of drawings and specifications but also on the "supervision" of the contract award and construction, which the architect did not do. The testimony of expert witnesses for the plaintiff as to the fair market value of the architectural services ranged from $24,000 to $37,000. But each expert testified that he did not know that the architect had only received some $13,000 for the original drawings and specifications. The judge found that it would be inconceivable for the developer to expect to pay more than the $13,000 he paid for the first set. As a result, the judge concluded that the reasonable value was $13,000. As the original plans had been revised, the judge deducted the cost of making the revision, as this was an expense the architect did not have to incur. This reduced the actual damages to $10,000.

The federal copyright law gives the claimant, in addition to actual damages, any profits of the in-

---

[9]*Community for Creative Non-Violence v. Reid*, 490 U.S. 730 (1989).

[10]*Aymes v. Bonelli*, 980 F.2d 857 (2d Cir.1992) (program created by computer programmer not a work for hire and programmer entitled to copyright); *Kunycia v. Melville Realty Co.*, 755 F.Supp. 566 (S.D.N.Y.1990) (architect retained by client entitled to copyright, as copyrightable drawings were not made pursuant to a work for hire).

[11]*Aitken, Hazen, Hoffman, Miller, P.C. v. Empire Constr. Co.*, 542 F.Supp. 252 (D.Neb.1982).

fringer that were not taken into account in computing actual damages. The copyright owner need present proof only of the infringer's gross revenues. The infringer must prove deductible expenses and elements of profit attributable to factors other than to copyrighted work. The developer made a gross profit of close to $60,000 on the project. After deducting a portion of the project's administrative and general overhead expenses, his net profit was some $17,000.

The developer sold the apartments to a third party, who was also a defendant in the infringement action. The judge concluded that the third party made *no* profit but actually suffered a loss.

Finally, the judge refused to award attorneys' fees to the plaintiff architect because the infringement had occurred before the architect had registered his copyright.

The final judgment awarded the architect $10,000 actual damages and the profits of $17,000 earned by the developer. The cost of a long and costly trial may have far exceeded the award. This should not be taken to mean that a copyright claim by a design professional will never produce a significant damage award. In a recent case, the architect recovered the fair market value of the plans ($11,968) plus the profit ($42,250) made by the party who infringed the architect's copyright in building a residence.[12] Nevertheless a careful calculation must be made of the likely recovery, because those against whom infringement claims are brought often raise a number of attacks to the validity of the claim.

One reason for design professionals seeking common law rather than statutory copyright protection was the presumed burden of depositing copyrighted works with the Copyright Office. Before and after the 1976 Act, registration could be made at any time prior to an infringement action being brought. However, under the 1976 Act, failure to register within three months of publication precludes recovery of attorneys' fees or statutory damages.[13] But registration need not be difficult or expensive. Even the pre-1976 copyright regulations

permitted substitutions for the original materials themselves if they were bulky or if it would be expensive to require deposit.

The 1976 Act gives authority to the Register of Copyrights to specify by regulation the nature of copies required to be deposited. These regulations can allow deposit of identifying material rather than copies. One rather than two copies can be permitted.[14]

E. Advice to Design Professionals

If copyright protection is to be sought, design professionals should comply with the statutory copyright requirements. First, they should be certain that they have not assigned their right to copyright ownership. They no longer need a contract clause giving ownership rights to the plans and specifications to the design professional. But see Section 12.11.

Second, design professionals should comply with the copyright notice requirements. The word *Copyright* can be written out, or the notice can be communicated by abbreviation or symbol. The authorized abbreviation is "Copr," and the authorized symbol is the letter "C" enclosed within a circle (©). The year of first publication should be given, and the name of the copyright owner or an abbreviation by which the name can be recognized or generally known can be used. The notice must be affixed to the copies in such a manner and location as to give reasonable notice of the copyright claim. The Register of Copyrights is given authority to specify methods by which copyright notice can be given.[15]

If the design professional wishes to take advantage of the statutory damage award and to recover attorneys' fees, she should register the copyrighted work within three months of publication. Methods are available to minimize this burden. The Copyright Office should be consulted.

SECTION 16.05 Moral Rights of Artists

The moral rights of artists were recognized by the enactment of the Visual Artist Rights Act of 1990,

---

[12]*Eales v. Environmental Lifestyles, Inc.,* 958 F.2d 876 (1992 9th Cir.), cert. denied sub nom *Shotey v. Eales,* 113 S.Ct. 605 (1992).
[13]17 U.S.C.A. § 412.

[14]17 U.S.C.A. § 408(c).
[15]As indicated, 17 U.S.C.A. §§ 405 and 406 provide relief if the copyright notice is omitted or erroneously made.

which was also designed to bring America into line with the Berne Convention.[16] Two kinds of rights are protected: the right of attribution and the right of integrity. Works protected are limited to graphic, sculptural, and photographic works. While generally this topic is beyond the scope of this treatise, one aspect should be mentioned.

The right of integrity consists of the artist's right to prevent certain distortions, mutilations, or other modifications of her work as well as the right to prevent destruction of work of recognized stature. This protection relates only to acts that will be prejudicial to the author's honor or reputation.

The act has a number of limitations, one of which relates to visual art incorporated in buildings. If an artist has consented to the incorporation of her visual art in a building, the owner of the building can remove the art even if it cannot be removed without destroying it or otherwise violating the artist's right of integrity. Consent that is given after June 1, 1991, must be evidenced by a written instrument signed by the artist and the owner and must specifically give the owner the right to remove even if removal destroys the work of art.

If the incorporated art can be removed *without* destruction, the owner of the building may still avoid infringing on the artist's moral rights by notifying the artist and giving her a chance to remove the art and pay for its removal within ninety days of receiving notice. The owner who is unable to notify the artist despite a good-faith attempt to do so can proceed to remove the incorporated visual art. The Register of Copyrights provides a method of recording names and addresses of artists and permits building owners to record their efforts to locate an artist.

SECTION 16.06 Patents: Some Observations and Comparisons

A. Scope of Coverage

A design professional who wishes to institute legal action for infringement of a patent or a copyright will retain an attorney. Although the steps for perfecting a copyright are simple, an inventor who

wishes to obtain a patent must secure the services of a patent attorney. A patent attorney is needed to guide the inventor through the maze of patent law and the complexities of a patent search. Perfection of a copyright as a rule will not require the services of an attorney.

B. Patent and Copyright Compared

The subjects of patents generally are products, machines, processes, and designs.[17] Copyright generally protects writings.

The principal protection accorded by copyright law is the exclusive right to reproduce, prepare derivative works, or distribute the copyrighted material. Copyright law does not protect against someone who, without knowledge of the copyrighted work or access to it, creates a similar work. A patent gives the patent holder a monopoly. The patent holder can exclude anyone from the field covered by the patent even if the same invention has been developed independently and without any knowledge of the patented device.

The most important difference between patent and copyright law is the higher degree of creativity required for issuance of a patent. Copyright law requires only that the work have some originality and be the independent labor of the author. Patent law requires that the work be original, inventive, useful, novel, and not obvious from the prior art in the particular field.

Protecting an invention begins with the issuance of a patent, which is supposed to presumptively establish that the patent is valid. In the bulk of patent infringement cases, the defendant attacks the validity of the patent. The defendant has a good chance of establishing that the patent is not valid.

One study covering patent cases back to 1920 demonstrates the difficulty of sustaining a patent in the courts.[18] From 1920 to 1973, the Supreme Court found 18% of the patents it reviewed to be valid. The Court of Appeals found 35% valid, while the trial courts found validity in 45% of the cases. Even more ominous for patent holders, the recent

---

[16]Title VI 101-650, 104 Stat. 5089, 5128–5133 (December 1990). See *DRATLER*, supra, note 8, at Section 6.01(6).

[17]For an instructional design patent case in an architectural context, see *Blumcraft of Pittsburgh v. Citizens & Southern Nat'l Bank of South Carolina*, 407 F.2d 557 (4th Cir.1969).
[18]Baum, *The Federal Courts and Patent Validity: An Analysis of the Record*, 56 J.Pat.Off.Soc. 758 (1974).

tendencies have been to find patents invalid, probably reflecting the prevailing hostility toward monopoly. The difficulty of sustaining a patent's validity often induces those who have developed industrial data to keep the information secret rather than publish it and seek patent protection. (Trade secrets are discussed in Section 16.07.)

Patent infringement suits are lengthy, complicated, and expensive. The plaintiff can recover damages and, in some cases, a compulsory royalty for patent infringement. In some flagrant infringement cases, the plaintiff can recover treble damages and attorneys' fees.

A patent's duration is seventeen years. This is substantially less than copyright protection. The limited protection accorded copyright and the monopoly protection accorded a patent are probably the reason the patent protection is more limited in duration.

Generally, patent protection is harder to acquire than copyright but once acquired is worth much more.

SECTION 16.07 Trade Secrets

A. Definition

The Restatement of Torts has defined a trade secret as

> . . . any formula, pattern, device, or compilation of information which is used in one's business, and which gives him an opportunity to obtain an advantage over competitors who do not know or use it. It may be a formula for a chemical compound, a process of manufacturing, treating or preserving materials, a pattern for a machine or other device . . . A trade secret is a process or device for continuous use in the operation of the business. Generally it relates to the production of goods, as, for example, a machine or formula for the production of an article.

> * * *

> The subject matter of a trade secret must be secret. Matters of public knowledge or of general knowledge in an industry cannot be appropriated by one as his secret. Matters which are completely disclosed by the goods which one markets cannot be his secret. Substantially, a trade secret is known only in the particular business in which it is used. It is not requisite that only the proprietor of the business know it. He may, without losing his protection, communicate it to employees in-

volved in its use. He may likewise communicate it to others pledged to secrecy.

The Restatement sets forth the following factors that are considered in determining whether particular information is a trade secret:

1. The extent to which the information is known outside of the proprietor's business.
2. The extent to which it is known by employees and others involved in the proprietor's business.
3. The extent of measures taken by the proprietor to guard the secrecy of the information.
4. The value of the information to the proprietor and her competitors.
5. The amount of effort or money expended by the proprietor in developing the information.
6. The ease or difficulty with which the information could be properly acquired or duplicated by others.[19]

B. Context of Trade Secret Litigation

Trade secret litigation can arise when an employee leaves an employer either to go into business or to work for a new and frequently competing, employer. If the former employee has commercial or technical information, the prior employer may seek a court decree ordering the former employee not to disclose any trade secrets "belonging" to the prior employer and a decree ordering the new employer not to use the secret information. Such a court order can be justified by a confidential relationship between the prior employer and the former employee or the breach of an employment contract between the prior employer and the former employee.

Trade secret litigation can result when the proprietor of a trade secret learns that someone to whom a trade secret has been disclosed on a basis of confidentiality intends to make or has made unauthorized use of the information. For example, the developer of a new product may give technical information relating to the product to the contractor building the plant in which the product is to be manufactured or to the manufacturer who is to build the machinery needed to make the product.

---

[19]Restatement (First) of Torts § 757 comment b. The Second Restatement of Torts omitted this topic. Although less authoritative, the First Restatement still provides a useful summary.

Trade secret litigation can also result if the person to whom the disclosure has been made intends to make or has made unauthorized use of the information. Similarly, a confidential disclosure of the information may be made to a manufacturer by an inventor who seeks to interest the manufacturer in a process or product developed by the inventor. The unauthorized use of such a precontract disclosure can lead to trade secret litigation.

Developers of technology sometimes try to recover research costs by licensing others to use the data. To protect the secrecy of the technology and to enable them to sell the data to others, developers usually obtain a promise from the licensee not to disclose the data to anyone else. Breach or a threatened breach of such a nondisclosure promise may cause the proprietor of the trade secret to seek a court decree forbidding any unauthorized use or disclosure.

C. Contrast to Patents: Disclosure vs. Secrecy

Patent law requires public disclosure of the process, design, or product that is the subject of a patent. In exchange for this public disclosure, the patent holder obtains a seventeen-year monopoly. Trade secret protection, on the other hand, requires that the data asserted to be a trade secret be kept relatively private and nonpublic.

A patent requires an invention to be novel, unique, useful, and not obvious from the prior art. The trade secret need not meet these formidable requirements.[20] Although the courts are not unanimous on the point, it seems clear that the person who asserts ownership of a trade secret must show that she has made some advance on what is generally known. If the information is generally known or generally available, the information is not a trade secret.

Roughly, trade secret protection has the same relationship to patent protection that common law copyright had to statutory copyright. Both the doctrine of common law copyright and the doctrine of trade secrets are predicated on extending legal protection to creative people by giving them the right to determine when, how, and if the fruits of their

intellectual labor should be made generally available. Patent and statutory copyright are predicated on disclosure. Trade secret protection is accorded by state law and suffers from the same lack of uniformity that common law copyright had. On the other hand, patent law is governed by federal law, resulting in general uniformity throughout the United States.

D. Adjusting Competing Social Values

The doctrine of trade secrets, like many other legal doctrines, must consider and adjust various desirable, yet often antithetical, objectives. This can be shown by examining these objectives from the points of view of the various persons affected.

Those who seek trade secret protection—primarily inventors and research-oriented organizations—want to be rewarded economically for their creativity. Restricting others from using the information and technology that they have developed can make their information more valuable. Without adequate economic incentives, scientific and industrial progress is likely to be impeded. Protection of trade secrets can discourage industrial espionage and corruption.

Trade secret protection can restrain the freedom of choice and action for research employees. Creative employees can be prevented from making the best economic use of their talents. An employer's failure to consider or develop an employee's research ideas can destroy the employee's creativity. Many hi-tech industries developed when creative people banded together to start new companies. Had they been tied to an older established company unwilling to engage in experimental research, many of these industries might not have developed or might have taken considerably longer to do so.

Overzealous protection of trade secrets can hamper commercial, scientific, and industrial progress. To a great extent, such progress is made possible by the free dissemination of technical and scientific information. Dissemination of such information can avoid costly duplication of research efforts.

Overprotection of trade secrets can also have an anticompetitive effect. Protection of trade secrets can give the developer a virtual monopoly that can hinder the development of competitive products and can result in higher prices to consumers.

---

[20]*Univ. Computing Co. v. Lykes-Youngstown Corp.*, 504 F.2d 518 (5th Cir.1974); *Raybestos-Manhattan, Inc. v. Rowland*, 460 F.2d 697 (4th Cir.1972).

Trade secret law has had to consider and adjust all these competing objectives—not an easy task.

E. Availability of Legal Protection

Duty Not to Disclose or Use: Confidential Relationship and Contract. The circumstances surrounding the disclosure and the nature of the information disclosed are relevant in determining whether a duty exists not to use or disclose the information. If the disclosure is accompanied by an express promise not to use or disclose, a general duty exists not to disclose. It may still be necessary to interpret the agreement to determine what cannot be disclosed, to whom disclosure is prohibited, and the duration of the restraint.

Suppose a licensing agreement exists by which the licensor permits the licensee to use technological data disclosed by the licensor to the licensee. Does the restraint on disclosure include information that the licensee knew before the disclosure? Does it include information developed by the licensee from the disclosed information? Does the restraint include parts of the technological data disclosed that are known at the time of disclosure or become generally known? Can disclosure be made to an affiliated or successor company? Is there a continuing obligation for either or both parties to communicate new technology? These questions should be and usually are covered in the licensing agreement. If not, courts must interpret the agreement and, if necessary, imply terms.

In some circumstances, there is no express provision prohibiting unauthorized use or disclosure. The method by which the information is acquired will often determine whether the disclosure is made in confidence and whether the person to whom it is disclosed obligates herself not to disclose it to others. This is similar to the process by which the law implies certain promises between contracting parties not expressed in the written contract. The communication may be part of a contractual arrangement. For example, the possessor of the information may communicate it to a consulting engineer who has been retained to advise the possessor on the type of machinery to be used in the process. If a written contract exists, the possessor will usually require a promise by the consulting engineer not to divulge certain specified information. Even without such an express promise not to disclose, the law would probably imply such a promise, based on surrounding facts and circumstances. The same is true if the disclosure is made to the manufacturer of the machine or to a building contractor.

Circumstances exist where there is no contractual relationship between the possessor of the information and the person to whom it is disclosed. For example, an inventor may disclose information to a manufacturer in order to interest the manufacturer in buying the information. It is possible for the inventor to obtain a promise from the manufacturer not to disclose the information. Even without such a promise, if it is apparent from the surrounding facts and circumstances that the disclosure is made in confidence, any disclosure of the information by the manufacturer would be a breach of the confidence and would give remedies to the inventor.

Nature of Information. If the person to whom the information is disclosed—whether a contractor hired to build a plant, a manufacturer hired to build a machine, or a consulting engineer hired to furnish technical services—knows that the information is not generally known in the industry, this is likely to persuade a court that a confidential relationship was created or that a nondisclosure promise should be implied.

The nature of the information will also determine the legal remedy for a breach of confidence or a breach of contract. Under American law, the normal remedy for a breach of contract is a judgment for money damages. Only if that remedy is inadequate will the law specifically order that a defendant do or not do something. This is crucial in trade secret cases. Typically, if the information is truly valuable and not generally known, the most important remedy is the court decree ordering that the person who has the information not disclose it to anyone else. Violation of such an order is punishable by a fine, or even imprisonment, under the contempt powers of the court. Such a decree puts the plaintiff in a good position to demand a substantial royalty or settlement price if the defendant needs to use the trade secret information.

To obtain such an extraordinary remedy, the plaintiff must show that irreparable injury would occur without such a court order and, as mentioned, that a judgment for money damages would be inadequate. In a trade secret case, the plaintiff seeks to show that irreparable economic harm would be suffered if the information claimed to be a trade secret is broadly disseminated. The plaintiff usually asserts that such broad disclosure will enable competitors to "catch up" despite the plaintiff's research expenditure to turn out a better product or develop a better process. The plaintiff will also claim that it is difficult, if not impossible, to establish the actual damages suffered by general dissemination of the secret information. A court concluding that the information is a trade secret usually gives injunctive relief.

The principal defense in trade secret cases is that the information was not secret.[21] Often defendants point to the existing literature in a given scientific or technical area, with a view towards showing that a person diligently searching for this information could put it together and arrive at the process independently. Courts have not been particularly receptive to this defense. Usually the defendant has not gone through the literature to ferret out the secret. The information is often obtained from an employee of the trade secret possessor, paid for its disclosure by virtue of a licensing agreement, or received through a confidential, limited disclosure. Although the defense has occasionally worked, on the whole it has not been successful. Part of the difficulty in arguing for this defense is that sometimes information and data are available, but not in a collected, organized, convenient, and usable form. These factors are the principal advantages of the trade secret. Sometimes the data are collected and organized in readily accessible form, but most people in the industry are unaware of this fact or unable to locate the material easily.

Employee Cases. There are special aspects to the cases where the information has been learned or developed by an employee and that employee goes into business herself, joins in a venture with others, or is hired by an existing or potential competitor of the prior employer. In addition to using the confidentiality theory, the former employer often points to an employment contract under which the employee agreed not to disclose the information after leaving the employment. Sometimes the lim-

itation on disclosure far exceeds what is reasonable. Often the employee has little bargaining power in deciding whether to sign such an agreement. Some courts have recognized the adhesive (nonbargain) nature of such agreements and have refused to give these agreements literal effect. However, the employer who has an agreement by the employee not to divulge information is in a better position to obtain a court decree ordering the employee not to disclose particular information.

In addition to recognizing the take-it-or-leave-it nature of most employment contracts, some courts feel that agreements under which employees cannot practice their trade or profession or use the information that is their principal means of advancement are unduly oppressive to employees. Such courts are not likely to be sympathetic to claims for trade secret protection. On the other hand, other courts manifest great concern with immorality and disloyalty on the part of employees and look on employee attempts to cash in on information of this type as morally indefensible. These courts are likely to deal harshly with employees in trade secret cases.

To sum up, a former employee will be restrained from using confidential information if that restraint is reasonable,[22] taking into account the legitimate needs of the former employer, the former employee, and the public.

F. Scope of Remedy

A trade secret claimant can recover damages suffered, profits made by the infringer resulting from the infringement, and a court decree prohibiting her from using or divulging the information.[23] The injunctive relief usually does not exceed the protection needed by the plaintiff. An injunction may be only for a period of time commensurate with the advantage gained through the technological in-

[21]In *ILG Ind., Inc. v. Scott*, 49 Ill.2d 88, 273 N.E.2d 393 (1971), the court held that the possibility of "reverse engineering" (starting from the finished product and working backwards) was not a defense when the process of doing so was quite time consuming.

[22]An interesting case examining these factors and restraining the former employee is *B.F. Goodrich Co. v. Wohlgemuth*, 117 Ohio App. 493, 192 N.E.2d 99 (1963).
[23]One court held that a plaintiff cannot receive both its losses and profits of the defendant. *Sperry Rand Corp. v. A-T-O, Inc.*, 447 F.2d 1387 (4th Cir.1971). Another stated that losses occur only as a practical matter when the trade secret is destroyed. *Univ., Computing Co. v. Lykes-Youngstown Corp.*, supra note 20. That case explored formulas for determining gain to defendant.

formation improperly acquired or used.[24] If the defendant could have ascertained the information within a designated period, the court decree may require that she not use the information for that period of time. Unless the defendant has made the information public, the court order for nondisclosure will apply only until the information is generally known. Some courts take a more punitive attitude and will order that the trade secret not be used even if it becomes generally known.[25] Generally, the more reprehensible the conduct by the defendant, the broader the injunction.

G. Duration of Protection

The trade secret is protectible as long as it is kept relatively secret. This unlimited time protection has caused some to advocate protecting trade secrets for a limited period of time by according a patent-like monopoly to the developer of a trade secret.

Some trade secrets are patentable. Unlimited duration of protection for a trade secret can frustrate the seventeen-year patent monopoly policy. To the extent that states, through protection of trade secrets, frustrate patent law, such trade secret protection may be unconstitutional.

---

[24]*ILG Ind., Inc. v. Scott*, supra note 21 (for 18 months); *Sperry Rand Corp. v. A-T-O, Inc.*, supra note 23 (for two years).
[25]A. TURNER, LAW OF TRADE SECRETS 437–438 (1962).

H. Advice to Design Professionals

Design professionals who invent processes, designs, or products should, wherever possible, use contracts to give them protection against the possibility that persons to whom they divulge the information may disclose the information to others or use it themselves.

Design professionals who occupy managerial positions in companies where trade secrets are important should use all methods possible to keep the information secret. Only those who have an absolute need to use the information should be given access to it, and these persons should expressly agree in writing not to disclose the information. Management should also realize that employee loyalty is probably the best protection against the loss of trade secrets. Reasonable treatment of employees is likely to be a better method of preserving trade secrets than litigation.

Design professionals who are technical employees and who wish to take their technological information to start their own business, join in a business venture, or work for a competitor of their present employer should realize that their departure under these circumstances may result in litigation, or at least the threat of litigation. Legal advice should be sought to examine the legality of any asserted restraints and to determine the scope of risk involved to the employee who chooses to leave her present employment.

Planning the Project: Compensation and Organization Variations

SECTION 17.01 Overview

A. Some Attributes of the Construction Industry

Please review Chapter 8, with special attention devoted to contractors. Industry characteristics are central to this chapter.

B. Owner's Objectives

In the design phase, quality, price, and completion date are interrelated. An owner who wishes the highest quality may have to trade off quantity, price, and completion date. If early completion is crucial, the owner may have to sacrifice price and probably quality and quantity.

After these choices have been made, the owner may have to make *additional* choices when it determines *how* it will select a contractor, the contractor or contractors it *will* select, and the type of construction contract or contracts. These choices should be made in a way that maximizes the likelihood that the owner will receive quality that complies with the contract documents and on-time completion at the lowest *ultimate* cost. Compromises may be needed. The contractor who will do the highest quality work is not likely to be cheapest and quickest. The quickest contractor may not be the one who will provide the best quality. The importance the owner attaches to these objectives will affect the process the owner uses to select the contractors and how construction contracts are organized. This chapter looks at pricing and organizational variations. Chapter 18 focuses on competitive bidding.

C. Blending Business and Legal Judgments

This treatise examines law in the *context* of the construction industry. A sharp line cannot always be drawn between business and legal considerations.

Clearly, choices as to the pricing of a construction contract and how the project is to be organized must seek to achieve the owner's objectives. Choices made in these crucial matters should seek to obtain the best on-time work at the best price. To a significant degree, this will depend on each participant's knowing what it is supposed to do and being able to do so in the most efficient way. Modern methods designed to bring efficiency to what can be a chaotic process are described in Section 17.04. Although Chapter 8 describes some characteristics of the construction industry, one characteristic—operational inefficiency—is relevant to this chapter. Section 17.05 looks at internal efficiency—mainly authority and communication.

An empirical study dealt with "the pervasive and distressing inefficiency" in construction work. One writer stated that the study indicated that only 32% of the total time spent on a construction site involved actual work on the project.[1] He quoted the study as showing that the remainder of the work was divided in the following manner:

1. 7% for equipment transportation delays
2. 13% for travelling on the job site

---

[1] Foster, *Construction Management and Design-Build/Fast Track Construction: A Solution Which Uncovers a Problem for the Surety,* 46 Law and Contemp.Probs. 95, 116 n. 119 (1983).

3. 29% consumed by waiting delays
4. 8% for late starts and early quits
5. 6% for receiving instructions
6. 5% for personnel breaks

Quoting these statistics, the writer concluded that much inefficiency is due to the many contractual relationships and parties all working on the same structure, each "under a different management and each marching to the beat of a different drummer."[2]

Choices of the type described in this chapter that deal with these matters are to a great extent best made by those who know design and construction. What role does the law play? The answer can be divided into two categories: direct and indirect legal controls.

Direct legal controls involve such matters as registration and licensing laws, legal controls dealing with how a construction contract is awarded, the standard to which a contracting party is held, and the way in which the law will treat claims by a party who has suffered losses that it seeks to transfer to someone else.

Indirect legal considerations are often as important. For example, where there are blurred lines of authority and unclear allocations of risk, the law will frequently be called on to pass on claims. Similarly, inefficiency and other matters that cause losses are likely to lead to claims that, though usually settled, are done so against the backdrop of what the law would provide in the event the dispute ended up in the courthouse.

Increasingly, it is being recognized that one of the overhead costs in construction work incurred by all participants is making, avoiding, preparing for, and resolving claims. Choices of the type described in this chapter must be made carefully and intelligently. Failure to do so can only increase the cost of construction.

D. Public vs. Private Projects

An important criterion is the status of the owner—whether it is a private party or a public entity. A private party who wishes to build can choose any type of compensation plan it can persuade a contractor to accept. It can award a contract in any way it chooses. It can make one contract with a prime contractor or a number of separate contracts with individual contractors.

Public entities may be limited by laws and regulations in making these choices. Commonly, public entities must award their construction contracts by competitive bidding under which contractors are all given a chance to submit a bid for particular work. That design professionals for public projects do not—at least at the first round—compete on the basis of price[3] causes problems when a public entity wishes to use the design/build system as noted in Section 17.04(F).

Some states and cities require that separate contracts be used for certain types of public work. Often limitations are placed on cost contracts used in public works. A public entity may be controlled in the way it resolves disputes with its contractors. Early in this century, public entities frequently took the position that they could not arbitrate disputes, as this would be delegating power to private arbitrators. Increasingly, by law or regulation, public entities are being required to use arbitration or some other method of resolving disputes.

Those working on public contracts must first examine statutes and regulations applicable to them to determine the restraints placed on them in awarding or organizing construction contracts. This chapter largely assumes that there are no restraints and that the owner and contractor can make any type of contract they wish and the owner is not limited in the way in which it chooses to organize participants contractually and administratively.[4]

SECTION 17.02 Pricing Variations

Selecting a compensation system must take into account the responsibility for certain risks. Though many variations are possible, each major category deals with risk allocation.

A. Fixed-Price or Lump-Sum Contracts: Some Variations

In American usage, unlike that in continental Europe,[5] fixed-price and lump-sum contracts are used

---

[2]Ibid.

[3]Refer to Section 11.03.
[4]For an instructional case study on planning for construction of a hospital, see Macomber, *You Can Manage Construction Risks*, Harv.Bus.Rev., March–April 1989, at 155.
[5]There it is much harder to obtain a price increase under a fixed-price contract than under a lump sum. The ameliorating doctrine of *rebus sic stantibus* does not apply to the former.

interchangeably. Under such contracts, the contractor agrees to do the work for a fixed price. Almost all of the performance risks—that is, events that make performance more costly than planned—fall on the contractor. For example, in *American Casualty Co. v. Memorial Hospital Association*,[6] no relief was given the contractor despite the actual costs being twice those anticipated.

Only if the contract itself provides a mechanism for increasing the contract price can be contractor receive more than the contract price. Although not common in *ordinary* American contracts (English and international contracts often use fluctuations clauses), some contracts have price escalation clauses under which the contract price is adjusted upward (and sometimes downward), depending on market or actual costs of labor, equipment, or materials. Such a provision protects the contractor from the risks of any unusual costs that play a major part in its performance.[7] It usually requires the contractor to use its best efforts to obtain the *best* prices.

Many construction contracts contain changed conditions clauses that allow a price increase if conditions under the ground or in existing structures are discovered that are substantially different from those anticipated by the parties.

Sophisticated procurement systems sometimes provide *variant* fixed-price formulas even in fixed-price contracts. For example, one is the federal fixed-price incentive firm (or FPIF) contract used in *negotiated* contracts. Before contract finalization, owner and contractor negotiate the following items:

1. Target cost—against which to measure final costs.
2. Target profit—a reasonable profit for the work at target cost.
3. Ceiling price—the total dollar amount for which the owner will be liable.
4. Sharing formula—the arrangement for establishing final profit and price.

After the work is completed, the contractor and the owner negotiate the final costs of the contract,

sharing the overruns or underruns according to the agreed formula. To illustrate, suppose the target cost for a contract is one million dollars, the target profit is $100,000, the price ceiling is $1,180,000, and the sharing formula is 75% (owner) and 25% (contractor). Under the formula, the contractor would keep 25% of every dollar saved. To earn a total profit of $120,000, it would have to reduce costs by $80,000 below target cost. Because there is no profit ceiling, profit would continue to increase indefinitely as the amount of underrun increased. Conversely, the contractor would have to overrun the target cost by $80,000 to reduce its profit to $80,000. If it overran by more than $180,000, it would lose money, as there is no minimum profit guaranteed in this contract type. Regardless of the final cost to the contractor, the contractor must meet the contractual specifications, and the owner's liability cannot exceed the ceiling price of $1,180,000.

This form of contract has the advantage of establishing a price ceiling similar to the guaranteed cost to the owner under the fixed-price contract. By penalizing the contractor for cost overruns above the target estimate (which is always something less than the price ceiling) and by rewarding it for cost savings, this type of contract provides financial motivation to the contractor to perform at the most economical cost.

A contract of this type is appropriate where the owner's plans are not sufficiently detailed to allow fixed-price bidding without excessive provision for contingencies yet are sufficiently advanced that a reasonably accurate target estimate can be made. Such a contract can also, if desired, include monetary incentive provisions to the contractor for early completion.

A fixed-price contract has the obvious advantage of letting the owner and those providing funds for the project know in advance what the project will cost. It works best when clear and complete plans and specifications are drawn. Incomplete contract documents are likely to cause interpretation questions that can lead to cost increases that under a cost-plus-overhead-and-profit changes clause can convert what appears to be a fixed-price contract into a cost contract.[8] The fixed-price con-

[6]223 F.Supp. 539 (E.D.Wis.1963).
[7]*RCI Northeast Services Div. v. Boston Edison Co.*, 822 F.2d 199 (1st Cir.1987) (retrospective insurance premium allowable cost as parties intended full pass-through of costs).

[8]*Rudd v. Anderson*, 153 Ind.App. 11, 285 N.E.2d 836 (1972) (fixed price of $35,000 became time and materials contract for $58,000).

tract is used most efficiently when a reasonable number of experienced contractors are willing to bid for the work. This is less likely to be the case where the design is experimental or where there is an abundance of work.

Another advantage to the fixed-price contract is that the owner need not be particularly concerned with the contractor's record keeping. If changed work or extra work is priced on a cost basis, there may have to be some inquiry into the contractor's cost. On the whole, the fixed-price contract avoids excessive owner concern with cost records of the contractor. Conversely, such a contract is attractive to the contractor, who need not expose its cost records, something as seen in (B) that occurs in a cost contract.

The fixed-price contract has come under severe attack because of its inherent adversarial nature. A contractor who reduces its costs increases its profits. As long as the cost reduction does not come at the expense of the owner's right to receive performance specified in the construction contract, the owner cannot object.

In construction, however, the performance required and whether such performance has been rendered are often difficult to establish. In this gray area of compliance, the interests of owner and contractor can clash. Unless the contractor is interested in its reputation for quality work or values goodwill, it is likely to perform no more than is demanded by the contact. This has led to some of the variations described in Section 17.04, particularly the cost contract with a guaranteed maximum price (GMP), sometimes called a guaranteed maximum cost (GMC).

Another disadvantage to the fixed-price contract is that the risk of almost all performance cost increases falls on the contractor. A prudent contractor will price these risks and include them in its bid. However, in a highly competitive industry, the prudent contractor might not receive the award because there may be others who are more willing to gamble with a low price and either hope that problems will *not* develop or recoup any losses by asserting claims for extras and delays. Even if a prudent contractor does take these risks into account and does receive the award, if the risks do not materialize, the owner may be paying more than it would have had the risk been taken out of the contractor's bid.

B. Cost Contracts

When prospective contractors cannot be relatively certain of what they will be expected to perform, or where they are uncertain as to the techniques needed to accomplish contractual requirements, they are likely to prefer to contract on a cost basis. For either reason, it is likely that projects that involve experimental design, new materials, or work at an unusual site or those in which the design has not been thoroughly worked out are likely to be made on a cost basis.[9]

Usually a cost contract allows the contractor to be paid its costs *plus* an additional amount for overhead and profit. This should be distinguished from what is sometimes called a time and materials contract, which at least in one case was held to preclude recovery by the contractor of overhead on direct labor costs.[10]

The cost contract has two principal disadvantages. First, the owner does not know what the work will cost at the time it engages the contractor. Second, as a general rule, a cost contract does not give sufficient incentive to the contractor to reduce costs. These have led to variants on a pure cost contract developed in both public and private contracting systems.

Cost contracts often contain provisions that require the contractor to use its best efforts to perform the work at the lowest reasonable cost. Often provisions are included that require the contractor who has reason to believe that the cost will overrun any projected costs to notify the owner or its representative and give a revised estimate of the total cost. Sometimes these provisions state that failure to give notice of prospective cost overruns will bar recovery of any amounts over any cost estimates that have been given.[11]

---

[9]See Rosenfeld & Geltner, *Cost-plus and Incentive Contracting: Some False Benefits and Inherent Drawbacks*, 9 Constr. Management & Economics 481 (1991) (argues that at a macro level, widespread use of cost-plus contracts on the average and over time contributes to adverse selection, blunts incentives for production efficiency, leads to higher costs and prices in the industry, and helps mediocre contractors win jobs against efficient ones).

[10]*Colvin v. United States*, 549 F.2d 1338 (9th Cir.1977).

[11]In *Jones v. J.H. Hiser Constr. Co., Inc.*, 60 Md.App.671, 484 A.2d 302 (1984), the court in a cost plus a fixed fee contract implied an obligation by the contractor "to know of, keep track of, and advise the owner that actual costs were substantially exceeding estimates."

Another method of keeping costs down is to include provisions in the contract stating that a fiduciary relationship has been created between owner and contractor that requires that each use its best efforts to accomplish the objectives of the other and to disclose any relevant information to the other.[12]

Even without specific provisions designed to protect the owner from excessive costs, the design professional and the contractor should keep the owner informed of costs based either on the covenant of good faith and fair dealing[13] or on an obligation inherent in the fiduciary relationship created by such a contract.[14]

One type of cost contract gives the contractor *cost plus a percentage of cost* for overhead and profit. Obviously, such a contract not only creates little incentive to cut costs but also grants a reward for increasing costs. For this reason, it is not used in federal procurement. However, sometimes it is used to price changed work in private contracts. It provides a readily accepted guideline for determining the percentage in contrast to the less readily definable negotiated fixed fee—its alternative for compensating overhead and profit. With a contractor of the highest integrity, this type of cost contract is useful.

Another type is the *cost plus a fixed fee* contract. The parties agree that the contractor will be reimbursed for allowable costs and paid a fee that is fixed at the time the contract is made. The fee is normally not affected when actual cost exceeds or is less than the estimated cost. However, if the scope of the work is substantially changed, sometimes the fee is renegotiated. Because the contractor's fee is not affected by cost savings, the contractor has no compensation incentive to reduce costs. For this reason, in federal procurement, this type of contract has largely been superseded by cost contracts that create incentives to reduce costs.

---

[12]AIA Doc. A111, ¶ 3.1, its predecessor cited in *Jones v. J.H. Hiser Constr. Co., Inc.*, supra note 11, states that the contractor accepts a relation of trust and confidence and will further the interests of the owner.

[13]See Section 19.02(D).

[14]*Williams Eng'g, Inc. v. Goodyear, Inc.*, 496 So.2d 1012 (La.1986) (engineer responsible for cost overrun because he did not update estimates, did not advise of other types of contracts, and did not hire a cost estimator).

One incentive cost contract used in federal procurement is the *cost plus award fee contract*, sometimes referred to as a CPAF contract. Under a CPAF contract, the owner reimburses the contractor for its actual allowable costs and the contractor is paid a base fee that is negotiated before contract award and usually is a low percentage of the agreed *estimated* total cost of the work.

The contractor is given an opportunity to earn through superior performance an additional specified award fee that may be two or three times the amount of the base fee. The award fee is determined by the owner and is based on the owner's evaluation of the contractor's performance. The basis for this evaluation is set forth in the contract and focuses on those goals the owner considers most important. The award fee is designed to give the contractor incentive for high-quality performance.

A CPAF contract states the performance factors and the weights assigned to them that will be used in the owner's periodic evaluation of the contractor's performance. If the owner's goals change, the contract can be amended by mutual agreement to revise the weights assigned or add new factors. In an average construction program, evaluation factors might be control and reduction of costs, quality of construction, and maintenance of schedules.

During performance, the owner pays only the *basic* fee each month when that month's portion of the work is completed. Periodically, the owner's designated representatives evaluate the contractor's performance for a given period and give it a numerical grade in those factors set forth in the contract. This overall grade determines the portion of the maximum attainable award fee that the contractor has earned and will be paid for the period being graded. Although the owner under such a contract has the sole right to evaluate performance, good owner-contractor relations usually require that the owner discuss evaluations with the contractor so that the contractor knows what the owner expects and how its performance is being rated.

Another type of incentive cost contract pioneered by the federal government is the *cost plus incentive fee* contract, sometimes called the CPIF contract. It is used when firm bids cannot be made in advance of performance but where the owner wishes to give the contractor profit motivation to reduce costs. In a CPIF contract, before award, owner and contractor negotiate the target cost, target fee, minimum and maximum fee, and fee adjustment formula. The formula determines the amount of fee payable to the contractor based on a comparison between the negotiated target cost and the final total allowable cost.

After the work is completed, contractor and owner negotiate the final fee in accordance with the fee adjustment formula. The formula can provide, for example, that the contractor would be penalized 25% of actual cost overruns above the target estimate and rewarded by 25% of the underruns. In each case, the formula is subject to the previously agreed-upon minimum and maximum fees. Such a contract can have separate incentive provisions for early completion.

Such an incentive contract took a disastrous twist for the contractor in *Koppers Co., v. Inland Steel Co.*[15] The contractor agreed to design, procure materials for, and construct an industrial plant for the owner. At the time of the award, the contractor estimated that the project would cost $267 million. Because of the likely number and scope of changes, a fixed price was not feasible. As a result, the contract contained targeted costs and an adjustment provision for agreed changes. The contract also provided a bonus under which the contractor would receive 50% of any cost underrun up to $6.3 million. However, in the event of an overrun, the contractor would refund up to $4.8 million.

The project cost nearly $444 million. The owner brought suit, charging that the contractor should be liable for much of the excess cost. Although there had been numerous design changes, the owner alleged that the actual cost far exceeded that which should have resulted from these changes because of the contractor's failure to use reasonable care in design and construction.

The contractor contended that the incentive bonus/penalty provision was the owner's sole remedy for cost overruns, regardless of why or how they occurred. The issue was whether a distinction should be drawn between increased costs attributable to errors and omissions by the contractor and those that occurred for other reasons. The court held that only clear language can deprive the contracting party of the normal rights that would be accorded by law. The incentive bonus/penalty provision was not intended, according to the court, to supplant the owner's remedies for any failure by the contractor to use due diligence in its performance. As a result, the owner was awarded some $64 million in damages from the contractor.

Another method of providing cost reduction incentives is value engineering (discussed in (C)).

These methods of seeking to keep costs down, although sometimes successful, still do not accomplish the objective of letting the owner or anyone supplying funds for the project know that the costs will not exceed the particular designated amount. To deal with this problem, owners sometimes insist that the contractor give a guaranteed maximum price (GMP) or that there be an "upset" price included in the contract. This is designed to give some assurance that the project will not cost more than a designated amount. This should be differentiated from any cost estimates given by the contractor, although there is always a risk that any cost figures discussed will end up being a GMP.[16]

Construction managers are frequently asked to give a GMP if *they* engage the specialty trade contractors or perform some of the work with their own forces. For that reason, discussion of a GMP is postponed until Section 17.04(D). However, a GMP may not be worth much if the design is quite incomplete at the time the GMP is given. If costs exceed the GMP, a claim is likely to be made by the contractor that there has been such change in the scope of the work that the GMP no longer applies.[17]

The owner has additional administrative costs in a cost contract. Usually the design professional will seek a higher fee than for a fixed-price contract because there are many more changes made in a cost contract as the work progresses. The design professional may have additional responsibilities for checking on the amount of costs incurred by the contractor and for ensuring that the costs claimed actually went into the project and were required under the contract. The determination of costs in-

---

[15]498 N.E.2d 1247 (Ind.App.1986).

[16]*J.E. Hathman, Inc. v. Sigma Alpha Epsilon Club*, 491 S.W.2d 261 (Mo.1973) (earlier AIA Doc. A111 with *no* blank for a guaranteed maximum price).
[17]*C. Norman Peterson v. Container Corp. of America*, 172 Cal.App.3d 628, 218 Cal.Rptr. 592 (1985).

volves not only often exasperating problems of cost accounting but also the creation of record management and management techniques for determining just what costs have been incurred.

Innumerable variations of allowable costs exist. Usually there is no question on certain items, such as material, labor, rental of equipment, transportation, and items of the contractor's overhead directly related to the project.[18] However, sometimes disputes arise over such matters as whether the cost of visits to the project by the contractor's administrative officials, the cost of supervisory personnel employed by the contractor, and the preparatory expenses or delay claims by subcontractors are allowable costs.

The drafter must try to anticipate all types of costs that can relate directly or indirectly to the project. A determination should be made as to which will be allowable costs for the purposes of the contract. Some of the troublesome areas can be highlighted by comparing Art. 7 and Art. 8 of AIA Doc. A111 which deal with cost-plus arrangements. Art. 7 lists reimbursable costs, and Art. 8 specifies certain costs that are not to be reimbursed. The years of experience of federal procurement have generated complicated allowable cost rules. Yet problems still arise in this troublesome area.

C. Value Engineering

Owners are always seeking methods of reducing costs in both fixed-price and cost contracts. *Value engineering,* a method developed by the federal procurement system, attempts to provide an incentive to the contractor to analyze each contract item or task to ensure that its essential function is provided at the lowest overall lifetime cost. The federal method states that an owner who accepts a value engineering change proposal initiated and developed by the contractor grants the contractor a share in any decrease in the cost of performing the contract and in any reduced costs of ownership.

While it is difficult to quarrel with the concept and objectives of value engineering, the actual operation of the system has generated a significant amount of litigation. Three federal procurement cases are instructive. The first, *John J. Kirlin, Inc. v.*

United States,[19] involved a contract for the renovation of the Pentagon's heating, ventilating, and air-conditioning system. The contractor was required, among other things, to replace certain minimum dampers but was not required to replace the "maximum" outside air dampers.

One month after the contract had been awarded, the contractor submitted a proposal suggesting that the maximum outside air dampers be replaced. The contractor asserted this would reduce the government's annual energy costs by $1.6 million. This proposal was rejected, the contracting officer believing that the estimate of energy savings was not valid.

Nevertheless, about three months after completion of the contract, the agency awarded another contract to a different contractor. The original contractor then asserted that it was entitled to share in the energy and maintenance cost savings that the agency would realize.

The U.S. Claims Court denied recovery, concluding that the value engineering proposal must be for the purpose of changing any requirement of the contract and does not deal with work beyond the scope of the contract that could not be accomplished by change order. Because the contractor could not be ordered to perform such a drastic change, the court rejected the claim.

The U.S. Circuit Court of Appeals agreed that value engineering does not cover changes that would be a cardinal change, one beyond the scope of the project. Yet it did not find it necessary to determine whether this would have been a cardinal change. It concluded that the contractor's claim required that its proposal be accepted. It rejected the contractor's contention that the proposal had been constructively accepted when the government agency decided to do the work after the contract had been completed. The contractor had no right to savings based on a subsequent contract to which the contractor was not a party. The court would not follow a decision by the Armed Services Board of Contract Appeals (ASBCA), which had granted recovery on the theory of an implied contract, in effect a conclusion that denying the claim would create unjust enrichment.[20]

[18]As to insurance, see *RCI Northeast Services Div. v. Boston Edison Co.,* supra note 7.

[19]827 F.2d 1538 (Fed.Cir.1987).
[20]*Alan Scott Indus.,* ASBCA 24729, 82-1 BCA ¶ 76,852.

The second case, *H&S Corp.*,[21] demonstrated that a value engineering proposal carries risks. In this case, the contractor had proposed a different type of material. Its proposal was accepted, and it received 55% of the savings. However, the material did not function properly, and the contractor had to pay for the cost of correction.

The third case, *ICSD Corp. v. United States*,[22] involved a contract to supply night vision field sites. The value engineering clause provided that the contractor would receive 50% of the government "contract" savings, that is, those realized on future purchases of essentially the same item as that to be acquired under the contract. The contractor would receive only 20% of the savings, however, for cost reductions considered "collateral," that is, reductions of operation costs and government-furnished property costs. Under ICSD's contract for night vision gunsights, batteries, though a major component of the sites, were not to be manufactured or delivered to the army.

ICSD submitted a value engineering change proposal (VECP) to substitute alkaline batteries for the mercury batteries specified. The contracting officer determined that the savings were over $1,000,000 per year and awarded ICSD a share of the savings. However, the contracting officer concluded that these savings were collateral to the contract and that the amount owed ICSD was 20%, not the 50% that it had sought. Also, the contracting officer made no award for savings based upon the increased safety of alkaline batteries, reduced disposal costs, reduced logistics costs, and the elimination of the need for a cold-weather adaptor.

While ICSD's proposal was under consideration, another government contractor submitted a similar VECP. The government evaluated both proposals and decided that ICSD's was superior in three areas while the other contractor's was superior in one. This led the contracting officer to split the savings share award, giving 75% to ICSD and 25% to the other contractor.

The court held that to be considered contract savings, there must be a reduction in the cost of the item or essentially the same item as that to be acquired under the contract. But because batteries were not acquired under the ICSD contract, the proper award was 20% of the government's savings. As to the other collateral savings not recognized by the government, the court found that the savings were not measurable, documentable, or ascertainable and that ICSD's estimates were neither reasonable nor credible. Finally, the court held that splitting savings awards between contractors is permissible.

The court rejected the argument that granting only 20% thwarts the policy of the incentive clause, noting that while the 50% award would provide greater incentives, less incentive to make proposals is needed where the reduced cost does not relate to items furnished under the contract.

In 1990, the Federal Acquisition Regulations (FAR), which regulate federal procurement, adopted value engineering principles and procedures for architect/engineer contracts. When authorized by the contracting officer, such studies will be performed after completion of 35% of the design stage or such other stages as the contracting officer determines. However, unlike value engineering for contractors, there is no cost sharing of savings between the government and the contractor. Instead, the design professional provides a fee breakdown schedule for the services relating to value engineering activities, and when approved, these services will be compensated.[23]

D. Unit Pricing

One risk of a fixed-price contract relates to the *number* of units of work that will have to be performed. This risk can be removed by the use of unit pricing. The contractor is paid a designated amount for each unit of work performed.[24]

A number of factors must be taken into account in planning unit pricing. First, the unit should be clearly described. The cost of a unit should be capable of accurate estimation. Best unit pricing involves repetitive work in which the contractor has achieved skill in cost predicting.

Second, it must be clearly specified whether the unit prices include preparatory work such as cost

---

[21]ASBCA 29156, 87-2-BCA ¶ 100,005.
[22]934 F.2d 313 (Fed.Cir.1991).

[23]48 C.F.R. § 52.248-2 (1992).
[24]This subsection of an earlier edition of this treatise cited in *Gottschalk Bros. v. City of Wausau*, 56 Wis.2d 848, 203 N.W.2d 140, 142 (1973).

of mobilizing and demobilizing apparatus needed to perform the particular work unit.

Third, it is important to decide whether the invitation to bidders will include an "upset" or maximum price or whether the invitation will include a minimum price. The reason for an upset price is obvious. However, the reason for a minimum price involves understanding the use of an unbalanced bid. Under such a bid, the contractor does not actually base its unit bid price on its prediction of the cost that will be incurred for performing that unit. Commonly, the contractor bids high for unit work that will be performed early and low for work that will be performed later. In a competitive bid evaluation, such a contractor suffers no disadvantage by this distortion. The award is based on the total unit prices multiplied by the total estimated quantities. As a result, such a bidder takes no *apparent* risk when it unbalances the bid. (But suppose units bid high are deleted or reduced.)[25]

The reasons for an unbalanced bid, sometimes called "pennying," were revealed in a New Jersey decision, *Boenning v. Brick Township Municipal Utilities Authority.*[26] The case involved a bidding invitation that set minimum unit prices for certain portions of work connected with the installation of a municipal sewerage system. The issue was whether the municipality could set *minimum* unit prices.

The court first distinguished *fixed* from *discretionary* items. It described a fixed unit as one that can be measured with certainty by reference to the plans and specifications, such as sewer pipe whose size and length are shown on the plans. (This type forms the basis of what the English call measured contracts, with the price based on so many board feet of lumber, doors, windows, etc.)

Discretionary items (an unfortunate term)[27] are a type of work whose quantities cannot be accurately measured or ascertained in advance of the work itself being performed. The contract before the court involved two discretionary items: underground and restoration work. The underground work required laying sewer pipes in trenches ranging from 6 to 20 feet in depth. Although test borings had been taken before design, the soil conditions that would be encountered could not be determined with certainty. Some soil conditions would require specified material or a concrete cradle to support the pipe. Timber or steel shoring left in place might be needed to shore up the trench. Similarly, it could not be determined in advance how much restoration work, such as asphalt paving or landscaping, would be required after the sewers had been installed.

In what the court called discretionary work, engineers for the awarding authority usually estimate units that will be required based on the borings, their knowledge of local conditions, and their experience. (The problem of mistaken estimates is taken up later in this subsection.) An unbalanced bid—or what the court referred to as penny bidding—can encourage the contractor who has pennied certain discretionary items to challenge an engineer's orders because the contractor may not want to incur the expenses of doing work for which it made a nominal bid. This can increase the number of disputes and interfere with timely completion of the project. Yet according to the court, it had become common practice in New Jersey bidding for utility work to penny units of discretionary work.

The court then dealt with how a pennying contractor can compete with one who does not penny units. The former may simply gamble that such work will not be needed or may believe that the engineer has overestimated the need for such items. Such a contractor will gain a competitive advantage by pennying and can afford to increase its bid on items that are certain, which can give it a windfall.

Another, more unsavory, reason for pennying may be the prospect of collusion between the contractor and the engineer, the latter finding the pennying units low and the excessively priced units high. Even without a specific "deal," the contractor may believe that the prospective gains might make it worthwhile to attempt to influence or even bribe the engineer.

The contractor may submit an unbalanced bid to obtain progress payments more quickly, called "front-end loading." This can mean a greater chance of late-finishing subcontractors and suppliers not being paid.

---

[25]*Gregory & Reilly Assoc., Inc.*, FAACAP 65-30, 65-2 BCA ¶ 4918 (1965).

[26]150 N.J.Super. 32, 374 A.2d 1214 (1977).

[27]Most disputes involve excavation. The units are *not,* as a rule, discretionary or fixed. The issues usually are overruns or underruns (discussed later in this subsection).

The court noted that the minimum prices were fixed below the estimated cost so that a contractor cannot reap a windfall if there is an overrun on any high-priced item. Establishing a minimum at a figure less than the likely cost will encourage competition among bidders. The minimum ensures that a contractor who has bid that price will be paid at least a part of its cost and can minimize gambling by contractors and potentially unnecessary disputes. It can also inhibit front-end loading by use of the unbalanced bid.

The court rejected an argument that if the contractors gamble and lose, their sureties will complete the job and the local authority will receive a cheaper price. The court held that the local authority could use the minimum bid for units inasmuch as it could, if it chose, reserve the right to reject unbalanced bids.[28]

The second legal issue, and perhaps one that arises most frequently, relates to inaccurate estimates of units to be performed. These usually involve excavation cases in which the actual units substantially overrun or underrun the estimates. Pricing the unit work usually assumes that there will not be a substantial deviation from the estimates. If the actual units substantially underrun, the cost cannot be spread over the number of units planned and will cost on a unit basis more than the contractor expected. If the unit is overrun, the contractor may be expected to perform more unit work in the same period of time, another factor that can increase planned costs.

Often contracts specifically grant price changes if costs overrun or underrun more than a designated amount. For example, the federal procurement system grants an equitable adjustment if costs overrun or underrun more than 15%.[29]

AIA Doc. A201 deals with this problem in a more limited way. Paragraph 7.1.4 grants an equitable adjustment to either party if the quantities are changed *by change order or construction change directive* and application of the unit prices will cause "substantial inequity." This does not grant an automatic adjustment for overruns or underruns.

Many contracts provide equitable adjustments if the subsurface conditions actually encountered vary from the conditions stated to exist or from those that are usually encountered (see Chapter 25).

Some of the more difficult cases have involved claims where the owner simply gives an estimate that turns out to be inaccurate (most commonly grossly inaccurate) without any showing of negligence chargeable to the owner or any contractual adjustment method.

As a rule, the mere fact that there is a variation does not entitle the contractor to additional compensation or a change in the unit price. Such a result would be based on the contractor's having assumed the risk or, occasionally, on an express provision in the contract stating that estimates cannot be relied on.[30] For example, in *Costanza Construction Corp. v. City of Rochester*,[31] the drawings showed 20 cubic yards of rock, and the specifications estimated 100 cubic yards. The actual amount of rock to be removed was 600 cubic yards. The contractor claimed that it bid below cost because it expected to encounter only a small amount of rock. It tried to avoid the unit price by claiming that the actual amount of rock found constituted a cardinal change that altered the essence of the contract. However, it was not granted relief, because the contract included a disclaimer by the city of responsibility as to the accuracy of the estimate and required that the contractor make its own inspection. The court brushed off a claim that there was no time for an independent inspection and that the cost would have been excessive.

The dissent stated that normal variations in quantities can be dealt with by a unit price. Here, however, the difference was so great (the contract price was $936,000, the actual cost for excavation was alleged to be $800,000, and the contractor's es-

---

[28]See also *Department of Labor and Industries v. Boston Water and Sewer Comm'n,* 18 Mass.App. 621, 469 N.E.2d 64 (1984) (agency could not refuse a bid that had pennied some items, especially where bid was not unbalanced, front-end loaded, or otherwise inflated).

[29]48 C.F.R. § 52.212.11 (1992). This was applied in *Victory Constr. Co., Inc. v. United States,* 206 S.Ct.Cl. 274, 510 F.2d 1379 (1975). The equitable adjustment was based on the cost differential directly attributable to the volume deviation greater than the "upset" amount and not the average cost of the entire unit work.

---

[30]*Zurn Eng'rs v. California Dept. of Water Resources,* 69 Cal.App.3d 798, 138 Cal.Rptr. 478 (1977).

[31]147 A.D.2d 929, 537 N.Y.S.2d 394, appeal dismissed, 74 N.Y.2d 715, 541 N.E.2d 429, 543 N.Y.S.2d 400 (1989).

timate was $2,500) that the estimates became meaningless. Application of the unit price, the dissent argued, could bring economic ruin to a good faith bidder through no fault of its own.

Reasonable reliance on the estimate, however, may justify reforming the contract to grant a price readjustment. For example, in *Peter Kiewit Sons' Co. v. United States*,[32] the contractor estimated its unit prices for three types of work of varying difficulty but then learned that the specifications required it to submit a combined unit price bid. It computed and submitted a unit price bid by averaging the three types of excavation and based its bid on the units estimated by the government.

When the work was in progress, the government ordered a reduction in one excavation type, a change that, because of the averaging used, the contractor claimed would increase the composite unit cost of the work. The claim was denied based on the contention that the estimates did not bind the government and could not be relied on, pointing to language to that effect in the specifications.

One basis for the contractor's claim was that each party had made a mutual mistake of fact as to the quantity of work that would be required. The Court of Claims upheld the claim, stating that the contract was intended not to be speculative but to be capable of proper computation and conservative bid. The composite bid being required meant that a great variation in the actual quantities performed either could be ruinous to the contractor or could cause the government to pay far more than the work was worth. The court concluded that the government did not intend to contract on such an irrational basis. If the parties each believe the estimates to be accurate, a mutual mistake to that effect would be a basis for reforming the contract.

As to language stating that estimates are not guaranteed, the court first pointed to language granting an equitable adjustment in the event of changed conditions, noting that such a provision was in conflict with the one stating that estimates are not guaranteed. Perhaps more important, the court stated that the disclaimer did not mean "that

all considerations of equity and justice are to be disregarded, and that a contract to do a useful job for the Government is to be turned into a gambling transaction."[33]

Additionally, this case demonstrates the not uncommon conflict between clauses that grant relief under certain circumstances and those that seek to place risks closely related to the former on the contractor (explored in greater detail in Chapter 25). The case also reflects a recognition of the contractual doctrine of mutual mistake that is sometimes applied in extreme cases despite the presence of language that appears to place the risk of unexpected quantities on the contractor.

Davidson & Jones, Inc. v. North Carolina Department of Administration[34] employed the mutual mistake doctrine in a case involving a claim for delay damages. The contract stated that the unit price for rock excavation would apply to "all rock removed above or below these quantities." The contractor was required to include 800 cubic yards of rock excavation in its basic bid. However, early in the excavation it was clear that this estimate was grossly inaccurate. Actually the contractor excavated 3,714 cubic yards. It was also delayed six months while doing this excavation. It did not seek an adjustment in its unit price but sought delay damages.

The court rejected the owner's argument that the contractor assumed the risk of delay and stated:

The trial court found that the plaintiff had inspected the site as required in the bidding documents and had seen nothing to indicate the presence of such an excess. The court further found that it was neither customary nor reasonable for a contractor to order his own subsurface investigation. Contractors customarily relied upon the State's figures; plaintiff here had actually relied upon them. Some variation was to be expected. Based upon the evidence presented at the trial, the trial court found that 10–15 percent was a reasonable variation. The court's findings of fact were based upon competent evidence and may not be disturbed on appeal.[35]

---

[32] 109 Ct.Cl. 517, 74 F.Supp. 165 (1947). See also *Timber Investors, Inc. v. United States*, 218 Ct.Cl. 408, 587 F.2d 472 (1978) (reformation allowed if estimated quantities "grossly inaccurate" and reliance reasonable, such as inability to verify).

[33] 74 F.Supp. at 168. For a similar result based on misrepresentation that induced a particular composite bid, see *Acchione & Canuso, Inc. v. Pennsylvania Dept. of Transp.*, 501 Pa. 337, 461 A.2d 765 (1983).
[34] 315 N.C. 144, 337 S.E.2d 463 (1985).
[35] 337 S.E.2d at 468.

Bidding mistakes are discussed in Section 18.04(E), but one case that involved a unit price clerical error merits comment here. *Pozar v. Department of Transportation*[36] involved a bid for highway construction. One unit item was "Binder," a dust palliative. The estimated quantity was ninety tons. Pozar's unit price per ton was $20. The total price should have been $1,800, which would have been the low total bid for the work. But by mistake, the total binder unit price was $18,000.

The bidding information stated that in the case of discrepancy between per-unit price and unit price totals, the per-unit price prevails unless it is "ambiguous, unintelligible or uncertain."

The awarding authority wanted to use the per-unit price and award the contract to Pozar. But the legal adviser, noting the estimate for binder was $300 a ton (this was, evidently, an unbalanced bid), concluded the bidder must not have intended to bid $20 a ton.

Pozar asked the court to review the award made to another bidder. The court held that the $20-per-ton unit bid was not "ambiguous, unintelligible or uncertain" and ordered the award be made to Pozar.

Ironically, as seen in Section 18.04(E), the legal adviser must have been worried about the possibility of a palpable bidding error giving Pozar a *right* to refuse to enter the contract. But here Pozar *wanted* the contract. But suppose its bid had been *much* lower than all others and it wanted to *withdraw* its bid. Very likely, it could have done so unless the unbalanced bid barred relief for a clerical error. This appears to give the bidder the best of both worlds.

E. Cash Allowance

Sometimes the owner wishes to select certain items after the prime contract has been awarded. For example, it may wish the right to select particular hardware or fixtures after a contractor is selected. To deal with this, a cash allowance can be specified. This allowance should cover the cost of the items to be selected. If the cost of the items ultimately selected varies from the cash allowance, an appropriate adjustment in the contract sum is made.

Disputes sometimes arise relating to what is encompassed within the allowance. The contract should specify whether the cash allowance encompasses only the net cost of the materials and equipment delivered and unloaded at the site including taxes or whether it also includes handling costs on the site, labor, installation costs, overhead and profits, and other expenses. Commonly, the allowance includes only the direct cost of the items selected.[37]

F. Contingencies

Contingencies are often included in contracts to deal with uncertainties. For example, in a bid contract, the contractor might be told to include in its bid a specific sum to cover the cost of contingencies (defined as items necessary or desired to complete the project in accordance with the owner's wishes). The items might be related to latent conditions discovered after work begins, items omitted from the documents, or items that the owner feels are necessary for a complete project. Under a contingency clause, if the contractor runs into no unexpected problems and completes the job in accordance with the plans and specifications, it receives the contract price but not the contingency amount. If it encounters unexpected expenses or conditions, the contingency amount is expected to be spent on them.[38]

As shall be seen in Section 17.04(D), contingencies are often included in cost contracts with a guaranteed maximum price (GMP).

PROBLEMS

1. Your client is about to secure a contractor for the construction of a ten-story commercial office building. It wishes to know whether it should use a single or separate contract system. It also wants to know whether it should use a fixed-price or cost contract. It asks you how "fixed" is a fixed-price contract and whether it should set aside a reserve above any fixed-price figure in the construction contract. It asks whether it should conduct a com-

[36]145 Cal.App.3d 269, 193 Cal.Rptr. 202 (1983).

[37]AIA Doc. A201, ¶ 3.8.
[38]*Godfrey, Bassett & Kuykendall, Architects, Ltd. v. Huntington Lumber & Supply Co., Inc.*, 584 So.2d 1254 (Miss.1991) (architect liable to contractor for contingency amount when he mistakenly told the contractor that the amount was being deleted from the contract).

petitive bid. What would be your advice? What would be the considerations that would bear on any advice you would give?

2. A fraternity house was heavily damaged by a fire, a portion of the building totally destroyed. Because of the importance of making the house habitable as soon as possible, the fraternity employed an architect who prepared plans and specifications and started negotiations with a particular contractor.

The work was begun immediately, even though final plans and specifications had not been prepared. Plans and specifications were prepared and submitted during construction, when a written contract was prepared by the fraternity and agreed upon by the contractor. The contract was prepared on a printed form that stated that it was to be used when the cost of the work plus a fixed fee was a basis for payment. It also stated that the contractor was to provide all labor and material and do all things necessary to complete construction according to the plans and specifications. The contract stated that any maximum cost would be adjusted in accordance with change orders. A typed-in provision stated, "Estimated maximum cost of this work is $300,000." The contract also stated that the contractor would be reimbursed for all costs necessarily incurred for the proper execution of the work, with particular items included and others excluded.

a. What sources would be used to determine whether the contractor has a legal right to recover for costs incurred that exceed $300,000?

b. What is likely to have been the intention of the contractor? The fraternity?

c. What if the intentions were different and neither party knew or should have known of the other's intention?

SECTION 17.03 Traditional Organization: Owner's Perspective

This chapter looks at compensation and organizational aspects of the Construction Process mainly from the owner's perspective. It is generally assumed that contractors who engage in the process are individual legal entities, usually corporations. However, some projects may be beyond the capac-

ity of an individual contractor, who may seek to associate other contractors with it through the use of a joint venture. This organizational method, roughly a partnership for one project, was discussed in Section 3.07.

A. Traditional System Reviewed

The traditional system separates design and construction, the former usually performed by an independent design professional and the latter by a contractor or contractors. The sequencing used in this system is creation of the design followed by contract award and execution. (The commonly used method of awarding construction contracts through competitive bidding is discussed in detail in Chapter 18.)

The contractor to whom the contract is awarded is usually referred to as the prime or general contractor. It will perform some of the work with its forces but is likely to use specialized trades to perform other portions of the work. The prime contractor is both a manager of those whom it engages to perform work and a producer in the sense that it is likely to perform much or some of the work with its own forces. The traditional system used in the United States usually gives certain site responsibilities to the design professional who has created the design (discussed in detail in Section 12.08).

The principal advantages of this system are as follows:

1. The owner can select from a wide range of design professionals.
2. For inexperienced owners, an *independent* professional monitoring the work with the owner's interest in mind can protect the owner's contractual rights.
3. Not awarding the contract until the design is complete should enable the contractor to bid more accurately, making a fixed-price contract less likely to be adjusted upwards except for design changes.
4. Subcontracting should produce highly skilled workers with its specialization of labor and should create a more competitive market because it takes less capital to enter.

B. Weaknesses

The modern variations to the traditional system described in Section 17.04 developed because, despite its strengths, the traditional system developed weaknesses. It is important to see these weaknesses as a backdrop to Section 17.04.

Separation of design and construction deprives the owner of contractor skill *during* the design process, such as sensitivity to the labor and material markets, knowledge of construction techniques, and their advantages, disadvantages, and costs. A contractor would also have the ability to evaluate the coherence and completeness of the design and, most important, the likely costs of any design proposed.

Sequencing the work in the traditional system not only precludes work from being performed while the design is being worked out but also deprives the contractor of the opportunity of making forward purchases in a favorable market. Although the leisurely pace that often accompanies the traditional Construction Process at its best can produce high-quality work, it is almost inevitable that such a process will take more time than one that allows construction to begin while the design is still being completed.

The frequent use of subcontractors selected and managed by the prime contractor causes difficult problems. Subcontractors complain that their profit margins are unjustifiably squeezed by prime contractors demanding they reduce their bids and, more important, reducing the price for which they will do the work after the prime contractor has been awarded the contract. Subcontractors also complain that contractually they are not connected to the owner—the source of authority and money. They also complain that prime contractors do not make sufficient effort to move along the money flow from owner to those who have performed work and that they withhold excessive amounts of money through retainage.

The traditional system tends to keep down the number of prospective prime contractors who could bid for work and thereby reduces the pool of competitors. The traditional system, with its emphasis on a fixed-price contract and competitive bidding, also can create an adversarial relationship between owner and contractor (described in Section 17.04(A)).

Also, the traditional system with its linked set of contracts—owner-design professional, owner-contractor, contractor-subcontractors—did not generate a collegial team joining together with a view

toward accomplishing the objectives of all the parties. Some of the variations noted in Section 17.04 emphasize better organization, the creation of a construction team, and the need for all involved to pull together to accomplish the objectives.

Another weakness of the traditional system relates to the role of the design professional during construction. For the reasons outlined earlier,[39] modern design professionals seek to exculpate themselves from responsibility for the contractor's work and to limit their liability exposure. Perhaps even more important, many design professionals lack the skill necessary to perform these services properly.

Under the traditional method, the managerial functions of the prime contractor may not be performed properly. The advent of increasing, pervasive, and complex governmental controls over safety often found many contractors unable to perform in accordance with legal requirements. In addition, the managerial function of scheduling, coordinating, and policing took on greater significance as pressure mounted to complete construction as early as possible and as claims for delay by those who participate in the project proliferated. A good prime contractor should be able to manage these functions, as its fee is paid to a large degree for performance of these services. Not all prime contractors were able to do this managerial work efficiently. Some owners believed the managerial fee included in the cost of the prime contract could be reduced.

The division between design and construction, although at least in theory creating better design and more efficient construction, had the unfortunate result of dividing responsibility. When defects develop, as shall be seen in Chapter 24, the design professional frequently contends that such defects were caused by the contractor's failure to execute the design properly, while the contractor asserts that the design was defective. This led to bewildered owners not being certain who was responsible for the defect as well as to complex litigation.

---

[39]Refer to Section 12.08.

SECTION 17.04   Modern Variations

A. Introductory Remarks

Generalizations are avoided in this section. Although separate contracts have been used for a substantial period of time, construction management (CM) and design/build (D/B) as well as phased construction (fast track) have only recently been the center of attention. As a result, crystallized and accepted practices do not exist.

Much of the standardization in the traditional system is attributable to general acceptance of construction contract forms published by the American Institute of Architects (AIA) and endorsed by the Associated General Contractors (AGC) that relate to that system, such as A101/201. (The AIA does not seek endorsement for B141, its contract for design services.) These associations have developed well-understood and accepted practices.

Until 1992, the AIA and AGC published competing CM documents, the former emphasizing the CM as agent and the latter emphasizing the CM as constructor. In 1992, the AIA published a set of CMa documents (A101/CMa, A201/CMa, B141/CMa, and B801/CMa) that are recommendations for how the CM fits into the contract system when he acts as an adviser to the owner. At the same time, the AIA published A121/CMc, and the AGC published AGC 565, a joint product to be used when the CM acts as constructor. In 1993 the AIA published B144/ARCH-CM, an amendment to B141 to be used when the architect also performs CM services. These documents may tend to create trade usages that bring some order to the at-present irregular CM system. However, the modern variations discussed in this section are more likely to occur in large-scale projects. They use tailor-made contracts developed for the special needs of the parties. Yet the joint effort is surely a welcome one.

Owners who commission projects of widely varying types often have different ideas of the most efficient way of building a project. As a result, a wide variety of contracts are used. Descriptions must be general and tentative.

As to legal conclusions, this section raises more questions than it can answer. Any observations will have to be tentative, not only because of the lack of crystallized practices but also because the law must deal with a series of new problems that cannot be easily solved by use of simple reference to earlier precedents or statutes. Law functions to a large degree through crude categories aided by analogies, largely for administrative convenience. Law often lags behind organizational and functional shifts in the real world. Variations of the type to be described in this section will inevitably create a period of temporary disharmony. It is hoped that predictable solutions will emerge.

Many legal rules are premised on the traditional system, as described in Section 17.03. Modern variations change this system. The use of separate contracts (multiple primes) (to be discussed in (C)) shifts responsibility for coordination from a prime contractor to someone else, often an independent professional adviser retained by the owner or a principal (not a prime) contractor who has no direct contracts with those contractors whose work it must, as a manager, coordinate.

Phased construction (fast tracking, discussed in (B)) allows construction to proceed during design. When the spotlight is placed on construction management (in (D)), a new adviser or coordinating contractor is engaged (not, as a rule, after the design is completed but during design) by the owner, who also plays a significant monitoring role during construction. Here the owner, through at least some of the permutations of construction management, has a more active role in managing the project and is not simply turning over the design to the contractor, a businessperson engaged in the venture to earn a profit.

In turnkey and design/build (discussed in (E) and (F)), the owner has not engaged an independent adviser to prepare the design but has turned over everything to a businessperson who represents that it has paid or can pay for the skill to both design and build, and in the case of a turnkey contract even more such as providing financing, furnishing interior furnishings, and providing a computer system. Even more difficult, the D/B owner can, as noted in (F), span a wide range of owners.

B. Phased Construction (Fast Tracking)

Phased construction, or what has come to be known as fast tracking, is not an organizational variation. It differs from the traditional method in that construction can begin before design is completed. There is no reason it cannot be used in a traditional single contracting system, although it is likely to mean that the contract price will be tied to costs. However, it has received greatest attention since emphasis has been placed on modern variations on the traditional method discussed in this section.

Construction can begin while design is still being worked out. Ideally this means that the project should be completed sooner.[40] If a contractor is engaged during the design phase and knows which material and equipment it will have to use, it can make purchases or obtain future commitments earlier and often cut costs.

Another advantage to the owner is the early activation of the construction loan. In traditional construction, an owner, particularly a developer, does not receive loan disbursements until the construction begins. Yet it must pay for investigation and acquisition of the project site. It may have had to pay either the purchase price for the land or installments of ground rent, as well as insurance premiums, real estate taxes, design service fees, or legal expenses. If it must make these expenditures before it starts to receive construction loan funds, it will have to draw on its personal unsecured credit line, necessitating monthly interest payments and reducing its credit line, which could be used for other purposes. If construction can begin through fast tracking, the owner need not use its personal funds but can employ loan disbursements.

The principal disadvantage of fast tracking is the incomplete design. The contractor will be asked to give some price—usually after the design has reached a certain stage of completion. Very likely the contract price will be cost plus overhead and profit, with the owner usually obtaining a guaranteed maximum price (GMP).[41] But the evolution of the design through a fast-tracking system is likely to generate claims. Completing the design

and redesigning can generate a claim that the contract has become one simply for cost, overhead, and profit and that any GMP has been eliminated.[42] The owner may be constrained in making design changes by the possibility that the contractor will assert that the completed drawings must be consistent with the incomplete drawings.[43]

Two other potential disadvantages need to be mentioned here. First, there is greater likelihood that there will be design omissions—items "falling between the cracks." Needed work is not incorporated in the design given *any* of the specialty trade contractors or subcontractors.[44] Although omissions of this type can occur in any design, they are more likely to occur when the design is being created piecemeal rather than prepared in its entirety for submission to a prime contractor.

Second, there is a greater likelihood in fast tracking that one participant may not do what it has promised and adversely affect the work of *many* other participants. For example, an opinion of a federal contract appeals board discussing fast tracking in the context of construction management and multiple primes stated:

> "Phased" construction has been analogized to a procession of vehicles moving along a highway. Each vehicle represents a prime contractor whose place in the procession has been pre-determined. The progress of each vehicle, except that of the lead vehicle, is dependent on the progress of the vehicle ahead. The milestone dates have been likened to mileage markers posted along the highway. Each vehicle is required to pass the mileage markers at designated times in order to insure steady progress.[45]

When any of the vehicles do not pass the mileage markers assigned to them, claims for delays and complex causation problems are likely to result.[46]

---

[40]Some recent studies have indicated that these ideals do not always mature. One study indicates that very low percentages of design completion before commencement of construction may result in considerable construction delays. See Laufer and Cohenca, *Factors Affecting Construction-Planning Outcomes*, 116 J. of Constr.Eng'g. & Mgmt. 135 (1990). Similarly, another study states that fast track is of much less value than it appears to be. The benefit of early completion, it stated, is more than half offset by the consequent shift in the timing of construction expenditures. It not only may wipe out any benefits gained by an early start but also may lead to the accumulation of losses. *Rosenfeld & Geltner,* supra note 9.

[41]Refer to Sections 17.02(B), 17.04(D), and (F).

---

[42]*Armour & Co. v. Scott,* 360 F.Supp. 319 (W.D.Pa.1972), affirmed 480 F.2d 611 (3d Cir.1973); see *C. Norman Peterson Co. v. Container Corp. of Am.,* supra note 17.

[43]*City Stores Co. v. Gervais F. Favrot Co.,* 359 So.2d 1031 (La.App.1978) (completed drawings must be consistent with incomplete drawings).

[44]*Grinnell Fire Protection Systems Co. v. W.C. Ealy & Assoc., Inc.,* 552 S.W.2d 747 (Tenn.App.1977).

[45]*Pierce Assoc., Inc.,* GSBCA 4163, 77-2 BCA ¶ 12,746, citing *Paccon, Inc. v. United States,* 185 Ct.Cl. 24, 399 F.2d 162 (1968), at 61, 942.

[46]The use of construction management often accompanied by fast tracking in federal projects led to an inordinate number of delay claims. See *Federal Construction: Use of Construction Management Services,* U.S. General Accounting Office GAO/GGD-90-12 (January 1990).

C. Separate Contracts (Multiple Primes): *Broadway Maintenance v. Rutgers*

Separate contracts developed because of owner concern over the subcontracting process briefly described in Section 17.03(B). Separate contractors give the owner greater control in specialty contracting and avoid the subcontractor complaints that they are a contract away from the source of power and funds. Separate contracts were used if owners were not confident of prime contractor management skill. Their use was also spurred by successful legislative efforts by subcontractor trade associations to require that state and sometimes local construction procurement use separate contracts.

Another factor, though of less significance, was the hope that separate contracts could develop lower contract prices. Breaking up the project into smaller bidding units allows more contractors to bid. Owners hoped the managerial fees could be reduced by taking this function from the prime contractor. These hoped-for pricing gains can compensate for the additional expense of conducting a number of competitive bids.

In the traditional contracting system, the linked set of contracts determines communication and responsibility. If subcontractors have complaints, they can look to the prime contractor. If the prime has problems, it can look to the subcontractors or the owner.[47]

Separate contractors are required to work in sequence or side-by-side on the site. But they do *not* have contracts with one another. Disputes between them must be worked out by the participant performing the coordination function, something that is not as neat in the separate contract system as in the single contract system. To be sure, even in the traditional system, this problem can arise if there are disputes between subcontractors who have no contract with each other. The traditional system handled this with relative clarity by requiring that these matters be dealt with by the prime contractor as part of its managerial function and the owner remove itself from these problems.

Who performs the managerial function? Who has legal responsibility? Under the construction management system, these often fall to the construction manager (CM). But the managerial function can be performed by the design professional if he has the skill and willingness to do so, by a staff representative of the owner, or by a managing or principal separate contractor. If it is not *clear* who will coordinate, police, and be responsible in *any* system, the project will be delayed and claims made. A case that generated all these problems is reproduced here.

---

[47]Even in the traditional system, lines were becoming blurred as owners and subcontractors sued each other. See Sections 28.05(B), 28.07(H), and 28.08(B). AIA prime contracts gave the owner power to intervene to a degree in the prime-subcontractor relationship. See AIA Doc. A201, ¶¶ 5.2.2, 5.3.1, 9.3.1.2, 9.6.2, and 9.6.3.

BROADWAY MAINTENANCE CORP. v. RUTGERS

Supreme Court of New Jersey, 1982. 90 N.J. 253, 447 A.2d 906.
[Ed. note: Footnotes renumbered and some omitted.]

SCHREIBER, J.

Two contractors engaged in the construction of the Rutgers Medical School in Piscataway Township sought to recover damages from Rutgers, The State University (Rutgers), for its failure to coordinate the project and to compel timely performance by a third contractor. Rutgers asserts that it allocated the sole duty to coordinate the work of the prime contractors on the project and ensure their timely performance to Frank Briscoe Co., Inc. (Briscoe), its general contractor. Plaintiffs deny their right to sue Briscoe as third party beneficiaries of Briscoe's contract with Rutgers. We must determine first, whether plaintiffs were intended to be beneficiaries of that contract. Even if they were, we must determine whether Rutgers, as owner, retained any duty to coordinate or supervise which could give rise to a cause of action in plaintiffs. Finally, we must determine whether Rutgers is excused from any such liability by an exculpatory clause in the contract.

On October 31, 1966, Rutgers signed contracts for general construction work with Briscoe for $7,392,000, electrical work with plaintiff, Broadway Maintenance Corp. (Broadway), for $2,508,650, and plumbing and fire protection with plaintiff, Edwin J. Dobson, Jr., Inc. (Dobson), for $998,413. Six other contracts, also entered into, covered precast concrete; structural steel; elevators; heating, ventilating and air conditioning; laboratory furniture and independent inspection and testing.[48] In contrast with construction projects in which the owner contracts with a general contractor who undertakes the entire project and coordinates operations of subcontractors, Rutgers entered into contracts with each of several prime contractors. In Rutgers' agreement with Briscoe, Briscoe agreed to act as the supervisor on the job and coordinator of all the contractors.

After the work was finished, at a date well beyond the scheduled time for completion, Dobson and Broadway filed separate complaints against Rutgers in the Superior Court, Law Division, asserting a variety of claims, including damages due to delays and disruptions caused by Rutgers' failure to coordinate the activities of the various contractors on the site. Rutgers filed third party complaints seeking indemnification from Briscoe and its surety. The two actions were consolidated for trial with a pending third suit brought by Briscoe against Rutgers for money due under the Rutgers-Briscoe contract. Dobson and Broadway never added Briscoe as a party defendant in their actions. The suit between Rutgers and Briscoe was settled before trial, except for two claims for indemnification. Briscoe is not a party to this appeal.

The non-jury trial proceeded for 43 days. Rutgers produced no evidence and rested at the end of plaintiffs' case. The trial court in an extensive written opinion granted plaintiffs judgments for some of their claims, but denied recovery against Rutgers for failure to coordinate the activities of the prime contractors, including Briscoe. 157 *N.J.Super.* 357, 384 *A.*2d 1121

(Law Div.1978). Dobson and Broadway appealed to the Appellate Division, which affirmed. 180 *N.J.Super.* 350, 434 A.2d 1125 (App.Div.1981).

Each plaintiff petitioned for certification. We granted both petitions to consider three questions: (1) in a multi-prime contract, is each prime contractor liable to the other, (2) in such a contract, does the owner have a duty to coordinate the work of the contractors, and (3) does the exculpatory clause in the prime contracts at issue here shield Rutgers from liability for damages due to delay. The Mechanical Contractors Association of New Jersey, Inc. and the Building Contractors Association of New Jersey were granted leave to file briefs as *amici curiae* ["friends of the Court" allowed to present their views by court permission].

[The court noted that jurisdictions had reached different results when faced with the question of whether one separate contractor could sue another. It concluded that the record supported the findings of the trial court that held that all parties had agreed that the separate contractors could maintain legal action against each other for damage due to unjustifiable delay. In determining this question, the court held that the terms and conditions of the contract should be examined to determine the intent of the parties.]

The existence of a third party claim does not necessarily extinguish all claims between the parties to the contract. Here plaintiffs argue that Rutgers breached its agreement with them and is liable irrespective of any third party claims they may have against the general contractor, Briscoe. The plaintiff's contention is sound at least with respect to those matters that Rutgers had contractually obligated itself to do and for which it would be responsible. The trial court did award damages to the plaintiffs against Rutgers for certain contractual breaches. Rutgers has not appealed from those determinations.

The narrow questions before us on this appeal are whether Rutgers had agreed to synthesize the operations of the prime contractors, including Briscoe, and, if so, whether Rutgers breached that duty and would be liable for the delay flowing from that breach. The order granting the petitions for certification was limited to the subject matter of coordination of the operations of all the prime contractors including Briscoe. Damages flowing therefrom involve only delay on the job and its consequential costs.

Plaintiffs urge various claims that they relate to delay such as additional expenses incurred because of lack of elevators and stairs. These particular claims were disallowed by the trial court, which pointed out, among other things, that Briscoe's delay did not cause the asserted damages. Though these items are not before us on this appeal, we have reviewed the record and are satisfied that the trial court's findings have adequate factual support.

---

[48]Rutgers apparently believed that it was subject to N.J.S.A. 52:32-2, which requires the letting of multiple contracts in connection with the construction of public buildings. This Court has since determined, however, that the statute does not apply to Rutgers. *Rutgers v. Kugler*, 58 N.J. 113, 275 A.2d 441 (1971), aff'g 110 N.J.Super. 424, 265 A.2d 847 (Law Div.1970).

Plaintiffs also continue to press before us matters such as Rutgers' alleged default in not withholding funds due Briscoe in order to satisfy plaintiff's claims for delay against Briscoe, and Rutgers' non-fulfillment of its supposed duty to place the site in condition so that plaintiffs could proceed. On this appeal we are concerned solely with the general supervision over the contractors. What duties Rutgers had in this respect depends upon what obligations were imposed upon it by the contract.

In the absence of any compelling public policy, an owner has the privilege to eliminate a general contractor and enter into several prime contracts governing the construction project.[49] In that event the owner could engage some third party or one of the contractors to perform all the coordinating functions. Where all the parties enter into such an arrangement the owner would have no supervisory function. The situation would be analogous to one where a general contractor had been engaged to construct the project on a turnkey basis. Surely the subcontractors would have no claim against the owner for failure to coordinate.

If no one were designated to carry on the overall supervision, the reasonable implication would be that the owner would perform those duties. In so doing, the owner impliedly assumes the duty to coordinate the various contractors to prevent unreasonable delays on the project. See, *e.g., Born v. Malloy*, 64 *Ill.App.*3d 181, 184, 21 *Ill.Dec.* 117, 120, 381 *N.E.*2d 52, 55 (1978); *Carlstrom v. Independent School District No. 77, Minn.*, 256 *N.W.*2d 479 (1977). That is a reasonable assumption because the contracting authority has the power to use its superior position and to invoke its contractual rights to compel cooperation among contractors. *Shea-S & M Ball v. Massman-Kiewit-Early*, 606 *F.*2d 1245, 1251 (D.C.Cir.1979). The owner is impliedly obligated to act in good faith and to do that which it reasonably can to ensure that the other contractors adhere to the time schedules established for the project. *Paccon, Inc. v. United States*, 399 *F.*2d 162, 169–70 (Ct.Cl.1968). An owner's failure to take action in the face of unnecessary and unreasonable delays by one of the contracting parties would ordinarily evidence bad faith and constitute a breach of its implied duty to coordinate.

[49]Use of multiple prime contracts assumes the owner will benefit from savings that will accrue from elimination of the overhead and profits of the general contractor. Provisions for construction of any public buildings by the State provide for separate bids for different major aspects of the job and bids for all work in one contract, the award to be made to whichever method results in a lower cost. N.J.S.A. 52:32-2.

It is of course also possible that the owner might fractionalize those supervisory functions. Where the owner has chosen to engage several contractors directly, the bottom line is to ascertain who the parties agreed would orchestrate and harmonize the work. The answers may be found in the contract language as illuminated by surrounding circumstances.

The complications that arise in this case are due in part to the unclear nature of who was to perform what supervisory function. Briscoe had agreed to supervise the job generally. It was entrusted with the "oversight, management, supervision, control and general direction" of the project. It was to control "the production and assembly management of the building construction process." Briscoe was obligated to have sufficient executive and supervisory staff in the field so as to handle these matters efficiently and expeditiously. The General Conditions also recited that Rutgers relied upon Briscoe's management and skill to supervise, direct and manage Briscoe's own work and the efforts of the other contractors; indeed, the agreements stated that the contractors also relied upon Briscoe's supervisory powers to deliver the building within the scheduled time.

However, Rutgers also designated its Department of New Facilities to represent it in technical and administrative negotiations with the contractors and the Department could stop the work if necessary. Rutgers also selected a critical path method consultant whose progress chart was to be followed, but could be amended with the consultant's approval. Further, Rutgers engaged an architect. The architect agreed to "[s]upervise the construction of work by periodic inspections sufficient to verify the quality of the construction and the conformity of the construction to the plans and specifications . . . and by which supervision the architect shall coordinate the work of the various contractors and expedite the construction of the Project." If a contractor were delayed in completing the work whether due to the owner, any other contractor, the architect, or other causes beyond the contractor's control, the architect alone was authorized to determine extensions of time to which a prime contractor would be entitled. In this respect the trial court found that the architect acted as an impartial and independent umpire, and as such was not the agent of either Rutgers or the contractor.

Plaintiffs contend that Rutgers retained supervisory control because Briscoe had not been given the power to enforce its coordinating authority. They argue that Rutgers had the economic weapons of terminating the contracts and withholding payments, so that only Rutgers could effectively cause the prime contractors, particularly Briscoe, to keep up to schedule. Though this contention has some surface appeal, it fails to account for the entire contractual structure governing the project. Rutgers' power to terminate a contract or withhold funds did not alter its expressed intent to have someone else supervise the work. Rutgers had delegated that overall responsibility to Briscoe. Rutgers never intervened to coordinate the operations. It never assumed control.

The plaintiffs also claim that Rutgers retained coordination and supervision over the job because of the roles of the Department of New Facilities and critical path method consultant. These contentions are misconceived. The Department's functions were primarily quality control and only incidentally affected time of performance. The consultant plotted actual progress as against the scheduled performance, but did not supervise and coordinate the work. Nor does the architect's supervisory role support plaintiffs' position. First, the architect's agreement was not incorporated in the contractors' contracts. Second, Rutgers' delegation of coordinating functions to the architect confirms, if anything, that Rutgers itself was not engaged in any such undertaking. Lastly, plaintiffs rely on an indemnity clause in each contract that provided that if a contractor or subcontractor sued Rutgers, the contractor would defend, indemnify and save Rutgers harmless. However, that provision simply confirms the intent that Rutgers was not to be responsible for defaults of other prime contractors, including Briscoe.

When viewed in its entirety, the contractual scheme contemplated that if a contractor were adversely affected by delays, it could maintain an action for costs and expenses against the fellow contractor who was a wrongdoer. Furthermore, a contractor had a right to obtain extensions of time to complete the work when delayed by "any act or neglect" of any other contractor. Other than those remedies, the contractor could not look to Rutgers for recourse because of its failure to coordinate the work.

[The court reviewed the enforceability of a "no-damage" (no pay for delay) clause and concluded that it would be given effect in this case. The court rejected an argument that the clause be given effect only when the delays are reasonable and concluded that Rutgers did not actively interfere with the progress of the work nor did it act in bad faith.]

In summary then, we hold that a third party beneficiary may sue on a contract when it is an intended and not an incidental beneficiary. Resolution of that issue depends upon examination of the contractual provisions and the attendant circumstances. In a construction project where the owner directly hires the various contractors, the owner may engage a separate contractor to coordinate and supervise the project and agree with the several contractors that the general supervision will be carried out in that manner. Lastly, the owner may exculpate itself from liability for damages to the extent delineated in the contract, in the absence of any public policy reasons to the contrary.

The judgment is affirmed.

For affirmance—Justices PASHMAN, CLIFFORD, SCHREIBER, HANDLER and O'HERN—5.

For reversal—None.

The *Rutgers* decision exposes some of the administrative and legal difficulties inherent in the separate contracting system. Injection of a new "entity" into the process—in that case, a CPM consultant (perhaps a limited CM)—without a well-defined role or clearly delineated responsibility can create problems. The case demonstrates the administrative difficulties of giving one entity managerial responsibilities without the tools to effectively manage—one reason the *Rutgers* decision has been criticized.[50]

If managerial functions are given the managing contractor, what effect does this have on any owner responsibility? For example, had that power been given to the architect or even a CM, it is likely that the owner would be responsible if either had failed to coordinate properly.

The AIA in its frequently used general conditions, A201[51] ¶ 6.1.3, requires the owner to provide coordination. Each separate contractor agrees to cooperate. A separate contractor can claim another has damaged or destroyed its work or property. Most important, a separate contractor can claim that another has failed to perform in accordance with the latter's contractual obligation and adversely affected the former's scheduling and performance.

If each separate contractor has reciprocal obligations (¶ 6.1.1 requires that separate contracts be identical or substantially similar to A201), under

---

[50]Bynum, *Construction Management and Design-Build/Fast Track Construction from the Perspective of a General Contractor*, 46 Law & Contemp.Probs. 25, 33 (1983).

[51]This is not substantially different in A201 CMa ¶ 6.1.3 (1992) and B801 CMa, ¶ 2.3.6 (1992), its CM agency documents.

¶ 6.2.5, disputes between separate contractors are handled by ¶ 4.3, a complicated disputes process found in A201. The process will deal with two disputes: the claimant separate contractor against the owner and the owner against the separate contractor whose acts were alleged to have generated the claim.

Many cases have involved who can sue whom when separate contracts or multiple prime contracts are used. Sometimes the owner, as in the *Rutgers* case, can persuade the court that the separate contract system insulates the owner from any claims by a separate contractor. As demonstrated in that case, much depends on the language of the contract and any right that one separate contractor has to maintain a direct action against the other. Although one case held that the owner is insulated from responsibility,[52] another did not.[53] Even if the owner is not absolved from responsibility, it is not clear whether the owner's obligation is "strict" or whether its obligation is simply to use its best efforts, in effect a negligence standard.[54] Finally, as to the right of one separate contractor to maintain a direct action against the other, though Illinois re-

cently held that one cannot sue the other,[55] Tennessee, in *Moore Construction Co., Inc. v. Clarksville Dept. of Electricity*,[56] came to a contrary conclusion in a case involving AIA Doc. A201. The court reviewed the cases and pointed to what it saw as an emerging trend. It stated:

> Unless the construction contracts involved clearly provide otherwise, prime contractors on construction projects involving multiple prime contractors will be considered to be intended or third party beneficiaries of the contracts between the project's owner and other prime contractors. They have been permitted to recover when the courts have found that their fellow prime contractor assumed an obligation to the owner to them (the "duty owed" test) or that their fellow contractor assumed an independent duty to them in their own contract with the owner (the "intent to benefit" test). The courts have generally relied upon the following factors to support a prime contractor's third party claim: (1) the construction contracts contain substantially the same language; (2) all contracts provide that time is of the essence; (3) all contracts provide for prompt performance and completion; (4) each contract recognizes the other contractor's rights to performance; (5) each contract contains a non-interference provision; and (6) each contract obligates the prime contractor to pay for the damage it may cause to the work, materials or equipment of other contractors working on the project.[57]

The court found these requirements present in the A201 to which both separate contractors had assented.

Suppose one separate contractor can sue the other for the latter's breach of contract. Should this have any bearing on the separate contractor's right to sue the owner? Under A201, ¶ 6.1.1, the owner provides coordination. If it does not do so, the separate contractor has a claim against the owner. Suppose, however, that an agreed schedule provides for coordination but one separate contractor does not perform as agreed. Under the *Moore* case, the separate contractor whose work has been disrupted can sue the separate contractor whose fail-

---

[52]*Hanberry Corp. v. State Bldg. Comm'n.*, 390 So.2d 277 (Miss.1980).

[53]*Shea-S & M Ball v. Massman-Kiewit-Early*, 606 F.2d 1245 (D.C.Cir.1979).

[54]The Court of Claims was not able to resolve this issue squarely. Compare *Fruehauf Corp. v. United States*, 218 Ct.Cl. 456, 587 F.2d 486 (1978) (warranty based on suspension of work clause), with *Paccon, Inc. v. United States*, 185 Ct.Cl. 24, 399 F.2d 162 (1968) (not liable if owner takes reasonable steps). In *Blinderman Constr. Co. v. United States*, 695 F.2d 552 (Fed.Cir.1982), only best efforts were required. See also *Shea-S & M Ball v. Massman-Kiewit-Early*, supra note 53. In *Amp-Rite Elec. Co. v. Wheaton Sanitary Dist.*, 220 Ill.App.3d 130, 580 N.E.2d 622 (1991), appeal denied, 143 Ill.2d 635, 587 N.E.2d 1011 (1992), the owner was held liable to two multiple primes, despite one of the multiple primes being required to coordinate the work. Although the court did not find that the owner had warranted (strict liability) that delays would not occur, the owner owed an implied duty to coordinate. This was breached when the owner actively created or passively permitted to continue a situation over which it had control. It alone had the power to control cooperation by withholding payments or terminating or taking over the work. This demonstrates that whatever pains the owner takes to relieve itself from responsibility for coordination, the owner is likely to be held responsible because of its other powers. See Goldberg, *The Owner's Duty to Coordinate Multi-Prime Construction Contractors, A Condition of Cooperation*, 28 Emory L.J. 377 (1979), for an exhaustive treatment.

[55]*J.F. Inc. v. S.M. Wilson & Co.*, 152 Ill.App.3d 873, 504 N.E.2d 1266 (1987), appeal denied, 115 Ill.2d 542, 511 N.E.2d 429 (1987).

[56]707 S.W.2d 1 (Tenn.App.1985), affirmed March 24, 1986.

[57]Id at 10.

ure to perform properly has caused the disruption. But can the separate contractor also sue the owner? Can it assert that the owner has impliedly warranted that each separate contractor's work would be properly coordinated with the work of others? If so, the separate contractor has claims against the owner and the other separate contractor. Under A201, these claims would then have to be processed through ¶ 4.3 dealing with disputes.

Wisconsin blends these problems by a statute regulating state public contracts. The state is not liable to a separate contractor for delay caused by another if the state took reasonable steps "to require the delaying prime contractor to comply with its contract." If the state is exonerated, the prime delayed can sue the delaying prime.[58]

Suppose one separate contractor damages the work or property of the other separate contractor? Under the *Moore* case, a properly drafted construction contract would give the separate contractor harmed a legal claim against the separate contractor causing the harm. But does the former also have a claim against the owner? Such a claim could be based on the owner's breach of an express or implied provision to indemnify a separate contractor if its property is damaged by another separate contractor.

Finally, can a separate contractor maintain a direct action against the surety of the other separate contractor? New Jersey faced this in a state public contract.[59] State contracts usually require a bond be furnished to protect unpaid subcontractors, suppliers, or laborers. In this case, each separate contractor furnished a performance bond (one not required by statute) and a payment bond (one required by statute). The statute specifically gave rights to persons who perform work under the contract of the contractor who furnished the bond, such as suppliers an subcontractors, but not to those who perform under their own contracts, such as other separate contractors. Consequently, New Jersey did not allow the separate contractor to sue

on the bond. Also, it held that only the owner could maintain an action on the performance bond. As a result, the separate contractor was left with a claim against a bankrupt separate contractor. Although New Jersey stated that it was following the majority rule, much depends on the language of any statute compelling that a bond be furnished, on the bond language itself, and on any language in the contract requiring the bond, as well as on the common law.

Owners contemplating a separate contract system should consider these problems. Clearly, a separate contract system ought not to be used unless the managerial function can be better performed and at a lower cost than if it is performed by a prime contractor under a single contract system. If the separate contract system is selected, all contracts should make clear who has the administrative responsibility and who has the legal responsibility if a contractor has been unjustifiably delayed. More particularly, if this responsibility is to rest solely in the hands of the persons who are given the responsibility for managing the contract, all contracts should make clear whether the owner is exculpated.

D. Construction Management

Before seeking to define construction management, it is important to outline the reasons for its development. Principally, it developed because of perceived weaknesses and inefficiencies in the traditional Construction Process with special reference to the inability of design professionals and contractors to use efficient management skills. Design professionals were faulted because of their casual attitude toward costs, their inability to predict costs, and their ignorance of the labor and materials market, as well as costs of employing construction techniques. Owners were also concerned about the tendency of design professionals to take less responsibility for quality control, policing schedules, and monitoring payments. Contractors also came in for their share of blame. Some lacked skills in construction techniques and the ability to work with new materials. Others did not have the infrastructure to comply with the increasingly onerous and detailed workplace safety regulations.

Construction management instead would be an *efficient* tool for obtaining higher quality construction at the lowest possible price and in the quickest

---

[58]West Ann.Wis.Stat. § 16.855 (14b). Presumably, if the state is not exonerated, the state can still sue the delaying prime contractor.
[59]*MGM Constr. Corp. v. N.J. Educ. Facilities Auth.*, 220 N.J.Super. 483, 532 A.2d 764 (1987). For a case that came to a different conclusion, see *Hanberry Corp. v. State Bldg. Comm'n.*, supra note 52.

possible time.[60] A limited study by the U.S. General Accounting Office of the use of Construction Management by certain federal agencies records some problems in the past, mainly due to delay claims in fast-track projects.[61] It notes the difficulty of using CM concepts in federal construction and reports that projects using the CM concept experienced time delays exceeding six months more often than projects that did not use the concept. From a cost standpoint, the CM projects less often had cost increases that exceeded 10%. Obviously, there are many ways of accounting for these differences other than simply the CM process. In any event, the spotlighting of the CM process should lead to more studies of its effectiveness.

One reason for emphasizing construction management rather than construction managers is the great variety of approaches taken to achieve these objectives. All that can be done in this treatise is to briefly describe some of these techniques. The advantages and disadvantages have been covered elsewhere.[62]

Sometimes a CM is part of the design professional's organization and the design professional in his contract with the owner agrees to provide CM services. As noted in Section 17.04(A) in 1993 the AIA published B144/ARCH-CM, an amendment to B141 to be used when the architect also performs CM services. More commonly, the CM is separately retained, in addition to a design professional, by the owner. (There is rarely a contract between the design professional and the CM.) The design professional still furnishes a design and interprets the contract documents. But the CM advises the owner and the design professional as to design and takes over many of the site services performed by the design professional in the traditional system. As noted in Section 17.04(A), in 1992, the AIA published a new set (CMa) of CM documents for a CM as adviser. Also, it jointly published documents with the AGC for a CM who is a constructor (CMc).

In addition to containing the contracts between the participants, the CMa material includes change orders, applications and certificates of payment, certificates for substantial completion, construction change directives, and other G series administrative documents. The method of compensating the CM when acting as professional adviser can be any of the various methods used to compensate a design professional, such as a percentage of construction costs, a personnel multiplier, cost plus a fee, or a fixed price.[63]

One of the problems that has made construction management so difficult to fit into traditional legal concepts relates to those arrangements under which CMs themselves engage specialty contractors. Sometimes they do this as agent for the owner, with the contractual relationship then being between the owner and the specialty trades. Such CMs look like professional advisers. Alternatively, CMs may contract on their own with specialty trades, a system that makes them look very much like prime contractors.

CMs who contract on their own usually guarantee a maximum price (GMP)—a cost contract with a "cap." The courts have taken a strict attitude toward the GMP. Rejecting a contention by the CM that the GMP was a "target figure," an Indiana court held the CM to the GMP, despite the fact that one of the trade contractors gave a price quotation substantially higher than planned and the CM decided to do the work itself.[64] In such arrangements, CMs may also perform some construction with their own employees. As noted, in 1992, the AIA and the AGC jointly drafted CM documents in which the CM acts as constructor.[65]

When CMs prepare a GMP, they usually agree to give a GMP after the design has been worked out to a designated portion of finality, say 75%. They protect themselves by getting as many fixed-price contracts from specialty trades as they can and usually build enough into their GMP to take into account not only their fees but also contingen-

[60]For an admirable analysis of bonding in CM, see Foster, *Construction Management and Design Build/Fast Track Construction: A Solution Which Uncovers a Problem for the Surety*, 46 Law & Contemp.Probs. 95 (1983).
[61]*Federal Construction*, supra note 46.
[62]For a series of articles on construction management, see 46 Law & Contemp.Probs. Winter 1983. See also Note, 8 U. Bridgeport L.Rev. 105 (1987) (job-site safety).

[63]For a case describing functions and compensation of a CM, see *Gibson v. Heiman*, 261 Ark. 236, 547 S.W.2d 111 (1977).
[64]*TRW, Inc. v. Fox Dev. Corp.*, 604 N.E.2d 626 (Ind.App.1992).
[65]The AIA version is A121/CMc, and the AGC version is AGC 565.

cies that may arise from both design changes and unforeseen circumstances.[66]

Despite this, one of the risks inherent in a GMP is that the specialty trades or the CM who gives a GMP will claim that there have been drastic scope changes that eliminate any contractual price commitments made. Usually, though not invariably, a construction management system will use fast track, a principal justification for construction management, to complete the project in a shorter time. Yet it is *possible* to have CM without having a fast-track system. The dangers of fast tracking—sloppier work and uncontrolled costs—may persuade an owner who wishes to use the construction management system not to fast track the job.

The legal issues have at their core whether the CM is more like a design professional or an entrepreneurial contractor. For example, must the CM be registered or licensed by the state, and if so, which type of registration or license is needed?[67] This may depend on which form of CM is used, whether the CM is engaged solely as a professional adviser, or whether the CM undertakes to perform some construction himself.

Must awards for CM services in public contracts be made in the same manner as other professional services?[68] The role of the CM in awarding construction contracts also raises the issue of analogies. For example, one case held that the CM represented the owner when he awarded a contract, thereby allowing the trade contractor to demand arbitration with the owner.[69] If he is simply a professional adviser who, unlike most design professionals, has been given the authority to award contracts, the contractor has made a contract with the owner *through* the CM. Here he looks more like a design professional than a contractor.[70] (Usually

the design professional *does not* have authority to award contracts.) However, if he contracts on his own with specialty trade contractors, he looks like a prime contractor.[71]

Liability problems also involve analogies. For example, it has been held that a contractor can bring a negligence claim against the CM just as a negligence claim can be instituted against a design professional.[72] If the CM is analogized to a design professional or performs services usually performed by a design professional,[73] will the CM be given the benefit of the professional standard described in Section 14.05 or quasi-judicial immunity noted in Section 14.09(D)? The professional standard has its greatest application when the design professional is performing design rather than administrative services.[74] Inasmuch as the services performed by a CM are connected more to administration than to design, it is possible that the *ordinary* but not *professional* negligence standard will be applied, meaning that no expert testimony will be required. But if the CM is simply an adviser performing administrative tasks for the owner, he looks *less* like a design professional. As a result, the CM may be able to defend himself more successfully if sued by other participants by asserting that he was simply advising the owner and not acting as an independent professional.

In some states, persons in charge of the construction work are strictly liable for violations that injure workers. Those in charge are more commonly prime contractors, although it is possible to include design professionals or CMs if they have broad powers over the process.[75] If CMs look like prime contractors, either because they are managing the work of the specialty trades and have overall safety responsibility or because they are

---

[66]For a case involving a D/B contract and the permissible design changes, see *City Stores Co. v. Gervais F. Favrot Co.*, 359 So.2d 1031 (La.App.1978).

[67]See Bynum, supra note 50 at 27–28; Lunch, infra note 81 at 86–87.

[68]Compare *Attlin Constr., Inc. v. Muncie Community Schools*, 413 N.E.2d 281 (Ind.App.1980), with *City of Inglewood—L.A. County Civic Center Auth. v. Superior Court*, 7 Cal.3d 861, 500 P.2d 601 103 Cal.Rptr. 689 (1972).

[69]*Seither & Cherry Co. v. Illinois Bank Bldg. Corp.*, 95 Ill.App.3d 191, 419 N.E.2d 940 (1981).

[70]See *Bagwell Coatings, Inc. v. Middle South Energy Inc.*, 797 F.2d 1298 (5th Cir.1986) (contractor cannot sue CM who was owner's agent).

[71]But see *Turner Constr. Co.*, ASBCA 25602, 81-1 BCA ¶ 15,070 (trade contractor engaged by CM could sue public agency who maintained significant direct involvement in award and performance despite disclaimer of any contractual relationship between public agency and trade contractor).

[72]*Gateway Erectors Div. v. Lutheran Gen. Hosp.*, 102 Ill.App.3d 300, 430 N.E.2d 20 (1981).

[73]*Pierce Assoc., Inc.*, supra note 45 (CM interprets contract and judges performance).

[74]Refer to Section 14.05(A).

[75]*Kenny v. George A. Fuller Co.*, 87 A.D.2d 183, 450 N.Y.S.2d 551 (1982) (also liable as agent of owner).

performing some of the construction with their own forces, they may be found to be contractors with contractorlike liability.[76]

In some states, being a "contractor" CM can be advantageous in the event of worker claims. For example, such a CM may claim immunity as a statutory employer.[77] This was unsuccessful where the CM did not construct anything but was simply hired to manage and inspect,[78] more like a design professional than a contractor. A "design professional" CM with safety functions may find himself exposed to claims by injured workers.[79] Similarly, one who has broad powers over the Construction Process and does none of the construction work may also expose the owner to liability.[80]

Those planning to engage a CM or those about to engage in CM work must also determine whether the insurance coverage of the CM will be adequate. The CM who is principally a contractor will find that his comprehensive (increasingly called commercial) general liability coverage will not include design, while a CM who is essentially a design professional may find that his professional liability insurance excludes any coverage that relates to the Construction Process or the work of the contractor.[81]

These legal uncertainties demonstrate the difficulty of placing the CM in traditional categories in the Construction Process, a task made even more difficult because there has yet been no agreement on the proper function of the CM. Perhaps this is inevitable during a shakedown period while the industry struggles to come to some consensus based on a sufficient experimentation time. The future may see lawmakers, such as legislators and judges, revise existing laws to take this new actor into account if he cannot be fit within the existing categories.

E. Turnkey Contracts

The discussion in this subsection must be taken together with the material in (F) dealing with design/build. Both subsections involve the owner contracting with an entity who agrees to both design and build. However, because turnkey contracts have other functions, they should be taken separately, both on their own and as an introduction to design/build.

There are a great variety of turnkey contracts. At its simplest, the contract is one in which the owner gives the turnkey builder some general directions as to what is wanted and the turnkey builder is expected to provide the design and construction that will lead to the satisfaction of the express needs of the client. In theory, once having given these general instructions, the owner can return when the project is completed, turn the key, and take over.

Many turnkey projects are not that simple. The instructions often go beyond simply giving a general indication of what is wanted. They can constitute detailed performance specifications.[82] Also, the obligation to design and build may be dependent upon the owner furnishing essential information or completing work upon which the turnkey contractor relies to create the design and to build. Finally, the owner who has commissioned a turnkey project is not likely to remain away until the time has come to turn the key. As in design/build, the owner may decide to check on the project as it is

---

[76]*Bechtel Power Corp. v. Secretary of Labor,* 548 F.2d 248 (8th Cir.1977) (OSHA liability for injury to CM employee); *Lemmer v. IDS Properties, Inc.,* 304 N.W.2d 864 (Minn.1980) (liable as "possessor of land"). As to OSHA, see *Skidmore, Owings & Merrill,* 5 O.S.H. Cas. (BNA) 1762 (1977). See Note, 32 Loy.L.Rev. 447 (1986) (concerned that CM will be held liable for site safety beyond assumed undertakings).

[77]See Section 31.02.

[78]*Brady v. Ralph Parsons Co.,* 308 Md. 486, 520 A.2d 717 (1987), affirmed 327 Md.275, 609 A.2d 297 (1992).

[79]*Caldwell v. Bechtel, Inc.,* 631 F.2d 989 (D.C.Cir.1980); *Phillips v. United States Eng'rs. & Constructors, Inc.,* 500 N.E.2d 1265 (Ind.App.1986); *Riggins v. Bechtel Power Corp.,* 44 Wash.App. 244, 722 P.2d 819 (1986), review denied 107 Wash.App. 1003 (1986). See Note, 8 U. Bridgeport L.Rev. 105 (1987), which assumes expanded tort liability for CMs and suggests protection by use of the professional standard and indemnity from the contractor.

[80]*Everette v. Alyeska Pipeline Service Co.,* 614 P.2d 1341 (Alaska 1980).

[81]See Lunch, *New Construction Methods and New Roles for Engineers,* 46 Law & Contemp.Probs. 83, 93–94 (1983).

[82]Another definition is given in *Smithco Eng'g, Inc. v. Int'l Fabricators, Inc.,* 775 P.2d 1011 (Wyo.1989), citing 1 B.J. MCBRIDE and I. WACHTEL, GOVERNMENT CONTRACTS § 9.120[12] (1980) as follows: A turnkey project is one in which the developer builds in accordance with plans and specifications of his own architect subject to performance specifications for quality and workmanship, and with limited guidance for features as style of house, number of bathrooms, etc.

being built and is almost certain to be making progress payments while the project is being built. The most important attribute of the turnkey project is that one entity both designs and builds, a system discussed in greater detail in (F).

Some turnkey contracts require the contractor not only to design and build the building but also to provide the land, the financing, and interior equipment and furnishings.

A turnkey contract looks more like a sale than a contract for services. As a result, one court held that a turnkey contract created warranties that made the seller-contractor responsible for any defects.[83] This is something that may also be found in any design/build contract.

F. Combining Designing and Building (D/B)

Transactions in which the architect or a contractor both designs and builds have become an important variation from the traditional method of organizing for construction. D/B (design/build) can encompass at one extreme the homeowner building a single-family home patterned on a house that the builder has already built and at the other a large engineering company agreeing to both design and build highly technical projects, such as petrochemical plants. The former is likely to be a builder with an in-house architect on its staff or one that engages an independent architect where it is required by law that design be accomplished by a registered architect or engineer. At the more complex extreme, the designer/builder may employ a large number of construction personnel and licensed architects and engineers in-house to offer a total package for projects such as power plants, dams, chemical processing facilities, and oil refineries. Between these extremes, D/B is often used for less technical repetitive work such as warehouses or small standard commercial buildings.

Organizationally, D/B has variations. One method is an architect promising to design and build and employing a contractor to execute the design. Because of the capital needed, a more common technique is the D/B to be a contractor who engages a design professional to create the design. Finally, a D/B can be a joint venture between a design professional and a contractor. As shall be noted, each form raises licensing and insurance issues.

Even more than in construction management, the design/build system has generated a large number of legal problems that must be confronted before the choice can be made to use such a system. In public projects, the principal problem is the usual requirement that construction work be competitively bid by the contractor and that the design professional be chosen either entirely on the basis of demonstrated competence and experience or in a two-stage process where the design professional competes with others as to competence and, once that has been determined, enters into a negotiation with the public entity.[84]

In the traditional competitive bidding process by which contractors are selected (discussed in greater detail in Chapter 18), the design is completed and given to potential contractors to bid for the work. The lowest responsible bidder is usually selected. Bids of this sort cannot be made unless there is a relatively worked out design.

Often a fast-track system is used when the design/build system is employed. This means that prospective bidders cannot be given a detailed set of plans and specifications upon which to make their bids. While the D/B system can be used competitively,[85] on the whole it does not fit with the requirement that the competitive bidding process be used to award construction work. As a result, there are many barriers to the use of design/build projects for public work. Some states have not allowed this method to be used, while in other states, it can be used, usually because it has been authorized for specific projects or because the state has enacted legislation that allows—or even requires—the design/build system to be used for certain types of construction.

Another problem in public contracts is that the design for a design/build project may be made by a designer who has not been selected in accordance with the state laws for selecting design professionals for public works. Florida met this problem in 1989 by enacting legislation that sets forth spe-

[83]*Mobile Housing Environments v. Barton & Barton*, 432 F.Supp. 1343 (D.Colo.1977).

[84]Refer to Section 11.03.
[85]*Ogden Dev. Corp. v. Federal Ins. Co.*, 508 F.2d 583 (2d Cir.1974).

cific standards when a public agency seeks to use the D/B system.[86] These standards include a requirement that the agency engage a licensed professional to prepare a design criteria package. Also, the agency must select no fewer than three D/B firms and specify criteria procedures and standards for evaluation of the proposals. In evaluating proposals, the agency must consult with the retained design criteria professional concerning the evaluation of bids.

A public entity wishing to use design/build must pilot through the shoals of the competitive bidding laws and the laws under which design professionals are selected before such a method can be used.[87]

In private contracting, the principal problem in design/build again relates to licensing laws. Under Missouri law, a contractor who did not have an architect's license could not recover for work performed under a D/B contract.[88] As noted, the design/builder may be a business organization that will either furnish the design services needed in-house or retain an independent design professional to perform these design functions.[89] In addition, many states bar business corporations from performing architectural services. As a result, design/build contracts have been challenged, even if the design portion was to be fulfilled by the engaging of a licensed design professional.[90] While the trend has been toward permitting such contracts despite the absence of a registered independent design professional, legal opinion should be sought if the design/build system is to be employed.

Finally, the liability insurance of a design professional usually excludes construction and work performed under a joint venture, while the liability insurance of a contractor usually excludes design. An entity planning to engage in a D/B venture must check insurance carefully.

Despite the legal obstacles to the design/build system, D/B has clearly been found to be useful. The principal advantage of D/B (called in England a "package" job), usually associated with fast tracking, is speed. One entity replaces *different* entities who design and build with the inevitable delay that using two entities whose work intersects will cause. Those who design and build frequently do repetitive work and acquire specialized expertise. Perhaps most important from the standpoint of an unsophisticated owner, the system concentrates responsibility on the designer/builder. Owners are often frustrated when they look to the designer who claims that the *contractor* did not follow the design, with the latter claiming that the problem was poor design.

Another impetus for the use of design/build occurred when the AIA, at first on a provisional basis, decided that its members could engage in business without running afoul of ethical constraints. The AIA did this because it felt that economic realities in the construction industry dictated that architects be given the opportunity of engaging in design/build ventures. After a study of how this was working, the AIA approved its members engaging in design/build work and published a series of design/build documents: A191 (owner and design-builder), A491 (design-builder and contractor), and B901 (design-builder and architect).

D/B has weaknesses. The absence of an *independent* design professional selected by the owner can deprive the owner of the widest opportunities for good design. An unsophisticated owner often lacks the skill to determine whether the contractor is doing the job well or as promised. This can reflect itself not only in substandard work but also in excessive payments being made early in the project or slow payment or nonpayment of subcontractors.

Owners, such as developers, do not like to make cost-plus contracts with designer/builders. Clear and complete drawings and specifications protect a developer who uses a cost-plus contract. Although the price is not defined, unless there is an effective GMP, the project is defined. In design/

[86]West Ann.Fla.Stat. § 287.055.
[87]For a thorough study, see Buesing, *The Law Struggles to Keep Pace with the Trend of State and Local Government Experience with Design/Build*, 11 Constr.Lawyer No. 4, October 1991, p. 22. See also Halsey & Quatman, *Design/ Build Contracts: Valid or Invalid?* 9 Constr.Lawyer No. 3, August 1989, p. 1 (focusing on Private Contracts); Comment, 17 U. Dayton L.Rev. 109 (1991).
[88]*Kansas City Community Center v. Heritage Industries, Inc.,* 972 F.2d 185 (8th Cir.1992).
[89]Such a contract was upheld in New York in a 4–3 decision in *Charlebois v. J.M. Weller Assocs.*, 72 N.Y.2d 587, 531 N.E.2d 1288, 535 N.Y.S.2d 356 (1988). The dissent worried about the effect on the public and the registration laws when the design professional is engaged by the D/B.
[90]See note 88 supra.

build, however, the owner may not even know what is to be built when it enters into the contract.

The contractor is reluctant to give a fixed price on a design/build project. It cannot know with any certainty what it will be expected to build. If the design has not been worked out at the time the contract is made, the design must be completed later. Very likely there will be redesign. As a result, the owner will prefer a fixed price, while the contractor would like an open-ended, cost contract.

To ameliorate this possible impasse, the owner may find it helpful to prepare a set of *performance* specifications. Even if the specifications cannot define the elements of the building in detail, they may be able to prescribe intelligent criteria for performance in advance.

Another useful technique is to prepare a budget for each phase of the work and designate the budget estimate as a target price. If the actual cost is greater or less than the estimate, the contract price can be adjusted.

Progress payments also provide problems in design/build contracts. Under traditional methods, the architect certifies progress payments. Although some complain that the architect cannot be impartial when he does this, the architect's professional stature gives hope that he will be impartial.

In design/build, payment certificates issued by an architect retained by the contractor may not be reliable. Owners who use design/build often take control of the progress payment process, a method that can operate to the disadvantage of the contractor. As a result, it may be useful for each party to agree on an independent certifier for progress payments. The system works best in repetitive work of a simple type, such as warehouses and uncomplicated residential construction, although it is also used in high-technology work.[91]

Care must be taken not to lock D/B to fast tracking. One advantage of combining designing and building is the ability to fast track the project. However, the two need not go together. One entity who *both* designs and builds may create the complete design before starting construction.[92]

If the design professional must be independent and retained by the D/B contractor, legal problems may develop. For example, the owner with whom the design professional has no contract may wish to bring a claim against the design professional based upon negligence.[93] For example, in *Nicholson & Loup, Inc. v. Carl E. Woodward, Inc.*,[94] the owner recovered a judgment against the design/build contractor, the soils firm that had produced a defective soils report for the contractor, and the architect who had been an employee of the design/builder. The in-house architect's name appeared on the contract as the person who would prepare the plans and specifications. The architect argued that he was not responsible for design defects because he was not involved in selecting the soils engineer and had no input on the structural design or the selection of piles or pile lengths. While the architect's seal appeared on all pages of the plans and specifications except the foundation plans and those marked for electrical and mechanical plans, the court pointed to a notation on the first page of the documents where the architect stated that he had prepared them or that they were prepared under his close personal supervision. This meant, according to the court, that the architect assumed responsibility for all the documents in the packet, including the foundation plans, the electrical plans, and the mechanical and plumbing plans. By law, the architect could not affix his seal to any document not prepared by him or under his responsible supervision.

Similarly, the D/B contractor may be giving sellerlike warranties to the purchaser-owner but may be able to hold an independent design professional only to the professional standard unless a more specific standard is set forth in a contract between builder and design professional.

On the other hand, suppose the designer is an *employee* of the D/B contractor. The owner may believe that an independent design professional will perform the traditional services that such a person performs when retained by the owner. Employees who work for a design/build entity need not look out for the interest of the owner as they would

---

[91] Advantages and disadvantages are explored in materials cited supra note 62.

[92] See Dibner, *Construction Management and Design-Build: An Owner's Experience in the Public Sector*, 46 Law & Contemp.Probs. 137, 143–144 (1983).

[93] This was permitted in *Keel v. Titan Constr. Corp.*, 639 P.2d 1228 (Okl.1981).

[94] 596 So.2d 374 (La.App.), writ denied, 605 So.2d 1098 (1992).

were they retained by the owner. This may raise problems if the employee is a registered architect or engineer who may be expected to look out for the public interest rather than simply being an employee of a contractor.

An owner building a single-residence home, a warehouse, or simple commercial building relies entirely on the contractor for design and construction. Here warranties of fitness are appropriate. The owner is buying not services but a finished product. The sophisticated owner who employs D/B and has within its own organization the services of skilled professionals to control design or monitor performance has bought not a product but rather expert services to prepare and execute the design it chooses. Such an owner should not receive a sale-of-goods-like warranty.

The differentiation between the two types also reflects itself in pricing. It is likely that the homeowner commissioning a builder to design and build a house will contract on a fixed price. The sophisticated owner building a petrochemical plant will contract on a cost basis with a GMP, at least until the design is put in final form.[95] Such projects create immense exposure for consequential damages. It is likely that the D/B will limit its obligation to correction of defects.

G. Partnering

"Partnering" has been receiving increasing attention in the Construction Process world. While it is not strictly speaking a form of organization, it is an approach that can remedy some of the stiffness and adversary tendencies that are so commonly part of the Construction Process.

Partnering is designed to create harmonious relationships needed in a successful construction project. The partnering process usually is initiated when the principal project participants are assembled. It can also be used to steer a troubled project back on course. These relationships are to be structured upon trust and dedication to the smooth operation of the project. Those who advocate this concept hope that it will accomplish the project successfully and minimize the likelihood of claims. Partnering, then, is a method of building a closeness that transcends organizational boundaries. This closeness should be premised upon a shared culture and the desire to maximize the likelihood of attaining the common goals.

While the objectives of partnering can be sought and achieved on an informal basis, focus upon this concept has led to the designation of a facilitator and development of an agreement designed to highlight the goals of partnering and the methods by which they are to be achieved.

Partnering can suffer unless the entities involved in the effort are willing to make sacrifices to keep the team intact. This is particularly a problem in dealing with design professionals, clients often expressing concern when design professionals with whom they have been working are assigned to other work. But this can also be a problem in large engineering firms and the organizations that employ them. Contacts between persons working at different operational levels of the various entities must be developed, and a sense of confidence and trust among these workers must be built up and maintained.

Partnering arrangements can be formalized by contracts, which, at the basic level of fulfilling operational needs, should seek to make concrete the parties' often nebulous commitments. What is important is that the parties express as clearly as possible what each party expects and what each party is willing to undertake. Clearly, expressions of this sort are difficult to make sufficiently specific and concrete. But some commitments, such as confidentiality and the agreement not to profit at the other "partner's" expense, may be worth expressing in the contract.

It is likely that resort to litigation will be rare in partnering arrangements. Judicial intervention will almost never take the form of ordering people to continue such arrangements or compensating one party—particularly as to acts in the future—because of the other's refusal to continue. There may, however, be resort to legal action if one party has breached confidentiality or has stolen business opportunities that should go to the other. For this reason, it is useful to deal with this in the contract. Finally, legal action may be taken if one party has incurred expenses that it feels the other should bear or at least share. Again, these are matters that may be appropriate for contractual resolution.[96]

[95]See Foster, supra note 60 at 118–119.

[96]See Roberts & Parisi, *Partnering: Prescriptions for Success*, 13 Constr.Litig.Rep. 298 (1992) (contains suggestions and cites relevant literature).

H. Teaming Agreements

Teaming agreements have been used when the federal government has embarked in programs to fulfill procurement needs, mainly in the high-tech defense area. A teaming agreement is a contract between a potential prime contractor and another firm that agrees to act as a potential subcontractor. The prime, sometimes with the assistance of the potential subcontractor, draws up a proposal for a federal agency contract in competition with other prospective contractors. Typically, the prime promises that if it is awarded the prime contract, it will reward its "teammate" with an implementing subcontract in return for the often unpaid support given it in making the bid.

Most of the litigation has involved claims by aggrieved teammates against the prime when they have not been awarded the implementing subcontract. Such agreements generally use such phrases as "best efforts" and "reasonable efforts" as well as other nonspecific terminology. Often the question of whether the prime is actually committed to award the implementing subcontract to the potential subcontractor is left vague and uncertain. These arrangements resemble in some ways the normal prime contractor-subcontractor relationship discussed in Chapter 28. As seen in Section 28.02, the primes seek to retain as much flexibility as possible while still keeping their teammate interested in supporting the arrangement. Yet the prime would like to ensure that any commitments made by the teammate can be legally enforced if necessary.

Moreover, similar to the prime-subcontract relationship discussed in Chapter 28, the terms that would govern the relationship if the prime wins the award and awards the subcontract to its teammate frequently have not been fully worked out. Sometimes such agreements include express covenants of good faith and fair dealing, but such terms do not necessarily mean that the two parties will always agree on more substantive questions. Also, the twilight area between unenforceable agreements to agree and enforceable agreements to negotiate in good faith can create difficult legal questions.[97]

I. Build/Operate/Transfer (BOT)

In the international field, some countries are not able to finance large infrastructure projects. As a result, they may make an agreement after competitive bidding with a consortium under which the latter operates the project and then ultimately transfers it back to the state. The consortium that seeks to enter into such a contract usually must provide the financial backing to build the project, get it operating, and keep it running so that it can be transferred (BOT). BOT projects are used in large infrastructure projects such as energy generation, bridges, and road systems. Most BOT projects are turnkey contracts, which makes financing more secure and easier to obtain. However, cost overruns are not unknown.

The parties involved in a simple BOT project are as follows:

1. The granting authority (usually a sovereign state).
2. The project sponsors (the entity that has put together the principal actors).
3. The financiers.
4. The users of the project.

The granting authority wants to be certain that the project is built properly, since its citizens are likely to use the project directly or indirectly and since it will be the recipient of the project at the end of the designated period. For their part, the project sponsors and financiers—who are operating in a highly regulated environment—want to ensure that the project finances are sufficient through assurances from the granting authority that there will be no rate decreases for electricity or tolls to please its citizen users.

In addition to having the need to coordinate a large number of participants with different interests, a BOT contract carries the normal risks of any international transaction that involves bringing together entities from different legal systems into a complex legal arrangement.[98]

J. Summary

The modern variations described in this section developed because of weaknesses in the traditional

---

[97]For a comprehensive analysis and concrete suggestions, see Note, 91 Colum.L.Rev. 1990 (1991).

[98]See Merna et al. *Benefits of a Structured Concession Agreement for Build-Own-Operate-Transfer ("BOT") Projects*, 10 Int'l.Constr.L.Rev. Pt. 1, 32 (1993).

system, particularly leisurely performance, divided responsibility, and complaints of subcontractors. However, as these new systems become used, they will become time tested and court tested. In the process, deficiencies will be revealed. This does not in any way diminish the utility of these variations, although it does impose a serious burden on those who wish to engage in them to anticipate the problems in deciding whether to use a variant from the traditional process and to plan in such a way so as to minimize the difficulties, both administrative and legal.

SECTION 17.05 Administrative Problems

A. Overview

This chapter has examined how a construction project can be organized, mainly the problems of selecting the type of contract (fixed-price or cost), the type of contract system to use (single, separate, D/B), and methods to bring efficiency to the management of the process (prime contractor, design professional, or construction manager). Throughout this chapter, reference has been made to the need for efficient organization of any project, with emphasis principally on the coordination of the activities of the various participants.

Yet another factor essential to construction efficiency must be examined. A successful construction project recognizes the importance of clear and efficient lines of communication among the main participants. It is particularly important to know who has authority to bind the owner (with the hope that such a person can be identified and found), how communications (and there are many) are to be made, to whom they should be directed, and what the rules are for determining their effectiveness (effective when posted or when received?).

Communications are essential elements in the responsibility for defects (Chapter 24) and delays (Chapter 26) as well as equitable adjustments for unanticipated subsurface conditions (Chapter 25). (When those topics are discussed, the problems of communications will be addressed.) After a contract has been awarded and before performance begins, the participants must address these administrative issues openly and seek to make clear at the outset who speaks for whom, to whom and how communications are to be directed, and

what the rules are for their implementation. Usually they are addressed at a preconstruction meeting attended by the representatives of the major participants in the process. This section deals with some of these problems.

B. Authority: Special Problems of Construction Contracts

Were the construction contract an ordinary one, it would be sufficient simply to refer to Chapter 4 dealing with agency and authority. However, special characteristics of the construction project make it essential to devote additional discussion to that topic. One relates to the multifaceted roles of an independent design professional engaged to design and monitor performance, to act as agent of the owner, to interpret and judge performance, and in general to be the central hub around which the process revolves. When a construction manager is used, the problems do not become easier, since the process revolves around two participants of often unclear responsibilities.[99]

In construction projects that will justify it, the owner may have a full-time site representative who observes, records, and reports the progress of the work. The representative (called a clerk of the works, project representative, resident engineer, or otherwise) can be a regular employee of the owner, an employee of the design professional or CM and paid for either by the employer or by the owner, or an independent entity retained to perform this service. The representative's permanent presence invites difficulties, with the contractor often contending that he directed the work, knew of deviations, accepted defective work, or knew of events that would be the basis for a contractor claim for a time extension or additional compensation.[100]

Communication is vital to construction. Problems often develop that should be brought to the attention of the appropriate party. For example, any design difficulties should be reported to the design professional. If there is intervention by public authorities, clashes between participants, material shortages, or strikes—to name only a few

---

[99]Even more complex was an arrangement with two CMs, one called an "Overall CM." See *DiSalvatore v. United States*, 456 F.Supp. 1079 (E.D.Pa.1978), 472 F.Supp. 816 (E.D.Pa.1979).

[100]See Section 21.04(H).

problems—the owner and design professional or construction manager should be made aware of them. Early discussion of problems or potential difficulties should effect prompt corrective action, efficient readjustments, and gathering of necessary information for the ever-present "claim." A successful construction project requires clear lines of communication and their use.

Looking first at the traditional organization (CMs are discussed later), the design professional, though an agent of the owner, has limited authority;[101] the authority of a project representative is even *more* limited.[102] Which acts of the design professional or project representative will be chargeable to the owner? Before looking at that question, it is important to differentiate between *acts* of the design professional that bind the owner by principles of agency from those *decisions* made by the design professional as interpreter of the contract and judge of its performance. Although courts sometimes blur this distinction, this section concentrates on the first. Chapter 29 examines the second.

Generally, the owner retains a design professional to provide professional advice and services, *not* to make or modify contracts or issue change orders. Although design professionals may enter into discussions with prospective contractors or even assist in the administration of a competitive bid, participants in the construction industry recognize that design professionals do not have authority to make contracts, modify them, or issue change orders on behalf of owners.

This does not mean that a design professional will never bind the owner. First, a design professional may be given express authority to perform any of these functions. For example, the design professional is frequently given authority to make *minor* changes in the work.[103] In addition, the design professional may be vested with certain authority that by implication includes other authority. For example, an architect who had the power to require that a contractor post a payment bond was authorized to represent to a subcontractor that such a bond would be required.[104]

Similarly, the owner may, by its acts, cloak the architect with apparent authority. For example, suppose the architect directs major changes for which he does not have express authority. If the owner stands by and does not intervene to deny such authority or pays for the changes, either may represent to the contractor that the architect has apparent authority to order that work be changed.[105]

Care must be taken to determine whether the principal—that is, the owner—has authority that can be exercised through its agent, the design professional. For example, suppose the engineer directs the contractor as to *how* the work is to be performed where this choice by contract is given the contractor. The *owner* would not have the authority to direct the methods. Any directions given by the engineer would not have to be followed any more than any direction by the owner. This is not an agency issue but one of contract interpretation.

Accepting defective work raises problems that are discussed in Section 24.05. A few points need to be made in this chapter, however. First, the design professional, as interpreter and judge of performance, decides whether work complies. If he decides that the work is proper, he has not accepted defective work. Of course, his determination may be reversed by arbitrators or a court. However, if he determines that the work has not conformed, any acceptance by him would be changing the contract or waiving the owner's right to proper performance, neither of which he has authority, at least in the ordinary case, to do.

Because of the failure to make this distinction, it is more likely that a design professional's acceptance of the work, even defective work, will be con-

[101]*Crown Constr. Co. v. Opelika Mfg. Corp.*, 480 F.2d 149 (5th Cir.1973); *Metro. Sanitary Dist. v. Anthony Pontarelli & Sons, Inc.*, 7 Ill.App.3d 829, 288 N.E.2d 905 (1972); *Kirk Reid Co. v. Fine*, 205 Va. 778, 139 S.E.2d 829 (1965). But an actual employee may have more authority. *Grand Trunk Western R. Co. v. H.W. Nelson Co.*, 116 F.2d 823 (6th Cir.1941) (employee design professional authorized to make important contract).
[102]*Lemley v. United States*, 317 F.Supp. 350 (N.D.W.Va.1970), affirmed 455 F.2d 522 (1971); *Missouri Portland Cement Co. v. J.A. Jones Constr. Co.*, 323 F.Supp. 242 (M.D.Tenn.1970), affirmed 438 F.2d 3 (6th Cir.1971); *Samuel J. Creswell Iron Works, Inc. v. Housing Auth. of the City of Camden*, 449 F.2d 557 (3d Cir.1971); *Acoustics, Inc. v. Trepte Constr. Co.*, 14 Cal.App.3d 887, 92 Cal.Rptr. 723 (1971); *Savignano v. Gloucester Housing Auth.*, 344 Mass. 668, 183 N.E.2d 862 (1962); *Hunt v. Owen Bldg. & Inv. Co.*, 219 S.W. 138 (Mo.App.1920).

[103]AIA Doc. A201 ¶ 7.4.1.
[104]*Bethlehem Fabricators, Inc. v. British Overseas Airways Corp.*, 434 F.2d 840 (2d Cir.1970).
[105]See Section 21.04(C).

sidered to bind the owner. To preclude this result, many contracts provide that acts of the design professional, such as issuing certificates, will not be construed as accepting defective work.[106]

As indicated earlier in this section, the Construction Process requires many communications among the participants. Clearly, a contract can provide that a notice can or must be given to the design professional.[107] If this is the case, the notice requirement has been met if notice has been given to the design professional. The notice need not come to the owner's attention. In the absence of specific contract language of this type, will a notice that should go to the owner be effective if it is delivered to the design professional?

Because of the close relationship between owner and design professional and the tendency of the law generally to be impatient with notice requirements if they would appear to bar a meritorious claim, there is a strong likelihood that a notice given to the design professional will be *as if* given to the owner. For example, in *Lindbrook Construction, Inc. v. Mukilteo School District Number 6*,[108] the contractor discovered unanticipated subsurface conditions substantially at variance with the contract documents. It notified the architect and claimed an equitable adjustment in the price and additional time. The architect denied that the contractor was entitled additional compensation and directed it to proceed. The contractor completed the work and sued the owner.

The issue before the appellate court was whether the contract requirement that notice be given to the *owner* had been satisfied by notice being given to the *architect*. The court held that the architect was the only representative of the owner with whom the contractor had contact. It noted that the architect had complete knowledge of the changed condition and that the contractor was going to claim additional compensation. There was no evidence of a breakdown in communication between the architect and the owner, and the court noted that it would be unbelievable to assume that the architect did not notify the owner. The court held that the notice to the architect was imputed to the school district.

The dissenting judges argued that the architect's knowledge of the extra work could not be imputed to the owner because the architect had no authority to modify the contract that required that notice be given to the *owner*. The architect was therefore, according to the dissenters, not operating within the scope of his authority, and the school district should not be bound. The dissenters also pointed to specific language stating that the architect could not commit the district to cost allowances and noted the importance of protecting public funds from unlawful expenditure.

Sometimes the notice problem is dealt with specifically by statute, such as provisions in mechanics' lien laws that state that notices can be given to the architect.

Are facts known to the design professional *as if* they were known by the owner? For example, one case held that the notice of a limited warranty given to a mechanical engineer consultant bound the owner-buyer.[109] Similarly, a design professional's knowledge of the construction industry custom was held to be chargeable to the owner.[110]

Yet language in one federal trial court opinion seems to point in another direction.[111] The case involved a claim by an owner for costs incurred to detect and remove asbestos contained in fireproofing material sold by the defendant. One issue was whether the claim had been barred by the passage of time. The asbestos fireproofing material had been installed in 1971, but representatives of the owner testified that they did not know that the fireproofing contained asbestos until January 1989. If the statute of limitations commenced in 1971, the claim would have been barred, but if the period for commencing the statute of limitations was 1989, the claim would still be valid.

The manufacturer contended that the architect was the agent of the owner and that, as such, if he (the architect) knew or should have known of the hazards of asbestos, this knowledge should be imputed to the owner. The court rejected this, noting that whether the architect is an independent con-

---

[106]AIA Doc. A201, ¶¶ 9.6.6, 4.3.5.
[107]Id. at ¶ 8.3.2 (time extensions).
[108]76 Wash.2d 539, 458 P.2d 1 (1969).

[109]*Trane Co. v. Gilbert,* 267 Cal.App.2d 720, 73 Cal.Rptr. 279 (1968).
[110]*Fifteenth Ave. Christian Church v. Moline Heating and Constr. Co.,* 131 Ill.App.2d 766, 265 N.E.2d 405 (1970).
[111]*Blue Cross and Blue Shield of South Carolina v. W.R. Grace & Co.,* 781 F.Supp. 420 (D.S.C.1991). After the trial court's opinion, the parties settled the claim.

tractor or an agent depends upon the particular function performed. It noted that the architect is an independent contractor in matters concerning the preparation of plans and specifications and that the selection of materials was part of that preparation. However, the court also noted that the architect is the agent of the owner during "supervision," his actual authority depending upon the contract and upon any apparent authority the acts of the owner granted him.

The court rejected the manufacturer's contention that the architect could have issued a change order regarding materials during the supervision phase and therefore was the owner's agent during construction. The manufacturer made this contention to establish that any knowledge of the architect regarding materials is imputed to the owner once construction begins, but the court found that this contention would undermine the legal distinction between the different roles of the architect.

In any event, according to the court, there was no evidence that the architect knew anything of the hazards of asbestos or that the fireproofing material contained asbestos when it was specified in 1971.

A more bizarre use of the imputation doctrine was unsuccessful in a claim by an owner against its engineer based on conspiracy with the contractor to prepare and submit false estimates and vouchers. One contention made by the defendant engineer was that his knowledge of the conspiracy was imputed to the *owner*. This unusual use of the imputation doctrine was rejected, the court noting that imputation applies only when the agent had knowledge that he has a duty to communicate to his principal. Where he had the motive to conceal, knowledge cannot be imputed to the principal.[112]

Clearly, if the design professional has limited authority, the project representative has even less. Usually the project representative is simply authorized to observe, keep records, and report.[113] Use of a project representative is often accompanied by a document that describes these functions and limits the authority of any project representa-

tive.[114] There is always a risk that a project representative will become overactive and seek to perform unauthorized activities, such as directing or accepting work. Although these are clearly unauthorized and the contractor should realize it, the contractor may submit and then later claim that these acts bind the owner. Usually, such assertions are not successful.[115]

This section has assumed the traditional construction delivery system. However, the problem of authority may have to take into account the way in which an architect or engineer, in a traditional system, may differ from a CM. Again, great caution must be exercised in making generalizations because of the fluidity and imprecision of construction management.

Nevertheless, it is likely that the difference in professional status between a design professional and a CM may reflect itself in the extent to which acts of the CM may be held to bind the owner. In many instances, it may be difficult to distinguish the activities of regular employees of a sophisticated owner, such as members of an engineering department, from the activities of a CM. Even if the CM is an independent entity, much of what the CM does, though resembling what design professionals were expected to do in the traditional system, is more geared toward the efficiency expected from the owner's own employees. As a result, it may be easier to establish apparent authority of a CM than of a design professional.[116]

To sum up, independent advisers such as design professionals, CMs, and project representatives, though agents of the owner, do not have actual authority to make contracts, modify existing contracts, accept defective work, or waive contract requirements on behalf of the owner. Yet doctrines such as apparent authority or ratification may, in a proper case, justify a conclusion that acts of these professional advisers will be chargeable to the owner.[117]

---

[112]*Metro. Sanitary Dist. v. Anthony Pontarelli & Sons, Inc.,* supra note 101.
[113]But in *Town of Winnsboro v. Barnard & Burk, Inc.,* 294 So.2d 867 (La.App.1974), certiorari denied 295 So.2d 445 (1974), knowledge of a project representative was chargeable to the architect.

[114]AIA Doc. B352. A new edition was published in 1993.
[115]See note 102, supra.
[116]*Turner Constr. Co.* ASBCA 25602, 81-1 BCA ¶ 15,070 (owner bound by notice to CM).
[117]See Section 21.04(C).

C. Communications

A communication can be made personally, by telephone, or by a written communication. It is advisable to specify in the contract that a facsimile (fax) or electronic mail (e-mail) is a written communication. Generally, written communications should be required where possible. If it is not possible for a written communication to be made, at the very least a person making the oral communication should give a written confirmation as soon as possible. If the contract provision that deals with communications of notices does not state how communications are to be made, the communication can be made in any reasonable manner.

Section 17.05(B) discussed the problems of authority and noted that it is essential to have designated persons with authority for each party to make decisions and to take responsibility. With regard to communications, each party should know to whom it should direct any particular communication. The authorized person should be designated by name in the contract and his address given. It is important to notify the other party if there has been a change in personnel and notices are to be sent to someone else.

If the contract does not designate time requirements for notices, any notice or communication required must be given within a reasonable time. Commonly, the construction contract will specify time requirements. For example, AIA Doc. A201, ¶ 4.3.3, requires a notice "within 21 days after the occurrence of the event giving rise to such claim or within 21 days after the claimant first recognizes the condition giving rise to the Claim, whichever is later."

It is also important to designate whether days are working days or calendar days. Working days raise problems because of the interposition of weekends and holidays of various sorts and the not unlikely possibility of delays due to inclement weather. It is usually best, as AIA document A201, ¶ 8.1.4, suggests, to specify that the days are calendar days.

The absence of a base point, such as simply stating that a twenty-day notice must be given, can create difficulties. Does the notice period begin to run when the event occurred, when the contractor found out about its occurrence, when it could have found out about its occurrence, or when the event had sufficient impact on the work to cause the delay? Such a dispute would be best resolved by concluding that the period of time for giving notice began when it was reasonably clear that a delay would occur.

Suppose the owner justifiably notifies the contractor that the contract will be terminated unless the contractor pays certain subcontractors within ten days. When does the ten-day period begin? Suppose the letter was dated on September 1, mailed on September 2, and received on September 4. Under such circumstances, when does the time period begin?

It is likely that the time would begin when the letter is received. Doubts will be resolved against the party sending the letter because the latter could have clarified any possible uncertainty. The receiver would probably expect the time period to begin to run when the letter was received unless the receiver knew that there had been an inordinate delay in the mail.

At the other end of the notice, suppose a contractor is requested to respond to a communication within ten days. The contractor posts the letter on the tenth day but the letter is not received until the twelfth. Has the contractor complied?

American law generally follows what is called the mailbox, or dispatch, rule. Posting a reply within the time period will be effective unless it is made clear that actual receipt must be had within the time period. The mailbox rule will apply if the communication is sent by a reasonable means of communication. The mailbox concept, though developed in cases involving formation of contract, would very likely be applied to construction project communications.[118]

Many of the difficulties discussed can be avoided by careful contract drafting. It is best to make absolutely clear when periods begin, when they conclude, and whether receipt will be effective when placed in the means of communication or whether a communication must be actually received.

Systems should be developed that make it likely that communications will be received and that

---

[118]*Palo Alto Town & Country Village, Inc. v. BBTC Co.*, 11 Cal.3d 494, 521 P.2d 1097 113 Cal.Rptr. 705 (1974) (mailbox rule applied to the exercise of an option).

proof exists that communications were sent and received. Although false claims of dispatch and receipt are probably rare, a carefully conceived and administered construction project should consider the importance not only of proving that notices were sent or received but also of knowing when these events occurred.

Communications received and copies of communications dispatched should be logged and kept in readily accessible files. Such records should be kept for a substantial time after completion of the project to deal with the possibility of long-delayed claims being asserted.

Contracts frequently set forth formalities to avoid difficult proof problems. During the course of contract administration, parties often dispense with formal requirements. If they do, there is a serious risk that the formal requirements have been eliminated by the conduct of the parties. For this reason, formal requirements should be complied with throughout contract administration. If the for-

mal requirements must be dispensed with, it is important for the party who wishes to rely on the formal requirements at a later date to notify the other party that the dispensation of the requirements on this occasion will not operate as an elimination of the formal requirements for the balance of the contract. Sometimes general conditions contain provisions stating that waiver of formal requirements in one or more instances will not operate to eliminate formal requirements in the future. These clauses may not mean much if the parties, by their conduct, show an intention to generally dispense with formal requirements.

Sometimes legal rights, such as the right to a mechanics' lien, may require compliance with statutory provisions for notices. Many of the problems that have been discussed in this section can arise when statutory notices are required. Particularly in the area of mechanics' liens, it is necessary to obtain and follow legal advice regarding notice requirements.

PROBLEMS

1. The statute in a state requires that all contracts for over a designated amount of money be awarded by competitive bidding, "except those that involve professional services." A public agency in that state is planning a massive public building program. It intends to hire independent architects, technical engineers, construction managers, and critical path method (CPM) consultants. Which, if any, of these persons can be selected by the public agency without competitive bidding?

2. O owned unimproved land on which he wished to build a large suburban office building. He was inexperienced in construction and decided to engage PM as project manager. The contract between O and PM authorized, among other things, PM to enter negotiations with specialty trade contractors based on contract documents that would be prepared by an architect O would engage. The drawings and specifications were partially completed, and PM started negotiations with a number of specialty trade contractors. PM's negotiations with E, an electrical contractor, were successful, and E agreed to do the work for $5 million as long as the completed contract documents did not differ ma-

terially from those that had been the basis of their negotiations. PM wanted to close the deal, as he believed he had made a very good contract for O. O, however, was out of the country and could not be reached. PM decided to make a binding contract with E. PM prepared AIA Document A101 CMa and attached to it an A201 CMa. He inserted O's name as owner in the first part of the form, typed it in at the end, and under that signed his name as project manager.

When O returned from his trip, he conferred with PM regarding contract negotiations. PM told him that he had made a contract with E. O became infuriated at PM's having exceeded his authority. Is O bound to this agreement? If he is bound, does he have a claim against PM? What additional facts would you need to answer his questions? Assuming you develop these facts, what would be your advice to him? What if PM had been designated as a construction manager with the "usual" construction manager functions?

3. A construction contract between O and C stated that completion of the project would be no later than December 1. A provision of the contract stated that O could accelerate the completion date, not to exceed sixty days, if he notified C of this in writing no later than April 1.

C's main office was in Chicago; O's main office was in New York; the construction project was in Toledo, Ohio. On April 1, O drafted a letter requesting a forty-five-day time acceleration. The letter was mailed on April 1 and was directed to C's office in Chicago. The normal time for expedited mail between New York and Chicago is one day and two days by normal mail. By mistake the clerk sent the letter by normal mail, and the letter did not arrive in Chicago until April 3. The employee receiving the letter at C's Chicago office did not know that it pertained to the Toledo job, and for that reason, actual notice to the Toledo employees of C was not received until April 5. Does O have the legal right to accelerate the construction date forty-five days? What would your answer be if, on April 1, an employee of O called the project supervisor of C at Toledo and told him to accelerate the performance forty-five days and then sent the letter?

Competitive Bidding: Theory, Realities, and Legal Pitfalls

SECTION 18.01 Basic Objectives Reconsidered

The owner wishes to obtain a completed project that complies with the contract requirements as to quality, quantity, and timeliness at the lowest possible cost. Just as the organizational method and pricing discussed in Chapter 17 must take this into account, so must the method selected to designate the contractor.

In looking at cost—the factor most commonly emphasized in competitive bidding—attention must be directed not only to the bid price but also to other costs. For example, it is likely to cost more to conduct a competitive bid than to negotiate a contract. Similarly, the ultimate cost must take into account any administrative costs (increasingly these days called "claims overhead") incurred to obtain the promised performance and resolve disputes. These are likely to be greater under competitive bidding because the low bidder's performance and administration can increase the cost of monitoring performance and resolving disputes as well as create a greater likelihood of claims increasing the ultimate contract price because of extra work.

The ultimate cost of a project should take into account the cost of maintenance and the durability of the project. Methods exist that take these factors into account.[1]

Although many of these costs are difficult to quantify, any owner who can *choose* to have competitive bidding should take them into account. Techniques within the competitive bidding system can reduce some of these risks.[2]

SECTION 18.02 Competitive Bidding: Theories and Some Pitfalls

According to an Ohio court, competitive bidding

> . . . gives everyone an equal chance to bid, eliminates collusion, and saves taxpayers money. . . . It fosters honest competition in order to obtain the best work and supplies at the lowest possible price because taxpayers' money is being used. It is also necessary to guard against favoritism, imprudence, extravagance, fraud and corruption.[3]

This is a tall order. This section assumes that goods or services requested can be objectively evaluated or compared, preferably before award, or at least after. An award of pencils of a standardized type is an illustration. All pencils would be comparable, and all that would be needed would be to compare prices. If a number of competing sellers will bid, the price should be as low as can be obtained.

Even here, performance uncertainties exist. Will the party awarded the pencil contract deliver as promised? But if it does not, replacement pencils

---

[1]Vickrey and Nicol, *Total-Cost Bidding—A Revolution in Public Contracts?* 58 Iowa L.Rev. 1 (1972) (bidders bid initial cost, guarantee maintenance costs, and give a price to repurchase goods: variables totaled and bid made to lowest *total* cost bidder).

[2]Comment, 130 U.Pa.L.Rev. 179 (1981) (competitive negotiation).
[3]*United States Constructors & Consultants, Inc. v. Cuyahoga Metro. Housing Auth.*, 35 Ohio App.2d 159, 163, 300 N.E.2d 452, 454 (1973).

could be obtained and the excess procurement cost charged to the contractor or its surety. For example, California held that a public award for a construction manager must be made competitively,[4] while Massachusetts held that competitive bidding was not required in the purchase of an existing vessel for the use as a ferry boat.[5] The Massachusetts court noted that competitive bidding does not work well where the various properties offered for purchase are not identical, as in the sale of real property. This noncomparable factor is a reason given for holding that turnkey contracts for public housing are not subject to the requirement for competitive bidding.[6] However, the same argument can be made in favor of permitting a negotiated contract for a construction manager. Yet the California court was more impressed with the need for competition as a method of reducing costs.

A procurement for the award of police cars that would meet designated standardized performance specifications would fit the competitive bidding model. Either before or after award, testing can be done to determine whether the bidders have prototype vehicles that will comply or have complied with the specifications. Again, there can still be the problem of predicting whether all the cars will conform and be delivered on time.

The principal objective of the awarding agency will be frustrated if the performance standards are not met. Time has been lost and the objective not yet accomplished. Even if the successful bidder is financially solvent or has furnished a surety bond, the objective has still not been achieved. Collecting for nonperformance certainly was not the objective of the procurement.

The difficulty in predicting whether a prospective bidder will be able to meet the performance standards is an important reason that competitive bidding is rarely used in research and development contracts. In these procurements, the agency must find a contractor who has the technological skill to perform properly. Competitive bidding, unless it is accompanied by some preselection process, may not be the best method to select a contractor in such a procurement.

Although most construction work is not as sophisticated or experimental as building nuclear submarines or space capsules, there are still elements in construction work that do not make it perfectly suitable for competitive bidding. First, what is the likelihood that prospective bidders have the willingness and the ability to do the work they promise? Second, how difficult will it be to determine whether they have done the work they have promised? Third, what is the likelihood that the actual price will substantially exceed the contract price for reasons other than owner-directed changes or circumstances over which neither party had control?

The competitive bidding process can be structured to minimize some of these risks. The invitation may state that the contract will be awarded to the lowest *responsible*[7] or lowest and *best* bidder.[8] Preselection can screen out bidders who do not have the requisite competence or capacity. Nevertheless, despite these provisions, the likelihood of success depends on the integrity and ability of the contractor, which are often difficult to measure in competitive bidding where the tendency is to look solely at price.

Competitive bidding also assumes that there are free bids and true competition. If there is collusion among the bidders to "take turns" or submit fictitious bids, antitrust laws are violated and competitive bidding cannot accomplish its objective of obtaining the lowest price. Although construction is, on the whole, a fiercely competitive business, collusion between competitors in the competitive bidding process is not unknown.

Anticompetitive devices can be found in competitive bidding. If product specifications do not provide for alternative products and a viable method for substitutes, competitive pricing may be unduly restricted. Also, specifying a particular product may create a warranty of commercial availability by the awarding authority to the bid-

[4]*City of Inglewood—L.A. County Civic Center Auth. v. Superior Court*, 7 Cal.3d 861, 500 P.2d 601, 103 Cal.Rptr. 689 (1972). But refer to Section 17.04(D).

[5]*Douglas v. Woods Hole, Martha's Vineyard, & Nantucket S.S. Auth.*, 366 Mass. 459, 319 N.E.2d 892 (1974).

[6]*United States Constructors & Consultants, Inc. v. Cuyahoga Metro. Housing Auth.*, supra note 3. Refer to Section 17.04(E).

[7]*City of Inglewood-L.A. County Civic Center Auth. v. Superior Court*, supra note 4.

[8]*Cedar Bay Constr. Inc. v. City of Fremont*, 50 Ohio St. 3d 19, 552 N.E.2d 202 (1990).

der.[9] New Jersey prohibits specifications in local public contracts that limit free and open bidding, such as specifying brand names without allowing equivalent products to be substituted.[10]

Another weakness of competitive bidding is the difficulty of involving the contractor in the design process. This can be alleviated by the use of a construction manager[11] or, in the extreme case, the use of a design/build contractor.[12] But the traditional contract system, where the owner supplies a design prepared by the design professional and submits it to the contractors for competitive bidding, generally fails to employ any design skills of the contractor.

Some competitive bidding systems permit *alternative bidding*. The design professional prepares a design for one method of conventional construction and, in addition, prescribes parameters and other specific configurations or finish requirements for the alternate. Usually, performance specifications for allowed alternates are included in the contract documents. The contractor can select the alternative design or bid on the design supplied. However, this method is limited to those projects where the design professional is aware of another viable design solution. Another limitation to alternative bidding is the greater likelihood of disputes as to responsibility and liability.

Prebid design, or what federal procurement calls "two-step" formal advertising (discussed in Section 18.03(D)), is another method of involving the contractor in the design. This employs a competition first on design, then on price, similar to the Brooks Act, under which design services are procured for federal projects.[13]

To sum up, the competitive bidding system is presently, and is likely to continue to be, the major method of obtaining construction contractors. However, competitive bidding has pitfalls, and in the proper circumstances, serious thought should be given to another method of obtaining a contractor.

SECTION 18.03　The Competitive Bidding Process

A. Objectives

Competitive bidding should result in contract awards made impartially at the lowest price. Nonconforming bids are usually disregarded because conformity is needed for a proper comparison of bids and to give each bidder an equal opportunity. The competitive bidding system cannot function properly unless honest and capable bidders have enough confidence in the fairness of the system to submit bids. Submitting a bid proposal is an expensive and time-consuming operation. The bidders are entitled to reasonable assurance that they will be treated fairly and that the owner will follow its own rules.

No effort will be made to cover every aspect of the competitive bidding process. Most public agencies have standard forms for the competitive bidding process, and they are often regulated by statutes, regulations, and ordinances. Private owners often have engaged in substantial construction work and have also developed their own forms and methods. Design professional associations have developed standard or recommended forms for bidding documents.[14] They are usually available to the design professional. The principal objective of the following subsections is to present an overview of the process.

B. Invitation to Bidders

The initial step in conducting a competitive bid is the Invitation to Bidders. In federal procurement, this is known as the Request for Bids (RFB). Frequently, public agencies are required to invite bids by public advertising. By using the broadest dissemination, such as trade newspapers, professional journals, and government publications, the largest number of bidders can participate. This gives all bidders a chance and should obtain the lowest price.

Although inviting the maximum number of competitors should result in a lower price, dangers exist in having too many bidders. Good bidders may be discouraged if the large number of bidders makes their chances of winning quite remote.

---

[9]*Edward M. Crough, Inc. v. Dept. of Gen. Serv. of Dist. of Columbia*, 572 A.2d 457 (D.C.App.1990).
[10]N.J.Stat.Ann. 40A:11–13.
[11]Refer to Section 17.04(D).
[12]Refer to Section 17.04(F).
[13]Refer to Section 10.03.

[14]AIA Doc. A701 (Instructions to Bidders).

C. Prequalification

Devices are available to avoid the competence risks inherent in the price-oriented competitive bid. In addition to a statement that the award will be made, if at all, to the lowest responsible or the lowest and best bidder, a prequalification system can be used. The owner selects a group of bidders, all of whom are likely to have the capability of doing a competent job. The design professional or construction manager may request information from specific contractors on prior jobs completed, capital structure, machinery and equipment, and personnel (including supervisory personnel). After evaluation of this information, the owner's professional adviser usually determines which contractors should be permitted to receive an invitation to bid.[15]

The advantage to prequalification is that there is a better chance of finding a good contractor. However, there will be administrative time and expense expended in making this preselection or prequalification, and the competitive aspect is likely to be diminished. Such a method should be used where the project is of great magnitude or involves new construction techniques and where cost is less important than ensuring the quality of performance.

Some states have prequalification systems. For example, New Jersey requires that all school construction be done by persons certified by the state board of education.[16]

D. Two-Step Process

The federal procurement regulations sometimes permit a two-step formal advertising procedure.[17] This is used where definite specifications cannot be prepared and offered for fixed-price competitive bids. As a first step, unpriced technical proposals are solicited to meet certain specified requirements that are set forth in the solicitation for proposals. Typically, these requirements are for an end-result product. All proposals received are evaluated, and those that are considered to be within the range of acceptability are discussed with the proposers to obtain clarification and more detailed definition.

The second step is a request for price bids from all those whose first-step proposals met the criteria specified in the original solicitation for proposals. Award is then made in accordance with the procedures for award of a fixed-price contract, with each bidder pricing its own technical proposal previously approved as having met the specified criteria.

E. First Article

In the procurement of products or equipment, the federal government can protect itself from having to wait for a product that may not be adequate by inserting a "first article" clause that requires that the contractors deliver for government testing a preproduction model of what the contractor has agreed to furnish.

F. Deposits

In public procurement and occasionally in private procurement, the invitation requires the bidder to deposit a bid bond, a cashier's check, or a certified check to provide security in the event that the bidder to whom the award is made does not enter into the contract. The amount is either a stated *percentage* of the bid or a *fixed* amount determined by a designated percentage of the estimated costs made by the design professional.

The invitation should also state how long the securities will be held, as this is a matter of importance to bidders. After award, the owner should release the securities of all but the lowest three or four bidders. The bidder to whom the award is made may not enter into the contract. It still may be possible under some circumstances to hold the next lowest bidder. However, there is no justification for holding the securities of bidders who are not likely to be awarded the contract. After the successful bidder *signs* the contract, all securities should be released.

Usually the bidders must make a small monetary deposit when they request information to study a possible proposal. Sometimes this has the effect of discouraging persons who are not serious about making a bid proposal from obtaining the plans and specifications merely out of curiosity. The deposit is usually refunded when the bidding documents are returned.

---

[15]For a discussion of competitive negotiated contracts in state procurement, see note 2 supra. See AIA Doc. A305 (Contractor's Qualification Statement).

[16]The method is discussed in *Donald F. Begraft, Inc. v. Borough of Franklin Bd. of Educ.*, 133 N.J.Super. 415, 337 A.2d 52 (1975).

[17]See *Wheelabrator Corp. v. Chafee*, 455 F.2d 1306 (D.C.Cir.1971).

G. Alternates

Sometimes larger projects can be divided into designated stages. The owner may decide not to build the entire project if it does not have the money or if bids are too high for certain portions of the work. For this reason, the Invitation to Bidders can be divided into the stages or project alternates. But alternates can mean that favoritism can be accomplished by the award. For example, there may be an Invitation to Bidders involving a hospital, housing for nurses, and a parking structure. One alternate could be the hospital alone. A second could be the hospital with the parking structure, and a third the hospital with the nurses' housing. One bidder may be low on the total bid. Another bidder may be low on the first alternate, another on the second, and another on the third. The determination of which alternate is to be awarded could be based on favoritism to one of the bidders. To avoid this difficulty, the Invitation to Bidders should state the preferred alternate choices.

The same type of award manipulation is possible if the alternates consist of different methods of construction or materials. A list of preferences within price limits should avoid suspicion of possible favoritism.

H. Information to Bidders

Information to Bidders—called in federal procurement the Information for Bids (IFB)—accompanying the invitation usually consists of drawings, specifications, basic contract terms, general and supplementary conditions, and any other documents that will be part of the contract. Sometimes soils test reports are included. Alternatively, the Information to Bidders may state that designated reports are available in the office of a particular geotechnical engineer or the design professional for examination by the bidder. Subsurface information is discussed in Section 25.03.

Bidders should be given adequate opportunity to study the bidding information, to make tests, to inspect the site, and to obtain bids from subcontractors and suppliers.[18] Even when there is adequate time, bidders often wait to complete their bids until the bid closing deadline is imminent.

Generally, this is due to reluctance on the part of subcontractors to give subbids to bidders until shortly before the deadline for bid submissions. (The reasons for this reluctance are explored in Section 28.02.) Even if bidders do not make proper use of the time available to them, having allowed reasonable time for bid preparation can be helpful to the owner if a dispute arises over claimed computation errors or unforeseen subsurface conditions.

The drawings and specifications should be detailed and complete so that bidders can make an intelligent bid proposal. Imprecise contract documents and much discretion to the design professional may discourage honest bidders from submitting bids and encourage bidders of doubtful integrity who make low proposals in the hope that they will later be able to point to ambiguities and make large claims for extras.

The Information to Bidders can specify that any uncertainties observed by the bidder must be resolved by a written request for a clarification to the design professional before bid opening. Any clarification issued should be in writing and sent to all persons who have been invited to bid.

The law increasingly requires that the party conducting the competitive bid disclose information in the bidding documents under certain circumstances (discussed in Section 18.04(B)).

I. Bid Proposals: Changes

Bid proposals should be submitted on forms provided by the owner. To properly compare bids, all bidders must be proposing to do the same work under the same terms and conditions. The bid proposal form should include or make reference to any important disclaimers contained in the bidding information. Such disclaimers include denying responsibility for subsurface information and specifying any rules that seek to govern the rights of the bidder to withdraw the bid, such as limiting withdrawal to clerical errors or setting forth other requirements such as a deadline for claiming mistake.

The bid proposal should be signed by an authorized person. If the bidder is a partnership, the entire name and address of the partnership should be given. If the bidder is a corporation, the bid should be signed by appropriate officers and the corporate seal should be attached. It may be desir-

---

[18]Under AIA Doc. A201, ¶ 1.2.2, the contractor warrants that it has visited the site and checked local conditions.

able to attach to the bid a resolution of the board of directors approving the bid.

Invitations can preclude bids from being changed, corrected, or withdrawn after submission. However, there is a tendency toward more flexibility. More commonly, invitations permit changes, corrections, and withdrawals of submitted bids before bid opening.

Permitting changes can help the owner by permitting and encouraging reduction of bids in a changing market. However, overliberality in permitting changes, corrections, and withdrawals may cause bidders to lose confidence in the honesty of the bidding competition.

Where the right to change is given, it may be limited to reducing the bid. If changes, corrections, or withdrawals are to be permitted, the invitation can limit such a right to a designated time period and require that any change, correction, or withdrawal be expressed in writing and received by the owner by a designated time.

The owner should keep all the records connected with the bidding process. Time logs should be kept that show exactly when the bidder obtained the bidding information. Changes in bidding information should be sent to all persons who have picked up bidding information, with copies of the changes kept for future reference. Changes after the bidding information is disseminated should be kept to a minimum.

J. Bid Opening

Bid proposals are sent in a sealed envelope to the owner or the person designated to administer the competitive bidding, such as a design professional or construction manager. In public contracts, bids are opened publicly at the time and place specified in the Invitation to Bidders. Usually the person administering the process announces the amount of the bids and the bidders. The invitation specifies a designated period of time in which the owner can evaluate the bids. At bid opening, care must be taken to avoid any impression that the low bidder has been awarded the contract. Often the person administering the process states that a particular bidder is the "apparent low bidder."[19]

In private competitive bids, the invitation can state that bids need *not* be opened in public. When the bid opening is private, the bidders, at least at that time, are not able to compare their bids with the others. This minimizes the likelihood that the low bidder will begin to suspect a bidding error if its bid is much lower than the others. One case involved the successful bidder not finding out that its bid had been much lower until it was well into the project. At that time, it threatened to walk off the project unless it was given a price adjustment, claiming that the owner must have known its bid was erroneous. Although the court did not grant relief to the contractor, it is likely that its determination to grant the contractor relief on another theory was affected by the belief that the owner should have notified the contractor of the wide discrepancy between its bid and the others.[20] This may reflect the law beginning to impose on contracting parties, or even those in the process of making a contract, an obligation of good faith and fair dealing, such as drawing attention to a possible mistake. A private bid opening may tempt the owner not to notify a bidder when it is apparent that a mistake has been made.

K. Evaluations of Bids

Some legal problems relating to bid evaluations and awards are discussed in Sections 18.04(E) and (F). This section outlines some steps that should be taken when evaluating bids.

First, the owner or its representative should follow the procedures set forth in the Invitation to Bidders. The bids should be checked to see whether they conform. A nonconforming bid is a proposal based on performance not called for in the invitation or information or that offers performance different from that specified. Nonconformity may relate to the work, the time of completion, the bonds to be submitted, or any requirements of the contract documents. Bidding documents sometimes provide that bidding technicalities can be waived by the owner. This allows the owner to accept the lowest bid despite minor irregularities. However, advance notice that tech-

---

[19]In *McCarty Corp. v. United States*, 204 Ct.Cl. 768, 499 F.2d 633 (1974), the contracting officer stated that the plaintiff was the "apparent low bidder." This, among other facts, precluded a finding that the plaintiff's bid

had been accepted. See also *Scheckel v. Jackson County*, 467 N.W.2d 286 (Iowa App.1991) (assistant engineer telling low bidder "he was to be awarded the job" not valid acceptance).
[20]*Paul Hardeman, Inc. v. Arkansas Power & Light Co.*, 380 F.Supp. 298 (E.D.Ark.1974).

nicalities can be waived can encourage careless proposals and may also create the impression that favoritism may be shown to certain bidders. If a bid is not in conformity and the defect cannot be waived, the bid must be rejected.

Second, the owner[21] should examine the Invitation to Bidders to determine whether *any* bid must be accepted. Typically, the invitation reserves the right to reject all bids. The invitation must be reviewed to determine the *basis* for making an award if one is made. Commonly, the award is to be made to the lowest responsible or lowest and best bidder.

To determine the lowest responsible bidder, the owner can take into account the following factors as well as any others that bear on who would be the bidder most likely to do the job properly:

1. expertise in type of work proposed
2. financial capability
3. organization, including key supervisory personnel
4. reputation for integrity
5. past performance

These illustrations relate to the basic objectives of obtaining a contractor who is likely to do the job properly at the lowest price and with the least administrative cost to the owner.

In contracts for the purchase of machinery, bid evaluation should take into account the following factors:

1. length and extent of warranty
2. availability of spare parts
3. service and maintenance
4. cost of replacement parts
5. cost of installation
6. durability

Gary Aircraft Corp. v. United States[22] illustrates the methods of evaluating a bidder used by the Department of the Air Force. The contract contemplated a three-year fixed-price contract. Two bids were received, but the awarding authority doubted the capability of the low bidder. The contracting officer first screened information regarding the two bidders. This information included credit reports; Defense Department records; reports from custom-

ers, suppliers, and bankers; financial data; and current and past production records.

Because the information created doubts regarding the low bidder, an advisory preaward survey was made. The survey of the low bidder took four days, and the report following it recommended that the award *not* be made to the low bidder because of deficiencies in the following areas:

1. quality control system
2. past performance
3. meeting required schedules
4. property control system
5. plant facilities and equipment

The award was *not* made to the low bidder.

Although the owner should evaluate criteria other than price in selecting a contractor, there is a substantial risk of a lawsuit challenging the award if the bid is awarded to someone other than the low bidder. Sometimes bid awards in public contracts can be judicially challenged. This should not deter even a public owner from awarding the contract to someone other than the low bidder. However, compelling reasons for not awarding it to the low bidder must exist and be documented.

Third, if bidders were asked to bid separately on project alternates, the owner should determine whether any bidders had made "all-or-nothing" bids. Such a bid indicates an unwillingness on the part of the bidder to perform any part of the project except the entire project. An all-or-nothing bid is permitted unless the invitation specifically precludes it.

L. Notification to Bidders

When the successful bidder is selected, the successful and unsuccessful bidders should be officially notified. However, sometimes with or without legal justification, the successful bidder will not enter into the contract. For this reason, the notification to unsuccessful bidders should state that their bids still remain available for acceptance by the owner for the period of time specified in the invitation to bid. This is done to preclude any later contention that awarding the contract released the unsuccessful bidders.

M. Post-Award Changes

If the owner changes the contract terms after the bid has been awarded and the change makes the obligation less burdensome to the successful bid-

---

[21]"Owner" includes design professional or construction manager advising the owner.
[22]342 F.Supp. 473 (W.D.Tex.1972).

der, this can be unfair to the other bidders. For this reason, changes of this type should be done with caution. (See *Conduit and Foundation Corp. v. Metropolitan Transportation Authority*, reproduced in Section 18.04(D).)

N. Signing the Formal Contract

The successful bidder is sent to the formal contract for signature or requested to meet at a specified time and place to sign the formal documents. Records of all correspondence should be kept in the event that subsequent disputes arise between the owner and the contractor relating to the bidding process and awarding of the contract.

The successful bidder must enter into the formal contract awarded unless legal grounds exist for refusal (discussed in Section 18.04(E)). The effect of signing the formal contract, as well as such signing's effect on bidding documents and the function of the formal contract, is discussed in Section 18.04(G). Since the date the contract has been made can measure the time commitment, this is discussed in Sections 26.02 and 26.04.

O. Readvertising

Sometimes the bids are all too high and the owner decides to readvertise the project. In public contracts, there are procedures for readvertising, which require time and administrative expense. Some feel that readvertising without reducing the scope or quality of the project is unfair because it is an attempt to beat down the bids of the contractors, often resulting in deficient workmanship and substandard materials. Readvertising frequently means higher bids from all bidders because the less skillful bidders have seen what the better bidders have bid.

P. Special Rules for Public Contracts

Owners generally seek a construction contract that will give them the balance they choose between quality, timely completion, and cost. On the basis of these criteria, the owner looks for a contractor who will do the best work at the best price.

Looking only at the project and these goals, the owner will not be concerned with the racial or gender characteristics of the contractor's labor force, the wages the contractor pays its employees, and the source of the supplies. Obviously, this uncon-

cern is not absolute. The owner could be concerned with the wages if it felt that workers who were not paid the prevailing rate would not perform properly. The source of the supplies could be important if the owner felt that supplies from certain sources were higher quality. But on the whole, the owner who wishes to obtain the best price through competitive bidding or through negotiation must give broad latitude to the contractor in these matters.

Public contracts, however, involve billions of taxpayer dollars. Elected officials frequently look to the procurement process and public spending as ways of accomplishing goals that go beyond the best quality at the best price. In doing so, they often respond to interest groups who also see procurement as a means of obtaining their objectives. The federal government has been the pioneer in using the procurement process to achieve social and economic objectives. State and local public agencies and even some private owners have also begun to see procurement in this light.

This treatise cannot deal in detail with the many rules that seek to achieve a variety of objectives through the public contract process. However, some goals sought to be achieved by the public contract process are as follows:

1. Providing employment opportunities to disadvantaged minorities, women, handicapped persons, or disabled war veterans.
2. Setting aside certain procurement awards for small businesses or disadvantaged minorities.[23]
3. Favoring contractors, suppliers, or workers who reside in a particular state or city. (One city enacted an ordinance giving preference to suppliers with a designated child-care program for employees.)
4. Awarding contracts to bidders located in economically depressed areas.
5. Ensuring that workers are paid at the local prevailing wage rate.[24]
6. Avoiding corruption in procurement.

---

[23]The leading case is *City of Richmond v. J.A. Croson Co.*, 488 U.S. 469 (1989). Among the many recent articles dealing with this case and its ramifications, see Comment, 40 U. Kan.L.Rev. 257 (1991); Note, 1991 B.Y.U.L.Rev. 1633; Note, 67 Ind.L.J. 169 (1991).

[24]See *Lusardi Constr. Co. v. Aubry*, 1 Cal.4th 976, 824 P.2d 643, 4 Cal.Rptr.2d 837 (1992) (requirement that employees on public projects be paid a prevailing wage applies even if not included in the contract).

Clearly, such rules have their costs. In addition to higher bid prices, administrative costs are likely to be incurred by all participants to see that these rules are followed. These costs may be worth incurring if they accomplish the objectives sought. However, attempts to use the public contract process for these objectives have generated intense controversy, both as to the fairness of the programs and as to whether the objectives can be accomplished in a different way at a lower cost.

Every aspect of awarding public contracts, especially competitively bid awards, must avoid even the appearance of impropriety. For example, public officials who make procurement decisions are expected to have the interests of their agency in mind and avoid conflict of interest. This is demonstrated in the *Conduit* case reproduced in Section 18.04(D).

As a result of public concern over contract awards, public officials often are rigid and unbending in such matters as bidding irregularities and withdrawal of bids.

SECTION 18.04 Some Legal Aspects of Competitive Bidding

A. Invitation to Bidders

Ordinarily, the invitation is not an offer and does not create a power of acceptance in the bidders. It is a request that bidders make offers to the owner that can be accepted or rejected. But the invitation is an important document. It sets up the ground rules for the competitive bid. If a contract is formed with one of the bidders or if the successful bidder refuses to enter into the contract, the invitation may have legal significance. Of course, the invitation can be modified by the formal agreement between bidder and owner, but often the formal contract simply memorializes what has been agreed to in the invitation and bid proposal.[25]

B. Duty to Disclose

The early common law rarely obligated a contracting party to disclose information that the other party would want to know. Contracting parties were generally expected to look out for themselves.

They could not make deliberate misrepresentations but could conceal matters within their knowledge, even if they knew that the other party would want to know them.[26] But currently the common law is likely to require disclosure of vital information that the other party is not likely to discover.[27]

This has been reflected in an increasing tendency to require that the party conducting a competitive bid disclose certain information to bidders. The Court of Claims has held that a federal procurement agency must disclose procurement plans of other federal agencies of which it knows and which may affect the pricing assumption of bidders.[28] Similarly, it has held that the procuring agency must disclose technical information that it possessed relating to the manufacturing process that it knew the contractor intended to use.[29] The court noted that on many occasions each party has an equal opportunity to uncover the facts. But where one party knew much more than the other and that the latter was proceeding in the wrong direction, it could not betray the contractor into a "ruinous course of action by silence." State courts have also recognized a duty to disclose.[30]

Yet contractors cannot rely too heavily on the possibility of legal protection. Generally parties who *can* protect themselves *must* do so.[31] Contracting parties, although expected to cooperate, do not owe each other fiduciary obligations to look out for each other.[32]

---

[25]See Section 18.04(G).

[26]*Swinton v. Whitinsville Sav. Bank*, 311 Mass. 677, 42 N.E.2d 808 (1942).

[27]*Obde v. Schlemeyer*, 56 Wash.2d 449, 353 P.2d 672 (1960).

[28]*J.A. Jones Constr. Co. v. United States*, 182 Ct.Cl. 615, 390 F.2d 886 (1968) (Corps of Engineers knew of Air Force plans); *Bateson-Stolte, Inc. v. United States*, 145 Ct.Cl. 387, 172 F.Supp. 454 (1959) (must prove Corps of Engineers knew of AEC plans). See also *Hardeman-Monier-Hutcherson v. United States*, 198 Ct.Cl. 472, 458 F.2d 1364 (1972) (duty to disclose weather and sea conditions after contractor request).

[29]*Helene Curtis Indus., Inc. v. United States*, 160 Ct.Cl. 437, 312 F.2d 774 (1963).

[30]*Welch v. California*, 139 Cal.App.3d 546, 188 Cal.Rptr. 726 (1983) (records of earlier attempt to repair that documented tidal difficulties); *Pennsylvania Dept. of Highways v. S.J. Groves & Sons Co.*, 20 Pa.Commw. 526, 343 A.2d 72 (1975) (another contractor would occupy site after access given).

[31]*H.N. Bailey & Assoc. v. United States*, 449 F.2d 376 (Ct.Cl.1971) (information obtainable elsewhere).

[32]See Section 19.02(D).

C. Bid Proposal

The bid proposal is an offer and creates a power of acceptance in the owner. The owner has a right to close a deal and bind a bidder without a further act of the bidder. Subject to many exceptions,[33] offers are generally revocable even if stated to be irrevocable for a specified period of time. This raises an issue that does not usually surface in construction litigation. Suppose either *before* bid opening or *after*, but *before* the formal contract is signed, the bidder revokes its bid. It is generally assumed, at least in public contracts, that despite the common law rule of revocability the bid is irrevocable for the period stated in the invitation unless the invitation *permits* withdrawal before bid opening. What is the justification for such an assumption?

Sometimes public contract competitive bidding is authorized by statutes or regulations that state bids to be irrevocable. This may supersede the common law rule of revocability.[34] Sometimes the bid is revocable but any deposit made with it is forfeited.[35] One possibility is that the bidder receives the benefit of having the bid considered by the owner.[36] The bid may be irrevocable if the owner has justifiably relied on the bid. The owner may have given up the opportunity of negotiating a contract having relied on the bidders' stating that their bids would be irrevocable or, at the very least, expended considerable time and money in conducting the competitive bid process.[37] If the transaction is one covered by the Uniform Commercial Code, the bid may be irrevocable as a "firm offer."[38] If the procurement involves the sale of goods with installation incidental, the Code would govern.[39] Even if it does not, it may be possible to use the Code by analogy.[40]

Once the bid has been opened, it is generally assumed that it has become irrevocable and can be withdrawn only if legal grounds exist for doing so, usually a mistake in preparing the bid.[41]

Can the owner hold bidders to whom the bid has *not* been awarded for the period specified in the Invitation to Bidders? The not uncommon refusal of the bidder to whom the contract has been awarded to enter into the contract makes it imperative that no expectation be created that the unsuccessful bids are no longer binding when an award has been made. Of course, the period of time cannot be extended, but it should not be shortened simply because the award appears to have been made to someone else.

D. Award and Waiving of Irregularities: *Conduit and Foundation Corp. v. Metropolitan Transportation Authority*

Usually the owner promises to award the contract to the lowest responsible or lowest and best bidder, reserving the right to reject *all* bids. While these two standards for determining to whom a contract should be awarded are often used as if they dictate the same outcome, it can be seen by a careful study of the language that this is not always the case. Being the lowest *responsible* bidder would not be the same as being the lowest and *best* bidder. The California Supreme Court emphasized that "responsible" simply means that the low bidder has the quality, fitness, and capacity to satisfactorily do the required work. In other words, according to the court, a contract must be awarded to the low bidder unless it has been found that it is not responsible, that is, "not qualified to do the particular work under consideration."[42]

To make this decision, the owner must evaluate the bids. This is done between bid opening and award, if any. Yet this evaluation period is not open-ended. Bids are irrevocable for the period specified in the Invitation to Bidders. This is illustrated in *Hennepin Public Water District v. Petersen Con-*

---

[33]Refer to Section 5.06(B).
[34]*Powder Horn Constructors, Inc. v. City of Florence*, 754 P.2d 356 (Colo.1988).
[35]1 A.CORBIN, CONTRACTS § 46–47 (1963).
[36]Rejected in *Sooy v. Winter*, 188 Mo.App. 150, 175 S.W.132 (1915).
[37]*Drennan v. Star Paving Co.*, 51 Cal.2d 409, 333 P.2d 757 (1958) (subcontractor bid irrevocable because prime relied).
[38]U.C.C. § 2-205.
[39]*Bonebrake v. Cox*, 499 F.2d 951 (8th Cir.1974).
[40]*Transatlantic Financing Corp. v. United States*, 363 F.2d 312 (D.C.Cir.1966).

---

[41]*Elsinore Union Elementary School Dist. v. Kastorff*, 54 Cal.2d 380, 353 P.2d 713, 6 Cal.Rptr. 1 (1960); *A.J.Colella, Inc. v. City of Allegheny*, 391 Pa. 103, 137 A.2d 265 (1958).
[42]*City of Inglewood—L.A. County Civic Center Auth. v. Superior Court*, supra note 4, 500 P.2d at 604. 103 Cal.Rptr., at 692.

struction Co.,[43] where the invitation had stated that the awarding authority would have to obtain financing within sixty days from bid opening. There was a tentative acceptance before the awarding authority obtained financing. Then sixty-seven days after bid opening, the awarding authority forwarded the formal contract to the successful bid-

---

[43]54 Ill.2d 327, 297 N.E.2d 131 (1973).

der. The court held the acceptance too late and the bidder released.

Before examining legal constraints on the awarding authority when it considers bids, the following case reviews some of the purposes of competitive bidding in the public sector discussed in Section 18.02 and passes on the conditions under which the awarding authority can reject all bids and rebid the project.

CONDUIT AND FOUNDATION CORP. v. METROPOLITAN TRANSPORTATION AUTHORITY

Court of Appeals of New York, 1985. 66 N.Y.2d 144, 485 N.E.2d 1005, 495 N.Y.S.2d 340.
[Ed. note: Footnotes omitted.]

JASEN, Judge.

The narrow issue presented on this appeal is whether the evidence on the record supports a holding that the Metropolitan Transportation Authority and the New York City Transit Authority acted unlawfully in rejecting all bids submitted in an initial round of bidding on a public works project.

The relevant facts are not in dispute. Respondents, the Metropolitan Transportation Authority and the New York City Transit Authority (Transit Authority), solicited bids for a public works contract for the massive rehabilitation of part of New York City's subway system. Three bids were received, including that of petitioner, a joint venture, whose bid was the lowest and fell within the advertised estimated cost between $120 [million] and $140 million. Shortly after opening the bids, the Transit Authority met with each of the three bidders for the stated purpose of determining why the prices submitted were as high as they were and whether the project costs might be reduced. Although the bidders were met separately, each was informed that the meetings with the others were taking place.

Subsequent to the meetings, the second lowest bidder advised the Transit Authority by letter that significant cost reductions were possible and suggested another meeting to discuss the possibility of preparing revised contract documents. Several days thereafter, it was petitioner that met again with the Transit Authority. Petitioner was informed that its bid, though the lowest, was far in excess of the revised estimated cost and that, therefore, all the bids might be rejected. Nevertheless, petitioner was told that it was deemed technically and financially qualified to undertake the project and, indeed, was afforded the opportunity to

reduce its bid for consideration. Within the next two weeks, petitioner once again met with the Transit Authority and offered to reduce its bid price by two million dollars, a proportionately small amount. The reduction was deemed insufficient by the Transit Authority's Chief Engineer who, on the following day, recommended that all the bids be rejected and that the contract, with modifications, be readvertised for a second round of bidding. The recommendation was adopted by the President of the Transit Authority and the three bidders were notified. Thereafter, the Transit Authority circulated a notice soliciting new bids for the project and indicating an estimated cost range between $100 [million] and $120 million, $20 million lower than that originally advertised.

Petitioner commenced this article 78 proceeding seeking an injunction against a second round of bidding and a judgment directing respondents to award the contract to petitioner as the lowest responsible bidder. Special Term granted the petition concluding that the Transit Authority's postbid communications with the three bidders, its revision of the advertised cost estimate and its purpose of obtaining bid prices lower than that originally received rendered the decision to reject all first round bids arbitrary and capricious. The Appellate Division agreed with Special Term, two justices dissenting, that the contract should be awarded to petitioner, but modified the judgment to reduce the contract price by two million dollars on the basis of petitioner's prior offer to the Transit Authority to so lower its bid. The court held that the record demonstrates an "appearance of impropriety" on the part of the respondents which "might have" a detrimental effect "on the broad interest of the public in the entire public bidding process". (111 A.D.2d 230, at p. 234, 489 N.Y.S.2d 265.) The two dissenters argued that the mere "appearance" of impropriety is not sufficient ground to disturb the decision of the Transit Authority absent a showing of actual favoritism fraud or similar evil which competitive bidding is intended to prevent. We agree and now reverse the order below.

As this court has stated on prior occasion, the purpose of the laws in this State requiring competitive bidding in the letting of public contracts is "to guard against favoritism, improvidence, extravagance, fraud and corruption." (*Jered Contr. Corp. v. New York City Tr. Auth.*, 22 N.Y.2d 187, 193, 292 N.Y.S.2d 98, 239 N.E.2d 197; . . .

These laws were not enacted to help enrich the corporate bidders but, rather, were intended for the benefit of the taxpayers. They should, therefore, be construed and administered "with sole reference to the public interest". (10 McQuillin, Municipal Corporations § 29.29, at 302 [3d rev ed]; . . . This public interest, we have noted, is sought to be promoted by fostering honest competition in the belief that the best work and supplies might thereby be obtained at the lowest possible prices. . . .

Indeed, this policy is reflected explicitly in the statutory provisions which mandate that public work contracts be awarded "to the lowest responsible bidder". (Public Authorities Law § 1209[1]; General Municipal Law § 103[1]; *see also*, General Municipal Law § 100-a.) Dishonesty, favoritism and material or substantial irregularity in the bidding process, which undermines the fairness of the competition, impermissibly contravene this public interest in the prudent and economical use of public moneys. . . .

Nevertheless, where good reason exists, the low bid may be disapproved or, indeed, all the bids rejected. Neither the low bidder nor any other bidder has a vested property interest in a public works contract . . . and statutory law specifically authorizes the rejection of all bids and the readvertisement for new ones if deemed to be "for the public interest so to do" (Public Authorities Law § 1209[1]; . . .

Although the power to reject any or all bids may not be exercised arbitrarily or for the purpose of thwarting the public benefit intended to be served by the competitive process . . . the discretionary decision ought not to be disturbed by the courts unless irrational, dishonest or otherwise unlawful. . . .

Moreover, it cannot be gainsaid that a realistic expectation of obtaining lower contract prices upon a second round of bidding constitutes a reasonable, bona fide ground for rejecting all first round bids and serves the public's interest in the economical use of public moneys. . . .

Nor is the decision to reject all bids rendered arbitrary and capricious solely on the basis of nondiscriminatory postbid changes in the contract specifications . . . or postbid communications with individual bidders for the bona fide purpose of ascertaining how contract costs might be reduced. . . .

Only upon a showing of actual impropriety or unfair dealing—i.e., "favoritism, improvidence, extravagance, fraud and corruption" (*Jered Contr. Corp. v. New York City Tr. Auth., supra*, 22 N.Y.2d at p. 193, 292 N.Y.S.2d 98, 239 N.E.2d 197)—or other violation of the statutory requirements, can the decision to reject all bids and readvertise for a second round of bidding be deemed unlawful. . . .

Consequently, where the party challenging the decision does not satisfy the burden of making such a demonstration, that decision should remain undisturbed. . . .

Here, where the court below found only that the postbid activity of the Transit Authority created an "appearance of impropriety" and "cast a doubt" upon its fair dealing with the bidders, the respondents' decision to seek a rebid ought not to have been disturbed. It may be true, as the majority at the Appellate Division found, that it would have been wiser for the Transit Authority to meet with all the bidders at the same time, instead of separately, in order to avoid the possible appearance of unfair dealing. Likewise, other bid interactions between the Transit Authority and certain bidders may have been more discreet. Nevertheless, as noted in the dissenting opinion of Justice Niehoff absent some finding that the respondents had, in fact, engaged in some unfair or unlawful practice tainting the impartiality of the competitive bidding process, it was error to grant the petition.

The Transit Authority clearly was empowered to reject all the first round bids, and its expectation of obtaining lower bid prices upon readvertisement was a rational basis for deciding to do so. . . .

Accordingly, the order of the Appellate Division should be reversed, with costs, and the petition dismissed.

WACHTLER, C.J., and MEYER, SIMONS, KAYE, ALEXANDER and TITONE, J.J., concur.

The U.S. Court of Claims has held that a bidder is entitled to honest consideration of its bid. To show that the bid was not properly considered, the bidder must establish that the government acted arbitrarily and capriciously and that there was no reasonable basis for the government's decision.[44]

On the other hand, California rejected a bidder's contention that the awarding authority must consider bids in good faith.[45] The court stated:

> Such a promise would be contrary to a well established rule which allows a public body where it has expressly reserved the right to reject all bids, to do so for any reason and at any time before it accepts a bid. If the entity so decides, it may return all bids unopened.
>
> The courts have consistently refused to interfere with the exercise of a public body's right to reject bids, however arbitrary or capricious.[46]

The court also rejected a claim based on misrepresentation because of immunity granted the public entity. But the awarding authority cannot solicit a bid without any intention of considering it.[47] However, California allowed the low bidder to recover its bidding expenses when the contract was improperly awarded to another bidder.[48]

Although the California rule gives maximum discretion to the awarding authority, there is an undeniable trend toward holding government agencies accountable for their acts.

Procedures vary where the low bidder is rejected. For example, New Jersey precludes the local awarding authorities from rejecting the high bid for property or low bid for work without giving a hearing to the disappointed bidders.[49] California, however, does not require a full-fledged courtlike hearing if the low bid is rejected. Rather, the awarding authority must give the low bidder access to any evidence that has reflected on its responsibility received from others or produced as a result of an independent investigation. The bidder must be afforded an opportunity to rebut such adverse evidence and present evidence that it is qualified to perform the contract.[50]

States vary as to the requirement for a hearing and its nature. But again, with the emphasis on greater accountability, it is likely that low bidders who are not awarded the contract can at least demand that reasons be given and that they be given an opportunity to rebut adverse evidence.

As for specific reasons for rejecting a bidder, New Jersey held that disputes over a previous job were insufficient to deny award to the low bidder.[51] But Mississippi upheld a provision in the invitation that allowed the awarding authority to reject a bidder for being in arrears on an existing contract or in litigation with the awarding authority for having defaulted on a previous contract.[52] The bidder's wholly owned subsidiary was suing the awarding authority on another matter, and on this basis the bidder's bid was rejected. The court recognized the provision as possibly coercing contractors from asserting their legal rights but held that the standard of rejecting bidders was one within the discretion of the authority. This allows an awarding authority to use the procurement process for an improper purpose.

Massachusetts upheld rejection of a subcontractor who had had a dispute with the prime contractor on an earlier job and had made misstatements of his previous work experience.[53] A federal court applying Mississippi law upheld rejection of a bidder because of the bidder's poor reputation for quality work.[54] Generally, public entities can reject bidders who intend to use nonunion workers.[55]

---

[44]*Keco Indus., Inc. v. United States*, 428 F.2d 1233 (Ct.Cl.1970). Absence of fraud or palpable abuse of discretion is required in Mississippi. *Warren G. Kleban Eng'g Corp. v. Caldwell*, 361 F.Supp. 805 (N.D.Miss.1973), vacated on other grounds, 490 F.2d 800 (5th Cir.1974). New Jersey requires a bona fide judgment. *Mendez v. City of Newark*, 132 N.J.Super. 261, 333 A.2d 307 (1975).

[45]*Universal By Products, Inc. v. City of Modesto*, 43 Cal.App.3d 145, 117 Cal.Rptr. 525 (1974).

[46]117 Cal.Rptr. at 529. See also *Pacific Architects Collaborative v. California*, 100 Cal.App.3d 110, 166 Cal.Rptr. 184 (1979) and *Commercial Indust. Constr. Inc. v. Anderson*, 683 P.2d 378 (Colo.App.1984).

[47]Supra notes 45, 46.

[48]*Swinerton & Walberg Co. v. City of Inglewood—L.A. County Civic Center Auth.*, 40 Cal.App.3d 98, 114 Cal.Rptr. 834 (1974). But see *Universal By-Products, Inc. v. City of Modesto*, supra note 45.

[49]*Mendez v. City of Newark*, supra note 44; *D. Stamato & Co. v. Township of Vernon*, 131 N.J.Super. 151, 329 A.2d 65 (1974).

[50]*City of Inglewood—L.A. County Civic Center Auth. v. Superior Court*, supra note 4.

[51]*D. Stamato & Co. v. Township of Vernon*, supra note 49.

[52]*M.T. Reed Constr. Co. v. Jackson Mun. Airport Auth.*, 227 So.2d 466 (Miss.1969).

[53]*Kopelman v. Univ. of Massachusetts Building Auth.*, 363 Mass. 463, 295 N.E.2d 161 (1973).

[54]*Warren G. Kleban Eng'g Corp. v. Caldwell*, supra note 44.

[55]*Building and Construction Trades Council of the Metro Dist. v. Associated Builders and Contractors of Massachusetts/Rhode Island, Inc.*, ___ U.S. ___, 113 S.Ct. 1190 (1993). See also *Image Carrier Corp. v. Beame*, 567 F.2d 1197 (2d Cir.1977). But see *Wittie Elec. Co. v. New Jersey*, 139 N.J.Super. 529, 354 A.2d 659 (1976). Much can depend on the language of the statute or local ordinance that expresses the power of the agency to reject bids as well as the judicial attitude toward the trade union movement.

An awarding authority or owner conducting a competitive bid should be able to reject a bidder if a good-faith determination has been made that the bidder is not likely to be able to complete the required performance. As the reason for rejection departs from evaluation of the bidder's experience, financial ability, integrity, and availability of facilities necessary to perform the contract, it is more likely that the rejection has not been made in good faith.[56] Bidder evaluation was discussed earlier.[57]

The remedy accorded a bidder who can establish that its bid was not properly considered or rejected has tended to be limited. Where the substantive claim can be upheld, the bidder will likely be able to recover its bidding expenses but not any profits it would have made had it been allowed to perform the contract.[58] The difficulty of establishing a substantive claim and the limited remedy where the contract has been awarded by a public entity reflects judicial hesitance to interfere with the operations of a public entity.[59]

Suppose there are irregularities in the proposal or process. The awarding authority may believe them minor and wish to waive them and accept the bid. The legal issues have been whether the irregularity in question is minor and whether waiver would encourage carelessness, create opportunity for favoritism, and operate unfairly to other bidders.

The awarding authority's discretion to waive a nonresponsive bid depends on the importance of the deviation and possible prejudice to other bidders.[60] The awarding authority's past conduct may have led a bidder to reasonably believe nonconformity would be disregarded. In such cases, the bid may not be rejected where the awarding authority had authority to waive the irregularity.[61]

Many cases involve bid submission deadlines. *H.R. Johnson Construction Co. v. Board of Education*[62] held that an awarding authority could not accept a bid made one minute late. The court was not persuaded that one minute could not give a bidder a competitive advantage over other bidders. The court felt that if the awarding authority is given the discretion to allow a one-minute deviation, it could stretch this power to even fifteen minutes or an hour.

In *William F. Wilke, Inc. v. Department of Army*,[63] the bids were to be opened at 3:00 p.m. At that time, a representative of the low bidder was in the room, but he neglected to put the sealed bid in the receptacle designated for that purpose. At 3:04, the contracting officer gathered the receptacle and started to sort out the bids. During the sorting, the low bidder's representative added his sealed bid to the box of yet unopened bids. The awarding authority awarded the bid to the low bidder, and the next low bidder complained.

The court concluded that the bid should not have been accepted, as it had been deposited four minutes late. But because it did not appear that the low bidder had obtained any competitive advantage, the court would not order that the bid be awarded to the next low bidder but limited the disappointed bidder to recovery of its bidding expenses. Other cases have allowed the awarding agency discretion to waive such irregularity, especially if the bidding information permitted this.[64]

E. Withdrawal or Correction of Mistaken Bids: *Sulzer Bingham Pumps Inc., v. Lockheed Missiles and Space Co.*

Sometimes a bidder will seek to withdraw or correct a bid. Such a request usually occurs just after bid opening or, more rarely, after the formal

---

[56]*D. Stamato & Co. v. Township of Vernon*, supra note 49.
[57]Refer to Sections 18.03(K) and 18.04(D).
[58]*Heyer Products Co. v. United States*, 135 Ct.Cl. 63, 140 F.Supp. 409 (1956); *Swinerton & Walberg Co. v. City of Inglewood—L.A. County Civic Center Auth.*, supra note 48.
[59]See Section 18.04(I) on standing to contest judicially an award.
[60]*Albano Cleaners, Inc. v. United States*, 455 F.2d 556 (Ct.Cl.1972). The court held that a substantial deviation affects price, quality, or quantity. In *Rossetti Contracting Co. v. Brennan*, 508 F.2d 1039 (7th Cir.1975), the court held that nonconformity as to affirmative action hiring was not correctable. Here the court was facing a substantial deviation but one that the bidder offered to correct. The court was fearful that allowing "correction" would give the bidder an "option" exercisable after seeing the other bids.

[61]*Albano Cleaners, Inc. v. United States*, supra note 60.
[62]16 Ohio Misc.99, 241 N.E.2d 403 (1968).
[63]485 F.2d 180 (4th Cir.1973).
[64]*William M. Young & Co. v. West Orange Redeveloping Agency*, 125 N.J.Super. 440, 311 A.2d 390 (1973) (two minutes late preceded by a telephone call from the bidder stating he would be a few minutes late because of inclement weather); *Gostovich v. City of West Richland*, 75 Wash.2d 583, 452 P.2d 737 (1969) (three *days* late due to mail mixup).

award. Usually the basis for the request is computation errors, such as omitting a large item, making a mathematical miscalculation in determining an item price or making an error in adding bid price components.[65]

Early cases would not relieve a bidder for these mistakes.[66] A reason given for denying relief was that the mistake was "unilateral," one made only by the bidder and not shared in by the owner. There was, as a rule, negligence in computing the bid. The fear of false claims and the integrity of the bidding system were other reasons for denying relief.

Some courts began to moderate the strictness of this doctrine. They pointed to the rule that a person to whom an offer has been made cannot accept the offer if she knows or should know that the offer was made by mistake. For example, if the owner or design professional knew or should have known that an entire item had been left out of the bid or that there had been a mistake in adding the total, it would be unfair to accept the proposal and seek to bind the bidder.

The courts usually focus upon the knowledge or constructive (what it should have known) knowledge of the owner before the formal award and execution of the formal contract, most commonly at the time the bids are open. Some courts have required that there be actual knowledge on the part of the owner,[67] while others appear to require constructive knowledge that the mistake was known or ought to have been known.[68] In practice, the difference between actual and constructive knowledge will rarely control the outcome of a case. The typical case involves the low bidder claiming that it had made a mistake at the time of bid opening or shortly thereafter. This "snap-up" doctrine is demonstrated in *Santucci Construction Co. v. County*

of Cook,[69] in which the awarding authority had estimated that the cost of drain work would be $1.9 million.[70] The cost of the drainpipe, including labor, was expected to cost $1.4 million. Santucci submitted a total bid of $1.1 million, with $775,000 for the drainpipe, including labor. Three other contractors submitted total bids between $1.7 million and $1.8 million, with their bids for the drainpipe, including labor, being between $1.2 million and $1.4 million. The engineer for the awarding authority had thought that Santucci's bid was "cheap, low." A day after bid opening, Santucci claimed a clerical error and sought to withdraw his bid. The request was refused, and Santucci would not enter the contract. The awarding authority retained Santucci's bid deposit. Santucci brought legal action to recover it.

Noting that ultimately the work was let for $1.6 million, the appellate court affirmed the finding of the trial court that Santucci had made a mistake and that the awarding authority should have known of this mistake. It rescinded Santucci's bid and ordered that the deposit be returned to him. The court emphasized that Santucci's bid was $600,000 less than the next lowest bidder and over $800,000 less than the awarding authority's estimate for the project.

The "snap-up" concept used by the court in the *Santucci* case has been the vehicle for bidder relief where courts have thought it appropriate. Yet an appraisal of the cases and an awareness of the immense variation of the bids for many types of construction work leads inescapably to the conclusion that it is not the fact that the awarding authorities *should* have known of the mistake but the unfairness of holding a bidder who has made an honest mistake when the next bid can still be accepted.[71]

---

[65]Occasionally, claims for relief are based on a sub-bidder's refusal to contract at the price proposed or for an error of judgment relating to performance cost.

[66]*Steinmeyer v. Schroeppel*, 226 Ill. 9, 80 N.E. 564 (1907).

[67]*Westinghouse Elec. Corp. v. New York City Transit Auth.*, 735 F.Supp. 1205 (S.D.N.Y.1990) (interpreting *Iversen Constr. Corp. v. Palmyra-Macedon Central Sch. Dist.*, 143 Misc.2d 36, 539 N.Y.S.2d 858 (1989) as requiring actual knowledge).

[68]*Bromley Contracting Co. v. United States*, 219 Ct.Cl. 517, 596 F.2d 448 (1979).

[69]21 Ill.App.3d 527, 315 N.E.2d 565 (1974).

[70]Amounts are approximations.

[71]Cases holding that a right to rescind existed and that the claimant met the standards required for rescission are *Marana Unified School Dist.*, 144 Ariz. 159, 696 P.2d 711 (1984) (bidder did not need to forfeit bond despite statute); *Power Horn Constructors, Inc. v. City of Florence*, supra note 34 (reversed intermediate court's requirement that bidder show it was not negligent). See Jones, *The Law of Mistaken Bids*, 48 U.Cin.L.Rev. 43 (1979); Cavico, *Relief for Unilateral Mistake in Construction Bids*, 10 Thurgood Marshall L.Rev. 1 (1985). The many cases are collected in Annot., 2 A.L.R.4th 991 (1980).

Some jurisdictions deny relief.[72] These jurisdictions emphasize the need to protect the process from possible favoritism that may accompany the power to allow withdrawal. Such decisions reflect skepticism that the fact-finding process can determine whether honest mistakes have been made. Even in these jurisdictions, relief can be granted if the facts appear to make it inequitable to hold the bidder to its bid.[73]

Most jurisdictions will relieve the bidder if the mistake is clerical rather than an error of judgment and involves a substantial portion of the total bid or a large amount of money and if the owner has not relied to its detriment on the mistaken bid.

---

[72]*Alaska International Constr. Inc. v. Earth Movers of Fairbanks, Inc.*, 697 P.2d 626 (Alaska 1985) (no loss on contract; strong dissent); *Anco Constr. Co. v. City of Wichita*, 233 Kan. 132, 660 P.2d 560 (1983); *Nelson Inc. of Wisconsin v. Sewerage Comm'n.*, 72 Wis.2d 400, 241 N.W.2d 390 (1976) (despite statute). Care must be taken to distinguish cases that refuse relief because the elements are *not* established from those that hold that no relief *can* be granted in any case.

[73]Pennsylvania holds that a bidder cannot revoke after bid opening and must forfeit the bond. But to forfeit the bond, the bidder must give *formal* notice of withdrawal

of the bid or actually refuse to execute the contract documents presented. In *Travelers Indem. Co. v. Susquehanna County Comm's.*, 17 Pa.Commw. 209, 331 A.2d 918 (1975), the court refused to forfeit the bid bond because the bidder had not actually withdrawn his bid but had simply requested to do so and the awarding authority had failed to present the bidder with contract papers for execution. On this technicality, the bidder obtained relief. Likewise, a federal district court, though stating that the bidder probably should have been relieved, felt precluded because Arkansas law denied relief because the contractor was negligent. Yet the court granted a large recovery to the plaintiff contractor on the grounds that the owner's termination was not made in good faith. *Paul Hardeman, Inc. v. Arkansas Power & Light Co.*, 380 F.Supp. 298 (E.D.Ark.1974) (a thorough discussion).

The Federal Acquisitions Regulations (FAR) permit the awarding agency to correct clerical mistakes before an award is made.[74] The Regulations provide a systematic method of dealing with bidding mistakes. A correction may be made if the bidder requests permission to correct a mistake and if there is "clear and convincing evidence" that there was a mistake and of the "bid actually intended." However, if the correction would displace one or more bids, correction "shall not be made unless the existence of the mistake and the bid actually intended are ascertainable substantially from the invitation and the bid itself."[75] If the evidence of the mistake is clear and convincing but the bidder does not provide clear and convincing evidence of the bid actually intended, the public agency may determine that the bidder be allowed to withdraw the bid. Suppose the bidder wishes to withdraw the bid rather than correct it. If the evidence is clear and convincing, both as to the existence of a mistake and as to the bid actually intended, and if the bid as corrected would be the lowest received, the agency may correct the bid and not permit its withdrawal.

The Regulations govern the relationship between bidders and the federal agency conducting the bidding competition. But they may also control bidding mistakes in disputes between a subcontractor and a prime contractor. If the subcontract selects federal procurement law as governing, the FAR will be applied.[76]

State legislation increasingly regulates attempts by those who bid on public contracts to withdraw bids because of mistake after bid opening. In 1945, California adopted legislation that allowed a bidder to be relieved from a bid if it established to the satisfaction of the court that

1. A mistake was made.
2. It gave the department written notice within five days after the opening of the bids of the mistake, specifying in the notice in detail how the mistake occurred.

---

[74]48 CFR, § 14.406-2 (1992).

[75]Id. at 14.406-3(a)(b). The Federal Procurement Regulations are discussed in Rudland, *Rationalizing the Bid Mistake Rules*, 16 Pub. Contract L.J. 446 (1987), and Hagberg, *Mistake in Bid, Including New Procedures Under Contract Disputes Act of 1978*, 13 Pub. Contract L.J. 257 (1983).

[76]*Sulzer Bingham Pumps, Inc. v. Lockheed Missile & Space Co., Inc.*, 947 F.2d 1362 (9th Cir.1991) (reproduced in part later in this subsection).

3. The mistake made the bid materially different than it intended it to be.

4. The mistake was made in filling out the bid and not due to error in judgment or to carelessness in inspecting the site of the work or in reading the plans or specifications.[77]

This statute as originally enacted applied only to certain state public agencies.

In 1982, California incorporated this approach in its Public Contract Code, which applies to all public entities—state, county, and local as well as other public entities.[78]

California case decisions had developed standards for determining whether a bidder could withdraw its bid, based upon unilateral mistake, that were flexible and emphasized the unfairness of the public entity holding the bidder to its mistaken bid.[79] It was held, however, that the exclusive method by which a California public bidder obtains relief from a mistaken bid is under the state statute.[80] This leaves the earlier flexible standards available for private contracts.

The statute reflects a more mechanical method of dealing with the problem, particularly in barring relief unless notice is given within five days after bid opening specifying in detail how the mistake occurred. There is no requirement under the statute that holding the bidder would be unconscionable or that the city have knowledge before the bid was accepted that there had been a clerical mistake making it unjust and unfair for the city to take advantage of the bidder's error. Nor does the statute require any evaluation of the degree of negligence.

The evolution of legal rules dealing with bidding mistakes reflects a slow process, starting first in those transactions where it is determined that the awarding authority knew or should have known of the mistake but limiting mistake to cler-

ical errors.[81] Yet suppose a mistake has been made that is not a *simple* clerical arithmetic error. Cases have differed as to whether relief is confined to *clerical* arithmetic mistakes. It may be useful to look at cases refusing to draw this distinction. For example, *Balaban-Gordon Co. v. Brighton Sewer District Number 2*[82] involved separate general construction and specialty trade contracts. The bidder by mistake included equipment in its *plumbing* bid that should have been in its *general* construction bid. It was awarded only the *general* construction contract. It was relieved from its bid. The court refused to draw a line between clerical and judgmental errors, noting the mistake to have been objectively discoverable.

In an astounding case, *White v. Berenda Mesa Water District*,[83] the court granted relief when the mistake related to the amount of hard rock that would have to be excavated. The bids ranged from $427,000 to $721,000, and the bid next lowest to that of White was $494,000. When the bids were opened, White reviewed his bid. He learned that the soil report on which he relied was in error but that the project specifications were accurate. Despite the fact that White *knew* that soil reports are prepared before detailed specifications and that the specifications and not the soil report control, and despite White's having a copy of the specifications that indicated that the owner did not warrant the accuracy of the soil report, White was granted relief. The court felt that a line between clerical and judgmental errors was too fine a line to draw.

Courts refusing to bar relief for judgment errors seem persuaded that if proof *exists* of the mistake and the owner can simply accept the next low bid, the owner has not been harmed. To these courts it seems unconscionable to force the bidder to pay the difference between its bid and the next low bid, often a substantial amount.

[77]Cal.Stat.1945, ch. 118, p. 501, now West Ann.Cal.Pub.Cont. Code § 10202. A bidder relieved cannot bid on the same job. For an interpretation of this provision, see *Colombo Constr. Co. v. Panama Union School Dist.*, 136 Cal.App.3d 868, 186 Cal.Rptr. 463 (1982) (bidder barred).
[78]See West Ann.Cal.Pub.Cont. Code § 5103.
[79]*Elsinore Union Elementary School Dist. v. Kastorff*, 54 Cal.2d 380, 353 P.2d 713, 6 Cal.Rptr. 1 (1960).
[80]*A&A Elect., Inc. v. City of King*, 54 Cal.App.3d 457, 126 Cal.Rptr. 585 (1976).

[81]See also *Osberg Constr. Co. v. City of The Dalles*, 300 F.Supp. 442 (D.Or.1969); *Elsinore Union Elementary School Dist. v. Kastorff*, supra note 79; *Powder Horn Constructors Inc. v. City of Florence*, supra note 34; *Missouri State Highway Comm'n. v. Hensel Phelps Constr. Co.*, 634 S.W.2d 168 (Mo.1982).
[82]41 A.D.2d 246, 342 N.Y.S.2d 435 (1973). See also *Wil-Fred's Inc. v. Metro. Sanitary Dist.*, 57 Ill.App.3d 16, 372 N.E.2d 946 (1978) (relief despite error as to how work was to be done).
[83]7 Cal.App.3d 894, 87 Cal.Rptr. 338 (1970).

Yet another method of relieving from large forfeitures exists, though it is often ignored. It looks at the *security* that bidders are usually requested to deposit, such as a certified check, a cashier's check, or a bid bond, based on a designated percentage of the bid. Suppose a bidder who has posted security seeks to withdraw its bid based on an asserted clerical or mathematical error of the type discussed in this section. Usually, one of the reasons to allow withdrawal is a great variation between the bid submitted and the bid intended to be submitted or the other bids. Suppose, however, that the owner sought only to keep the deposit rather than attempt to hold the bidder to its bid and seek damages based on the difference between the low bid and the next low bid. The owner would contend that the bidder has simply breached the obligation to *enter into* the construction contract and not the construction contract itself. The stipulated damage for such a breach is the amount of the deposit. In such a case, the owner would claim the right to retain the deposit but not damages based on the difference between the bid and the next low bidder.

This would seem a fair solution to the vexatious bidding mistake cases. The contractor's exposure would be the deposit. Although a few courts have drawn this distinction,[84] most generally have held that the contractor is relieved from its performance *entirely* if the requirements for relief are established. In such a case, the contractor is entitled to recover the deposit. Although the deposit may be only 5% to 10% of the bid, in large jobs, this amount is substantial. Where the facts are sufficient

to allow the mistake doctrine to be applied, most courts would prefer to relieve the bidder entirely.[85]

Courts generally have been reluctant to allow a bidder to *correct* a mistake, particularly if the claim to reform the contract is made after performance has begun or been completed.[86] Yet some cases, particularly in the federal procurement system, do permit the bidder to correct its mistake.[87] Some courts outside the federal procurement system have also permitted correction through the equitable doctrine of reformation.[88] Yet clearly rescission (cancellation) is easier to obtain than reformation.[89]

A recent case, which involves the application of the Federal Acquisition Regulations, deals with what appears to be a mistaken bid created by errors of judgment and employs an equitable solution when the contract has been partly performed. This instructive case is reproduced here.

---

[84]*Triple Contractors, Inc. v. Rural Water Dist. No. 4*, 226 Kan. 626, 603 P.2d 184 (1979). The Canadian Supreme Court has used a similar analysis. See *The Queen in Right of Ontario v. Ron Eng'g and Constr. (Eastern), Ltd.*, [1981] S.C.R. 111.

[85]*Marana Unified School Dist.*, supra note 71; *Power Horn Constructors, Inc. v. City of Florence*, supra note 34 (court might have done so had clause liquidated damages).
[86]*Lemoge Elec. v. County of San Mateo*, 46 Cal.2d 659, 297 P.2d 638 (1956); *Midway Excavators, Inc. v. Chandler*, 128 N.H. 654, 522 A.2d 982 (1986); *Anco Constr. Co. v. City of Wichita*, supra note 72.
[87]*United States v. Hamilton Enters., Inc.*, 711 F.2d 1038 (Fed.Cir.1983) (contractor must establish claim by clear and convincing proof of clerical or arithmetic error or misreading of specifications); *Bromley Contracting Co. v. United States*, supra note 68 (U.S. "overreached" and rescission no longer possible); *North Landing Line Constr. Co.* Comp.Gen.Dec. B-239662 (July 20, 1990, 90-2 CPD ¶ 60); *W.H. Hussey & Associates*, Comp.Gen.Dec. B-237207 (February 1, 1990, 90-1 CPD ¶ 137), request for reconsideration denied, May 2, 1990. Much depends upon the language in the FAR. Refer to notes 74, 75 supra.
[88]*Nat Harrison Assocs., Inc. v. Louisville Gas & Elec. Co.*, 512 F.2d 511 (6th Cir.1975) (dictum), cert. denied, 421 U.S. 988 (1975); *Iversen Constr. Co. v. Palmyra-Macedon Central School Dist.*, 143 Misc.2d 36, 539 N.Y.S.2d 858 (1989).
[89]*Liebherr Crane Corp. v. United States*, 810 F.2d 1153 (Fed.Cir.1987).

SULZER BINGHAM PUMPS, INC., v.
LOCKHEED MISSILES & SPACE COMPANY, INC.

United States Court of Appeals, Ninth Circuit, 1991. 947 F.2d 1362.

Before KILKENNY, GOODWIN and SCHROEDER, Circuit Judges.

SCHROEDER, Circuit Judge.

This appeal arises out of an unusual dispute between a major government contractor and a subcontractor providing components for the United States Navy's Trident II nuclear submarines. The contractor, Lockheed Missiles & Space Company, awarded a subcontract to the low bidder for the subcontract, the appellee Sulzer Bingham Pumps, Inc. Sulzer Bingham's bid, however, was in fact millions of dollars lower than it would have been if Sulzer Bingham had not committed a series of errors in preparing the bid. According to the findings of the respected district judge who heard the evidence, Lockheed doubted that Sulzer Bingham could perform the contract at that price, but nevertheless awarded the contract for the bid price, resulting in an unconscionably low price. The district court concluded that Lockheed's conduct amounted to overreaching in violation of basic contractual principles, and that its failure to ask the subcontractor to verify its bid violated the terms of the subcontract. The district court declined to rescind the contract. The district court did, however, award Sulzer Bingham some equitable relief in the form of the actual costs it incurred above the contract price, in a total amount which was not to exceed the next lowest bid. Lockheed appeals.

* * *

FACTUAL BACKGROUND

The facts as determined by the district court are not seriously disputed on appeal. They can be summarized as follows.

Lockheed is the prime contractor for the Navy's Trident II nuclear submarines. The subcontract at issue involved the production of ballast cans. When the submarines are not carrying nuclear missiles, they need ballast cans for stability. Each ballast can weighs 64,000 pounds and stands 15 feet high.

In 1988, Lockheed sent out a request for quotation to potential subcontractors, seeking bids for the manufacturer of 124 ballast cans. In February 1989, Lockheed received eight bids, including one from Sulzer Bingham. Sulzer Bingham was the lowest bidder at $6,544,055. The next lowest bid was $10,176,670, and the bids ranged up to $12,940,540, with one high bid at $17,766,327. Lockheed estimated that the job would cost about $8.5 million.

Lockheed's employees were shocked by Sulzer Bingham's bid and thought it was surprisingly low. Price extensions, submitted to Lockheed by Sulzer Bingham, revealed no arithmetic errors. Lockheed then asked Sulzer Bingham to verify that its bid included shipping charges and First Article Compatibility Testing, but did not ask for verification of the entire bid. Sulzer Bingham informed Lockheed that its bid was complete. Lockheed then inspected Sulzer Bingham's Portland facility to evaluate Sulzer Bingham's technical capabilities. The inspection revealed that Sulzer Bingham would have to make many modifications to its existing facility in order to complete the contract. The turntable Sulzer Bingham anticipated using for machining and assembling the ballast

cans was inadequate, and an entirely new lead pouring facility needed to be constructed. None of these shortcomings were revealed to Sulzer Bingham by Lockheed.

At no time did Lockheed notify Sulzer Bingham that it suspected a mistake in Sulzer Bingham's bid. Lockheed did not inform Sulzer Bingham that its bid was significantly lower than the next lowest bid, and lower than Lockheed's own estimate of the cost of the job as well. Lockheed never informed Sulzer Bingham that it suspected that Sulzer Bingham would not be able to complete the contract at the bid price.

Sulzer Bingham made a variety of errors in its bid. It had underestimated the number of hours the job would require, and had used hourly labor rates that were below cost. Sulzer Bingham had not realized that its existing facilities were inadequate for the job, and had overlooked certain costs of the job. Sulzer Bingham's bid broke down to $30,707 per ballast can. The next lowest bid broke down to $58,137 per can, and Lockheed estimated at least $40,000 per can.

In late February 1989, Lockheed accepted Sulzer Bingham's bid and Sulzer Bingham started work. In November 1989, Sulzer Bingham revised its estimate of the cost of the job and asked Lockheed for additional $2,111,000 in compensation. Lockheed rejected Sulzer Bingham's request for additional compensation.

* * *

The district court . . . concluded that under section 14.406-3 of the Federal Acquisition Regulations, Lockheed had a duty to notify Sulzer Bingham when it suspected a mistake in Sulzer Bingham's bid. The district court further concluded that Lockheed breached this duty, and Sulzer Bingham was therefore entitled to equitable relief.

The district court denied rescission because Sulzer Bingham had delayed its request for rescission and because Sulzer Bingham had already completed about half of the contract. The district court then concluded that it could award Sulzer Bingham other equitable relief because Lockheed had breached its duty to verify Sulzer Bingham's bid, and the resulting contract was unconscionable. The district court required Sulzer Bingham to complete the contract, and ordered that Sulzer Bingham could "recover its actual costs only, including a reasonable amount for depreciation and overhead. Under no circumstances may [Sulzer Bingham's] recovery exceed the amount of the next lowest bid."

THE APPLICABILITY OF FEDERAL ACQUISITION REGULATIONS REGARDING BID VERIFICATION

Lockheed argues that the district court erred in holding that the Federal Acquisition Regulations ("FAR") imposed specific verification duties on Lockheed. [Ed. note: The Court held that the subcontract imposed these duties on Lockheed.]

Section 14.406-3 of the Federal Acquisition Regulation requires a contracting officer to take the following steps if a bidding mistake is suspected:

1. The contracting officer shall immediately request the bidder to verify the bid. Action taken to verify bids must be sufficient to reasonably assure the contracting officer that the bid as confirmed is without error, or to elicit the allegation of a mistake by the bidder. To assure that the bidder will be put on notice of a mistake suspected by the contracting officer, the bidder should be advised as appropriate—
   i. That its bid is so much lower than the other bids or the Government's estimate as to indicate a possibility of error;

   * * *

   iv. Of any other information, proper for disclosure, that leads the contracting officer to believe that there is a mistake in bid.

48 C.F.R. § 14.406-3(g)(1). The district court held that Lockheed breached its duty to Sulzer Bingham by not notifying Sulzer Bingham that it suspected a mistake, and not informing Sulzer Bingham that the bid was much lower than all other bids.

* * *

The district court correctly concluded that FAR governed the parties' conduct during the bid acceptance and award period, and that Lockheed breached its duty under FAR by failing to notify Sulzer Bingham that it suspected a mistake in the bid.

THE APPROPRIATE REMEDY FOR LOCKHEED'S BREACH

Lockheed argues that even if FAR applies and Lockheed breached its duty to properly verify Sulzer Bingham's bid, nevertheless Sulzer Bingham is not entitled to relief. In support of this argument Lockheed relies on cases holding that the terms of a substantially performed contract may not be changed through reformation if the bid mistake is not attributable to an arithmetic or clerical error.

It is apparently well settled government contract law that reformation, based upon a mistake in bid-

ding, is available to correct only "clear cut clerical or arithmetical error, or misreading of specifications." *Aydin Corp. v. United States,* 229 Ct.Cl. 309, 314, 669 F.2d 681, 685 (1982) (quoting *Ruggiero v. United States,* 190 Ct.Cl. 327, 335, 420 F.2d 709, 713 (1970)). According to the district court's findings, errors in judgment predominated in this case. It is apparently well settled government contract law that such errors in judgment do not, by themselves, justify the award of equitable relief in the form of a change in the contract price. . . . Such a conclusion is consistent with the understanding that contracting entities should live up to their contractual obligations when bidding errors are based on economic misjudgment, and not attributable to the other contracting party's conduct.

In this case, however, the district court did not fashion equitable relief solely on the basis of the bidder's economic misjudgments. Rather, the district court awarded relief because Lockheed, in failing to follow contractual provisions requiring bid verification, accepted an unconscionably low bid. The verification procedures Lockheed ignored were designed to ensure that such unconscionably priced contracts would not be awarded.

None of the authorities relied upon by Lockheed involve such a situation. In *Aydin Corp.,* 229 Ct.Cl. at 318, 669 F.2d at 687, for example, the disparity in bids was not sufficient to put the contracting officer on constructive notice of any mistake in the bid, whereas in this case, Lockheed had actual notice. In *Hamilton Enterprises,* 711 F.2d at 1045, the bidder underestimated the number of hours needed to perform the contract, and the government failed to adequately verify the bid. The bidder defaulted on the contract. The court denied the claims of both parties, describing the case as one of "mutual fault to the extent that neither party is entitled to recover on the claims asserted against the other." There was no finding of unconscionability. In this case, unlike *Hamilton Enterprises,* Lockheed is reaping the rewards of Sulzer Bingham's performance at an unconscionably low price. The Defense Department's own adjudicatory arm has itself recognized in contract disputes that equitable principles do apply to prevent the enforcement of an unconscionable contract.

* * *

AFFIRMED.

Can an *owner reduce* the contract price because of a claimed mistake made by the bidders? In *Edward D. Lord, Inc. v. Municipal Utilities Authority,*[90]

two sub-bidders agreed to cross-excessive bids, with each bidding to prime bidders at agreed excessive prices. Ultimately, this was discovered, and the bidder to whom the contract had been awarded cancelled the subcontract and rebid the job for much less than the cross-excessive bids.

---

[90]133 N.J.Super. 503, 337 A.2d 621 (1975).

New Jersey held that the awarding authority could reform the contract because of mutual mistake. Both bidder and awarding authority assumed that the bidding was fair and not rigged and the price agreed on reflected this basic assumption. However, the assumption turned out to be incorrect because of the collusive conduct of the subcontractors. Because public funds were involved, the court concluded that the awarding authority would be granted reformation reducing the contract price to take into account the excessive cross bids.

Suppose the Invitation to Bidders states that bidders will not be released for errors. One court interpreted this language to cover errors of judgment and not clerical errors.[91] Another refused to employ this interpretation technique to give relief for a clerical error.[92] The former approach is preferable. If the invitation *clearly* covers clerical errors, the risk of even clerical errors should be placed on the contractor. But it is more likely that such language will not tie the hands of courts to grant relief for such mistakes if the enforcement of the mistaken bid would be unconscionable.

Mirroring increased judicial activism in other fields, constitutional principles have been invoked in bid mistake cases. In *Midway Excavators, Inc. v. Chandler*,[93] the court rejected the claim by the bidder that the failure by the public entity to develop guidelines determining the type of technical mistakes that would justify relief violated the bidder's constitutional right of not being deprived of property without due process of law.

F. Bid Deposit

The bidders are usually requested to submit deposits with their proposals. Is the deposit a security deposit out of which the owner can take whatever damages it has incurred? Does the payment limit the damages to a specified figure that still obligates the owner to prove damages up to that figure? Is the deposit submitted in an attempt to set damages in advance by agreement of the parties?

Suppose Bidder A submits a proposal for one million dollars and the Invitation to Bidders requires it to submit a bid bond for 5% of its bid. A bid bond for $50,000 is deposited. The bids are opened, and A is lowest. The next lowest bidder has submitted a bid of $1.1 million. A is offered the contract but without any legal justification declines to enter into it. The contract is offered to the next lowest bidder, who bid $1.1. Is the owner entitled to $100,000 in damages, with $50,000 of it as a security deposit out of which it can assure itself that it will be able to collect at least a part of its damages? Or is the owner limited to $50,000 because this is what the parties have agreed will be the actual damage amount whether the actual damages are higher or lower?

Suppose the next lowest bidder had been $1,025,000 instead of $1.1 million. In such a case, can the owner keep the entire $50,000 or be limited to $25,000?

To a certain extent, the parties are free by their contract to determine whether the amount submitted or the bid bond deposited liquidates (agrees in advance on the amount of) damages, is a security deposit, or is a limitation of liability. A security deposit is an amount of money deposited with one party out of which the latter can satisfy whatever damages to which it is entitled. If the 5% deposit is merely a security deposit, the owner can retain this amount and sue for any balance to which it is entitled, or it must return the excess of the deposit over damages.

If the deposit is a valid liquidated damages clause, the parties have agreed in advance that whatever the amount of the actual damages, the breaching party will pay the amount stipulated in the clause. The owner could retain the $50,000 whether the damages were $200,000 or one dollar.

A valid liquidated damages clause requires that it be difficult to ascertain the damages at the time the contract is made and that the amount agreed be a genuine preestimate of the potential damages.

An amount disproportionate to the actual or anticipated damage chosen merely to coerce performance is a penalty and unenforceable. For example, if the deposit were 50% of the bid and if it were most unlikely that there would be damages approaching this amount, the clause would be a penalty and unenforceable. The owner would be entitled only to actual damages without regard for the amount of the deposit, and it would have to refund the excess of the deposit over its actual damages.

---

[91] *M.F. Kemper Constr. Co. v. City of Los Angeles*, 37 Cal.2d 696, 235 P.2d 7 (1951). See also *Jobco, Inc. v. County of Nassau*, 129 A.D.2d 614, 514 N.Y.S.2d 108 (1987) (disclaimer in bid bond did not preclude rescission for mistake).

[92] *City of Newport News v. Doyle & Russell, Inc.*, 211 Va. 603, 179 S.E.2d 493 (1971).

[93] 128 N.H. 654, 522 A.2d 982 (1986).

Legislation, state or local, may give the public agency damages based on the difference between the low bid and the next lowest bid if the low bidder does not enter into the contract awarded to it. In such a case, the deposit is for security and is not an attempt to liquidate damages. A liquidated damages clause *establishes* the damages.

Legislation sometimes provides that the public agency may retain the amount deposited but only to the extent of the difference between the defaulting bidder's bid and the amount for which the contract is ultimately awarded. For example, suppose the amount deposited was $50,000 or 5% of the one-million-dollar bid and the next bidder was awarded the contract at $1,025,000. In such a case, the owner would be entitled to retain $25,000. Such statutes set up a liquidation of damages that will apply only if actual damages are greater than the amount deposited. If actual damages are less, the deposit is simply security.

Is it desirable to liquidate damages? From the owner's standpoint, the chances of collecting an amount in excess of the deposit from the contractor are remote. In addition, the owner would like to retain the amount deposited without having to show actual damages. For these reasons, it is preferable to liquidate damages rather than use the deposit solely as security. The pure security deposit does allow the owner to seek to recover an amount beyond the deposit. This would occur if the discrepancy between the defaulting bidder's bid and the next bidder is *more* than the deposit. In such a case, there is a strong likelihood of a mistake that would permit the bidder to withdraw its bid. Many contractors would not be able to satisfy a large court judgment. This is a reason for providing the security deposit. Liquidated damages protect the contractor from the risk of excessive damages and guarantee the owner a reasonable amount of collectible damages.

A properly drafted liquidated damages clause is likely to be enforced if created by legislation that specifically permits the awarding authority to forfeit the deposit.[94] In the absence of such legislation,

courts divide. Some enforce such a clause;[95] others do not.[96]

The uncertainty of the amount of damages must exist at the time the contract is made and at the time of the deposit. Some courts ignore this and seem to look at whether the amount of damages can be easily ascertained at the time of breach. These courts seem unwilling to forfeit an amount in excess of actual damages. This approach does not take note of the long-haul aspects of denominating the forfeiture clause as stipulated damages. Over the long haul of many competitive bids, the losses to the public agency probably average per competition the amount stipulated in each competitive bid. Though the long haul may seem unfair to the *particular* bidder who must lose more than what it *appears* the agency has been damaged in *this* competitive bid, the particular bidder is relieved from any risk *beyond* the deposit amount.

There are administrative costs when the next lowest bidder is selected. Admittedly, that loss often seems much less than the amount forfeited. But again, the parties *expect* the amount to be deposited to be forfeited, and the bidder is relieved from the risk of loss beyond the deposit. As long as the amount selected is reasonable, the forfeiture clause should be considered a valid liquidated damages clause.

The clause should be *clearly* enforceable if it is necessary to rebid the entire project. Rebidding entails substantial additional administrative expense.

Suppose both low bidder *and* the next lowest bidder unjustifiably refuse to enter into the contract? Can the owner retain the deposit by both bidders? Although it may seem unfair to retain both bidders' deposits, it is not logically indefensible. Each bidder has breached, and each has been to some degree relieved from the risk of excessive damages by the use of an agreed damage provision. However, goodwill and the avoidance of litigation may necessitate some solution, such as

[94]*A & A Elec., Inc. v. City of King*, 54 Cal.App.3d 457, 126 Cal.Rptr. 585 (1976). But see *Petrovich v. City of Arcadia*, 36 Cal.2d 78, 222 P.2d 231 (1950) (did not seem willing to enforce such a clause). Cf. *Powder Horn Constructors, Inc. v. City of Florence*, supra note 34.

[95]*Jobco, Inc. v. County of Nassau*, supra note 91 (dictum); *Bellefonte Borough Auth. v. Gateway Equip. & Supply Co.*, 442 Pa. 492, 277 A.2d 347 (1971); *City of Fargo v. Case Dev. Co.*, 401 N.W.2d 529 (N.D.1987) (failure to develop property).
[96]*Petrovich v. City of Arcadia*, supra note 94; *Ogden Dev. Corp. v. Federal Ins. Co.*, 508 F.2d 583 (2d Cir.1974), held that the clause forfeiting the deposit was a penalty.

retaining one half of each rather than trying to retain both deposits.[97]

G. The Formal Contract

The culmination of a successful competitive bidding process is the award by the awarding authority to the successful bidder. Usually, a formal contract is forwarded or given to the successful bidder for its execution.

Suppose the award is made *before* the expiration of the period during which the bid is irrevocable but the formal contract is *not* executed within that period. Although one case discussed earlier held that a tentative acceptance within the period was not a sufficient acceptance,[98] two Wisconsin cases held that the validity of the contract did not require execution of the contract. In one,[99] the awarding authority had voted to accept the bid in the presence of the bidder. In the other,[100] approval by a federal regulatory agency that conditioned the award was not received until the morning of the final day of the period during which the bid was irrevocable. On that morning (a Friday), the engineer notified the contractor that the formal contracts would be in the mail that day. The following day (Saturday), the contractor wrote that it was withdrawing its bid, since the forty-five-day period during which its bid was irrevocable had expired. The contracts were received on the following Monday. The court could have held that the acceptance took place on the forty-fifth day—when the contracts were put in the mail.[101] The court held in favor of the awarding authority. The contracts had been formed, and the formal contracts merely memorialized the agreement that had already been made.

Suppose the formal contract is not consistent with earlier communications exchanged between the successful bidder and the awarding authority.

The Court of Claims held that the formal contract, while typically superseding all previous negotiations, documents, etc., is merely a reduction to form of the actual agreement made by the advertisement, bid, and its acceptance.[102] This is another recognition of the formal contract often being simply a memorial of the agreement that has been made. However, the date of the formal contract *may* set into motion any time commitment of the contractor.

H. Bidding Documents

The culmination of the complex competitive process usually is assent by both parties to the construction contract. This process has generated a series of many long and complex writings, such as the Invitation to Bidders, Information to Bidders, and the Bid Proposal. Included among these are other materials, such as plans, specifications, general conditions, supplemental conditions, addenda, and the agreement forms. Chapters 19 and 20 deal with the problems generated by this wealth of written material. But what about the materials generated by the bid process itself?

AIA Doc. A201, ¶ 1.1.1, seeks to deny the bidding materials *any* legal effect by excluding bidding requirements as contract documents. Bidding materials have been superseded by execution of the construction contract. It is important to avoid contradiction in the voluminous contract documents. But a provision can state that in the event of conflict, the construction contract takes precedence over the bidding documents. Although implementation of provisions of this type is not as simple as it appears, the solution AIA has selected—*excluding bidding documents*—is undesirable.

Most participants in the process believe that the bidding documents do have legal efficacy.[103] Information is given in the bidding documents that is relied on by the bidders. For example, subsurface information is frequently included in the Information to Bidders. Under AIA Doc. A201, ¶ 4.3.6, unless this information is found in the specifications,

---

[97]West Ann.Cal.Pub.Cont. Code §§ 10181–10182 permit forfeiture of the security of lowest, second lowest, *and* third lowest if none will enter into the contract.

[98]*Hennepin Public Water Dist. v. Petersen Constr. Co.*, supra note 43, discussed in Section 18.04(D).

[99]*Nelson Inc. of Wisconsin v. Sewerage Comm'n*, supra note 72.

[100]*City of Merrill v. Wenzel Brothers, Inc.*, 88 Wis.2d 676, 277 N.W.2d 799 (1979). See also *Citizens Bank of Perry v. Harlie Lynch Constr. Co.*, 426 So.2d 52 (Fla.App.1983) (oral acceptance by owner's board valid acceptance).

[101]Refer to Section 5.06(C).

[102]*Dana Corp. v. United States*, 200 Ct.Cl. 200, 470 F.2d 1032 (1972).

[103]*Jack B. Parson Constr. Co. v. State Dep't of Transp.*, 725 P.2d 614 (Utah 1986) (effect given to bidding information).

it cannot be the basis of any claim by the contractor for an equitable adjustment for subsurface conditions different from those usually encountered or disclosed by the contract documents. Similarly, AIA Doc. A701 (Instructions to Bidders) imposes many contractual terms intended to survive, such as liquidating damages for failure to enter into the construction contract and requiring that a particular type of surety bond be used. If the award is made and the construction contract documents are signed *without* these provisions, if taken literally, A201 would discharge any obligations the contractor may have that are expressed in the Instruction to Bidders. There may be factual material in the bidding instructions or information that the owner may wish to point to at some later date if a claim has been made.[104]

The bidding material was useful to the contractor in *Village of Turtle Lake v. Orvedahl Constr. Inc.*[105] Here the bidding material *was* a contract document since the arbitration clause included disputes relating to the contract documents, the bidder was given a chance to submit its claim of mistake to arbitration even if it could not meet the requirements of the statute regulating bidding mistakes in public contracts.

Attempts to deny any legal effectiveness to the bidding documents may not be in the owner's best interest and may frustrate the reasonable expectations of both parties.

I. Judicial Review of Agency Action

Public procurement encompasses a vast number of social and economic goals. The result has been a complicated set of statutes and regulations with the increasing likelihood of irregularities.

To challenge an agency decision, the challenger must have "standing."[106] Many states allow taxpayer suits to challenge official actions sometimes done in the name of the disappointed bidder or someone acting on its behalf. The challenge may also be made by a taxpayer who simply believes that the agency action will cost the taxpayers

money.[107] Massachusetts grants a disappointed bidder the right to challenge the agency action.[108] Pennsylvania allows *only* the low bidder to challenge the agency decision.[109] Much depends on state statute, but at the state and local level, standing does not seem to have been the problem it has been at the federal level.

Before 1970, disappointed bidders were not granted standing to challenge federal agency decisions based on *Perkins v. Lukens Steel Co.*,[110] which held that the government has the sole right to choose with whom and under what terms it will contract. A disappointed bidder has only a privilege and not a right to do business with the government. In addition to this wooden logic, the court was concerned that judicial interference with government procurement would cause delay and involve the courts in decision making beyond their competence.

But in *Scanwell Laboratories, Inc. v. Shaffer*,[111] a federal appeals court cited a statute passed after *Perkins v. Lukens Steel Co.* and held that a disappointed bidder had standing to seek a judicial order stopping an allegedly invalid procurement award.

The *Scanwell* case seemed to open some federal courthouse doors to disappointed bidders seeking federal contracts. This would involve courts in difficult procurement problems. A year later, the same court had second thoughts. It held that judicial interference with a procurement award required a demonstration by the challenger that there was no "rational basis" for the award. In addition, trial courts were given broad discretion to refuse to interfere with the procurement.[112] It is easier for a disappointed bidder to recover damages that are

[104]*D.A. Collins Constr. Co. v. New York*, 88 A.D.2d 698, 451 N.Y.S.2d 314 (1982) (information regarding possible delay barred contractor delay claim).
[105]135 Wis.2d 385, 400 N.W.2d 475 (App.1986).
[106]Standing was also discussed in Section 9.14.
[107]But in *Lynch v. Devine*, 45 Ill.App.3d 743, 359 N.E.2d 1137 (1977), plaintiff had no standing as bidder *or* taxpayer.
[108]*Grant Constr. Co. of R.I. v. City of New Bedford*, 1 Mass.App. 843, 301 N.E.2d 463 (1973).
[109]*Pullman Inc. v. Volpe*, 337 F.Supp. 432 (E.D.Pa.1971).
[110]310 U.S. 113 (1940).
[111]424 F.2d 859 (D.C.Cir.1970). Not all federal circuit courts give the disappointed bidder standing. But the Second Circuit chose to follow Scanwell. *B.K. Instrument, Inc. v. United States*, 715 F.2d 713 (2d Cir.1983) noted with approval in 51 Brooklyn L.Rev. 191 (1984).
[112]*M. Steinthal & Co. v. Seamans*, 455 F.2d 1289 (D.C.Cir.1971).

usually limited to bidding expenses *if* it can establish a defect in procurement procedures.[113]

J. Illegal Contracts

The imposing number of legal controls on public contracts, the interest generated by public projects, and the staff inadequacies in small public entities all create a substantial risk that an award may be made illegally. Suppose the party to whom the award had been made partly or fully performed.

Clearly, recovery cannot be made under an illegal contract. But two other issues can arise. First, and most frequently, can the contractor who has performed under an illegal contract recover for the work it has performed based on restitution? Second, can any payments that have been made to a contractor be recovered by the awarding authority?

These issues of unjust enrichment, like attempts by bidders to withdraw their bids, generate sharp differences of opinion. Those who would deny recovery or even require repayment stress the importance of an honest competitive bidding system and the need to protect public funds. Although a recognition exists of the occasional unfairness of denying a contractor recovery for work it has performed because of technical irregularities, those who take a hard line cite the difficulty of making these judgments and the importance of not allowing any loopholes in the laws regulating public contracts.

Those who take a softer approach are willing to concede that there should not be recovery or that there should even be repayment where there is venality or corruption, but they draw a distinction between those cases and ones that do not involve serious criminal misconduct. Where corruption is pervasive they would concede that only harsh and unremitting punishment has a chance of deterring such corruption. On the other hand, in some jurisdictions, inefficiency is common and corruption rare. In those jurisdictions, there should be greater willingness to allow payment to a contractor where the award was not tainted with bad faith, fraud, or corruption.

Generally, illegally awarded contracts cannot be the basis for restitution. But in 1971, restitution *was* allowed in *Blum v. City of Hillsboro*.[114] Immediately after acceptance of the $47,000 contract, council members of the awarding authority asked the contractor if it would be willing to do additional work for a designated price. The contractor stated it would, and the awarding authority, through its mayor and city council, specified the work to be done and drew up an amendment to the original contract.

The contractor performed the additional work, which increased the amount of the contract to $154,000. The awarding authority paid $82,000 but refused to pay the balance of $72,000. The contractor sought to recover the balance, and the awarding authority counterclaimed for the amount it paid in excess of the original contract price. This claim was based on the failure to follow state law, which required that the additional work be competitively bid.[115]

The court held that recovery of restitution based on benefit conferred (though a minority view) was justified in this case. According to the court, failure to allow profits would be sufficient deterrence. If recovery is not granted, a claim will very likely be made on the municipality to use its discretionary power to pay moral claims. The court limited recovery to actual costs, including overhead not exceeding actual benefit, but denied profit. Nor could recovery exceed the unit cost of the original contract that had been properly awarded.

In 1977, the Minnesota Supreme Court followed the *Blum* case in a well-drilling contract that was illegally awarded because it violated the competitive bidding statute. Whether there had been a violation was a close question, inasmuch as the awarding authority thought there had been a sufficient emergency that granted it an exemption

---

[113]*Keco Indus., Inc. v. U.S.*, 428 F.2d 1233 (Ct.Cl.1970). Lost profits were not allowed in *Armstrong & Armstrong, Inc. v. United States*, 514 F.2d 402 (9th Cir.1975); *Swinerton & Walberg Co. v. City of Inglewood*, 40 Cal.App.3d 98, 114 Cal.Rptr. 834 (1974); *Paul Sardella Constr. Co. v. Braintree Housing Auth.*, 329 N.E.2d 762 (Mass.App.1975). See Rosengren & Librizzi, *Bid Protests: Substance and Procedure on Publicly Funded Construction Projects*, 7 Constr. Lawyer No. 1, January 1987, p. 1.

[114]49 Wis.2d 667, 183 N.W.2d 47 (1971). Relief was also granted in *Bd. of Comm'rs v. Dedelow, Inc.*, 159 Ind.App. 563, 308 N.E.2d 420 (1974) (need to start work despite challenge and contractor's lack of culpable misconduct).
[115]See Annot. 40 A.L.R.4th 968 (1985).

from the competitive bidding requirements. Yet because the contractor had not actually found water, the court denied recovery because there had been no benefit to the awarding authority.[116]

Sloppy administration surfaced in *McCuistion v. City of Siloam Springs*.[117] The contract was illegally awarded, as it had never been formally authorized and approved by the city council. Additionally, the contractor faced a difficult problem of establishing that there had *ever* been a written contract made. The contractor testified that he signed the contract in the office of the city attorney and that he either saw executed copies or was told by the city attorney that the copies had been executed. He had never been furnished an executed copy. All the parties seemed to *assume* that a written contract had been made, but *no one* was ever able to find the contract or a record of it. The court held that the contractor could recover.

Yet other jurisdictions have taken a much harder line. *Manning Engineering, Inc. v. Hudson County Park Commission*[118] involved pervasive corruption in the awarding of contracts in Jersey City, New Jersey. The court not only denied the engineering company recovery for work that it had performed but also indicated that the awarding authority would have had a good claim had it sought repayment of funds that had been paid. It cited a New York case[119] that had involved a contractor who had been convicted of conspiring to violate state bribery laws through a kickback system. When the contractor sued for the unpaid balance, the city successfully defended the claim and recovered payments that it had paid.

Gerzof v. Sweeney[120] demonstrated not only the difficulty of fashioning an appropriate remedy but also how the "tough" New York court can be persuaded to relax its harsh rules. In this case, the Village of Freeport (New York) had advertised for bids for a 3,500-kilowatt generator. Enterprise bid $615,000, and Nordberg bid $674,000. After an advisory committee had recommended acceptance of Enterprise's bid, a new village election was held at which a new mayor and two new trustees were elected. Shortly thereafter, Nordberg's *higher* bid was accepted.

Enterprise obtained a court order setting the award aside. The board of trustees then drew up *new* specifications for a 5,000-kilowatt generator with the active participation of Nordberg. The specifications were so rigged that only Nordberg could comply. As expected, Nordberg was the only bidder, and its bid of $757,000 was accepted. Nordberg installed the generator and was paid.

After a court declared the *second* award invalid, the trial court held that the village should retain the generator and recover the $757,000 from Nordberg. The intermediate appellate court modified that judgment by providing that Nordberg could retake the machine upon posting a bond for $357,000 to secure the village against damages from removal and replacement of equipment. New York's highest court, the Court of Appeals, first emphasized the importance of protecting the public against corruption and collusion between public officials and bidders.

In the *normal* case it would make no difference whether the village was defending a claim brought by Nordberg or was seeking to recover the money paid Nordberg. But this was *not* a normal case. Granting recovery of payments made would cost Nordberg three quarters of a million dollars, and the village would have its generator. Motivated by the enormity of the forfeiture, the court awarded a remedy different from that awarded by the trial court *or* the intermediate appellate court. The Court of Appeals stated that the award should have been made to Enterprise. Had this been done, the village would have had a 3,500-kilowatt generator for $615,000. The court awarded judgment against Nordberg based on the difference between the $757,000 paid Nordberg and the $615,000 that the village would have paid Enterprise. To this was added $37,000, the difference between what it cost the village to install the Nordberg generator and what it would have cost to install the one offered by Enterprise. In addition, the village was awarded interest.

Suppose the contract should have been awarded to X but was awarded illegally to Y. Suppose X seeks the profits from Y that Y made on the contract. Although some cases have denied recovery,[121]

[116]*Layne Minnesota Co. v. Town of Stuntz*, 257 N.W.2d 295 (Minn.1977).
[117]268 Ark. 148, 594 S.W.2d 233 (1980).
[118]74 N.J. 113, 376 A.2d 1194 (1977).
[119]*S.T. Grand, Inc. v. City of New York*, 32 N.Y.2d 300, 298 N.E.2d 105, 344 N.Y.S.2d 938 (1973).
[120]22 N.Y.2d 297, 239 N.E.2d 521, 292 N.Y.S.2d 640 (1968).

[121]*Savini Constr. Co. v. Crooks Bros. Constr. Co.*, 540 F.2d 1355 (9th Cir.1974); *Royal Services, Inc. v. Maintenance, Inc.*, 361 F.2d 86 (5th Cir.1966).

a federal court decision applying Iowa law employed unjust enrichment to award a bidder who *should* have been awarded the contract the profit of the contractor who had been awarded the contract improperly.[122]

SECTION 18.05 Subcontractor Bids

The relationship between prime and subcontractor is discussed in greater detail in Section 28.02(B).

However, one aspect of that relationship should be mentioned briefly here. A legal problem that has surfaced frequently relates to the right of a prime contractor to hold a subcontractor to its bid after the former has used that bid in computing its own bid and submitting it to the owner. Although the cases are by no means unanimous,[123] the clear trend is toward holding the subcontractor's bid irrevocable after it has been used by the prime contractor.[124]

---

[122]*Iconco v. Jensen Constr. Co.*, 622 F.2d 1291 (8th Cir.1980), cert. denied 474 U.S. 848 (1985). See also *Tectonics, Inc. of Fla. v. Castle Constr. Co.*, 753 F.2d 957 (11th Cir.1985). See also *Integrity Management Int'l Inc. v. Tombs & Sons, Inc.*, 836 F.2d 485 (10th Cir.1987) (SBA did not preempt state law).

[123]*Home Elec. Co. of Lenoir, Inc. v. Hall & Underdown Heating & Air Conditioning Co.*, 86 N.C.App. 540, 358 S.E.2d 539 (1987), affirmed 322 N.C. 107, 366 S.E.2d 441 (1988), noted in 10 Campbell L.Rev. 293 (1988).
[124]The leading case is *Drennan v. Star Paving Co.*, 51 Cal.2d 409, 333 P.2d 757 (1958). See Section 28.02(B).

PROBLEMS

1. O decided to construct a building and to obtain competitive bids. She selected five contractors and sent them invitations to bid as well as bidding information. The invitation stated that bids would be irrevocable until thirty days after bid opening. It also stated that the bid would be awarded to the lowest responsible bidder. Bids were to be submitted no later than 2:00 p.m. on June 10. Each bidder was to accompany the bid with a certified check or bid bond for 10% of the bid. This amount would be forfeited if the bidder were awarded the contract but did not sign it.

When the bids were opened, they disclosed that four of the bids ranged from one million dollars to $1.4 million. The fifth bid by Acme Construction was for $840,000. The owner asked Acme if there had been some mistake. Acme said that it was certain that this was the correct bid. The owner then told Acme that it had the job and that she did not have to evaluate Acme's responsibility because she knew that Acme had an excellent reputation.

The next day, Acme came in and stated that its bid was underbid by $120,000. There was a computation error of $80,000 caused by Acme's bookkeeper, who had not added the columns correctly. Also, the electrical subcontractor had claimed that it left a 40,000-dollar item out of its bid figures, and it canceled its bid.

a. Would you release Acme?

b. Would you let it "correct" its bid?

c. Would a court permit Acme to cancel? To correct?

d. Can the owner accept the next lowest bid if Acme refuses to perform?

e. If Acme unjustifiably refuses to perform, can the owner retain her damages for the difference in bids, or is she limited to the 10% deposit?

2. O wanted to build a residence on a lot she owned. She asked three contractors to give her a fixed-price bid on the project. None would do so, because experimental methods of construction were called for and because the architect was considered difficult to get along with. O asked C to make a fixed-price bid. She did not tell C that three other contractors had refused to do so. C gave a written bid of $100,000, which O accepted. Three days later, C found out that the three other contractors had refused to bid on the job. It claims O should have told it of this. Does C have a legal defense if it refuses to go through with the contract? Explain.

3. O wanted to build a hospital. It engaged A to prepare the design and assist in obtaining a contractor. A made some informal telephone calls and invited six bidders to submit bids. O wished to avoid using an attorney to prepare the bidding documents. A volunteered to draft the Instruction to Bidders. Among the instructions were the following provisions:

> In awarding or rejecting bids, the owner reserves the following rights:
>
> 1. Identity of successful bidder will not be determined at bid opening. The owner reserves the right to obtain legal advice from its attorney on the legality and sufficiency of all bids.
> 2. The owner at its sole discretion may reject all bids and waive any irregularities in any bid.
> 3. Time is of the essence in this project. In awarding the contract, consideration will be given both to the bid price and to the number of calendar days to achieve substantial completion.
> 4. Providing that the bid does not exceed the amount of funds estimated by the owner available to finance the project, the contract will be awarded to the responsible bidder submitting the lowest, responsive bid and earliest date for substantial completion of the work.

The bids ranged from $4.3 million to $4.8 million, with A's estimate at $5 million. C submitted the lowest bid of $4.3 million. The next lowest bid was $4.45 million. C also specified the shortest period for completion—300 days. Other bidders submitted completion dates of 365, 395, 425, and 441 days. At bid opening, there was a startled reaction from the other bidders both to C's price and, more particularly, to C's time commitment. A thought that C had lowballed the bid. O and A conferred. They were concerned because they had received information that C was having some difficulty on other projects and that C's superintendent had not had a prior hospital construction experience. They also felt that 300 days was an unrealistic time to complete the project.

As a result, A suggested to O that the project be awarded to the next lowest bidder. After its own evaluation, O agreed and sent letters to this effect to C and to the next lowest bidder.

C brought legal action asking that it be awarded the construction contract and, if this is not possible, to be awarded bidding expenses and its lost profits. If O loses, will it have a valid claim against A? (Review Section 12.07.) C also sued A for having advised O not to award the contract to C. (As to this claim, review Section 14.08(F).) Evaluate the validity of C's claim.

Sources of Construction Contract Rights and Duties: Contract Documents and Legal Rules

SECTION 19.01 Contract Documents

An idea of the complexity of construction contract documents can be seen in Figure 19.01a.

A. Bidding Documents

Bidding documents were discussed in Section 18.04(H).

B. Basic Agreement

The basic agreement culminates competitive bidding or negotiation. As an illustration, AIA Doc. A101 set forth in Appendix B identifies the parties and the architect and contains provisions dealing with the work to be performed, time of commencement and completion, the contract sum, and provisions for progress payments and final payment. In essence, the basic agreement sets forth the principal incentives of each contracting party. The owner seeks a project completed on schedule, and the contractor seeks agreed compensation for its performance.

The construction project is a complex undertaking. The basic agreement forms are but one part of the total package of construction documents. See Figure 19.01a. Frequently, the other documents are incorporated by reference in the basic agreement. For example, A101 includes as contract documents the general and supplementary conditions, drawings, specifications, addenda issued prior to execution of the basic agreement, and all modifications issued subsequently. Article 9 provides space to list the documents incorporated by reference.

Frequently, contracts incorporate industry standards. The parties can even incorporate a docu-

ment not yet in existence. For example, in *Randolph Construction Co. v. Kings East Corp.*,[1] the basic agreement incorporated plans that had not yet been completed. But according to the court, if the completed plans are substantially different than anticipated, no contract exists unless the parties agree to the completed plans.

Incorporation by reference—a technique for giving legal effectiveness to writings not physically attached to the contract—should be differentiated from *referring* to other writings. Often reference is made to other writings for informational purposes without any intention of making them part of the contract obligations. Classifying a writing referred to in the contract as simply providing information can be a technique to avoid binding a party to a writing of which it was unaware. Although this can occur when an owner signs an A201, it is more of a problem for subcontractors. Frequently, reference is made in subcontracts to the prime contract, and there can be serious questions as to whether the prime contract provisions are incorporated into the subcontract as well as to reconciling contrary provisions in prime and subcontracts. For that reason, incorporation into subcontracts is discussed ahead.[2]

C. Drawings (Plans)

The drawings graphically depict the contractor's obligations. Together with the other contract doc-

---

[1]165 Conn. 269, 334 A.2d 464 (1973).
[2]See Section 28.04.

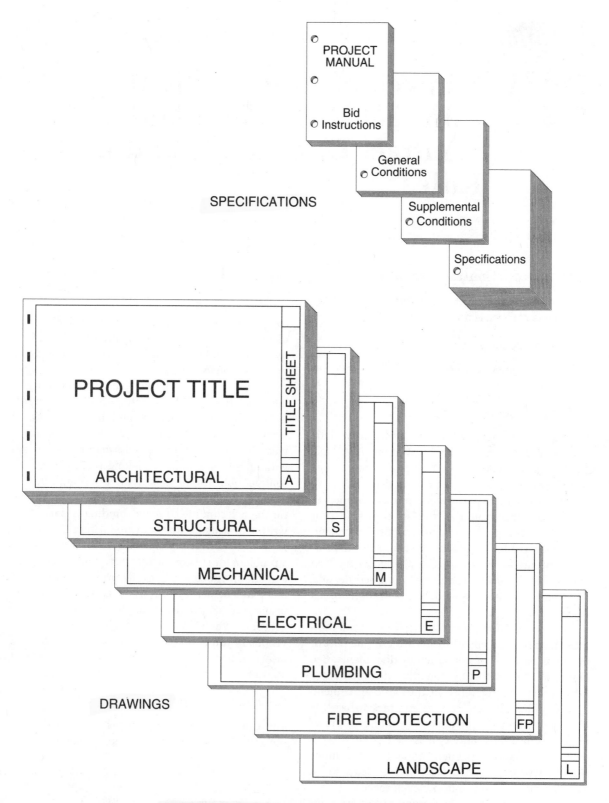

PROJECT MANUAL

Bid Instructions

General Conditions

SPECIFICATIONS

Supplemental Conditions

Specifications

PROJECT TITLE

TITLE SHEET

ARCHITECTURAL   A

STRUCTURAL   S

MECHANICAL   M

ELECTRICAL   E

PLUMBING   P

DRAWINGS

FIRE PROTECTION   FP

LANDSCAPE   L

CONSTRUCTION DOCUMENTS

uments, particularly the specifications, they define and measure the contract obligation.

Drawings are of great importance to the Construction Process. Compliance with them usually relieves the contractor if the project is unsuccessful or is not in accord with the owner's expectations.[3] Incomplete drawings and those that are inconsistent with the specifications almost always generate increased construction costs. Defective drawings or drawings that do not fit with the specifications usually result in liability to the design professional.[4]

A common Construction Process problem is inconsistency between the drawings and specifications (dealt with in Section 20.03).

D. Specifications and *Fruin-Colnon v. Niagara Frontier*: Work Preservation Clauses

Specifications use words to describe the required quantity and quality of the project.[5] They also provide information that will help the contractor plan its price and performance, such as subsurface conditions, site access for heavy equipment, and availability of temporary power. They should be clear and complete and should "fit together." If not, there are likely to be defects, disputes, increased costs, and litigation.

Specifications are classified by type. Most important are *design, performance,* and *purchase description.*

Design Specifications (sometimes called Materials and Methods or Detail Specifications) state precise measurements, tolerances, materials, con-

struction methods, sequences, quality control, inspection requirements, and other information. They tell the contractor *in detail* the material it must furnish and how to perform the work.

Performance Specifications state the performance characteristics required; for example, the pump will deliver fifty units per minute, a heating system will heat to 70°F within a designated time, or a wall will resist flames for a designated period. Design and measurements are not stated or considered important as long as the performance requirements are met.

Under a pure performance specification, the contractor accepts responsibility for design, engineering, and performance requirements, with general discretion as to how to accomplish the goal. Sometimes the contract documents give the contractor suggestions for foundation work. They may be clearly labeled as indicative of "general requirements" or accompanied by a statement that the foundation shall be redesigned to suit subsurface conditions.

Performance specifications are more common in large-scale industrial work where the contractor agrees to design and build a plant that will turn out a designated number of units of a particular quality in a specified period of time. However, performance specifications may also be used in residential or commercial work.

The following case excerpt differentiates performance from design specifications and indicates how a contract should be interpreted to determine whether performance or design specifications have been created. The case involved tunnel construction and a claim by the contractor that it was entitled to additional compensation. The owner asserted that the contractual requirement of watertightness was a performance specification.

---

[3]See Section 24.02.
[4]See Section 14.03.
[5]See Sections 23.05(E) and (F) for more on specifications.

FRUIN-COLNON CORP. ET AL v. NIAGARA FRONTIER TRANSPORTATION AUTHORITY

Supreme Court, Appellate Division, 4th Department, 1992. 180 A.D.2d 222, 585 N.Y.S.2d 248.

[The contractor agreed to construct twin subway tunnels, each approximately two miles long, as a part of the Buffalo Light Rail Rapid Transit System.]

* * *

A performance specification requires a contractor to produce a specific result without specifying the particular method or means of achieving that result. . . . Under a performance specification, only an objective or standard of performance is set forth, and the contractor is free to choose the materials, methods and design necessary to meet the objective or standard of performance. ... Concomitant with control over the choice of design, materials and methods is the corresponding responsibility to ensure that the end product performs as desired. ... In other words, the contractual risk of nonperformance is upon the contractor. ... That is in contrast to a design specification, where the owner specifies the design, materials and methods and impliedly warrants their feasibility and sufficiency.... A contractor must follow a design specification without deviation and bears no responsibility if the design proves inadequate to achieve the intended result. ... In that instance, the contractor's guarantee, even if framed in "absolute" terms, is limited to the quality of the materials and workmanship employed in following the owner's design. ... Whether a provision is a performance specification or a design specification depends upon the language of the contract as a whole. ... Other factors to consider include the nature and degree of the contractor's involvement in the specification process, and the degree to which the contractor is allowed to exercise discretion in carrying out its performance under the contract.

* * *

In arguing that the watertightness requirement was a performance specification that plaintiff assumed the responsibility of meeting, defendant relies primarily on the language of article 3.12 (§ 03300) of the contract, the watertightness clause. Read in isolation, article 3.12 appears to be a performance specification; it specifies the end objective (watertightness) and the standards for measuring that objective, but does not specify the methods of achieving watertightness. Nevertheless, the language and structure of the contract as a whole, as well as the parties' usage and course of performance under the contract, support the conclusion that a design specification was created.

Although the watertightness clause itself does not set forth a particular method for achieving watertightness, the contract as a whole establishes complex and exacting standards for design and construction of the tunnel. Plaintiff was to construct an unreinforced, cast-in-place concrete liner of precise dimension. The type and mix of the concrete was precisely specified, as were detailed requirements for placing, curing, protecting, and finishing the concrete. Plaintiff was given no discretion to deviate from those specifications, whether for the purpose of waterproofing or otherwise. For example, plaintiff had no discretion to install an impermeable outer limit to resist the hydrostatic pressure that both parties knew would exist following completion of construction.

Other provisions of the contract contemplate that waterproofing would be accomplished by means of fissure grouting, which also was to be carried out pursuant to detailed specifications. Additionally, the payment and warranty provisions of the contract support the conclusion that, as a whole, it created a design specification. The contract explicitly provides that "all measures necessary for achieving the degree of watertightness specified in ... article 3.12, including remedial treatments to stem leaks", would be paid for at the contract unit prices. It is unlikely that defendant would have agreed to pay plaintiff on a per unit basis if, as defendant contends, plaintiff had assumed the responsibility of achieving watertightness. Further, unlike the general warranty set forth in the contract, the extended watertightness warranty did not provide that plaintiff would remedy water leaks at its own expense, as it would have if plaintiff had assumed the responsibility of achieving watertightness.

The parties' course of dealing prior to construction also supports the inference that a performance specification was not intended. Bidders had no input into the design of the tunnel, nor did plaintiff exercise any independent design judgment after it was awarded the contract. Defendant relies heavily on plaintiff's March 5, 1980 letter submitted in support of its Value Engineering Change Proposal (VECP), a type of bilateral change order. In that letter, plaintiff asserted, as a reason why it should be permitted to change the design of the tunnel liners to a "full circle pour", that plaintiff bore the risk of meeting the watertightness requirement. In relying on the language of that letter, defendant overlooks the context in which it was sent. If the contract had created a performance specification with respect to watertightness, it would have been unnecessary for plaintiff to obtain defendant's approval for a design change, and it would have been improper for defendant to withhold such approval.

Similarly, the parties' course of dealing during and following construction illustrates that the contract did not establish a performance specification. As evidence that plaintiff undertook the responsibility of waterproofing, defendant cites the fact that plaintiff, on its own initiative, developed and carried out a course of chemical grouting, a method not mentioned in the contract and for which plaintiff did not bill defendant. It is far more significant, however, that defendant routinely denied plaintiff's numerous requests to implement various other waterproofing methods. Even before the concrete liners were put in place, and while the water diversion system was operating, plaintiff repeatedly requested permission to grout the numerous water-bearing fissures that were present in the exposed rock. Those requests consistently were denied based on defendant's erroneous view that fissure grouting was not intended to achieve watertightness. After the tunnel was constructed and the water diversion system turned off, plaintiff unsuccessfully requested permission to fissure grout, plug the deep wells and piezometer testing holes, and construct a permanent dewatering system. Those requests were denied by defendant, which maintained an uncooperative and obstructive attitude. Plaintiff nonetheless proceeded to fissure grout under protest, based on its proper interpretation of the contract. It would have been unnecessary for plaintiff to seek defendant's consent for such measures, and contractually impermissible for defendant to withhold such approval, if plaintiff were in fact responsible for achieving watertightness and had discretion to choose the means to achieve that objective.

* * *

For the foregoing reasons, the court properly awarded judgment to plaintiff on its watertightness claim. Because the contract did not create a performance specification, plaintiff was not contractually responsible for making the tunnels watertight at its own expense. Consequently, plaintiff is entitled to reimbursement for the work performed in an attempt to achieve watertightness.

* * *

Accordingly, the judgment and subsequent order of the court should be affirmed.

Judgment unanimously affirmed without costs.

Purchase Description Specifications designate the product or equipment required by manufacturer, trade name, and type. Sometimes the contractor can select from an approved list. These specifications usually increase contract cost, since they limit the contractor's ability to use materials or equipment that may be just as good as that specified and may cost less. For this reason, public contracts and many private contracts frequently create a method by which the contractor can seek approval to use alternative products or materials.

When such methods are used, a number of legal problems can arise. First, does use of substitute materials or products transfer the risk of design failure?[6] Second, does unreasonable delay in passing on such a request expose the design professional to liability to the owner or contractor?[7] Third, will the contractor have a claim for intentional interference with its contract if the specifications and methods for approving alternates are rigged by the design professional in favor of a particular manufacturer?[8]

The proposal to use an alternate is made most commonly before bid opening but occasionally during performance. When such proposals are made, the design professional should be prompt and fair in passing on whether the alternate product or material is the equal or equivalent of the brand specified. If the design professional has a reputation for intransigence or unreasonably delayed decisions, a contractor is likely to assume that alternates will not be available and bid accordingly. Unreasonable delay in passing on requests for alternates can result in liability to the design professional.[9]

Specifications should state that the determination of whether a proposed alternate is the equivalent of that specified can take into account not only function and performance but also aesthetics, manufacturer's warranty, and the reputation of the manufacturer for servicing the product and supplying spare parts.

Another question that can arise with purchase description specifications, particularly those that do not provide for authorized substitutions, is whether the owner who uses such specifications makes any warranties regarding the commercial availability of the designated materials or equipment. While such a warranty has been found to have been created, as a rule, it is a very limited one. The owner warrants that the supplier is capable but does not warrant its willingness to meet time requirements, nor does the owner warrant that the supplier designated will accede to the

---

[6]See Section 24.03.
[7]*E.C. Ernst, Inc. v. Manhattan Constr. Co. of Texas*, 551 F.2d 1026, 5th Cir: 1977 rehearing denied in part and granted in part, 559 F.2d 268 (5th Cir.1977) cert. denied sub nom *Providence Hospital v. Manhattan Constr. Co. of Texas*, 434 U.S. 1067 (1975) (claim by owner). As to contractors, refer to Section 14.08.
[8]*Waldinger Corp. v. CRS Group Eng'rs Inc.*, 775 F.2d 781 (7th Cir.1985).

[9]*E.C. Ernst, Inc. v. Manhattan Constr. Co. of Texas*, supra note 7.

terms and conditions insisted upon by the contractor. Finally, the warranty does not include an assurance that the supplier will perform. In the case where this limited warranty was recognized, the court simply concluded that it would not change the normal rights and responsibilities attendant upon the use of subcontractors and suppliers.[10]

Specification writing has many legal ramifications. Design professionals can find themselves enmeshed in antitrust litigation if a manufacturer contends that the design professional, through specification writing, participated in a scheme to illegally restrain trade. The design professional's role in the specification drafting process can make the design professional a target for high-pressure tactics or bribery by a manufacturer who wishes to limit competition by the specification process.[11]

Specification writing can play an important part in relations between contractors and any labor union representing employees. Traditionally, most of the labor that goes into a construction project is performed at the site. The relatively high wages paid to unionized craft workers, the frequent disruption caused by labor disputes at the site,[12] and delays caused by weather conditions have encouraged systems under which more work would be performed in factories.

Construction specifications increasingly require integrated units or prefabricated materials designed to reduce the amount of labor performed at the site.[13] But reduction of work performed at the site affects job security of those in the construction craft trades. As a result, many craft unions have won what are called "work preservation" clauses in their collective bargaining agreements with employers under which the employers, mainly subcontractors, agree not to handle products on which labor is performed at the factory when that labor traditionally was performed at the site. Under these work preservation clauses, unions are given the right to strike if there is a violation.

Enterprise Association of Steam, etc., Local Union Number 638 v. NLRB involved the construction of the Norwegian Home for the Aged in New York City. The prime contractor prepared the specifications, which required that climate control units manufactured by a designated supplier be installed as integrated units. These units contained factory-installed internal piping. If the units were installed complete with factory prepiping, the supplier guaranteed all units for a year.

The subcontractor who had agreed to install the heating, ventilating, and air conditioning had had a collective bargaining agreement for many years with the plumbers local that contained a provision requiring the subcontractor employer to preserve certain cutting and threading work for performance at the job site by its own employees. That agreement would have required the internal piping already installed in the climate control units to be cut and threaded at the job site.

When the units arrived at the job site, the union's business agent inspected them and informed both prime contractor and subcontractor that the union employees would not install them. The dispute delayed completion, and the prime contractor filed an unfair labor practice charge with the National Labor Relations Board (NLRB), alleging that the union could not instruct members to refuse to handle the units, as their object would have been to force a "neutral"—the prime contractor—to cease using the supplier's products.

In dealing with work preservation cases, the NLRB and some of the federal circuit courts of appeals had adopted a "right to control" test. If the struck employer (here the subcontractor) had the right to control work assignments, the strike is legal. The pressure must be on the employer in such a case, as it can control work assignment. But if it does not have the right to control, the pressure must be on a "neutral"—the prime contractor. Then the strike is illegal.

---

[10]*Edward M. Crough, Inc. v. Dept. of Gen. Serv. of Dist. of Columbia*, 572 A.2d 457 (D.C.App.1990).
[11]*George R. Whitten, Jr., Inc. v. Paddock Pool Builders, Inc.*, 508 F.2d 547 (1st Cir.1974), involved an antitrust action between two competitors where design professionals were courted with gifts to draft specifications in a certain way.
[12]See Section 23.05(B).
[13]The recent subject matter of work preservation clauses illustrates some of the technologically motivated decisions being made by those drafting specifications. For example, cases have involved precast walls, *NLRB v. Carpenters Dist. Council of Kansas City and Vicinity*, 439 F.2d 225 (8th Cir.1971); a prepiped sink unit, *Associated General Contractors of California, Inc. v. NLRB*, 514 F.2d 433 (9th Cir.1975); prefabricated fireplaces, *Western Monolithics Concrete Products, Inc. v. NLRB*, 446 F.2d 522 (9th Cir.1971); and prepiped heating and ventilating controls discussed in the text of this subsection.

Applying the right to control test to the *Local 638* case, the NLRB concluded that the subcontractor never had the power to assign the disputed piping work to its union employees. This, according to the NLRB, made the subcontractor a neutral. The union's principal target was the prime contractor, inasmuch as the subcontractor had no right to control the work assignment. Pressure was being put on the subcontractor to force the prime contractor to stop buying the units and restore the work. Such pressure was a secondary boycott and an unfair labor practice.

However, the Federal Circuit Court of Appeals for the District of Columbia,[14] following opinions in other federal circuit courts, refused to follow the right to control test and concluded that the subcontractor was not a neutral, inasmuch as the union was simply attempting to enforce its lawful work preservation clause with the employer subcontractor. The court also suggested that the pressure could have been placed on the subcontractor to negotiate a compromise or to terminate the contract with the prime contractor. Because the core of the union's grievances was with the subcontractor, the court held the strike lawful. However, the U.S. Supreme Court reversed the circuit court and concluded that this was an illegal secondary boycott.[15]

Design professionals drafting specifications must take work preservation clauses into account. Craft unions will oppose specifications that reduce their work. Various steps, some legal and some probably illegal, are likely to be taken by the craft union if substantial work traditionally performed by their members will be performed by others. This does not mean that the best and cheapest technology should never be specified. However, in taking into account the advantages of specifications that involve prefabricated products or sealed-at-the-factory components, the disadvantages of potential work stoppages and legal battles must be considered.

E. Conditions: General and Supplementary

Construction is a complex undertaking and a dispute-prone activity. Guidelines are needed to spell out clearly and completely the rights and duties of the parties. For these reasons, most construction documents include general conditions (often supplemented by supplementary conditions) of the contract. They are the ground rules under which the project will be constructed. They are often lengthy and deal with the following subjects:

1. scope of contract documents and resolution of conflicts between them
2. roles and responsibilities of the principal participants in the project
3. subcontractors and separate contractors
4. time
5. payments and completion
6. protection from and risk of loss to persons and property
7. changes
8. corrections
9. termination
10. disputes
11. insurance

A differentiation should be made between a legal condition, general conditions, and supplementary conditions. The first is a legal classification. It is an event that must occur or be excused before an obligation to perform arises. Whether a particular event—sometimes all or part of a promised performance and sometimes an event not within the control of a contracting party—is a condition depends on any contractual language manifesting this conclusion, the probable intentions of the parties, and elements of fairness. However, this is not the sense in which the term *conditions* is used in this subsection.

General conditions are usually expressed in standardized prepared printed contract forms often published by professional associations such as the American Institute of Architects (AIA) or the Engineers Joint Contract Documents Committee (EJCDC). Usually the court seeks to determine how terms should be interpreted by ascertaining the intention of the parties. Because forms prepared by the AIA or EJCDC are in essence prepared by third parties—not the contracting parties—they raise difficult interpretation questions.[16]

---

[14]521 F.2d 885 (D.C.Cir.1975).
[15]*NLRB v. Enterprise Ass'n of Steam, etc., Local Union Number 638*, 429 U.S. 507 (1977).

[16]See Section 20.02.

Sometimes general conditions are prepared by owners who are "repeat players" in the world of construction. These owners may enter into many similar transactions. As shall be seen in Section 20.02, unclear general conditions are likely to be interpreted against the owner that has prepared them.

Although frequently called general conditions, the provisions within it are almost never automatically considered legal conditions, although there is nothing that prevents language within a general condition creating a legal condition. General conditions are prepared in a way that allows them to be used in many types of transactions, either by different contract makers or by a single contract maker.

Yet any individual construction contract may have attributes that make it necessary to have supplementary conditions. For example, the indemnity provisions in general conditions prepared for national use may not be enforceable or desirable in states that have specific statutes regulating indemnification. Similarly, the frequent existence of specialized statutes dealing with arbitration may make it essential to add supplementary conditions if those requirements are to be met. Finally, because of the individualized aspects of insurance, specific insurance requirements will always be found in supplementary conditions. Not all contract makers wish to use general conditions drafted by others, such as the AIA. As a result, these contract makers may use AIA Doc. A201 but either modify it or attach their own supplementary conditions and make those conditions take precedence over the general conditions.

Two further observations must be made. First, some owners or design professionals may lull contractors into a false sense of security by prescribing *general* conditions with which the contractors are familiar and comfortable but may make many changes in the *supplementary* conditions that destroy some of the protection accorded contractors by the general conditions.

Second, the law will frequently be called on to sort out inconsistencies between general conditions and supplementary conditions.[17] Although it is best to delete from general conditions any provisions that are supplemented by the supplementary conditions, often this requires more work than the

attorney wishes to expend. As a result, the attorney may simply state that the supplementary conditions take precedence over the general conditions. This often requires a judge or arbitrator to seek to reconcile apparently conflicting language. This problem is an endemic one to construction with its wealth of contract documents.

The role of the design professional in drafting or suggesting that particular general conditions be used was discussed in Section 12.07.

F. Site Subsurface Test Reports

Frequently, Invitations to Bidders contain subsurface test reports or site data or a reference that such data are available in the office of a geotechnical consultant engaged by the owner. This information and data are of great importance in the construction project. The legal effect of furnishing this information is discussed in Chapter 25.

G. Prior Negotiations and the Parol Evidence Rule

The parol evidence rule relates to the provability of oral agreements made before or at the same time as a written contract. This rule is based on the concept of "completeness of writings." If the written agreement and other written documents incorporated by reference or attached to it are the complete agreement of the parties, prior or contemporary oral agreements are not binding on the parties. Even if they were made, they were integrated, or merged, into the written, complete document.[18]

Most construction contracts are complete. Nevertheless, a possibility exists that oral agreements have been made before or at the same time as the final written agreement. If the written agreement contains an integration or merger clause—one that specifies that the written agreement is the complete and final agreement—it will be difficult for either party to prove a prior oral agreement.[19] It will not be impossible, however, because doctrines that attack the validity of the contract, such as fraud, permit proof of an oral agreement even if there are

---

[17]See Section 20.03(B).

[18]Refer to Sections 11.04(E) and 12.03(C) for further discussion.

[19]*Lower Kuskokwin School Dist. v. Alaska Diversified Contractors, Inc.,* 734 P.2d 62 (Alaska 1987), denied motion for reconsideration, 778 P.2d 581 (Alaska 1989), cert. denied, 493 U.S. 1022 (1990).

provisions specifying that the written agreement is the complete document. If no contract provision deals with the question of completeness, testimony claiming an oral agreement generally will be admitted.

It is important to incorporate the entire agreement into the writing. Even if the writing is prepared and ready for execution, the design professional, and certainly the owner's attorney, should insist on incorporating any changes or additions into the writing itself. The entire document need not be retyped if time does not permit (word processing has made this easy). If the oral agreement is not included in the writing, a substantial risk exists that it cannot be proved.

Suppose there is a dispute over meaning of terms used in the writing. If the words chosen by the parties are ambiguous, the court can look at the surrounding facts and circumstances to determine how the parties used those particular words. These circumstances include the setting of the transaction, the contracting parties' objectives in making the contract, and any conversations they may have had. Even if a writing is considered complete, antecedent or subsequent conversations may be admissible to interpret the writing.

Two cases are instructive, both involving the parol evidence rule in the design and construction context. The first, *Godfrey, Bassett & Kuykendall, Architects, Ltd. v. Huntington Lumber & Supply Co., Inc.,*[20] was discussed in Section 11.04(E). The case noted the closeness of the relationship between the contracting parties as one basis for determining whether one party could claim that there had been an antecedent agreement not expressed in the writing, despite a failure by that party to have read the agreement. In the case, the court looked carefully at the background and relationship between the parties to determine whether failure to include an asserted promise precluded the party from testifying as to its existence.

Before discussing the second case, some background is in order. There are many expectations to the parol evidence rule. Many scholars and important judges have expressed hostility toward the rule's continued existence. This has led some to believe that in no case will a party be prevented from

testifying of an asserted antecedent oral agreement despite its not having been included in the writing.

These beliefs would certainly have been supported by cases that were decided in California in the 1960s.[21] However, *Brinderson-Newberg Joint Venture v. Pacific Erectors, Inc.*[22]—a construction contract dispute decided in 1992 by the Ninth Federal Circuit Court applying California law—shows that there is still vitality in the parol evidence rule.

The case involved a contract to build a coal-fired power plant. Brinderson entered into negotiations with Pacific, one of its prospective subcontractors, who submitted the low bid for certain erection work. Negotiations took place between Brinderson and Pacific on three occasions over what work was included in Pacific's bid. Brinderson later claimed that Pacific agreed to erect all of the flue gas system (FGS), while Pacific contended that it would erect only "pick and set" of the FGS components.

After one of the preliminary meetings, Brinderson drafted a contract and sent it to Pacific for review. At a subsequent meeting, representatives of the parties "reviewed the contract line by line and negotiated a number of changes or clarifications concerning Pacific's scope of work."[23] No change was made in Brinderson's proposal. Brinderson stated that it expected Pacific to erect all the FGS components, while Pacific contended that no change was made, because it relied on Brinderson's assurance that the language only required picks and sets be erected and the scope of work would be limited to jobs Pacific customarily performed. Pacific also contended that Brinderson claimed it did not want to take the time to write the pick-and-set agreement into the contract. The contract contained a clause stating that the contract was a completely integrated agreement.

The parties admitted that the contract was completely integrated, but Pacific contended that evidence of the negotiations and Brinderson's alleged promise should be used to interpret the contract. However, the court did not wish to eviscerate the

[20]584 So.2d 1254 (Miss.1991).

[21]*Masterson v. Sine,* 68 Cal.2d 222, 436 P.2d 561, 65 Cal.Rptr. 545 (1968); *Pacific Gas & Elec. Co. v. G.W. Thomas Drayage & Rigging Co.,* 69 Cal.2d 33, 442 P.2d 641, 69 Cal.Rptr. 561 (1968).
[22]971 F.2d 272 (9th Cir.1992), cert. denied, 113 S.Ct. 1267 (1993).
[23]Id. at 275.

parol evidence rule. It required that there be reasonable harmony between the parol testimony offered into evidence and the integrated contract. The court concluded that the testimony would be permitted only if the written contract is susceptible to the meaning that Pacific gave it. The court concluded that the language of the contract was not reasonably susceptible to the proposed interpretation put forth by Pacific. It concluded by stating that if such an oral agreement had been made, Pacific did so at its own peril, because the completely integrated contract negates any such agreement.

When the *Godfrey* case and the *Brinderson* case are compared, it can be seen that the surrounding facts and circumstances and closeness of the relationship between the parties can have a significant effect on whether the parol evidence is admitted.

H. Modifications

Modifications should be differentiated from changes. Changes occur frequently during a construction project and are discussed in detail in Chapter 21. They are generally governed by carefully drawn and complete provisions under which the owner has the right to order changes and the contractor must perform them, with appropriate contract language dealing with compensation for deletions or additions caused by the changes. A modification is a change agreed upon by both parties in the basic obligation not based on any contractual provision giving the owner right to order changes.

In addition to agreement, a valid modification of a service contract requires consideration. To demonstrate, suppose there is a construction project for $100,000. The obligations of the contractor are expressed in the contract documents. Suppose that during the term of the agreement the parties mutually agree that the contract price will be increased to $110,000. The owner must receive something in exchange for the additional compensation. If, in exchange for this, the contractor agrees to an increase in the quality or quantity of the building or to shorten the time for completion, there is an appropriate exchange and the modification agreement is binding. Difficulties arise when the modification agreement encompasses an increase in price without any change in the contractor's obligation. Such an agreement can be invalidated by the preexisting duty rule. Unless one of the many exceptions applies, the agreement is not binding if

the party—in this case the contractor—is obligating itself to do no more than it was previously obligated to perform under the original contract.[24]

This preexisting duty rule is criticized. It limits the autonomy of the parties by denying enforceability of agreements voluntarily made. Implicit in the rule is an assumption that an increased price for the same amount of work is likely to be the result of expressed or implied coercion on the part of the contractor, as if the contractor is saying, "Pay me more money or I will quit and you will have to go to court to get damages." However, suppose the parties have voluntarily arrived at a modification of this type. There is no reason for not giving effect to their agreement. There are a number of exceptions that can relieve against the sometimes harsh effect of the preexisting duty rule.

In the construction contract, minor changes in the contractor's obligation have been held sufficient to avoid the rule even where the increase in price was not commensurate with the change in obligation on the part of the contractor. This approach may permit a contract modification to be enforced if the parties have had enough foresight or legal knowledge to provide for some minor and relatively insignificant change in the contractor's obligation as a means of enforcing the increased price.

An exception that developed in construction contracts enforces the modification if it is fair and equitable in view of the circumstances not anticipated by the parties when the contract was made.[25] As this exception (frequently referred to as the unforeseen circumstances exception) has occurred mainly in the area of subsurface conditions, it is discussed in greater detail in Section 25.01(D).

Another indication of dissatisfaction with the requirement of consideration for a modification is manifested by Sec. 2-209 of the Uniform Commercial Code dealing with goods transactions. The section states that modifications are valid without consideration. However, a modification obtained in bad faith is not valid.

---

[24]*Crookham & Vessels, Inc. v. Larry Moyer Trucking, Inc.*, 699 S.W.2d 414 (Ark.App.1985); *Hiers-Wright Assoc., Inc. v. Manufacturer's Hanover Mortgage Co.*, 182 Ga.App. 732, 356 S.E.2d 903 (1987).
[25]*Linz v. Schuck*, 106 Md. 220, 67 A. 286 (1907) followed in the Restatement (Second) of Contracts § 89 (1981).

Curiously, despite the movement toward aboli-tion of the consideration requirement for a modi-fication, the rule is applied vigorously in federal public contracts. Unless the government receives something for an increase in price or a reduced price for a deletion, the modification is not valid. This rigid adherence to the consideration require-ment may be justified as a means of ensuring that there be no gifts of public funds and preventing corruption and collusion between contractor and public official. Giving some advantage to a con-tractor without the government's getting anything in return may be unfair to the other bidders.

Contracts frequently state that modifications must be in writing. Such provisions generally were not enforced by the common law.[26] In enforcing oral modifications despite such clauses, the courts have stated that by making a subsequent oral mod-ification, the parties have changed the agreement requiring that the modification be in writing. How-ever, under Sec. 2-209 of the Uniform Commercial Code, oral agreements modifying the written agreement are not effective if the written agree-ment contains a provision requiring that modifi-cations be in writing. There are some exceptions to this, but the Code expresses a policy that such con-tractual provisions requiring a writing as a condi-tion to enforcement of an asserted modification agreement should be given more effect than the courts have given them in the past. It remains to be seen whether this change in the law relating to the sale of goods will have any impact on judicial thinking in other types of contracts.

SECTION 19.02 Judicially Determined Terms

A. Necessity to Imply Terms

When making a contract, parties frequently do not consider all problems that may arise. Some matters may have been considered, but the parties believed the resolution of the matters to be so obvious that contract coverage would be unnecessary. Matters have been discussed by the parties during negoti-ations, but the parties could not agree on a contract solution to the problem. Yet these parties may in-tend to have a binding contract despite their in-ability to resolve all problems during negotiation. In such cases, the parties may state in the contract that they will agree in the future on certain less important contract matters or omit the matter from the contract entirely.

Courts may be asked to fill in the gaps not cov-ered by the contract or decide matters left for fu-ture agreement where the parties cannot agree. Courts will be more likely to perform these func-tions if convinced that the parties intended to make an enforceable agreement, especially where per-formance has commenced.

Until recently, common law judges hesitated to imply terms. These judges saw themselves as en-forcing contracts made by the parties rather than making contracts for the parties.

One court stated:

1. The implication must arise from the language used or it must be indispensable to effectuate the in-tention of the parties.

2. It must appear from the language used that it was so clearly within the contemplation of the parties that they deemed it unnecessary to express it.

3. Implied covenants can be justified only on the grounds of legal necessity.[27]

4. A promise can be implied only where it can be rightfully assumed that it would have been made if at-tention had been called to it.

5. There can be no implied covenant where the sub-ject is completely covered by the contract.[28]

These requirements reflect judicial reluctance to imply terms. In addition, courts during this period hesitated to provide a term where the parties had agreed to agree but did not.[29]

Modern judges seem less insecure and more re-alistic about their role in contract disputes. They

---

[26]*Prince v. R.C. Tolman Constr. Co., Inc.,* 610 P.2d 1267 (Utah 1980); *United States v. Klefstad Eng'g Co.,* 324 F.Supp. 972 (W.D.Pa.1971). A similar provision in an AIA document was not binding in *Fishel & Taylor v. Grifton United Methodist Church,* 9 N.C.App. 224, 175 S.E.2d 785 (1970). Four states, most notably California, require that modifications of written contracts must be in writing. There are many exceptions.

[27]This means that courts will imply a promise if it is nec-essary in order to have a valid contract and the parties have so intended.
[28]*Stockton Dry Goods Co. v. Girsh,* 36 Cal.2d 677, 681, 227 P.2d 1, 3–4 (1951).
[29]Refer to Section 5.06(F).

are beginning to recognize that they "make" contracts for the parties. Their recognition of the proliferation of standard form contracts often made in an adhesion context makes them more willing to imply terms to redress unequal bargaining than were courts a quarter century ago.

The law should exercise restraint in implying terms where the contract has been negotiated. In such cases, it is likely that most of the major problems were considered and the absence of a promise may be deliberate. Implying a term in such cases could frustrate the bargain.

Even in negotiated contracts it may be necessary to imply terms that were so obvious that the parties did not think it necessary to express them. A court should "complete" the deal by filling in gaps where parties intended to make a contract and have left minor terms for future agreement but have not been able to agree.

Again, a word of caution. The existence of express contract terms generally precludes terms on that subject being implied.[30] Also, custom takes precedence over implied terms. First, the contract must be examined. If the contract deals with subject matter that relates to the proposed implication, a court is less likely to imply terms. However, sometimes the effect of implication is achieved by interpreting particular contract terms in a way consistent with the implication.

One court discussing the owner's duty to furnish a site stated:

> Each party to a contract is under an implied obligation to restrain from doing any act that would delay or prevent the other party's performance of the contract. . . . A party who is engaged to do work has a right to proceed free of let or hindrance of the other party, and if such other party interferes, hinders or prevents the doing of the work to such an extent as to render the performance difficult and largely diminish the profits, the first may treat the contract as broken and is not bound to proceed under the added burdens and increased expense.[31]

The court here speaks of the conduct of the owner that may delay or prevent the contractor from performing the work. Implied obligations can go further and, under certain circumstances, require that the owner perform positive acts to assist the contractor in performance. However, the law will be more reluctant to impliedly require affirmative acts of cooperation than to preclude negative acts of hindrance or prevention. Positive acts are more likely to have been thought about if they were important, and failure to express them in the contract may indicate that neither expected them to be done. For example, *C.A. Davis, Inc. v. City of Miami*[32] involved a claim by a defaulting contractor that the city had failed to cooperate and thereby hindered the contractor's performance. Evidently the contractor contended that the city had a duty to secure utility company cooperation with regard to underground obstructions encountered on the project. The trial court judge refused the contractor's request that the jury be instructed that the city had a duty to cooperate. The appellate court concluded that the trial judge had been correct in refusing to give this instruction, as there had been no evidence that the city had failed to cooperate. The court stated:

> The evidence before the court demonstrated that the City repeatedly and routinely cooperated in an effort to keep the project going. In fact, the City had no implied duty to do [the contractor's] work, but only a duty not to hinder or impede that work.[33]

This language indicates that implied terms of the type under discussion must often be correlated with contract interpretation. Had the contractor been obligated to secure utility company cooperation, it would have been difficult to imply a term obligating the city to do so.

Any duty to cooperate should not require undertaking heavy burdens of cooperation that would very likely frustrate the contractual allocations of responsibility made by the parties. However, if one party can assist the other party's performance at a minimal cost, such cooperation should be required.

Some specific illustrations of implied terms in the construction contract have been mentioned ear-

---

[30]*Weber v. Milpitas County Water Dist.*, 201 Cal.App.2d 666, 20 Cal.Rptr. 45 (1962).

[31]*United States v. Guy H. James Constr. Co.*, 390 F.Supp. 1193, 1206 (M.D.Tenn.1972), affirmed without opinion, 489 F.2d 756 (6th Cir.1974); *Lewis-Nicholson, Inc. v. United States*, 550 F.2d 26 (Ct.Cl.1977). As to subcontracts, see *Elte, Inc. v. S.S. Mullen, Inc.*, 469 F.2d 1127, 1132–1133 (9th Cir.1972).

[32]400 So.2d 536 (Fla.Dist.Ct.App.1981).
[33]Id. at 539.

lier, such as the standard of the design profession-al's performance[34] and the obligation of the owner to coordinate work of separate contractors or stand behind the obligation of another entity given that responsibility.[35]

There are other illustrations. The owner impliedly promises that the site will be ready for the contractor to commence performance.[36] (Often, as seen in Section 26.02, this is dealt with by not starting the obligation to commence performance until a notice to proceed has been given.) The owner impliedly promises to obtain the necessary easements or rights to enter the project site or land of another necessary for the contractor's performance.

The prime contractor impliedly promises the owner to use proper workmanship and materials and to complete the project free and clear of liens.[37] A contractor in a cost contract promises to inform the owner of prospective overruns of cost estimates.[38] One court held that the prime contractor promised the subcontractors that the work would be coordinated and that the prime contractor would process subcontractor claims to the owner.[39] But another court refused to imply a promise that the prime contractor would create and maintain an efficient schedule when this was neither customary nor bargained for.[40]

The owner impliedly promises that it and its designated representative, usually the design professional, will perform in such a way as to reasonably expedite the contractor's performance. For example, the owner impliedly promises that the design professional will give contract interpretations and pass on sufficiency of shop drawings within a reasonable time.[41]

The law usually implies a promise by the owner to supply adequate drawings and specifications (discussed in greater detail in Section 23.05(E)).

However, that such terms will be implied does not eliminate the necessity of focusing more closely on the nature of the implication. For example, suppose the owner designates specified material. That will very likely mean that if the contractor uses that material, the contractor will not be responsible if the material proves to be unsuitable.[42] However, does specification of a material imply that the material is available? Does it imply that the material is stocked by local suppliers? Such questions expose the basic function of implying terms—that of allocating risk and responsibility.

A federal appeals board held that the owner does not warrant that a product specified will be in stock inasmuch as parties do not usually guarantee the performance of third parties.[43] Such a result may also be based on the likelihood that owners know no more about the stock of suppliers than does the contractor.

Usually the owner reserves certain powers during the contractor's performance to protect its interests. This should not be automatically converted into an affirmative duty for the benefit of the contractor or subcontractors. For example, it has been held that the power to monitor contractor performance and stop the work if necessary did not imply that the owner had *promised* to direct the contractor's workers.[44]

Courts are sometimes asked to imply completion times to construction contracts. A specified time for completion is usually given in the contract. If there is no express provision dealing with this question, courts hold that the contractor must complete performance within a reasonable time. Reasonableness will be determined in the light of all the surrounding facts and circumstances.

Courts should not imply a reasonable time for performance if performance has not yet commenced and if the failure to agree on a time for performance indicates that the parties have not yet intended to conclude a contract. The absence of agreement on such an important question may mean that the parties are still in a bargaining stage.

---

[34]Refer to Section 14.05(A).

[35]Refer to Section 17.04(C).

[36]*North Harris County Junior College Dist. v. Fleetwood Constr. Co.*, 604 S.W.2d 247 (Tex.Ct.App.1980).

[37]But see AIA Doc. A201, ¶ 9.3.3 (work previously paid for free of liens as far as contractor knows).

[38]*Jones v. J.H. Hiser Constr. Co., Inc.*, 60 Md.App. 671, 484 A.2d 302 (1984).

[39]*Citizens Nat'l Bank of Orlando v. Vitt*, 367 F.2d 541 (5th Cir.1966).

[40]*Drew Brown Ltd. v. Joseph Rugo, Inc.*, 436 F.2d 632 (1st Cir.1971).

[41]See AIA Doc. A201, ¶¶ 4.2.7, 4.2.11.

[42]See Section 24.02.

[43]*James Walford Constr. Co.*, GSBCA 6498, 83-1 BCA ¶ 16,277. See *Edward M. Crough Inc. v. Dept. of Gen. Serv. of Dist. of Columbia*, supra note 10 (discussed in Section 19.01(D)).

[44]*Nat Harrison Assoc., Inc. v. Louisville Gas & Elec. Co.*, 512 F.2d 511 (6th Cir.1975).

B. Custom

Customary practices are important in determining rights and duties of contracting parties, particularly in complex transactions such as construction, where not everything can be stipulated in the contract. Mention has been made in Section 12.11 of the frequent claim by design professionals that customarily they retain ownership of drawings and specifications. Custom plays other significant roles in construction. In placing such a heavy emphasis on customary practices when *interpreting* contract terms, courts often state that parties contract with reference to existing customs. In that sense, courts can be said to be simply giving effect to the actual intention of the parties. However, a contracting party may be held to those customs that it knew or should have known. The law places a burden on contracting parties to learn those customs that apply in the type of transaction in which they are about to enter and in the place where they contract.

In any event, an established custom can be a more convenient and proper method of filling gaps than a court determination of what is reasonable. For example, if customarily the contractor obtains a building permit, the court is likely to place this responsibility on the contractor, but only where there is a contract gap on this point.[45] If the contract does not deal with the matter, custom would be preferable to a judge's determination of who should obtain the permit.

Custom can assist the court when the terms must be interpreted or conflicts exist in the contract documents. For example, in *Fifteenth Avenue Christian Church v. Moline Heating & Construction Co.*,[46] the owner brought a legal claim against the heating contractor claiming that the contractor was obligated to install a "one hour fire resistive" ceiling in the boiler room of the church because provisions of the local building code required it. The specifications made no reference to the walls or ceiling of the boiler room. After completion and final payment, the city fire marshal called the owner's attention to the fact that the ceiling did not meet code

requirements. About a year later, the owner demanded that the ceiling be replaced, but the heating contractor refused.

The general conditions required the contractor to comply with all local laws and to notify the engineer if the specifications were at variance with local laws. Technical provisions included in the contract documents also gave precedence to code regulations over specifications.

Witnesses called by the contractor stated that provisions and specifications requiring that the contractor comply with all local laws were standard provisions and referred to the work *specifically* included in the contract. They testified that if the plans and specifications did not specifically provide for the installation of a particular type of ceiling in the boiler room, the *heating* contractor would not be obligated to install such a ceiling. They further testified that the local codes referred to in the specification related to the code rules relating to the type of work to be done in the specifications and that the installation of ceilings in a boiler room fell within the general category of work done by the general construction trades and not the plumbing and heating trades.

The appellate court held that it was proper to receive evidence of custom to interpret the language in the contract as long as the custom is known or should have been known to the parties and there is no specific stipulation in the contract to the contrary. In affirming the decision for the contractor, the court noted that the engineers who represented the owner were professionals and must be presumed to know the customs and usages of the trade and that this knowledge was imputed to the owner.

The *Fifteenth Avenue* decision is instructive. Custom *is* important.[47] Also, the owner will be held responsible for what the design professional knew or should have known, which can ultimately mean enlarged responsibility for the design professional. Blanket recitals that the work will be in compliance with local codes should not be taken to enlarge the contractor's obligation when it has bid on specifications that did not include disputed work and

---

[45]In *Weber v. Milpitas County Water Dist.*, supra note 30, custom that the *owner* procured these permits was not relevant because there was an *express* provision in the contract requiring the *contractor* to obtain such permits.
[46]131 Ill.App.2d 766, 265 N.E.2d 405 (1970).

[47]In *Chicago Bridge & Iron Co. v. Reliance Ins. Co.*, 46 Ill.2d 522, 264 N.E.2d 134 (1970), the court admitted custom that subcontractors execute lien waivers despite not having received payment.

when that disputed work is customarily performed by a different contractor.

Custom is an important part of construction contracts. However, parties should never rely on custom when there are express provisions to the contrary. Custom can be a trap for a contractor who performs a project in a locality where customs may be different from those where it customarily works. Custom can often be difficult to establish. In the absence of an express provision controlling the dispute, the party seeking to establish custom will be given an opportunity to do so.

C. Building Laws

Building codes and land use controls play a pervasive role in construction. Frequently, it is implied or expressed in the contract that the contractor will comply with applicable laws. Further analysis is needed to separate several related but distinct problems.

Where design and construction are separate, the former is the responsibility of the owner and is usually accomplished by the design professional. Suppose the design violates building code requirements. The contractor may contend that its job is to build the design, but the contractor cannot ignore violations of law. It should direct the design professional's attention to any *obvious* code violations in the design.

Sometimes a building code prescribes that installation must be made in a particular way. In *Aldridge v. Valley Steel Construction, Inc.*,[48] the plaintiff was injured when electrical machinery installed by an independent contractor exploded. The independent contractor followed the plans provided for installation of the equipment. However, the Alabama Supreme Court pointed to an affidavit submitted by an expert for the plaintiff. The expert was willing to testify that the installation violated the National Electric Code and that a prudent contractor would have known that installation in accordance with the plans created a dangerous situation that might result in an injury-causing accident. This meant that there was a genuine issue of fact precluding the entry of a summary judgment (without a trial) in favor of the contractor. (In many construction contracts, the plans do not prescribe particular means and methods of executing the design. Instead, the contract is likely to require the contractor to comply with all building laws.)

The contractor's responsibility in design code violations is dealt with by AIA Doc. A201, ¶ 3.7.3, which states that it is not the contractor's responsibility to make certain that the design accords with legal building requirements but that *if* the contractor observes variances, it must notify the architect in writing. Under ¶ 3.7.4, the contractor's performance of work *knowing* it to be contrary to such laws without notifying the architect places full responsibility on the contractor.[49]

This solution effectuates a sound middle ground between putting an unreasonable responsibility on the contractor and allowing the contractor to close its eyes in the face of danger. A contractor who does not comply with the requirements of ¶ 3.7.3 is exposed to liability, not only for the cost of correction and consequential damages that the owner may suffer but also for losses suffered by third parties.

As noted, the contractor is not usually responsible for design when that design is furnished to it by the owner. But suppose the contractor both designs and builds. In *Quedding v. Arisumi Brothers, Inc.*,[50] the contractor supplied the design, which clearly specified only *vertical* rebars. Local building codes required *both vertical* and *horizontal* rebars. The court held that the existing law was part of the contract unless stipulated to the contrary. As a result, the contractor breached when it violated the code even if it followed the specifications. Here the court properly protected the reasonable expectations of the owner that the contractor would build in accordance with the local building laws. The owner would not be expected in such a transaction to know that the design violated the building codes.

In this case, there was no provision requiring the contractor to comply with local building law, yet the court implied this obligation. Had the contract provided that the contractor would comply with local building codes, there would have been a conflict between that provision and the specifications. Although it is usually held that the specific provi-

---

[48]603 So.2d 981 (Ala.1992).

[49]Refer to Section 14.05(D), dealing with design that violates local building codes.
[50]661 P.2d 706 (Hawaii 1983).

sion in a contract takes precedence over a general provision, in such a case, an unsophisticated owner unaware of the conflict should have its expectations protected by giving preference to the provision requiring that the contractor comply with building laws.

Yet in many instances, no *direct* clash exists between building codes and the design. In such instances, the owner and the design professional expect the contractor to follow building codes. It is in this sense that the contractor must build in conformity with legal requirements. Essentially, this is a gap-filling instrumentality. Additionally, such code compliance provisions are intended to obtain compliance with worker safety rules.

Other legal requirements can consist of permits issued by public officials at various stages of the performance. These permits are not limited to building and occupancy permits. Frequently, permits must be obtained from public utilities to connect the utilities of the project to public utility lines. Drainage rights of way may be needed as well as permits from state highway officials for certain types of construction.

Who must obtain permits required by law? Construction documents should cover these matters.[51] If the contract does not, custom may allocate responsibility for obtaining permits. In the absence of an express contract provision or accepted custom, the law will be likely to imply that the owner will obtain the more important permanent permits, such as land use control permits,[52] but the contractor should get operational permits, such as building permits[53] and occupancy permits. The contractor is likely to be required, by implication of law, to obtain permits from public officials that are associated with facilities and equipment. The contractor should obtain utility hookup permits and permits for temporary construction. The deter-

mination of who should obtain particular construction permits will depend on the extent of experience the particular owner has had in construction projects.

Suppose the contractor is required to obtain certain permits but does not. In *Lew Bonn co. v. Herman*,[54] the electrical subcontractor sued the prime contractor for work that had been performed. The prime contractor contended that the contract was unenforceable because the plaintiff subcontractor had failed to file a copy of the plans with the city building inspector as required by law. The court held that this slight violation of the statute would not make the contract unenforceable and allowed the subcontractor to recover for his work. The court emphasized that the public was not placed in jeopardy, and the subcontractor's failure should not entitle the prime contractor to a windfall. Generally, the subcontractor's failure to comply with any requirement to obtain a work permit or submit the plan does not affect its right to recover compensation.[55] However, it is dangerous to rely on a court's subsequently determining that the violation was technical.

D. Good Faith and Fair Dealing

The common law did not generally hold contracting parties to obligations of good faith and fair dealing. Refusal to do so reflected the common law's belief in the sanctity of the written contract, the notion that contracting parties should take care of themselves, and the imprecision of good faith.

Beginning in the mid-twentieth century, however, and heavily influenced by the Uniform Commercial Code,[56] American contract law began to hold contracting parties to the covenant of good faith and fair dealing. One party cannot be expected to guarantee that the other party will receive the benefit it expects from a contract. Nor must a contracting party make unreasonable sacrifices for the other party. However, each party should not only avoid deliberate and willful frustration of the other party's expectations but also extend a helping hand where to do so would not be unreasonably burdensome. Contracting parties, though not partners in a legal sense, must recognize the interdependence of contractual relationships.

---

[51]AIA Doc. A201, ¶ 3.7.1 (contractor obtains building permit).
[52]*COAC, Inc. v. Kennedy Eng'rs*, 67 Cal.App.3d 916, 136 Cal.Rptr. 890 (1977) (Environmental Impact Report).
[53]In *Drost v. Professional Bldg. Service Corp.*, 153 Ind.App. 273, 286 N.E.2d 846 (1972), the contract documents required the contractor to furnish "all necessary . . . drawings for construction of this facility as described for the use in obtaining the necessary building permits." The court held that this language placed a duty on the contractor to obtain these permits.

[54]271 Minn. 105, 135 N.W.2d 222 (1965).
[55]See Annot., 26 A.L.R.3d 1395 (1969).
[56]U.C.C. § 1-203.

Some applications of the doctrine are simply recognition of implied terms discussed in (A). But more expansive use of the doctrine emphasizes the unspoken objectives of the contracting parties, the spirit of the contract itself, and the need for elemental fairness. This more expansive use can even dictate an outcome contrary to the literal interpretation of the written contract. Finally, some breaches of this covenant, particularly in contracts of insurance and employment, violate public policy. When they do, they can be the basis for tort remedies, including punitive damages.

Because of the pervasiveness of the doctrine and of the particular interdependence of the Construction Process, illustrations of the covenant of good faith and fair dealing are sprinkled throughout this treatise.[57] Yet it may be useful to note some illustrative cases here.

In *EKE Builders, Inc. v. Quail Bluff Association*,[58] the owner engaged the contractor to construct an apartment complex. The agreement was conditioned on the owner's obtaining financing. The lender advised the owner that the plaintiff was unacceptable as a contractor unless it obtained performance and payment bonds from a bonding company approved by the lender. However, the owner did not communicate the lender's conditions but told the plaintiff that the lender had refused to approve him under *any* circumstances. The owner terminated the contract. Yet that same day, the owner entered into a construction agreement with a second contractor for $113,000 less than the original contract price.

The court held that the owner not only breached its agreement with the contractor but also committed the tort of deceit, which under certain circumstances can be the basis for an award of punitive damages. Most important, the court held that the owner's power to terminate if financing was not obtained would not be interpreted as permitting arbitrary or unreasonable conduct that would prejudice the contractor's reasonable expectations. Also, the court held that the owner had an obligation to act in good faith to obtain financing. The owner must deal fairly with the contractor and not use a subterfuge to deny him the benefit of its contract.

During the Construction Process, the owner makes progress payments based on certificates issued by the design professional. To protect each party from abuse of power, the design professional must exercise his power in good faith.[59] Similarly, a contract clause empowering a dissatisfied prime contractor to correct defective work at the subcontractor's expense must also be exercised in good faith.[60]

Maier's Trucking Co. v. United Construction Co. illustrates good faith and fair dealing in the context of a deductive change.[61] The prime contractor suggested a modification that would essentially eliminate a subcontractor's work. Although the majority held that this did not breach the covenant of good faith and fair dealing, two judges strongly dissented. They would have ordered a trial to determine whether this suggestion had been made in bad faith.

Another element of good faith and fair dealing can be seen in the need for communication. For example, AIA Doc. A201, ¶¶ 9.4.1 and 9.5.1, require that an architect who cannot certify the full amount requested notify the owner and contractor and give his reasons. Similarly, AIA Doc. A312, ¶ 3.1, states that the surety's obligation arises after the owner has notified the contractor and surety that it considers defaulting the contractor and "has requested . . . a conference with the contractor and surety . . . to discuss methods of performing the Construction Contract."

Both elements illustrate *specific* requirements of obligations that would very likely be encompassed into the covenant of good faith and fair dealing.

The settlement process and the pressures attendant to it are fertile ground for claims of bad faith. One case employed the concept of duress to void a settlement.[62] Another held that a prime contractor who was empowered by contract to negotiate the settlement of a delay claim on behalf of its sub-

---

[57]See Sections 14.04(B), 26.03(B) and *Broadway Maintenance Corp. v. Rutgers*, reproduced at Section 17.04(C).
[58]714 P.2d 604 (Okla.App.1985).

[59]*City of Mound Bayou v. Roy Collins Constr. Co.*, 499 So.2d 1354 (Miss.1986).
[60]*United States v. Mountain States Constr. Co.*, 588 F.2d 259 (9th Cir.1978) (not a clear holding).
[61]237 Kan. 692, 704 P.2d 2 (1985).
[62]*Rich & Whillock, Inc. v. Ashton Devt., Inc.*, 157 Cal.App.3d 1154, 204 Cal.Rptr. 86 (1984), reproduced in Section 27.13.

contractors was required to do so in good faith.[63] Still another case held that a termination by the owner in accordance with the powers given it under the contract was improper, since it had not been made in good faith.[64]

Even this sampling shows the range of breaches. Some are considered tortious, some illustrate that contract language does not necessarily insulate a party from its obligation to act in good faith and to deal fairly, and some are based on implied terms. The outpouring of cases dealing with this doctrine demonstrates that the doctrine will be an important component of the construction contract obligation.[65]

E. Unconscionability

The common law generally allowed contracting parties to decide the terms of their contracts with relatively minimal judicial intervention. But again, as noted in (D), the Uniform Commercial Code regulating transactions in goods led to the broadening of the unconscionability doctrine that had been used in the courts of equity.[66] The obligation of good faith and fair dealing placed limits on the conduct of contracting parties. The unconscionability doctrine authorizes an inquiry into the circumstances under which the contract was made and permits a judge to refuse to enforce unconscionable clauses or contracts.

The doctrine has rarely been applied in construction cases.[67] Clauses where it might have some utility, such as those of exculpation or indemnification, have been regulated by requiring a high degree of specificity or by statute.[68] Yet the unconscionability doctrine is likely to be increasingly invoked in the future.

---

[63]*T.G.I. East Coast Constr. Co. v. Fireman's Fund Ins. Co.*, 534 F.Supp. 780 (S.D.N.Y.1982).
[64]*Paul Hardeman, Inc. v. Arkansas Power & Light Co.*, 380 F.Supp. 298 (E.D.Ark.1974). Cf. *Darwin Constr. Co., Inc. v. United States*, 811 F.2d 593 (Fed.Cir.1987), cert. denied 484 U.S. 1008 (1988).
[65]See Section 18.04(E) for another illustration—that of requiring an owner to notify a bidder if it appears a bidding mistake has been made.

[66]U.C.C. § 2-302.
[67]*Arcwel Marine, Inc. v. Southwest Marine, Inc.*, 816 F.2d 468 (9th Cir.1987) (exculpatory clause); *Curtis Elevator Co. v. Hampshire House, Inc.*, 142 N.J.Super. 537, 362 A.2d 73 (1976) (strike clause). Both decisions upheld the clauses attacked as unconscionable.
[68]See Sections 32.05(D) and (E).

C H A P T E R   T W E N T Y

Contract Interpretation: Chronic Confusion

SECTION 20.01 Basic Objectives

The basic objective in contract interpretation is to determine the intention of the parties. However, many problems lurk within this relatively simple standard, some of which follow:

1. What can be examined to ascertain the intention of the parties?
2. Once the relevant sources are examined, how is the intention of the parties determined?
3. What if each of the parties has different intentions?
4. What if one party knows of the other party's intention?
5. What if the parties have no particular intention about the matter in question?
6. Can the court go beyond these presumed intentions of the parties and interpret in accordance with what the court thinks the parties would have intended had they thought about it?
7. Can a court disregard the intention of the parties and base determination of the rights and duties of the parties on judicial notions of proper allocation of risk?

The problems can be even more complicated when the language has been selected not by any or all of the parties but by a third party such as the American Institute of Architects (AIA). Should the intention of AIA personnel who selected the language be examined if it can be ascertained? If it cannot, what should the law do when neither party had any intention whatsoever when it agreed to use AIA documents, not an uncommon phenomenon as to certain types of contract clauses.

The preceding list is given merely to indicate that phrasing the test as the process of ascertaining the intention of the parties is deceptively simple, often hiding difficult interpretation problems.

SECTION 20.02 Language Interpretation: *Newsom v. U.S.*

Words have no inherent meaning, yet they develop meanings because people who use them as tools of communication attach meanings to them. A judge asked to interpret contract terms could decide all interpretation questions simply by using a dictionary and choosing the dictionary meaning that seemed most appropriate. However, even dictionary meanings are not exclusive. The choice of which dictionary meaning to be selected can be a formidable task. For these reasons, interpretation of contract terms should take into account the setting and function of the transaction and other matters not found in the contract or the dictionary. Courts seek to put themselves in the position of the contracting parties and determine what the contracting parties must have meant or intended when they used the language in question.

Yet judges and juries do not have absolute freedom to determine the meaning of words. This constraint may be traceable to the fear that juries, and sometimes trial judges, will be unduly sympathetic to a hard-luck story and too inclined to protect the party in the weaker bargaining position. Underlying this reluctance is the skepticism the law has toward the ability of the trial process to separate truth from falsehood, especially when parties differ in their testimony as to what transpired during the negotiations.

Reluctance to give absolute discretion to the fact finders may also be traceable in part to the fear that juries, and sometimes trial judges, may not realize the importance to the commercial world of attaching consistent, commercially accepted, meanings to terms.

The plain meaning rule has been the method employed to limit interpretation powers of trial judges and juries. A judge must first determine whether the words used by the parties had a plain meaning. If so, the judge cannot look beyond the document itself to determine what the parties meant when they used those particular words.

If the judge determines that language cannot be interpreted solely by looking at the writing, other evidence can be examined. A judge will look at the surrounding facts and circumstances that led to the making of the contract, the preliminary negotiation and statements of the parties, any written codes the parties may have adopted, custom and usage, and any interpretation the parties may have placed on the words by their own acts.

Perhaps even judges who state that the meaning is plain on its face have already formally or informally reviewed evidence outside the writing and determined that those outside sources were not helpful.

Despite the plain meaning rule, courts will look at evidence of custom and usage even though the custom and usage seems at variance with the apparently plain meaning of the terms in the contract. For example, suppose a contract specified that meat to be delivered must not have a fat content exceeding 50%. Yet custom in the industry could be introduced that sellers are allowed to deliver meat with a fat content of 50.5%.[1]

The plain meaning rule applies only in litigation. For example, an arbitrator could examine evidence outside the writing without first having to determine that the language was susceptible to more than one meaning.

Appellate courts have stressed that rarely is language sufficiently plain in meaning to preclude extrinsic evidence from being introduced. As a result, modern trial judges seem hesitant to invoke the plain meaning rule and preclude extrinsic evidence. Often the result is a lengthy, expensive trial that does not seem to have advanced the inquiry beyond the language.[2]

Surrounding facts and circumstances often determine how language is interpreted. The relevant surrounding facts and circumstances typically are those that existed at the time the contract was made. However, sometimes courts look at circumstances that existed at some time during *performance* to interpret contract language.[3]

Evidence of the surrounding facts and circumstances that courts will *not* examine is any *undisclosed* intentions of the parties they claim existed at the time they made the contract. If these intentions are made known to the other party, they may be relevant.

Disclosed intention was relevant in *United States v. F. D. Rich Co.*,[4] a dispute between a subcontractor and prime contractor. The contract documents on which the subcontractor prepared its bid were inconsistent as to certain work. The specifications appeared to require that the work be done, although the drawings did not. The subcontractor met with the prime contractor before submitting the bid. The subcontractor testified that at the meeting, the contract documents were examined and discussed. The subcontractor testified that it made it clear that its bid was based on not doing the work in question.

During performance, the public agency insisted that the disputed performance was required. The prime contractor ordered the subcontractor to do the work, which was done under protest. Later, the subcontractor sued for the additional costs of doing the disputed work.

In sustaining a finding by the trial court that the work did not fall within the subcontract, the appellate court stressed that the prime contractor *knew* that the subcontractor's bid did not include the disputed work.

---

[1]*Hurst v. W. J. Lake & Co.*, 141 Or. 306, 16 P.2d 627 (1932).

[2]*Metro. Paving Co. v. City of Aurora, Colorado*, 449 F.2d 177 (10th Cir.1971) is an illustration.
[3]*Contracting & Material Co. v. City of Chicago*, 20 Ill.App.3d 684, 314 N.E.2d 598 (1974). The holding was reversed, not on the admissibility of the evidence but on finding the evidence irrelevant because of a strict interpretation given to the contract clause in question. See 64 Ill.2d 21, 349 N.E.2d 389 (1976).
[4]434 F.2d 855 (9th Cir.1970).

Statements and, more important, acts of the parties *before* the dispute arose may indicate how the parties interpreted the language. Courts often invoke and give considerable weight to those acts under what is called practical interpretation. The practices of the parties often indicate their intention at the time the contract was made. For example, making two progress payments without a showing that the work complied with the contract manifested an intention that the contractor was entitled to progress payments despite noncompliance.[5] Likewise, the prime contractor periodically billing the owner in accordance with certain unit prices and the owner paying these billings indicated that the unit prices were correct.[6] The contractor doing what the owner had directed without complaint indicated the contractor's acquiescence in the owner's interpretation.[7] But the doctrine of practical interpretation applies only to language susceptible to more than one interpretation.[8] This rarely is a difficult obstacle, however.

Not infrequently the surrounding circumstances and the predispute conduct of the parties provide little assistance. When this occurs, courts sometimes resort to secondary assistance, called "canons of interpretation." These canons are interpretation guides. Some are used frequently. One, *expressio unius est exclusio alterius,* excludes an item from relevance when there is a list of items and the item in question is not expressed. For example, suppose a party were excused from performance in the event of strikes, fire, explosion, storms, or war. Under the *expressio* guide, the occurrence of an event not mentioned, such as a drought, would not excuse performance. Where the parties have expressed five justifiable excuses, they must have intended to exclude all other excuses for failure to perform. This guide is sometimes harshly applied and does not give realistic recognition of the difficulty of drafting a complete list of events.

Another guide, *ejusdem generis,* states that the meaning of a general term in a contract is limited by the specific illustrations that accompany it. Any-

thing not specifically mentioned must be similar in meaning to those things that are. For example, damages payable for harm to crops, trees, fences, and premises probably would not include depreciation in market value of the land. The particular item in question is too unrelated to those items listed to be covered by the contract terms.

As to other guides, one court stated:

> . . . where one interpretation makes a contract unreasonable or such that a prudent person would not normally contract under such circumstances, but another interpretation equally consistent with the language would make it reasonable, fair and just, the latter interpretation would apply.[9]

This does not give the court the power to rewrite the language. Nor does it preclude the contracting parties from agreeing to language that a court might consider unreasonable. However, for an unreasonable interpretation to be selected, the language must make clear that this is what the parties intended.

Another court required that the language be interpreted "so as not to put one side at the mere will or mercy of the other."[10] Another stated that a particular interpretation should be rejected because it "would certainly be both unconscionable and inequitable" and the recognized rule is that the interpretation that makes a contract fair and reasonable will be preferred to one leading to a harsh and unreasonable result.[11]

Another important interpretation guide, *contra proferentem,* interprets ambiguous language against the party who selected the language or supplied the contract. Usually this guide is applied not to negotiated contracts but to those mainly prepared in advance by one party and presented to the other on a take-it-or-leave-it basis.

One basis for this guide is to penalize the party who created the ambiguity. Another and perhaps more important rationale is the necessity of protecting the reasonable expectations of the party

---

[5]*Giem v. Searles,* 470 S.W.2d 327 (Ky.1971).

[6]*Berry v. Blackard Constr. Co.,* 13 Ill.App.3d 768, 300 N.E.2d 627 (1973).

[7]*Bulley & Andrews, Inc. v. Symons Corp.,* 25 Ill.App.3d 696, 323 N.E.2d 806 (1975).

[8]*Dana Corp. v. United States,* 470 F.2d 1032 (Ct.Cl.1972).

[9]*Elte, Inc. v. S.S. Mullen, Inc.,* 469 F.2d 1127, 1131 (9th Cir.1972).

[10]*Contra Costa County Flood Control & Water Conservation Dist. v. United States,* 206 Ct.Cl. 413, 512 F.2d 1094, 1098 (1975).

[11]*Glassman Constr. Co. v. Maryland City Plaza, Inc.,* 371 F.Supp. 1154, 1159 n. 3 (D.Md.1974).

who had no choice in preparing the contract or choosing the language. This rationale had its genesis in the interpretation of insurance contracts. Insureds frequently were given protection despite what appeared to be language precluding insurance coverage. This preference recognizes that insurance policies are difficult to read and understand and the insured's expectations as to protection are derived principally from advertising, sales literature, and salespeople's representations. Giving preference to the insured's expectations may also rest on the judicial conclusion that insurance companies frequently exclude risks that should be covered.

The *contra proferentem* guide can be a tie breaker when all other evidence is either inconclusive or unpersuasive. It can be mentioned by the court to bolster an interpretation that has already been determined by other evidence.

This guide has been used in the construction context, such as disputes between contractor and owner[12] and between prime and subcontractor.[13]

Design and construction work frequently are performed after parties have assented to a standard prepared contract form created by associations such as the American Institute of Architects (AIA) or the group of engineering associations who created the Engineers Joint Contract Document Committee (EJCDC). Sometimes such printed contracts are agreed to without careful consideration of the language. On other occasions, all or some of the language is considered carefully by the parties before entering into the contract. Sometimes the parties have dealt with and understood the language, and on other occasions, one party may be unfamiliar with the terminology and concepts employed in the standard contract.

A look at a few cases in the construction context may be instructive. *Durand Associates, Inc. v. Guardian Investment Co.*[14] construed an AIA standard contract against the engineer who had supplied it despite the fact that the owner was an investment company about to build a medical clinic that later was changed to an apartment complex. Although the owners in the case seemed to be experienced

businesspersons, evidently the court felt that their knowledge and experience did not equip them to carefully appraise the language of the AIA document. Yet the owners *read* the document and made some revisions that were accepted by the engineer. Perhaps the court's construing the document "strictly" against the engineer was based more on the court's reluctance to compel the owner to pay a fee when the project was abandoned because of excessive cost.

Other cases have involved disputes between owners and contractors, with varied outcomes. One case noted that the contractors were sophisticated businesspersons and that the owner was legally represented were indications that the document should not be construed against either party.[15] Another pointed to the contract's having been the result of arm's-length bargaining.[16] A third interpreted A201 against the owner.[17]

Although the contract involved was not one published by a professional society, *W. C. James, Inc. v. Phillips Petroleum Co.*[18] is instructive. The plaintiff, a large pipeline contractor, contracted with Phillips, a large gasoline supplier. The contractor contended that the contract had been preprepared, or boiler-plated, and presented to bidders on a take-it-or-leave-it basis. The trial court noted that the clause in question—one waiving damages for delay—was common in construction work. The trial court stated that the contractor entered into the contract voluntarily, intelligently, and knowingly. The trial court upheld the clause, stating that it was not unfair and that the contract was not one of adhesion.

The Federal Circuit Court of Appeals affirmed the trial court, noting that the pipeline contractor was of sufficient size, even in relationship to Phillips, so that it could not seek relief on the grounds of an adhesion contract.

It is likely that the pipeline contractor in the *James* case had no choice. Yet if the contractor was aware that it was taking the risk of delay damages, it could

---

[12]Ibid.

[13]*United States v. Klefstad Eng'g Co.*, 324 F.Supp. 972 (W.D.Pa.1971).

[14]186 Neb. 349, 183 N.W.2d 246 (1971).

---

[15]*Robinhorne Constr. Corp. v. Snyder*, 113 Ill.App.2d 288, 251 N.E.2d 641 (1969), affirmed 47 Ill.2d 349, 265 N.E.2d 670 (1970).

[16]*Cree Coaches, Inc. v. Panel Suppliers, Inc.*, 384 Mich. 646, 186 N.W.2d 335 (1971).

[17]*Osolo School Buildings v. Thorlief Larson and Son, Inc.*, 473 N.E.2d 643 (Ind.App.1985).

[18]485 F.2d 22 (10th Cir.1973).

adjust its price and take this risk into account. It is not in the same position as a consumer, who is usually not in the position to adjust to the risk that the contract language seeks to place on her.

The preceding discussion has demonstrated the need for an analysis that will recognize some of the particular problems of dealing with standard contracts published by the AIA or EJCDC. First, it is important to look at the surrounding facts and circumstances that led to the use of the standard agreement. It should not always be assumed that the architect dictates that an AIA document be used for either design or construction services. Nor should it be assumed that an owner dictates the use of an AIA document for construction services. It is not inconceivable that an AIA document is selected by each party to the contract because of the reputation of the document for fairness or familiarity. This is more likely to be the case when AIA construction documents are used unless the owner has dealt with design services before.

Second, changes are frequently made in AIA documents, principally in the area of payments, changes, indemnification, arbitration, and the responsibilities of the design professional. It would be administratively inconvenient to use different standards of interpretation when substantial changes have taken place. Where there have been substantial changes, it is likely that the parties have considered the entire document, in some cases even jointly negotiating the agreement. Where the latter is the case, the agreement should be interpreted neutrally. Where one party has used some clauses from a standard agreement that favor it and then incorporated other clauses more favorable than those contained in the standard agreement, clearly the language should be interpreted against that party.

Third, it is possible to interpret any B-series agreements—those that deal with design services—in favor of the client, as the AIA receives no input from other organizations for these documents and is likely to be drafting with the best interests of architects in mind.[19] However, the AIA obtains endorsement and approval for some of its construction services documents, such as A201, from the

Associated General Contractors (AGC). This can lead to an assumption that the agreement has been jointed negotiated by representatives of owners and contractors and should be interpreted neutrally. However, although in some respects the AIA can be said to be thinking of the interests of the owner, there is too much in A201 that is done with the interests of the architect in mind, and there is no direct owner representation. It would be a mistake to treat the agreement as if it had been jointly worked out by representatives of owners and contractors.[20]

Fourth, courts usually seek to find the intention of the contracting parties as the lodestar of contract interpretation. However, it is not uncommon for neither party to an AIA document to have any intention whatsoever as to certain clauses and their function when they have used an AIA document as the basis for their agreement. Does that mean that parties should be allowed to introduce any evidence of the intention of AIA representatives, if any can be obtained? This is analogous to the use of legislative history by courts in interpreting statutes. Inasmuch as there is currently no official AIA publication that seeks to explain the reasons for language the AIA has selected, any such evidence would have to consist of informal letters or affidavits by AIA officials.[21] Such evidence is untrustworthy at best, and the parties would have no opportunity to question those who have sought to describe the intention of the Institute. It is probably best for courts to ignore this evidence and decide intention questions based on common law principles and the hypothetical intention of the parties— what they would have intended.

A Florida court faced this problem when it was asked to interpret the meaning of an applicable

---

[19]Two cases in which this has been done are *Malo v. Gilman*, 177 Ind.App. 365, 379 N.E.2d 554 (1978), reproduced in Section 12.03, and *Kostohryz v. McGuire*, 298 Minn. 513, 212 N.W.2d 850 (1973).

[20]For an exploration of how AIA documents are prepared, see Sweet, *the American Institute of Architects, Dominant Actor in the Construction Documents Market,* 1991 Wis.L.Rev. 317.

[21]In 1988, the AIA issued R671, B141 Commentary, which along with a similar commentary on A201, is found in the AIA Handbook of Professional Practice published in 1988. The commentary is directed at practicing architects. It contains instructions for preparing B141, furnishing more detail than B141a, the instruction to users of B141. It also contains brief provision-by-provision discussion. It cannot be regarded as a full-fledged comprehensive commentary. For the author's commentary, see J. SWEET, SWEET ON CONSTRUCTION INDUSTRY CONTRACTS: MAJOR AIA DOCUMENTS (2d ed. 1992).

building code. Expert testimony was submitted by both parties. The architect's expert had served as manager of engineering of the Southern Building Code Congress and had been responsible for building code changes, hearings, plan reviews, and code interpretations. In effect, that expert testified as to what the drafters had in mind when they created the code. The design, he stated, was consistent with the intent of the code. However, the court concluded that the most important evidence to interpret the code was the interpretation given by the code official who had authority to implement the code.[22]

Courts often look at the particular clause in question to determine whether to enforce it or how to interpret it. Courts give indemnification clauses careful scrutiny, either invalidating them entirely or construing them strongly against anyone seeking to be indemnified for its own negligence. Such scrutiny is based on a desire to redress an unequal bargaining position and to avoid an interpretation that could encourage carelessness.

The competitive bidding process has developed special rules. Prospective bidders are given a complex set of construction documents that have been prepared over a long period of time. Bidders are asked to review the materials within the short period of time available to them before submitting a bid. They are also asked to report any errors or inconsistencies that they have observed. This laudable attempt to catch errors and invoke the knowledge and skill of the contractor often generates problems because many errors are not actually detected until after the work is in progress, mainly through the review of shop drawings.

One way of dealing with interpreting construction contracts in this context is to accept the interpretation that would have been given by a reasonably prudent bidder.[23] However, courts have had to struggle, sometimes painfully, with the question of whether the contractor should have sought clarification before bidding because the ambiguity was patent and glaring.[24] The following case explores this problem.

---

[22]*Edward J. Seibert, A.I.A., Architect and Planner, P.A. v. Bayport Beach & Tennis Club Assn., Inc.*, 573 So.2d 889 (Fla.Dist.Ct.App.1990).

[23]*Corbetta Constr. Co. v. United States*, 198 Ct.Cl. 712, 461 F.2d 1330 (1972).
[24]*Zinger Constr. Co., Inc. v. United States*, 807 F.2d 979 (Fed.Cir.1986) (design would not produce an operational system; duty to inquire); *Brezina Construction Co. v. United States*, 196 Ct.Cl. 29, 449 F.2d 372 (1971). See also *Wickham Contracting Co. v. United States*, 212 Ct.Cl. 318, 546 F.2d 395 (1976) (scale of drawings). New Jersey adopted the patent ambiguity rule in *D'Annunzio Bros., Inc. v. New Jersey Transit Corp.*, 245 N.J.Super. 527, 586 A.2d 301 (1991).

NEWSOM v. UNITED STATES

United States Court of Claims, 1982. 676 F.2d 647.
[Ed. note: Footnotes renumbered and some omitted.]

SMITH, Judge.

This case is an appeal by petitioner, George E. Newsom, of a decision of the Veterans Administration Board of Contract Appeals (board). The board found that certain parts of the contract for hospital improvements were patently ambiguous and that, having failed to consult with the contracting officer about the ambiguities, petitioner was barred from recovering for work done beyond that required under petitioner's interpretation of the contract. We affirm the decision of the board.

On August 28, 1978, the Veterans Administration (VA) issued an invitation for bids for building medi-prep and janitor rooms in the VA hospital at Knoxville, Iowa. Drawings and specifications for the work to be done were supplied to the prospective bidders.

Paragraphs 4, 5, and 6 of the specifications described, respectively, buildings 81, 82, and 85. Each paragraph had two parts: the first described the first floor of the building and referenced page 7 of the drawings; the second described the second floor of the building and referenced page 8 of the drawings. Conversely, the caption block on page 7 of the drawings indicated that it described work for all three buildings, 81, 82, and 85. However, page 8 of the drawings indicated only building 85. Petitioner at no time inquired about this discrepancy.

As a consequence, petitioner included in his bid the costs of the second floor of building 85 only. He was the low bidder and the contract was awarded to him on October 13, 1978. It was not until March 29, 1979, that the parties realized that there was a discrepancy between what the VA had intended and what petitioner had understood. Petitioner then did the work as intended by the VA at an additional cost of $14,600, and he appealed the decision of the contracting officer denying relief to the Veterans Administration Board of Contract Appeals. The board held against petitioner on the ground that the error on page 8 of the drawings was a patent ambiguity which imposed upon the contractor a duty to inquire about it. Petitioner now appeals that finding to this court under the Contract Disputes Act.

The doctrine of patent ambiguity is an exception to the general rule of *contra proferentem* which requires that a contract be construed against the party who wrote it. If a patent ambiguity is found in a contract, the contractor has a duty to inquire of the contracting officer the true meaning of the contract before submitting a bid.[25] This prevents contractors from taking advantage of the Government; it protects other bidders by ensuring that all bidders bid on the same specifications; and it materially aids the administration of Government contracts by requiring that ambiguities be raised before the contract is bid on, thus avoiding costly litigation after the fact. It is therefore important that we give effect to the patent ambiguity doctrine in appropriate situations.

The existence of a patent ambiguity is a question of contractual interpretation which must be decided de novo by the court. This determination cannot be made upon the basis of a single general rule, however. Rather, it is a case-by-case judgment based upon an objective standard. In coming to our decision, we are bound neither by the legal conclusions of the board, nor by the subjective beliefs of the contractor, subcontractors, or resident engineer as to the obviousness of the ambiguity.

The analytical framework for cases like the instant one was set out authoritatively in *Mountain Home Con-*

tractors v. United States.[26] It mandated a two-step analysis. First, the court must ask whether the ambiguity was patent. This is not a simple yes-no proposition but involves placing the contractual language at a point along a spectrum: Is it *so* glaring as to raise a duty to inquire? Only if the court decides that the ambiguity was not patent does it reach the question whether a plaintiff's interpretation was reasonable. The existence of a patent ambiguity *in itself* raises the duty of inquiry, regardless of the reasonableness . . . of the contractor's interpretation. It is crucial to bear in mind this analytical framework. The court may not consider the reasonableness of the contractor's interpretation, if at all, until it has determined that a patent ambiguity did not exist.[27]

Examining the contract itself, we find that a patent ambiguity existed. Two parts of the contract said very different things: the specification required construction on the second floors of buildings 81, 82, and 85, whereas the drawings required construction on the second floor of only building 85. It is impossible from the words of the contract to determine what was really meant. The contractor speculated that it meant that part of the project had been dropped along the way. Looking at the same language, the Government can insist that it was clearly a drafting error. We do not consider which interpretation is correct; at this stage we determine only whether there was an ambiguity. What is significant about the differing interpretations is that neither does away with the contractor's ambiguity or internal contradiction. There is simply no way to decide what to do on the second floors of buildings 81 and 82 without recognizing that the contract also indicates otherwise.

Mountain Home, discussed above, involved a very similar ambiguity. The specifications ordered inclusion of kitchen fans in certain housing units, but the drawings appeared to indicate that kitchen fans were not to be installed. There is a crucial difference between that case and this, however. In *Mountain Home* the indication on the drawing was that the fans were to be under an alternate bid. Thus, the drawings indicated that the Government was reserving the option of either including the fans in the main contract or ordering them separately. The drawings did not state that the fans were simply not to be included. The

---

[25]*Beacon Constr. Co. v. United States*, 161 Ct.Cl. 1, 6, 314 F.2d 501, 504 (1963); *Blount Bros. Constr. Co. v. United States*, 171 Ct.Cl. 478, 495–96, 346 F.2d 962, 971–972 (1965).

[26]*Mountain Home Contractors v. United States*, 192 Ct.Cl. 16, 20–21, 425 F.2d 1260, 1263 (1970).
[27]If the court finds that a patent ambiguity did not exist, the reasonableness of the contractor's interpretation becomes crucial in deciding whether the normal *contra proferentem* rule applies.

Mountain Home contract, therefore, was susceptible of an interpretation which did not leave significant ambiguities or internal contradictions. Here, even petitioner's interpretation acknowledges that the contract is not internally consistent. Petitioner's interpretation explains the reason for the inconsistency but does not eliminate it.

We recognize that the instant case does not represent a difference in kind from the *Mountain Home* facts, but this area of the law involves a case-by-case determination of placement along a spectrum. In our opinion, this case is closer to *Beacon Construction*[28] than to *Mountain Home.* In *Beacon Construction,* the specifications stated only that "weatherstrip shall be provided for all doors," while the drawings describing the weatherstripping clearly indicated weatherstripping around the windows as well. The conflict be-

tween the specifications and the drawings was direct, as in the instant case. And the court was not swayed by the mere fact that the contractor was able to come up with a highly plausible interpretation of the ambiguity. No interpretation could in *Beacon Construction,* or can in the instant case, eliminate the substantial, obvious conflict between the drawings and the specifications.

Finally, we emphasize the negligible time and the ease of effort required to make inquiry of the contracting officer compared with the costs of erroneous interpretation, including protracted litigation. While the court by no means wishes to condone sloppy drafting by the Government, it must recognize the value and importance of a duty of inquiry in achieving fair and expeditious administration of Government contracts.

Accordingly, upon consideration of the submissions, and after hearing oral argument, the decision of the Veterans Administration Board of Contract Appeals is

AFFIRMED.

---

[28]*Beacon Constr. Co. v. United States,* supra note 25, 161 Ct.Cl. at 4–5, 314 F.2d at 502–03.

Note the difference between the drawings and the specifications in the *Newsom* case. Often contracts seek to deal with this problem by incorporating a precedence of documents clause, which gives more weight or precedent to one document than to others. This method of dealing with ambiguity in the construction context is discussed in Section 20.03.

Inasmuch as many of the disputes that have dealt with the duty of the contractor to draw attention to defective design have, such as in the *Newsom* case, come before a federal contracting agency appeal board, it may be useful to look at criteria that have been developed by those boards to determine whether a defect should have been brought to the attention of the owner. Some of these criteria are as follows:

1. Did other bidders discover the error and seek clarification?
2. Did skilled professionals for the agency detect the error?
3. Were the construction documents so complicated that a small detail could have been easily missed?
4. Was the cost of correcting the defect a relatively small part of the contract price?
5. Did the defect occur in one item out of many items involved in the bid?

6. Was there a single prime contractor or many (a "multi-prime" job)?
7. Will the contractor make a profit from a failure to inquire?[29]

Interpretation guides are only guides. Courts look principally at the language, the surrounding facts and circumstances, and the acts of the parties. Nevertheless, interpretation guides are useful in close cases and may provide the court with a rationalization for a result already achieved.

SECTION 20.03 Resolving Conflicts and Inconsistencies

A. Within the Written Agreement

The difficult process of contract interpretation requires a court to put itself in the position of the parties and determine the parties' intentions at the time of the agreement. This process is complicated by the modern tendency for longer agreements.

---

[29]For agency decisions dealing with this issue, see *Pathman Constr. Co.,* ASBCA 22343, 81-1 BCA ¶ 15,010; *George Hyman Constr. Co.,* ENG BCA 4506, 81-2 BCA ¶ 15,363; *Sam Bonk Uniform & Civilian Cap Co.,* ASBCA 23592, 81-1 BCA ¶ 14,840; *R&C Corp.,* GSBCA 6041, 81-2 BCA ¶ 15,369; *Linde Constr. Co.,* GSBCA 5840, slip op. (1981); *S&W Contracting Co.,* IBCA 1307-10-79, 81-1 BCA ¶ 15,133.

Often, long agreements are not carefully examined to avoid conflicts and inconsistencies. This is especially true where a set of standardized general conditions or contract terms is appended to a letter agreement that expresses the basic elements of the contract.

Courts generally interpret the contract as a whole and, wherever possible, seek to reconcile what may appear to be conflicting provisions. It is assumed that every provision was intended to have some effect. Yet this process of reconciliation may not accomplish its objective, and it may be necessary to prefer one provision over another.

Where one clause deals generally with a problem and another deals more specifically with the same problem, the specific takes precedence over the general. The specific clause is likely to better indicate the intention of the parties. For example, suppose a building contract specified that *all* disputes were subject to arbitration but also stated in a different paragraph that disputes as to aesthetic effect were to be resolved by the design professional whose decision was to be final. It is likely that the parties intended the specific clause dealing with nonreviewability of specific decisions to control the general clause.

If parties have expressed themselves *specifically*, their failure to change the general clause is likely due to a desire to avoid cluttering up the contract with exceptions and provisos. Also, the parties may not have noticed the discrepancy.

Later clauses take precedence over earlier ones. Although this is rarely used, it is premised on the assumption that parties sharpen their intentions as they proceed, much like an agreement made Tuesday displaces one made the preceding Monday.

Operative clauses take precedence over "whereas" clauses that seek to give the background of the transaction. As to inconsistency between printed, typed, and handwritten provisions, handwritten will be preferred over typed and typed preferred

to printed. These preferences are premised on the assumption that the parties express themselves more accurately when they take the trouble to express themselves by hand. Likewise, specially typed provisions on a printed form are more likely to be an indication of the intention of the parties than the printed provisions.

Some of these rules may have to be reevaluated in the light of the frequent use of contracts printed by word processors. Such agreements may appear to have been typed particularly for this transaction but may have been prepared for a variety of transactions just as a printed contract form. The test in seeking to reconcile conflicting language within a written document should be whether some language was specially prepared for this transaction and indicated with greater accuracy the intention of the parties.

These priorities are simply guides to assist the court in resolving these questions. Guides to the meaning of language *within* a document are similar to the canons of interpretation that were explored in Section 20.02. Each party can usually point to guides that support its position. What must be done is to focus on Section 20.01 and its emphasis on the intention of the parties. Which meaning is most likely to accomplish those objectives without grossly distorting the language?

B. Between Documents: *Unicon Management Corp. v. U.S.*

Construction contracts present particularly difficult interpretation questions because of the number and complexity of contract documents and the frequent incorporation of bulky specifications by reference. Suppose work is called for by one document but not specifically required in another. Suppose one document sets up procedures for changes different from those set forth in another document. At the outset, it may be instructive to reproduce a case dealing with this problem.

UNICON MANAGEMENT CORP. v. UNITED STATES

United States Court of Claims, 1967. 375 F.2d 804.
[Ed. note: Footnotes renumbered and some omitted.]

DAVIS, Judge.

In March 1959 the contractor agreed with the Corps of Engineers to construct, for a fixed price, two

phases of the Missile Master Facilities near Pittsburgh. The current claim is that the Government required plaintiff to install in one room a steel-plate flooring which was not called for by the plans and specifications. The contracting officer and the Armed Services Board of Contract Appeals (65-1 BCA para. 4775) refused the demand for an equitable adjustment and this suit was brought. The problem arises because the most pertinent specification, if read alone, could be said to contemplate a wholly concrete rather than a partial steel-plate floor, while the most pertinent drawings, if read alone, direct the steel-plate covering. The contractor resolves the difficulty by relying, mainly, on the contractual clause that "in case of difference between drawings and specifications, the specifications shall govern." The Board and the defendant invoke another provision that "anything mentioned in the specifications and not shown on the drawings, or shown on the drawings and not mentioned in the specifications, shall be of like effect as if shown or mentioned in both."[30] Since the question is one of contract interpretation, we are free to decide the matter for ourselves.

The room was Equipment Room No. 1 (also designated as Room 117) in one of the buildings being erected by plaintiff. To secure heavy equipment, this space was to have pallets along the floor with parallel cable trenches for bringing electric current to the machines. Steel beams were to be used in and along the floor in connection with these pallets. It is agreed that the top cover of the floor, apart from the portions devoted to the pallets and the trenches, was to be a resilient floor tile. There was also to be a concrete base. The dispute is whether the tile was to be laid directly on this concrete or over a quarter-inch steel plate above the concrete. The specifications, in a section of the Technical Provisions on miscellaneous metalwork, contain a paragraph dealing with the floor of this

room and with the pallets (TP 17-23).[31] By itself, this provision can be interpreted as implicitly envisioning an all-concrete floor; there is no reference to steel plate but there are references to "the concrete floor", "a concrete floor", "pouring of the floor", "pouring of the floors", and steel beams being "embedded" in the concrete floor.

The relevant drawings, by themselves, give a very different impression. The most significant sketch—a cross-section of the floor of Equipment Room No. 1— shows the trench covered by a removable steel cover plate, with a depression into which the plate's handle is fitted flush with the floor level; the open-floor portion of the drawing shows the top of two steel beams covered by $1/4$" steel plate and the resilient floor tile on top of this plate. This is a specific directive to use steel plate immediately beneath the tile. To the same effect, another detail drawing of the trenches and beams in the room shows steel plate on top of an "I" beam and the resilient tile above the steel plate. The other drawings are uninformative or unclear on the point, but they do not contradict the sketches explicitly showing the steel plate. The result is that the drawings as a whole affirmatively support the Government's understanding as to the nature of the floor of Equipment Room No. 1.[32]

We agree with the Board that these plans need not, and should not, be construed as in conflict with specification TP 17-23 footnote [31], supra but, instead, as supplementing the latter. The parties directed in the contract that "anything . . . shown on the drawings and not mentioned in the specifications, shall be of like effect as if shown or mentioned in both" (footnote [30] supra). This rule is peculiarly applicable here be-

[30]These sentences are from Article 2 of the General Provisions which reads as follows:

"The Contractor shall keep on the work a copy of the drawings and specifications and shall at all times give the Contracting Officer access thereto. Anything mentioned in the specifications and not shown on the drawings, or shown on the drawings and not mentioned in the specifications, shall be or like effect as if shown or mentioned in both. In case of difference between drawings and specifications, the specifications shall govern. In any case of discrepancy either in the figures, in the drawings, or in the specifications, the matter shall be promptly submitted to the Contracting Officer, who shall promptly make a determination in writing. Any adjustment by the Contractor without this determination shall be at his own risk and expense. The Contracting Officer shall furnish from time to time such detail drawings and other information as he may consider necessary, unless otherwise provided."

[31]17-23 *Equipment Room No. 1:* AAOC Main Building. Floor shall have steel beams embedded in the concrete floor so that they run transversely to the equipment pallet lengths. The flange surfaces shall be flush with the finished cement. Two beam runs shall be used, one at each end of the pallets, as shown on the drawings. These beams will be used to anchor the equipment pallets, to level the equipment pallets and can be used as references for cement finishing tools during pouring of the floor. These beams shall be connected to the building ground system so that they will serve as a grounding means for the equipment pallets. The steel beams shall be on one continuous length, with portions removed to provide clearance for the cable troughs. The cross section area of the beams or jumpers, at the points where metal is removed shall provide at least as much conductivity as the ground cables used to connect the beams to the ground system. The grounding system must be installed prior to the pouring of the floors, as specified in Section 47, 'Electrical Work.' "
[32]Ed. note: Figure 20.01 is a diagram of the disputed floor.

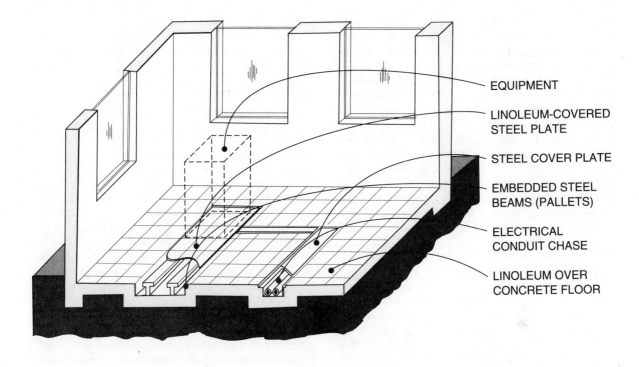

EQUIPMENT

LINOLEUM-COVERED
STEEL PLATE

STEEL COVER PLATE

EMBEDDED STEEL
BEAMS (PALLETS)

ELECTRICAL
CONDUIT CHASE

LINOLEUM OVER
CONCRETE FLOOR

UNICON

cause TP 17-23 obviously does not cover the entire subject of floor construction. For instance, it does not mention the important item of resilient floor tile, or the metal plates for the trenches, nor does it even describe fully the construction of the steel beams running transversely to the pallets. When the contract is viewed as a whole, the function of the paragraph is seen as calling attention to the steel beams set in the floor to support the pallets—the references to concrete are in that connection and in that special light—not as drawing together all of the requirements on the makeup of the floor. Certainly, as the Board pointed out, the specification does not provide in terms or by necessary implication that the resilient tile is to rest directly on concrete; nothing is said about the tile or its placing. It is therefore proper to read TP 17-23 as open to complementation by the drawings (or by other specifications) insofar as these cover items or aspects other than the features of the steel beams discussed in the former provision. Plaintiff seems to insist that, if the specification *can* be read as conflicting with the drawings, that reading must be adopted even though a more harmonious interpretation is also reasonably available. The rule, however, which the courts have always preferred is, where possible, to interpret the provisions of a contract as coordinate not contradictory. . . . Contractors, too, have long been on notice that in reading contract documents they should seek to find concord, rather than discord, if they properly can.

This brings us to another reason why plaintiff's position is weak. There is no evidence as to the view

actually taken by the company's estimators before it submitted its bid (plaintiff did not call them to the stand in the administrative proceeding). But assuming . . . that they read TP 17-23 the way plaintiff now does, if they examined the plans and specifications carefully they could not have helped notice the drawings which specifically embodied the contrary requirement. If they were not aware of this fact they should have been. The contract provided that "in any case of discrepancy either in the figures, in the drawings, or in the specifications, the matter shall be promptly submitted to the Contracting Officer, who shall promptly make a determination in writing. Any adjustment by the Contractor without this determination shall be at his own risk and expense" (footnote [30] supra). A warning of this kind calls upon the bidder to bring to the Government's attention any serious or patent discrepancy of significance, of which he is or should be cognizant. . . . The discrepancy here—if TP 17-23 was then thought to mean what plaintiff now contends—surely met that standard of importance. Yet there is no suggestion that plaintiff brought it to the contracting officer or sought the required guidance before bidding. If, on the other hand, plaintiff did not study the plans and specifications before bidding, it cannot complain that the Board and this court strive, in accordance with the established canon, to read the relevant contract provisions together rather than at odds.

The plaintiff is not entitled to recover. Its motion for summary judgment is denied and the defendant's is granted. The petition is dismissed.

Other decisions have dealt with conflicts in documents. First, as in the *Unicon* decision, the court attempts to reconcile the apparently conflicting language.[33] When this cannot be done, the court will often base its decision on contract language that seeks to resolve potential conflicts.[34]

Another technique for dealing with potential conflict is seen in AIA Doc. A201, ¶ 1.2.3, which states:

---

[33]*Warren G. Kleban Eng'g Corp. v. Caldwell*, 361 F.Supp. 805 (N.D.Miss.1973), vacated on other grounds, 490 F.2d 800 (5th Cir.1974); *Contracting & Material Co. v. City of Chicago*, supra note 3.
[34]*Graham v. Virginia*, 206 Va. 431, 143 S.E.2d 831 (1965) (specifications over general conditions); *Dunlap v. Warmack-Fitts Steel Co.*, 370 F.2d 876 (8th Cir.1967) (specifications over bid proposal). AIA Doc. B141, ¶ 9.2, requires terms be given same meaning as A201.

The intent of the Contract Documents is to include all items necessary for the proper execution and completion of the Work by the Contractor. The Contract Documents are complementary, and what is required by one shall be as binding as if required by all; performance by the Contractor shall be required only to the extent consistent with the Contract Documents and reasonably inferable from them as being necessary to produce the intended results.

This paragraph warns the contractor to assume that work called for under *any* document will be required even if apparently omitted from other documents. The paragraph requires work not specifically covered if that work is necessary to produce the intended results. Undoubtedly, such a provision gives considerable power to the architect inasmuch as she may be called on to determine whether particular work is reasonably inferred though not specifically required. A cautious contractor must take this into account when submitting a bid. Even though there are limits to the architect's power, the contractor may find itself at the mercy of an architect who wishes to cover up for poor document drafting. Understandably, a provision is needed to take into account that it is not possible to specifically include *every* aspect of work that should be done. However, such a provision can generate performance difficulties if the architect does not exercise this power fairly.

The AIA has chosen not to include a precedence-of-documents clause discussed in the *Unicon* case. It believes that such a clause can be a disincentive for contractors to report errors, omissions, or inconsistencies in the design documents. Certainly, if such a clause is read to eliminate the need to make inquiry at the bidding or negotiations stage, the AIA's position can be justified. The AIA also believes that inclusion of such a clause can lead to "wooden" interpretation decisions that ignore the documents as a whole. As seen in the *Unicon* case, the reconciliation process is not always an easy one. The AIA also believes that there is no standardized precedence system that can be the basis for a provision in a nationally used contract form. It suggests that any parties that wish to include such a provision can simply add it.

Other arguments against a precedence-of-documents clause can be made. The common law has developed a method of reconciling apparently inconsistent clauses within a writing and among writings, which may do as good a job as a precedence-of-documents clause, particularly where the language can be easily supplemented by evidence of industry custom and a course of dealing between the parties. As can be seen in the Court of Claims decisions reproduced in this section; not only is it often difficult to determine whether a precedence clause diminishes or eliminates the obligation to inquire, but it also requires mental gymnastics to distinguish differences within a document from those differences that appear between documents. A precedence-of-documents clause does not always deal well with problems where the evidence of a lesser ranked document is strong while the evidence of a more highly ranked document is weak.

Yet the frequent inclusion of precedence-of-documents clauses must indicate that such clauses have their value. In the often insoluble interpretation problem, they may provide some method of arriving at a solution. It is likely that this is the principal reason they are used despite their weaknesses.

To conclude this subsection, it may be instructive to examine a case that demonstrates the complexity that can be generated by the wealth of contract documents.

William F. Klingensmith, Inc. v. United States[35] involved a contract dispute between the contractor and the U.S. General Services Agency (GSA) over the base layers of certain roadways and paved areas. The contractor asserted that it was required only to use "bank run gravel" for a particular base layer. However, the GSA took the position that macadam was required.

The specifications supported the GSA's position, although the drawings contained details that supported the contractor's position. However, the general conditions expressly provided that "in case of difference between drawings and specifications, the specifications shall govern." The court concluded that because there was no ambiguity in the specifications, there was no need to invoke the rule that ambiguities are resolved against the drafter.

Yet the contractor had another arrow in its quiver. The contract had incorporated certain provisions of the Maryland State Roads Commission's specifications dealing with materials, highways, and bridges. The Maryland specifications stated that in the event of discrepancy between plans and specifications, the *plans* controlled. However, this arrow did not reach the target. The court concluded that the *specific* precedence-of-documents language in the contract *itself* took precedence over precedence language of specifications *incorporated* by reference. The specific controls the general.

A number of other contractor arrows did not reach the mark. The contractor's contention that macadam had not been used as a base course in the locality in recent years was rejected. Likewise, its showing that the government's estimator also assumed gravel could be used was not persuasive.

---

[35]205 Ct.Cl. 651, 505 F.2d 1257 (1974).

(Evidently the government estimator also relied on the drawings.)

The contractor's estimator, having noticed the conflict between drawings and specifications, called the architect-engineer who had been employed by GSA but was told that the matter was no longer in its hands and suggested that the estimator seek clarification from GSA. On requesting such a clarification from GSA, the estimator was advised that GSA could not respond to an inquiry on the telephone "since such clarifications were required to be made through a formal addendum issued to all prospective bidders." This was impossible, as bid opening was only three hours away.

The contractor claimed that this conversation showed that the government *knew* the contractor's interpretation and was bound by it. However, the court rejected this, pointing to provisions of the invitation for bids stating that requests for clarifications or interpretations must be submitted within a specified period of days from bid opening and the request was too late. The court concluded by stating that resolving ambiguities against the drafter cannot be done in federal government contracts where the contractor did not seek a clarification of potential conflicts.

The *Klingensmith* case demonstrates the immense complexity in construction document interpretations. It also shows the importance of reading specifications carefully before preparing bids and allowing enough time to seek clarifications.

SECTION 20.04 Reformation of Contracts

Sometimes parties to a contract reduce their agreement to a writing, and for various reasons, the writing does not correctly express the intention of the parties. The remedy of reformation granted by equity judges rewrites the contract to make the writing conform to the actual agreement of the parties. Because there were no juries in equity cases, courts felt freer to go into the difficult questions of intention and mistake.

Most cases have involved an improper description of land in a deed. These descriptions were complicated and often taken from old deeds and tax bills. It was not unusual for the person copying the old description to make a mistake. If it could be shown by reliable evidence that a mistake had been made, the court would reform the contract.

This would be a judicial declaration that the contract covered the particular land that the parties actually intended to buy and sell rather than the land covered in the description.

In addition to the cases of mutual mistake in description, other types of mistake would justify reformation. Usually these mistakes would involve the process by which the actual agreement was reduced to writing.

The reformation doctrine has been given an interesting application by the U.S. Court of Claims in federal procurement contracting.[36] In *National Presto Industries, Inc. v. United States*,[37] the contractor under a fixed-price contract sought more money when its costs were substantially increased because a method that it had intended to use was not feasible for mass production. The court found mutual mistake in that the contractor and the government each assumed that the particular process in which the United States had interest could be used. Usually the remedy for mutually mistaken fundamental assumptions is to relieve the performing party—in this case the contractor—from the obligation to perform. But the Court of Claims extended reformation beyond its normal correction of mistakes function to change the fixed-price contract to one of a joint enterprise in which each joint enterpriser would share the unforeseen expenses.

Such a doctrine can impair the certainty and risk assumption features of procurement. As a result, two years later, the Court of Claims seemed to have had second thoughts about its decision in *National Presto* and sought to make clear that the reformation remedy awarded in *National Presto* could be justified only where there is a joint enterprise experimental situation in which neither party has assumed the particular risk, a great concern on the part of the government in the process and not merely the end product and a distinct benefit to the government from the contractor's period of trial and error.[38] Nevertheless, in exceptional circum-

---

[36]This court was replaced by the Claims Court, a trial court, and the U.S. Circuit Court for the Federal Circuit, an appellate court. The former is currently called the U.S. Court of Federal Claims.

[37]167 Ct.Cl. 749, 338 F.2d 99 (1964), cert. denied 380 U.S. 962 (1965).

[38]*Natus Corp. v. United States*, 178 Ct.Cl. 1, 371 F.2d 450 (1967).

stances, reformation may be the vehicle to redistribute risks at least in federal procurement.[39]

Reformation can correct writings that do not reflect the actual agreement of the parties,[40] but the requirements for invoking this equitable remedy are formidable. Parties obviously cannot rely on the possibility that the law will correct their mistakes. They must make every effort to make the writing conform to their actual agreement.

---

[39]*Dynalectron Corp. v. United States*, 207 Ct.Cl. 349, 518 F.2d 594 (1975).
[40]*Timber Investors, Inc. v. United States*, 218 Ct.Cl. 408, 587 F.2d 472 (1978) (mistaken estimates in quantities). Refer to Section 17.02(D).

PROBLEMS

1. O owned a large house. She entered into negotiations with C regarding the painting of the house. They signed a written contract under which it was agreed that C would receive $2,000. C drafted the contract. In part the contract stated:

> C will do a first-class job of painting O's house. (Garage not included.)

A dispute arose over the number of coats of paint to be used and whether C was to paint the concrete steps leading to the house. C claimed two coats were adequate, while O stated that three coats were needed for a first-class job. C claimed the contract did not include the concrete steps. O states that she pointed to the steps when they inspected the house and that she said that the steps needed a coat of paint. C admitted O said this, but C contends that when it was not included in the contract, she assumed she would not have to do it. How should a judge decide this dispute?

2. A contract for the construction of a major water delivery pipeline required the contractor to excavate a trench, place bedding material in the bottom of the trench, install the pipe, and place backfill material around and above the pipe. The contract specified four zones of bedding and backfill material—Zones 1, 2, 3, and 4. Zone 3 backfill material was to be placed at certain points depending on whether steel or concrete pipe was used. Also, depending on the type used, all or a portion of the backfill was to be compacted.

The contract provided that backfill used in Zones 1 and 2 should contain no material larger than three quarters of an inch. The dispute involves the size limitation, if any, of Zone 3 material. The specific provision dealing with Zone 3 backfill reads:

> 12.2.44 *Zone Backfill Material.* Zone 3 backfill material shall consist of selected material from the trench excavation, free from frozen material and lumps or balls of clay, organic, or other objectionable material. When compaction of Zone 3 backfill is called for, the material shall be well graded and easily compacted throughout a wide range of moisture content. Alternatively, if flooding, jetting and vibration are to be used for placing and compaction, the material shall meet the additional requirements specified in paragraph *Zone 1 and Zone 2 Bedding Material* for material to be placed and compacted by flooding, jetting and vibration. The maximum size shall pass a 2-inch U.S. Standard Series sieve.

The contractor contended that no size limitation existed whatsoever on Zone 3 backfill material unless compaction was by flooding, jetting, or vibration, in which case a two-inch limitation existed. The owner contended that all Zone 3 backfill material must meet the two-inch test.

a. If this dispute came before a court, should it look at evidence outside the writing?

b. What evidence would be relevant if the court were willing to look beyond the writing?

c. Assuming that all the relevant evidence is examined, how should the clause be interpreted?

Changes: Complex Construction Centerpiece

SECTION 21.01 Definitions and Functions of a Changes Clause: *Watson Lumber Co. v. Guennewig*

After award of a construction project, the owner may find it necessary to order changes in the work. The contract documents are at best an imperfect expression of what the design professional and owner intend to be performed by the contractor. Circumstances during the Construction Process develop that may make it necessary or advisable to revise the drawings and specifications.

Design may prove to be inadequate. Methods specified become undesirable. Materials designated become scarce or excessively costly. From the owner's planning standpoint, program or budget may change. Natural events may occur that necessitate changes. For any of many reasons, it often becomes necessary to direct changes after the contract has been awarded to the contractor.

"Changes" problems can arise in a number of different contexts. Most commonly, they involve claims by contractors that they have gone beyond the contract requirements and are entitled to additional compensation, sometimes called *disputed* changes. These are primarily matters of contract interpretation, and the implementation of any system under which the design professional resolves disputed questions. While to some degree this chapter examines certain aspects of the changes process that bear upon such disputes, the principal focus of this chapter is on the power of the owner to order a change, the process by which the change is made, and the effect upon the contract price and time of the change.

Less commonly, disputes may arise when a contractor defends a claim brought against it by the owner for noncompliance with the contract requirements by asserting that the original contract requirements were changed and that the contractor has complied with the contractual obligations as changed.

Preliminarily, a differentiation should be made between *changes, extras,* and *deletions,* though the AIA encompasses all of them in AIA Doc. A201, Art. 7, dealing with changes. However, reference shall be made in this chapter to "extras," which, for example, is the way the court classified the problem involved in the *Watson* case to be reproduced in this section. This *type* of change highlights the difficulties that a long post-completion list of extras can present to the owner or those furnishing funds for the project. Extras usually involve additional or more expensive items than those in the original contract documents. Conversely, a deletion, depending on how it is priced, can adversely affect planning. Although technically both are changes, additions or deletions that affect contract price or time are sensitive areas.

A slight variant of this differentiation involves a comparison between *additional* work and changed work. This can relate to whether or not the owner *must* order changes from the contractor (discussed in Section 21.02).

Changes must be contrasted with *modifications* and *waivers.* Modifications are two-party agreements in which owner and contractor mutually agree to change portions of the work. They are discussed in various contexts in Sections 5.11(C),

19.01(H), and 25.01(D). (One way to avoid formal requirements in a changes clause is to conclude that the work in question was a modification agreed upon by the parties, *not* a change. See Sections 21.03(A) and 21.04(H).)

A change is the term used in construction contracts that allows the owner to unilaterally direct that changes be made *without* obtaining consent to perform the work by the contractor.

A waiver is generally based on the owner's acts that either manifest an intention to dispense with some of the contractual requirements or lead the contractor to reasonably believe that the owner is giving up its right to have required work performed. In such a case, the contractor can recover the full contract price despite not complying with the contract documents (discussed in Section 22.06(E)).

Reference has been made to the changes process and to the changes clause. In 1987, the AIA published a new edition of A201 that not only moved changes from Article 12 to Article 7 but also introduced new terminology. What had been referred to as a change order—that is, the exercise of a contractual power to unilaterally direct changes in the work—has been designated under ¶ 7.3 as a Construction Change Directive (CCD). Under the current A201, a change order (CO) is the wrap-up paperwork after there has been agreement as to price and time adjustment. What A201 calls a change order looks very much like a contract modification.

This change in terminology was designed to expedite the changes process, but it remains to be seen whether it will be adopted by other standard contracts. This chapter uses traditional terminology, such as the changes process and the changes clause. However, when the AIA documents are discussed, AIA terminology is used.

The beginning of this section noted the reason that changes are common in construction projects. Yet an understanding of the changes process requires that the process be looked at from the perspective of the principal parties to the construction contract.

While the contractor recognizes that some changes are inevitable in construction work, it is fearful that it will not receive adequate compensation for the changed work or for the unchanged work that is affected by the change or a proper

time extension and compensation for delay or disruption caused by having to do work out of order. It will also be concerned that poor administrative practices relating to changes will impede its cash flow and place an undue burden on its financial planning.

Another concern expressed by contractors is that unexpected changes may place a drain upon their resources, divert capital that they would like to use on other projects, and require technical skills beyond those they possess.

Contractors also complain about the unilateral nature of the changes clause. They must perform before they know how much they are going to receive in price and time adjustments. While usually changes clauses provide a pricing mechanism through initial decisions by the design professional, the contractor may be fearful that fair compensation and time adjustment will not be granted by the design professional especially when the change will reflect upon the professional competence of the design professional.

To the owner, the changes process looks quite different. As mentioned earlier, the changes process is needed to make those design changes made necessary by subsequent experience and events. But the owner is also fearful that the changes process can expose it to large cost overruns that may seriously disrupt its financial planning and capacity. In that sense, the changes process is an important element of cost control. Owners and, perhaps more important, lenders fear a "loose" changes mechanism. They are fearful that bidders of questionable honesty and competence will bid low on a project with the hope that clever and skillful post-award scrutiny of the drawings and specifications will be rewarded by assertions that requested work is not *required* under the contract and generate claims for additional compensation. A changes clause, which gives either too much negotiating power to the contractor or too much discretion to the design professional or arbitrator, will convert a fixed-price contract into an open-ended cost type tied to a generous allowance for overhead and profit. This is the reason owners and lenders want tight, complete specifications, a mechanical pricing provision, such as unit pricing, and limits on overhead and profit. Their horror is the prospect that the end of the project will witness a long list of claimed extras which, if paid, will substantially increase the ultimate construction contract payout.

The following case explores the changes mechanism mainly from the latter perspective.

WATSON LUMBER COMPANY v. GUENNEWIG

Appellate Court of Illinois, 1967. 79 Ill.App.2d 377, 226 N.E.2d 270.
[Ed. note: Footnotes renumbered and some omitted.]

EBERSPACHER, Justice.

The corporate plaintiff, Watson Lumber Company, the building contractor, obtained a judgment for $22,500.00 in a suit to recover the unpaid balance due under the terms of a written building contract, and additional compensation for extras, against the defendants William and Mary Guennewig. Plaintiff is engaged in the retail lumber business, and is managed by its president and principal stockholder, Leeds Watson. It has been building several houses each year in the course of its lumber business.

* * *

[Ed. note: The project was a four-bedroom, two-bath house with air-conditioning for a contract price of $28,206. The total amount claimed as extras and awarded by the trial court was $3,840.09.]

The contractor claimed a right to extra compensation with respect to no less than 48 different and varied items of labor and/or materials. These items range all the way from $1.06 for extra plumbing pieces to $429.00 for an air-conditioner larger than plaintiff's evidence showed to be necessary, and $630.00 for extra brick work. The evidence, in support of each of these items and circumstances surrounding each being added, is pertinent to the items individually, and the evidence supporting recovery for one, does not necessarily support recovery for another.

* * *

Most of the extras claimed by the contractor were not stipulated in writing as required by the contract. The contractor claims that the requirement was waived. Prior to considering whether the parties, by agreement or conduct dispensed with the requirement that extras must be agreed to in writing, it should first be determined whether the extras claimed are genuine "extras." We believe this is an important area of dispute between these parties. Once it is determined that the work is an "extra" and its performance is justified, the cases frequently state that a presumption arises that it is to be paid for. . . .

No such presumption arises, however, where the contractor proceeds voluntarily; nor does such a presumption arise in cases like this one, where the contract makes requirements which any claim for extras must meet. . . .

* * *

The law assigns to the contractor, seeking to recover for "extras," the burden of proving the essential elements. . . . That is, he must establish by the evidence that (a) the work was outside the scope of his contract promises; (b) the extra items were ordered by the owner, . . . (c) the owner agreed to pay extra, either by his words or conduct, . . . (d) the extras were not furnished by the contractor as his voluntary act, and (e) the extra items were not rendered necessary by any fault of the contractor. . . .

The proof that the items are extra, that the defendant ordered it as such, agreed to pay for it, and waived the necessity of a written stipulation, must be by clear and convincing evidence. The burden of establishing these matters is properly the plaintiff's. Evidence of general discussion cannot be said to supply all of these elements.

The evidence is clear that many of the items claimed as extras were not claimed as extras in advance of their being supplied. Indeed, there is little to refute the evidence that many of the extras were not the subject of any claim until after the contractor requested the balance of the contract price, and claimed the house was complete. This makes the evidence even less susceptible to the view that the owner knew ahead of time that he had ordered these as extra items and less likely that any general conversation resulted in the contractor rightly believing extras had been ordered.

In a building and construction situation, both the owner and the contractor have interests that must be kept in mind and protected. The contractor should not be required to furnish items that were clearly beyond and outside of what the parties originally agreed that he would furnish. The owner has a right to full and good faith performance of the contractor's promise, but has no right to expand the nature and extent of the contractor's obligation. On the other hand, the owner has a right to know the nature and extent of his promise, and a right to know the extent of his liabilities before they are incurred. Thus, he has a right to be protected against the contractor voluntarily going ahead with extra work at his expense. He also has a right to control his own liabilities. Therefore, the law required his consent be evidenced before he can be charged for an extra . . . and here the contract provided his consent be evidenced in writing.

The amount of the judgment forces us to conclude that the plaintiff contractor was awarded most of the extra compensation he claims. We have examined the record concerning the evidence in support of each of these many items and are unable to find support for any "extras" approaching the $3,840.09 which plaintiff claims to have been awarded. In many instances the character of the item as an "extra" is assumed rather than established.[1] In order to recover for items as "extras," they must be shown to be items not required to be furnished under plaintiff's original promise as stated in the contract, including the items that the plans and specifications reasonably implied even though not mentioned. A promise to do or furnish that which the promisor is already bound to do or furnish, is not consideration for even an implied promise to pay additional for such performance or the furnishing of materials. The character of the item is one of the basic circumstances under which the owner's conduct and the contractor's conduct must be judged in determining whether or not that conduct amounts to an order for the extra.

The award obviously includes items which Watson plainly admits "there was no specific conversation". In other instances, the only evidence to supply, even by inference, the essential element that the item was furnished pursuant to the owner's request and agreement to pay is Mr. Watson's statement that Mrs. Guennewig "wanted that". No specific conversation is testified to, or fixed in time or place. Thus it cannot be said from such testimony whether she expressed this desire before or after the particular item was furnished. If she said so afterward, the item wasn't furnished on her orders. Nor can such an expression of desire imply an agreement to pay extra. The fact that Mrs. Guennewig may have "wanted" an item and said so to the contractor falls far short of proving that the contractor has a right to extra compensation.

* * *

Many items seem to be included as "extras" merely because plaintiff had not figured them in the original cost figures.

It is clear that the contractor does not have the right to extra compensation for every deviation from the original specification on items that may cost more than originally estimated. The written contract fixes the scope of his undertaking. It fixed the price he is to be paid for carrying it out. The hazards of the undertaking are ordinarily his. . . .

* * *

"If the construction of an entire work is called for at a fixed compensation, the hazards of the undertaking are assumed by the builder, and he cannot recover for increased cost, as extra work, on discovering that he has made a mistake on his estimate of the cost, or that the work is more difficult and expensive than he anticipated." 17A C.J.S. Contracts § 371(6), p. 413.

Some so called "extras" were furnished, and thereafter the owner's agreement was sought.[2] Such an agreement has been held to be too late. . . .

The judge, by his remarks at the time of awarding judgment, shows that the definition of extras applied in this case was, indeed, broad. He said,

"substantial deviation from the drawings or specifications were made—some deviations in writing signed by the parties, some in writing delivered but not signed, but nevertheless utilized and accepted, some delivered and not signed, utilized and not accepted, some made orally and accepted, some made orally but not accepted, and some in the trade practice accepted or not accepted".

While the court does not state that he grants recovery for all extras claimed, he does not tell us which ones were and were not allowed. The amount of the judgment requires us to assume that most were part of the recovery. It can be said with certainty that the extras allowed exceeded those for which there is evidence in the record to establish the requirements pointed out.

---

[1] We cite as some examples: An extra charge was made for kitchen and bathroom ceilings, concerning which Watson testified that he was going to give these as gifts "if she had paid her bill." According to the testimony, the ceilings were lowered to cover the duct work. We consider it unlikely that the parties intended to build a house without duct work or with duct work exposed. Likewise an "extra" charge was made for grading, although the contract clearly specifies that grading is the contractor's duty. An "extra" charge is sought for enclosing the basement stairs, although the plans show the basement stairs enclosed. An extra charge is sought for painting, apparently on the basis that more coats than were provided in the contract were necessary.

[2] The drain tile around the foundation of the house, according to the evidence, was already in place when it was disclosed to the owner that it was more expensive material. It was only then that the contractor secured the owner's consent to pay for one-half the cost of the more expensive material.

Mere acceptance of the work by the owner as referred to by the court does not create liability for an extra.... In 13 Am.Jur.2d 60, "Building & Cont." § 56, it is stated, that: "The position taken by most courts considering the question is that the mere occupancy and use do not constitute an acceptance of the work as complying with the contract or amount to a waiver of defects therein". Conversation and conduct showing agreement for extra work or acquiescence in its performance after it has been furnished will not create liability.... More than mere acceptance is required even in cases where there is no doubt that the item is an "extra"....

The contractor must make his position clear at the time the owner has to decide whether or not he shall incur extra liability. Fairness requires that the owner should have the chance to make such a decision. He was not given that chance in this case in connection with all of these extras. Liability for extras, like all contract liability, is essentially a matter of consent; of promise based on consideration....

The Illinois cases allow recovery for extra compensation only when the contractor has made his claim for an extra, clear and certain, before furnishing the item, not after. They are in accord with the comments to be found in 31 Ill.Law Rev. 791 (1937). There the author, after reviewing the cases, makes the following analysis:

"The real issue in these cases is whether or not the contractor has, at the time the question of extra work arises, made his position clear to the owner or his agents and that would seem to be the true test in situations where a written order clause is sought to be disregarded. If he does expressly contend that work demanded is extra, the owner certainly cannot be said to be taken unawares, and if orders are given to go ahead it is with full knowledge of the possible consequences."

The contractor claims that the requirement of written stipulation covering extras was waived by the owner's conduct. The defendants quite agree that such a waiver is possible and common but claim this evidence fails to support a waiver of the requirement. There are many cases in which the owner's conduct has waived such a requirement.... In all the cases finding that such a provision had been waived thus allowing a contractor to collect for extras, the nature and character of the item clearly showed it to be extra. Also, in most cases the owner's verbal consent of request for the item was clear beyond question and was proven to have been made at the time the question first arose while the work was still to be finished. The defendants' refusal to give a written order has in itself been held to negative the idea of a waiver of the contract requirements for a written order.... We think the waiver of such a provision must be proved by clear and convincing evidence and the task of so proving rests upon the party relying upon the waiver....

[Ed note: The court ordered a new trial, stating that the contractor could recover only for those extras he could prove were ordered as such by the owner in the proper form, unless he could show that the owner waived the requirements of a writing by clear and convincing evidence.]

Undoubtedly, the court is correct in emphasizing the owner's right to know whether particular work will be asserted as extra. However, in a small construction project, the Illinois court's requirements would place an inordinate administrative burden on the contractor. It would not only have to make clear its position that particular work was extra but also obtain the written change order executed by the owner. If it did not obtain a written change order, it *might* be able to assert the doctrine of waiver if it could persuade the court that the work was extra and that the owner was made aware of this and allowed the work to proceed.

A contractor in a small project, such as that involved in the *Watson* case, might better add a contingency in the contract price to cover small extras that are very likely to be requested rather than comply with the excessively formal requirements set forth by the court in the *Watson* case.

The *Watson* case shows that courts often ignore the cost of complying with rules of law. Undoubtedly, larger projects will bear the administrative costs of doing things correctly and "according to the book." But smaller jobs may not permit "by the book" contract administration.

The changes process involves the following:

1. The exercise of a power to order a change or a direction the contractor contends is a change in the work.
2. Methods for the contract or the parties to price the change and its effect on time requirements.
3. A residuary provision that controls the price in the event the parties do not agree.

SECTION 21.02 Shifts in Bargaining Power

To appreciate the centrality of the changes process to construction, the shifts in bargaining power because of changes must be appreciated. Some of them have been noted in the preceding section. But it is important that these factors be underlined. When a dispute develops, the resolution of the dispute, whether by design professional, arbitrator, or court, may be influenced in some way by the bargaining power of the parties in structuring the changes process or at the time the dispute develops. In addition, the bargaining power is important at the time the owner must decide whether to issue a change order.

When preparing to engage a contractor, the owner, as a rule, has superior bargaining power. The hotly competitive construction industry and the frequent use of competitive bidding usually allow the owner to control many aspects of the construction contract terms.

A few contractors will deliberately bid low and drive out more prudent and experienced contractors with the hope that they will be able to demand and receive additional compensation by pointing to design ambiguities and amassing large claims toward the end of the project.

The contractor who is performing moves into a much stronger bargaining position. This is clearly so if the owner *must,* for either practical or legal reasons, order any additional work from the contractor. It would be in an even stronger bargaining position if it could refuse to execute the change unless there were a mutually satisfactory agreement on the effect of the change on price or time. Yet any bargaining advantages to the contractor by being in the position to refuse to do the work until there is an agreement are usually tempered by contract provisions that require the contractor to do the work even if there is no agreement on the price or time. (To counter this, a contractor can assert that the direction is a cardinal change discussed in Section 21.03(A).) This can be made even worse by the dominance some owners have over the changes process through their control over the purse strings. This power, exercised either directly or through the design professional, to withhold payment until the contractor agrees on price and time can exert immense pressure on the contractor to accept whatever the owner or design professional is willing to pay.

The changes mechanism can operate adversely to the contractor if the owner makes many small changes but is niggardly in its proposals for adjusting the price. (This can backfire, however, generating claims by the contractor, particularly in a losing contract, that the cumulative effect of any changes has created a cardinal change.)

To sum up, the changes mechanism on the whole favors the owner except if it is dealing with a clever claims-conscious contractor who uses the changes mechanism to extract large amounts of money at the end of the job. Judicial resolution of disputes that involve changes may take into account the owner's strong position, particularly if the owner seems to have abused its power.

SECTION 21.03 Types of Changes

A number of classifications in addition to those noted in Section 21.01 can be made that can help one understand the changes process.

A. Cardinal Change

While Section 21.04(B) discusses the limits on the owner's power to order changes, it is important to look at the cardinal change, a concept developed in federal procurement law that relates to the power to order a change. The cardinal change originated not because of potential abuses created by a changes clause but because of jurisdictional aspects of federal procurement law. Before 1978, a dispute between a contractor and a federal procurement agency would have to first go before an agency board of contract appeals if it arose *under the contract.* To bypass the board and bring the dispute before the U.S. Court of Claims (now the U.S. Court of Federal Claims), the contractor had to show a *breach of the contract.* To accomplish this, the contractor would seek to show a course of conduct—usually a large number of changes, drastic changes, or other wrongful agency conduct that it could establish as a breach of contract.

Despite legislation enacted in 1978 that allows the contractor to choose which route to take,[3] the

---

[3]41 U.S.C.A. §§ 601 et seq.

concept still has utility in both federal procurement and private disputes.

Clearly, any changes clause must be interpreted. Not *any* change can be ordered. A direction that goes beyond the scope of the work, if that is the standard, need not be obeyed. Such an order shall be called a "one-shot" scope change, to distinguish it from the more common "nibbling" or "aggregate of changes," currently the most common claim.

Suppose the contractor complies. Now there has been an agreement. But does it fall within the jurisdiction of the changes clause with its procedural and pricing mechanisms? The facts may indicate that the contractor, knowing it could not be compelled to perform because the order went beyond the scope of the work, agreed to waive any "beyond the scope" defense and allow the work to be governed by the changes clause. In the absence of such evidence, the parties have simply made a new agreement. If the parties cannot agree on price, the contractor is entitled to reasonable compensation.[4]

A cardinal change is considered a breach largely because of its origin in federal procurement jurisdictional issues. Can the contractor choose to treat it as a serious material breach and terminate any further contractual obligations it owes the owner?[5]

The real nature of the direction or order can now emerge. It is simply a proposal, an order, that the contractor can *choose* to accept. It is *not* a breach of contract unless the owner asserts that it will *not* proceed further unless it *is* accepted. If so, this is a contract repudiation, and the contractor can terminate and recover any damages it may have suffered.

The other type of cardinal change does not occur all at once, as in the one-shot direction, but occurs throughout the performance of the contract. It consists of many changes,[6] drastic changes,[7] or other conduct that has gone beyond the reasonable expectations of the contractor and made the transaction different from what the parties had in mind

when they made the contract.[8] This *is* a breach. But what are the remedies?

Here two scenarios emerge. One involves the contractor's deciding in the *middle* of the project that it has had enough, that it feels that it can walk off the job because the total effect of the owner's conduct constitutes a material breach.[9] If upheld, the contractor can recover the reasonable value of its work based on restitution[10] or damages.[11] (The former is selected in losing contracts.) The second, and more common, scenario involves *completion* of the work by the contractor and a claim for the reasonable value of the services and materials, along with an allowance for overhead and profit, bypassing any contract price or guaranteed maximum price (GMP).[12]

B. Constructive Change

Again, reference must be made to federal procurement law. Claims for breach of contract could be taken *only* to the Court of Claims before 1978. A claim had to go before any procuring agency appeals board if it arose "under the contract." Relief had to be provided by statute or by a clause in the contract. To keep claims within the appeals board, the constructive change developed. Had a change order been issued, any claim clearly came *under* the contract.

---

[4]*Nat Harrison Assoc., Inc. v. Gulf States Util. Co.*, 491 F.2d 578 (5th Cir.1974).

[5]See Section 34.04(A).

[6]*Wunderlich Contracting Co. v. United States*, 240 F.2d 201 (10th Cir.1957) (6,000 changes).

[7]*Saddler v. United States*, 152 Ct.Cl. 557, 287 F.2d 411 (1961) (doubling excavation in small contract).

[8]*Allied Materials & Equip. Co. v. United States*, 215 Ct.Cl. 406, 569 F.2d 562 (1978). *Wunderlich Contracting Co. v. United States*, 173 Ct.Cl. 180, 351 F.2d 956 (1965) (cumulative effect of magnitude and quality of changes). But see *Hensel Phelps Constr. Co. v. King County*, 57 Wash.App. 170, 787 P.2d 58 (1990). In this case, the painting subcontractor's claim for a cardinal change entitling it to *quantum meruit* was based upon a drastic acceleration, a redoing of its work, and the stacking of trades (more than one trade working at the same time). The court pointed to the contractor having been compensated under the changes clause for its additional expenses and to other clauses indicating the contractor had taken certain risks. The court also pointed to the fact that there had been no fundamental alteration of the project itself, such as the shape or the square footage of the surfaces painted.

[9]See Section 34.04(A).

[10]See Section 27.02(E).

[11]See Section 27.02(D).

[12]*C. Norman Peterson Co. v. Container Corp. of Am.*, 172 Cal.App.3d 628, 218 Cal.Rptr. 592 (1985) (cardinal change eliminated GMP); *Rudd v. Anderson*, 153 Ind.App. 11, 285 N.E.2d 836 (1972). See also Section 27.02(F).

In many cases, the contractor claimed that a direction or order was a change. The contracting officer acting for the procuring agency refused to issue a change order because he asserted that the work ordered was *within* the contract. The fiction of the constructive change allowed the boards of appeal to take jurisdiction by concluding that a change *should* have been issued if it were to later agree with the contractor. It is then *as if* a change order *had* been issued. The doctrine was also used if specifications were defective, requiring the contractor to do additional work. Even though the doctrine has no *jurisdictional* significance because of 1978 federal legislation,[13] it still merits comment.

Federal procurement disputes clauses require the contractor to keep working pending resolution of a dispute. If a contractor wishes to stop work, it may assert that its claim is based on breach and does not arise *under* the contract. Here the agency may assert that there has been a constructive change and the contractor must continue *under* the contract. The changes clause precludes claims after final payment. Post-final payment claims may be asserted to be based on *breach,* not "under the contract." In private contracts, the problem can still arise, although not in a jurisdictional sense.

Suppose that there is a dispute between design professional and contractor over disputed work. The design professional contends that the work is called for under the contract, and the contractor claims it is not. Suppose the contractor performs the work but makes clear that it considers the position of the design professional unjustified and that it intends to claim additional compensation.[14]

Here, as in so many other aspects of the Construction Process, it is important to recognize the design professional's power to interpret the document and his quasi-judicial role. If the design professional determines that the work falls within the contract, many construction contracts make his decision final unless it is overturned by arbitration or litigation. If the contractor later requests additional compensation for the work and is met with the contention that no change order has been issued, the absence of a written change order should not bar the contractor's claim as long as it made clear that it intended to claim additional compensation. However, the claim should be denied *if* the design professional's decision is binding and has *not* been overturned. A contractor dissatisfied with the design professional's decision should invoke any process under which that decision can be appealed. If it later is determined that the design professional had been incorrect and the decision is overturned, the absence of a written change order should not bar the claim. Although no change order had been issued, there was a *constructive* change order—one *should* have been issued.[15] (Some problems of this type are handled under "waiver," largely because the party directing the change had no authority to do so.[16])

Another type of constructive change developed in federal procurement law—the constructive acceleration (discussed in Chapter 26 dealing with time).

C. Deductive Change (Deletion)

Changes clauses usually permit the owner to delete a portion of the work, sometimes known as a deductive change. While as a rule, a deductive change is lumped together with changes that *add* to the contractor's contractual commitment, it raises special problems. First (as seen in Section 21.04(B)), the power to change the work typically is limited to those changes that fall within the scope of the work. It is clear that a deductive change is not susceptible to be measured by that standard.

Similarly, problems can develop over whether deleted work reduces the overhead and profit that are usually added in the event of additional work (discussed in Section 21.04(I)).

Finally, a problem similar to that discussed in Section 21.04(E)—the duty to order additional work from the contractor—can arise when there is a deductive change. Suppose the owner wishes to delete part of the work and offer it to another contractor. Unless the owner had good reason to do so, or at least acted in good faith, this would be a breach of the contract.

---

[13]See note 3 supra. Before 1968, constructive changes were also used to get around the *Rice* doctrine—*United States v. Rice,* 317 U.S. 61 (1942)—having barred recovery for additional expense of *unchanged* work.

[14]See 48 C.F.R. § 52.243-1(c) (1992); AIA Doc. A201, ¶ 4.3.7.

[15]*Chris Berg, Inc. v. United States,* 197 Ct.Cl. 503, 455 F.2d 1037 (1972).

[16]*Weeshoff Constr. Co. v. Los Angeles County Flood Control Dist.,* 88 Cal.App.3d 579, 152 Cal.Rptr. 19 (1979).

D. Minor Change

It sometimes becomes necessary to make minor changes in the drawings and specifications that are not intended to affect the contract price or completion date. AIA Doc. A201, ¶ 7.4.1, gives the architect authority to execute a written change order for such changes so long as the changes are consistent with the intent of the contract documents. (Even if the architect is given authority to make minor changes, it is best not to make changes when the owner is available.)

Suppose the contractor believes the change is *not* minor and demands additional time or compensation. Paragraph 7.4.1 requires that the contractor carry out the order and dispute its propriety later. The contractor should make its intention to claim additional compensation clear at the time it performs the work.[17]

E. Tentative Change

As mentioned in Section 21.01, the classic changes clause requires the contractor to comply with the change order before any agreement is made on price. Contractors generally are not happy with this arrangement. But owners may not wish to be obligated to pay for the change without knowing what it will cost. For that reason, either by contract provisions or by practice, some owners will issue a tentative change, obtain a price from the contractor, and then decide whether to implement it. If time does not permit the processing of such a tentative change order, the owner can direct the change unilaterally, and the contractor must comply.

One problem that sometimes arises when a tentative changes process is invoked is the contractor's claim that it should be compensated for its expenses in preparing price quotation if the change is not ordered. Owners usually resist such claims, contending that the cost of preparing a price quotation is similar to the overhead costs of preparing a bid at the outset.

SECTION 21.04 Change-Order Mechanism

A. Judicial Attitude Toward Changes Mechanism

From a planning standpoint, a changes mechanism is essential. Design flexibility and cost con-

trol cannot be accomplished without a system for changing work.

Yet judicial attitude toward the changes mechanism often determines how courts will interpret contract language and how quickly courts will find that the changes mechanism has been waived.

Judicial attitude is reflected by language in opinions that must pass on these questions. For example, Massachusetts held that the contractor could not recover for extra work and concluded:

> Although it seems a hardship for the [contractor] not to be able to recover for the extra work which apparently it performed in good faith, yet such failure results from its not obtaining from the architect or his agents written authority to perform the work. . . .[18]

A federal court granted recovery to a subcontractor despite the absence of formal requirements by pointing to the fact that "[f]rom the beginning of the contract work the parties to the subcontract ignored the provisions as to written orders and proceeded with the work with little or no regard for them."[19] It was obvious that the court thought the formal requirements simply technicalities that should not preclude a contractor from recovering for work performed beyond the contract requirements.

In a decision denying recovery for extras ordered orally by the city engineer, the Pennsylvania Supreme Court seemed unwilling to force the City of Philadelphia to pay for work not properly ordered and that could not be returned by the city.[20] Here, control over public funds predominated. Likewise, the West Virginia Supreme Court denied recovery for work performed at the direction and with the knowledge of the county court president and other board members because of statutory requirements that all proceedings of the county court be entered into the record books of such court. In justifying its decision, the court stated:

> So the requirements of the statute and the decision that county courts must enter of record its orders for the expenditure of the funds must not be construed as meaningless, but must be enforced for the benefit of the whole public and not for the benefit of any partic-

---

[17]AIA Doc. A201, ¶ 4.3.7.

[18]*Crane Constr. Co. v. Commonwealth,* 290 Mass. 249, 195 N.E. 110, 112 (1935).
[19]*Ross Eng'g. Co. v. Pace,* 153 F.2d 35, 49 (4th Cir.1946).
[20]*Montgomery v. City of Philadelphia,* 391 Pa. 607, 139 A.2d 347 (1958).

ular individual who may suffer on account of a mistaken reliance upon invalid acts of individual officers, however unfortunate, harmful or deceptive any such acts may have been, unless there is a predominant reason not to do so. To extend the doctrine of liability in every instance because of unjust enrichment is to open the door to all such claims as have not been properly authorized. . . . Too often is it apparent that the expenditures first authorized under a contract are enlarged by so-called "extras" without proper authorization, either intentionally or otherwise, and any such practice should certainly not be given judicial sanction. Nor should sloven management of county affairs be approved. In the absence of special reason, based on competent evidence, why the paramount consideration inherent in the legislative act requiring formal orders by county courts should not be made effective, there should be no digression therefrom as to its enforcement.[21]

It is more difficult to grant recovery to a contractor where there has not been compliance with formal requirements in contracts for public works. However, as seen in the *Watson Lumber* case (reproduced in Section 21.01), courts are also protective of inexperienced owners who are building their first home. Such owners must be made aware of price increases for proposed changed work.

The judicial attitude toward formal requirements must take into account the contract language, the experience of the parties in construction, the need to protect public funds, the reasonable expectations of the owner, and the potential unjust enrichment that can result if the contractor is unable to recover for admittedly extra work.

B. Limitation on Power to Order Changes

While a changes clause is essential to construction projects, it is, as suggested earlier in this chapter, amenable to abuse. While a contractor can expect some changes, the changes clause should not be a blank check for the owner to order the contractor to do anything it wishes, compel it to perform before there has been an agreement on price, and take care of the compensation later. This can be even more abusive if the owner seeks to avoid any increased cost of performing unchanged work or any cost incident to disruption of the contractor's method of performance.

To determine whether a proposed change can be ordered, the overall character of the work must be considered. More particularly, the change must permit the work to retain its specific character, must be similar in nature to the work already contracted for, must not change the extent of the total performance in an intolerable manner, and must be within the capability of the contractor, taking into account its technical skill and its financial capability.

Almost all construction contracts place some limit on the owner's power to direct changes unilaterally. Sometimes a limit is based upon a certain percentage of the contract price. The difficulty with this is that the limitations may not be known until the contract is close to full performance. More commonly in the United States, the power to direct changes is limited to the scope of the work. For example, an owner may be able to order a 10% increase in the floor space of a residential home or that a carport be built or even a swimming pool but not that a beach house be built twenty miles away.

As mentioned in Section 21.03(C), the scope limitation is difficult to apply when the owner wishes to delete part of the work. Here the only suitable guideline is the reasonable expectation of the contractor.

Suppose the owner points to the changes clause and directs that the contractor accelerate its performance. Must the contractor comply? For example, if the contract time is 500 days after the Notice to Proceed, can the owner direct that the work be done in 400 days?

AIA Doc. A201 does not address this question directly. Paragraph 7.3.1 simply provides that the owner may "order changes in the Work within the general scope of the Contract consisting of additions, deletions or other revisions." Work is defined in ¶ 1.1.3 as "the construction and services required by the Contract Documents," language contemplating the physical structure and not the time during which it must be completed. In the light of the drastic nature of acceleration, A201 should not empower the owner to direct an acceleration. A201 should be interpreted to allow only quantity and quality changes.[22]

---

[21]*Earl T. Browder, Inc. v. County Court*, 143 W.Va. 406, 102 S.E.2d 425, 432–433 (1958).

[22]*Mobil Chemical Co. v. Blount Bros. Corp.*, 809 F.2d 1175 (5th Cir.1987) (acceleration order assumed to be a breach of contract). Similarly, see *Hensel Phelps Constr. Co. v. King County*, supra note 8.

Can the owner order a delay or a ''stretch-out'' by a change? For example, suppose the owner directs the 500-day time be increased to 600 days. It is unlikely that ¶ 7.3 would permit such a stretch-out. Yet ¶ 14.3 does allow the owner to *suspend* performance for any reason. If so, the contractor receives a price adjustment if the suspension increases its cost, plus profit on the increased cost.

It can be contended that if the owner can suspend, it can stretch out. And if it can stretch out, it should be able to speed up. But there is a difference between a suspension, with its generous remedy, and a stretch-out and certainly a speed-up, with the indeterminate remedy (some consider it generous) of ¶ 7.3.6.

The federal procurement system speaks directly to this point, giving the federal agency the power to order accelerations.[23] Even if a pure acceleration *were* permitted, there is still a scope limit. The amount of the acceleration cannot exceed what could be reasonably expected and that for which a reasonable price adjustment can be made.

Can the owner through the changes process take over control of means, methods, and techniques by which the contract is to be performed? Under AIA Doc. A201, ¶ 3.3.1, this power belongs to the contractor. Such a change would not be within the owner's power. The changes process is limited to quantity and quality changes.

A limitation sometimes ignored is the duration of the power to order changed work. As seen in Section 26.04, some contracts use substantial compliance as the benchmark for determining compliance with the contractor's time commitment. It is at that point that the owner can make effective use of the project even though some things still need to be completed. It is at that point that the owner's power to direct changes unilaterally should terminate. Otherwise, the contractor might have to perform for a period of time not contemplated at the time the contract was made.

Another limitation that a contractor with strong bargaining power would like is that the contractor can demand that the owner show evidence that it is in the position to pay for additional work that the owner may wish to have performed. The AIA gives the contractor, at the time the contractor

makes the contract, and promptly from time to time thereafter, the power to determine whether a strong likelihood exists that the contractor will be paid for the work it performs.[24] A contractor without this contractual protection either at the beginning of the project or when substantial changes are ordered will have to take the owner's ability to pay into account at the time it decides the price for which it is willing to do the work.

Even if such protection is not found in the contract, if the owner directs a substantial change, the obligation of good faith and fair dealing should require that the owner satisfy the contractor that it is in the position to pay for the work that it has ordered.

C. Authority to Order Change

In the absence of any contract clause dealing with the question of authority to make changes, the question of who can order a change is controlled by doctrines of agency. The owner can order changes. Members of the owner's organization may also have the authority to order changes expressly, impliedly, or by the doctrine of apparent authority. Usually, carefully thought-out construction contracts specify which members of the owner's organization have the power to order changes.

Neither the design professional nor, surely, the project representative has inherent authority by virtue of his position to direct changes in the work. Yet sometimes the ultimate outcome is based on a contractor's claim that a *direction* had been given by someone clearly *without* the authority to direct the change when this unauthorized act was known by authorized officials. For example, *Chris Berg, Inc. v. United States*[25] involved a contract awarded under federal procurement rules. The contractor painted a stairway clearly not required under the contract to be painted. A field memo had been executed by a project engineer and a resident engineer, neither of whom had authority to issue a change order, which impliedly directed the contractor to paint the stairway in question.

Responsible officials in the agency *knew* that the plaintiff was doing work not called for under the contract. The plaintiff had requested a ruling on

---

[23]48 C.F.R. § 52.243-4 (1992).

[24]AIA Doc. A201, ¶ 2.2.1.
[25]Supra note 15.

whether the stairway should be painted and that it be advised of the government's interpretation of the field memo. The court stated that the agency had ample opportunity to warn the plaintiff that it was painting a stairway not called for under the contract. The court concluded that this was a *constructive change* and the contractor was entitled to an equitable adjustment. In addition, the constructive change was based on the contractor's having painted certain areas after having received the tacit and undoubted oral approval of the project engineer.

Similarly, in *Weeshoff Construction Co. v. Los Angeles County Flood Control District*,[26] the contractor performed additional work in a road construction contract largely under the direction of the agency's site inspector. The court noted the problem of the site inspector's lack of authority to waive a written change-order requirement. However, it concluded that the change order had already been issued, inasmuch as the procuring agency had threatened to do the work with its own forces if the contractor did not do it. The court concluded that the contractor was justified in relying on the site inspector's statements.

These two decisions demonstrate that directions by project representatives coupled with other facts that would make denial of recovery unjust may be the basis for the contractor's receiving additional compensation. (Excusing formal requirements is discussed in Section 21.04(H).)

Construction contracts frequently specify who has authority to issue change orders. As an illustration, AIA Doc. A201, ¶ 7.2.1, states that a change order is prepared by the architect and signed by the architect, the contractor, and the owner. Paragraph 7.3.1 states that the construction change directive is again prepared by the architect but signed only by the architect and the owner. Finally, ¶ 7.4.1 requires a written order signed only by the architect for a minor change.

This raises an issue that is rarely addressed. All documents connected to the changes process are prepared by the architect. But ¶ 7.1.2 requires the architect's agreement, as well as that of the owner, for change orders and construction change directives. Can the architect prevent a change either by

refusing to prepare the change order or refusing to sign either the change order or the construction change directive?

Suppose AIA Doc. B141 is used. Paragraph 2.6.13 requires the architect to prepare change orders and construction change directives "for the Owner's approval." (Under ¶ 3.3.3, services in connection with these documents are additional services.) Is agreement by the architect a purely ministerial act, one that does not give the architect the power to refuse to sign a change order or constructive change directive?

As seen in the *Duncan* case reproduced in Section 10.04(B), the law places professional responsibility on the architect because he has been registered and qualified by the state, a responsibility that can take precedence over contract obligations. Surely the architect could refuse to sign if the change clearly would violate law. In addition, suppose his refusal to sign is based on his belief that his professional reputation would be ruined by the finished project. Would this justify his refusal?

To be sure, the architect who refuses to sign a change order or (using AIA current terminology) a construction change directive authorized by the owner and agreed to by the contractor may lose a client. Yet the owner should take seriously any objection the architect may have to signing these documents. This would be part of the covenant of good faith and fair dealing the owner owes the architect. As noted, the architect would not be required to sign a change order or construction change directive that clearly violates building laws. Yet even if the matter involved only aesthetics, the architect's professional standing should be taken into account by the owner. Admittedly, the money to be spent is that of the owner, and the performance is to be that of the contractor. Yet the owner should meet and confer with the architect and give serious consideration to the architect's professional judgment. In the end, the owner may decide to go ahead with the change. But its decision to do so must be made in good faith. If the architect cannot continue without either violating the law or compromising his professional integrity, the contractual relationship should be terminated. Neither party has breached. But the architect should be paid for work he has performed for which he has not been paid based on restitution.

Just as the owner can give the design professional express authority, the owner may cloak the

---

[26]Supra note 16.

design professional with apparent authority.[27] The owner's acts can reasonably lead the contractor to believe that the design professional has the authority to change the work. Suppose the owner knows that the design professional is executing change orders and makes no objection or, even further, pays based on change orders issued by the design professional. This activity could reasonably lead the contractor to believe that the design professional has authority to change the work. The apparent authority doctrine is less likely to be successfully invoked by the contractor if the project involves public work.[28]

Fletcher v. Laguna Vista Corp.[29] illustrates enlarged authority. Changes were required to be ordered in writing by owner and architect. However, in concluding that the architect had authority to execute a written change order, the court stated:

> . . . the manner in which the parties themselves have interpreted the contract through their course of dealings is of utmost importance. The record in this case is filled with testimony to the accord that both [owners] and [contractor] had relied on architect Frye to make adjustments in the contract sum and had abided by his decision. [Owners] knew that there would be at least a slight overage in the sums spent by [contractor] for overhead but had never objected. [Owners] accepted decreases in the cost of millwork which were incorporated into a change order signed only by architect Frye and the contractor. The parties themselves have interpreted the contract to allow an increase and a decrease in the contract sum with only the written signature of architect Frye. Even if the contract does not grant this authority to architect Frye, the parties through their course of dealings have interpreted and modified the document so as to place in the hands of architect Frye the final authority to authorize increases and decreases in the contract sum.[30]

It is possible for the owner to give the architect this authority. However, if the basis for enlarging the architect's normal authority is the architect's power to interpret the contract documents, the case is incorrect. The court failed to distinguish the normal power given to the design professional to interpret the contract documents from the power to order changes.

Sometimes there is insufficient time to obtain authorization from the owner for work needed immediately because of impending danger to person or property. The contractor should have authority to do such emergency work. Frequently, contracts provide that extra work in emergencies can be performed without authorization.[31] Even without express or implied authority, the contractor should be able to recover for emergency work based on the principle of unjust enrichment.

D. Misrepresentation of Authority

If the contractor performed the additional work at the order of the design professional and this order was beyond the latter's actual authority, what recourse does the contractor have?

First, the contractor would seek to establish that the design professional had *apparent* authority to order the work. But in the absence of the owner's having led the contractor to believe that the design professional had this authority, the contractor will not be successful. Likewise, any claim that the owner has been unjustly enriched by the work would not be successful. The contractor would be considered a volunteer and denied recovery.

Next, the contractor would look to the design professional. The threshold question would be whether the contractor reasonably relied on the misrepresentation of authority. In many cases, reliance would not be reasonable, because design professionals typically are not given this authority and because the contractor should have checked with the owner. But the reliance element is often downgraded, and in any event, suppose compliance with the order had been reasonable under the circumstances. In a case of misrepresentation of authority, the contractor can recover from the design professional.[32] The recovery would be the cost of performing the unauthorized work and any cost of correction made necessary to make the work con-

---

[27]See Section 4.06.
[28]This is discussed in greater detail in Section 21.04(H).
[29]275 So.2d 579 (Fla.Dist.Ct.App.1973).
[30]Id. at 580–581 (the court's footnotes omitted).

[31]AIA Doc. A201, ¶ 10.2.1 (contractor can act in emergency affecting safety of persons or property).
[32]*Brown v. Maryland Cas. Co.*, 246 Ark. 1074, 442 S.W.2d 187 (1969) (dictum).

form to the contract documents. It would be unlikely that the design professional held to have misrepresented his authority could recover from the owner on the theory of unjust enrichment.

Suppose the contractor *had* recovered from the owner because the latter's acts cloaked the design professional with apparent authority. It must be remembered that the design professional acted without authority even though the acts of the owner may have created apparent authority. Suppose the owner seeks to transfer this loss to the design professional because of the latter's misrepresentation of authority. Passing by the question of whether the owner had suffered any real loss other than having work it did not choose, the loss was caused *both* by the design professional's misrepresentation of authority and by the owner's acts making it appear that the design professional *had* the authority. Because the more culpable act appears to be the misrepresentation of authority, this loss should be borne by the design professional.

Yet if the misrepresentation resulted in the value of the property having been increased, the owner has received a benefit. Although the law does not explicitly divide losses in such cases, it would be fair to do so here. Another method of sharing the loss would be to give the owner any difference between what the owner had to pay the contractor and the enhanced value of the project because of the unauthorized work.

At this point, a distinction should be made between the contractor's recovery having been based on implied authority or on apparent authority. In the former case, there has been no wrongdoing by the agent, who was authorized to act even though authority was not expressly given. For example, the architect's representation to the subcontractor in the *Bethlehem* case noted in Section 28.07(J) was authorized implicitly though not explicitly. Had the architect's statement been *beyond* his authority—express or implied—recovery against the owner would have had to have been based on apparent authority. If so, the owner would have had a claim against the architect. In the *Bethlehem* case, there was no unjust enrichment of the type present in the preceding discussion. As a result, if the subcontractor's recovery against the owner had been based on an unauthorized representation by the architect, the owner should be reimbursed for what it paid the subcontractor.

E. Duty to Order Change from Contractor

The changes clause clearly gives the owner the power to order that work within the general scope of the contract be performed by the contractor. Suppose the owner asserts the right to award work within the general scope of the contract to a third party or perform that work with its own forces. The power to do this would give the owner some bargaining advantage in negotiating the price for the changed work with the original contractor.

Most changes clauses do not deal with this question specifically. The contractor could contend that giving the owner this power would be unfair, putting the contractor at the mercy of a changes clause without the compensating advantage of being able to perform all work within the general scope of the contract. Giving the contractor the right to do the changed work, it could assert, would not put the owner at a disadvantage. Usually there are pricing formulas that apply to the changed work if the parties do not agree on a price. Although as a practical matter the owner may find it inexpedient to have the original contractor working side by side with a substitute or its own employees, the express language of most changes clauses does *not* give the contractor the power to perform changed work.

In *Hunkin Conkey Construction Co. v. United States*,[33] the contractor had entered into a contract with the Corps of Engineers for the construction of a dam. During performance, subsurface problems developed that necessitated a significant change in design. The contractor and the Corps discussed alternative designs but were unable to agree on a price for an alternative design desired by the Corps. As a result, the Corps negotiated a contract with a third party to perform the alternative design work at a cost of $200,000 *less* than that proposed by the original contractor.

After completion of the project, the contractor contended that the work should have been awarded to it. The Court of Claims was not persuaded that the power to direct changes meant that the government had a duty to order those changes from the original contractor. But the court directed its attention to another provision of the contract. This provision allowed the government to undertake or award other contracts for "additional

---

[33]198 Ct.Cl. 638, 461 F.2d 1270 (1972).

work" and required the original contractor to co-operate with other contractors or government employees. This, according to the Court of Claims, had to be read together with the changes clause. When so done, it was clear that the government was not obligated to award the additional work to the contractor.

Although the conclusion that the work in question was additional can be debated, the line of demarcation may be useful both from a practical, administrative basis (it would be difficult for the *two* contractors in such a dispute to work side by side) and from the normal expectations of the owner. The owner may not want to be "locked in" to the contractor for "extra" work. Conversely, though, the contractor may assert an "expectation" that its price was predicated on a "monopoly" on "extras."

F. Formal Requirements

As noted in (C), construction contracts generally require that change orders be written and signed by the person or persons authorized to execute change orders. Obviously, it is *best* to have the change order issued and price agreed on *before* the work is started. If price cannot be agreed on before the work is begun, issuance of a change order does assure the owner that the changed work will be compensated in accordance with a changes clause formula if no agreement is reached and assures the contractor that the owner will not contend that the work in question falls within the contract requirements.

Is it best to require that the change order be issued before the work is begun? If the change order *is* issued after the work is begun or completed, issuance generally forecloses any question of whether the work was within the contract requirements. However, *not* requiring that the change order be issued *before* the work is begun invites issuance of oral change orders with the assurance by the owner that a written change order will follow. Although this sometimes may be necessary under certain circumstances, suppose the owner *denies* issuing an oral change order. (In many cases, the owner admits giving a particular direction but contends it was not a change.) In such cases, the owner will insist that no additional compensation should be paid, and the contractor will contend that a requirement for a writing has been dispensed with by the issuance of an oral change order.

It is advisable to require that the change order be issued *before* the start of work even if this means some delay while the written change order is issued.[34]

Under AIA Doc. A201, ¶ 7.3, the sequence is the issuance of a written construction change directive, commencement of the work with a simultaneous *attempt* to agree on price and time, and the residual power in the architect to decide price or time changes if the parties cannot agree.

Some contracts do *not* require a written change order in *advance* of the work being done. The contractor can recover if a *subsequent* written change order is issued. Here the obvious advantage is speed, but the disadvantage is the possibility of subsequent disputes over whether the work was extra. At the very least, the issuance of a written change order eliminates this difficulty. Selection of a less rigorous changes mechanism may reflect lack of confidence that a more structured clause will be effective. This may be attributed to the belief that courts will not deny compensation to the contractor despite the absence of a written change order if it appears that the owner has directed that work be performed that the court decides was beyond the contract documents.[35]

Another problem that surfaces with regularity relates to the interaction between a changes clause with *its* formal requirements and other provisions of the contract that may have *different* formal requirements. This arises most frequently when the contractor encounters subsurface conditions different from those specified or normally expected and seeks to receive an equitable adjustment. (Inasmuch as this is a special problem, it is discussed in Chapter 25.)

---

[34]In *Uhlhorn v. Reid*, 398 S.W.2d 169, 175 (Tex.Ct.App.1965), the clause stated that "No extra work or charges under this contract will be recognized or paid for unless agreed to in writing before the work is done or the charges made."

[35]As examples, see *Universal Builders, Inc. v. Moon Motor Lodge, Inc.*, 430 Pa. 550, 244 A.2d 10 (1968), and *Ecko Enter., Inc. v. Remi Fortin Constr., Inc.*, 118 N.H. 37, 382 A.2d 368 (1978). See also Section 21.04(G).

G. Intention to Claim a Change

A lengthy list of extra work submitted after completion is a sad, though not infrequent, part of construction work. Section 21.03(B), which treated constructive changes, noted that a vehicle for such claims is the contention that particular direction or instructions given by owner or design professional were changes entitling the contractor to extra compensation. One method of minimizing this problem is to require the contractor to give a written notice within a designated number of days after the occurrence of any event that the contractor will claim as the basis for an increase in the contract price. Paragraph 4.3.7 of AIA Doc. A201 requires such a notice as specified in ¶ 4.3.3. Except in an emergency, the notice must be given before beginning the work.[36]

A similar provision is used by the California Department of Public Works. A general contractor doing work for the state must submit a written protest to the state architect within thirty days after receiving a *written* order from the state architect to perform any disputed work. Failure to do so precludes compensation for the work.[37] The California provision also requires that the protest notice specify in detail how requirements were exceeded and the resulting appropriate change in cost. This content requirement goes considerably beyond ¶ 4.3.7, which simply requires written notice of an intention to make a claim in the contract price.

H. Excusing Formal Requirements

Not uncommonly, the change-order mechanism is disregarded by the parties. This may be because of unrealistically high expectations by the drafters, time pressures, or unwillingness by the parties to make and keep records. When contractor claims for extras are denied by the owner because of the absence of a written change order or intention to claim an extra, contractors frequently assert that the requirements have been waived. Waiver questions can be divided into three issues:

1. Is the requirement waivable?
2. Who has the authority to waive the requirement?
3. Did the facts claimed to create waiver lead the contractor to reasonably believe that the requirements have been eliminated or indicate that the owner intended to eliminate the requirements?

Except where the waiver concept cannot be applied to public contracts,[38] formal requirements can be waived. They are not considered to be an important element of the exchange and are often viewed as simply technical requirements.[39]

As a rule, only parties who have authority to order changes have authority to waive the formal requirements.[40] Usually, only the owner or its authorized agent can waive the formal requirements, although a design professional with authority to order changes should have authority to waive the writing requirement.

Acts that can be the basis for waiver generally are conduct by the owner that indicated no written change order would be required[41] or oral orders by the owner or its authorized representative.[42] Generally, if it appears that the owner orally ordered the changed work, it will be assumed that the writing requirement has been waived.

Unfortunately, courts do not differentiate between cases where the owner *admits* ordering the work but claims it was within the contract requirements from owner admissions that the work was extra and was ordered. One purpose for having a requirement for a written change order is to obviate the question of whether the work was extra. The absence of a writing indicates at the very least that the owner did not consider the work extra.

---

[36]*Pioneer Roofing Co. v. Mardian Constr. Co.*, 152 Ariz. 455, 733 P.2d 652 (1986) (oral order by prime to subcontractor in emergency).

[37]This provision was interpreted in *Acoustics, Inc. v. Trepte Constr. Co.* and *Weeshoff Constr. Co. v. Los Angeles County Flood Control Dist.*, discussed in Section 21.04(H).

[38]In *Delta Const. Co. of Jackson v. City of Jackson*, 198 So.2d 592 (Miss.1967), the court held that waiver could operate against a municipality but not where the formality in question was a supplemental agreement on price. In *Metro. Sanitary Dist. of Greater Chicago v. Anthony Pontarelli & Sons, Inc.*, 7 Ill.App.3d 829, 288 N.E.2d 905 (1972), the court held that a provision giving the public agency the right to recover illegal and excessive payments meant that there was no way that the required approval by the board of trustees could be waived.

[39]See Section 21.04(A).

[40]See Section 21.04(C).

[41]*Huang Int. Co. v. Foose Constr. Co.*, 734 P.2d 975 (Wyo.1987).

[42]*City of Mound Bayou v. Roy Collins Constr. Co.*, 499 So.2d 1354 (Miss.1986); *T. Lippia & Son, Inc. v. Jorson*, 32 Conn.Supp. 529, 342 A.2d 910 (1975).

However, courts seem to disregard this factor. If they determine that the work was extra and the contractor was ordered to perform it, absence of a written change order is not likely to prevent the contractor from recovering. But in a public contract, recovery could be denied even in such a case unless a *constructive* change is found.

Another factor that must be taken into account is whether the owner or design professional knew of the facts that would be the basis for a claim of a change by the contractor. For example, in *Moore Construction Co. v. Clarksville Department of Electricity*,[43] one issue was whether one separate contractor could claim delay damages caused by another separate contractor. However, the claimant did not give a notice of an intention to claim additional compensation as required by the predecessor of AIA Doc. A201, ¶ 4.3.7. The court held that this notice could be waived. It noted that all the participants knew that the claimant was being delayed and that the delays were not the fault of the claimant. The claimant had been given a time extension through a job site memo, but no format change order had been issued.

The court noted that the claimant could have reasonably believed that the owner would not demand that it submit a written notice of a claim for additional compensation because it had been granted a time extension. In addition, throughout this project the requirement that the contractor submit written notices had not been followed. (In fact, most of the formal requirements appear to have been abandoned.) In addition, the court concluded that the owner had not shown that it had been prejudiced by the failure of the contractor to give the required written notice.

Owner payment based on oral change orders by the design professional or conceivably by a resident engineer or project representative can be a waiver. Payment can lead the contractor to reasonably believe that the design professional has the authority to order changes orally.[44] If payment is accompanied by a notice that makes clear to the contractor that the formal requirements are *not* being waived, no waiver should be found.

It may be useful to examine language in some waiver cases. In *Rivercliff Co. v. Linebarger*,[45] the court found a waiver of the writing requirement, stating:

> For a second ground, appellant contends that . . . the trial court should not have made any allowance to the contractor because the extra work was not authorized in accordance with the terms of the contract. This contention appears to be supported by the terms of the contract, which provides that extras must be approved in writing prior to execution. This provision was not complied with but it does not constitute a defense available to appellant, because, as we hold, a strict compliance with this provision of the contract was waived by appellant in this instance. It is not disputed that the extra excavation was done with the knowledge and at the direction of Smith who was not only the architect supervising the work for Rivercliff but was also a part owner of the appellant corporation. From his testimony we gather that he refused to approve an allowance for extras mainly because he did not think the contractor was entitled to anything as a result of the changed method of constructing the foundation. It appears that other changes in construction had been made and paid for where no written change order had been previously issued. Although it was shown that several such changes had been made and paid for during the construction of the four buildings, yet Mr. Smith testified that only one written change order had been made.[46]

Another court stated:

> Several situations may form the basis for waiver: (1) when the extra work was necessary and had not been foreseen; (2) when the changes were of such magnitude that they could not be supposed to have been made without the knowledge of the owner; (3) when the owner was aware of the additional work and made no objection to it; and (4) when there was a subsequent verbal agreement authorizing the work.[47]

Finally, in a subcontractor context, a court stated:

> . . . the court finds that plaintiff is entitled to recover for several of the jobs which it performed that were not incidental to the building of a cofferdam and that were

[43]707 S.W.2d 1 (Tenn.App.1985), affirmed March 24, 1986.
[44]*Oxford Dev. Corp. v. Rausauer Builders, Inc.*, 304 N.E.2d 211 (Ind.App.1973).

[45]223 Ark. 105, 264 S.W.2d 842 (1954).
[46]264 S.W.2d at 846.
[47]*Nat Harrison Assoc., Inc. v. Gulf States Util. Co.*, 491 F.2d 578, 583 (5th Cir.1974).

required to be done due to the insistence or inaction of James. Where a party is aware that extra work is being done without proper authorization but stands by without protest while extra work is being incorporated into the project, there is an implied promise that he will pay for the extra work. See *United States v. Klefstad Engineering Co.*, 324 F.Supp. 972 (W.D.Pa.1971). In *Klefstad*, the prime contractor was given the right to recover for work it performed which was the responsibility of the subcontractor; whereas in the present case the subcontractor, E & R, is entitled to recover for work it performed which was the responsibility of the prime contractor, James. Whether the theory of recovery is considered as quasi-contract, implied-in-fact, or promissory estoppel, a subcontractor is entitled to be compensated for extra work it performed as the result of inducing statements and conduct by the prime contractor.[48]

The difficulty in predicting when a court will find that formal requirements have been excused is demonstrated by two California cases. In *Acoustics, Inc. v. Trepte Construction Co.*,[49] the court held that compliance with contractual provisions for written orders is indispensable and denied recovery to a contractor who had been *verbally* ordered to perform changed work. The court noted that the state inspector who had ordered the changes had no authority to waive the formal requirement and that the contractor erred in expecting payment without a written change order from the state architect as required in the contract.

Eight years later, in *Weeshoff Construction Co. v. Los Angeles County Flood Control District*,[50] which involved a similar claim with the only differentiation being the apparent knowledge by the awarding agency that a site inspector was directing that changes be made, the court came to the opposite result based on, though not explicitly, the *constructive* change rationale.

Another way of avoiding the formalities of the changes clause is to conclude that the work was not a "change." Section 21.03(A) noted that work may be considered outside the changes clause. Likewise, work made necessary because of errors of the de-

sign professional must be paid for despite the absence of the formal requirements set forth in the changes clause. Such work can be regarded as a remedy for defective work given to the contractor.

I. Pricing Changed Work

Pricing changed work is an important part of the changes mechanism. A tightly drawn pricing formula, along with clear and complete contract documents, can discourage a deliberately low bid made with the intention of asserting a long list of claims for extra work.

Where possible, work should be compensated by any unit prices specified in the agreement. If there is no specifically applicable unit price specified in the contract, compensation should be based upon an analogous unit item, taking into account the difference between it and the required work. It is important, however, to take into account the possibility of great variations in units of work requested and the effect on contractor costs. (Refer to Section 17.02(D).)

In 1987, the AIA amplified its changes clause by requiring a mechanism be used to effectuate agreement on price and time adjustments. For example, A201, ¶ 7.3.1, requires the construction change directive to state "a proposed basis for adjustment" of time and price. Paragraph 7.3.3 provides formulas that can be used as the basis of an agreement between the owner and the contractor. When the contractor receives the directive under ¶ 7.3.4, it must proceed with the change in the work and indicate whether it agrees or disagrees with the proposal for adjusting price and time. If it agrees, under ¶ 7.3.5, it signs the construction change directive. If it does not respond promptly or disagrees with the proposed method of adjustment, the adjustment will be determined by the architect "on the basis of reasonable expenditures and savings . . . including, in the case of an increase in the Contract Sum, a reasonable allowance for overhead and profit." Similarly, in the absence of an agreement on time, ¶ 7.3.8 requires that it be referred to the architect for determination.

Some contracts provide that if the parties cannot agree, the contractor will be paid cost plus a designated percentage of cost in lieu of overhead and profit. Pricing extra work in this fashion can discourage the contractor from reducing costs. However, it may be difficult to arrive at a fixed

---

[48]*United States v. Guy H. James Constr. Co.*, 390 F.Supp. 1193, 1223 (M.D.Tenn.1972), affirmed 489 F.2d 756 (6th Cir.1974).
[49]14 Cal.App.3d 887, 92 Cal.Rptr. 723 (1971).
[50]Supra note 16.

fee for overhead and profit when the nature of the extra work cannot be determined until it is ordered. For this reason, it is likely that cost plus a percentage of cost will continue to be used to price changed work.[51]

When a cost formula is used, the design professional must examine the cost items to see whether they are reasonable and required under the change order. The changes clause can specify that only certain material, equipment, and labor costs or direct overhead costs are to be included as cost items.[52]

Public contracts—federal and state—frequently provide that if the parties cannot agree, there will be an "equitable" adjustment of the price.[53]

Suppose the change *reduces* the work. Does deductive change require that overhead and profit also be deducted for deleted work? Paragraph 7.3.6 of AIA Doc. A201 permits the architect to add a reasonable allowance for overhead and profit *only* for an increase in the work but appears to preclude reduction if work is deleted. On the other hand, a federal procurement decision held that overhead and profit would be deducted when work is deleted.[54]

Suppose a changes clause specifies that the parties will agree on compensation for changed work and the parties cannot agree. Early cases concluded that such an agreement was not enforceable as simply "an agreement to agree."[55] Today, courts would very likely determine a reasonable price for

the work where the parties do not agree.[56] One case involving changed work on a subcontract held that where the parties could not agree on a price, the subcontractor could receive cost plus overhead and profit.[57]

A changes clause that specifies that the parties will agree should also provide an alternative if the parties do not agree, such as a pricing formula or a broadly drawn arbitration clause.

Earlier federal procurement cases seemed to prohibit contractors from recovering for added costs of doing *unchanged* work caused by the change order.[58] However, federal procurement regulations have been changed to permit the contractor to recover additional costs of performing unchanged work.[59]

In addition to claims involving the cost of performing unchanged work, an excessive number of change orders of the type that can be considered cardinal changes are, along with other acts of the owner, often the basis for the now frequent delay damage claim (discussed in Section 26.10).

SECTION 21.05 Effect of Changes on Performance Bonds

When sureties were usually uncompensated individuals, courts held that any changes in the contract between the principal and the obligee would discharge the surety. This would be unjust where a professional surety bond company is used, especially as changes and modifications are common in construction contracts. Most surety bonds provide that modifications made in the basic construction contract will not discharge the surety. Some changes clauses permit changes up to a designated percentage of the contract price without notifying the surety.

[51]Federal procurement regulations have developed weighted guidelines for determining profit. The factors taken into account are degree of risk, relative difficulty of the work, size of job, period of performance, contractor's investment, assistance by government, and subcontracting. Each of these factors is weighted depending on the particular procurement. An illustration of how these guidelines are used can be found in *Norair Eng'g. Corp.*, ASBCA 10856, 67-2 BCA ¶ 6619.
[52]AIA Doc. A201, ¶ 7.3.6.
[53]48 C.F.R. § 52.243-1(b) (1992). For a municipal contract containing a similar clause, see *Fattore Co. v. Metro. Sewerage Comm.*, 454 F.2d 537 (7th Cir.1971).
[54]*Algernon Blair, Inc.*, ASBCA 10738, 65-2 BCA ¶ 5127.
[55]See Section 5.06(F).

[56]*Purvis v. United States*, 344 F.2d 867 (9th Cir.1965).
[57]*Hensel Phelps Constr. Co. v. United States*, 413 F.2d 701 (10th Cir.1969).
[58]*United States v. Rice*, supra note 13.
[59]48 C.F.R. § 52.243-1(b) (1992).

PROBLEMS

1. A was the architect for a construction project being built by C for O. The specifications called for

soundproof tile for certain room ceilings. C wanted to put in tile of a certain brand that was represented by the manufacturer to be soundproof. A insisted that another brand, which was more expensive, be used. C claimed that this would be an extra and demanded a change order. A claimed that as judge of performance, he was ruling that the tile that the contractor wanted to use did not meet the contract requirements. C used the tile specified by A and, after completion of performance, demanded that he be paid extra for the tile he had used. A instructed O not to pay because the tile used was required under the contract and because the contract required that a written change order be issued before there could be extra payment for any work. Is A correct in his position? (Assume that no tiles are actually soundproof but that the tile demanded by A resisted sound better and cost more than the tile that C wanted to use.)

2. C contracted with O to build a five-story commercial building under a changes clause allowing changes in "the general scope of the work." Must C perform an order to

a. Add an underground garage?

b. Build a sauna bath on the roof?

c. Substitute toilets with a lower water tank capacity for those designated?

Must C delete a library taking up one half of one floor if ordered to do so? If the library *were* deleted, how should the price be reduced?

Payment: Money Flow as Lifeline

SECTION 22.01 The Doctrine of Conditions

This chapter deals principally with the process by which the contractor is paid for performing work required by the construction documents. The rules that control this process are derived principally from the contract documents, most notably the basic agreement and the general conditions. An important backdrop to these contract provisions is that part of the legal doctrine of conditions that deals with the order of performance.

This important legal doctrine seeks to protect the actual exchange of performance specified in the contract. A party should not have to perform its promise without obtaining the other party's promised performance. For example, suppose a contract is made under which a supplier agrees to deliver supplies to a small manufacturer. The supplies arrive by truck at the warehouse. However, the seller's truck driver refuses to unload the supplies until payment is made. The buyer's employee refuses to pay until the supplies are unloaded and placed on the buyer's receiving dock. Obviously, such a dispute could have been dealt with in the first instance by appropriate contract language dealing with the question of whether delivery precedes payment. When the contract does not deal with this question, the law must determine the sequence of performance.

Before proceeding to the legal resolution of this question, it should be noted that had the seller unloaded the supplies and not been paid, the seller would have had a valid claim for payment. Conversely, had the buyer paid but the supplies not been unloaded, the buyer would have had a valid

claim for the value of the supplies or return of the money. However, neither party wishes to exchange its actual performance for a legal claim. Each would prefer to receive the other's performance before rendering its own.

Likewise, buyer and seller of real estate wish to avoid performance without obtaining the other's performance. The seller does not wish to transfer ownership by deed without obtaining the money. Conversely, the buyer does not wish to pay without receiving the deed. Protecting the desire of each is usually accomplished in a real estate purchase by use of a third party or escrow holder. The seller will transfer the deed, and the buyer will deliver the money to the third party. Each believes that the third party will effectuate the exchange.

In the absence of a third-party system or specific contract clause dealing with sequence of performance, the law must determine whether performance or payment must come first. The common law required that the performance of services *precede* the payment for those services.[1] This protected the party receiving the services by permitting that party to withhold payment until *all* the services were performed. Not only did this rule allow the paying party to avoid the risk of paying and then not receiving performance, but it also allowed the paying party to dangle payment before the performing party, a powerful incentive for rendering performance.

However, the "work first and then be paid" rule is disadvantageous to the performing party. The latter must finance the entire cost of performance.

---

[1] E.A. FARNSWORTH CONTRACTS 614 (2d ed. 1990).

The party performing services must take the risk that *any* deviation, however trivial, would enable the paying party to withhold money greatly in excess of the damages caused by the deviation. The performing party must assume the risk that the paying party might *not* pay after complete performance, leaving the performing party with a legal claim.

The doctrine of conditions is central to the discussion of progress payments (discussed in Section 22.02) and the right of the contractor to recover despite noncompliance (discussed in Section 22.06). Though less important, it is also relevant to retainage and final payment, discussed in Sections 22.03 and 22.05, respectively.

SECTION 22.02 Progress Payments

A. Function

As indicated in the preceding section, the doctrine of conditions requires that work be performed before any obligation to pay arises. Such a rule in a construction context places severe financial obligations on the contractor and creates a substantial risk of nonpayment. To avoid these problems, construction contracts generally provide for periodic progress payments made monthly or at designated phases of the work. This section examines the process by which progress payments are made and some common problems involved in this process.

B. Schedule of Values

To facilitate the computation of the amount to be paid under progress payments in a fixed-price contract, the contractor is generally required to submit to the design professional a schedule of values before the first application for payment. This schedule, when approved, constitutes an agreed valuation of designated portions of the work. The aggregate of the schedule should be the contract price. Contracts sometime permit adjustments as work proceeds.

C. Application for Payment Certificate

The contractor submits an application for a progress payment. This application is generally submitted a designated number of days before payment is due. The application is usually accompanied by documentation that supports the contractor's right to be paid the amount requested. If payment is requested for materials or equipment stored on or off the site, AIA Doc. A201, ¶ 9.3.2, requires the contractor to submit bills of sale or comply with other procedures that establish the owner's title to such property or equipment. If the material is stored off site, ¶ 9.3.2 requires, in addition to owner approval, the contractor to show that the material has been properly insured and will be transported to the site.

Allowing the contractor to retain possession of material for which the owner has paid can create legal problems, mainly claims by creditors of the contractor or the trustee in bankruptcy if the contractor is declared bankrupt.[2] To avoid this, some owners require that material be stored only in bonded warehouses.

Some contracts require the contractor to give assurance or proof that it has paid its subcontractors and suppliers when progress payments applications are made. Alternatively, such contracts allow the prime contractor to submit documents executed by subcontractors or suppliers that give up their rights to any mechanics' liens.

Neither the standard contract published by the American Institute of Architects (AIA) nor the one published by the Engineers Joint Contracts Documents Committee (EJCDC) requires such assurance or proof of payment or submission of lien waivers at the time *progress* payments are made. This is justified in part by the formidable administrative requirement that can result in large projects if partial lien waivers or evidence of partial payment is required. These steps may not be necessary if other techniques exist that protect against such risks. One function of a retainage (discussed in Section 22.03) is to protect against liens. Also, ¶ 9.3.3 in AIA Doc. A201 requires the contractor to warrant to the best of its knowledge, information and belief that all work and materials will be free of liens. This will not preclude liens but may create a right against the contractor or surety under a payment bond.

Allowing design professionals to withhold progress payments when they learn that subcontrac-

---

[2]See *Traveler's Indem. Co. v. Ewing, Cole, Erdman & Eubank*, 711 F.2d 14 (3d Cir.1983), cert. denied 464 U.S. 1041 (1984) (architect not liable for not warning owner of bankruptcy risk in paying for materials stored off site).

tors are not being paid or that liens have been or are likely to be filed may be sufficient owner protection. Usually such information is communicated quickly to the design professional.

The AIA has acceded to requests by subcontractor associations to include more protection for subcontractors in A201. For example, ¶ 9.3.1.2 bars the contractor from including in the payment application amounts it does not intend to pay a subcontractor or supplier "because of a dispute or other reason." Similarly, ¶ 9.6.3 requires the architect, on request, "if practicable" to furnish information to a subcontractor regarding the percentages of completion or amounts applied for by the contractor and action taken thereon on account of work performed by the subcontractor.

Also, the AIA recognizes the increasingly difficult problems of cash flow for the contractor. Paragraph ¶ 9.3.1.1 allows the contractor to include in its application amounts that had been authorized by construction change directives though not formally included in change orders. Similarly, ¶ 7.3.7, dealing with changes, allows the contractor to include in payment applications "amounts not in dispute."

D. Observations and Inspections

Before issuing a payment certificate, the design professional visits the site to determine how far the work has progressed. This is the basis on which progress payments are made.

The inspection should uncover whatever an inspection principally designed to determine the *progress* of the work would have uncovered. Each new edition of A201 has sought to limit the architect's responsibility during the progress payment certification process. Currently, under ¶ 9.4.2, issuance of a certificate is "based on the Architect's observations at the site and the data comprising the Application for Payment." Paragraph 9.4.2 also states that the certificate warrants only that the work complies with the contract requirements "to the best of the Architect's knowledge, information and belief." The certificate is

> subject to an evaluation of the Work for conformance with the Contract Documents upon Substantial Completion, to results of subsequent tests and inspections, to minor deviations from the Contract Documents correctible prior to completion and to specific qualifications expressed by the Architect.

Finally, ¶ 9.4.2 states that the issuance of a certificate does not represent that the architect has made exhaustive or continuous on-site inspections, reviewed the construction methods, reviewed copies of requisitions from subcontractors and suppliers, or has made any examination to determine how previous payments have been used.

Some owners, particularly developers influenced by their lenders, require more intensive inspections and place more responsibility on the architect. To accomplish this, these owners require the architect to give warranties that go beyond the experience of the design professional and the services she has contractually committed to perform. To deal with this, the AIA in its B141, ¶ 4.11, requires that the architect receive for her "review and approval" certificate language "at least fourteen days prior to execution" and prevents the owner from requiring certifications "that would require knowledge or services beyond" the scope of B141.

E. Amount Certified for Payment

The amount certified for payment depends on the pricing provisions in the construction contract. In fixed-price contracts, the pricing benchmark is the contract price. In cost contracts, the principal reference point for determining payments is the allowable costs incurred by the contractor. In unit-priced contracts, the progress payments are based on the number of units of designated work performed.

Each pricing provision commonly uses a retainage system under which a designated amount is withheld from progress payments to provide security to the owner. (Retainage is discussed in Section 22.03.)

The computation of the amount to be paid is facilitated by the use of a schedule of values discussed in (B). Such an agreed schedule determines the extent to which the work has progressed. Smaller contracts without a schedule of values sometimes provide that specified payments are to be made at designated phases of the work with an appropriate allowance for retainage where used.

Despite a properly prepared schedule of values, measurement problems can develop. A significant amount of time elapses between the contractor's ordering material and equipment and incorporating that material and equipment into the work. The contractor prefers payment as early as possible.

The owner prefers not to have to pay until the material and equipment have been incorporated into the project. Paragraph 9.3.2 of AIA Doc. A201 generally requires that the work be incorporated before payment is to be made. However, as noted in (C), it allows for payment for materials and equipment not incorporated but stored on and off the site. Also, as noted in (C), payment prior to incorporation into the project raises legal and insurance questions.

Construction contracts allow the design professional to make *partial* certification for payment. Partial certificates authorize payments of an amount less than requested by the contractor. Usually such certificates are based on the design professional's determination that the work has not progressed to the extent claimed by the contractor or on a determination that the work does not meet the requirements in the contract documents.[3] If an architect intends to withhold the certificate either in whole or in part, ¶ 9.4.1 of AIA Doc. A201 requires the architect to notify the contractor of such a decision and to give reasons for this action. Under ¶ 9.5.1, failure by the contractor and architect to agree on payment amount allows the architect to issue a certificate for payment for the amount "for which the Architect is able to make such representations to the Owner."

Construction contracts frequently give the design professional the power to revoke a previously issued certificate. Revocation can be effectuated by a partial certificate or by withholding a certificate for work that has been performed. One reason for revocation is the discovery of defective work that had been the basis of a previously issued certificate.

Paragraph 9.5.1 of AIA Doc. A201 gives the architect power to make payment adjustments.[4] These adjustments are designed to protect the owner from losses that have occurred or may occur in the future. Losses can relate to nonconforming work, nonpayment of subcontractors and suppliers, or claims made by other contractors or other third parties against the owner for which the contractor may be responsible.

Although the law would give the owner certain offset rights under these circumstances without a contractual right of offset, it is advisable to expressly recognize this right. Doing so avoids the necessity of establishing a right to offset and can exceed the protection given by law. However, the design professional should not be unreasonable or arbitrary in determining when and how much to withhold from payment amounts earned by the contractor.[5] Issuing partial certificates or, more important, withholding certificates can result in contractor default. For this reason, language that requires design professional and contractor to discuss and negotiate on these matters is useful.

Suppose the contractor in a public contract demands the right to a hearing before the design professional withholds progress payments. In *Signet Construction Corp. v. Borg*,[6] the court held that constitutional requirements of due process did not require a formal hearing before withholding. Informal hearings are sufficient. The court was concerned that such a requirement could cause delay and unreasonably burden the public entity. Increasingly, contractors invoke constitutional rights in their disputes with state entities.[7]

Federal procurement policy allows reduction or suspension of progress payments despite the absence of any present contract breach by the contractor where the latter's financial position makes future nonperformance likely.[8]

F. Time of Payment

Increasingly, the federal government[9] and the state legislatures have been responsive to complaints by

---

[3]*Berry v. Blackard Constr. Co.*, 13 Ill.App.3d 768, 300 N.E.2d 627 (1973), held that compliance with contract documents did not condition contractor's right to recover progress payments. This is justified only if there had been a course of conduct of payment despite noncompliance. Normally, the architect should not certify for payment work that the architect knows is not in compliance with the contract documents.

[4]See *Howard S. Lease Constr. Co. v. Holly*, 725 P.2d 712 (Alaska 1986) (need not notify contractor before withholding). But see AIA Doc. A201, ¶¶ 9.4.1, 9.5.1.

[5]*City of Mound Bayou v. Roy Collin Constr. Co.*, 499 So.2d 1354 (Miss.1986) (altering requests without proof of poor or incomplete work); *S.I.E.M.E., S.r.1.*, ASBCA 25642, 81-2 BCA ¶ 15,377 (improper to withhold $10,000 when $1,000 would have been sufficient to complete the few remaining items).

[6]775 F.2d 486 (2d Cir.1985).

[7]See Sections 18.04(E) and 34.03(J).

[8]This power was exercised in *National Eastern Corp. v. United States*, 477 F.2d 1347 (Ct.Cl.1973).

[9]Federal Prompt Payment Statutes 31 U.S.C.A. § 3901 et seq.

contractors and subcontractors that payments are not made promptly to them. For example, California legislation establishes deadlines for making both progress and retention payments, places a cap on withholding where amounts are disputed, and imposes penalties for noncompliance.[10]

Care must be taken to check for legislation dealing with the payment process. Some such legislation covers public contracts, some private contracts, and some both. Also, such legislation should be checked to see whether it provides that parties can, by contract, change the legislative rules.

G. Passage of Title

As work proceeds, the materials and equipment are procured and installed in the building or attached to the land. Problems can develop that relate to when title passes from the contractor to the owner. There may be a preliminary question as to when title passes between the vendor of the product and the purchaser, in the construction context, the contractor or subcontractors. But for purposes of the construction contract, the principal question relates to when the materials and equipment belong to the owner.

Passage of title is important. If the materials and equipment still belong to the contractor, the contractor has the risk of loss, and the creditors of the contractor can seize the property or equipment in payment for the contractor's debts. Conversely, once the title has passed to the owner, the owner has the risk of loss, and its creditors may have rights in the property.

The risk of loss is usually dealt with by insurance. Either the contractor or, more commonly, the owner takes out insurance on all materials and equipment that reach the site or, in the case of materials and equipment stored, at an earlier point. But typically, the owner would like to have title pass to it as soon as possible to avoid the possibility that creditors of the contractor will claim that they can seize the property of the contractor in payment

for the contractor's debts or because of unsatisfied court judgments.

By law, title passes when the materials and equipment are incorporated into the project. Other solutions can be expressed in the contract, such as when materials are delivered to the site or when payment is made.

H. Assignment of Payments

Contractors or subcontractors often must borrow funds to operate their businesses. Lenders commonly require collateral to secure them against the possibility that the borrower will not repay the loan. Sometimes collateral consists of funds to be earned under specific construction contracts. Lenders often seek information from owners regarding the construction contracts whose payments are to be used as security and assurances by the owner that payments will be made to the lender or that checks will be issued jointly to lender and contractor.[11]

Rather than rely on a promise, more commonly, lenders demand that assignments be made to them of the payments. An assignment transfers the right to receive payment and effectuates a change of ownership in the rights transferred. It is a more substantial security than a promise. The party making the transfer is the *assignor*. The party to whom ownership is transferred is the *assignee*. The party owing the obligation being transferred is the *obligor*. Using the fact pattern of the prime contractor seeking a loan, the prime is the assignor, the bank the assignee, and the owner the obligor.

Such assignments were difficult if not impossible to accomplish early in English legal history. But modern law not only makes such assignments possible but also encourages them. However, these assignments should not put the obligor in a substantially worse position.

---

[10]West Ann. Cal.Bus. & Prof.Code § 7108.5 (applies to payments by contractors on private and local public contracts); West Ann. Cal.Civ.Code § 3260.1 (applies to private contracts); West Ann. Cal.Pub.Cont.Code § 10261.5 (applies to state agencies); Id. at 10202.5 (applies to state public works agencies.) See also Va. Ann. Code § 11.62-1 et seq. (public contracts).

[11]Sometimes subcontractors seek to borrow funds, and information or assurances are sought from the prime contractor. Whether giving information and assurances is legally enforceable by the lender is often difficult to determine. *Central Nat'l Bank & Trust Co. of Rockford v. Consumers Constr. Co.*, 5 Ill.App.3d 274, 282 N.E.2d 158 (1972), held a prime contractor liable to a lender who lent to a subcontractor in reliance on letters sent by the prime contractor. Yet another Illinois decision held that a lender could *not* recover from the owner. *Bank of Marion v. Robert "Chick" Fritz, Inc.*, 9 Ill.App.3d 102, 291 N.E.2d 836 (1973).

Encouragement of assignments has gone to the extreme of invalidating contract clauses that prohibit assignment of such rights where the assignments are given as collateral to obtain loans.[12] Terms in contracts precluding assignment of payments for such purposes are invalid.

Although the validity of such assignments no longer raises serious questions, other legal problems exist. They center principally on the extent to which the original contract obligation can be changed because a third party (the assignee) now owns contract rights.

An obligor prime contractor was held liable to the assignee bank who had lent to an assignor subcontractor when the prime contractor did not see to it that the funds were paid to the bank *despite* the prime's assent to the assignment's being conditioned on its being given no greater burden.[13]

Normally, the obligor must pay the assignee after it has received notice of the assignment. But sometimes the obligor, such as the owner, can pay to the assignor contractor to enable it to complete the contract.[14]

Modification of existing contracts whose rights have been assigned can be made "in good faith and in accordance with reasonable commercial standards" as long as payments assigned have not as yet been earned.[15] The obligor (owner) can offset against the assignee (lender) amounts owed it by the assignor (prime contractor).[16]

I. Lender Involvement

The assumption in this section has been that the owner is making payments to its prime contractor. Sometimes lenders involve themselves directly in the payment process. However, this can create the type of problem that was presented in *Davis v. Nevada National Bank*.[17] The lender was held liable for paying directly to a prime contractor despite warning by the owner that there were structural defects. The owner successfully pursued a claim against the contractor, but the contractor's bankruptcy wiped out that claim. As a result, the owner brought a claim against the lender. Recognizing that this was an unusual decision and seeking to reassure lenders that it would not expose them to unreasonable risk, the court noted:

> Nothing we have said should be interpreted beyond the comparatively narrow confines of the instant case. Specifically, under usual construction loan terms and conditions, no lender should consider itself at risk if it elects not to generally inspect the progress of the construction of a project financed by the lender. Nor is a lender to consider itself at risk if it volitionally elects to inspect and does so negligently or ineffectively. A lender also has no duty, under our instant holding, either to withhold payment at borrowers' requests or to inspect upon such requests, for construction deficiencies or omissions of a type that inevitably will occur in all projects and that commonly are remedied by a contractor as part of a "punch list" prior to project completion and the release of retained funds. Stated otherwise, lender liability may arise under a construction loan when: (1) the lender assumes the responsibility or the right to distribute loan proceeds to parties other than its borrower during the course of construction; (2) the lender is apprised by its borrower of substantial deficiencies in construction that affect the structural integrity of the building; (3) the borrower requests that the lender withhold further distributions of loan proceeds pending the satisfactory resolution of the construction deficiency; (4) the lender continues to distribute loan proceeds in complete disregard of its borrower's complaints and without any bona fide attempt to ascertain the truth of said complaints; and (5) the borrower ultimately is damaged because the substance of the borrower's complaints was accurate and the borrower is unable to recover damages against the contractor or other party directly responsible for the construction deficiencies.[18]

J. Joint Checks

One of the principal aims of the payment process is to make certain that payments go to those who

---

[12]*Mississippi Bank v. Nickles & Wells Constr. Co.*, 421 So.2d 1056 (Miss.1982); *Aetna Cas. & Sur. Co. v. Bedford-Stuyvesant Restoration Constr. Corp.*, 90 A.D.2d 474, 455 N.Y.S.2d 265 (1982). See U.C.C. § 9-318(4).

[13]*Bank of Yuma v. Arrow Constr. Co.*, 106 Ariz. 582, 480 P.2d 338 (1971).

[14]*Fricker v. Uddo & Taormina Co.*, 48 Cal.2d 696, 312 P.2d 1085 (1957).

[15]U.C.C. § 9-318(2).

[16]Id. at § 9-318(1).

[17]737 P.2d 503 (Nev.1987).

[18]Id. at 506.

have provided labor or materials. Those who have provided labor or materials for private projects, such as contractors, subcontractors, and suppliers, usually have the right to assert a mechanics' lien on the property they have improved when they are not paid. This gives them a security interest in the property and a right to have the property sold to pay their claims.

To avoid this and also to ensure that those who are doing the work will continue to have the incentive to do the work, those who make payments, such as owners and prime contractors, seek to avoid diversion of the funds from those to whom the funds should go. One method sometimes used to avoid diversion is to have a payment issued by a joint check. For example, an owner may issue a joint check to a prime contractor for work that includes work by a subcontractor or materials furnished by a supplier. The names of both payees appear on the check. Each will have to endorse the check for it to be converted into cash. This should avoid the possibility that the prime contractor will take the funds and not pay those whose services or materials have provided the basis for the payment. Similarly, prime contractors sometimes issue payments to subcontractors by using a joint check to ensure that the subcontractor pays sub–subcontractors or suppliers.

This method, though apparently simple, creates legal problems. The first is whether the contract allows the payor to make payments by the use of a joint check. Using the owner as payor as an illustration, the power to pay by joint check initially depends upon whether the prime contract allows this method of payment. The imposition of the joint check method may generate opposition by the prime contractor who does not wish to have to obtain cooperation from the joint payee to cash the check. (Of course, this is the purpose for the joint check method.) Without a contractual power giving the owner to use the joint check method, its unilateral use would be a breach of contract.[19] The contractor can claim with justification that the contract requires payment to it and that any attempt to interject a third party interferes with its management prerogatives. Where this is discussed in advance and jointly resolved, it is likely that the

clause will allow the owner to use the joint check process if the contractor is in default in paying subcontractors or suppliers.

Another contractual solution is to bar the issuance of joint checks if the contractor requests the owner not to do so because of a good-faith dispute it has with a subcontractor or supplier. The more difficult problem is whether the owner would impliedly have the power to use the joint check process if it had reasonable grounds to believe that the payee will not pay those to whom payments should be made. It is likely that the obligation of good faith and fair dealing would give the owner this power under such circumstances, but the outcome of such a dispute would be difficult to predict.

Even if the payor has the power to pay by joint checks, problems can develop if the process misfires, principally because the payee obtains an endorsement from the copayee but the latter does not actually receive payment or the amount owed the copayee and the amount of the payment are not the same. In a California case,[20] the owner drew a check payable jointly to the contractor and a supplier, after which the supplier waived its mechanics' lien rights. Both parties endorsed the check. Proceeds were paid to the prime contractor, who gave its personal check to the supplier. Unfortunately for the supplier, the check was dishonored because the contractor did not have sufficient funds in its account.

The supplier then sued the contractor and received a judgment for the amount of the debt but could satisfy only a portion of it from the assets of the prime contractor. It then brought a lawsuit against the surety of the prime contractor for the balance. The court held that the unpaid supplier did not lose its right to claim on the bond despite its having waived its lien and endorsing the check. Yet it is clear that the joint check process failed because the supplier did not insist on being paid at the time it endorsed the check but relied upon a personal check issued by the prime.

Special problems develop in the implementation of the joint check system if the check is a progress payment. This issue was addressed by the Arizona Supreme Court in 1990 in a dispute between a sup-

---

[19]*Piedmont Eng'ng & Constr. Corp. v. Amps Elec. Co.*, 162 Ga.App. 564, 292 S.E.2d 411 (1982).

[20]*Ferry v. Ohio Farmers Ins. Co.*, 211 Cal.App.2d 651, 27 Cal.Rptr. 471 (1963).

plier and an owner over a lien asserted by the former.[21] Initially, the court, citing cases from a number of jurisdictions, adopted the joint check rule. It stated:

> ... when an owner or general contractor makes a materialman and a subcontractor joint payees on a check that includes payment for labor and materials furnished, and no other agreement exists between the materialman and the owner or general contractor as to the allocation of the proceeds, the materialman, by endorsing the check, will be deemed to have been paid the money due him, up to the amount of the joint check.[22]

Under the joint check rule, the unpaid materialman cannot recover or assert a lien (it may be able to recover on the bond), as it is conclusively presumed that it has taken all the money of that payment and applied it to *any* debts owed it by the copayee even if it has received only part or none of the money owed.

The court held that when the payment is a progress payment, the implied intention of the parties is that the proceeds of the payment are allocated based upon the labor or materials furnished by the joint payee that are the basis for *that* payment. This modifies the general joint check rule under which the materialman in this case would be held to have received *all* payments owed it under the contract up to the amount of the payment of the check. The court also held that if the general contractor still owes money on the particular supply contract in question, the joint check rule is inapplicable and the materialman (if it is unpaid) may, to that extent, pursue its lien claim.

A case decided by the highest Massachusetts court in 1991 also demonstrates the complexity of the joint check process.[23] In this case, the general contractor issued a joint check to a subcontractor and its supplier for materials for a private project on which there was a payment bond. However, the subcontractor did not owe the supplier for some of the delivered materials as of the date of the issuance of the check, as the subcontractor had been given thirty days to make payment after delivery. The supplier took payment only for the materials then due, even though substantially more had been delivered, the payment for which was not due for thirty days.

When the subcontractor did not pay for the remaining materials, the supplier brought an action against the surety. Even though the total amount paid by the subcontractor was more than enough to cover the cost of the delivered materials, the court held that the mere fact that the joint check was issued for an amount greater than the value of the delivered materials was not conclusive as to whether the supplier was precluded from seeking payment from the surety. The court stated that the amount due the supplier at the time of payment was determinative—with only that amount being deemed to have been paid and any additional amount being deemed to have been unpaid. This could be defeated only by an explicit agreement as to what was to be paid with a joint check or by the establishment of a customary practice in the industry to the effect that the amount deemed to have been received was the amount allocable to the materials that had been delivered, not simply the amount that had been due and owing.

K. Surety Requests That Payment Be Withheld

Payment bonds require the surety to pay unpaid subcontractors and suppliers. To recover any losses caused by having to make such payments, sureties often demand that the owner stop paying the prime contractor when the latter has defaulted in its payments to subcontractors or suppliers.

In public work, unpaid subcontractors and suppliers do not have lien rights. As a result, the owner may decide to pay the prime contractor to enable the latter to continue performance and complete the work. In doing so, the surety's right to reimbursement can be affected adversely. Cases decided in the U.S. Court of Claims, now the U.S. Court of Federal Claims, have given the contracting officer broad discretion to pay earned progress payments to the contractor despite some minor defaults.[24] The court has recognized the government's interest in obtaining a completed project by giving

---

[21]*Brown Wholesale Elec. v. Beztak,* 163 Ariz. 340, 788 P.2d 73 (1990).
[22]788 P.2d at 76.
[23]*Suffolk Builders Supply, Inc. v. Tocci Building Corp.,* 410 Mass. 151, 571 N.E.2d 377 (1991).

[24]*Argonaut Ins. Co. v. United States,* 434 F.2d 1362 (Ct.Cl.1970).

discretion to make payments even though payment can harm the surety.

Although the contracting officer is given considerable discretion, a subsequent opinion by the Court of Claims requires that that discretion be exercised ''responsibly'' and that the surety's interest be considered.[25] In one case, a summary judgment had been granted to the government by the trial court based solely on the contractor's having been on schedule and having completed 91% of the project. The appellate court held that the summary judgment in favor of the government was improper where the surety contended that progress payments and retainage should not have been released when it notified the government of the contractor's impending default.[26] Other factors must be examined to determine whether the government exercised reasonable discretion in disbursing the funds.

Suppose the owner accedes to the surety's requests or demands. In such a case, the unpaid prime contractor may, in addition to claiming a right to the payment withheld, assert a claim against the surety for wrongful interference with the contract that the prime contractor has made with the owner.[27]

L. Remedies for Nonpayment

For convenience, the discussion will assume an unpaid prime contractor. However, any conclusions expressed in that context are likely to apply to unpaid first- or second-tier subcontractors. The law, as noted in Section 22.02(F), recognizes the importance of prompt payment. Yet under certain circumstances, drastic remedies for nonpayment may not be appropriate. Sometimes delay in payment is unavoidable. Delay may not harm the contractor. Care must be taken to avoid an unpaid party's using minor delay as an excuse to terminate an unprofitable contract and causing economic dislocation problems. Despite these possibilities, on the whole nonpayment is—and should be—considered a serious matter.

At the outset, there must be a determination that failure to make payment is a breach of contract. Because construction contracts are detailed and contract procedures often ignored, those accused of not paying in accordance with the contract often assert that prompt payment has been waived.[28]

If payment is not made as promised, the contractor is entitled to interest on the payment. Because as a rule the payments are liquidated or relatively certain in amount, the interest runs from the date payment was due.

AIA documents state that late payments are paid at the ''legal rate'' of interest in the absence of a specified rate in the contract.[29] This term generates confusion. Generally, the legal rate is the amount specified by law that is paid on unpaid court judgments. Yet a significant number of states hold that the legal rate is the highest rate that could be lawfully exacted for the particular transaction. Until the explosive inflation of the 1970s, the interest rate on unpaid judgments ran roughly between 5% and 8%. During that inflationary period, the actual market rate for a commercial loan was between 15% and 20%. In states that used the unpaid judgment rate, owners were often tempted to delay payment and pocket the difference between the two rates.

Yet inflation subsided in the 1980s, and some states increased the rate of interest on unpaid judgments. For that reason, generalizations at any given time as to the differential between these two rates can be perilous. To add complexity, in many states, commercial loans have no limits as to interest. Contracting parties should agree on an interest rate for late payments related to the market rate rather than leave it to the legal rate.

As noted in Section 22.02(F), state prompt payment statutes increasingly provide stiff penalties for violations. For example, in California, failure to pay a subcontractor triggers a penalty of 2% of the amount due per month for each month the payment is not made in addition to whatever interest is provided by law or by contract.[30] Also, if the

---

[25]*United States Fid. & Guar. Co. v. United States*, 475 F.2d 1377 (Ct.Cl.1973).
[26]*Balboa Ins. Co. v. United States*, 775 F.2d 1158 (Fed.Cir.1985).
[27]Such a claim was unsuccessful in *Gerstner Elec. Inc. v. American Ins. Co.*, 520 F.2d 790 (8th Cir.1975), the court failing to find the requisite malice or bad faith. See Section 14.08(F).

[28]*Bart Arconti & Sons, Inc. v. Ames-Ennis, Inc.*, 275 Md. 295, 340 A.2d 225 (1975) (waiver by accepting late payments 50% of the time); *Silliman Co. v. S. Ippolito & Sons*, infra note 35 (no waiver).
[29]AIA Docs. A101, ¶ 7.2; A201, ¶ 13.6.1; A401, ¶ 15.2; and B141, ¶ 11.5.2. See Section 6.08 for further discussion of interest.
[30]West Ann. Cal.Bus. & Prof.Code § 7108.5.

subcontractor must resort to legal action and it prevails, it is entitled to its attorneys' fees and costs.[31] An unpaid prime contractor can recover the 2% penalty, but this is in lieu of any other interest it would be entitled to recover. It can also recover attorneys' fees if it prevails.[32] Under a Virginia law enacted in 1990, there is a 1% per month penalty for delayed payment on public contracts.[33] Again, local law must be consulted.

Contractors *unable* to borrow money may suffer other losses, such as the ability to bid on other projects, loss of key personnel who leave when they are not paid, or eviction caused by nonpayment of rent. If such losses are reasonably foreseeable at the time the contract is made and could not have been reasonably avoided by the contractor, they can be recovered by the contractor if they can be proven with reasonable certainty. The obstacles to recovering for losses of this sort are formidable, and as a rule, interest is all that can be recovered for delayed payment or nonpayment.

A formidable weapon in the event of nonpayment is the power to suspend work. In addition to placing heavy pressure on the owner, suspension avoids the risk of further uncompensated work. The availability of a mechanics' lien is a pale substitute when there is a substantial risk of nonpayment. Until recently, the law was not willing to recognize a remedy short of termination for such breaches. Some cases now hold that the contractor can suspend work if it is not paid.[34]

Nonpayment certainly should not automatically give the right to suspend performance. Shutting down and starting up a construction project is costly. It would be unfair to allow the contractor to shut down the job simply because payment is not made absolutely on schedule.

Under AIA Doc. A201, ¶ 9.7.1, seven days after a progress payment should have been made, the contractor can give a seven-day notice of an intention to stop work unless payment is made. Failure

to pay after expiration of the notice period permits suspension. Paragraph 9.7.1 also states that the contractor shall receive a price increase for its reasonable costs of shutdown, delay, and startup. Undoubtedly, the contractor should be able to suspend work when it is not paid after a reasonable period of time. However, the two seven-day periods may be too short.

Can an unpaid contractor terminate its obligation to perform? Suspension is temporary. Termination relieves the contractor from the legal obligation of having to perform in the future. In the absence of any contract provision dealing with this question, termination is proper if the breach is classified as "material."

Although materiality of breach is discussed in greater detail in Section 34.04, it may be useful to look at some factors that relate to nonpayment as a material breach. The most important are effect on the contractor's ability to perform and the likelihood of future nonpayment. Persistent nonpayment may indicate that the problem is a serious one. A clear statement that performance will *not* be made is a repudiation and clearly gives the right—and probably the obligation—to stop performance. Termination is an important decision. The law looks at *many* factors before deciding whether there has been a material breach. Many cases, however, have concluded that a failure to pay gives the contractor the right to terminate its obligation to perform under the contract.[35]

---

[31]Ibid.

[32]West Ann. Cal.Civ.Code § 3160.5.

[33]Va. Ann. Code § 11-62.11.

[34]*Hart & Son Hauling Inc. v. MacHaffie*, 706 S.W.2d 586 (Mo.App.1986) (can suspend until assurance of payment); *Watson v. Auburn Iron Works, Inc.*, 23 Ill.App.3d 265, 318 N.E.2d 508 (1974) (facts indicated more than simple nonpayment); *Zulla Steel, Inc. v. A & M Gregos*, 174 N.J.Super. 124, 415 A.2d 1183 (1980); *Aiello Constr. Inc. v. Nationwide Tractor Trailer, etc.*, 413 A.2d 85 (R.I.1980). See also Restatement (Second) of Contracts § 237 (1981).

[35]*Guerini Stone Co. v. P.J. Carlin Constr. Co.*, 248 U.S. 334 (1919); *United States v. Western Cas. & Surety Co.*, 498 F.2d 335 (9th Cir.1974); *St. Paul-Mercury Indem. Co. v. United States*, 238 F.2d 917 (10th Cir.1956) (subcontractor); *United States v. Premier Contractors, Inc.*, 283 F.Supp. 343 (N.D.Me.1968) (supplier); *Integrated, Inc. v. Alec Fergusson Elec. Contractors*, 250 Cal.App.2d 287, 58 Cal.Rptr. 503 (1967) (many factors in addition to nonpayment); *Silliman Co. v. S. Ippolito & Sons, Inc.*, 1 Conn.App. 72, 467 A.2d 1249 (1983) (subcontractor); *Berry v. Blackard Constr. Co.*, supra note 3 (owner's refusal based on contention that work did not conform); *Leto v. Cypress Builders, Inc.*, 428 So.2d 819 (La.App.1983) (nonpayment made work precarious); *Zulla Steel, Inc. v. A & M Gregos, Inc.*, supra note 34 (protracted suspension converted to termination); *Shapiro Eng'g Corp. v. Francis O. Day Co.*, 215 Md. 373, 137 A.2d 695 (1958); *Aiello Constr. Inc. v. Nationwide Tractor Trailer, etc.*, supra note 34 (substantial underpayment for prolonged period); *Tennessee Asphalt Co. v. Purcell Enter., Inc.*, 631 S.W.2d 439 (Tenn.App.1982) (subcontract: substantial delay required); *Darrell J. Didericksen & Sons, Inc. v. Magna Water & Sewer Improvement Dist.*, 613 P.2d 1116 (Utah 1980); *H.E. & C.F. Blinne Contracting Co.*, ENG BCA 4174, 83-1 BCA ¶ 16,388.

Yet the decision to walk off the job is one that carries serious consequences, a conclusion demonstrated by *Keyway Contractors, Inc. v. Leek Corporation, Inc.*[36] The sub–subcontract provided that during performance of the work, all disputes had to first be submitted to the subcontractor for a decision. This decision would be binding on the sub-subcontractor unless the subcontractor commenced an arbitration within thirty days. The contract also provided that the sub–subcontractor would not stop work because of any dispute or controversy. This contractual system seeks to avoid a dispute being the basis for one of the disputing parties walking off the job.

That goal, however, did not work in this case. The sub–subcontractor ceased work and walked off the project. The subcontractor then brought a lawsuit against the sub–subcontractor because of its "walk-off." The sub–subcontractor defended and counterclaimed, stating that it had not been paid as provided for under the contract. A trial court decision in favor of the subcontractor was upheld by the court of appeals. The court determined that the parties had agreed to a method for resolving such disputes. Walking off the job rather than using that method was an unjustified breach of contract.

The decision to walk off the job was based on the erroneous conclusion that a failure to pay in accordance with the contract meant that the sub-contractor could not invoke the contract provisions requiring that the dispute be submitted to it for resolution and ultimately to arbitration.

Emphasis in the discussion and in the cases has been on the effect of nonpayment on the contractor's ability to perform. This can operate to the disadvantage of a financially sound contractor. Another approach, and perhaps a better one, is to permit termination if a changed cash flow will cause the contractor to finance the project to a larger degree than anticipated. If, however, the facts clearly demonstrate that a particular contractor was selected for, among other factors, its capacity to absorb payment delay, this *may* show an intention by the contractor to accept the risk of substantial alteration of financing the work.

Suppose the contractor continues performance. Continued performance may indicate that nonpayment was not sufficiently serious to constitute termination. Continued performance may manifest an intention to continue performance that is relied on by the owner. However, continued performance should not invariably preclude nonpayment from justifying termination. An unpaid contractor may *choose* to continue work for a short period while awaiting performance.[37] This would be especially true if the contractor clearly indicated that if payment were not forthcoming, work would cease.

AIA Doc. A201, ¶ 14.1.1.3, as noted earlier in this subsection permits the contractor to terminate if the work has been suspended for thirty days for nonpayment. The contractor must first give a seven-day written notice that it intends to terminate. Does this express termination provision affect any common law power to terminate for nonpayment? That depends on whether the express power to terminate is exclusive. Generally, the law hesitates to make specified remedies in a contract exclusive unless there is a clear indication that this is the intention of the parties.[38] This has been codified in A201, ¶ 13.4.1, which states that remedies specified are in addition to any remedies otherwise imposed or available by law.

Can the owner preclude termination by paying all unpaid progress payments with interest during the seven-day period? Is the notice period designed to allow the owner to cure past defaults?[39] As termination is a drastic remedy, common sense and the obligation of good faith and fair dealing support the conclusion that the owner can cure past defaults during the notice period. Remedies to a contractor who has justifiably ceased performance or who has been wrongfully removed from the site are discussed in Section 27.02. Some courts allow the contractor to recover only for work performed and not for lost profits.[40] However, the general tendency is to treat this breach as no different from any other.[41]

---

[36]189 Ga.App. 467, 376 S.E.2d 212 (1988).

[37]*Darrell J. Didericksen & Sons, Inc. v. Magna Water & Sewer Improvement Dist.*, supra note 35. But see *Drew Brown Ltd. v. Joseph Rugo, Inc.*, 436 F.2d 632 (1st Cir.1971).
[38]*Glantz Contracting Co. v. General Elec.*, 379 So.2d 912 (Miss.1980); *Bender-Miller Co. v. Thomwood Farms, Inc.*, 211 Va. 585, 179 S.E.2d 636 (1971).
[39]See Section 34.03(F).
[40]*Palmer v. Watson Constr. Co.*, 265 Minn. 195, 121 N.W.2d 62 (1963).
[41]*Leto v. Cypress Builders, Inc.*, supra note 35.

The effect of owner nonpayment on the right of a subcontractor to recover for its performance is treated in Section 28.06.

M. Payment as Waiver of Defects

Sometimes contractors contend that making progress payments waives any deviations from contract document requirements. Generally, payment in and of itself does not waive defects.[42] Suppose the owner knows or should know of the defect and makes payment. This may lead the contractor reasonably to believe that the owner intends to pay despite the defect. If so, minor defects are waived.

To preclude waiver in such cases, construction contracts frequently contain provisions stating that progress payments do not waive claims for defects. For example, ¶ 9.6.6 of AIA Doc. A201 states that progress payments are not an acceptance of work that is not in compliance with the contract documents. The payment certificate should make clear that payment does not constitute waiver.

Acts that can constitute waiver of defects more commonly arise when the project is accepted or completed and final payment is made. Detailed discussion of this problem is found in Section 24.05.

N. Progress Payments and the Concept of Divisibility

Divisibility, sometimes called severability, is a multipurpose legal concept sometimes applied to a contract where performance is made in installments or is divided into designated items. As an illustration of the first, the contract for the sale of goods could allow twelve monthly deliveries with payment at the time of each delivery. An illustration of the second would be a contract for the sale of goods that consisted of five different items to be delivered with different prices for each item.

If a contract is considered divisible, it would be as if there were twelve separate contracts in the first illustration and five in the second. Each partial performance, whether a monthly delivery or delivery of less than all the five items, would be considered the equivalent of the money promised for each installment or item. Without describing the many legal issues that can sometimes depend on the divisibility classification, the progress payment mechanism should not result in the construction contract being considered divisible. The amounts certified are approximations and not agreed valuations for the work as it proceeds. The amounts can be—and frequently are—adjusted at the time of final payment.[43]

SECTION 22.03 Retainage

Retainage is a contractually created security system under which the owner retains a specified portion of earned progress payments to secure itself against certain risks. For example, suppose the schedule of values establishes that the contractor is entitled to a progress payment of $50,000. Construction contracts commonly provide that a portion of this $50,000 will be retained by the owner and paid at the end of the project or at some later date. The purpose of retainage is to provide money out of which claims that the owner has against the contractor can be collected without the necessity of a lawsuit. Retainage is *not* an agreed damage or damage limit.[44]

Retainage has become controversial. Contractors and subcontractors increasingly request legislators to help them. They contend that retention of money that they have earned is unfair to them and costly and unnecessary to the owner. They assert that the reasons for creating the security, such as defective work or the prime contractor not paying subcontractors or suppliers, are dealt with by the contractor's furnishing performance and payment bonds. Contractors and subcontractors also contend that financing costs that are inevitably incurred because of retention are ultimately transferred to the owner through the contract price. They then argue that the owner is paying twice for the same risk—once through financing costs included in the contract price and once by the cost of surety bond premiums.

---

[42]See Annot., 66 A.L.R.2d 570 (1959). Waiver generally is discussed in Section 22.06(E).

[43]*Dravo Corp. v. Litton Systems, Inc.,* 379 F.Supp. 37 (S.D.Miss.1974); *Shapiro Eng'g Corp. v. Francis O. Day Co.,* supra note 35; *Kirkland v. Archbold,* 113 N.E.2d 496 (Ohio App.1953).
[44]Earlier edition of treatise cited and followed in *Van Knight Steel Erection, Inc. v. Housing & Redevelopment Auth. of the City of St. Paul,* 430 N.W.2d 1, 3 (Minn.App.1988).

This argument equates the efficacy of retainage with bonds. Obviously, it is better for the owner to actually have funds within its control that it can use to secure the owner against these risks rather than have to deal with or send unpaid subcontractors or suppliers to the surety. Also, owners can earn interest on the funds retained.

Early finishing subcontractors, such as excavating subcontractors, make another attack on retainage. They would like line-item retention, in essence decoupling their work from that of the other contractors on the project and granting them the right to the retainage for their work when their work is finished.

As one purpose of retainage is to pay lien claimants, subcontractors and suppliers can benefit from retainage, something that will be shown by state statutes discussed later in this section.

Contractors also complain that they should be able to substitute other securities for the retainage and obtain the use of the retainage as quickly as possible. They also assert that owners unfairly earn interest on the retainage.

Finally, contractors and subcontractors may worry about the safety of the retainage. Will the owner have the money to pay the retainage at the end of the project?

There has been some tendency to reduce retainage, and some contracts eliminate it entirely. Sometimes retainage is limited to the first 50% of the work.

Complaints of contractors led to a decision by the Office of Federal Procurement Policy[45] to effectuate a uniform government-wide retainage policy. Retainage must not be used as a substitute for good contract management. The agency cannot withhold funds without good cause. Determinations concerning the use of retainage should be based on an assessment of the contractor's past performance and the likelihood that such performance will continue. It is suggested that retainage not exceed 10% and that it be adjusted downward as the contract approaches completion, particularly if there is better than expected performance or alternate safeguards. Once all contract requirements have been completed, all retained amounts should be paid promptly to the contractor.

Contractor complaints, as noted, have increasingly led to legislation. Statutes in some states recognize the problems described earlier in this section and attempt to deal with them. Texas, for example, has long allowed a mechanics' lien on retainage. In contracts for which liens by subcontractors and suppliers may be filed, the owner must retain a 10% retainage as security for lien claimants. In addition, lien claimants are given a preference on the retainage that may entitle them to payment ahead of the prime contractor. The owner's failure to comply with its requirement for retainage gives lien claimants preferential mechanics' liens on the property itself.[46]

Michigan prohibits state owners from comingling retainage with other funds. With some exceptions for federally funded projects, retainage must be placed in an interest-bearing account. The contractor receives the retainage and the interest it earned at final payment. After 94% of the work has been completed, the contractor can post an irrevocable letter of credit and receive the retainage.[47]

Tennessee requires the owner, public or private (or contractor who holds retainage for subcontractors), in contracts over $500,000 to deposit retainage in escrow. The escrow holder must give security for the retainage. The interest earned by the retainage goes to the contractor or subcontractor. Rights granted by the statute cannot be waived.[48] Tennessee allows the prime contractor to obtain the retainage held by local entities by posting security.[49]

In 1988, California enacted legislation that requires local entities to include provisions in any invitation to bid, with certain exceptions, that allow the contractor to substitute certain securities for retainage held by the agency or an escrow agent. When this system is invoked, the owner makes full progress payments to the contractor from which retainage normally would have been withheld. The owner can draw on the securities in the event of a default by the contractor by notifying the escrow

---

[45]OFPP Policy Letter 83-1, 48 Fed.Reg. 22,832 (1983). This has been implemented by the General Services Administration. See 48 Fed.Reg. 37,997 (1983).

[46]Tex.Ann.Prop.Code § 53.101-105.
[47]Mich.Ann.Comp.Laws § 125.1563.
[48]Tenn.Ann.Code § 66-11-144.
[49]Id. at § 12-4-108.

agent who converts the securities to cash and distributes the cash as instructed by the owner. Alternatively, at the written request of the contractor, the owner makes payments of the retainage directly to the escrow agent, who invests it on behalf of the contractor. Upon satisfactory completion of the contract, the escrow agent distributes to the contractor the securities and any payments made to the escrow agent by the owner. The contractor must pay each subcontractor no later than twenty days after receipt of this payment the respective amount of interest earned attributable to funds withheld from each subcontractor. All expenses of the escrow arrangement are paid by the contractor.[50]

Another indication of legislative activity can be seen by legislation enacted in Idaho in 1990.[51] The legislation covers private works of improvement. Under the statute, the retainage must not exceed 5% of any payment and the total withheld can never exceed 5% of the contract price if the contractor or subcontractor posts a performance bond. The 5% maximum does not apply to contracts for the improvement of residential property consisting of from one to four units (one of which is occupied by the owner).

Thirty-five days after substantial completion, the amount retained must be reduced to whichever is less: 150% of the estimated value of the work to be completed or the amount that has been retained by the owner—again not to exceed 5%. Within thirty-five days from final completion, the retainage must be released except in the event of a dispute. If there is a dispute, the amount that can be withheld from the final payment cannot exceed 150% of the estimated value in dispute.

The prime contractor must pay its subcontractors any retainage plus interest that is received within ten days from the time of receipt. The prime contractor can deduct from the interest paid a 1% fee for administration. However, the retainage need not be paid to the subcontractor if a bona fide dispute exists between the prime and the subcontractor. The amount withheld in such case must not be more than 150% of the estimated value of the work to be completed or amount in dispute. If the retainage is not paid within the time limits required, the payments will be subject to an additional charge over and above interest of $1\frac{1}{2}\%$ per month. Provisions also exist for segregated accounts created by the owner or lender under which funds are not mixed with other funds of the owner. Retained funds are deposited into these segregated accounts when withheld by the owner, and the accounts bear interest. The interest earned shall be paid to the prime contractor, but the owner can deduct from the interest an amount to cover administration. Most important, the Idaho legislation states that the parties cannot waive any provisions of this statute.

It can be seen that increased legislative activity in the field of payments with regard to their promptness and with regard to retainage has meant that parties who are about to engage in a construction contract must check local laws to see whether these laws can be varied by contract and to ensure that these laws are followed. Particular reference must be paid to whether the laws are limited to public contracts, are limited to private contracts, or cover both types. Typically, a regulation of this sort begins in public contracts and often is extended to private contracts. However, as seen in the Idaho statute, smaller projects are sometimes exempt.

One problem incident to retainage can develop at the end of the project. Suppose the retained amount is $50,000 and the owner is entitled to take $20,000 from it to remedy defects in the work. A solvent contractor will be paid the balance. However, in the volatile construction industry, it is not uncommon for a contractor to run into financial problems. In such a case, a horde of claimants descends on the owner and demands the money. The claimants can be lenders who have received assignments of contract payments as security for a loan, sureties who have had to discharge the obligations of the contractor, taxing authorities who claim the funds when the contractor has not paid its taxes, and, in the event the contractor has gone bankrupt, the trustee in bankruptcy.

The scramble for funds usually is dealt with by the owner's initiating what is called an "interpleader" action. The owner pays the disputed funds into court, files a lawsuit in which it names as parties all claimants to the funds, and withdraws

---

[50]West Ann.Cal.Pub.Cont.Code § 22300.
[51]Idaho Code § 29-115.

CHAPTER 22 / PAYMENT: MONEY FLOW AS LIFELINE

from the fray. The court must unscramble the claims.[52]

Although the possible multiple claimants' confusion at the end of the job may not in itself be a reason to limit or eliminate retention, the troublesome disputes that relate to retainage ownership should be taken into account in determining whether and to what extent retainage should be used.

SECTION 22.04 Substantial Completion

Sometimes construction contracts provide that a payment will be made to the contractor on substantial completion. Perhaps more important, the "end of the job" process, which begins with the contractor's representation that it has substantially completed the project and concludes with a certificate of final completion, is one about which a number of legal issues cluster (described in this section and in Section 22.05).

Although time problems are discussed in Chapter 26, it should be noted that under AIA Doc. A101, ¶ 3.2, and A201, ¶ 8.2.3, compliance with the contractor's time obligation is measured by *substantial* completion.

The process for determining substantial completion under A201, ¶ 9.8.2, begins with a submission to the architect by the contractor of a list of items to be completed or corrected, followed by an inspection by the architect to determine whether she should issue a certificate of substantial completion. Such inspections are more carefully made than are inspections for ordinary progress payments. More important issues are involved. For example, this may be the point at which the owner retakes possession of the site and begins to use the project.[53] This can have important insurance consequences. Also, statutes of limitation that cut off liability a designated number of years after completion frequently use substantial completion as the event that triggers the commencement of the period.[54]

In 1987, the AIA added ¶ 9.9.1 to A201. Paragraph 9.9.1 permits the owner to occupy or use a portion or all of the project prior to substantial completion.

SECTION 22.05 Completion and Final Payment

The AIA process begins with an act by the contractor, a notice that the work is ready for final inspection and acceptance, and an application for final payment. The architect inspects the work to determine whether the certificate should be issued—an inspection that should be undertaken with great care.

This stage in the Construction Process is a benchmark that has serious implications for major participants in the process. If the owner has not taken possession of the project at the time of substantial completion, it will certainly do so at this point.

Possession by the owner involves legal responsibilities that rest on the possessor of land.[55] Perhaps more important, the end of the job often has an effect on allocation of risks, such as those that relate to defects, and the continued existence of any claim. Because completion is sometimes equated with acceptance—an important legal doctrine that relates to defects—it is discussed in Sections 14.09(B) and 24.05.

Contractors also have claims that may be affected by completion and final payment. As noted in Section 27.13, the law looks for benchmarks that can put an end to disputes. In that section the benchmark under discussion—payment for changed work—reflected some of the pressure contractors face when seeking final payment.

The effect of final payment accompanied by a release and its effect on contractor claims surfaced in *Mingus Constructors, Inc. v. United States*.[56] Mingus substantially completed his work on July 30, 1982. In July and August, he sent letters to the contracting officer stating that he intended to file a claim and giving some indication as to the basis of the claim.

On October 28, 1982, the contracting officer inquired about the status of the claims release form it had sent to Mingus. This was part of the final

[52]For one of the many articles dealing with this issue, see Mungall, *The Buffeting of the Subrogation Rights of the Construction Contract Bond Surety by United States v. Munsey Trust Co.*, 46 Ins. Couns. J. 607 (1979).
[53]AIA Doc. A201, ¶ 9.8.2
[54]See Section 23.03(G).

[55]See Section 7.08.
[56]812 F.2d 1387 (Fed.Cir.1987).

payment process. Mingus was advised that he could "except" in the space provided any claims against the government and that the exception would not hold up processing of his final payment. On the following day, he executed the claims release form, which stated that he released all claims. In the space provided for exceptions, Mingus stated that pursuant to correspondence, he would file a claim, the amount of which was undetermined. He was paid on November 2, 1982. On November 5, 1982, he wrote to the government seeking confirmation of his understanding that by inserting the statement in the form he had not waived his claims. The contracting officer replied on November 10, 1982, stating that he could "except" any claims in the space provided without waiving his rights.

Mingus finally submitted his claim on January 5, 1984. When the claim was returned to Mingus, Mingus brought legal action in the Claims Court. The government took the position that Mingus had not protected his claim when he executed the release. The Claims Court held in favor of the government. This was affirmed by the federal appellate court based on a contractual provision barring claims after final payment. This, according to the court, can be avoided only by specifically excepting claims at the time of the general release.

Mingus was barred because of his release. Releases are strictly construed against the contractor. The exception Mingus included in the release, even if supplemented by his prior letters, was a "blunderbuss exception," which "does nothing to inform the government as to the source, substance, or scope of the contractor's specific contentions."[57] The court felt it was important that releases put an end to controversies. Although the contractor need not include a final certified claim when it executes the release, merely stating the claim in general terms does nothing to preserve the claim.[58]

Failure to provide a contractually required notice at the time of final payment was held to have waived not only the public contractor's claims against the public entity but also its negligence claim against the public entity's engineer. This was based upon the acceptance of the final payment

form stating that acceptance released the owner and others for any claims or any liability to the contractor arising out of the work.[59]

A201 deals with release of contractor claims in ¶ 9.10.4. To protect its claim at the time of final payment, the contractor must have made the claim previously in writing and must have identified that claim as unsettled at the time of the application for final payment. Again, the needed specificity of the previous written claim can generate disputes. For the owner to be aware of the nature of the contractor's claim—both to investigate it and to have some idea of the contingencies that must be set aside to deal with it[60]—the contractor should give as much information as is reasonably available. As noted in the *Mingus* case, blunderbuss reservations of rights may not be sufficient.

Mechanics' liens can surface at the end of the job. Although they are discussed in greater detail in Chapter 28 dealing with subcontracts, this section looks briefly at liens from the vantage point of the owner, with particular reference to the end of the job.

Owners employ a variety of techniques to avoid liens. Some relate to the payment process. In some contracts, though not those of the AIA, owners seek to avoid liens by requiring evidence that subcontractors and suppliers who have lien rights either have been paid or have executed lien waivers before the contractor receives *progress* payments. Even if this is not required for a progress payment, it is almost certainly required for final payment. For example, AIA Doc. A201, ¶ 9.10.2, conditions final payment on the contractor's submitting an affidavit that it has paid all its bills and gives the owner the right to require other data "such as receipts, releases and waivers of liens . . . arising out of the Contract, to the extent and in such form as may be designated by the Owner." Because of concern that lien waivers or—even worse—"no lien" contracts[61] (under which the prime contractor

---

[57]Id. at 1394.

[58]See also *John E. Fisher Constr. Co., Inc. v. Onondaga*, 115 A.D.2d 993, 497 N.Y.S.2d 557 (1985), appeal denied, 67 N.Y.2d 609, 494 N.E.2d 114, 502 N.Y.S.2d 1028 (1986).

[59]*McKeny Constr. Co., Inc. v. Town of Rowlesburg*, 187 W.Va. 521, 420 S.E.2d 281 (1992).

[60]*Bornstein v. City of New York*, 94 A.D.2d 683, 463 N.Y.S.2d 198 (1983) (contractor had filed claim before accepting final payment).

[61]O.C.G.A. §44-14-366 enacted in Georgia effective January 1, 1992 makes lien waivers made in advance of furnishing labor and materials null and void and sets forth statutory interim and final lien waiver forms. Similarly, See West Ann.Wis.Stat. § 779.035 (1m)(a).

waives all liens in advance for itself, its subcontractors, and suppliers) may be unfair to subcontractors and suppliers, legal controls have been enacted by the states that regulate these methods.[62] Where this is the case, as noted in (J), owners may choose to issue final payments as well as progress payments through joint checks.

Although these legislative activities have been designed to protect subcontractors and suppliers, other legislative activity protects owners from liens. For example, an Iowa statute provides that the owner need not make final payment until ninety days after completion of the project if the project would be subject to a lien.[63] This does not apply if the prime contractor has filed a payment bond or supplied signed receipts or lien waivers. Generally, liens must be filed a designated number of days after substantial or final completion.[64] One way of avoiding liens is to delay final payment until the time for filing liens has expired.

Owner protection does not eliminate the requirement that contracting parties be fair to one another. For example, a case involved a contract with lien avoidance conditions before final payment. Yet the contractor was awarded final payment *without* complying because the owner had insisted that the contractor waive its delay damage claim before final payment would be made. The court found this tactic to be unconscionable.[65]

SECTION 22.06 Payment for Work Despite Noncompliance

A. The Doctrine of Conditions

The promise by the owner to make any payments is conditioned on the contractor's compliance with the contract documents. During the course of the work, the contractor is not entitled to be paid unless, as discussed in Section 22.02(A), provisions are made for progress payments. At the end of the project, the doctrine of conditions in its strictest sense requires absolute compliance with the contract documents before the contractor is entitled to any unpaid part of the contract price.

Although the owner would be permitted to hold back all the money until it receives what has been promised, in construction contracts, strict application of the doctrine can cause hardship to the contractor. During performance, progress payments are essential to finance the job and avoid the risk of going unpaid for work. At the end of the job, a strict application of the doctrine of conditions can cause a loss to the contractor disproportionate to the loss caused by nonperformance. Such loss can result in unjust enrichment where the owner is occupying the project. This section outlines some of the legal doctrines that have developed to relieve the contractor when performance has not been in *exact* compliance with contract documents.

The doctrine of conditions can apply to a contractor who fails to complete the project on time. A subcontractor's promise to complete by a designated time was held to be a promise and not a condition to the prime contractor's obligation to pay the contract price.[66] The court noted that conditions are not favored, because they can create forfeiture and unjust enrichment. The contractor can recover the unpaid balance, but it must pay any damages caused by delay.

In the absence of a valid liquidation of damages clause, delay damages are very difficult to prove. The ultimate outcome is likely to be that the contractor will collect for the work despite its late performance. However, were timely performance a condition to final payment—unlike defects of quality that can be cured—the prime contractor would have the benefit of the subcontractor's work without paying fully for it.

B. Substantial Performance: *Plante v. Jacobs*

The substantial performance doctrine developed early in English law and made its way into American law. While the dimensions of the doctrine will be sketched later in this subsection, in its simplest form, it requires the owner to pay the balance of the construction contract price if the contractor has

---

[62]West Ann.Cal.Civ.Code § 3262. See note 61 supra and Section 28.07(D).

[63]Iowa Ann.Code § 572.13.

[64]West Ann.Cal.Civ.Code § 3116 (ninety days after completion or cessation of work or thirty days after notice of cessation or completion filed by owner).

[65]*North Harris County Junior College Dist. v. Fleetwood Constr. Co.*, 604 S.W.2d 247 (Tex.Ct.App.1980).

[66]*Landscape Design & Constr., Inc. v. Harold Thomas Excavating, Inc.*, 604 S.W.2d 374 (Tex.Ct.App.1980).

substantially performed, leaving the owner a claim for damages based upon failure of the contractor to perform strictly in accordance with the contract requirements.

Most substantial performance cases involve construction contracts. Construction is a complex undertaking, with detailed contract requirements. A strong likelihood exists that minor deviations will surface at the end of the job. Work performed by prime contractor employees may escape the scrutiny of the prime's superintendent. Much of the work is performed by subcontractors, and the exact quantity and quality may be difficult to determine while the work is being performed.

The complexity of contract documents frequently generates interpretation questions, and it is often difficult to get an authorized interpretation by the design professional while work is being performed. Likewise, it may be difficult to find someone with authority to direct a change or approve substituted materials or products claimed by the contractor to be equal or equivalent to those specified.

Unlike in other types of contracts, the breaching party—that is, the contractor—cannot take back its performance. The contractor's performance has become incorporated into the owner's land, and very frequently the owner has taken possession of the project.

With some limited exceptions,[67] substantial performance issues usually arise at the end of the job. The prime contractor contends that the owner has *essentially* what it bargained for and often points to the owner's occupying and using the project. One court defined substantial performance as

> . . . when construction has progressed to the point that the building can be put to the use for which it was intended, even though comparatively minor items remain to be furnished or performed in order to conform to the plans and specifications of the completed building.[68]

Another case added that the defects must not "so pervade the whole work that a deduction in

damage will not be fair compensation."[69] Yet another case concluded that there had not been substantial performance when there was an accumulation of small defects, even though the owner was occupying the premises.[70]

The ratio of defect correction costs to contract price and the determination of whether the building has met its essential purpose are relevant.[71] But one court affirmed a judgment of substantial performance despite cost of completion and correction of defective work being 31% of the contract price.[72]

Jacob & Youngs, Inc. v. Kent,[73] a leading case, found that there had been substantial performance when it was discovered *after completion* that one brand of pipe had been substituted for the one specified. The court noted that the pipe substituted was of equal quality to that designated.

Kirk Reid Co. v. Fine[74] involved the installation of an air-conditioning system. The owner proved that the following deviations existed:

1. The system was 46 to 50 tons short in capacity.
2. The primary air unit was of a lower rating by 7,850 cubic feet per minute.
3. The condenser water pipe was 120 gallons per minute short of capacity.

Yet the court held that the contractor had substantially performed, stating:

> It appears from the . . . finding that "there was a workable air conditioning plant installed and operating by May 1, 1959," the time called for by the contract. It further appears from the record that the system so installed and operating has been used, since that time, by the defendant without complaint by him that it is insufficient to perform its task.[75]

A case dealing with construction of a church held that the contractor had substantially performed when the deviation consisted of a two-foot lower ceiling, windows shorter and narrower than

---

[67]*Nordin Constr. Co. v. City of Nome,* 489 P.2d 455 (Alaska 1971) (owner sued to recover payments).
[68]*Southwest Eng'g Co. v. Reorganized School Dist. R-9,* 434 S.W.2d 743, 751 (Mo.App.1968).

[69]*Jardine Estates, Inc. v. Donna Brook Corp.,* 42 N.J.Super. 332, 126 A.2d 372, 375 (1956).
[70]*Tolstoy Constr. Co. v. Minter,* 78 Cal.App.3d 665, 143 Cal.Rptr. 570 (1978).
[71]*Stevens Constr. Corp. v. Carolina Corp.,* 63 Wis.2d 342, 217 N.W.2d 291 (1974). See also 3A A. CORBIN, CONTRACTS § 706 (1960).
[72]*Jardine Estates, Inc. v. Donna Brook Corp.,* supra note 69.
[73]230 N.Y. 239, 129 N.E. 889 (1921).
[74]205 Va. 778, 139 S.E.2d 829 (1965).
[75]139 S.E.2d at 837.

called for, and seats narrower than designated by the specifications.[76] Another court held that the contractor substantially performed when it was discovered at the completion of the project that the roof shingles were discolored.[77] The court noted that the shingles could not be seen from the street and did not constitute a functional defect.

Finally, another court concluded that the contractor had substantially performed when it buried pipe six and one-half feet into the ground instead of the required seven.[78]

Where the contractor leaves the project *before* completion, it is more difficult, though not impossible, to establish substantial performance.[79]

Early cases required the contractor to show that it was free from fault and that the breach was not willful, such as by establishing that the deviation was caused by a subcontractor or was not done knowingly.[80] However, recent cases have reflected ambivalence on this issue. One case held that substantial performance could be used despite a finding that the contractor knowingly deviated from the contract.[81] Other cases have held that the reason for the breach must be taken into account with other factors to determine whether there has been substantial performance.[82]

Although substantial performance is designed to avoid forfeiture and unjust enrichment, it would be difficult to conclude there was unjust enrichment if a contractor testified that it had substituted one brand of pipe for another because "it was just as good" or because it did not feel like going to the warehouse.[83]

It has been stated that explicit contract language precludes substantial performance from determining the contractor's rights to the contract price.[84] Where such statements are made, they are likely to be in cases where the contractor has been found to have substantially performed.[85] The dilemma posed by this question requires that the essential nature of the substantial performance doctrine be determined. If the doctrine merely reflects a common intention that minor deviations will not preclude recovery, a contract clause negating this intention should be controlling. However, if the doctrine is based on the avoidance of forfeiture by the contractor and corresponding unjust enrichment of the owner, an express contract clause should not bar use of this doctrine. Resolving this difficult question will be aided by an evaluation of *Plante v. Jacobs* (reproduced ahead).

Following this case will be a further critique of the substantial performance doctrine and its relationship to contract language dealing with strict compliance.

Usually the contractor must prove it has substantially performed.[86] However, once substantial performance has been established, as a rule, the owner must prove its damages.[87] Substantial performance, though entitling the contractor to recover the balance of the contract price, is still a breach. Because damages reduce the amount to which the contractor is entitled, the law places the burden of proving damages on the owner once substantial performance has been established. It is likely that the *extent* of damages will be an important factor in the owner's attempt to establish that there has not been substantial performance.

A few states place the burden of establishing damages on the *contractor*, an anomalous result in the light of the first line of attack by the contractor usually being that it has *fully* performed.[88]

[76]*Pinches v. Swedish Evangelical Lutheran Church*, 55 Conn. 183, 10 A. 264 (1887).
[77]*Salem Towne Apartments, Inc. v. McDaniel & Sons Roofing Co.*, 330 F.Supp. 906 (E.D.N.C.1970). But in *O.W. Grun Roofing & Constr. Co. v. Cope*, 529 S.W.2d 258 (Tex.Ct.App.1975), the court held that streaks in a structurally sound roof precluded substantial performance and required the roof to be replaced.
[78]*All Seasons Water User Assn. v. Northern Improvement Co.*, 399 N.W.2d 278 (N.D.1987).
[79]In *Watson Lumber Co. v. Mouser*, 30 Ill.App.3d 100, 333 N.E.2d 19 (1975), though, a trial court conclusion of substantial performance was affirmed.
[80]*Jacobs & Youngs, Inc. v. Kent*, supra note 73.
[81]*Kirk Reid Co. v. Fine*, supra note 74.
[82]*Hadden v. Consol. Edison Co. of N.Y., Inc.*, 34 N.Y.2d 88, 312 N.E.2d 445, 356 N.Y.S.2d 249 (1974); *Nordin Constr. Co. v. City of Nome*, supra note 67; *Watson Lumber Co. v. Mouser*, supra note 79 (requiring performance be in good faith and breach not be willful).
[83]*O.W. Grun Roofing & Constr. Co. v. Cope*, supra note 77.
[84]*Jacob & Youngs, Inc. v. Kent*, supra note 73.
[85]Ibid. But see *Winn v. Aleda Constr. Co., Inc.*, 227 Va. 304, 315 S.E.2d 193 (1984) (language barred substantial performance).
[86]*A.W. Therrien Co. v. H. K. Ferguson Co.*, 470 F.2d 912 (1st Cir.1972).
[87]*Maloney v. Oak Builders, Inc.*, 224 So.2d 161 (La.App.1969); *Hopkins Constr. Co. v. Reliance Ins. Co.*, 475 P.2d 223 (Alaska 1970).
[88]*Vance v. My Apartment Steak House of San Antonio, Inc.*, 677 S.W.2d 480 (Tex.1985). See Note, 22 Hous.L.Rev.1093 (1985), attacking Texas rule.

Damages are sometimes measured by the cost of correction of the defect, sometimes by the diminished value of the project, and sometimes by a combination of the two.[89] The interrelation of various damage measurements can be seen in *Plante v. Jacobs*, reproduced here.

---

[89]These damage measurements are discussed in Section 27.03(D). In *Jacob & Youngs, Inc. v. Kent*, supra note 73, the measure of recovery for substitution of pipe was the diminished value of the project that the court concluded

was nominal. But in *Salem Towne Apartments, Inc. v. McDaniel & Sons Roofing Co.*, supra note 77, the diminished value of the project was found to have been about one half the cost of correction. But see *Kaiser v. Fishman*, _____ A.D.2d _____ , 590 N.Y.S.2d 230 (1992) (approved award of almost $215,000 as the cost of correction, despite what court called a quick-fix correction work of $60,000).

PLANTE
v. JACOBS

Supreme Court of Wisconsin, 1960. 10 Wis.2d 567, 103 N.W.2d 296.

Suit to establish a lien to recover the unpaid balance of the contract price plus extras of building a house for the defendants, Frank M. and Carol H. Jacobs, who in their answer allege no substantial performance and breach of the contract by the plaintiff and counterclaim for damages due to faulty workmanship and incomplete construction.... After a trial [before the judge without a jury] judgment was entered for the plaintiff in the amount of $4,152.90[90] plus interest and costs, from which the defendants, Jacobs, appealed and the plaintiff petitioned for a review....

The Jacobs, on or about January 6, 1956, entered into a written contract with the plaintiff to furnish the materials and construct a house upon their lot in Brookfield, Waukesha county, in accordance with plans and specifications, for the sum of $26,765. During the course of construction the plaintiff was paid $20,000. Disputes arose between the parties, the defendants refused to continue payment, and the plaintiff did not complete the house. On January 12, 1957, the plaintiff duly filed his lien.

The trial court found the contract was substantially performed and was modified in respect to lengthen-

ing the house two feet and the reasonable value of this extra was $960. The court disallowed extras amounting to $1,748.92 claimed by the plaintiff because they were not agreed upon in writing in accordance with the terms of the agreement. In respect to defective workmanship the court allowed the cost of repairing the following items: $1,550 for the patio wall; $100 for the patio floor; $300 for cracks in the ceiling of the living room and kitchen; and $20.15 credit balance for hardware. The court also found the defendants were not damaged by the misplacement of a wall between the kitchen and the living room, and the other items of defective workmanship and incompleteness were not proven. The amount of these credits allowed the defendants was deducted from the gross amount found owing the plaintiff, and the judgment was entered for the difference and made a lien on the premises....

HALLOWS, Justice. The defendants argue the plaintiff cannot recover any amount because he has failed to substantially perform the contract. The plaintiff conceded he failed to furnish the kitchen cabinets, gutters and downspouts, sidewalk, closet clothes poles, and

---

[90]Ed. note: The computation of the judgment must have been:

Original contract price	26,765.00	
Extras	960.00	
Contract price as adjusted		27,725.00
Payments received by plaintiff	20,000.00	
Balance unpaid		7,725.00
Less: 1) Amount conceded by plaintiff for omissions (kitchen cabinets, gutters and downspout, sidewalk, closet clothes poles, and entrance seat)	1,601.95	
2) Cost of correction for patio wall ($1550), patio floor ($100), cracks in living room and kitchen ceiling ($300), and credit balance for hardware ($20.15).		3,572.10
		$4,152.90

entrance seat amounting to $1,601.95. This amount was allowed to the defendants. The defendants claim some 20 other items of incomplete or faulty performance by the plaintiff and no substantial performance because the cost of completing the house in strict compliance with the plans and specifications would amount to 25 or 30 percent of the contract price. The defendants especially stress the misplacing of the wall between the living room and the kitchen, which narrowed the living room in excess of one foot. The cost of tearing down this wall and rebuilding it would be approximately $4,000. The record is not clear why and when this wall was misplaced, but the wall is completely built and the house decorated and the defendants are living therein. Real estate experts testified that the smaller width of the living room would not affect the market price of the house.

The defendants rely on Manitowoc Steam Boiler Works v. Manitowoc Glue Co., . . . for the proposition there can be no recovery on the contract . . . unless there is substantial performance. This is undoubtedly the correct rule at common law. . . . The question here is whether there has been substantial performance. The test of what amounts to substantial performance seems to be whether the performance meets the essential purpose of the contract. In the Manitowoc case the contract called for a boiler having a capacity of 150 per cent of the existing boiler. The court held there was no substantial performance because the boiler furnished had a capacity of only 82 percent of the old boiler and only approximately one-half of the boiler capacity contemplated by the contract. In Houlahan v. Clark, . . . the contract provided the plaintiff was to drive pilings in the lake and place a boat house thereon parallel and in line with a neighbor's dock. This was not done and the contractor so positioned the boat house that it was practically useless to the owner. Manthey v. Stock, . . . involved a contract to paint a house and to do a good job, including the removal of the old paint where necessary. The plaintiff did not remove the old paint, and blistering and roughness of the new paint resulted. The court held that the plaintiff failed to show substantial performance. The defendants also cite Manning v. School District No. 6, . . . However, this case involved a contract to install a heating and ventilating plant in the school building which would meet certain tests which the heating apparatus failed to do. The heating plant was practically a total failure to accomplish the purposes of the contract. See also Nees v. Weaver, . . . (roof on a garage).

Substantial performance as applied to construction of a house does not mean that every detail must be in strict compliance with the specifications and the plans. Something less than perfection is the test of spe-

cific performance unless all details are made the essence of the contract. This was not done here. There may be situations in which features or details of construction of special or of great personal importance, which if not performed, would prevent a finding of substantial performance of the contract. In this case the plan was a stock floor plan. No detailed construction of the house was shown on the plan. There were no blueprints. The specifications were standard printed forms with some modifications and additions written in by the parties. Many of the problems that arose during the construction had to be solved on the basis of practical experience. No mathematical rule relating to the percentage of the price, of cost of completion or of completeness can be laid down to determine substantial performance of a building contract. Although the defendants received a house with which they are dissatisfied in many respects, the trial court was not in error in finding the contract was substantially performed.

The next question is what is the amount of recovery when the plaintiff has substantially, but incompletely, performed. For substantial performance the plaintiff should recover the contract price less the damages caused the defendant by the incomplete performance. Both parties agree. Venzke v. Magdanz, . . . states the correct rule for damages due to faulty construction amounting to such incomplete performance, which is the difference between the value of the house as it stands with faulty and incomplete construction and the value of the house if it had been constructed in strict accordance with the plans and specifications. This is the diminished-value rule. The cost of replacement or repair is not the measure of such damage, but is an element to take into consideration in arriving at value under some circumstances. The cost of replacement or the cost to make whole the omissions may equal or be less than the difference in value in some cases and, likewise, the cost to rectify a defect may greatly exceed the added value to the structure as corrected. The defendants argue that under the Venzke rule their damages are $10,000. The plaintiff on review argues the defendants' damages are only $650. Both parties agree the trial court applied the wrong rule to the facts.

The trial court applied the cost-of-repair or replacement rule as to several items, relying on Stern v. Schlafer, . . . wherein it was stated that when there are a number of small items of defect or omission which can be remedied without the reconstruction of a substantial part of the building or a great sacrifice of work or material already wrought in the building, the reasonable cost of correcting the defect should be allowed. However, in Mohs v. Quarton, . . . the court held when the separation of defects would lead to confusion, the rule of diminished value could apply to all defects.

In this case no such confusion arises in separating the defects. The trial court disallowed certain claimed defects because they were not proven. This finding was not against the great weight and clear preponderance of the evidence and will not be disturbed on appeal. Of the remaining defects claimed by the defendants, the court allowed the cost of replacement or repair except as to the misplacement of the living-room wall. Whether a defect should fall under the cost-of-replacement rule or be considered under the diminished-value rule depends upon the nature and magnitude of the defect. This court has not allowed items of such magnitude under the cost-of-repair rule as the trial court did. Viewing the construction of the house as a whole and its cost we cannot say, however, that the trial court was in error in allowing the cost of repairing the plaster cracks in the ceilings, the cost of mud jacking and repairing the patio floor, and the cost of reconstructing the nonweight-bearing and non-structural patio wall. Such reconstruction did not involve an unreasonable economic waste.

The item of misplacing the living room wall under the facts of this case was clearly under the diminished-value rule. There is no evidence that defendants requested or demanded the replacement of the wall in the place called for by the specifications during the course of construction. To tear down the wall now and rebuild it in its proper place would involve a substantial destruction of the work, if not all of it, which was put into the wall and would cause additional damage to other parts of the house and require replastering and redecorating the walls and ceilings of at least two rooms. Such economic waste is unreasonable and unjustified. The rule of diminished value contemplates the wall is not going to be moved. Expert witnesses for both parties, testifying as to the value of the house, agreed that the misplacement of the wall had no effect on the market price. The trial court properly found that the defendants suffered no legal damage, although the defendants' particular desire for specific room size was not satisfied. . . .

On review the plaintiff raises two questions: Whether he should have been allowed compensation for the disallowed extras, and whether the cost of reconstructing the patio wall was proper. The trial court was not in error in disallowing the claimed extras. None of them was agreed to in writing as provided by the contract, and the evidence is conflicting whether some were in fact extras or that the defendants waived the applicable requirements of the contract. The plaintiff had the burden of proof on these items. The second question raised by the plaintiff has already been disposed of in considering the cost-of-replacement rule.

It would unduly prolong this opinion to detail and discuss all the disputed items of defects of workmanship or omissions. We have reviewed the entire record and considered the points of law raised and believe the findings are supported by the great weight and clear preponderance of the evidence and the law properly applied to the facts.

Judgment affirmed.

Returning to the question posed before *Plante v. Jacobs,* to what extent can contract language preclude use of the substantial performance doctrine? *Plante v. Jacobs* stated:

> Substantial performance as applied to construction of a house does not mean that every detail must be in strict compliance with the specifications and the plans. Something less than perfection is the test of specific performance *unless all details are made the essence of the contract.* [Editor's emphasis.] This was not done here. There may be situations in which features or details of construction of special or of great personal importance, which if not performed, would prevent a finding of substantial performance of the contract.

A distinction must be drawn between different methods by which the contract itself might preclude use of the doctrine. First, a clause could simply state that strict compliance with all requirements of the contract documents is required. Such a clause is likely to be found among the many clauses included in the general conditions of the contract.

Second, the contract could create, as does AIA in A201, a system dealing with minor defects discovered when an inspection is made to determine whether there has been substantial completion and a mechanism for dealing with these defects between substantial and final completion.

Third, interpretation of specific language in the contract documents and the surrounding facts and circumstances may indicate that exact compliance with certain contract requirements was expected. For example, suppose in *Plante v. Jacobs* that the contractor knew or should have known that the Jacobs had purchased expensive furniture that could not be used because the living room wall had been misplaced. Alternatively, suppose the Jacobs had informed Plante that for aesthetic reasons the living room *must* comply in all respects to the floor plan dimensions and *no* deviation would be allowed. In either case, it could be argued that the substantial performance should not be used, as the owners expected strict performance and this was communicated to the contractor at the time the contract was made.

If the doctrine of substantial performance is based on the implication that "close to perfect" compliance is all the owner could reasonably expect, the doctrine should not be applied to contracts using any of the methods previously outlined. Any method could indicate that the owner expected full compliance and that the contractor was aware of this responsibility. Perhaps a clause of the first type would be less persuasive, as it could be part of boilerplate, preprinted clauses and would not sufficiently bring to the contractor's attention the need for strict compliance. Even here the contractor should be aware of the importance of strict compliance.

A misplaced wall could be troublesome even if the contract appeared to negate use of the substantial performance doctrine. This is the very type of defect for which the doctrine is designed. If so, another objective of the doctrine is to avoid economic waste, a goal that originally helped generate the doctrine.

In *Plante v. Jacobs,* correction of the misplaced wall would have cost $4,000, but the misplaced wall was found not to have diminished the value of the house. Although the court did not face a contract method for avoiding substantial performance, it is doubtful that the court would have either ordered the contractor to correct the mistake or permitted the owner to take a 4,000-dollar deduction from the final payment for damages. Actual replacement of the wall either by court order or by allowing the Jacobs to reimburse themselves out of the final payment after they corrected the work themselves would cause or endorse an uneconomic expenditures of $4,000 that would not increase the value of the house.

Suppose the Jacobs were allowed to deduct the $4,000 *without* the need to show they were suffi-

ciently concerned to correct the mistake. Would the Jacobs simply take the money and not make the correction? In the likely event they would not make the correction, they would be given a windfall. But this windfall possibility assumes that the Jacobs built their house principally as an investment and not as a place to live. Even assuming that the Jacobs would not invest the $4,000 to correct the mistake, some recognition should be given to the possibility that the Jacobs might have been damaged in some difficult-to-measure way by having to live in a house with a living room one foot shorter than they had planned even though the misplaced wall gave them an extra foot in the kitchen. Although some might doubt that a mistake of this kind could cause emotional distress, as was seen in Section 6.06(H), the law is slowly moving toward the recognition that certain contracts are made not simply for commercial reasons but for reasons of personal solicitude. Were this the case here, such a loss could have been taken into account in determining the damages caused by the misplaced wall.

The substantial performance doctrine straddles concepts of contract and unjust enrichment. There *can* be risk assumption by contract, though economic waste and windfall *cannot* be excluded. As evidence becomes clear that the risk was specifically assumed by the contractor, there may be less room for the doctrine. But as it becomes apparent that there will be economic waste or windfall, the doctrine has more scope despite contract language that might seem to preclude its use.

Plante v. Jacobs illustrates yet another aspect of the interrelationship between the contract and substantial performance. The court stated:

> In this case the plan was a stock floor plan, no detailed construction of the house was shown on the plan. There were no blueprints. The specifications were standard printed forms with some modifications and additions written in by the parties. Many of the problems that arose during construction had to be solved on the basis of practical experience.

In this type of transaction, the contract documents, such as they were, may have only been starting-out points from which the parties work together toward a particular solution. In such a case, it would be unfair to hold the contractor to the original requirements of the contract because contractor and owners were essentially designing the house as it was being built.

Similarly in this type of transaction, there is not likely to be a detailed set of general terms and conditions that provide a method for determining when there has been substantial completion and ultimately final completion. As a result, it is much easier to insert the common law substantial performance doctrine because there is no contractual method to deal with disputes where the owner has received largely what it had bargained for but had not received everything.

Plante v. Jacobs also illustrates possible abuse of the doctrine. The defects at the end of the job consisted of the contractor's failure to furnish the kitchen cabinets, gutters and downspouts, sidewalk, closet clothes poles, and entrance seat. The estimated cost of furnishing these items was $1,610.95. The amount retained was $7,725. Superficially, it would seem as if Plante had substantially performed and should be entitled to the outstanding balance less the cost of correction. In theory, the deduction from the outstanding balance would compensate the owner for the omissions. In actuality, the legal measure of recovery does not accomplish this. There are additional costs, such as obtaining a substitute contractor, the delay in effectuating the necessary corrections, and the risk of a substitute contractor not performing, which would be difficult to deduct from the outstanding balance.

Even if the contractor *had* removed its workers from the project, it would not impose any serious burden on it to return and complete the job as promised. Would it be unreasonable for the owner to withhold the entire $7,725 as an unashamedly coercive device to get the contractor to finish the work if the owner chose this route rather than engage a substitute contractor? If the amount withheld appears excessive, it could be reduced to two or three times the projected cost of correction (often statutes use 150%). But the amount retained must be enough to make it worthwhile for the contractor to finish the job properly. If not, some contractors will refuse to perform and invoke the substantial performance doctrine to recover the outstanding balance less the amount deducted as damages. In such a case, it is likely that a court would conclude that there had been substantial performance. The court noted that the house had been decorated and

occupied by the owners even though the owners were dissatisfied with the contractor's work. Certainly the house, with the omissions, was fit for its primary purpose, and close to strict completion had been accomplished.

The above criteria were generated from cases that clearly involved economic waste or windfall because the defects were discovered at the end of the project and would have involved substantial redoing of the project at expenditures greatly in excess of the diminished value caused by the deviation. However, where these elements are not present, the substantial performance doctrine should not be a device by which contractors can walk away without having completed the project and expect to be paid the balance less the damages allowed by law. Yet one court expressed this view:

> The doctrine of substantial performance is a necessary inroad on the pure concept of freedom of contracts. The doctrine recognizes countervailing interests of private individuals and society; and, to some extent, it sacrifices the preciseness of the individual's contractual expectations to society's need for facilitating economic exchange. This is not to say that the rule of substantial performance constitutes a moral or ethical compromise; rather, the wisdom of its application adds legal efficacy to promises by enforcing the essential purposes of contracts and by eliminating trivial excuses for nonperformance.[91]

This court views the doctrine as one to protect contractors from owners who seek an excuse to avoid paying. This attitude can lead a contractor to walk away from a commitment and be able to force the court to perform an "accounting job," compute the balance owed by deducting the cost of correction or completion, or use the diminished value, a result *not* bargained for by the owner. It can promote a "just as good" philosophy and encourage sloppy, incomplete work. Whether it promotes promise making is, at best, debatable.

C. Divisible Contract

For various reasons, the law sometimes artificially treats one contract as if it were composed of a number of separate contracts. Section 22.02(N) dis-

---

[91]*Bruner v. Hines,* 295 Ala. 111, 324 So.2d 265, 269–270 (1975).

cussed whether progress payments in construction contracts make the contract divisible.

One function that can be served by the divisibility fiction is to enable a contractor to recover despite not having fully performed the contract. For example, suppose a construction project has five well-defined phases and designated payments for each phase. Suppose the contractor completes three phases but omits the remaining two. In such a case, classifying this essentially single contract into a series of five contracts would enable the contractor to recover for the three phases completed, with the owner being left with a deduction for the damages caused by the failure to complete the final two.

The presence of progress payments does not mean that the parties have agreed that each segment of work is an agreed equivalent for the amount of the progress payment being made. The progress payments are approximations and not intended as agreed figures. Although it is not inconceivable that a court would classify a construction contract as divisible in order to permit the contractor to recover despite default, the doctrine of substantial performance, discussed in Section 22.06(B), and the doctrine of restitution, to be discussed in Section 22.06(D), are more appropriate vehicles for giving relief to a contracting party who has not fully performed yet has benefited the other party.

D. Restitution: Unjust Enrichment

The defaulting building contractor may not have performed sufficiently to take advantage of the substantial performance doctrine. Another concept sometimes available that can enable the contractor to recover for any net benefit that its performance has conferred on the owner is restitution, a concept sometimes called quasi-contract, or *quantum meruit.*

Restitution, unlike its substantial performance counterpart, has not received uniform acceptance. Early cases were not sympathetic to defaulting parties, and many held that a party in default could not recover for work performed.[92] However, there has been some relaxation of this attitude, and in many jurisdictions, contractors are able to recover

restitution despite their own breach in order to avoid unjust enrichment of the owner.[93]

Suppose there is a construction contract in which a promise to pay $100,000 is exchanged for the promise to construct a building. Suppose the contractor unjustifiably leaves the project during the middle of performance. The defaulting contractor may have conferred a net benefit despite its having breached by abandoning the project. For example, suppose the defaulting contractor has received $20,000 in progress payments but the work has been sufficiently advanced so that it can be completed by a successor contractor for $75,000. Absent any delay damages, the defaulting contractor has conferred a net benefit of $5,000 on the owner and should be entitled to this amount.

Some of the same problems that have been discussed in relation to substantial performance apply when the contractor seeks to use restitution to recover for the benefit that has been conferred. Such recovery may be precluded if the contractor's breach is considered "willful" and similar problems of measuring the damages caused by the breach arise when restitution rather than substantial performance is the basis of recovery.

Restitution cases do not arise as frequently as do "end of the job" substantial performance disputes. Part of this may be due to the unlikeliness of a contractor's walking off a project that would be profitable. Only in these cases is there likely to be a net benefit to the owner that exceeds progress payments that have been made.

E. Waiver

During performance or at the end of the project, the owner may manifest a willingness to pay the full contract price despite the contractor's not having fully complied with the contract documents. Such manifestation is often described as a waiver.

---

[92]*Kelley v. Hance,* 108 Conn. 186, 142 A. 683 (1928); *Steel Storage & Elevator Constr. Co. v. Stock,* 225 N.Y. 173, 121 N.E. 786 (1919).

[93]*American Sur. Co. of N.Y. v. United States,* 368 F.2d 475 (9th Cir.1966); *United States v. Premier Contractors, Inc.,* supra note 35; *Starling v. Housing Auth. of City of Atlanta,* 162 Ga.App. 852, 293 S.E.2d 392 (1982); *R.J. Berke & Co. v. J.P. Griffin, Inc.,* 116 N.H. 760, 367 A.2d 583 (1976); *Nelson v. Hazel,* 91 Idaho 850, 433 P.2d 120 (1967); *Herbert W. Jaeger & Assoc. v. Slovak-American Charitable Assn,* 156 Ill.App.3d 106, 507 N.E.2d 863 (1987) (no net benefit); *Sunset Indus., Inc. v. Bartel,* 84 Or.App. 537, 734 P.2d 897 (1987) (dictum); *Kreyer v. Driscoll,* 39 Wis.2d 540, 159 N.W.2d 680 (1968). See Annot., 76 A.L.R.2d 805 (1961).

If the deviation is an important aspect of the contract, any promise by the owner to pay despite noncompliance would have to be supported by consideration. Consideration, as discussed in Section 5.08, usually consists of something given in exchange for accepting the deviation, such as a price reduction. Alternatively, the waiver of even important matters would be enforced if the contractor had reasonably relied on the statement. Examples are stating ''Do the extra work, and I won't insist on a written change order'' or misleading the contractor by a course of conduct such as paying despite noncompliance.[94]

Waivers of less important matters are effective despite the absence of any consideration or reliance. All that is needed in such cases is evidence that an intention to waive or give up the requirement was communicated to the other party.[95]

A waiver can be directly expressed (You told me you would pay even though I omitted the doorstops). It can be expressed indirectly by implication (When you made the final payment knowing I omitted the doorstops, I assumed you were waiving the defect). Waiver is more likely to be found when the defect is small (doorstops) or technical (failure to obtain a written change order). It is more likely to be found where it is directly *expressed* (I'll pay anyway, or Change orders are a bother, and I know you did the extra work). It is less likely to be found when expressed by implication. (When you paid, I thought you were waiving those defects of which you were aware, or When you didn't complain, I assumed you waived the defects). These factors recognize that small things can be given up if the evidence is clear that the party entitled to them was willing to do so and that acts are often ambiguous.

The act frequently contended to indicate waiver of defects is the owner's accepting by using or occupying the project; particularly those defects apparent at the time the owner used or occupied the project will be asserted to have been waived. Although this topic is treated in greater detail in Section 24.05, it should be noted that court decisions are not consistent, though the trend is against acceptance as waiver of a claim.[96]

General conditions of construction contracts sometimes seek to eliminate the possibility of waiver. For example, ¶ 4.3.5.2 of AIA Doc. A201 states that payment is not an acceptance of nonconforming work. Sometimes clauses more generally state that when one party does not insist on its full contract rights, it shall not be precluded from insisting on these rights in the future. Clauses of this type, though supportive of a conclusion that there has been no waiver,[97] are not absolutely waiver proof. Where it appears that there has been a course of conduct that deviates from contract requirements or clear evidence that a defect has been given up, waiver can be found. The clause is removed by concluding that the nonwaiver clause itself has been waived.

Waiver, unlike substantial performance or restitution, allows recovery of the entire contract balance without any deduction for damages. Waiver usually deals with technical requirements such as written change orders[98] or minor defects in the work.

The person asserted to have waived contract rights must have been authorized to do so. Authority is determined by agency concepts and is often dealt with in the contract.[99] However, it is likely that it would take less authority to waive a minor contract defect than to make a contract in the first instance or to modify an existing contract. Waiver is less likely to be applied in public contracts.[100]

---

[94]See Section 21.04(H).

[95]*Texana Oil Co. v. Stephenson*, 521 S.W.2d 104 (Tex.Ct.App.1975).

[96]Although moving into the house and signing a form of acceptance for a federal agency was held to waive defects in *Cantrell v. Woodhill Enter., Inc.*, 273 N.C. 490, 160 S.E.2d 476 (1968), *Honolulu Roofing Co. v. Felix*, 49 Hawaii 578, 426 P.2d 298 (1967), held that mere occupancy is not a waiver, particularly when the owners had to move in because they had vacated their own home. See also *Aubrey v. Helton*, 276 Ala. 134, 159 So.2d 837 (1964), which held that occupancy alone would not waive defects; only where there is accompanying conduct clearly indicating that there has been acceptance will occupancy and use constitute acceptance.

[97]*Armour & Co. v. Nard*, 463 F.2d 8 (8th Cir.1972).

[98]See Section 21.04(H) and Section 26.08 dealing with waiving formalities to obtain time extension.

[99]On the question of authority, see Section 17.05(B).

[100]See Section 11.01(B). But in *Kenny Constr. Co. v. Metro. San. Dist. of Greater Chicago*, 52 Ill.2d 187, 288 N.E.2d 1 (1972), the court treated waiver in a public contract much as if it had been a private contract.

SECTION 22.07 The Certification and Payment Process: Some Liability Problems

The design professional's role in the certification and payment process can generate a variety of claims, such as failure to discover defects, issuing certificates without determining whether the contractor has paid subcontractors and suppliers or whether liens have been or will be filed, and failure to issue certificates within a reasonable time. But this section looks at the design professional's liability for certifying an incorrect amount.

As discussed earlier,[101] one of the multifaceted aspects of the design professional's performance is to interpret the contract, judge performance, and decide disputes. The design professional's role can be analogized to that of a judge when performing some of these functions. When the design professional looks like a judge, in some states she will be accorded quasi-judicial immunity—being liable for only corrupt or dishonest decisions or those not made in good faith. Quasi-judicial immunity where granted by common law is not likely to be asserted as a defense when the design professional has been sued for improper certification. In fact, the landmark English decision reversing earlier decisions granting immunity was based on a claim of improper certification.[102]

Quasi-judicial immunity can also be created by contract. For example, quasi-judicial immunity was granted to the architect by A201, ¶ 2.2.10, published in 1976 for all interpretations and decisions made in good faith. The AIA does not view the architect's role in the certification process as creating quasi-judicial immunity. If the amount certified is disputed by either owner or contractor, the dispute must be submitted to the architect under ¶ 4.3 before it can proceed to arbitration under ¶ 4.5 or to litigation. But currently, A201 does not create

quasi-judicial immunity for the architect's participation in the disputes process.[103]

The principal legal issue has been the extent to which persons other than the owner can pursue a claim for improper certification. Potential claimants are the contractor, subcontractors, or sureties. Overcertification shrinks the retainage and can cause harm to the owner. It can also cause harm to the surety. The surety looks to the retainage for security if the contractor defaults and the surety must take over. Undercertification can harm the contractor and the surety as well. It can adversely affect the contractor's capacity to continue or complete performance.

Case decisions have tended to permit third parties such as contractors and sureties to maintain legal action against the design professional for improper certification.[104] Those cases that have protected the architect have pointed to a lack of privity.[105] The principal protection against such claims in transactions that involve AIA documents is the multitiered disputes process under which the architect's certificate, if challenged, must (ironically) go back to the architect and ultimately to arbitration. Any mistakes can be corrected by the arbitrators. As to the finality of the decision if challenged in court, see Section 29.09.

---

[103]In 1987, quasi-judicial immunity was limited to interpretations and decisions. See A201, ¶ 4.2.12. The AIA believes this will provide such immunity for decisions made under ¶ 4.3. Yet it would have been better to put immunity in ¶ 4.3, dealing with claims and disputes.

[104]*Aetna v. Hellmuth, Obata & Kassabaum, Inc.*, 392 F.2d 472 (8th Cir.1968); *Palmer v. Brown*, 127 Cal.App.2d 44, 273 P.2d 306 (1954) (immunity not discussed); *State v. Malvaney*, 221 Miss. 190, 72 So.2d 822 (1954); *Westerhold v. Carroll*, 419 S.W.2d 73 (Mo.1967). *Designed Ventures, Inc. v. Auth. of the City of Newport*, 132 B.R. 677 (D.R.I.1991).

[105]*Peerless Ins. Co. v. Cerny & Assoc., Inc.*, 199 F.Supp. 951 (D.Minn.1961); *Engle Acoustic & Tile, Inc. v. Grenfell*, 223 So.2d 613 (Miss.1969) (sharply limiting *State v. Malvaney*, supra note 104). See Section 14.08(C).

---

[101]See Section 14.09(D).

[102]*Sutcliffe v. Thackrah*, [1974] A.C. 727.

PROBLEMS

1. A construction contract called for oak panelling in certain rooms. Because oak was unavailable, a subcontractor installed birch panel. The prime did not notice this deviation. Oak and birch panelling

cost the same. When the owner saw the birch panelling, she told the prime contractor she liked it. The next day the owner told the prime contractor she wanted oak panelling installed. Is the contractor obligated to replace the panelling? If she does not, what, if anything, can the owner recover?

2. On February 1, C was entitled to a progress payment of $10,000 based on a certificate issued by the architect. The owner informed the contractor on February 2 that payment would be delayed for a week because the proceeds of the construction loan had run out and the owner was in the process of getting a new loan. What are the contractor's rights at this point? What additional facts would be helpful in resolving this question?

CHAPTER TWENTY-THREE

Expectations and Disappointments: Some Performance Problems

SECTION 23.01 Introduction to Chapters 23 through 28

Each party to a construction contract (the contract between owner and contractor will be illustrative) has expectations when the contract is created. The owner expects to receive the project specified in the construction contract at the time promised. It also expects to pay the contract price as adjusted for change orders.

The contractor's expectations are more complex. Its performance entails a variety of tasks dependent on cooperation by many entities as well as on weather conditions. The contractor expects that it will be able to purchase labor and materials at or within certain prices and that work will proceed without excessive slowdowns or stoppages due to weather, labor difficulties, acts of public authorities, or failure by the owner to cooperate. The contractor does not anticipate that construction of the work will be impeded or destroyed by fire, earthquake, or vandalism. It expects to be paid as it performs.

Yet either or both may be disappointed. The owner may not receive the project it expected for a variety of reasons, explored in Chapter 24, dealing with defects, and in Chapter 25, dealing with subsurface problems. The owner may not receive the completed project at the time promised, a disappointment discussed in Chapter 26. Chapter 27 explores the methods of measuring any claims the owner may have against the contractor for not performing.

The contractor's disappointments that relate to failure to receive progress payments treated in Chapter 22 are explored in greater detail in Chap-

ter 27. This chapter deals with the disappointments that relate to obstacles to its performance or insecurity as to the performance by the owner. The special problems of increases in the cost of performance due to unexpected subsurface conditions are treated in Chapter 25. The contractor's disappointed expectations as to the pace with which it will work are treated in Chapter 26.

Disappointments can encompass losses caused by subcontractors, whose special problems are discussed in Chapter 28.

This chapter introduces performance problems discussed in Chapters 24 through 28. Initially, it describes and briefly illustrates the essential attributes of legal doctrines central to performance problems. In addition, the chapter notes some performance problems that do not justify full chapter treatment.

SECTION 23.02 Affirmative Legal Doctrines: The Bases for Claims

A. Introduction: The Shopping List

This section summarizes in mostly nontechnical terms the legal theories that can be the basis for construction-related claims between owner and contractor. Unfortunately, it has become common for claimants to assert a variety of claims based on assorted legal theories, with the law generously permitting, at least up to a point in a lawsuit, the use of such a shotgun approach. This chapter deals only with claims by contract-connected parties. As will be seen, the variety of doctrines is astounding.

488

B. Fraud

Fraud is the intentional deception of one contracting party by the other. The deception induces the former to make a contract, one that it would not have made had it not been deceived. Usually the deception consists of giving false information or half-truths or occasionally concealing information. The deception must relate to important matters. The party defrauded relies by making the contract. For example, an owner may defraud a contractor who is concerned about the owner's resources by falsely telling the contractor it has received a loan commitment. The contractor may defraud the owner by falsely representing it has access to equipment needed to perform particular work. In most states, fraud can consist of making a promise to induce the making of a contract without intending to perform it.

The defrauded party can rescind the contract and recover any benefit it has conferred on the other party based on restitution. It can affirm the contract, cease or continue performance, and recover damages. Punitive damages are generally available.

Fraud should not play a significant role in construction performance disputes. Although it can on rare occasions be committed during performance, such as deceiving an owner by telling it that certain work has been performed or has been approved by a city inspector, it usually consists of pre-performance conduct. Performance disputes over unanticipated subsurface conditions sometimes involve allegations that an owner deliberately gave false information that the contractor relied on.

Fraud is easy to allege, particularly in a free-swinging construction dispute. Although fraud claims have a nuisance and settlement value, they are rarely successful.

C. Concealment of Information

Concealing information, though not always fraudulent, can be the basis for a claim. The early common law placed no duty on a contracting party to disclose information to the other party during negotiation. But as noted in Section 18.04(B), there is an increasing tendency to require a party to disclose information to the other party that the latter would want to have known and that it is not likely to discover on its own.

Concealing information, like fraud and misrepresentation (to be discussed in (D)), relates mainly to subsurface problems, as discussed in Chapter 25. Failure to disclose when disclosure should have been made permits the innocent party to rescind the contract and recover for any benefit it may have conferred on the other party based on restitution. Alternatively, it can be the basis for a claim for damages, based on either the additional expenses incurred or the difference between the expenses that should have been incurred and those that were incurred.

D. Misrepresentation

Misrepresentation (discussed in Section 7.07) is intentionally, negligently, or innocently furnishing inaccurate information relied on by the other party in deciding to make the contract and its terms—principally price and time. Although, like fraud and concealment, it is principally a preperformance claim, it can be based on conduct during the performance of the contract, such as misrepresenting the value or the quality of work done. But like fraud and concealment, misrepresentations occurring before making the contract are usually discovered during performance. Again, this arises mainly, though not exclusively, in subsurface disputes.

E. Negligence

Negligence claims generally involve parties not connected by contract, such as a claim by a person delivering materials to a site against the contractor based on the latter's negligent conduct. But as seen in Section 14.05(D), negligence can be the basis of a claim by one party to a contract against the other party. A negligence claim requires a duty owed to the claimant, a violation of that duty, and a suffering of harm caused factually and proximately by the negligent party.[1]

Negligence claims have become more common in construction *contract* disputes that relate to defects or delays, usually with the hope of involving the contractor's public liability insurer, avoiding exculpatory clauses in the contract, or obtaining a more expansive remedy than one that could be obtained for breach of contract.[2]

---

[1]See Section 7.03.
[2]See Sections 14.05(E) and 24.08.

Negligence claims require that the claimant establish that the defendant's conduct did not live up to the standard required by law, something not required in a simple breach-of-contract claim. Negligence claims are barred after a shorter period than breach-of-contract claims.

The remedy for negligent conduct related to construction is usually the additional expense incurred. Although it is *roughly* similar to damages for breach of contract, certain defenses used in breach-of-contract claims, such as foreseeability,[3] offer less protection to a defendant when the claim is based on negligence. Negligence as a tort can be more easily the basis for a claim for emotional distress.[4] If wrongdoing goes beyond simple negligence and involves intentional misconduct or recklessness, the wrongful party cannot defend on the basis that the loss that was incurred was not reasonably foreseeable, and punitive damages may be recovered.

F. Strict Liability

As noted in Section 7.09, strict liability has its greatest modern use in claims against manufacturers who market defective products. It does not require proof of negligence. In that sense, it is like a breach of contract or its near relative, breach of an implied warranty. Although it is often the basis of claims by those who buy homes or lots built by developers,[5] it is rarely used as the basis for a claim between owner and contractor in the ordinary construction contract.

G. Breach of Contract

To determine whether a contract breach has occurred, the following must be interpreted:

1. The written contract, if any.
2. Antecedent negotiations not superseded by a written contract.
3. Terms implied by law.

The contract may prescribe conditions that relate to claims, set forth exculpatory provisions, or control remedies. Breach does not require fault, although doctrines described in Section 23.04 may provide a defense. Remedies were discussed generally in Chapter 6 and are discussed more particularly in Chapter 27. Basically, the claimant can recover gains prevented and losses incurred that are established with reasonable certainty, were caused substantially by the breach, could not have been reasonably avoided, and were reasonably foreseeable at the time the contract was made.

H. Express Warranty

A warranty is an assurance by one party relied on by the other party to the contract that a particular outcome will be achieved by the warrantor. For example, a manufacturer may expressly warrant that a tank it promises to build will travel one hundred miles on one hundred gallons of fuel, will not be inoperable more than ten days a month, and will hit a moving target ten miles away while the tank is traveling at twenty-five miles per hour.

Failure to accomplish the promised objectives is a breach. Fault is not required. (Defenses, though rare, are discussed in Section 23.03.) Frequently the *method* in which the promised outcome is to be achieved is determined by the warrantor. Courts have been reluctant to hold the warrantor to its warranty if it has not been given the freedom to determine the method by which the warranted outcome is to be achieved.

Express warranty for various purposes has been considered a *special* form of contract breach. It *can* be considered simply an outcome promise by the warrantor. Perhaps express warranty continues to be used to emphasize its nonfault nature. It is less susceptible to contract doctrines that can *excuse* nonperformance by a contracting party. In some jurisdictions, breach of a contract to perform services must be based on negligence.[6] Warranty, express or implied, in such jurisdictions may simply bring the law back to the normal nonfault contract standard. (Express warranty can create rights for a third party. It was used before the modern law of product liability allowed a user to sue a manufacturer directly without privity and based on nonfault conduct in manufacturing the product.[7]) Its

---

[3]See Section 6.06(C).
[4]See Sections 6.06(H) and 27.09.
[5]See Sections 7.09(K) and 24.10.

[6]*Samuelson v. Chutich*, 187 Colo. 155, 529 P.2d 631 (1974).
[7]See *Baxter v. Ford Motor Co.*, 168 Wash. 456, 12 P.2d 409 (1932). It is still used when the product was not defective. *Collins v. Uniroyal, Inc.*, 64 N.J. 260, 315 A.2d 16 (1974).

use in construction relates mainly to warranty or guarantee clauses discussed in Section 24.09. Its remedy is the same as for breach of contract.

I. Implied Warranty

This is another "subspecie" of contract breach that developed historically as a separate basis for a claim. It is like express warranty in that it must be relied on and an outcome is warranted. Liability is strict. No negligence need be shown. Remedy is the same as for breach of an express warranty.

In sale of goods, implied warranty usually relates to merchantability or fitness for the buyer's known purposes. Implied warranty developed as an offshoot of modern tort law, mainly to allow a claimant who had suffered harm—usually personal harm—to maintain an action against a manufacturer with whom it was not in privity, without the necessity of establishing negligence. It was also used to avoid real property doctrines that often barred an occupier of a home from maintaining a legal claim against a builder who had sold the home to it or to a predecessor owner, again, with no need to show negligence.

In construction contract disputes, as shall be seen,[8] a contractor often claims that the owner *impliedly* warrants the sufficiency of the design. This could have been described as an implied term, that the owner "promises" that executing the design will achieve the expected outcome of both owner (a successful project) and contractor (performance in a reasonably efficient manner).

Where the construction contract does not specify the quality of the contractor's services,[9] the law usually implies a warranty that the contractor will perform its services in a workmanlike manner, free of defects of workmanship.[10] In operation, this implied warranty will closely resemble the negligence standard. However, because warranty is more re-sult oriented, application of a warranty is likely to place the burden on the contractor where the work has not measured up to the ordinary expectation of the owner.

Although sometimes implied terms are justified as simply reflecting the intention of the parties,[11] warranties are often implied because of "fairness," a conscious judicial allocation of risk. In sales transactions the express warranty may be one that is intended to limit or negate the warranties that would be provided by the law in the absence of contrary provisions in the contract. Such a *disclaimer* warranty (it takes away when it appears to give) must be clear and conspicuous to be effective[12] and even if clear and conspicuous may not be enforced if such a warranty is unconscionable.[13] An implied warranty is preempted (displaced) by an express warranty, just as any express contract term takes precedence over or may bar an implied term.

An implied warranty is not only preempted by an express warranty but may not have been relied upon by the other party, an essential element in a claim for breach of warranty. It is an issue in Chapter 25 (which deals with subsurface claims), with the owner often giving information but disclaiming responsibility for the information's accuracy.

J. Consumer Protection Legislation

Most states have enacted consumer protection legislation. Such laws bar misleading or fraudulent business practices in consumer transactions. They provide a list of forbidden activities and enlarge the remedies available to consumers, such as permitting treble damages and awarding attorneys' fees.

In some states, these statutes have been used against contractors who build for consumers. As illustrations, the following acts have been found to constitute unfair trade practices:

1. Using substandard material.[14]
2. Failing to complete work within a specified period of time.[15]

---

[8]See Sections 23.05(E) and 24.02.

[9]See AIA Doc. A201, ¶ 3.5.1, for a warranty as to labor.

[10]*New Zion Baptist Church v. Mecco, Inc.*, 478 So.2d 1364 (La.App.1985). But see *Milau Assoc. v. North Avenue Dev. Corp.*, 42 N.Y.2d 482, 368 N.E.2d 1247, 398 N.Y.S.2d 882 (1977), which held that there was no implied warranty of fitness in a service contract. The case dealt principally with the implication of a successful result in terms of the expectations of the owner, not the method of measuring whether work executed by the contractor meets a contractual standard.

[11]See Section 19.02(A).

[12]U.C.C. § 2-316.

[13]U.C.C. § 2-302.

[14]*New Mea Constr. Corp. v. Harper*, 203 N.J.Super. 486, 497 A.2d 534 (1985).

[15]*Watson v. Bettinger*, 658 S.W.2d 756 (Tex.Ct.App.1983).

3. Failing to adequately supervise the construction project.[16]
4. Intentionally supplying low estimates.[17]
5. Providing defective workmanship.[18]

Texas, whose Deceptive Trade Practices Act has had a significant impact on construction litigation, has applied its strict liability provisions to professional services.[19]

Sometimes deceptive trade practices acts can be applicable to disputes between prime contractors and subcontractors. For example, the Connecticut Unfair Trade Practices Act was violated when a prime contractor made an oral contract with a subcontractor, listed the subcontractor as a disadvantaged business enterprise in its bid with a public entity, and then sought to renegotiate the contract. In addition to being awarded lost profits, the subcontractor was awarded its attorneys' fees under the Act.[20]

As noted in Section 10.11, consumer dissatisfaction sometimes results in the charge that the licensed contractor should be disciplined because of its failure to perform properly. Such complaints have generated another method of consumer protection, structuring an informal complaint process, with the contractor's license at stake.

SECTION 23.03 Defenses to Claims

A. Introduction: Causation and Fault

On the whole, the doctrines described in this section relate to the occurrence or discovery of a supervening event that one contracting party asserts should relieve it from its obligation to perform. A party asserting such a defense must show that the event has seriously disrupted performance planning or performance itself. If the occurrence of the event could have been prevented or if the effect of

the event on performance could have been minimized or avoided, the occurrence of the event does not provide a defense. For example, suppose laborers struck because of illegal conduct by the contractor who employed them. The contractor will not receive relief, because its conduct has caused the disruptive event. Likewise, if work were shut down because of a court order issued because the owner did not comply with land use controls, the owner will not be given relief if a contractor makes a claim.

Similarly, a party who does not appear to be "innocent" is less likely to be given relief, either because its conduct has caused the event or because its fault denies it the right to obtain relief from its contractual obligations.

On the whole, the doctrines described in this section deal with events that affect performance that cannot be said to have been the fault of either party.

B. Contractual Risk Assumption

Some of the doctrines discussed in this section were developed by the common law.[21] They involve claims for relief based on doctrines developed by courts to deal with *drastic* changes of circumstances. The willingness or even the power of courts to use these common law doctrines is often affected negatively by *specific* contract clauses. For example, an owner may defend against a contractor's claim for additional compensation for delays caused by wrongful acts of the owner by pointing to a clause under which the contractor is limited to a time extension.

Similarly, the common law doctrine of impossibility is not likely to be employed if there is a *force majeure* provision, which grants relief under specified circumstances to the performing party. For example, contract clauses frequently allow for a time extension if certain events occur. The contractor may defend a claim by the owner that the contractor has not performed on time by pointing to contract language that justified a time extension.

A fixed-price contract itself is a contractual assumption of risk that in many instances bars the contractor from receiving additional compensation in the event its performance costs have increased

---

[16]*Building Concepts, Inc. v. Duncan,* 667 S.W.2d 897 (Tex.Ct.App.1984).
[17]*Hannon v. Original Gunite Aquatech Pools, Inc.,* 385 Mass. 813, 434 N.E.2d 611 (1982) (dictum); *East Lake Const. Co. v. Hess,* 102 Wash.2d 30, 686 P.2d 465 (1984).
[18]*Jim Walter Homes, Inc. v. White,* 617 S.W.2d 767 (Tex.Ct.App.1981).
[19]*White Budd Van Ness Partnership v. Major-Gladys Drive Joint Venture,* 798 S.W.2d 805 (Tex.Ct.App.1990).
[20]*Bridgeport Restoration Co., Inc. v. A. Petrucci Constr. Co.,* 211 Conn. 230, 557 A.2d 1263 (1989).

[21]See Section 1.07.

because of events over which it may have had no control.

Contractual risk assumption is not limited to express contract terms. Courts often hold that the owner warrants the "sufficiency" of its design. The law will protect a contractor who follows the design if the owner asserts a claim for a defect by concluding that the owner impliedly warranted the design.

As contracts become more detailed, the potential for judicial intervention may narrow. Yet the more open recognition of the adhesive nature of many construction contracts may tempt a court to judge the *fairness* of contract clauses dealing with these claims for relief.[22]

C. Mutual Mistake

Each contracting party, particularly the party agreeing to perform services, has fundamental assumptions often not expressed in the contract. For example, often the contracting parties assume that the subsurface conditions that will be encountered will not vary greatly from those expected by the design professional or the contractor. Although some deviation may be expected, drastic deviation that has a tremendous effect on the contractor's performance may be beyond any mutual assumption of the parties. In such a case, the contractor may be relieved from further performance (discussed in Section 25.01(E)).

Gevyn Construction Corp. v. United States[23] involved a contract with the U.S. Post Office Department. The specifications referred to a letter sent by Michigan state officials to the post office that stated that the contractor could tap a drain into a designated storm sewer of the state of Michigan. Relying on this, the contractor made its bid. But the state of Michigan refused permission, and the contractor was forced to connect the drain to a more distant outlet. Although the court to some degree emphasized that the U.S. Post Office Department invited the contractor to rely on the letter, the main basis for relief was the parties' having based their agreement on the false assumption that the drain could be tapped into the state's storm sewer.

Another illustration of the occasional use of mutual mistake as a basis for relief is seen in Section 23.05(B), dealing with disruptive labor activities.

D. Impossibility: Commercial Impracticability

Despite the generally harsh attitude taken by nineteenth-century common law courts toward claims of impossibility, in that century contracting parties could be relieved if it was concluded that their performance had become impossible.[24] However, it soon became recognized that true impossibility is rare, and what is usually asserted as a defense by a performing party is that performance cannot be accomplished without excessive and unreasonable cost. This recognition emphasized "commercial impracticability" rather than actual impossibility.[25]

Despite the apparently increased willingness to grant relief to a contracting party, the need for commercial certainty meant that relief would be granted sparingly. First, something unexpected must occur. Second, the risk of this unexpected occurrence must not have been allocated to the performing party by agreement or by custom. The occurrence of the event must have rendered performance commercially impracticable.[26] These are formidable requirements, and relief requires extraordinary circumstances.

In addition to the differentiation between impossibility and impracticability, other differentiations are useful. First, *objective* impossibility is sometimes differentiated from *subjective* impossibility. To grant relief, it must usually be shown not simply that the contracting party could not perform (subjective impossibility) but that other contractors similarly situated (objective impossibility) would not have been able to do so.

Second, *temporary* impossibility should be differentiated from *permanent* impossibility. For example, suppose there is a severe material shortage that makes it impossible for the contractor to continue performance. This shortage may be relieved

---

[22]See Sections 5.04(C) and 19.02(E).
[23]357 F.Supp. 18 (S.D.N.Y.1972).

[24]*Taylor v. Caldwell*, 122 Eng.Rep. 309, 3 Best & S. 826 (Q.B.1863) (theatre owner exonerated when theatre burned).
[25]U.C.C. § 2-615(a).
[26]*Transatlantic Financing Corp. v. United States*, 363 F.2d 312 (D.C.Cir.1966) (charterer not relieved when Suez Canal closed in 1956).

by elimination of the event that caused the shortage, such as a transportation strike. In such a case, impossibility is only temporary. Unless the contractor is held to assume these risks, the only relief it should obtain is a time extension, not termination. However, if it appears that the event will continue an indefinite time or for so long a time that it will drastically affect the cost of performance, termination may be appropriate.

Third, *partial* impossibility must be differentiated from *total* impossibility. Suppose a material shortage is caused by a strike. As a result, a supplier who has contract and customer commitments of 1,000 units has only 500 in stock and cannot obtain others. It will be allowed relief if it makes a good-faith allocation of its available supply to all its customers.[27] Suppose it has a contractual commitment to deliver ten units to a contractor. If it would be an exercise of commercial good faith to supply all customers—those with contracts and those with commercial commitments—alike, the contractual obligation would be discharged if the supplier delivered five units to the contractor.

E. Frustration

Frustration of purpose developed early in the twentieth century. It looked at the effect of subsequent events not on performance but on desirability of performance. For example, the leading English case[28] giving rise to this doctrine involved a contract under which the plaintiff let rooms to the defendant from which the defendant would have a good view of the coronation of King Edward VII. The coronation was postponed because the King became ill. The defendant was relieved from his obligation to pay for the rooms. Clearly this did not involve impossibility. The coronation parade postponement made the contract much less *attractive* than it was originally. This doctrine has great similarity to mutual mistake. Each party probably assumed that the coronation would take place as originally scheduled, and each was equally mistaken.

It can be seen that relief under this doctrine must be awarded sparingly or contracts would lose much of their effectiveness. The leading American case, *Lloyd v. Murphy*,[29] involved a claim by a commercial tenant at the onset of World War II that it should be relieved of its obligation to pay rent on premises from which it intended to sell new cars that became unavailable because of the war. The court held that relief under this doctrine required that the value of the contract be almost totally destroyed by an event that was not reasonably foreseeable by the party seeking relief. The court held that the tenant did not meet these requirements.

Suppose a law were passed during construction of a racetrack that made horse racing illegal in the state. Suppose the owner sought to relieve itself of its contractual obligation by pointing to this event. If the project can be used only for an activity that is now illegal, it can be argued that the owner should be freed from its construction contract. However, much would depend on whether the legislative action were reasonably foreseeable and whether the project could be used for other purposes, such as a go-cart track, tennis court, commercial exhibitions, a nine-hole golf course, or auto racing.

While frustration claims are rare in construction projects, the Massachusetts Supreme Judicial Court employed this doctrine in a case that involved a contract to resurface and improve a public road.[30] The contract required the contractor to replace a grass median strip with precast concrete barriers. The prime contractor entered into subcontracts under which a designated subcontractor would provide the concrete barriers for the median strips. After work began, a citizens' group filed a lawsuit protesting the removal of the grass median strip. The public entity settled the case by agreeing not to install any additional barriers and deleting the concrete barriers from the project. As a result, the prime contractor cancelled its subcontract with the subcontractor to whom the work had been awarded, generating a claim for lost profits by the cancelled subcontractor. (The subcontractor had been paid at the contract price for the barriers it had delivered.)

The court held that the prime contractor was not responsible for the elimination of the barriers and that the risk of elimination of the barriers was not allocated by the subcontract. The prime contractor

---

[27]U.C.C. § 2-615(b).

[28]*Krell v. Henry*, 2 K.B. 740 (C.A.1903).

[29]25 Cal.2d 48, 153 P.2d 47 (1944).

[30]*Chase Precast Corp. v. John J. Paonessa Co., Inc.*, 409 Mass. 371, 566 N.E.2d 603 (1991).

was excused because of the frustration of purpose rule, because neither party to the subcontract anticipated that the public entity would cancel a major portion of the job. The court's analysis did not draw any significant distinction between frustration and commercial impracticability, unlike that done in earlier cases and commentaries.

F. Acceptance

Sometimes the contractor claims a defense when a claim is asserted against it for defective work based on the owner's having accepted the project. Inasmuch as this defense is principally related to defects, it is discussed in Section 24.05.

G. Passage of Time: Statutes of Limitation

Sometimes a claimant is met with a defense that it did not begin legal action on the claim within the time required by law. Though this is not strictly speaking a defense based on the contract or common law doctrines, it has generated immense complexity and must be looked at in any treatment of claims generally.

Statutes of limitations (discussed elsewhere[31]) deal with the effect of the passage of time on the maintainability of claims. They are designed to protect defendants from false or fraudulent claims that may be difficult to disprove if not brought until relevant evidence has been lost or destroyed and witnesses become unavailable. In addition, entirely apart from the fault aspect of delay in bringing legal action, the law can also seek to promote certainty and finality in transactions, especially commercial transactions, by terminating contingent liabilities at specific points in time. This second function is sometimes accomplished by a related statute dealing with time, sometimes called a "statute of repose."[32] The statute can bar a claim of which the claimant was never aware.

Statutes of limitations also encompass what are called completion statutes (discussed later in this section).

Although these statutes are necessary, their implementation has generated great difficulty in construction performance claims, claims often discovered long after the breach by the contractor or completion of the work. Such statutes frequently are unclear as to the time the statutory period begins. In construction, the period can begin when the wrongful act occurred, when the contract performance had been completed, when the defective work had been discovered, when the owner knew of the defect and its cause, or when the owner knew or should have known all the elements of its claim, including against whom the claim can be made. For example, suppose a contract is made to build a commercial building in 1980. Work proceeds during that year. The contractor does not use proper workmanship in the application of adhesive materials in the construction of the roof. The building is accepted and occupied in 1980. In 1990, severe roof leaks occur that are traceable to improper workmanship by the contractor.

Suppose the applicable statute of limitations states that a claim for breach of contract must be begun within four years from the time the "cause of action accrued." If the cause of action accrued at the time of the improper workmanship or the acceptance of the building, the claim has been barred by the passage of time. If the claim accrued at the time of *damage*, should the period begin at the time of completion (at that time the improper workmanship could not have been corrected) or at the time the poor workmanship was discovered? If the cause of action accrued in 1990, legal action can be brought until 1994. If so, the contractor will be forced to defend a claim based on events that occurred fourteen years earlier. Its defense may be hampered by its inability to produce witnesses and documentary evidence. Yet if the earlier time is selected, the owner has lost a claim before it becomes aware it had one unless an objective standard is used and a reasonable inspection would have discovered the poor workmanship during the time of performance.[33]

Participants in the construction industry persuaded all states but Iowa and Vermont to enact completion statutes that cut off liability a designated number of years after substantial completion (also called statutes of repose). Some statutes

---

[31]See Sections 2.03 and 14.09(C).
[32]*Gates Rubber Co. v. United States Marine Corp.*, 508 F.2d 603 (7th Cir.1975).

[33]See *D.J. Witherspoon v. Sides Constr. Co.*, 219 Neb. 117, 362 N.W.2d 35 (1985) (where architect merely designs, period begins when design documents delivered; where he supervises, it begins at completion of his services); *Holy Family Catholic Congregation v. Stubenrauch Assoc.*, 136 Wis.2d 515, 402 N.W.2d 382 (App.1987) (commences when contractor loses significant control and owner can occupy or use it for its intended purpose; not determined by architect certificate or when contractor stopped work).

dealt only with claims based on breach of contract, while others included tort claims.[34] Most differentiated between latent and patent defects, the former being defects not discoverable by reasonable inspection and the latter being defects that were reasonably discoverable. The statutes varied as to the persons who could take advantage of them. Typically, those protected were design professionals and contractors, with protection not accorded owners or suppliers. The time period selected by the legislatures also varied.

The statutes were attacked in many states, principally based on the unfairness of depriving a party of the claim that it did not realize it had. More technically, statutes were attacked based on constitutional grounds, mainly as violating constitutional equal protection of the law. Some persons were given legislative protection while others were not.

Most courts upheld such statutes. Some were held unconstitutional. Those held unconstitutional were usually redrafted to meet constitutional objections. Even some redrafted were found to be unconstitutional. A number of such statutes have an uncertain status, neither being held constitutional nor unconstitutional.[35] In addition, the statutes generated a great amount of litigation, not only passing on their constitutionality but also seeking to integrate them with the normal statutes of limitations of the state.[36] As this treatise cannot provide a detailed description and analysis of these statutes, local law must be consulted.

One observation that relates to risk management (discussed in greater detail in Chapter 32) merits mention at this point. Racing to the legislatures may have distracted from another approach to the long-delayed claim. It is possible for the contracting parties to regulate this problem. Although contractual regulation will not affect claims of third parties, such as those who may be injured if a building collapses years after it is completed, the bulk of the exposure in this area usually involves claims by the owner against the contractor or design professional.

Parties in a construction contract can regulate both the period of time in which the claim can be made and when that period begins. AIA Doc. B141, ¶ 9.3, and A201, ¶ 13.7.1, seek to control the latter. As to the former, § 2-725(1) of the Uniform Commercial Code which governs transactions in goods and can, by analogy, be applied to service transactions creates a four-year period of limitation but allows a contractually created period of not less than one year but not longer than the four-year statutory period.

Sometimes contractors contend that any warranty or guarantee clause is a private period of limitations (discussed in Section 24.09(B) dealing with these clauses).

---

[34]See *Martinez v. Traubner*, 32 Cal.3d 755, 653 P.2d 1046, 187 Cal.Rptr. 251 (1982) (did not apply to tort claim for personal injury). See also *Stoneson Dev. Corp. v. Superior Court*, 197 Cal.App.3d 178, 242 Cal.Rptr. 721 (1987) (applied to claim based on strict liability).

[35]For a list of the statutes and their status see A/E Legal Newsletter, Special Supplement No. 1, April 1992 and Legislation Division Update, 12 Constr. Lawyer, No. 4, Nov. 1992, p. 36. The leading case refusing to uphold such a statute is *Skinner v. Anderson*, 38 Ill.2d 455, 231 N.E.2d 588 (1967) (statute subsequently revised). The leading case upholding such a statute is *Freezer Storage, Inc. v. Armstrong Cork Co.*, 234 Pa.Super. 441, 341 A.2d 184 (1975). As to recent cases, see *Turner Constr. Co. Inc. v. Scales*, 752 P.2d 467 (Alaska 1988) (challenge by owner: held unconstitutional); *American Liberty Ins. Co. v. West & Conyers*, 491 So.2d 573 (Fla.Dist.Ct.App.1986) (upheld); *Shibuya v. Architects Hawaii, Limited*, 65 Hawaii 26, 647 P.2d 276 (1982) (unconstitutional, despite having been redrafted after an earlier decision holding the statute unconstitutional); *State Farm Fire & Cas. Co. v. All Elec. Inc.*, 660 P.2d 995 (Nev.1983) (upheld over strong dissent); *Daugaard v. Baltic Cooperative Bldg. Supply Assoc.*, 349 N.W.2d 419 (S.D.1984) (held unconstitutional), noted at 30 S.D.L.Rev. 157 (1984); *Sun Valley Water Beds of Utah, Inc. v. Herm Hughes & Son, Inc.*, 782 P.2d 188 (Utah 1989) (violated open court provision of Constitution and an arbitrary and unreasonable means of achieving the statutory objective; statute redrafted in 1991). Michigan law

is discussed in Baiers, *Statutes of Limitations Applicable to Architects, Engineers and Contractors in Michigan,* 1989 Det.C.L.Rev. 25. For a discussion of Pennsylvania law see Comment, 30 Duq.L.Rev. 681 (1992); For a discussion of Vermont law see Note, 9 Vt.L.Rev. 101 (1984). Care must be taken not to rely on a court decision holding the statute unconstitutional. The legislature may have subsequently amended the statute to meet the objections of the court decision.

[36]This topic was dealt with extensively in *Northern Indiana Pub. Service Co. v. Fattore Constr. Co.*, 486 N.E.2d 633 (Ind.App.1985). The court held that a special statute takes preference over a general statute, especially if the former has been more recently enacted. It also held that if either of two statutes can apply, the court is to apply the longer statute.

H. Release

In some disputes, the party against whom a claim is made asserts that the claim has been settled. Section 27.13 discusses this defense, and Section 22.05 discusses it in the context of final payment.

I. Sovereign Immunity

Although sovereign immunity has been abolished in many states, some states use immunity to provide a defense to a public owner. Despite the many cases abolishing immunity, one case recently stated that a statute abolishing immunity must be strictly construed so as not to undermine sovereignty.[37] A number of cases with varied outcomes have recently passed on whether a public entity waives its immunity by entering into a construction contract.[38]

SECTION 23.04 Restitution

Some of the doctrines described in Section 23.03 must be looked at in connection with restitution. If a contractor is relieved because of common law doctrines described in Section 23.03, it may seek affirmatively to recover for the benefit that it has conferred before the time that its obligation to perform was terminated. One aspect of this, destruction of the work in progress, is discussed in Section 23.05(C). For two reasons, this problem has not generated much difficulty in construction disputes. First, the contractor is usually being paid as it performs, and any restitutionary claim it might have would be reduced by these progress payments. This means that the amount at stake is likely to be small. Second, courts have used these relief doctrines sparingly in construction cases. However, in a series of Massachusetts cases, subcontractors were granted broad restitution remedies when the prime contract was found to have been illegally awarded.[39]

---

[37]*De Fonce Constr. v. State*, 198 Conn. 185, 501 A.2d 745 (1985).

[38]*G.M. McCrossin, Inc. v. W. Va. Bd. of Regents*, 355 S.E.2d 32 (W.Va.1987); *C&M Constr. Co., Inc. v. Commonwealth*, 396 Mass. 390, 486 N.E.2d 54 (1985); and *Architectural Woods, Inc. v. State of Washington*, 92 Wash.2d 521, 598 P.2d 1372 (1980). But see *Ramie Constr. Co. v. Apache Tribe of the Mescalero Reservation*, 673 F.2d 315 (10th Cir.1982).

[39]*Albre Marble & Tile Co. v. John Bowen Co.*, 338 Mass. 394, 155 N.E.2d 437 (1959); *Boston Plate & Window Glass Co. v. John Bowen Co.*, 335 Mass. 697, 141 N.E.2d 715 (1957); *M. Ahern Co. v. John Bowen Co.*, 334 Mass. 36, 133 N.E.2d 484 (1956).

Restitution, however, has another use. Section 23.02 discussed claims and their bases. Sometimes a party wishes to measure its recovery for its claim not by the other party's promised performance but by the benefit it has given the other party (discussed in greater detail in contractor claims analyzed in Section 27.02).

SECTION 23.05 Specific Applications of General Principles

A. Increased Cost of Performance

In the course of performance, the contractor may find that its cost of performance has risen dramatically. Illustrations can be drastic increases in the cost of materials, equipment, supplies, utilities, labor, or transportation. These risks are generally assumed by the contractor under a fixed-price contract.

The common law provided little protection to a contractor who performs under a fixed price. To be sure, extraordinary and unanticipated events may occur that drastically affect the cost of the contractor's performance and may be the basis for a claim for relief. Here again the remedy sought may be influential if not determinative. Additional time is easiest to obtain, although this depends on the language of the clause that grants time extensions. Additional compensation and termination, if sought, are more difficult to obtain.

There are exceptions, most relating to specific contract provisions or terms implied into the contract. Examples of the former are equitable adjustment provisions that relate to subsurface work, as discussed in Section 25.06; escalation clauses that can grant price increases under certain market conditions; and time extension provisions, as discussed in Section 26.08. Examples of the latter are implied warranties by the owner that the design can be constructed economically and efficiently as discussed in (E).

B. Labor Disruptions: The Picket Line

The labor dispute is one of the most disruptive events in a construction project. Such disputes can cause frequent and lengthy work stoppages. Craft unions sometimes dispute who has the right to perform certain work. Not uncommonly, employees of unionized contractors refuse to work on a site with nonunion employees of another contractor. Workers, as noted in Section 19.01(D), sometimes refuse to work on a project because prefabricated units are introduced on the site in violation of a "work preservation" clause in a subcontract or collective bargaining agreement. Any of these situations can result in a strike, a picket line, or both. Workers frequently refuse to cross a picket line. Such refusals can shut down a job directly because no workers are willing to work on the site or indirectly through the refusal of workers to deliver materials to a picketed site.

Pressure and even coercion are common tactics in the struggles between employers and those groups attempting to organize the workers as well as represent them in negotiating wages, hours, job security, and working conditions. Each of the disputants—employer and union—not only seek to persuade but also use economic weapons to obtain a favorable outcome.

Disputants direct pressure at each other by use of economic weapons such as strikes or lockouts. Even these direct primary pressures affect neutrals not involved in the labor dispute. A strike affects nonstriking workers, the families of the strikers, and those who deal with the struck employer, such as those who supply it goods or purchase its products or services. Likewise, a lockout by the employer affects neutrals.

Economic warfare often expands to include pressure and coercion on third parties important to the employer, such as suppliers or customers. The pressure can be direct through communicated coercive threats or indirect by the use of a picket line whose signs describe the union's reason for its grievance against the employer and asks that certain activity or nonactivity be taken. A simple illustration may be helpful in understanding how economic warfare in labor relations can spread.

Suppose there is a strike between a manufacturer of kitchenware and a union seeking to organize its plant workers or obtain better wages, hours, or working conditions. Initially, the economic warfare might consist of a strike at the manufacturer's plant that would be accompanied by a picket line at the entrance designed to discourage union drivers or drivers sympathetic to the union's cause from delivering supplies to the manufacturer.

In addition, the union might broaden its attack by picketing places that sell the kitchenware made by the manufacturer, such as hardware stores or supermarkets. The picket signs might appeal to union drivers not to deliver goods to the hardware store. The picket signs might also be directed to those who might buy in the hardware store. Such signs might ask patrons not to buy kitchenware products made by the manufacturer and might even to go the extent of asking that patrons not purchase *anything* at the hardware store because the store sells the manufacturer's kitchenware.

The purpose for such picket lines would be to reduce the sales of kitchenware items by the manufacturer, either directly by encouraging people not to buy them or indirectly by coercing hardware store outlets for the manufacturer's products into canceling orders for the manufacturer's goods.

As these tactics broaden the field of economic warfare, they may begin to seriously harm neutrals and become what are called secondary boycotts. Such impermissible activities are unfair labor practices, giving the party injured by such activity the right to damages and, more important, the right to obtain an injunction ordering that such activities cease or be modified.

The U.S. Supreme Court has stated that the law reflects

the dual congressional objectives of preserving the right of labor organizations to bring pressure to bear upon offending employers in primary labor disputes and of shielding unoffending employers and others from pressures and controversies not their own.[40]

In addition, the court recognized that the law was concerned not only with pressure brought to bear on the other disputants but also with pressure brought to force third parties "to bring pressure on the employer to agree to the union's demands."[41]

The line between permitted primary and prohibited secondary boycotts is difficult to draw in industrial collective bargaining warfare. It is even more difficult in construction work because of the transient nature of the workers, the seasonal nature of the work, the proliferation of craft unions, the frequent occupation of the construction site by employees of many bargaining entities such as contractors and subcontractors, and the special rules

[40]*NLRB v. Denver Building & Constr. Trades Council*, 341 U.S. 675 (1951).
[41]*NLRB v. Local 825, International Union of Operating Eng.*, 400 U.S. 297 (1971).

that govern collective bargaining in the construction industry.

One of the above-described factors merits additional comment. As indicated, the construction project usually involves work by a number of contractors sometimes at different stages of the work but often at the same time. Some of the employers may have collective bargaining agreements with a union, while others may not. Some employers may have collective bargaining agreements with one union, while others may have collective bargaining agreements with a different one. For example, there may be a nonunion prime contractor working alongside union subcontractors. Sometimes there are union subcontractors working on a site with employees of a subcontractor who does not have a collective bargaining agreement with any union. A union engaged in an economic struggle with an employer can maximize its bargaining power if it can shut down the entire project by putting up a picket line that no union workers will cross. The very effectiveness of common situs picketing (picketing an entrance used by all workers) is the reason it can also enmesh many neutrals and cause frequent and costly work stoppages.

The law limited common situs picketing by allowing the prime contractor to set up separate gates, one for the employees involved in the labor dispute and the other for those not involved in the dispute. This two-gate system prevents the project from being totally shut down. Picketing can be done at the first gate but not at the second. In *N.L.R.B. v. Int. Union of Elevator Constructors,*[42] an employee of a neutral subcontractor refused to use the neutral gate when another gate at the site was being picketed as a result of a labor dispute involving another subcontractor. The union attempted to enforce a contractual picket line clause that protected its employees from discipline when they refused to use a neutral gate. The Court of Appeals upheld a decision of the National Labor Relations Board that the union was guilty of a violation of the National Labor Relations Act when it sought to enforce such a clause, as the employee's refusal to work during picketing by another subcontractor's employees was not a protected activity.

But suppose a labor dispute does shut down or curtail a project. What effect will this have on the construction contract obligations of participants in the project?

Labor disputes, like damage to a partially completed project, can involve three separate but related questions:

1. Is the contractor entitled to a time extension for the period that work is shut down or curtailed?
2. Can the contractor collect additional compensation for expenses caused by the shutdowns or curtailment of work?
3. Can the contractor whose work is severely disrupted by a labor dispute to which it is not a party terminate its obligation to perform under the contract?

As for the first, construction projects typically provide time extensions for work disrupted by labor difficulties. For example, ¶ 8.3.1 of AIA Doc. A201 awards the contractor a time extension for delays caused "by labor disputes."[43] Without such a provision granting relief, a contractor will have a difficult time obtaining a time extension in the event of labor difficulties. Generally, the contractor assumes the risk of events that would increase the cost and time of performance.

Moving to the second question, it is generally difficult for the contractor to recover additional compensation when its costs are markedly increased because of unusual labor disruptions. *McNamara Construction of Manitoba, Limited v. United States*[44] denied a claim for additional compensation by a contractor who completed performance on a 21-million-dollar project in the midst of labor problems that caused 303 days of work stoppages in a contract with a completion date of three years. The work stoppage resulted from seemingly never-ending jurisdictional disputes, featherbedding practices of the unions, and what appeared to be the gross abuse of union power in matters relating to employee discipline. The contractor was given time extensions but sought an additional $6 million

---

[42]902 F.2d 1297 (8th Cir.1990).

[43]New Jersey held a strike clause used in all elevator installer contracts not unconscionable. *Curtis Elevator Co. v. Hampshire House Inc.*, 142 N.J.Super. 537, 362 A.2d 73 (1976).
[44]509 F.2d 1166 (Ct.Cl.1975).

that it claimed were costs caused by the unusual labor conditions.

To recover additional compensation, the contractor contended that both contractor and government labored under a mutual mistake of fact when they made the contract, each "assuming" that labor unions would behave in a responsible manner. In view of how things turned out, the contractor contended that this was a mutually mistaken assumption. To justify the recovery of additional compensation, the contractor claimed the contract should be "reformed"[45] and that it should be awarded additional expenses.

Over a strenuous dissenting opinion arguing that the contractor faced not ordinary but abnormally difficult labor difficulties, the court concluded that there was no mutual mistake. Labor problems, according to the majority of the court, are inherent in the vast majority of construction contracts and are a fact of modern commercial life. Rejecting the contention that the parties assume the risk of only normal strikes, the court emphasized that contractors in a fixed-price contract assume the risks of unexpected costs, a matter that a contractor should have taken into account when pricing the work.

The third question, that of termination, will depend principally on the construction contract. Interestingly, the AIA standard documents, though providing for time extensions, do not provide for termination. The absence of a clause specifically allowing termination should not necessarily preclude termination from being a proper remedy under certain circumstances.[46] Some courts will hold that a contract that specifically mentioned labor problems and did not make them the basis for termination indicated that the parties contemplated continued performance, with time extensions being the only remedy.

It may be useful to examine a case that involved a defense of impossibility asserted to be grounds for terminating performance obligations where the contract did not treat labor difficulties. In *Mishara*

Construction Co. v. Transit-Mixed Concrete,[47] a prime contractor brought legal action against a supplier who claimed it could not deliver because of a picket line at the site. The court rejected the prime contractor's contention that a labor dispute that makes performance more difficult *never* constitutes an excuse for nonperformance. The court instead held that whether a labor dispute excuses the supplier's obligation to perform would be a fact question that must be determined by a jury. The court noted that a picket line might constitute a mere inconvenience and not make performance impracticable. Also, according to the court, a contract made in the context of an industry with a long record of labor difficulties shows that the parties assumed the risk of labor disputes. For example, if the supplier knew it was agreeing to deliver concrete to an employer who had had chronic and bitter labor difficulties, its having made a contract *without* providing for contractual protection could indicate that it was assuming this risk. The court stated:

> Where the probability of a labor dispute appears to be practically nil, and where the occurrence of such a dispute provides unusual difficulty, the excuse of impracticability might well be applicable.[48]

The court concluded by noting that the tendency has been to recognize strikes as an excuse for nonperformance. Although the court indicated that under certain circumstances, absence of protective language might be considered an assumption of risk, the Uniform Commercial Code, with its emphasis on commercial impracticability, might not find the absence of a clause *conclusive* on the question of whether a particular risk was assumed.[49]

C. Partial or Total Destruction of Project: Insurance and Subrogation Waivers

During project construction, the project may be partially or totally destroyed by circumstances for which neither party is chargeable. The work may

---

[45]See Section 20.04, dealing with reformation as a method adopted by the old Court of Claims to deal with unanticipated expenditures made in the performance of a fixed-price contract.

[46]See AIA Doc. A201, ¶ 13.4.1, which states contractual remedies are not exclusive. See also Section 34.03(A), which discusses this in the context of termination.

[47]365 Mass. 122, 310 N.E.2d 363 (1974).

[48]310 N.E.2d at 368.

[49]New Jersey stated that even without a clause excusing performance hindered by labor disputes, the doctrine of commercial impracticability might give the performing party relief. *Curtis Elevator Co. v. Hampshire House, Inc.*, supra note 43.

be destroyed by fire, unstable subsurface conditions, or violent natural acts such as earthquakes or hurricanes. Although destruction is often discussed in terms of impossibility or assumption of risk, it is essential to recognize that at least three issues can arise when such events occur:

1. Are the parties relieved from any further obligation to perform?
2. Must the contractor repair and restore damage to the work?
3. Is the contractor entitled to be paid for work incorporated into the structure prior to destruction?

As for the first question, the performance obligation is not terminated in the absence of a clause relieving the contractor from its performance obligation. Similarly, with regard to the second question, the contractor has the responsibility for repair to and restoration of damaged work. Courts have stated that a party who in unqualified terms promises to perform is itself at fault when it does not expressly protect itself from these contingencies.[50] If the contract is for repairs or additions to an existing structure or for *part* of a new structure and the work is destroyed in whole or in part without the fault of either party, each party is relieved from further contract obligations.[51]

As to the third question—that of recovery for work performed *prior* to destruction—the result depends, as a rule, on the termination issue. If the contractor is not discharged from its obligation to perform, it cannot recover for work performed. If it is discharged, such as where it had agreed to build only *part* of the structure or to repair an existing structure, generally the contractor can recover for work performed prior to destruction.[52]

Usually each participant in the Construction Process insures *its* property from loss of the type discussed in this subsection. Typically, the owner insures the work in progress while the contractor insures its equipment and other property that will not go into the project. When the project is destroyed during construction, the owner receives proceeds from the insurance company. It holds these proceeds as fiduciary for those who have suffered property damage covered by the proceeds. Because much of the work has been paid for by progress payments, the proceeds that are held for the contractor are typically work for which progress payments have not yet been received and work for which progress payments were received but from which the retainage has been deducted. A well-planned property insurance system should reimburse the parties for the losses suffered. If this occurs, an argument can be made for the rule that does *not* discharge the parties from their obligation to perform further. This result follows whether the construction was for an entire structure or part of one or for the repair of an existing structure.

This appears to be the solution under AIA Doc. A201. Article 14, dealing with the termination of the contract, does not include destruction of the project as a ground for termination. Even more, A201, ¶ 11.3.1, requires the owner to purchase property damage insurance, sometimes known as builders risk insurance. If there is a loss, the owner under ¶ 11.3.10 adjusts and settles with the insurers unless parties with an interest in the proceeds object within five days. Most importantly, ¶ 11.3.9 states that "replacement of damaged property shall be covered by appropriate Change Order." This seems to assume that the parties must continue their performance, that the damaged property will be replaced, and that a change order will effectuate a time extension.

If a substantial period of time has elapsed from the making of the contract until the destruction of the project, it would be unfair to require each party to begin over. Even if property losses have been reimbursed through insurance, the original bid was based on prices and conditions existing at the time bids were made. The prime contractor might not be able to hold subcontractors to the original subcontracts in states where those who build only part of a whole structure are relieved from further performance obligations. In these states, new subcontracts would have to be negotiated. Where it would be an essentially different contract than originally made, holding that the contractor must still perform would be unfair. Likewise the owner's perspective may be different after the project has been

---

[50]*Stees v. Leonard,* 20 Minn. 494 (1874). Although the principal cases articulating this rule were decided in the nineteenth century, the rule was applied in *Dravo Corp. v. Litton Systems, Inc.,* 379 F.Supp. 37 (S.D.Miss.1974).

[51]*Fowler v. Ins. Co. of North Am.,* 155 Ga.App. 439, 270 S.E.2d 845 (1980).

[52]Annot., 28 A.L.R.3d 788 (1969).

destroyed while in the process of construction.

New information may bear on the economic viability of the project, and it might be unreasonable to compel the owner to continue as if nothing has happened.

It is advisable to incorporate a provision in the construction contract that would give either party the right to terminate if the project is totally or nearly destroyed after a designated period of time has expired from start of performance.

As indicated throughout this treatise, legislatures are increasingly enacting laws that regulate construction mainly on public but increasingly on private construction projects. For example, in 1990, California enacted a law that requires that public contracts

> not require the contractor to be responsible for the cost of repairing and restoring damage to the work, which damage is determined to have been proximately caused by an Act of God, in excess of 5% of the contracted sum. . . .[53]

The statute provides this protection only if the contractor has built in accordance with accepted building standards and the project's plans and specifications. However, the contracts may contain provisions for terminating the contract. The statute defines Acts of God to include only earthquakes that exceed 3.5 on the Richter scale and tidal waves.

Although discussion has emphasized destruction for which neither party can be held accountable, another issue often arises over responsibility. The most frequent cause of destruction—destruction by fire—is usually an insured risk. Fire losses are often traceable to human failings. When the insurer pays, it often looks for reimbursement from a party other than the insured who it can claim negligently caused the fire.

Assignment and the equitable doctrine of subrogation are methods insurers use to obtain reimbursement. Insurance policies often provide that on being paid for the loss, the insured assigns to its insurer its claims against third parties who caused the loss. Subrogation allows the insurer to "step into the shoes" of the insured. The insurer can assert any claims the insured may have against third parties (not its insured) who caused the loss. But construction projects, with their wealth of entities, usually provide ample third parties for the insurer to pursue.

Construction contracts frequently use techniques to bar subrogation. One is to require that the property insurance designate *all* the active participants as insureds. Another, often used together with the former method, is to require all active participants to waive any rights they have against other participants. The AIA uses such waiver of subrogation clauses in all AIA documents.[54] To illustrate, suppose a fire has been caused by the prime contractor and the insurer pays the owner for the damage. The insurer will not be able to pursue the prime for negligently causing the fire if the owner has waived its rights of subrogation against the prime. This method seeks to place the entire loss on the insurer and frees the participants in the construction project or, more realistically, their public liability insurers from responsibility.

Some insurers do not look favorably on such waivers. To them, waivers give away their claims, much like a motorist admitting he was wrong at the scene of an auto accident. An insured lost its coverage by waiving subrogation, held to be a violation of the failure to cooperate provision in the insurance policy.[55] For that reason A201, ¶ 11.3.7, gives the insurer notice by requiring that the waivers of subrogation be included as an endorsement on the policy.

Insurers defend subrogation claims by contending that as between the blameless insurance company and a negligent third party, the latter or its liability insurer should pay the loss. In addition, they suggest that relieving all the parties from negligent conduct can encourage carelessness. Waivers also create an externality, a cost borne by others that should be allocated to the party whose activity has caused the loss.

Those who oppose subrogation claims point to their transaction costs, a second lawsuit often needed to complete the process. They also state that subrogation recoveries are insurer windfalls,

[53]West Ann.Cal.Pub.Cont.Code § 7105.

[54]See AIA Docs. A201, ¶¶ 5.3.1, 11.3.7, A401, ¶ 13.5, and B141, ¶ 9.4.
[55]*Liberty Mutual Ins. Co. v. Altfillisch Constr. Co.*, 70 Cal.App.3d 789, 139 Cal.Rptr. 91 (1977).

as they are too uncertain to be included in the data used to determine premiums. Finally, those who structure a waiver of subrogation system feel that such a system best accords with the intention of all the parties to the construction project.

The controversial nature of subrogation waivers has led to some decisions reflecting hostility to subrogation.[56] Others see subrogation as a useful method of loss distribution.[57]

D. Governmental Acts

Government interference can take different forms. A law may be passed after the contract has been made that makes performance of the contract illegal. For example, suppose a law is passed prohibiting the construction of any nuclear facility. Certainly enactment of such legislation would terminate any contract to build a nuclear plant. Performance would require an illegal act. The contractor's right would be limited to recovery for any work performed before the enactment of the legislation. Such recovery would be restitutionary and based on unjust enrichment.

Another form of governmental interference would be the issuance of a judicial or administrative order shutting the project down. For example, a project might be shut down because of improper construction methods, an invalid building permit, or a design or use that violated land use controls. If the project is shut down for an appreciable period of time, the party not responsible for the shutdown should be relieved from further performance obligations and may have a cause of action against the other party. For example, if a shutdown is due to improper design or failure to comply with land use controls, the contractor may have a valid claim

for damages as well as be relieved from further performance obligations.[58] If the shutdown is due to poor construction methods or failure to comply with the contract documents, the owner should have a valid legal claim against the contractor for breach as well as be able to terminate any further obligations owed the contractor. The mere fact that performance was stopped by a government official does not necessarily absolve a party from contract breach even if the work is shut down. If the shutdown was due to the unexcused nonperformance by one of the parties, there can be breach as well as a possible termination.

Despite the generalization made in the preceding paragraph, that government intervention may be traceable to the fault of one of the parties or someone for whose acts the parties are responsible, the difficulty of placing clear responsibility on one of the parties for acts of government intervention makes it unlikely that government intervention in the ordinary case will be chargeable to one of the contracting parties.

Suppose the governmental acts that stop performance are not the responsibility of *either* contracting party. The site may have been condemned by the state's exercise of its power of eminent domain. In wartime, the state may have drafted workers, requisitioned essential materials, or commandeered all transportation facilities. Such acts would relieve the contractor from further contractual obligations. (They might also discharge the owner's obligations through frustration of purpose. Refer to Section 23.03(E).) Any claims for work performed by the contractor prior to shutdown would be restitutionary.

Suppose a work stoppage results because important equipment being used by the contractor has been repossessed by the owner of the equipment or someone with a security interest in it. Repossession is often accomplished by court order. However, such a court order is a risk clearly assumed by the contractor. That the actual act that interferes with performance is an order by a public

---

[56]*Tokio Marine & Fire Ins. Co. v. Employers Ins. of Wausau*, 786 F.2d 101 (2d Cir.1986) (purpose of subrogation waiver to place loss on first-party insurer to avoid disruptions and disputes among parties); *Blue Cross of Southwestern Virginia and Blue Shield of Southwestern Virginia, v. McDevitt & Street Co.*, 234 Va. 191, 360 S.E.2d 825 (1987) (consensus by parties to exempt themselves and their liability insurers from liability and shift risk to property insurer). For a fuller discussion, see J. SWEET, SWEET ON CONSTRUCTION INDUSTRY CONTRACTS § 23.24 (2d ed. 1992).

[57]*United States Fire Ins. Co. of Ammala*, 334 N.W.2d 631 (Minn.1983) (subrogation against subcontractor permitted despite subcontractor being an insured).

[58]In *Gordon v. Indusco Management Corp.*, 164 Conn. 262, 320 A.2d 811 (1973), the failure of the owner to obtain a building permit did not give the owner a defense when sued by the contractor, because the defendant owner had assured the contractor that the building permit could be obtained.

official should not relieve the contractor from its obligation to perform.

Suppose the job was shut down because a vital piece of *subcontractor* equipment was seized by court order. The result would be the same as if the equipment were owned by the prime contractor. The prime contractor would not be entitled to a time extension either because this was an assumed risk, as in the preceding paragraph, or because the job was not really shut down by the state.

Suppose legislation is enacted that would bar a method of performance contemplated at the time the contract was made. It is likely that the performing party will not receive relief if the legislation was reasonably foreseeable and did not increase costs astronomically.[59]

E. Defective Specifications: *U.S. v. Spearin*

Contractor allegations that the specifications were defective are common in construction performance disputes. Unfortunately, such allegations often produce more heat than light and obscure the real issues in the dispute. Before looking at these issues, it is useful to reproduce the fountainhead case that is often cited to support such an allegation.

_____

[59]*Levine v. Rendler*, 272 Md. 1, 320 A.2d 258 (1974) (change in requirements for composition of road in a residential development).

UNITED STATES v. SPEARIN

Supreme Court of the United States, 1918. 248 U.S. 132.
[Ed. note: Footnotes renumbered.]

BRANDEIS, Justice.

Spearin brought this suit in the Court of Claims, demanding a balance alleged to be due for work done under a contract to construct a dry-dock and also damages for its annulment. Judgment was entered for him. . . .

First. The decision to be made on the Government's appeal depends upon whether or not it was entitled to annul the contract. The facts essential to a determination of the question are these:

Spearin contracted to build for $757,800 a dry-dock at the Brooklyn Navy Yard in accordance with plans and specifications which had been prepared by the Government. The site selected by it was intersected by a 6-foot brick sewer; and it was necessary to divert and relocate a section thereof before the work of constructing the dry-dock could begin. The plans and specifications provided that the contractor should do the work and prescribed the dimensions, material, and location of the section to be substituted. All the prescribed requirements were fully complied with by Spearin; and the substituted section was accepted by the Government as satisfactory. It was located about 37 to 50 feet from the proposed excavation for the dry-dock; but a large part of the new section was within the area set aside as space within which the contractor's operations were to be carried on. Both before and after the diversion of the 6-foot sewer, it connected, within the Navy Yard but outside the space reserved for work on the dry-dock, with a 7-foot sewer which emptied into Wallabout Basin.

About a year after this relocation of the 6-foot sewer there occurred a sudden and heavy downpour of rain coincident with a high tide. This forced the water up the sewer for a considerable distance to a depth of 2 feet or more. Internal pressure broke the 6-foot sewer as so relocated, at several places; and the excavation of the dry-dock was flooded. Upon investigation, it was discovered that there was a dam from 5 to $5\frac{1}{2}$ feet high in the 7-foot sewer; and that dam, by diverting to the 6-foot sewer the greater part of the water, had caused the internal pressure which broke it. Both sewers were a part of the city sewerage system; but the dam was not shown either on the city's plan, nor on the Government's plans and blue-prints, which were submitted to Spearin. On them the 7-foot sewer appeared as unobstructed. The Government officials concerned with the letting of the contract and construction of the dry-dock did not know of the existence of the dam. The site selected for the dry-dock was low ground; and during some years prior to making the contract sued on, the sewers had, from time to time, overflowed to the knowledge of these Government officials and others. But the fact had not been communicated to Spearin by anyone. [Spearin] had, before entering into the contract, made a superficial examination of the premises and sought from the civil engineer's office at the Navy Yard information concerning the conditions and probable cost of the work; but he had made no special examination of the sewers nor special enquiry into the possibility of the work being flooded thereby; and had no information on the subject.

Promptly after the breaking of the sewer Spearin notified the Government that he considered the sewers under existing plans a menace to the work and that he would not resume operations unless the Government either made good or assumed responsibility for the damage that had already occurred and either made such changes in the sewer system as would remove the danger or assumed responsibility for the damage which might thereafter be occasioned by the insufficient capacity and the location and design of the existing sewers. The estimated cost of restoring the sewer was $3,875. But it was unsafe to both Spearin and the Government's property to proceed with the work with the 6-foot sewer in its then condition. The Government insisted that the responsibility for remedying existing conditions rested with the contractor. After fifteen months spent in investigation and fruitless correspondence, the Secretary of the Navy annulled the contract and took possession of the plant and materials on the site. Later the dry-dock, under radically changed and enlarged plans, was completed by other contractors, the Government having first discontinued the use of the 6-foot intersecting sewer and then reconstructed it by modifying size, shape and material so as to remove all danger of its breaking from internal pressure. . . .

The general rules of law applicable to these facts are well settled. Where one agrees to do, for a fixed sum, a thing possible to be performed, he will not be excused, or become entitled to additional compensation, because unforeseen difficulties are encountered. *Day v. United States*, 245 U.S. 159; *Phoenix Bridge Co. v. United States*, 211 U.S. 188. Thus one who undertakes to erect a structure upon a particular site, assumes ordinarily the risk of subsidence of the soil. *Simpson v. United States*, 172 U.S. 372; *Dermott v. Jones*, 2 Wall. 1. But if the contractor is bound to build according to plans and specifications prepared by the owner, the contractor will not be responsible for the consequences of defects in the plans and specifications. *MacKnight Flintic Stone Co. v. The Mayor*, 160 N.Y. 72; *Filbert v. Philadelphia*, 181 Pa.St. 530; *Bentley v. State*, 73 Wisconsin, 416. See *Sundstrom v. New York*, 213 N.Y. 68. This responsibility of the owner is not overcome by the usual clauses requiring builders to visit the site, to check the plans, and to inform themselves of the requirements of the work, as is shown by *Christie v. United States*, 237 U.S. 234; *Hollerbach v. United States*, 233 U.S. 165, and *United States v. Utah & c. Stage Co.*, 199 U.S. 414, 424, where it was held that the contractor should be relieved, if he was misled by erroneous statements in the specifications.

In the case at bar, the sewer, as well as the other structures, was to be built in accordance with the plans and specifications furnished by the Govern-

ment. The construction of the sewer constituted as much an integral part of the contract as did the construction of any part of the dry-dock proper. It was as necessary as any other work in the preparation for the foundation. It involved no separate contract and no separate consideration. The contention of the Government that the present case is to be distinguished from the *Bentley Case, supra,* and other similar cases, on the ground that the contract with reference to the sewer is purely collateral, is clearly without merit. The risk of the existing system proving adequate might have rested upon Spearin, if the contract for the dry-dock had not contained the provision for relocation of the 6-foot sewer. But the insertion of the articles prescribing the character, dimensions and location of the sewer imported a warranty that, if the specifications were complied with, the sewer would be adequate. This implied warranty is not overcome by the general clauses requiring the contractor, to examine the site,[60] to check up the plans,[61] and to assume responsibility for the work until completion and acceptance.[62] The obligation to examine the site did not impose upon [the contractor] the duty of making a diligent enquiry into the history of the locality with a view to determining, at his peril, whether the sewer specifically prescribed by the Government would prove adequate. The duty to check plans did not impose the obligation to pass upon their adequacy to accomplish the purpose in view. And the provision concerning contractor's responsibility cannot be construed as abridging rights arising under specific provisions of the contract.

* * *

The judgment of the Court of Claims is, therefore,

---

[60]"271. *Examination of site.*—Intending bidders are expected to examine the site of the proposed dry-dock and inform themselves thoroughly of the actual conditions and requirements before submitting proposals."

[61]"25. *Checking plans and dimensions; lines and levels.*—The contractor shall check all plans furnished him immediately upon their receipt and promptly notify the civil engineer in charge of any discrepancies discovered therein. . . . The contractor will be held responsible for the lines and levels of his work, and he must combine all materials properly, so that the completed structure shall conform to the true intent and meaning of the plans and specifications."

[62]"21. *Contractor's responsibility.*—The contractor shall be responsible for the entire work and every part thereof, until completion and final acceptance by the Chief of Bureau of Yards and Docks, and for all tools, appliances, and property of every description used in connection therewith. . . ."

Before attempting to break down the imprecise term *defective specifications* into more workable categories, it is important to note the different legal issues affected by the quality of specifications. The risk of building defects may depend on whether the defect was caused by the design. If so, the entity who supplied design specifications (materials and methods)—usually the owner—will not be able to transfer the cost of correction to the contractor who has executed the design. (Whether it can transfer this loss to the design professional was discussed in Chapter 14.) Put another way, the contractor who executes the required design is not liable for the cost of correction, as seen in *United States v. Spearin*.

Suppose the contractor claims that it has expended more than it anticipated because of defective specifications. Chapter 25 deals with this in the context of subsurface conditions. This subsection looks briefly at this issue in other contexts.

A contractor who cannot comply with performance specifications, particularly when the contractor has been required to follow detailed specifications, sometimes asserts that the specifications were "impossible"[63] (discussed in (F)).

It may be helpful to start with the attributes of good or high-quality specifications. This requires an understanding of what specifications include and what they are designed to accomplish.

In the sense that specifications measure the contractor's contract obligations, the specification should make clear to the contractor what it will be expected to do. A design specification tells the contractor what it is to do and how it is to do it. A purchase description specification is even more limiting, telling the contractor that the materials and equipment it *must* furnish will be made by a particular manufacturer and be of a designated model or type. A performance specification describes a specific outcome. When the performance specification is combined with a design specification, the owner has sought to tell the contractor what it must do, how it must do it, and the result it must achieve. As shall be seen in Section 24.03, this can cause problems.

In addition, specifications as seen in *United States v. Spearin* can provide relevant information that the contractor needs and uses to determine whether it will enter into the contract, what will be its price, and how it expects to perform. In that case, the United States did not tell the contractor that there was a dam in the 7-foot sewer. Put another way, it failed to provide complete information. It described the sewer but did not inform the contractor that there was a dam in it or that there had been prior flooding problems. Information can relate to the site, its subsurface characteristics, its accesses, and any conditions that would be helpful in planning performance.

High-quality specifications clearly inform the contractor what it will be expected to do and the conditions under which it will perform. They enable the contractor to plan its performance and price with the expectation of predictable and efficient work sequences. Similarly, a subcontractor relies on information supplied by the owner or the prime.[64] From the owner's standpoint, high-quality specifications will advance the owner's anticipated objectives, particularly those objectives expressed in or implied by the contract documents.

Defective specifications do not accomplish these objectives. More detailed classification helps understand some of the legal issues that surround this term, one that hides a multitude of sins.

Erroneous specifications contain factual errors of the type described in *United States v. Spearin* and subsurface data errors, to be discussed in Chapter 25. They also include *legal* errors, such as noncompliance with applicable laws such as building and housing codes and environmental regulations.[65] Usually legal errors are the responsibility of the owner, with ultimate responsibility belonging to the design professional.[66] Only if the contractor

---

[63]Sometimes specifications are impossible in the sense that they are not coherent, that they do not "fit." Compliance with one part makes compliance with another part an "impossibility." In (F), impossibility means beyond the state of the art.

[64]See *W.F. Magann Corp. v. Diamond Mfg. Co.*, 775 F.2d 1202 (4th Cir.1985) (prime did not disclose information it received from owner to subcontractor).
[65]*St. Joseph Hosp. v. Corbetta Constr. Co.*, 21 Ill.App.3d 925, 316 N.E.2d 51 (1974) (panelling violated code); *Atlantic Nat. Bank of Jacksonville v. Modular Age, Inc.*, 363 So.2d 1152 (Fla.Dist.Ct.App.1978) (wall violated code).
[66]But see *Green v. City of New York*, 283 A.D. 485, 128 N.Y.S.2d 715 (1954), and *Quedding v. Arisumi Bros., Inc.*, 661 P.2d 706 (Hawaii 1983), where contractors who designed and built were liable where specified materials violated code because of express or implied promise to comply with the law.

knew of the errors may this risk have been shifted or shared.[67]

Another erroneous specification is mechanical. An illustration is physical problems with the project, such as water intrusion, foundation settling, insufficient heating or cooling, or a partial or total collapse of the structure. Although such problems are often classified as defects (the subject of Chapter 24), the party responsible for defects of this type that are caused by the design is the party who created the design or had the design created for it, usually the party with a preponderance of expertise in the element of design that failed. Typically, this is the owner, unless the owner has transferred this risk to the contractor.

Errors can be functional, such as a project failing to accomplish the owner's desired objectives. This can result from bad business judgment or changed circumstances—both owner risks. The owner may be able to transfer this risk to the design professional if the latter warranted a successful outcome or if the design professional's failure to perform as other design professionals would have performed caused the project to fail.

Defective specifications can "fail" as a method of communication. For example, specifications that do not describe what will be demanded or that fail to give sufficient design detail to enable the contractor to accomplish the desired objective are "incomplete."

The complexity of determining who is responsible for an incomplete specification can be illustrated by *Tibshraeny Brothers Construction, Inc. v. United States*.[68] The case involved a contract between the Department of the Air Force and the plaintiff for modifying existing fuel facilities and constructing a new fuel system at a National Guard facility. The Air Force had hired an architect/engineer to design and prepare the system based upon design drawings prepared by the Air Force. The architect/engineer did what he was hired to do, and his plans and drawings became part of the bid documents.

During construction, a dispute arose over who had the responsibility for providing control wiring diagrams. The Air Force and the architect/engineer took the position that the diagrams were to be prepared by the plaintiff. The plaintiff contracted with its electrical subcontractor to prepare such diagrams, and the latter retained the services of an experienced electrical engineer and instructed him to design operable control diagrams for the system.

After a number of rejected proposals for diagrams, the diagrams prepared by the electrical engineer retained by the electrical subcontractor were finally accepted. The plaintiff claimed its additional expenses resulting from the preparation of the diagrams and the delays created by the dispute.

The claims court had no difficulty in concluding that the design was incomplete without the control wiring diagrams. But the question it next faced was who was responsible for the diagrams' preparation. The plaintiff contended that the specifications did not specifically require that it prepare the diagrams and that since the diagrams were part of design, they should be prepared either by the government or by its architect/engineer. The plaintiff also contended that where the specifications were ambiguous as to who was to prepare the diagrams and in the absence of a clear statement shifting responsibility to the plaintiff, the diagrams were part of the design function for which it was not responsible.

The court concluded that at the very least, the contract was ambiguous. Since the contract was reasonably susceptible to more than one interpretation, both parties' interpretations could be said to be reasonable and even plausible, but the interpretation favoring the nondrafting party—the contractor—prevailed.

The court rejected any evidence that the Air Force intended to have the contractor prepare the diagrams, since such intention was not communicated by the bid documents or by any other method until the dispute had arisen. Nor did the court find the language so apparently unclear as to require that the plaintiff inquire. It buttressed its conclusion by pointing to trade usage, indicating that the contractor is not generally required to design and prepare the diagrams. It also noted that there had been prior dealings between the plaintiff and the Air Force that showed that the Air Force had furnished similar diagrams to the plaintiff after construction began. The court found that the plaintiff could rely upon this course of dealing.

Finally, the court pointed to the contract between the Air Force and its architect/engineer un-

---

[67] AIA Doc. A201, ¶ 3.7.4.
[68] 6 Cl.Ct. 463 (1984).

der which the latter agreed to design a fully operational system that included enough information from which a complete and operable system could be built. This indicated to the court that it was clear that the architect/engineer was to do the design work and that where the plans and specifications were not sufficiently complete, the architect/engineer was to prepare the required diagrams.

Sometimes the specifications are confusing or contradictory, again a problem of failing to communicate the performance that will be demanded.

In *Jasper Construction, Inc. v. Foothill Junior College District*,[69] the specification stated:

> 11. CONSTRUCTION JOINTS
>
> A) Locations and details of construction joints shall be as indicated on the structural drawings, or as approved by the Architect. Relate required vertical joints in walls to joints in finish. In general, approved joints shall be located to least impair the strength of the structure.[70]

The locations of the construction joints were not shown on the drawings. However, Jasper contended that certain structural drawings indicated *to him* that the steel was from "floor to floor" and therefore the concrete would be poured in the same manner. He began to pour the basement "floor to floor," but the architect informed him that the joints would have to be "wall to wall." The contractor complained but did the work as directed and made a claim.

The jury was instructed that a public entity that issues plans and specifications *impliedly warrants* that they are *free of defects,* and *complete,* and will, if followed, result in the project intended. [Emphasis added.]

The appellate court found this jury instruction erroneous. The court recognized that warranties do attach to owner-supplied specifications. However, the court limited the implied warranty doctrine to *affirmative misrepresentations* or concealment of material facts that misled the contractor. The court concluded that the contractor had not been misled. It could have cleared up any ambiguities or incompleteness in advance by seeking a clarification from the architect.

The result in the *Jasper* case in no way undermines the importance or existence of implied warranties. Warranties are implied either where the issue is who bears the risk of inaccurate information, such as in the subsurface cases, or in disputes that involve the outcome of compliance with specifications. In either case, an implied warranty *may* be found it if is more equitable to make the owner responsible for inaccurate information relied on reasonably by the contractor or to make the owner pay for defects caused by the design or unanticipated expenses of complying with that design. Implied warranties do not depend on fault. They are found if this is what the parties are very likely to have intended or, more commonly, if this is the fairest way of allocating the risk for particular losses.

The result in the *Jasper* case was correct if it can be assumed at best that the specifications were unclear and Jasper should have sought a determination. But its limiting of the implied warranty doctrine to affirmative misrepresentation unduly limited the utility of the implied warranty doctrine.

Most implied warranty cases involve owners who were public entities. There is no reason to differentiate public from private owners. Private owners who have the same resources and expertise of public entities should be held to similar implied warranties of quality specifications. If a private entity does *not* have these resources and expertise (and there may be cases where public entities do not either) and relies on the contractor, the owner may be the beneficiary of an implied warranty by the contractor.

F. "Impossible" Specifications

Performance specifications require the performing party to accomplish a designated objective.[71] For example, an aircraft company might agree to manufacture an airplane that will fly twice the speed of sound or a machinery manufacturer might agree to build a system for a plant that would turn out 1,000 units of a particular quality per hour. Suppose the airplane manufacturer or machinery maker fails. Ordinarily, the failure to comply with

---

[69]91 Cal.App.3d 1, 153 Cal.Rptr. 767 (1979).
[70]153 Cal.Rptr. at 769.

[71]For a cavalier judicial treatment of performance specifications, see *Kurland v. United Pacific Ins. Co.*, discussed in Section 24.03.

performance specifications is a breach of contract.[72] But suppose the party promising to meet these specifications asserts that it was "impossible" to do so. Does proof of this provide a defense?[73] Does such proof entitle the performing party to reimbursement for the expenses incurred while seeking to meet the performance specification?

The answers to these questions in the first instance depends on the contract. The contract can place such risks on one or both parties. Often contracts are not clear on this point, and the law must determine the answers to these difficult questions.

Unfortunately, such problems are classified as "impossible" specification cases because the performing party, who will be referred to as the contractor, claims that it was "impossible" to meet these performance specifications. But the word *impossible* has a number of subtle shadings that often complicate cases and make prediction uncertain.

Another difficulty with the "impossible" label is that it can obscure the crucial issue of risk assumption. Clearly, a party can promise to do the impossible, although evidence that such a foolish risk was taken should be clear.[74] Conversely, the fact that it is not physically possible—that is, beyond the state of the art—should not invariably resolve the matter in favor of the contractor.

Foster Wheeler Corp. v. United States,[75] a Court of Claims opinion, dealt at length with an impossible specifications problem. Foster Wheeler Corporation (FWC) entered into a fixed-price supply contract under which it agreed to design, fabricate, and deliver within thirteen months two boilers and perform a "dynamic shock analysis" study called a DDAM (dynamic design-analysis method) that would demonstrate that the boilers could withstand shock up to certain designated intensities set forth in the contract specifications. The total contract price was $280,000. The boilers were ultimately to be installed in naval ships. Because the boilers could not be subjected to actual shock testing, the DDAM—a mathematical model to represent a piece of equipment and the use of dynamic inputs to substitute for physical stresses and failure criteria—was to substitute.

After many months of design work and creation of mathematical models, the contractor ceased performance and sought an equitable adjustment of $192,000, claiming "impossibility" specifications. This claim was based on the impossibility of meeting the performance specifications of "shock hardness." Given other design requirements, this could not be demonstrated by the DDAM.

The court recognized that the term *impossibility* does not require *absolute* impossibility but encompasses impracticability, a type of commercial impossibility caused by extremely unreasonable expense to perform. *Absolute* impossibility, in the sense of requiring performance beyond the state of the art, would entitle the contractor to recover its cost in attempting to perform unless it assumed this risk. The court concluded that demonstrating the boiler to be shock hard by the DDAM method was *both* "commercially and absolutely impossible."

After giving facts that supported a conclusion of absolute impossibility, the court went on to the more controversial "commercial impossibility," stating:

> Under this theory, it is contended that the construction of shock-hard boiler, even if ultimately possible, could not be accomplished without commercially unacceptable costs and time input far beyond that contemplated in the contract. To design a shock-hard boiler by means of a mathematical model and dynamic analysis could . . . take an infinite amount of time. . . . The evidence shows . . . that the . . . contract contained specifications which were impossible to meet, either commercially or within the state of the art.[76]

---

[72]*Gurney Indus., Inc. v. St. Paul Fire and Marine Ins., Co.*, 467 F.2d 588 (4th Cir.1972). One court intimated that the doctrine of substantial performance (discussed in Section 22.06(B)) might be applicable to performance specifications. *Votaw Precision Tool Co. v. Air Canada*, 60 Cal.App.3d 52, 131 Cal.Rptr. 335 (1976). Under this approach, coming "close" might be sufficient to justify receiving the unpaid balance of the contract price, less damages caused by the breach.

[73]The discussion in this subsection emphasizes attempts by the performing party to be reimbursed for its efforts, as this has been the principal issue in those cases raising this problem. If there were a sufficient degree of impossibility to justify reimbursement, it seems obvious that the performing party would have a defense if sued by the other party for not complying with the performance specifications. The contractor was given a defense when the plans were considered "impossible" in *City of Littleton v. Employers Fire Ins. Co.*, 169 Colo. 104, 453 P.2d 810 (1969).

[74]*J.C. Penney Co. v. Davis & Davis, Inc.*, 158 Ga.App. 169, 279 S.E.2d 461 (1981).

[75]206 Ct.Cl. 533, 513 F.2d 588 (1975).

[76]513 F.2d at 598.

This did not end the matter. The government argued that FWC assumed the contractual responsibility for performing the impossible. To determine who should assume this risk, the court examined which party had the greater expertise in the subject matter of the contract and which party took the initiative in drawing up specifications and promoting a particular method or design. The court ruled for FWC.

G. Weather

An empirical study reports that weather is "one of the most important causes of delay in construction, its influence being felt through lost or non-productive working days, idle equipment, spoiled materials, contingency overheads and consequently higher prices."[77] Generally, the weather is considered to be a risk borne by the contractor. Time extensions will be granted only if the weather is severe and abnormal for the time and place.[78] Even if the time extension is justified because of weather conditions, in the absence of a changed condition clause that covers weather, weather conditions do not justify additional compensation.[79] The owner does not warrant that the weather will not interfere unduly with a contractor's performance.

H. Financial Problems

After the construction contract is made and before completion of each party's performance, either party can suffer severe financial reverses. Such reverses may manifest themselves in difficulties that range from short delay in paying bills to insolvency and even bankruptcy.

Financial reverses of this sort can raise two related but separate legal questions. First, the party in financial difficulty may contend that it should be relieved from further performance because it does not have the financial capacity to continue performance. Second, one party may be unwilling to continue performance if it appears that the other party's financial difficulties will make it unlikely that the latter will perform as promised. Under certain circumstances, the law allows a party who has legitimate concern over the other party's ability to perform to refuse to continue performance until it is reassured that the other party will perform its promise.[80]

AIA, in its documents relating to both design and construction services, gives the performing parties—the architect in B141[81] and the contractor in A201[82]—the right to demand certain information regarding the owner's financial resources.

Suppose the contractor asserts that the owner has made a *promise* to supply adequate information and that its failure to do so constitutes a breach. Such assertion would be based on the furnishing of information not only being a *condition* to the contractor's obligation to execute the contract but also being a *promise* by the owner that it will furnish this information if requested. If so, the contractor is entitled to damages. At the very least, this should encompass its expenses incurred in negotiating or submitting a competitive bid. However, the contractor should not recover lost profits. To avoid this, it is likely that such a provision would be held to create a *condition* rather than a *promise*.

Suppose the required information is not provided. Can the contractor suspend performance until the information is furnished? Can it terminate? In the absence of any contractual provision dealing with this issue, it is likely that the contractor can demand adequate assurance and cease its performance until it is furnished.[83] Failure to furnish the information will constitute a repudiation of the contract. A201, ¶ 14.1.1.5, gives the contractor the right to terminate its obligation in the event it has ceased performance for thirty days because the owner has not furnished the requested financial information. By implication, this gives the contractor a right to suspend work if the information is not furnished.

Some owners may object to such a provision. (Such a clause can give the contractor an opportunity to get out of the contract if it finds that it has bid too low.) They may justify their refusal by noting that contractors in *private* construction have a right to mechanics' liens in the event they are not

---

[77]Laufer & Cohenca, *Factors Affecting Construction-Planning Outcomes*, 116 J. of Constr. Eng'g & Mgmt, 135, 147 (1990).
[78]See Sections 26.07 and 26.08.
[79]*Hardeman-Monier-Hutcherson v. United States*, 198 Ct.Cl. 472, 458 F.2d 1364 (1972).

[80]Restatement (Second) of Contracts §§ 251, 252 (1981).
[81]AIA Doc. B141, ¶ 4.3.
[82]AIA Doc. A201, ¶ 2.2.1.
[83]Restatement (Second) of Contracts § 251 (1981).

paid. However, the contractor would prefer a battery of weapons to deal with financial insecurity and nonpayment.

Suppose an A201 is used but that portion that gives the contractor this power is deleted. Would a court use common law doctrines that might otherwise be available to the contractor in the event it has reasonable insecurity as to the owner's ability to pay? Deletion would probably indicate that the parties did not intend that this "gap filler" be part of this transaction.

Suppose the contractor's power is limited to that period between award and execution of the formal contract. Would this preclude a court from using any common law financial insecurity doctrines that might otherwise be available during performance?[84] Does a contract that deals with a particular problem signal a court that the law should not imply terms? The detailed treatment of this problem would make legal intervention unlikely.

Financial problems encountered by the contractor are not likely to provide it with any justification with refusal to continue performance. However, construction contracts usually grant the owner certain remedies in the event the contractor runs into financial difficulties.[85] This should make it unnecessary for the law to employ any common law doctrines dealing with financial insecurity. This is particularly likely in the light of the frequent use of surety bonds to provide financial security for the owner in the event the contractor does not perform as promised.

I. Asbestos and Other Hazardous Materials

Section 9.13 described some of the contours of the environmental movement and the increasing liability exposure for those who are involved in any way with hazardous materials. While focus of this subsection is upon the discovery of such condition during construction and the effect of this upon the contractual obligations of the parties to a construction contract, it may be useful to direct attention to the liability exposure of a contractor when it is later determined that it had come in contact with hazardous materials during its performance.

In *Kaiser Aluminum and Chemical Corp. v. Catellus Dev. Corp.*,[86] the city had purchased property from the owner and then hired an excavation contractor to excavate and grade the land for a proposed public housing project. While excavating, the contractor spread some of the displaced soil over other parts of the property. Unknown to the contractor, this excavated soil contained hazardous chemical compounds.

The city sued the owner from whom it had purchased the land to recover part of the costs of removing the contaminated soil from the property. The prior owner asserted a legal claim against the contractor alleging that it had made the contamination worse when it extracted the contaminated soil and spread it over uncontaminated areas of the property. The court held that the contractor could be liable to the prior owner because of federal legislation that extends liability to anyone who operates a facility at which hazardous substances are disposed of. The court held that the excavation contractor had authority to control the cause of the contamination and that the dispersal of contaminated soil constituted disposal of hazardous substances. The court concluded that disposal encompasses placing hazardous wastes on any land and is not limited to the initial introduction of the waste onto the property. The contractor could also be held as a transporter of hazardous waste. It excavated it and spread it over uncontaminated property. When liability is not based upon negligence it is strict. Excavation contractors who believe they may face such a problem should either obtain an indemnification from the party hiring it to do the work or be certain it is covered by special endorsement to its liability insurance policy.

Suppose asbestos or other hazardous material is discovered during construction. How does this affect the obligation of the contractor to continue performance or to resume performance? AIA Doc. A201, ¶ 10.1.2, allows cessation of work if asbestos or PCB (other contracts add a list of other hazardous substances) is found and has not been rendered harmless. After an investigation and treatment, work will be resumed only by written agreement between the owner and the contractor. In the absence of such agreement, resumption requires a decision by the architect not taken to arbitration.

---

[84]See Section 34.04(B).
[85]As to bankruptcy, see Section 34.04(C).

[86]976 F.2d 1338 (9th Cir.1992).

The difficulty with the AIA's solution is that it does not take into account the increasing public control over such work. If asbestos or PCB is discovered, it is almost certain that state law will require that only contractors licensed to do asbestos abatement work will be allowed to work on that portion of the premises. The AIA treats this problem as if public controls over this activity did not exist.

Also, there has been criticism of the AIA's having limited hazardous materials (in AIA Document A201) to asbestos and PCB. Responding to this criticism, in 1989 the AIA published A512, which provides language that users can incorporate into supplementary conditions. While not defining hazardous materials, as is done so frequently in other contracts, A512, ¶ 10.1.5, states that the contractor must stop work and report to the owner and architect in writing "if reasonable precautions will be inadequate to prevent foreseeable bodily injury or death to persons resulting from a material or substance encountered on the site." The AIA thought that this was superior to seeking to define hazardous materials and toxic substances.

A512, ¶ 10.1.6, also proposes as a supplementary condition language that requires the owner to obtain the services of a licensed laboratory to "verify a presence or absence of the material or substance reported by the Contractor" and to "verify that [the substance] has been rendered harmless".

In 1990, three years after the AIA published its current A201, the Engineers Joint Contracts Documents Committee (EJCDC) published Document No. 1910-8. The document contained sharper definitions, a more sensible procedure when hazardous conditions are encountered, and a better system for price and time adjustments.[87] It is useful to consult this document when planning any construction contract where there is any possibility that hazardous materials will be encountered.

Cessation and resumption of work raise other legal issues. Suppose the work is stopped, an asbestos abatement contractor performs asbestos removal, and the work is resumed. Who must pay for the cost of asbestos abatement and removal? Can the contractor receive an equitable adjustment if discovery of asbestos or other hazardous material has had an adverse effect on its performance cost? If its work has been drastically affected, does the contractor have the power to terminate the contract? Similarly, if the project looks much different to the owner after the high cost of abatement, can the owner use impossibility or frustration to terminate its obligation to proceed further under the contract? Finally, if the owner chooses not to resume the work, has it breached its contract with the contractor?

These questions should be addressed in the contract. If the contractor encounters such materials, it should receive an equitable adjustment if it can continue to perform any work. The owner should be allowed to terminate the contract if in good faith it believes that it would not be economical to continue working under such conditions.

Another problem that is beginning to develop in construction relates to the discovery of **radon,** either during construction or after construction has been completed. Increasingly, radon will create legal problems, both of a public and of a private nature.[88]

[88]Note, 15 Seton Hall Legis.J. 171 (1991); Note, 37 Wash.Univ.J. of Urb. & Contemp.Law 135 (1990).

[87]¶¶ 1.4, 1.21, 1.30, 1.32, 4.5.

PROBLEMS

1. P was the prime contractor for the construction of a five-story office building. The project was destroyed by fire after approximately 50% was completed. All participants were paid for the work they had performed either by earlier progress payments or by insurance proceeds turned over to them by the owner. Are the participants, the owner, the prime contractor, and the subcontractors obligated to begin performance again? Should they be?

2. A prime contractor failed to comply with safety rules. This was reported to a union representing the employees. The union shop steward ordered the workers to leave the job until the safety rules were corrected. Should the contractor be given a time extension or additional compensation under a contract that includes a clause granting time extensions ''for labor disputes''?

3. O is a highly paid executive for a computer company. He made a construction contract with C un-der which C agreed to build a luxurious home and O promised to pay $350,000. During construction, O was fired because of a financial scandal in which he was implicated. C is concerned that O will not be able to make progress payments. What are his legal rights under AIA Doc. A201? Under the common law?

Defects: Design, Execution, and Blurred Roles

SECTION 24.01 Introduction: The Partnership

The sad but not uncommon discovery that the project has defects often generates a claim by the owner against parties it holds responsible for having caused the defect. Claims against the design professional were discussed in Chapter 14. This chapter concentrates on owner claims against the contractor.[1]

Defects in a house can include a leaky roof, a sagging floor, structural instability, and an inadequate heating or plumbing system. A commercial structure can include these defects as well as an escalator that is unsafe or inefficient or that requires excessive repairs. An industrial plant can include the preceding defects as well as inadequate space to install machinery or the inability of the computer system to operate the assembly line.

A differentiation must be made between temporary and permanent work. Usually defects deal with *permanent* work—the finished product that the owner intends to use. Under AIA Doc. A201, ¶ 3.3.1, *how* the permanent work is to be accomplished, such as means, methods, and sequences, including methods for accomplishing the *temporary* work, such as temporary shoring, bracing, formwork, or coffer dams, is usually the responsibility of the contractor.[2]

It is important to differentiate the different types of owners discussed in Section 8.02(A). The most important is a differentiation between an owner who supplies a design—usually by an independent design professional—and one who hires a contractor to both design and build. (Design/build (D/B) was discussed in Section 17.04(F).)

It is also important to review certain types of specifications noted in Chapters 19 and 23. They are *design* specifications, *performance* specifications, and *purchase description* specifications. The first requires the contractor to use designated methods and materials. The second gives the contractor particular goals, with the contractor often but not always able to determine how the goals are to be achieved. The third designates materials and equipment by name of manufacturer, trade name, and type. Sometimes purchase description specifications allow the contractor to request authorization to substitute something that is equal to or the equivalent of the designated product. Specifications can combine types, such as design and performance. Also, they vary as to completeness of materials and methods requirements.

Often it is difficult to determine what causes a defect. A defect can be caused by the design, by workmanship, by extraneous factors such as weather, or by a combination of factors. Perhaps the greatest difficulty involves the defects traceable to materials (discussed in Section 24.02).

Looking ahead, one *rough* classification is to charge the owner with defects caused by design and hold the contractor accountable for defects caused by failure to follow the design or by poor workmanship. This can be deceptive unless account is taken of the increasingly *blurred* roles related both to design and to its execution.

Although design is principally the responsibility of the owner (usually acting through the design professional), the contractor plays a role in design,

---

[1]For claims against public liability insurers, see Section 24.08.

[2]Currently AIA Doc. A201, ¶ 1.1.3, defines "work" as any activity performed by the contractor. This can include *temporary* work, such as shoring, bracing, or formwork, activity not included in the definition of "work" in 1976. One architect stated that the architect could be responsible for temporary work because of her site visit responsibilities, a result clearly *not* intended by the AIA. See McCrary, *An Individual Look at the New A201*, Architecture, October 1987, p. 50.

particularly a contractor retained because of its specialized skill in certain work or because it retains subcontractors with those skills. Illustrations of this shared responsibility can be those contracts that require the contractor to submit drawings that indicate how it proposes to do the work (often prepared by specialized subcontractors) along with the requirement that the submittals be approved by the design professional. (This was discussed in Section 12.08(C).)

Sometimes bidders are asked to provide design alternates or do so voluntarily. Even more important, at the bidding stage, contractors are frequently required to examine the contract documents and report any errors they observe. Before or after award or during performance, the contractor may request to substitute different equipment or material than that specified. Although approval by the design professional is usually required, approval often is based on representations or even warranties by the contractor that the proposed substitution will be at least as good as that specified or will accomplish the desired result. The contractor, though clearly subordinate to the design professional, plays an important role in design.

Similarly, the design professional frequently monitors the contractor's performance during the work and at the end of the job. Some design professionals may even direct *how* the work is to be done, although this is typically not within their power or responsibility.

There is a rough "partnership" between owner and contractor that makes it difficult to neatly divide responsibility into design and its execution.

SECTION 24.02 Basic Principle: Responsibility Follows Control

As a basic principle, responsibility for a defect rests on the party to the construction contract who essentially controls and represents that it possesses skill in that phase of the overall Construction Process that substantially caused the defect. Usually defects caused by design are the responsibility of the owner in a traditional construction project and the contractor who both designs and builds. Control does not mean simply the *power* to make design choices. Usually every owner has the power to determine design choices. For example, an owner may require a particular tile to be used, a power within its contract rights. But the control needed to invoke the basic principle means a skilled choice, either one made by an owner who has professional skill in tile selection or an adviser such as an architect with those skills.

This principle recognizes that the owner who supplies the design is responsible for design that does not accomplish the owner's objective and yet may not have a claim against the design professional. Usually the standard to which the design professional is held is whether she would have performed as would have other design professionals similarly situated.

Under this principle, the owner, though faultless, bears the cost of correcting defects. The owner has the principal economic stake in the project and will benefit from a successful project. There is no reason why it cannot be responsible for project failures even though it is blameless and cannot transfer the loss.

Many cases have held that the contractor who follows the design is not responsible for a defect unless it warrants the design or was negligent.[3] This establishes the principle that the owner is responsible for any design it has furnished. Similarly, the owner impliedly warrants the accuracy of specific information it furnishes that is reasonably relied on by the contractor.[4] It also warrants that any required materials, design features, or construction methods will create a satisfactory end product within the completed time and without extraordinary unanticipated expense.[5]

---

[3]For the most recent of the many cases, see *Larry Smith v. George M. Gilmer*, 488 So.2d 1143 (La.App.1986); *John Grace & Co., Inc. v. State Univ. Constr. Fund*, 99 A.D.2d 860, 472 N.Y.S.2d 735 (1984); *Burke County Public Schools Bd. of Educ. v. Juno Constr. Corp.*, 50 N.C.App. 238, 273 S.E.2d 504 (1981); affirmed 304 N.C. 159, 282 S.E.2d 779 (1981); *Reiman Constr. Co. v. Jerry Hiller Co.*, 709 P.2d 1271 (Wyo. 1985). But see *United States Fid. and Guar. Co. v. Jacksonville State Univ.*, 357 So.2d 952 (Ala.1978), discussed in Section 24.03. It held the contractor responsible, the result based on an apparent guarantee of a successful outcome given by the contractor. As to the legal effect of the contractor following a defective design, see Section 24.06. See also the recent U.S. Supreme Court decision granting immunity to a federal government contractor who followed approved U.S. specifications and who warned the federal agency of possible danger. *Boyle v. U.S. Technologies Corp.*, 487 U.S. 500 (1988).

[4]See Section 25.03.

[5]See Section 23.05(E).

Cases supporting this principle usually involve traditional construction in which the owner furnishes and monitors the design but is not primarily responsible for its execution. Risk allocation is based on the probable intention of the parties, the greater skill possessed or supplied by the owner, the contractor's lack of discretion, and the owner's being in the best position to avoid the harm, as well as the owner's ability to spread, absorb, or shift the risk to the design professional. Similarly, contractors are generally held to have impliedly warranted the quality of their workmanship.[6]

Contractors who *both* design and build usually warrant that the finished product will be fit for the purposes of the owner of which the contractor knew or should have known for the same reasons that were given to place the risk on the owner in a traditional method. Also, these warranties are similar to those placed on *sellers* of goods. For example, one court found an implied warranty[7] where:

1. The contractor holds itself out, expressly or by implication, as competent to undertake the contract; and the owner
2. has no particular expertise in the kind of work contemplated;
3. furnishes no plans, designs, specifications, details or blueprints; and
4. tacitly or specifically indicates its reliance on the experience and skill of the contractor, after making known to him the specific purposes for which the building is intended.

Defective material generates the most difficult problems. The huge variety of available materials (often untested), the pressure to cut costs or weight, and the inability to test or rely on manufacturers all combine to make this a prime cause of defects.

Materials specified may be *unsuitable* and will *never* accomplish the purpose for which they have been specified. As this is part of design, responsibility for such materials falls on the person who controls the design or to whom the risk is transferred.[8]

Suppose the materials were suitable but were faulty goods that came off the assembly line on a particular day that did not measure up to the quality requirement of the manufacturer. This problem is addressed in greater detail in Section 24.09(C), dealing with the AIA warranty clauses. A few observations can be made at this point.

Arguments can support placing the risk of faulty materials on either owner or contractor. As to putting the risk on the owner, it is often difficult to determine whether the materials were unsuitable or faulty. It is more efficient to place responsibility for *any* specified materials on the party in control of the design—the owner in the case of traditional construction and the contractor who both designs and builds. It is unfair to use purchase description specifications and then seek to hold the contractor accountable. When a purchase description specification is used, the contractor is not a seller but is simply a procurer of goods ordered by someone else. It should not be held to the warranties of sellers or manufacturers.

Cogent arguments can be made for holding the contractor responsible for faulty materials. Although it may be difficult to determine whether a defect is caused by unsuitable or faulty materials, it is also difficult to determine whether the defect is caused by materials *at all* or by the contractor's failure to install them properly. Very likely a contractor is responsible for faulty materials if it knew or should have known they were faulty. If so, it is administratively more efficient to make the contractor responsible for faulty materials, particularly as the contractor is likely to have a better claim

---

[6]*Northern Pac. Ry. Co. v. Goss*, 203 F. 904 (8th Cir.1913); *Trahan v. Broussard*, 399 So.2d 782 (La.App.1981); *Smith v. Erftmier*, 210 Neb. 486, 315 N.W.2d 445 (1982). But see *Samuelson v. Chutich*, 187 Colo. 155, 529 P.2d 631 (1974) (fault required).

[7]*Dobler v. Malloy*, 214 N.W.2d 510, 516 (N.D.1973). See also *Rosell v. Silver Crest Enter.* 7 Ariz.App. 137, 436 P.2d 915 (1968); *Barraque v. Neff*, 202 La. 360, 11 So.2d 697 (1942); *Robertson Lumber Co. v. Stephen Farmers Coop. Elevator Co.*, 274 Minn. 17, 143 N.W.2d 622 (1966). But see *Milau Assoc. v. North Ave. Dev. Corp.*, 42 N.Y.2d 482, 368 N.E.2d 1247, 398 N.Y.S.2d 882 (1977) (no implied warranty of fitness in service contract).

[8]*Trustees of Indiana University v. Aetna Cas. & Sur. Co.*, 920 F.2d 429 (7th Cir.1990); *Teufel v. Wienir*, 68 Wash.2d 31, 411 P.2d 151 (1966). The *Aetna* case is discussed in great detail by Reynolds, *What Is a Contractor's Warranty Responsibility for Owner-Specified Materials—Trustees of Indiana University v. Aetna Casualty & Surety*, 11 Constr.Lawyer No. 4, October 1991, p. 1. While the author represented Aetna in the case and admits that he might be biased, the article is a careful and perceptive treatment of the problem.

against the supplier or manufacturer than the owner.

This deceptively simple problem is one that *should* be handled specifically in any contract with design or purchase description specifications.

SECTION 24.03 Displacing the Basic Principle: Unconscionability

Autonomy (freedom of contract) generally allows contracting parties to determine how particular risks will be borne. As a rule, the owner is the party who usually seeks to take advantage of autonomy.

Displacement should be differentiated from clauses that clarify or augment the basic principle. A warranty clause (discussed in Section 24.09) can transfer design risks to the contractor. Such a clause would displace what would otherwise be the common law rule. However, such a clause may make *clear* that the contractor is responsible for poor workmanship and specify a remedy. The latter warranty clause seeks not to displace but to augment the basic principle.

Any clauses that seek to displace the basic principle will be scrutinized carefully to determine both whether the contracting party on whom the risk is placed was *aware* of the risk allocation and the *fairness* of displacement of the basic principle. Although cases do not always openly recognize this, the courts are more willing to determine the fairness of such clauses.

W.H. Lyman Construction Co. v. Village of Gurnee[9] involved a specification that required that the contractor use a particular manhole base and seal. The specifications also stated that the contractor assumed the risk of complying with infiltration limits. If the contractor thought the design would be inadequate, it was to direct attention to this in writing at the time it submitted its bid.

The court held that this was an "impermissible attempt on the part of the Village to shift the responsibility for the sufficiency and adequacy of the plans to the contractor, without providing the contractor the corresponding benefit of something to say about the plans that he is strictly bound to follow."[10] After noting the "possible" unconsciona-

bility of such a clause, the court concluded that the clause would not shield the village from its negligence because of public policy. Shielding the village would discourage bidders, and the public interest would suffer in the long run.

Kurland v. United Pacific Insurance Co.[11] demonstrates the unwillingness of courts to simply *apply* language that seeks to displace this basic principle. It involved a claim by an owner against a subcontractor's surety. The subcontractor had undertaken to install an air-conditioning system in an apartment building. The plans and specifications designated equipment to be used and required the contractor to meet performance standards for cooling and heating. The subcontractor did as required, but the air-conditioning system did not function as required.

In affirming a judgment for the surety, the court noted that the plans and specifications had been prepared by the architect. This was an *owner* design choice, perhaps a negligent one. The court held that it could not be *reasonably* concluded that the subcontractor would assume the responsibility for the adequacy of the plans and specifications. The court concluded that warranty–like language was simply an undertaking by the subcontractor that it would work as effectively as possible to achieve the desired result. But it was *not* a warranty of a successful outcome. Nevertheless, courts do not *always* relieve the contractor if it follows the plans and specifications dictated by the owner. In *United States Fidelity & Guaranty Co. v. Jacksonville State University*,[12] the contractor did as it was required by the specifications. But the court held the contractor responsible for wall leaks caused by unsuitable sealing material, as language in the contract was found to be a guarantee by the contractor.

The *Kurland* and *Jacksonville* cases, along with *Teufel v. Wienir*,[13] raise a problem that relates to subcontracting. Often the prime contract contains

---

[9]84 Ill.App.3d 28, 403 N.E.2d 1325 (1980).
[10]403 N.E.2d at 1332.

[11]251 Cal.App.2d 112, 59 Cal.Rptr. 258 (1967). Also see *Fanning & Doorley Const. Co. v. Geigy Chem. Corp.*, 305 F.Supp. 650 (R.I.1969) and *Wood-Hopkins Contracting Co. v. Masonry Contractors, Inc.*, 235 So.2d 548 (Fla.Dist.Ct.App.1970), which relieved contractors who simply installed what they were told.
[12]357 So.2d 952 (Ala.1978).
[13]Supra note 8.

language that requires the prime contractor to obtain a warranty from the subcontractor of its work. As seen in the *Kurland* case, such warranties may not mean much if purchase description specifications are used.

Subcontractor warranties raise another issue reflected in the *Jacksonville* and *Teufel* cases. Do these warranties affect any *prime* contractor obligation? Commonly, contract language makes the prime contractor responsible for the subcontractor's work. On the other hand, the contract may indicate that the owner was exclusively relying on the *subcontractor's* warranty. This exonerates the prime contractor.

That a subcontractor can be or has been sued should not change the legal obligation of owner and prime contractor unless the owner, by either demanding or accepting the subcontractor's warranty, manifests an intent to release the prime. For example, the trial court judge gave the prime contractor a defense in *Teufel v. Wienir* because the specifications indicated that the owner agreed to look only to the subcontractor.[14]

Suppose the design is created by the *subcontractor*. As between owner and prime, responsibility falls on the contractor.[15]

Another problem of risk shifting results from proposals by the prime contractor for substitutions of materials or equipment for that specified. Although substitutions must be approved by the design professional, to protect herself and to recognize that she may be relying on representations of the contractor, the design professional often obtains a guarantee by the contractor.

The effectiveness of the contractor's warranty will depend on the language of the warranty and surrounding facts and circumstances. Although a few court decisions have upheld warranties in the context of a substitution request,[16] the outcome of such a dispute cannot be easily predicted.[17]

Another method of displacing the basic principle is the frequent inclusion in construction contracts of provisions requiring that the contractor study and compare the design documents and report any observed errors as well as any violation of building laws that it observes.[18] These provisions should be strictly interpreted. If given effect, they displace the basic principle set forth in Section 24.02 by relieving the owner and ultimately the design professional from responsibility for design. The contractor should be required to use whatever design skills it possesses as long as doing so does not place unreasonable burdens on the contractor. Yet design in the traditional method of construction is still the responsibility of the owner. These objectives can be accommodated by requiring the contractor who has breached to share the cost of correction and any consequential damages with owner. This is discussed in Section 24.06.

Varying results show that language will not *automatically* displace the basic principle of control outlined in Section 24.02. Yet the outcome can often be predicted. Which party really made the design choice? Which party had the greater experience and skill? Which party relied on the other? The answers do not always implicate the owner, even in a traditional system. Even with these questions answered, account must be taken of the different attitudes toward the use of contract language to achieve what appears to be an unconscionable result.

SECTION 24.04 Good Faith and Fair Dealing: A Supplemental Principle

Parties who *plan* to enter into a contract are not, as a general rule, expected to look out for each other. Although there are exceptions, this is still a strong principle in American contract law.

Once a contract has been made, the law, led by the Uniform Commercial Code (U.C.C.) § 1-203 dealing with certain commercial transactions, increasingly expects parties to act in good faith and to deal fairly with one another.[19] In construction contracts, either party—though most commonly the contractor—must warn the other when the other is proceeding in a way that will cause failure. Sometimes this is reflected in contract clauses that

---

[14]The appellate court affirmed a judgment for the contractor by finding that the prime contractor simply followed the design. *Teufel v. Wienir,* supra note 8.

[15]*Stevens Constr. Corp. v. Carolina Corp.,* 63 Wis.2d 342, 217 N.W.2d 291 (1974).

[16]*Urania v. M.P. Constr. Co.,* 492 So.2d 888 (La.App.1986); *New Orleans Unity Soc. v. Standard Roofing,* 224 So.2d 60 (La.App.1969), cert. denied, 254 La. 811, 227 So.2d 146 (1969).

[17]A warranty whose meaning was not clear did not shift the risk to the contractor in *Habenicht & Howlett v. Jones-Allen-Dillingham,* Cal. Court of Appeals, 1 Civ. 46449 (1981) (an unpublished opinion that cannot be cited in California).

---

[18]AIA Doc. A201, ¶¶ 3.2.1, 3.7.3, and 3.7.4.

[19]Refer to Section 19.02(D).

require that the contractor bring to the attention of the owner or design professional design or other problems that can adversely affect the project.[20]

Suppose there is no specific contract obligation? What does the law demand of the contractor? Must it

1. Examine the contract documents, but only to prepare its bid, with design errors being none of its business?
2. Examine the contract documents principally to prepare its bid but also to note and report any errors observed?
3. Examine the contract documents, both to prepare its bid and to check for errors, it being held for errors it should have observed?

Even making allowance for factual variations, the cases display diverse results. This is common when a new doctrine of a vague nature limits the powerful principle of autonomy. A few appear to permit the contractor to ignore any design errors it may even observe, with its responsibility simply to build and not design.[21] The better reasoned cases and, as indicated, the AIA require the contractor to warn the owner if the contractor believes a suitable result cannot be obtained from the design.[22]

Is the contractor's conduct measured objectively? Is it responsible for an examination being what it *should* have discovered and what it *should* have reported?

Cases that have given the contractor who has followed the design a defense have stated that this defense will be lost if the contractor has been negligent, an apparently objective standard.[23] However, the AIA has used a subjective standard, a fairer result. The owner is paying for whatever design skill the contractor possesses and uses. The principal purpose for the contractor's reviewing the contract documents is to prepare its bid. The owner should expect attention to be drawn only to those errors that the contractor *does* discover.

The owner often has years to prepare the design, contrasted to the thirty days or so given the contractor to review the design and prepare its bid.[24] An objective approach may operate in a way as to place design risks unfairly on the contractor.

The contractor who *does* notify the owner of errors is not charged for defects that result despite its warnings and is entitled to be paid for what it has done.[25] The contractor who does *not* report obvious errors should not be given advantage of the "following the plans" defense.[26] If it performs work knowing that it violates legal requirements or technical competence, it will not be able to recover for the work it performed.

If failure to comply with its requirement causes a loss, as it will undoubtedly do, should the contractor be responsible for the *entire* loss? Sharing responsibility rather than placing it all on the contractor is fairer, particularly if the contractor did not have actual knowledge of the error. This is discussed in Section 24.06.

SECTION 24.05 Acceptance of Project

If the owner communicates a *clear* intention to relinquish a claim for obvious defects, the claim is barred.[27] Acts frequently *asserted* to have indicated such an intention are payment, particularly the final payment, and taking possession of the project.

Standard forms increasingly make clear that these acts do *not* constitute a waiver of any claim

---

[20]AIA Doc. A201, ¶¶ 3.2.1, 3.7.3.
[21]*Lewis v. Anchorage Asphalt Paving Co.*, 535 P.2d 1188 (Alaska 1975); *Luxurious Swimming Pools, Inc. v. Tepe*, 177 Ind.App. 384, 379 N.E.2d 992 (1978); *Rubin v. Coles*, 142 Misc. 139, 253 N.Y.S. 808 (Cty.Ct.1931).
[22]*Lebreton v. Brown*, 260 So.2d 767 (La.App.1972); *Hutchinson v. Bohnsack School Dist.*, 51 N.D. 165, 199 N.W. 484 (1924); *Home Furniture, Inc. v. Brunzell Constr. Co.*, 84 Nev. 309, 440 P.2d 398 (1968).
[23]See supra note 3.

[24]*Southern New England Contracting Co. v. State*, 165 Conn. 644, 345 A.2d 550 (1974); *Pittman Constr. Co. v. Housing Auth. of New Orleans*, 169 So.2d 122 (La.App.1964). See also *T.H. Taylor, Inc.*, ASBCA 27699, 86-2 BCA ¶ 18,743, which required the contractor to report only obvious errors.
[25]Architects who pointed out design difficulties can recover if the owner still proceeds. *Greenhaven Corp. v. Hutchcraft & Assoc.*, 463 N.E.2d 283 (Ind.App.1984) (design violated fire code); *Bowman v. Coursey*, 433 So.2d 251 (La.App.1983), certiorari denied 440 So.2d 151 (1983) (design below acceptable professional practice).
[26]See dicta in *Davis v. Henderlong Lumber Co.*, 221 F.Supp. 129 (N.D.Ind.1963), cited in *Snider v. Bob Heinlin Concrete Constr. Co.*, 506 N.E.2d 77 (Ind.App.1987).
[27]*A.H. Sollinger Constr. Co. v. Illinois Bldg. Auth.*, 5 Ill.App.3d 554, 283 N.E.2d 508 (1972); *Maloney v. Oak Builders, Inc.*, 224 So.2d 161 (La.App.1969); *Salem Realty Co. v. Batson*, 256 N.C. 298, 123 S.E.2d 744 (1962); *Stevens Constr. Corp. v. Carolina Corp.*, supra note 15.

for defective work.[28] The universal presence of warranty (guarantee) clauses makes it quite unlikely that such a claim will be lost because of the occurrence of these acts.[29]

Generally, the owner by payment[30] or by taking possession of the project[31] does not waive its claim for defective work. Either act alone does not *unambiguously* indicate the owner's intention to give up a claim for work to which it was entitled and for which it paid. Possession may be taken for reasons other than satisfaction with the work. Such holdings are frequently supported by contract clauses denying that such acts have waived claims for defects. Cases that have *found* waiver have been ones that have involved disputes over particular work followed by some act of the owner, such as payment or taking possession, which communicated satisfaction with the work and an intention not to assert any claim.[32]

SECTION 24.06 Owner Claims and Divided Responsibility

If proof establishes the cause of the defect and that cause is clearly the responsibility of one of the parties to the construction contract, the loss is chargeable to that party. But in a venture as complicated as construction, it is not uncommon for a construction defect to be traceable to multiple causes. For example, suppose the owner asserts a claim against the contractor based on the contractor's not having followed the plans and specifications. Suppose the contractor admits having breached the contract but points to other possible causes, such as abnormal weather conditions or third parties for whom the contractor is not responsible. Legal responsibility for breach of contract does not require that the claimant eliminate all causes except acts or omissions by the party against whom the claim has been made. The defendant's breach must only be a *substantial factor* in causing the harm.[33] If other conditions or actors played a minor or trivial part in causing the loss, the contractor is responsible for the entire loss.

Multiple causation becomes more complicated when the defect is traceable both to the party against whom the claim has been made and to the claimant itself. For example, Section 24.02 stated the basic principle that in the traditional construction project, the owner is responsible for the design and the contractor for its execution. Yet as noted in Section 24.01, these activities are not watertight compartments. The owner plays a role in execution because of the design professional's site responsibilities. The contractor plays a role in the design because through a specialty subcontractor, it may supply the design or it may be obliged to study and compare the contract documents and report any errors observed.

Another complicating factor is that the cause of the defect may be traceable to wrongful acts of the design professional—some of which may be chargeable to the owner. This section deals with defect claims by the owner against the contractor, with the defect having been caused in whole or in part by acts of the owner or someone for whom the owner is responsible. More commonly in construction, the owner asserts a claim for defective work against both the contractor and a third party, usually the design professional. This claim is discussed in Section 27.14.

Before analyzing shared responsibility, a threshold question must be addressed. Can the contractor

[28]AIA Doc. A201, ¶¶ 4.3.5 and 9.6.6. See Section 22.02(M). A federal regulation that stated that taking possession or use is not acceptance was influential in concluding that the project had not been accepted in *M.C. & D. Capital Corp. v. United States*, 948 F.2d 1251 (Fed.Cir.1991).

[29]AIA Doc. A201, ¶¶ 3.5.1 and 12.2.

[30]*Metro. Sanitary Dist. of Greater Chicago v. Anthony Pontarelli & Sons, Inc.*, 7 Ill.App.3d 829, 288 N.E.2d 905 (1972); *Parsons v. Beaulieu*, 429 A.2d 214 (Me.1981); *Handy v. Bliss*, 204 Mass. 513, 90 N.E. 864 (1910); *Burke County Public Schools Bd. of Educ. v. Juno Constr.*, supra note 3; *Quin Blair Enterprises, Inc. v. Julien Constr. Co.*, 597 P.2d 945, 955 (Wyo.1979) (citing earlier edition of treatise).

[31]*M.C. & D. Capital Corp. v. United States*, 948 F.2d 1251 (Fed.Cir.1991) (U.S. not precluded from terminating for default despite the fact that it had taken possession); *Coastal Modular Corp. v. Laminators, Inc.*, 635 F.2d 1102 (4th Cir.1980) (by implication); *Aubrey v. Helton*, 276 Ala. 134, 159 So.2d 837 (1964); *Honolulu Roofing Co. v. Felix*, 49 Hawaii 578, 426 P.2d 298 (1967); *Kangas v. Trust*, 110 Ill.App.3d 876, 441 N.E.2d 1271 (1982); *Hemenway Co., Inc. v. Bartex, Inc. of Tex.*, 373 So.2d 1356 (La.App.1979); *Bismarck Baptist Church v. Wiedemann Indus.*, 201 N.W.2d 434 (N.D.1972); *Hurley v. Kiona-Benton School Dist. No. 27*, 124 Wash. 537, 215 P. 21 (1923).

[32]*Saldal v. Jacobsen*, 154 Iowa 630, 135 N.W. 18 (1912); *Grass Range High School Dist. v. Wallace Diteman, Inc.*, 155 Mont. 10, 465 P.2d 814 (1970); *Cantrell v. Woodhill Enter., Inc.*, 273 N.C. 490, 160 S.E.2d 476 (1968). See generally, Sweet, *Completion, Acceptance and Waiver of Claims: Back to Basics*, 17 Forum 1312 (1982).

[33]*Krauss v. Greenbarg*, 137 F.2d 569 (3d Cir.), cert. denied 320 U.S. 791 (1943).

who does *not* follow the plans and specifications point to defective design? A few cases have precluded such a contractor from pointing to a design defect.[34] This conclusion is too punitive and does not take into account the role the design professional played in causing the loss. A federal court applying Arkansas law refused to bar the contractor's claim against the owner for a defective design simply because the contractor failed to execute the design.[35]

Where defects can be traced to both the owner and the contractor, the law has taken two approaches. Some cases have held that the loss will be shared, with owner's claim being reduced by a rough formula comparison such as that used in those states that use comparative negligence as part of their tort law.[36] This takes into account the complexity of causation and the desire to avoid all-or-nothing outcomes.

Yet other cases have not been willing to apply comparative fault in a contract claim for economic losses.[37] They seek to preserve a sharp line between contract and tort claims, a difficult objective to attain in the light of the possibility that the owner in some jurisdictions may be able to elect whether to maintain its claim in contract or tort.[38]

The apportionment problem can be controlled by contract language. AIA Doc. A201 provides an illustration. If the contractor does not fulfill its obligation under ¶ 3.2.1 to report errors that it discovers, that paragraph states that the contractor "shall assume appropriate responsibility . . . and shall bear an appropriate amount of the attributable costs for correction." This would appear to invite a shared responsibility.

Yet if the contractor does not fulfill its obligation under ¶ 3.7.3 to report any portions of the design that violate building laws, ¶ 3.7.4 states that it shall "assume full responsibility for such Work and shall bear the attributable costs." This would appear to place the *entire* responsibility on the contractor. Such an outcome would relieve the owner and ultimately the architect from any responsibility for supplying a design that did not comply with building laws. Admittedly the intervening cause of the contractor's failing to report errors of which it is aware is an *intentional* act, which can cut off responsibility for a negligent one. Yet this exculpates the owner (and the architect) from its responsibility for the design.

SECTION 24.07   Third-Party Claims

This chapter has emphasized the discovery of defects that cause harm to the owner. But defects caused either by design, for which the owner is responsible, or by execution, for which the contractor is responsible, or by a combination of the two can cause harm to third parties. As seen in Section 14.08, third-party claims create complex legal issues.

Claims for personal harm are discussed in Chapter 31. As to damage to the property of a third party, a tort claim will be dealt with as would any tort claim.[39] To illustrate, as noted in Section 7.03(D), a contractor who negligently breached its contract causing property damage to a third party contended that heavy rains were a superseding cause of the loss.[40] In rejecting this defense, the court stated that the heavy rains that caused the flooding and damaged the plaintiff's property were a reasonably foreseeable peril and not an exculpatory intervening cause.

Suppose a third party suffers property damage that can be traceable to defective design. The complexity of third-party tort claims is demonstrated by *Cincinnati Riverfront Coliseum, Inc. v. McNulty Co.*[41] Five years after completion of construction, an elevated walkway for pedestrian traffic suffered extensive damage and deterioration. The owner repaired the structure and brought legal action against eleven defendants. All settled except the consulting engineer who designed the walkway and the city, which had obligated itself to maintain it.

[34]*Valley Constr. Co. v. Lake Hills Sewer Dist.*, 67 Wash.2d 910, 410 P.2d 796 (1965); *Robert G. Regan Co. v. Fiocchi*, Ill.App.2d 336, 194 N.E.2d 665 (1963), cert. denied 379 U.S. 828 (1964).
[35]*Arkansas Rice Growers Coop. Ass'n v. Alchemy Indus., Inc.*, 797 F.2d 565 (8th Cir.1986).
[36]*Grow Constr. Co. v. State*, 56 A.D.2d 95, 391 N.Y.S.2d 726 (1977); *Circle Elec. Contractors, Inc.*, DOT CAB 76-27, 77-BCA ¶ 12, 339.
[37]*Broce-O'Dell Concrete Products, Inc. v. Mel Jarvis Constr. Co.*, 6 Kan.App.2d 757, 634 P.2d 1142 (1981) (claim by subcontractor against concrete supplier).
[38]See Section 14.05(E).

[39]See Section 7.03.
[40]*Diamond Springs Lime Co. v. American River Constructors*, 16 Cal.App.3d 581, 94 Cal.Rptr. 200 (1971).
[41]28 Ohio St.3d 333, 504 N.E.2d 415 (1986).

The jury found that 40% losses were attributable to the negligence of the consulting engineer and 5% to the city. The consulting engineer claimed that he was not responsible for construction defects, because the structure had not been constructed in accordance with the design, and that his tort liability was cut off by the intervening acts of the contractor.

The court stated that the engineer is responsible for the foreseeable consequences of his failure to exercise reasonable care. The jury could have found that it was reasonably foreseeable that the contractor would alter the design without approval of the engineer and fail to execute the design properly. The engineer can escape responsibility if the deviations are material and were the "proximate cause" of the loss. The other cause, the breach by the contractor, must independently break the causal connection between the negligent design and the damage. If so, it supersedes the negligent design as the cause. The issue of whether the contractor's failure to properly execute the design was an independent intervening cause was properly submitted to the jury. The jury's determination against the engineer was based on sufficient evidence. (The contractor was one of the defendants that had settled. Very likely it compensated the owner in part for the cost of correction and other attendant losses.)

As noted in Sections 14.08(C), (D), and (E), if the losses suffered by third parties are purely economic losses unconnected to personal harm or damage to property, some jurisdictions do not allow tort claims to be maintained.

The *Cincinnati Riverfront* case dealt with one party's liability for negligence being cut off by the acts of another. Other courts have looked on design negligence and failure to follow the design as *concurrent* and not *independent* causes. If so, the claimant (in defect cases, the owner) can contend that *both* caused the loss and can recover an entire, indivisible loss from either party, the *ultimate* responsibility dealt with by indemnity or contribution.[42]

SECTION 24.08 Claims Against Liability Insurer

Commonly, the contractor is required to carry comprehensive (increasingly called "commercial") general liability (CGL) insurance. CGL insurance indemnifies the insured contractor against claims by third parties who assert that they have suffered losses because the insured contractor has not acted in accordance with tort law. Usually claimants are workers on the job who are not employees of the contractor (the latter are usually covered by workers' compensation) or members of the public. Claims covered are those that involve personal harm or property damage.

In the 1970s, contractors began to assert claims against their CGL insurers based on "public" liability for defects as "property damage." (This was and often is the reason for an owner claim to be based on negligence.) Such an approach is advantageous to the contractor. The insurer, unlike a surety, cannot claim against its insured.

Generally, CGL policies are not intended to cover ordinary business losses, defective work being an example. They are intended to cover unusual, unexpected losses, such as a wall collapsing, destroying a Porsche automobile parked nearby. As a result, CGL policies usually exclude work products, property in the control of the insured (this should be handled by property insurance), or work performed by or on behalf of the contractor. (This excludes subcontractor work.)

A minority of jurisdictions have concluded that defective workmanship is covered under the CGL policies, usually based upon the court's conclusion that the language is ambiguous and therefore requires coverage.[43] However, most states that have faced this issue have denied the contractor coverage where there are allegations of faulty or defective workmanship.[44] This is usually based upon the language in the standard exclusions and the concept that poor workmanship is a business expense.

One commentator sought to justify refusal to cover poor workmanship by noting that repair and

---

[42]*Northern Petrochemical Co. v. Thorsen & Thorshov, Inc.,* 297 Minn. 118, 211 N.W.2d 159 (1973). See Section 27.14.

[43]*Colard v. American Family Mut. Ins. Co.,* 709 P.2d 11 (Colo.App.1985) (despite work product exclusion); *Emcasco Ins. Co. v. L&M Dev., Inc.,* 372 N.W.2d 908 (N.D.1985) (implied warranty exception to exclusion).

[44]*Reliance Ins. Co. v. Mogavero,* 640 F.Supp. 84 (D.Md.1986); *Maryland Cas. Co. v. Reeder,* infra note 46; *Wechsler v. Beall & Assoc. Roofing Corp.,* 442 So.2d 1104 (Fla.Dist.Ct.App.1987); *Vernon Williams & Son Constr. Inc. v. Continental Ins. Co.,* 591 S.W.2d 760 (Tenn.1979). For general discussion and particular emphasis upon coverage for work of subcontractors and construction managers, see Mack and Mack, *Construction Claims Under the Comprehensive General Liability Policy,* 40 S.C.L.Rev. 1003 (1989).

replacement costs, typically the measure of recovery for defective work, are generally or at least to some degree within the control of the insured. These costs can be minimized by careful purchasing, inspection of materials, quality control, and hiring policies. If these costs were to be covered by the CGL policy, the incentive to exercise care or make repairs at the least possible cost would be minimized, since the insurance company would be paying for the repair and replacement. The losses attributable to poor workmanship tend to occur more frequently than losses through injuries to other property, but the losses are relatively limited, since the cost of total replacement would be the greatest cost. It is advantageous for the insured to stand these losses, as it can then purchase insurance more cheaply, since the insurer is freed from administering many small claims. The insured can receive a reduced rate for the more unusual risk of injury to property other than the work performed by the contractor.[45]

Another problem relates to duration of coverage. The *standard* CGL policy terminates coverage after final payment. This would bar coverage for defects discovered after that point. But it is possible to obtain a broad form property damage endorsement, including completed operations, to cover such claims. This endorsement is sometimes obtained to provide the prime contractor coverage if the defective work has been performed by subcontractors. Some cases have held that this endorsement provides the prime contractor with coverage for work done by subcontractors, while others have not.[46] Insurance problems raised by such claims is beyond the scope of this treatise.[47]

SECTION 24.09 Warranty (Guarantee) Clauses

A. Relationship to Acceptance

Acceptance of the project is an important benchmark in the relationship between owner and contractor. Contractors would like to know when the law can no longer call on them to perform further work or to respond to a legal claim. For that reason, they frequently contend that acceptance of the project manifests owner satisfaction and the owner cannot make any further complaints about the work.

The law has been reluctant to find that there has been sufficient "acceptance" to bar any future claims for latent or nondiscoverable defects against the contractor.[48] Undoubtedly, this reluctance is traceable to the difficulty of discovering defects at the time the project is turned over. Many defects will not be apparent until the owner has taken over the project and used it. One method of dealing with the risk that acceptance will be found and claims barred is to insert a provision making clear that liability does not end on the project's being turned over to the owner.[49] The exact nature of that liability is discussed in (B).

B. Purposes: *St. Andrew's Episcopal Day School v. Walsh Plumbing Co.*

The deceptively simple warranty clause obscures a variety of possible purposes. Before discussing the many purposes, a case involving such a clause is reproduced.

---

[45]Macaulay, *Justice Traynor and the Law of Contracts*, 13 Stan.L.Rev. 812, 825–826 (1961).

[46]Compare *Knutson Constr. Co. v. St. Paul Fire & Marine Ins. Co.*, 396 N.W.2d 229 (Minn.1986) (broad form property damage endorsement does not change loss from business cost to insurance risk), and *Blaylock & Brown Constr., Inc. v. AIU Ins. Co.*, 796 S.W.2d 146 (Tenn.Ct.App.1990), appeal denied (July 23, 1990) (did not cover negligence of subcontractor because prime still in overall control), with *Fireguard Sprinkler Sys., Inc. v. Scottsdale Ins. Co.*, 864 F.2d 648 (9th Cir.1988) (prime may have little control over sub's work, and broad form property damage endorsement covers this risk), and *Maryland Cas. Co. v. Reeder*, 221 Cal.App.3d 961, 270 Cal.Rptr. 719 (1990) (followed interpretation given by insurance industry).

[47]See *Chemstar, Inc. v. Liberty Mut. Ins. Co.*, 797 F.Supp.

1241 (C.D.Cal.1992) (defective batch of lime causing damage to a number of homes—a single occurrence); *McDonald v. State Farm Fire & Cas. Co.*, 119 Wash.2d 724, 837 P.2d 1000 (1992) (coverage excluded where negligent construction, excluded from coverage, is the efficient proximate cause of damage unless a covered loss otherwise ensues).

[48]See Section 24.05.

[49]AIA Doc. A201 ¶ 12.2.6.

ST. ANDREW'S EPISCOPAL DAY SCHOOL v. WALSH PLUMBING CO.

Supreme Court of Mississippi, 1970. 234 So.2d 922.

ROBERTSON, Justice.

The appellant, St. Andrew's Episcopal Day School, a charitable corporation, brought suit in the Chancery Court of the First Judicial District of Hinds County, Mississippi, against Appellee Walsh Plumbing Company, the contractor of the mechanical work, and Appellee The Trane Company, the manufacturer of the major portion of the air conditioning system installed in the new Day School building. The suit was one for breach of warranty or guaranty. The chancellor, after a full trial, dismissed the bill of complaint, and the Day School appeals from this judgment.

On April 8, 1965, appellant entered into a contract with Walsh Plumbing Company whereby, in consideration of $149,420, Walsh agreed to:

> furnish all of the materials and perform all of the work shown on the Drawings and described in the Specifications entitled: Item II, Mechanical Construction, St. Andrews Episcopal Day School, Old Canton Road, Jackson, Mississippi.

The General Conditions of the Contract provided in Article 20:

Correction of the Work After Substantial Completion

The Contractor shall remedy any defects due to faulty materials or workmanship and pay for any damage to other work resulting therefrom, which shall appear within a period of one year from the date of Substantial Completion as defined in these General Conditions, and in accordance with the terms of any special guarantees provided in the Contract. The Owner shall give notice of observed defects with reasonable promptness. All questions arising under this Article shall be decided by the Architect subject to arbitration, notwithstanding final payment.

The Construction Specifications, in Paragraph 22 of Mechanical Construction, required:

> This contractor shall guarantee each and every part of all apparatus entering into this work to be *the best of its respective kind and he shall replace within one year from date of completion all parts which during that time prove to be defective and he must replace these parts at his own expense.*
>
> He shall guarantee to install each and every portion of the work in strict accordance with the plans and specifications and to the satisfaction of the owner.
>
> Guarantee to include replacement of refrigerant loss from air conditioning and refrigerant system. [Emphasis added.]

On August 19, 1966, Lomax, North and Beasley, consulting engineers, by letter, advised Biggs, Weir, Neal & Chastain, architects, that the mechanical contractor should furnish "as built drawings which locate all underground piping and clean-outs, framed operating instructions in the boiler rooms, CFM figures for all air units, and certification that all safety valves and devices have been tested." The engineers ended their letter with this comment:

> Other than the above, we feel that the mechanical work is ready for final certification, subject to contract guarantee provisions.

On August 30, 1966, the engineer wrote the architects:

> Final inspections of the subject project have been completed and we recommend final certification of the mechanical contract, subject to contract guarantee provisions.

In accordance with the requirements of Paragraph 22 of the Construction Specifications, on August 16, 1966, Walsh Plumbing Company wrote St. Andrew's Episcopal Day School:

> We hereby guarantee all work performed by us on the above captioned project to be free from defective materials and workmanship for a period of one (1) year, unless called for in the specifications to be a longer period of time."

Between August 16, 1966, and July 17, 1967, Appellee Walsh was called on several times to remedy defects in the air conditioning system, and Walsh always responded promptly. On July 17, 1967, the air conditioning system broke down completely and ceased to function. The headmaster, the Reverend James, immediately tried to contact Ray Walsh, only to find that he was out of town. Mrs. Walsh suggested that the School call somebody else.

During the first two weeks of July, 1967, James E. Davis, Jr., operator of Davis-Trane Service Agency and also a salesman and representative of The Trane Company, was contacted by John B. Walsh and together with Mr. Walsh attempted to determine why the air conditioning system was not cooling. Mr. Davis described the meeting that took place on July 18, 1967, in these words:

> [T]he meeting that we had on a particular day at the school, in which Mr. Nicholson was there, and Mr. Ray Walsh was there, and Mr. Nicholson said, 'We want to get

this thing fixed,' and *Mr. Walsh told me to fix it.* He said, 'You have the people, the personnel and the know-how, *you go ahead and do it* and I will just stay out of it,' so Mr. Nicholson said, 'Okay, Davis-Trane Service Agency, go ahead and fix this machine.' " [Emphasis added.]

Forrest G. North, the mechanical engineer, testified that the failure of two safety devices, the flow control switch and the freeze protection thermostat caused the copper tubes inside the chiller shell and the shell itself to rupture. The purpose of the flow control switch was to prevent the operation of the machine when there was no circulation of water in the chiller shell. The flow control switch was installed by Walsh outside the chiller and was not Trane equipment. The freeze protection thermostat was a Trane part, and was installed inside the chiller unit by Trane at its factory.

The repairs made by Davis-Trane Service Agency pursuant to Ray Walsh's instructions to Davis to go ahead and fix it amounted to $6,813.05. One of the major items of expense was a new chiller unit purchased from the Trane Company. The Trane Company and Davis-Trane billed the appellant, and the appellant paid Trane separately for the new chiller unit, and Davis-Trane for all the repairs.

The appellant is in the business of running a Christian day school; it is not in the air conditioning business. Appellant does not profess to have any knowledge or expertise about air conditioning systems or equipment. That was the main reason for the provision in the Construction Specifications that the mechanical contractor "shall guarantee each and every part of all apparatus entering into this work to be *the best of its respective kind,* and he shall replace within one year from date of completion all parts which during that time prove to be defective and he must replace these parts at his own expense."

Walsh was a reputable, responsible and knowledgeable contractor of mechanical work; and when Lomax, North and Beasley, consulting engineers for the Day School, recommended to the appellant that the bid of Walsh Plumbing Company be accepted, the duty and responsibility was placed squarely on Walsh's shoulders to purchase and properly install the best air conditioning system on the market. Not only was Walsh to purchase and install, he was to guarantee the system and its installation for one year. This was what Walsh contracted to do, and this was what the appellant paid Walsh to do. Not knowing anything about air conditioning systems, the appellant employed experts in this field and reposed full confidence in these experts to look after its interests.

The chancellor was correct in finding that the breakdown of the air conditioning system occurred "within the time of the warranty by Walsh," and that the repairs were made necessary to properly repair the system.

The chancellor was in error in holding that:

"St. Andrews never gave any written notice or made any demand on Walsh to comply with his warranty to fix the machine, which would be as I hold a condition precedent to hiring someone else to do the work."

The sole purpose of notice is to give the contractor who selected, purchased and installed the system the first opportunity to remedy the defects at the least possible expense to him. Appellee Walsh was given this opportunity.

The evidence is undisputed that Walsh was at the Day School building on July 18, 1967, the day after the breakdown, with James E. Davis, Jr., of Davis-Trane Service Agency and John W. Nicholson of the Day School. Davis, called as an adverse witness by the appellant, testified that Ray Walsh said at that time:

You have the people, the personnel and the know-how, you go ahead and do it and I will just stay out of it ". . . ." [Emphasis added.]

Walsh was afforded the opportunity of doing the work himself or employing somebody else to do it. With full knowledge and full notice he chose to employ Davis to go ahead and remedy the defects and make the necessary repairs.

Appellee Trane was the major supplier of items and equipment going into the air conditioning system and Trane was well paid for these. It is unfortunate that Trane's one-year guaranty to Walsh had run out at the time of the complete breakdown of the air conditioning system. Trane guaranteed the items and equipment furnished by it for one year from the date of shipment; these parts and equipment were shipped in March, 1966. The complete breakdown did not occur until July, 1967. The chancellor was correct in holding that Trane's warranty had expired.

The chancellor found that $6,813.05 was "a reasonable amount to make the repairs" and put the air conditioning system back in operation.

The judgment of the chancery court is affirmed as the Appellee The Trane Company, but the judgment is reversed as to Appellee Walsh Plumbing Company, and judgment is rendered against Walsh Plumbing Company on its warranty for $6,813.05, together with 6% interest from July 17, 1967.

Judgment affirmed as to the Trane Company, but reversed and rendered as to Walsh Plumbing Company.

GILLESPIE, P.J., and JONES, BRADY, and INZER, J.J., concur.

From the owner's vantage point, one of the problems in the traditional contracting system is the possibility of being "whipsawed" between the design professional and the contractor when a defect is discovered. The design professional contends that the defect resulted from improper execution of the design. The contractor contends that the defect existed because of design inadequacy. To make matters worse, each or both may contend that the defect was caused by conditions over which neither had control, such as unusual weather, misuse, or poor maintenance by the owner. One way of dealing with divided responsibility is to hire one entity to both design and build.

Another method of dealing with this problem is to incorporate a provision under which one entity is responsible for defects *however* and *by whomever* caused. The entity on whom the owner is likely to place this risk is the contractor. For example, the owner could require that the contractor give a full warranty on the roof that would cover defects *however* they may be caused.

Despite the undoubted advantage in placing the risks of defects *however caused* on the contractor, this is not often done. Contractors may respond to such a roof guaranty by drastically increasing the contract price to take into account the possibility of design errors, the additional cost of checking on the quality of the design, the cost of any special endorsement that will have to be obtained to cover design risks, or the cost of a roofing bond. Nevertheless, the owner may decide that these additional costs are worth the advantage of centralizing responsibility. Although enforcement requires very clear language (even *that* may not be sufficient), such clauses have been enforced.[50]

Another use of a warranty clause relates to the nonfault nature of express warranty. In some states, one who contracts to perform services need only perform nonnegligently.[51] In those states, a warranty clause as to workmanship can avoid the burden of having to establish that the contractor performed negligently.

Many issues lurk in a simple warranty clause. Earlier discussion emphasized the *owner's* purpose in having such a clause. The *contractor* is also concerned with how such a clause will function. Although it would prefer to be relieved from responsibility on acceptance, it may be willing to accept a warranty clause if it believes that expiration of the period will terminate its obligation. Similarly, the contractor may be concerned with exposure created by such a clause even within the warranty period. Failure to take immediate steps to deal with a defect may increase the loss. The cost of correcting a defect is likely to be greater if correction is done by someone other than the contractor. For that reason, contractors sometimes contend that the clause is for their benefit, that its function is to give the contractor notice as quickly as possible to enable the contractor to cut the losses and repair the defect as inexpensively as possible.[52] On the other hand, the owner may contend that the clause is for its benefit. If so, it need *not* call back the contractor (discussed in (C)).

Another possible purpose is to bar the contractor from contesting *how* the owner has chosen to correct the defect if the contractor refuses to attempt to correct the work when requested to do so. The clause may place the burden on the warrantor to establish that *other* causes, such as abnormal weather or owner misuse, caused the defect.

In the consumer context, warranties can mislead the consumer. The consumer may believe that what looks like a very broad warranty has been so hedged about with restrictions that it may not amount to very much. To deal with this problem, the Congress enacted the Magnuson-Moss Warranty Act, which carefully regulates consumer war-

---

[50]*Potler v. MCP Facilities Corp.*, 471 F.Supp. 1344 (E.D.N.Y.1979); *Bryson v. McCone*, 121 Cal. 153, 53 P. 637 (1898) (construction of industrial plant); *New Orleans Unity Soc. v. Standard Roofing Co.*, 224 So.2d 60 (La.App.1969), cert. denied 254 La. 811, 227 So.2d 146 (1969); *St. Andrew's Episcopal Day School v. Walsh Plumbing Co.*, reproduced in this section; *Burke County Public Schools etc. v. Juno Constr.*, supra note 3 (agreement to maintain roof for five years); *Shuster v. Sion*, 86 R.I. 431, 136 A.2d 611 (1957); *Shopping Center Mangement Co. v. Rupp*, 54 Wash.2d 624, 343 P.2d 877 (1959); cf. *Pinellas County v. Lee Constr. Co. of Sanford*, 375 So.2d 293 (Fla.Dist.Ct.App.1979).

[51]*Samuelson v. Chutich*, supra note 6.
[52]See *St. Andrew's Episcopal Day School v. Walsh Plumbing Co.*, reproduced in this section. If it is clear that the contractor will not correct the defect, giving the notice is excused. *Orto v. Jackson*, 413 N.E.2d 273 (Ind.App.1980).

ranties.[53] The Act also specifies remedies for breach of such a warranty and allows the consumer who prevails in a claim based on the Act to recover its attorneys' fees.

While the Act was thought principally to deal with consumer products, a case held that it also applies to a contract to roof a house.[54] In this case, the court looked at the legislative history and the desire to protect consumers and concluded that the Act was not limited solely to products (the award of attorneys' fees exceeded the damages).

Moving back to the warranty clause as putting a time limit on liability, although a few decisions have held the clause to have created a private (created by the contract) period of limitation,[55] most have held that such a provision is *not* intended to cut off liability at the end of the warranty period.[56] Clearly, the AIA has *not* made its warranty clause a period of limitation.[57] The law will not bar a claim by a contractual provision of this type unless the clause *clearly* shows this intention.[58]

Does the clause affect the remedy? Usually the clause states that the contractor will correct any defect within the warranty period. As discussed in (C), will this affect the issuance of a court decree *specifically ordering* the contractor to come and correct the work? If the contractor fails to correct the work, the owner can do so and charge the contractor. This recognizes the adequacy of the money award, usually a bar to judicially ordered specific performance. A warranty clause should not affect specific performance.

Remedy can be approached another way. Suppose there were *no* warranty clause. The owner can recover the cost of correction unless it would be disproportionately high in relation to the diminished value caused by the defect. If the latter, the measure is the difference in value between the project as built and as it *should* have been built. In addition, the owner may be able to recover consequential damages, such as lost use or profits if reasonably foreseeable and proved with reasonable certainty.

A warranty clause is a promise by the contractor to correct defects within the warranty period. Does this make correction cost the only remedy? Is diminished value barred? Does the clause limit liability to cost of correction?

Very likely the clause was not intended to deal with any of these subtle issues. The law should not give the clause *any* remedial effect.[59]

Although historically such clauses may have been justified as a method to avoid acceptance barring claims, is there a current justification for inclusion of a warranty clause?

An unfortunate problem that often plagues the owner is the unwillingness of the contractor to return and correct defects for which it is responsible. Including a specific provision under which the contractor promises to return and correct defects if notified to do so within a designated period may persuade a contractor that it must do as it has promised.

Exhortation, of course, may not be the *sole* function of such a clause or even *a* function of a particular clause. A warranty clause may be designed to shift design risks to a contractor or to affect the legal standard of conduct. Nevertheless, on the average, *modern* warranty clauses are principally exhortations to get defects corrected quickly before they generate greater losses. This does not mean that the drafter may not want a *different* purpose.

[53]15 U.S.C.A. § 2301 et seq.

[54]*Muchisky v. Frederic Roofing Co., Inc.*, 838 S.W.2d 74 (Mo.App.1992).

[55]*Cree Coaches, Inc. v. Panel Suppliers, Inc.*, 384 Mich. 646, 186 N.W.2d 335 (1971); *Indep. Consol. School Dist. No. 24 v. Carlstrom*, 277 Minn. 117, 151 N.W.2d 784 (1967).

[56]*First Nat'l Bank of Akron v. Cann*, 503 F.Supp. 419 (N.D.Ohio 1980), affirmed 669 F.2d 414 (6th Cir.1982) (citing earlier edition of treatise); *Norair Eng'g Corp. v. St. Joseph's Hosp., Inc.*, 147 Ga.App. 595, 249 S.E.2d 642 (1978); *Bd. of Regents v. Wilson*, 27 Ill.App.3d 26, 326 N.E.2d 216 (1975) (citing and following earlier edition of treatise); *Michel v. Efferson*, 223 La. 136, 65 So.2d 115 (1953); *Newton Housing Auth. v. Cumberland Constr. Co.*, 5 Mass.App.Ct 1, 358 N.E.2d 474 (1977); *All Seasons Water Users Assn. v. Northern Improve. Co.*, 399 N.W.2d 278 (N.D.1987); *City of Midland v. Waller*, 430 S.W.2d 473 (Tex.1968).

[57]AIA Doc. A201, ¶ 12.2.6.

[58]*Glantz Contracting Co. v. General Elec.*, 379 So.2d 912 (Miss.1980); *Bender-Miller Co. v. Thomwood Farms, Inc.*, 211 Va. 585, 179 S.E.2d 636 (1971) (remedy not exclusive).

[59]*United States v. Franklin Steel Products, Inc.*, 482 F.2d 400 (9th Cir.1973) (consequential damages recoverable); *Oliver B. Cannon & Son, Inc. v. Dorr-Oliver, Inc.*, 336 A.2d 211 (Del.1975); *New Zion Baptist Church v. Mecco, Inc.*, 478 So.2d 1364 (La.App.1985). But see *Leggette v. Pittman*, 268 N.C. 292, 150 S.E.2d 420 (1966). This holding was, though, limited in *Salerno Towne Apartments, Inc. v. McDaniel & Sons Roofing Co.*, 330 F.Supp. 906 (E.D.N.C.1970).

If so, the objective should be expressed clearly in the contract. Even so, clauses will always need interpretation as to coverage.[60]

C. Warranty and Correction of Work Under A201

A201, ¶ 3.5.1, creates an express warranty as to the quality of the contractor's work. It must be read together with ¶ 12.2, which deals with correction of work that does not meet the quality warranty.

Paragraph 3.5.1 creates three warranties from the contractor to the owner and to the architect:

1. "[M]aterials and equipment . . . will be of good quality and new unless otherwise required or permitted."
2. "The Work will be free from defects not inherent in the quality required or permitted."
3. "The Work will conform to the requirements of the Contract Documents."

Section 24.02 discussed the difference between unsuitable and faulty (defective) materials. Suppose purchase description specifications that require the contractor to use particular materials or equipment by manufacturer and model number are *not* used. Paragraph 3.5.1 makes the contractor responsible for faulty materials or equipment. Suppose, though, purchase description specifications *are* used. If the contractor did as ordered, would the contractor be liable under ¶ 3.5.1 for defects caused by faulty materials? It failed to supply good quality. But it can be contended that the contractor is relieved because it installed what was "required or permitted." This contention would be supported by the strong body of case law that relieves the contractor when it does as it has been ordered.[61] That legal doctrine does not apply if the contractor has given a warranty or if it violated its obligation of good faith and fair dealing, such as by failing to communicate information that would have warned the architect not to proceed. Yet ¶ 3.5.1 is not sufficiently *specific* to create a warranty that can overcome the basic rule that the contractor is relieved if it does what it has been required to do.

In 1987, the AIA added language that exculpates the contractor from its warranty if the defect is caused by "abuse, modifications not executed by the Contractor, improper or insufficient maintenance, improper operation, or normal wear and tear under normal usage." The loss of control and the possibility of such exculpatory events are reasons contractors seek to be relieved from *any* responsibility after expiration of the warranty period.

If there is a breach of the quality requirement under ¶ 3.5.1, ¶ 12.2 deals with correction of work. But ¶ 12.2 deals with the mechanism for correcting defective work, whether discovered during performance or within the one-year period measured by substantial completion.

Although the Associated General Contractors (AGC) has sought to cut off the contractor's liability for patent (discoverable) defects after the expiration of the warranty period, the AIA has made clear in ¶ 12.2.6 that expiration of the period does not terminate the contractor's liability, which is terminated only by the expiration of the appropriate statute of limitation. Other issues that have been discussed in this subsection are not so clearly resolved.

Does the contractor have "first crack" at defective work that is discovered within the first year?[62] The clause does state that the contractor shall correct the defect promptly after receipt of written notice and that the owner shall give notice promptly after discovery of the condition. Also, ¶ 12.2.4 gives the owner the right to correct *if* the contractor does not.

If this clause is intended for the owner's benefit, the owner need not give the contractor the first chance to make the corrections.[63] Yet applying the obligation of good faith and fair dealing, it would not be fair for the owner to have the absolute right to determine whether to give the contractor the first chance to correct the work. It can refuse *only* if the owner has had good reason to lose faith in the contractor, such as prior assurances by the contractor having proved to be unreliable.

The AIA approach highlights the possibility that a defect can be discovered during performance, within the one-year period, or after expiration of the warranty period but before the claim would be

---

[60]*Sheldon v. Ramey Builders, Inc.*, 156 Ga.App. 670, 275 S.E.2d 743 (1980) (warranty as structural flaws included flooding caused by failure to waterproof).
[61]See Section 24.02. See also *Teufel v. Wienir*, supra note 8.

[62]*St. Andrew's Episcopal Day School v. Walsh Plumbing Co.*, reproduced in Section 24.09(B).
[63]*Baker Pool Co. v. Bennett*, 411 S.W.2d 335 (Ky.1967).

barred by law. Article 12 and the sections to which it refers dealing with remedy contemplate the discovery of defective work *during* performance. As a result, some of the mechanisms, such as those that involve the architect, would not be applicable for defects discovered during the one-year period. The architect is no longer "on board," and any role she might have can no longer be fulfilled during the one-year period. (But in 1987, ¶ 4.2.1 allows the owner to recall the architect during the one-year period.)

What if the defect is discovered *after* the expiration of the warranty period? Must the owner notify the contractor? Notification provisions should apply *only* to defects discovered within one year. Yet ¶ 12.2.2 covers *either* defects discovered during the year period or those not barred by the statute of limitations. One court troubled by this problem pointed to ¶ 7.4.1 (now ¶ 4.3.9), which requires either party who suffers injury or damage to make a claim in writing to the other party within a reasonable time.[64] This was very likely intended to apply only to harm suffered during performance and not to place a notice requirement as a condition precedent for *any* claim.

The one-year correction of work requirement expressed in ¶ 12.2.1 can be expanded if there is "an applicable special warranty required by the Contract Documents." This recognizes that there may be some specific warranties that exceed the one-year period. For example, roof warranties are often given for lengthy periods. If such a warranty is considered a specific warranty required by the contract, the one-year correction of work period should be extended to the period covered in the specific warranty clause.[65]

Does the warranty clause control the remedy? It is true that ¶ 12.2.4 allows the owner to *correct* the work if the contractor fails to do so, apparently whether failure occurs during the contractor's performance and after the project has been turned over to the owner, whether within the year period or not. Yet it is unlikely that this was intended to bar any diminished value measure of damages or

limit the owner to correction cost, barring other losses resulting from the defect.

AIA takes the position that the purpose of the warranty clause is *remedial*, to give the owner a right to specific performance (judicial order to the contractor to correct the defect).[66] It supports this by pointing to its language in ¶ 12.2.1 stating that the contractor has a specific obligation to correct the work. It is not likely that a court would accord the language any control over the remedy, the money award, or the specific performance.

SECTION 24.10 Implied Warranties in the Sale of Homes: Strict Liability

A. Home Buyers and Their Legal Problems

Project success in the preceding section assumes construction documents that set forth the contractor's obligations and are the basis for judging the design professional's performance. Yet persons often buy partially or totally completed homes without the protection of plans that must be met before final payment is made. They purchase homes from developers or builder-vendors, each of whom builds homes for immediate sale to home buyers. Suppose after the purchase the basement leaks, the walls crack, or the heating system malfunctions. What recourse does the buyer have?

Before the mid-1960s, home buyers who faced such problems had little recourse against the sellers, or for that matter against anyone. Part of this was traceable to legal rules designed to avoid uncertainties in real property transfers and to transfer certain risks by ownership change. In addition, the common law expected buyers to protect themselves, a concept expressed in the maxim "caveat emptor," or "let the buyer beware." Even if these barriers were surmounted, the homeowner discovering these defects could only, as a rule, pursue the party who had sold her the house. Recovery could not be obtained from those more responsible, such as the builder, the designer, or the developer who had not sold the house to the person discovering the defect. In a fast turnover market, such require-

[64]*First Nat'l Bank of Akron v. Cann,* 669 F.2d 415 (6th Cir.1982).
[65]*Hillcrest Country Club v. N.D. Judd Co.,* 236 Neb. 233, 461 N.W.2d 55 (1990).

[66]3 Architect's Handbook of Professional Practice, A201 Commentary p. 69 (1987). But *Gerety v. Poitras,* 126 Vt. 153, 224 A.2d 919 (1966) refused to order specific performance of a guaranty clause.

ments of privity often left the discoverer of the defect with a claim only against the person who had sold the discoverer the house.

B. The Implied Warranty Explosion of the 1960s

Dissatisfaction with the traditional denial of protection meshed with the general consumer movement that began in the 1960s. Consumers of goods began to demand products that worked, and home buyers demanded similar protection. In response to this demand, courts recognized an implied warranty of habitability in the sale of new homes.[67] Some courts have gone beyond implied warranty and have held that a builder of mass-produced homes can be held strictly liable for injuries caused by the conditions in the home.[68]

C. Current Problems

The breakthrough consisted of overthrowing the old rules that had denied any form of recovery. With the exception of a few jurisdictions,[69] most courts that have recently faced the question have done this. But the courts and legislatures must still work out details. Courts have articulated similar but somewhat variant formulas for describing the nature of the implied warranty.[70]

The exact nature of the warranty will depend on the price of the house, the customary standards of the community, and the reasonable expectations of the buyer. Perhaps it would be useful to examine a few cases to see what specific facts have constituted breach of the implied warranty.

A number of cases have involved sewerage, water, and moisture damage. One case involved an overflowing septic tank,[71] another, a leaky basement,[72] still another, water and fill seepage,[73] and yet another, basement seepage.[74] One unfortunate home buyer found water seepage in the basement, surface waters flowing in from the outside, and a septic tank that backed up.[75]

Yet not all water disappointments fall within the implied warranty. For example, Connecticut refused to imply a warranty in a contract to sell a house and dig a well that the water supplied would be potable, quoting the trial judge as having stated that only the Lord can guarantee the quantity or quality of water from a well.[76]

Other defects that have justified a claim that the implied warranty has been breached were structural defects, such as a collapsing stairway[77] or cracked walls.[78] Lead-based paint violated the implied warranty of habitability.[79] Finally, claims have been based on a defective heating system,[80] a defective air-conditioning system,[81] and a malfunctioning fireplace.[82]

The preceding list of defects seems to require that the developer or vendor-builder furnish basic shelter from the elements and reasonable comfort. Obviously, whether these general requirements have been met will depend on the contract of purchase and the circumstances surrounding the particular transaction.

For whose protection does the implied warranty doctrine exist, and who is subject to its requirements? Clearly, a buyer can sue the seller. Beyond that, the law varies in the different jurisdictions. Some courts allow a subsequent purchaser to main-

---

[67]One of the most interesting of the many opinions is *Humber v. Morton*, 426 S.W.2d 554 (Tex.1968).
[68]*Schipper v. Levitt & Sons, Inc.*, 44 N.J. 70, 207 A.2d 314 (1965). See Comment, 10 Ohio No.U.L.Rev. 103 (1983).
[69]*Druid Homes, Inc. v. Cooper*, 272 Ala. 415, 131 So.2d 884 (1961), held that there was no implied warranty for improvements on real property in Alabama.
[70]*Hartley v. Ballou*, 20 N.C.App. 493, 201 S.E.2d 712 (1974) (suitable for habitation); *Padula v. Deb-Cin Homes, Inc.*, 111 R.I. 29, 298 A.2d 529 (1973) (fit for human habitation); *Gable v. Silver*, 258 So.2d 11 (Fla.Dist.Ct.App.1972) (fit and merchantable); *Crawley v. Terhune*, 437 S.W.2d 743 (Ky.1969) (major structural features constructed in a workmanlike manner with suitable materials) Ann.Conn.Gen.Stat. § 47-121 (built in accordance with building codes). Implied warranties are discussed in detail in Note, 42 S.C.L.Rev. 503, 514–524 (1991).

[71]*Rutledge v. Dodenhoff*, 254 S.C. 407, 175 S.E.2d 792 (1970).
[72]*Crawley v. Terhune*, supra note 70.
[73]*Wawak v. Stewart*, 247 Ark. 1093, 449 S.W.2d 922 (1970).
[74]*Elmore v. Blume*, 31 Ill.App.3d 643, 334 N.E.2d 431 (1975).
[75]*Norton v. Burleaud*, 115 N.H. 435, 342 A.2d 629 (1975).
[76]*Bertozzi v. McCarthy*, 164 Conn. 463, 323 A.2d 553 (1973).
[77]*Rogers v. Scyphers*, 251 S.C. 128, 161 S.E.2d 81 (1968).
[78]*Oliver v. City Builders, Inc.*, 303 So.2d 466 (Miss.1974).
[79]*City of Philadelphia v. Page*, 363 F.Supp. 148 (E.D.Pa.1973).
[80]*Kriegler v. Eichler Homes, Inc.*, 269 Cal.App.2d 224, 74 Cal.Rptr. 749 (1969).
[81]*Gable v. Silver*, supra note 70.
[82]*Humber v. Morton*, supra note 67.

tain an action,[83] while some do not.[84] One case held that the infant son of a tenant could bring an action against the developer.[85]

As to other potential defendants, successful actions have been brought against a lender who took more than a normal lender's role in creating the development[86] and an engineer who conducted soil studies.[87]

What types of property carry with it implied warranty protection? Such protection has been held to include, at one extreme, a condominium,[88] and at the other, a sale of a lot.[89] Between these two extremes California would not apply strict liability (much akin to implied warranty) to a builder who built two houses at different times and in different locations. The court found that this reflected occasional construction and not mass production.[90]

Courts seem hesitant to imply a warranty in the sale of an existing home not built for immediate sale.[91] Here it is more difficult to justify warranty implication as part of the normal enterprise risk of developers or builder-vendors. The caveat emptor doctrine, with the modern qualification that sellers disclose serious defects of which they know and that the buyer would not likely be able to discover, seems proper. In this regard, the California Supreme Court, in holding builders and vendors to an implied warranty that the completed structure was designed and constructed in a reasonably workmanlike manner, stated:

> In the setting of the marketplace, the builder or seller of new construction—not unlike the manufacturer or merchandiser of personality—makes implied representations, ordinarily indispensable to the sale, that the builder has used reasonable skill and judgment in constructing the building. On the other hand, the purchaser does not usually possess the knowledge of the builder and is unable to fully examine a completed house and its components without disturbing the finished product. Further, unlike the purchaser of an older building, he has no opportunity to observe how the building has withstood the passage of time. Thus he generally relies on those in a position to know the quality of the work to be sold, and his reliance is surely evident to the construction industry.[92]

Most plaintiffs seek damages based on either the cost of correction of the defect or the diminished value of the property.[93] A host of structural defects and incurable water problems should enable the buyer to call the deal off and receive any money paid less any benefit received by occupying the premises. Like cars, houses can be lemons.

Can the seller disclaim the implied warranty by a contract clause? Express contract language usually takes precedence over implied warranties. It is likely that exculpatory clauses will not be effective where the plaintiff has suffered personal harm. Exculpation may be effective where there is damage to property or other economic loss if the exculpatory language was brought to the buyer's attention in such a way as to *clearly* indicate that the buyer assumed this risk.[94] Marketing methods of most developers and vendor-builders are not likely to employ this open approach.

---

[83]*Wells v. Clowers Const. Co.*, 476, So.2d 105 (Ala.1985); *Kriegler v. Eichler Homes, Inc.*, supra note 80; *McMillan v. Brune-Harpenau-Turbeck Builders*, 8 Ohio St.3d 3, 455 N.E.2d 1276 (1983) (negligence needed); *Keyes v. Guy Bailey Homes*, 439 So.2d 670 (Miss.1983).

[84]*Redarowicz v. Ohlendorf*, 92 Ill.2d 171, 441 N.E.2d 324 (1982); *Barnes v. MacBrown & Co.*, 264 Ind. 227, 342 N.E.2d 619 (1976); *Evans v. Mitchell*, 77 N.C.App. 598, 335 S.E.2d 758 (1985) review denied 316 N.C. 376, 342 S.E.2d 89 (1986). See Note, *Gupta v. Ritter Homes, Inc.*, 35 Baylor L.Rev. 670 (1983).

[85]*Schipper v. Levit & Sons, Inc.*, supra note 68.

[86]*Connor v. Great Western Sav. & Loan Assn.*, 69 Cal.2d 850, 447 P.2d 609, 73 Cal.Rptr. 369 (1968). See Ferguson, *Lender's Liability for Construction Defects*, 11 Real Est.L.J. 310 (1983). Lender liability is also discussed in Note, S.C.L.Rev. 503, 510–512, 519–520 (1991).

[87]*Avner v. Longridge Estates*, 272 Cal.App.2d 607, 77 Cal.Rptr. 633 (1969).

[88]*Gable v. Silver*, supra note 70.

[89]*Avner v. Longridge Estates*, supra note 87.

[90]*Oliver v. Superior Court*, 217 Cal.App.3d 86, 259 Cal.Rptr. 160 (1989).

[91]However, a reconditioner seller, the FHA, was held in an implied warranty case in *City of Philadelphia v. Page*, supra note 79.

[92]*Pollard v. Saxe & Yolles Dev. Co.*, 12 Cal.3d 374, 525 P.2d 88, 91, 115 Cal.Rptr. 648, 651 (1974).

[93]These measures of recovery are discussed in more detail in Section 27.03.

[94]See *Melody Home Mfg. Co. v. Barnes*, 741 S.W.2d 349 (Tex.1987) (cannot disclaim implied warranty of good workmanship). See also Anderson, *Disclaiming the Implied Warranties of Habitability and Good Workmanship in the Sale of New Homes*, 15 Tex.Tech.L.Rev. 517 (1984).

D. Insurance Protection

The preceding discussion may have over-emphasized the value of the implied warranty to buyers. Although the 1960s did see an explosion of protection for home buyers, the law in any given jurisdiction may still be unclear, especially as to the particular nature of the warranty. Often the defects complained of are not large, and instituting individual legal action to enforce the warranty may cost more than what can be recovered in court. This is a reason for increased use of the class action in which a number of buyers "similarly situated" join together in one lawsuit. In any event, there are often long delays in litigating these cases.

Developers and vendor-builders are engaged in a hazardous business and often are either out of business or unable to pay for damages by the time a claim is brought. As a result, private industry has been offering homeowner's warranties to cover the types of defects that have been discussed in this section.

E. Deceptive Practices Statutes

For a review of these statutes, see Section 23.02(J).

SECTION 24.11 A Suggestion: Defect Response Agreements

This section is directed to owners who are about to engage in projects of a substantial nature, particularly those with the likelihood of defects developing. It is an attempt to outline a skeletal proposal for dealing with defects.

Dealing with defects involves three stages. Stage I, the *diagnostic* stage, involves correcting the defect. Stage II is the *preparation* to resolve the allocation of legal responsibility for the losses caused by the defect. Stage III, the actual trial, involves *judicial* resolution of legal responsibility.

Although Stage III is most dramatic (and costly), Stages I and II are the most crucial. The defect must be fixed for the owner to receive the project for which it has paid and to avoid large losses. Stage II, the gathering of information that can enable the parties to try to voluntarily settle a dispute, involves a large expenditure of time and money before the parties are in the position to settle the dis-

pute. Stage III is rarely reached. Regardless of how expensive Stages I and II are, the expenses of Stage III with a trial of from one to six months add immense direct and indirect costs and often generate unsatisfactory outcomes. As a result, most disputes and large projects never go beyond Stage II, with the dispute being settled out of court.

Before the parties are in the position to seriously discuss settlement, they must have some idea of what caused the dispute, who is legally responsible, and how the losses should be apportioned. To prepare for serious negotiations, much expense must be incurred, such as expenses of lawyers, experts, testers, and the often ignored indirect expense of officers and key employees of the major participants having to spend time that does not earn their employers any revenue. In disputes of this nature, the stakes are sufficiently high so that no participant feels that it can "cut costs."

When all the major participants in the dispute—and there are many—have sufficient information or feel under sufficient pressure to seriously discuss negotiations, it is likely that there is enough blame to spread around, enough technical uncertainty and disagreements among experts, and uncertainty as to the legal outcome. As a result, the major participants—the owner, contractors, design professionals, major suppliers, manufacturers, sureties, insurers, and funding agencies, to name the most visible—each decides it will pitch some money into a settlement "pot."

How do these participants determine how much they are willing to pitch into the pot? It will very likely result from hard bargaining with all the variable factors that make negotiation an art, such as willingness or reluctance to continue the fight, financial strength and weakness, appraisal of the strength and weakness of each participant's position, and a prediction of how the matter will be resolved, whether decided by arbitration or litigation.

All of this takes place after a large expenditure of time and money needed to amass the information that each party feels it needs before it can enter into meaningful negotiations. Is there any way that this can be avoided? One method is to simulate in a rough way the type of negotiations that result at Stage II but to do it *before* the project begins. This can be accomplished by a Defect Response Agreement (DRA) made between *all* the major participants. To accomplish this, the major participants enter into the negotiation as they "sign on" the project, all under the direction and supervision of the owner. The DRA deals solely with the response to defects, how to correct the defects, and who is responsible for the "immediate" response as well as with the formula for determining the ultimate loss distribution and possibly the security to back up any such formula.

Because all the major parties are signatories, there would be none of the increasingly complex third-party problems. The formula itself would be determined by negotiation, influenced very likely by the same matters that affect any negotiation—a calculation of risk and profits. Fear of the open-ended consequential damages would lead to each

participant's giving up claims it would have for such damages.

A DRA would have to be imposed by a strong private owner[95] that recognizes that *it* bears the large burden of the wasteful cost incurred in the present system. It can "persuade" other participants that they will benefit in the long run by a system that will avoid the worst aspect of the current dispute resolution process.[96]

_____

[95]Public owners would be too fearful of accountability requirements to try such a no-fault approach.
[96]Another suggestion is to use a property damage policy that would cover defects, however caused, with *all* participants waiving subrogation claims and consequential damages. Defects would then be an insurance problem.

PROBLEMS

1. The City Museum engaged the services of an architect to prepare plans and specifications for the construction of a one-story, brick and block warehouse building for the storage of paintings and other works of art. The owner furnished the architect with a topographic survey and test boring reports previously prepared by a geotechnical engineer employed by the owner. The parties then executed AIA Document B141, "STANDARD FORM OF AGREEMENT BETWEEN OWNER AND ARCHITECT."

The architect delivered the plans and specifications to the owner with an estimated cost of one million dollars. Upon approval by the owner, invitations to bid were advertised, and the lowest acceptable bid submitted by a prime contractor resulted in the execution of AIA Document A201, "GENERAL CONDITIONS OF THE CONTRACT FOR CONSTRUCTION." (See Appendix C.)

The prime contractor engaged the services of several subcontractors for the various specialties of the project. Contractual arrangements were created by the use of AIA Document A401, "STANDARD FORM OF AGREEMENT BETWEEN CONTRACTOR AND SUBCONTRACTOR." (See Appendix E.)

After construction began, the prime contractor notified the architect that there was disturbed earth where the concrete foundation footings were to be

placed at the northeast corner of the lot. Rather than undergo the expense of a change in the foundation as originally designed, the architect revised her plans to provide for reinforced poured concrete grade beams and additional steel at the point where the earth was disturbed. The boring report provided by the owner did not show any water at this location.

The architect further revised the plans to provide for a specific brand of waterproofing materials to be applied from the bottom of the grade beam to a point well above the grade.

The waterproofing materials were a new product and had not previously been used by the architect or contractor, but the manufacturer was a reputable company that had a long history of producing acceptable building materials and had tested the product and recommended its use to the architect. The architect did not make any independent test of the materials before specifying their use after the job condition developed.

During one of her periodic visits with her consulting engineer, the architect determined that the work was not being performed according to the plans as revised with regard to the installation of steel at the northeast corner by the subcontractor who had been employed by the contractor to perform this work. The contractor and the subcontractor complained that the objection was frivolous and that the work was being performed according to the revised plans and specifications.

Following an exchange of recriminations between the parties, the architect rejected the work and then directed the contractor to remove a portion of the beam already in place and to use more and a different type of steel in accordance with the plans. The contractor finally agreed to these demands, stating that it did so under protest, and instructed the subcontractor to make the changes.

On completion of the building, the architect issued a certificate of final payment to the contractor and the contractor was paid in full.

A month later, the museum found it necessary to move a number of valuable paintings into the warehouse and placed them on the floor at the northeast corner of the building. The following night, a torrential downpour caused an accumulation of water deep enough to destroy and damage a number of valuable paintings.

Subsequent investigation revealed that the beam in the northeast corner of the warehouse had cracked and had settled to a point where a gap was exposed between the beam and the concrete floor. It was further found that the subcontractor had failed to install the additional concrete beams and steel as demanded by the architect and instructed by the contractor. The investigation also revealed that the waterproofing materials had been torn as a result of the settlement and the flooding, but testing showed that the waterproofing material might not have repelled all of the moisture in any event.

QUESTIONS

1. Does the owner have a claim against the architect for the water damage to the building based on design error? (Review Chapter 14).

2. Is the architect liable for selecting defective waterproofing materials? (Review Chapter 14.)

3. Would the architect be liable had she not specified the type of waterproofing materials and the materials had been selected by the contractor?

4. Is the architect liable for failure to "monitor" the construction work? (Review Chapter 14.)

5. Is the contractor liable to the owner for the subcontractor's failure to comply with the architect's instructions? (Consult Chapter 28.)

6. May the contractor defend against the owner's claim for damages on the grounds that the architect's plans were defective? (Review Chapter 23.)

7. Does the owner have a claim against the architect based on the issuance of the final certificate of payment? (Review Chapters 14 and 22).

8. Does the owner's claim for damages extend to the damaged and destroyed paintings resulting from the flooding? (Review Chapter 6.)

9. How should the owner's claim for damages be apportioned between architect and contractor?

10. Had the contractor not made the discovery of the soil condition, would it have been liable for damages resulting from defective plans? (Consult Chapter 25.)

PROBLEMS

Your state is considering enacting a law that will prescribe certain warranty periods that builders must give to those to whom they sell houses. The required warranty periods are as follows:

1. Defect in pipes, including central heating—2 years.

2. Penetration of water or moisture from the roof, walls, or basement—3 years.

3. Defect in machines, motors, or boilers—3 years.

4. Peeling of wall coverings in the stairwells—3 years.

5. Sinking of floor tiles on the ground floor, parking places, sideways, pathways, and all areas of the building—3 years.

6. Cracks in the walls and ceilings, noticeable peeling in the exterior coverings—5 years.

7. Any other defect—1 year.

The liability prescribed by the warranty clause only deals with contractual liability and does not bar claims based on negligence.

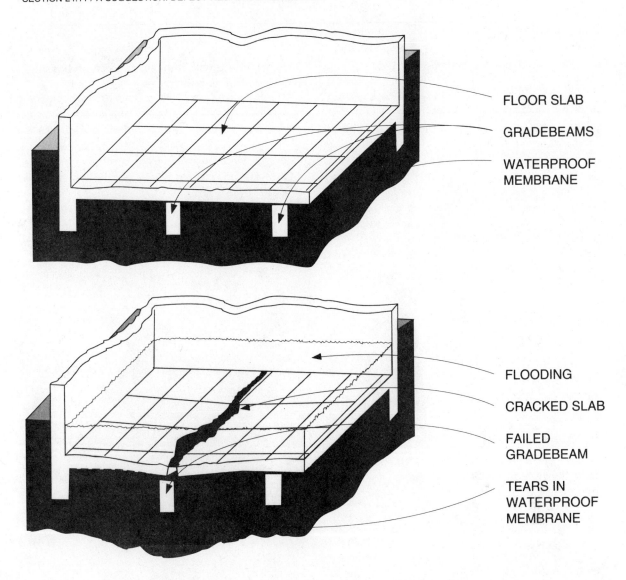

HYPOTHETICAL

QUESTIONS

1. How should these periods be selected?

2. Will this improve quality?

3. What effect will it have on cost of housing? Number of units built?

C H A P T E R   T W E N T Y - F I V E

Subsurface Problems: Predictable Uncertainty

SECTION 25.01 Discovery of Unforeseen Conditions

A. Effect on Performance

For reasons to be explored in (B), the discovery of unforeseen subsurface conditions is not unusual in the Construction Process. When such conditions are discovered, they usually have an adverse effect on the contractor's planned performance and prediction of performance costs. Sometimes subsurface materials encountered are more difficult to excavate or extract, with many cases involving the discovery of hardpan or more rock than expected. When this occurs, performance is likely to take more time and cost more.

Sometimes the subsurface conditions encountered generate a great increase or decrease in the quantities to be excavated. This also can affect the time and costs, particularly if the contractor has bid a composite unit price that may be adversely affected by finding that certain work overruns and other work underruns. (This was discussed in Section 17.02(D), dealing with unit pricing.) Materials (borrow pits) that the contractor expects to use for fill or compaction may turn out to be unsuitable. If so, the contractor must obtain materials from a site more distant than planned or more costly to extract. When these conditions are discovered, the contractor may request more time and more money.

B. Causes

Why are unexpected subsurface conditions frequently encountered? Soil testing is expensive, and only a limited number of borings are usually made.

As a result, the data reported may not reflect subsurface conditions throughout the entire area. Subsurface conditions even within small areas may vary greatly.

Geotechnical engineers may be exposed to great liability if the information they report is incorrect or their suggestions do not accomplish the owner's expectations. Although this chapter concentrates mainly on subsurface conditions encountered *during* construction, problems such as structural instability, settling, or cracking can develop later. They are often traceable to the subsurface, often leading to claims against geotechnical engineers. As a result, geotechnical engineers may not perform services where liability exposure is greatly in excess of anticipated profit, or they may sharply increase their fees. This can eliminate testing in smaller projects or make less extensive testing more likely.

Information, however reliable, must be gathered and used by designers and contractors. When what is encountered varies from what is anticipated, the risk will have to be borne by someone. This chapter examines contractor claims for additional compensation when contractors' costs are more than expected because of unforeseen subsurface conditions.

C. Two Models

Again it is important to recognize the different methods by which construction is accomplished. First, consider construction done by a contractor who is given a site and asked to both design and build a particular project. *Stees v. Leonard,*[1] cited

---

[1] 20 Minn. 494 (1874), cited in *Dravo Corp. v. Litton Systems, Inc.,* 379 F.Supp. 37 (S.D.Miss.1974).

many times for the proposition that the contractor bears all the risks of subsurface conditions, involved a contract under which the contractor both designed and built a house for the owner. In such cases, it is easy to place the risk of success on the contractor.[2]

The traditional American Construction Process *divides* design and construction, with the owner engaging a design professional to design, a geotechnical engineer to gather data, and a contractor to build. This makes it more difficult to determine who will bear the risk of unforeseen subsurface conditions.

Although there are abundant factual variations, typically the owner—private or public—calls for bids on a construction project that will involve subsurface excavation. Building specifications may be furnished, but methods of excavation and construction will probably be left to the contractor.

Depending on owner identity and project size, soil tests are taken by an independent geotechnical engineer for purposes of cost estimation, design, and scheduling. The reports will be made available to the design professional and the owner. Although the information is likely to be available to the contractor, owners take different routes as to whether they will stand behind the accuracy of the information.

Whichever approach is taken, the bidder will usually be warned that it must inspect the site and, under some contracts, conduct its own soil testing. However, often the bidder will not make independent soil tests because the profit potential may be too small to justify such expenditure or because the bids are due in a relatively short period of time. As a result, contractors frequently bid without knowledge of actual conditions that will be encountered. Bids are calculated on the basis of expected conditions, and if the unexpected is encountered, the actual cost will vary widely from that anticipated. The focus of this chapter is on who will bear this risk.[3]

D. Enforceability of a Promise to Pay More Money

When unforeseen subsurface conditions are discovered, the owner may promise to pay additional compensation. The enforceability of such a promise depends on the application of the preexisting duty rule, discussed in Section 5.11(C). The traditional application of this rule would deny enforceability of such a promise because the owner is getting for its promise nothing more than it was entitled to get under the contract.[4] However, special rules have developed in subsurface cases that can in some jurisdictions permit enforcement of the promise. Sometimes a promise is enforced because the contractor gave up its right to rescind the contract for a mutual mistake as consideration for the promise.[5] More commonly, the law enforces such a promise so long as "the modification is fair and equitable in view of circumstances not anticipated by the parties when the contract was made."[6]

E. Supervening Geotechnical Conditions and Mistake Claims

Most problems involve encountering unexpected *preexisting* subsurface conditions. Additional costs can result from a change in geological conditions because of unexpected weather (unseasonable rains or frost) or third-party interference (flooding from adjacent lands). A distinction can be drawn between the types of risks involved. The risk of a preexisting geological condition can be eliminated if the parties wish to spend enough time and money. On the other hand, the likelihood of a future occurrence, whether from third-party conduct or atmospheric conditions, can rarely be predicted with any degree of accuracy.

---

[2]See Section 24.02.

[3]For a comparative study, see Wiegand, *Allocation of the Soil Risk in Construction Contracts: A Legal Comparison*, 6 Int'l.Constr.L.Rev., Pt. 3, 282 (1989).

---

[4]For a case indicating that the traditional rule still retains its vigor, see *Crookham & Vessels, Inc. v. Larry Moyer Trucking, Inc.*, 16 Ark.App. 214, 699 S.W.2d 414 (1985). Cf. *Hiers-Wright Assoc. Inc. v. Manufacturer's Hanover Mort. Corp.*, 182 Ga.App. 732, 356 S.E.2d 903 (1987). For a collection of cases, see Annot., 85 A.L.R.3d 259, 315–336 (1978).

[5]*Healy v. Brewster*, 251 Cal.App.2d 541, 59 Cal.Rptr. 752 (1967).

[6]*Angel v. Murray*, 113 R.I. 482, 322 A.2d 630, 636 (1974), quoting what is now Restatement (Second) of Contracts § 89(a). See also *Linz v. Schuck*, 106 Md. 220, 67 A. 286 (1907).

Supervening changes of geological conditions are not discussed in this chapter. Usually when such events occur, the contractor seeks a time extension but not additional compensation. The differing site conditions clause used by the federal government and discussed in Section 25.06(A) generally does not apply to atmospheric difficulties or third-party interference.[7]

If contracting parties share a fundamental mistake as to material assumptions at the time the contract is made, the parties may be discharged because of mutual mistake unless the risk has been allocated by contract or by custom to the party whose performance has been adversely affected. For example, if both parties have fundamental assumptions as to particular subsurface characteristics, the discovery of substantially different subsurface characteristics could entitle the party adversely affected, usually the contractor, to terminate its obligation under the contract and be paid for what it has performed based on restitution.

Most subsurface disputes involve claims by the contractor for additional compensation, not for the right to terminate its obligation. As a rule, these claims are based on misrepresentation, as discussed in Section 25.03, or on contractual provisions, as discussed in Section 25.06. As a result, mutual mistake does not play a significant role in this chapter.

Mistake can be relevant in two instances. As seen in Section 25.01(D), mistaken assumptions may be part of the basis for enforcing a subsequent promise to pay additional compensation. Also, the differing site conditions method under which the contractor receives additional compensation as discussed in Section 25.06 can be substituted for any common law right to terminate because of mutual mistake (discussed in that section).

SECTION 25.02 Common Law Rule

Unless the owner has furnished site or subsurface information on which the contractor can rely, the risk of added costs of performance falls on the contractor. In *United States v. Spearin* (reproduced in Section 23.05(E)), the court stated that a contracting party who agrees to perform something possible for a fixed sum will not be excused or entitled to additional compensation because unforeseen difficulties make performance more costly than anticipated.

Similarly (though not the focus of this chapter), if the completed project slides, settles, or is destroyed because of subsurface conditions, generally the contractor is responsible for redoing the work or for the diminished value of the premises unless the owner has designated the excavation method to be used or the contractor has relied on misrepresentations made by the owner. As an illustration, Mississippi held a builder responsible for severe cracking because the builder was found to have been negligent in not warning the buyer of potential soil problems and not recommending that soil tests be made.[8] The court predicated its holding in favor of a subsequent purchaser on the builder's duty to use its technical knowledge to build a house on yazoo clay that would protect users from settling. The dissenting judge would have exonerated the builder because he concluded that there was no such duty imposed upon the builder and because the builder had built according to the plans and specifications required by the Veterans Administration.

As a general proposition, then, the contractor will bear the risk of unforeseen subsurface conditions unless (1) it can establish that it has relied on information furnished by the owner (discussed in Sections 25.03 and 25.05), (2) the contract itself provided protection (discussed in Section 25.06), (3) the owner did not disclose information that it should have disclosed,[9] or (4) the cost of performance was extraordinarily higher than could have been anticipated, a fact that would justify the ap-

[7]*Hardeman-Monier-Hutcherson v. United States*, 198 Ct.Cl. 472, 458 F.2d 1364 (1972). It can apply if representations are made to the contractor regarding sea or climate conditions.

[8]*George B. Gilmore v. Garrett*, 582 So.2d 387 (Miss.1991). See also *Stees v. Leonard*, supra note 1.
[9]See *P.T. & L. Constr. Co., Inc. v. State of New Jersey Dep't of Transp.*, 108 N.J. 559, 531 A.2d 330 (1987). See also *U.S. Home Corp, Rutenberg Home Div. v. Metro. Property & Liability Ins. Co.*, 516 So.2d 3 (Fla.Dist.Ct.App.1987) (seller's possession of superior knowledge changes latent defect into patent defect). See also Section 18.04(B).

plication of mutual mistake or impossibility as discussed in Sections 23.03(C) and (D).

The number of exceptions to the basic rule should not convey the impression that the policy of placing these risks on the performing party no longer exists or does not reflect, at least to some courts, an important legal principle in fixed-price contracts. For example, in *W. H. Lyman Construction Co. v. Village of Gurnee*,[10] a contractor had been engaged to perform a sanitary sewer project. One basis for a claim against the public entity was that the sewer had to be constructed through subsurface soil that was for the most part waterbearing sand and silt rather than clay, as indicated by the soil-boring log shown on the plans. A high groundwater table was also discovered. This required the contractor to install numerous dewatering wells.

It is important to note the judicial attitude in the *Lyman* case, which dealt with the claim by the contractor that the public entity impliedly warranted that the plans and specifications would enable it to accomplish its promised performance in the manner anticipated. In rejecting the contractor's claim, the court stated:

> It is well settled that a contractor cannot claim it is entitled to additional compensation simply because the task it has undertaken turns out to be more difficult due to weather conditions, the subsidence of the soil, etc. To find otherwise would be contrary to public policy and detrimental to the public interest.[11]

The court looked on the common law rule as expressing a principle of great importance, one needed to protect public entities and public funds. It is not clear that the court would have felt as strongly as it did were the contract a private one. Yet, owners and those who supply funds for the project are also greatly concerned with the ultimate cost of the project and rely heavily on the contract price in their planning.

SECTION 25.03 Information Furnished by Owner

Owners who intend to build substantial projects often make subsurface and soil reports available to prospective contractors. One exception to the general allocation of unexpected costs to the contractor relates to the owner furnishing information that is relied on by the contractor. The owner may make this information available but state that it is *not* responsible for its accuracy. This section emphasizes the effect of providing this information with the full realization that the contractor is likely to rely on it and without any technique to shield the owner from responsibility.

At the outset, differentiation must be made between the types of information that may be made available to the contractor. Usually reports of any tests that have been taken are among the information made available. The reports may also contain opinions or inferences that the geotechnical engineer may have drawn from observation and tests taken. The information may include estimates as to the type and amount of material to be excavated or needed for fill or compaction. (This information may also be included in the specifications.) Though less common, the report can recommend particular subsurface operational techniques.

Misrepresentation is the basic theory on which the contractor bases its claim. A threshold question involves what constitutes a misrepresentation. Differentiation must be made between facts and opinions. Reporting the result of tests is clearly a factual representation, while professional judgments that seek to draw inferences from this information may be simply opinions.

A misrepresentation claim may be based on improperly selected test sites, the inference being that a contractor may believe that the test sites selected will generally represent the site. Misrepresentation can consist of a combination of providing some information but not disclosing all the information that qualifies the information given. Half-truths can be just as misleading as complete falsehoods. This is distinguished from simply failing to disclose *any* information or information that the owner knows would be valuable to the contractor and that the contractor is not likely to be able to discover for itself.

Representations may be fraudulent, that is, made with the intention of deceiving the contractor. Fraud claims, though having a heavy burden of proof, are most valuable to a contractor. The contractor's negligence in failing to check the data gen-

[10]84 Ill.App.3d 28, 403 N.E.2d 1325 (1980).
[11]403 N.E.2d at 1328.

erally does not bar recovery.[12] Recovery would be barred only if the contractor relied on information *it* gathered. Fraud gives the contractor a variety of remedies. The contractor can rescind the contract and refuse to perform further, raise fraud as a defense if sued for nonperformance, or—most important—complete the contract and recover additional compensation in a claim for damages.

More commonly, though, misrepresentations are not made with the intention of deceiving the contractor. If the misrepresentations were made negligently, some added complexities develop. Very likely the negligence is that of the geotechnical engineer, and owner recovery against the geotechnical engineer may not always be available.[13] Clearly, if the negligence had been that of the owner, the contractor can recover any losses it suffers that could not have been reasonably avoided.

What is the responsibility of the owner for any *negligent* representations in the soil information generated by the geotechnical engineer? If the geotechnical engineer is an independent contractor—which is likely to be the case—the owner would not be chargeable with the latter's negligence. However, it is very likely that the owner will be chargeable either through one of the exceptions to the independent contractor rule[14] or for breach of contract (that the breach is caused by a person whom it engages to perform services does not relieve it of responsibility),[15] or implied warranty of the accuracy of the information supplied by the owner that is relied on by the contractor.[16]

The least culpable conduct is that of innocent misrepresentation. In such cases, there is neither intention to deceive nor negligence. Generally, innocent misrepresentation allows the contracting party misled to rescind the contract. Some states allow a restitutionary damage remedy—the difference between what was received and what was paid out[17]—an ineffective remedy in subsurface cases.

Actual rescission in these subsurface cases is rare. Many legal and factual issues may make it difficult to predict whether the right to rescind is available. Walking off the job under these conditions exposes the contractor to liability. Even if its rescission were determined to be justified, the contractor can recover only the reasonable value of its services. This should, at least in theory, put the contractor in the position it was in at the time it made the contract.

It may be difficult to recover lost overhead and profit using this restitutionary remedy.[18] More likely, the contractor will continue performance and later claim additional compensation. Because innocent misrepresentation is not a contract *breach*, however, contractors will shift to implied warranty to justify their claims for additional compensation. The principal problem with the implied warranty theory is the owner's attempt to exonerate itself by the use of contract language negating the necessary reliance.[19]

SECTION 25.04 Risk Allocation Plans

One of the principal planning decisions that those who prepare construction contracts must make relates to subsurface problems broadly defined. As seen in Section 25.02, in the absence of any contractual risk assumption system incorporated into the contract, the contractor will bear the risk of additional expenses attributable to having to perform work under subsurface conditions that are different than anticipated. This means that the contractor will have to bear the cost of any additional expenses and cannot be relieved from any further obligation to perform.

---

[12]*Seeger v. Odell*, 18 Cal.2d 409, 115 P.2d 977 (1941).
[13]See Section 14.05. Compare *Texas Tunneling Co. v. City of Chattanooga*, 329 F.2d 402 (6th Cir.1964) (claim denied), with *M. Miller Co. v. Central Contra Costa Sanitary Dist.*, 198 Cal.App.2d 305, 18 Cal.Rptr. 13 (1961) (claim allowed). The latter has been followed in the Restatement (Second) of Torts, § 525 (1977).
[14]See Section 31.05(C).
[15]*Harold A. Newman Co. v. Nero*, 31 Cal.App.3d 490, 107 Cal.Rptr. 464 (1973); *Brooks v. Hayes*, 133 Wis.2d 228, 395 N.W.2d 167 (1986).
[16]Restatement (Second) of Torts § 552 (1977). The remedy is the difference in value between what was received and what was paid out—a "restitutionlike" remedy.

---

[17]See Section 34.01.
[18]See Section 27.02(E).
[19]See Section 25.05.

The exceptions to this basic risk allocation were noted in Section 25.02. But those who plan risk allocation for such contracts must keep in mind that generally the law will permit the contract to distribute the risks in any way chosen by the parties. That it is usually the owner who makes this determination and that the only choice the contractor has is to enter into the contract or not rarely affects the law's respect for the contract provisions.

Assuming that the owner is the person to make the risk distribution choices, there are essentially three options:

1. Develop no information and let the contractor proceed based upon its own evaluation.
2. Gather the necessary information, make it available to the contractor, but seek to disclaim any responsibility for its accuracy (disclaimer system).
3. Gather the information, make it available to the contractor, and promise to give the contractor an equitable adjustment in the contract price if the actual conditions turn out to be different from those represented or different from those anticipated (differing site conditions (DSC) system).

The preceding techniques are discussed in Sections 25.05 and 25.06. In those sections, however, the assumptions are that a particular risk distribution choice has been made and how the law deals with that choice. In this section, the emphasis is upon the advantages and disadvantages of the three systems.

Emphasis in this section will be not upon the first system but upon the second and the third. In construction projects of any size, gathering subsurface information is crucial for design choices, and it will be a rare project where the owner simply allows the contractor to deal with the problem. Contractors are not always geared to gather the information and make these choices, and any attempt to use the first system runs a great risk of construction failure.

With the important exceptions noted in Section 25.06, most private and many state and local public owners make information available to the bidders but use contractual disclaimers in an attempt to relieve themselves from any responsibility for the accuracy of the information.

Those who adopt the disclaimer system recognize the possibility or even the likelihood that the contractor will encounter physical conditions different from those represented or anticipated. They expect that the contractor will calculate the risks and include contingencies in the bid price, which takes this risk into account. Those who prefer this system also think that it will encourage contractors to be more careful in evaluating the information furnished, and in making site inspections if the contractor will have to pay for the added expenses for corrective work. Advocates of this system want the pricing of such uncertainty to go into the contract at the front end rather than at the back end through claims. To sum up, the principal justification for the disclaimer system is the need many public entities and many private owners have to know at the outset what the project will cost.

But the disclaimer system has drawbacks. First, as shall be seen in Section 25.05, the disclaimer does not always work. A study by the American Society of Civil Engineers concluded that despite owners seeking to place these risks on the contractor through disclaimer language, contractors will make claims that in the end will be compensated by the owner. The claims will be based upon other contract clauses or different legal theories.[20]

Second, if the disclaimer system does not protect the geotechnical engineer who furnishes the information, the contractor may be able to make a successful claim against the geotechnical engineer based upon negligence. This will frustrate the risk allocation system by giving the contractor a windfall if it has priced the contingency in its contract price and can still recover against the engineer. This uncertainty may also force the engineer to protect himself by demanding higher compensation or indemnification.

Third, the ruthlessly competitive construction market may mean that contractors do not include contingencies for subsurface conditions into their bid prices. While this may appear to be beneficial to the owner, the contractor who loses money is likely to make a claim and may win it; in any event, all the parties will suffer extensive claims overhead.

---

[20]*Managing Unforeseen Site Conditions,* 113 J. of Constr. Eng'g & Mgmt., Vol. 113, No. 2 (June 1987), quoted by Jones, *The U.S. Perspective on Procedures for Subsurface Ground Conditions Claims,* 7 Int'l Constr.L.Rev., Pt. 2, 169 (1990).

Despite these undoubted disadvantages, the need for certainty, particularly when budgets are fixed and tight, motivates many owners to seek to use the disclaimer system.

The third system—the differing site conditions (DSC) method, assures the contractor that it can bid on what it believes it is likely to encounter and that it will receive an equitable adjustment if actual conditions do not turn out that way. This system is supposed to generate lower bid prices because the contractor need not attempt the often difficult task of providing a contingency in its pricing for the subsurface uncertainties and need not incur extensive costs connected with its own testing if it is not certain that the information furnished by the owner under the disclaimer system is accurate.[21]

Obviously, the DSC method has disadvantages. The most important is the uncertainty of the ultimate contract price. Over the long run—and this is the point made by many owners who are repeat players—the prices will be lower. Also, the cost of administering the DSC system can be formidable, and a contractor who is not convinced that it will be treated fairly may include a contingency price anyway. The determination of the amount of the equitable adjustment can generate difficult problems and expensive claims overhead.

Another argument sometimes made against using the DSC system is the uncertainty as to whether it actually induces lower bids. Some contend that bids are based on workload, the desirability of keeping the work force together, and the prospects of a well-administered construction project. The presence of the DSC system may be a relatively insignificant item in determining the actual bid price. Also, the bid price may not be reduced unless the contractor is confident that obtaining an equitable adjustment will not be extraordinarily difficult or administratively expensive.

Finally, another factor in favor of using the DSC system is that it should provide an incentive for the owner to furnish the best subsurface information by hiring competent soil testers and giving them sufficient time and compensation to develop accurate information.

It can be seen that no system is perfect, and those who must plan the risk allocation face difficult choices. There is no doubt that there is an increased tendency, as demonstrated in Section 25.06, to use the DSC system. But it is likely that in the greater number of construction projects, the disclaimer system will still be employed.

SECTION 25.05 Disclaimers—Putting Risk on Contractor

Owners use a variety of techniques to relieve themselves of responsibility for the accuracy of subsurface information they have obtained and made available. Although emphasis is on making the information available and then seeking to disclaim responsibility for it, attention should preliminarily be directed to the possibility that the owner will choose simply not to make this information available. Owners may seek to do this if they believe that no sure technique exists that will relieve them of the responsibility for the information's accuracy. Owners are increasingly expected to disclose information that would be valuable to the contractor if it is likely that the contractor would not be able to discover this information for itself. Yet an opinion of the Pennsylvania Commonwealth Court[22] held that the owner's failure to include information on subsurface conditions was not constructive fraud. Even more, an opinion of the Post Office Board of Contract Appeals held that the public entity was not required to make a subsurface investigation if it did not wish to do so.[23]

In these cases, other factors made it difficult for the contractor to recover. For example, in the first case, the court noted that the contractor was experienced and had access to public documents describing the mine water levels in the project area. In the second case, the court noted that the contractor was aware of the existence of subsurface rock, that before bidding, the contractor anticipated that some rock would be encountered in excavation, and that the amount of rock actually encountered was not unusual for the area. These factors might have been sufficient to enforce any dis-

---

[21]See *Foster Constr. C.A. & Williams Bros. Co. v. United States*, 193 Ct.Cl. 587, 435 F.2d 873, 887 (1970), for an articulation for the reasons for a differing site conditions clause.

---

[22]*Tri-County Excavating, Inc. v. Borough of Kingston*, 46 Pa.Commw. 315, 407 A.2d 462 (1979).

[23]*Wyman Constr. Inc.*, PSBCA. 611, 80-1 BCA ¶ 14,215.

claimer were the information given and responsibility for it disclaimed.

More commonly, the owner needs the information, commissions it, and makes it available to the contractor. A number of techniques are employed by some owners to avoid responsibility for the accuracy of the information. Sometimes owners place the responsibility on the contractor to check the site and make its own tests. As seen in *United States v. Spearin* (reproduced in Section 23.05(E)), these generalized disclaimers are not always successful.

Some owners use a different approach. They do not include subsurface data in the information given bidders but state where such information can be inspected by the contractor.

Another approach is to give the data but state that it is information and is not intended to be part of the contract. This is another way of stating that the contractor cannot rely on the accuracy of the information, with the hope being that the owner is shielded from any claim based on misrepresentation or warranty.[24] Attempts to use these techniques generate varied outcomes, a reflection of the difficulties courts have in deciding whether a party can use the contract to shift a risk to the other party.

Courts should proceed cautiously before upsetting contractual disclaimer systems, particularly in public contracts. Judicial ambivalence can be demonstrated by comparing *Wiechmann Engineers v. State Dep't of Public Works*[25] with *Stenerson v. City of Kalispell*.[26] The *Wiechmann* case involved a road construction contract. The contractor made a claim for additional compensation based on its having uncovered subsurface boulders along the length of the project that drastically increased its expenses. The state caused test holes to be drilled. It made the information developed available in the office of

the district engineer but did not include the information in the bid package. The information noted that in thirteen of the eighteen tests, boulderous conditions were found. The contract required bidders to carefully inspect the site, the plans, and the specifications and stated that the bidder was representing that it was aware of the conditions to be encountered. The contract also indicated that the information that was available was made only for the purpose of design and study and was not a part of the contract. The state disclaimed all responsibility for the sufficiency or accuracy of the investigations made. The risk of subsurface conditions that differed materially from those represented or expected was to be borne by the contractor.

The court noted that there was no deliberate concealment by the state or no misleading information furnished. It distinguished an earlier case granting additional compensation to the contractor by noting that in the case before the court, there were no representations as to subsurface conditions, the court relying on the statement that the information was not to be the basis of the bid. Also, the contractor had as much access to information as did the state. Finally, the disclaimers clearly put this responsibility on the contractor, the court noting that the boulders were actually visible in the job area—something known to the contractor before the bid was submitted.

The court stated that fairness to bidders required that the successful bidder should not recover additional compensation when it was imprudent or careless in its investigation, that a careless contractor cannot convert its carelessness into a claim of fraudulent concealment, and that the

> State of California is not the guardian of every contractor who seeks to perform services for the public and at public expense. Such a concept would be grossly unfair to the prudent and careful contractor who is frequently underbid by a careless competitor.[27]

This should be compared with the *Stenerson* case, which involved a contract for rough grading and other work on a municipal golf course. Included in the bid materials was a booklet of specifications with a map entitled "Rough grading plans" with ground elevations on which the con-

---

[24]*City of Columbia v. Paul N. Howard Co.*, 707 F.2d 338 (8th Cir.1983). This case is discussed in Section 25.06(B).

[25]31 Cal.App.3d 741, 107 Cal.Rptr. 529 (1973). See also *Cook v. Oklahoma Bd. of Public Affairs*, 736 P.2d 140 (Okla.1987). The contractor was denied recovery even where the owner knew of the condition, which it did not mention in bid information. A reasonably prudent bidder would have known of the condition.

[26]629 P.2d 773 (Mont.1981). See also *Jack B. Parson Constr. Co. v. State*, 725 P.2d 614 (Utah 1986) (bidder could rely reasonably on specific misrepresentation despite disclaimer).

[27]107 Cal.Rptr. at 536.

tractor relied. The work was awarded on a lump-sum basis, with a unit price for additional work. The contractor made a claim based on its having moved some 27,000 cubic yards of earth more than it expected when it prepared its bid. The city refused to pay based on language in the contract indicating that calculations on the grading plan were approximate and that the contractors were to make their own determination as to the amount of top soil and grading work to be done.

The trial court treated the statement of estimated quantities as a representation that was reasonably relied on by the contractor in making its bid. The contractor testified that it relied on this information, that it made on-site inspections and saw no discrepancies, and that those in its business rely on the expertise of the people who make such estimates. It testified that it had no reason to believe the figures were inaccurate and had no way of checking the accuracy of the figures without surveying. The contractor testified that it had bid on jobs some 200 times and always used the plans and specifications as the basis for its bid. It also indicated that in the business of earth moving it was not customary to double-check the information by resurveying.

The court held that the trial court judge's determination that the contractors had justifiably relied on the erroneous information was reasonable and upheld a judgment for the contractor.[28]

The *Wiechmann* and *Stenerson* decisions can be reconciled. The contractor in the *Wiechmann* case had the opportunity to examine the information but did not. It also observed the site and knew that boulders were present. It would be difficult to justify a claim for additional expenses under these conditions. The public entity made available all it knew, and the information was readily accessible to the contractor. The contractor lost its claim because it did not *rely* on the information. This invoked the common law rule that places the risk of performance that proves to be more expensive than anticipated on the contractor.

In the *Stenerson* case, the court concluded that the public knew or should have known that the information would be relied on and that any on-site inspection would not have revealed any information that was not on the plans. The *Wiechmann* case condemned the contractor because it did not perform as the court felt contractors *should* perform. The court in the *Stenerson* decision concluded that the contractors did as other contractors would have done. They relied on the expertise of the geotechnical engineers.

The fact remains that the *Wiechmann* decision reflects the court's unwillingness to deprive the public entity of the opportunity of disclaiming responsibility and placing the risk on the contractor. The *Stenerson* decision, on the other hand, showed an unwillingness to allow the public entity to exonerate itself when it knew that the contractor did rely on the information, with reliance in the long run being beneficial to the public entity because it generated a lower price.[29]

Although all judges are likely to give great autonomy to contracting parties to apportion risks as they choose, they often differ when one party has dictated a risk allocation plan. To be sure, a contractor can adjust its bid price to take into account the risks that it is being asked to bear.

The law operates through judges who may view the disclaimer process differently. Those judges who are unwilling to allow owners to furnish information and disclaim responsibility for its accuracy (the issue here is not fraud or negligent misrepresentation) seem more influenced by the apparent unfairness of allowing the owner to place the risk on a party often less in a position to distribute that risk or shift it. This is even more persuasive if the owner knows that the information will be relied on and derives a benefit through lower bid prices when the contractor does not make its own tests. They are also influenced by the belief that the contracting parties owe each other the duties of good faith and fair dealing. They believe autonomy should not be used as a means of placing subsurface risks on the contractor when the contractor did as others and relied. Judges favoring broad autonomy are less influenced by these con-

---

[28]But see *Costanza Constr. Corp. v. City of Rochester*, 147 A.D.2d 929, 537 N.Y.S.2d 394 (1989), in which the contractor under a unit-price contract was held to have taken the risk of extremely different quantities of concrete because of disclaimers of responsibility as to the accuracy of the estimate. The case is discussed in Section 17.02(D).

---

[29]See Parvin & Araps, *Highway Construction Claims*, 12 Pub.Cont.L.J. 255 (1982).

siderations. They seek only to determine *how* the risk was apportioned by the contract.

SECTION 25.06 Contractual Protection to Contractor

A. Public Contracts: The Federal Approach

In the absence of contract protection, misrepresentation, or breach of warranty, the contractor must bear the risk of unforeseen site and subsurface conditions. If disclaimer language is chosen carefully and the bidder given a reasonable opportunity to observe and test, it is likely that the disclaimer will be effective. Even where this is not so, the owner, at best, will be held responsible for factual representations, not opinions or inferences.

The owner may find that this risk shift to the contractor is not in the owner's best interest. In such a case, the owner may choose to accept the risk of unforeseen site and subsurface conditions. This is accomplished in federal construction through what was formerly called the changed conditions clause and is now known as the differing site conditions (DSC) clause.

The differing site conditions system plays a significant part in federal procurement. In 1982, an EPA study identified DSC claims as representing 50% of the dollar amount in contract modifications.[30] A 1984 study showed about 34% of all federal contract modifications were based on the DSC.[31]

The federal DSC clause states:

> (a) The Contractor shall promptly, and before the conditions are disturbed, give a written notice to the Contracting Officer of (1) subsurface or latent physical conditions at the site which differ materially from those indicated in this contract, or (2) unknown physical conditions at the site, of an unusual nature, which differ materially from those ordinarily encountered and generally recognized as inherent in work of the character provided for in the contract.
>
> (b) The Contracting Officer shall investigate the site conditions promptly after receiving the notice. If the conditions do materially so differ and cause an in-

crease or decrease in the Contractor's cost of, or the time required for, performing any part of the work under this contract, whether or not changed as a result of the conditions, an equitable adjustment shall be made under this clause and the contract modified in writing accordingly.

> (c) No request by the Contractor for an equitable adjustment to the contract under this clause shall be allowed, unless the Contractor has given the written notice required; provided, that the time prescribed in (a) above for giving written notice may be extended by the Contracting Officer.
>
> (d) No request by the Contractor for an equitable adjustment to the contract for differing site conditions shall be allowed if made after final payment under this contract.[32]

A DSC (differing site condition) creates two methods of obtaining an equitable adjustment, conditions different from those represented (Type I) and unanticipated conditions (Type II).

A Type I DSC requires that there be an actual physical (subsurface or latent) condition encountered at the site that differs materially from the conditions represented in the contract documents. Initially, the contract documents must be examined. In federal procurement, such documents are broadly defined and include the invitation for bids, drawings, specifications, soil-boring data, representations of the type of work to be done, and the geographical area of construction. Included among the contract documents are not only specific information as to the subsurface characteristics but also the description of the nature of the project and the physical conditions that relate to the work. Physical conditions include the details of excavation or construction work, as they may be representations of the physical conditions. The contractor can draw reasonable inferences as to the physical conditions, and there need not be express representations of them. The important thing is whether the representations are sufficient to provide a reasonably prudent contractor with a sufficient basis upon which to rely when the contractor prepared its bid.

Economic, governmental, or political conditions do not constitute physical conditions. Moreover, natural conditions, such as abnormally heavy rainfall, rough seas, strong winds, hurricanes, severe

---

[30]Report of Internal and Management Audit, Environmental Protection Agency, Office of the Inspector General.
[31]Contract and Change Order Summary Reports, Naval Facilities Engineering Command.

[32]48 C.F.R. § 52.236-2 (1992).

temperatures, frozen ground, or flooding by themselves do not constitute a differing site condition. However, an abnormal weather condition may interact with a latent physical condition in such a way as to create a DSC. Physical conditions can also include artificial or manufactured conditions, such as underground electric power lines, sewer lines, or gas pipelines that are unexpectedly encountered at the site.

The condition must be subsurface or latent. Most claims involve encountering something different than anticipated in the subsurface materials or structure—usually physical or mechanical properties, behavioral characteristics,[33] quantities, etc. But a DSC need not always be a subsurface condition; it can also be a latent physical condition at the site, such as undisclosed concrete piles, the thickness of a concrete wall, or the height of ceilings above a suspended ceiling.

The DSC must be encountered at the site, which usually is the place where construction will be undertaken but can include off-site pits from which soil is to be borrowed for fill or disposal sites if the contractor is directed to obtain or dispose of materials off site.

The physical conditions encountered must be materially different from those indicated. Was there a substantial variance from what a reasonably prudent contractor would have expected to encounter, based upon its review of the contract documents and what it would have encountered had it complied with the site investigation clause in the contract? Even without such a clause, a contractor would be expected to make at least a minimal inspection to familiarize itself with the site. Typically, site inspection clauses obligate the contractor to inspect and familiarize itself with the conditions at the site, but the contractor is obligated only to discover conditions apparent through a reasonable investigation. It is not expected to perform burdensome, extensive, or detailed tests or analysis. The contractor can rely on information received from the owner, particularly when it does not have adequate time or opportunity to conduct a thorough investigation. What other bidders did or what they thought they would encounter will bear heavily on

the question of whether the site investigation was properly performed.

A Type II DSC requires a variance between the site condition actually encountered and that which would be reasonably expected at the time the contract was made. Expectations look at the information furnished to the contractor or information acquired from other sources. Were the conditions encountered common, usual, and customary for that geographical area? The contractor is expected to be aware of conditions under which the work will be performed. For example, a contractor who is working in winter in a mountainous area where snow is common should expect to encounter wet conditions.

To obtain an equitable adjustment, the contractor must notify the contracting officer promptly in writing. This generates the inevitable difficulty over substantial compliance and waiver.[34] Courts seem more willing to waive strict notice requirements if it appears that the owner knew that the contractor had encountered unforeseen subsurface conditions and that it was likely that a claim would be made.[35] In *Kenny Construction Co. v. Metropolitan Sanitary District of Greater Chicago*,[36] the contractor notified the engineer, who told the contractor to go ahead and they would figure costs later. Despite the absence of a written order by the engineer, the court granted recovery, because the contractor relied on the engineer's statement and the public entity was estopped to assert the condition.

In *Centex Construction Co.*,[37] the written notice requirement was waived because the project representative was aware of the oil dump initially at the prebid site investigation and again when the con-

---

[33]See Smyth, *Behavioral Differing Site Conditions Claims*, 11 Constr.Litig.Rep. 158 (1990) (a tunneling contract).

[34]*Moorhead Constr. Co. v. City of Grand Forks*, 508 F.2d 1008 (8th Cir.1975); *Metro. Paving Co. v. City of Aurora, Colorado*, 449 F.2d 177 (10th Cir.1971) (waiver: owner denied claim as condition anticipatable); *Acchione & Canuso, Inc. v. Commonwealth Dep't of Transp.*, 501 Pa. 337, 461 A.2d 765 (1983).
[35]*Brinderson Corp. v. Hampton Roads Sanitation Dist.*, 825 F.2d 41 (4th Cir.1987) (actual or constructive notice of condition and an opportunity to investigate waives formal notice); *Roger J. Au & Son, Inc. v. Northeast Ohio Regional Sewer Dist.*, 29 Ohio App.3d 284, 504 N.E.2d 1209 (1986) (claim depends on whether owner already knew and was not prejudiced).
[36]52 Ill.2d 187, 288 N.E.2d 1 (1972). A subsequent opinion is found in 56 Ill.2d 516, 309 N.E.2d 221 (1974).
[37]ASBCA 26830 through 26849, 83-1 BCA ¶ 16,525.

tractor orally advised him of the condition when work commenced. The Board of Appeals also noted that failure to provide written notice did not prejudice the government's position. But as in other cases dealing with notices, outcomes will vary.

References have been made to cases outside the federal procurement system. The use of a DSC is becoming more common in public contracts made by entities other than the federal government. This can create difficulties. For example, one case involved a clause that incorporated a Type I DSC but not a Type II DSC.[38] Similarly, confusion can develop because a local public entity may be compelled to use a DSC because the project is being partially funded with federal money granted by the Environmental Protection Agency (EPA). But the local entity making the contract may have had a practice of using disclaimers, and these disclaimers are sometimes still included in the contract documents. One case faced with this question concluded that the DSC provision took precedence.[39]

Similarly, in 1989, DSC clauses were required to be incorporated into road-building contracts for federally aided highway programs under which roads are constructed by private contractors under contract with state highway agencies. However, under the applicable regulations, the DSC provisions need not be included if the state law requires a different method of risk distribution.[40] Also, statutes can mandate slight deviations from classic federal DSCs. For example, in 1989, California enacted a statute dealing with public works contracts of local entities that involve digging trenches or other excavations that extend deeper than four feet below the surface. Such contracts are required to include a truncated Type I DSC and a Type II DSC clause.[41] The California statute states that conditions that involve hazardous waste justify a change order.

If the necessary substantive requirements are met under the DSC, the contractor or the owner receives an equitable adjustment. The legal systems that have had to deal with such claims, particularly those in federal procurement, have had to struggle with a variety of measurement techniques—all of which tend to employ approximations rather than precise measurements. Using the contractor's claim as an illustration, the best proof is item-by-item proof of the difference between the cost it would have incurred had it not encountered the differing site condition and what it actually incurred. However, often it is impossible to establish those additional costs with anything resembling precision, and the contractor is likely to seek to employ a crude formula such as total cost, modified total cost, or jury verdict. These are discussed in greater detail in Section 27.02(F).

Sometimes contractors claim that encountering the subsurface conditions required that they performed the work out of sequence or that it delayed the completion of the project. In such cases, contractors sometimes seek additional compensation in addition to time extensions. As shall be seen in Section 26.10(A), claims for additional compensation may be barred by a "no damage" (no pay for delay) clause. Although the law has had considerable difficulty when such clauses are attacked, the current trend seems to be toward increased scope for such exculpatory clauses.

Finally, the law has struggled with contractor claims that it is also entitled to additional overhead and profit.[42]

Suggestions have been made for contract provisions that will deal in some way with the amount of the equitable adjustment. One commentator suggests the possibility of a contractual provision that requires the contractor to include a per-diem calculation of delay damages.[43] The contractor would include within the stipulated amount field overhead, home office overhead, idle labor and equipment, and any loss of efficiency or impact on later performance. These figures would be in-

[38]*Metro. Sewerage Comm'n v. R. W. Constr., Inc.*, 72 Wis.2d 365, 241 N.W.2d 37 (1976).

[39]*Fattore Co. v. Metro Sewerage Comm'n*, 505 F.2d 1 (7th Cir.1974). See also *Andrew Catapano Co., Inc. v. City of New York*, 116 Misc.2d 163, 455 N.Y.S.2d 145 (S.Ct.1980); *Baltimore Contractors Inc. v. United States*, 12 Cl.Ct.328 (1987).

[40]23 C.F.R. § 635.109 (1992).

[41]West Ann.Cal.Pub.Cont.Code § 7104.

[42]*Kenny Constr. Co. v. Metro. Sanitary Dist. of Greater Chicago*, 52 Ill.2d 187, 288 N.E.2d 1 (1972) (allowed as part of changes process); *Fattore Co. v. Metro. Sewerage Comm'n*, supra note 39 (profit on unperformed work denied).

[43]Ashcraft, *Avoiding and Managing Risk of Differing Site Conditions*, a chapter in DIFFERING SITE CONDITIONS CLAIMS (1992).

cluded in the bid. Determining the low bidder would also involve an assumed number of delay days appropriate to the project multiplied by the bidder's per-diem daily delay price, which would be added to the bid.

Another method is to include a provision under which the contractor agrees that certain items would not be included in any DSC claim and that any dispute resolution system would not include in its award payment for profit, labor and efficiencies, cost of vital equipment, or project and home office overhead.

It is likely that these attempts to structure remedies for a DSC could be used only in less regulated public contracts or in private contracts.

Does the right to receive an equitable adjustment preclude the contractor from terminating its obligation if the condition discovered would have granted it that power? In addition to taking this potentially disastrous risk away from the contractor, the clause is designed to keep the job going. It should substitute for any common law rights that the contractor might be accorded.[44]

B. AIA Approach: Concealed Conditions

In 1937, the AIA adopted a similar approach to unknown subsurface conditions. AIA Doc. A201, ¶ 4.3.6, now uses language identical to the federal DSC clause, with its Type I and II substantive bases for additional compensation.

Whether the occasional owner who builds once in a lifetime is better off with such a provision is debatable. Similarly, it is questionable whether lenders will want a provision that can substantially expand the ultimate cost of the project.

In AIA Doc. A201, ¶ 1.2.2, the contractor represents that it has visited the site and familiarized itself with local conditions under which the work is to be performed. Conditions that could have been observed at a normal site visit cannot justify additional compensation.

This raises an issue often ignored. Suppose the contractor *should* have seen certain site conditions or actually *saw* site conditions that varied from the conditions *indicated* by the contract documents. If relief should be accorded only when the conditions are concealed, a contractor who knows or should know what it will encounter because of access to the site should not receive a Type I equitable adjustment under ¶ 4.3.6.

In 1987, the AIA substantially expanded the administrative features of ¶ 4.3.6. The observing party must give a notice "to the other party promptly before conditions are disturbed and in no event later than 21 days after first observance of the conditions."

Although many notices required by AIA documents can be given to the architect, this notice must go to the owner if the contractor is the observing party. The architect can be considered the owner's agent for this purpose, and a notice to the architect may be sufficient.[45] The law has been hesitant to give literal effect to notice provisions.[46]

A201 is honeycombed with notice provisions. For example, if the contractor wishes to receive additional compensation, it must give a written notice under ¶ 4.3.7. If it wishes additional time, it must give a written notice under ¶ 4.3.8.1. If it has suffered injury to property, it must give a written notice under ¶ 4.3.9. Each of the three previously mentioned sections requires a written notice. Yet ¶ 4.3.6 simply requires that notice be given. Certainly, it is safer for a contractor discovering such conditions to give notices under ¶¶ 4.3.6, 4.3.7, 4.3.8.1, and 4.3.9. However, if an oral notice were given in accordance with ¶ 4.3.6, failure to give the other notices might be waived if the owner knew from the notice given that claims for additional compensation, additional time, and even damage to property are likely to be made.

Returning to the substantive basis for a claim, the Type I claim under ¶ 4.3.6 requires conditions be encountered "which differ materially from those indicated in the Contract Documents." AIA Doc. A201, ¶¶ 1.1.1, excludes bidding requirements from the contract documents. (A101, ¶ 9.1.7, recognizes that some may wish to include them.)

---

[44]*North Harris County Junior College Dist. v. Fleetwood Constr. Co.*, 604 S.W.2d 247 (Tex.Ct.App.1980), appears to allow termination. Actually, the court held that a refusal by the owner to acknowledge that there was a changed condition could have justified the contractor's terminating. AIA Doc. A201, ¶ 13.4.1, makes contractual remedies nonexclusive.

[45]See Section 17.05(B).
[46]See Section 21.04(H).

Because specifications are contract documents, the contractor will not face any obstacles if the information is included in the specifications.[47] But suppose the information is included in the instructions to bidders or is simply made available at the geotechnical engineer's office. A literal interpretation of ¶¶ 1.1.1 and 4.3.6 could bar the contractor's claim.

However, two contentions could be made that would overcome this unfortunate conclusion. First, ¶ 1.1.1 speaks of "bidding requirements." This could be interpreted to include only that material that deals with the rules for the competitive bidding process. To be sure, ¶ 1.1.1 does give illustrations of bidding requirements and includes the "advertisement or invitation to bid, instructions to Bidders," and other materials. Yet these illustrations can be interpreted as excluding only the materials found in them that deal with the rules for the competitive bid. This contention can be supported by ¶ 1.1.7, which defines the Project Manual as including the bidding requirements. This would indicate that information that continued to have relevance after award, such as subsurface information, would continue to have legal significance after the contract is awarded.

Second, it can be contended that inclusion of a DSC, yet giving the contractor relevant information in something *other* than a contract document, would be at least a violation of the covenant of good faith and fair dealing and possibly fraud. Put another way, assuring the contractor that it will receive an equitable adjustment if it encounters something unexpected and then refusing to grant an equitable adjustment in a Type I claim, the most common basis for equitable adjustments, would violate the obligation of good faith and fair dealing.[48]

Labeling the subsurface data as solely for information or in some way indicating that they are not part of the contract is an indirect and often uncertain method of relieving the owner from responsibility. In *City of Columbia v. Paul N. Howard Co.*,[49] the contract involved the construction of a public

sewer improvement and contained a DSC clause, with the contractor being entitled to reimbursement of cost resulting from conditions differing materially from those indicated in the contract.

The contractor received a document entitled "Specifications and Documents" with an appendix that contained test boring logs. Its claim was based on subsurface conditions differing from those indicated in the logs. The city claimed the logs were not part of the contract and pointed to a disclaimer on the cover page of the test boring logs that stated:

> These reports are for reference only and are not part of the contract documents.[50]

The trial court ruled for the city.

The Court of Appeals looked at other parts of the contract and not simply at the disclaimer on the test boring logs. It noted that the contract documents were defined as items listed in the table of contents. The logs were listed in the table of contents under "Appendix." In addition, a supplementary condition stated:

> Test hole information represents subsurface characteristics to the extent indicated, and only for the point location of the test hole. Each bidder shall make his own interpretation of the character and condition of the materials which will be encountered between test hole locations.[51]

This supported, according to the court, the city's argument that it was not responsible for unanticipated conditions between the *test holes*, but it also supported the contractor's argument that it could rely on logs at least as to the *actual* conditions at the location of the test holes.

The court stated that the trial court should have reconciled *all* these provisions and concluded that the various parts could be interpreted to mean that the contractor could rely on the logs only for their accuracy and not for conditions between the test hole locations.

The court conceded the city's argument that the appendix was not an item listed in the table of contents and therefore was not part of the contract. But the court held that the equitable adjustment depended not on the logs being part of the contract but solely on conditions differing materially from

---

[47]*North Harris County Junior College Dist. v. Fleetwood Constr. Co.*, supra note 44 (reference to information in specifications).
[48]See Section 19.02(D).
[49]707 F.2d 338 (8th Cir.1983). See also *Jack B. Parsons Constr. Co. v. State*, supra note 26.

[50]707 F.2d at 339. (Note the conflict with the DSC.)
[51]Id. at 339–340.

those indicated in the contract. The court concluded by noting that though the logs were not part of the contract, they were *indicated* in the supplementary conditions, at least enough to make the differing site conditions provision available.

A201 creates a complicated system to determine the existence and extent of the equitable adjustment. The architect investigates. If he determines that the contractor is entitled to an equitable adjustment, he so recommends. If he determines the contractor is *not* entitled to an equitable adjustment, he notifies the owner and contractor in writing giving his reasons. If either party objects to a determination that no equitable adjustment be granted, that party can invoke the disputes process under ¶ 4.3, and again the matter is submitted to the architect. Although the language is by no means clear, it appears that if the architect recommends an equitable adjustment, and if either party is dissatisfied, that party can again have the matter resubmitted to the architect under ¶ 4.3, the disputes clause. If either party is dissatisfied with the architect's decision under the disputes clause, that party can demand arbitration under ¶ 4.5. A201 gives no guidance as to the amount of the equitable adjustment, but presumably the vast federal jurisprudence can be looked to, especially in the light of the identical language used.

Some suggestions have been made for modification of the DSC in A201. The modifications relate to inclusion of language dealing with asbestos and other hazardous substances, which one commentator suggests be an automatic DSC.[52]

As noted in Section 25.06(A), the delay that may be caused by the discovery and treatment of the DSC may be compensated by the use of a liquidated damages clause. Some have suggested language limiting the remedy to which the contractor is entitled.

C. EJCDC

The Engineers Joint Contracts Documents Committee (EJCDC) employs a more complex methodology in its No. 1910-8. It divides coverage into subsurface and underground facilities physical

conditions. Section 4.2.1 states that information will be given in the Supplementary Conditions. Under ¶ 4.2.2, the contractor can rely on "the general accuracy of the 'technical' data" even though the data are not contract documents. But ¶ 4.2.2.1 states that the contractor cannot rely on the completeness of the information for purposes of execution or safety nor under ¶ 4.2.2 upon "other data, interpretations, opinions and information."

Paragraph 4.2.3 precludes reliance on its conclusions as to interpretation or conclusions from any data, including the technical data. This is a sensible line of demarcation and fits the DSC system.

To claim an equitable adjustment, a written notice must be given under ¶ 4.2.3 describing the technical data the contractor claims it intends to rely upon and why the data are materially inaccurate or stating that the condition differs materially from that shown or indicated in the contract documents. The Type II language of ¶ 4.2.3 is similar to that in the federal procurement system and the AIA.

After a claim is made, the engineer can review and test. He advises the owner of his findings and conclusions. If he concludes a change is required, he issues a work change directive. But this does not guarantee price or time changes in the contract. Under ¶ 4.2.6, these are made only to the extent that the condition "causes an increase or decrease in the cost or time." The inability of the parties to agree on the amount of adjustment in price or time is a dispute that falls into the disputes articles, Articles 11 and 12, under which the engineer plays roughly the same role as does the architect under A201.

Underground facilities are treated differently. Information is given by the owners of these facilities. Under ¶ 4.3.1.2, neither the owner nor the engineer is responsible for its accuracy or completeness. The contractor must review and check such information. But if uncovering reveals an underground facility not shown or indicated or reasonably expected, Paragraph 4.3.2 gives the contractor an increase in price (decreases are also allowed) "to the extent that they are attributable to the existence of any Underground facility that was not shown or indicated that could not have been reasonably expected." Again claims are handled under Articles 11 and 12.

EJCDC documents are more detailed than those of the AIA, largely because of the type of engineering projects for which they are designed. But

[52]O'Leary, *Negotiating the General Construction Contract* in DESIGN AND CONSTRUCTION CONTRACTS: NEW FORMS, NEW REALITIES, 249, 279 (1987).

their classifications and detail would be useful in many building contracts.

D. FIDIC Approach

While emphasis in this treatise is on domestic construction contracts and not on those made in an international context, occasionally it is useful to look at a particular problem from the vantage point of a standardized contract used in international transactions. The most commonly used standard contract is the one generally referred to as the FIDIC contract, published by the Federation Internationale des Ingenieurs-Conseil (International Federation of Consulting Engineers). The fourth edition of the FIDIC published in 1987 has a relatively simple provision for dealing with unexpected subsurface conditions. Captioned "Adverse Physical Obstructions or Conditions," ¶ 12.2 states that if the contractor encounters

> physical obstructions or physical conditions, other than climatic conditions on the Site, which obstructions or conditions were, in his opinion, not foreseeable by an experienced contractor, the Contractor shall forthwith give notice thereof to the Engineer, with a copy to the Employer. On receipt of such notice, the Engineer shall, if in his opinion such obstructions or conditions could not have been reasonably foreseen by an experienced contractor, after due consultation with Employer and Contractor . . . grant a time extension and price adjustment.

It is not likely that such a provision would be attractive to most American owners because of both the imprecision of the language and the power given to the engineer. However, where there is a good relationship between the owner and the contractor and where both have confidence in the technical ability and sense of fairness of the design professional, such a system has much to recommend it.

SECTION 25.07 Some Advice to Courts

Undoubtedly, arguments can be made for a disclaimer system or a system that uses a DSC clause. Courts should effectuate *whichever* choice has been made. If it appears clear that the contractor has, happily or not, accepted the risk of unforeseen subsurface conditions, the court should effectuate that risk allocation and not seek to destroy it by tortured interpretation generated by a belief that it is unconscionable for the owner to place these risks on the contractor. Only if it is clear that the contractor was not made aware of the risk or if the imposition of the risk violates the obligation of good faith and fair dealing should the court entertain a result that does violence to the loss distribution selected and expressed in the contract.[53]

Similarly, courts should *not* allow a contractor to maintain a tort action against a geotechnical engineer if it is clear that the contractor was expected to assume this risk. Allowing the contractor to transfer this risk to the geotechnical engineer through a tort action frustrates the efficiency of the system selected.

---

[53]An illustration would be deliberately misleading the contractor or not giving the contractor sufficient time to examine the data, visit the site, or take its own tests. Whether this would encompass using a disclaimer system knowing that the contractor does rely is unclear. See *Stenerson v. City of Kalispell*, discussed in Section 25.05.

PROBLEMS

1. The state decided to pave a portion of an interstate route. It furnished all prospective bidders with a bound set of documents. The documents were to be used as a basis for the bids. Sheet 2B showed the location and boundaries of two borrow pits labeled "prospect 1" and "prospect 2." The sheet described the contents as "limestone ledge rock" and reported tests run on the materials in the pits to determine whether they met the contract specifications.

Another document, sheet 44, described the prospects in detail and included the statement that each had been used previously for base and surface course on interstate projects. Although bidders could request permission to use other sources, these two prospects were the only economically feasible sources of paving materials in the general area of the project. All of those who bid on the project specified these prospects as their proposed source.

The state standard specifications for road and bridge construction require bidders to visit the prospective materials site and visually inspect it. Two employees of C, the ultimately successful bidder, visited the pits. They concluded that the materials were consistent with the test data and would be suitable.

C submitted the low bid and was awarded the contract. After it crushed some materials from prospect 2 and laid some of the asphalt, inspection revealed that the asphalt did not meet contract specifications because of the poor quality of the crushed rock. C adjusted the crushing process and added blend sand, but to no avail.

Some four months after starting work, C shut down the operation, insisting on changes in the contract, which the state would not make.

C's claim was based on misrepresentation by the state. However, the state contended that sheet 2B was not part of the plans and specifications, as it did not fall within the technical definition of the word *plans* set forth in the state's standard specifications. C contended that sheet 2B as a *practical* matter was part of the plans and specifications. Sheet 2B had been furnished to the bidders to provide foundation materials for the bid and was something on which all bidders were entitled to rely. It must be scrutinized for accuracy.

C also contended that the information on sheets 2B and 44 was taken from tests conducted nine years before its bid and that all the tested materials had been removed from the sites long ago. At the time C made its bid, materials of a similar quality were found in the prospects, only under a substantial layer of overburden.

In addition, C contended that the state had made more recent tests showing that the materials might be marginal, but the tests were not listed in sheet 2B. C also established that the state knew but did not disclose that after the date of the earlier tests, a contractor had experienced difficulty in using prospect 2 and had recommended that that prospect not be relied on in the future.

The state claimed that two of C's employees conducted a physical inspection of the prospect that included photographing all exposed cuts, feeling rock samples for texture, and performing scratch tests. It claimed that the employees believed that this inspection confirmed the accuracy of the information furnished and chose to rely on it rather than conduct additional laboratory work.

The state also contended that a reasonably prudent contractor would have gained sufficient information from a properly conducted site visit, along with lab tests, to anticipate the need to either remove some of the overburden or somehow handle and dispose of it during the crushing operation to avoid contaminating the asphalt.

How would you resolve this dispute?

2. C was awarded a contract to do construction work on the renovation of a fish hatchery for a state conservation agency. The project included reworking existing ponds, building a concrete drain and harvest kettle structures, and installing new fill and drain lines.

C attended neither of the two prebid conferences that were held. At those conferences, subsurface water conditions were discussed. C made no attempt to investigate the subsurface conditions as required by the contract because he felt he could rely on the soil borings furnished by the state with the plans and specifications, and these soil borings did not indicate wet site conditions. C sent his son to inspect the site six days after the prebid conference. The inspection consisted merely of driving through the location in the company of the hatchery manager. Other bidders testified they were able to discover the wet soil condition by looking at the site or observing water in the bottom of existing lagoon cells.

C began work, complained about the wet soil conditions, and refused to continue further performance. C made a claim, stating that the state had misrepresented the moisture content and had failed to disclose the wet soil conditions. He also asserted that the state had failed to disclose the underground aquifer of which it knew. Finally, the soil boring showed no evidence of moisture, and C contended that it was customary to show moisture content by borings.

The contract required that C remove all surface water and groundwater; that he visit the site and fully inform himself of all existing conditions, including physical characteristics; and that he investigate the subsurface conditions and determine for his information the character and proportionate quantity of soils, rocks, and other subsurface materials that may be encountered. C was also to inform himself by independent research of difficulties to be encountered and judge for himself factors affecting the cost of doing the work or time needed. The contract concluded with a provision stating that the owner assumed no responsibility for physical characteristics at the site, and C agreed he would make no claim for additional compensation or time "because of his failure to fully ac-quaint himself with all conditions relating to work."

How would you resolve C's claim for additional compensation and lost profits?

CHAPTER TWENTY-SIX

Time: A Different but Important Dimension

SECTION 26.01 An Overview

The law has *not* looked at time as part of the *basic* construction contract exchange—that is money in exchange for the project. This may be because of the frequency of delayed construction projects.[1] Timely completion depends on proper performance by the many participants as well as optimal conditions for performance, such as weather and anticipated subsurface conditions.

Delayed performance is less likely to be a valid ground for termination.[2] Delayed performance is also less likely to create legal justification for the owner's refusal to pay the promised compensation, with the owner left to the often inadequate damage remedy.[3] Similarly, delayed payment by the owner is less likely to automatically give the contractor a right to stop the work or terminate its performance.[4] Finally, a strong possibility exists that a performance bond that does not expressly speak of delay will not cover delay damages.[5]

Computation of damages is *also* different. If the owner does not pay or the contractor does not build properly, measuring the value of the claim is, relatively speaking, simple. If the owner does not pay, at the very least, contractors are entitled to interest. If the contractor does not build properly, the owner is entitled to cost of correction or diminished value of the project.[6]

Delay creates serious measurement problems. The owner's *basic* measure of recovery for unexcused contractor delay is lost use of the project. The contractor's *basic* measure of recovery for owner-caused delay is added expense. Lost use is difficult to establish in noncommercial projects. Added expense is even more difficult to measure. Because of measurement problems, each contracting party—whether it pictures itself the potential claimant or the party against whom a claim will be made—would like a contractual method to deal with delay claims, either to limit them or to agree in advance on amount. This does not mean that time is not important in construction. The desire to speed up completion to minimize financing costs and accelerate revenue-producing activities was a large factor in leading to techniques intended to generate efficient methods of organizing construction (discussed in detail in Section 17.04). Time is another dimension, however.

SECTION 26.02 Commencement

The very nature of construction sometimes makes complicated what in other contracts is simple. In

---

[1]*Johnson v. Fenestra, Inc.*, 305 F.2d 179 (3d Cir.1962), rejected a contention that delays are calculated risks and not breaches of contract.

[2]*Hartford Elec. Applicators of Thermalux, Inc. v. Alden*, 169 Conn. 177, 363 A.2d 135 (1975) (forty-day delay in furnishing site not material breach).

[3]*Landscape Design & Constr., Inc. v. Harold Thomas Excavating, Inc.*, 604 S.W.2d 374 (Tex.Ct.App.1980).

[4]Refer to Section 22.02(L).

[5]*American Home Assur. v. Larkin General Hosp.*, 593 So.2d 195 (Fla.1992). AIA Doc. 312, the AIA's most recent surety bond, specifically provides for delay damages.

---

[6]See Section 27.03(D).

an ordinary contract, such as an employment contract, or in a long-term contract for the sale of goods, as a rule, the commencement dates are simple to establish. In construction, the date when the performance can begin *in earnest* (procurement can precede site access) is usually the date when site access is given. This cannot always be precisely forecast. The owner may need to obtain permits, easements, and financing before the contractor can be given site access. To avoid responsibility for site access delay and to measure the contractor's time obligation fairly, the commencement of the time commitment period, when measured in days, is often triggered by the contractor's being given access to the site, usually by a Notice To Proceed (NTP). When an NTP is used, the contractor assumes the risk of ordinary delays in site access but not those that go beyond the reasonable expectations of the contracting parties.

Commencement raises other problems. For example, must the contractor actually begin work at the site when an NTP is given? Is the actual commencement date of the NTP a date specified in the NTP or the date the NTP is received? The contract should deal with these issues but frequently does not. What are reasonable expectations of the parties? The custom in the industry? What is fair?

If the NTP does not expressly specify the commencement date, the date should begin when the NTP is received. Any date specified in the NTP should be effective only if it meets the standard of good faith and fair dealing. For example, the notice should not specify a date that the owner knows the contractor cannot meet or that fails to take into account realistic commencement requirements.

If an NTP system is employed, the owner will know when the contractor has commenced performance. However, if it is not employed, the contractor could conceivably commence performance before the commencement date. This could operate to the disadvantage of the lender who wishes to perfect its security interest before work begins in order to have priority over any mechanics' liens. Similarly, early commencement before insurance is in place can create a coverage gap.

With this in mind, in the absence of an NTP, AIA Doc. A101, ¶ 3.1, states the contractor must notify the owner in writing not less than five days before beginning work to allow time for filing security interests. Similarly, AIA Doc. A201, ¶ 8.2.2,

bars the contractor from premature commencement prior to the effective date of insurance.[7]

Suppose there is a delay between bid opening, bid award, and execution of a formal contract. These issues were involved in two instructive cases. *Quin Blair Enterprises v. Julien Construction Co.*[8] involved a competitively bid contract to build a motel. Julien's bid stated that he would complete 240 days from the date the contact was signed. Julien was awarded the contract and was asked to prepare the agreement. He used AIA Doc. A101 and filled in the blank with 240 days, without anything specific as to when the period began. But A101 stated that the contract was made on October 8, 1971.

AIA Doc. A201, ¶ 8.1.2, in effect at that time stated that the date of commencement, if there is no notice to proceed, begins on the date of the agreement "or such other date as established therein."

Blair signed on October 22 and Julien on October 25. Julien could not start until Blair cleared the site. That was not completed until November 18. When did the 240-day period commence? On October 8? On October 25? On November 18?[9] The trial court chose October 8, but the Supreme Court disagreed. Recognizing the ambiguities and seeking to harmonize all the writings, the court concluded that the time began on October 25.

Because no date was specified in A101, the court referred to A201. Because there was no NTP, the commencement date should have been the date of the agreement. But was that date October 8, the date on the agreement, or October 25, when the agreement was signed by the contractor? Because A201 states the date of the agreement, one would think it was the date on the agreement, not the date when it was made. According to the court, each party agreed that the bid was part of the contract. The bid stated that time would begin on *signing* the contract, and although parties *can* designate a retroactive date, there was nothing to indicate that "either party ever intended . . . [a] retroactive date."[10]

---

[7]For a neater and sharper method of determining commencement of contract time, see EJCDC No. 1910-8, ¶ 2.3.
[8]597 P.2d 945 (Wyo.1979).
[9]Julien's failure to submit the required notice of an intention to ask for an extension barred a time extension.
[10]597 P.2d at 951.

Bloomfield Reorganized School District No. R-14, Stoddard County v. Stites[11] did find a retroactive date. The contract was dated August 8, 1955, and provided that the contract was to be substantially completed in 395 calendar days. The architect mailed the contract to the contractor on August 17, 1955. The contractor signed the contract and returned it to the architect, who delivered it to the school superintendent for execution by the school board. School board officials signed the contract, and the superintendent mailed it to the architect on September 14. The latter forwarded the signed copy to the contractor on September 22, *six weeks* after the contract date. Yet the court looked *solely* at the language of the contract, which stated that the agreement had been made on August 8, 1955.

SECTION 26.03 Acceleration

A. Specific: The Changes Clause

One method to accelerate the completion date is a specific directive by the owner that the contractor must complete in a time shorter than that originally agreed. Power to accelerate is usually determined by the changes clause. See Section 21.04(B).

B. Constructive Acceleration

Constructive acceleration originated in federal procurement law. Although the original jurisdictional basis for its development is no longer applicable,[12] it can be applied to all construction contracts.

Constructive acceleration is based on the owner's *unjustified* refusal to grant a time extension. It requires that a cause exist that would have justified a time extension, a request for a time extension, denial of that request, demand (express or implied) that performance be completed on time, and an actual acceleration.[13]

The modern justification for constructive acceleration is that denial of a deserved time extension *can* force additional expenses when work is not performed in the order planned. Suppose, though, that the contractor continues to perform as it would have performed had an extension been granted. This will very likely lead to untimely completion. If the time extension *should* have been granted and it is done so *later* (by agreement, by an arbitrator, or by a court), any attempt by the owner to recover actual or liquidated damages would be unsuccessful. The constructive acceleration doctrine gives the contractor the option of speeding up its performance and recovering any additional expenses it can establish or using the wrongful denial of the time extension as a defense to any claim that might be brought against the contractor by the owner for late completion.

C. Voluntary: Early Completion

Delays are so common in construction that attention is rarely paid to the legal effect of the contractor's completing early or claiming that it would have completed early had it not been delayed by the owner.

Early completion may be desirable to some owners. This can be evidenced by a penalty-bonus clause.[14] On the other hand, early completion may, if unexpected, frustrate owner plans. For example, suppose a contractor building a factory finishes substantially earlier than planned. The owner may have to take possession before it can install its machinery. Early completion can require payments in advance of resource capabilities. It can be as disruptive as late completion.

AIA Doc. A101, Art. 3, requires the contractor to substantially complete the project "not later than" a specified date. This appears to give the contractor the freedom to complete early even if this were to disrupt owner plans.

Construction contracts of any magnitude usually have schedules. It is unlikely that the owner will be greatly surprised by early completion. Yet even awareness during construction that performance will be completed earlier than required may not enable the owner to make the adjustments needed to avoid economic losses.

The obligations of good faith and fair dealing require that a contractor notify an owner if it intends to finish much earlier than expected or when

---

[11]336 S.W.2d 95 (Mo.1960).

[12]Like constructive changes, constructive acceleration was a claim based on the contract and not its breach. This gave jurisdiction to the agency appeals board. Since 1978, a claimant can choose to bring a claim before either an appeals board or the U.S. Court of Federal Claims. See Section 21.04(B).

[13]*M.S.I. Corp.*, GSBCA 2429, 68-2 BCA, ¶ 7377.

[14]See Section 26.09(B).

it appears that this is likely to be the case.[15] If this is done or the owner is aware of that prospect, the contractor should receive additional compensation if the owner interferes with any realistic schedule under which the contractor would have completed earlier than required by the contract.[16]

SECTION 26.04 Completion

Generally, construction contracts provide a method for measuring compliance with the time commitment of the contractor. For example, as discussed in Section 22.04, the AIA has selected substantial completion as the benchmark for determining compliance with the time commitment. Other construction contracts will use actual or final completion.

Completion is an important benchmark in the history of a construction project similar to agreement as to changed work or final payment. The effect of completion often linked to final payment and acceptance was discussed in Section 24.05.

SECTION 26.05 Schedules: Simple and Critical Path Method (CPM)

A project schedule is a formal summary of the planned activities, their sequence, and the time required and the conditions necessary for their performance. A schedule alerts the major participants of the tasks they must accomplish to keep the project on schedule. It can reduce project cost by increasing productivity and efficiency, facilitates monitoring of the project, and can support or disprove delay claims.

The schedule for a *very* simple project, such as the construction of a garage, may simply be starting and completion dates. A somewhat more complex project, such as a residence, may add designated stages of completion, mainly as benchmarks for progress payments. When construction moves upscale, for example, from a simple commercial structure to a nuclear energy plant, the schedule will take on more complex characteristics. Until the past twenty-five years, schedules in anything but the simplest project would be a bar chart, sometimes referred to as a Gantt Chart, after its

inventor. One such bar chart[17] is shown in Figure 26.05a.

Bar charts have deficiencies. They provide no logical relationship between work packages. There are limits to the number of work packages that can be represented in a bar chart—perhaps thirty to fifty—until the level of detail becomes unwieldy. Rates of progress within a package may not be uniform. According to two authors, however:

> . . . the visual clarity of the bar chart makes it a very valuable medium for displaying job schedule information. It is immediately intelligible to people who have no knowledge of CPM[18] or network diagrams. Its familiarity breeds confidence. It affords an easy and convenient way in which to present information developed from a CPM study. It is far less expensive than CPM. For these reasons, bar charts undoubtedly will continue to be widely used in the construction industry.[19]

Construction contract scheduling received great attention in the 1970s traceable to the following:

1. Greater use of management techniques and computers.
2. Roaring inflation and high cost of money.
3. Increased willingness of the law to compensate the contractor when the owner disrupted the construction schedule.

Some of these factors led to the development of new construction specialties, such as construction management and CPM consultants, as well as construction techniques such as fast tracking.

A variety of legal issues have surfaced. They are not new issues, but the now common delay damage claim, sometimes known as an impact or ripple claim, necessitates greater emphasis on the schedule, from both an operational and a legal standpoint. This section first notes the AIA's handling of scheduling, describes the CPM method, and concludes with some description of legal issues involved.

A. AIA

The approach taken by AIA is reflected in AIA Docs. B141 and A201. B141, ¶ 2.2.6, disclaims any

---

[15]See Section 19.02(D).
[16]*BECO Corp.*, ASBCA 27090 (1983).

[17]M. CALLAHAN & M. HOHNS, CONSTRUCTION SCHEDULES 38 (1983).
[18]Critical path method (CPM) is discussed in (B).
[19]M. CALLAHAN & M. HOHNS, CONSTRUCTION SCHEDULES 39 (1983).

AREA	1968 F	M	A	M	J	J	A	S	O	N	D	1969 J	F	M	A	M	J	J	A	S	O	N	D	1970 J	F
FOUNDATION	■	■	■																						
STRUCTURAL STEEL				■	■	■																			
CONCRETE					■	■	■	■																	
ENCLOSURE STONE							■	■	■	■	■														
MASONRY							■	■	■	■	■														
WINDOW FRAMES							■	■	■	■	■														
JAIL EQUIPMENT								■	■	■	■	■	■	■	■	■	■	■							
FINISH WORK														■	■	■	■	■	■	■	■	■	■		

architect responsibility for sequences or submissions of contractors. A201, ¶ 3.10,1, requires the contractor to submit its construction schedule for the information of the owner and the architect. The schedule must provide "for expeditious and practicable execution of the Work." The contractor's failure to conform to the most recent schedule constitutes a breach under ¶ 3.10.3.

Neither details nor schedule type is required. All is left to the contractor. Of course, a contractor is also interested in completing the project as promised. It *should* develop and meet a schedule that will accomplish that objective. But should responsibility be put *solely* in the hands of the contractor? Failure to meet the completion date, though giving a claim against the contractor, is not getting an on-time project. Of course, payments are keyed to work progress[20] and can be an incentive to move the work along. Failure to supply enough skilled workers may be grounds for termination,[21] though

rarely will this be sought. Taken as a whole, A201 seems to ignore delay claims and contains few levers to obtain timely completion.[22]

---

[20]Section 22.02(E).
[21]AIA Doc. A201, ¶ 14.2.1.1.

---

[22]Much more is required in the Construction Management Documents. In 1992, the AIA published a new set of CM documents, the principal ones being A101/CMa, A201/CMa, B141CMa, and B801CMa. The documents use a CM system under which the CM is an adviser to the owner. Also, the AIA responded to industry pressures for documents under which the CM acts as constructor, and may guarantee a maximum price (GMP). The AIA now also publishes A121/CMc, which is used with 1987 editions of A201 and B141. Worked out jointly with the Associated General Contractors (AGC), under the A121/CMc, and its corresponding AGC 565, the construction manager contracts directly with subcontractors, is compensated on a cost plus a fixed fee, and provides the owner with a GMP. For more on the CM system, see Section 17.04(B). The Construction Management Association of America (CMAA) published a set of CM documents in 1993 under which the CM acts as an advisor. They consist of CMAA Docs. A-1, A-2, A-3, and A-4. The CMAA has prepared a draft of documents under which the CM acts as a constructor and guarantees a maximum price, CMP-1, GMP-2 and GMP-3.

Contracts prepared by experienced public or private owners, particularly private owners under the influence of their lenders, usually prescribe much greater detail and take greater control over the contractor's schedule. This can manifest itself in language requiring that the schedule be on a form approved by the owner or the owner's lender; that each monthly schedule specify whether the project is on schedule (and if not, the reasons therefore); that monthly schedule reports include a complete list of suppliers and fabricators, the items that they will furnish, the time required for fabrication, and scheduled delivery dates for all suppliers; and that the contractor hold weekly progress meetings and report in detail as to schedule compliance.

Similarly, the Engineers Joint Contract Documents Committee (EJCDC) takes progress much more seriously than does the AIA. For example, the EJCDC's No. 1910-8, ¶ 2.6, requires the contractor to submit within ten days after the effective date of the contract an estimated progress schedule, a preliminary schedule of values, and a preliminary schedule of submittals. The finalized schedule must be acceptable to the engineer, and ¶ 6.6 requires the contractor to submit for acceptance adjusted progress schedules.

B. Critical Path Method (CPM)[23]

This system has generated burgeoning literature.[24] The description here must be simple, its goal

---

[23]This discussion owes much to student research done by James K. Graves and Jeffrey M. Chu.

[24]M. CALLAHAN & M. HOHNS, CONSTRUCTION SCHEDULES (1983); K. CUSHMAN, CONSTRUCTION LITIGATION 153 (1981); Wickwire, *The Use of Critical Path Method Techniques in Contract Claims*, 7 Pub.Cont.L.J. 1 (1974). The latter seminal paper was updated by Jon M. Wickwire and his colleagues. See Wickwire, Hurlbut, & Lerman in *Critical Path Method Techniques in Contract Claims: Issues and Developments*, 1974–1988, 18 Pub.Cont.L.J. 338 (1989). For a more skeptical look at CPM, see Laufer & Tucker, *Is Construction Project Planning Really Doing Its Job? A Critical Examination of Focus, Role and Process*, 5 Constr.Mgmt. & Econ. 243 (1987). Laufer and Tucker note that the CPM system is being used increasingly as an administrative and legal tool rather than a planning instrument, and they state that the success of CPM/Pert networks has been limited. Laufer and Tucker point to a study indicating that only 15% of large construction companies have found CPM very successful and another in which only 43% use CPM effectively. They note that in small projects, only 10% attempt to use

mainly to point out the essential characteristics of the process and note the effect of float or slack time. The CPM process also relates to measuring claims (discussed in Section 27.03(E)).

To show how a CPM schedule operates, a very simple construction project will be used as illustration, without all the complexities of arrow diagrams, precedence diagrams, and nodes.

First, the contractor divides the total project into different activities or work packages. A major project may have thousands of activities, with each subcontractor generally performing a different activity.

Next, the contractor determines the activities that must be completed before other activities can be started. These constraints are the key to the CPM schedule. For example, usually excavation must be completed before foundation work can be begun. Conversely, plumbing and electrical work can usually be performed at the same time, as neither is dependent on the other. Subcontractors performing this work can work side by side.

Finally, the contractor estimates how long it will take subcontractors to complete their activities. This estimate is made after the contractor consults with its subcontractors and analyzes the design drawings. These data influence the number of days allocated to each activity.

In the sample project, the following are activities and their respective durations and constraints:

Activity	Duration	Constraint
1. Excavation	7 days	None
2. Formwork	5 days	Excavation
3. Concrete pour	5 days	Formwork and plumbing
4. Plumbing	4 days	Excavation
5. Electrical	2 days	Excavation
6. Roof	4 days	All

---

the CPM system. They state that CPM refers to technological constraints, while limitations of resources are barely considered. Even where technological relationships are stressed, CPM may be suitable for sequential operations but not for bulk operations, where detailed sequencing of activities is often irrelevant or unimportant. Laufer and Tucker find CPM difficult to apply in practice, as the majority of activities on the site are overlapping and these are difficult to express accurately in an activity network. Finally, Laufer and Tucker criticize the network model as presupposing that interferences and variability occur rarely when uncertainty is not a brief intrusion but exists in construction everywhere.

Constraints dictate the form of the CPM schedule. Because excavation has no constraints, it can be performed first. Once it is completed, the formwork, plumbing, and electrical activities can be performed. The concrete pour activity cannot be performed until the formwork and plumbing are completed. The roof cannot be installed until the concrete pour and electrical have been completed. The CPM schedule for this project is illustrated in Figure 26.05b. The total project under this schedule should be completed in twenty-one days.

The critical path, the longest path on this simple schedule, consists of those activities that will cause a delay to the *total* project if *they* are delayed. In the above example, excavation, formwork, concrete pour, and roofwork are on the critical path. A delay to any of these activities will delay the entire project.

In contrast, plumbing and electrical activities are not on the critical path. Their delay, *up to a point*, will not delay the total project. If electrical work is delayed seven days, the total project will not be delayed. The number of days each noncritical path activity can be delayed before the total project is affected is called float or slack time. In the illustration, plumbing and electrical work have one day and eight days of float, respectively.

If a noncritical path activity is delayed beyond its float period, it becomes part of the critical path. Moreover, some activities that were previously on the critical path will no longer be there. Suppose there is a three-day delay to plumbing. Originally, plumbing had one day of float. Now the CPM must be adjusted as shown in Figure 26.05b.

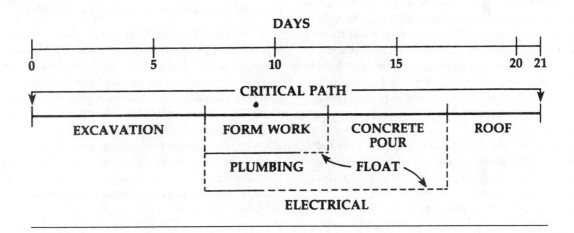

Figure 30.05c

The total project has now been delayed two days. Plumbing has become part of the critical path, and formwork has moved off the path.

The use of this method can be illustrated by *Morris Mechanical Enterprises v. United States.*[25] The contractor was to deliver and install a chiller within 120 days. The contractor delivered the chiller 231 days late. As a result, the government withheld $23,100 as liquidated damages from the final payment.

In ruling for the contractor, the court pointed to the CPM. The schedule showed that delivery and installation of the chiller were originally on the critical path. However, they were later taken off because of delays to other activities for which the contractor was not responsible. The chiller was to be installed in an equipment room. Another contractor who was responsible for completing the equipment room had difficulty procuring materials. When the chiller was actually delivered, the room had not yet been completed. Even though the chiller was delivered 231 days late, the contractor's breach did not delay the total project inasmuch as its performance was no longer on the critical path. The court relieved the contractor by concluding that it should have been given a time extension, precluding the agency from deducting from the unpaid contract balance.

Requiring the contractor to *construct* and *maintain* a CPM schedule has at least three advantages.[26] First, it should require the contractor to work more efficiently. Second, it gives the owner notice of the actual progress of the work. Third, from a litigation standpoint, requiring the contractor to maintain a CPM schedule helps prove or disprove the impact of an owner-caused delay. (Schedules are more persuasive evidence if they are actually used during the construction of the project.)

A CPM schedule has disadvantages. First, it will increase the total contract price. Such schedules are expensive to create and maintain. Second, the contractor may believe such a requirement an unnecessary intrusion into its work. In such a case, the contractor's creation and maintenance of the schedule during construction may be haphazard.

Authoritative commentators state that the following questions must be asked in evaluating a delay claim in which CPM is going to be employed:

1. How was the project actually constructed?
2. What are the differences between the project as planned and as constructed with respect to activities, sequences, durations, manpower, and other resources?
3. What are the causes of the differences or variations between the project as planned and the actual performance?
4. What are the effects of the variances in activities, sequence, duration, manpower, and other resources as they relate to the costs experienced, both by the contractor and by the owner?[27]

As discussed earlier, float is the number of days a noncritical path activity can be delayed before it becomes part of the critical path. Both owner and contractor prefer that the project have float periods. The float periods reflect that every project has some flexibility. A noncritical path activity can be started a few days after it theoretically can begin without delaying the project. By contrast, a critical path activity must start immediately once the preceding critical path activity has been completed. Furthermore, a noncritical path activity does not have to be completed, as shown in the *Morris* decision, by the date it was scheduled for completion. Again, by contrast, the total project will be delayed only if a critical path activity is delayed. The only constraint on a noncritical path activity is that it cannot be delayed longer than its float period.

A number of cases have explored the problem of "who owns the float."[28] Suppose activity A has

---

[25]1 Cl.Ct. 50, 554 F.Supp. 433 (1982).

[26]It is misleading to speak of *a* CPM schedule. In any project where CPMs are used to prove delay claims, it is likely that the claimant will have to show a reasonable "as planned" CPM, an "as-built" CPM reflecting all delays—government, contractor, and excusable—and an adjusted CPM to establish completion of the project absent government delays. See Wickwire, Hurlbut, & Lerman, supra note 24.

[27]Wickwire, Hurlbut, & Lerman, supra note 24, at 341, analyze these issues in detail in their paper.
[28]Compare *Brooks Towers Corp. v. Hunkin-Conkey Constr. Co.,* 454 F.2d 1203 (10th Cir.1972), with *Natkin & Co. v. George A. Fuller Co.,* 347 F.Supp. 17 (W.D.Mo.1972). See generally, M. CALLAHAN & M. HOHNS, CONSTRUCTION SCHEDULES § 4–6 (1983). Wickwire, Hurlbut, & Lerman, supra note 24; Finke, *The Burden of Proof in Government Contract Delay Claims,* 22 Pub.Cont.L.J. 125 (1992).

a float period of thirty days and is the only activity delayed on the project. If either the owner or the contractor causes activity A to be delayed for fewer than thirty days, neither is responsible for delay damages. The delay did not result in delaying the total project. The project can still be completed by the contract completion date. Conversely, if one of the parties causes activity A to be delayed beyond thirty days, that party is liable for delay damages because activity A has become part of the critical path and the project has been delayed.

Suppose contractor and owner each cause a twenty-day delay to activity A. The total project will be delayed ten days, because originally this activity had a thirty-day float. But who is liable for the ten-day delay? The project would not have been delayed at all if either owner or contractor had not delayed the activity.

If the contractor owned the float, the owner could not charge the contractor for project delay and would be liable for any contractor delay expenses. The contractor's argument that it owns the float can stress that *it* is responsible for scheduling the activities of the project. In a sense, it created the float because it created the CPM schedule.

But what if the owner requires a CPM schedule or demands ownership of the float? The contractor could change the float period by changing the critical path. Also, owner and contractor rarely bargain for the ownership of the float. Usually the contract provides that the contractor must complete by a certain date and the owner must pay a designated sum of money for the completed project. Implicit is that the owner will not interfere with the contractor's performance. If there were interference, the contractor would not be able to complete the project by the completion date (this seems to be the AIA hands-off approach). If the owner takes part of the float period, it interferes with the contractor's performance. If interference causes the contractor to incur delay damages, the owner should be liable for the time-related added costs. These are contractor arguments to claim the float.

Some owners take float ownership by contract.[29] Where this is done, the contractor does not receive

a time extension even if delayed by the owner. In a noncritical path activity, if the delay can be absorbed, completion is not delayed.[30]

A responsible contractor will inflate its bid price to cover its potential liability for a delayed project. Also, it must absorb its delay costs. At the bidding stage, the contractor cannot estimate the number of days of float the owner will use. Depending on the complexity of the project, the total increase in the bid can be substantial. In addition, initial enforcement of float ownership clauses may be uncertain, principally because of the owner's superior bargaining position.

The *project* can own the float using one of two methods. First, the party actually causing the delay to the project is liable. Suppose, in the above example, the owner caused the first twenty-day delay. Then the contractor caused another twenty-day delay. The contractor is liable for the ten-day delay of the total project. The owner's delay had used up twenty days of the total thirty-day float, leaving ten days of float remaining. The contractor's delay used up the balance of that period but then caused the activity to be delayed beyond its float period.

This method works fairest if both parties innocently or negligently caused delays to the same activity. However, if the owner's delay had been caused intentionally or willfully, the owner rather than the contractor should be liable.

The second method is borrowed from tort law. Liability for the delayed project depends on the comparative fault of the parties. In the above example, because both parties had caused an equal delay to activity A, each is liable for one half of the other's time-related losses. This method can become complicated if liquidated damages are used (see Section 26.09(B)).

Each of these solutions has advantages and disadvantages. Giving *either* the contractor or the owner the float provides some certainty. A party is exonerated for delays that use up the float of a noncritical path activity if *it* exclusively owns the float.

---

[29]For example, "... Contractor shall be entitled to an extension of time only with regard to the completion requirements affected by such delay and only by the amount of time he is actually delayed thereby in the per-

formance of the work ..." is language used by a strong industrial owner. Float is discussed in detail by Wickwire, Hurlbut, & Lerman, supra note 24, at 361–362.

[30]*E. C. Ernst, Inc. v. Manhattan Constr. Co. of Tex.*, 387 F.Supp. 1001 (S.D.Ala.1974), affirmed 551 F.2d 1026, rehearing denied in part, granted in part. 559 F.2d 268 (5th Cir.1977).

In complex projects, it may be best to agree that neither owns the float.[31] In such projects, it is equally likely that both parties will cause delay to the project. The owner will anticipate that it is likely to delay the project but plan on using some of the float. Nevertheless, the contract price may increase if the *project* owns the float. A responsible contractor will increase its bid when some of its previously recognized rights are taken away but not as much as if the *owner* exclusively owns the float.

Where the project is simple, it may be best that the contractor own the float. This should develop the lowest contract price. Fewer delays by owner and contractor are likely to occur. If the owner does cause a delay and uses part of the float, the contractor's delay damages may be less than the addition to the contract price that would result if the owner took the entire float for itself.

Generally, the CPM system is practical for only major construction projects. Contractors that bid for these projects already appreciate the usefulness of the schedule. The amount of money spent to construct and maintain the schedule is nominal in comparison to the total cost of the project. Liability exposure in such projects can be vast.

C. Some Other Legal Issues

Sometimes contractors, mainly subcontractors, contend that they are entitled to have their work scheduled properly. One court would not imply an obligation by the prime contractor that it would supply a schedule for subcontractor work because this was not customarily done and the subcontractor did not ask for it in negotiations.[32] The court also observed that the subcontractor had not made a showing that scheduling was important in that particular project. However, the increasing use of obligations of good faith and fair dealing may be the basis for implying that there would be a schedule, particularly if this were customarily done.[33]

An issue often ignored is whether failure to comply with a progress schedule is a breach of contract. AIA Doc. A201, ¶ 3.10.3, appears to require that the contractor conform to the most recent schedule. Similarly, EJCDC No. 1910-8 ¶ 6.29 requires a contractor to adhere to the progress schedule, even though there are disputes or disagreements. To underline the importance of this, ¶ 15.2.1 states that the owner can terminate if the contractor does not adhere to the current schedule.

Although schedules are important and the contractor's performance can be terminated for consistent failure to meet the progress schedule, schedule dates must be more flexible than completion dates. The contractor must be given some latitude both in creating and in maintaining the schedule. Yet changing the schedule, not complying with it, and even abandoning it can be serious enough to be a breach, perhaps even a material one.

SECTION 26.06 Causation

Delays can be caused by a variety of participants and events. Delay can be caused by acts of the owner or someone for whose acts the owner is responsible, such as the design professional. Sometimes delays are caused by the contractor or someone for whose acts the contractor is responsible, such as a subcontractor. Sometimes delays are caused by events not chargeable to either owner or contractor, such as nonnegligent fires, unpreventable labor difficulties, or unforeseeably extreme weather conditions. Delays can be caused by third parties, such as a union shutting down the project over a labor dispute.

The multiple causes for delay create a number of legal problems. For example, delays caused by events *not* chargeable to either party will not justify additional compensation unless there is a warranty by one party that those events will not occur. If a delay is caused by both owner and contractor, it may be extremely difficult to determine which portion of the delay is caused by either party. This can mean that neither party will be able to recover losses it suffers because of the delay[34] or that any

---

[31]See Wickwire, Hurlbut, & Lerman, supra note 24, at 361–362. The authors state that in significant public procurements, contract clauses are included that provide that the float is not for the exclusive benefit of either party to the project. Such clauses permit the individual that gets to the float first to gain the benefit of the float.
[32]*Drew Brown Ltd. v. Joseph Rugo, Inc.*, 436 F.2d 632 (1st Cir.1971).
[33]See Section 19.02(D).

---

[34]*Westinghouse Elec. Corp. v. Garrett Corp.*, 601 F.2d 155 (4th Cir.1979) (discretion in trial judge to deny *both* damages); *Armour & Co. v. Scott*, 360 F.Supp. 319 (W.D.Pa.1972) (owner received restitution for overpayments), affirmed 480 F.2d 611 (3d Cir.1973); *Hartford Elec. Applicators of Thermalux, Inc. v. Alden*, supra note 2 (breach by *both* converts completion date to a reasonable time).

clause liquidating damages will not be upheld or that the contractor will be limited to a time extension.[35]

Frequently, the cause or causes of delay are not easy to trace. This, plus the not uncommon multiple causes of delay, makes delay a particularly complicated legal problem.

SECTION 26.07 Allocation of Delay Risks

At the outset, it is important to divide risk allocation relating to delay into two categories. First, the party whose performance has been delayed may seek relief from its obligation to perform by a particular time. For example, a contractor who has agreed to complete the project by a given date may justify its failure to do so by pointing to the causes that it claims excuses its obligation to complete by the time specified. Second, this contractor may claim additional compensation based on an increase in its anticipated spending because events occurred that delayed its performance. As a general rule, the law has been more willing to excuse performance than to grant additional compensation. Because of this differentiation, speaking simply of risk allocation of events that impede performance can be misleading. This shall be demonstrated in this section.

A party who has agreed to perform by a specific time generally assumes the risks of most events that may delay its performance. In the absence of any common law defenses, such as impossibility or mutual mistake, a contractor will not be relieved of its obligation to perform as promised, let alone receive any additional compensation. Relief, if any, must come from the contract.

The contractor normally does not assume the risk that it will be unreasonably delayed by the owner or someone for whom the owner is responsible. However, under certain special circumstances, a contractor may have assumed even this impediment to its performance. For example, suppose an owner constructs an addition to a functioning plant. The contract provides that the owner can order the contractor off the site if the owner's manufacturing operation requires that be done.

The contractor has assumed the risk of these delays, though not delays that are beyond those normally contemplated, such as constant or excessive delays or those caused by the owner's bad faith.

Weber Construction Co. v. State[36] involved a railroad construction contract that required work by the contractor to be coordinated with that of the railroad. The contract provided that the contractor should take into account possible delay or interruption caused by the operation or maintenance of the railroad at the time it determined its contract price. Although the contractor received time extensions when the railroad delayed its performance, it was not given additional compensation.

The law can take into account the blameworthiness of the party causing the delay. For example, in *Broome Construction, Inc. v. United States*,[37] the Court of Claims held the government not liable for delay in making a worksite available where it sought to do so in good faith. The contractor assumed the risk of *this* delay but would not have assumed the risk of *negligently* caused delay. The contractor's request for additional compensation requires a clause expressly warranting that the government was making itself *strictly* liable for its failure to furnish the site.

Fault can also play a role when a party who has been delayed seeks to be relieved from its responsibility. For example, in *J.D. Hedin Construction Co. v. United States*,[38] the Court of Claims held that a contractor was entitled to a time extension because of a cement shortage. The court criticized findings by the Agency Appeals Board that the delays were foreseeable and were the fault of the contractor. The court stated that the contractor could not be expected to show prophetic insight but need only resort "to the usual and long-established methods employed by the commercial world in general."[39]

Many events can occur that cause delay in construction. The common law placing almost all these risks on the contractor has led to the frequent use of *force majeure* clauses, which single out specific

---

[35]*Hartford Elec. Applicators of Thermalux, Inc. v. Alden,* supra note 2. But see Section 26.09(B).

[36]37 A.D.2d 232, 323 N.Y.S.2d 492 (1971), affirmed 30 N.Y.2d 631, 282 N.E.2d 331, 331 N.Y.S.2d 443 (1972).
[37]203 Ct.Cl. 521, 492 F.2d 829 (1974). But in *Hartford Elec. Applicators of Thermalux, Inc. v. Alden,* supra note 2, a 40-day delay in furnishing the site was found to be an owner breach but not a material breach allowing termination.
[38]187 Ct.Cl. 45, 408 F.2d 424 (1969).
[39]408 F.2d at 429.

events and general causes as justifying relief to the contractor. As an illustration, AIA Doc. A201 provides the following:

> 8.3.1 If the Contractor is delayed at any time in the progress of the Work by any act or neglect of the Owner or the Architect, of an employee of either, or of a separate contractor employed by the Owner, or by changes ordered in the Work, or by labor disputes, fire, unusual delay in deliveries, unavoidable casualties, or any causes beyond the Contractor's control, or by delay authorized by the Owner pending arbitration, or by other causes which the Architect determines may justify delay, then the Contract Time shall be extended by Change Order for such reasonable time as the Architect may determine.

In 1987, the AIA moved weather conditions from ¶ 8.3.1, dealing with time extensions, to ¶ 4.3.8.2, dealing with claims. The latter requires the contractor to document "by data substantiating that weather conditions were abnormal for the period of time and could not be reasonably anticipated" before it can receive a time extension for weather conditions. In addition, ¶ 4.3.8.2 requires that the contractor document that the weather conditions "had an adverse effect on the scheduled construction."

These requirements can make it difficult for a contractor who does not keep detailed weather records to claim a time extension for adverse weather conditions. They reflect a belief by the AIA that weather generally is a risk assumed by the contractor and that only in extraordinary circumstances should weather be the basis for a time extension.[40]

Because a contractor will frequently contend that severe weather should be the basis for a time extension, it is useful to look at *Fortec Constructors v. United States*.[41] The contractor contended that it should have been given a fifty-seven-day time extension because of unusually severe weather. The contract had provided that the contractor would be relieved if it were delayed for causes "other than normal weather," including but not restricted to severe weather. The court then stated:

> Unusually severe weather is adverse weather which at the time of year in which it occurred is unusual for the place in which it occurred. Proof of unusually severe weather is generally accomplished by comparing previous years' weather with the weather experienced by the contractor. In the present case, contract provision 1A-06(b), a meteorological chart of past weather averages, established the usual weather conditions to be expected during contract performance. Notwithstanding the occurrences of unusually severe weather, however, a plaintiff is only entitled to an extension of contract time if such unusually severe weather has an adverse impact on the construction being performed. . . . On a daily basis, the contractor completed a Daily Inspection Report (DIR) and the Government completed a Quality Assurance Report (QAR). These reports record daily rainfall, temperature extremes, and a rating of how the weather affected work that day. While certain testimony indicated that the parties did not always complete the reports on a daily basis, the Court believes that the DIRs and QARs represent the most reliable documents presented regarding both the actual weather at the job site and its effect on job performance. Accordingly, the Court has utilized these documents in assessing the merits of the plaintiff's weather claims. [Citations omitted.][42]

The court compared the DIRs and the QARs for each month against the contract's chart of bad weather conditions to see whether the contractor had experienced more days of severe weather than expected. The court concluded that for some months, the contractor had experienced less severe weather than expected and awarded it a one-day time extension.

Returning to ¶ 8.3.1, the residual clause "beyond the Contractor's control" would be subject to the *ejusdem generis* guide.[43] Events claimed to fall within the general clause must be *similar* to the specific events. For example, the clause may not encompass an extreme rise in labor wages due to the outbreak of war but may encompass delay in transportation of workers to the site caused by a bankruptcy of a transportation company on which the contractor relied.

This catchall can invoke the *expressio unius est exclusio alterius* guide.[44] This interpretation guide

---

[40]See Annot., 85 A.L.R.3d 1085 (1978).
[41]8 Ct.Cl. 490 (1985), affirmed, 804 F.2d 141 (Fed.Cir.1986). This case was also considered a seminal one for determining the effect of a CPM schedule.

[42]8 Ct.Cl., at 492–93.
[43]See Section 20.02.
[44]Ibid.

asserts that anything not included in the catalog of events must have been intended to be excluded. Suppose there is an abrupt and sharp rise in the cost of material. This has not been included. This would normally exclude it, except for the catchall. But *ejusdem generis* must be applied. Is this event "like" the events listed?

Suppose the prime contractor seeks to excuse its responsibility for delay by pointing to the delay having been caused by a subcontractor. Unlike some contracts, ¶ 8.3.1 does *not* expressly include subcontractor-caused delay among those that justify a time extension. It does give the architect the right to grant a time extension for any cause that justifies the delay. However, failure to specifically include a common cause of delay, such as subcontractor-caused delay, should mean that the subcontractor-caused delay should not be encompassed within the catchall or within the broad grant of power given the architect.

Another reason for not granting a time extension is the single contract system's objective of centralizing administration and responsibility in the prime contractor. Only if subcontractor-caused delay is *specifically* included should it excuse the prime contractor.

The independent contractor rule, though subject to many exceptions, relieves the employer of an independent contractor for the losses wrongfully caused by the latter.[45] Contractors sometimes assert that the subcontractor is an independent contractor inasmuch as the subcontractor is usually an independent business entity and can, to a large extent, control the details of how the work is performed. Even so, the independent contractor rule does *not* relieve the employer of an independent contractor when the independent contractor has been hired to perform a contract obligation and the party who suffers the loss caused by the independent contractor is the party to whom the contract obligation was owed.[46] In the construction contract context, the owner usually permits the prime contractor to perform obligations through subcontractors. This does not usually mean that the prime contractor is relieved of its obligation to the owner unless the

owner specifically agrees to exonerate the prime contractor.

The residuary power granted to the architect to grant time extensions has been criticized as leading to a deterioration of any fixed completion date. It does have the virtue of not forcing the drafter to think of every possible event and include it in the "catalog of events" justifying a time extension.

Delays caused by a separate contractor to another separate contractor were discussed in Section 17.04(C).

SECTION 26.08 Time Extensions

Construction contracts usually provide a mechanism under which the contractor will receive a time extension if it is delayed by the owner or by other designated events such as those described in Section 26.07. This section examines the time extension process.

Ideally, owner and contractor should agree on the issuance and extent of a time extension. In many construction contracts, the resolution of these issues, at least in the first instance, is given to the design professional or construction manager. The finality of her decision in the event of subsequent arbitration or litigation is discussed ahead.[47] Whether a time extension should be granted usually requires that the *force majeure* clause be applied to the facts asserted to constitute justification for time extension.

Suppose a time extension is justified. How is the amount of time extension determined? To some degree, this was discussed in Section 26.05, dealing with scheduling. Depending on the existence of float and who can take the benefit of it, delay in the performance of a particular activity may not justify any time extension.

Suppose abnormal weather conditions not normally anticipatable precluded work from October 1 to October 14. Suppose that period contained ten working days. Unless the contractor can show that it would have worked on other than normal working days, the time extension should be ten days.[48]

The extent of time extension need not necessarily, however, be the same as the number of days

---

[45]See Section 31.05(B).

[46]*Harold A. Newman Co. v. Nero*, 31 Cal.App.3d 490, 107 Cal.Rptr. 464 (1973); *Brooks v. Hayes*, 133 Wis.2d 228, 395 N.W.2d 167 (1986). See also Section 31.05.

[47]See Section 29.09.

[48]*Missouri Roofing Co. v. United States*, 357 F.Supp. 918 (E.D.Mo.1973), appears to support this conclusion.

of delay. Suppose a fourteen-day delay caused the contractor to work during a period when the weather was more rigorous than during the period of delay. Under such circumstances, a time extension of fifteen days is proper when the contractor was actually precluded from working for only ten days.

One court faced the imprecision of measuring the exact impact of delay. That court arbitrarily granted a time extension of 65 days for a 131-day delay because this amount was "as accurate an estimate as can be made from the actual resulting delay."[49] Whoever determines the amount of delay will be given considerable latitude. However, it is vital for both owner and contractor to keep careful and detailed records. (See Sections 26.10(E) and 27.05.)

Usually time extension mechanisms provide that the contractor must give notice of the occurrence of an event that is to be the basis for a time extension claim and its probable effect.[50]

Some courts seem to consider such notices as technicalities. For this reason, these courts seem quick to find that the notice condition has been waived if it appears that the owner—or someone with actual or apparent authority—knew of the delay-causing event and was not harmed by failure to give the notice. Likewise, the requirement will be waived if in any way the owner has misled the contractor into believing that the notice will not be required.[51]

Notice conditions, however, serve a useful function. In dealing with the requirement that a home buyer give notice of a breach of warranty within a reasonable time after discovery of the breach, a court stated:

> The requirement of notice of breach is based on a sound commercial rule designed to allow the defendant opportunity for repairing the defective item, reducing damages, avoiding defective products in the

future, and negotiating settlements. The notice requirement also protects against stale claims.[52]

Likewise, in the context of a time extension mechanism, the notice of an intention to claim a time extension serves a number of useful functions. The notice informs the design professional or owner that persons for whom it is responsible, such as other separate contractors, consultants, or the design professional, are delaying the contractor. This can enable the owner or design professional to eliminate the cause of the delay and minimize future delays or damage claims. A timely notice should permit the design professional to determine what has occurred while the evidence is still fresh and witnesses remember what actually transpired.

The notice shows that the contractor has been adversely affected and can eliminate long-delayed, sometimes spurious contractor claims made after completion of the work. If the owner or design professional knew of the event causing the delay, the impact of the event on the contractor's performance, and that it was quite likely that the contractor would make a time extension request, waiver may be proper.[53] However, the value of the notice requirement is substantially reduced if the requirement is too frequently ignored by courts.[54]

SECTION 26.09 Unexcused Contractor Delay

Delay as justification for termination is discussed in Section 34.04(A). This section deals with the recovery of damages for delay. Although measurement of claims owners and contractors have against each other is discussed in greater detail in Chapter 27, this section deals with damage liquidation—the contractual method used most frequently to deal with contractor delay.

[49]E. C. Ernst, Inc. v. Manhattan Constr. Co. of Tex., supra note 30.
[50]AIA Doc. A201, ¶¶ 4.3.3, 4.3.8.1.
[51]Travelers Indem. Co. v. West Georgia Nat'l Bank, 387 F.Supp. 1090 (N.D.Ga.1974) (provision that HUD approve time extension waived). For additional discussion on waiver of technical requirements, see Section 21.04(H).

[52]Pollard v. Saxe & Yolles Dev. Co., 12 Cal.3d 374, 525 P.2d 88, 92, 115 Cal.Rptr. 648, 652 (1974).
[53]Southwest Eng'g Co. v. Reorganized School Dist. R-9, 434 S.W.2d 743 (Mo.App.1968).
[54]Strictly enforced in Herbert & Brooner Constr. Co. v. Golden, 499 S.W.2d 541 (Mo.App.1973), and Quin Blair Enter., Inc. v. Julien Const. Co., supra note 8. Narrowly construed in Hartford Elec. Applicators of Thermalux, Inc. v. Alden, supra note 2.

A. Actual Damages

The damage formula applied most frequently by the common law to delayed contractor performance is the value of the lost use of the project caused by the delay. For example, delayed completion of a residence to be occupied by the owner would very likely be measured by the lost rental value.[55] Although the owner may have suffered other losses, such as the inconvenience of living in a motel or with relatives or having to transport a child to a more distant school, such consequential damages would generally be difficult to recover.

In commercial construction, the owner will very likely be able to recover the lost use value in the event of unexcused delay by the contractor.[56] Even in projects that have readily ascertainable commercial value, losses are often suffered that may be difficult to recover. This is even a greater problem when the project is built for a public entity that intends to use it as a school, an office building, or a freeway. Although some public projects have a readily ascertainable use value, most do not.

B. Liquidated Damages: *Bethlehem Steel Corp. v. Chicago* and *Rohlin Constr. Co. v. City of Hinton*

Because proof of delay damages is very difficult, particularly in public projects, construction contracts commonly include provisions under which the parties agree that certain types of unexcused delay will result in damages of a specific amount. They are usually known as liquidated damages clauses. A court stated:

> There was a time when the courts were quite strong in their view that almost every contract clause containing a liquidated damage provision was, in fact, a forfeiture provision which equity abhorred, and therefore, nothing but actual damages sustained by the aggrieved party could be recovered in case of contract breach caused by delay past the proposed completion date. But, in modern times, the courts have become

more tolerant of such provisions, probably because of the Anglo-Saxon reliance on the importance of keeping one's word, and have become more strongly inclined to allow parties to make their own contracts and to carry out their own intentions, free of judicial interference, even when such non-intervention would result in the recovery of a prestated amount as liquidated damages, upon proof of a violation of the contract, and without proof of actual damages.[57]

The early common law also felt that contractually stipulated damages could be used unconscionably by parties possessing strong bargaining power. The common law saw its role in awarding contract damages as compensating losses and not punishing or effectuating abuse of power.

The earliest use of such clauses was penal bonds, a precursor of surety bonds explored in Chapter 33. Penal bonds were *absolute* promises by the maker to pay a designated sum (the penal sum) followed by a condition that this promise would be null and void *if* certain things occurred, such as the maker, usually a debtor or a surety, paying the full amount of the debt. Frequently, the penal sum greatly exceeded the harm caused. Equity courts would not enforce penalties. (This is the historical basis for differentiating liquidated damages clauses from penalties.) In addition, English courts felt that damages were the exclusive province of the courts. They would not enforce clauses that would "oust the court of jurisdiction."

Despite the nineteenth-century common law courts' belief in liberty and autonomy, the common law courts continued what equity courts had started by placing strict limits on enforcement of these clauses. Courts were willing under *limited* circumstances to enforce liquidated damages clauses. This would be done if there were a showing by the party seeking to enforce the clause that the damages were extremely difficult to ascertain at the time the contract was made and that the amount selected was a genuine preestimate of the damages likely to occur as a probable result of the breach. In addition, some courts would not enforce these

---

[55]*Muller v. Light,* 538 S.W.2d 487 (Tex.Ct.App.1976) ($100-a-day clause held a penalty). See also *Miami Heart Institute v. Heery Archs. & Eng'rs,* 765 F.Supp. 1083 (S.D.Fla.1991) (delay claim by owner against architect) (discussed in detail in Section 12.14(B)).
[56]*Ryan v. Thurmond,* 481 S.W.2d 199 (Tex.Ct.App.1972).

[57]*Sides Constr. Co. v. City of Scott City,* 581 S.W.2d 443, 446 (Mo.App.1979) annotated in 12 A.L.R.4th 891 (1982) collecting many cases.

clauses unless it were shown that the parties intended the clause to compensate and not to punish. The amount selected must be the sole money award. The proponent should not have the option of choosing liquidated or actual damages.

Modern courts, particularly in public contracts, recognize the difficulty of proving damages generally, particularly those relating to delay and the certainty that these clauses can provide *both* parties. As indicated, the law is more willing to be relieved of the burden of measuring damages.[58] Although the results are by no means unanimous, clauses liquidating damages for construction delay are generally enforced if they are reasonable as

judged by the circumstances existing at the time the contract was made.[59] Often a comparison is made between the amount stipulated and the contract price.[60] Some states even require that such clauses be inserted in public construction contracts.[61] A case that typifies the modern judicial attitude toward such clauses is reproduced here.

---

[58]*Sides Constr. Co. of City of Scott City,* supra note 57 (avoids laborious item-by-item damage recitations); *Osceola County v. Bumble Bee Constr., Inc.,* 479 So.2d 310 (Fla.Dist.Ct.App.1985) (city tourist information center).

---

[59]*Dahlstrom Corp. v. State Highway Comm'n of State of Mississippi,* 590 F.2d 614 (5th Cir.1979); *Pembroke v. Gulf Oil Corp.,* 454 F.2d 606 (5th Cir.1971); *Dave Gustafson & Co. v. State,* 83 S.D. 160, 156 N.W.2d 185 (1968). Cases are collected in 12 A.L.R.4th 891 (1982). The Uniform Commercial Code (UCC) enforces them if reasonable in the light of *actual* or *anticipated* damages. See UCC § 2-718.
[60]*Dahlstrom Corp. v. State Highway Comm'n of State of Mississippi,* supra note 59.
[61]*Westinghouse Elec. Corp. v. County of Los Angeles,* 129 Cal.App.3d 771, 181 Cal.Rptr. 332 (1982).

BETHLEHEM STEEL CORP. v. CHICAGO

United States Court of Appeals, Seventh Circuit, 1965. 350 F.2d 649.
[Ed. note: Footnotes omitted.]

GRANT, District Judge.

Plaintiff-Appellant (Bethlehem) brought this action to recover an item of $52,000.00 together with certain items of interest, etc., withheld by the Defendant (City), as liquidated damages for delay in furnishing, erecting, and painting of the structural steel for a portion of the South Route Superhighway, now the "Dan Ryan Expressway", in the City of Chicago.... [T]he District Court concluded that Plaintiff's claims on the items in controversy should be denied and entered judgment accordingly. We agree and we affirm.

The trial court's findings included the following uncontroverted facts:

* * *

"The work which Bethlehem undertook was the erection in Chicago of structural steel for a 22-span steel stringer elevated highway structure, approximately 1,815 feet long, to carry the South Route Superhighway from South Canal Street to the South Branch of the Chicago River. Bethlehem's work was preceded and followed by the work of other contractors on the same section.

"The 'Proposal and Acceptance' in the instructions to bidders required the bidders to '... complete ... within the specified time the work required....' Time was expressly stated to be the essence of the contract and specified provisions were made for delivery of the steel within 105 days thereafter, which was to be not later than 15 days from notification. The successful bidder was to submit to the Commissioner of Public Works a 'Time Schedule' for his work and if 'less than the amount ... specified to be completed' were accomplished 'the City may declare this contract forfeited....' The work had to be completed irrespective of weather conditions.

"The all important provision specifying $1,000 a day 'liquidated damages' for delay is as follows:

'The work under this contract covers a very important section of the South Route Superhighway, and any delay in the completion of this work will materially delay the completion of and opening of the South Route Superhighway thereby causing great inconvenience to the public, added cost of engineering and supervision, maintenance of detours, and other tangible and intangible losses. Therefore, if any work shall remain uncompleted after the time specified in the Contract Documents for the completion of the work or after any authorized extension of such stipulated time, the Contractor shall pay to the City the sum listed in the following schedule for each and every day that such work remains uncompleted, and such moneys shall be paid as liquidated damages, not a penalty, to partially covered losses and expenses to the City.

'Amount of Liquidated Damages per Day . . . $1,000.00.

"The City shall recover said liquidated damages by deducting the amount thereof out of any moneys due or that may become due the Contractor. . . .'

"Provision was made to cover delay in a contractor's starting due to preceding contractor's delay. Unavoidable delays by the contractor were also covered, and extensions therefor accordingly granted."

Bethlehem's work on this project followed the construction of the foundation and piers of the superhighway by another contractor. Bethlehem, in turn, was followed by still another contractor who constructed the deck and the roadway.

Following successive requests for extensions of its own agreed completion date, Bethlehem was granted a total of 63 days' additional time within which to perform its contract. Actual completion by Bethlehem, however, was 52 days after the extended date, which delay the City assessed at $1,000.00 per day, or a total of $52,000.00 as liquidated damages.

Bethlehem contends it is entitled to the $52,000.00 on the ground that the City actually sustained no damages. Bethlehem contends that the above-quoted provision for liquidated damages is, in fact, an invalid penalty provision. It points out that notwithstanding the fact that it admittedly was responsible for 52 days of unexcused delay in the completion of its contract, the superhighway was actually opened to the public on the date scheduled.

In other words, Bethlehem now seeks to re-write the contract and to relieve itself from the stipulated delivery dates for the purposes of liquidated damages, and to substitute therefor the City's target date for the scheduled opening of the superhighway. This the Plaintiff cannot do.

In Wise v. United States, . . . the Supreme Court said:

". . . [T]he result of the modern decisions was determined to be that . . . courts will endeavor, by a construction of the agreement which the parties have made, to ascertain what their intention was when they inserted such a stipulation for payment, of a designated sum or upon a designated basis, for a breach of a covenant of their contract.

. . . When that intention is clearly ascertainable from the writing, effect will be given to the provision, as freely as to any other, where the damages are uncertain in nature or amount or are difficult of ascertainment or where the amount stipulated for is not so extravagant, or disproportionate to the amount of property loss, as to show that compensation was not the object aimed at or as to imply fraud, mistake, circumvention or oppression. *There is no sound reason why persons competent and free to contract may not agree upon this subject as fully as upon any other, or why their agreement, when fairly and understandingly entered into with a view to just compensation for the anticipated loss, should not be enforced.*

"*. . . The later rule, however, is to look with candor, if not with favor, upon such provisions in contracts when deliberately entered into between parties who have equality of opportunity for understanding and insisting upon their rights, as promoting prompt performance of contracts and because adjusting in advance, and amicably, matters the settlement of which through courts would often involve difficulty, uncertainty, delay and expense. . . .*

". . . It is obvious that the extent of the loss which would result to the Government from delay in performance must be uncertain and difficult to determine and it is clear that the amount stipulated for is not excessive. . . .

"The parties . . . were much more competent to justly determine what the amount of damage would be, an amount necessarily largely conjectural and resting in estimate, than a court or jury would be, directed to a conclusion, as either must be, after the event, by views and testimony derived from witnesses who would be unusual to a degree if their conclusions were not, in a measure, colored and partisan." (Italics supplied.)

* * *

Affirmed.

The *Bethlehem* decision illustrates increased judicial cordiality toward liquidated damages clauses in the construction context, particularly public contracts. The law is more willing to enforce such contracts if they have been bargained for and if actual damages would be difficult to establish in court. Although the amount stipulated is rarely the result of bargaining in a competitive bid contract, the contractor can adjust its bid to take this risk into account.[62]

---

[62]See Sweet, *Liquidated Damages in California*, 60 Ca-lif.L.Rev. 84, 118–123 (1972). But see *Spacemaster International, Inc. v. City of Worcester*, 940 F.2d 16 (1st Cir.1991), to which reference was made in the *Rohlin* case repro-duced later in this section. The contract involved modular classrooms, and delay meant that children had to have classes in hallways, gymnasiums, auditoriums, and li-braries. Educational programs were forced to be compro-mised, and morale suffered. These facts would certainly have been good enough to allow damages that could be liquidated. The court pointed to the fact that the amount

Yet in another road-building contract, the Su-preme Court of Iowa expressed concern with the arbitrary way in which the liquidated damages amount was selected. That case is reproduced at this point.

---

selected was not the result of arm's-length bargaining but was simply put out to competitive bidding. It used this fact to support its holding that a summary judgment should not be granted. It did not recognize the difficulties of negotiating a liquidated damages clause in competitive bidding for public work. See Rives, *Liquidated Damages in the Construction Industry,* 12 Constr. Lawyer No. 3, (Au-gust 1992) p. 6.

ROHLIN CONSTRUCTION CO., INC., v. CITY OF HINTON

Supreme Court of Iowa, 1991. 476 N.W.2d 78.

Considered by HARRIS, P.J., and SCHULTZ, CARTER, LAVORATO and SNELL, JJ.

SCHULTZ, Justice.

The issues in this appeal center on the question of whether the liquidated damage clauses in three road construction contracts were actually penalty clauses and therefore void. The trial court refused to impose liquidated damages on the contractor. Our court of appeals affirmed. We affirm both courts.

In the spring of 1988, defendants Plymouth County (county) and the City of Hinton (city) entered a joint project for resurfacing certain county and city roads. Defendants jointly planned and advertised for bids on one city and two county resurfacing projects. Plaintiff Rohlin Construction Co., Inc. (Rohlin) was the suc-cessful bidder and entered into two contracts with the county for total prices of $221,588.39 and $251,696.99, and one contract with the city for a price of $37,957. The two county contracts were dated May 31, 1988, and the city contract was dated June 16, 1988.

All three contracts contained a provision requiring that work be completed within forty "working days." The contracts specified a completion date of Septem-ber 2, 1988, but did not specify a starting date. The contract language was prepared by the Iowa Depart-ment of Transportation on "proposal forms" which contained the following language:

> If this bid is accepted, Bidder agrees . . . to either com-plete the work within the contract period or pay liquidated damages, which shall accrue at the daily rate specified below, for each additional working day the work remains uncompleted.

Each contract established $400.00 per day as the amount of liquidated damages.

Rohlin commenced work on September 17 and completed the project in less than thirty days. Rohlin completed the project 25.5 days past the September 2 deadline on the city contract and 27.5 days and 28 days late on the county contracts. The city withheld $10,200 and the county withheld $22,200 as liquidated damages for late completion of the project.

Rohlin then commenced separate law suits against the city and county seeking judgments against both defendants for the sums withheld, plus interest and attorney fees. Following trial, the court ruled in favor of Rohlin, allowing recovery for the amount of the withholdings plus interest at eight percent from No-vember 15, but denying attorney fees.

* * *

In the past, we disfavored the use of liquidated damage clauses and favored interpretation of con-tracts that make stipulated sums penalties. . . . Later, we relaxed this penalty rule and recognized that par-ties may fix damages by contract when the amount of damages is uncertain and the amount fixed is fair. . . . This change in contractual interpretations is con-sistent with the trend of favoring liquidated damage clauses.

* * *

We often turn to Restatements of the Law and believe it is appropriate to do so in this case. The American Law Institute adopts a more conservative approach as follows:

Damages for breach by either party may be liquidated in the agreement but only at an amount that is reasonable in the light of the anticipated or actual loss caused by the breach and the difficulties of proof of loss. A term fixing unreasonably large liquidated damages is unenforceable on grounds of public policy as a penalty.

Restatement (Second) of Contracts § 356(1) (1981). However, the American Law Institute shows no hostility toward liquidated damages by stating:

The parties to a contract may effectively provide in advance the damages that are to be payable in the event of breach as long as the provision does not disregard the principle of compensation. The enforcement of such provisions for liquidated damages saves the time of courts, juries, parties and witnesses and reduces the expense of litigation. This is especially important if the amount in controversy is small. However, the parties to a contract are not free to provide a penalty for its breach. The central objective behind the system of contract remedies is compensatory, not punitive. Punishment of a promisor for having broken his promise has no justification on either economic or other grounds and a term providing such a penalty is unenforceable on grounds of public policy.

Id. comment a. This Restatement section also sets out the test for a penalty:

Under the test stated in Subsection (1), two factors combine in determining whether an amount of money fixed as damages is so unreasonably large as to be a penalty. The first factor is the anticipated or actual loss caused by the breach. The amount fixed is reasonable to the extent that it approximates the actual loss that has resulted from the particular breach, even though it may not approximate the loss that might have been anticipated under other possible breaches. Furthermore, the amount fixed is reasonable to the extent that it approximates the loss anticipated at the time of the making of the contract, even though it may not approximate the actual loss. The second factor is the difficulty of proof of loss. The greater the difficulty either of proving that loss has occurred or of establishing its amount with the requisite certainty, the easier it is to show that the amount fixed is reasonable. To the extent that there is uncertainty as to the harm, the estimate of the court or jury may not accord with the principle of compensation any more than does the advance estimate of the parties. A determination whether the amount fixed is a penalty turns on a combination of these two factors. If the difficulty of proof of loss is great, considerable latitude is allowed in the approximation of anticipated or actual harm. If, on the other hand, the difficulty of proof of loss is slight, less latitude is allowed in that approximation. If, to take an extreme case, it is clear that no loss at all has occurred, a provision fixing a substantial sum as damages is unenforceable.

Id. comment b (citations omitted). We believe that application of these principles is appropriate to our determination in this case.

A review of the record reveals uncertainty of how the sum of $400 per day for liquidated damages was derived. This liquidated damage amount was placed in the specifications as a result of the county engineer's consultation with someone in the office of the Department of Transportation (DOT). A DOT construction manual contained a schedule of suggested rates for liquidated damages based strictly on the engineer's estimate of the contract price. According to the schedule in the manual, the city's contract called for $100 per day in liquidated damages and the two county contracts called for liquidated damages of $200 and $300 per day respectively, for a total of $600. As noted previously, there were three separate contracts but the resurfacing project was a joint city-county project involving connected highways. Thus, if the total amount rather than the individual amounts of the three contracts is used, then the manual's suggested rate is $400 per day in liquidated damages.

The county engineer indicated that the reason for deviating from the manual's suggested rates was due to the city and county's desire for a completion date prior to the increased traffic that would accompany the start of school and the grain-hauling season in Hinton. In addition, the engineer stated that "we wanted the liquidated damage amount to be sufficient to make the contractor aware that we need that project completed."

There is no valid justification for the individual liquidated damage amounts contained in each of the three contracts. Under the record of this case, the person who set the $400-per-day amount in each contract is unknown and was not called as a witness. Additionally, no witness was called to justify the suggested liquidated damage amounts contained in the DOT manual schedule. The county engineer did not conduct studies or present any other data suggesting that defendants anticipated that the government entities and the public could sustain damages equivalent to the $400-per-day liquidated damage amount contained in each of the three contracts. Furthermore, plaintiff called the school superintendent as a witness to give evidence that the school experienced no problems due to the road work. The county engineer also indicated that a Hinton grain elevator company had not complained that delayed completion of the road work caused the company or its patrons any damages or losses.

Plaintiffs seem to contend that Rohlin's delayed completion of the project caused no damages. The county did sustain damages, however, due to erosion because it could not seed the highway shoulders because of the delay. Undoubtedly, there was some inconvenience to school bus drivers and to grain haulers because of the late completion.

We agree with the trial court that the provisions for liquidated damages in the three contracts were penalties rather than reasonable amounts for liquidated damages. "Liquidated damages must compensate for loss rather than punish for breach...." *Space Master Int'l, Inc. v. City of Worcester*, 940 F.2d 16, 18 (1st Cir.1991). We recognize that proving the amount of loss with any degree of certainty is difficult; nevertheless, the amount of liquidated damages set in each contract appears to be unreasonably large and goes far beyond the anticipated loss caused by delay in performance of the contract. The road project would inevitably cause some inconvenience to the parties and the public for a certain number of days regardless of when performed. Therefore, we conclude that the $400-per-day liquidated damage clause contained in each of the three contracts is an unrealistic amount and is therefore a penalty that should not be enforced.

The court of appeals affirmed the trial court's denial of an award of attorney fees to plaintiff. We agree that this determination is a matter within the discretion of the trial court and it did not abuse its discretion in denying plaintiff attorney fees.

Costs on appeal are taxed two-thirds to Plymouth County and one-third to the City of Hinton.

DECISION OF COURT OF APPEALS AND JUDGMENT OF DISTRICT COURT AFFIRMED.

The court's concern over the absence of evidence indicating why the particular amount was selected, with particular reference to the DOT manual schedule, will place unreasonable obstacles to the enforcement of such clauses. Difficulty of amassing proof of this sort is one of the reasons why parties liquidate and why modern courts as exemplified in the *Bethlehem* decision are more inclined to enforce such clauses. Had the public entity introduced testimony indicating why it felt it necessary to deviate from the manual schedule, the outcome might have been different. But if all the bidders had taken into account the likely enforcement of the liquidated damages clauses and had seen the importance the public entity attached to timely completion, they should have included in their bids an amount to take care of the possibility that additional resources would have to be used to accomplish timely completion. If Rohlin did not, and thereby obtained the contract, this would not have been fair to the other bidders who bid rationally. If Rohlin had included these pricing contingencies, it unjustly enriched itself in this contract. To be sure, there would not have been unjust enrichment if Rohlin had signed the contracts upon the assumption that the liquidation of damages clause would never be enforced.

The court's decision also reflects its belief that if a party's delayed performance cannot be said to have caused monetary losses, the liquidation of damages clause should not be enforced. It is clear that establishing actual losses in such cases cannot be done at the time the contract is made or even after it is performed. This is the reason for the use of liquidation of damages clauses.

Despite greater willingness to enforce such clauses, additional legal issues reflect some judicial ambivalence. For example, courts generally will not enforce liquidated damages clauses unless there is a rational apportionment of that portion of the damage caused by the contractor and the owner. Usually this is accomplished by a clause granting time extensions for certain delay.[63] In addition, jurisdictions vary as to whether the proponent of a liquidation clause must bear the burden of establishing its enforceability. Jurisdictions hostile to liquidation place the burden on the proponent,[64] but the modern tendency has been to enforce such clauses unless the opponent can show invalidity.[65]

Despite increased enforceability of reasonable liquidated damages clauses, traditional suspicion

---

[63]*E. C. Ernst. v. Manhattan Construction Co. of Tex.*, supra note 30 (enforced); *Jasper Constr. v. Foothill, Junior College District*, 91 Cal.App.3d 1, 153 Cal.Rptr. 767 (1979) (enforced); *Hartford Elec. Applicators of Thermalux, Inc.*, supra note 2 (not enforced).
[64]*Utica Mutual Ins. Co. v. DiDonato*, 187 N.J.Super. 30, 453 A.2d 559 (1982).
[65]*Commercial Union Ins. v. LaVilla Indep. School Dist.*, 779 S.W.2d 102 (Tex.Ct.App.1989) (party opposing liquidation of damages must prove actual damages to show liquidation amount not a reasonable approximation); A modern California statutory revision puts the burden on the party attacking the clause. West Ann.Cal.Civ.Code § 1671(b).

of these clauses makes it important to consider particular issues that surface when such clauses, though clearly enforceable, must be interpreted.

Should a liquidation clause be applied if the contractor *abandons* the project? It can be contended that parties drafting such clauses are thinking principally of *delayed* completion by the contractor and not *abandonment*, particularly if followed by abandonment of the project by the *owner*.[66] Another reason for not applying the clause is the fear that the owner may be tempted to delay completion and thereby increase the liquidated damage amount.

Although application can appear to create open-ended liability, the clause can be applied for a reasonable period of time, not for infinity. Another way of dealing with this problem is to contractually "cap" the liquidation amount and thereby avoid the open-ended liability that can generate a forfeiture vastly disproportionate to the actual damages.[67]

Unjustified abandonment by the contractor or proper termination by the owner can cause the owner *two* harms: additional cost to complete the project and additional cost caused by delayed completion. Damage liquidation can apply to the second, delay in completion determined by the actual completion by a successor or when the project could have been reasonably completed by a successor contractor.[68]

Another problem relates to the difference between substantial and final completion of the project.[69] This issue was treated in *Hungerford Construction Co. v. Florida Citrus Exposition, Inc.*[70] The contract involved an exhibition center for the citrus industry. One important feature of the project was a concrete dome 170 feet in diameter that was to operate as the center's roof. The roof was to be waterproof without independent waterproof covering. Completion time was 180 calendar days, and the liquidated damages were specified to be $200 per calendar day of delay.

The project was completed within 180 days, and the owner moved into the project. However, from the beginning, the roof leaked, and it was necessary to do corrective work. The corrective work did not require that the owner leave the premises, but it did preclude the owner from making the premises available on a rental basis to those who might want to use it for exhibition purposes. The court referred to this use as a secondary use. Even after the corrective work, the secondary use was diminished because there was unsightly discolored plaster across the roof caused by the leaking and correction.

The court held that the liquidate damages clause would not be applied in this case. The court stated that the building was available to the owner for its primary use and that the loss of secondary use was entirely speculative. The court stated:

> One auto show may have been lost but there is no evidence as to the amount of rent which would have been realized out of this transaction or whether the loss occurred during the period in suit. The only other loss of use claimed was in the form of a daily admission charge to the public to see the building and its contents. The proof demonstrated only that this use did not rise above a suggested use. In any event, such loss of secondary use as there may have been is capable of proof. The proof that was offered was entirely disproportionate to the sum of $200.00 . . . per day. It is thus clear that the claim for liquidated damages was far in excess of such compensatory damages as would be indicated from the slight deprivation of secondary use claimed by the owner.[71]

To deal with this problem, the contract can create a two-tier liquidation system, with one tier dealing with substantial completion and the other with final completion. This would avoid the *Hungerford* case result, in which it is likely that the owner will be unable to establish losses with sufficient certainty for the reduced use value of the

---

[66]*Town of North Hempstead v. Sea Crest Constr. Corp.*, 119 A.D.2d 744, 501 N.Y.S.2d 156 (1986).

[67]*Reed & Martin, Inc. v. Westinghouse Elec. Corp.*, 439 F.2d 1268 (2d Cir.1971).

[68]*Construction Contracting & Mgmt., Inc. v. McConnell*, 112 N.M. 371, 815 P.2d 1161 (1991). Compare *City of Boston v. New England Sales & Mfg. Corp.*, 386 Mass. 820, 438 N.E.2d 68 (1982), and *Austin Griffith, Inc. v. Goldberg*, 224 S.C. 372, 79 S.E.2d 447 (1953) (limited to abandonment after completion date), with *Continental Realty Corp. v. Andrew Crevolin Co.*, 380 F.Supp. 246 (S.D.W.Va.1974) (refusal to apply clause).

[69]AIA Doc. A101, ¶ 3.2, measures the time commitment by substantial completion.

[70]410 F.2d 1229 (5th Cir.1969).

[71]Id. at 1232.

structure while corrective work was being performed.

Applicability of the clause also was an issue in *Northern Petrochemical Co. v. Thorsen & Thorshov, Inc.*[72] After completion, serious structural defects were discovered. This necessitated large-scale redesign and reconstruction. Correction of the defects took eight months. The court affirmed an award based on lost profits because of delayed occupancy and excess operating costs while awaiting occupancy. The court rejected application of the contractual liquidated damages clause, noting that it was intended to apply to *normal* delay in the "rate of construction," not to an extraordinary eight months' delay due to redesign and reconstruction.

The court's conclusion would limit liquidation to normal, expectable delays where there is *completion*. But suppose the proof of actual damages was difficult and the liquidation amount high. The owner would then have sought to enforce the clause. Theoretically, such an attempt would fail.

Suppose the defects took one month to correct. Delays caused by defects discovered and corrected during performance and before completion are within the clause. There is no reason to draw a line at completion. But a line *can* be drawn, as did the court, at *extraordinary* delay whether caused by correction of defects during or after completion. To be sure, this does inject an element of uncertainty. But the law should respect the intention of the parties, and the intention was not to deal with extraordinarily long delays.

The use of fast tracking[73] can generate particular problems, especially if the project is being built by a number of separate contractors or multiple prime contractors. In *Casson Construction Co.,*[74] the contractor, though finishing on time, failed to complete a phase of its work that delayed a follow-on contractor. The follow-on contractor was ordered to accelerate. It did so by the use of overtime and double shifts for several weeks. This acceleration cost $174,000.

The breaching contractor claimed that its liability was limited to the liquidation of damages amount of $240 a day contained in *its* contract. The

cost of acceleration was $644 a day. The Board pointed to another provision in the contract that stated that the contractor would indemnify the public agency for acceleration payments made to other contractors. The court held that the latter clause controlled and that the same breach can invoke different clauses. The Board would not apply the liquidated damages clause to milestone date delays, only to delay in completing the *entire* contract performance. The Board noted that the standardized language used in the contract was adopted long before the advent of fast-track construction.

Subcontracting raises special problems. Sometimes prime contractors seek to charge all subcontractors a ratable amount of any liquidated damages that the prime contractor must pay the owner. This avoid the often difficult question of which subcontractors have caused the delay. Yet it requires a subcontractor who has not caused the delay to share the liquidation loss.[75] As a result, AIA Doc. A401, ¶ 3.3.1, expressly provides that assessment of liquidated damages against the subcontractor are assessed "only to the extent caused by the Subcontractor" or those for whom it is responsible. Unfortunately, this language could justify a claim by the prime contractor against a subcontractor who has *caused* the delay even if the subcontractor has not *breached* its contract.

Another difficulty is shown by *P & C Thompson Brothers Construction Co. v. Rowe.*[76] The subcontract liquidated damages at $250 a day. The prime contractor deducted from the subcontractor's final payment its *actual* damages—the liquidated sum and a pro rata portion of the liquidation of damages it had to pay the owner. The court would not enforce the clause, as it would be more than the prime contractor would have to pay the owner. The court found this to be unconscionable.

[72]297 Minn. 118, 211 N.W.2d 159 (1973).
[73]See Section 17.04(B).
[74]GSBCA 4884, 78-1 BCA ¶ 13,032.

[75]*United States v. Arundel Corp.,* 814 F.2d 193 (5th Cir.1987).
[76]433 So.2d 1388 (Fla.Dist.Ct.App.1983). See also *Hall Constr. Co., Inc. v. Beynon,* 507 So.2d 1225 (Fla.Dist.Ct.App.1987) (clause interpreted to cover both amount prime paid owner and prime's actual damages). Sometimes the subcontractor can take advantage of the liquidation clause in the prime contractor because of the flow-through clause. See *Industrial Indem. Co. v. Wick Constr. Co.,* 680 P.2d 1100 (Alaska 1984). For a discussion of flow-through clauses, see Section 28.04.

Actual damages suffered by the prime contractor itself can be separated from amounts that the prime contractor must pay the owner.[77] There is no reason why the liquidated damages clause cannot encompass *both*.[78] Although sometimes the amount that the prime contractor may have to pay the owner can be ascertained at the time the contract is made (the stipulated amount in the owner-prime contract), the *other* damages that the prime may suffer *in addition* may be difficult to establish. Two provisions *can* be included: one making clear that the subcontractor will reimburse the prime for any amounts it must pay the owner attributable to the subcontractor's breach and another to liquidate damages for other harm suffered by the prime. Alternatively, the contract can specify that the liquidation amount applies only to the liquidation amounts that must be paid by the prime contractor to the owner, not precluding the recovery of actual damages suffered by the contractor as well. But the parties should be able to include *both* in one clause.

How important is enforceability? If the clause is unenforceable, the owner can still prove actual damages. For this reason, some drafters insert a stiff liquidation amount, with actual damages still recoverable if the clause is found to be a penalty. This is not desirable. It gives up the possibility of a *sure* liquidation. Actual delay damages are extremely difficult to prove. (But an unforceable penalty may still get on-time completion.)

Project completion dates can be "hard" or "soft," depending on the importance attached to timely completion. When a high liquidation amount is selected, the contractor should increase its contract price either to ensure timely completion (double time, expedited deliveries, more workers) or to pay damage liquidation. Coupling a soft completion date with a stiff damage liquidation amount will generate timely completion at a high cost when timely completion is not crucial.

When the completion date is hard, the amount selected must be sufficiently high (but not too high so as to risk finding that it is a penalty) to make it more profitable for the contractor to finish on time than to delay and pay liquidated damages. Select-ing an amount to accomplish this objective requires a sophisticated understanding of contractor costs. It involves awareness of the increased administrative costs of such a clause. Invariably, there will be more requests for time extensions. Any stiff clause should be accompanied with a contractual justification for the amount, as in the *Bethlehem* case.

This advice is inconsistent with the stated requirement that liquidated damages be based on a genuine preestimate of actual damages that will result in the event of delay. In many construction contracts—mainly public but also private—the amount selected does not in reality represent an attempt to estimate damages. In such contracts, it is almost impossible to estimate the economic loss caused by delay. For example, it is likely that the $1,000 a day selected in the *Bethlehem* case did not reflect the city of Chicago's judgment as to the economic loss that it or its citizens would suffer in the event the project was delayed. It is more likely that the amount was selected to make it more economical for the contractor to perform on time than to delay and pay damages. Although courts do not *overtly* concede that they are doing so, their enforcement of liquidated damages clauses for a delay in many construction contracts amounts to enforcing reasonable penalties.

There are possible variations to a per-diem formula. For example, a contract made by the San Francisco Golden Gate Bridge District for repair of the Golden Gate Bridge specified that for each ten minutes of delay, the liquidation amount would be $1,000. This was done to obtain the promised performance by the contractor of finishing before the morning rush hours.[79]

Another variant was observed in *Grenier v. Compratt Construction Co.*,[80] which involved a blasting contract. The work had to be done before constructing the roads in a subdivision. The contract stated that damages were liquidated at $1,500 at the end of the first week, an additional $2,000 at the end of the second week, an additional $2,500 at the end of the third week, and $3,000 for each

---

[77]*United States v. Foster Constr. (Panama) S.A.*, 456 F.2d 250 (5th Cir.1972).

[78]*Hall Constr. Co., Inc. v. Beynon*, supra note 76 (dictum).

[79]San Francisco Sunday *Examiner & Chronicle*, Feb. 19, 1984, p. B-2. The English Department of Transportation charges repair contractors who are late a "lane rental" fee of up to $14,000 a day. The Economist, 1–7 Sept. 1984, p. 27.

[80]189 Conn. 144, 454 A.2d 1289 (1983).

additional week of unexcused delay. Despite a "penalty" label, the court enforced it.

Why did the amount go up for each week's delay? Perhaps the developer would incur escalating liability or lost income as the delay continued. It is more likely that the developer was seeking to coerce performance by making it increasingly expensive for the contractor not to complete on time.

Drafting also must take into account the difference between a damage liquidation and a limitation of liability. As indicated earlier, a valid liquidated damages clause establishes the damages in advance. The stipulated amount is recoverable whether actual damages, if they are established, are higher or lower. A clause limiting damages requires that actual damages be proved but limits the amount recoverable. In *Burns v. Hanover Insurance Co.*,[81] the clause read:

> The Owner will suffer financial loss if the Project is not completed by the above date. The Contractor shall be liable for and paid to the Owner on an actual expense basis as established by receipts, not more than $1,000 for packing and storage of furnishings and $30 per day for temporary accommodation.

The majority held this to be a valid liquidated damages clause. The dissent correctly stated that the clause simply put a limitation of liability but required that actual losses be established.

Sometimes liquidation clauses are joined with bonus provisions. Under such clauses, the contractor forfeits a designated amount for each day of unexcused delay but gains a designated daily bonus if it completes the project in advance of the completion date. Although enforceability does not require that a bonus be attached to a damage liquidation, it may be tactically desirable to do so. It may help to enforce the liquidation clause, with its "mutuality" attractiveness. Its use may also make it appear that the amount had actually been bargained by the contractor and owner. However, a bonus clause should not be used unless it is very important to obtain performance in advance of the completion date.

Liquidated damage clauses should specifically give the owner the right to deduct the amount from the final payment or any retainage even if it may be possible to do so without such a provision.

Damage liquidation for unexcused contractor delay is usually desirable even though claims based on liquidated damage clauses are often traded away in a final settlement. Liquidated damage claims must be drafted carefully. First, the applicable law must be determined. Second, the clause must be tailored to the particular type of delay to which it is expected to apply. Third, the amount selected should take into account the importance of timely completion, the likely lost-use value, and the likelihood that the amount selected will actually achieve the objective.

SECTION 26.10 Owner-Caused Delay

A. Assumed Contractual Risk: No-Damage Clauses

Increasingly, contractors make large claims for delay damages. As a result, it is becoming even more common for clauses in the contracts—particularly public contracts and private ones drafted largely with the interests of the owner in mind—to attempt to make the contractor assume the risk of owner-caused delay.

Public entities are often limited by appropriations and bond issues. As a result, they must know in advance the ultimate cost of a construction project. To do this, many public entities use the disclaimer system for unforeseen subsurface conditions described in Section 25.05. Similarly, they wish to avoid claims being made at the end of the project based on allegations that they have delayed completion or required the contractor to perform its work out of sequence. These entities recognize that barring claims may cause higher bids, but they would prefer to see bidders increase their bids to take this risk into account rather than face claims at the end of the job.

Before looking at the troublesome question of the validity of no-damage clauses, it is important to note other methods that can be used to eliminate or reduce the likelihood of such claims. The contract may specify that the owner has the right to delay the contractor and that the interference is not a contract breach. Another technique is to contend that any time-extension mechanism is the *exclusive* remedy. Generally, the availability and use of the

---

[81]454 A.2d 325 (D.C.App.1982).

time-extension mechanism does not impliedly preclude delay damages.[82] The AIA's time-extension mechanism specifically states that it does *not* bar the contractor from recovering delay damages.[83]

An indirect technique for limiting delay claims is to require written notice by the contractor to the owner or design professional if events have occurred that later will be asserted as justifying delay damages. Often this notice is stated to be a condition precedent to any right to delay damages. Although such clauses are not looked on favorably by courts and are frequently found to have been waived,[84] noncompliance with a notice provision can be the basis for barring a claim for delay damages.[85]

Many public and some private construction contracts meet the delay damage problem head on. Rather than setting up notice conditions or specifying what can be recovered, these contracts contain no-damage or no-pay-for-delay clauses. Such clauses attempt to place the entire risk for delay damages on the contractor and to limit the contractor to time extensions. Generally, such clauses are enforced. However, exculpation clauses of this type are rarely attractive to courts, and in special circumstances, they will not be given effect. Like all contract clauses, they must be interpreted to determine what they are intended to cover. Using the process of interpretation, no-damage clauses do not preclude recovery for delay damages if the delay was not the type contemplated, if the delay was an abandonment by the owner, or if the delay was caused by bad faith or active interference.[86]

New York law illustrates the difficulty courts have had in determining whether such clauses will be given effect. In 1983, the New York Court of Appeals in *Kalisch-Jarcho, Inc. v. City of New York*[87] held that no-damage clauses bar recovery of delay damages unless the owner has committed fraud, malice, or bad faith or has shown a reckless indifference to the rights of others, a standard apparently requiring actual malice or conduct approaching intentional wrongdoing. The court based its unwillingness to interfere on its description of such public contracts as having been made at arm's length by sophisticated contracting parties.

Yet in 1986, the court in *Corinno Civetta Construction v. City of New York*,[88] a consolidated opinion dealing with a series of cases, appeared to step back *slightly* from the hands-off position it expressed in 1983. It added illustrations of facts to those given in the *Kalisch-Jarcho* case and noted that damages for delay could be recovered if any of the following is true:

1. The delay is not contemplated by the parties.
2. The delay is so unreasonable as to constitute an intentional abandonment of the contract by the owner.
3. The delay results from the owner's breach of a fundamental obligation of the contract.

These cases had involved delays caused by a variety of causes, including the city's moratorium on street openings, unanticipated subsurface conditions, a dispute over the type of pipe joint to be used, omission of design data, incompatability of equipment supplied by other separate prime contractors, and the city's failure to coordinate the work of the separate prime contractors. The city presented evidence that these delays were contemplated or the city attempted to resolve the problems indicating that there was no abandonment or breach. The court concluded that a motion for a summary judgment granted the city had not been

---

[82]*Selden Breck Constr. Co. v. Regents of Univ. of Michigan,* 274 F. 982 (E.D.Mich.1921). See *Glantz Contracting Co. v. General Elec.,* 379 So.2d 912 (Miss.1980); *Bender-Miller Co. v. Thomwood Farms, Inc.,* 211 Va. 585, 179 S.E.2d 636 (1971) (remedy not exclusive).

[83]AIA Doc. A201, ¶ 8.2.3.

[84]*Montgomery Ward & Co. v. Robert Cagle Bldg. Co.,* 265 F.Supp. 469 (S.D.Tex.1967).

[85]*Tuttle/White Constructors, Inc. v. State,* 371 So.2d 1096 (Fla.Dist.Ct.App.1979).

[86]*Peter Kiewit Sons Co. v. Iowa Southern Utilities Co.,* 355 F.Supp. 376 (S.D.Iowa 1973) (citing many authorities). In 1984, California added § 7102 to its Public Contract Code, stating that no-damage clauses do not apply to unreasonable delays or those not contemplated by the parties. See Cal.Stat. 1984, ch. 42. Similarly, see Or.Rev.Stat. § 279.063 enacted in 1985. In Virginia, a contractor cannot waive damages for unreasonable delay in public contracts. Va.Ann.Code § 11-56.2 (Michie 1992). Similarly,

Colorado outlawed no-damage clauses in public contracts. Colo.Rev.Stat. § 24-91-103.5(1)(a) (West 1989). See Parvin and Araps, *Highway Construction Claims,* 12 Pub.Cont.L.J. 255, 277–281 (1982); Vance, *Fully Compensating the Contractor for Delay Damages in Washington Public Works Contracts,* 13 Gonz.L.Rev. 410 (1978); Comment, 28 Loy.L.Rev. 129 (1982).

[87]58 N.Y.2d 377, 448 N.E.2d 413, 461 N.Y.S.2d 746 (1983).

[88]67 N.Y.2d 297, 493 N.E.2d 905, 502 N.Y.S.2d 681 (1986).

proper. A trial would be needed to determine whether the clause would be upheld.

If the *Corinno* case moderated the hands-off attitude expressed in *Kalisch-Jarcho,* the message did not seem to reach one intermediate appellate court. In *Buckley & Company v. City of New York,*[89] the delay was due principally to unexpected subsurface conditions. The contract granted an equitable adjustment in the event of unanticipated subsurface conditions. This, according to the court, indicated that the parties were aware that these conditions might delay completion of the project. Also, the court did not find that the city had been grossly negligent or had maliciously interfered but had been simply negligent in its evaluation of the subsurface conditions and its design of the cofferdam. The *Buckley* case did not read the *Corinno* case as retreating from the strictness of the *Kalisch-Jarcho* case.

New Jersey has also tended to enforce such clauses. *Broadway Maintenance Corp. v. Rutgers*[90] (reproduced in part in Section 17.04(C)) enforced such a clause. The court stated:

> No damage provisions are a part of the economic package upon which the parties agree. The contractor who chooses to accept these risks will reflect the accompanying responsibility in his price.[91]

It is important to look at the intention of the parties. Such inquiry determines whether the type of delay that occurred was the type the parties intended to exclude as a basis for a delay claim.

The contractor had complained of backfill problems, slow-pouring concrete, and lack of temporary heat. The court affirmed a finding by the intermediate court that the situations were *not* sufficiently exceptional to fall *outside* the no-damage clause. Such situations are ordinary and usual types of damage that most contractors frequently encounter.

Another court[92] held that such a clause would bar *delay* damages, which it defined as time lost when work cannot be performed because materials have not been delivered or preliminary work done. But it does *not* bar a claim based upon *hindering* the work, such as failure to coordinate the work or

supply temporary heat. This may only illustrate *active* interference. Another illustration of the exception for active interference was based upon the owner's testimony that he would "break" the contractor before he would pay.[93]

The preceding discussion has assumed a clause that bars any damages for owner-caused delay. Some contracts contain provisions that limit the contractor's remedy to a time extension and to only designated damages, such as reimbursement for its reasonable additional costs resulting from owner-caused delay, excluding recovery for any additional profit or increasing the fee. Another modification on the "no-pay" or "no-damage" provision is to provide that the contractor can recover delay damages only after a designated number of days of delay caused by the owner.

Some contract clauses permit *some* delay damages but avoid large claims for diminished productivity implemented by open-ended total cost formula. One contract provided that the contractor could recover delay damages but only for premiums paid on its bond and for wages and salaries of workers needed to maintain the work, the plant, and the equipment during the delay.[94]

Yet not all courts have been as unwilling as those of New York and New Jersey to tamper with such clauses. For example, Pennsylvania held that such clauses will not be enforced if there is affirmative or positive interference or a failure to act in some essential matter necessary to the prosecution of the work.[95] The court held that the failure to keep a lake in a drained condition actively interfered with a contractor's lake excavation contract. The court looked at the facts rather liberally from the perspective of the contractor and concluded

---

[89]121 A.D.2d 933, 505 N.Y.S.2d 140 (1986).
[90]90 N.J. 253, 447 A.2d 906 (1982).
[91]447 A.2d at 914.
[92]*John E. Green Plumbing & Heating Co., Inc. v. Turner Constr. Co.,* 742 F.2d 965 (6th Cir.1984).

[93]*Newberry Square Dev. Corp. v. Southern Landmark, Inc.,* 578 So.2d 750 (Fla.Dist.Ct.App.), case dismissed, 584 So.2d 999 (Fla.1991).
[94]*Calumet Constr. Corp. v. Metropolitan Sanitary Dist.,* 222 Ill.App.3d 374, 581 N.E.2d 206 (1991), appeal denied, 143 Ill.2d 636, 587 N.E.2d 1012 (1992).
[95]*Coatesville Contractors & Eng'rs, Inc. v. Borough of Ridley Park, Pa.,* 509 P.2d 553, 506 A.2d 862 (1986). Similarly, see *Carrabine Constr. Co. v. Chrysler Realty Corp.,* 25 Ohio St.3d 222, 495 N.E.2d 952 (1986); *McDevitt & Street Co. v. Marriott Corp.,* 713 F.Supp. 906 (E.D.Va.1989), affirmed in part, 911 F.2d 723 (4th Cir.1990) (clause upheld unless delay unreasonable, intentional, or fraudulently caused), upon remand as to issue of interest, 754 F.Supp. 513 (E.D.Va.), affirmed in part and revised in part without opinion, 948 F.2d 1281 (4th Cir.1991).

that this was not the type of interference contemplated by the parties. Similarly, the bad-faith or active-interference exception was applied in *United States Steel Corp. v. Missouri Pacific Railroad Co.*[96] The claim of active interference was based on the owner's directing the contractor to proceed before requisite work of an earlier contractor had. been completed. Access to the work was denied during completion of the earlier contractor's work, causing a delay of 175 days. The owner granted a time extension for this delay but, based on a no-damage clause, refused to pay any delay damages.

The court noted that the owner had an implied obligation to refrain from anything that would reasonably interfere with the contractor's opportunity to proceed with its work and to allow the contractor to carry on that work "with reasonable economy and dispatch." The court concluded that issuance of the notice to proceed was an affirmative, willful act and that the owner's bad faith was demonstrated by its knowledge of circumstances that would prevent the contractor from proceeding timely with its work.

The popularity of delay or disruption claims by contractors will force courts to pass on the validity of such exculpatory clauses. Undoubtedly, case decisions in different jurisdictions will apply different tests and will come to different outcomes. Essentially, such clauses will be given effect if the delays or disruptions encountered were ones that could have been expected in the particular contract in question and that could not have been reasonably avoided by the owner.

Whether such a clause should be included in a contract depends both on the likelihood of its being enforced in a proper case and on the willingness to take costs "at the front end" rather than through claims at the end of the work. However, the use of these clauses in a highly competitive construction market can encourage some bidders to bid low and take their chances.

B. Subcontractor Claims

Delays caused by the owner not only harm the prime contractor but also may harm subcontractors. The absence of a contract between subcon-

tractors and the owner generally precludes direct legal action and, as a rule, direct negotiations between subcontractors and owners over delay claims. Often the subcontractor's claim is processed by the prime contractor. This processing is dealt with in Section 28.08(B).[97]

C. Liquidated Damages

Until recently, it was rare for owners to liquidate damages for delays they caused. Public owners usually protected themselves, or at least hoped to, by the use of no-damage clauses. However, it is becoming more common to liquidate damages as an alternative to exposure to the often open-ended and difficult to establish or to disprove delay damage claims. Although no judicial analysis of these clauses has been found, very likely courts will enforce them. The process of establishing the additional costs incurred by the contractor for delay is even *more* complex and *more* difficult than establishing lost use by owners, even public owners. This should encourage enforcement, provided the amount bears some reasonable relationship to the anticipated or actual damages.

D. Measurement

The measurement of the value of the contractor's claim when without excuse the owner delays the contractor's performance is dealt with in Section 27.02(F).

E. Records

In delay disputes, the party with the best records has a great advantage. Each party should keep job records such as the site representative's daily field reports, correspondence, memoranda, photographs or video recordings, and change orders. These records should include data on labor, equipment, and materials used for each activity and should document the cause and impact of every delay. See Section 27.05.

---

[96]668 F.2d 435 (8th Cir.1982).

[97]Sometimes the prime and the subcontractor use liquidating agreements under which the prime confesses liability to the subcontractor for owner-caused delay and the subcontractor releases the prime for all other liability. Under such agreements, the subcontractor is relegated to whatever delay damages the prime can recover from the owner (discussed in Section 28.08(B)).

SECTION 26.11 Third-Party Claims

Delayed completion can harm third parties, such as tenants who expect to move into the completed project, tenants who are displaced for an unreasonable period during construction, or even others, such as restaurants, who expect to earn revenue from tenants of the completed project or their employees or customers. Suppose these third parties seek to transfer their losses to those they claim responsible for the delay, such as owner, design professional, or contractor.

Claims against a contract-connected party, such as a claim by a tenant against its landlord-owner would be decided by reference to the contract, contract law, and contract remedies. Claims by a noncontract-connected party, such as a tenant against the contractor or design professional, are more complex. The issue arose in *J'Aire Corp. v. Gregory*[98] (discussed in Sections 14.08(E) and 14.11(C)). The court allowed a tenant to assert a claim against a contractor for negligent delay in completing a renovation project that adversely affected its restaurant business. In essence, the tenant contended that the contractor *negligently* interfered with its restaurant business. Usually the law requires that interference with contractual relations or prospective advantages be *intentional*.[99] But the court, using an ''easy to satisfy'' foreseeability test, held that the tenant should be allowed to show that the contractor's negligence caused it economic harm. States that do not permit tort claims for purely economic losses would not permit such a claim to be brought. However, in such jurisdictions, a claimant may try to establish that it is an intended beneficiary of a contract to which it is not a party.[100] For example, a developer who contracted with the city asserted, unsuccessfully, that it was an intended beneficiary of the contract between the city and a contractor the city had hired who had not performed in accordance with its time commitment.[101]

Finally, such clauses can be used *defensively* by third parties in some states. For example, an engineer defended on the basis of a no-damage clause in the construction contract to which she was not a party.[102]

As noted earlier,[103] injection into a series of planned transactions—the contract between owner and contractor and owner and tenant—of tort concepts can destroy contract planning. Yet claims against major construction participants based on tort or assertions that the claimant is an intended beneficiary are growing. Risk management planning, such as realistic pricing, contract language, and other devices, is becoming essential. It was discussed in Chapter 15 in the context of the design professions. It is also the subject of Chapters 32 and 33, dealing with indemnification and suretyship.

---

[98]24 Cal.3d 799, 598 P.2d 60, 157 Cal.Rptr. 407 (1979).
[99]Restatement (Second) of Torts §§ 766(B)(C) (1979).

[100]See Section 14.08(B).
[101]*Chard Realty, Inc. v. City of Shakopee,* 392 N.W.2d 716 (Minn.App.1986).
[102]*Bates & Rogers Constr. Co. v. Greeley & Hansen,* 109 Ill.2d 225, 486 N.E.2d 902 (1985).
[103]See Sections 14.05(E) and 14.11(C).

PROBLEMS

1. C and O contract for C's construction of a five-story office building for $500,000. The completion date is December 1. The contractor provides that for every day of unexcused delay, C shall be chargeable with liquidated damages of $3,000 a day. The building was completed twenty days late, and the delay was not excused by any events set forth in the *force majeure* clause. What can O recover for the delay? What added facts would be helpful in answering this question?

Suppose the agreed damage figure had been $1,000 a day and C abandoned the project on December 30. O obtained a replacement who completed the project twenty-five days later. The substitute contractor cost $60,000 more than the contract price. What can O recover from C?

2. You have been asked to design and monitor construction of a luxury single-family home for O. The contract price is $500,000. You know that O is a particularly demanding client. She wants things "just so." Also, she is likely to make many design changes, and she likes to visit the job site. Should you advise her attorney to include a no-damage clause? What would be your reasons?

3. Your client wants to use an AIA Doc. A201, but she is concerned about the completion point for determining time commitments being substantial completion. She asks if this is fair. How would you answer her question?

Claims: By-Products of Construction Process

SECTION 27.01 Introduction

Review Chapter 6, the basic building block dealing with remedies for breach of contract.

The law seeks to compensate the contracting party who has suffered losses because of the other party's breach of contract. Generally, it does not punish the party that has breached the contract.[1] The court judgment seeks to put the injured party in the position it would have reached had there been *no* breach (expectation) or seeks to restore the party to the position it occupied before performance began (restitution).

The law developed conventional formulas to implement the compensation objective. At times, a choice can be made between different formulas. Judicial attempts to determine what *would* have happened can lead to unsupported speculation and guesswork. The complexity of construction performance, especially in large projects, makes it difficult to determine what *has* happened.

As a result, claims measurement uses rough approximations that may be all that is available. This chapter reveals a constant tension between demanding specific and solid proof and the willingness to use formulas or approximations.

Measurement of claims and remedies for breach of the construction contract have generated a sea of reported case decisions from the fifty American jurisdictions. The differing judicial attitudes toward claims, the multitude of American jurisdictions, and the inevitable factual variations make generalizations treacherous, except at the most abstract levels. Although the basic principle of compensation is the foundation of claims measurement (except for the increasing use of punitive damages), implementation of the compensation objective can invoke different formulas. This chapter describes the varying rules, discusses the reasons behind them, provides illustrations, and notes trends.

SECTION 27.02 Measurement: Contractor vs. Owner

A. Illustrations

Claims by contractors against owners can arise in many contexts. The principal claims are based on the following:

1. Refusal of the owner to permit the contractor to commence performance after the contract has been awarded.
2. Wrongfully terminating the contractor's right to perform during performance.
3. Committing acts that justify the contractor in ceasing performance.
4. Failing to pay the contractor for work performed under the contract.
5. Committing acts that increase the contractor's cost of performance.

B. Cost Contracts

This section does not deal with claims by the contractor for additional compensation under cost contracts. Usually, such claims are based on assertions

---

[1]See Section 6.04.

that costs were incurred and that the owner's payment did not cover those costs. Those questions generally involve factual determinations of whether particular work was done or whether it was called for under the contract and the value of the work performed. Although the contractor must establish the particular work that was done and its cost, the owner has the burden of showing that the cost of particular work was unreasonable.[2] The purpose of the cost contract is to place the risk of costs incurred on the owner.

C. Project Never Commenced

Although there may be consequential damages asserted when the contractor is not given the opportunity to commence performance, general damages in such cases are the lost profits on the contract. Because lost profits arise when the contractor has commenced performance and is then wrongfully terminated or has justified grounds for ceasing performance, lost profits are discussed in the next subsection.

D. Project Partially Completed: Damages

The damages award should place the contractor in the position it *would* have been in had the owner performed in accordance with the contract. Three possible formulas, which for convenience are referred to as Formulas I, II, and III, can determine the amount of recovery. Formula I determines what *would* have happened by awarding the contractor the contract price. This is the amount it *would* have had on completion. But from this, the cost of completion must be deducted.[3] This is an expense saved by the breach. Progress payments received also must be deducted. Completion cost is what a reasonably prudent contractor in the contractor's position would have had to spend to complete the work.[4]

A contractor who finds it difficult to prove what its costs *would have* been can use the costs of a suc-

cessor, provided those costs were reasonable.[5] If the owner can establish that the contractor would have sustained a loss had it completed the project, this loss will be deducted from the recovery.[6] (As seen in (E), a deduction need not be made if the claim brought by the contractor is based on restitution.)

Formula II is the contractor's expenditures in part performance, including preparation, and, if the contractor can establish them, profits on the entire project.[7] To illustrate Formulas I and II, assume that a project has the following figures:

1. Contract price—$100,000.
2. Expenditures in part performance—$60,000.
3. Cost of completion—$30,000.
4. Progress payments received—$50,000.

Under Formula I, the contractor would receive

$100,000	(contract price)
− 30,000	(cost of completion)
$ 70,000	
− 50,000	(progress payments received)
$ 20,000	(net recovery)

Under Formula II, the recovery would be

$ 60,000	(expenditures in part performance)
+ 10,000	(profits—$100,000 contract price less $90,000, total cost of project)
$ 70,000	
− 50,000	(progress payments received)
$ 20,000	(net recovery)

The result using Formulas I and II is identical.

But suppose expenditures in part performance were $90,000. Had the contract been completely performed, the contractor would have lost $20,000. Using Formula I, the contractor would still receive $20,000, inasmuch as the two principal items of the formula are not changed. However, the contractor has already expended $90,000 and has received $70,000 through progress payments ($50,000) and the money award ($20,000). This would create the

[2]*Sloane v. Malcolm Price, Inc.*, 339 A.2d 43 (D.C.App.1975).
[3]*Tull v. Gundersons Inc.*, 709 P.2d 940 (Colo.App.1985). See Kirksey and Smedley, *Protecting the Contractor's Expectation Interest After the Owner's Substantial Breach*, 6 Constr. Lawyer No. 1 (Oct. 1985) p. 1.
[4]*Watson v. Auburn Iron Works, Inc.*, 23 Ill.App.3d 265, 318 N.E.2d 508 (1974).

[5]*Carchia v. United States*, 485 F.2d 622 (Ct.Cl.1973). But in *Edward Elec. Co. v. Metro. Sanitary Dist.*, 16 Ill.App.3d 521, 306 N.E.2d 733 (1973), the contractor's attempt to base the cost of completion on the completed work was unsuccessful.
[6]*Watson v. Auburn Iron Works, Inc.*, supra note 4.
[7]Restatement (Second) of Contracts § 347, Comment D (1981).

same loss of $20,000 that it would have suffered had it completed the contract.

Formula II, that is, expenditures in part performance coupled with profits or less losses, would compute recovery as follows:

$90,000	(expenditures in part performance)
− 20,000	(loss that would have been suffered)
70,000	
− 50,000	(progress payments received)
$20,000	(net recovery)

Under Formula II, the contractor would be in the same position as under Formula I. The contractor expended $90,000 and received $70,000, leaving the same $20,000 loss.

Formula III entitles the contractor to such proportion of the contract price as the cost of the work done bears to the entire cost of completing performance and, for the remaining portion of the work, the profit that would have been made as to that work.[8] Formula III should produce the same recovery on profitable contracts but will produce a different result in losing contracts. However, as Formula III is rarely used, detailed illustrations of its application will not.be given.

Compensation can involve the difficult question of establishing lost profits. Some contracting systems make a sharp differentiation between profits on performed and unperformed work. For example, federal procurement policies rarely give recognition to the profit on unperformed work.[9] However, the ordinary contract measurement formulas that have been set forth do not make this differentiation. In such cases, the principal problem has been the standard of certainty required to recover lost profits.

Early case decisions took a negative view toward lost profits as part of contract damages. Many cases held that profits could not be established for a new business.[10] Even lost profits by someone in an existing business were closely scrutinized. Was the profit reasonably foreseeable by the contract parties at the time the contract was made? In contracts for the sale of goods, this often excluded recovery of unusual resale profits.

In construction contracts, however, the principal problem has been certainty.[11] How does the contractor establish not only the profits on work performed but also the profits on unperformed work? If it uses Formula I—that of contract price less cost of completion—lost profits need not be established directly. However, what it *would* have cost to complete the project can often be difficult to show. As a result, more commonly, contractors use Formula II—expenditures in part performance coupled with profits. Although the former can be difficult to establish if the contractor has not kept good records, it is generally easier to prove than cost of completion. However, the contractor will have to establish profits.

E. Project Partially Completed: Restitution[12]

The law generally permits a contracting party who has *not* fully performed to use an alternative measure of recovery where the other party has committed a serious breach.[13] Breaches sufficiently

---

[8]*Kehoe v. Borough of Rutherford*, 56 N.J.L. 23, 27 A. 912 (1893).

[9]*General Builders Supply Co. v. United States*, 409 F.2d 246 (Ct.Cl.1969). A weighting formula developed by the Chief of Engineers for determining profit on performed work was cited and applied in *Norair Eng'g. Corp.*, ASBCA 10856, 67-2 BCA ¶ 6619.

[10]D. DOBBS, REMEDIES. 153–155 (1973).

[11]See Section 27.04.

[12]One confusing aspect of restitution is the frequent mention of the term *quantum meruit*. Often courts state that a particular action is brought on a *quantum meruit*. Early in English legal history, a party seeking relief had to bring itself within the language of writs issued by public officials commanding persons to appear before one of the King's courts. One frequently used writ was *Assumpsit*. This writ was broken down into a number of subwrits, one of which was called *General Assumpsit*. Because of common usage of certain subwrits such as *General Assumpsit*, many of them became known as Common Counts. One of the Common Counts was given the name of *quantum meruit*, which usually dealt with a method of recovering for the reasonable value of services furnished the defendant at his request. It was used as a method of recovering for breach of contract and could be used when the claim was based on unjust enrichment. This treatise refers to restitution rather than *quantum meruit*, even though some courts still use that term in describing the process by which the plaintiff recovers based on benefit conferred rather than damages.

[13]*Kass v. Todd*, 362 Mass. 169, 284 N.E.2d 590 (1972) (dictum). The seriousness is similar to that needed for a power to terminate. See Section 34.04(A). But see *United States v. Mountain States Constr. Co.*, 588 F.2d 259 (9th Cir.1978) (Washington law), which required a tortious breach. This is not in the mainstream.

serious to justify restitution have been failure to make progress payments,[14] excessive changes,[15] and failure to perform those acts during performance that would allow the contractor to perform in the most expeditious way.[16]

Expectation (discussed in (D)) looks forward and seeks to determine what would have happened had the parties completed performance. Restitution looks backward to the position the parties were in at the time they entered into the contract. In some contracts, this is accomplished by a party's returning any performance the other party may have conferred on it. In construction, actual restitution is ordinarily impracticable, as the work has been attached to the owner's land. Instead, the contractor is entitled to receive the value that its performance has benefited the owner, generally the reasonable value of the materials and labor it contributed to improving the owner's land.

Restitution should be measured objectively, with the contractor recovering an amount that equals what the owner would have had to pay to purchase the materials and services from one in the contractor's position at the time and place the services were rendered.[17] Ordinarily, the contractor seeks to introduce evidence of its actual costs incurred in performing its contractual obligations. Although this is not conclusive, it is likely to be accepted by the court unless the owner can establish that they were not market rate costs or that they were costs incurred to correct deficiencies chargeable to the contractor. One case affirmed a judgment for a subcontractor in its claim against the prime contractor by pointing to the opinion testimony of the subcontractor as to the direct costs and the court's examination of the site.[18] Another court looked at the bids submitted by unsuccessful bidders in determining the reasonable value of the services.[19]

The contractor should be able to recover its overhead costs, as they are incurred by the contractor in improving the owner's land and should be considered as benefiting the owner.[20]

Clearly, the contractor who seeks restitution cannot recover profit on unperformed work. It has been allowed to recover profit on work that has been performed.[21] This would be included in what other contractors would have charged.[22]

The most difficult problem relates to the effect of the contract price on the contractor's recovery. Should the law ignore the contract price, use it as evidence of the value of the benefit conferred, use the pro rata contract price to actually measure the recovery, or use the contract price to put a limit on the contractor's restitutionary claim? Although the general tendency has been to use the contract price only as evidence of value,[23] some cases have limited the contractor to the contract price.[24]

The effect of contract price on recovery has stirred great debate. The majority rule—that of giving the contract price only evidentiary effect—can allow a contractor who has entered into a losing contract to at least break even. One argument given to justify this result has been that the restitutionary principle, that is, looking backward and avoiding unjust enrichment, has as much validity as expectation, that is, looking forward to what would have happened had the contract been performed properly. Another justification for the majority rule is that an owner who has breached a contract should not be able to take advantage of the contract price. A reason frequently given is that the expenses most likely exceed the contract price because of breaches by the owner. Perhaps—most persuasive—restitution not limited to the contract price is relatively easy to administer. If expectation (discussed in (D)) were used, losses would be deducted. This com-

---

[14]*United States v. Algernon Blair, Inc.*, 479 F.2d 638 (4th Cir.1973).

[15]*Glassman Constr. Co. v. Maryland City Plaza, Inc.*, 371 F.Supp. 1154 (D.Md.1974); *Rudd v. Anderson*, 153 Ind.App. 11, 285 N.E.2d 836 (1972).

[16]*Leo Spear Constr. Co. v. Fidelity & Cas. Co. of New York*, 446 F.2d 439 (2d Cir.1971).

[17]*United States v. Algernon Blair, Inc.*, supra note 14.

[18]*Leo Spear Co. v. Fidelity & Cas. Co. of New York*, supra note 16.

[19]*Paul Hardeman, Inc. v. Arkansas Power & Light Co.*, 380 F.Supp. 298 (E.D.Ark.1974).

[20]*Leo Spear Constr. Co. v. Fid. & Cas. Co. of N.Y.*, supra note 16.

[21]Ibid. See also *C. Norman Peterson Co. v. Container Corp. of Am.*, 172 Cal.App.3d 628, 218 Cal.Rptr. 592 (1985) (overhead and profit).

[22]*W.F. Magann Corp. v. Diamond Mfg. Co.*, 775 F.2d 1202 (4th Cir.1985) (profit not recoverable per se but can be recovered if relevant to reasonable value).

[23]*United States v. Algernon Blair, Inc.*, supra note 14; *Paul Hardeman, Inc. v. Arkansas Power & Light Co.*, supra note 19; *Kass v. Todd*, supra note 13.

[24]*United States v. Mountain States Constr. Co.*, supra note 13; *Johnson v. Bovee*, 40 Colo.App. 317, 574 P.2d 513 (1978).

putation would require a determination of cost of completion, often at best a guess. If the contract price is not controlling, all that is needed is to determine the reasonable value of the work performed.

Those who support using the contract price either to measure the value or to limit recovery point to the anomaly that can result if the contractor *does* complete performance. Restitution *cannot* be used if the contractor has completed performance and all that is left is for the owner to pay the balance of the contract price. If the owner's breach has caused the contractor to lose money, the contractor must establish such losses. (As shall be seen, various crude formulas are sometimes allowed, making proof easier. See (F).) Barring the owner from using the contract price because it has breached does not take into account the possibility or even the likelihood that the question of breach has been a close one, with much fault attributable to both parties. Why take away the benefit of the bargain made by the owner because it is ultimately determined that the owner has breached?

The difficulty with the majority rule is that it can encourage a contractor in a losing contract to claim that it has adequate grounds for termination and to stop work. If it continues to perform, it will almost certainly lose money and face difficult problems if it seeks to recover compensation based on breaches by the owner. If it stops and claims the right to terminate, it runs the risk that it will be found to be in default and have to pay heavy damages. These damages may be no greater, however, than the contractor will suffer if it continues to perform a losing contract when there is very little hope of turning things around. In the confusion of a construction dispute in court, a good chance exists that the contractor may be exonerated.

This section has discussed restitution as a method for measuring the recovery for breach of contract. In *C. Norman Peterson Co. v. Container Corp. of America*,[25] restitution, including an allowance for overhead and profit, was granted a contractor based on an "implied" agreement to abandon the original cost contract with its guar-

anteed maximum price coupled with the subsequent "implicit" understanding by the parties that the parties would proceed on a *quantum meruit* basis. These implications resulted from an incomplete design and vast number of changes.

F. Project Completed: *C.B.C. Enterprises v. U.S.*, *Amp.-Rite Elec. Co. v. Wheaton San. Dist.*, and *New Pueblo Constructors v. State*

A contractor who has fully performed but has not been paid the balance of the contract price cannot measure its recovery by the reasonable value of its services.[26] It recovers only the unpaid balance of the contract price and any other losses it can prove.

Limitation on the restitutionary remedy is sometimes justified on historic grounds. However, the modern justification is the ease of using the contract price as a measure compared with making the more complicated factual determination of the reasonable value of the services. Here, unlike the past performance discussed in (E), there is no need to determine value.

Most complex construction contract disputes involve claims by the contractor who has completed the project and not, as a rule, for the unpaid balance of the contract price (it has probably been paid), except for the additional costs it incurred for which it claims compensation. The contractor asserts that defective specifications required it to perform in an inefficient and costly manner, and, most important, that unexcused delays caused by the owner increased its cost of performance.

The problems of defective specifications and unforeseen subsurface conditions, to the extent that they do *not* involve delay, cause less difficulty. It is not difficult, comparatively speaking, to establish the value of additional services and materials that were necessary because of defective specifications or unforeseen subsurface conditions. Even more, in the case of claims based on unforeseen subsurface conditions, the justification for additional compensation is usually based on a differing site conditions clause, which either may contain a formula for measuring the additional compensation or has developed legal rules that provide a solution for measuring compensation. (Where delay does result, the crude formulas for measuring damages

---

[25]Supra note 21. See also *Hensel Phelps v. King City*, 57 Wash.App. 170, 787 P.2d 58 (1990) (*quantum meruit* not allowed where contract exculpated owner or provided relief such as equitable adjustments and time extensions).

[26]*Kass v. Todd*, supra note 13.

such as total cost, jury verdict, and extended home office overhead may have to be applied.) However, the greatest difficulty arises in those cases where the contractor claims that it had to perform work in an inefficient manner mainly because of excessive changes by the owner or administrative incompetence of the owner or someone for whose conduct it is responsible. To set the scene, a court facing such a claim stated:

> We note parenthetically and at the outset that, except in the middle of a battlefield, nowhere must men coordinate the movement of other men and all materials in the midst of such chaos and with such limited certainty of present facts and future occurrences as in a huge construction project such as the building of this 100 million dollar hospital. Even the most painstaking planning frequently turns out to be mere conjecture and accommodation to changes must necessarily be of the rough, quick and *ad hoc* sort, analogous to ever-changing commands on the battlefield. Further, it is a difficult task for a court to be able to examine testimony and evidence in the quiet of a courtroom several years later concerning such confusion and then extract from them a determination of precisely when the disorder and constant readjustment, which is to be expected by any subcontractor on a job site, become so extreme, so debilitating and so unreasonable as to constitute a breach of contract between a contractor and a subcontractor. This was the formidable undertaking faced by the trial judge in the instant case and which we now review on the record made by the parties before him.[27]

Walter Kidde Constructors, Inc. v. Connecticut[28] involved a contractor (Kidde-Briscoe) claim in a dispute over a project that finished almost 900 days later than planned. Much of the responsibility fell on the public owner. The owner failed to make certain parts of the site available as promised. It issued many hold orders that substantially interfered with and disrupted construction while plans were being redesigned. It was late in processing change orders and approving shop drawings while still insisting that the contractors accelerate their work under threat of imposing a 250-dollar-per-day penalty for not completing within the time require-

ments of the contract. In describing the effect of these delays, the court stated:

> The long delay in completing construction had a devastating impact on Kidde-Briscoe. They were not only required to provide labor and materials far in excess of and different from that called for in the original contract but to perform work out of sequence at an accelerated rate and in a manner not planned and not utilized under normal conditions. The unanticipated almost two and a half years of delayed construction extended through a period of increasing inflation and escalation in labor rates and material costs, through two winters with adverse building conditions and into a time when Kidde-Briscoe was faced with an 89 day regional strike by the sheet metal workers and 137 day regional strike by the plumbers union.
>
> A computation of the consequent damages to Kidde-Briscoe involves damages of two sorts: (a) delay damages due to the extended periods of field and home office overhead and (b) damages due to disruption, loss of productivity, inefficiency, acceleration and escalation.

In addition to the items noted in the *Kidde* decision, contractors often claim and are sometimes awarded the following:[29]

1. Idleness and underemployment of facilities, equipment, and labor.
2. Increased cost and scarcity of labor and materials.
3. Utilization of more expensive modes of operation.
4. Stopgap work needed to prevent deterioration.
5. Shutdown and restarting costs.
6. Maintenance.
7. Supervision.
8. Equipment and machinery rentals and cost of handling and moving.
9. Travel.
10. Bond and insurance premiums.
11. Interest.

---

[27]*Blake Constr. Co. v. C.J. Coakley Co.*, 431 A.2d 569, 575 (D.C.App.1981).

[28]37 Conn.Supp. 50, 434 A.2d 962, 977 (1981).

[29]For an opinion that illustrates a claim involving a large number of these items, see *Contracting & Material Co. v. City of Chicago*, 20 Ill.App.3d 684, 314 N.E.2d 598 (1974). The opinion was reversed by the Supreme Court for failure by the contractor to comply with a contractual condition precedent of working double shifts, 64 Ill.2d 21, 349 N.E.2d 389 (1976), but the facts provide a good illustration.

In the *Walter Kidde* decision just discussed, the court stated that damages could include loss of productivity and inefficiency. *Luria Bros. & Co., Inc. v. United States*[30] discussed a claim by a contractor that the defendant's delay had caused lost productivity of its workforce. The workers were required to work during severe winter conditions as well as under adverse water conditions as a result of constant revisions in the contract drawings that resulted in confusion and interruption of the orderly progress of the work.

The court noted that loss of productivity can rarely be proven by books and records and that it almost always has to be proven by opinions of expert witnesses. But any expression of opinion will be looked at skeptically if there is no further evidence to support that opinion or provide a sufficient basis for making a reasonable approximation of damages.

The plaintiff presented testimony of an expert witness. Though the court found the witness was competent and well qualified to express an opinion, it did note that the witness had been a former employee of the contractor and that he might have "a certain predilection for his old employer"[31] and might have "wanted to 'help them out' all he could."[32] The witness testified that productivity was reduced, since men had to work outside on trench excavations and foundation construction in winter weather. This required the men to wear gloves and warmer clothing and work on frozen or extremely wet ground. The witness estimated that the loss of productivity during this period was $33^1/_3\%$, and as to other periods, he estimated the loss of productivity between 25% and 20%.

The court, while noting that the testimony had not been rebutted, observed that these percentages were merely estimates based upon observation and experience. The court reduced the estimates from $33^1/_3\%$ to 20%, from 25% to 10%, and from 20% to 10%.

Other approaches can be taken that are sometimes successful in proving lost productivity or, for that matter, other types of losses. One is to present evidence that measures productivity at a time other than when it was impacted adversely by job conditions and compare it to the actual productivity during the period when work was adversely affected. This is related to the total cost theory (discussed later in this subsection).

Another approach is to apply impact factors developed by trade associations that, though highly subjective, may be based upon extensive research into the question of decreased productivity. There are methods other than presenting expert testimony, as suggested in the *Luria* case.

The *Luria* case also involved an award for excess home office overhead. This has been one of the more complicated questions in measuring damages for owner breach, particularly in the field of federal procurement. Also, it is a method of measuring damages that is largely in the hands of accountants employed as expert witnesses and requires careful understanding of the difference between job site overhead, home office overhead, fixed overhead costs, and variable overhead costs.

When a contractor bids for construction work, it will usually take into account not only job site overhead but also home office overhead. Many decisions of the federal boards of appeal and the old Court of Claims have dealt with the question of home office overhead costs incurred after the original contract completion date and caused by compensable delays during the project. Home office overhead includes costs that are incurred to the mutual benefit of all contracts and cannot be tied to a specific project. One commentator stated:

> Costs which are normally included are: executive and clerical salaries, outside legal and accounting expense, mortgage expense, rent, depreciation, property taxes, insurance, utilities/telephone, auto/travel, professional and trade licenses and fees, employee recruitment, relocation, training and education, photocopying, data processing, office supplies, postage, books and periodicals, miscellaneous general and administrative expenses, advertising, interest on borrowing and other financial costs, entertainment, contributions and donations, bad debts, losses on other contracts, and bid and proposal costs.[33]

Under various federal regulations, some of these costs cannot be included in home office overhead.

---

[30]177 Ct.Cl. 676, 369 F.2d 701 (1966).
[31]369 F.2d at 713.
[32]Ibid.

[33]Long, *Extended Home Office Overhead Damages*, VI KC News, No. 2, June 1989, p. 1.

Illustrations are advertising, interest on borrowing, entertainment costs, and contributions and donations.

But even after the excluded items are eliminated, there are difficult allocation problems, as demonstrated by the following case, which involved what is sometimes called the Eichleay formula for allocating extended home office overhead as part of the contractor's damages when work is not completed within the contract time.

C.B.C. ENTERPRISES, INC., v. THE UNITED STATES

United States Court of Appeals, Federal Circuit, 1992. Rehearing and a suggestion for Rehearing En Banc Declined Dec. 1, 1992. 978 F.2d 669.

Before RICH, Circuit Judge, COWEN, Senior Circuit Judge, and CLEVENGER, Circuit Judge.

CLEVENGER, Circuit Judge.

C.B.C. Enterprises, Inc. (CBC) appeals from the September 18, 1991 decision of the United States Claims Court granting summary judgment to the United States. *C.B.C. Enters., Inc. v. United States,* 24 Cl.Ct. 187 (1991). The Claims Court held that CBC was not entitled to use the Eichleay formula to calculate extended home office overhead on a 24-day extension of contract performance. We affirm.

* * *

I.

In September, 1989, CBC contracted with the United States Navy to build certain improvements to a Marine Corps Air Station building at Cherry Point, North Carolina. The original contract price was $927,300 and CBC originally was to complete the improvements by July 11, 1990. During the course of construction the Navy modified the contract several times. On certain modifications which extended the work performance period, the parties agreed to calculate the additional home office overhead at 13.94 percent of direct costs, the rate at which home office overhead was fixed in the basic contract. The present appeal concerns the Navy's unilateral modification numbered P00003. That modification called for work with direct costs of $10,846.00 and extended CBC's work performance period by 24 days, during which time CBC's work was not suspended, delayed or disrupted. All aspects of this modification were mutually agreeable except for the amount necessary to compensate CBC for home office overhead expenses. The Navy awarded CBC $1,512.00 for those expenses, that is 13.94 percent of the modification's direct costs.

CBC contended that the Navy was required to award it home office overhead for the modification calculated at a daily rate derived from a formula known in the building arts as the Eichleay formula. The Eichleay formula was devised to calculate reimbursable home office overhead costs in the event of suspension of work on a contract, when the suspension decreases the stream of direct costs against which to assess a percentage rate for reimbursement. In the extreme, when direct costs are decreased to naught, employment of a fixed percentage rate absurdly denies recovery of any otherwise demonstrably reimbursable extended home office overhead costs. The Eichleay formula provides a method of constructively calculating daily extended home office overhead, using contract billings, total billings for the contract period, total overhead, days of contract performance, and days of delay. *See, e.g., Capital Elec. Co. v. United States,* 729 F.2d 743, 747 (Fed.Cir.1984). Using the Eichleay formula, CBC figured that its total extended home office overhead for modification P00003 was $15,317.54, or $13,805.54 more than the Navy awarded CBC. CBC submitted a claim for that amount to the contracting officer who denied the claim in November, 1990. He affirmed the Navy's position that it was impermissible to use Eichleay to calculate extended home office overhead when additional work, not suspension of work, extends the contract performance period.

* * *

The Claims Court read our decision in *Capital Electric* as a general rule that restricts use of the Eichleay formula to situations involving suspensions of work, when direct costs have been greatly reduced or eliminated. *C.B.C.,* 24 Cl.Ct. at 191. Citing *Savoy Constr. Co.,* 85-2 B.C.A. (CCH) ¶ 18,073 (1985), the Claims Court however discerned an exception to that general rule, stating that "there may be instances where use of the Eichleay formula . . . is justified in unusual situations where work performance extension is involved. . . ." *C.B.C.,* 24 Cl.Ct. at 193. The Claims Court granted summary judgment to the government, reasoning that the performance extension in this case was not "so unusual or so unreasonable" as to warrant extending use of the Eichleay method beyond work suspension situations. *Id.*

II.

CBC argues that this court, the Court of Claims and the Claims Court have consistently held that a contractor is entitled to use the Eichleay daily rate formula to recover extended home office overhead for delays, suspensions or extensions of contract performance caused by the government. Accordingly, CBC asserts that the Claims Court erred in confining use of the Eichleay formula to situations involving work suspension generally and exceptionally to contract extensions exhibiting "unusual circumstances." In CBC's view, using the Eichleay formula to calculate home office overhead for extended contract performance periods should be the rule, with only two exceptions: added work not extending the performance period and performance extensions involving added work equal to or greater than the original contract's daily rate of direct costs. In those circumstances, use of the parties' agreed percentage overhead rate would be appropriate because the direct cost stream has not been diminished. *Amicus curiae* [Friends of the Court] Associated General Contractors of America urges us to adopt CBC's arguments.

By contrast, the government contends that the Eichleay formula is never appropriately applied to mere extensions of contract performance occasioned by contract modifications adding work to be performed. From the government's perspective, the general rule requires recovery of extended home office overhead under agreed percentage rates. The Eichleay formula, the government says, is an exception to that rule, and may only be applied where a contractor incurs extended overhead expenses as a result of government-caused delay, disruption or suspension of work.

III.

This case thus requires us to decide which of two methods should be used to reimburse a contractor for home office overhead costs incurred by reason of a contact extension to perform additional work. Those two methods are: (i) a fixed percentage mark-up of the direct costs incurred, in this case 13.94 percent; and (ii) a constructive daily rate derived using the Eichleay formula.

Contractors dealing with the government are entitled to recover in the contract price a portion of their home office overhead. 48 C.F.R. §§ 31.203–.205 (1990). Home office overhead includes the cost of such items as weekly payrolls, Davis-Bacon reports, checks,

W-2's, 941's and other required tax forms, cost records, review submittals from subcontractors, weekly and monthly progress reports to the government, salaries, dues and subscriptions, auto and travel, telephone and photocopying. *Capital Elec.*, 729 F.2d at 746. Conventional wisdom has it that such costs cannot, by their nature, be specified and traced to any particular contract. Were such specification and tracing possible, such charges should properly be direct costs of a contract. . . . For this reason, extended home office overhead costs are approximated.

The Eichleay method of calculating extended home office overhead has a long history. As we shall see, contractors have been permitted to use this method to calculate extended home office overhead in situation where disruption, delay or outright suspension [has] cast a cloud of uncertainty over the length of the performance period of the contract.

[Ed. note: The Court described the pre-Eichleay cases that began to employ a proportional amount of total home office overhead.]

In 1960, the Armed Services Board of Contract Appeals devised a similar daily rate formula to estimate extended home office overhead. *Eichleay Corp.*, 60-2 B.C.A. (CCH) ¶ 2688 (1960), *aff'd on reconsid.*, 61-1 B.C.A. (CCH) ¶ 2894 (ASBCA 1961).* Citing *Fred R. Comb*, the Board stated that the contractor need not prove a specific amount of extended overhead, but instead need only assert a fair allocation to compensate for the government delays which caused suspension of approximately one-half the work under the contract. *Id.* at 13,573, 13,575. The Board ultimately concluded that its formula, destined to be titled the "Eichleay" formula, was the best way to "allocate home office suspension of work" when "it was . . . not

---

*

The Eichleay Formula
(as approved by the Federal Circuit)

First, calculate:

$$\frac{\text{contract billings}}{\text{total billings for contract period}} \times \begin{array}{c}\text{total overhead}\\\text{[incurred during}\\\text{the] contract}\\\text{period}\end{array} = \begin{array}{c}\text{overhead}\\\text{allocable}\\\text{to the}\\\text{contract}\end{array}$$

Second, calculate:

$$\frac{\text{allocable contract overhead}}{\text{[actual] days of [contract] performance}} = \begin{array}{c}\text{daily}\\\text{contract}\\\text{overhead}\end{array}$$

Third, calculate:

$$\text{daily contract overhead} \times \begin{array}{c}\text{[number}\\\text{of] days}\\\text{[of] delay}\end{array} = \begin{array}{c}\text{amount}\\\text{recoverable}\end{array}$$

Capital Elec., 729 F.2d at 747.

practical for the contractor to undertake the performance of other work which might absorb them." *Id.* at 13,573. The Board reasoned that:

> overhead costs, including the main office expenses involved in this case, cannot ordinarily be charged to a particular contract. They represent the cost of general facilities and administration necessary to the performance of all contracts. It is therefore necessary to allocate them to specific contracts on some fair basis of proration.

Id. at 13,574. On reconsideration, the Board clarified the type of suspension at issue in *Eichleay Corp.*:

> Performance of the contracts was at no time completely suspended and the delays were not continuous. One item of work would be suspended, then another and so on through an extended series of suspensions. The partial suspensions were lifted at innumerable, varying intervals over a prolonged period of time with the issuance of the numerous modifications providing for changes in the contracts.

Eichleay Corp., 61-1 B.C.A. (CCH) ¶ 2894, at 15,117 (ASBCA 1961).

Following the *Eichleay Corp.* decision, the Court of Claims approved award of overhead costs prorated on a daily basis in cases where the government caused disruption, suspension or delay during performance of the contract. *See, e.g., Luria Bros. & Co. v. United States*, 369 F.2d 701, 706–07, 711, 177 Ct.Cl. 676 (1966) (subsoil investigations meant the contractor was ordered to stop all pourings of footings so that the entire project substantially fell to a halt); *J.D. Hedin Constr. Co. v. United States*, 347 F.2d 235, 242, 245, 171 Ct.Cl. 70 (1965) (no productive work accomplished for 11 weeks while pile driving operations were suspended pending decision of the Veterans Administration).

Then in 1984, this court refused to accede to the government's request to jettison use of the Eichleay formula to calculate such delay damages. *Capital Elec. Co.*, 729 F.2d at 747. Instead, this court approved the use of the Eichleay formula to calculate extended home office overhead under the suspension of work clause provided that compensable delay occurred, and that the contractor could not have taken on any other jobs during the contract period. The contractor having met this burden, the government may only preclude use of the Eichleay formula if it can somehow show that the contractor would suffer no loss by using a fixed percentage mark-up formula. *Id.* at 745–46. The facts underlying our decision in *Capital Electric* tie use of the Eichleay formula to the uncertainty caused by suspension, disruption or delay of contract performance because, although its work was never completely stopped, Capital Electric's work was "un-

reasonably delayed and disrupted" for 302 days such that it "was never able to man the job as planned." *Capital Elec. Co.*, 83-2 B.C.A. (CCH) ¶ 16,548, at 82,302 (GSBCA 1983), *rev'd on other grounds*, 729 F.2d 743 (Fed.Cir.1984).

This same element of uncertainty, engendered by the fact of disruption, suspension or delay of contract performance, has been present whenever the courts or the Boards of Contract Appeals have permitted extended home office overhead to be calculated under the Eichleay formula.

[Ed. note: Here the Court cited cases supporting this proposition and noting their facts.]

In all of these cases when disruption, suspension or delay caused by the government has reduced the stream of direct costs in a contract, it is appropriate to use the Eichleay formula to calculate extended home office overhead instead of the fixed percentage rate formula because the latter would not adequately compensate the contractor for extended home office overhead.

IV.

CBC and *amicus curiae* argue that use of Eichleay should be permitted in any instance in which a contract modification results in an erosion of direct costs because a percentage mark-up of the decreased additional direct costs will not allocate a fair proportion of home office overhead to the contract. This desire to extend availability of the Eichleay formula to pure contract extensions would likely transform use of the formula from an exception to a rule, making the formula applicable to nearly every contract. CBC maintains that extended home office overhead expenses would not be calculated using the Eichleay formula for the majority of pure contract extensions. CBC, however, can only articulate two occasions when use of the Eichleay formula would be inappropriate, *i.e.*, would not reduce the stream of direct costs: contract modifications that add direct costs but do not extend the time for performance, and contract modifications that, over an extended period of time, add direct costs equal to or greater than the amount of direct costs per day under the original contract. Moreover, CBC offers only attorney argument, not empirical data, to support its contention that the Eichleay formula will thus be inapplicable for the majority of contract extensions. We are not persuaded. It appears more likely that the greater part of all contract modifications would qualify for the Eichleay rule under CBC's contention.

In our view, CBC seeks a drastic shift in the circumstances under which the Eichleay formula has been available. Such an extension of the formula's use is no less drastic than the government's urging in the *Capital Electric* litigation that Eichleay be abandoned altogether. As we noted in *Capital Electric,* we believe that Congress, not the courts, should make that decision. *Capital Elec.,* 729 F.2d at 747.

We decline the invitation to stand availability of the Eichleay formula on its head. The *raison d'etre* of Eichleay requires at least some element of uncertainty arising from suspension, disruption or delay of contract performance. Such delays are sudden, sporadic and of uncertain duration. As a result, it is impractical for the contractor to take on other work during these delays. *See George Hyman Constr. Co. v. Washington Metro. Area Transit Auth.,* 816 F.2d 753, 756–58 (D.C.Cir.1987). By contrast, CBC negotiated a change order with the government which extended contract performance for a brief known period of time. CBC experienced no suspension of work, no idle time and no uncertain periods of delay during the agreed upon extended contract performance period. Where no element of uncertainty is imposed on the contractor, use of the Eichleay formula to calculate extended home

office overhead is not permissible. Such a limitation on the use of the Eichleay formula is reasonable because, after all, the Eichleay formula only roughly approximates extended home office overhead. *See generally 5 The Nash & Cibinic Rep.,* ¶ 62, at 166 (Nov. 1991). Thus, computation of extended home office overhead using an estimated daily rate is an extraordinary remedy which is specifically limited to contracts affected by government-caused suspensions, disruptions and delays of work. Absent these circumstances, the Claims Court properly recognized that it is inappropriate to use the Eichleay formula to calculate home office overhead for contract extensions because adequate compensation for overhead expenses may usually be calculated more precisely using a fixed percentage formula. *C.B.C.,* 24 Cl.Ct. at 192.

For these reasons the judgment of the Claims Court is

AFFIRMED.

Courts have not been uniform in their treatment of methods by which a contractor can establish its damages for extended home office overhead.[34] Some are strict in demanding accurate proof of the loss caused by the delay and are unwilling to allow formulas to be employed in place of strict proof requirements.[35] Others are more willing to employ formulas such as the Eichleay formula, recognizing the difficulties of proof and being willing to give the contractor the benefit of the doubt.[36] Strict application of certainty rules can make it difficult if not impossible for the contractor to establish the exact amount of its loss.[37] To counter this, some courts have been willing to relax these rules through the use of formulas that, though not perfect, provide a rough way of measuring an undoubted but difficult-to-establish loss.

To illustrate, some contractors seek to use an even rougher formula than those that have been noted in this section. A contractor faced with the almost insurmountable obstacle of establishing actual losses knows that it has lost money and feels that its losses were attributable not to its own poor estimating or inefficient performance but to acts of the owner or to those for whom the owner is responsible. As a result, the contractor will sometimes seek to use what is called the *total cost* method, a comparison of the actual costs of performance with what the contractor contends should have been the cost of the project.

Some courts have not been willing to accept the *total cost* formula because it assumes that the defendant's breaches caused all the expenditures in excess of the original estimate, that the original estimate accurately represented the cost of performance, and that the contractor efficiently performed its work.[38] Although the Court of Claims allowed this method in *WRB Corporation v.*

---

[34]For an extensive discussion, see Note, 17 Ga.L.Rev. 761 (1983). See also McGeehin, *A Farewell to Eichleay,* 14 Pub.Cont.L.J. 276 (1984).

[35]*W.G. Cornell Co. v. Ceramic Coating Co.,* 626 F.2d 990 (D.C.Cir.1980). *Berley Industries, Inc. v. City of New York,* 45 N.Y.2d 683, 385 N.E.2d 281, 412 N.Y.S.2d 589 (1978).

[36]*Southern New England Contracting Co. v. State,* 165 Conn. 644, 345 A.2d 550 (1974), followed in *Walter Kidde Constructors, Inc. v. Connecticut,* supra note 28. *Broward County v. Russell, Inc.,* 589 So.2d 983 (Fla.Dist.Ct.App.1991).

[37]See Sections 6.06(B) and 27.04.

[38]*Namekagon Dev. Co. v. Bois Forte Reservation Housing Auth.,* 395 F.Supp. 23 (D.Minn.1974); *Huber, Hunt & Nichols, Inc. v. Moore,* 67 Cal.App.3d 278, 136 Cal.Rptr. 603 (1977). The current status of the latter case is doubtful in the light of *C. Norman Peterson Co. v. Container Corp. of Am.,* supra note 21, and *State Dept. of Transp. v. Guy F. Atkinson Co.,* 187 Cal.App.3d 25, 231 Cal.Rptr. 382 (1986).

United States, it placed limits on possible abuse by requiring proof that

(1) The nature of the particular losses make it impossible or highly impracticable to determine them with a reasonable degree of accuracy;
(2) The plaintiff's bid or estimate was realistic;
(3) Its actual costs were reasonable; and
(4) It was not responsible for the added expenses.[39]

Pennsylvania will apply the total cost *only* if there are no other means of determining damages and the claimant has presented reasonably accurate evidence of the various costs incurred.[40]

The *modified* total cost method seeks to avoid the crudeness of the total cost method. It focuses on the impacted work activities[41] and adjusts the original estimate to remove mistakes, inaccuracies, and work items not affected.

Yet the often insurmountable proof problems have led an increasing number of courts, often in road construction disputes, to recognize total cost where appropriate limitations are observed.[42]

Another rough measurement sometimes employed either in conjunction with a total cost theory or by itself is the jury verdict. This formula recognizes the inherently imprecise nature of the proof that can be produced in such cases. However, a jury verdict can be used only where there is a clear proof of injury, there is no more reliable method for computing damages and the evidence is sufficient to make a fair and reasonable approximation of damages.[43] Inasmuch as these crude formulas depend heavily upon the facts in a particular case, understanding their use requires that case decisions be examined. The following two cases, reproduced in part, involve these crude formulas and come to the conclusion that use of these formulas was appropriate.

---

[39]183 Ct.Cl. 409, 426 (1968).

[40]*John F. Harkins Co., Inc. v. School Dist. of Philadelphia*, 313 Pa.Super. 425, 460 A.2d 260 (1983) (contractor did not meet burden).

[41]*Servidone Constr. Corp. v. United States*, 931 F.2d 860 (Fed.Cir.1991) (reduced award because bid too low and substituted a reasonable bid). *E.C. Ernst v. Koppers Co.*, 626 F.2d 324 (3d Cir.1980); *Seattle Western Indus., Inc. v. David A. Mowat Co.*, 110 Wash.2d 1, 750 P.2d 245 (1988) (deducted cost for which subcontractor claimant responsible).

[42]*Moorehead Constr. Co. v. City of Grand Forks*, 508 F.2d 1008 (8th Cir.1975); *C. Norman Peterson Co. v. Container Corp. of Am.*, supra note 21; *Dep't of Transp. v. Hawkins Bridge Co.*, 457 So.2d 525 (Fla.Dist.Ct.App.1984); *Glasgow, Inc. v. Commonwealth of Pa., Dep't of Transp.*, 108 Pa.Commw. 48, 529 A.2d 576 (1987); *State Highway Comm'n of Wyoming v. Brasel & Sims Constr. Co.*, 688 P.2d 871 (Wyo.1984). For a comprehensive analysis, see Aaen, *The Total Cost Method of Calculating Damages in Construction Cases*, 22 Pac.L.J. 1185 (1991) (noting that majority of appellate cases accept total cost and illustrating various formulas).

[43]*Dawco Constr. Co. v. United States*, 930 F.2d 872 (Fed.Cir.1991) (jury verdict improperly employed in a case that should have been one based upon total cost); *Fattore Co. v. Metropolitan Sewerage Comm'n.*, 505 F.2d 1 (7th Cir.1974) (approved jury verdict but warned that such formulas must require clear proof that the contractor has suffered a loss and that there was no more reliable method for computing damages); *Meva Corp. v. United States*, 206 Ct.Cl. 203, 511 F.2d 548 (1975); *State Department of Transp. v. Guy F. Atkinson Co.*, supra note 38 (approved decision by arbitrator who arbitrarily reduced award 35% from the highest claim estimates submitted by contractor); *Telex Corp. v. International Business Machines Corp.*, 367 F.Supp. 258 (N.D.Okla.1973) (humanly impossible to trace, find, and specify in detail and quantity the numerous circumstances that cause or contribute to financial consequences) (affirmed in part, reversed and remanded on other grounds, 510 F.2d 894 (10th Cir.1975); *State Highway Comm'n of Wyoming v. Brasel & Sims Constr. Co.*, supra note 42.

AMP-RITE ELEC. CO., INC. v. WHEATON SAN. DIST.

Appellate Court of Illinois, 1991. 220 Ill.App.3d 130, 580 N.E.2d 622, appeal denied, 143 Ill.2d 635, 587 N.E.2d 1011 (1992).

[Ed. note: Amp-Rite was one of three prime contractors hired by the Sanitary District to perform electrical work. Because the work was not in a state of readiness, Amp-Rite could not perform as planned. These delays increased its cost and were the basis for a claim against the Sanitary District.

The jury ruled for Amp-Rite. The Sanitary District's post-trial motion for a directed verdict (it claimed the judge erred in allowing the jury to use the total cost method) was based upon the jury's permitted use of the total cost method of computing damages. The motion was denied. The Sanitary District appealed. The next part of the court's opinion dealt with the appropriateness of the total cost method in this case.]

The District next contends Amp-Rite's proof failed, as a matter of law, to entitle it to use the "total cost" method of calculating its damages, as opposed to the "discrete" method, and that judgment notwithstanding the verdict should have been entered on count I. Of the total amount of damages claimed by Amp-Rite, alternately calculated amounts of either $203,600 or $154,814 were claimed for additional labor expense. As noted previously, the larger figure was calculated with reference to Amp-Rite's own bid; the lesser figure was calculated with reference to the third lowest bid received by the District for "reasonableness-of-the-bid" comparison purposes.

Unlike the "usual" total cost approach which is based on total contract expenditures ... Amp-Rite's total cost method of calculating damages compared only the total number of labor hours actually expended in constructing the project (10,103) with the total number of labor hours Amp-Rite estimated would be required to construct the project (3,806). Additional supervision hours were then subtracted out as a separate item, and the remainder was multiplied by the man-hour rate being charged to the District for all contract extras and the overhead fee provided in the contract. The damages thusly calculated totaled $203,616. In contrast, the "discrete" method of calculating damages would isolate either specific items of work which were delayed or time periods of lower workers' productivity and then identify the overrun hours specifically attributable thereto.

The only Illinois case discovered which has discussed the total cost method came out of this district. We found there, though, that we did not need to determine whether the total cost method was improperly used in that case to calculate delay damages where we found the contracting authority had not breached its contract with the claimants. (*Gust K. Newberg, Inc. v. Illinois State Toll Highway Authority* (1987), 153 Ill.App.3d 918, 929, 106 Ill.Dec. 858, 506 N.E.2d 658.) Consequently, the parties necessarily have resorted to Federal cases which have either allowed, or not, that the total cost method was appropriate under the facts presented. These Federal cases are lengthy and detailed. The total cost method guidelines which emerge therefrom are consistent, however.

The total cost method is premised on the principle that, where a contractor is entitled to an adjustment,

the contracting entity should not be relieved of its liability for same merely because the contractor is unable to prove its increased costs within a mathematical certainty.... The total cost method is not favored, however.... It is not a preferred means of calculation because it is imprecise ... and because it assumes the plaintiff's costs were reasonable, that plaintiff was not responsible for any increases in cost, and that the plaintiff's bid was accurately computed, which is not always the case.... Whether the total cost method of determining damages will be deemed acceptable in a given case depends on the plaintiff proving "that (1) the nature of the particular losses make [*sic*] it impossible or highly impractical to determine them with a reasonable degree of accuracy; (2) the plaintiff's bid or estimate was realistic; (3) its actual costs were reasonable; and (4) it was not responsible for the added expenses."

* * *

The District argues here that Amp-Rite's proof in each of these regards was insufficient. Specifically, it first contends Amp-Rite failed to prove that the damages it sustained as to its additional labor expense were impossible or highly impractical to determine with a reasonable degree of accuracy by a method other than total cost. The District claims more than ample records were available from which Amp-Rite could have calculated its damages using the discrete, or direct, method. It points to minutes of the construction progress meetings which showed that Amp-Rite recognized as early as August 1, 1983, which was two months into the project, that its work was being impacted by the delayed performance of the other prime contractors and that Amp-Rite continued to complain of such delays in subsequent meetings. Coupled with this early "claim recognition," the District further asserts Amp-Rite had the ability to, and did, keep contemporaneous records of delays and extra work, and it recorded on a daily basis the type of work the workers were performing. As such, the District concludes Amp-Rite's expert's opinion was entitled to no weight. As noted earlier, Dr. Adrian's opinion was that if it were not for the use of the total cost method, someone with his industrial engineering background would have had to conduct an on-site video camera "analysis of arm and leg movement loss because of the impact and the added distances that [the workers] walked" in order to arrive at some kind of computation of damages.

Amp-Rite responds that, although it could easily recount the hours required to do work which was not anticipated, the problem it encountered here was the performance of base contract work done under unanticipated conditions. That is, the project, rather than proceeding in an orderly progression from west to east, spanned the entire one-third-mile length. The project was extended from 14 to 22 months and included two winters, whereas Amp-Rite anticipated it would perform work indoors through one winter based on the contract specifications. Additionally, the physical site was "like a bomb hit it," due to the numerous excavations throughout the site, making storage of equipment and travel back and forth exceedingly difficult. Amp-Rite tried to calculate a specific amount of delay and a specific dollar value attributable to each item of delay but found it was impossible. Amp-Rite's project manager testified it would have required having someone follow workers around the jobsite as the work was being done in order to compare the amount of time it was taking to what the time would be under normal circumstances. The "ripple effect" caused by the delays—one delay leading to another—compounded the problem of identifying the impact on an individual item-of-delay basis. Further, Amp-Rite's expert's opinion that the total cost method was appropriate in this case was premised on the fact the delays caused by the other prime contractors impacted the productivity of the workers in terms of "waiting, weather, job site conditions, traveling, and difficult working conditions."

In our opinion, such impact is not "discrete," which term means separate and distinct, not attached to others, unrelated or discontinuous. (Webster's New World Dictionary 402–03 (1972).) Rather, it pervaded the project, making the nature of Amp-Rite's losses in terms of additional labor expense "impossible or highly impractical" to determine with a reasonable degree of accuracy. Where the pervasive nature of the problems encountered renders an exact calculation of losses impossible or highly impractical, the total cost method is applicable. . . . The District next argues the total cost method was not appropriate here because the testimony showed that not all the labor overrun hours were due to causes attributable to it. As in all breach of contract cases, the costs must be tied in to fault on the defendant's part. . . . For example, it asserts Amp-Rite's total cost claim included hours for these causes: (1) the flood in the sand filter building; (2) acts of God; (3) damage caused by other contractors to Amp-Rite's work; (4) problems with materials furnished by other prime contractors; and (5) problems with Amp-Rite's subcontractor, Allan Engineering. Citing *Boyajian v. United States* (1970), 423 F.2d 1231, 1244, 191 Ct.Cl. 233, the District argues that because there was no evidence whereby the jury could isolate and remove these impermissible hours from its award, Amp-Rite's total cost claim must fail.

Initially, we note the District argues its nonresponsibility for these causes as a "corollary" to the requirement that before the total cost method may be utilized, the plaintiff must prove that its own ineffi-

ciencies were not responsible for the added expenses. It is uncontradicted, however, that no inefficiencies of Amp-Rite were responsible for the added labor expenses. The District's eventual grant of a 15-week extension of the contract time to November 15, 1984, under the terms of article 12 of the contract and its subsequent implied extension of time to January 17, 1985, bespeak Amp-Rite's lack of fault.

* * *

Of the numerous factors shown to have had an impact on Amp-Rite's job progress here, not all were claimed to have increased its labor hours. Its evidence clearly tied the labor overrun hours specifically to the District's failure to coordinate the work of the contractors according to the general construction schedule set forth in the contract documents. The District's failure to enforce coordination was alleged to be the cause of the hopscotching, the starting and stopping, the waiting, the gearing up to do a job and not being able to do it, and the need to work outdoors in less productive winter weather conditions. Close examination of the events the District claims were not attributable to it shows either that delay damages were applicable under the contract provisions or that the event was not a labor-hour increasing one. Article 12 of the contract did not preclude damages for delay caused by acts of God (such as the flood in the sand filter building due to heavy rains), or for delays caused by acts or neglect by any separate contractor employed by it (such as the second flooding of the sand filter excavations, damage to Amp-Rite's work by other subcontractors, and problems with materials furnished by other subcontractors).

As to problems with Amp-Rite's subcontractor, Allan Engineering, there was evidence presented by the District that Amp-Rite withheld payments to Allan and that this delayed the completion of the supervisory control system which, in turn, delayed final completion of the project. That evidence was introduced by the District in relation to Amp-Rite's claim in count II that the District breached its contract in assessing liquidated damages against it for failing to complete the project on time. No problems with Allan were claimed by Amp-Rite to have increased its labor hours. We conclude the District's argument in this regard fails to establish the inapplicability of the total cost method on this basis.

The District next argues Amp-Rite's bid was "unrealistic" in that it unrealistically assumed (1) the west end buildings would be completed before the east end buildings were started; (2) the multiple prime contractor delivery system would have no bearing on the number of hours required to do the work; (3) Northwestern's underground piping excavation and installation would be done without interruption; and (4) the conditions of installation warranted double downward factoring of the labor hours in its bid to reflect easiest-level installation conditions.

Amp-Rite responds that, in preparing its bid, it was entitled to rely on the contract documents . . . including the general construction schedule therein, which every witness, including the District's own expert, understood to show a west-to-east progression of the project work. Voluminous evidence was adduced concerning the manner in which Amp-Rite's bid was prepared by Messrs. Voss, Williamson and Adaska. Its reasonableness was not overwhelmed by the District's evidence of the unfactored, "straight-NECA [National Electrical Contractors Association] takeoff" bid prepared by its expert which was shown to be fatally flawed in its unrealistic conclusion that the number of labor hours which should have been estimated for the project was 9,433.

Viewing the evidence in the light most favorable to Amp-Rite, its bid was not unreasonable. It is clear the general construction schedule in the contract documents was revised more than once to accommodate Wil-Freds[44] major concrete construction work, to make up for Wil-Freds' and Northwestern's[45] delays and to stay on schedule. Although these revisions may have been in the District's best interest, they prejudiced the benefit of the bargain Amp-Rite had struck with the District. . . . The District openly acknowledged it knew the revised schedule prepared in November would cause problems for Amp-Rite. Amp-Rite reasonably need not have anticipated the extensive delays which occurred from the very start of the project. It was entitled to rely on the general construction schedule in its bid preparation.

Although hindsight on this project shows that the possibility there might be multiple prime contractors on the project should perhaps have been a bidding consideration . . . there was no way for Amp-Rite to know at the time it bid the project whether a single [bid], a combination [of], or multiple bids would be accepted by the District. Thus, the NECA manual's caution concerning increasing labor hours depending on the known effectiveness or experience of "the person who has the overall responsibility to manage the construction" reasonably could not have been heeded. Further, the recommended labor units in the NECA manual themselves were based on the assumption, as was Amp-Rite's labor estimate, that the construction project would be "effectively" managed. Amp-Rite was entitled to assume that the District would reason-

ably use its punitive powers under the contract to keep offending contractors on schedule.

It is true that the general construction schedule did not set forth with any specificity *when* Northwestern would install its underground piping so that Amp-Rite could follow along. However, Northwestern's contract with the District required that it also follow the general construction schedule, and, despite the District's belated suggestion that Amp-Rite use an alternate trenching route for a portion of the conduit paralleling Spring Brook Creek at its own expense, the plans and specifications showed that Amp-Rite's electrical conduit was to be placed in the same location as Northwestern's underground piping. Consequently, it was not unreasonable for Amp-Rite to assume when making its bid that Northwestern would effectively prosecute its work in line with the general construction schedule, thus allowing Amp-Rite seasonably to follow along. Although some delays are inevitable in any project, Amp-Rite could not know during its bid preparation that Northwestern would fall six months behind schedule in its work or that the delays experienced by Wil-Freds and the accelerated schedules devised in response would so substantially prejudice it. A "reasonable" bid is just that; it need not be prepared along "worst scenario" lines.

As to the double downward factoring of the labor hours in Amp-Rite's bid, the District argues this was not justified based on its own expert's testimony that the instant wastewater treatment project, which was equivalent in terms of difficulty of construction to that of a chemical plant, warranted a bid based on 9,433 labor hours. This compares with Amp-Rite's 3,806 estimated hours and its 10,103 actual hours.

The District's expert's estimate admittedly did not take into consideration any aspects of the project which would cause an increase or decrease in the estimated hours, however, and Amp-Rite's cross-examination and rebuttal exposed numerous labor hour computations regarding prewired control systems and panels and multiple conduit runs that would have justified downward factoring of the labor hours. Additionally, the District's expert was neither an electrician nor a builder and had no experience in using the NECA manual. The second downward factoring of Amp-Rite's bid was performed by its president, Ron Adaska, who based that second reduction in labor units on the high productivity of workers in Du Page County as experienced by Amp-Rite in previous projects.

The District's final contention as to the applicability of the total cost method is that Amp-Rite did not prove its actual costs were reasonable. The District notes it does not contend that Amp-Rite did not accurately record the labor overruns. Rather, it believes

---

[44]Ed. note: general construction contractor in charge of scheduling.

[45]Ed. note: plumbing, heating, and ventilating separate contractor.

that the overrun shows either that Amp-Rite had the correct number of workers on the project to do the work as it became available, in which case Amp-Rite's estimated hours were too low, or [that] Amp-Rite kept an inordinate number of men available to do work which was not ready to be done, in which case Amp-Rite's overrun hours are unreasonable.

We agree with Amp-Rite that the District's argument in this regard stands in "rather astounding juxtaposition" to its preceding argument that Amp-Rite's labor hours were estimated unreasonably low and should have been more in the neighborhood of 9,433 hours without even considering any of the adverse conditions which were actually experienced during the construction. If the District's position is that 9,433 labor hours would have been a "reasonable" estimate under normal conditions, can it seriously argue that total actual labor hours of 10,103 (which includes

6,297 labor overrun hours) under the adverse and extenuated conditions which occurred here are unreasonable? We do not think so.

We further note that the District's belated extensions of time for completion of the contract as requested by Amp-Rite, and Wil-Freds' "Just watch where I go" approach to project coordination, placed Amp-Rite in the position of having to keep men on the job to do whatever work became available in order to try to meet the scheduled completion date.

In sum, we conclude the evidence, viewed in its aspect most favorable to Amp-Rite, did not so overwhelmingly favor the District that no contrary verdict could ever stand. Consequently, the court did not err in denying the District's motion for judgment notwithstanding the verdict.

* * *

NEW PUEBLO CONSTRUCTORS, INC., v. STATE OF ARIZONA

Supreme Court of Arizona, En Banc, 1985. 144 Ariz. 95 696 P.2d 185.

[Ed. note: A catastrophic storm and unusually heavy rainfall caused extensive damage to road work. The contractor (NPC) performed corrective work and sued the state (ADOT) to recover its added expenses under a clause that relieved the contractor for "Acts of God."]

* * *

III. THE MEASURE OF DAMAGES

The jury awarded damages of approximately $201,000 in actual damages (without overhead) including the following amounts:

1. Cleaning and refurbishing medians and ditches — $ 3,500
2. Replacing eroded material in the Anamax pit lost in the flood, including costs of hauling substitute material from the Duval pit to the Anamax pit — 28,500
3. Restoring eroded and lost special backfill material — 8,000
4. Restoring mineral aggregate from new area after flooding of the Agua Linda pit — 102,000
5. Additional stripping of the Duval Mineral aggregate pit — 19,000
6. Dewatering of the Duval and Agua Linda mineral aggregate pit — 40,000

$201,000

The contractor kept records of the actual costs of the entire project but did not keep separate records of the actual costs of the rebuilding and rework caused by the weather, except for item #6.

* * *

When additional work is performed on construction projects, there are traditionally at least three ways of proving costs. Specifically:

1. *Actual cost.* Keeping separate records of the actual costs of additional work is the most reliable method of qualifying costs.

* * *

2. *Jury verdict.* Where the contractor cannot prove actual costs, the contractor may present evidence of the cost of additional work to the finder of fact including any actual cost data, accounting records, estimates by law and expert witnesses, and calculations from similar projects.

* * *

3. *Total cost.* Under certain circumstances, the contractor may subtract the estimated cost or bid of the entire project from the final cost of the entire project. The resulting figure is the amount claimed as damages.

* * *

Courts have resorted to the total cost method only under exceptional circumstances and then only as a last resort method.... This method of measuring costs suffers from many defects. By simply subtracting the bid estimate from the cost of the overall project, the total cost method:

(a) presumes that the bid estimate was realistic;
(b) can pass along costs to the state which might have been incurred despite the act of God or the other party's breach;
(c) can reward the contractor's inefficiency, managerial ineptitude, financial difficulties and other failings by passing these costs along to the state.

* * *

To avoid these defects of the total cost method, NPC proposes a fourth method of measuring costs in construction contract cases.

4. *Modified total cost (or cost variance).* The original bid for a particular item of work is subtracted from the actual costs for this item of work, though both the bid and actual costs are limited and adjusted in the following manner. Damages are limited to only certain time intervals wherein the work was adversely affected by the weather. The actual costs are totaled only for these work activities affected by the weather and to only those cost accounts recording such work. The employee and machinery time used for a particular item of work is determined by reference to the superintendent's reports describing the weekly work performed, engineer's diaries, labor distribution reports, and equipment distribution reports. Actual costs which were otherwise compensated or unrelated to such work activity are eliminated. The original bid estimate is redetermined in the light of the actual unit costs for the month before and after the months adversely affected. The unit cost overrun for a particular work item during the rainy months would be multiplied by the total actual work performed to arrive at the cost of rework. The adjusted estimate of the particular items of work in dispute is subtracted from the actual adjusted cost of this work only if: (1) the nature of the particular losses makes it impossible or highly impracticable to determine them with a reasonable degree of accuracy; (2) the contractor's bid was realistic; (3) his actual costs were reasonable; (4) the added expense was not caused by the contractor or by some cause for which he assumed the risk, but was proximately caused by the unforeseen circumstance or the other party's breach.

It is misleading to refer to this as a separate ''method'' for determining damages. Most courts re-

fuse to apply the total cost method in the absence of the aforementioned circumstances.... Additionally, the acceptability of such a determination of damages involves a factual question as to how carefully the total cost method was modified to restrict its deficiencies. The trial judge must play an active role in this fact-bound inquiry in determining that the measure of damages is appropriate to the nature of the harm involved and that the specific estimates have been appropriately adjusted to avoid recovery of unrelated costs by the contractor.

* * *

First, a sufficient foundation for the use of the cost variance method has been established in this case. Most of the additional expenses were caused by an invisible, rising subterranean water table, so that segregating and precisely recording rework costs and original work costs was impracticable. Other cases have applied a similarly modified total method as a last resort method on similar facts. *See J.D. Hedin Construction Co. v. United States,* 171 Ct.Cl. 70, 85–88, 347 F.2d 235, 246–47 (1965) (contractor suffered weather-related damage during extended period of work caused by faulty and changed government specifications); *State Highway Comm'n v. Brasel & Sims Const., supra,* 688 P.2d at 878–79 (contractor suffered increased expenses and delay due to failure of a state to supply adequate water and inferior quality of state supplied gravel); *Moorhead Construction Co. v. City of Grand Forks,* [508 F.2d 1008 (8th Cir.1975).] (excess moisture and lack of soil compaction caused additional expense and are different than represented); *Thorn Construction Co. v. Utah Dept. of Transp.,* 598 P.2d 365 (Utah 1979) (additional expense is caused contractor by inferior quality of road material and it is necessary to excavate other materials at another pit).

Because the additional work was best performed concurrently with the principal contract work, it was not feasible in this situation to quantify the actual costs of rework.

* * *

Nor has ADOT shown that NPC's adjusted estimates of the costs of rework were unrealistic. NPC submitted the low bids on both the Tubac and Carmen projects and increased these estimates to reflect the costs actually experienced on the project. The jury obviously concluded that the weather was the cause of the contractor's cost overruns during the months in question. NPC did keep some actual cost records where it was feasible to do so, namely, for item #6 of the damages, the dewatering of the mineral aggregate pits.

* * *

NPC used the jury verdict method to determine damages for ... additional stripping of the Duval aggregate pit. The use of the jury verdict method is appropriate when (1) the state is liable for a changed condition that increases the contractor's expenses, (2) due to circumstances beyond his control, the claimant cannot feasibly prove specific damages by a more reliable method, and (3) when there is sufficient evidence in the record to provide a reasonable basis for approximating the damages.

* * *

We also believe that there is sufficient foundation for the use of the jury verdict method. Courts have approved this method of damages on similar facts. *Metro Sewerage Comm'n v. R.W. Constr., supra*, 78 Wis. at 466, 255 N.W.2d at 302 (contractor suffered increased costs due to dewatering of tunnel caused by artesian water not shown on specifications); *Foster Constr. v. United States*, 193 Ct.Cl. 587, 435 F.2d 873 (1970) (contractor caused additional expense of uncertain amount by misrepresented subsurface conditions in constructing pier). We believe that the use of the jury verdict was appropriate in this case for much the same reasons that the use of the modified total cost method was warranted. We are far from unqualifiedly endorsing either the jury verdict or the modified total cost measure of damages. The availability of both measures of damage must be proven by the contractor. Neither measure of damages can be used where

there is no excuse for the failure to keep track of actual costs. *Pickard's Sons Co. v. United States*, 209 Ct.Cl. 643, 532 F.2d 739 (1976); Appeal of Soledad Enterprises, Inc. ASBCA 20376, 77-2 BCA 12757 (1977).

We do not believe that this is a case wherein the contractor inexcusably failed to keep track of costs because this was not feasible due to circumstances beyond his control. The case at bar is quite different from the cases cited in this respect. *Id.*

Nor can the contractor use the modified total cost method when there is sufficient evidence to use the jury verdict method.

* * *

The contractor cannot use the jury verdict method if he can prove only that the state caused part of the damages and cannot make a reasonable approximation of those damages. *See Electronic & Missile Facilities, Inc. v. United States*, 189 Ct.Cl. 237, 256–57, 416 F.2d 1345, 1357–58 (1969) (contractor was caused additional expense to remove contaminated gravel, but cannot reasonably approximate government causation of damages). Even on facts where the use of both these measures of damages is appropriate, they are subject to close judicial scrutiny to insure that the contractor does not receive a windfall. We find no error, however, on the facts of this case.

SECTION 27.03 Measurement: Owner vs. Contractor

A. Illustrations

The principal measurement problems relate to the contractor's unjustified failure to start or complete the project, to complete the project as specified, or to complete the project on time.

B. Project Never Begun

Damages are determined by subtracting the contract price from the market price of the work. This is usually based on the best competitive price that can be obtained from a successor contractor for the same work. Suppose the contract price were $100,000 and the successor cost $120,000. The owner would be entitled to $20,000. This would protect its contract bargain.

Claims for lost bargains do not receive the protection that claims based on out-of-pocket losses do. Very likely in a case of this type it would be

concluded that there was no difference between contract and market price, no lost bargain, and no recovery for that element of the breach.

In addition to the expectation interest (loss of bargain), the owner would be entitled to any losses caused by the delay (discussed in (E)).

C. Project Partially Completed

Suppose the contractor ceases performance unjustifiably or is properly terminated by the owner. Restitution is one way of measuring the owner's claim. This measure seeks to restore the status quo as it existed at the time the contract was made. This can be accomplished by restoring to the owner any progress payments it has made to the contractor. From this is deducted the benefit to the owner of the partially completed project. However, the difficulty of measuring the benefit conferred may lead a court to conclude that restitution is not proper. For example, one case denied the owner restitution of its payments in a well-drilling contract in which the well produced 80% of the required amount of water.[46] In such a case, diminished expectation would be more appropriate. The owner would be entitled to the cost of performance that would generate the promised performance by the contractor less the payments it had made.

The owner was allowed to recover progress payments when a fallout shelter leaked and finally caved in.[47] To restore the status quo, the contractor would be entitled to deduct the extent to which its work added value to the land. The more likely event is that it added no value, with the owner being entitled to recover the cost of removing the shelter and restoring the land to its precontract condition.[48]

Restitution is rarely used to measure owner claims in construction contract disputes. More commonly, failure to complete generates two types of damages. First, the owner very likely will have to pay more than the balance of the contract price to have the work completed by a successor. In many cases, the work will have been performed defectively, and this too may justify additional compensation (discussed in (D)). There is likely to be delayed completion, with delay damages being discussed in (E).

D. Defective Performance: Correction Cost or Diminished Value?

Claims relating to defective performance may arise when a contractor has performed improperly, never having reached the level of substantial performance.[49] More commonly, defective performance cases involve attempts by the contractor to recover the balance of the contract price by alleging either substantial or full performance, the secondary issue being the deduction available to the owner if the work was only substantially performed.

The law seeks to place the owner in the position that it would have been in had the contractor performed properly. One way of doing this is by giving the owner the *cost of correction*—the amount necessary to correct the defective work or complete the work to bring it to the state required under the contract.

Another method of giving the owner what it was promised is to award the *diminished value*—the difference between the value of the project as defectively built and what it would have been worth had it been completed as promised. This measure gives the owner's balance sheet what it would have had by full performance. This is achieved by combining the value of the project as it sits with the diminished value measure of recovery because of less than complete performance.

Each method presents problems. Suppose cost of correction is used and the correction costs greatly exceed the value correction would add to the property. The likelihood that the owner will not use the money for this purpose presents windfall possibilities. The award would put the owner in a *better* position than if the contractor had performed properly. If the owner *did* correct defective work or complete the work when it would not be economically sound to do so, this would waste scarce societal resources.

But awarding diminution of value deprives the owner of the performance bargained for and forces the owner to take an amount that is usually determined by the testimony of expert witnesses. Such testimony is expensive to procure and may not take into account the subjective expectations of the owner when it made the contract. In addition, this evidence is "softer" than the evidence of the cost of correcting the work. Finally, if the award is less than the cost of correction and the owner elects to correct or complete, the award will be inadequate.

Predictably, different jurisdictions will go off in different directions; there may even be contradictions within a particular jurisdiction. A few states employ only the cost of completion or correction. Some hold that diminution in value is the only proper measure. Some states hold that the owner can elect to invoke either measure. A few hold that the proper measure of recovery is cost of correction or diminution in value, whichever is less. As shown in *Plante v. Jacobs*,[50] (reproduced in Section

---

[46]*Village of Wells v. Layne-Minnesota Co.*, 240 Minn. 132, 60 N.W.2d 621 (1953).

[47]*Economy Swimming Pool Co., Inc. v. Freeling*, 236 Ark. 888, 370 S.W.2d 438 (1963).

[48]*Bourgeois v. Arrow Fence Co.*, 592 So.2d 445 (La.App.1991) (addition so defective so as to be irreparable permitted owner to recover payments made and cost of demolition), writ denied, 596 So.2d 214 (1992). Compare *Mayfield v. Swafford*, 106 Ill.App.3d 610, 435 N.E.2d 953 (1982), to be discussed in (D).

[49]See Section 22.06(B).

---

[50]10 Wis.2d 567, 103 N.W.2d 296 (1960).

22.06(B)), one formula can be applied to some items and the other to other items. If cost of correction will diminish the value, both can be applied.[51] Most states prefer the cost of correction measure unless it would generate "economic waste."[52] One author stated:

> "Economic waste" is primarily a result-oriented concept, not a fiscal one. Economic waste comes into play in those cases in which the defective building is still serviceable and useful to society. If repairs are possible but would completely destroy a substantial portion of the work, damage or injure good parts of the building, impair the building as a whole, or involve substantial tearing down and rebuilding, then that is "economic waste."[53]

The choice of an appropriate measure may also depend on whether either party, particularly the contractor, has acted in good faith. If there does not appear to have been a good reason for the deviation, there is a stronger likelihood that a measure will be selected that is preferable to the owner.[54]

This topic was dealt with in Section 22.06(B), which examined substantial performance. If the contractor has substantially performed, the owner can offset against the balance of the contract price the damages caused by the contractor's failure to perform the contract as promised. The offset sometimes is the cost of correction and sometimes the diminution in value or a combination of the two. Please review Section 22.06(B).

The most difficult cases are those that involve a residence constructed for the use of the owner. In one case, the contractor had installed roofing shingles that were discolored. The cost to correct this properly functioning roof by replacing the roofing material was estimated at between $14,000 and $25,000. The contractor contended that the weathering of the shingles would ultimately lead to a uniform color and that the roof was not visible to passersby. The court concluded the measure of recovery to be $7,500, based on diminution in value.[55]

Another case involving the installation of a shingled roof concluded that the contractor had *not* substantially performed, because the roof had yellow streaks, though it was structurally sound and would protect the owner from the elements. The court concluded that in matters relating to homes and their decoration, taste or preference "almost approaching whimsy" may be controlling with the homeowner so that variations that under other circumstances might be considered trifling would preclude a finding that there had been substantial performance.[56] This gave the owner a defense to any contractor claim for the balance and to be the basis for an owner claim measured by the cost of replacing the roof.

As noted earlier, a combination of the two measures is appropriate. For example, some deviations can be measured by diminution in value, while others can employ cost of correction.[57] The two formulas can be applied together if the cost of correction will not achieve the outcome promised under the contract. For example, in one case, there had been settling and slanting due to a failure to sink piles far enough. The court held that the owner could recover not only the cost of doing as well as could be done to correct the problem but also the diminished value, as the cost of correction could never accomplish the contractual requirements.[58]

---

[51]*Italian Economic Corp. v. Community Eng'rs., Inc.*, 135 Misc.2d 209, 514 N.Y.S.2d 630 (1987) (structural repairs needed to comply with Code reduced floor space and windows).

[52]The many cases are collected in Annot., 41 A.L.R.4th 131 (1985). For some recent cases, see *Granite Constr. Co. v. United States*, 962 F.2d 998 (Fed.Cir.1992) (though material did not meet technical requirements of the specifications, it exceeded by 20 the safety factor for the overall project: diminution in value applied), cert. denied, 113 S.Ct. 965 (1993); *City of Charlotte v. Skidmore, Owings, & Merrill*, 103 N.C.App. 667, 407 S.E.2d 571 (1991) (sidewalk deteriorated because of negligent design by architect; replacement value applied). See also Abney, *Determining Damages for Breach of Implied Warranties in Construction Defect Cases*, 16 Real Est. L.J. 210 (1988); Chomsky, *Of Spoil Pits and Swimming Pools: Reconsidering the Measure of Damages for Construction Contracts*, 75 Minn.L.Rev. 1445 (1991) (suggests that measure exceed diminution of market value if cost to complete too burdensome; wants all factors considered that relate to owner's loss and uses highest level consistent with fairness to the contractor).

[53]Abney, op. cit supra note 52 at 218 (footnotes omitted).

[54]In *Kaiser v. Fishman*, 187 A.D.2d 623, 590 N.Y.S.2d 230 (1992), the court affirmed an award for cost of correction that involved physically lifting the residence off its pilings and performing substantial corrective work, mainly because the contractor had deliberately failed to comply with the contract terms.

[55]*Salem Towne Apartments, Inc. v. McDaniel & Sons Roofing Co.*, 330 F.Supp. 906 (E.D.N.C.1970).

[56]*O.W. Grun Roofing & Const. Co. v. Cope*, 529 S.W.2d 258 (Tex.Ct.App.1975).

[57]*Plante v. Jacobs*, reproduced in Section 22.06(B).

[58]*Kahn v. Prahl*, 414 S.W.2d 269 (Mo.1967). See also *Northern Petrochemical Co. v. Thorsen & Thorshov, Inc.*, 297 Minn. 118, 211 N.W.2d 159 (1973); *Italian Economic Corp. v. Community Engineers, Inc.*, supra note 51.

When cost of correction is used, several other factors must be considered. First, the cost of correction cannot be recovered if the corrective work includes work not called for under the original contract. In one case, replacement of leaky gutters and downspouts with galvanized gutters was not the proper measure of recovery when aluminum had originally been called for.[59] The measure should have been the cost of installing aluminum gutters that would have stopped the leaking, assuming this was the contractor's responsibility.

This generalization may not apply if the basis for the claim is an implied warranty of habitability. For example, in *Parsons v. Beaulieu,*[60] the defendant had contracted to build a septic tank for the plaintiff. The completed work did not comply with the legal requirements. The plaintiff replaced the tank with a larger one that did comply. The court held that the defendant contractor had to pay for the cost of the larger septic tank because the original one would not function in a way that would meet the contractor's implied warranty. The court noted that the owner had installed the least expensive one that would meet the legal requirements. In the *Parsons* case, the owner was not being put in a better position, as it had contracted for a tank that met the legal requirements.

Suppose the correction takes place a year after completion. The installation of new materials and equipment can be said to put the owner in a better position than that bargained for. The owner now has new materials and equipment to replace materials and equipment that had used up some of its useful life and had become depreciated. Generally, the law takes into account any extended life that the replacement may give the owner.[61] But in *Dickerson Construction Co. v. Process Engineering Co.,*[62] the court refused to deduct any amount for extending the life of the building, inasmuch as the defects had been discovered only six months after completion.

At what point is the cost of correction determined? Usually it is the time of the breach. This assumes that the owner knows of the defect and is in the position to have the defective work corrected. However, some cases have used the time of *trial* as the benchmark for determining when the cost of correction should be computed. These cases had looked on the highly inflationary period of the 1970s and noted that an award based on time of breach rendered years after the work has been completed would not put the owner in the position it would have been had the work been done right in the first place.[63]

A similar problem can arise when the issue is diminution in value. A Florida case is instructive on this and other points. *Grossman Holdings Limited v. Hourihan*[64] involved a claim by the owners against a contractor for failure to build a house as promised. The owners contracted with the developer to purchase a house to be built in a planned development. Both the model and the office drawings showed the house with a southeast exposure, and the contract stated that the developer would construct the house "substantially the same" as in the plans and specifications or as in the model. A short time later, a new model and map went on display that showed the owners' lot and to-be-built house facing the opposite direction from that which they expected and wanted. The owners brought this discrepancy to the attention of the developer and demanded the construction of the house that they had contracted for. (They wanted the house facing a particular direction for aesthetic and energy-saving reasons.) The developer refused to change the plans and began constructing the house facing in the direction opposite of that expected by the owner.

The trial court refused to award any damages, finding that the cost of turning the house around would have been economically wasteful and out of proportion to the good to be attained. It also noted

---

[59]*St. Joseph Hosp. v. Corbetta Const. Co.,* 21 Ill.App.3d 925, 316 N.E.2d 51 (1974); *Steinbrecher v. Jones,* 151 W.Va. 462, 153 S.E2d 295 (1967).

[60]429 A.2d 214 (Me.1981). See also *Kangas v. Trust,* 110 Ill.App.3d 876, 441 N.E.2d 1271 (1982) (better material needed because contractor's breach made it more costly to repair).

[61]*Freeport Sulphur Co. v. S.S. Hermosa,* 526 F.2d 300 (5th Cir.1976).

[62]341 So.2d 646 (Miss.1977). Similarly, see *Five M. Palmer Trust v. Clover Contractors, Inc.,* 513 So.2d 364 (La.App.1987).

[63]*Anchorage Asphalt Paving Co. v. Lewis,* 629 P.2d 65 (Alaska 1981); *Corbetta Constr. Co. v. Lake County Public Bldg. Comm'n,* 64 Ill.App.3d 313, 381 N.E.2d 758 (1978) (contractor denied breach and refused to correct).

[64]414 So.2d 1037 (Fla.1982).

that the value of the house had risen *above* the contract price since the date of the contract.

The intermediate court of appeals applied a different measure of damages.[65] That court held that the economic waste doctrine does not apply to residential construction. It also found that the developer's willful and intentional failure to perform as promised barred it from using the substantial compliance doctrine. The court held that the proper damages would have been an amount necessary to reconstruct the dwelling to make it conform to the plans and specifications.

The Florida Supreme Court did not agree. First, it noted that the proper remedy for breach of a construction contract is usually the reasonable cost of correction if this is possible and does not involve unreasonable economic waste. Where it does, the measure is the difference in value between the product contracted for and the value of the performance that has been received. The court saw no reason to separate residential buildings from other construction. The court concluded that repositioning the house would result in economic waste but held that the trial court had incorrectly applied the proper formula—diminution in value. The court stated that the measurement is to be determined at the date of breach. Fluctuations in value after breach do not affect the measure of recovery. (The house would have been worth even more had it been built properly.)

The *Grossman* case also reflects the difference between objective and subjective measurements of value. The impersonal marketplace may not have found any difference in the house facing in one direction or the other. But the Hourihans did.

There is something to be said for the intermediate appeals court decision that refused to use the economic waste argument in residential construction.[66] Balance sheet gains or losses are appropriate for certain types of commercial ventures but hardly make sense when a couple is buying a home in which they expect to live.

The case also rejects, or at least seems to reject, the statement made in other cases that allows the economic waste measurement to be used only if the breach is not willful. Here the owners complained as soon as they noticed that the developer was starting to turn the house around, and the developer simply refused to make any change. To be sure, the developer might have wanted all the houses facing in a particular way and might have thought it had sufficient discretion to make that sort of change inasmuch as the contract stated that the house that was built would be *substantially* the same as the model and house as shown on the maps. Yet this could be classified as a willful and deliberate breach.

Another illustration of the difficulty of applying the diminished value rule can be seen in *Mayfield v. Swafford.*[67] The contract in question involved the construction of a swimming pool for $7,000. Within a short period of time, substantial defects were discovered that made the pool useless. The cost of correcting the defects would have been $11,000, or about 50% more than the contract price. As a result, the court first held that cost of correction cannot be used, because its cost was clearly disproportionate to the probable loss in value to the owner.

How should diminution in value be measured? The pool can be looked at as part of the real estate, and a measurement can be taken of the diminished value of the entire real estate, pool included. Alternatively, the diminished value can be what the *pool* is currently worth compared with what it should have been worth had it been built properly.

The pool in its present condition has no use, function, or utility to the owner. Its presence detracts from the value of the property. The diminished value of either property or pool should reflect the cost of removal of the pool and restoration of the site to its original condition.

The measurement first looks at the value that a properly constructed pool would have added to the *property,* an amount the court determined to be the 7,000-dollar contract price. This is compared with the value of the property with no pool at all, which must also take into account the cost of removing the pool and restoration of the site, less any salvage obtained in the removal process. Suppose the cost of removal and restoration less salvage is $3,000. There is a negative enhancement that would make the diminution in value a net $10,000. The

---

[65]396 So.2d 753 (Fla.Dist.Ct.App.1981).
[66]*Edenfield v. Woodlawn Manor, Inc.* 62 Tenn.App. 280, 462 S.W.2d 237 (1970).

[67]106 Ill.App.3d 610, 435 N.E.2d 953 (1982).

court held that the owners are entitled to the judgment for the lesser of the two amounts to be compared: the cost of repair or the diminution in the value of the owner's property.

E. Delay

Chapter 26 discussed the problems of delay, and Section 26.09(B) looked at contractual attempts to agree on damages in advance, liquidating damages. Delay was noted in Section 27.02(F), dealing with a contractor's claims that it was not allowed to perform its work in the most economic and efficient manner. This section looks briefly at owner claims for unexcused contractor delay, delay claims that were not liquidated in the contract.

The basic measurement formula used for delay is lost use, measured as a rule by rental value.[68] The difficulty arises when the owner makes claims that go beyond lost use value—usually added finance charges[69] and lost rentals. Clearly, the owner cannot recover both the interest it had to pay on a loan used to build the project and the lost rental value. One court stated:

> This is not a proper element of damage where damages are sought for the reasonable market rental value of the building in question. The rental income the owner received for his building is the interest he receives from his investment. If he has to borrow money to build the building, the interest he pays is part of his expense in furnishing the building. It is the loss of the rental income from the building that makes up the element of his damage, if any.[70]

In another case, the owner was able to be compensated for a loan extension fee that had to be paid to the lender after the contractor had been forty-five days in default.[71]

Suppose the owners made leases with tenants who could not be put into possession because of delayed construction. One case allowed the owner to recover for such rentals that were not paid.[72]

Other cases have concluded that no damages were suffered when the leases were *extended* to the full lease period as long as the tenants made no claims against the owner for the delay.[73]

The uncertainties over the lost rental value and other related expenses provide an incentive to liquidate damages for contractor-caused delay.

SECTION 27.04 Certainty

Please review Section 6.06(B) for background on the certainty requirement. Construction litigation can require adequate proof of the following:

1. Contractor cost of performing the work.
2. Owner or contractor projected cost of completing the work.
3. Additional cost incurred by contractor because of owner disruption or delay.
4. Contractor profit on performed work, unperformed work, or other consequential damages.
5. Owner lost profit for delayed performance or faulty work.

As to (1), a contractor who kept time records but who did not segregate them by jobs was able to prove how much should have been attributed to the defendant's project by testimony of his foreman.[74] Similarly, subcontractors are sometimes relieved from strict certainty proof.[75]

On the other hand, a prime contractor was limited to the amount that it could prove by time cards for labor work performed after the defendant subcontractor had defaulted.[76] Similarly, an estimate by the owner as to the cost of correction was held inadequate to prove damages.[77]

As to (2), this is usually established by introduction into evidence of estimates by other contractors or, even better, contracts for completion and correction.

[68]*Ryan v. Thurmond*, 481 S.W.2d 199 (Tex.Ct.App.1972). See *Miami Heart Institute v. Heery Architects & Engineers, Inc.*, 765 F.Supp. 1083 (S.D.Fla.1991), discussed in Section 12.14(B), in which the court employed lost use value in a claim by an owner against its architect.
[69]*Ralph D. Nelson Co. v. Beil*, infra note 73.
[70]*Ryan v. Thurmond*, supra note 68.
[71]*Herbert & Brooner Constr. Co. v. Golden*, 499 S.W.2d 541 (Mo.App.1973).
[72]Ibid.

[73]*Ryan v. Thurmond*, supra note 68; *Ralph D. Nelson Co. v. Beil*, 671 P.2d 85 (Okl.App.1983) (owner may not *keep* property for full term of twenty-five-year lease).
[74]*Don Lloyd Builders, Inc. v. Paltrow*, 133 Vt. 79, 330 A.2d 82 (1974).
[75]*St. Paul-Mercury Indem. Co. v. United States*, 238 F.2d 917 (10th Cir.1956); *McDowell-Purcell, Inc. v. Manhattan Constr. Co.*, 383 F.Supp. 802 (N.D.Ala.1974), affirmed 515 F.2d 1181 (5th Cir.1975); *Certain-Teed Products Corp. v. Goslee Roofing & Sheet Metal, Inc.*, 26 Md.App. 452, 339 A.2d 302 (1975).
[76]*Welch & Corr Constr. Corp. v. Wheeler*, 470 F.2d 140 (1st Cir.1972).
[77]*Gross v. Breaux*, 144 So.2d 763 (La.App.1962).

As to (3), see Section 27.02(F).

As to (4), although a court refused to award profits in a construction contract claim because of the lack of experience, personnel, equipment, and background of the contractor,[78] the principal questions have been the type of proof that will support a claim for lost profits. One court allowed a subcontractor to establish general profitability of the job.[79] Another submitted lost profits to the jury based on testimony of the president of the plaintiff contractor as to profits on other similar work, his method of estimating profits, and his expectation as to profits.[80] In the latter case, another witness had testified as to a range of profits. However, the appellate court reduced the judgment to the smallest profit that could be supported by the testimony.[81]

Court decisions are not always consistent as to the amount of certainty required to establish a loss. Some courts will not hold the claimant to a high standard of certainty if they are convinced that a loss has occurred, that it has been caused by the defendant's breach, and that the claimant has marshalled the best available evidence. But other courts, particularly in cases in which the issue of breach is a close one and it appears that the defendant has performed in good faith, will use the certainty requirement to either limit or bar recovery.

In arbitration or other nonjudicial systems, certainty will be less of an obstacle to obtaining the amount claimed. Nonjudicial dispute resolution involves practical people aware of the realities of record keeping and more willing to guess, compromise, or use rough formulas in these intractable proof problems.

SECTION 27.05 Records

A. During Performance

The certainty requirement for establishing damages discussed in Section 27.04 demonstrates the importance of record keeping during performance of the contract. The contractor who seeks to prove the reasonable value of the labor and materials it has furnished will be in a substantially better position if it can produce accurate, detailed records of what it spent in performing the work.[82] If it seeks to prove additional expenditures that it claims resulted from wrongful acts of the owner, it will be in a substantially better position if it can establish those costs that resulted from the breach by the owner. If this cannot be done, it may be able to take advantage of crude formulas, such as total cost, modified total cost, or the jury verdict discussed in Section 27.02(F). But the party that goes into settlement negotiations, arbitration, or litigation with the most complete and detailed records is in the best position to obtain an optimal settlement, award, or court judgment.

In addition, the contractor should keep and maintain its records on *other* contracts, particularly records that establish its worker productivity, scheduling, and profit margins.

A contractor who wishes to obtain a time extension for bad weather should maintain careful records of the weather conditions at the place of performance. This is particularly important to those who use AIA documents. A201, ¶ 4.3.8.2, requires the contractor to produce records relating to weather and establish the effect weather had on its performance.

Record keeping is a prime motivation for having a full-time site representative.[83]

B. After Dispute Arises

After a dispute has arisen, each party seeks to augment its records with the records of the other party to the dispute. Usually this is handled through discovery prior to trial.[84]

Those who deal with public owners may be able to invoke Freedom of Information Laws[85] or con-

[78]*Elec. Service Co. of Duluth v. Lakehead Elec. Co.*, 291 Minn. 22, 189 N.W.2d 489 (1971).

[79]*Constr. Limited v. Brooks-Skinner Bldg. Co.*, 488 F.2d 427 (3d Cir.1973).

[80]*Natco, Inc. v. Williams Bros. Eng'g Co.*, 489 F.2d 639 (5th Cir.1974).

[81]See Section 27.02(D) for formulas that include profit. For profits unconnected to *this* contract, see Section 27.06.

[82]In *Huber, Hunt & Nichols, Inc. v. Moore*, 67 Cal.App.3d 278, 136 Cal.Rptr. 603 (1977), the court refused to admit computer costs records, as they did not segregate costs incurred because of the breach.

[83]See Sections 12.08(D) and 17.05(B).

[84]See Section 2.09.

[85]*De Maria Bldg. Co. v. Dep't of Management and Budget*, 159 Mich.App. 729, 407 N.W.2d 72 (1987). But see *Sea Crest Const. Corp. v. Stubbing*, 82 A.D.2d 546, 442 N.Y.S.2d 130 (1981), which held that the correspondence between the public agency and its consultants were, with some exceptions, exempt and did not have to be disclosed.

stitutional due process rights as a basis for obtaining relevant records of the public owner.[86]

SECTION 27.06 Consequential Damages

A. Defined

Direct losses can occur in any construction contract. They are usually measured by conventional objective formulas, such as lost use for unexcused contractor completion delay,[87] cost of correction or diminished value for other contractor defaults,[88] interest for owner-delayed payment,[89] the balance of the purchase price for nonpayment, and the additional costs caused by owner disruption or delay.[90]

Sometimes a claimant claims consequential damages, those that go beyond *direct* losses. Consequential damages are abnormal, related to the particular transaction, and more subjective. The legal doctrines relevant to consequential damages are foreseeability, discussed in Section 6.06(C), and certainty, discussed in Sections 6.06(B) and 27.04. Consequential damages are more difficult to establish than direct losses.

B. Owner Claims

Generally, owners' consequential damage claims are for lost profits caused by delay in completing the project. Traditionally, recovery of lost profits was denied if the business in which the owner intended to engage was a new one, an inflexible application of the certainty rule.[91] However, the

modern tendency has been to treat this issue as any other issue and regard it as a question of fact that should be resolved based on relevant evidence.[92]

C. Contractor Claims

Generally, contractor claims are based on the loss of other business opportunities or business goodwill in general.[93] Many claims are for lost profits because the owner's conduct has reduced or impaired the contractor's bonding capacity. Most cases have denied these claims as being too speculative.[94] Some claims have been allowed in cases where the contractor could clearly establish a history of making consistent profits.[95] The contractor must also establish that these damages were reasonably foreseeable by the parties at the time the contract was made.[96]

D. Tort and Contract

The interrelationship of claims based on breach of contract and those based on tort was discussed earlier.[97] *R.E.T. Corp. v. Frank Paxton Co., Inc.*,[98] demonstrates the increasing fusion of tort and contract claims and reviews some of the material discussed in Section 27.03(D).

The owner built two apartment complexes. The second complex had problems because tenants complained of cold apartments and high heating bills. An investigation revealed that the contractor had not complied with the insulation requirements of the contract. Corrections were attempted that af-

[86]*Zurn Eng'rs v. California State Dept. of Water Resources*, 69 Cal.App.3d 798, 138 Cal.Rptr. 478 (1977) (due process requires that the State Engineer notify the contractor of the factual basis for his decision and give him a chance to rebut or supplement it before hearing). This decision and earlier ones on which it was based relied heavily on the power granted the State Engineer to resolve disputes. This was subsequently changed by statute in California to provide for arbitration and the discovery rights granted ordinary civil litigants. See West Ann.Cal.Pub.Cont.Code § 10240.11.
[87]See Section 26.09.
[88]See Section 27.03(D).
[89]See Sections 6.08 and 22.02(L).
[90]See Section 27.02(F).
[91]*Drs. Selike & Conlon, Ltd. v. Twin Oaks Realty, Inc.*, 143 Ill.App.3d 168, 491 N.E.2d 912 (1986); *Exton Drive-In v. Home Indem. Co.*, 436 Pa. 480, 261 A.2d 319 (1969), cert. denied 400 U.S. 819 (1970).

[92]*Mechanical Wholesale, Inc. v. Universal-Rundle Corp.*, 432 F.2d 228 (5th Cir.1970) (recovery not automatically excluded for new businesses); *Melvin Grossman v. Sea Air Towers, Ltd.*, 513 So.2d 686 (Fla.Dist.Ct.App.1987) (lost rents are reasonably certain in nature).
[93]*Larry Armbruster & Sons, Inc. v. State Public School Bldg. Auth.*, 505 A.2d 395 (Pa.Commw.1986) (not established).
[94]*Indiana & Michigan Elec. Co. v. Terre Haute Indus., Inc.*, 507 N.E.2d 588 (Ind.App.1987); *Hirsh Elec. Co. v. Community Services, Inc.*, 133 A.D.2d 140, 518 N.Y.S.2d 818 (1987) modified on reargument, 145 A.D.2d 603, 536 N.Y.S.2d 141 (1988).
[95]*Tempo, Inc. v. Rapid Elec. Sales and Service, Inc.*, 132 Mich.App. 93, 347 N.W.2d 728 (1984); *Lass v. Montana State Highway Comm'n*, 157 Mont. 21, 483 P.2d 699 (1971) (contractor had earned profit for past twenty-two years).
[96]*Texas Power & Light Co. v. Barnhill*, 639 S.W.2d 331 (Tex.Ct.App.1982) (owner aware of existence of collateral contracts).
[97]See Section 14.05(E).
[98]329 N.W.2d 416 (Iowa 1983).

ter some initial difficulties seemed to cure the problems, and tenant complaints dropped. But by the time the corrections were made, almost three years after completion, the complex was in serious financial trouble. The tenants forced management to offer rent inducements and adjust the rents. Vacancies increased, and rents could not be increased sufficiently to cover increasing costs. The second apartment complex dragged down the first, the business went into receivership, and the property ultimately was sold at a substantial loss.

The claim against the builder was, as is customary, based on a number of theories. The trial court found that the builder had breached its contract, breached express and implied warranties, and negligently installed the insulation. It awarded damages of $105,000 for repair costs, $237,000 for lost rents, and $650,000 as the difference between the fair market value of the property (the amount the sale should have realized) and the reduced price (the amount that was finally obtained). The total damages were almost one million dollars.

The Iowa Supreme Court held that the trial court had been correct in using both cost of correction and diminished value, because no single formula would sufficiently compensate the plaintiff. However, the reviewing court found that the trial court had erroneously concluded that the damages are the same for breach of contract and for negligence. The court held that foreseeability is a limitation if the claim is based on breach of contract but not if based on tort. The court held that the cost of repairs and lost rentals could well have been contemplated by the parties when the contract was made, as they are routine elements of a damage recovery for breach of a construction contract. However, the court concluded that parties would not normally contemplate diminution in the value of the complex and its subsequent distress sale. This would be recoverable in a claim based on tort but not based on contract. Yet the court concluded that the error was harmless and affirmed the judgment.

The court dealt with another issue raised in Section 27.03(D). The trial court had determined the diminution in value as of 1978, the time the building was sold, rather than in 1974–1975, the time the complex was built. The owner argued that the injury was a continuing one that extended until the final distress sale. The court concluded that these were special facts that militated against using the

usual rule that diminution is determined as of the time of completion. The court concluded that the injury had no impact on the plaintiff so as to become final until the complex went into receivership and was sold.

SECTION 27.07 Avoidable Consequences: The Concept of Mitigation

The general dimensions of this doctrine were discussed in Section 6.06(D). Some defendants have invoked this concept in construction disputes. In *C.A. Davis, Inc. v. City of Miami*,[99] a defendant contractor sought to reduce the recovery by the owner by asserting that the owner spent more than was necessary to correct the contractor's work. The court held that the contractor could challenge the completion cost only if it could show waste, extravagance, or lack of good faith. It will be unusual for a court to conclude that expenses incurred by the claimant to correct defective work by the contractor were out of line and not recoverable. Yet one court did reduce an award because it concluded that overtime was not necessary and the claimant's refusal to allow the designer-builder to provide free engineering services was not justified.[100]

In *Great Lakes Gas Transmission Co. v. Grayco Constructors, Inc.*,[101] a breaching construction contractor contended that the owner, a public utility, could pass its loss to its customers through rate increases. In addition to the uncertainty as to whether this *could* be done, it seems as if denial of this assertion was based in part on the "chutzpah" of the contract breaker in seeking relief by putting such a burden on the utility and its customers.

This doctrine was invoked, strangely, in *S.J. Groves & Sons Co. v. Warner Co.*, by a subcontractor in default who claimed the prime contractor should have fired him.[102] The prime contractor sought damages against the subcontractor who had failed to supply ready-mixed concrete.

Trouble developed almost at the outset. The owner was forced repeatedly to reject the concrete

---

[99]400 So.2d 536 (Fla.Dist.Ct.App.1981).
[100]*First Nat'l Bank of Akron v. Cann*, 503 F.Supp. 419 (1980), affirmed 669 F.2d 415 (6th Cir.1982).
[101]506 F.2d 498 (6th Cir.1974).
[102]576 F.2d 524 (3d Cir.1978).

supplied by the subcontractor. In addition, the subcontractor frequently failed to make deliveries in accordance with the prime contractor's instructions. The prime contractor considered using other sources but felt it had no real alternatives. It would cost too much to build its own plant. The only other concrete source had not been certified to do state work, and its price was higher than that of the subcontractor. In addition, the only alternative source had limited facilities and trucks. Despite the difficulties, the subcontractor continued to assure the prime contractor that things would improve.

Despite these promises, the subcontractor's performance continued to be erratic, and the public entity ordered all construction halted until the subcontractor's performance could be discussed at a conference. Again after renewed assurances that things would improve, the public entity allowed work to resume. For succeeding months, the subcontractor's performance continued to be uneven and unpredictable.

During performance, the prime contractor approached the alternate source, which by then had been certified by the state. The alternate source agreed to reduce its price to the same price as the subcontractor, but the prime contractor continued to use its subcontractor as its sole supplier.

Had the prime contractor acted reasonably in continuing to use the subcontractor despite its poor performance? The trial court concluded that the prime contractor had not, but the appellate court did not agree.

After noting that the burden of proving that the losses could have been avoided by reasonable effort was on the breaching party, the court looked at the alternatives available to the prime contractor. One alternative was to simply terminate the subcontractor, an alternative the court did not find realistic. Another alternative was for the prime contractor to set up its own cement batching plant, an alternative the court found impractical because of time and expense. Another alternative was to accept the subcontractor's assurances that performance would be satisfactory in the future, the alternative selected. Another alternative was to use the alternate supplier as a supplemental source or as a substitute.

The appellate court concluded that *all* the alternatives had their drawbacks. Even if the alternate supplier had been engaged as a supplemental source, there was still no guarantee that the sub-

contractor would perform properly. The use of two suppliers might raise other problems, and there was a question as to whether the alternate supplier would have been able to perform.

The court concluded that confronted with these choices, the prime contractor's decision to stay with the subcontractor may not have been the best choice. The test is, however, whether the course chosen was *reasonable,* not whether it was necessarily the best. The court was not willing to engage in hypercritical examination of the choice made. It concluded that staying with the subcontractor may have been not only reasonable but also the best choice under the circumstances. The court noted that the breaching subcontractor, who sought to second-guess the choice made by the prime contractor, could also have engaged a supplemental supplier and that, where each party had the equal alternative to reduce the damages, the defendant was in no position to contend that the plaintiff failed to mitigate.

SECTION 27.08 Collateral Source Rule

This doctrine was described generally in Section 6.06(F). Some illustrations have been seen in construction disputes. For example, in *Industrial Development Board v. Fuqua Industries, Inc.,*[103] a local development board contracted for the construction of a plant that was to be used by a manufacturer. The development board planned to execute a long-term lease of the plant to the manufacturer. The development board made the contract to enable the manufacturer to obtain better financing terms. In effect, the plant was designed by the manufacturer for the manufacturer's use. The defects in the plant were corrected by the manufacturer, and this, according to the court, precluded the development board from having any contract claim against the contractor. (However, the court would allow the manufacturer to recover from the contractor to preclude unjust enrichment.)

New Foundation Baptist Church v. Davis[104] came to a different conclusion. A church sued the contractor for defects in the sanctuary floor that caused collapse during a funeral three years after the church was completed. A church member carpen-

[103]523 F.2d 1226 (5th Cir.1975).
[104]257 S.C. 443, 186 S.E.2d 247 (1972).

ter donated his labor and repaired the damage for a total cost to the church of $3,000. Yet the jury's award of $6,500 to the church was upheld. The court noted that the church member had no wish to benefit the contractor, and it would be unfair for the contractor to receive the advantage of the church member's generosity.

The issue arose in *Huber, Hunt & Nichols, Inc. v. Moore,*[105] where a contractor claimed against the owner and the architect. The contractor settled with the owner before trial. One of the reasons given for denying recovery against the architect was that the contractor had been paid substantial amounts by the owner. This was a tort claim, there being no contract between contractor and architect. Although this would make the payments come from a collateral source that cannot be used to reduce the claim, the court held that the owner and the architect were sufficiently associated with each other to preclude the owner from being considered a collateral source.

SECTION 27.09 Noneconomic Losses

The difference between those losses that can be measured in the marketplace and those that cannot were discussed generally in Section 6.06(H). In looking at claims for emotional distress in construction disputes, care must be taken to differentiate the substantive basis for the claim. Such a claim may be based on a breach of contract, negligent conduct, or intentional conduct. Particular cases are instructive.

Randa v. United States Homes, Inc.[106] upheld a jury verdict for *intentional* infliction of emotional distress against a contractor. There were many defects, causing the wife of the buyer of a 160,000-dollar home to spend time in the hospital for a nervous breakdown. She was "petrified" when told she could not put mirrors in the bedroom. Filing of liens shocked the buyers, with the wife getting "sick" and starting "to cry."

Other decisions have allowed recovery for noneconomic losses without the need to establish intentional infliction of emotional distress. *B & M Homes,*

Inc. v. Hogan[107] involved a claim by the Hogans against Morrow (the owner of B & M Homes) based on the contract under which the Hogans agreed to buy a lot and a house to be built by Morrow. During construction, Mrs. Hogan discovered a hairline crack in the concrete slab. Morrow told her such cracks were common and not to worry. After the Hogans moved in, the crack widened and caused extended damage. Morrow was notified and dealt with the damage but did not attempt to repair the slab himself. The claim was based in part on mental anguish that the Hogans suffered. The Hogans introduced evidence that they were concerned over their safety because they believed the house to be structurally defective, that the condition of the house might cause gas and water lines to burst, and that they were forced to live in a defective house because they could not afford to move.

The court noted that as a general rule, damages for mental anguish are not recoverable for breach of contract. However, an exception is made for contracts that involve "mental concern or solicitude." The court held that this contract fell into that category and that it was reasonably foreseeable that faulty construction of a house would cause the homeowners to suffer severe mental anguish. The court noted homes as the largest single investment for most families and one that places the family in debt for many years. The court pointed to an earlier decision that had allowed recovery for mental anguish when a builder performed improperly under a contract to build a home, emphasizing the homeowner's view of her home as her castle and a place to protect her against the elements and to shelter her belongings.

Orto v. Jackson[108] allowed recovery for "aggravation and inconvenience." In upholding the award, the court stated:

> . . . for over two years the Jacksons have suffered partial loss of use and enjoyment of their basement, which still leaks after heavy rains, and the total loss of use and enjoyment of their backyard due to the cesspool located there. . . . The aggravation and inconvenience suffered by the Jacksons certainly arose naturally from the builders' breach of the contract and it could hardly

---

[105]Supra note 82.
[106]325 N.W.2d 905 (Iowa App.1982). See also *Alsteen v. Gehl,* 21 Wis.2d 349, 124 N.W.2d 312 (1963) (dictum).

[107]376 So.2d 667 (Ala.1979). See Annot., 7 A.L.R.4th 1178 (1981).
[108]413 N.E.2d 273, 278 (Ind.App.1980).

be said that these damages were not in the parties' minds at the time they entered the contract.

These cases do not suggest that noneconomic losses will be freely awarded in construction contract disputes. They do suggest that recovery for such losses will no longer be automatically denied. The law, as part of consumer protection, may compensate those who suffer emotional distress when homes built for them are defective and the builders refuse to stand behind their work. Although such claims are susceptible to abuse, the *possibility* of recovery may cause builders to take their contractual obligations seriously.

SECTION 27.10 Punitive Damages

Punitive damages were discussed in the contract context in Section 6.04 and in the tort context in Section 7.10(C). This chapter deals mainly with claims for breach of the construction contract. The award of punitive damages, almost unheard of in construction contract disputes, has begun to play an important role. To justify punitive damages the conduct must go beyond a simple or even negligent breach of contract. It must be at least *grossly* negligent, outrageous, or malicious.

Despite the windfall aspects of punitive damages, both contract law and tort law have begun to look at punitive damages as a way of deterring outrageous conduct, particularly in that part of the construction contract world that involves ordinary consumers rather than knowledgeable business persons.

Again, particular cases are instructive. In *F.D. Borkholder Co. v. Sandock,*[109] the contractor deviated from the plans. The court justified the punitive damages award by concluding that the defendant was guilty of intentional and wrongful acts that constituted fraud, misrepresentation, deceit, and gross negligence in its dealings. The court stated:

> The Court of Appeals cited our decision in *Hibschman Pontiac, Inc. v. Batchelor,* 266 Ind. 310, 362 N.E.2d 845 (1977) for the proposition that punitive damages are recoverable in breach of contract actions only when a separate tort accompanies the breach of tort-

like conduct mingles in the breach. Here, prior to the execution of the contract, Sandock representatives expressed their concern about moisture on the walls. Under the terms of the contract, they were to pay $200 for plans to be drawn up by Borkholder's architect. The contract provided that all labor and material would be furnished in accordance with specifications. Sandock was given a copy of the plans. However, contrary to these plans, the top and bottom courses of block forming the one wall were not filled with concrete, thus constituting latent variances. Furthermore, the roofline was shortened which represented an additional deviation from the plans. There was testimony that the cut-off roofline enabled water to leak down into the top of the block wall. Other evidence indicated that the wetness problem resulted from this water percolating down through the inside of the wall, collecting at the bottom, and then rising again by capillary action. Sandock made numerous complaints but was constantly reassured by several Borkholder representatives that the problem was caused by simple condensation, a theory ultimately disproved by an on-site test conducted by the Borkholder firm. Sam Sandock testified that Freeman Borkholder, president of the company, promised that the situation would be remedied whereupon Sandock tendered all but $1,000 of the contract price. The problem was never corrected. The Borkholder people knew, of course, that the blocks in the wall were not filled with concrete. Also, Borkholder himself conceded that the roofline adjustment increased the likelihood of water running down into the core of the wall.

We believe that there is cogent and convincing proof that the Borkholder firm engaged in intentional wrongful acts constituting fraud, misrepresentation, deceit, and gross negligence in its dealings with Sandock. *Hibschman Pontiac, Inc.,* supra. Accordingly, we agree with the Court of Appeals that the trial court could have concluded that separate torts accompanied the breach. Next, relying on *Hibschman,* the Court of Appeals attempted to identify the public interest to be served by imposing punitive damages. However, the majority could not perceive any such interest and refused to let the award stand. We disagree. As Judge Garrard stated in his dissent:

"I have no problem identifying the public interest to be served in requiring that the builders of public buildings be deterred from fraudulently disregarding building code requirements or those contained in the plans and specifications they have agreed to comply with." *Sandock v. Borkholder, supra,* at 959.

---

[109]274 Ind. 612, 413 N.E.2d 567, 570–571 (1980).

The purpose of punitive damages generally is to punish the wrongdoer and to deter him and others from engaging in similar conduct in the future. . . . An award of such damages is particularly appropriate in proper cases involving consumer fraud. . . .

The building contractor occupies a position of trust with members of the public for whom he agrees to do the desired construction. Few people are knowledgeable about this industry, and most are not aware of the techniques that must be employed to produce a sound structure. Necessarily, they rely on the expertise of the builder. Here, the builder has been found to have engaged in fraudulent or deceptive practices by constructing a building with latent deviations from the plans which resulted in damage to the owner. Further, the builder has attempted to disclaim responsibility for such damage when it may be inferred that it knew or should have known that its work was the cause. Under these circumstances, certainly the imposition of punitive damages furthers the public interest.

In *Jeffers v. Nysse*,[110] a purchaser made a claim against a developer-builder. The jury had found that the builder had misrepresented the insulation and heating costs when he sold the house, knowing those misrepresentations to be untrue and with the intention of deceiving the buyers. Yet the jury did not find that the seller's conduct had been malicious or vindictive. Nevertheless, the court affirmed a judgment of punitive damages because the conduct had been wanton, willful, or reckless despite the absence of malice. The court concluded that a requirement of malice

would shield the defendants from any liability beyond the costs of compensating the [buyers] for their costs in putting the house in the condition it was represented to be in originally. But "putting the cookies back in the jar" when caught is not enough. If that result were reached, sellers could make any misrepresentation necessary to make a sale. If it was not discovered, or was discovered or not pursued, the seller would make a windfall gain. If the fraud were discovered and successfully proven, the seller would only be liable to make good on his representations. He would suffer no punishment nor would he be deterred from similar conduct in the future.

In *Joseph v. Bray*,[111] punitive damages were awarded where the contractor deliberately installed wrong materials, worked when the owner was not on the premises, concealed defective work, and promised the owner to make repairs but did not. These actions constituted fraud, which would justify punitive damages.

The cases described have involved performance problems. However, *Brant Construction Co. Inc. v. Lumen Construction Inc.*[112] involved a smoke screen to hide evasion of the Minority Business Enterprise (MBE) requirements of a public contract. The prime contractor deceived the subcontractor and used it as a "front" to obtain award of a federally funded contract that required good-faith effort toward employing a designated percentage of minority contractors.

The MBE program is designed to give experience to minority contractors. The prime contractor entered into a contract with a minority subcontractor but for much of the contract period used another subcontractor. When it finally gave some work to the designated subcontractor, the prime did not offer the required assistance and prevented it from performing properly. This attempted circumvention of the MBE program justified awarding punitive damages to the subcontractor to deter other primes from similar conduct by making an example of the prime.

But noting cases in which punitive damages *were* awarded should not be taken to mean that they are common in construction contract disputes. For example, one case refused to award punitive damages when there had been a wrongful termination because there had been no showing of malicious or oppressive conduct. The court did not want to open the claims floodgates in the wake of bitterly disputed construction contracts.[113]

Similarly, another court refused to award punitive damages for gross negligence in performance of a contract.[114]

---

[111]182 Ga.App. 131, 354 S.E.2d 878 (1987).

[112]515 N.E.2d 868 (Ind.App.1987).

[113]*Indiana & Michigan Elec. Co. v. Terre Haute Indus.*, supra note 94.

[114]*Jim Walter Homes, Inc. v. Reed*, 711 S.W.2d 617 (Tex.1986). Punitive damages were not awarded where the issue of breach was a close one—*Quedding v. Arisumi Bros., Inc.*, 661 P.2d 706 (Hawaii 1983)—or where there was no evidence the contractor intended not to perform when the contract was made—*Centuries Properties, Inc. v. Machtinger*, 448 So.2d 570 (Fla.Dist.Ct.App.1984). See Annot., 40 A.L.R.4th 110 (1985).

---

[110]98 Wis.2d 543, 297 N.W.2d 495 (1980).

Finally, punitive damages were denied in a breach of contract case involving only delay or nonperformance.[115] In that case, the intentional breach of an underpriced contract was held not to subject the breaching party to punitive damages. The court was unwilling to force a contractor faced with a substantial financial loss to perform when it made the sound business judgment not to perform, knowing it would be subject only to compensatory damages.

Although punitive damages will still be rare in the ordinary construction contract dispute, the increased willingness of courts to award such open-ended damages determined by a jury means that punitive damages will play an increasingly important role in construction disputes, particularly where contractors prey on unsuspecting owners and where ordinary compensatory damages will not be sufficient to deter such conduct.

SECTION 27.11 Cost of Dispute Resolution: Attorneys' Fees

See Section 6.07.

SECTION 27.12 Interest

See Section 6.08.

---

[115]*Construction Contracting & Mgmt, Inc. v. McConnell,* 112 N.M. 371, 815 P.2d 1161 (1991).

SECTION 27.13 Disputes and Settlements: *Rich & Whillock v. Ashton Dev.*

When important benchmarks in a transaction are reached, the law must deal with the effect of those benchmarks on claims. Illustration of such benchmarks in the Construction Process are the issuance of progress payment certificates, the issuance of certificates of substantial and final completion, and the acceptance of the project by the owner.

The law seeks to put an end to disputes. As a result, the common law generally discharged any claims if the claimant accepted a tendered payment with the knowledge that the tendered payment was made in full and final settlement of any claims relating to the incident giving rise to the payment. This was accomplished through the doctrine of accord and satisfaction.

Yet the owner may be taking advantage of the desperate condition of the contractor to force an unfair settlement knowing the contractor simply cannot resist. Because of financial pressure and the belief of the contractor that it takes only that to which it is entitled, the contractor may take the payment. To deal with the effect of payment on its claim, the following case employed economic duress to set aside a settlement that had been made between a prime and a subcontractor.

RICH & WHILLOCK, INC. v. ASHTON DEVELOPMENT, INC.

California Court of Appeal, 1984. 157 Cal.App.3d 1154, 204 Cal.Rptr. 86.
[Ed. note: Footnotes omitted.]

WIENER, Associate Justice.

Ashton Development, Inc. and Bob Britton, Inc. appeal from the judgment awarding Rich & Whillock, Inc. $22,286.45 for the balance due under a grading and excavating contract. Following a nonjury trial the court entered judgment after it found a settlement agreement and release signed by Rich & Whillock, Inc. were the products of economic duress and thus provided no defense to its contract claim. We conclude substantial evidence supports the court's finding and affirm the judgment.

FACTUAL AND PROCEDURAL BACKGROUND

On February 17, 1981, Bob Britton, president of Bob Britton, Inc., signed a contract for grading and excavating services to be provided by Rich & Whillock, Inc. at a price of $112,990. The work was to be done on a project by Ashton Development, Inc. Bob Britton, Inc. was general contractor on the project and the agent for Ashton Development, Inc. in all dealings with Rich & Whillock, Inc. Work began the day the contract was signed.

In late March 1981 Rich & Whillock, Inc. encountered rock on the project site. A meeting was held at the site to discuss the problem. In attendance were Greg Whillock and Jim Rich, president and vice-president of Rich & Whillock, Inc., Bob Britton, Berj Aghadjian, president of Ashton Development, Inc., and a man from a blasting company. Everyone agreed the rock would have to be blasted. The $112,990 contract price expressly excluded blasting. The contract also stated "[a]ny rock encountered will be considered an extra at current rental rates." In response to Britton's inquiry, Whillock and Rich estimated the extra cost to remove the rock would be about $60,000, for a total contract price of approximately $172,000. They also emphasized, however, the estimate was not firm and the actual cost could go much higher due to the unpredictable nature of rock work.

Britton directed Whillock and Rich to go ahead with the rock work and bill him for the extra costs and said they would be paid. Rich & Whillock, Inc. proceeded accordingly, submitting invoices and receiving payments every other week. The invoices separately stated the charges for the regular contract work and the extra rock work and were supported by attached employee time sheets. Toward the end of April Whillock asked Britton if he had any questions and told Whillock to continue with the rock work because it had to be done.

By June 17, 1981, after receiving payments totaling $190,363.50, Rich & Whillock, Inc. submitted a final billing for an additional $72,286.45. After consulting with Aghadjian, Britton refused to pay. When Whillock asked why, Britton explained he and Aghadjian were short on funds for the project and had no money left to pay the final billing. Up until he received that billing, Britton had no complaints about the work done or the invoices submitted by Rich & Whillock, Inc. and had never asked for any accounting of charges in addition to that already provided. Whillock told Britton he and Rich would "go broke" if not paid because they were a new company, the project was a big job for them, they had rented most of their equipment and they had numerous subcontractors waiting to be paid. Britton replied he and Aghadjian would pay them $50,000 or nothing, and they could sue for the full amount if unsatisfied with the compromise.

On July 10, 1981, Britton presented Rich with an agreement for a final compromise payment of $50,000. The agreement provided $25,000 would be paid "upon receipt of this signed agreement," to be followed by a second $25,000 payment on August 10, 1981 "upon receipt of full and unconditional releases for all labor, material, equipment, supplies, etc., purchased, acquired or furnished for this contract up to and including August 10, 1981." Rich repeated Whillock's earlier statements about the probable effects of nonpayment on their business. Britton replied: "I have a check for you, and just take it or leave it, this is all you get. If you don't want this, you have got to sue me." Rich then signed the agreement and received a $25,000 check after telling Britton the agreement was "blackmail" and he was signing it only because he had to in order to survive. Rich & Whillock, Inc. received the second $25,000 payment on August 20, 1981, at which time Whillock signed a standard release form.

In December 1981 Rich & Whillock, Inc. filed this action for damages for breach of contract. The court found Ashton Development, Inc. and Bob Britton, Inc. were liable for the $22,286.45 balance due under the contract, and that the July 10 agreement and August 20 release were unenforceable due to economic duress. On the latter point the court found Britton and Aghadjian "never really disputed the amount of plaintiff's charge in that they never asked for an accounting nor documentation concerning the extra work." The court also stated it disbelieved Britton when he testified Rich & Whillock, Inc. had agreed to do the extra work for a sum not to exceed $90,000. By disbelieving Britton and finding no dispute about the actual amount owed, the court impliedly found Britton and Aghadjian acted in bad faith when they refused to pay Rich & Whillock, Inc.'s final billing and offered instead to pay a compromise amount of $50,000. Based upon its finding of bad faith, the court concluded the July 10 agreement and August 20 release were signed "under duress in that plaintiff felt they would face financial ruin if they did not accept the lesser sum and that defendants, knowing this, threatened no further payment unless plaintiff accepted the lesser sum."

DISCUSSION

"At the outset it is helpful to acknowledge the various policy considerations which are involved in cases involving economic duress. Typically, those claiming such coercion are attempting to avoid the consequences of a modification of an original contract or of a settlement and release agreement. On the one hand, courts are reluctant to set aside agreements because of the notion of freedom of contract and because of the desirability of having private dispute resolutions be final. On the other hand, there is an increasing recognition of the law's role in correcting inequitable or unequal exchanges between parties of disproportionate bargaining power and a greater willingness to not enforce agreements which were entered into under coercive circumstances." (*Totem Marine T. & B. v. Alyeska Pipeline, Etc.* (Alaska 1978) 584 P.2d 15, 21, fn. omitted.)

California courts have recognized the economic duress doctrine in private sector cases for at least 50 years. (*Young v. Hoagland* (1931) 212 Cal. 426, 430–432, 298 P. 996.) The doctrine is equitably based (*Burke v. Gould, supra,* 105 Cal. at p. 281, 38 P. 733) and represents "but an expansion by courts of equity of the old common law doctrine of duress." (*Sistrom v. Anderson* (1942) 51 Cal.App.2d 213, 220, 124 P.2d 372.) As it has evolved to the present day, the economic duress doctrine is not limited by early statutory and judicial expressions requiring an unlawful act in the nature of a tort or a crime. (Civ.Code, § 1569, subd. 2; . . . .

Instead, the doctrine now may come into play upon the doing of a wrongful act which is sufficiently coercive to cause a reasonably prudent person faced with no reasonable alternative to succumb to the perpetrator's pressure. . . . The assertion of a claim known to be false or a bad faith threat to breach a contract or to withhold a payment may constitute a wrongful act for purposes of the economic duress doctrine. . . .

Further, a reasonably prudent person subject to such an act may have no reasonable alternative but to succumb when the only other alternative is bankruptcy or financial ruin. . . .

The underlying concern of the economic duress doctrine is the enforcement in the marketplace of certain minimal standards of business ethics. Hard bargaining, "efficient" breaches and reasonable settlements of good faith disputes are all acceptable, even desirable, in our economic system. That system can be viewed as a game in which everybody wins, to one degree or another, so long as everyone plays by the common rules. Those rules are not limited to precepts or rationality and self-interest. They include equitable notions of fairness and propriety which preclude the wrongful exploitation of business exigencies to obtain disproportionate exchanges of value. Such exchanges make a mockery of freedom of contract and undermine the proper functioning of our economic system. The economic duress doctrine serves as a last resort to correct these aberrations when conventional alternatives and remedies are unavailing. The necessity for the doctrine in cases such as this has been graphically described:

"Nowadays, a wait of even a few weeks in collecting on a contract claim is sometimes serious or fatal for an enterprise at a crisis in its history. The business of a creditor in financial straits is at the mercy of an unscrupulous debtor, who need only suggest that if the creditor does not care to settle on the debtor's own hard terms, he can sue. This situation, in which promptness in payment is vastly more important than even approximate justice in the settlement terms, is too common in modern business rela-

tions to be ignored by society and the courts." (Dalzell, *Duress by Economic Pressure II* (1942) 20 N. Carolina L.Rev. 340, 370.)

Totem Marine T. & B. v. Alyeska Pipeline, Etc., supra, 584 P.2d 15, presents an example of economic duress remarkably parallel to the circumstances of this case. Totem, a new corporation, contracted with Alyeska to transport pipeline construction materials from Houston, Texas to a port in southern Alaska, with the possibility of one or two cargo stops along the way. Totem chartered the equipment necessary to perform the contract. Unfortunately, numerous unanticipated problems arose from the outset which impeded Totem's performance. When Totem's chartered tugs and barge arrived in the port of Long Beach, California, Alyeska caused the barge to be unloaded and unilaterally terminated the contract. Totem then submitted termination invoices totalling somewhere between $260,000 and $300,000. At the same time, Totem notified Alyeska it was in urgent need of cash to pay creditors and that without immediate payment it would go bankrupt. After some negotiations, Alyeska offered to settle Totem's account for $97,500. In order to avoid bankruptcy, Totem accepted Alyeska's compromise offer and signed an agreement releasing Alyeska from all claims under the contract. (*Id,* at pp. 17–19).

About four months after signing the release agreement Totem sued Alyeska for the balance due under the contract. The trial court entered summary judgment for Alyeska based on the release agreement. (*Totem Marine T. & B. v. Alyeska Pipeline, Etc., supra,* 584 P.2d at p. 19.) The Supreme Court of Alaska reversed, explaining:

"[W]e believe that Totem's allegations, if proved, would support a finding that it executed a release of its contract claims against Alyeska under economic duress. Totem has alleged that Alyeska deliberately withheld payment of an acknowledged debt, knowing that Totem had no choice but to accept an inadequate sum in settlement of that debt; that Totem was faced with impending bankruptcy; that Totem was unable to meet its pressing debts other than by accepting the immediate cash payment offered by Aleyska; and that through necessity, Totem thus involuntarily accepted an inadequate settlement offer from Alyeska and executed a release of all claims under the contract. If the release was in fact executed under these circumstances, we think that under the legal principles discussed above that this would constitute the type of wrongful conduct and lack of alternatives that would render the release voidable by Totem on the ground of economic duress." (*Id.,* at pp. 23–24, fn. omitted.)

Here, Britton and Aghadjian acted in bad faith when they refused to pay Rich & Whillock, Inc.'s final billing and offered instead to pay a compromise amount of $50,000. At the time of their bad faith breach and settlement offer, Britton, and through him, Aghadjian, knew Rich & Whillock, Inc. was a new company overextended to creditors and subcontractors and faced with imminent bankruptcy if not paid its final billing. Whillock and Rich strenuously protested Britton's and Aghadjian's coercive tactics, and succumbed to them only to avoid economic disaster to themselves and the adverse ripple effects of their bankruptcy on those to whom they were indebted. Under these circumstances, the trial court found the

July 10 agreement and August 20 release were the products of economic duress. That finding is consistent with the legal principles discussed above and is supported by substantial evidence. Accordingly, the court correctly concluded Ashton Development, Inc. and Bob Britton, Inc. were liable for the $22,286.45 balance due under the contract.

DISPOSITION

Judgment affirmed.

COLOGNE, Acting P.J., and STANIFORTH, J., concur.

As noted, the accord and satisfaction doctrine generally bars claims. Yet there have been inroads on this venerable doctrine, such as case decisions emphasizing economic duress as in the *Rich & Whillock* case,[116] judicial outrage at unconscionable conduct by the stronger party,[117] a minority of cases holding that the Uniform Commercial Code eliminated accord and satisfaction under certain conditions,[118] and state legislation.[119] Although a

person who accepts tender of an amount under these circumstances still runs a substantial risk that its claim will be barred, increasingly the law is willing to allow the claim to be pursued despite the claimant's having accepted the tendered payment with the restrictive conditions.[120]

SECTION 27.14 Claims Against Multiple Parties

Chapter 24—particularly Sections 24.06 and 24.07—noted the not uncommon phenomenon of defects that can be traceable to design and to failure by the contractor to execute the design properly. That chapter looked at disputes between owner and contractor. In this section, emphasis is on a claim made by the owner against both its independent design professional and the contractor. How is recovery measured when there are multiple causes?

The problem is best illustrated by *Northern Petrochemical Co. v. Thorsen & Thorshov, Inc.*[121] The owner contracted with an architect to design its new headquarters, which contained offices, a ware-

[116]See also *Centric Corp. v. Morrison-Knudsen Co.*, 731 P.2d 411 (Okla.1986) (CM took advantage of the trade contractor's precarious financial condition; latter's release incident to final payment set aside). But economic duress claims are not easy to sustain. In *Turner v. Low Rent Housing Agency of the City of Des Moines*, 387 N.W.2d 596 (Iowa 1986), failure to show an improper motive (bad faith) precluded a finding of duress despite the owner's taking advantage of the contractor's necessitous condition. Also, the contractor failed to show it had no viable alternative. See also *Selmer Co. v. Blakeslee-Midwest Co.*, 704 F.2d 924 (7th Cir.1983), in which a subcontractor did not succeed in setting aside a settlement on the basis of its being in financial difficulty. This *alone* will not be sufficient.
[117]*City of Mound Bayou v. Roy Collins Constr. Co.*, 499 So.2d 1354 (Miss.1986); *North Harris County Junior College Dist. v. Fleetwood Constr. Co.*, 604 S.W.2d 247 (Tex.Ct.App.1980).
[118]*Horn Waterproofing Corp. v. Bushwick Iron & Steel Corp.*, 66 N.Y.2d 231, 488 N.E.2d 56, 497 N.Y.S.2d 310 (1985) (citing cases and authorities). In 1990 the UCC changed § 1-207 to make clear that it did not eliminate the common law accord and satisfaction doctrine.
[119]West Ann.Cal.Pub.Cont.Code § 7100 (barred in state contracts). In 1987, California drastically reduced the effectiveness of accord and satisfaction in the check-cashing situation. See West Ann.Cal.Civ.Code § 1526. A creditor can strike out a "paid in full" notation, cash the check, and still preserve a *disputed* claim.

[120]*John Grier Constr. Co. v. Jones Welding & Repair, Inc.*, 238 Va. 270, 383 S.E.2d 719 (1989) (subcontractor's claim not barred despite its endorsement of a check marked "paid in full" because its lack of knowledge meant no meeting of the minds).
[121]297 Minn. 118, 211 N.W.2d 159 (1973). Similarly, see *City of Reno v. Ken O'Brien & Assoc., Inc.*, an unpublished opinion of the U.S. Court of Appeals for the Ninth Circuit, No. 74-2094, March 23, 1978 (design and construction combined to create an indivisible loss).

house, and a manufacturing plant. It also contracted with a general contractor under a fixed-price contract. The construction began in the fall of 1967 and was virtually completed in April of 1968. At that time, it became clear that there were major structural flaws in the building. Large cracks were developing. Walls and columns were out of plumb. An investigation led to remedial action, but the deterioration continued. It became apparent that the building was moving in fits and starts. An agreement was made by the major participants to perform corrective work.

Until actual reconstruction began, it was believed that the sole reason for the building's failure was a structural defect in a support wall. When corrective work began, it was determined that the reinforcing steel for the concrete flooring had not been imbedded in the concrete and the underlying fill had not been compacted as required. These omissions left large voids under the floor that caused cracking and uneven settling. Correcting these faults required a massive reconstruction process that took approximately eight months. The trial court concluded that the owner's damages were approximately $750,000. It determined that some of the problem was traceable to design, some to faulty construction, and the balance to both.

Interestingly, the court treated this as a tort case, finding that both the architect and the contractor had been negligent. It then shifted to tort principles applicable when there are co-wrongdoers and there is no difference in the culpability of wrongdoing. It concluded that where it is not reasonably possible to make a division of the damage caused by the separate acts of negligence closely related in point of time, the negligent parties, even though they acted independently, are jointly and severally liable. Each was responsible for the entire indivisible loss, even though they did not act together. The court adopted a rule that puts the burden on either architect or contractor to limit its liability by providing a method of apportionment. In other words, either the architect or the contractor must prove that its negligence caused a particular harm for which it is responsible. In the absence of a rational method of apportionment, each is responsible for the entire loss. (The rights and duties between the co-wrongdoers are discussed in Section 32.03.)

In the 1970s, the issue of joint and several liability became controversial. If two or more defendants are responsible for an indivisible loss, one may be much more at fault than the other. For example, if one is 90% at fault and the other 10%, and the party 90% at fault cannot pay a court judgment, the entire loss may fall on the party who was 10% at fault.

This possibility led the Iowa legislature to enact legislation that allows joint and several liability only if a party is at least 50% negligent. Under such a statute, a case involving a claim against a contractor and a design professional concluded that the contractor was 86% responsible and the design professional 14%. However, the contractor was judgment proof; that is, it could not respond to a judgment. As a result, the owner desperately sought to show that the design professional was 50% responsible so that it could saddle the design professional with the entire loss. This was not successful, and the design professional was responsible for only 14% of the loss.[122]

In another case, the city sued both the architect and the contractor when a decorative sidewalk designed by the architect was not adequate. The architect showed that the contractor had deviated from the plans. However, the court concluded that this did not insulate the architect from liability. It only shifted to the contractor the burden of proving that its deviation did not damage the owner. If the contractor is successful in convincing the judge or jury that its breach did not substantially cause the harm, it is relieved from responsibility, and the entire loss will fall upon the design professional.[123]

The increased use of comparative negligence and shared responsibility can lead to another approach. The loss can be divided by comparing negligence or, in the absence of a tort standard being appropriate, the level or intensity of wrongdoing in a nontort sense.[124]

---

[122]Eventide Lutheran Home v. Smithson Elec. & Gen. Constr. Inc., 445 N.W.2d 789 (Iowa 1989), applying West Ann. Iowa Code § 668.4.

[123]*City of Charlotte v. Skidmore, Owings & Merrill,* 103 N.C.App. 667, 407 S.E.2d 571 (1991). See also *White Budd Van Ness Partnership v. Major-Gladys Drive Joint Venture,* 798 S.W.2d 805 (Tex.Ct.App.1990) (architect's breach was substantial factor in total loss and could not receive credit for damage caused by the contractor), cert. denied, 112 S.Ct. 180 (1991).

[124]*Shepard v. City of Palatka,* 414 So.2d 1077 (Fla.Dist.Ct.App.1981) (dictum).

Suppose the architect *were looked on* as having causing 60% of the loss and the contractor 40%. This would place 60% of the loss on the architect and 40% on the contractor. Here cause would be replaced by "fault." This could also bypass complicated contribution and indemnity issues. But it would expose the owner to a greater likelihood of being uncompensated if either defendant cannot pay the judgment.

SECTION 27.15 Security for Claims

A. Owner Claims

The owner can secure its claim against a contractor by withholding payment of funds for work performed[125] or refusing to release retainage.[126] Another method is to require that the contractor furnish either individual guarantees, unsecured or backed up security interests in property, or a performance or warranty bond.

B. Prime Contractor Claims

The contractor indirectly can secure its claim against the owner by the pressure of a threatened or actual stoppage or even termination until payment is made. It can file a mechanics' lien against

any private property that the prime contractor has improved.[127]

C. Subcontractor Claims

In addition to the methods described in (B), in public projects, the subcontractor can secure its claim by looking to any payment bond that the prime contractor has been required to furnish. In some states, stop notice rights can be effective to obtain payment.[128]

D. Summary

Claims related to the construction project are often worth very little unless methods have been used to collect them. Obtaining a court judgment in many instances does not provide actual reparation for the loss. For that reason, participants in the construction project should plan their transactions to give them as much security as they can obtain by their contracts and use every effort to perfect any security that is given to them under the contract, such as a surety bond, or perfect security provided by law, such as a mechanics' lien or stop notice.

---

[125]See Section 22.02(E).
[126]See Section 22.03.

[127]See Section 28.07(D).
[128]See Section 28.07(E). See also Sections 28.07(F), (G), (H) and (J) for additional methods available to subcontractors.

PROBLEMS

1. O and C entered into a construction contract under which C was to construct a commercial office building and receive payment of $100,000 from O. During performance, O wrongfully terminated the contract and ordered C to leave the project. C's costs before termination were $60,000. It would have cost C an additional $60,000 to complete the work. C has been paid progress payments of $50,000. How much should C be able to recover from O?

2. O and C entered into a construction contract under which C was to build a commercial building

in accordance with plans and specifications drafted by A, an architect retained by O. The contract stated that for every day of unexcused delay O would deduct $2,000 as a "penalty." The contract price was $250,000. C was twenty-five days late. Can O recover $50,000? (Review Section 26.09.) If the clause is invalid, what are O's damages?

3. O hired DB to design and build a warehouse. Six months after completion and final payment, the building developed cracks and leaks in the roof. How would O's claim against DB be measured? (Assume that DB is responsible for the defects.)

The Subcontracting Process: An "Achilles' Heel"

SECTION 28.01 An Overview of the Process

At the risk of oversimplification, the subcontracting process, as used in this chapter, is the method of construction organization under which the prime contractor is allowed to perform some or even much of its contract obligations through other contracting entities. The latter contracting entities are first-tier subcontractors. Likewise, the process in a large construction project can involve first-tier subcontractors performing their contract obligations through other contracting entities called second-tier subcontractors, or sub-subcontractors.

Frequently, other business entities furnish equipment, machinery, products, supplies, or materials incorporated into the project or used to construct the project. These entities—collectively called suppliers—usually make contracts with prime contractors or subcontractors. Although the line between subcontractors and suppliers is sometimes a difficult one to draw, for purposes of the discussion, subcontractors are defined as persons who perform significant services at the site.[1]

What results in the subcontracting process is a chain of contracts that runs from owner to prime contractor or separate contractors, from prime or separate contractors to subcontractors, and from subcontractors to sub-subcontractors. Likewise, there are direct contract lines between contractors that for purposes of discussion include prime contractors and subcontractors and their suppliers.

Some legal problems generated by the subcontracting process have been discussed, and others are discussed in subsequent portions of the treatise. But subcontracting is the legal Achilles' heel of the Construction Process. It generates many legal problems and merits a separate chapter.

The principal advantage of a subcontracting system is improved efficiency, accomplished by breaking down work into categories that require a small number of related skills and the development of those skills by repetition. Those who perform services, whether they be laborers working on the site, cost estimators making bid proposals, or managers making procurement decisions, should become more skilled as they repeat performance of these specified services.

The subcontracting process, if working properly, can reduce costs not only by more efficient work but also by the creation of many competitive prime contractors and subcontractors. Entry into the prime contract field is facilitated by allowing contractors to conserve capital by relieving them of investment and financial burdens to the extent that subcontractors are used. The subcontracting system enables prime contractors to shift over much of the contract risks to subcontractors. If these risks involve the particular skill of the subcontractor and are ones over which the subcontractor has direct control, this can be an efficient allocation of risk.

As for subcontractors, the subcontracting system should encourage many smaller, highly competitive subcontractors to enter the construction field. The subcontracting system may be one of the reasons the construction industry, unlike the automobile industry, is made up of many contractors with specialized talents who operate mainly in lim-

---

[1]*U.S. Industries v. Blake Constr. Co.*, 671 F.2d 539, 543 (D.C.Cir.1982) (citing earlier edition of treatise).

ited localities. This has meant vigorous competition for work that, though it has the disadvantage of economic instability, frequent financial failures, and disputes, should, through competition, reduce construction cost.

An often ignored advantage of the subcontracting system is that many subcontracting entities can create social mobility and avoid the rigidity of more closed class structures. This is accomplished by permitting individuals or small business units to enter the field with a minimum amount of capital.

These undoubted potential and actual advantages have their cost. The subcontracting system generates a large share of construction legal problems.

The principal subcontract problem deals with payments. As shall be seen in Section 28.06, the subcontracting system creates risk of delayed payment and nonpayment to subcontractors. Typically, the prime contractor on any substantial project is paid monthly as work progresses, with a customary retainage of from 5% to 10%. This creates a delay of cash flow from prime contractor to those to whom it owes payment. Even in no-retainage contracts, the prime contractor often has cash-flow problems caused by the lag between payments and obligations. When there are retainages, the cash-flow problem is more serious. To the extent of payment delay, the prime contractor is providing financing services for the owner that it seeks to transfer to subcontractors.

Prime contractors generally use subcontract payment provisions to minimize cash-flow problems. These provisions frequently permit the prime contractor to delay paying subcontractors until the former have been paid by the owner. Such delayed payment provisions may squeeze subcontractors who have to pay for their labor and supplies. Even those with credit face financial hardship when credit is withdrawn or limited.

Cash-flow problems are increased as a greater percentage of work is performed by subcontractors. Increased subcontracting means a greater likelihood that nonperformance by one subcontractor will delay payment to others who have performed properly.

The construction industry is composed of many small businesses with limited financial capacity and credit. This is particularly true with regard to subcontractors, many of whom started as tradesmen and many of whose businesses are family owned. A short delay in the flow-through of funds from the owner to those contractors performing services on the site can cause financial hardship that may deprive a contractor, especially one at the lowest tier, of the funds needed to continue performance. The subcontracting system heightens the financial stress inherent in the flow-through process because it increases the distance of the money flow.

The extent of work subcontracted also plays a role in creating legal problems. Subcontractors frequently assert that prime contractors are merely assemblers or brokers for the services of others. Prime contractors deny this and claim that they supply much of the materials and services themselves. Prime contractors and subcontractors agree that the *amount* of work subcontracted depends on the *type* of construction. Whether prime contractors or subcontractors are correct on this controversial question, it is clear that the greater the percentage of subcontracted work, the greater the likelihood of severe cash-flow problems. A prime contractor who does not have much money tied up in the project is less concerned about the swiftness of the cash flow. This is especially true if the prime contractor's obligation to pay subcontractors is conditioned on receiving payments from the owner.

Moving from cash flow delay to nonpayment problems, the volatility of the construction industry must again be emphasized. Business failures and bankruptcies frequently occur in the construction industry. The higher the proportion of work subcontracted, the greater the risk that there will be unpaid subcontractors who will seek some type of legal relief when their work has benefited the owner. One type of relief often sought by unpaid contractors and suppliers is a mechanics' lien. Valid liens usually result in owners paying lien claimants to remove the liens. To avoid liens, owners create payment structures to minimize payment diversion by the prime contractors. Owners—private and especially public—frequently require payment bonds to give an effective remedy to unpaid subcontractors and suppliers. Bond requirements inject the complexity of surety bonds and additional parties into the already complicated construction structure.

Problems of delayed payment and nonpayment highlight another significant feature of the subcontracting process. The subcontractor is "a contract away" from the major source of power and control over the project—the owner. As it has no direct contractual relationship with the owner, it must look to the prime contractor for payment, for dealing with disputes over performance, and to process claims against the owner. This can create a sense of powerlessness in subcontractors, leading to friction and lawsuits at worst and poor communication at best.

The subcontracting process compounds the already difficult problems caused by the multitude of construction documents that regulate the relationships in the Construction Process. One principal cause of legal problems is the wealth of potentially conflicting documents that regulate the relationship between owner and contractor. Add first- and second-tier subcontracts, which frequently refer *generally* to contract provisions in contracts above them on the contract chain, and additional ingredients for disputes are added.

The subcontracting process generates a potentially large number of construction contractors—all working on the same site often at the same time. For purposes of this discussion, contractors include prime contractors, separate contractors, and the various tiers of subcontractors. Contractors may disagree on who will have access to a particular part of the site at a particular time. One contractor's performance often depends on another contractor's work being completed or at least in a certain state of readiness. One contractor's work may be disturbed or ruined by another contractor's workers, and disputes may develop over which contractor is responsible. One contractor may employ workers who belong to a construction trade union, while another uses nonunion workers. One contractor's employee may be injured or killed, and the responsibility may be asserted against or shared by a number of other contractors. The sheer number of different contractors at work can create immense administrative and, ultimately, legal problems.

The illustrations given do not exhaust the legal problems incident to subcontracting. Additionally, there is frequent bargaining disparity present in construction contracts and especially in subcontracts. Generally, the dominant bargaining strengths parallel the money flow. Lenders can exact terms from owners because the total number of borrowers seek more money than lenders have to lend. The owner who has funds or can borrow them seeks a contractor from among the many willing to perform the work. This gives the owner substantial bargaining advantage over the prime contractor. Although the prime contractor awarded the contract does not as yet have funds, it has contract rights. These contract rights generally give the prime contractor bargaining advantage over subcontractors, many of whom are looking for work. As a result, at this stage, the prime contractor usually has the bargaining power to demand, and often obtain, favorable terms from subcontractors, sometimes terms that are more favorable than those in the prime contract. Likewise, subcontractors frequently exert parallel bargaining advantage over sub-subcontractors.

While lawyers do not play a significant role drafting and reviewing construction contracts generally, this is particularly the case with subcontractor contracts. Lawyers are frequently not brought into the picture until the need for legal action becomes imminent. This can mean extraordinarily one-sided contracts forced on subcontractors, who are generally unaware of legislation or case decisions that might protect them.

The bargaining position of suppliers and contractors cannot be so easily generalized. Suppliers range from large manufacturing companies with strong financial positions to small distributors with limited financial capacity. As a result, some suppliers are in the position to dictate terms to subcontractors and even to prime contractors. Yet even suppliers in this advantageous position must sell and must often extend credit to do so. Often their extension of credit is predicated on statutory lien rights or the existence of payment bonds.

In the typical construction project, one finds the lowest tier subcontractor with the weakest bargaining power. This bargaining pattern is illustrated by the frequent passing down of increasingly harsh and one-sided indemnity agreements. In addition to the strong bargaining pressures from prime contractors or higher tier subcontractors, lower tier subcontractors also face strong bargaining power of large suppliers and construction trade unions. Subcontractors, especially those at the lowest tier, often find themselves squeezed on all sides because of their poor bargaining position.

Institutional bargaining disparities generate legal problems. Harsh terms exacted at the bargaining table may be resisted by the weaker party when disputes develop. The law looks with disfavor on harsh terms, even though the legal system generally allows parties to write their own contracts. As a result, the uncertainty of enforcement of harsh terms exacted by the dominant party generates legal disputes that often require judicial resolution.

Those in vulnerable bargaining positions may seek other avenues of relief. For example, industry associations sometimes obtain legislative enactments such as anti-indemnity legislation to overturn contract clauses exacted from them by parties in a stronger bargaining position. Injection of legislative rules in an area regulated largely by private contracts adds an additional complicating feature to construction problems.

Associations composed of members who are often in a weak bargaining position often seek to remedy this weakness by participation in the process by which standard construction contracts such as those published by the American Institute of Architects (AIA) are created. For example, prime contractors, through the Associated General Contractors (AGC), seek to persuade the AIA to incorporate or eliminate language in AIA Standard Documents that they might not otherwise be able to do in dealing with an owner. Correspondingly, associations representing subcontractors frequently request the AIA to incorporate provisions in the AIA prime contract that require that certain rights be accorded subcontractors when these rights might not be obtainable in normal bargaining between prime contractors and subcontractors. When the prime contract deals with subcontracting and subcontractors, an increased likelihood exists of conflicting contract documents and a more generally complicated prime contract.

Subcontractor associations have sought protective language not only in prime contract documents published by the AIA but also in the AIA's Standard Subcontract, A401. Just as prime contractors prefer an AIA document over a construction contract drafted by the owner, subcontractors are likely to prefer a subcontract drafted by the AIA to one prepared by the prime contractor.

One final aspect of the subcontracting process adds additional complexity. Contracts for the performance of design and construction services are basically regulated by common law rules, that is, rules that have evolved through court decisions. On the other hand, contracts for the sale of goods are governed by the Uniform Commercial Code (U.C.C.), a comprehensive statutory regulation. When a subcontractor, as is usually the case, provides both goods and services, which rules apply— the common law or U.C.C.? Generally, the test is whether goods or services are the predominant aspect of the transaction.[2] This can mean that a particular subcontract or a severable part of a particular subcontract is governed by the U.C.C., while the rest may be governed by the common law.

To sum up, although the subcontracting system has undoubted advantages, it generates a host of legal problems. With this in mind, some particular subcontract problems are addressed.

SECTION 28.02 Subcontractor Bidding Process

A. Use of Sub-Bids

In technical projects requiring specialized skills, the prime contractor may do very little of the work itself. In such projects, the cost to the prime contractor of work to be done by subcontractors is not likely to be known until subcontractors submit sub-bids. The prime contractor cannot bid until hearing from all the prospective subcontractors, something that does not usually occur until close to time for submitting the bid to the owner. The prime contractor relies on the subcontractors' bid in submitting its own.

The prime contractor's reliance on subcontractors is most acute in the mechanical specialty trades (e.g., electrical, plumbing, air conditioning). Prime contractors utilize a large number of subcontractors from the nonmechanical specialty trades (e.g., masons, roofers). This activity can be bid with relative accuracy by the prime contractor, with the sub-bids providing downside protection. These subcontractors primarily provide the prime contractor with a skilled work force, one that the prime contractor

---

[2]*Bonebrake v. Cox*, 499 F.2d 951 (8th Cir.1974), followed in *Pittsburgh-Des Moines Steel Co. v. Brookhaven Manor Water Co.*, 532 F.2d 572 (7th Cir.1976). See also *Arango Constr. Co. v. Success Roofing, Inc.*, 46 Wash.App. 314, 730 P.2d 720 (1986) (U.C.C. did not apply).

could not possibly afford to keep employed between projects.

Prime contractors generally use the bids given by subcontractors in computing their bids. Suppose a subcontractor withdraws its bid (usually because it contends that its bid had been inaccurately computed or communicated). Can the prime contractor hold the subcontractor to its bid by contending that it had relied on the sub-bid in making its bid?

B. Irrevocable Sub-Bids

The early cases dealing with this problem used traditional contract analysis to allow the subcontractor to revoke its bid. Generally, the prime contractor does not make a contract with the subcontractor, conditioned on the prime contractor's being awarded the bid. The prime contractor wishes to preserve maximum freedom to renegotiate with the low bidder or negotiate with other bidders. Ignoring the prime contractor having relied on the sub-bid in making its own bid, *James Baird Co. v. Gimbel Brothers*[3] did not hold the subcontractor to its bid.

However, contract law has expanded reliance as a basis for making an offer irrevocable. This carried over into subcontractor bid cases, the leading case being *Drennan v. Star Paving Co.*[4] The subcontractor had submitted the lowest sub-bid for the paving portion of the work, a sub-bid the prime contractor used in computing its overall bid. The prime contractor listed the defendant on the owner's bid form as required by statute.[5] The prime contractor

was awarded the contract and stopped by the subcontractor's office the next day to firm up the "subcontract." Upon arrival, the subcontractor immediately informed the prime contractor that it had made a mistake in the preparation of its bid and would not honor it. The prime contractor brought suit and was awarded the difference in cost between the subcontractor's sub-bid and the cost of a replacement.

The court concluded that the use of the sub-bid did not create a bilateral—or two-sided—contract between the plaintiff and the defendant.[6] But the court held the subcontractor to its bid because the bid was a promise relied on reasonably by the prime contractor when it submitted its bid. The possible uncertainty of subcontract terms (discussed later in this section) was brushed aside.

Reliance making the bid irrevocable created a one-sided contract despite submission of the sub-bid inviting acceptance by the prime contractor.[7] Subcontractors do not wish to subsidize the prime contractor's costs of bid preparation. They want the subcontract.

A survey by the American Bar Association Public Contract Law Section in 1992 reported that of the fifty states, the District of Columbia, and Guam, twenty-six jurisdictions hold that subcontractors are bound to their quotations, four reject promissory estoppel making the subcontractor's bid irrevocable, and the remaining twenty-two have so far left the issue open.[8] In addition to the question of

---

[3]64 F.2d 344 (2d Cir.1933).

[4]51 Cal.2d 409, 333 P.2d 757 (1958), followed in the Restatement (Second) of Contracts, § 87(2) (1981). Cases following the *Drennan* case are *Preload Technology, Inc. v. A.B. & J. Constr. Co., Inc.*, 696 F.2d 1080 (5th Cir.1983); *Alaska Bussell Elec. Co. v. Vern Hickel Constr. Co.*, 688 P.2d 576 (Alaska 1984); *Mead Assoc., Inc. v. Antonsen*, 677 P.2d 434 (Colo.App. 1984); and *Arango Constr. Co. v. Success Roofing, Inc.*, supra note 2. But *Drennan* was not followed in *Home Elec. Co. of Lenoir, Inc. v. Hall and Underdown Heating & Air Conditioning Co.*, 86 N.C.App. 540, 358 S.E.2d 539 (1987) (rejected as one-sided), affirmed 332 N.C. 107, 366 S.E.2d 441 (1988) noted in 10 Campbell L.Rev. 293 (1988).

[5]As to Listing Laws, see Section 28.03. In *Southern California Acoustics Co. v. C. V. Holder, Inc.*, 71 Cal.2d 719, 456 P.2d 975, 79 Cal.Rptr. 319 (1969), the court held that an improper substitution created a claim based on the statutory violation against the prime contractor in favor of the improperly substituted subcontractor. This approxi-

mates binding *both* parties. But listing the subcontractor to comply with the statute generally does not create a bilateral contract. *Holman Erection Co. v. Orville E. Madsen & Sons, Inc.*, 330 N.W.2d 693 (Minn.1983).

[6]For a case holding that use of the bid is not an acceptance, see *Mitchell v. Siqueiros*, 99 Idaho 396, 582 P.2d 1074 (1978). Similarly, see *Four Nines Gold, Inc. v. 71 Constr., Inc.*, 809 P.2d 236, 239 (Wyo.1991) (citing earlier edition of treatise).

[7]For a justification of a one-sided contract, see *Holman Erection Co. v. Orville E. Madsen & Sons, Inc.*, supra note 5. But see *Home Elec. Co. of Lenoir, Inc. v. Hall & Underdown Heating & Air Conditioning Co.*, supra note 4, which did not find the sub-bid irrevocable, as that would bind the sub but not the prime.

[8]State and Local Contracting Committee of the Public Contract Law Section of the American Bar Association, ARE SUBCONTRACTOR QUOTES BINDING? (1992). This monograph also reprinted four articles from the Construction Lawyer dealing with subcontractor bids.

whether the subcontractor's bid is irrevocable, many cases have raised questions of whether the Statute of Frauds applies to promises (certain transactions requiring a written memorandum) enforced because of reliance.[9]

C. Bargaining Situation: Shopping and Peddling

The *Drennan* rule improves the prime contractor's already powerful bargaining position. Although the prime contractor under the *Drennan* rule is not free to delay its acceptance or to reopen bargaining with the subcontractor and still claim a right to accept the original bid, it can at least for a short period seek or receive lower bid proposals by other subcontractors.

Under the *Drennan* rule, until the prime contractor is ready to sign a contract with the subcontractor whose bid it has used, the subcontractor is not assured of getting the job. Before the award of the prime contract, the plurality of competing prime contractors' bidding on a project tends to diffuse their bargaining power over subcontractors. This competition before the award of the contract should result in lower sub-bids and consequently lower overall bids by the prime contractors, a definite benefit to the owner. Although subcontractors often wait until the last minute to submit their sub-bids in an effort to minimize the prime contractor's superior bargaining position, a substantial amount of competition still exists among the subcontractors themselves.

After award of the contract, the relative bargaining strengths of the successful prime contractor and the competing subcontractors change drastically. The prime contractor now has a "monopoly" and a substantially superior bargaining position over the subcontractors under the *Drennan* rule. The sub-bids used provide the prime contractor with a protective ceiling on the cost of the work with no obligation to use the subcontractors. The prime contractor is therefore free to look elsewhere for yet a better price and is able to increase its profits by engaging in postaward negotiations.

Postaward negotiations have become controversial. Sometimes they are called "bid shopping"; the prime contractor uses the lowest sub-bid to "shop around" with the hope of getting still lower sub-bids. "Bid peddling" is the converse, with other subcontractors attempting to undercut the sub-bid to the prime, in essence engaging in a second round of bidding. Subcontractors often refer to these postaward negotiations as "bid chopping" and "bid chiseling." (These tactics can also be used *before* prime bids are submitted.)

Both subcontractors and owners have reasons to condemn postaward competition. The subcontractors assert that the preparation of a bid involves considerable expense. Subcontractors who "bid peddle" may not even prepare their own bids, saving overhead expenses. The subcontractor who went to the expense of calculating a bid subsidizes the bid-peddling subcontractor's overhead costs as well as the prime contractor's costs of bid preparation.

Subcontractors who fear bid shopping often wait until the last minute to submit their sub-bids to the prime contractor to give the prime contractor as little time as possible to bid shop. This last-minute rush is the cause for many mistakes by both subcontractors and prime contractors. Some subcontractors simply refrain from bidding on jobs where bid shopping is anticipated to save the expense of preparing a bid. To that extent, competition among subcontractors is diminished and higher prices can result.

Subcontractors feel that they must pad their bids to make allowance for the eventual postaward negotiations. This "puffing" raises the cost to the owner, as the inflated bid is the bid the prime contractor uses to compute its overall bid. Any subsequent negotiations that result in a reduction of the price inure to the prime contractor alone.

The superior bargaining position of a successful prime contractor spurs cutthroat competition among subcontractors, resulting in lost profits that can upset industry stability. Prime contractors respond by noting that sub-bids are often unresponsive to the specifications and require further clarification and negotiation. This may be especially true when prime contractors are dealing with subcontractors with whom they have never dealt.

---

[9] *Allied Grape Growers v. Bronco Wine Co.*, 203 Cal.App.3d 432, 249 Cal.Rptr. 872 (1988), found fourteen jurisdictions that recognized estoppel as an exception to the Statute of Frauds and only four that did not. The court followed the majority in concluding that the Statute of Frauds did not apply to a promise enforced because of reliance. See also infra note 19.

They must investigate the subcontractor's reputation and work experience before making a firm contract.

Prime contractors state that bids are often requested for alternative proposals, and they lack the time to evaluate all the alternatives in the short time available between receipt of the sub-bids and bid closing. They assert that negotiations are sometimes required to decide on the specific alternative to be chosen. Prime contractors also justify post-award negotiation by stating that estimating a job is a normal cost of overhead in the construction industry.

D. Avoiding *Drennan*

Subcontractors can avoid their sub-bids' being firm offers that bind them and not the contractor. They can denominate their bids as "requests for the prime to make offers" to them or call them "quotations" given only for the prime contractor's convenience.[10] They may also try to annex to their sub-bids language that states that the use of the bid constitutes an acceptance that ties the prime contractor to them.[11] They may refuse to submit bids unless they receive a promise by the prime contractor to accept the bid if it is low and the prime contractor is awarded the contract.[12] They may refuse to begin work without receiving a letter of intent from the prime contractor that provides that they will be paid if the parties cannot agree on the terms of the subcontract.[13] Finally, one subcontrac-

tor notified the public *owner* that it had made a mistake in its bid to the prime, and the owner cancelled the procurement.[14]

The difficulty with most of these methods of avoiding the *Drennan* rule is that either the subcontractors do not have the bargaining power to implement them or the process does not make it convenient for them to use them.

The *Drennan* rule can also be avoided if many crucial areas have been left for further negotiation.[15] Finally, though unsuccessful in the *Drennan* case, a case upheld the subcontractor's claim of mistake.[16]

E. Bid Depositories

Whenever individuals are weak at the bargaining table they seek alternative approaches. Subcontractors have organized themselves and created bid depositories (sometimes called bid registries). A typical bid depository is a facility established and operated by a trade association that receives bids from the subcontractors for the supplying of construction services for large construction projects. The depository collects the bids from the subcontractors and delivers them to the prime contractors who intend to bid on that particular project. Bid depositories are used most frequently in the mechanical specialty trades, these subcontractors most vulnerable to the detrimental effects of bid shopping and bid peddling.

The subcontractors must submit their bids to the depository before the cutoff time, typically four hours before bid closing. The bids are submitted in sealed envelopes, one copy addressed to the prime contractor and the other to the depository. To facilitate compilation of the bids, special bid forms are often provided. After the cutoff time, no bids may be received, and none received may be revoked or amended.

At closing time, the bids addressed to the prime contractors are distributed to whom they are ad-

---

[10]*Leo F. Piazza Paving Co. v. Bebek & Brkich*, 141 Cal.App.2d 226, 296 P.2d 368 (1956); *Cannavino & Shea, Inc. v. Water Works Supply Corp.*, 361 Mass. 363, 280 N.E.2d 147 (1972).

[11]*Sharp Bros. Contracting Co. v. Commercial Restoration, Inc.*, 334 S.W.2d 248 (Mo.App.1960) (bid must be accepted in ten days). But see *S. M. Wilson & Co. v. Prepakt Concrete Co.*, 23 Ill.App.3d 137, 318 N.E.2d 722 (1974) (protection lost by continuing to deal after expiration of deadline).

[12]*Elec. Constr. & Maintenance Co., Inc. v. Maeda Pacific Corp.*, 764 F.2d 619 (9th Cir.1985) (sub sued prime when latter used a different sub). Similarly, the subcontractor recovered lost profits when there had been an oral agreement that the subcontractor was to be awarded the subcontract if the prime contractor was awarded the prime contract, the subcontractor was listed in the prime's bid as a disadvantaged business enterprise, and then the prime contractor sought to renegotiate after it was awarded the contract. *Bridgeport Restoration, Co. Inc. v. Petrucci Constr. Co.*, 211 Conn. 230, 557 A.2d 1263 (1989).

[13]*Quake Constr. Inc. v. American Airlines, Inc.*, 141 Ill.2d 281, 565 N.E.2d 990 (1990), discussed in Section 5.06(H).

---

[14]*Four Nines Gold, Inc. v. 71 Constr., Inc.*, supra note 6.

[15]*Preload Technology, Inc. v. A. B. & J. Constr. Co., Inc.*, supra note 4 (dictum). But the court rejected this in *Arango Constr. Co. v. Success Roofing, Inc.*, supra note 2, pointing to the sub's reason for not making the contract relating to price and not other terms.

[16]*Tolboe Constr. Co. v. Staker Paving & Constr. Co.*, 682 P.2d 843 (Utah 1984). See Section 18.04(E).

dressed. The prime contractors have four hours before bid closing to prepare their own bids. The other copy of each bid is retained by the depository. After bid closing, the depository may publish either all the sub-bids, only the lowest sub-bids, or none at all. In any event, at least the depository knows the identity and price of the lowest subbidder who used the depository. Further renegotiation with the subcontractor who submitted the low bid becomes highly suspect and may subject the prime contractor to a disciplinary action by the depository.

Depositories are generally open to any subcontractor in the applicable trade. Membership is usually required before the subcontractor may use a depository's facilities. Prime contractors can be members, but normally an agreement to abide by depository rules is all that is required of the prime contractor.

Most depository rules state the following:

1. Subcontractors must use the depository exclusively, if at all.
2. Prime contractors may not accept bids from subcontractors except through the depository.
3. The prime contractor must agree to accept the lowest bid from the depository or not use the depository at all. (For a public contract version, see the "prefiled" bidding system noted in Section 28.03.)
4. Bid splitting (combining different parts of different sub-bids together to get a yet lower aggregate bid) is prohibited.
5. Violation of the depository rules results in a forfeiture of a filing fee, imposition of a fine, or both.

Some bid depositories with tight controls have been found to violate laws against price fixing.[17] They eliminate bid shopping, a form of free-market competition.

Those who feel that free competition will achieve the best result see all negotiation as not only useful but also desirable. To subcontractors, negotiation from weakness means harsh terms.

Subcontractors also look to the owner—public or private—to assist them, contending that the owner is also disadvantaged by this negotiation. Involving the owner through an approval process is discussed in Section 28.03.

F. Uniform Commercial Code

Offers generally at common law are revocable even if stated to be irrevocable. Dissatisfaction with this rule has generated solutions taking many forms, including the *Drennan* rule making the bids irrevocable. One method adopted was § 2-205 of the Uniform Commercial Code (U.C.C.). Section 2-205 recognizes firm offers and enforces them if certain requirements are met.

The U.C.C. does not apply to service contracts or those that involve interests in land. The U.C.C. has also had limited use in making sub-bids by suppliers irrevocable. Section 2-205 requires a written offer stating that the bid will be held open. Bids are often communicated by telephone.

The U.C.C. is not a viable solution to the problem of sub-bids for another reason. Rarely will all the needed terms be expressed in a telephone conversation. Even if the bid is communicated by a writing with all the terms, rarely do prime contractors intend to be bound to those terms, choosing to either use terms used before or dictate terms later. The firm offer of § 2-205 does not fit the subbid, another reason why the *Drennan* rule, a substitute for the U.C.C. firm offer, was incorrectly decided.

California revised its version of the U.C.C. in 1980. Under § 2205(b) of its Commercial Code, a written or oral sub-bid for *goods* made to a licensed contractor that the bidder knows or should know will be relied on is irrevocable for twenty days after the prime contractor is awarded the contract but not later than ninety days after the bid. Oral bids of over $2,500 must be confirmed in writing within forty-eight hours. The bid can limit the duration of the offer.

Again the problem of terms can arise. The statute does not preclude the sub-bidder from asserting that the terms needed to cure any completeness requirement were not included in the offer, particularly if it were oral. Although the statute as revised expands the enforcement of oral sub-bids for

[17]*Oakland-Alameda County Builders' Exchange v. F.P. Lathrop Constr. Co.*, 4 Cal.3d 354, 482 P.2d 226, 93 Cal.Rptr. 602, (1971) (invalid). But for a more permissive attitude, see *Cullum Elec. & Mechanical, Inc. v. Mechanical Contractors Assn. of South Carolina*, 436 F.Supp. 418 (D.S.C.1976) (used peaceable persuasion and no intent to monopolize; bid peddling not essential to pure price competition), affirmed 569 F.2d 821 (4th Cir.1978), cert. denied 439 U.S. 910 (1978).

goods, it has the same defects as the *Drennan* rule (noted in (G)).

G. A Suggestion

The *Drennan* rule, now firmly in the saddle in most American jurisdictions, was a laudable attempt to avoid the common law rule of revocability and expand reliance as a method of enforcing *promises that should be enforced.* It is singularly inappropriate for construction sub-bids. First, it gives advantage to the prime contractors that already have a strong bargaining position. Second, it creates a binding offer before there has been agreement on important terms such as bonds, payment, indemnification, and dispute resolution.[18] Third, its nature has raised technical difficulties.[19] The traditional common law rule expressed in the *Baird* case[20] is more appropriate. One-sided contracts should be found only when this is clearly intended or where absolutely necessary. Neither occurs here.

Use of the bid should be an acceptance of the offer and a contract concluded *if* there are sufficient terms agreed on by *both* parties. This can be based on well-established custom or a course of dealings between the parties supplying the ancillary terms in addition to the work to be performed and payment. When this does not exist, neither party

should be bound until both parties work out the terms needed to make a binding contract.[21]

H. Teaming Agreements

Teaming agreements are, as a rule, informal arrangements between prime contractors who wish to obtain federal procurement high-technology contracts and subcontractors who assist in devising proposals designed to help the prime contractor win the competition. The informality of these arrangements can often lead to litigation if the prime contractor wins the competition and does not award subcontract work to its "teaming partner." This was discussed in Section 17.04(H).

SECTION 28.03 Subcontractor Selection and Approval: The Owner's Perspective

Owners wish to have competent contractors building their projects. They want subcontractors treated fairly and given an incentive to perform the work expeditiously. Some owners contract directly with the specialty trades to avoid the prime contractor as an "intermediary" between owner and specialized trades. Some contract directly with the specialized trades and assign those contracts to a main or prime contractor. Some owners dictate to the prime contractor which subcontractors will be used, known in England as the nominated subcontractor system.

The American version used in some public contract systems is called "pre-filed" bids. Prospective subcontractors file sub-bids with the public entity. The successful prime bidder must contract with the lowest subcontract bidders.[22]

These methods of direct intervention are not common in the traditional contracting system. Some owners leave subcontracting exclusively to the prime contractor. Others take a role that gives them *some* control but does not involve the owner

---

[18]This problem was recognized in *Saliba-Kringlen Corp. v. Allen Eng'g Co.,* 15 Cal.App.3d 95, 92 Cal.Rptr. 799 (1971). But too much emphasis on this would frustrate the *Drennan* rule, an outcome the court was not willing to endorse. Similarly, see *Arango Constr. Co. v. Success Roofing, Inc.,* supra note 2. As to certainty, compare *Debron Corp. v. National Homes Construction Corp.,* 493 F.2d 352 (8th Cir.1974), with *C.H. Leavell & Co. v. Grafe & Assoc., Inc.,* 90 Idaho 502, 414 P.2d 873 (1966). The former was unimpressed with the lack of certainty argument, while the latter was receptive. Refer to *Preload Technology, Inc. v. A. B. & J. Constr. Co., Inc.,* supra note 4.

[19]Is it a contract for Statute of Frauds purposes? See *N. Litterio & Co. v. Glassman Constr. Co.,* 319 F.2d 736 (D.C.Cir.1963), and note 9 supra. Is the claim "at law" or "in equity" an issue that controls the requirement of a jury? See *C&K Eng'g Contractors v. Amber Steel Co.,* 23 Cal.3d 1, 587 P.2d 1136, 151 Cal.Rptr. 323 (1978) (equitable). Does breach by anticipatory repudiation apply to the prime before it makes a subcontract? See *Alaska Bussell Elec. Co. v. Vern Hickel Constr. Co.,* supra note 4 (dictum stating it does not).

[20]See note 3 supra.

---

[21]This approach has been suggested in *Loranger Constr. Corp. v. E. F. Hauserman Co.,* 376 Mass. 757, 384 N.E.2d 176 (1978). See, generally, Closen & Weiland, *The Construction Industry Bidding Cases,* 13 J.Mar.L.Rev. 565 (1980).

[22]This is described in *J. F.White Contracting Co. v. Dep't. of Public Works,* 24 Mass.App. 932, 508 N.E.2d 637 (1987).

directly with the subcontractor. Using the prime contractor as a buffer is done for administrative and legal reasons. The subcontracting system requires a well-defined organizational and communication structure under which each participant knows what it must do and with whom it must deal. From a legal standpoint, the owner does not want to be responsible for subcontract work, does want the prime contractor to be responsible for defective subcontract work, and does not want to be responsible if subcontractors are not paid. .

One "part-way" control is to require that prime contractors list their subcontractors at the time they make their bid. Statutes in some states, called Listing Laws,[23] impose this on prime contractors in public projects. Although undoubtedly some of the impetus for such laws came from subcontractor trade associations, one reason for Listing Laws is to assure the awarding authority that only competent subcontractors will perform on the project.

Legislatures frequently justify listing statutes by condemning bid shopping and bid peddling that they assert cause poor quality of materials and workmanship to the detriment of the public and also deny

> the public . . . the full benefits of fair competition among prime contractors and subcontractors, and lead to insolvencies, loss of wages to employees, and other evils.[24]

Usually Listing Laws regulate substitution of listed subcontractors by providing specific justifications for substitutions and a procedure for determining the grounds for replacing one subcontractor with another. Complications have developed not only as to grounds for substitution but also as to who has a claim for violation of the statute.[25] From the subcontractor's viewpoint,

Listing Laws dampen any attempt by the prime contractor to reopen negotiations.

Listing is used by the American Institute of Architects (AIA) in a more limited way. AIA Doc. A201, ¶ 5.2.1, requires the contractor "as soon as practical after the award" to furnish the owner and architect the name of subcontractors and those who will furnish materials and equipment fabricated to a special design. Before 1976, the architect had to *approve* subcontractors. To make the architect's role more passive and to minimize the likelihood of liability, A201 currently states that the architect "will promptly reply to the Contractor in writing stating whether or not the Owner or the Architect, after due investigation, has reasonable objection to any proposed" subcontractor. Failure to "reply promptly shall constitute notice of no reasonable objection." If the architect has reasonable objection, ¶ 5.2.3 requires a substitution and an increase or decrease in the contract price. This has been criticized as encouraging a prime contractor to submit subcontractors who may not be satisfactory to get a price increase when the substitution is made. Before 1976, A201 allowed the architect or owner to revoke approval at any time. Currently, A201 bars rejection once performance begins.

Some control over subcontractors, though desirable, has negative features. Requiring that a subcontractor not be removed unless there is "due investigation" can invite a subcontractor who has been removed to assert a claim that the architect or owner has intentionally interfered with its actual or prospective contract. Although there may be defenses,[26] this does increase potential liability exposure.

---

[23]West Ann.Cal.Pub.Cont. Code §§ 4100 et seq.

[24]Id. at § 4101.

[25]See *Southern California Acoustics Co. v. C. V. Holder, Inc.,* supra note 5, giving an improperly delisted subcontractor a statutory claim only against the prime contractor. A prime who violates the Listing Law cannot bring a claim against the subcontractor who performed though it had not been properly listed. The contract was void. *R. M. Sherman Co. v. W. R. Thomason, Inc.,* 191 Cal.App.3d 559, 236 Cal.Rptr. 577 (1987). Generally, listing in accordance with bidding requirements does not constitute an accep-

tance. *Mitchell v. Siqueiros,* supra note 6. Where a designated percentage of subcontracts must be set aside for certain disadvantaged groups, listing a particular subcontractor to comply with this requirement may help to show that the parties have made an agreement. While the court in *Bridgeport Restoration Co., Inc. v. Petrucci Constr. Co.,* supra note 12, affirmed the trial court's conclusion that there had been an oral contract for subcontracted work, the court specifically pointed to the prime's having listed the sub as a disadvantaged business enterprise in response to a requirement of setting aside certain contracts for these entities. The court did not specifically conclude that listing the subcontractor acted as an acceptance, but undoubtedly this had some influence on the court's decision.

[26]*Commercial Indus. Constr. Co. v. Anderson,* 683 P.2d 378 (Colo.App.1984) (architect given adviser's privilege). See Section 14.08(F).

The Engineers Joint Contracts Documents Committee (EJCDC), in its No. 1910-8, ¶ 6.8.2, does not require that the identity of subcontractors be furnished. Only if the supplementary conditions require the identity of subcontractors and suppliers are their names to be submitted to the owner for acceptance by the owner and the engineer.

Suppose the contractor wishes to remove an approved subcontractor? AIA Doc. A201, ¶ 5.2.4, bars a contractor from doing so if the owner or architect makes a reasonable objection. But objections to proposed substitutions must be exercised carefully. In *Meva Corp. v. United States*,[27] the contracting officer denied the contractor permission to substitute, and the contractor was forced to keep the original subcontractor on. The subcontractor did a poor job, however, and the contractor recovered losses caused by the subcontractor from the public agency.

To sum up, owner intervention into the relationship between prime contractor and subcontractor may be essential, but it carries risks. If the risks are so great but the need for intervention so strong, the owner should consider a method other than the traditional contracting system. (The owner's attempt to keep subcontractors "on board" in the event the prime contract is terminated is discussed in Section 34.06.)

SECTION 28.04 Sources of Subcontract Rights and Duties: Flow-Through Clauses

The principal source of contract rights and duties between prime contractor and first-tier subcontractor is the subcontract itself. Similarly, the principal source of contract rights and duties between first- and second-tier subcontractors is the sub-subcontract, and so on down the subcontract chain. However, this source is not exclusive. As with any contract, the express terms must be interpreted by language often outside the writing. Express terms will be supplemented by terms implied judicially into the subcontract relationship.

Like other aspects of the subcontract system, each contract on the subcontract chain can be and frequently is affected by contracts higher up the chain. The subcontract relationship is usually affected and may be controlled by terms in the prime contract. Correspondingly, second-tier subcontract relationships are affected and may be controlled by both first-tier subcontract and prime contract provisions. For convenience, discussion focuses on the relationship between prime contractor and first-tier subcontractor and the effect of the prime contract to which the subcontractor is not a party on that relationship.

The discussion in this section centers principally on the subcontractor's being bound to provisions of the prime contract because the latter is referred to or incorporated in the subcontract. However, it has been contended that a series of interrelated contracts can create a single contract binding all parties to the entire series even though each party may not have signed each contract in the series. Although such a contention was rejected in the context of an architect-owner contract followed by an owner-prime contractor contract,[28] a California decision intimated that such a conclusion might be sustained in a case involving a subcontract followed shortly by a prime contract.[29]

There is a superficial attractiveness in *directly* binding all participants in the series of linked contracts to the terms of each contract in the series. It would certainly avoid some of the problems discussed later in this section. Such a fusion of parties, though, would frustrate the intention of the parties when they create the linked series of contracts such as that found in subcontracting. The contracts, though clearly interdependent for many purposes, are *not* intended to join all the participants in one contractual arrangement. The very purpose of setting up a linked though separate set of contracts is to create a hierarchical structure that establishes lines of authority and responsibility. That this attempt is not entirely successful for all purposes, especially where personal harm is involved[30] or a participant is not paid, does not mean that the sys-

---

[28]*C. H. Leavell & Co. v. Glantz Contracting Corp.*, 322 F.Supp. 779 (E.D.La.1971).
[29]*Varco-Pruden, Inc. v. Hampshire Constr. Co.*, 50 Cal.App.3d 654, 123 Cal.Rptr. 606 (1975). The language expressing this opinion was not necessary to the holding of the case and must be considered as "dictum." A subsequent court would not be required to follow such a conclusion.
[30]See Sections 14.08(G), and 28.07(J).

---

[27]206 Ct.Cl. 203, 511 F.2d 548 (1975).

tem should be scrapped entirely and all the participants involuntarily joined together in one contract arrangement.[31]

In addition, binding all parties to one contract, with the prime contract taking precedence, takes away the autonomy of prime contractor and subcontractor to make a contract with *different* terms. Although often the prime contractor can dictate subcontract terms, this is not inevitably the case. Autonomy provides flexibility that would be lost or diminished if the linked set of contracts were considered one contract. This approach provides many opportunities for conflicting language. Perhaps if there were one contract that would be signed on by each party as it comes onto the project, the system might work. However, considering the linked set as if it is one consolidated contract probably raises more problems than it solves.[32]

Returning to the series of linked individual contracts, which contract—prime or subcontract—should be considered the dominant contract in dealing with disputes between prime contractor and subcontractor when the contracts do not make this clear? There are arguments to prefer the subcontract. First, this is the writing directly assented to by prime contractor and subcontractor. Second, the references to or incorporation of prime contract documents may often be vague, and the subcontractor rarely has the time or ability to compare the two for inconsistency. Third, the law does not compel the subcontract and prime contract to "track together."

There are arguments for giving preference to the prime contract. Often the subcontractor sees only the prime contract bidding documents at the time a sub-bid is submitted. Suppose the subcontractor refuses to enter into the subcontract because the subcontract contains terms different from those in the prime contract. If its bid is irrevocable, it can contend that it should be able to perform under the terms of its firm offer that include the prime contract provisions that apply to the subcontract work. This is yet another reason the *Drennan* rule does not fit well with subcontractor bidding.

But the contractual system usually plans for this problem. The owner and the prime contractor generally seek to bind the subcontractors at least to the performance provisions of the prime contract. The prime contractor's objective is to ensure that the entity it has selected to perform its obligations is obligated to do so. The owner's reason is less obvious. It could simply bind the contractor and let the prime take care of ensuring that there is consistent performance obligation down the line. But the owner seeks proper performance and not claims against its prime contractor. It is most likely to obtain proper performance if the subcontractors down the line have committed themselves to do so. As an ancillary but much less important reason for obtaining such commitments from subcontractors, some owners might be seeking to perfect a claim against their subcontractors by asserting that they are intended beneficiaries of the subcontracts. This is discussed in Section 28.05(B).

To tie subcontractors, the prime contract will usually contain a "flow-through," or conduit, clause. Such a clause requires the prime contractor to tie the subcontractors to provisions of the prime contract that affect their work. But flow-through clauses are not self-executing, the objective being accomplished only if the prime contractor does incorporate the prime contract into the subcontract. It is crucial to be aware of the maze of different contract documents[33] and those administrative provisions, such as those that deal with disputes, that go *beyond* substantive performance obligations. To implement the objectives of owner and prime, the latter must obtain a promise by the subcontractor to perform the performance commitments of the prime and be bound to the administrative provisions, such as design professional decisions and arbitration. When a total system is used, such as AIA Docs. A101/201 and A401, this is done automatically. But if the prime uses its own subcontract or one negotiated with a subcontractor, the owner should check the ultimate subcontracts to be certain that the prime contractor has done what it has obligated itself to do in the prime contract.

Implementation usually is accomplished by incorporation by reference. This must be distinguished from simply referring to the prime

---

[31]See Section 24.11 for use of such a system to deal with defects.

[32]A similar argument can be made for combining AIA Docs. B141, A201, and A401.

---

[33]See Section 19.01.

contract. If there is merely a reference to the prime contract, the subcontract will be interpreted in the light of the prime contract, or the prime will take precedence in the event of a conflict.[34] Although the prime contract—principally the general and supplementary conditions and technical writings— need not be physically attached to the subcontract, the subcontract should clearly state that the prime contract is incorporated by reference. If properly incorporated, it is part of the subcontract.[35]

Except for interpretation of specifications or what work is included in the work of a particular specialty, debates over whether the subcontractor must comply with the performance aspects of the prime contract are rare. Most reported appellate decisions involve dispute resolution—principally the issue of whether particular disputes between prime and subcontractor must be resolved by arbitration.[36] The cases have struggled with conflicting arbitration clauses in prime and subcontract,[37] with transactions where arbitration is included in the prime but not in the subcontract,[38] and even with contracts that do not contain an arbitration clause and do not even refer to another contract that does contain one.[39]

A study of the reported appellate cases reveals apparent inconsistencies due to the different factual patterns and varying judicial attitude toward arbitration.[40] A look at a few of the many cases is instructive. But before doing so, it is important to look at a flow-through, or conduit, clause.

As has been noted, the *principal* reason for such a clause is to ensure that subcontractors commit themselves to the performance requirements of the prime contract. As an illustration, AIA Doc. A201, ¶ 5.3.1, requires that the contractor

> require each Subcontractor, to the extent of the Work to be performed by the Subcontractor, to be bound to the Contractor by the terms of the Contract Documents and to assume toward the Contractor all the obligations and responsibilities which the Contractor . . . assumes toward the Owner and Architect.

But this clause goes beyond that objective. Under it, the *benefits* given the prime in A201 flow down to the subcontractor. For example, ¶ 5.3.1 gives the subcontractor all the "rights, remedies and redress" against the contractor that the contractor has against the owner *unless* the subcontract specifically provides otherwise. For example, prime contracts will usually contain *force majeure* provisions that excuse delayed performance if certain events occur. Such a flow-through of benefits provision would give identical rights to the subcontractor if a claim is made against it by the prime contractor for delay.

A201's flow-through of benefits applies only if the subcontract does not specify otherwise. The flow-through provision may not be of much value to the subcontractor, as subcontracts frequently contain provisions less favorable to the subcontractor than the prime contractor has in its contract with the owner. Note that ¶ 5.3.1 requires the contractor to make available to proposed subcontractors, *before* execution of the contract, copies of the general conditions to which the subcontractor will be bound and "identify to the Subcontractor any terms and conditions of the proposed Subcontract which may be at variance with the Contract Documents." The AIA hopes to give the subcontractor some bargaining muscle when the prime contractor seeks to force provisions on the subcontractor that are harsher than those in the prime contract. Whether this will have any effect on the bargaining power in the actual subcontracts is debatable.

Flow-through benefit provisions raise problems. For example, AIA Doc. A201, ¶ 2.2.1, gives the prime contractor a right to inquire into the financial arrangements made by the owner. Does the flow-

---

[34]*Oxford Dev. Corp. v. Rausauer Builders, Inc.,* 158 Ind.App. 622, 304 N.E.2d 211 (1973) (conflict over work to be performed); *United States v. Foster Constr. (Panama) S.A.,* 456 F.2d 250 (5th Cir.1972) (conflict over terms and conditions).

[35]*West Bank Steel Erectors Corp. v. Charles Carter & Co.,* 248 So.2d 52 (La.App.1971), held that a subcontractor was required to perform in accordance with specifications in the prime contract.

[36]But see *George Hyman Constr. Co. v. Precision Walls, Inc. of Raleigh,* 132 A.D.2d 523, 517 N.Y.S.2d 263 (1987), which held that a forum selection clause in the prime contract was not incorporated into the subcontract. Only clear language will accomplish this.

[37]*Falcon Steel Co. v. Weber Eng'g. Co.,* 517 A.2d 281 (Del.Ch.1986).

[38]*Maxum Foundations, Inc. v. Salus Corp.,* 779 F.2d 974 (4th Cir.1985).

[39]*John Ashe Assoc. v. Envirogenics Co.,* 425 F.Supp. 238 (E.D.Pa.1977).

[40]*Beacon Constr. Co. v. Prepakt Concrete Co.,* 375 F.2d 977 (1st Cir.1967), refused to compel arbitration, but *J. S. & H. Constr. Co. v. Richmond County Hosp. Auth.,* 473 F.2d 212 (5th Cir.1973), describing arbitration in laudatory terms, came to a different conclusion.

through of benefits provision give a similar right to the subcontractor to inquire into the financial sources of the prime contractor? The prime contractor may with justification contend that this power constitutes a trap for unwary prime contractors who may not realize how the use of a boilerplated flow-through provision can compel it to disclose sensitive information to subcontractors.

The attempt to give negotiation power to the subcontractor to augment its weak position at the bargaining table has other difficulties. For example, A201 sets up a dispute resolution system under which initially disputes are decided by the architect subject to arbitration. If this is a *burden* under the prime contract, the subcontractor must accept this method of dispute resolution. If, however, it is a *benefit*, the subcontractor is entitled to it only if there is nothing specific in the subcontract to the contrary.

Structuring the flow-through of benefits provision to apply only if nothing to the contrary is found in the subcontract raises the inevitable problem of seeking to determine when a benefit has been specifically excluded by the subcontract. Not only that, this attempt to give bargaining power to the subcontractor seems to be dependent on a requirement in ¶ 5.3.1 that the contractor make available to proposed subcontractors copies of the prime contract and identify to the subcontractor "any terms or conditions" of the subcontract that may be "at variance" with the prime contract. Is there a difference between "at variance" and "specifically provided otherwise"?

(Some prime contractors complain that requiring them to identify to the subcontractor provisions in the proposed subcontract, which may vary from the prime contract, requires that they act as the subcontractor's attorney.)

This flow-through of benefits provision demonstrates some of the difficulties in construction contract drafting. It seems attractive to include a clause that helps the subcontractor in its often difficult negotiations, if there are any at all, with the prime contractor. However, often drafters do not think about the possible applications of a general clause both as to substance and procedures. This well-meaning attempt to aid the subcontractors may only create more problems in construction contract administration and in dispute resolution.

As to specific cases, *Pioneer Industries v. Gevyn Construction Corp.*[41] involved legal action by a subcontractor against the prime contractor and its surety for work performed before the prime contract was terminated by the public awarding authority. The prime contractor disputed the termination and sought arbitration in accordance with the prime contract arbitration provision. However, the subcontractor sought to have the dispute decided by the courts rather than by arbitrators. The prime contractor pointed to a provision in the subcontract that incorporated the arbitration clause of the prime contract into the subcontract. The court pointed to subcontract provisions, stating that any conflict between prime contract and subcontract would be controlled by the subcontract and that disputes arising out of termination would be arbitrated only if the dispute involved *less* than $3,500. Because the dispute involved in the lawsuit involved *over* $3,500, the court concluded that the dispute did not fall within the arbitration clause of the subcontract. This holding illustrates that the *specific* subcontract arbitration clause takes precedence over the *incorporation by reference* of the prime contract arbitration clause. The probably unintended result was an arbitration between prime contractor and owner and litigation between prime contractor and subcontractor.

Blanket incorporation by reference can create difficulty if the purpose of incorporation is to ensure that the performance obligations are assumed by the subcontractor. For example, in *Falcon Steel Co. v. Weber Engineering Co.*,[42] the arbitration clause in the subcontract was broad but the prime contract's arbitration provision was limited to $50,000. Despite a flow-through clause, the court held that the subcontract arbitration clause controlled. The flow-through provision incorporated only the scope-of-work clause to ensure that the subcontractor followed the plans and specifications. The court pointed to the subcontract arbitration clause as being plain, broad, and without exceptions. The court was not influenced by language stating that the prime contract took precedence over the subcontract.

The two cases described enforced the subcontract arbitration clause. However, the court in *John*

---

[41]458 F.2d 582 (1st Cir.1972).
[42]Supra note 37.

F. Harkins Co., Inc. v. The Waldinger Corp.[43] chose the arbitration clause in the prime contract. The subcontract contained a broad arbitration clause but also contained a clause stating that the subcontractor would be bound by all applicable provisions of the prime contract, the latter containing a much narrower arbitration clause.

After a dispute arose, the subcontractor sought to invoke the arbitration clause in the subcontract. The prime contractor contended it did not have to arbitrate, as the dispute was not arbitrable under the prime contract. The appellate court upheld a decision of the trial court that the dispute was not subject to arbitration.

The court pointed to preliminary negotiations that indicated that the subcontractor realized that the subcontract was favorable to the prime contractor and sought unsuccessfully to obtain language that would give it the benefit of provisions in the prime contract. Evidence also showed that both parties intended that the incorporation of the prime contract would not give the subcontractor greater rights against the prime than the prime had under the prime contract. This conclusion, over a vigorous dissent, came in the face of federal law generally extending great scope to arbitration clauses.

Two cases held that arbitration would be required even though the subcontract did not contain an arbitration clause. The first, *Maxum Foundations, Inc. v. Salus Corporation*,[44] involved AIA Doc. A201, which incorporated the general conditions into the subcontract and also contained a flow-through provision. The court held that this was sufficient to incorporate the arbitration clause of A201 into the subcontract. Despite the prime contractor not having "passed through" prime contract obligations, the court concluded that all the evidence indicated that the clause was intended to be part of the subcontract, even though not expressly included in it. The court was influenced by the policy of federal courts favoring arbitration.

The second case, *John Ashe Associates v. Envirogenics Co.*,[45] involved a purchase order subcontract that did not even refer to the prime general conditions with its arbitration clause. Yet the subcontractor was bound to arbitrate because

1. The prime normally sent its prime general conditions with an arbitration clause to all prospective subcontractors.
2. The specifications and special conditions of the specifications referred to the prime general conditions.
3. The subcontractor referred to the prime general conditions to prepare its bid.

These reasons were held sufficient to put the subcontractor on notice that the arbitration clause of the general conditions was incorporated into the purchase order executed by the subcontractor.

These cases reveal some of the chaos that can occur in dispute resolution because of the subcontract system. Although, as indicated earlier in this section, there is no requirement that the prime and subcontracts be perfectly parallel, the risk of contradictory dispute resolution provisions or the absence of such provisions in one contract and their presence in another generates complicated issues. This makes disputes between prime and subcontractors extraordinarily difficult to settle and resolve.

SECTION 28.05 Subcontractor Defaults

A. Claims by Contract-Connected Parties

A subcontractor's unexcused refusal or failure to perform in accordance with its contract obligations can damage those with whom the subcontractor has contracted, such as the prime contractor or a lower tier subcontractor. As shall be seen in (C), a prime contractor may be accountable to the owner for defaults of the subcontractor. This can mean that the extent of the claim and the method of resolving any dispute over it may be affected by provisions in contracts other than the subcontract.[46]

---

[43]796 F.2d 657 (3d Cir.1986), cert. denied. sub.nom. *TWC Holdings, Inc. v. John F. Harkins Co.*, 479 U.S. 1059 (1987).
[44]Supra note 38.
[45]Supra note 39.

[46]*United States v. Foster Constr. (Panama) S.A.*, supra note 34 (sub liable for prime's lost overhead and liquidated damages); *Reed & Martin, Inc. v. Westinghouse Elec. Corp.*, 439 F.2d 1268 (2d Cir.1971) (sub pays liquidated damages only if prime does). See also *P&C Thompson Bros. Constr. Co. v. Rowe*, 433 So.2d 1388 (Fla.Dist.Ct.App.1983), discussed in Section 26.09.

B. Claims by Third Parties

Tort claims against a subcontractor by those who suffer personal harm are discussed later.[47] This section looks briefly at claims against a subcontractor based upon economic losses suffered by third parties, principally the owner.

Owners employ three theories to recover losses they have suffered that were caused by a subcontractor breach. First, an owner may contend that the subcontractor has breached its contract with the prime contractor—the contract intended for the benefit of the owner. Second, the owner may assert that the subcontractor's breach was tortious in that it failed to live up to the legal standard of care.[48] Third, the owner may contend that the prime contractor was merely a conduit between owner and subcontractor—essentially a contention based on the prime contractor's contracting as an agent of the owner.[49]

Generally, attempts have met with mixed success, with some courts allowing the owner to assert a claim as an intended beneficiary[50] while others, taking a realistic look at the intention to benefit test, have not.[51] However, a prime contractor who began dealing with a sub-subcontractor because the subcontractor was having financial problems was allowed to sue the sub-subcontractor for defective work.[52]

Because of the unsettled state of the law, the third party's *principal* claim is against someone with whom it has a contractual relationship and that party institutes action against the subcontractor. For example, if the owner has suffered a loss, it is likely that the owner's action will be against the prime contractor and the prime contractor will bring an action against the subcontractor.

The third-party claim—that is, the direct claim against the subcontractor—may also be asserted. But the uncertainty of its success makes it ancillary to the principal claim unless the claim against the contract-connected party would be uncollectable.

C. Responsibility of Prime Contractor

The subcontracting process permits the prime contractor to discharge its obligations by the use of a subcontractor. However, this power to perform through another does not relieve the prime contractor of responsibility for subcontractor nonperformance.[53] Most standard construction contracts make the prime contractor responsible for subcontractor defaults.[54]

In *Norair Engineering Corp. v. St. Joseph's Hospital, Inc.,*[55] one issue involved the responsibility for defective work performed by a subcontractor who was hired at the owner's demand by the contractor who had been brought in by the bonding company to replace the original prime. The court concluded that the contractor was well aware that it was taking the risk of the subcontractor's default when it entered into the prime contract, that it was not without remedies against the subcontractor, and that it would never have been awarded the contract to complete the work unless it had agreed to the provisions under which the owner could select and approve subcontractors.

Owners can exercise a variety of controls over the selection of subcontractors.[56] That the owner approves a proposed subcontractor should not relieve the prime contractor of responsibility. Clearly,

---

[47]See Chapter 31.

[48]See cases cited in Sections 14.08(C) (D) and (E).

[49]*National Cash Register Co. v. UNARCO Industries, Inc.,* 490 F.2d 285 (7th Cir.1974) (dictum that owner could sue as principal but claim allowed under subrogation); *Seither & Cherry Co. v. Illinois Bank Bldg. Co.,* infra note 58.

[50]*Chestnut Hill Dev. Corp. v. Otis Elevator,* 653 F.Supp. 927 (D.Mass.1987) (ambiguity because clause that allowed the owner to assume the contractor's rights did not appear in the subcontract, but subcontract required sub to perform strictly in accordance with prime contract); *United States v. Ogden Technologies Laboratories, Inc.,* 406 F.Supp. 1090 (E.D.N.Y.1973); *Gilbert Financial Corp. v. Steelform Contracting Co.,* 82 Cal.App.3d 65, 145 Cal.Rptr. 448 (1978).

[51]*Vogel Bros. Building Co. v. Scarborough Constructors, Inc.,* 513 So.2d 260 (Fla.Dist.Ct.App.1987) (owner could not compel prime and sub to arbitrate); *Wheeling Trust & Savings Bank v. Trem Co., Inc.,* 153 Ill.App.3d 136, 505 N.E.2d 1045 (1987); *Vogel v. Reed Supply Co.,* 277 N.C. 119, 177 S.E.2d 273 (1970); *Manor Junior College v. Kaller's Inc.,* 507 A.2d 1245 (Pa.Super.1986).

[52]*White Constr. Co. Inc. v. Sauter Constr. Co.,* 731 P.2d 784 (Colo.App.1987).

[53]*Kahn v. Prahl,* 414 S.W.2d 269 (Mo.1967); *Waterway Terminals Co. v. P. S. Lord Mechanical Contractors,* 242 Or. 1, 406 P.2d 556 (1965) (prime responsible for subcontractors).

[54]AIA Doc. A201, ¶ 5.3.1. This is aided by requiring that the prime contractor obtain agreements from subcontractors that preserve and protect the rights of the owner "so that the subcontracting thereof will not prejudice such rights."

[55]147 Ga.App. 595, 249 S.E.2d 642 (1978).

[56]See Section 28.03.

this is a matter for negotiation, and the law should enforce the choice that the contracting parties have made. Where there is no clear evidence one way or the other, a strong argument can be made that the owner *dictating* a particular subcontractor should relieve the prime contractor of responsibility.[57] In such a case, the prime contractor may be acting as an agent of the owner to engage a particular subcontractor and should not be responsible unless it knew or should have known that the subcontractor was incompetent.[58] If the owner steps in and takes away the prime contractor's power to manage the subcontractors, the prime contractor cannot be held accountable for the work of the subcontractors. Its obligation is conditioned on its right to manage the job without unreasonable interference.[59]

Sometimes the prime contractor supports a contention that it should not be held responsible by pointing to the independent contractor rule. This rule, though subject to many exceptions, relieves the employer of an independent contractor from responsibility from the latter's improper performance.[60] However, this defense, where available, does not relieve the employer of the independent contractor when the independent contractor's failure to perform properly has damaged a party to whom the employer of the independent contractor owed a contract right of proper performance.[61]

This section has looked largely at judicial claims against the prime contractor based upon conduct of the subcontractor. However, increasing attention is being directed toward claims by public safety officials against the prime contractor under powers granted by the Occupational Safety and Health Act (OSHA), a federal law that regulates the workplace. An OSHA citation was issued against a prime contractor for defective work performed by its electrical subcontractor.[62] The citation was based upon the conclusion that the prime should have general knowledge of electrical requirements. While the prime was entitled to rely on the knowledge and expertise of its subcontractors, the OSHA Review Commission concluded that the prime did not show evidence of its reasonable reliance on the subcontractor. This may mean that a prime contractor will be accountable under OSHA for defective work performed by specialty trades unless it obtains some written acknowledgement from the subcontractors that requirements are being complied with and that the prime is relying on the subcontractor's expertise.

SECTION 28.06 Payment Claims Against Prime Contractor

One of the particularly sensitive areas in the subcontract relationship relates to money flow, with subcontractors frequently contending that they invest substantial funds in their performance and are entitled to be paid as they work and to be completely paid when their work is completed. This issue surfaces around two concepts: line-item retention and payment conditions. The subcontractors wish to divorce their work and payment for it from the prime contract. They argue *for* line-item retention and *against* payment conditions.

Line-item retention gives the subcontractor the right to be paid after it has fully performed. Delay usually occurs because the owner holds back retainage—a designated amount of the contract price for the entire performance, that of prime contractor and *all* subcontractors. When all the work is completed and the project accepted, the owner will pay the retainage. However, early finishing subcontractors may have to wait a substantial period of time after they have completed their work for the retention allocated to their contract because the entire project is not yet completed. They would like to disassociate their contract from the rest of the subcontracts and the prime contract.

A payment condition (pay when paid clause or more realistically, from the position of the prime, "pay *only* if paid") makes payment to the prime contractor a condition to the prime contractor's ob-

---

[57]Cf. *National Cash Register Co. v. UNARCO Industries, Inc.,* supra note 49. Here the owner was allowed to sue the subcontractor that it ordered the prime contractor to use. Very likely the prime would be relieved of responsibility.

[58]*Seither & Cherry Co. v. Illinois Bldg. Corp.,* 95 Ill.App.3d 191, 419 N.E.2d 940 (1981) (CM agent of owner in hiring contractor).

[59]See Section 19.02(A).

[60]See Section 31.05(B).

[61]*Harold A. Newman Co. v. Nero,* 31 Cal.App.3d 490, 107 Cal.Rptr. 464 (1973); *Brooks v. Hayes,* 133 Wis.2d 228, 395 N.W.2d 167 (1986) (quoting and following earlier edition of treatise).

[62]*Secretary of Labor v. Blount Int'l, Ltd.,* 15 O.S.H.Cas. (BNA) 1897 (O.S.H.Rev.Comm'n, Sept. 18, 1992).

ligation to pay the subcontractor. Prime contractors seek to create such a condition by including language in the subcontract stating that the prime contractor will pay "if paid by the owner," "when paid by the owner," or "as paid by the owner." An endless number of cases have interpreted this language. Does the language create a condition to payment or simply indicate that the payment flow contemplated some delay, with payment in any event being required after the expiration of a reasonable time?

All courts agree, or at least so they state, that the parties can make a payment condition under which the subcontractor assumes the risk that it will not be paid for its work. The legal issue has centered around the requisite degree of specificity needed to create such a condition. On the surface, the courts simply examine the language and seek to determine the intention of the parties. Other factors are operating, however. This section looks at what courts have done, discusses some reasons for such clauses, and concludes by looking briefly at the system used in the AIA Standard Documents.

In 1933, New York held that language of the type that appears to tie the subcontractor's right to payment to the contractor's receiving payment created a payment condition precluding the subcontractor from recovering when the prime contractor had not been paid.[63] In 1962, the leading case of *Thomas J. Dyer Co. v. Bishop International Engineering Co.*[64] came to a different conclusion. The court stated that performing parties usually expect to be paid for their work. Any result under which a party would suffer a forfeiture (that is, performing the work and not being paid) must be supported by clear and specific language that this risk has been taken. The court stated that the subcontractor, in addition to its mechanics' lien protection, contracts mainly on the basis of the solvency of the prime contractor. To change this normal credit risk, the contract should "contain an express condition clearly showing that to be the intention of the parties."[65] The court concluded that the language dealing with linkage of payment to payment to the prime contractor was designed "to postpone payment for a reasonable period of time after the work was completed, during which the general contractor would be afforded the opportunity of procuring from the owner the funds necessary to pay the subcontractor."[66]

Most recent cases have followed the reasoning in the *Dyer* case.[67] But some decisions have concluded that a payment condition has been created.[68] Courts here are not simply construing contract language. The language used indicates that the prime contractor need not pay *until* it is paid. Courts have tilted the balance toward the subcontractor because of the feeling that the subcontractor has little choice as to the contract language, that the subcontractor has no opportunity to evaluate the solvency of the owner, and that it would be unfair for subcontractors to have to bear this risk when the prime contractor is in the best position to evaluate the credit of the owner.

---

[63]*Mascioni v. I. B. Miller, Inc.*, 261 N.Y. 1, 184 N.E. 473 (1933).
[64]303 F.2d 655 (6th Cir.1962).
[65]303 F.2d at 661.

[66]Ibid.
[67]*Nicholas Acoustics & Specialty Co. v. H & M Constr. Co.*, 695 F.2d 839 (5th Cir.1983) (clause gives prime contractor opportunity to see whether subcontractor had completed its work); *Culligan Corp. v. Transamerica Ins. Co.*, 580 F.2d 251 (7th Cir.1978); *Midland Eng'g. Co. v. John A. Hall Constr. Co.*, 398 F.Supp. 981 (N.D.Ind.1975); *Yamanishi v. Bleily & Collishaw, Inc.*, 29 Cal.App.3d 457, 105 Cal.Rptr. 580 (1972); *Sasser & Co. v. Griffin*, 133 Ga.App. 83, 210 S.E.2d 34 (1974); *Chartres Corp. v. Charles Carter & Co.*, 346 So.2d 796 (La.App.1977); *D. K. Meyer Corp. v. Bevco, Inc.*, 206 Neb. 318, 292 N.W.2d 773 (1980); *Schuler-Haas Electric Co. v. Aetna Casualty & Surety Co.*, 40 N.Y.2d 883, 357 N.E.2d 1003, 389 N.Y.S.2d 348 (1976); *Grossman Steel & Aluminum Corp. v. Samson Window Corp.*, 78 A.D.2d 871, 433 N.Y.S.2d 31 (1980); *Elk & Jacobs Drywall v. Town Contractors, Inc.*, 267 S.C. 412, 229 S.E.2d 260 (1976). In addition, courts have found other ways to protect subcontractors. See *Excavators & Erectors, Inc. v. Bullard Eng'rs, Inc.*, 489 F.2d 318 (5th Cir.1973) (separate oral contract not governed by subcontract payment condition clause); *Pioneer Indus. v. Gevyn Constr. Corp.*, supra note 41 (payment clause violates Massachusetts payment bond statute). For a more complete discussion, cases, secondary authorities, and sample clauses, see J. SWEET, SWEET ON CONSTRUCTION INDUSTRY CONTRACTS, § 18.10 (2d ed. 1992).
[68]*Architectural Sys., Inc. v. Gilbane Bldg. Co.*, 760 F.Supp. 79 (D.Md.1991); *Brown v. Maryland Cas. Co.*, 246 Ark. 1074, 442 S.W.2d 187 (1969); *D. I. Corbett Elec., Inc. v. Venture Constr. Co.*, 140 Ga.App. 586, 231 S.E.2d 536 (1976); *Dorman Strahan v. Landis Constr. Co.*, 499 So.2d 417 (La.App.1986). Note that in this and the preceding footnote, cases are cited for opposite conclusions in the same jurisdiction. The result can depend on the specificity of the language and/or the cause for the delay. For example, in the *Chartres* and *Midland* cases, there was extremely long delay, in each case almost two years.

The *reason* for nonpayment is important. Two clear cases should be set aside. First, if the nonpayment to the prime contractor results from the *subcontractor's* failure to perform properly, the subcontractor is not entitled to recovery. Second, if the reason for nonpayment is improper performance by the prime contractor or an unwillingness to make reasonable efforts to obtain payment, the subcontractor is entitled to be paid. Even if a condition to payment *has* been created, prevention of its occurrence or not taking reasonable efforts to make it occur excuses the condition allowing the subcontractor to recover.

The more difficult questions involve the owner's nonpayment under circumstances for which neither the prime contractor nor the subcontractor can be held directly responsible. Suppose the nonpayment results from the design professional's refusal to issue a payment certificate or revocation of a previously issued certificate.[69] For example, suppose work by *another* subcontractor had been previously accepted but was discovered to be defective after payment to the subcontractor. In such a case, the latter's correction of the work will start the money flow again. Under these circumstances, the prime contractor should be required to pay the unpaid subcontractor who *had* performed properly. The prime contractor hired the subcontractor whose performance was improper and is in the best position to put pressure on *that* subcontractor to correct the defective work. Additionally, the prime contractor could have required the subcontractor whose work had been found to be defective to post a performance bond. This would have entitled the prime contractor to look to the surety for the nonperformance. Allowing the prime contractor to refuse to pay the unpaid subcontractor whose work was properly performed would also strengthen the already strong bargaining position the prime contractor is likely to have. This can give the latter additional leverage to force settlement of disputes between prime and unpaid contractor even for projects unaffected by the payment dispute.

Another difficult problem involves nonpayment due to owner insolvency or bankruptcy. Generally,

such risks should be borne by the prime contractor.[70] The prime has entered into the contract with the owner and is in the best position to make financial capacity judgments.[71] Although the subcontractor's willingness to perform work may have rested to some degree on its evaluation of the financial position of the owner, it would not be sufficient to conclude that the prime contractor has transferred the risk to its subcontractors in the absence of a clear showing that this was intended.

Other factors exist that may determine whether a payment condition has been created. If the prime contractor has invested substantial amounts of money into the project itself, it can more persuasively contend that it and the unpaid subcontractor should share the loss. If the prime contractor has invested very little of its own money, a stronger argument can be made for refusing to find a payment condition, with the prime contractor in such a case acting as a broker who should take the risk of the owner's insolvency.

Sometimes it is contended that the unpaid subcontractor has a right to a mechanics' lien if it has improved the owner's property. Although the laws relating to mechanics' lien are varied and often difficult to decipher, the right to a lien should be based on a legal right to payment from the prime contractor. If a payment condition has been created, the unpaid subcontractor should not be able to obtain a mechanics' lien. This strengthens its argument that a payment condition should not be found unless it is extremely clear that the subcontractor took this risk. To be sure, some court decisions look solely at the enrichment of owners through the improvement on their property as the basis for a mechanics' lien. At best, the uncertainty of the mechanics' lien points toward a conclusion that a payment condition generally has *not* been created.

The AIA has been buffeted by both sides of this controversy. The AGC has argued that matters between the prime contractor and the subcontractor

[69]*R. C. Small & Assoc., Inc. v. Southern Mechanical, Inc.*, 730 S.W.2d 100 (Tex.Ct.App.1986) (certificate not a condition precedent).

[70]*Pacific Lining Co. v. Algernon Blair Constr. Co.*, 819 F.2d 602 (5th Cir.1987) (payment to prime not condition precedent).
[71]This conclusion would be strengthened under AIA Doc. A201, ¶ 2.2.1. A contractor can require the owner to furnish the contractor reasonable evidence that the owner has·made financial arrangements to fulfill its obligation.

should be left to them and should not be regulated in A201, the AIA's standard general conditions. On the other hand, subcontractor associations have contended that the AIA should use A201 to ensure that subcontractors are properly treated.

As a result AIA Doc. A201, ¶ 9.6.2, is vague on this question, with the AIA leaving the ultimate determination to court decisions. Yet A401, the AIA's standard construction subcontract, not only seeks to make clear there is *no* payment condition but also creates line-item retention, with each subcontractor paid its retainage as it finishes.[72]

The struggle between primes and subcontractors in some states has moved to the legislative arena. For example, North Carolina and Wisconsin have barred the use of payment conditions (pay only if paid).[73] The Wisconsin legislation does not prohibit contract provisions that might delay a payment to a subcontractor until the contractor receives payment.

SECTION 28.07 Payment Claims Against Property Funds or Persons Other Than Prime Contractor

A. Court Judgments and Specific Remedies

Legal relief generally comes in the form of court-issued judgments. Most judgments simply state that the judgment creditor—the person who seeks and obtains the judgment—is entitled to a specific amount of money from the judgment debtor—the person against whom the judgment is issued. The judgment creditor is given methods of collecting on these judgments, but as mentioned in Section 2.13, such remedies are often cumbersome and ineffective. If the judgment debtor does not pay voluntarily, collection problems sometimes make the judgment worthless.

Specific remedies, on the other hand, are remedies that either command the defendant to do or not do a particular act (specific performance decrees or injunctions) or operate against specific property (liens against particular funds, goods, or property). Specific remedies are equitable decrees

backed up by the contempt power of the court. The judge can punish a person adjudged to be in contempt of court by a fine or imprisonment. See Section 6.03. They are more effective, as a rule, than the ordinary money award court judgment.

One reason unpaid subcontractors seek mechanics' liens[74] or other specific remedies is that they operate against particular property and are more effective than ordinary judgments, whether obtained against the prime contractor or third parties such as the owner. Sometimes even specific remedies are ineffective, but on the whole, they are preferable to ordinary court judgments.

B. Statutory and Nonstatutory Remedies

Most of the material discussed in this section, such as mechanics' liens, stop notices, and trust fund protection, have been created by the state legislatures. Many bonds are required on public work because of legislative enactments. The statutory nature of these remedies has added complications.

Reported appellate cases that seem contradictory are often traceable to the variant statutes before the courts. Statutes change frequently. Compliance is difficult, and the law becomes murky. Although state statutes follow broad patterns of similarity, variance in details places a heavy burden on those contractors who operate in different states and may in part account for the local character of the construction industry.

Much of the law in this area involves interpretation of these frequently complex state statutes. Such state statutes create many requirements for the creation of lien rights or stop-notice protection. Who will be accorded statutory protection, how such protection is achieved, and the nature of the protection are often resolved by reference to the complicated, almost unreadable statutes. Frequently, such interpretation questions are resolved by holding that the statutes are designed to prevent

---

[72]AIA Doc. A401, ¶¶ 11.3, 12.1.

[73]N.C.Gen.Stat. § 22C-2 (1991); West Ann.Wis.Stat. § 779.035 (1m)(c).

[74]The many mechanics' lien laws make generalizations perilous. For example, a mechanics' lien generally creates only a *security* interest in the property improved by the lien claimant. However, Florida permits a lien claimant a *personal* judgment against the owner even if there is no contract relationship between them. West Ann.Fla.Stat. § 713.75(1). Similarly, what are called "liens" against public buildings may only be liens against funds. See note 125 infra. Also, Texas gives a lien on the retainage. Ann.Tex.Prop.Code § 53.103.

unjust enrichment by protecting unpaid subcontractors, the intended beneficiaries of the legislative protection. Matters can become even more complicated in jurisdictions that state that the standards for perfecting a mechanics' lien will be *strictly* required, but once the lien has been perfected, the remedy will be administered *liberally*.[75] Strained interpretations and language distortion often result, yet construing these statutes to protect subcontractors does not invariably result in lien protection. The principal legacy of this approach is legal uncertainty and unpredictability.

The extensiveness of the statutory system can make it more difficult for subcontractors to assert nonstatutory claims. Where the subcontractor has been given statutory protection but did not take the necessary steps to obtain it, courts sometimes deny the *nonstatutory* claim.[76] Although the statutory systems are not exclusive, their existence can persuade courts that they offer sufficient protection to justify denial of a nonstatutory claim.

C. Public and Private Work

The nature of the remedy available may depend on whether the work is public or private. Mechanics' liens, to the extent that they permit foreclosure rights against public buildings, are not available. As a result, the federal government enacted the Miller Act, a compulsory prime bonding system. Many states have enacted comparable legislation for state construction projects.

Sometimes systems developed in public work spill over into private construction. For example, stop notices (discussed in Section 28.07(E)) were originally developed to compensate contractors for public work because they were not accorded lien rights. However, in those states that have stop notices, the stop notice is often available for private work.

D. Mechanics' Liens

Mechanics' lien laws are complicated and vary considerably from state to state.[77] For that reason, it would be inadvisable to attempt a summary of all aspects of these statutory protections accorded certain participants in the Construction Process. Instead, the discussion focuses on rationales for such protection and salient features and current criticisms of lien laws.

Participants in the Construction Process who can in various ways trace their labor and materials into property improvements of another are given lien rights against the property in the event they are not paid by the party who has promised to pay them. The most important lien recipients are prime contractors, subcontractors, suppliers, laborers, and design professionals. The remedy accorded a lien holder is the right to demand a judicial foreclosure or sale of the property and be paid out of the proceeds, including, in some states, the legal costs of perfecting the lien, such as attorneys' fees.[78]

Lien claimants are divided into two principal categories, those who have direct contract relations with the owner and those who do not. Typical illustrations of the first are design professionals and prime contractors. Illustrations of the second are subcontractors, laborers, and suppliers to prime contractors and subcontractors. Owners can avoid liens by paying their design professionals and prime contractors. But because the majority of the difficult problems are generated by lien claimants not connected by contract with the owner and because this chapter focuses on subcontractor problems, the discussion centers around the second class of lien claimants. The usual justification given for granting these liens is unjust enrichment. Those

---

[75]*Talco Capital Corp. v. State Underground Parking Comm'n,* 41 Ohio App.2d 171, 324 N.E.2d 762 (1974).
[76]*Engle Acoustic & Tile, Inc. v. Grenfell,* 223 So.2d 613 (Miss.1969); *Banks v. City of Cincinnati,* 31 Ohio App.3d 54, 508 N.E.2d 966 (1987). See, generally, Comment, *Mississippi Law Governing Private Construction Projects: Some Problems and Proposals,* 47 Miss.L.J. 437 (1976).

[77]The National Conference of Commissioners of Uniform State Laws has proposed a new "Uniform Construction Lien Act." One of its features would be to protect against double payment by the owner, a failing of many mechanics' lien laws. An earlier attempt at uniformity failed, no state adopting the proposed Act. The commissioners withdrew the earlier Act.
[78]However, Delaware held that the state statute allowing recovery of attorneys' fees by successful plaintiffs in mechanics' liens actions but not by successful defendants violates the constitutional guarantee of equal protection. *Gaster v. Coldiron,* 297 A.2d 384 (Del.1972).

whose labor or materials have gone into the property of another should have lien rights in the property when they are not paid as promised.[79]

The first mechanics' lien law was enacted in Maryland in 1791. One reason was to provide a quick and effective remedy for unpaid workers who cannot wait until a full trial to collect their wages. Quick and certain remedies can induce workers to work on construction by assuring them they will be paid. As an illustration, the Maryland lien laws were enacted at the urging of Thomas Jefferson and James Madison to stimulate and encourage the rapid building of the city of Washington.

Amplification of this inducement so vital to a developing country could and did lead to expansion of lien beneficiaries to include not only laborers but also all those who participate directly in the Construction Process. The state gives credit to prime contractors by granting subcontractors lien rights, which encourages persons to furnish labor and materials for construction. This state credit was especially needed to bolster an unstable construction industry composed of many contractors unwilling or unable to pay subcontractors and suppliers. This is probably the principal reason for giving lien rights today.

Expansion of lien laws is undoubtedly traceable to the realities of the political process. Once some participants in the Construction Process had received lien rights on a frequently asserted unjust enrichment theory, it was relatively easy to expand the list of lien beneficiaries. Those who might oppose lien expansion, such as owners, are often unrepresented as an organized group in the legislatures. This too may have accounted for expansion of lien beneficiaries and lien rights.

The desire by participants to expand mechanics' lien rights is understandable, since the mechanics' lien is a more effective remedy than a money award. However, legislatures that respond to such pressures often ignore the fact that these liens come at the expense of others: in the case of subcontractor liens, they come at the expense of the unsecured creditors of prime contractors; in the case of prime contractor liens, they come at the expense of the unsecured creditors of the owner.

Some salient characteristics of lien laws have been mentioned. A lien is a security interest in the property and can be foreclosed on by the lien claimant. Some states known as having New York type of statutes limit the amount of the lien to the unpaid balance owed the prime contractor by the owner. Some states known as having Pennsylvania type of statutes do not place a limit of this type on the lien. States with an open-ended lien permit the owner to limit the lien to the contract price by filing a copy of the contract and posting a bond—something rarely done.

Although owners typically set the lien-creating events into motion by contracting with prime contractors, tenants can create liens by hiring contractors. Property owners can avoid such liens by posting a notice of nonresponsibility within a designated time after the owners learn the improvement is being made.

Those entitled to liens are usually set forth in the statute, and the list typically is lengthy and expanding. Whether particular work qualifies for a lien is often unclear because of lien statutes. Some statutes use generic terms such as improvement, building, or structure. Others that attempt to be detailed do not always keep up with changes in the Construction Process. For example, liens have been denied where the lien claimant had placed engineering stakes and markers,[80] where a claimant had graded and installed storm and sanitary sewers, paving, curbing, and seating;[81] where a claim-

---

[79]The unjust enrichment rationale loses some of its attractiveness when ''double payment'' is considered. An owner who pays the prime contractor may have to pay *again* to an unpaid subcontractor if it wants to remove the lien. As noted in note 97 infra, states increasingly protect residential owners from double payment. See Ann.Mich.Comp.Laws § 570.1203. As shall be noted in (H), unpaid Michigan subcontractors are given a claim against a state-administered fund created by payments of licensed contractors.

---

[80]*South Bay Eng'g. Corp. v. Citizens Sav. & Loan Ass'n*, 51 Cal.App.3d 453, 124 Cal.Rptr. 221 (1975).

[81]*Sampson-Miller Assoc. Companies v. Landmark Realty Co.*, 224 Pa.Super. 25, 303 A.2d 43 (1973). The court reached its result reluctantly and urged that the legislature liberalize the statute to permit liens for work similar to that done by the claimant. The court noted a number of states where liens are available for preliminary work, such as California, Hawaii, Texas, and Illinois.

ant had performed demolition work;[82] where a claimant had installed a swimming pool;[83] where a claimant had performed electrical work on a modular home erected at a factory;[84] and where an entity had helped subcontractors assemble a work force, had dealt with payroll, and had advanced funds.[85] Most states deny liens to lessors of equipment used in construction unless the items are consumed in the process of use.[86] Not all work that would be considered part of the Construction Process can qualify for a lien.

Lien laws typically create an obstacle course of technical requirements for claimants. For example, a California subcontractor lien claimant must file a preliminary notice with the owner, the general contractor, and construction lender within twenty days after furnishing the materials.[87] The claimant must also record the claim of lien within ninety days of completion of the work of improvement or, if a notice of completion or a notice of cessation of the work is recorded, within thirty days of such notice.[88] The lien terminates unless an action to enforce it is begun within ninety days of the completion of the improvement[89] and such action is subject to discretionary dismissal if not brought to trial within two years.[90] Generally, any failure to comply with these requirements will invalidate the lien. Substantial compliance is insufficient.[91]

If the lien amounts exceed the value of the property after those with security interests that take priority are paid, all claimants are treated equally. In some states, liens of prime contractors are subordinated to other lien claimants, while in other states, laborers are sometimes given preference.

As between lien claimants and others with security interests in the property, such as the seller of the property who retains a security interest or a construction lender, the party who perfects its interest first takes priority. For this reason, lenders will not make construction loans if work has begun on the project, for fear that their security interest will not take priority over those who have already begun work. As a rule, lenders and other security holders in the land perfect their security interest *before* work begins. As a result, lien claims can become valueless if trouble develops and prior security holders foreclose on the property. This occurs frequently because of market imperfections. The lender is typically able to buy in at less than the amount owing on the construction loan because the liens of other claimants are not extensive enough to justify bidding in or they may not have sufficient funds to be able to compete with the lender.

To deal with this problem, New Hampshire enacted a statute that gives mechanics' lien claimants priority over construction lenders.[92] Additionally, one court found a way to grant lien claimants priority over construction lenders.[93]

Lien claimants must establish that they have performed work under the terms of a valid contract. As has been emphasized throughout this treatise, there has been a proliferation of legal controls on the Construction Process. (Licensing laws, land use controls, building and housing codes, and the controls imposed on projects built in part with public funds are illustrations.) As a result, a lien may be denied because of a technical violation of a law or regulation.[94]

---

[82]*John F. Bushelman Co. v. Troxell,* 44 Ohio App.2d 365, 338 N.E.2d 780 (1975). The court noted that it could not extend the right to demolition work but left the problem for legislative action.

[83]*Freeform Pools, Inc. v. Strawbridge Home for Boys, Inc.,* 228 Md. 297, 179 A.2d 683 (1962). The statute gave liens for buildings, and the court concluded that a swimming pool was not a building.

[84]*C & W Elec., Inc. v. Casa Dorado Corp.,* 34 Colo.App. 117, 523 P.2d 137 (1974). The court noted that no owner of real property had requested that the work be performed, and this was a requirement of the statute.

[85]*Primo Team v. Blake Constr. Co.,* 3 Cal.App.4th 801, 4 Cal.Rptr.2d 701 (1992).

[86]Annot. 3 A.L.R.3d 573, 578 (1965). In California, the seller of equipment who retains a security interest cannot obtain a mechanics' lien. *Davies Machinery Co. v. Pine Mountain Club, Inc.,* 39 Cal.App.3d 18, 113 Cal.Rptr. 784 (1974). But Louisiana granted a lien for nails, lumber, and plyform consumed in temporary work. *Slagle-Johnson Lumber Co. v. Landis Constr. Co.,* 379 So.2d 479 (La.1979).

[87]West Ann.Cal.Civ.Code § 3114.

[88]Id. at § 3116.

[89]Id. at § 3144.

[90]Id. at § 3147.

[91]*IGA Aluminum Products, Inc. v. Mfg. Bank,* 130 Cal.App.3d 699, 181 Cal.Rptr. 859 (1982).

[92]Ann.N.H.Rev.Stat. 447:12-a.

[93]*Security Bank & Trust Co. v. Pocono Web Press, Inc.,* 295 Pa.Super. 455, 441 A.2d 1321 (1982) (trade contractors given lien even if they did not begin work before lender lien filed despite their notice of lender's mortgage).

[94]Liens were upheld in *Excellent Builders, Inc. v. Pioneer Trust & Savings Bank,* 15 Ill.App.3d 832, 305 N.E.2d 273 (1973) (design violated setback, sideline, and sideyard requirements), and *M. Arthur Gensler, Jr. & Assoc. Inc. v. Larry Barrett, Inc.,* 7 Cal.3d 695, 499 P.2d 503, 103 Cal.Rptr. 247 (1972) (failure to obtain a new building permit when costs increased).

Mechanics' lien laws have been attacked as violating the U.S. Constitution. Attacks were based on decisions of the U.S. Supreme Court that had invalidated certain remedies granted before a full trial as depriving the defendants against whom the remedy was allowed the use of their property without due process of law.[95] Generally, such attacks have been unsuccessful.[96] Courts have held that the owner's deprivation of the use of its property is not substantial. Although marketability may be impaired to a degree, there is no interference with the owner's possession. Courts that have not sustained these attacks have emphasized that the work of the claimants has improved the value of the property and that the owner can force an expeditious adjudication of the lien claim within a short period.

The brief examination of some of the salient characteristics of mechanics' lien protection has revealed the weaknesses of the remedy. Most important, the statutes are complex and change frequently. Carelessness in compliance can result in the lien's being lost. On the other hand, strict compliance is costly—perhaps, in smaller jobs, more than the value of the lien. The lien is most important in construction projects that fail. It is in such situations that it is most likely that lien claimants will find their claims wiped out because prior security holders have foreclosed and the funds left over for lien claimants are nonexistent. These deficiencies have led subcontractors to seek other forms of legislative and judicial relief, the focal point of the balance of Section 28.07.

Discussion has centered on the weaknesses of mechanics' lien laws as protection for unpaid subcontractors and suppliers. However, another frequently made criticism of such laws is that they can compel an inexperienced owner to pay twice for the same work. The owner may pay the prime contractor and then have to pay an unpaid subcontractor to remove a lien. The possibility of double

payment has led to some legislative change to protect homeowners.[97]

Another obstacle to asserting a mechanics' lien is the presence in the prime contract of a provision under which the prime contractor gives up liens for itself and its subcontractors. In some states, such no–lien contracts are common. Whether a subcontractor will be precluded from asserting a mechanics' lien under such circumstances depends on whether the law requires that it consent to giving up its lien.[98]

Because of the circumstances under which such lien waivers are given, legislatures have started to regulate these transactions.[99]

E. Stop Notices

The ineffectiveness of mechanics' liens in private work and the unavailability in public work has led some states to supplement mechanics' lien protection by enactment of stop-notice laws. Like mechanics' lien laws, these statutes vary from state to state, but for simplicity, reference is made to the California stop-notice law.[100]

Those entitled to mechanics' liens can take advantage of stop-notice laws. Using a subcontractor as an illustration, in private projects, a subcontractor must file with owner and lender a preliminary notice twenty days after beginning work that de-

---

[95]The first of this long series of cases was *Sniadach v. Family Finance Corp.*, 395 U.S. 337 (1969).

[96]*Columbia Group, Inc. v. Jackson*, 151 Ariz. 76, 725 P.2d 1110 (1986), and *Connolly Dev., Inc. v. Superior Court of Merced County*, 17 Cal.3d 803, 553 P.2d 637, 132 Cal.Rptr. 477 (1976), upheld mechanics' lien laws, but *Barry Properties, Inc. v. Flick Bros. Roofing Co., Inc.*, 277 Md. 15, 353 A.2d 222 (1977) did not. See Note, 34 Wash. & Lee L.Rev. 1067 (1967).

[97]Ann.Mich.Comp.Laws § 570.1203. For cases see *Grier Lumber Co. v. Tryon*, 337 A.2d 323 (Del.Super.1975); *Ridge Sheet Metal Co. v. Morrell*, 69 Md.App. 364, 517 A.2d 1133 (1986); *Aztec Wood Interiors, Inc. v. Andrade Homes, Inc.*, 104 N.M. 45, 716 P.2d 236 (1986).

[98]*Pero Bldg. Co., Inc. v. Donald H. Smith*, 6 Conn.App. 180, 504 A.2d 524 (1986) (contractor cannot bind subs and suppliers). If a subcontractor knows the prime contract is a no-lien contract, it may be barred. *Baker Sand & Gravel Co. v. Rogers Plumbing and Heating Co.*, 228 Ala. 533, 154 So. 591 (Ala.1934).

[99]West Ann.Cal.Civ.Code § 3262 (bars owner or prime from waiving liens of others, except with their written consent, and differentiates between conditional release, which is given in the advance of payment, and unconditional release, given when payment has been made); Ill.Rev.Stat. Ch. 82, ¶ 1.1 (1992) (invalidates lien waivers given in anticipation of and in consideration of being awarded a contract or subcontract); West Ann.Wis.Stat. § 779.035 (1m)(a)(1992) voids provisions requiring a contractor or supplier to waive its lien or claim on payment bond before it has been paid).

[100]For a listing of such state statutes, see Blanton and Diak, *Legislation Division Update*, 12 Constr. Lawyer No. 1, Jan. 1992, at p. 31.

scribes the work to be done, the parties, the site, and a statement that if bills are not paid, the property being improved may be subject to a mechanics' lien.[101] In public work, a somewhat simplified preliminary notice must be filed twenty days after the work is begun.[102] In private work, a stop notice sent to a lender may be accompanied by a bond in the amount of 1.25 times the claim to protect the lender from damages if the claim is not ultimately established as valid. If a *nonbonded* stop notice is filed with a lender, the latter *may* withhold construction funds to pay the claim.[103] If a *bonded* stop notice is filed with a lender, the latter *must*—unless a payment bond has been filed—set aside construction loan funds to pay the claim.[104]

A stop notice sent to an owner—public or private—need not be accompanied by a bond.[105] Owners—public or private—must withhold funds from money owed to the prime contractor to answer the stop-notice claims unless a payment bond has been filed.[106]

In California, the claim can include only materials furnished or work performed.[107] The notice need not state that the claimant has not yet been paid.[108]

If the funds withheld are insufficient to pay all stop-notice claimants, each claimant shares pro rata in the funds without regard to when the stop notices were filed.[109] An owner, lender, or prime contractor who questions the amount of the notice can file a bond for 1.25 times the amount stated with the party served with a stop notice. When such bond is filed, the funds withheld must be released.[110] California statutes authorizing stop notices on public works specify a detailed, speedy procedure to deal with disputed stop notices.[111]

The stop notice is more effective than a mechanics' lien in obtaining payment. It can stop the flow of funds, which can jeopardize the project. Effectiveness for a subcontractor does not mean that the possible unfairness to the prime contractor should be ignored. An effective subcontractor remedy may not always be fair to prime contractors, especially where there is an honest dispute over the subcontractor's performance.

F. Trust Fund Legislation: Criminal Penalties

A number of states have enacted legislation designed to avoid prime contractor diversion by designating funds received for work performed by others as trust funds that the trustee prime contractor must pay to those whose work generated the funds.[112] Some trust fund statutes apply to public works, some to private works, and some to both.[113]

If the funds paid can be located—ordinarily a difficult task—the statutes typically give the claimant priority over bankruptcy trustees and creditors. To enforce the trust, the court can order an accounting, set aside as a diversion any unauthorized payments, award damages for breach of trust, or issue an order terminating or eliminating the authority of the contractor to apply trust assets.

Because such funds often disappear, the main effect of trust fund statutes is to provide harsh penalties for those who violate the trust. Sanctions for violation of the trust vary. Frequently, the breach of trust caused by diversion is a crime. For example, in New York, a diverting prime contractor is guilty of larceny[114] and faces a maximum sentence of seven years in prison. In New Jersey the contractor can be convicted of theft[115] and be imprisoned up to three years; the punishment can include a 1,000-dollar fine. Although the penal sanction is rarely used,[116] its existence should deter diversion and result in payments to subcontractors and suppliers.

New York law gives civil remedies to trust fund beneficiaries,[117] including the award of punitive

---

[101]West Ann.Cal.Civ.Code § 3097.
[102]Id. at § 3098.
[103]Id. at §§ 3083, 3162.
[104]Ibid.
[105]Id. at § 3103.
[106]Id. at §§ 3161, 3186.
[107]Id. at § 3161.
[108]Id. at § 3103.
[109]Id. at § 3167.
[110]Id. at § 3171.
[111]Id. at §§ 3197–3205.

[112]For a study of Alberta (Canada), see Ettinger, *Trusts in the Construction Industry,* 27 Alberta L.Rev. 390 (1989).
[113]One ignored aspect of this theory is whether the subcontractor is entitled to interest for the period the funds are held in trust. See Section 22.03.
[114]N.Y.—McKinney's Lien Law § 79-a. A 1987 amendment makes paying with a bad check an inference payment was not intended establishing an intent to commit larceny.
[115]Ann.N.J.Stat. 2C:20-9.
[116]*People v. Van Keuren,* 31 A.D.2d 711, 295 N.Y.S.2d 892 (1968).
[117]N.Y.—McKinney's Lien Law § 77(1).

damages.[118] Oklahoma allows the additional possibility of pursuing a civil action directly against certain officers of any diverting corporation.[119] Some states allow a civil remedy without express language in the statute creating such a remedy.[120]

Rather than employ the trust fund designation, some states directly designate diversion as a crime. For example, under certain circumstances, the California Penal Code makes willful diversion of amounts over $10,000 punishable by up to five years' imprisonment and a fine of up to $10,000. Diversion of amounts under $10,000 is a misdemeanor punishable by imprisonment not to exceed six months or a fine not exceeding $500 or both. The California Penal Code makes it a crime to submit a false voucher to obtain construction loan funds.[121]

California narrowly interpreted the statute, rendering it relatively ineffective.[122] Yet the owner may find it useful to insert the language of the statute in a contractor's application for payment form.

G. Michigan Homeowners Construction Lien Recovery Fund

In 1982, Michigan enacted legislation that avoids a residential owner paying twice (owner pays prime and then pays the unpaid subcontractor to remove the latter's lien).[123] At the same time, Michigan allowed a licensed subcontractor who could not receive a lien because the prime contractor had been paid to recover from a state-administered fund created by fees paid by all licensed contractors. The subcontractor who seeks to recover from the fund must establish that it would have had a lien, that payments were made to its prime, that the prime retained the funds, and that the subcontractor demanded payment from the prime. A cap of $75,000 was placed on claims against the fund for any residence.

H. Texas Trapping Statute

The Texas "trapping statute" is designed to aid subcontractors and suppliers. It allows an owner to withhold and pay a subcontractor directly if the latter notifies the owner that it has not been paid and the prime does not object.[124]

I. Compulsory Bonding Legislation

Reference has been made to the general unavailability of lien remedies against public structures.[125] To compensate for this and to encourage work on public projects, most public work—federal and state—requires that prime contractors post payment bonds to give protection to subcontractors and suppliers. Statutes and case decisions vary as to how far down the line such protection exists.

Like mechanics' lien laws, the compulsory bonding legislation has its pitfalls. Again, rules determine who must file notices,[126] to whom they must be sent, and what they must state, as well as requirements relating to when the lawsuit must be filed. On the whole, with their deficiencies, payment bonds on public work are a more effective remedy for subcontractors and suppliers than are mechanics' liens.[127]

J. Nonstatutory Claims

Subcontractors and suppliers who do not receive payment for work performed sometimes seek remedies against third parties not based on statutes of the type described earlier in the section. Instead or alternatively, they seek recovery based on a variety of theories against a number of parties other than the party with whom they have made the construc-

---

[118]*Sabol & Rice, Inc. v. Poughkeepsie Galleria*, 175 A.D.2d 555, 572 N.Y.S. 811 (1991). In this case, the owner diverted $28 million to its partners and did not pay its contractors. Even though punitive damages are not specified in the statute, they were allowed in this case. The court noted that the conduct would have been sufficient to justify criminal charges of larceny.
[119]Ann.Okl.Stat. tit. 42, § 153.
[120]*B. F. Farnell Co. v. Monahan*, 377 Mich. 552, 141 N.W.2d 58 (1966); *Hiller & Skoglund, Inc. v. Atlantic Creosoting Co.*, 40 N.J. 6, 190 A.2d 380 (1963).
[121]West Ann.Cal.Penal Code §§ 484b, 484c.
[122]*People v. Butcher*, 185 Cal.App.3d 929, 229 Cal.Rptr. 910 (1986).
[123]Ann.Mich.Comp.Laws §§ 570.1201, 570.1203.

[124]Applied in *Don Hill Constr. Co. v. Dealers Elec. Supply Co.*, 790 S.W.2d 805 (Tex.Ct.App.1990).
[125]Some states have what are called mechanics' lien statutes that apply to public work. See Ky.Rev.Stat. § 376-210. These statutes do not give the lien holder the right to foreclose on the public work. The remedy is against the funds for the project in the hands of the public agency similar to a stop notice.
[126]Generally, Miller Act notice requirements need not be complied with strictly. *United States v. Merle A. Patnode Co.*, 457 F.2d 116 (7th Cir.1972).
[127]See Note, 51 Miss.L.J. 351 (1980–81).

tion contract. To illustrate, an unpaid subcontractor may seek recovery for work performed from the owner, the lender, or a design professional. Similarly, a claim may be made by another party farther down on the subcontract chain, such as a second-tier subcontractor against all parties on the contract chain other than the first-tier subcontractor with whom the contract was made.

Nonstatutory claims can be further divided into specific remedies and ordinary court judgments. An illustration of a specific remedy is a claim against funds in the hands of the owner or lender earmarked for the project. An illustration of other remedies is a claim generally against the owner, lender, or design professional that would result in an ordinary judgment and not a claim against specific property.

Claims to specific funds such as funds in the hands of a lender or owner are usually based on the assertion of an equitable lien—an equitable remedy based on unjust enrichment. Such claims generally point to the claimant's having improved the property of the defendant owner or having improved the land in which the lender has a security interest.

Despite the availability of the statutory stop-notice remedy in California, until 1967, the equitable lien was successfully asserted by claimants against California construction lenders. Claimants were allowed to recover against the unexpended construction loan funds if the claimant, typically a subcontractor, had relied on the funds when it entered into the subcontract.[128]

In 1967, the California legislature enacted California Civil Code § 3264, which shut the door on equitable liens by making the stop-notice statutory remedy exclusive against construction loan funds.

Although California made the statutory remedies exclusive, some courts grant equitable liens on construction loan funds held by owners,[129] by lenders,[130] or on the retainage.[131]

Although the law is somewhat unsettled, it seems likely that claimants such as unpaid subcontractors or suppliers or those sureties that have had to pay such parties will be able to recover from the design professional if the latter did not perform in accordance with the professional standards established by the law when issuing payments to the prime contractor.[132] Alternatively, if the claimant can show that the design professional breached the contract with the owner and that this contract was in part for the claimant's benefit, recovery is possible. These burdens have been difficult to sustain.[133] Claims against owners and lenders are based on similar theories and have had some, but limited, success.[134]

Gee v. Eberle, 279 Pa.Super. 101, 420 A.2d 1050 (1980), granted recovery to a subcontractor against a lender on an equitable lien theory despite the recognition that this was a minority view. (This holding was sharply limited by the Pennsylvania Supreme Court in *D. A. Hill Co. v. CleveTrust Co. Investors*, 524 Pa. 425, 573 A.2d 1005 (1990). See generally Reitz, *Construction Lenders' Liability to Contractors, Subcontractors, and Materialmen*, 130 U.Pa.L.Rev. 416 (1981).

[131]*Williard, Inc. v. Powertherm Corp.*, 497 Pa. 628, 444 A.2d 93 (1982).

[132]See Section 22.07. But see Sections 14.08(B) through (F).

[133]*Engle Acoustic & Tile, Inc. v. Grenfell*, supra note 76, denied recovery against owner and architect. Similarly, a contractor sued an architect in contract and in tort and was unsuccessful. *C. H. Leavell & Co. v. Glantz Contracting Corp.*, supra note 28. On the other hand, an architect was held liable to a contractor for economic loss in *United States v. Rogers & Rogers*, 161 F.Supp. 132 (S.D.Cal.1958). The confusion in this area can be shown by *A. R. Moyer, Inc. v. Graham*, 285 So.2d 397 (Fla.1973), which held that a contractor could sue the architect in tort but not as the intended beneficiary of the owner-architect contract.

[134]Subcontractors were unsuccessful in *Helash v. Ballard*, 638 F.2d 74 (9th Cir.1980) (against owner); *Urban Systems Dev. Corp. v. NCNB Mortgage Corp.*, 513 F.2d 1304 (4th Cir.1975) (against lender); *United States Fid. & Guar. Co. v. United States*, 201 Ct.Cl. 1, 475 F.2d 1377 (1973) (against owner); *Stratton v. Inspiration Consolidated Copper Co.*, 140 Ariz. 528, 683 P.2d 327 (Ariz.App.1984) (against owner); *Sofias v. Bank of America N&TS Ass'n.*, 172 Cal.App.3d 583, 218 Cal.Rptr. 388 (1985) (sub's claim as intended beneficiary against lender unsuccessful); *Rochelle Vault Co. v. First National Bank*, 5 Ill.App.3d 354, 283 N.E.2d 336 (1972) (against owner); *Engle Acoustic & Tile Co. v. Grenfell*, supra note 76 (against owner); *Haggard Drilling, Inc. v. Greene*, 195 Neb. 136, 236 N.W.2d 841 (1975) (against owner); and *Seegers v. Sprague*, 70 Wis.2d 997, 236 N.W.2d 227 (1975) (against owner). In addition to the equitable lien cases cited supra at notes 129 and 130, *Votaw Precision Tool Co. v. Air Canada*, 60 Cal.App.3d 52, 131 Cal.Rptr. 335 (1976), seemed willing to grant recovery on the third-party beneficiary theory, but the facts did not justify a recovery in

---

[128]*McBain v. Santa Clara Sav. & Loan Ass'n*, 241 Cal.App.2d 829, 51 Cal.Rptr. 78 (1966).

[129]*Avco Delta Corp. v. United States*, 484 F.2d 692 (7th Cir.1973). Another case that found an equitable lien was *Active Fire Sprinkler Corp. v. United States Postal Service*, 811 F.2d 747 (2d Cir.1974).

[130]*Trans-Bay Eng's & Builders, Inc. v. Hills*, 551 F.2d 370 (D.C.Cir.1976) (prime contractor recovered from lender and HUD); *Bankers Trust Sav. & Loan Ass'n v. Cooley*, 362 F.Supp. 328 (N.D.Miss.1973) (sub recovered from lender).

One differentiation between claims brought against design professionals and those brought against owners is that claims against the latter can be based on unjust enrichment. Claimants assert that their work or materials have benefited the owner and that it would be unjust for the owner to retain this benefit without paying the claimants.

As a rule, such claims have not been successful because the owner can show that it has paid someone, usually the prime contractor, or that the retention of benefit was not unjust because the claimant could have protected itself by using the statutory remedies.[135] One case granted recovery where the owner had not paid anyone, with the prime contractor's having left town before being paid.[136]

The subcontractor's reliance on a promise made by the owner's authorized agent that the prime contractor would be required to file a surety bond was the basis for a successful subcontractor claim against the owner.[137]

With the exception of an occasional recovery based on reliance on a direct promise or unjust enrichment, claimants such as subcontractors and suppliers have had only limited success against owners. Subcontractors must use statutory remedies or bonds or they will be left to whatever claim they have against the party with whom they dealt directly.

Second-tier subcontractors sometimes, like first-tier subcontractors, look up the subcontract chain for a solvent party to pay for work when the party who has promised to do so does not. In *Friendly Ice Cream Corp. v. Andrew E. Mitchell & Sons, Inc.*,[138] an unpaid second-tier subcontractor asserted claims based on a number of theories against a prime contractor but was unsuccessful. But a third-tier subcontractor was allowed to recover against a first-tier sub based on the latter's promise to the prime contractor to pay all suppliers.[139]

The preceding discussion should indicate the importance of dealing with a party who has the desire and financial capability to pay for work that is ordered. Obviously, a subcontractor's principal concern would be the financial responsibility of the prime contractor with whom it has dealt. Yet the panoply of legislative protection and the occasional nonstatutory relief given indicates that such parties frequently are not paid and seek other methods to collect.

K. Joint Checks

One way owners and prime contractors seek to avoid liens and other claims is to issue joint checks. For example, the owner may issue a joint check to the prime and to the subcontractors, while the prime contractor may issue joint checks to subcontractors and their sub-subcontractors and suppliers. This was discussed in Subsection 22.02(J).

SECTION 28.08 Other Subcontractor Claims

A. Against Prime Contractor

Suppose a subcontractor asserts that its cost of performance was wrongfully increased because of acts

---

the case. Subcontractors were successful in a claim against a public corporation lender that did not require the prime contractor to post a required bond that would have protected the subcontractor. *New England Concrete Pipe Corp. v. D/C Systems of New England, Inc.*, 495 F.Supp. 1334 (D.Mass.1980), vacated on jurisdictional grounds, 658 F.2d 867 (1st Cir.1981). But see *Haskell Lemon Constr. Co. v. Independent School Dist.*, 589 P.2d 677 (Okl.1979). A special statute justified a recovery by a subcontractor against the owner in *Sweetman Constr. Co., Inc. v. South Dakota*, 293 N.W.2d 457 (S.D.1980). *Seither & Cherry Co. v. Illinois Bank Building Corp.*, 95 Ill.App.3d 191, 419 N.E.2d 940 (1981), allowed a subcontractor to recover against the owner by concluding that the construction manager acted as an agent of the owner. A subcontractor was successful in *Port Chester Elec. Constr. Corp. v. Atlas*, 40 N.Y.2d 652, 389 N.Y.S.2d 327, 357 N.E.2d 983 (1976), where owner and prime contractor were all controlled by one person. A subcontractor successfully sued the owner for the latter's negligence in *Berkel & Co. Contractors, Inc. v. Providence Hospital*, 454 So.2d 496 (Ala.1984). In *Bituminous Constr., Inc. v. Rucker Enter.*, 816 F.2d 965 (4th Cir.1987), the subcontractor recovered from the owner who had promised to issue joint checks but did not.

[135] *Garwood & Sons Constr. Co., Inc. v. Centos Assoc. Ltd. Partnership*, 8 Conn.App. 185, 511 A.2d 377 (1986); *Lee Bros. Contractors v. Christy Park Baptist Church*, 706 S.W.2d 608 (Mo.App.1986); *Seegers v. Sprague*, supra note 134. Similarly, an excavation subcontractor's claim against an adjacent landowner based on unjust enrichment was unsuccessful in *D. C. Trautman Co. v. Fargo Excavating Co.*, 380 N.W.2d 644 (N.D.1986).

[136] *Costanzo v. Stewart*, 9 Ariz.App. 430, 453 P.2d 526 (1969).

[137] *Bethlehem Fabricators, Inc. v. British Overseas Airways Corp.*, 434 F.2d 840 (2d Cir.1970) (promise made by architect). But see *Greenville Indep. School Dist. v. B&J Excavating, Inc.*, 694 S.W.2d 410 (Tex.Ct.App.1985).

[138] 340 A.2d 168 (Del.Super.1975). Similarly, see *Insulation Contracting & Supply v. Kravco, Inc.*, 209 N.J.Super. 367, 507 A.2d 754 (1986).

[139] *Aetna Cas. & Sur. Co. v. Kemp Smith Co.*, 208 A.2d 737 (D.C.App.1965). Similarly, a second-tier subcontractor successfully sued a prime contractor in *Merco Mfg. Inc. v. J. P. McMichael Constr. Co.*, 372 F.Supp. 967 (W.D.La.1974).

or omissions by the prime contractor or someone for whom the prime contractor is responsible. Generally, the law implies an obligation on the part of the prime contractor to take reasonable measures to ensure that the subcontractor can perform expeditiously and is not unreasonably delayed.[140] However, terms will not be implied if express provisions in the subcontract directly deal with this matter. Commonly subcontracts, amplified by relevant provisions in prime contracts incorporated into the subcontract, deal in some way with this problem. Express provisions can place such a responsibility on the prime contractor, such as the contract requiring that the site be ready by a particular date or that the work be in a sufficient state of readiness by a designated time to enable the subcontractor to perform specific work.

Conversely, contract provisions may indicate that the subcontractor has assumed certain risks regarding the sequence of performance. For example, the subcontract might require that the work be performed "as directed by the prime contractor," and such a provision would give the prime contractor wide latitude to determine when the subcontractor will be permitted to work.

Additional clauses in the prime contract to which the subcontract refers or that are incorporated into the subcontract are also relevant. The contract can specify that certain delay-causing risks were contractually assumed risks or grant only a time extension.[141]

In addition, contract clauses may control by denying recoverability of delay damages. For example, a no-damage clause that limited the subcontractor to a time extension in the subcontract, especially if tracked with a no-damage clause in the prime contract, would very likely preclude recovery of delay damages by the subcontractor against the prime contractor for delays caused by owner or prime contractor.[142]

The problems of tracking or parallelism in prime contracts or subcontracts are a complicating factor in these claims. As a general rule, prime contracts and subcontracts are parallel in terms of rights and responsibilities. For example, should a prime contractor be held liable for delay to the subcontractor for acts caused by the owner when the prime contractor is precluded from recovering from the owner because of a no-damage clause?

The bargaining situation sometimes permits the prime contractor to better its position in the subcontract. For example, the prime contractor might be able to include a no-damage clause in the subcontract when the prime contract allows the prime contractor its delay damages against the owner. Although a careful subcontractor might be able to preclude this, often time or realities of the process frustrate parallel rights. An abuse of prime contractor bargaining power can result in discontented subcontractors and poor performance. It is becoming increasingly common for owners to insist that prime contracts contain provisions requiring the prime contractor to give the subcontractor benefits parallel to those given the prime contractor in its contract with the owner. (See Section 28.04.)

Other disputes between prime and subcontractor can arise. For example, a case that sharply divided the Kansas Supreme Court involved a claim by a subcontractor against the prime based on the latter having suggested a modification in the contract with the owner that would have essentially eliminated all of the subcontractor's performance. The court over a dissent held that this suggestion did not violate the obligation of good faith and fair dealing owed by the prime to the subcontractor.[143]

B. Pass-Through Claims Against Owner: Liquidating Agreements

When a project is disrupted or delayed by the owner or other responsible entity, all participants incur expenses. Frequently, they seek to transfer these expenses to the owner. Because of the high costs of pursuing claims, subcontractors often will pool their claims with those of the prime, and the prime will present them against the owner. However, even where there is an agreement under which the subcontractors will take whatever the

---

[140]Annot., 16 A.L.R.3d 1252, 1254 (1967). See also Section 19.02(A).

[141]*Loughman Cabinet Co. v. C. Iber & Sons, Inc.*, 46 Ill.App.3d 873, 361 N.E.2d 379 (1977) (subcontractor denied delay damages where prime contractor granted only time extension).

[142]*McDaniel v. Ashton-Mardian Co.*, 357 F.2d 511 (9th Cir.1966) (subcontractor assumed the risk of government-caused delays). But in *J. J. Brown Co. v. J. L. Simmons Co.*, 2 Ill.App.2d 132, 118 N.E.2d 781 (1954), a subcontractor recovered delay damages from the prime contractor when the former was delayed by another subcontractor.

[143]*Meier's Trucking Co. v. United Constr. Co.*, 237 Kan. 692, 704 P.2d 2 (1985).

prime gets for them, the prime must pursue the subcontractors' claims in good faith.[144]

One mechanism for accomplishing this is a liquidating agreement. Under such an agreement, the prime confesses liability to the subcontractor for owner-caused delay, the subcontractor releases the prime from all other liability, and the subcontractor is relegated to whatever delay damages the prime can recover from the owner.

In addition to saving costs in authorizing the prime to handle the subcontractor's claim, such pass-through or liquidating agreements avoid the problem of the subcontractor not being in privity of contract with the owner. Usually, to make a claim against the owner, the prime must show that it has suffered a loss that it seeks to transfer to the owner. Usually this is done by the prime showing either that it has reimbursed the subcontractor for the latter's damages or that it remains liable for such reimbursement. Sometimes, as noted, the liquidating agreement is drafted so as to make the contractor liable to the subcontractor but only as, when, and to the extent that the contractor recovers from the owner based upon the subcontractor's claim. This is similar to the Severin doctrine discussed in (C).

Such pass-through claims have generated litigation. For example, in one case, the liquidating agreement obligated the contractor to pay the subcontractor 9.5% of the first twenty million dollars recovered and 5% of any excess. Combined with a partial payment, this formula left the contractor responsible for passing through less than the face amount of the subcontractor's claim. The court rejected the owner's attempt to limit its liability to the prime to the amount the prime would pay its subcontractor under the liquidating agreement. The court noted that the prime's liability to its subcontractor and therefore the prime contractor's right to recover on the subcontractor's claims are not connected to the agreed division of the recovery.[145]

While this should encourage such pass-through or liquidating claims, a decision by the New York Court of Appeals created grave doubts about the utility of liquidating agreements. In that case, the subcontractor sued a separate prime contractor for delays caused primarily by the owner, its engineer, other prime contractors, and bad weather. At the end of the subcontractor's case, the court granted the separate prime contractor's motion to dismiss, concluding that the subcontractor would not be able to recover damages from the prime and that the prime did not cause the loss over which it had no control.[146]

While this result is in no way unusual, it must be looked at in the light of liquidating agreements under which the prime confesses liability for owner-caused delay and the subcontractor releases the prime from all other liability and relegates itself to whatever delay damages the prime can recover from the owner. If the subcontractor cannot recover from its prime, it is difficult to see how the prime can "confess" liability to its subcontractor and, if so, how a prime can recover from an owner for subcontractor claims for which the prime is not independently responsible to the subcontractor. This again raises the problem of the Severin doctrine (discussed in (C)).

C. Federal Procurement and the Severin Doctrine

A significant but not insurmountable barrier to subcontract recovery in federal procurement is the Severin doctrine. Subcontractors cannot bring direct action against a federal agency owner because of an absence of a contractual relationship between subcontractor and the agency.[147] As a rule, subcontractor claims against federal agencies, usually for extra work, are brought by the subcontractor in the name of and with consent of the prime contractor. The Severin doctrine[148] precludes the prime contractor from bringing an action against the government for damages suffered by the subcontractor if the prime contractor is not liable to the subcontractor because the subcontract had absolved the prime contractor. However, the courts have not

---

[144]*T.G.I. East Coast Constr. Corp. v. Fireman's Fund Ins. Co.*, 534 F.Supp. 780 (S.D.N.Y.1982).

[145]*Frank Briscoe Co. v. County of Clark*, 772 F.Supp. 513 (D.Nev.1991).

[146]*Triangle Sheet Metal Works, Inc. v. James H. Merritt & Co.*, 79 N.Y.2d 801, 588 N.E.2d 69, 580 N.Y.S.2d 171 (1991).

[147]*Clifton D. Mayhew, Inc. v. Blake Constr. Co.*, 482 F.2d 1260 (4th Cir.1973).

[148]*Severin v. United States*, 99 Ct.Cl. 435 (1943), cert. denied 322 U.S. 733 (1944).

been comfortable with this limitation, and many exceptions have been developed that have enabled the subcontractor, through the prime contractor, to process claims against the government even though the prime contractor would not have to respond to the subcontractor unless the claim against the U.S. was successful.[149]

D. Against Third Parties

Suppose the costs of performance are increased because of acts of third parties such as owners or design professionals. In addition to making any claim that might be available against the prime contractor, suppose a subcontractor seeks legal relief directly against those third parties who it asserts caused the loss. For example, Section 28.07(J) stated that subcontractors sometimes seek payment against third parties such as owners or design pro-

fessionals when the prime contractor does not pay them. Claims for additional cost of performance are not treated as favorably as claims to be paid for work performed. Even the latter claims are difficult to sustain against third parties. Very likely, claims against third parties for increased cost of performance will meet more severe resistance. However, a subcontractor who can persuade a court that the third party did not live up to the legal obligation imposed by law and thereby committed a tort or that the third party promised to perform in a certain way to the prime contractor and that this promise was for the benefit of the subcontractor can recover against third parties.[150]

---

[149]*Keydata Corp. v. United States*, 205 Ct.Cl. 467, 504 F.2d 1115 (1974).

---

[150]The owner who had negotiated a contract with subcontractors and then assigned the contract to a prime contractor was held liable to a subcontractor for the owner's change of the critical path method. See *Natkin & Co. v. George A. Fuller Co.*, 347 F.Supp. 17 (W.D.Mo.1972). See also cases cited in notes 138–139 supra and Section 14.08(B).

PROBLEMS

1. Prime contractor and subcontractor enter into a contract under which the prime contractor would pay "as payments are received from the owner." The subcontractor completed its work, but the prime contractor was not paid because the owner went bankrupt. Should the prime contractor be obligated to pay the subcontractor?

2. Prime contractor and subcontractor enter into a subcontract that incorporated all provisions of the prime contract not directly inconsistent with the subcontract. Prime contractor and subcontractor disputed whether particular performance by the subcontractor was in conformity with the plans and specifications. The prime contractor pointed to a provision in the prime contract stating that the architect would be the judge of performance and her decisions would be final. No provision relating

to judging performance could be found in the subcontract. However, the subcontractor claimed that the architect should not judge, as she was biased, because the subcontractor's claim was that the specifications were unclear and this was the responsibility of the architect.

When the subcontractor had bid on the job, the prime contract general conditions were part of the bidding package, but the subcontractor did not read them. The reason given for not reading them at that time was that it expected to be given a copy before it entered into the subcontract but was told at the contract signing that it was unavailable. Should the prime contract provision making the architect the judge of performance control the dispute between prime contractor and subcontractor?

Design Professional as Judge: Tradition Under Attack

SECTION 29.01 Relevance

In many construction projects, the design professional makes decisions affecting the rights and duties of owners, contractors, and subcontractors. Usually the power to make these decisions is created by the contract between owner and contractor. For example, the design professional interprets the contract documents and determines whether performance complies. To do this, the design professional is given the right to make inspections and to uncover work. The design professional has the right to grant time extensions, approve substitution of subcontractors, and pass on the sufficiency of shop drawings submitted by the contractor. The design professional plays an important role in terminating the contractor's right to perform, having either the power to terminate or, even where the owner is given that power, the power to issue a certificate that cause exists for termination.[1]

Not all the illustrations given in the preceding paragraph apply to all construction contracts. Because of increased liability, there has been a slight tendency to reduce these powers in contract administration. For example, standard contracts published by the American Institute of Architects (AIA) and the Engineers Joint Contract Documents Committee (EJCDC) no longer give the design professional the power to order that the work be stopped. Generally, the design professional has a "judging" role in most construction projects. This chapter deals with the legal aspect of this role.

SECTION 29.02 Doctrine of Conditions

Parties to a contract can condition specific obligations on the occurrence of designated events. Normally, unless these events occur or are excused, the duty to perform does not arise or ceases to exist.

Commonly, parties to a construction contract give the design professional power to make certain decisions. Such third-party decisions can be expressed as conditions. For example, if a certificate issued by the design professional conditions payment, issuance of that certificate or the existence of facts that excuse issuance is necessary before the owner is obligated to pay.

Although a decision of a third party can be a valid condition, there are two aspects of the circumstances under which this is created in construction contracts that will bear on legal treatment of such conditions. First, the owner pays, and for all practical purposes selects, the design professional. Under these circumstances, one basic element of a decision-making process—that of an impartial judge—may not be present in the construction contract dispute process. Second, in many construction contracts, the contractor, and even more so the subcontractor, has no real choice but to accept the decision-making process under which the design professional is given broad powers.

As shall be seen,[2] the law accords a degree of finality to decisions made by the design professional. But the degree to which these two factors cast doubt on the impartiality of the design profes-

[1]AIA Doc. A201, ¶ 14.2.2.

[2]See Section 29.09.

sional may cause courts to scrutinize these decisions carefully.

SECTION 29.03   Excusing the Condition

Under certain circumstances, the condition can be excused. If so, any disputes between owner and contractor would have to be negotiated by the parties, submitted to arbitrators if the parties have agreed to use this process, or decided by a court.

The condition is excused if the design professional becomes unavailable or unable[3] to interpret the contract or judge its performance. The construction contract can provide for a successor design professional when the one originally designated cannot or will not perform this function. For example, AIA Doc. A201, ¶ 4.1.3, allows the owner to appoint a successor "against whom the Contractor makes no reasonable objection." Even if no successor mechanism is specified in the contract, the parties can agree to use a replacement design professional. But if there were neither an agreement nor a mechanism for a successor design professional, any obligation conditioned on the issuance of a certificate or the resolution of a dispute by the design professional would be unconditional and the condition no longer be part of the contract.

Other acts can excuse the condition. For example, one court excused the condition when the architect examined the work, found it satisfactory, but still refused to issue the certificate.[4] Similarly, another court held that the issuance of a payment certificate had been waived because the architect failed to reinspect the work and specify which terms remained to be corrected.[5] The condition was excused in a private contract for development of a road when the city engineer, whose approval was a condition, simply refused to act.[6]

Clearly, collusion between the design professional and either owner or contractor excuses the condition.[7] A condition can be excused if the parties agree to eliminate it.[8]

The condition can be excused if the party for whom the condition is principally inserted manifests an intention to perform its obligations despite nonoccurrence of the condition. Although one case held that the condition of a certificate is for the owner's benefit and the owner can waive it,[9] there is no reason why such a condition cannot be considered for the benefit of *both* parties, particularly where the contractor establishes that it entered into the contract on the assumption that a *particular* design professional would make these decisions.

SECTION 29.04   Design Professional as Judge: Reasons

To a continental European, the design professional as judge seems incongruous. With the exception of Great Britain, European construction administration might include the design professional giving his interpretation of design documents he drafted. He would not be given "a judging role." The close association between owner and design professional based on the former's selection and payment of the latter precludes this role being given to the latter.

Despite the close association between owner and design professional, most American construction contracts, both public and private, give the design professional broad decision-making powers.[10] A number of reasons can be given for the development and continuation of this system.

First, the stature and integrity of the design professions may give both parties to the construction contract confidence that the decisions will reflect technical skill and basic elements of fairness.[11]

---

[3]*United States v. Klefstad Engineering Co.*, 324 F.Supp. 972 (W.D.Pa.1971) (surveyor lost records); *Grenier v. Compratt Constr. Co.*, 189 Conn. 144, 454 A.2d 1289 (1983); *Manalili v. Commercial Mowing & Grading*, 442 So.2d 411 (Fla.Dist.Ct.App.1983).

[4]*Anderson-Ross Floors, Inc. v. Scherrer Constr. Co.*, 62 Ill.App.3d 713, 379 N.E.2d 786 (1978).

[5]*Hartford Elec. Applicators of Thermalux, Inc. v. Alden*, 169 Conn. 177, 363 A.2d 135 (1975).

[6]*Grenier v. Compratt Constr. Co.*, supra note 3.

[7]*Metro. Sanitary Dis. of Greater Chicago v. Anthony Pontarelli & Sons, Inc.*, 7 Ill.App.3d 829, 288 N.E.2d 905 (1972).

[8]*Steffek v. Wichers*, 211 Kan. 342, 507 P.2d 274, 281–82 (1973).

[9]*Halvorson v. Blue Mountain Prune Growers Co-op*, 188 Or. 661, 214 P.2d 986 (1950).

[10]The history is described in Dreifus, *The "Engineer Decision" in California Public Contract Law*, 11 Pub.Cont. L.J. 1 (1979).

[11]See *Zurn Eng'rs v. State Dep't of Water Resources*, 69 Cal.App.3d 798, 138 Cal.Rptr. 478 (1977). One reason for enforcing such a mechanism was the court's belief that the parties had mutually agreed to accept the expertise of the state engineer rather than the lay opinion of a judge or jury.

Second, the design professional's role in design before construction equips the design professional with the skill to make decisions that will successfully implement the project objectives of the owner. In a sense, the role as interpreter and judge is a continuation of design.

Third, owners are often unsophisticated in matters of construction and need the protection of a design professional to obtain what they have been promised in the construction documents. Implicit is the assumption that without a champion and protector, the owner might be taken advantage of by the contractor. Couple this with the owner's bargaining strength, and the present system results.

Fourth, even assuming that complete objectivity is lacking and that the contractor rarely has much choice, the alternative would be worse. Suppose the design professional did not act as interpreter and judge. Such matters would have to be resolved by owner and contractor which for many owners would require professional advice. If owner and contractor cannot agree, the complexity of construction documents and performance will necessitate many costly delays, because the alternative forums for owners and contractors who cannot agree—litigation and arbitration— still involve time and expense. The present system, with its deficiencies, is probably more efficient than any alternative.

Fifth, despite the dangers of partiality and conflict of interest (the design professional may overlook defective workmanship to induce the contractor not to press a delay claim or a claim for extras based on the design professional's negligence, or the design professional may find a defect due to poor workmanship rather than expose himself to a claim for defective design), the system seems to have worked well. Arbitrations in construction disputes are rare, and litigation is even more rare. This can mean genuine satisfaction with the system or the recognition that the alternatives are worse. Perhaps a quick decision that may at times be unfair is better than a more costly, cumbersome system that *might* give better and more impartial decisions.[12]

The potential for bias toward the owner[13] and decision making often involving the design professional's own prior design work are bound to reflect themselves in the judicial treatment of the design professional's decisions and whether the design professional will be held responsible for decisions (discussed in Section 14.09(D)).

SECTION 29.05 Jurisdiction of Decision-Making Powers

The design professional's interpreting and judging functions are created by the contract between the owner and the contractor. As a result, interpretation of the contract determines whether the design professional has the jurisdiction to resolve the dispute. The contract will also govern the finality of the design professional's decision (discussed in greater detail in Section 29.09).

Generally, jurisdictional grants to decide disputes are likely to be broader in matters that involve the experience and expertise of the design professional.[14] Jurisdictional grants are likely to be broad where an on-the-spot decision during the operational phases, as opposed to post-completion disputes, needs to be resolved.[15] Despite these basic guidelines, decisions are likely to vary, depending on the particular language of the contract and judicial attitude toward this decision-making process.

In 1987, the AIA drastically revised the part of A201 that deals with the role of the architect in interpreting the contract documents and resolving disputes. First, a line was drawn between the architect's interpreting and deciding matters concern-

---

[12]*Cofell's Plumbing & Heating, Inc. v. Stumpf*, 290 N.W.2d 230 (N.D.1980).

[13]To avoid creating the false impression that the cases generally involve the design professional's refusing to award a certificate, it should be noted that many cases involve the owner's refusal to pay despite the design professional's having issued a certificate.

[14]But see *Paschen Contractors, Inc. v. John J. Calnan Co.*, 13 Ill.App.3d 485, 300 N.E.2d 795 (1973) (architect had *no* jurisdiction, as the subcontractor's claim involved architect's work); *Edward Elec. Co. v. Metro. Sanitary Dis. of Greater Chicago*, 16 Ill.App.3d 521, 306 N.E.2d 733 (1973) (jurisdiction based on "engineering nature" included performance costs).

[15]*Cofell's Plumbing & Heating, Inc. v. Stumpf*, supra note 12. See also *County of Rockland v. Primiano Constr. Co.*, 51 N.Y.2d 1, 409 N.E.2d 951, 431 N.Y.S.2d 478 (1980).

ing performance and the architect's resolving claims and disputes. The architect's jurisdiction to interpret and decide performance matters under ¶ 4.2.11 relates to the requirements of the contract documents. This can be very broad in the light of the contract documents covering many matters, some technical, some administrative, and some legal. The architect's jurisdiction to pass on claims and resolve disputes under ¶ 4.3 is even more broad, covering almost any dispute related to the project that could involve the owner and the contractor. Under ¶ 4.3.2, disputes must first (subject to some exceptions to be discussed later) go to the architect.

To test these provisions, a look at the increasingly frequent contractor claims based on owner-caused disruption or delay is instructive. Typically, the claims are made after the contract performance has been completed.[16] Must such a claim first go to the architect before it can be taken to arbitration or litigation?

The broad language in ¶ 4.3.2 would appear to require that such a dispute be submitted first to the architect. But the principal reason for giving the architect the power to interpret and to resolve disputes is to ensure that work continues while the dispute is being resolved by a trusted third party familiar with the context in which the dispute arises.[17] This policy would not be advanced by requiring the architect to first pass on a post-completion dispute.[18] Finding the position of the architect to be vacant—an exception under ¶ 4.3.2—avoids this improper use of the architect's dispute resolution power for the purpose of delay.

A201, ¶ 4.2.1, assumes that the architect will provide administration until final payment is due. However, that paragraph also permits the owner to extend the architect's service through the one-year correction of work period. If the owner extends the architect's performance under ¶ 4.2.1 in order to delay arbitration or litigation by demanding that disputes be first submitted to the architect, this would violate the obligation of good faith and fair dealing.[19]

The conflict of interest noted briefly in Section 29.04 must be addressed. Architects should not be given the power to interpret, judge performance, or resolve disputes if the competence of their own work is a serious issue.[20] It is better practice for architects to withdraw from such judging roles if they would be placed under a serious conflict of interest. Where this is a *serious* issue, they should seek permission from *both* parties before proceeding to interpret, judge performance, or decide disputes. Because the design professional's design and administrative functions are central, it would not be wise to push this conflict of interest problem too far. If it were pushed to its extreme, it would almost eliminate the design professional's judging function. Because that function is deeply ingrained in American Construction Process practices and seems to have worked tolerably well, design professionals should remove themselves only if it is quite clear that their work is *directly* being challenged.

As noted, ¶ 4.3.2 provides exceptions to claims and disputes first being resolved by the architect. Most exceptions deal with there being no architect or the architect not resolving the dispute in accordance with the required time deadlines. However, the fifth exception involves claims that "relate to a mechanics' lien."

If *any* claim that, if successful, would entitle the contractor to a mechanics' lien would be exempt from the disputes process required by ¶ 4.3, almost every claim by a contractor could bypass the process. This would make the process largely useless. The AIA did not intend such an outcome. This exclusion was intended to relate to issues that involve the technical *validity* of a mechanics' lien, such as the effect of a no-lien contract or the validity of lien releases, not the underlying substantive claim. These are technical issues that the architect should not decide.

SECTION 29.06 Who Can Make the Decision?

The interpreting and judging powers given the design professional by the contract are important, and

---

[16]It is assumed that the notice requirement of ¶ 4.3.3 has been met or waived.
[17]AIA Doc. A201, ¶ 4.3.4.
[18]See cases cited at note 15 supra.
[19]See Section 19.02(D).

[20]*Paschen Contractors, Inc. v. John J. Calnan Co.*, supra note 14. But A201, ¶ 4.3.2, specifically requires that claims that allege an error or omission by the architect first go to the architect.

the parties—particularly the owner—can expect those powers to be exercised by the particular person or persons in whom the design professional has confidence. If a design professional partnership is named, any principal partner should be able to make such interpretations and judgments. However, the owner may expect that the particular partner with whom it dealt or the partner with principal responsibility for the design would be the appropriate and suitable person to interpret and judge. An individual design professional who is empowered to make such determinations cannot delegate this power without the permission of the contracting parties.[21]

Suppose AIA Doc. A201 is the general conditions of the contract. Paragraph 4.1.1 assumes the architect may have an authorized representative. But this should not permit the designated architect to delegate the responsibility of interpreting the contract or resolving disputes, even to a person in his firm. A representative may be appropriate for some administrative purposes, such as receiving notices or even executing change orders. But delegating the highly personal service of giving interpretations or resolving disputes requires consent of owner and contract. This should not be affected by ¶ 4.1.1.

SECTION 29.07 The Contract as a Control on Decision-Making Powers

In addition to jurisdiction, there are other limits to the decision-making powers of design professionals. For example, a contractual power to grant time extensions usually provides standards for determining whether a time extension should be granted. The clause may set forth the events that will justify a time extension and the impact that these events must have on performance. Sometimes, as in AIA Doc. A201, ¶ 8.3.1, the design professional is given a residual power to determine whether time extensions should be granted. Only the existence of this power gives the design professional the authority to enlarge the list of events that justify a time extension.

Similarly, the power to decide disputes does not give the design professional authority to change the specifications. For example, in *Northwestern Marble & Tile Co. v. Megrath*,[22] the power to interpret the contract did not permit the design professional to require a subcontractor to install galvanized iron when the specifications permitted the installation of either galvanized iron or mild steel pipe.

SECTION 29.08 Procedural Matters

A. Requirements of Elemental Fairness: A201

The design professional as interpreter or judge invites comparison with arbitration and litigation. Should the design professional conduct a hearing similar to that used in arbitration or litigation? Clearly, the formalities of the courtroom would be inappropriate and unnecessary. Even the informal hearings conducted by an arbitrator would not be required. Unless a clause that confers jurisdiction requires a hearing, no hearing at all is necessary.

Cogent reasons exist for some semblance of elemental fairness to both parties. First, continued good relations on the project necessitate a feeling on the part of the participants that they have been treated fairly. Each party should feel it has been given a fair chance to state its case and be informed of the other party's position. A fair chance need not necessarily include even an *informal* hearing. The design professional should listen to the positions of each party where feasible before making a decision.

The exact nature of what is fair will depend on the facts and circumstances existing at the time the matter is submitted to the design professional. Small matters and those that require quick decisions may not justify the procedural caution that would be necessary where large amounts of money are at stake or where an urgent decision is less important.

The second reason for elemental fairness is the likelihood that a decision made without it will not be accorded much finality. For example, *John W. Johnson, Inc. v. Basic Construction Co.*[23] involved a

---

[21]*Huggins v. Atlanta Tile & Marble Co.*, 98 Ga.App. 597, 106 S.E.2d 191 (1958). But see *Atlantic Nat'l Bank of Jacksonville v. Modular Age, Inc.*, 363 So.2d 1152 (Fla.Dist.Ct.App.1978), where a different architect judged the quality of modular units, the delegation issue not having been raised. An A201 was used.

---

[22]72 Wash. 441, 130 P. 484 (1913).
[23]292 F.Supp. 300 (D.D.C.1968), affirmed 429 F.2d 764 (D.C.Cir.1970).

dispute in which the architect had ordered a prime contractor to terminate a particular subcontractor. The trial judge stated:

> This amazing directive was issued by the architect's office without notice to the plaintiff and without giving the plaintiff any opportunity to be heard, orally or in writing, formally or informally. This action on the part of the architect's office was contrary to the fundamental ideas of justice and fair play. The suggestion belatedly made at the trial that it was not appropriate for the architect to maintain any contacts with subcontractors is fallacious in this connection. Any such principle as that did not bar the architect's representative from according a hearing to the subcontractor before directing that his subcontract be cancelled.[24]

Termination is a serious matter, and more process fairness in such matters can be expected. But the case also reflects judicial unwillingness to uphold rash, impetuous decisions.

The third reason relates to the immunity sometimes given to design professionals when they act in a quasi-judicial role (explored in Section 14.09(D)). The more design professionals act like judges or arbitrators, the more likely they will be given judicial protection from suit by someone dissatisfied with the decision.

In 1987, the AIA greatly amplified the provisions in A201 for resolving disputes.[25] Detailed procedural directions and time deadlines were given. This was done to give greater guidance to architects and to resolve disputes as they develop to avoid their being "blown up" at the end of the project. Yet it is likely that these provisions will not be followed in many transactions—either because parties become impatient with the burdens and paperwork or because the transaction does not involve enough money to make it worthwhile incurring the overhead expenses such a system would inevitably entail.

B. Standard of Interpretation

Section 29.09 deals with the finality to be accorded decisions by the design professional. This section emphasizes the process from the vantage point of the design professional. How should relevant contract language be interpreted?

This treatise has dealt extensively with methods by which contracts—especially construction contracts—are to be interpreted.[26] Were it not for the fact that design professionals are often called on to interpret construction documents that they either have prepared or have had a role in preparing, the earlier discussion would suffice. However, the complexity of interpreting and judging one's own work requires some additional comments in this section.

In addition to the limits imposed by the contract language being interpreted, contracts that give interpreting and judging powers to the design professional often specify general standards of interpretation to be used by the design professional. For example, AIA Doc. A201, ¶ 4.2.12, requires that all "interpretations and decisions of the Architect . . . be consistent with the intent of and reasonably inferable from the Contract Documents." Despite the limits and guides, the design professional is often faced with the formidable task of deciding disputes over the meaning of contract document language. The prior participation in the development of the contract documents makes this difficult task even more difficult.

N.E. Redlon Co. v. Franklin Square Corp.[27] held that the architect must not take into account the intention of the owner or contractor or his intention at the time he drafted the documents in question. According to the court, the architect should first look at the contract documents to see whether they have provided guidance in interpretation matters. In *Redlon*, the contract specifications stated that terms were to be used in their trade or technical sense. After applying any contractual guides given, the architect was to use an objective standard to determine the meaning of contract language.

If an objective standard is desired, design professionals should not consider their or the owner's intention.[28] The court may have gone too far when

---

[24]Id. at 304.

[25]For more detail, see J. SWEET, *SWEET ON CONSTRUCTION INDUSTRY CONTRACTS:* §§ 11.8–11.18 (2d ed. 1992).

[26]See Section 20.02.

[27]89 N.H. 137, 195 A. 348 (1937).

[28]The architect's intention was overcome by other evidence in *Alabama Society for Crippled Children & Adults v. Still Constr. Co., Inc.*, 309 So.2d 102 (Ala.App.1975).

it directed the architect to ignore any intention the *contractor* may have had when it made the contract.

The design professional should judge the contract documents from the perspective of an honest contractor examining them before bid or negotiation. The preparation of the contract documents by the owner through the design professional takes a long time, much longer than given to the contractor to examine and bid. Any ambiguities that the contractor should not have been expected to notice and to which attention should not have been directed prior to bid submission or negotiation should be resolved in favor of the contractor. Conversely, unclear language to which the contractor should have directed attention should be resolved against it.

This conclusion should not be eliminated by a printed clause in the general conditions or specifications that seeks to put the risk of all unclear language in the contract documents on the contractor unless it brings this to the design professional's attention. For reasons described elsewhere,[29] this is an unfair burden to place on the contractor.

Admittedly, a standard that looks at the honest contractor can place design professionals in a difficult position where the language in question was derived from the drawings and specifications they prepared.[30] Openly acknowledging that the specifications were unclear can be a confession of professional failure. The standard of interpretation suggested—that of favoring the contractor under certain circumstances—can inhibit the design professional from ever finding language unclear for fear that it would reflect on his work. An honest design professional should be fair to both owner and contractor despite this possibility.

Perhaps it will be expecting too much for the design professional to step back from his own work and judge it objectively. This possibility may be a reason to accord less finality to the decision. In any event, if unwilling to construe unclear language in favor of the contractor, the design professional will likely use or claim to use the objective standard required in the *Redlon* decision.

C. Form of Decision

The contract clause giving the design professional the power to interpret documents can require that a particular form be followed when a decision is made. AIA Doc. A201, ¶ 4.2.12, requires that interpretations be in writing or in the form of drawings. For this and other reasons, it is generally advisable for decisions to be made in writing and communicated as soon as possible to each party. Where it is not feasible to make a decision in writing on the spot, any oral decision should be confirmed in writing and sent by a reliable means of communication to each party. A written communication giving the design professional's interpretation of his decision need not give reasons to support the decision. The essential requirement is that the parties know that a decision has been made and that they know the nature of the decision. However, the process will work more smoothly if the participants are given a reasoned explanation for the decision. The decision need not be elaborate or detailed but should specify the relevant contract language and facts and the process by which the decision has been made.

If A201 is used, the architect can create a thirty-day time limit for demanding arbitration by stating in his written decision that the decision is final but subject to arbitration and that a demand must be made within thirty days after receipt of the final written decision.[31] However, that paragraph does not *require* that this mechanism be employed. If it is not, arbitration must be demanded within a reasonable time.[32]

The refusal to require such a statement was based upon a desire to discourage arbitration of decisions and judicializing the design professional decision-making process.

D. Appeal

Some construction contracts require arbitration, such as A201, ¶ 4.5.1, or some other dispute resolution process. These dispute resolution systems are discussed in Chapter 30.

---

[29]See Section 24.03.

[30]The discussion assumes that the language in question was not part of the basic contract or general conditions, writings that should have been drafted or supplied by an attorney. See Sections 12.07, 19.01(B), and 19.01(D).

[31]AIA Doc. A201, ¶ 4.5.4.1.

[32]Id. at ¶ 4.5.4.2.

E. Costs

Whether the design professional will be compensated for interpreting the contract and resolving disputes depends on the contract between the owner and the design professional. Generally, this is considered part of basic services and does not earn additional compensation. However, in 1987, AIA Doc. B141 made drastic changes in the demarcation between basic and additional services and added ¶ 3.3.7, making "evaluating an extensive number of claims" a contingent additional service. The Engineers Joint Contracts Document Committee (EJCDC) provides in its No. 1910-1, ¶ 3.2.6, that evaluating "an unreasonable claim or an excessive number of claims" is a required additional service justifying additional compensation above the basic fee. This is a more precise standard than that of AIA.

The costs incurred by the parties, such as transporting witnesses to any informal hearing, obtaining any expert testimony, or attending any informal hearings, will be borne by the parties who incur them unless the contract provides otherwise. As most dispute resolution done by the design professional is informal, costs of this sort are not likely to be comparable to those that will be incurred in an arbitration or litigation.

Sometimes judging performance requires the design professional to order that work be uncovered. Uncovering work is costly. Often contracts specify who will pay for the cost of uncovering and recovering. AIA Doc. A201 is an illustration. Under ¶ 12.1.1, if work had been *improperly* covered by the contractor, such as covering work despite a request by the architect that it *not* be covered, the cost must be borne by the contractor even if the work had been properly performed. Paragraph 12.1.2 deals with work *properly* covered. If the work is found "in accordance with the Contract Documents," costs are borne by the owner. If the work did *not* comply, the contractor must pay the cost of uncovering and recovering unless the deviation was caused by the owner or a separate contractor. Similarly, ¶ 13.5.3 places the entire cost of "testing, inspection or approval" on the contractor if these processes reveal that the contractor has not complied with the contract documents.

Should the contractor be required to pay the entire cost if *any* deviation is discovered? One purpose of having the architect visit the site

periodically is to observe work before it is covered. The assumption under ¶ 12.1.2 is that the architect did not see the defective work before it was covered or request the contractor not to cover it until he had a chance to inspect it. Under such circumstances, the architect may be fearful of a claim being made against *him* unless *some* defective work is found.

The all-or-nothing solution is justified only if there were major deviations or if the work was covered to hide defective work. It is better to share the cost of covering and uncovering if the defect is slight or inadvertent.

SECTION 29.09 Finality of Design Professional Decision

Parties to a contract can create a mechanism by which disputed matters or other matters that require judgment can be submitted to a third party for a decision. The finality of that decision—that is, whether it can be challenged and the extent of the challenge—can range from 0 to 100%. It can be *absolutely* unchallengeable. At the other extreme, a decision by a third party can be simply advisory.

Most impartial third party decisions have *some* finality but not the finality of an arbitral award[33] or a court decision. They are clearly more than advisory. Unless they are shown either to have been dishonestly made or to be clearly wrong, they are likely to be upheld if challenged in court.[34]

The preceding discussion has dealt with decisions by impartial third parties. A design professional cannot be said to be a disinterested third party. In discussing the reasons for limiting the effect of the refusal by an architect to issue a certificate for payment, a New York judge stated:

> The rule is based upon the fact that the architect, in contracts of this sort, rarely a disinterested arbiter, is usually the representative of the party, often the owner, who must ultimately bear the cost of the work.[35]

---

[33]See Section 30.14.
[34]Earlier edition of treatise cited and followed in *Bolton Corp. v. T.A. Loving Co.*, 94 N.C.App. 392, 380 S.E.2d 796, 801 (1989).
[35]*Arc Elec. Constr. Co. v. George A. Fuller Co.*, 24 N.Y.2d 99, 247 N.E.2d 111, 112–114. n.2, 299 N.Y.S.2d 129, 133 n.2 (1969).

This factor—the dubious impartiality of the decision maker—and others to be noted in this section make the law uncertain and apparently contradictory.

Disputes submitted to the design professional can range from purely factual (Did the work meet certain specific standards?) to matters that though sometimes called legal are really factual (How should this clause be interpreted?) to legal questions (Is the substantial performance doctrine applicable where the design professional is to judge performance?).[36]

The wide range of issues, along with the ambivalent status of the design professional and possible conflict of interest, has led to uncertainty over finality of the design professional's decision. Finality in the first instance depends on the language of the contract giving the design professional the power to make decisions. If it says *nothing* about conclusiveness or finality, the decision would be purely advisory. But generally, clauses giving this power state that the decision shall be "final and binding."

Suppose the architect's decision is final and binding unless arbitration is invoked. If the arbitration clause has been deleted or arbitration waived, the *court* will extend a considerable amount of deference to the architect's decision. (If arbitration is sought, the arbitrators need pay no attention to the architect's decision. This can encourage arbitration by the party that is dissatisfied with the architect's decision.)

As noted earlier in this section, however, language of finality does not actually mean that the decision is final. One court held the decision to be binding unless there was fraud or bad faith.[37] The Restatement (Second) of Contracts makes the decision binding so long as it is made honestly and not on the basis of gross mistake as to the facts.[38]

New York gives greater finality to a certificate that has been issued than to the absence of a certificate. In the absence of a certificate, the contractor can recover if it can show substantial performance, with refusal to issue the certificate being unreasonable.[39]

These varying, though related, degrees of finality make the decision conclusive if honestly made unless it is clear that the design professional made a serious mistake. But the standards do not tell the entire story. Other relevant factors determine the degree of finality.

The particular nature of the dispute is important. If the dispute is more technical and less legal, the decision will be given more finality.[40]

AIA Doc. A201, ¶ 4.2.13, states that architect decisions as to aesthetic effect "will be final if consistent with the intent expressed in the Contract Documents." In addition, ¶ 4.5.1 makes clear that such decisions cannot be arbitrated. The creation of an apparently unreviewable decision can lead to abuse. Certainly, the architect should not use this power to force the contractor to perform in a way not required by the contract, and the language of ¶ 4.2.11 appears to recognize this. But the very subjectivity of aesthetic effect and the power for abuse that such a clause can create will inevitably lead to a difference of opinion among courts as to the enforceability of such a clause.[41]

If a certificate is issued and the *owner* refuses to pay, it is likely that the certificate will be considered final.[42] Undoubtedly, this is a recognition that

---

[36]Earlier edition of treatise cited in *In the Matter of Dutchess Community College*, 57 A.D.2d 555, 393 N.Y.2d 77, 78 (1977). Conflict of interest is discussed in detail in J. SWEET, SWEET ON CONSTRUCTION INDUSTRY CONTRACTS § 11.11 (2d ed. 1992).

[37]*Laurel Race Course, Inc. v. Regal Constr. Co.*, 274 Md. 142, 333 A.2d 319 (1975).

[38]Restatement (Second) of Contracts § 227 comment c, illustrations 7 and 8 (1981). The First Restatement was applied in *James I. Barnes Const. Co. v. Washington Township*, 134 Ind.App. 461, 184 N.E.2d 763 (1962). Other somewhat variant standards are expressed in *Perini Corp. v. Massachusetts Port Auth.*, 2 Mass.App. 34, 308 N.E.2d 562 (1974)

(binding unless arbitrary or in bad faith); *City of Mound Bayou v. Roy Collins Constr. Co.* 499 So.2d 1354 (Miss.1986) (binding if made in good faith, an honest judgment after a fair consideration of the facts); and *E.C. Ernst, Inc. v. Manhattan Constr. Co. of Texas*, 387 F.Supp. 1001 (S.D.Ala.1974), affirmed 551 F.2d 1026, rehearing denied in part and granted in part 559 F.2d 268 (5th Cir.1977) (engineer must use good faith).

[39]*Arc Elec. Constr. Co. v. George A. Fuller Co.*, supra note 35.

[40]*Yonkers Contracting Co. v. New York State Thruway Auth.*, 25 N.Y.2d 1, 250 N.E.2d 27, 302 N.Y.S.2d 521. (1969); *John W. Johnson, Inc. v. J.A. Jones Constr. Co.*, 369 F.Supp. 484 (E.D.Va.1973).

[41]Compare *Baker v. Keller Constr. Corp.*, 219 So.2d 569 (La.App.1969) (reviewed decision), with *Mississippi Coast Coliseum Comm'n v. Stuart Constr. Co.*, 417 So.2d 541 (Miss.1982) (refused to review).

[42]*Hines v. Farr*, 235 S.C. 436, 112 S.E.2d 33 (1960).

the owner for all practical purposes has selected the design professional and should be given less opportunity to challenge the decision. If a subcontractor's rights are at stake, less finality will be given.[43] The subcontractor had even less of a role in selecting the design professional. Power to the design professional is often accomplished by incorporating prime contract general terms by reference into the subcontract. The subcontractor may not have had much opportunity to present its case to the design professional.[44]

The availability of arbitration may also play a role. If the decision can be appealed to arbitration, arguably more finality can be given.[45] Another factor that may bear on the degree of finality to be given design professional decisions is the process by which the decision is made. If it appears to have been made precipitously without elemental notions of fairness, less finality, if any, will be accorded.[46]

Another problem relates to the interaction between language giving finality to a design professional decision and language that bars acceptance of the project from waiving claims for defective work subsequently discovered.[47] One case held that the architect's power to judge performance did not give his decisions finality because of a clause stating that neither the issuance of a final certificate nor final payment relieved the contractor from responsibility for faulty materials or workmanship.[48] Yet another decision more sensibly reconciled these two clauses by concluding that the issuance of a certificate is conclusive where defects are patent or obvious but not where defects are latent, that is, not reasonably discoverable.[49]

SECTION 29.10　Finality: A Comment

If the parties to a contract *voluntarily* designate a third party to make certain determinations or decide certain disputes and agree that these deter-

minations or decisions shall be binding on both parties, the law should give effect to such an agreement. Parties who genuinely agree to abide by such third-party decisions and who are satisfied that the decision was honestly made are likely to perform in accordance with the decision. The third party may be better equipped in the view of the contracting parties to give a fair and quick decision. Failure to make such agreements final may discourage parties from agreeing to submit determinations and disputes to third parties or encourage them not to live up to such agreements. The result can be an increasing burden on an already burdened judicial system.

Contract language and courts often state that in making certain determinations, the design professional is acting in a quasi-judicial function and not as representative of one of the contracting parties. This does not change the realities. The design professional is *not* a neutral judge, and the contractor has little choice but to accept the current system.

Generally, determinations and decisions by the design professional are followed by the parties. This may be because the parties are satisfied, because it is too costly to arbitrate or litigate, or because of the need to retain the goodwill of the design professional. In most cases, this will mean that giving the design professional decision-making powers, *in the first instance,* will provide a quick method of handling construction disputes. According such decisions some degree of finality simply gives effect to the superior bargaining position the owner frequently enjoys at the time a construction contract is made.

In *Cofell's Plumbing & Heating, Inc. v. Stumpf,*[50] one issue related to the finality of an engineer's decision under a contract that gave the engineer "binding authority to determine all questions concerning specification interpretations in the execution of the contract." After completion, a dispute arose that related to how particular work should be priced. The project engineer had concluded that the work should be compensated under a particular provision in the contract. The owner refused to pay, but the trial court concluded that the contractor was entitled to be paid in accordance with the engineer's decision.

---

[43] *Walnut Creek Elec. v. Reynolds Constr. Co.,* 263 Cal.App.2d 511, 69 Cal.Rptr. 667 (1968).
[44] *John W. Johnson, Inc. v. Basic Constr. Co.,* supra note 23.
[45] *Roosevelt Uni. v. Mayfair Constr. Co.,* 28 Ill.App.3d 1045, 331 N.E.2d 835 (1975).
[46] *John W. Johnson, Inc. v. Basic Constr. Co.,* supra note 23.
[47] Section 24.05.
[48] *Flour Mills of America Inc. v. American Steel Bldg. Co.,* 449 P.2d 861 (Okl.1969).
[49] *City of Midland v. Waller,* 430 S.W.2d 473 (Tex.1968).

---

[50] Supra note 12.

The appellate court held that a line must be drawn between "disputes or specification interpretations which deal with work performance and require immediate resolution and those which deal with other contractual disputes, such as rates, method, or time of payment, and do not require prompt on-the-site determination."[51] The court held that the contract should be interpreted to give finality to the former but not to the latter. The project will continue whatever the determination by the engineer, at least in the view of the majority.

The dissenting judge wished to expand the finality provision to *both* types of decisions to prevent unnecessary delays in the Construction Process while disputes are settled. He contended that the majority's decision will invite delays by contractors who will be forced to stop work while resolving other types of disputes that the engineer, under the interpretation of the majority, cannot resolve with any degree of finality.

It is important to keep the project moving. Decisions needed to accomplish this may be given finality to accomplish this goal. Yet it is likely that the project can be expedited by getting a decision but still allowing a dissatisfied party to challenge it later.

Some American jurisdictions give the design professional immunity when acting as a judge.[52] In those jurisdictions, to both give the design professional immunity and accord substantial finality to his decisions concentrates too much power in the hands of the design professional. Because this power will occasionally be abused, in those jurisdictions very little, if any, finality should be given to the design professional's decision. Even where the design professional has *no* immunity, it is better to accord no finality to his decisions. The system will work without it, and a needless issue can be removed from construction litigation.

PROBLEM

Under the terms of a construction contract between O and C, A, the architect, was given the power to terminate C's performance if C was not making reasonable progress. A informed C that C's progress was not satisfactory and ordered C to leave the job. C tried to show A records of weather conditions to prove that unseasonably cold weather had cut down its efficiency, that there had been a jurisdictional strike between the unions of two subcontractors, and that A had taken too long to approve shop drawings and render interpretations. A refused to look at the evidence or to consider these contentions. He also refused to give any time to the contractor to catch up. C left the site. There was no arbitration provision.

C sued O and A, claiming that he had been terminated without just cause. Does C have a valid claim against O? Against A? If it has a valid claim against either, what should be its measure of recovery?

---

[51] 290 N.W.2d at 234.
[52] See Section 14.09(D).

C H A P T E R   T H I R T Y

Construction Disputes: Arbitration and Other Methods to Reduce Costs and Save Time

SECTION 30.01 Introduction

This chapter deals principally with a voluntary method of resolving disputes by submitting disputes to a third party and agreeing to be bound by that party's decision. Third-party resolution can be used to determine narrow issues such as the strength of concrete or the value of property. The former would be classified as third-party testing and the latter as third-party appraisal. The parties can go further and authorize a third party to decide *any* dispute that might arise between them in the performance of a contract. Such a general referral to the third party can encompass specific disputes narrow in character or broad disputes that can encompass any matter that might be resolved by a court. This latter system, frequently called arbitration, is the focal point of this chapter.

This chapter also looks at some methods that are implemented by the judicial system, such as the use of special referees or masters or a summary jury trial. In addition, the chapter looks at techniques by which a third party will facilitate an agreement by the disputing parties. What links these methods is that they all seek to avoid the traditional lawsuit as a means of resolving construction disputes.

While this chapter deals with the intervention by third parties in some form, it must be emphasized that most disputes in the Construction Process are resolved through negotiation. While it has become fashionable to speak of alternative dispute resolution (ADR) and its various methodologies, it cannot be overemphasized that the most important initial step toward the resolution of disputes is good-faith negotiations by the parties to try to

solve the problem. This means that the parties must gather information, look at the dispute from the other party's position as well as their own, and recognize the horrendous costs that can be incurred in terms of time, money, and irrational outcomes because of the intervention of third parties, either through ADR or through the litigation system.

SECTION 30.02 The Law and Arbitration

Until the 1920s, the law was openly hostile to arbitration. Although it would enforce an arbitral award *after* it was made, it frustrated agreements to arbitrate disputes that might arise in the future. Some courts found such agreements invalid, some allowed a party to revoke such an agreement prior to award, and some would give only nominal damages for breaching a contract to arbitrate future disputes. Arbitration could not thrive in such a legal system.

In the 1920s, commercial arbitration was greatly encouraged by statutory enactments in a substantial number of states, including commercially important states, which sought to remedy some of the deficiencies in the legal treatment of arbitration. Principally, these arbitration statutes accomplished the following:

1. Made agreements to submit future disputes to arbitration irrevocable.
2. Gave the party seeking arbitration the power to obtain a court order compelling the other party to arbitrate.
3. Required courts to stop any litigation where there had been a valid agreement to arbitrate a pending arbitration.

4. Authorized courts to appoint arbitrators and fill vacancies when one party would not designate the arbitrator or arbitrators withdrew or were unable to serve.
5. Limited the court's power to review findings of fact by the arbitrator and her application of the law.
6. Set forth specific procedural defects that could invalidate arbitral awards and gave time limits for challenges.

Almost all states and the federal government currently have modern arbitration statutes. Clearly, this has greatly encouraged the use of arbitration. Yet this has also created concurrent jurisdiction, where a party is often free to invoke either state or federal arbitration laws. If this does occur, one party may contend that the Federal Arbitration Act (FAA) preempts state arbitration law. This can slow down the arbitration process, as such an issue must be resolved in court. The more opportunities to seek judicial rulings, the less desirable it is to include an arbitration clause.

The more favorable attitude toward arbitration has been tempered by the modern judicial recognition that many agreements to arbitrate are forced on the weaker party in an adhesion contract. This can be demonstrated by two cases. The first case, *Spence v. Omnibus Industries*,[1] involved a dispute between a homeowner and a remodeling contractor over a remodeling contract. The contract—a standardized contract—included a clause requiring arbitration in accordance with the rules of the American Arbitration Association.

After the dispute arose, the homeowner brought a lawsuit against the remodeling contractor seeking damages of $37,000. The remodeling contractor filed a petition for arbitration. The petition was granted, and the court ordered that the homeowner pay the arbitration filing fee of $720. The homeowner was willing to arbitrate but appealed the court's decision requiring her to pay the filing fee.

The court reversed the decision regarding the filing fee and in doing so compared the cost of beginning an action in court with the cost of submitting the dispute to arbitration.

The court noted that the filing fee for commencing an action in court would have been $50.50, but

it would cost the homeowner $720 to submit the matter to arbitration. The court stated:

> The reason for this disparity is obvious. Courts are established and supported by the State in order to afford forums to which all, rich and poor alike, may present controversies at minimum cost to the parties. Arbitration is supported by the parties. If the parties are equal in bargaining power, arbitration is good. If the parties are not equal, arbitration may deny a forum to the weaker.[2]

After characterizing the contract as one of adhesion imposed by the party of greater bargaining power on the weaker,[3] the court noted that the contract was over 2,000 words jammed into a tightly printed jumble of "terms and conditions." The court noted that it was quite unlikely that one homeowner in one hundred would ever read the massive information on the reverse page. Despite the judicial policy favoring arbitration, the court pointed to the strong judicial policy to protect the weaker party to the bargain. The court felt that a 720-dollar fee could discourage a homeowner from presenting a claim against a builder.

Concluding that the homeowner waived her right to arbitrate by filing an action in court,[4] the court held that the contractor seeking arbitration became the initiating party and required the latter to pay the filing fee.[5] Had the homeowner wanted arbitration, according to the court, she would have had to pay the filing fee.

The holding in the *Spence* decision points to some of the inadequacies of the adhesion contract analysis. It is just as likely as not that the homeowner had stronger bargaining power. However, when faced with a pre-prepared, complicated form contract that includes matters beyond the understanding of the homeowner, the homeowner may become the weaker party, not because she lacks bargaining power but because she does not have the skill or ability to understand the terms agreed

---

[1] 44 Cal.App.3d 970, 119 Cal.Rptr. 171 (1975).

[2] 119 Cal.Rptr. at 172.
[3] See Section 5.04(C).
[4] See Section 30.05.
[5] A court less impressed with the adhesive aspects of the contract held that the party who commenced litigation would have to pay the initiating fee when the other party demanded arbitration. See *A.P. Brown Co. v. Superior Court, County of Pima*, 16 Ariz.App. 38, 490 P.2d 867 (1971).

to.[6] This problem becomes even more difficult as legislatures begin to bar arbitration clauses in adhesion contracts.[7]

The second case, *Player v. George M. Brewster & Son, Inc.,*[8] involved an arbitration clause in a subcontract. The work was to be performed in California, and the arbitration clause required that arbitration be governed by New Jersey law and be held in New Jersey. The arbitration clause specified that each party would select one arbitrator and the third would be picked by a New Jersey trial judge.[9] The prime contractor, though having its home office in New Jersey, performed much of its work in the western part of the United States.

The court interpreted the clause to determine whether it covered the particular dispute. Like the court in *Spence v. Omnibus Industries,* the court in this case noted that this was a contract of adhesion and should be construed in favor of the subcontractor. The court stated:

> As a whole it appears to be a "house attorney" prepared form intended by Brewster to be submitted to all of its subcontractors on a take-it-or-leave-it basis.[10]

The court then paid some tribute to arbitration stating:

> The law favors contracts for arbitration of disputes between parties. They are binding when they are openly and fairly entered into and when they accomplish the purpose for which they are intended.
>
> * * *
>
> Our trial courts are clogged with cases, many of them involving disputes between contracting parties. One of the principal purposes which arbitration proceedings accomplish is to relieve that congestion and to obviate the delays of litigation.[11]

The court was concerned that a strong prime contractor that made all of its subcontractors agree to such clauses can deprive subcontractors access to the courts in their states. The prime contractor would possess a powerful weapon enabling it to force the subcontractors to arbitrate thousands of miles away from the place of business of the subcontractor and have the third arbitrator, the neutral, be selected by a hometown judge where the prime contractor had its offices. The court concluded:

> A skepticism is born when we read paragraph 13 as to whether it was written as Brewster wrote it for the purpose of expeditious disposition of controversies with its subcontractors. Its plan, it is suggested, may have been designed to effectuate a more unilateral benefit to itself.
>
> * * *
>
> We think the courts would and should scan closely contracts which bear facial resemblance to contracts of adhesion and which contain cross-country arbitration clauses before giving them approval.[12]

Modern courts are beginning to look carefully at arbitration clauses in an adhesion context. Is the clause designed to frustrate the weaker party's right to make claims rather than provide a fair and efficient resolution of disputes? Is the clause one that is not brought to the attention of the weaker party and phrased in a way so as *not* to be understood? Is the clause, though expected and understood, one that essentially rigs the process in favor of one of the parties?[13] If the answer to any of these questions is yes, despite the favorable attitude toward arbitration[14] as an alternative dispute resolution system, the law may not enforce the clause.

---

[6]Suppose a client signs an AIA Doc. B141 with its arbitration clause?

[7]Ann.Iowa Code § 679A.1.

[8]18 Cal.App.3d 526, 96 Cal.Rptr. 149 (1971).

[9]By the time the matter went to court, the prime contractor was willing to have the arbitration heard in California but insisted that the third arbitrator be picked by the New Jersey trial court judge.

[10]96 Cal.Rptr. at 154.

[11]Ibid.

[12]96 Cal.Rptr. at 156. But see *Pacific Gas and Elec. Co. v. Superior Court,* 15 Cal.App. 4th 576, 19 Cal.Rptr.2d 295 (1993). It rejected a challenge to the arbitration clause based upon the assertion the contract was one of adhesion. The court noted that in "a commercial context" such a clause is valid unless it is structured or utilized unfairly by the party with the superior bargaining power. 19 Cal.Rptr.2d at 310.

[13]*Graham v. Scissor-Tail, Inc.,* 28 Cal.3d 807, 623 P.2d 165, 171 Cal.Rptr. 604 (1981) (arbitration clause in a contract between an experienced rock music promoter and a performer that required arbitration by a committee of the performer's union invalidated). Similarly, see *Naclerio Contracting Co., Inc. v. City of New York,* 116 A.D.2d 463, 496 N.Y.S.2d 444 (1986), affirmed 69 N.Y.2d 794, 505 N.E.2d 625, 513 N.Y.S.2d 115 (1987), barring the commissioner of a public agency from exercising a contractual power to determine all questions related to the contract and its performance. See Section 30.03(B).

[14]*La Stella v. Garcia Estates Inc.,* 66 N.J. 297, 331 A.2d 1 (1975).

Yet despite increased recognition of realities in the contract-making process, arbitration continues to be looked on favorably as a method to avoid the courthouse.[15]

SECTION 30.03 Agreements to Arbitrate and Their Validity

Arbitration is a voluntary system based on a valid contract to arbitrate. For that reason, analysis of the validity of a general arbitration clause involves the requisite elements for a valid contract as well as an awareness of the special treatment accorded agreements to arbitrate by the courts.

A. Manifestations of Mutual Assent: An Agreement to Arbitrate

An agreement to arbitrate is usually manifested by each party's signing a written contract that contains the arbitration clause or makes reference to arbitration in another document. However, in consumer transactions, even a signature may be insufficient to create a valid contract to arbitrate. In such transactions, it may be necessary to show that the consumer had her attention drawn to the arbitration clause and understood its import.[16]

In construction contracts, the principal problems have been incorporation by reference of arbitration clauses into subcontracts (discussed in Section 28.04).

Even in prime contracts, incorporation by reference can be a problem for an unsophisticated owner. An owner who uses AIA documents will sign the Basic Agreement, A101, which states nothing about arbitration but incorporates A201, the General Conditions, into the Contract Documents. To complicate the process, A201's arbitration clause[17] is among a number of provisions in Art. 4 captioned "Administration of the Contract." Even more difficult for the unsophisticated owner, ¶ 4.5.1 requires arbitration in accordance with Construction Industry Arbitration Rules of the American Arbitration Association (CI Rules).[18] These formi-

dable rules are not included in the Contract Documents.

Modern arbitration statutes generally require simply that the agreement be written. However, states sometimes enact special formal requirements for agreements to arbitrate.[19] Local statutes must be consulted.

B. Validity Attacks on Arbitration Clauses

Agreements to arbitrate future disputes were generally not enforceable at common law. This impediment has been removed in most states by enactment of modern arbitration statutes. For transactions involving interstate commerce, the Federal Arbitration Act makes agreements to arbitrate future disputes valid.[20]

In addition to the now decreasingly successful claims based on the invalidity of arbitration clauses, other attacks are being made. One frequently used attack by the party who does not wish to arbitrate is that the contract was fraudulently procured. Unquestionably, a contract procured by fraud is invalid. But who decides the disputed issue of fraud—the arbitrators or the court?

The clause should be no more enforceable than the contract that creates it. The claim of fraud is easily made and difficult to sustain. A judicial determination of fraud can take time and would suspend the arbitration hearing while lengthy hearings are held in court. As a result, a claim of fraud can have a tactical advantage for the party wishing to delay. In addition, dividing the dispute between those that should be resolved by the court and those by the arbitrator causes other problems. Suppose the fraudulent claim is *not* sustained by the court but is upheld by the arbitrator.

To encourage arbitration and discourage broad assertions of fraud that did not *specifically* attack the arbitration clause itself, a number of courts employ a legal fiction that artificially severs the arbitration clause from the rest of the contract. This divisibility concept permits the fraud issue to go to the arbi-

---

[15]See West Ann.Cal.Bus. & Prof.Code § 7085 (consumer complaints to California Contractors Licensing Board resolved by arbitration). See Section 10.11.

[16]*Spence v. Omnibus Industries,* supra note 1, discussed in Section 30.02.

[17]AIA Doc. A201, ¶ 4.5.

[18]A number of references to those rules are in this chapter and are referred to as the CI Rules (reproduced in Appendix F).

---

[19]But in 1987, Texas enacted Chapter 817, which *eliminated* the requirement that agreements contain a notice stating that the contract provides for arbitration. This was done to expedite arbitration. It is no longer required to rubber-stamp contracts with boilerplate arbitration language.

[20]9 U.S.C.A. § 2.

trator unless the arbitration clause itself was attacked as having been fraudulently obtained.[21] Other courts have refused to use this fiction and submitted issues of fraud to the court.[22]

In 1967, the U.S. Supreme Court adopted the divisibility concept in *Prima Paint Corp. v. Flood & Conklin Manufacturing Co.*[23] This made attacks of fraud less effective and strengthened arbitration. Although state courts deciding cases under state arbitration laws were not bound to follow this decision, most have.[24] Yet the court in its desire to remove impediments to arbitration may have encouraged contracting parties not to agree to general arbitration clauses for fear that matters as important as fraud will not go to a court but will go to an arbitrator.

Another attack is that the arbitration clause is unconscionable. California held that a contract between a rock music promoter and entertainers that included an arbitration clause giving the power to arbitrate disputes to a committee of the American Federation of Musicians was unconscionable.[25] Despite the court's conclusion that the promoter knew what he was doing, the court found the clause unconscionable. The cards were stacked in favor of the performers. Similarly, Texas recently enacted a statute precluding arbitration if the clause is unconscionable.[26]

Another attack on an arbitration clause is that it lacks mutuality—when one party is required to arbitrate but the other is not. After reviewing cases coming to different conclusions, New Jersey upheld a clause that gave only one party the right to arbitrate.[27]

Suppose one party terminates the contract for an asserted material breach by the other. This does not eliminate any jurisdiction that had been created by an arbitration clause in the original but now terminated agreement.[28] The terminating party or party being terminated can still seek arbitration. Termination was not an attack on the validity of the contract.

SECTION 30.04 Specific Arbitration Clauses: Jurisdiction of Arbitrator and Timeliness of Arbitration Requests

A. Jurisdiction Conferred by Clause

A frequently disputed issue relates to whether the arbitration clause covers the particular matter in dispute. Obviously, much depends on the language of the clause, and a broadly drafted arbitration clause can cover almost anything that relates to the contract between the parties.

As a general rule, close questions as to jurisdiction are likely to be resolved against the party that prepared the clause.[29] However, courts favorably disposed toward arbitration are unwilling to narrowly interpret clauses to deny arbitrability of particular disputes. This issue can arise at the time arbitration is challenged or after an arbitration award has been taken to court for confirmation. Often, one ground for refusing to confirm an arbitration award is that it exceeded the arbitrator's jurisdiction. One court noted that overtechnical judicial review of arbitration awards on the basis of the scope of the arbitrator's authority can frustrate the basic purposes of arbitration.[30] It is more difficult to attack jurisdiction *after* the arbitration has

---

[21]*Robert Lawrence Co. v. Devonshire Fabrics, Inc.*, 271 F.2d 402 (2d Cir.1959).
[22]*Lummus Co. v. Commonwealth Oil Refining Co.*, 280 F.2d 915 (1st Cir.1960).
[23]388 U.S. 395 (1967). The same concept was applied to a dispute that arose after a mutual termination of the contract in *Clifton D. Mayhew, Inc. v. Mabro Constr., Inc.*, 383 F.Supp. 192 (D.D.C.1974) (noting split in federal courts).
[24]For example, see *Security Constr. Co. v. Maietta*, 25 Md.App. 303, 334 A.2d 133 (1975); *Weinrott v. Carp*, 32 N.Y.2d 190, 298 N.E.2d 42, 344 N.Y.S.2d 848, (1973); *Ericksen, Arbuthnot, etc., Inc. v. 100 Oak Street*, 35 Cal.3d 312, 673 P.2d 251, 197 Cal.Rptr. 581 (1983). But see *City of Blaine v. John Coleman Hayes & Assoc.*, 818 S.W.2d 33 (Tenn.App.), appeal denied (1991) (allegations of fraud decided by court, not arbitrator).
[25]*Graham v. Scissor-Tail, Inc.*, supra note 13. Similarly, see *Naclerio Contracting Co. v. City of New York*, supra note 13.
[26]Ann.Tex.Rev.Civ.Stat. Art. 224.

[27]*Kalman Floor Co. v. Jos. L. Muscarelle, Inc.*, 196 N.J.Super. 16, 481 A.2d 553 (1984), affirmed 98 N.J. 266, 486 A.2d 334 (1985). Similarly, see *Sablosky v. Edward S. Gordon Co.*, 73 N.Y.2d 133, 535 N.E.2d 643, 538 N.Y.S.2d 513 (1989).
[28]*County of Middlesex v. Gevyn Constr. Corp.*, 450 F.2d 53 (1st Cir.1971); *Riess v. Murchison*, 384 F.2d 727 (9th Cir.1967).
[29]*Player v. George M. Brewster & Son, Inc.*, supra note 8.
[30]*Federal Commerce & Navigation Co. v. Kanematsu-Gosho, Ltd.*, 457 F.2d 387 (2d Cir.1972). See also *Muhlenberg Township School Dist. Auth. v. Pennsylvania Fortunato Constr. Co.*, 460 Pa. 260, 333 A.2d 184 (1975).

taken place. The clause can specify that it applies *only* to claims that do not exceed a certain amount or claims that involve *factual* as opposed to *legal* disputes.[31]

Whether the clause has conferred jurisdiction upon the arbitrator depends upon the language of the arbitration clause as well as judicial attitude toward arbitration. A few examples illustrate this. Courts have differed as to whether delay damages fall within an arbitration clause.[32] A court seeking to encourage arbitration would not limit arbitration to damages solely to person or property.[33] Another court less favorable to arbitration held that a general arbitration clause did not cover the owner's possible liability for water damage.[34] That court required that the clause be "crystal clear" before it would confer jurisdiction on the arbitrator.

Another court held that a general arbitration clause in a subcontract did not confer jurisdiction on the arbitrator to determine which portion of the funds the prime contractor received to train minority workers should go to the subcontractor.[35] That same decision refused to permit the architect to arbitrate a dispute between prime contractor and subcontractor when the subcontractor's principal claim was that the architect had committed design errors.

Courts have differed in their willingness to encompass implied terms under the arbitration clause.[36] Most courts hold that the arbitration clause will be applied to disputes that arise after the work is completed despite a provision in the arbitration clause stating that work will continue while the dispute is being arbitrated.[37]

The AIA has denied jurisdiction to an arbitrator to consider disputes over aesthetic effect.[38] Another jurisdictional issue relates to the arbitrability of tort claims.[39] Some arbitration clauses allow one party to arbitrate but not the other.[40] As noted earlier, judicial resolution of the jurisdictional question is likely to be influenced by the court's attitude toward arbitration, the relative bargaining power of the parties, and the apparent appropriateness of arbitration for a particular dispute.

Although arbitration typically involves performance problems, it has been held that a claim by a contractor that it should be relieved from a bidding mistake should be decided by arbitration.[41]

The increasing attempt of nonparties to enforce contracts by claiming that they are intended beneficiaries of contracts made by others has added to arbitration complexity. For example, a tenant who successfully maintained it was a third-party beneficiary of a contract between the architect and contractor was forced to use arbitration, as arbitration was included in the contract to which it claimed third-party rights.[42] Yet a liability insurer was held not to be a third-party beneficiary of a subcontract and could not force arbitration of the dispute between prime and subcontractor.[43] Cases such as these demonstrate the importance of contracting

---

[31]*Doyle & Russell, Inc. v. Roanoke Hosp. Ass'n*, 213 Va. 489, 193 S.E.2d 662 (1973).

[32]*Harrison F. Blades, Inc. v. Jarman Memorial Hosp. Bldg. Fund, Inc.*, infra note 36, held a delay damage claim beyond the scope of arbitration, while *Aberthaw Constr. Co. v. Centre County Hosp.*, 366 F.Supp. 513 (M.D.Pa.1973), held that arbitration was required.

[33]*Muhlenberg Township School Dist. Auth. v. Pennsylvania Fortunato Constr. Co.*, supra note 30.

[34]*Silver Cross Hops. v. S.N. Nielsen Co.*, 8 Ill.App.3d 1000, 291 N.E.2d 247 (1972).

[35]*Paschen Contractors, Inc. v. John J. Calnan, Co.*, 13 Ill.App.3d 485, 300 N.E.2d 795 (1973).

[36]*Roosevelt Univ. v. Mayfair Constr. Co.*, 28 Ill.App.3d 1045, 331 N.E.2d 835 (1975), and *Allentown Supply Corp. v. Hamburg Municipal Auth.*, 463 Pa. 167, 344 A.2d 477 (1975), held that arbitration encompassed implied terms, while *Harrison F. Blades, Inc. v. Jarman Memorial Hosp. Bldg. Fund, Inc.*, 109 Ill.App.2d 224, 248 N.E.2d 289 (1969), would not. Implied and express warranty came within the arbitration clause in *Harman Elec. Constr. Co. v. Consolidated Eng'g Co.*, 347 F.Supp. 392 (D.Del.1972).

[37]*Warren Bros. Co. v. Cardi Corp.*, 471 F.2d 1304 (1st Cir.1973); *Hudik-Ross, Inc. v. 1530 Palisade Ave. Corp.*, 131 N.J.Super. 159, 329 A.2d 70 (1974). A case to the contrary was *Hussey Metal Div. v. Lectromelt Furnace Div.*, 471 F.2d 556 (3d Cir.1972) (clause stating that no demand for arbitration could be made after final payment). See Section 34.03(I).

[38]AIA Doc. A201, ¶¶ 4.2.11, 4.5.1. For a review of cases involving AIA Doc. A201, see J. SWEET, SWEET ON CONSTRUCTION INDUSTRY CONTRACTS § 22.10 (2d ed. 1992).

[39]*Harman Elec. Constr. Co. v. Consolidated Eng'g Co.*, supra note 36; *Morton Levine & Assoc.*, Chartered v. Van Deree, 334 So.2d 287 (Fla.Dist.Ct.App.1976).

[40]See Section 30.03(B).

[41]*Village of Turtle Lake v. Orvedahl Constr. Co.*, 135 Wis.2d 385, 400 N.W.2d 475 (App.1986).

[42]*Dist. Moving & Storage Co. v. Gardiner & Gardiner, Inc.*, 63 Md.App. 96, 492 A.2d 319 (1985), affirmed 306 Md. 286, 508 A.2d 487 (1986).

[43]*Maryland Casualty Co. v. State Dep't of General Services*, 489 So.2d 57 (Fla.Dist.Ct.App.1986), review dismissed 494 So.2d 1151 (1986).

parties in this context excluding third-party claims by appropriate contract language and for courts to be cautious before granting a third party enforcement rights as an intended beneficiary.

Expiration of any time limit specified in the contract or statute for making the award generally ends jurisdiction of the arbitrator, and any awards made after expiration are invalid. This rule has been eroded where an award made after expiration of the time limit is enforced on the basis of waiver, such as proceeding with the arbitration without protest after expiration or failure to protest after a late award has been rendered. Waiver can also be predicated on a course of conduct that shows that the parties did not consider time to be of the essence, especially where no prejudice is shown. If the expiration period for making the award has been waived, the award must be made within a reasonable time.[44] One of the more difficult jurisdictional questions relates to the application of an arbitration clause to a disputed termination (discussed in Section 34.03(I)).

Jurisdiction problems highlight the variety of possibilities between *no* arbitration and a *general* arbitration clause. Drafting variations are discussed in Section 30.17.

B. Timeliness of Arbitration Demand

AIA Doc. A201 sets two standards for timeliness. Decisions by the architect that state that they are final and subject to appeal must be appealed to arbitration within thirty days.[45] If the decision does not contain such a statement, arbitration must be requested within a reasonable time.[46]

Milton Schwartz & Associates, Architects v. Magness Corp.[47] illustrates the difficulty of determining what is a reasonable time. It involved a dispute in September 1972 between the architect and a land developer over excess costs. Nonpayment was followed by a formal demand for payment on April 4, 1973, by the architect's lawyer in which the latter advised the developer that suit would be instituted if the defendant did not act promptly to ensure payment. On June 11, 1973, the developer's lawyer informed the architect's lawyer that the contract was terminated because of the architect's breach and that the developer was considering a counterclaim if the architect instituted legal action. On August 8, 1973, the architect brought legal action against the developer, and the latter responded immediately by demanding arbitration in accordance with the arbitration clause that required that the demand be made within a reasonable time after the dispute had arisen. The architect contended that eleven months had elapsed from the time the dispute developed in September of 1972 until the demand for arbitration on August 8, 1973, and this exceeded a reasonable period of time.

The court noted that Pennsylvania law, which governed the contract, looked with favor upon arbitration and concluded that the reasonable time standard depended on who sought arbitration—the claimant or the party opposing the claim. The court concluded that the opponent of the claim—in this case the developer—should not have the burden of considering an arbitration remedy *prior* to the time his potential adversary makes it clear that some form of enforcement proceedings will be forthcoming.

The formal demand was not made until April 4, 1973, and suit not brought until August 8, 1973. Whether the time for determining whether arbitration would be sought began at the time of the formal demand or at the time of the commencement of the lawsuit was irrelevant in the light of the court's conclusion that the defendant had sought arbitration within a reasonable time judged from either point.

It is preferable to have a fixed period of time rather than the method selected by A201 under which the architect is given the discretion to create a fixed thirty-day period.[48]

---

[44]Annot., 56 A.L.R.3d 815 (1974).
[45]AIA Doc. A201 ¶ 4.5.4.1. Without an *express* statement of the right to appeal, the thirty-day period does not apply. *Roosevelt Univ. v. Mayfair Constr. Co.*, supra note 36; *Niagara Mohawk Power Corp. v. Perfetto & Whalen Constr. Co.*, 52 A.D.2d 1081, 384 N.Y.S.2d 299 (1976).
[46]Id at ¶ 4.5.4.2. Two months after filing a mechanics' lien and one month after an action to foreclose was a reasonable time to demand arbitration in *Frederick Contractors, Inc. v. Bel Pre Medical Center, Inc.*, 274 Md. 307, 334 A.2d 526 (1975). See also *Chase Architectural Assoc. v. Moyers Corners*, 99 A.D.2d 646, 472 N.Y.S.2d 219 (1984) (demand made five years after completion reasonable, as final payment certificate never issued).
[47]368 F.Supp. 749 (D.Del.1974).

[48]EJCDC No. 1910-8, Exhibit GC-A, ¶ 16.2, uses a flat thirty-day period after the engineer's decision as the cutoff to demanding arbitration.

C. Who Decides Jurisdiction and Timeliness

One of the most troublesome questions relates to whether the court or the arbitrator will decide certain threshold issues. Section 30.03 dealt with attacks on the arbitration process. Suppose one of the parties attacks arbitration by claiming that the arbitrators have no jurisdiction to decide the particular dispute or that the process has not been requested within the time period specified in the arbitration clause. Both questions can involve contract interpretation, but the second in particular involves a factual determination that often involves applying the standard of reasonableness.

Some decisions have held that the threshold question should be decided by the judge, and others hold they can be decided by the arbitrator.[49] The uneasy partnership between arbitrator and judge and the complexity that can be generated by these preliminary procedural questions can be demonstrated by the *County of Rockland v. Primiano Construction Co.*[50] After performance had been completed, the prime contractor claimed that it had suffered delay damages because of acts of the public owner. The owner claimed that the demand for arbitration had come too late.

The New York Court of Appeals differentiated the role of arbitrator and judge in deciding these preliminary questions. The court held that the judge decides whether there was a valid agreement to arbitrate, including the frequently made contention that the dispute is one not arbitrable under the terms of the contract.

Next the court referred to what it called a second threshold issue, which also requires judicial determination, such as whether any preliminary requirements or conditions precedent have been met. As illustrations, the court gave clauses that require that disputes be first submitted to a third party for determination and that set specific contractual limitations to arbitration itself. It sharply distinguished them from what it called "procedural stipulations" that must be observed in the conduct of the arbitration proceedings itself, such as requirements that the demand for arbitration be made within a designated period of time. These, according to the court, are to be resolved by the arbitrator.

Although the court recognized that both can be classified as conditions precedent to the arbitration, it distinguished requirements that are "a prerequisite to entry into the arbitration process" from "a procedural prescription for the management of that process."[51] Putting time limits into the second category, the court concluded that the issue of whether the arbitration demand had been made within the requisite time was to be decided by the arbitrators.[52] Maryland held that jurisdiction and timeliness should be resolved by the judge.[53]

If, as in Maryland, the judge decides timeliness, it may be preferable not to use arbitration. The more avenues available to take a dispute to court, the less likely it is that arbitration will be a less costly and more expeditious method of resolving disputes.

SECTION 30.05 Waiver of Arbitration

Not infrequently, one of the parties to the contract contends that the other has waived the right to arbitrate. Such a contention is usually based on activity that appears to be inconsistent with an intention to arbitrate. For example, one party may institute legal action or file a mechanics' lien[54] or even go beyond these points and take preliminary steps before trial and *then* seek to arbitrate. In such a case,

---

[49]See Annot., 26 A.L.R.3d 604 (1969).

[50]51 N.Y.2d 1, 409 N.E.2d 951, 431 N.Y.S.2d 478 (1980).

[51]409 N.E.2d at 954, 431 N.Y.S.2d at 482.

[52]Similarly, *Boys Club of San Fernando Valley v. Fidelity & Deposit Co. of Md.*, 6 Cal.App.4th 1266, 8 Cal.Rptr. 2d 587 (1992), held that whether the statute of limitations in a surety bond bars the claim is to be decided by the arbitrator. See also *City of Lenexa v. C.L. Fairley Constr. Co.*, 15 Kan.App.2d 207, 805 P.2d 507 (1991), review denied (April 23, 1991) (arbitrator decides issue of timeliness of demand for arbitration).

[53]*Barclay Townhouse Assoc. v. Steven L. Messersmith, Inc.*, 67 Md.App. 493, 508 A.2d 507 (1986), affirmed, 313 Md. 652, 547 A.2d 1048 (1988).

[54]A number of cases have held that filing a lien does not waive arbitration. See *Pine Gravel, Inc. v. Cianchette*, 128 N.H. 460, 514 A.2d 1282 (1986), and *Sentry Eng'g & Constr., Inc. v. Mariner's Cay Dev. Corp.*, 287 S.C. 346, 338 S.E.2d 631 (1985). Undoubtedly, this is traceable to the need to file a mechanics' lien within a short period of time after completion of the work, at a time when the contractor may not have decided whether to arbitrate or litigate. Local state statutes must be consulted. For example, in California, filing a claim for lien does not waive arbitration if at the same time the lien claimant asks that the mechanics' lien action be stayed pending arbitration. See West Ann.Cal.Civ.Proc.Code § 1281.5.

the other party may contend that the prior activities were inconsistent with the desire to arbitrate and constitute a waiver.[55]

Two lines of authority have developed.[56] One examines the activities of the party alleged to have waived the arbitration clause to see whether they manifest an intention not to arbitrate. The other requires that the party claiming waiver show that it has been prejudiced by the activities of the other party. For example, one court found prejudice when the party claiming waiver expended a substantial amount of money or substantial preparation for trial.[57] Another court held that prejudice could consist of the party asserted to have waived the arbitration clause having discovered information through early litigation activity that would not have been available in arbitration.[58]

As has been seen in preceding sections, the inevitable question of who will decide disputed preliminary issues has played an important role in waiver cases. Will it be the arbitrator or the judge? Just as court decisions are not consistent as to the *existence* of waiver, they are inconsistent on the question of who will decide whether there has been a waiver.[59]

Although cases will continue to seem contradictory because of the many factual situations, the general trend may be indicated by a case that stated that arbitration was favored and that waiver would not be lightly inferred. The court noted the necessity for showing prejudice rather than simply showing inconsistent acts and indicated that recent waiver cases had involved demands for arbitration long after suits had been started and after discovery proceedings had taken place.[60]

The party wishing to protect its right to arbitrate takes risks if it engages in activity that can indicate an intention to forgo the arbitration process. It may be necessary to take such steps, but they should be taken only on advice of counsel and with the understanding that the risk is present that arbitration will not be available.

SECTION 30.06   Prehearing Activities

The party initiating arbitration usually files a notice of an intention to arbitrate and, under the CI Rules, must pay a fee based on the amount of the claim. The notice usually contains a statement setting forth the nature of the dispute, the amount involved, and any remedies sought.[61]

Once it is clear that arbitration will take place—and sometimes even before an arbitrator has been selected—one or both of the parties may wish to examine persons who might be witnesses for the other party and examine documents in the other's possession to evaluate the other party's case, to prepare for trial, and possibly to settle. In judicial proceedings, these activities are classified as "discovery."

An argument frequently made against arbitration is the absence of discovery.[62] Although discovery is still unavailable in many states, some states permit it,[63] and it is becoming more common for

---

[55]In *People ex rel. Delisi Constr. Co. v. Bd. of Educ.*, 26 Ill.App.3d 893, 326 N.E.2d 55 (1975).

[56]*Burton-Dixie Corp. v. Timothy McCarthy Constr. Co.*, 436 F.2d 405 (5th Cir.1971), stressed inconsistent conduct, while Hudik-Ross, Inc. v. 1530 Palisade Ave. Corp., supra note 37, stressed the necessity for showing prejudice. See Note, 1992 J.Disp.Resol. 178, annotating *Kramer v. Hammond*, 943 F.2d 176 (2d Cir.1992); Note, 13 Ariz.L.Rev. 479 (1971).

[57]*Commercial Metals Co. v. Int'l Union Marine Corp.*, 294 F.Supp. 570 (S.D.N.Y.1968).

[58]*Vespe Contracting Co. v. Anvan Corp.*, 399 F.Supp. 516 (E.D.Pa.1975).

[59]See note 49, supra. Many recent cases have involved the invocation of discovery. Most have held that this in and of itself is sufficient to waive arbitration, particularly if discovery is not available in arbitration and discovery uncovers information that would not otherwise be revealed. See *Christensen v. Dewor Developments*, 33 Cal.3d 778, 661 P.2d 1088, 191 Cal.Rptr. 8 (1983); *School Bd. of Orange County v. Southeast Roofing & Sheet Metal, Inc.*, 489 So.2d 886 (Fla.Dist.Ct.App.1986); *Joba Constr. Co. v. Monroe County Drain Comm'r*, 150 Mich.App. 173, 388 N.W.2d 251 (1986) (despite Construction Industry Arbitration Rule stating that judicial proceedings are not a waiver of arbitration). But see *Maxum Foundations, Inc. v. Salus Corp.*, 779 F.2d 974 (4th Cir.1985) (no waiver despite the party seeking to arbitrate having taken depositions and participated in pretrial conferences).

[60]*Gavlik Constr. Co. v. H.F. Campbell Co.*, 526 F.2d 777 (3d Cir.1975).

[61]CI Rules §§ 7, 9.

[62]For a debate (Hinchey arguing the affirmative and House and Corgan arguing the negative), see *Do We Need Special ADR Rules for Complex Construction Cases?* 11 Constr. Lawyer No. 3, August 1991, p. 1.

[63]M. DOMKE, COMMERCIAL ARBITRATION, § 27.01 (G. Wilner ed. 1984).

courts to order discovery under certain circumstances to aid the arbitration process.[64] In complicated arbitrations that involve stakes high enough to justify prehearing activities, it may be useful to permit or even suggest that the arbitrating parties submit in advance a written statement giving *in detail* the facts and legal justification for the contentions of the arbitrating parties. Such an advance submission can eliminate much irrelevant testimony at the hearing. In addition, advance statements can give each party some idea of what the other party will seek to assert. Like discovery, this can aid the parties to prepare for the hearing, can make the hearing more expeditious, and may lead to settlement.

SECTION 30.07 Selection of Arbitrators

Generally, the method for selecting arbitrators is specified in the arbitration clause. Sometimes a particular arbitrator or specific panel of arbitrators is designated by the parties in advance and incorporated in the arbitration clause. Advance agreement on the arbitrators or a panel of arbitrators should build confidence in the arbitration process. However, in construction contracts, such advance agreement is not common.

Generally, a procedure to select arbitrators rather than a designation of particular individuals is used. For example, the procedure can require each party to name an arbitrator, with the two-party appointed arbitrators designating a third or neutral arbitrator. Some arbitration clauses provide that each party will appoint an arbitrator and only if they cannot agree on the disposition of the dispute do they appoint a third arbitrator who makes the decision.

Some procedures permit the neutral administrator, such as the American Arbitration Association (AAA), to designate the arbitrator if the parties cannot agree. The AAA seeks to match the particular panel of arbitrators it proposes to the type of dispute. Parties are given an opportunity to strike names on the proposed panel. The AAA selects from those acceptable to both parties. Where they cannot agree, as stated, the AAA chooses under CI Rules § 13 the neutral arbitrators.

In a three-person arbitration, generally each arbitrator is expected to be neutral. Sometimes persons appointed by the parties are expected to be partisans for the party selecting them, and they advocate the position of that party. With the exception of arbitrators who are expected to be partisans, the process usually assumes that the arbitrators, especially the neutral, will render an impartial decision based on the evidence submitted as well as the expertise they bring to the dispute.

In *Commonwealth Coatings Corp. v. Continental Casualty Co.,*[65] the U.S. Supreme Court dealt with the extent to which the neutral arbitrator must disclose facts that may bear on the arbitrator's impartiality. It has stirred controversy and has had an impact on arbitration law.

The arbitration had involved a dispute between a subcontractor and the surety on a prime contractor's bond. The neutral member performed services as an engineering consultant for owners and building contractors. One regular customer was the prime contractor with whom the subcontractor had the dispute in question. That relationship was sporadic. The arbitrator's services were used only from time to time, and there had been no dealings between the arbitrator and the prime contractor for about a year before the arbitration. The prime contractor had paid fees of about $12,000 over a four- or five-year period. These facts were not revealed by the arbitrator until *after* the award had been made. When this was disclosed, the subcontractor sought to invalidate the award.

The court noted the resemblance between a judge and an arbitrator exercising quasi-judicial powers. Clearly, the judge must avoid any appearance of partiality. But the court felt it even more important that an arbitrator avoid any appearance of impartiality because of the limited review of the arbitrator's decision.

The court refused to confirm the arbitration award because the arbitrator had not disclosed this relationship. Although the court recognized that arbitrators cannot sever all their ties from the business world, it held that failure to disclose prior ac-

---

[64]*Burton v. Bush,* 614 F.2d 389 (4th Cir.1980). See also *Bigge Crane & Rigging Co. v. Docutel Corp.,* 371 F.Supp. 240 (S.D.N.Y.1973), noted in 44 U.Cin.L.Rev. 151 (1975).

[65]393 U.S. 145 (1968).

tivities of this type would create the impression of possible bias.

The concurring justices emphasized that arbitrators are people of affairs and part of the marketplace and cannot be compared strictly to judges. The concurring justices were concerned that too great a burden of disclosure would disqualify the best informed and most capable arbitrators. These justices felt that the arbitrator cannot be expected to provide the parties with a complete and unexpurgated business biography. Nevertheless, the concurring justices felt that in this case the relationship was *more* than trivial and for that reason agreed that the arbitration award should be set aside.

The dissenting justices felt that the losing party was simply grasping at straws and such a requirement of disclosure would discourage arbitrators from serving and would render arbitration less effective because it would be too easily challengeable.

Subsequent cases have reflected ambivalence toward the *Commonwealth Coatings* requirement of disclosure, especially in the context of a trade association or closely knit industry.[66] Yet the spirit of *Commonwealth* has been followed in other decisions.[67]

SECTION 30.08 Place of Arbitration

The arbitration clause need not specify where the arbitration is to take place. Any contractual designation of locale will be given effect as long as the selection is reasonable.[68] If hearing site selection is based on factors that will expedite the process and make it less expensive, it will certainly be considered reasonable. On the other hand, if the place is designated by the stronger party to frustrate the weaker party's right to have disputes heard, such a clause will not be given effect.[69] Section 11 of the CI Rules provides that if the parties cannot agree on the hearing location, the AAA shall determine the location and the decision shall be final and binding.[70]

In 1991, Wisconsin enacted legislation invalidating any provision in a contract for improvement of land in Wisconsin that requires that any litigation, arbitration, or other dispute resolution process occur in another state.[71] Because of this new statute, it would seem unlikely that any arbitration administrator would schedule an arbitration involving a Wisconsin construction contract outside Wisconsin.

SECTION 30.09 Multiple Party Arbitrations: Joinder and Consolidation

Frequently, a linked set of construction contracts contains identical arbitration provisions. For example, there are identical arbitration clauses in AIA standard contracts between owner and contractor and between owner and architect. Likewise, identical arbitration clauses are contained in the AIA standard contracts between architect and consulting engineer and between prime contractor and subcontractor.

Suppose there is a building collapse and the owner wishes to assert a claim. It may not be certain whether the collapse resulted from poor design or poor workmanship. The first would be chargeable primarily to the design professional and the second to the contractor. The owner may wish to arbitrate. Suppose there are identical arbitration clauses in the owner's contracts with the design professional and contractor. The owner *can* arbitrate separately with each, but there are disadvantages. Two arbitrations will very likely take longer and cost more than one. The owner may lose *both* arbitrations. (Although this may be unpalatable to

---

[66]*Garfield & Co. v. Wiest*, 432 F.2d 849 (2d Cir.1970), cert. denied, 401 U.S. 940 (1971) (stock exchange arbitration: waiver); *Baar & Beards, Inc. v. Oleg Cassini, Inc.*, 30 N.Y.2d 649, 282 N.E.2d 624, 331 N.Y.S.2d 670 (1972) (need for specialized knowledge and skill); *Reed & Martin, Inc. v. Westinghouse Elec. Corp.*, 439 F.2d 1268 (2d Cir.1971) (arbitrator need not provide complete and unexpurgated business biography); *William B. Lucke, Inc. v. Spiegel*, 131 Ill.App.2d 532, 266 N.E.2d 504 (1970) (bias too remote).
[67]*Sanko S.S. Co. v. Cook Indus., Inc.*, 495 F.2d 1260 (2d Cir.1973); *J.P. Stevens & Co. v. Rytex Corp.*, 41 A.D.2d 15, 340 N.Y.S.2d 933, affirmed 34 N.Y.2d 123, 312 N.E.2d 466, 356 N.Y.S.2d 278 (1974) (stating major burden of disclosure properly falls upon arbitrator). The cases are discussed generally in Annot., 56 A.L.R.3d 697 (1974). For CI Rules, see Appendix F.
[68]*Central Contracting Co. v. Maryland Cas. Co.*, 367 F.2d 341 (3d Cir.1966). Similarly, in an international context, see *Republic Int'l Corp. v. Amco Eng'rs, Inc.*, 516 F.2d 161 (9th Cir.1975).

---

[69]*Player v. George M. Brewster & Son, Inc.*, supra note 8.
[70]This provision was upheld in *Reed & Martin, Inc. v. Westinghouse Elec. Corp.*, supra note 66. See also *Aerojet-General Corp. v. Am. Arbitration Assoc.*, 478 F.2d 248 (9th Cir.1973).
[71]West Ann.Wis.Stat. § 779.035(1m)(b).

the owner, the result may be correct if the design professional performed in accordance with the standards required of design professionals and the contractor executed the design properly.)

Inconsistent findings may result. For example, one arbitrator may conclude that the design professional performed in accordance with the professional standard while the other did not. One may conclude that the contractor followed the design while the other did not. To avoid this and save time and costs, the owner may wish to consolidate the two arbitrations.

Suppose the owner demands arbitration only with the contractor. The contractor may believe that the principal responsibility for the collapse was defective design. In such a case, it might wish to add the design professional as a party to the arbitration or, as stated in legal terminology, "join" the design professional in the arbitration proceedings.[72]

A contractor can be caught in a similar dilemma if it felt that the responsibility for the owner's claim against it was work of a subcontractor. In such a case, the prime contractor may wish to seek arbitration with the subcontractor and consolidate the two arbitrations. Similarly, the architect may find it advisable to consolidate any arbitration she might have with her consulting engineer and any arbitration proceedings with the owner.

Similar problems surfaced in early American and English law when commercial transactions became more complex. The common law, or King's courts, refused to permit joinder and consolidation. The equity courts did so if claims resulted from the same transaction and the dispute could be handled efficiently in one lawsuit. Modern American procedural law ultimately adopted the view of the equity courts.

Modern courts have had to face this problem in the arbitration context. The first cases came to different conclusions on the issue of whether the trial court had the power to consolidate separate arbi-

trations.[73] Where it had been held that this power did not exist, some legislatures enacted legislation authorizing consolidation and joinder.[74] The tendency of the modern cases has been to consolidate existing arbitrations or to permit adding a party to an existing arbitration if that party has agreed to arbitrate.[75] As to this issue, local cases and statutes must be consulted.

Suppose the arbitration clause bars joinder without express consent of both parties. Would such a clause preclude a court from consolidating two arbitrations, one of which involved the design professional?

Current AIA Standard Documents do not permit consolidation of arbitrations between architect and owner and between owner and contractor without consent of all parties.[76] Consolidation of separate arbitrations between owner and prime contractor and prime contractor and subcontractor are permitted if a common question of fact or law requires resolution of the dispute.[77]

Why treat the two consolidation questions differently? One reason given is that an arbitration between architect and owner will involve a determination of whether the architect has lived up to the professional standard, while the issue between

---

[72]This assumes that contractor and design professional have agreed to arbitrate under the same rules as owner and contractor. Ordinarily, contractor and design professional do not have a contract, but they can agree to arbitrate their disputes. "Joining" the design professional as a party to the original arbitration is similar to but procedurally slightly different from consolidation. The latter merges two existing arbitrations into one.

[73]Compare *Grover-Diamond Assoc. v. Am. Arbitration Ass'n,* 297 Minn. 324, 211 N.W.2d 787 (1973) (power to consolidate), with *Stop & Shop Companies, Inc. v. Gilbane Bldg. Co.,* 364 Mass. 325, 304 N.E.2d 429 (1973) (no power to consolidate in absence of statute or specific clause permitting consolidation).

[74]See West Ann.Cal.Civ.Proc.Code § 1281; Ann.Mass.Gen. Laws c.251 § 2A.

[75]*Maxum Foundations, Inc. v. Salus Corp.,* 817 F.2d 1086 (4th Cir.1987); *Higley South, Inc. v. Park Shore Dev. Co., Inc.,* 494 So.2d 227 (Fla.Dist.Ct.App.1986) (AIA Doc. A201); *Manchester Township Bd. of Educ. v. Carney,* 199 N.J.Super. 266, 489 A.2d 682 (1985) (dictum); *Episcopal Housing Corp. v. Fed. Ins. Co.,* 273 S.C. 181, 255 S.E.2d 451 (1979). For a thorough canvassing of this issue, see Stipanowich, *Arbitration and the Multi-Party Dispute: The Search for Workable Solutions,* 72 Iowa L.Rev. 473 (1987).

[76]AIA Doc. A201, ¶ 4.5.5. The language was tortured to *permit* consolidation in *Garden Grove Community Church v. Pittsburgh-Des Moines Steel Co.,* 140 Cal.App.3d 251, 191 Cal.Rptr. 15 (1983).

[77]AIA Doc. A201, ¶ 4.5.5. Some cynics have observed that the reason for the differentiation is that for tactical reasons architects do not wish to participate in the same arbitration with contractors. Their interests are not so directly at stake when there is a consolidation between the arbitrations of owner and contractor and of contractor and subcontractor.

owner and contractor is whether the contractor has performed in accordance with the contract documents. Although there may be some *slight* difference in the legal standards, this is not sufficient justification to bar consolidation.

For reasons mentioned, consolidation and joinder are generally desirable. The possibility of confusion because of the potentially large number of parties in a consolidated arbitration can be handled by according the arbitrator the power to decide the number of parties and issues that would make consolidation or joinder too confusing. If so, a request to do so could be denied.

Robert Lamb Hart Planners and Architects v. Evergreen, Ltd.[78] demonstrated that while A201 could preclude consolidations of arbitrations without the architect's consent, there was a way to allow the contractor to pursue the owner's arbitration claim against the architect. A construction dispute led to the owner filing a demand for arbitration against the contractor and the contractor counterclaiming against the owner. During that arbitration, the contractor claimed that its problems stemmed from negligently designed plans and specifications by the architect. The owner and the architect also had a dispute, and the owner filed a demand for arbitration against the architect. The owner sought to consolidate its arbitration with the contractor and its arbitration with the architect, fearing inconsistent results from the two separate arbitrations. However, citing the language barring the architect from being added to another arbitration without his consent, the arbitrator refused the request for consolidation.

The owner and the contractor ultimately settled their dispute. As part of the settlement, the owner assigned to the contractor its rights against the architect. This allowed the contractor to pursue the owner's arbitration claim against the architect in the name of the owner.

Citing general principles under which most contract rights are freely assignable, the court permitted the contractor to whom the owner had assigned its rights to arbitrate the owner's claim against the architect. In doing so, it noted that arbitration is a favored measure of dispute resolution and that the general principles permitting free assignability of

rights favor permitting assignment to be used in such a situation. The court also noted that the architect had refused the owner's request for assistance in defending against the contractor's claim and had refused to participate as a witness in the owner's defense unless the architect was released from all liability.

Another method by which participants in the linked set of contracts can be involved in arbitration is what is called the "vouching in" system. This is a method by which a person involved in the arbitration—though not a party—is given the opportunity to participate and to seek to persuade the arbitrators to make an award in accordance with its contentions.[79]

SECTION 30.10 The Hearing

A. A Differentiation of Issues: Desirable vs. Required

Section 30.14 discusses attempts by the party satisfied with the award to have it confirmed in court or by the party dissatisfied with the award to have it vacated—that is, upset. That section refers principally to defects in the process that are of sufficient importance to justify not confirming or vacating the award. Even though complaints are made regarding the arbitration hearing that will not justify upsetting the award, examination of complaints might provide a blueprint for conducting a fair hearing. The arbitrator should consider not only what is compelled, which, as shall be seen, is relatively minimal, but also the type of hearing that will persuade the parties that they have been treated fairly.

The exact type of hearing will to a large degree depend on the intensity of feelings of the parties, the amount at stake, and the need for an expeditious decision. The discussion in this section assumes that a serious matter is brought before the arbitrators, one with sufficient economic importance to justify a careful and fair hearing. Because attorneys are frequently present in these arbitrations, unless otherwise indicated, it is assumed that the parties will be represented by legal counsel.

---

[78]787 F.Supp. 753 (S.D.Ohio 1992).

[79]Such a system is described in detail in *Perkins & Will v. Syska & Hennessy & Garfinkel, etc.*, 50 A.D.2d 226, 376 N.Y.S.2d 533 (1975).

This assumption does not negate the possibility or even likelihood that many arbitrations do not justify some of the steps suggested because of matters adverted to earlier in this paragraph. Nor does it ignore the important differentiation between arbitration and litigation as to speed. Over-judicialization of the arbitration process by giving to it those attributes of the legal process that led to arbitration in the first place is clearly undesirable.

B. Waiver

Parties can agree to submit the dispute to arbitration based solely on written statements, the contract documents, or any other written data the parties feel relevant. Such a paper submission can save time and could be valuable in minor disputes. However, such submissions are often deceptive and may provide insufficient information to make an award. Even if the parties have agreed that no hearing is necessary, an arbitrator can request a hearing in the presence of both parties if resolving the dispute requires it.

C. Time

The arbitrators should attempt to schedule the hearing as early as possible, but the arbitrators must take into account the time needed to prepare for the hearing. A complicated dispute may require the opinions or services of engineers, architects, accountants, attorneys, and photographers, among others. If adequate time is not allowed for preparation, continual requests will be made for recesses after the hearing process has begun.

Arbitrators should grant reasonable party requests to postpone a scheduled hearing or recess a hearing early. However, arbitrators must take into account that requests for delays and recesses are sometimes bargaining tactics used by the party who feels the other party's financial position will not tolerate delay. Arbitrators should avoid disrupting the process for their own convenience.

D. Proceeding Without the Presence of One of the Parties

Suppose one of the parties to the arbitration indicates that it will not participate in the hearings. Under such conditions, should the arbitrator proceed with the hearings, or can the party who *does* attend be awarded the amount of the claim? It is better

practice for the arbitrators to hear the evidence submitted by the party attending before making the award. Obviously, doubts will be resolved against the party who has chosen not to attend the hearings.

Paragraph 30 of the CI Rules states that the arbitration may proceed if a party without notice fails to be present or obtain a postponement, but an award under this paragraph cannot be made "solely on the default of a party." The party attending must present such evidence as the arbitrator may require.

E. The Arbitrators

At the outset, the parties should be permitted to question the arbitrators relating to any matters that could affect their impartiality. The arbitrators must disclose matters that could affect their impartiality.

The arbitrators should comply with any state arbitration laws requiring arbitrators to take an oath at the beginning of a hearing. Even if not required, it is good practice for arbitrators to take an oath that they will conduct the hearing and render their award impartially and to the best of their ability.

F. Rules for Conducting the Hearing

Generally, arbitration clauses do not prescribe detailed rules relating to the method of conducting the hearing. Certain arbitration statutes give general directives, such as requiring that the arbitrator permit each party to present its case and to cross-examine witnesses for the other party. Arbitration associations or trade groups that conduct arbitration often have simple rules relating to the conduct of the hearing. In the absence of rules, the arbitrator determines how the hearing is to be conducted.

G. Opening Statements

In a complicated arbitration, or even in matters that may not appear to be complicated, it is often helpful to permit the parties or their attorneys to make a brief opening statement. The statement can help the arbitrator determine which evidence is relevant to the dispute and can serve the function of reducing the number of issues by having the parties agree to certain facts and issues.

H. Production of Evidence: Subpoena Powers

Most states give arbitrators the power to issue a subpoena that compels witnesses to appear and testify and requires persons to bring in relevant records. Without such power, the arbitrator cannot compel witnesses to appear or documents to be produced.

Usually the arbitrating parties produce witnesses and supply whatever records are advantageous to their position. If issues can be resolved more easily if certain witnesses are produced or certain documents are presented, the arbitrator can resolve those questions against the party who refuses to produce the witnesses or records within its control. If the arbitrator indicates this to a reluctant party, the latter is likely to produce the witnesses or the records.

I. Legal Rules of Evidence

The rules of evidence applied in courts need not be followed by the arbitrator. Principally, the rules of evidence that can be dispensed with are those that relate to the *form* that evidence must take to be admissible. However, certain principles that are part of the legal rules of evidence should guide the arbitrator when determining whether material submitted by the parties should be considered. These principles relate to *relevance* and *administrative expediency*.

The arbitrator should not go into matters not germane to the dispute. This is often difficult to determine. At the outset, the arbitrator should not cut off a line of testimony that may not appear to be relevant at the moment it is presented. Perhaps this testimony will become relevant as the hearing proceeds. However, the arbitrator should ask the party presenting the evidence what it intends to establish. If it is then determined that what the party intends to establish is not germane to the dispute, the evidence should be disregarded and testimony cut off. The evidence presented should be relevant and should not be cumulative. Once a fact has been established, it is usually not necessary to establish it again.

J. Documentary Evidence

As stated, the many technical rules of evidence that relate to the admissibility of documentary evidence need not be followed by the arbitrator. Any documentary evidence submitted by the parties can be examined provided that it is relevant and not cumulative. The authenticity of the document can be taken into account. An excessive preoccupation with form, notarial seals, and witnesses to the document can slow down the hearing. Unless a party questions the authenticity of a document, the arbitrator should consider it.

K. Questioning Witnesses

One of the important constituents of a hearing is the testimony of witnesses. Failure to hear testimony of witnesses is likely to be procedural misconduct that can vitiate any award made by the arbitrator. Arbitrators should follow any state laws that require the witnesses be placed under oath. Even where the law does not require this, a simple oath adds to the dignity of the hearing and may induce truthful testimony.

Generally, legal rules of evidence prohibit an attorney from asking witnesses leading questions (questions that suggest the answer in the question). However, leading questions are permitted by the opposing attorney. This process, called cross-examination, is designed to test the credibility of the witness and expose dishonest or inaccurate statements.

Arbitrators need not follow these courtroom rules. However, cross-examination, whether by the opposing attorney or by the arbitrators, can be a useful device to test the veracity of the witness. Sometimes the narrative method permits the witness to testify in a logical and understandable fashion. However, the form of examining witnesses lies largely within the discretion of the arbitrators.

L. Visiting the Site

Generally, the hearing will be conducted in a hearing room or in an office. It is possible, and often helpful, to conduct a hearing at the site where the evidence is available to the arbitrator or arbitrators. Even when the hearing is not conducted at the site, the arbitrator—either on her own motion or when requested by a party or the parties—may view the premises. Preferably, the viewing of the premises should not be done without the presence of the arbitrating parties or their attorneys.

M. Ex Parte Communications

Ex parte communications are information or arguments communicated by one party to a dispute or by a third party to the person deciding the dispute without the knowledge of the other party. It would be improper for a judge to receive privately communicated information relative to a pending case from one of the parties, one of the attorneys for the parties, or a third party not connected with the case. The attorneys who represent the litigating parties should know what communications are being made to the judge in order to respond to them or to point out inaccuracies. This is one reason why arbitrators should notify the parties if they plan to view the premises and set a time that will enable the parties to be present.

Section 29 of the CI Rules prohibits any communication between the parties and the arbitrator except at the oral hearing.

N. Transcript

Normally, there is no requirement that testimony be transcribed or that a written transcript be made. Some arbitrators prefer to have the testimony transcribed and reproduced for their own use as well as for settling any disputes that may arise between the arbitrators and the parties over the exact testimony of witnesses.

O. Reopening Hearing

After the hearing has been closed, either party may request to reopen the hearing to introduce additional evidence. The arbitrator can determine whether to reopen the hearing. Newly discovered evidence usually will be a sufficient basis to reopen the hearing as long as the arbitrators have not yet made their award. If the evidence that is proposed to be introduced at an additional hearing could have been available for the original hearing or is merely cumulative, the arbitrator should not reopen the hearing. However, such matters are largely within the discretion of the arbitrator.[80]

SECTION 30.11 Substantive Standards

Sometimes the arbitration clause or the rules under which the arbitration is to be held give general guidelines to the arbitrator regarding standards by which to decide the dispute.

Some arbitration clauses or rules permit the arbitrator to do almost anything regardless of the language of the contract as long as what is done accords with justice or fairness. As a rule, however, arbitrators decide the dispute based on the evidence, and in a contract dispute, the contract terms play a central role.

Arbitrators need not follow case precedents. One of the reasons frequently given to arbitrate is that the arbitrator is free to make a proper decision *without* the constraint of earlier precedent or rules of law. However, in complicated arbitrations, attorneys may present legal precedents in an attempt to persuade the arbitrators. Although the arbitrators would be free to consider these precedents, clearly they would not be required to follow them.

Must the arbitrator apply legal rules that would provide a defense to the claim were it litigated? For example, suppose the claim would be barred because of the statute of limitations[81] or because the claimant is not licensed in accordance with state law.[82] Usually such defenses are asserted when the demand for arbitration is made. The party opposing arbitration goes to court contending that the contract including the arbitration clause no longer has any legal validity because the claim has been made beyond the period allowed by law. Similarly, the party may seek to oppose arbitration by contending that the party demanding arbitration cannot be allowed to recover because it does not have the requisite license.

If the matter has proceeded to arbitration and the claims have not been asserted as a *bar* to arbitration, the claims may have been waived. If asserted in the arbitration, arbitrators generally have discretion to apply them. Awards are generally confirmed even if based on mistake of law.[83] If the

---

[80]But see *Manchester Township Bd. of Educ. v. Carney,* supra note 75 (arbitration award overturned where arbitrators refused to hear rebuttal testimony).

[81]As to statutes of limitation, see Annot., 94 A.L.R.3d, 533 (1979).
[82]As to licensing, see *Merkle v. Rice Constr. Co.,* 271 So.2d 220 (Fla.Dist.Ct.App.1973) (issue for the arbitrator); *Parking Unlimited, Inc. v. Monsour Medical Foundation,* 299 Pa.Super. 289, 445 A.2d 758 (1982) (arbitrator upheld). See also *Starr v. J. Abrams Constr. Co., Inc.,* 16 Mass.App. 74, 448 N.E.2d 1311 (1983). The court affirmed an award despite the parties' having misled the FHA by failing to inform FHA of a special agreement as to costs. But see *Loving & Evans v. Blick,* 33 Cal.2d 603, 204 P.2d 23 (1949) (refused to confirm award, as party not licensed). (For a discussion of the rights of unlicensed parties, see Section 10.07(B).)
[83]See Section 30.14, which indicates that in some states, egregious errors of law may be the basis for vacating an arbitrator's award.

defenses appear technical rather than substantive, it is likely that the arbitrators will not bar the claim once the arbitration has proceeded.

SECTION 30.12 Remedies: Interim Relief

Commonly, the arbitration clause or rules under which the arbitration is being held states remedies that can be awarded. For example, § 43 of the CI Rules allows the arbitrator to grant any remedy for relief that is "just and equitable and within the scope of the agreement of the parties, including, but not limited to, specific performance of a contract."

Usually the arbitrator issues a money award, stating that one party owes the other party a designated amount of money. However, in a dispute between a developer and a builder, the arbitrator made an award that was not sought by either party. He ordered the builder to buy back the buildings it had built. There were many serious defects and numerous code violations. The buildings were of slight value and exposed the developer to liability.[84]

A Massachusetts case revealed a lacuna in the CI Rules.[85] As a result the CI Rules were amended to give the arbitrator the power to order a party to provide security for any award that may be issued. Paragraph 34, put into effect on January 1, 1993 (see Appendix F), states that

> . . . the tribunal may take whatever interim measures it deems necessary with respect to the dispute, including measures for the conservation of property.

The tribunal "may require security for the cost of such measures."

The arbitrator may also assess arbitration fees and expenses equally or in favor of any party and any administrative expenses in favor of AAA.[86] No specific provision allows the arbitrator to award attorneys' fees. However, *Harris v. Dyer*,[87] which involved AIA Doc. A201, awarded the winning party who had filed a mechanics' lien its attorneys' fees

incurred in arbitration. State law granted the prevailing party attorneys' fees connected to filing and foreclosing a mechanics' lien.

As a result of the proliferation of punitive damage claims in litigation, it is inevitable that the law would have to deal with whether an arbitrator can award punitive damages. While New York appeared to be unwilling to enforce a punitive damages award made by an arbitrator,[88] the federal courts have done so.[89] It has been suggested that arbitrators have this power to give them the opportunity to do complete justice in the case.[90] In any event, punitive damages awards in commercial construction will be rare.

Suppose the arbitrator specifically *orders* a party to perform in accordance with a contractual promise. Such an order cannot be legally enforced by the arbitrator, because the arbitrator does not have the power possessed by a judge to fine or jail a person who does not perform in accordance with a coercive court order. This absence of enforcement power on the part of the arbitrator is the reason arbitration awards are brought to court for enforcement when the parties do not comply.

Grayson-Robinson Stores, Inc. v. Iris Construction Corp.[91] involved a specific order made by an arbitrator in a dispute over a lease between a department store and shopping center developer. The developer did not build a shopping center, claiming it could not obtain funds. The arbitrator did not find this to be a defense and ordered the developer to build the shopping center. When the developer refused to comply, the department store sought confirmation of the award in court. Generally, courts do not specifically order contracts to be performed (see Section 6.03). Yet the court enforced the arbitrator's award, indicating that the arbitrator may have a remedial power not possessed by the judge.

Sometimes arbitrators order that the party perform and state in the award that failure to perform in accordance with the order will entitle the other party to a specified amount of damages. This tech-

---

[84]*David Co. v. Jim W. Miller Constr. Co.*, 444 N.W.2d 836 (Minn.1989).

[85]*Charles Constr. Co. v. Derderian*, 412 Mass. 14, 586 N.E.2d 992 (1992).

[86]CI Rules § 43. See also Annot., 57 A.L.R.3d 633 (1974).

[87]292 Or. 233, 637 P.2d 918 (1981).

[88]*Garrity v. Lyle Stuart, Inc.*, 40 N.Y.2d 354, 353 N.E.2d 793, 386 N.Y.S.2d 831 (1976).

[89]Raytheon Co. v. Automatic Business Systems, Inc., 882 F.2d 6 (1st Cir.1989).

[90]Stipanowich, *Punitive Damages in Arbitration: Garrity v. Lyle Stuart, Inc. Reconsidered*, 66 B.U.L.Rev. 953 (1986) (a thorough analysis).

[91]8 N.Y.2d 133, 168 N.E.2d 377, 202 N.Y.S.2d 303 (1960).

nique, where enforced, can give a method of enforcement without seeking court confirmation. However, such a technique is not likely to be effective unless the alternative damage claim is high enough to coerce performance. Although this may obtain enforcement of the award without going to court, California has held that its arbitration law did not authorize imposition of such an economic sanction. But it saw no legal impediment to enforcement of an agreement by the parties giving the arbitrator this power.[92]

The arbitrator's determination of whether damages or a specific order is appropriate should take into account whether the parties will continue to be able to work together after the arbitration. One reason that courts are hesitant to order parties to perform in accordance with promises is that by the time the matter has reached court, the parties will no longer cooperate with each other. The same can be true in arbitration. However, if there is a cooperative attitude between the parties and if the work continues to be performed during arbitration, it may be useful to specifically order that work be performed rather than to award damages.

SECTION 30.13 Award

Before making the award, the arbitrators review any documents submitted and listen to or read any transcription of the hearings. They consider any briefs that may have been submitted by the parties or their attorneys. Submission of briefs is uncommon except for disputes involving large amounts of money. The decision need not be unanimous unless the arbitration clause or the rules under which the arbitration is being held so require.

The form of the award can be simple. The arbitrator need not give reasons for the award. There are arguments for and against a reasoned explanation of the decision accompanying the award. An explanation may persuade the parties that the arbitrators have considered the case carefully. This may lead to voluntary compliance, which is obviously better than costly court confirmation.

A disadvantage of giving an explanation is the possibility that the dissatisfied party or parties may refuse to comply and seek to reopen the matter by objecting to the reasons given. Yet the scope of judicial review may be a reason *for* explaining an award. For example, one court refused to confirm an award that appeared to have ignored a contract clause even though the arbitrator had the *power* to refuse to enforce an unconscionable clause.[93] Had the arbitrator in his award stated that the clause had been disregarded because it was unconscionable, the award would have been sustained. An even greater argument against reasons accompanying the award is the additional time and expense entailed. Making the arbitration too much *like* a court trial can lose some of the advantages of arbitration.

Taking all of this into account, a short, reasoned explanation accompanying the award is advisable though not required.

The rules under which an arbitration is conducted often specify when the award must be made. For example, § 41 of the CI Rules requires an award be promptly made and, unless otherwise agreed or required by law, "no later than thirty days from the date of closing the hearing, or if the AAA's transmittal of the final statements and proofs to the arbitrator." Failure to make the award by the designated time may terminate the jurisdiction of the arbitrators, although more commonly, the parties agree to waive any time deadlines. More important, failure to make the award in the time required may deprive the arbitrators of any quasi-judicial immunity (discussed in Section 30.16). If there are no specific time requirements, the award must be made within a reasonable time.

SECTION 30.14 Enforcement and Limited Judicial Review

Failure to comply with an arbitration award may necessitate judicial involvement and review. A party wishing enforcement may have to go to court to obtain confirmation. The party seeking to challenge the award may go to court and ask that the award be vacated.

Most state arbitration statutes specify grounds for reviewing an arbitrator's award. As to grounds, § 12 of the Uniform Arbitration Act, enacted in

---

[92]*Luster v. Collins*, 15 Cal.App.4th 1338, 19 Cal.Rptr.2d 215 (1993) (arbitrator ordered wrongdoer to pay $50 a day for each day wrongful act continued).

[93]*Granite Worsted Mills, Inc. v. Aaronson Cowen, Ltd.*, 25 N.Y.2d 451, 255 N.E.2d 168, 306 N.Y.S.2d 934 (1969).

whole or with minor variations in a substantial number of states, permits an award to be vacated (set aside) where any of the following exists:

(1) The award was procured by corruption, fraud or other undue means;

(2) There was evident partiality by an arbitrator appointed as a neutral or corruption in any of the arbitrators or misconduct prejudicing the rights of any party;

(3) The arbitrators exceeded their powers;

(4) The arbitrators refused to postpone the hearing upon sufficient cause being shown therefor or refused to hear evidence material to the controversy or otherwise so conducted the hearing, contrary to the provisions of Section 5, as to prejudice substantially the rights of a party; or

(5) There was no arbitration agreement . . . and the party did not participate in the arbitration hearing without raising the objection;

But the fact that the relief was such that it could not or would not be granted by a court of law or equity is not grounds for vacating or refusing to confirm the award.

Section 13 of the Act allows modification or correction of an award within ninety days after delivery of a copy of the award where any of the following exists:

(1) There was an evident miscalculation of figures or an evident mistake in the description of any person, thing or property referred to in the award;

(2) The arbitrators have awarded upon a matter not submitted to them and the award may be corrected without affecting the merits of the decision upon the issues submitted; or

(3) The award is imperfect in a matter of form, not affecting the merits of the controversy.

Similar language is contained in the Federal Arbitration Act.[94] Grounds for vacating are limited and principally look to serious procedural misconduct on the part of the arbitrators.[95] Similarly, case decisions have employed language indicating a very limited judicial review of arbitration awards, mainly the manner of holding the arbitration. One court stated that the court will not inquire whether the determination was right or wrong.[96] Another stated that errors of fact or law are not sufficient to set aside the award.[97] Courts have held that an error of law was not reviewable unless the arbitrator gave a completely irrational construction to the provision in dispute.[98] Another stated that honest errors were not reviewable.[99] Another stated that arbitrators may apply their own sense of justice and make an award reflecting the spirit rather than the letter of the agreement.[100] Finally, California, pointing to the statutory grounds as exclusive and overruling earlier precedents, held that with certain exceptions, an arbitrator's decision is not reviewable for errors of fact or law whether or not such error appears on the face of the award and causes substantial injustice to the parties.[101]

The arbitrator's power on evidentiary matters was demonstrated in *Norwich Roman Catholic Diocesan Corp. v. Southern New England Contracting Co.*[102] The arbitrator did not examine the specifications that defined excavation materials that were designed to compute the unit prices for the work. The court held that because the arbitrator had ju-

---

[94]9 U.S.C.A. § 2.

[95]*Manchester Township Bd. of Educ. v. Carney*, supra note 75 (refusal to hear rebuttal evidence).

[96]*Drake-O'Meara & Assoc. v. American Testing & Eng'g Corp.*, 459 S.W.2d 362 (Mo.1970). See also *Seither & Cherry Co. v. Illinois Bank Bldg. Corp.*, 95 Ill.App.3d 191, 419 N.E.2d 940 (1981), noted in 22 A.L.R. 4th 356 (1983).

[97]*Mars Constructors, Inc. v. Tropical Enterprises Ltd.*, 51 Hawaii 332, 460 P.2d 317 (1969). See also *Clinton Water Ass'n v. Farmers Constr. Co.*, 163 W.Va. 85, 254 S.E.2d 692 (1979); *Lawrence v. Falzarano*, 380 Mass. 18, 402 N.E.2d 1017 (1980). But see *West v. Jamison*, 182 Ga.App. 565, 356 S.E.2d 659 (1987) (refused to confirm award because of *obvious* mistake of law).

[98]*Firmin v. Garber*, 353 So.2d 975 (La.1977) (award upheld, not grossly irrational but simply debatable); *Maross Constr. Inc. v. Central N.Y. Regional Transp. Auth.*, 66 N.Y.2d 341, 488 N.E.2d 67, 497 N.Y.S.2d 321 (1985) (upheld unless totally irrational). Similarly, noting the possibility that awards may be bizarre or completely irrational, *Pacific Gas and Electric Co. v. Superior Court*, supra note 12, held an award which was such an egregious mistake so as to be an arbitrary remaking of the contract would not be upheld. Such an award would be completely outside expectations of the parties. This court also would extend *more* judicial review, limited as it is, to clauses to arbitrate *future* disputes, as compared to submitting an *existing* dispute to arbitration.

[99]*Reith v. Wynhoff*, 28 Wis.2d 336, 137 N.Y.2d 33 (1965).

[100]*Matter of J.M. Weller Assoc., Inc.*, 169 A.D.2d 958, 564 N.Y.S.2d 854 (1991), appeal denied (June 4, 1991).

[101]*Moncharsh v. Heily & Blase*, 3 Cal.4th 1, 832 P.2d 899, 10 Cal.Rptr.2d 183 (1992), rehearing denied (Sept. 24, 1992).

[102]164 Conn. 472, 325 A.2d 274 (1973).

risdiction, it would not go into the merits of the dispute or evidentiary issues. The court pointed to CI Rules § 30 (currently § 31) under which the arbitration was being administered which stated that the arbitrator would be judge of the admissibility of the evidence.

Despite a pattern emerging under which there would be relatively little review of an arbitration award, the issue continues to divide the appellate courts. For example, this issue fragmented the New Jersey Supreme Court in *Perini Corp. v. Greate Bay Hotel & Casino, Inc.*[103] to the degree that no majority opinion could be written.

The dispute involved the contract between a casino in Atlantic City and a construction manager under which the casino would be renovated. The arbitration consumed sixty-four days of hearings, involved twenty-one witnesses, and resulted in almost 11,000 pages of transcript. The arbitrators (2 to 1) awarded the casino lost profits of $14.5 million against the construction manager. The CM's fee was $600,000 plus reimbursables.

The arbitration was followed by three and a half years of litigation, first at the trial level, then before the intermediate appellate court, and then before the Supreme Court of New Jersey. The litigation produced five judicial opinions (*excluding* the concurring opinion of sixteen double-columned pages) of over 150 pages. Except for two dissenting judges, the rest of the court was willing to affirm the arbitrator's award, but the three-judge plurality and the two-judge concurring opinion (of a seven-person court) strongly differed as to the proper scope of judicial review.

The plurality of three judges devoted an extensive part of its opinion to examining the legal issues that had been resolved in favor of the casino. While it recognized that a simple mistake of law is not a sufficient reason to overturn the award—that ground for reversal not being in the New Jersey statute—it was willing to extensively review the award to see whether it complied with New Jersey law. It noted that an award would be sustained unless the mistake or error of law or fact resulted from a failure of intent or error so gross as to suggest fraud or misconduct.

The concurring judges felt that this would open up too many arbitration awards to judicial scrutiny

and concluded that the only basis for overturning an arbitration award should be those reasons set forth in the arbitration statute (principally those that deal with corruption, fraud, or undue means).

The plurality stated that the purpose of arbitration is to get away from the "strictures and limitations of law".[104] It noted that the parties only expected honesty and "an attempt to reach a fair and just result, [based upon] the depth of their experience with the kinds of problems put before them.[105]

The concept of limited judicial review was undoubtedly designed to encourage arbitration by limiting the likelihood that an award can be overturned in court. However, limited judicial review almost to the point of no review can make contracting parties reluctant to use the arbitration process. As shall be seen,[106] one reason sometimes given for reluctance to enter into arbitration agreements is the absence of any meaningful review of the arbitrator's decision. Yet it is hard to escape the logic of the concurring opinion in the *Perini* decision. Even opening the door as little as is done by the plurality to include gross mistakes of law creates sufficient uncertainty and encourages appeals to the courts.

The limited review does not mean that arbitration awards are *never* upset. For example, a court that articulated the standard of complete irrationality nevertheless upset an arbitrator's decision by concluding that the words of the contract were so clear that there was nothing left to interpret.[107] For all practical purposes, however, an arbitrator's decision is final.

SECTION 30.15 Insurers and Sureties

Generally, a public liability insurance policy carried by a contractor or a professional liability insurance policy carried by a design professional indemnifies the insured if it incurs liability to a

---

[103]129 N.J. 479, 610 A.2d 364 (1992).

[104]610 A.2d at 396.

[105]Ibid.

[106]See Section 30.17.

[107]*O-S Corp. v. Samuel A. Kroll, Inc.*, 29 Md.App. 406, 348 A.2d 870 (1975). Similarly, an award was set aside because it would violate state public bidding law in *State of Conn. v. R.A. Civitello Co., Inc.*, 6 Conn.App. 438, 505 A.2d 1277 (1986).

third party. Insurers in fixing their rates must be able to predict their losses. The professional liability insurer assumes that its liability will be based on the normal professional activity performed by its insured. Similarly, an insurer who issues public liability insurance to a contractor expects to indemnify the insured only if accidents occur that arise out of its normal construction activities.

To standardize risks, insurers usually exclude liability assumed by contract. An insured who wishes to do so, such as a contractor who has contractually agreed to indemnify the owner or architect, must obtain a special endorsement. This gives the insurer the opportunity to examine the risk and decide whether to accept it, refuse it, or accept it with a premium adjustment.

Arbitration, though not *imposing* liability by contract, substitutes one form of dispute resolution for another. This is a more serious problem for professional liability insurers. Unlike claims against a contractor, claims against a design professional are likely to be made by parties with contracts (often containing arbitration clauses) against the insured design professional.

Generally, insurers are at least *wary* of arbitration. Although most do not specifically *exclude* arbitration, they may counsel that it not be used, suggest that only a certain clause be used, or, in rare cases, deny coverage for a design professional who intends to use or uses a general arbitration clause. If insurers are aware of such a clause at the time they insure or make no objection if they are asked their opinion, they have consented. Design professionals should check their policies, bring them to the insurers if they are in doubt as to the advisability of arbitration or its effect on coverage, and notify their insurer if they plan to submit an existing dispute to arbitration.

Sureties are discussed in greater detail in Chapter 33. The surety issues a bond to the prime contractor to protect the obligee—usually the owner—from the risk that any claim the owner will have against the prime contractor cannot be collected. Usually the surety bond either incorporates the construction contract by reference or refers to it.

Some cases have involved attempts by the surety to compel arbitration when the principal parties to the dispute, such as owner and prime contractor or prime contractor and subcontractor, do not wish to arbitrate. Usually the surety cannot compel arbitration, as it is not a party to the construction contracts that contain arbitration clauses.[108]

But the more difficult problems are attempts by sureties to distance themselves from the arbitration and then refuse to pay the arbitration award. This may be done simply to delay payment by invoking a technical defense. But it is also possible that the surety in good faith feel it cannot participate in the arbitration because it does not have sufficient information or records.

As a rule, the obligee (owner on a prime's bond or prime on a subcontractor's bond) prefers that the surety participate in the arbitration. Participation will ensure that the surety will pay or at least be liable for the award. This overcomes any disadvantage to the obligee of possibly having to face two opponents and two attorneys. Most often, though, the surety does not participate and then seeks to deny its responsibility to pay the award.

The surety must pay the award if it has involved itself in some way in the process, such as taking an assignment of the construction contract, being subrogated (placed in the contractor's position) to the contractor's rights, completing the project, or participating in the arbitration. It must pay the award if it has expressly promised to participate or be bound by the award.[109]

Generally, a surety will be obligated to pay the arbitration award if it had notice of the arbitration and an opportunity to defend. Even if the surety is not obligated to participate, its decision to forgo participation when it knows of the arbitration and

---

[108]*Aetna Cas. & Sur. Co. v. Jelac Corp.*, 505 So.2d 37 (Fla.Dist.Ct.App.1987). But see *J&S Constr. Co. v. Traveller's Indem. Co.*, 520 F.2d 809 (1st Cir.1975) (surety could invoke arbitration, as construction contract incorporated in bond).

[109]*Town of Melville v. Safeco Ins. Co.*, 589 So.2d 625 (La.App.1991). The Supreme Court sent the case back for further proceedings, as there were issues of fact related to personal defenses of the surety that may not be precluded by the arbitration award. 593 So.2d 376 (La.1992).

has had the opportunity to defend should result in its being bound to pay any arbitration award.[110]

To bind the surety to any arbitration award, the owner should incorporate the construction contract into the bond, notify the surety of any default, notify the surety of any arbitration, and invite the surety to participate. If the surety chooses not to participate, it should still have to pay the award.

Increasing legislative regulation of the settlement practices of sureties may make it more difficult for sureties to refuse to participate in any arbitration and use this as the basis for refusing to pay an arbitration award against its principal debtor.[111]

SECTION 30.16 Arbitrator Immunity

Some American states grant the design professional quasi-judicial immunity when acting as judge. More clearly, arbitrators are granted quasi-judicial immunity.[112] Arbitrators are even more like judges than are design professionals. However, *Baar v. Tigerman*[113] stripped an arbitrator of his quasi-judicial immunity when he did not make an award in accordance with the time requirements of the arbitration rules. In addition, the court refused

to grant immunity to the American Arbitration Association (AAA).

Although immunity was restored by subsequent legislation,[114] the case is a warning to arbitrators and those who run arbitral systems to set up reasonable deadlines for making the award and then to comply with them or obtain extensions.

SECTION 30.17 Arbitration and Litigation Compared

Arbitration is voluntary. Parties can *choose* to employ it. Choice involves a number of considerations, some of which are apparent and some not.

A lawsuit begins with the filing of pleadings by a lawyer. An arbitration process can start by the filing of a claim, which does not *require* a lawyer. However, the claimant may wish legal advice to decide whether arbitration or litigation is the better method to handle the dispute. Generally, it is quicker and less expensive to begin arbitration than litigation.

On the other hand, the filing fees for beginning litigation are modest compared with those required to initiate arbitration (see Section 30.02). The CI Rules employ a sliding scale of filing fees.[115] These fees apply to both claims and counterclaims, something common in construction disputes.

Arbitration and litigation differ in the ease with which basically one dispute involving a large number of claimants and claims resistors can be handled. Generally, joinder of parties and consolidation of disputes is more easily accomplished in a lawsuit than in arbitration (discussed in Section 30.09).

Another major difference is that in many states, arbitration does not allow for an effective method of discovery, the process by which attorneys in litigation can obtain evidence from the other party to prepare for the hearing. Although some states require it, an arbitration system can *contractually*

---

[110]*Raymond Intern. Builders, Inc. v. First Indem. of America Ins. Co.*, 104 N.J. 182, 516 A.2d 620 (1986). But see West Ann.Cal.Civ.Code § 2855, which appears to require that the surety be a party to the arbitration before it can be compelled to pay the award. It has been held that the surety *must* participate in the arbitration if the prime contract with an arbitration clause was incorporated into the bond. *United States Fid. & Guar. Co. v. West Point Constr. Co., Inc.*, 837 F.2d 1507 (11th Cir.1988). Similarly, see *Boys Club of San Fernando Valley, Inc. v. Fidelity & Deposit Co. of Md.*, 6 Cal.App.4th 1266, 8 Cal.Rptr.2d 587 (1992), and *St. Paul Fire & Marine Ins. Co. v. Woolley/Sweeney Hotel*, 545 So.2d 958 (Fla.Dist.Ct.App.), review denied, 559 So.2d 1666 (Fla.1989). In the latter case, the dissent noted that the surety would be bound to perform in accordance with the award but should not be required to arbitrate, a sensible distinction. Failure to participate would make the surety liable for any arbitration award. See generally Ruck, *Can a Contract Bond Surety Be Compelled to Arbitrate Claims Against It?*, 16 Forum 765 (1981).

[111]See Section 33.10(I).

[112]*Baar v. Tigerman*, 140 Cal.App.3d 979, 189 Cal.Rptr. 834 (1983). The case is noted in 67 Marq.L.Rev. 147 (1983).

[113]Supra note 112.

[114]West Ann.Cal.Civ.Proc.Code § 1280.1. The AAA was granted immunity under the statute. *American Arbitration Assn. v. Superior Court*, 8 Cal.App.4th 1131, 10 Cal.Rptr.2d 899 (1992), rehearing denied (September 11, 1992).

[115]See Appendix F.

compel it, and although the arbitrator can frequently compel the production of evidence at the hearing, none is as effective as discovery in a lawsuit.

Discovery availability demonstrates the balance between process fairness and a quick resolution of disputes. Its availability can make arbitration more desirable to attorneys. Yet its use can be costly, and its availability can slow down the process and allow one party to delay for bargaining purposes.

One of the advantages frequently given for arbitration is the informality and speed of the process. Although lawyers are generally required in litigation, arbitration does not require that the parties be legally represented. However, parties arbitrating important matters usually retain lawyers. This can hamper the arbitration, as there is no requirement that arbitrators be lawyers. Having nonlawyer arbitrators deciding legal questions argued by attorneys can be undesirable. In addition, difficult legal questions may have to be decided by nonlawyer arbitrators.

The skill with which a judge or arbitrator conducts a hearing not only determines how quickly the hearing can be completed but also affects a party's perception of the fairness of the process. Because of the greater hearing experience possessed by judges generally, the hearing process in court will be conducted by a person with greater hearing experience. Although some arbitrators who work permanently or frequently can develop the skill to conduct good hearings, in the construction industry hearing expertise is rarer than in other classes of arbitration. Arbitrators who decide construction disputes rarely make their living at arbitration. Not only does this result in less experienced arbitrators, but the part-time and, even more important, volunteer arbitrators must often take lengthy recesses in the hearings because of having to work at their profession or business.

One of the advantages of arbitration in terms of hearing speed is the freedom the arbitrator has to move the hearing along unhampered by technical rules of evidence that, when accompanied by contentious attorneys, can make court hearings excessively long. The arbitrator, unlike the judge, need *not* make a transcript of the hearing which, though helpful, is costly and time consuming.

Looking at all aspects of this problem, arbitration is likely to be quicker than judicial dispute res-

olution. Speedy hearings should not only result in faster decisions but also avoid the indirect costs of lengthy hearings, such as the unproductivity of witnesses who must attend the hearing, and direct costs, such as travel expense and attorneys' fees.

Hearing location is likely to be different. A court hearing is likely to take place at the principal city in the county where the lawsuit was commenced and whose court has jurisdiction over the dispute. This does not necessarily mean that the trial will be close to the project or convenient for witnesses. Often lawyers seek to gain tactical advantages by having the matter heard in a particular court. On the other hand, the arbitrator has more flexibility to schedule the hearing at a place that is more convenient for the witnesses. However, sometimes the arbitration clause can be deliberately designed to make it inconvenient for one of the parties to demand arbitration.

Another differentiation relates to the public nature of the hearing. Although arbitration hearings are private, the judicial process is public. Sometimes trials are reported to local newspapers. The privacy of the arbitration not only may avoid unwanted publicity but also can be a more sympathetic setting for witnesses.

One differentiation is often ignored. The courtroom and the services of judge and clerk are furnished free to the litigants. The room for the hearing and the arbitrator's fee are expenses that must be paid by the parties.

Viewing the project, often a helpful activity, is more easily accomplished in arbitration.

Quality comparisons are difficult. Arbitration has frequently been supported because of the expertise (often absent in judges) that arbitrators bring to the disputes. Doubtless, construction experience is useful. On the other hand, the dispute resolver who is *too* expert may conceive of her role not as providing a hearing and *then* making a decision but as simply deciding the dispute based on her own knowledge. On the assumption that experience in construction is likely to produce a better decision, it is worth comparing contextual experience in the two processes.

Judges have legal training and as a rule have spent ten to twenty years practicing law before they are appointed or elected to the judiciary. This experience *may* have involved construction matters, but on the average, it is likely that most judges have had little construction experience before becoming a judge. They often learn on the job. A trial judge who has been on the bench for five or ten years in a jurisdiction that has a wide range of cases may have gathered enough experience to be knowledgeable about construction matters. On the whole, though, trial judges do not bring great expertise to the resolution of construction disputes. Yet in some states, special masters or referees can be appointed to handle certain parts of the lawsuit. Often they are experienced.

Often arbitrators possess experience in the types of matters being arbitrated before them, particularly those who arbitrate under collective bargaining agreements or who arbitrate highly specialized disputes, such as disputes between members of the stock exchange, the diamond industry, or textile trade associations. Most disputes in these industries or between members of these associations are repetitive, deal with technical matters, and are handled well through arbitration.

On the other hand, the construction industry arbitration panels, though made up of persons with experience in construction such as attorneys, architects, engineers, and contractors, do not have a sufficient volume of cases to justify full-time arbitrators or even part-time arbitrators who work with sufficient regularity to develop specialized knowledge. The construction industry, though having some elements in common, is highly specialized, and an arbitrator's experience in electrical contracting may not prepare the arbitrator to handle a dispute involving road building. On the whole, though arbitrators are more likely to have had more experience in construction matters than have judges.

It would be beyond the scope of this treatise to attempt to do a detailed comparison from a substantive standpoint between arbitration awards and court judgments. However, two issues are frequently raised that merit brief comment.

It has been asserted that a dispute that runs through the arbitral process is less likely to be decided by the plain meaning of contract language than would be one that proceeds through the judicial process. To some degree, this is inherent in the concept of arbitration, which is supposed to be less formal and technical than the judicial process. Also, the arbitral process is generally not designed to create binding precedents, something that can be

created if a dispute culminates with a written opinion by an appellate court. Anecdotal evidence suggests that a party who seeks to base its claim or defense upon language of the construction contract is more likely to be successful in a judicial proceeding than in an arbitration.

The second criticism made of arbitration is that arbitrators often refuse to decide wholly on the merits but seek to give an award that will give each party something so that no one should walk away from the arbitration feeling that she has been demolished.

While opinions may differ as to the desirability of accomplishing the latter objective, again anecdotal evidence indicates that lawyers generally prefer arbitrators to decide strictly on the merits, and lawyers are not pleased when the arbitrator "splits the difference." That this is a concern of those who administer arbitration is reflected by a recent study conducted by the American Arbitration Association that sought to determine whether this is the case. The study reviewed all construction arbitrations administered by the AAA in 1990 and concluded that arbitrators decide clearly in favor of one party or the other in the vast majority of cases. According to this study, in only 12% of the cases was the award split between the parties within the 40%–59% range.

Another comparison of great importance is the degree of finality of the decision. A decision by a trial judge can be appealed to an appellate court. Although appeals on the whole have a low probability of success, a litigant who appeals will be successful if it can show that the trial judge has made an error of law or that factual findings were not supported by the evidence. On the other hand, for all practical purposes, an arbitrator's decision is final. To some, this is a great advantage of arbitration, as it ends the dispute quickly. To others, it is a disadvantage because of the immense power possessed by the arbitrator, including the power to make completely wrong decisions.

Comparing arbitration and litigation should not assume that the only alternatives are a general arbitration clause or none at all. Variations can be employed in the proper case that may be preferable to either extreme. Although the variations suggested are by no means exclusive, they do demonstrate that an arbitration clause can be "tailored" to make it preferable to either a general arbitration clause or none at all. Variations can include the following:

1. Limiting arbitration to factual disputes such as those involving technical performance standards or eliminating arbitration of other types of more "legal" disputes such as termination.
2. Specifying the place of arbitration.
3. Providing a designated person or persons as arbitrator or arbitrators.
4. Limiting arbitration to claims not exceeding a designated amount or percentage of the contract price.[116]
5. Limiting disputes to those that occur while the work is proceeding with an expedited one-person panel. (See CI Rules §§ 53–57).
6. Permitting consolidation of separate arbitrations.
7. Providing a right to discovery.
8. Limiting the award to the most fair of the last proposals or an amount between the two final proposals of the parties.
9. Eliminating the use of attorneys.
10. Making the award "nonbinding."

Obviously, the more variables, the greater the cost of obtaining agreement and expressing it. This probably inhibits specialized arbitration clauses, at least where one party cannot dictate the clause to the other. A look at the variables and attendant complexity may persuade an exasperated drafter or negotiator that it may be simpler where possible to agree to waive a jury and have all disputes tried before the judge.

Choosing a general arbitration clause may be influenced by those who issue surety bonds and professional liability insurance. If the insurer or surety will not cover work done under a contract with a general arbitration clause, this can be a significant factor in the choice.

Whether to agree to a general arbitration clause at a particular time and place requires that the choice go beyond simply comparing abstract models of dispute resolution. For example, if the choice is between taking a dispute to an efficient and competent legal system or an unknown panel of AAA arbitrators, the former is preferable. On the other hand, an inefficient court system with questionable judgment is much less preferable than a highly skilled arbitration system.

Many state dispute resolution systems have developed techniques that have eliminated the worst aspects of incompetence and delay that can cause parties to choose arbitration.[117]

The *Perini* case noted in Section 30.14 demonstrated that arbitration is not always a simple, expeditious, or inexpensive method of adjudicating commercial controversies. Any system will have its "horror story" cases, and perhaps the quality of the arbitrator's award could conceivably have been better than what would have resulted in court. It is unlikely, however, that arbitration in this case saved time or money.

Another important aspect of arbitration choice that favors arbitration relates to the likelihood that the parties will accept the arbitrator's decision without repeated and costly trips to the courthouse. This is becoming even more complicated by the expansion of jurisdiction under the Federal Arbitration Act and the increasing likelihood that arbitration may be sought in either state or federal court and under either a state arbitration statute or the federal act. This will undoubtedly add to the complexity of legal issues and may in the long run discourage arbitration.

This section has pointed to advantages and disadvantages of arbitration. The contracting parties and their attorneys should carefully consider such factors when choosing contract language dealing with dispute resolution. The choice between a general arbitration clause, a limited arbitration clause, or no arbitration clause is an important decision that should be made with care.

SECTION 30.18 Private Systems

This section lists briefly some methods that have been suggested or actually used for resolving disputes through the intervention of third parties by agreement of the disputants. Undoubtedly there are others.

A. AAA Complex Arbitrations

The American Arbitration Association (AAA), working with the newly renamed National Con-

---

[116]The Engineers Joint Contract Documents Committee limits arbitration to claims not exceeding $200,000 exclusive of interest and costs in its engineer-owner agreements. EJCDC No. 1910-1, ¶¶ G8.6.6, G8.6.7.

[117]See Section 30.19.

struction Dispute Resolution Committee, has developed *Guidelines for Expediting Larger, Complex Construction Arbitrations.* The AAA recommends that these guidelines be used in cases involving $250,000 or more. Among the key features is an administrative conference conducted by an AAA official early in the case to explore the ramifications of the dispute. Another option is a preliminary hearing run by the arbitrator to shape the proceedings and establish a schedule for exchange of information and to establish precise dates for the hearing.

B. AAA Expedited Procedures

In 1984, the CI Rules created an alternative method to expedite arbitration. Section 9 requires the use of the expedited procedures as defined in §§ 53–57 in any claim in which the claim of any party does not exceed $50,000 exclusive of interest or arbitration costs. The parties can agree to use the expedited procedures in claims that exceed that amount. The AAA can determine that those procedures shall not be applied to claims under $50,000.

The principal features of this process are informal notices as well as streamlined methods for selecting arbitrators and time and place of hearing. Perhaps most interesting, the hearing must be completed within one day unless the arbitrators for good cause schedule an additional hearing to be held in five days. The award must be made no later than five business days from the close of the hearing.

C. Mediation: Mediation-Arbitration

Mediation involves the introduction of a neutral third party, usually an individual, but occasionally a team of two co-mediators. Mediators must be knowledgeable in the subject matter and appreciate the subtleties needed to persuade each party to see the strengths and weaknesses of its respective position. Also, mediators need the trust and confidence of the disputants. The mediator does not render a decision and has no power to compel a party to submit any proposal or counterproposal, but seeks to be a catalyst to try to induce the parties to reach a voluntary settlement. When this is done successfully, the parties themselves negotiate a settlement that allows them to control the terms.

The parties may use the Construction Mediation Rules of the AAA, or they can informally select a mediator and suggest ways in which the mediator should operate. A number of private organizations offer mediation services. Mediation has become one of the favored third-party intervention methods.

Another method that has been proposed is mediation-arbitration, which has been developed in some collective-bargaining contexts and places the third party in two roles. The third party seeks to mediate disputes between the parties by aiding in the negotiation process with a view toward settlement, but if parties do not settle a dispute, the third party can use normal arbitration power to decide the dispute.

Because the third party must possess a variety of skills and would have considerable power, it is likely that this mediator-arbitrator would be selected in advance by the parties. For this reason, the system contemplates the use of the mediator-arbitrator throughout the performance period. Although this is technically available in ordinary arbitration, experience has shown that most disputes during performance are likely to be decided by the design professional. A mediator-arbitrator would very likely be called on more frequently during performance and may in practice supplant the dispute resolution role of the design professional.

Mediation-arbitration is rarely used in construction disputes. One difficulty is finding a third party in whom the contracting parties have confidence and who has not only the skill but also the time to play an ongoing role during construction. Opposition can come from design professionals or construction managers who see the injection of yet another major participant as a threat to their status and as an additional complicating factor in an already complicated system.

D. Minitrials

The minitrial has received attention in the popular press and is beginning to receive attention in the scholarly journals.[118] Although there can be many

---

[118]For discussions of minitrials, see Green, Marks, & Olson, *Settling Large Case Litigation: An Alternative Approach,* 11 Loy.L.A.L.Rev. 493 (1978); Nilsson, *A Litigation Settling Experiment,* 65 A.B.A.J. 1818 (1979); Anderson & Snipes, *Stretching the Concept of Mini-Trials: The Case of Bechtel and the Corps of Engineers,* 9 Constr. Lawyer No. 2, April 1989, p. 3; Klitgaard & Mussman, *High Technology Disputes: The Mini-Trial as the Emerging Solution,* 8 Santa Clara Computer & High Tech.L.J. 1 (1992).

variations (the process being essentially a private one made by a contract), attention has been directed toward a particular minitrial and the process it used.

After a lengthy process of negotiation, the disputants worked out an agreement for a minitrial. The parties agreed on a judge, or a neutral adviser. They were allowed an expedited discovery procedure and exchanged briefs. They were given a designated time to present their positions before executives of each disputant not *directly* connected with the dispute who have authority to settle. The hearing itself was limited to two days, moderated by the adviser. If the executives were unable to reach a settlement after the hearing, the adviser would issue a *nonbinding* opinion. The executives would meet again with the hope of settling. If they cannot, the parties can go to court, with any admissions made or the tentative opinion of the adviser *not* admissible in any subsequent trial.

E. Dispute Review Board

The dispute review board was developed in tunneling contracts. The contracting parties each appoint a member of the board, and the two select a third member. Each member pays the fees and costs connected with the board's operations. Essentially, such a board is a committee that meets periodically, talks to the active participants, walks the site, and is made aware of problems as they develop. Such a system can be expensive to operate, but if the parties are picked properly and their roles are carefully delineated, the system may save a great deal of claims overhead.

F. Claims Master

Recognizing the difficulties of litigation and current Alternative Dispute Resolution systems, one commentator has suggested the development of a claims master.[119] Such a person would be a neutral expert hired by all of the parties. This expert would be expected to identify the responsibilities of and problems caused by each party and to enforce an allocation of responsibility for time and cost overruns should they develop. Such a "master" is not

to be an advocate or an expert for either of the parties. She would be a party to the contract, and her fees would be preset and paid by all the parties based on the number of settled disputes. She would have to be given significant authority to be effective, according to the proponent. If litigation is required, the claims master would forfeit a portion of her fees. A bonus could be given if there is no litigation. Such a cost could be built into the project costs as a substitute for the ever-increasing claims overhead.

According to its proponent, the claims master would have certified credentials and licenses. The proponent also suggested that universities and professional organizations create a new curriculum for initial certification.

G. Expert Determination Process

Another system proposed would require the appointment of an expert as a substitution for arbitration. The expert's opinion would be final, as would any arbitration award. Material would be submitted by the parties to the expert, after which the expert would arrange a conference between the parties. After the conference at which the parties could present their evidence, the expert would answer the questions submitted and notify the answers in writing to the parties. Each party would bear its own costs of the process and share the cost of the expert.[120]

H. Defect Response Agreements

See Section 24.11.

SECTION 30.19 Adjuncts of Judicial Systems

Just as outsiders saw problems in the judicial system, those operating the system were well aware of the system's flaws. While wholesale restructuring was not considered possible, some judicial systems at federal and state court levels developed

---

[119]Long, *A New Role: The Claims Master*, 7 K.C.News No. 3, September 1990, p. 2.

[120]For details, see Burke & Chinkin, *Expert Determination as a Viable Alternative to Arbitration and Litigation*, 6 Int'l.Constr.L.Rev. pt. 4, 401 (1989). The rules can also be found in J. SWEET, SWEET ON CONSTRUCTION INDUSTRY CONTRACTS § 22.2 (2d ed. 1992) (supplement).

techniques to reduce the time and expense of providing a method of resolving private disputes. Two of these systems are noted in this section.

A. Special Masters and Referees

As courts have struggled with better methods of dealing with construction disputes, some have appointed individuals—called masters or referees—who are given certain powers to expedite construction litigation. Sometimes these individuals are given authority to set rules for the deposition process, which has become a costly method of obtaining information. Sometimes masters or referees informally act as mediators with a view toward persuading the litigants to settle the dispute. Finally, such individuals are sometimes authorized by the parties to resolve the dispute and are sometimes authorized by the judge to make findings of fact and conclusions of law, which can then be passed upon and adopted by the judge if the parties have given up their rights to a jury trial. Sometimes a presiding judge designates a referee or master as a temporary judge or a judge pro tem. Where the determination by the master or referee has been approved by the court or where the referee or master is a temporary judge or judge pro tem, the judgment is final and subject to the same appeal as an ordinary trial court judgment.

B. Summary Jury Trials

Summary jury trials have been developed in the federal trial courts. Although the Seventh Federal Circuit Court of Appeals has limited the trial judge's authority to compel that such a method be used,[121] it is likely that this process will be used increasingly by agreement or in those federal circuits that do not agree with the Seventh.

A jury is selected in the normal manner. The judge informs the jury that the parties have agreed in advance to an abbreviated procedure to save time and money. The jury is not told until after the proceedings that its determinations are not binding.

Each attorney makes what in a real trial would be a combined opening and closing statement. Attorneys may use charts, graphs, or other visual aids that would be used in a normal closing argument. However, no witnesses offer testimony. The other attorney responds, and the first attorney is given a limited amount of rebuttal time. The time allotted for the arguments runs between one-half and one full day. After the presentation is made by the attorneys, the jury is instructed by the judge as it would be at the conclusion of a normal jury trial. After instructions, the jury retires and then presents its verdict. Because the verdict is not binding, either party can demand a regular trial. But if the parties believe that the verdict is very likely what a real jury would determine, the mock verdict should encourage settlement. The obvious advantages to such a method are avoiding the expensive marshaling of documents, preparing a lengthy pretrial order, participating in a pretrial conference, and preparing and presenting the witnesses. If the parties take their obligation to present their cases in good faith and then negotiate in good faith after the jury verdict, this method should save time and money.

However, some believe that each jury is so different that what an advisory jury concludes may not be what a real jury would determine. Also, those who are not in favor of the summary jury trial emphasize the great difference between hearing summaries of the type that would be presented in a summary jury trial and actually seeing witnesses and listening to their testimony.

SECTION 30.20 Public Contracts

Public construction contracts involve considerations not found in private contracts. Many statutes, rules, and regulations govern the award of contracts, and the resolution of disputes under those contracts. This treatise cannot examine the details of these legal restraints because of both their complexity and their variations. However, the increased and often intense spotlighting of disputes in construction work necessitates some observations regarding dispute resolution in the context of public construction contracts.

A. Federal Procurement Contracts

Before 1978, the disputes process in federal procurement contracts was based on the disputes

---

[121]*Strandell v. Jackson County, Ill.*, 838 F.2d 884 (7thCir.1988), noted in 40 Case W.Res.L.Rev. 491 (1989–1990). See Metzloff, *Reconfiguring the Summary Jury Trial*, 41 Duke L.J. 806 (1992).

clause, which gave the contracting officer of the federal agency awarding the contract the power to decide disputed questions that arose during performance or thereafter. The contracting officer is usually a high administrative official of the agency awarding the contract whose decisions were conclusive unless appealed to the head of the agency within thirty days from the decision.

The agency appeals boards are appointed by the head of the agency. Their hearings and decisions are very much like those of a court. Before 1978, contractors had to go before the agency appeals boards if the dispute arose under the contract. They could appeal to the then Court of Claims if their claims were based on a breach of contract. Alternatively, in claims under $10,000, they could appeal to a federal district court. They could appeal from board decisions, but a board's decision was final on questions of fact if supported by substantial evidence. The Court of Claims could make its own determination on legal questions. Many federal procurement doctrines were developed that had as their objective either to keep a dispute before an agency appeals board or to allow the board to be bypassed in favor of the Court of Claims.

In 1978, Congress enacted the Federal Contract Disputes Act,[122] which gave *legislative* authorization for a disputes process that up to that time had been created solely by contract. In addition, changes were made. Appeals can no longer be taken to the federal district courts for claims under $10,000. Also, Section 605 requires contracting officers to issue decisions within sixty days on claims of $50,000 or less. Claims over that amount require a decision within sixty days or a notification of the time within which a decision will be issued. Failure to issue a decision within the time required permits the contractor to bypass the contracting officer and go directly to higher authorities. The contractor is given sixty days from the contracting officer's decision to appeal to the agency appeals board. Section 609 allows a contractor to bring an action directly to the Court of Claims.

Congress made drastic changes in 1982 in the Federal Courts Improvement Act.[123] It created a new U.S. Court of Appeals for the Federal Circuit and a new U.S. Claims Court[124] (to be distinguished from the earlier Court of Claims). Claims are initially presented to the Claims Court, and appellate functions go to the new Court of Appeals for the Federal Circuit.[125]

B. State and Local Contracts

A detailed treatment of the many state and local laws and regulations cannot be given in this treatise. It is important, though, to recognize the importance of complying with specialized requirements for disputes under such public contracts of this type. Some states have created special courts to deal with these claims. Court decisions in some states precluded arbitration of public contract disputes because this would place in the hands of private parties the power to decide public matters and expend public funds.[126] However, hostility has been fading, and contracts to arbitrate future disputes have been upheld. Some states even require the arbitration of these disputes.[127]

California's experience is interesting, and although it may not be typical, it demonstrates the continued interest and concern over the disputes resolution process. Before 1978, California state agencies followed the traditional pattern of having disputes resolved initially by a high official of the agency—the state engineer. The state engineer could issue a decision that had a substantial

---

[122]41 U.S.C.A. §§ 601 et. seq.
[123]96 Stat. 25, Pub.L. 97-164. See Anthony and Smith, *The Federal Court Improvement Act of 1982,* 13 Pub. Contract L.J. 201 (1983); Miller, *The New United States Claims Court,* 32 Clev.St.L.Rev. 7 (1983–84).

[124]This court is now called the United States Court of Federal Claims.
[125]For an article dealing with how the federal procurement agencies can use new methods of ADR, see Crowell & Pou, *Appealing Government Contract Decisions: Reducing Cost and Delay of Procurement Litigation with Alternative Dispute Resolution Techniques,* 49 Md.L.Rev. 183 (1990).
[126]*Pathman Constr. Co. v. Knox County Hosp. Ass'n,* 164 Ind.App. 121, 326 N.E.2d 844 (1975); *E.E. Tripp Excavating Contractor, Inc. v. County of Jackson,* 60 Mich.App. 221, 230 N.W.2d 556 (1975); *City of Madison v. Frank Lloyd Wright Found.,* 20 Wis.2d 361, 122 N.W.2d 409 (1963).
[127]North Dakota requires arbitration of highway construction contract disputes. The statute was upheld in *Nelson Paving Co., Inc. v. Hjelle,* 207 N.W.2d 225 (N.D.1973). Arbitration is also required for certain public contracts in Rhode Island. See *Sterling Eng'g & Constr. Co. v. Town of Burrillville Housing Auth.,* 108 R.I. 723, 279 A.2d 445 (1971). Likewise, Pennsylvania requires that public contracts contain arbitration clauses. See *U.S. Fid. & Guar. Co. v. Bangor Area Joint School Auth.,* 355 F.Supp. 913 (E.D.Pa.1973).

amount of finality, being final and conclusive unless fraudulent, capricious, arbitrary, or so grossly erroneous as to necessarily imply bad faith, a standard much like that applied to decisions by independent design professionals.

In 1977, a California appeals court decided a case that led to a drastic overhaul of the dispute resolution system. In that case, the contractor appealed to the trial court after an adverse decision by the state engineer. The trial court found that errors had been committed, disregarded the decision of the state engineer, and retried the case.

The Court of Appeals agreed that the state engineer committed procedural errors[128] but did not set aside the decision as had been done by the trial court. It sent the claim back to the state engineer for reconsideration. In doing so, the court emphasized that the parties had "by voluntary contract" agreed that the disputes would be resolved by a designated person. This conclusion was disputed by state contractors who stated that they had no choice but to agree to such a provision.

The furor caused by this case led to an executive order and later to a statute that required major state procuring agencies to use a judicialized arbitration.[129] Disputes would be decided by the persons certified as competent arbitrators by an arbitration committee composed of representatives of the state agencies and the construction industry. Those approved were placed on the State Construction Arbitration Panel. Those who wished to be placed on the panel had to submit information that indicated their education and experience and set a rate at which they were willing to serve. The disputants selected from this certified list.

Discovery is required. The arbitrator can utilize expert technical or legal advisers, depending on whether the arbitrator is an attorney or a technically trained person. Consolidation and joinder are permitted. Hearings are open to the public. The award must contain findings of fact, a summary of the evidence, and reasons underlying the award as well as conclusions of law. The award can be vacated (set aside) if it is not supported by substantial evidence or is based on an error of law.

California continued its innovation in 1990, enacting legislation that regulates disputes between contractors and local public entities for claims of $375,000 or less. This process uses informal conferences, nonbinding judicially supervised mediation, and judicial arbitration. The agencies were required to incorporate this process in plans and specifications that could give rise to a claim.[130]

SECTION 30.21 International Arbitration

In sections 2.15 and 8.09 mention was made of the difference between domestic construction contracts and those that involve parties who are nationals of different countries, particularly contracts made by American construction companies requiring that they build projects in a foreign country. Transactions of this type generate some issues of minimal or no importance in domestic contracts.

The contracts themselves may be expressed in more than one language, often generating problems that result from imprecise translation. They may also involve payment in currency that varies greatly in value. Contractors in such transactions may often have to deal with tight and often changing laws relating to repatriation of profits and import of personnel and materials. Perhaps most important, neither party may trust the legal system of the other, and the contractor may believe that it will not obtain an impartial hearing if it is forced to bring its disputes to courts in the foreign country, particularly if the owner is, as is so common in lesser developed countries, an instrumentality of the government.

Contractors making these contracts commonly insist on international arbitration to resolve disputes. Such arbitrations are usually held in neutral countries or in centers of respected commercial arbitration. Because of the complexities generated by international arbitration, this treatise does not discuss the subject in detail. Yet the emphasis on dispute resolution worldwide is a justification for

[128]*Zurn Eng'rs v. State Dep't of Water Resources,* 69 Cal.App.3d 798, 138 Cal.Rptr. 478 (1977), cert. denied 434 U.S. 985 (1977).
[129]See West Ann.Cal.Pub.Cont.Code §§ 10240 et seq. See also West Ann.Cal.Civ.Code § 1670, which requires local entities to either arbitrate or litigate. Employees of the agency cannot, as before, decide the dispute.

[130]West Ann.Cal.Pub.Cont.Code § 20104. This law will remain in effect only until January 1, 1994, unless it is reenacted by the legislature, which indicates the experimental nature of the system.

some information regarding systems used for international transactions.[131]

The most commonly used contract for international engineering is the one published by the Federation Internationale des Ingenieurs-Conseils (International Federation of Consulting Engineers). The federation's contract, generally known as the FIDIC contract, for civil engineering construction provides for international arbitration. However, subclause 67.2 states that if either party gives a notice of an intention to arbitrate, the arbitration shall not begin "unless an attempt has first been made by the parties to settle such disputes amicably." Unless the parties agree otherwise, arbitration can be commenced on or after the fifty-sixth day after which the notice of intention to commence arbitration is given "whether or not any attempt at amicable settlement thereby has been made." This was designed to ensure that the requirement of an attempted amicable settlement has not upset the timetable for arbitration.[132]

Another system has been proposed by the World Bank. In its instruction to bidders, contained in its Sample Bidding Documents for smaller (under 10 million dollars) projects, ¶ 35.1 provides that the owner will propose that a particular person be appointed as an adjudicator at a designated hourly fee. If the bidder disagrees, it must so state in its bid and made a counterproposal. If there is disagreement, the adjudicator will be appointed by a designated appointing authority. The rules under which the adjudicator would operate are specified in ¶ 25 of the general conditions of the contract. The adjudicator must give her decision within twenty-eight days. The costs are divided. Either party may refer a decision of the adjudicator to an arbitration within twenty-eight days of the adjudicator's written decision. In essence, the World Bank suggests that one role that has typically been performed by the consulting engineer—that of initially resolving disputes—be given to a different entity whose decision can be taken to arbitration.[133]

---

[131]See Paterson, *Canadian Developments in International Arbitration Law: A Step Beyond Mauro Rubino-Sammartano's International Arbitration Law,* 27 Willamette L.Rev. 573 (1991); Tiefenbrun, *A Comparison of International Arbitral Rules,* 15 B.C. Int'l & Comp.L.Rev. 25 (1992).
[132]For an evaluation of this provision, see Hollands, *F.I.D.I.C.'s Provision for Amicable Settlement of Disputes,* 6 Int'l Constr. L.Rev. Pt. 1 34 (1989).

[133]This is derived from the New Engineering Contract issued by the Institution of Civil Engineers in England.

PROBLEMS

1. The arbitration clause in a contract between O and C provided that the parties agree to arbitrate all disputes under this contract. C refused to perform under the contract because she claimed that O defrauded her into making the contract by telling her that he would award two other construction contracts to C when he knew he had already awarded them to another contractor. C claimed that she relied on the promise by making a much lower bid than she would have ordinarily made. O denied making the promise and demanded arbitration. C refused to arbitrate and started a court action for fraud. O insists on arbitration. Must this dispute be resolved by arbitration? Who decides?

2. O and C entered into a construction contract that contained a general agreement to arbitrate all future disputes. A dispute arose. O terminated the

contract. C brought legal action. C sought and used the discovery process permitted by state law and examined three of O's officials under oath. Shortly after the depositions were completed, C demanded arbitration. O does not wish to arbitrate. Should O be compelled to arbitrate?

3. O and A had a dispute relating to what compensation O is required to pay for A's architectural services. A has demanded $10,000, while O has offered to pay $5,000. Under state law, all claims for under $15,000 must go to an arbitrator appointed by the court whose decision is *not* binding on the parties. A has suggested to O that their dispute be handled under the American Arbitration Association expedited procedures (set forth in the CI Rules in Appendix F). If you were O, would you prefer to arbitrate or litigate?

Construction-Related Accidents: Interface Between Tort Law and Workers' Compensation

SECTION 31.01 An Overview

Statistics demonstrate the hazardous nature of the construction industry. The Bureau of Labor Statistics reported that in 1984 there were 22.4 deaths for every 100,000 construction workers. In 1985, they recorded 30.8 deaths for each 100,000 construction workers, the highest of any industry that year. (The national rate for all occupations is 6.2 deaths.) Similarly, a recent study by the National Institute of Occupational Safety and Health reports construction to be the most dangerous industry in the United States. It noted that more than 20% of all traumatic occupational deaths occurred in construction. That report noted that there were 23 deaths per 100,000 workers, making construction second only to mining in the rate of fatal injuries.

Looking at accidents from the perspective of total project costs, one study found that the total accident costs for insured and uninsured risks were as high as 3% of the total project costs, amounting to 10% of labor costs.[1]

The law early recognized that construction was a hazardous activity. For example, even before the enactment early in the twentieth century of its workers' compensation law, Illinois enacted its Structural Work Act (called the Scaffold Act). Its purpose was to protect workers who work above ground level as well as persons who might be hit by materials falling from construction. The original

act covers all devices that elevate workers and materials that are used to support them.

But reference will also be made in this chapter to persons who suffer physical harm *after* the project has been completed. Except for injuries to workers in the scope of their employment, which are accidents that fall within the jurisdiction of workers' compensation laws, tort law dominates construction accidents. Even accidents that fall within workers' compensation are affected by tort law. As shall be seen, third-party lawsuits—those by injured workers against persons other than their employers or others immune from such actions—have become significant as a means of compensation for injured construction workers.

Many of the doctrines discussed in this chapter were introduced in Chapter 7, which laid the groundwork for this chapter. This chapter spotlights the Construction Process and applies the general principles described in Chapter 7.

Chapter 17, which described the various ways in which construction projects are organized, also bears heavily on this chapter. That chapter noted the attributes of the traditional system of construction, such as divided design and construction, prime contractor coordination and organization, and design professional review of project progress. Much of that chapter noted modern variations on the traditional methods that bear heavily on the three principal issues described in this chapter:

1. *Statutory violations* and their effect on liability.
2. *Vicarious liability*, the liability of one party for the failure of another to perform as required by tort law.

---

[1]Levitt et al., *Improving Construction Safety Performance: The User's Role*, Technical Report No. 260, Stanford University (1981).

3. *Strict liability*, imposed on a party solely because of its status, as owner, as "person in charge," or as statutory employer.

These issues all involve control—the power to determine what is to be done and how it is to be done. In a traditional construction organization, the owner turns over the design and site to a prime contractor and allows the prime contractor to control the site, select and coordinate subcontractors, police their work, and act as a conduit or buffer between owner and subcontractors. In this model, the prime contractor has charge of the Construction Process, even though the owner, through the design professional, may reserve the right to monitor the work as it proceeds. This control is particularly emphasized in matters that relate to construction technique and the proliferating legal controls designed to protect workers and others from physical harm during construction and, on occasion, afterward.

Other methods of organization, such as separate contracts (multiple primes), coordinated and monitored by the owner often through a construction manager, and design/build, affect this central assumption as to control. In the former, the owner has taken a more active role, effectively replacing the prime as the party in control. In design/build, a *sophisticated* owner may still, through its infrastructure, have either control or the right to control the process. At the other extreme of design/build the owner may not even monitor the work through a design professional as in the traditional system, simply turning over everything to the design/build contractor.

Liability may depend on whether the owner is a large, sophisticated organization with its engineering, insurance, and legal departments that totally dominate the project or an unsophisticated owner, inexperienced in construction, that puts its trust and money in the hands of others, such as the owner's retaining an architect to design and a contractor to build or a contractor who does both.

The sophisticated owner has not only a big economic stake in the project but also the capacity to protect itself and actively control the process. The inexperienced owner is often a passive participant. This is demonstrated by developments in New York. Section 241 of the New York Labor Law requires all owners and contractors to provide a safe workplace. This statute was the basis for holding owners strictly liable for injuries to workers.[2] In 1980, this statute was amended to exempt owners who have one- or two-family homes built for them if they do not direct or supervise construction. (Would the use of an architect deny an owner this exemption?)[3]

The New York Labor Law was drafted originally not to place strict tort liability on *all* owners for injury to those working on their projects. However, a court reading the statute literally saw no way to avoid this result. Although it is possible to justify such a result even in the case of an unsophisticated owner by pointing to the possibility of obtaining insurance, it is clear that the legislature in 1980 chose not to place strict liability on such owners.

This chapter must be read in the light of the observations made in Chapter 7 that tort law emphasizes compensating those who suffer physical harm with less emphasis on the need to establish moral culpability. To be sure, not all claimants recover from all those against whom they assert claims. As a rule, however, if workers' compensation recovery is taken into account, claimants will recover from someone and will rarely go totally uncompensated. The principal issue is whether they can go beyond workers' compensation to receive more generous tort awards.

SECTION 31.02 Worker Claims: Workers' Compensation Legislation

Although tort law dominates this chapter, it is important to see workers' compensation laws (described in Section 7.04(C)) as a backdrop to worker claims based on tort law. First, workers' compensation usually provides a quick, sure recovery for a worker injured in the course of employment but one that does not provide as much compensation as a tort recovery. That factor is often influential in tort claims. Some judges view this as a reason for expanded tort recovery,[4] while others regard a

---

[2]*Allen v. Cloutier Constr. Corp.*, 44 N.Y.2d 290, 376 N.E.2d 1276, 405 N.Y.S.2d 630 (1978).

[3]In 1981, the N.Y. statute was amended to exclude design professionals who do not direct or control the work from strict liability. See also *Sherwood v. Omega Constr. Co.*, 657 F.Supp. 345 (S.D.N.Y.1987) (project representative who did not supervise not strictly liable).

[4]*Funk v. General Motors Corp.*, 392 Mich. 91, 220 N.W.2d 641 (1974) discussed in Section 31.05(B).

workers' compensation recovery as adequate.[5] Second, workers' compensation law, with some exceptions, immunizes the injured worker's employer from any tort liability.

As third-party claims have proliferated, states have made legislative changes designed to reduce those parties against whom such claims can be brought. Sometimes co-employees or those in a common employment are immune. As seen in Section 14.09(E), a substantial number of states have granted immunity to the design professional for third-party claims relating to site services unless the design professional has contracted to oversee safety or undertook to do so. The complexity of the construction process and the large number of entities involved have generated difficulty when such immunity statutes have come before the courts.[6] (As Section 32.03(D) shows, the employer is often a cross-defendant through indemnification.)

The frequent creation of statutory employers in workers' compensation laws complicates the immunity issue. Many states recognize the fragmentation and financial instability of the construction industry and seek to make a party farther up the employer chain from the injured worker the statutory employer of the injured worker. The statutory employer, though not in reality employing the

worker, is considered the worker's employer for workers' compensation purposes. This increases the likelihood that an injured worker will receive a compensation award. The creation of statutory employers was designed to enlarge the injured worker's right to a compensation award, with third-party claims (tort claims against all those not immune) not being common at that time. The proliferation of third-party claims has generated assertions by a party who might otherwise be liable in a tort claim that it is a statutory employer and entitled to the same immunity as the actual employer.

Predictably, in the light of the variant statutory language and often different attitudes toward these third-party claims, different results occur when the issues are whether the owner is a statutory employer and whether a statutory employer receives immunity.[7]

SECTION 31.03 Privity and Duty

Before determining whether the defendant has lived up to the standard of conduct required by tort law, two preliminary concepts, though diminishing in importance, must be examined.

The tort law requirement that persons conduct themselves in a particular way does not extend its liability protection into infinity. To avoid potentially crushing responsibility that can have an adverse effect on professional, commercial, and industrial activities, lines must be drawn to limit liability. Two concepts that can draw such a line relate to the relationship between plaintiff and defendant. One, called duty, asks whether the defendant owes a duty *to the plaintiff* to conduct itself in accordance with the tort standard of conduct. Another way of expressing this is to ask whether there is a privity between plaintiff and defendant. Usually this means the existence of a contract between the two. It also can encompass other close relationships that, though perhaps contractual in a sense,

[5]*Franklin Privette v. Superior Court*, 93 C.D.O.S. 5492 (Cal. Supreme Court, July 19, 1993); *Zueck v. Oppenheimer Gateway Properties*, 809 S.W.2d 384 (Mo.1991); *Evans v. Hook*, 239 Va. 127, 387 S.E.2d 777 (1990).
[6]*Wenzel v. Boyles Galvanizing Co.*, 920 F.2d 778 (11th Cir.1991) (architect/construction manager retained by owner not granted immunity despite Florida statute granting immunity to the employer's safety consultant); *Bergen v. Fourth Skyline Corp.*, 501 F.2d 1174 (4th Cir.1974) (owner and multiple primes immune from claim by injured worker under Virginia law but a factual question remains as to immunity of concrete supplier); *Halter v. Waco Scaffolding & Equip. Co.*, 797 P.2d 790 (Colo.1990) (architect immune as agent of owner); *Employers Ins. of Wausau v. Abernathy*, 442 So.2d 953 (Fla.1983) (under Florida law subcontractor not immune from claim by employee of prime contractor); *Nichols v. VVKR, Inc.*, 241 Va. 516, 403 S.E.2d 698 (1991) (under Virginia law, architect hired by transit agency not immune as agency not in construction business). For a survey of Virginia law, see Barnhill, *Annual Survey of Virginia Law: Construction Law*, 25 U.Rich.L.Rev. 699 (1991). For a general discussion of immunity under workers' compensation laws, see Note, 96 Harv.L.Rev. 1641 (1983). Since workers' compensation statutes are changed frequently and the holding of any case cited may be based upon a particular statute in effect at that time, case precedents are not always reliable.

[7]Compare *Hattersley v. Bollt*, 512 F.2d 209 (3d Cir.1975) (immunity), with *Laffoon v. Bell & Zoller Coal Co.*, 65 Ill.2d 437, 359 N.E.2d 125 (1976) (immunity denied). See also *Brady v. Ralph Parsons Co.*, 308 Md. 486, 520 A.2d 717 (1987) (CM who does not perform work not a statutory employer). But see *Evans v. Hook*, supra note 5 (architect immune as statutory employer of worker).

are created more by status than by contract. An illustration of such relationships is employer and employee, parent and child, hotel keeper and guest, or common carrier and passenger. The important element is whether one person has a duty to avoid exposing the other to an unreasonable risk of harm.

Drawing lines based on relationship and requiring either privity or a duty once had great importance in protecting defendants. Much of their effectiveness resulted from the existence of privity or a duty being a legal question to be decided by judge and not by jury. As seen in Section 7.09, dealing with liability for defective products, and to a lesser degree in Section 7.07, dealing with misrepresentation, these limiting factors have lost much of their effectiveness. Yet they still play a minor role in construction accidents.

The objective of compensating accident victims is often accomplished by allowing a wide range of persons who *may* be held accountable. The availability of insurance and the desire that enterprises should bear their legitimate responsibility for losses they cause have virtually eliminated privity and duty as barriers to liability where the plaintiff seeks compensation for personal harm. Elimination of these barriers when harm was caused by defective products began in 1916.[8] In construction accidents, the barriers began to fall in the 1950s.[9]

Court decisions have either minimized or largely eliminated the privity requirement in cases of personal harm. Similarly, any contention that the defendant owed no duty to the person suffering physical harm has had only limited success. As a result, the worker or visitor injured during construction generally is given the chance to bring legal action against most direct participants in the Construction Process. For example, an injured worker will very likely be able to bring legal action against any contractor other than his own employer or statutory employer,[10] the owner,[11] any design professional,[12] a distributor or manufacturer of defective products,[13] possibly the lender,[14] and, where immunity does not exist, public officials whose negligent conduct may have played an important part in causing the injury.[15]

Still, the privity doctrine has had some success. For example, provisions frequently contained in policies that insure a contractor for public liability or workers' compensation give the insurer the right to inspect premises and activities of its insured as part of its loss-prevention system. Workers have sought, with spotty success, to recover against these insurance companies if they can establish that the insurance company owed them a duty because the insurer *undertook* to perform a duty owed workers by their employer, the insured under the policy. If a duty *is* found because the insurer undertook the duty of its insured, the workers can show the insurer did not conduct an inspection when it should have or did not take necessary steps after having conducted one. One writer, after discussing the cases and legislation dealing with this problem, outlined several key factors as summarized in the following paragraphs.

Focusing first on the reaction of the insured to the inspection, reliance and duty are more likely to be found when the insured discontinued his safety program after the insurer began inspecting. The same result is probable if the insured did not completely discontinue his program but simply reduced it by a sufficient degree. The reaction of the insured is the most important factor in determining the applicability of the reliance theory.

Receiving equal attention will be the conduct of the insurer. An initial factor will be the purpose of the in-

---

[8]*MacPherson v. Buick Motor Co.*, 217 N.Y. 382, 111 N.E. 1050 (1916).

[9]*Hanna v. Fletcher*, 231 F.2d 469 (D.C.Cir.1956), cert. denied 359 U.S. 912 (1956); *Dow v. Holly Mfg. Co.*, 49 Cal.2d 720, 321 P.2d 736 (1958); *Inman v. Binghamton Housing Auth.*, 3 N.Y.2d 137, 699, 143 N.E.2d 895, 164 N.Y.S.2d 699 (1957); *Macomber v. Cox*, 249 Or. 61, 435 P.2d 462 (1967).

[10]*Schroeder v. C.F. Braun & Co.*, 502 F.2d 235 (7th Cir.1974).

[11]*Associated Eng'rs, Inc. v. Job*, 370 F.2d 633 (8th Cir.1966).

[12]*Miller v. DeWitt*, 37 Ill.2d 273, 226 N.E.2d 630 (1967); *Evans v. Howard R. Green Co.*, 231 N.W.2d 907 (Iowa 1975). But as noted in Section 14.08(G), a differentiation is made between design and site services which, in some jurisdictions, has led to the reimposition of something like a privity defense.

[13]*Morgan v. Stubblefield*, 6 Cal.3d 606, 493 P.2d 465, 100 Cal.Rptr. 1 (1972).

[14]*Connor v. Great Western Sav. & Loan Ass'n*, 69 Cal.2d 850, 447 P.2d 609, 73 Cal.Rptr. 369 (1968) (property damage to tract home).

[15]*Holman v. State*, 53 Cal.App.3d 317, 124 Cal.Rptr. 773 (1975). In *Morris v. County of Marin*, 18 Cal.3d 901, 559 P.2d 606, 136 Cal.Rptr. 251 (1977), an injured worker was allowed to bring legal action against the county that in violation of state law had issued a building permit to the worker's employer without requiring a workers' compensation certificate be filed.

spection. If the inspection is solely for internal purposes and on a limited basis, a duty is less likely to be found. Another important element will be whether open and obvious conditions were inspected. If so, a court might restrict liability to harm following from those conditions.

A final and perhaps most important consideration deals with the character of the inspection. If the insurer makes representations as to the efficiency of its program, coupled with high safety credentials of its inspectors, liability will be more likely. The authority to enforce recommendations will also help establish the existence of a duty. By the same token, the absence of these considerations will also lead to a finding of no duty.[16]

Those who enter another's land are not always entitled to insist that the possessor of land make the land reasonably safe for them. Here the relationship between the person who enters the land and the one who possesses (controls) it may *limit* the duty by requiring *less* of the possessor of land than is required of everyone else.

One complication of limited duty is the difficulty of determining the status of the construction worker who works on the site or at a plant while performing construction work. Some cases hold that the worker is an invitee entitled to a warning or to have the premises made reasonably safe for him.[17] Others conclude that the worker is a licensee with less protection, mainly to be warned of concealed dangers.[18] As jurisdictions begin to move toward one standard of conduct, these classification problems should be less important.[19]

The limited duty sometimes placed on the possessor of land requires a preliminary determination of *who* is the possessor of land. The owner, before construction, is responsible for the condition of the land. As has been seen, that obligation may not be delegable.[20] In a traditional construction project—that is, where the owner turns over the site along with construction documents to the contractor and asks for a particular result—the owner's reservation of certain rights during the construction phase does not preclude a conclusion that the contractor has taken possession of the land. This transfer of responsibility is permitted only if the contractor is an independent contractor (discussed in Section 31.05(B)), or if one of the many exceptions to the independent contractor rule does not apply. By and large, the question of who possesses the land for these purposes is determined by the resolution of these issues.

Another duty problem that arises in construction accidents relates to the employee-employer relationship. Frequently, the law requires that an employer furnish a safe workplace or safe tools for employees. Because of the multitude of entities present on a construction project, it is often difficult to determine *who* is the employer for these purposes in both traditional and nontraditional contracting methods. Here the duty concept can be invoked to determine whether a particular person responsible for furnishing a safe place of employment or tools has a duty other than to its *own* employees. For example, in the traditional process, there are prime contractors and various tiers of subcontractors. Does the prime contractor owe the employees of a subcontractor the obligation of furnishing safe tools or workplace?[21] In a nontraditional system with more owner control, does primary responsibility for these matters fall on the owner?[22] In the nontraditional system, does the

---

[16]Comment, 66 Ky.L.J. 910, 922–3 (1978). See also *Stark v. Commercial Union Ins. Co.*, 501 So.2d 1214 (Ala.1987) (no general duty to the *public*); *Jansen v. Fidelity & Cas. Co.*, 79 N.Y.2d 867, 589 N.E.2d 379, 581 N.Y.S.2d 156 (1992). Florida has immunized workers' compensation insurers for employers from claims by injured workers. See West Ann.Fla.Stat. § 440.11(3) (1993).

[17]*Hogge v. United States*, 354 F.Supp. 429 (E.D.Va.1972) (as to portions of site where he was authorized to go); *Elder v. Pacific Tel. & Tel. Co.*, 66 Cal.App.3d 650, 136 Cal.Rptr. 203 (1977) (but no duty to sub's employee where prime contractor not "operatively present"); *Crotty v. Reading Indus.*, 237 Pa.Super. 1, 345 A.2d 259 (1975) (over strong dissent); *Ferguson v. R. E. Ball & Co.*, 153 W.Va. 882, 173 S.E.2d 83 (1970) (dictum).

[18]*Nagler v. United States Steel Corp.*, 486 F.2d 794 (7th Cir. 1973); *Epperly v. City of Seattle*, 65 Wash.2d 777, 399 P.2d 591 (1965); *Daniel Constr. Co. v. Holden*, 266 Ark. 43, 585 S.W.2d 6 (1979).

[19]See Section 7.08(G).

[20]See Section 7.08(H). See also *Weber v. Northern Illinois Gas Co.*, 10 Ill.App.3d 625, 295 N.E.2d 41 (1973).

[21]*Caswell v. Lynch*, 23 Cal.App.3d 87, 99 Cal.Rptr. 880 (1972), held that it did. See also *Macomber v. Cox*, supra note 9 in which the dissenting judge viewed the prime contractor not as a builder but as an assembler and overseer of subcontractors and the participant most visible and responsible for the overall safety of the employees.

[22]*United States v. English*, 521 F.2d 63 (9th Cir.1975) (owner an employer under California Labor Code, as it controlled the premises).

prime contractor owe a duty to employees of the owner?[23]

These questions *could* be resolved by the statutory language. However, often these statutes are drafted without construction project organization in mind. But like the question of who possesses the land, the issue of who is an employer for these purposes is often dominated by a determination of whether the prime contractor or subcontractor is an independent contractor. For example, suppose a prime contractor's employee is injured because of the employer's statutory violation as to work conditions. If that worker *can* assert a legal claim against the owner, can the claim be based on the statutory violation? Similar problems can arise if the injured worker is an employee of a subcontractor or even of the owner. (These questions are discussed in Section 31.05.)

Where the issue is faced directly, such as primary responsibility for compliance with the Occupational Safety and Health Act (OSHA), it is likely that the prime contractor will be given primary responsibility.[24] The prime contractor is the participant to whom possession is given and the participant to whom the owner looks for overall safety responsibility. This is often evidenced by prime contract provisions placing this responsibility on the prime contractor. Where the owner in a nontraditional process has more overall control and operative presence, the principal responsibility may rest on it.[25] It may, as seen in Chapter 32, attempt to shift any loss it suffers by obtaining indemnification from the prime contractor. This does not divest it from responsibility to those injured.

Sometimes the existence of a duty is determined by the contract that the defendant has made. Do its provisions indicate that the defendant owed a duty to the person injured? For example, prime contracts frequently contain provisions placing worker safety in the hands of the prime contractor. Suppose a subcontractor's employee is injured because of a failure of the prime contractor to comply with these provisions. The contract provisions may create a duty to the injured worker if made for the worker's benefit. Courts have reached different conclusions.[26]

Similarly, sometimes the contract is used to *deny* any duty to the injured worker. Professional societies of architects and engineers have included language in their standardized construction contracts seeking to *negate* any duty on the part of the design professional to monitor work methods to protect workers. This is done to shield the design professional from any liability for accidents caused workers that are principally due to improper construction methods.[27]

SECTION 31.04 Standard of Conduct

A. Negligence and Strict Liability

The negligence standard—that of avoiding unreasonable risk of harm to others—dominates construction accidents.[28] A few exceptions exist, however. Claimants against those who have made or supplied defective products need not establish negligence. On the whole, construction is considered to be furnishing services rather than supplying goods. Yet even there, there are some exceptions to the usual requirement that negligence be established. Certain construction activities as noted in Section 7.04(A) can be classified as abnormally dangerous or ultrahazardous, dispensing with the negligence requirement. A few states have held developer-builders to a standard of strict lia-

---

[23]*Valdez v. J. D. Diffenbaugh Co.*, 51 Cal.App.3d 494, 124 Cal.Rptr. 467 (1975) (contractor owed duty to owner's employee).
[24]29 U.S.C.A. § 651. For the view that the prime contractor should have primary responsibility, see *Funk v. General Motors Corp.*, supra note 4. Louisiana held that the federal agency making a grant for a bridge across the Mississippi River and the state entity to which the grant was made were not obligated to enforce OSHA regulations. The duty fell solely on the prime contractor engaged by the state entity. *Walters v. Landis Constr. Co., Inc.*, 522 So.2d 1306 (La.App.1988). (See Sections 14.08(G) and 31.09.)
[25]*Funk v. General Motors Corp.*, supra note 4.

[26]Compare *J.D. Williams v. Fenix & Scisson, Inc.*, 608 F.2d 1205 (9th Cir.1979) (not intended beneficiary of owner-engineer contract), and *West v. Morrison-Knudsen Co.*, 451 F.2d 493 (9th Cir.1971) (not intended beneficiary of owner-contractor contract), with *Dunn v. Brown & Root, Inc.*, 455 F.2d 717 (8th Cir.1972) (intended beneficiary of prime contract).
[27]See Section 14.08(G).
[28]*Belcher v. Nevada Rock & Sand Co.*, 516 F.2d 859 (9th Cir.1975) (rejected strict liability in action against contractor); *La Rossa v. Scientific Design Co.*, 402 F.2d 937 (3d Cir.1968) (rejected strict liability in action against design professional).

bility.[29] Some states place strict liability on the party in charge of the work[30] or on owners and contractors.[31] For the most part, claims for physical harm related to construction require proof of negligence.

B. Negligence Formulas: Inadvertence and Conscious Choice

Whether a person has exposed another to the unreasonable risk of harm requires an evaluation of the magnitude of the risk, the utility of the conduct, and the burden of eliminating the risk. Some negligent acts are simple inadvertence where the actors either forget to do what they know they should do or do it carelessly. For example, cases have involved acts of carelessly covering a dangerous hole[32] and placing a ladder where it could cause injury.[33]

Negligence can be based on the actor's having exercised a conscious choice, with the issue being whether that choice was reasonable under the circumstances. *Stanley v. United States*[34] involved a fatal fall by a painting subcontractor's employee that resulted in an action against the U.S. owner. The worker's widow contended that a Navy antenna tower from which the worker had fallen had been defectively designed. The court described the tower as follows:

> The tower in question, No. S-0, is an elongated triangular prism, its three legs arising 979 feet from the apexes of a triangular base 12 feet on a side. The sides are skeletal, with a horizontal beam about every 12 feet, with a single diagonal cross of tie-rods, or windrods, so called, in the space between. The inside is empty, except that every 70 feet there is a triangular grating, or platform, 12' × 12' × 12' with a 2' × 3' rectangular aperture, or cut-out, to afford access to a

ladder which runs down the side to the platform next beneath. These apertures are staggered from platform to platform, as the ladder shifts from one side of the tower to another as one progresses from one platform to the next. Although there was no testimony on the subject, this seems an obvious safety feature, so that the platform at the base of the ascending ladder would be solid. The effect of this, of course, is to require everyone ascending or descending the tower to dismount the ladder and cross over at each platform. The principal purpose of the platform, apart from providing strength to the structure, is to serve as a place of rest. There are no guardrails around the cut-out, and none around the outside of the platform.[35]

Stanley was twenty years old and without experience in high work. Because of his inexperience, he was assigned to work on the platforms, which were thought to be safer. At the time of the accident, he was using a paint roller with a 6-foot rod that required both hands. He seemed to be having no difficulty but shortly thereafter fell 700 feet and died of a crushed skull.

The evidence showed negligence on the part of the deceased worker's employer in permitting the worker to work without a safety belt. Negligence by the government was based on a failure to supply guardrails to the ladder apertures or cutouts. To establish this, an expert witness testified for the plaintiff that good safety practice called for guardrails around the apertures. The expert conceded that the apertures were obvious dangers, although he stated that people engaged in painting might have their attention diverted and forget the existence of the apertures. The expert conceded that the absence of a guardrail did not constitute a danger to persons climbing the tower who could be expected to pause on the platform only for a rest and would not be as likely to be distracted from the existence of the aperture.

The government safety expert testified that guardrails would be dangerous to those servicing the antenna and cables who would be carrying equipment and who might hang the equipment up on any rail. In evaluating testimony of the design choice, the court stated:

> The platform, without rails, was a dangerous place for a painter, particularly an inexperienced painter, to

[29]*Schipper v. Levitt & Sons, Inc.*, 44 N.J. 70, 207 A.2d 314 (1965). But see *Macomber v. Cox*, supra note 9. For a general discussion, see Comment, 10 Ohio N.U.L.Rev. 103 (1983).
[30]Ann.Ill.Stat. ch. 48, ¶¶ 68, 69. See Section 31.04(F).
[31]See Section 31.01.
[32]*E.L. Jones Constr. Co. v. Noland*, 105 Ariz. 446, 466 P.2d 740 (1970).
[33]*Hill v. Lundin & Assoc., Inc.*, 260 La. 542, 256 So.2d 620 (1972). The contractor was relieved in this case because a third party moving the ladder operated as a superseding cause.
[34]476 F.2d 606 (1st Cir.1973).

[35]Id. at 607.

work. It was perhaps more dangerous than such a painter might appreciate. On the other hand, as to all other persons, who manifestly use the tower far more frequently, and who would have to traverse apertures 26 times for a single assent, guardrails were not only not needed, but were to some degree contra-indicated. Furthermore, the special danger applicable to painters could, as the court found, be obviated by the use of safety belts with a tag line attachment.[36]

In reversing a judgment against the United States, the court stressed that the design choice had been reasonable in the light of the circumstances.

The design choice arose in a different context in *George v. Morgan Construction Co.*[37] A worker in a steel plant was injured when a hot cobble was ejected from a rolling mill machine. Cobbles are ejected because the steel is defective or because of operational errors by the steel company. The worker recovered a workers' compensation award and then brought legal action against the designer and builder of the machine.

The worker contended that the injury could have been avoided by installation of cobblescreens as a protective guard. However, the machine designer contended that a cobblescreen would have blocked the view of employees controlling the mill from a glass-enclosed pulpit 20 feet above the floor that housed operational controls. But, stated the court:

The mill could have been so designed so that the cobblescreens could have been erected without blocking the pulpit, by putting the pulpit on the drive side of the mill instead of the working side. . . .

The judge then quoted Hume:

'. . . did I show you a house or palace, where there was not one apartment convenient or agreeable; where the windows, doors, fires, passages, stairs, and the whole economy of the building were the source of noise, confusion, fatigue, darkness, and extremes of heat and cold; you would certainly blame the contrivance, without any further examination. The architect would in vain display his subtlety, and prove to you that if this door or that window were altered, greater ills would ensue. What he says, may be strictly true: the alteration of one particular, while the other parts of

the building remain, may only augment the inconveniences. But still you would assert in general, that if the architect had had skill and good intentions, he might have formed such a plan of the whole, and might have adjusted the parts in such a manner, as would have remedied all or most of these inconveniences.'

A good architect would have designed the house in such a way as to avoid these disadvantages, so that one would not have to choose between a design that was bad and one that was worse. And in the same way, a non-negligent Construction Company would have designed the steel mill in such a way as to avoid requiring the Steel Company to choose between the Scylla of no cobblescreens and the Charybdis of an obstructed view from the pulpit.[38]

C. Common Practice: Custom

Deviation from and conformity to common practice are admissible but not conclusive on the question of negligence.[39] Such evidence can assist the finder of fact to determine what was feasible and what the defendant knew or should have known.[40]

D. Res Ipsa Loquitur

This concept permits negligence to be shown by circumstantial evidence—evidence of the accident coupled with inferences that the cause of the accident is more likely than not to have been the negligence of the defendant. The principal use of this doctrine is to provide sufficient evidence to submit the question of negligence to the jury.

The doctrine has uneven application in construction litigation mainly because of the many participants who were or may have been working

---

[36]Id. at 609.
[37]389 F.Supp. 253 (E.D.Pa.1975).

[38]Id. at 257. The Hume quotation was taken from *Dialogues Concerning Natural Religion*, Part XI, Norman Kemp Smith edition 204 (1935).
[39]*Fireman's Fund Ins. Co. v. AALCO Wrecking Co.*, 466 F.2d 179 (8th Cir.1972); *George v. Morgan Constr. Co.*, supra note 37; *Rivera v. Rockford Machine & Tool Co.*, 1 Ill.App.3d 641, 274 N.E.2d 828 (1971). In judging professional conduct, the common standard can be the legal standard. See Section 14.05(A).
[40]*Grant v. Joseph J. Duffy Co.*, 20 Ill.App.3d 669, 314 N.E.2d 478 (1974). The industry standards considered were the American Safety Code for Building Construction, the State of Illinois Health and Safety Act, and Manual of Accident Prevention of the Associated General Contractors.

on the project at the time the negligence occurred. As a result, it is often difficult to connect the negligence to a particular participant.

New York refused to apply the doctrine where there had been an injury from falling bricks where nineteen contractors had been working on the site.[41] Similarly, Wisconsin refused to apply the doctrine where a stone fell, injuring a subcontractor's employee.[42] The work was being performed on a state-owned building. Refusal was predicated on the failure to show that the contractor had exclusive right to control the work, as the architect and the state's representative also had control rights.

That the architect and the representative of the owner had certain powers during construction should not have defeated the plaintiff's attempt to use res ipsa loquitur. However, had other employees of the state been working in the area from which the stone had fallen, it would have been difficult to infer that the accident had been caused by negligence of the contractor.

The uncertainty of the application of this doctrine in construction work can be demonstrated by contrary conclusions in two cases, each of which involved injury to a worker while working at the site of a single-family home and defective stairs.[43] Yet there are cases where it has been successfully employed by the plaintiff.[44]

As tort law continues to emphasize compensating accident victims, res ipsa loquitur may be used

more frequently in construction accidents. This is more likely to be the case in smaller construction projects or construction projects largely accomplished by one contractor. As the project becomes more complex, it will be difficult to connect the accident with the negligence of any particular participant.

E. Prior Accidents

Sometimes it is necessary for the plaintiff to establish that the defendant knew of a dangerous condition and continued to allow it to exist. Evidence of this can be established by testimony of prior accidents of a reasonably similar nature.[45]

F. Departures from and Conformity with Statutory Standards

The Construction Process is honeycombed with laws regulating worker safety. The effect of statutes generally on liability was discussed generally in Section 7.03(C). This section examines this important question within the context of the Construction Process.

Some statutes determine liability directly. For example, workers' compensation laws determine the employer's liability for employment accidents.[46] Similarly, as noted in Section 31.04(A), some statutes create strict liability. For example, the Illinois Structural Work Act, known as the Scaffold Act, defines the standards and places liability on the party "in charge" who willfully violates the statute.[47] Similarly, the New York Labor Law places

---

[41]*Wolf v. American Soc'y Tract Soc.*, 164 N.Y. 30, 58 N.E. 31 (1900). Similarly, the doctrine was not used in *Hardie v. Charles P. Boland Co.*, 205 N.Y. 336, 98 N.E. 661 (1912), where a chimney fell and the cause could have been either the contractor or the architect. But see *Schroeder v. City & County Sav. Bank*, 293 N.Y. 370, 57 N.E.2d 57 (1944) (scaffold accident in which the doctrine was applied against the contractor who had erected the scaffold and another who was working nearby at the time of the accident).

[42]*Goebel v. General Bldg. Service Co.*, 26 Wis.2d 129, 131 N.W.2d 852 (1965).

[43]*Moore v. Denune & Pipic, Inc.*, 26 Ohio St.2d 125, 269 N.E.2d 599 (1971), held that it could not be used, while by dictum, *Ferguson v. R.E. Ball & Co.*, supra note 17, stated that it could.

[44]*Smith v. General Paving Co.*, 24 Ill.App.3d 858, 321 N.E.2d 689 (1974) (roadwork that required that drains be kept free followed by flooding); *Bedford v. Re*, 9 Cal.3d 593, 510 P.2d 724, 108 Cal.Rptr. 364 (1973) (decorative wall collapsed).

[45]*Grant v. Joseph J. Duffy Co.*, supra note 40.

[46]See Section 7.04(C).

[47]Supra note 30. Illinois has struggled with the issue of who is in charge. Compare *McGovern v. Standish*, 65 Ill.2d 54, 357 N.E.2d 1134 (1976) (architect not in charge), with *Emberton v. State Farm Mut. Automobile Ins. Co.*, 71 Ill.2d 111, 373 N.E.2d 1348 (1978) (architect in charge). *Sasser v. Alfred Benesch & Co.*, 216 Ill.App.3d 445, 576 N.E.2d 303 (1991) (architect with duty to inspect, administer, and stop certain work raised factual question as to whether he was in charge); *Coyne v. Robert H. Anderson & Assoc., Inc.*, 215 Ill.App.3d 104, 574 N.E.2d 863 (1991) (architect with duty to monitor for compliance with safety regulations in charge).

strict liability on the owner and the contractor.[48] Such direct statutory liability is relatively rare. Even the Federal Occupational Safety and Health Act of 1970 (OSHA), though relevant on the negligence issue, does not *grant* a civil remedy to employees against violators of the statute.[49] This statute, as most safety statutes, is designed to impose public law liability on violators and a public law duty to avoid harm to workers.

Sometimes statutes state a *general* standard of conduct (such as requiring a safe place) owed by one participant, usually the contractor or owner, to other participants, usually workers or visitors to the site. More important are statutes that indirectly but significantly affect liability by providing a *detailed* standard that plays an important role in determining liability. To have any significant effect on negligence, the statute must have been designed to protect the person injured and must have dealt with the type of risk that caused the harm. Though less important, the violation must not have been excused.

Statutory safety rules are generally for the protection of those who work on the site. For example, the Illinois Structural Work Act is designed to protect construction workers but not employees on the maintenance staff of the owner.[50] Courts on occasion have drawn an unreasonably narrow line in protection matters. For example, one court held that the statute requiring an employer to provide a safe place for its employees protected only employees of the employer and not those of an independent contractor hired to do work in the employer's plant.[51]

California, on the other hand, has given such statutes broad scope. For example, California has held that safety regulations imposed on a prime contractor were for the benefit of employees of a subcontractor,[52] for the benefit of a physician who entered the site to perform emergency medical treatment for two injured workers,[53] and for the protection of a sixteen-year-old trespasser who was taking a shortcut over the site.[54]

Expanded tort application of safety statutes gives effect to legislative concern for safety. It can aid courts in resolving the often difficult question of whether the defendant has exposed the plaintiff to an unreasonable risk of harm. Greater application of these safety statutes should induce compliance.

If the preceding requirements are met, most states hold that the violation is negligence per se. The trial court judge *must* take away any negligence question from the jury and rule that the plaintiff has conclusively proved negligence. For example, *Bowman v. Redding & Co.* held that violations of local safety rules requiring that the entrance to a shaft be provided with a substantial door gate or hinged bar and that a hoist signal system be employed were negligence per se when a

---

[48]New York has also struggled with the issue of who is strictly liable because of the statute. As to owners and architects, see Section 31.01. As to other participants, see *Russin v. Louis N. Picciano & Son*, 54 N.Y.2d 311, 429 N.E.2d 805, 445 N.Y.S.2d 127 (1981) (separate contractor not liable unless given authority to supervise and control portion of work where accident occurred); *Kenny v. Fuller*, 87 A.D.2d 183, 450 N.Y.S.2d 551 (1982) (construction manager *both* contractor and agent of owner); *Carollo v. Tishman Constr. & Research Co.*, 109 Misc.2d 506, 440 N.Y.S.2d 437 (1981) (rejected construction manager's argument that it was merely an expediter and held it to be a contractor). However, CMs have been able to protect themselves by indemnification agreements with the subcontractors (*Carollo v. Tishman Constr. & Research Co.*, supra, and *Kenny v. Fuller*, supra). The statute does not place strict liability on a project representative who does not supervise. *Sherwood v. Omega Constr. Co.*, supra note 3.

[49]*Jeter v. St. Regis Paper Co.*, 507 F.2d 973 (5th Cir.1975); *Russell v. Bartley*, 494 F.2d 334 (6th Cir.1974). *Tenenbaum v. City of Chicago*, 60 Ill.2d 363, 325 N.E.2d 607 (1975), held that the violation of a city safety ordinance does not create a civil remedy. As to an OSHA violation as *evidence* of negligence, see *Indust. Tile, Inc. v. Stewart*, 388 So.2d 171 (Ala.1980), cert. denied 449 U.S. 1081 (1981). Increasingly, state officials are bringing criminal charges against employers based upon unsafe workplace conditions. See *People v. Chicago Magnet Wire Corp.*, 126 Ill.2d 356, 534 N.E.2d 962, cert. denied sub nom *Asta v. Illinois*, 493 U.S. 809 (1989) (OSHA federal law does not preempt state criminal law).

[50]*Bitner v. Lester B. Knight & Assoc., Inc.*, 16 Ill.App.3d 857, 307 N.E.2d 136 (1974).

[51]*Olds v. Pennsalt Chemicals Corp.*, 432 F.2d 1033 (6th Cir.1970), cert. denied 401 U.S. 1010 (1971). Similarly, regulations of a prime contractor held not for the benefit of the subcontractor's employee in *Jones v. Indianapolis Power & Light Co.*, 158 Ind.App. 676, 304 N.E.2d 337 (1973).

[52]*Alber v. Owens*, 66 Cal.2d 790, 427 P.2d 781, 59 Cal.Rptr. 117 (1967).

[53]*Solgaard v. Guy F. Atkinson Co.*, 6 Cal.3d 361, 491 P.2d 821, 99 Cal.Rptr. 29 (1971).

[54]*Cappa v. Oscar C. Holmes, Inc.*, 25 Cal.App.3d 978, 102 Cal.Rptr. 207 (1972).

worker fell to his death.[55] Similarly, violation of a state law requiring shields over certain types of moving machinery was held to be negligence per se where a worker lost his balance and thrust his hand into the workings of a nearby postal conveyor belt.[56] Some jurisdictions hold that a statutory violation is evidence of negligence but not conclusive on this question.[57] One court in so concluding noted that the statute in question, one that required guardrails for steps in single-family residences, had been enacted after the home had been built and was not strictly enforced.[58]

Generally, compliance with a statutory standard is evidence that legal standards have been met but not conclusive on this issue.[59]

Contributory negligence or assumption of risk by the injured party as a defense to a negligence claim is discussed in Section 31.08.

SECTION 31.05 Vicarious Liability and the Independent Contractor Rule

Perhaps the most frequently contested issue in construction accident litigation is whether one person's negligence can be charged to another. Often the cause of the harm can be traced easily to one participant, but that participant (such as a contractor) is immune from tort action or, even if not immune, may be a small enterprise unable to pay any court judgment and may lack insurance coverage. For those reasons, the injured person often seeks to charge a person's negligence to another who can be sued and can respond directly or indirectly to a court judgment. For example, an injured employee of a subcontractor may be precluded from suing his employer and perhaps the prime contractor as well, the persons most closely connected to the accident. As a result, the injured employee often institutes legal action against the prime contractor (if the prime contractor is not a statutory employer) and the design professional. In addition, the em-

ployee may assert that the negligence of any contractor, whether immune or not, or the design professional is chargeable to the owner.

A. Vicarious Liability

One party can be vicariously responsible for the acts of another. The most common illustration is the responsibility of the employer for the acts of its employees committed in the scope of their employment. It must initially be determined whether the negligent person was an employee of the employer or acted as an independent contractor. There is an employment relationship if the employer controlled or had the right to control the details of how the negligent party performed its services (examined in greater detail in (B), dealing with the independent contractor rule).

B. The Independent Contractor Rule

Suppose an injured worker makes a claim against an owner. A number of issues can surface. First, the claimant may assert that the owner was itself negligent. Has the owner been negligent in engaging the contractor whose negligence caused the injury?[60] Usually this relates to the competence of the person engaged. However, it has been held that an owner who hires a financially irresponsible contractor is itself negligent.[61]

The next issue is whether the negligent party is an employee or an independent contractor. As noted in (A), the test is control or the right to control the *method* of performance, beyond the *result* to be accomplished. If the owner possesses or exercises this power, the negligent party is its employee. If the owner does not, simply asking for a result without dictating how that result is to be accomplished, the owner can assert that it engaged an independent contractor and is not responsible for the contractor's negligence.

Those supporting the rule contend that the person who has used proper care in hiring an independent contractor and has turned over the work to it should not be responsible. Those who oppose the rule assert that it allows enterprises to insulate themselves from enterprise risks. In response, de-

---

[55]449 F.2d 956 (D.C.Cir.1971).

[56]*O'Neill v. United States,* 450 F.2d 1012 (3d Cir.1971).

[57]*Fireman's Fund Ins. Co. v. AALCO Wrecking Co.,* supra note 39; *Tenenbaum v. City of Chicago,* supra note 49; *Johnson v. Salem Title Co.,* 246 Or. 409, 425 P.2d 519 (1967) (architect's violation of building code justified submitting question of negligence to jury).

[58]*Cooper v. Goodwin,* 478 F.2d 653 (D.C.Cir.1973).

[59]*Spangler v. Kranco, Inc.,* 481 F.2d 373 (4th Cir.1973).

---

[60]*Smith v. United States,* 497 F.2d 500 (5th Cir.1974).

[61]*Becker v. Interstate Properties,* 569 F.2d 1203 (3d Cir.1977), cert. denied 436 U.S. 906 (1978).

fenders say that the independent contractor is usually required to carry liability insurance and the cost of insurance is transferred to the employer of the independent contractor through the contract price. Yet, assert attackers, some independent contractors do not insure, and if the employer will pay indirectly, why not compel it to pay directly by denying protection of the rule, and so the debate rages.

Undoubtedly, the inconsistent cases and many exceptions to the rule reflect this controversy. Should the same rule be applied to a major addition made by General Motors to one of its plants[62] and to a farmer who has furnished materials and a truck to a one-person contractor who has agreed to build a chicken house?[63] In addition to the differing factual patterns, judges vary in their willingness to impose liability vicariously.

Considered swallowed up by exceptions[64] and concepts of enterprise liability, the independent contractor rule has displayed considerable modern vitality in many states. A large number of cases have exonerated the owner from responsibility for the negligence of the contractor despite the various powers the owner, usually through the design professional, can exercise during performance.[65] Such powers can include approval of subcontractors, inspection of the work, coordination of the work, condemnation of defective work, checking on the maintenance of schedules, and holding safety meetings. Cases that have given defenses based on the rule have involved both traditional methods of contracting and nontraditional methods, such as contract awards made by sophisticated owners with technical staff competence and work often performed within the physical area of an existing facility.[66] Similarly, an owner has been exonerated for negligence of its architect,[67] and prime contractors have been relieved from responsibility for subcontractors.[68]

Any language in the contract between employer and the party doing the work stating the latter to be an independent contractor has been considered relevant but certainly is not conclusive and often not persuasive.[69] The issue is for the jury. However, the independent contractor defense may be lost if the employer of the independent contractor assumes the latter's responsibilities.[70]

Without resort to the many exceptions to the rule, some cases have been unwilling to classify the negligent party as an independent contractor for whom the employer is not responsible. Most of these cases have been projects that have involved sophisticated, institutional owners that have never actually turned over exclusive control of the site or

---

[62]*Funk v. General Motors Corp.*, supra note 4 (discussed later in the section).

[63]*Smith v. Jones*, 220 So.2d 829 (Miss.1969).

[64]One court noted that there were so many exceptions that the rule applied only when there was not good reason for departing from it. *Walker v. Capistrano Saddle Club*, 12 Cal.App.3d 894, 896, 90 Cal.Rptr. 912, 914 (1970). But see *Franklin Privette v. Superior Court*, supra note 5 which breathed some life into the independent contractor rule in 1993 by limiting one of the exceptions. An earlier court stated that the rule existed primarily as a preamble to the catalog of its exceptions. *Pacific Fire Ins. Co. v. Kenny Boiler & Mfg. Co.*, 201 Minn. 500, 277 N.W. 226 (1937). Earlier edition of treatise cited in *Shannon v. Howard S. Wright Constr. Co.*, 181 Mont. 269, 593 P.2d 438, 441 (1979).

[65]See, e.g., *Jeter v. St. Regis Paper Co.*, supra note 49; *Foster v. Nat'l Starch & Chemical Co.*, 500 F.2d 81 (7th Cir.1974); *Campbell v. United States*, 493 F.2d 1000 (9th Cir.1974); *Jeffries v. United States*, 477 F.2d 52 (9th Cir.1973); *Stanley v. United States*, supra note 34; *Wolfe v. Bethlehem Steel Corp.*, 460 F.2d 675 (7th Cir.1972); *Fisher v. United States*, 441 F.2d 1288 (3d Cir.1971); *Olds v. Pennsalt Chemicals Corp.*, supra note 51; *Williams v. Gervais F. Favret Co.*, 499 So.2d 623 (La.App.1986), hearing denied 503 So.2d 19 (La.1987).

[66]*Jeter v. St. Regis Paper Co.*, supra note 49; *Foster v. Nat'l Starch & Chemical Co.*, supra note 65; *Wolfe v. Bethlehem Steel Corp.*, supra note 65, to name a few.

[67]*Milton J. Womack, Inc. v. State House of Representatives*, 509 So.2d 62 (La.App.1987) (negligent design); *Cutlip v. Lucky Stores, Inc.*, 22 Md.App. 673, 325 A.2d 432 (1974) (design and site services); *S.S. Kresge Co. v. Port of Longview*, 18 Wash.App. 805, 573 P.2d 1336 (1977) (design).

[68]*West v. Guy F. Atkinson Constr. Co.*, 251 Cal.App.2d 296, 59 Cal.Rptr. 286 (1967); *Kemp v. The Bechtel Constr. Co.*, 720 P.2d 270 (Mont.1986) (two dissenting); *Cafferkey v. Turner Constr. Co.*, 21 Ohio St.3d 110, 488 N.E.2d 189 (1986) (three dissenting).

[69]*Martin v. Phillips Petroleum Co.*, 42 Cal.App.3d 916, 117 Cal.Rptr. 269 (1974); *Reith v. General Tel. Co. of Illinois*, 22 Ill.App.3d 337, 317 N.E.2d 369 (1974); *Scrimager v. Cabot Corp.*, 23 Ill.App.3d 193, 318 N.E.2d 521 (1974).

[70]*C.H. Leavell & Co. v. Doster*, 233 So.2d 775 (Miss.1970). Similarly, *Hargrove v. Frommeyer & Co.*, 229 Pa.Super. 298, 323 A.2d 300 (1974), held the owner for defective design that injured an employee of the plaintiff when the plans and specifications were drawn by the owner's design department.

the project to the prime contractor.[71] *Funk v. General Motors Corp.*[72] is typical of those cases that have not applied the rule. Funk was an employee of a subcontractor and was seriously injured in a fall on a plant construction job. He recovered workers' compensation benefits from his employer and brought an action against the prime contractor and the owner of the plant, General Motors. After stating that primary responsibility for job safety belonged to the prime contractor, the Michigan Supreme Court stated that the owner of a building under construction is not ordinarily liable to construction workers for job safety. The court continued:

> In contrast with a general contractor, the owner typically is not a professional builder. Most owners visit the construction site only casually and are not knowledgeable concerning safety measures. . . . Supervising job safety, providing safeguards, is not part of the business of the typical owner.[73]

General Motors retained a significant amount of control of the project. Its internal division drafted the building plans, wrote the contractual specifications, and acted as architectural supervisor. It directly hired several of the contractors, including the employer of the plaintiff, and later assigned those contracts to the prime contractor. A General Motors representative was the architect-engineer superintendent for one of GM's divisions. He could order the prime contractor to terminate any prime or subcontractor's performance within twenty-four hours. The representative was at the job site daily and interpreted the contract specifications and plans for the prime. He made on-the-spot inspections, including inspections to determine whether safety requirements were being met.

The court concluded that an owner, such as GM here, will not be absolved from its failure to require observance of reasonable safety precautions when it in effect is the superintendent and has knowledge of the high degree of risk faced by construction workers. Similarly, some courts have held prime contractors for the negligence of their subcontractors.[74]

Courts refusing to allow the independent contractor defense seem motivated by hostility to an enterprise seeking to avoid responsibility for harms caused by the enterprise. Also, they wish to see victims compensated and are willing to search for a "deep pocket."[75]

C. Exceptions to Independent Contractor Rule

One court stated that the rule is "primarily important as a preamble to the catalog of its exceptions."[76] Although there are many exceptions, it is important to determine the extent of their application.

The preceding subsection noted that retention of a significant amount of control may mean that the party who has been negligent is not an independent contractor making the party engaging the negligent party vicariously liable. Retention of control can also be considered as an exception to the independent contractor rule. For example, the Restatement (Second) of Torts § 414 states that an employer who entrusts work to an independent contractor but who "retains the control of any part of the work" is liable for physical harm to those for whose safety the employer owes a duty to exercise reasonable care, if that harm is caused by its failure to exercise its control with reasonable care. Technically, this should not be considered an exception to the independent contractor rule, as the employer is liable for its own negligence. Nevertheless, it is still classified as an exception to the independent contractor rule, as the section assumes that the employer has retained an independent contractor.

Usually, a general right reserved to employers (this is usually applied to general rights retained by owners) does not mean that the contractor is

---

[71]*Nagler v. United States Steel Corp.,* supra note 18; *Magill v. Westinghouse Elec. Corp.,* 327 F.Supp.1097 (E.D.Pa.1971); *Weber v. Northern Illinois Gas Co.,* supra note 20; *Moore v. Denune & Pipic, Inc.,* supra note 43; *Kuhns v. Standard Oil Co. of California,* 257 Or. 482, 478 P.2d 396 (1970).

[72]See note 4, supra.

[73]220 N.W.2d at 646.

[74]*Smith v. United States,* supra note 60; *Summers v. Crown Constr. Co.,* 453 F.2d 998 (4th Cir.1972). Although some negligence seems to have been found against the prime contractor in the *Summers* case, the facts do not seem materially different from *West v. Guy F. Atkinson Constr. Co.,* supra note 68, which came to a contrary conclusion.

[75]Very little attention is paid to the possibility of the independent contractor's carrying liability insurance, passing that cost to the employer through the contract price, and creating double insurance.

[76]See note 64, supra.

controlled as to its methods of work or its operative detail.[77]

Yet in a claim by the employee of a subcontractor against a prime contractor employer, Arizona held that the employer need not control the day-to-day operations to be liable.[78] The court noted that contracts published by the American Institute of Architects (AIA) put the responsibility on the contractor for safety precautions and programs and that it must take reasonable precautions for the safety of *all* employees on the work.

It does not seem likely that the independent contractor rule will be a defense to the prime contractor if the contract clearly allocates the responsibility of safety to the prime contractor.[79]

One of the best known exceptions is for work "that the employer should recognize as likely to create during its progress a peculiar risk of physical harm to others unless special precautions are taken."[80] More commonly, courts have referred to this particular exception as work that is inherently or intrinsically dangerous or that involves a peculiarly high degree of risk.[81]

Although most jurisdictions recognize this exception, they vary in their willingness to apply it.[82] Even where the activity is considered inherently or intrinsically dangerous, not all accidents result in liability to the employer of the independent contractor. For example, one case held that the existence of the exception is not relevant unless a direct causal relationship exists between the peculiar risk inherent in the work and the plaintiff's injuries. The injury in this case was caused by a defective tractor/scraper/earth-moving machine. The court held the employer of the independent contractor not responsible.[83]

Large institutional owners with safety expertise and a high degree of control are likely to be responsible for the negligence of their independent contractors because of a judicial determination that the work performed was inherently dangerous or involved a peculiar risk of physical harm.[84] Courts unwilling to use this "exception" are often motivated by a desire to preserve the independent contractor rule in construction accidents.[85]

Even jurisdictions willing to recognize an exception for inherently dangerous work have difficulty coming to a common conclusion regarding the availability of this exception to employees of the independent contractor. For example, suppose that the owner hires a prime contractor to erect high-voltage electrical wires and that this is high-risk, inherently dangerous work. Suppose the prime contractor is negligent and one of the prime's employees suffers physical harm. The employee cannot bring legal action against the prime contractor because of workers' compensation immunity granted to the employer. If the employee asserts an action against the employer of the independent contractor—that is, the owner—some jurisdictions will permit the inherently dangerous exception to apply,[86] while most will not.[87]

Courts unwilling to extend the inherently dangerous exception to employees seem persuaded that the latter have an adequate remedy through workers' compensation.[88] Those that believe otherwise are likely to grant an injured worker the benefit of the inherently dangerous work exception to the independent contractor rule.[89]

---

[77]Restatement (Second) of Torts § 414, Comment C.

[78]*Louis v. N.J. Riebe Enters., Inc.*, 170 Ariz. 384, 825 P.2d 5 (1992).

[79]The *Louis* case is discussed and analyzed and other authorities are noted in 13 Constr.Litig.Rep. 183 (June 1992).

[80]Restatement (Second) of Torts § 416 (1965).

[81]W. PROSSER AND P. KEETON, TORTS, 512 (5th ed.1984).

[82]For an analysis of this exception, see Comment, 67 Iowa L.Rev. 589 (1982). The inherently dangerous work and peculiar risk exceptions as they apply to employees of the independent contractor are analyzed in cases collected in 13 Constr.Litig.Rep. 152–53 (May 1992).

[83]*Holman v. State*, 53 Cal.App.3d 317, 124 Cal.Rptr. 773 (1975).

---

[84]*McDonough v. United States Steel Corp.*, 228 Pa.Super. 268, 324 A.2d 542 (1974).

[85]*Musgrave v. Tennessee Valley Auth.*, 391 F.Supp. 1330 (N.D.Ala.1975).

[86]*Lindler v. District of Columbia*, 502 F.2d 495 (D.C.Cir.1974); *McDonough v. United States Steel Corp.*, supra note 84.

[87]*Campbell v. United States*, supra note 65; *Olson v. Red Wing Shoe Co.*, 456 F.2d 1299 (8th Cir.1972); *Reber v. Chandler High School Dist. No. 202*, 13 Ariz.App. 133, 474 P.2d 852 (1970); *Franklin Privette v. Superior Court*, supra note 5 (reversing prior decisions); *Zueck v. Oppenheimer Gateway Properties, Inc.*, 809 S.W.2d 384 (Mo.1991); supra note 5.

[88]*Franklin Privette v. Superior Court*, supra note 5. *Olson v. Red Wing Shoe Co.*, supra note 87; *Zueck v. Oppenheimer Gateway Properties, Inc.*, supra note 5.

[89]Undoubtedly, this played a significant part in the court's decision in *Alber v. Owens*, supra note 52.

The exception for inherently dangerous work has been applied vigorously in some jurisdictions but very cautiously in others. Similarly, other exceptions, such as for nondelegable duties[90] and violation of building codes,[91] have been recognized, but jurisdictions vary in their willingness to actually *apply* an exception to the independent contractor rule in a particular case.

The *existence* of many exceptions to the rule has by no means obliterated the rule. The rule still stands as a significant barrier to recovery in construction accidents.

SECTION 31.06 Cause in Fact

To transfer the loss from the victim to the defendant, the victim must prove that the defendant's conduct has substantially caused the loss.[92] One exception relates to harm caused by two negligent persons but the plaintiff cannot prove which person caused the loss.[93]

SECTION 31.07 Proximate Cause

Sometimes the law relieves a negligent party who has caused the loss because the actual loss was greatly out of proportion to what could have been expected, the loss occurred to an unforeseeable plaintiff, the loss occurred in a substantially unanticipated manner, or the loss was caused by an intervening agency that should be considered a supervening cause releasing the original wrongdoer.

A few applications of proximate cause in construction accidents have been made, mainly the effect of intervening causes. For example, an architect responsible for a negligently designed wall that collapsed was not relieved by the approval of his work by a building inspector.[94] Similarly, a defendant contractor who had left a trench across a street filled with gravel was not relieved from liability when a third party placed a thin layer of blacktop across the street that ultimately wore out, leaving holes.[95] The court concluded that the intervention by the third party was reasonably foreseeable. But an architect was relieved because of the absence of proximate cause because other individuals, the project manager, the site superintendent, and the masonry subcontractor, were all aware of the unsafe nature of the scaffold before the accident.[96] In addition, the building inspector had visited the site at least daily but had failed to notice the scaffold in its unsafe condition. These, according to the court, were sufficient to break the causal connection between any alleged negligence on the part of the architect and the worker's injury. The contractor who negligently left a ladder in a position where it might have caused an injury was relieved from liability when the ladder was unexpectedly moved by a third party.[97]

Sometimes the acceptance of the work that contains a patent defect absolves the contractor from responsibility.[98] The justification for this is that the owner deprives the contractor of the opportunity to rectify its wrong by occupying and resuming possession of the work. The owner is presumed to have made a reasonably careful inspection and know of its defects, and if the owner takes it in this condition, it accepts the defects and the negligence that caused them as its own. If it accepts the work that is in a dangerous condition, the owner has an immediate duty to make it safe, and if it fails to do so, and a third person is injured, its negligence is the proximate cause of the injury.

---

[90]*Reith v. General Tel. Co. of Illinois,* supra note 69. The employer of the independent contractor—a public utility—needed a franchise and permit from state authorities. This made its duty nondelegable and deprived it of the right to use the independent contractor rule. Similarly, in *Weber v. Northern Illinois Gas Co.,* supra note 20, the court held that the employer's duty to provide a safe work site to employees of a subcontractor was nondelegable.

[91]*Johnson v. Salem Title Co.,* supra note 57 (architect held liable for the negligence of his consulting engineer).

[92]A court seemed to relax this burden in *Fireman's Fund Ins. Co. v. AALCO Wrecking Co.,* supra note 39.

[93]*Bowman v. Redding & Co.,* 449 F.2d 956 (D.C.Cir.1971). See Section 7.03(D).

---

[94]*Johnson v. Salem Title Co.,* supra note 57. See Section 24.07.

[95]*Adams v. Combs,* 456 S.W.2d 288 (Ky.App.1971).

[96]*Belgum v. Mitsuo Kawamoto & Assoc.,* 236 Neb. 127, 459 N.W.2d 226 (1990).

[97]*Hill v. Lundin & Assoc., Inc.,* supra note 33.

[98]*Slavin v. Kay,* 108 So.2d 462 (Fla.1958). By dictum, this was affirmed in an opinion that divided the Florida Supreme Court in *Ed Ricke & Sons, Inc. v. Green,* 609 So.2d 504 (Fla.1992), rehearing denied (Jan. 5, 1993).

SECTION 31.08 Some Defenses

A. Assumption of Risk and Contributory Negligence

Sometimes those against whom claims have been made point to conduct by the claimant that they assert bars or at least reduces the extent of the claim. The defenses that relate to conduct by the claimant are assumption of risk and contributory negligence. Assuming a risk is a voluntary exposure to danger with full knowledge and appreciation of its existence—or taking one's chances. Contributory negligence, on the other hand, is conduct that does not rise to the standard of conduct required by tort law and that plays a substantial part in causing the injury.

The two concepts can overlap. The claimant may have chosen voluntarily to encounter the risk, and this choice may be conduct that falls below that required by law. Yet a party can reasonably assume a risk. Before briefly looking at a case that sought to make this differentiation, it is important to recognize the difference between a claim based on common law negligence and one based on the violation of a statute. Claims based on common law negligence can be defeated if it is determined that the claimant has expressly or impliedly assumed the risk.[99] Similarly, where contributory negligence by the claimant *bars* the claim (an increasingly small number of states do so), a claim based on common law negligence will be barred.[100] Complications develop when the claim is essentially based on the violation of a statute.

In *Fonseca v. County of Orange*,[101] the plaintiff was a construction worker who fell from a bridge that was being constructed by the worker's employer over a river for Orange County. He slipped on some wet excess cement as he was doing finish work on the concrete deck of the bridge at a height of at least 20 feet above the dry riverbed. No scaffolding or railing had been installed around the perimeter of the bridge for the protection of workers

as required by law. After recovering workers' compensation benefits from his employer, the plaintiff brought a third party tort claim against Orange County based upon the latter's vicarious responsibility for the contractor's violation of state safety orders requiring installation of railings where workers must work at heights in excess of $7^1/_2$ feet above ground level and scaffolding at heights above 15 feet.

One assertion made by Orange County was that the plaintiff had been contributorily negligent. The plaintiff had had considerable experience in cement work, and his experience should have motivated him to refuse to start work when he saw that there were safety violations. The plaintiff had testified that he told the county engineer that protective railings should be installed, but in any event, the plaintiff proceeded to work on the deck from which he fell and was injured.

After defining contributory negligence, the court noted that some jurisdictions have eliminated it as a bar to recovery in actions arising from violations of safety regulations but that in California it is available as a defense.[102] On the other hand, assumption of risk does *not* bar a worker's tort claim against a third party, particularly where there has been a violation of safety statutes or orders. (Clearly it would *not* bar a workers' compensation claim.) In other words, if the plaintiff's conduct was characterized as assumption of risk, it would not bar the claim, but if characterized as contributory negligence, at least at that time, it would bar the entire claim.

The court gave as a reason for this differentiation the weak bargaining position of the employee and the economic pressure on him to work in unsafe places or with an unsafe appliance.

Noting the difference between the two doctrines and the possible overlap between them, the court concluded that the plaintiff had assumed the risk because he knew of the danger and its magnitude and voluntarily encountered the risk. The court concluded that the plaintiff did not perform unreasonably when he continued to work despite the violation of the safety orders. This convenient conclusion—finding that the conduct did not bar the claim—reflects the tendency of some courts to give wide scope to worker third-party actions.

---

[99]*Mitchell v. Young Refining Corp.*, 517 F.2d 1036 (5th Cir.1975). But see *Gray v. Martindale Lumber Co.*, 515 F.2d 1218 (5th Cir.1975).
[100]*Mundt v. Ragnar Benson, Inc.*, 61 Ill.2d 151, 335 N.E.2d 10 (1975). But see *Funk v. General Motors Corp.*, supra note 4.
[101]28 Cal.App.3d 361, 104 Cal.Rptr. 566 (1972).

[102]See Section 7.03(G).

Injured workers in other jurisdictions have not fared as well as Fonseca. Courts, when they bar the claim, have held that a worker who stays on the job when asked or ordered to perform work under dangerous conditions has either assumed the risk or been contributorily negligent.[103] These courts have not been sympathetic to the economic compulsion arguments made by workers—that they had to work or would be sent home.

Admittedly, some arguments can be made for barring recovery in these cases. The worker *will* very likely receive workers' compensation. Barring recovery may motivate workers to refuse to work under dangerous conditions, and this may induce employers to make their work conditions reasonably safe. It can be argued that a worker who is a member of a construction trade union will often be able to find assistance there.[104]

The realities of the workplace and the economic pressures on workers support the position taken by the California court in the *Fonseca* decision. Admittedly, there is an element of paternalism in concluding that workers must be protected from their own foolhardiness and from the pressures of the workplace. Perhaps in the more individualistic nineteenth century this would be appropriate, but it does not seem to be in tune with notions of modern social welfare concepts that have become predominant in tort law. In this regard, the modern movement toward comparative negligence would mean that conduct considered as unreasonable (and Fonseca's conduct could fall under this category) would only lower recovery rather than bar the entire claim. It will be much more difficult for the defendant to establish that a *worker's* violation of a safety statute would bar or reduce the worker's claim.[105]

Procedural rules make successful use of these defenses difficult for defendants even when they are permitted to assert them. First, the defendants must plead and prove that the injured party has assumed the risk or has been contributorily negligent. Second, the determination of whether the defendant has proved this by a preponderance of the evidence is typically made by juries.[106] In addition, contributory negligence is slowly disappearing in most American states and is not likely to prove to be an absolute bar. The emphasis on occupational safety is likely to make successful use of assumption of risk and contributory negligence even rarer in the future.

B. Sovereign Immunity: Design Liability

The broad outlines of sovereign immunity, past and present, were given in Section 7.06(D). Several aspects of immunity that relate more directly to the Construction Process merit brief comment. The increased participation of governmental authorities in land use and the Construction Process usually makes it necessary to obtain a building permit before construction and an occupancy permit after construction. Issuing a building permit often involves checking of drawings and specifications,[107] while an occupancy permit usually necessitates an inspection to ensure compliance with building code standards.[108]

Public entities that issue such permits usually seek immunity from injuries or losses that may occur because of faulty design or construction. Sometimes this is accomplished by making this activity immune either by statute or by case decision. Some entities that issue permits seek to relieve themselves from liability by disclaimers they make at the time that such permits or approvals are given.[109]

The allocation of responsibility between design professionals who draft plans and specifications and the local building department that must review and approve them was an issue that faced the Louisiana Appellate court in *Sunlake Apartment Res-*

---

[103]*Mitchell v. Young Refining Corp.*, supra note 99; *Demarest v. T.C. Bateson Constr. Co.*, 370 F.2d 281 (10th Cir.1966); *Ralston v. Illinois Power Co.*, 13 Ill.App.3d 95, 299 N.E.2d 497 (1973).

[104]*Holman v. State*, supra note 15.

[105]*Morgan v. Stubblefield*, 6 Cal.3d 606, 493 P.2d 465, 106 Cal.Rptr. 1 (1972) (worker must have actual knowledge of the safety order before it will affect his claim).

[106]*Gantt v. Mobil Chemical Co.*, 463 F.2d 691 (5th Cir.1972); *Kirsch v. Dondlinger & Sons Constr. Co.*, 206 Kan. 701, 482 P.2d 10 (1971); *Bennett v. Young*, 266 N.C. 164, 145 S.E.2d 853 (1966).

[107]See *Manors of Inverrary XII Condominium Assoc. v. Atreco-Florida*, 438 So.2d 490 (Fla.Dist.Ct.App.1983) (no immunity).

[108]As to inspections, see Note, 18 U.Rich.L.Rev. 809 (1984).

[109]See Comment, 58 Wash.L.Rev. 537 (1983). See also *Morris v. County of Marin*, supra note 15.

idents v. Tonti Dev. Corp.[110] After a fire that destroyed an apartment building, a claim had been brought against a number of parties, including the architect, to recover for the fire damage. The claims were settled, but the architect and others then brought an action against state and local agencies based upon their alleged failure to detect the absence of fire stops and draft stops in the design as required by the building code.

The court held that the duty owed by public agencies was to the public. The lack of a duty to private parties exempted the agencies from liability for the negligence of its employees. The court gave another reason to reject the claim. It noted that responsibility for a failure to detect design errors would impose tremendous liability upon public agencies. In effect, liability would in whole or in part relieve architects, contractors, and building owners of their liability for their negligence and would make the public treasury an insurer for their negligence. In an astounding admission, the court stated that it is not uncommon for local building inspectors to issue permits for buildings that violate the building code. Since people in the local building department are poorly prepared and overworked, they must often rely upon the design professionals who sign the drawings.

A look at the California Design Immunity Legislation and its subsequent history is useful both as a review of design problems generally and as an illustration of the difficulties that develop because of the sharply differing policies reflected in immunity law.

Design liability has features that separate it from other forms of liability. First, it almost always involves conscious choices and trade-offs between quality, quantity, time, and cost as well as between function and aesthetics. Second, some aspects of design involve technical considerations sometimes thought to be beyond the competence of a jury. Third, events that occur *after* initial design can require reappraisal of the design. The state of the art changes, and a strong danger exists that those being charged will be held to a standard that existed at the time of the accident or even at the time of the trial rather than at the time the design decision was initially made.

In 1961, the California Supreme Court abolished unlimited sovereign immunity.[111] In 1963, the legislature adopted a comprehensive statutory scheme that sought to accommodate the various interests involved. One section[112] provides:

> Neither a public entity nor a public employee is liable under this chapter for an injury caused by the plan or design of a construction of, or an improvement to, public property where such plan or design has been approved in advance of the construction or improvement by the legislative body of the public entity or by some other body or employee exercising discretionary authority to give such approval or where such plan or design is prepared in conformity with standards previously so approved, if the trial or appellate court determines that there is any substantial evidence upon the basis of which (a) a reasonable public employee could have adopted the plan or design or the standards therefor or (b) reasonable legislative body or other body or employee could have approved the plan or design or the standards therefor.

In 1967, California granted immunity in *Cabell v. State of California*, a case where accidents had occurred before the one causing injury to the plaintiff and the claim had been made that the public entity *now* knew that the design had been improper.[113] The court emphasized the importance of judging the design at the time it was made and not at some later date. Yet four and one half years later, the California court overruled its holding, stating that it was convinced that the legislature did not intend to permit public entities to shut their eyes to the operation of a plan or design once it had been transferred from "blueprint to blacktop."[114]

SECTION 31.09 OSHA Violations

The emphasis in this chapter has been upon construction accidents and the use of the judicial process to compensate victims. But those active in the construction process may increasingly be subject to the Occupational Safety and Health Act (OSHA) or

---

[110]602 So.2d 22 (La.App.1992).

[111]*Muskopf v. Corning Hosp. District*, 55 Cal.2d 211 359 P.2d 457, 11 Cal.Rptr. 89 (1961).
[112]West Ann.Cal.Gov't Code § 830.6. As to road design immunity, see Note, 58 Temp.L.Q. 543 (1985).
[113]67 Cal.2d 150, 430 P.2d 34, 60 Cal.Rptr. 476 (1967).
[114]*Baldwin v. State of California*, 6 Cal.3d 424, 491 P.2d 1121, 99 Cal.Rptr. 145 (1972).

its state counterparts. These statutes were designed to make the workplace reasonably safe. While violations of OSHA rules do not create civil liability to an injured worker, the battery of enforcement authority given to the regulatory agencies has made OSHA liability exposure an important factor in the Construction Process.

Under OSHA, the Department of Labor can seek restraining orders and civil fines for violations. In addition, criminal penalties can be imposed in the event death results from a violation, including "a fine of not more than $10,000 or . . . imprisonment for not more than six months, or both" for a first violation. For successive violations, the fine may be doubled.[115] As a result, it is often crucial to determine who is an employer for purposes of deciding whether certain entities involved in the Construction Process can be subject to civil and criminal penalties.

Penalties generally are imposed upon employers who are defined as those who perform construction work or substantially supervise construction. The entity most responsible will be the prime contractor.[116] As a rule, design professionals who engage in traditional on-site activities are not considered employers for OSHA purposes.[117]

SECTION 31.10 Passage of Time: Statutes of Limitations

Some claims by those who have suffered personal harm are made many years after the Construction Process has been completed. This problem in the context of owner claims was discussed in Section 23.03(G). Reference should be made to that section, because both ordinary and special completion statutes discussed there can affect claims by those who have suffered personal harm.

PROBLEM

W was an employee of S, an electrical subcontractor in a project in which P was constructing an office building for O. The site and a set of construction documents drawn by A were turned over to P. W was working on a scaffold alongside the fourth floor of a partially constructed six-story building. Safety regulations required that W wear a safety belt. On the day of the accident, W decided not to wear a belt because it was hot. A heavy plank fell from the roof, striking W, knocking him off the scaffold, and sending him to his death. The plank came from an area in which two sub-subcontractors, S-1 and S-2, had been working.

W's widow has brought a lawsuit against S-1, S-2, P, O, and A, as well as S's workers' compensation carrier. The claim against S-1 and S-2 was based on the negligence of their employees, which caused the plank to fall. The claim against P was based on P's knowledge that safety orders were being violated by S. The claim against O was based on the work's being "inherently dangerous" and thereby depriving O of the independent contractor defense. The claim against A was based on his failure to insist on safe construction methods when he knew that S had a poor reputation in worker safety. The claim against the workers' compensation carrier was based on its unexercised power to conduct a loss-prevention program and to provide inspection. What is the likelihood of W's widow succeeding against any of these defendants? Would your answer have been any different had W asked for safety belts on the date of the accident, was told there were none, and was told that if he did not wish to work that day he could go home?

---

[115]29 U.S.C.A. § 666.

[116]*Secretary of Labor v. Blount International, Ltd.*, 15 O.S.H. Cas. (BNA) 1897 (O.S.H. Comm'n 1992) (prime responsible for failure to exercise reasonable diligence to discover a dangerous electrical panel distribution box installed by the subcontractor).

[117]*Secretary of Labor v. Skidmore, Owings & Merrill,* 5 O.S.H. Cas. (BNA) 1762 (O.S.H. Comm'n 1977) (architect); *Secretary of Labor v. Simpson, Gumpertz & Hegers, Inc.,* 15 O.S.H. Cas. (BNA) 1851 (O.S.H. Comm'n 1992) (consulting structural engineer), but a construction manager was held liable under OSHA, as the owner depended on him for safety despite the fact that he could not order subcontractors to work with proper regard for the safety of the workplace. *Secretary of Labor v. Kulka Constr. Management Corp.*, 15 O.S.H. Cas. (BNA) 1870 (O.S.H. Comm'n 1992). (As noted in Section 14.08(G), the Secretary of Labor has appealed the *Simpson* decision.)

Indemnification and Other Forms of Shifting and Sharing Risks: Who Ultimately Pays?

SECTION 32.01 First Instance and Ultimate Responsibility Compared

Chapter 31 demonstrated the increased likelihood that those who suffer losses incident to the Construction Process will be compensated, especially where those losses involve personal harm. Although claimants will not always receive judicial awards, the tendency has been to expand liability to ensure that those who suffer losses are compensated.

Lawsuits today generally begin with claims against a number of defendants, something that is legally permissible and relatively inexpensive. Typically, these defendants make claims against each other as well as claims against parties who have not been sued by the original claimant. The result is a multiparty lawsuit that generally involves as many as a half dozen interested participants as well as an almost equal number of sureties and insurers. For example, the injured employee of a prime contractor is likely to sue all contractors who are not immune, such as other separate contractors, and subcontractors, the owner, the design professional, any construction manager, and, depending on the facts, those who have supplied equipment or materials. Each of the defendants will very likely bring claims against the others based on indemnification. If there is a building defect, such as the failure of an air-conditioning system, the owner is likely to assert claims against the design professional, the contractor, and the manufacturers and sellers of the system. Again, those against whom claims have been asserted are likely to assert claims against each other.

In these lawsuits, there are two levels of responsibility. There is *first instance* responsibility to the original claimant, such as the injured worker or the owner. Resolution of this issue depends on whether any of the defendants or someone for whom they are responsible has failed to live up to the standard required by the contract or by the law. After this determination is made, the next level, that of *ultimate* responsibility must be addressed. Who among those responsible will ultimately bear the loss? This inquiry is the focal point of this chapter and one that has developed unbelievably complex and costly legal controversies.

SECTION 32.02 Contribution Among Wrongdoers

Suppose A and B, *acting together* is pursuance of a common plan or design, injure C. C can recover its loss from either A or B if each has committed a wrong. Because each has committed the wrong and because they have acted together, neither A nor B can receive contribution from the other if either has paid more than one half of the total judgment. In this particular instance, A and B are joint wrongdoers. For example, if the design professional and owner acted together to destroy the contractor's business or reputation, the contractor could sue either or both and recover its loss from either or both. In such a case, neither design professional nor owner would have a claim against the other if either paid more than one half of the judgment.

But a number of defendants may be sued in the same legal action even though they have not acted together. They are codefendants because each may have played a substantial role in causing the injury and for procedural convenience, all the claims are decided in one lawsuit. The defendants are concurrent wrongdoers, not in the sense that their wrongdoing occurred at the same time but in the sense that each played a substantial role in causing an individual loss to the claimant.

Suppose three defendants are held liable to the plaintiff and one defendant pays the award. Can the one defendant seek reimbursement or "contribution" from the other defendants? The basis for contribution is the unfairness of one defendant paying the entire judgment when all are responsible. Most American courts do not require contribution among wrongdoers. But one half of the states have created contribution by statute. The statutes vary considerably, but where they exist, two factors have reduced the effectiveness of contribution statutes. First, some states require that there be a joint judgment against the defendants before contribution can be compelled. This has obvious limitations, as it inhibits settlement. Second, and more important for purposes of construction accidents, contribution is frequently denied against a party who was immune from the original claim. For example, it would not be available against the employer (actual or statutory, as described in Section 31.02) of the injured party if the employer were immune from liability because of workers' compensation laws. To encourage settlement, some states have enacted legislation under which a defendant who settles with the claimant and seeks to be relieved from any claim that may be made by other claimants can ask the court to determine whether the settlement was made in good faith. If the court so determines, the settling defendant is released from any claims that may be made against it by other defendants.[1]

Where contribution does exist, either by judicial rule or legislation, each wrongdoer generally contributes a pro rata share. Where there are three wrongdoers, each would contribute one third. An increasing number of jurisdictions now compare wrongdoing quantitatively when determining the amount of contribution required.[2]

Sometimes qualitative comparisons between wrongdoers are made. But this comparison, accomplished through indemnity, can result in a total shift of the loss. Contribution, on the other hand, requires a sharing of ultimate responsibility.

SECTION 32.03 Noncontractual Indemnity[3]: Unjust Enrichment

A. Basic Principle: Unjust Enrichment

Wrongdoing can encompass a spectrum from innocent conduct to conduct intended to harm. Illustrations of the *least* culpable conduct are liability imposed on manufacturers for defective products, even though there has been no showing that they were negligent, and liability placed on the employer for the negligent acts of its employees' committed in the scope of employment. Similarly, as seen in Section 31.05(C), the employer of an independent contractor may under certain circumstances be liable for the negligence of that independent contractor.

An enlarged and qualitatively wide range of liability creates the opportunity to compare wrongdoing of defendants, either original or those added by cross actions brought by original defendants. For example, suppose a worker employed by a prime contractor is injured because of deliberate safety violations by her employer. Under certain circumstances, the worker can recover from the owner.[4] Liability in such a case may be based on the nondelegable statutory duty of the owner to furnish a safe workplace. Alternatively, liability may be based on the failure to determine whether

---

[1]West Ann.Cal. Civ.Proc. § 877.6.

[2]See *Skinner v. Reed-Prentice Div. Package Machinery Co.*, 70 Ill.2d 1, 374 N.E.2d 437 (1977); *Bielski v. Schulze*, 16 Wis.2d 1, 114 N.W.2d 105 (1962).

[3]This categorization separates indemnification based on *express* agreement from *other* forms of indemnification. The latter have various designations, such as common law indemnity, quasi-contractual indemnity, and equitable indemnity. Rather than attempt to sort out these terms and their implications, indemnification is divided into noncontractual and contractual.

[4]See Section 31.05.

safe practices were being followed or on the failure to discharge the contractor after becoming aware of the violations. Some of these reasons for owner liability would be less morally objectionable than the conduct of the prime contractor. In many states, a wrongdoer who pays the claim or who would be responsible under a court judgment for the claim can receive indemnification from the other wrongdoer. Suppose the owner seeks indemnification from the contractor. Indemnification would require a comparison of the *degree* of wrongdoing between the two parties. If the owner's moral culpability is less than that of the contractor, the latter would indemnify the former. The contractor would be unjustly enriched and the owner unjustly impoverished if the owner had to pay the claim.

Similarly, suppose the prime contractor violated safety orders and a claim was made against the design professional based on the latter's failure to detect the violation or to exercise corrective power given by the contract. The contractor's conduct can be considered active if it orders an employee to work under dangerous conditions. On the other hand, the conduct of the design professional can be considered passive, her liability based on failure to act. Although judicial determination of active and passive conduct is sometimes at variance with ordinary meaning (a point to be explored later in this section), courts sometimes make this differentiation the basis for giving the design professional indemnity from the contractor or from someone else more directly connected with the injury or loss.[5] Indemnity in such a case is based on the unjust enrichment of the more culpable wrongdoer that would result if it were not required to bear this loss.

B. Noncontract and Contract Indemnity Differentiated

Sections 32.04 and 32.05 treat indemnification implied from or expressed in a construction contract. As indicated, noncontractual indemnity is based principally on unjust enrichment and the concept that losses should be shifted from one wrongdoer to another based on qualitative comparative fault.

For historical reasons, however, courts have tended to treat noncontractual indemnity as more analogous to a contract claim than to a tort claim. For this reason, courts tend to classify this form of indemnity as quasi-contractual (something *like* a contract) even though it is not based on consent.[6]

C. Some Classifications

Courts have articulated various tests to determine whether noncontractual indemnification will be awarded. Although one court recognized an indemnification obligation in order to impose an ultimate burden on one who was the "active delinquent" in bringing about the injury rather than the "lesser delinquent,[7] most courts have employed the primary-secondary or passive-active differentiation, singly or together. For example, one court stated indemnity would be granted

> to a person who, without active fault on his own part, has been compelled, by reason of some legal obligation, to pay damages occasioned by the initial negligence of another, and for which he himself is only secondarily liable.[8]

The court stated that vicarious liability—that is, liability for the wrongs of another—that arises out of a positive rule of common or statutory law,[9] and liability imposed for failure to discover or correct a defect or remedy a dangerous condition caused by the act of the one primarily responsible are illustrations of *secondary* liability. The passive-active differentiation, perhaps used more frequently, looks to similar factors. Again, some positive acts create liability, while sometimes negative or pas-

---

[5]In *Owings v. Rosé*, infra note 6, an architect was given indemnity against a consulting engineer when the architect was held liable essentially for the negligence of the engineer.

[6]See *Ohio Cas. Ins. Co. v. Ford Motor Co.*, 502 F.2d 138 (6th Cir.1974); *Owings v. Rosé*, 262 Or. 247, 497 P.2d 1183 (1972). But see *J. L. Simmons Co. v. Fid. & Cas. Co.*, 511 F.2d 87 (7th Cir.1975).

[7]*Miller v. De Witt*, 37 Ill.2d 273, 226 N.E.2d 630, 642 (1967).

[8]*Builders Supply Co. v. McCabe*, 366 Pa. 322, 77 A.2d 368, 370 (1951).

[9]For example, in *Wrobel v. Trapani*, 129 Ill.App.2d 306, 264 N.E.2d 240 (1970), the prime contractor was held liable for having violated the Illinois Structural Work Act because he was in charge of the project. He sought indemnity from a subcontractor who had violated the statute by not having the subcontractor's employee use care in putting up and using a ladder. The court held that the prime contractor's liability was passive, the subcontractor's was active, and the former would be indemnified by the latter.

sive inaction, though creating liability, is less morally objectionable though differential enough to justify indemnification.

It may be useful to look at a few cases that have dealt with noncontractual indemnity. In *Adams v. Combs*,[10] a road contractor sought indemnity from the city for whom the road work had been performed. An accident occurred two years after completion of the contract. The court denied the claim, however, holding that any negligence on the part of the city in not maintaining the road was secondary and passive, while the contractor's negligence was primary and active.

Architects sought indemnity in *St. Joseph Hospital v. Corbetta Const. Co.*[11] and *Owings v. Rose.*[12] In the first case, the architect was *denied* indemnity from a supplier of defective tile because the architect knew the tile was defective. In the second case, the architect was *granted* indemnification against a negligent consulting engineer when the architect who was *not* negligent was liable because he had a nondelegable duty. Where each party has violated the *same* duty, indemnity is usually denied. For example, in *Harris v. Algonquin Ready Mix, Inc.*,[13] the electric company and the contractor it had hired each failed to warn employees of certain dangers. The absence of any *qualitative* difference in their liability precluded indemnity.

It is generally stated that failure to discover or remedy the defect caused by another is merely passive negligence.[14] However, in *Becker v. Black & Veatch Consulting Engineers*,[15] an engineer's failure to inspect was held to be active negligence when the engineer was employed for this express purpose. This precluded the engineer from receiving indemnity from owner or contractor, as all were considered actively negligent.

Contribution would have been permitted if such right existed and, generally, if *all* could have been held liable. If the contractor was immune because the injured person was its employee, however, the contractor could not be compelled to contribute to the judgment unless there was contractual indemnity. The judgment would be shared by owner and engineer unless there was contractual indemnity.

The perils of generalizations in this area are emphasized in an Illinois decision which, in discussing active and passive liability, stated:

> Determination of this question is not a matter of proceeding according to the usual dictionary definitions of the words "active" and "passive." These words are terms of art and they must be applied in accordance with concepts worked out by courts of review upon a case-by-case basis. Under appropriate circumstances, inaction or passivity in the ordinary sense may well constitute the primary cause of a mishap or active negligence. . . . It has been appropriately stated that "mere motion does not define the distinction between active and passive negligence." [16]

Noting that certain terms have become words of art is a signal that a particular doctrine has developed difficulty, and it is therefore important to read between the lines of judicial opinions dealing with that doctrine. It is also an indication of considerable uncertainty in the law and the importance of using contract language to specifically deal with the problem rather than leave the matter to the vagaries of court decisions when either vague concepts must be applied[17] or legal terminology varies from ordinary meaning.

One commentator, reviewing some of the criteria used for determining whether noncontractual indemnity would be granted, stated:

> Probably none of these is the complete answer, and, as is so often the case in the law of torts, no one explanation can be found which will cover all of the cases. Indemnity is a shifting of responsibility from the shoulders of one person to another; and the duty to indemnify will be recognized in cases where community opinion would consider that in justice the responsibility should rest upon one rather than the other. This may be because of the relation of the parties to one another, and the consequent duty owed; or it may be

---

[10]465 S.W.2d 288 (Ky.1971).

[11]21 Ill.App.3d 925, 316 N.E.2d 51 (1974).

[12]Supra note 6.

[13]59 Ill.2d 445, 322 N.E.2d 58 (1974).

[14]See *Builders Supply Co. v. McCabe*, supra note 8.

[15]509 F.2d 42 (8th Cir. 1974). See *Associated Eng'rs, Inc. v. Job*, 370 F.2d 633 (8th Cir.1966).

---

[16]*Moody v. Chicago Transit Auth.*, 17 Ill.App.3d 113, 117, 307 N.E.2d 789, 792–93 (1974).

[17]In *Santisteven v. Dow Chemical Co.*, 506 F.2d 1216 (9th Cir.1974), the court stated that quasi-contractual indemnity is based on whether it is "just" to shift liability.

because of a significant difference in the kind or quality of their conduct.[18]

D. Employer Indemnification

Sometimes the propriety of shifting liability depends on substantive law policies that tend to either protect a particular person from this responsibility or place it on her. Many noncontractual indemnity claims are brought against employers of injured persons. For example, suppose a subcontractor's employee is injured and the primary responsibility for the injury was the employer's failure to comply with safety rules. Because the injured employee cannot sue her own employer, suppose she institutes legal action against the prime contractor. Suppose the prime contractor can be sued as a third party; liability against the prime contractor is based on its responsibility as the employer to provide safe working conditions, and this duty cannot be delegated. This would be a classic case for granting the prime contractor noncontractual indemnity. The prime contractor's liability is based on a nondelegable duty, and clearly the subcontractor's negligence is active and primary.

Nevertheless, allowing indemnity would place *ultimate* liability on the subcontractor. Would this deny the subcontractor immunity from tort action granted by workers' compensation law? One of the trade-offs in workers' compensation was granting immunity to the employer from tort liability in exchange for denying the employer rights possessed prior to the enactment of workers' compensation statutes. Indemnification can, by indirection, frustrate this.

But there are cogent arguments for allowing indemnity in these cases. The third party, here the prime contractor, received nothing in the "trade" and arguably should not have existing rights taken away. This argument, however, ignores the likelihood that third parties are also employers who are part of some workers' compensation legislative trade. In addition to this argument, it would appear to be unjust to exonerate the more culpable employer when liability against the third party, such as a prime contractor or owner, may be vi-

carious, based on passive negligence or a statutory violation. It might also be argued that denial of indemnity may encourage carelessness by the employer.

The difficulty of this question is reflected in the case law. It has been considered the most evenly balanced issue in workers' compensation law.[19] But now it appears that a slight majority precludes *noncontractual* indemnity against an employer.[20] Those that grant indemnity usually find a separate duty owed by the employer to the third party seeking indemnity.[21]

E. Comparative Negligence

Many states have moved to comparative negligence as a method of avoiding the all-or-nothing aspects of the contributory negligence rule.[22] Suppose an injured worker claims successfully against owner and design professional and those two defendants assert a successful cross claim against the worker's employer based on noncontractual indemnity. Jurisdictions that grant contribution would give each of the two defendants in the original action and the additional defendant in the crossclaim contribution from the others if any party paid more than one third of the award.

Generally, no *quantitative* comparison is made when a claim for noncontractual indemnity is successful. There is a *qualitative* comparison, such as passive-active or primary-secondary. But this comparison is made to determine whether the *entire* loss will be shifted from, say, a passively negligent defendant to an actively negligent one.

---

[18]*W. PROSSER, TORTS,* 313 (4th ed. 1971). A subsequent edition made marginal modifications. See W. PROSSER and P. KEETON, TORTS, 344 (5th ed. 1984).

[19]See 2A A. LARSON, *WORKMEN'S COMPENSATION* § 76.11 (1983). Compare *Dole v. Dow Chemical Co.,* 30 N.Y.2d 143, 282 N.E.2d 288 (1972), 331 N.Y.S.2d 382, which allowed indemnity, with *Galimi v. Jetco, Inc.,* 514 F.2d 949 (2d Cir.1975), dealing with the Federal Employees Compensation Act, which did not. Larson, the leading scholar in this field, advocates abolition of noncontractual indemnification against the employer. See Larson, *Third-Party Action Over Against Workers' Compensation Employer* [1982] Duke L. J. 483.
[20]See *A. A. Equipment, Inc. v. Farmoil, Inc.,* 31 Conn.Supp. 322, 330 A.2d 99 (1974). California has barred noncontractual indemnity against the employer of the claimant by requiring an express written indemnification. West Ann.Cal.Lab.Code § 3864. Note that *express* indemnification is permitted.
[21]*Santisteven v. Dow Chemical Co.,* supra note 17; *Dole v. Dow Chemical Co.,* supra note 19.
[22]See Section 7.03(G).

To deal with this all-or-nothing approach and to eliminate the difficult passive vs. active and primary vs. secondary comparisons, New York, in *Dole v. Dow Chemical Co.,*[23] adopted quantitative comparative fault for defendants.[24] In this case, an employee sued the manufacturer of a chemical claiming the package was not properly labeled and users were not warned. The manufacturer brought a cross action against the employer of the injured employee claiming that the employer did not take proper precautions, used untrained personnel, and did not follow instructions. The intermediate appellate court struck down the cross action because the manufacturer's conduct had been active. However the Court of Appeals reversed, noting the inadequacy of both passive-active and primary-secondary classification (though preferring the latter) and holding that it is more just to *share* responsibility than to either *transfer* or deny transfer of the entire loss.

New York expressed reservations a year later, however, holding the *Dole* rule did not apply where a statute placed absolute liability on an apartment owner for another's negligence.[25] The apartment owner had made a contract with an elevator operating company that had been negligent. The court held that comparative fault would not apply where one party was liable vicariously or because the law had placed a nondelegable duty on them—classic cases of passive or secondary liability. The court also noted that contracts between the defendants can distribute the ultimate loss between them.

F. Preemption

Although express indemnification is discussed in detail in Section 32.05, one problem relating to indemnity clauses must be mentioned here. Does the presence of an express indemnification clause bar noncontractual indemnity?

If noncontractual indemnification is based on the likely intention of the parties, the matter should be resolved simply by interpreting the language of the indemnity clause. If no language deals specifically with preemption, it is likely that the parties, by focusing on indemnification, have demonstrated an intention to exclude other forms of indemnification. If noncontractual indemnification is based on unjust enrichment, however, the problem becomes more difficult. On the one hand, it can be contended that there should be no inquiry into unjust enrichment where the parties have dealt specifically with the issue. Yet unjust enrichment can have a life of its own, based on what the law considers fair and just.

The difficulty is reflected in case decisions that have passed on this problem. Most cases have concluded that the presence of an express indemnification provision bars indemnification based on a comparison of culpability.[26] But another case, *E. L. White, Inc. v. City of Huntington Beach,*[27] not only came to a different conclusion but also demonstrated some of the complexity raised by indemnification. For that reason, it may be useful to look at the facts in that case.

The city of Huntington Beach contracted with White to build a drain sewer. Under the contract, White agreed to indemnify the city. Problems developed after completion. White hired a subcontractor to make repairs. One employee of the subcontractor was killed and another injured in a cave-in caused by failure to shore or slope the trenches. The action by the workers resulted in judgments against the city and White. Before the legal claim was filed by the workers, the city filed a separate claim against White and its insurer asking, among other things, that a declaration be made that the city be entitled to indemnification from White under the indemnification clause contained in the prime contract. California law does not grant

---

[23]Supra note 19. California followed the *Dole* case in *Am. Motorcycle Ass'n v. Superior Court,* 20 Cal.3d 578, 578 P.2d 899, 146 Cal.Rptr. 182, (1978), despite the availability of statutory contribution.

[24]Quantitative comparisons seem more popular with contemporary courts than qualitative ones, especially all-or-nothing ones. See *Cooper Stevedoring Co. v. Fritz Kopke, Inc.,* 417 U.S. 106 (1974); *U.S. v. Seckinger,* 397 U.S. 203 (1970).

[25]*Rogers v. Dorchester Assoc.,* 32 N.Y.2d 553, 300 N.E.2d 403, 347 N.Y.S.2d 22 (1973). In 1974, New York enforced a noncontractual indemnity claim based on primary and secondary liability brought by a prime contractor against a subcontractor when a worker was injured and brought a claim against the prime. See *Kelly v. Diesel Constr. Div.,* 35 N.Y.2d 1, 315 N.E.2d 751, 358 N.Y.S.2d 685 (1974).

[26]*Southern Pac. Co. v. Morrison-Knudsen Co.,* 216 Or. 398, 338 P.2d 665 (1959); *Wyoming Johnson, Inc. v. Stag Indus., Inc.,* 662 P.2d 96 (Wyo.1983); *Covert v. City of Binghamton,* 117 Misc.2d 1075, 459 N.Y.S.2d 721 (1983).

[27]21 Cal.3d 497, 579 P.2d 505, 146 Cal.Rptr. 614 (1978). Similarly, see *Fairbanks North Star Borough v. Roen Design Assoc. Inc.,* 727 P.2d 758 (Alaska 1986).

indemnity when the contractual indemnity clause is written in general terms and the party seeking indemnification is *actively* negligent. The city was found to have been *actively* negligent, barring its indemnification claim.

After the judgment had been rendered, White and its insurer brought a legal action against the city seeking *equitable* indemnity against the city for amounts that it paid to satisfy the judgments.

The city contended that equitable indemnity was barred by the contractual indemnification clause, but the court did not agree. The court held that the two forms of indemnification—contractual and equitable—are separate bases for transferring a loss. If, however, the express contractual provision does not apply to the factual setting before the court, equitable indemnification can come into play.

The indemnity clause must be interpreted to see whether it has expressed *any* intention by the parties that *all* equitable indemnity is to be eliminated. The indemnification clause protected only the *city*. The court permitted equitable indemnity when a claim was made by the *contractor*, as the clause did not preclude equitable indemnity.

G. Settlements

As noted in Section 32.02, when there are a number of joint wrongdoers, the law seeks to provide incentives to settle. Some states have enacted statutes that allow a settling defendant to petition the court to conclude that the settlement was made in good faith as a means of relieving the settling defendant from claims by other defendants.

SECTION 32.04   Implied Indemnity

As mentioned earlier, one form of contractual indemnity rests not on express language but on the presumed intention of the contracting parties that one party will respond for a loss it causes to the other party. For example, suppose the prime contractor negligently left a hole uncovered. A mail deliverer who entered the site during construction with permission is injured. Suppose the latter successfully asserts a claim against the owner based on the owner's obligation as a possessor of land to keep the premises reasonably safe. Even in the absence of a specific indemnity clause in the prime contract, it can be contended that the prime con-

tractor has impliedly promised the owner that it would indemnify the owner if its negligent conduct harmed a third party who asserted a claim against the owner and obtained a court judgment.

This form of implied indemnity has had a complicated history in admiralty law where longshoremen have been injured. The relatively uniform presence of indemnification clauses and the expansive use of noncontractual indemnity, however, have made this form of indemnification relatively unimportant in modern construction disputes.[28]

SECTION 32.05   Contractual Indemnity

A. Indemnification Compared to Exculpation, Liability Limitation, and Liquidated Damages

This subsection compares indemnification clauses to other clauses that seek to distribute losses. Distinct differences exist between the various loss distributing clauses, but there is also much overlap. Indemnification—the principal topic in this chapter—has become controversial and highly regulated because it can be exculpatory; that is, it can relieve one party of liability that would otherwise exist. For that reason, it is instructive to look carefully at both exculpation and indemnification.

One of the most simple yet effective ways of relieving a party from responsibility is by a contractual exculpation under which a party who may suffer losses agrees that it will not pursue the party who is legally responsible for such losses. For example, on being admitted to a hospital, suppose a patient signs a form under which she agrees that the hospital will not be responsible if she is harmed as a result of the hospital's negligence. The hospital would be seeking to relieve itself from liability. Although the law accords considerable freedom to contracting parties to make their own rules and distribute losses as they wish, the hospital admissions room is hardly the place for the contract process. For that reason, it is likely that such an

---

[28]But in *Kemper Architects, P.C. v. McFall, Konkel & Kimball Consulting Eng'rs,* 843 P.2d 1178 (Wyo.1992), an architect's counterclaim against its consulting engineer was based upon implied indemnity inasmuch as there was no written contract between them. The court employed the professional standard to measure the duty of the consulting engineer. (See Section 14.05(A).)

agreement would be invalid.[29] Similarly, some courts and legislatures do not allow exculpation in a residential lease[30] or one in which the tenant is a small business.[31]

On the other hand, suppose racehorse owners stable their horses at a racetrack and sign an agreement under which they exculpate the racetrack from any loss or damage to the horses even if caused by negligence of the racetrack. Such an exculpatory clause was upheld in *Rutter v. Arlington Park Jockey Club*[32] when horses were killed in a barn fire. The court held that exculpatory clauses will be upheld if it is clear from the contract that the parties intended to accomplish this result and the evidence indicates that the parties were of relatively equal bargaining strength. After concluding that the language was sufficiently clear to include negligence by the racetrack, the court stated:

> This was a contract between businessmen and it reflected good business judgment to place the risk of loss upon the party who could least expensively insure against it. It is unlikely that the Club would provide stable facilities free of charge if by doing so it incurred liability for damage to horses of substantial but unknown value. The cost of insurance would be greater for the Club than for the horse owners because the Club could not accurately predict the number and value of the horses that would be housed in its facilities from time to time.[33]

The court was not impressed with the argument that enforcing such an exculpatory clause would encourage the racetrack to be careless.

The court, although enforcing the exculpatory clause, insisted that the language of exculpation be clear, exculpatory clauses requiring a high degree of specificity.[34]

Exculpatory clauses deny the right of one party to use the legal process when it would otherwise have the opportunity of doing so. Indemnification, on the other hand, does not preclude first *instance* liability. It deals with ultimate responsibility by shifting the loss from one party to another. For example, if an injured worker recovers against the owner, permitting the owner to shift the loss to the contractor by indemnification in no way precludes the injured worker from recovering. A shift of ultimate responsibility from owner to contractor occurs. If the owner has in any way been at fault, however, risk shifting through indemnification resembles exculpation. Because of this similarity, where one party is being indemnified despite its negligence, the judicial attitude toward indemnity clauses, both in terms of interpretation and in terms of enforcement, is likely to be similar to the judicial attitudes toward exculpatory provisions.[35]

Another contract clause that can be compared with an indemnity clause is one that seeks to limit the legal remedy. For example, sellers of machinery sometimes seek to limit their liability to repair and replace defective parts. Similarly, some design professionals seek to limit their liability to their client to a designated amount of money or a certain percentage of the fee.[36] If the actual damages in the latter illustration are *less* than the specified amount, only actual damages can be recovered. But a liability limitation sets a ceiling on the damages.

Many of the same considerations that have been discussed with regard to exculpatory clauses apply to liability limitations. Where they are determined in a proper setting by parties of relatively equal bargaining power and where the language clearly expresses an intention to limit the liability, they are given effect.[37]

Contract clauses sometimes stipulate the amount of damages in advance. As seen in Section 26.09(B), such clauses are frequently used for unexcused time delay. They are generally given effect as long as they are reasonable. But some of the same considerations that relate to bargaining power and appropriateness of advance agreement

---

[29]*Tunkl v. Regents of the Univ. of California,* 60 Cal.2d 92, 383 P.2d 441, 32 Cal.Rptr. 33 (1963).
[30]*Crowell v. Housing Auth. of the City of Dallas,* 495 S.W.2d 887 (Tex.1973).
[31]*McLean v. L.P.W. Realty Corp.,* 507 F.2d 1032 (2d Cir.1974).
[32]510 F.2d 1065 (7th Cir.1975).
[33]Id. at 1068. A case in the construction context emphasizing similar considerations is *New England Tel. & Tel. Co. v. Central Vermont Pub. Serv. Corp.,* 391 F.Supp. 420 (D.Vt.1975).
[34]*Cincinnati Gas & Elec. Co. v. Westinghouse Elec. Corp.,* 465 F.2d 1064 (6th Cir.1972).

[35]*Di Lonardo v. Gilbane Bldg. Co.,* 114 R.I. 469, 334 A.2d 422 (1975).
[36]See Section 15.03(D).
[37]*Delta Air Lines, Inc. v. Douglas Aircraft Co.,* 238 Cal.App.2d 95, 47 Cal.Rptr. 518 (1965).

will be taken into account. Liquidated damage clauses, unlike indemnity clauses, do not shift losses.

If the surrounding circumstances justify enforcement of an exculpatory clause or one that limits liability, those same circumstances may justify enforcement of an indemnification clause that has an exculpatory element.

B. Indemnity Clauses Classified

This section uses the prime contract as an illustration. The prime contractor is the *indemnitor,* that is, the party promising to indemnify. The owner is the *indemnitee,* that is, the party to whom indemnification has been promised. The analysis in this section also applies to a subcontract where the prime contractor is the *indemnitee* and the subcontractor the *indemnitor.*

Looking first at a claim as it relates to the *indemnitor's* conduct, the indemnity clause can be "work-related." Such a clause covers a broad variety of claims that may be asserted against the indemnitee owners by third parties such as injured workers or adjacent landowners. These claims may be predicated on fault of the indemnitor prime. However, the only indemnification requirement is that the claim "arise out of," "be occasioned by," or "be due to," to use common indemnity phrases, the work or activity of the prime.[38]

The clause can be more limited. It may cover only claims based on wrongful conduct of the prime. This is employed by AIA Doc. A201, ¶ 3.18 (discussed in Section 32.05(G)).

Focusing on the *indemnitee,* the claim usually asserts one or a number of bases for owner liability. The claimant may point to acts or failures to act by the owner itself or someone for whom it is responsible, such as design professional or prime contractor. This can raise issues of passive v. active or primary vs. secondary (discussed in Section 32.03).[39] The claim may be based on *status,* such as the owner's being the possessor of land or the employer with common law or statutory responsibilities often strict in nature. Also, the claim against the indemnitee may be based *in whole* or *in part* on the indemnitee's own wrongful acts. Clauses that cover claims solely based on the negligence of the indemnitee are called "broad form," while those that cover claims based in part on the negligence of the indemnitee are called "intermediate form" indemnity clauses.

The exculpatory feature of many clauses, along with work-related clauses that do not require any legal wrong by the indemnitor prime, has caused courts to scrutinize such clauses. Also, the bargaining power that often enables apparently one-sided clauses to be passed "down the line" from owner to prime, from prime to subcontractor, and so on, has also generated judicial concern, making litigation predictability hazardous.

C. Functions of Indemnity Clauses

The hostility the law has shown toward clauses under which one party has promised to indemnify another often ignores the important function served by the indemnification process. As an illustration, suppose an owner plans to make a construction contract with a prime contractor. The owner recognizes the increased likelihood that claims will be made against it that are based on the contractor's performance, sometimes but not always negligent. With a recognition that the law increasingly makes the owner responsible to third parties or at least that the law has made it more likely that claims may be made by third parties against the owner, the owner may say to the contractor:

> I have turned over the site to you. It is your responsibility to see to it that the building is constructed properly. You must not expose others to unreasonable risk of harm. The increasing likelihood that I will be sued for what you do makes it fair that you relieve me of ultimate responsibility for these claims by your agreeing to hold me harmless or to indemnify me.

Alternatively, a reassuring proposal may come from the contractor, who may say to the owner:

> I know you may be concerned about the possibility that a claim will be made against you by a third party during the course of my performance and that you will have to defend against that claim and either negotiate a settlement or even pay a court judgment. I always con-

---

[38]*Doloughty v. Blanchard Constr. Co.,* 139 N.J.Super. 110, 352 A.2d 613 (1976).

[39]California treats passive and active negligence differently when faced with the required amount of specificity to enforce such clauses. See *Rossmoor Sanitation, Inc. v. Pylon, Inc.,* 13 Cal.3d 622, 532 P.2d 97, 119 Cal.Rptr. 449, (1975), and *MacDonald & Kruse, Inc. v. San Jose Steel Co.,* 29 Cal.App.3d 413, 105 Cal.Rptr. 725 (1972). But see *Morton Thiokol, Inc. v. Metal Bldg. Alteration Co.,* infra note 56.

duct my work in accordance with the best construction practices, and I have promised in my contract to do the work in a proper manner. I am so confident that I will do this that I am willing to relieve your anxiety by holding you harmless or by indemnifying you if any claim is made against you by third parties relating to my work. You will have nothing to worry about, as I will stand behind my work. If you are concerned about my ability to pay you, I will agree to back it up by public liability insurance coverage.

In this context, indemnification acts to seal a deal when one party is anxious. The same scenario can be played but in a slightly different way if the architect asks the owner to obtain indemnification for her through the prime contract in the manner accomplished by AIA Doc. A201, ¶ 3.18. The architect may be saying to the contractor through the owner:

The law may hold me accountable for injury to your workers or to employees of your subcontractors because they may connect their injury with something they claim I did or should have done. You are being paid for your expertise in construction methods and your knowledge of safety rules. These are not activities in which I have been trained or in which I claim to have great skill or experience. For that reason, if a claim is made against me for conduct that is your responsibility, I want you to hold me harmless and indemnify me.

In addition, either owner or architect may back up its request for indemnification from the prime contractor by noting that it is exposed to potentially open-ended tort liability if claims are made by employees of the contractor or other subcontractors while the actual employer, either the prime contractor or subcontractor, who is most directly responsible, need only pay the more limited liability imposed on it by workers' compensation law.

Insurance plays a significant part in the indemnification process, as the promise to indemnify may be worthless unless it is backed up by a solvent insurer. Usually the indemnitor, such as the prime contractor, may have to secure a special endorsement from its insurer to cover contractually assumed liability. This will increase the cost of insurance, a cost that will ultimately be passed on to the owner. But the owner may believe that it is cheaper in the long run to pay this additional cost than to take the risk and insure against it. Distribution of risks through indemnification can facili-

tate insurance at the cheapest possible cost.[40] (An architect who is asked to indemnify, as noted in Section 15.04, may not be persuaded of this rationale if her professional liability insurance has a high deductible.)

Insurance facilitation assumes that there will be some accidents and seeks an efficient method of distributing the ultimate loss to an insurer. Yet indemnification can be looked on from the perspective of worker safety and avoidance of accidents.[41] Here the emphasis is on the exculpatory aspects of indemnification. This is particularly a problem with an indemnification provision, such as a broad form or intermediate form, which can exculpate the indemnitee from her own wrongdoing. Even more, a prime contractor who extracts indemnification provisions from all of its subcontractors may have little incentive to control project operations in a way that minimizes the likelihood of injuries to workers.

A federal judge, focusing on safety, stated that indemnification seeks to place the ultimate responsibility on the party who can most cheaply avoid the harm.[42] An application of this principle would extend broad scope to clauses under which subcontractors (closest to the actual operation) indemnify prime contractors and prime contractors (overall control) indemnify owners and design professionals. Despite this, some legislatures, as seen in (D), and some courts, as seen in (E), have not permitted full contractual freedom to the parties to decide how risks should be allocated by indemnification. Undoubtedly, much of this unwillingness is influenced by the belief that indemnification freely permitted will encourage carelessness. It is unlikely that unregulated indemnification (enforcing clauses as written) would be a disincentive for creating and monitoring a safety program.

[40]See *Hicks v. Ocean Drilling & Exploration Co.*, 512 F.2d 817 (5th Cir.1975); *Guy F. Atkinson Co. v. Schatz*, 102 Cal.App.3d 351, 161 Cal.Rptr. 436 (1980) (citing earlier edition of treatise); *Buscaglia v. Owens-Corning Fiberglas*, 68 N.J.Super. 508, 172 A.2d 703 (1961), affirmed 36 N.J. 532, 178 A.2d 208 (1962); *Di Lonardo v. Gilbane Bldg. Co.* 114 R.I. 469, 334 A.2d 422 (1975).
[41]*Ft. Wayne Cablevision v. Indiana & Michigan Elec. Co.* 443 N.E.2d 863 (Ind.App.1983); *Bosio v. Branigar Org., Inc.*, 154 Ill.App.3d 611, 506 N.E.2d 996 (1987).
[42]*McMunn v. Hertz*, 791 F.2d 88 (7th Cir.1986) (per Judge Posner).

The factor that bears most heavily on how the contractor performs is its insurance rates, which may in part be predicated on the number of claims that are made against it. But the insurance premium is more likely to be predicated on the type of work, the locality, and the volume of the prime contractor's business. Other factors can deter careless performance, such as the possibility that public officials will take action against the prime contractor or the contractor's license will be in jeopardy. In any event, the law allows parties to produce insurance even though the existence of insurance can encourage carelessness. Indemnification is a form of insurance, and there is no reason why the law should be hostile to this process. Undoubtedly, some of the hostility is also generated by the way in which indemnification is forced on the weaker party who must usually accept such a clause on a take-it-or-leave-it basis. Again, if the risk is clearly brought to the attention of the weaker party and that party can insure against that risk and pass the cost of insurance on to the stronger party in its contract price, hostility is not justified.[43]

D. Statutory Regulation

The concern over indemnification that has been noted in (C) has led to frequent statutory regulation and, as shall be seen in (E), to court decisions that in effect regulate indemnification clauses. Regulation does not view indemnification as a method of reassuring a nervous contract maker, obtaining insurance at the best possible cost, or placing the risk on the party that can avoid the harm most cheaply. Those who relate indemnification look principally at the exculpatory aspects of all-or-nothing indemnification and the means by which the stronger party obtains indemnification from the weaker party.

Legislative intervention initially resulted from a struggle between the American Institute of Architects (AIA) and the Associated General Contractors (AGC), more realistically a struggle between professional liability insurers of design professionals and public liability insurers of primes and subcontractors. In 1966, the AIA included an indemnification provision in A201 for the first time. The

AGC, using the muscle of public liability insurers, fought this and ultimately obtained a modification in 1967. But the turf extended to state legislatures. Contractors' associations were instrumental in obtaining legislation that limited enforceability of construction contract indemnification clauses. Even after the associations settled their dispute in 1967, demands continued to be made on legislatures for limitation of indemnification. Finally, in the 1980s, long after the battle between the AIA and the AGC was over, some legislatures were persuaded to eliminate the exculpatory features of such clauses with their all-or-nothing solutions.

Although a detailed examination of such statutes is beyond the scope of this treatise, a look at some state anti-indemnification laws will review the shape of legislative regulation of such clauses.[44]

Statutes initially simply barred broad form indemnity clauses, forbidding clauses that operated even if the indemnitee—the person to whom the indemnification was being made—was solely negligent.[45] Compliance with these statutes was relatively simple and usually accomplished. In any event, most claims for which indemnification was sought did not result solely from the activity or inactivity of the indemnitee. The interdependent nature of the construction project was the factor that led to claims being attributed to the negligence of both the indemnitee and the indemnitor.

In the 1980s, a number of states eliminated the exculpatory feature of all-or-nothing indemnification clauses by barring clauses under which the indemnitor would have to pay for a part of the loss that had been caused by the negligence of the indemnitee.[46]

---

[43]*Di Lonardo v. Gilbane Bldg. Co.,* supra note 40.

[44]The statutes are categorized and discussed by Claiborne and Smith, Legislative Division Update, 12 Constr.Lawyer No. 2, Apr. 1992, p. 30.
[45]Ann. Mich.Comp.Laws § 691.991 is an illustration. New Jersey in 1981 had barred all indemnity clauses in construction contracts, but in 1983 amended its law to bar only broad form clauses—those that operate even if the indemnitee was solely negligent. See Ann.N.J.Stat. § 2A:40A-1, which was applied retroactively in *Miller v. Hall Bldg. Corp.,* 210 N.J.Super. 248, 509 A.2d 316 (1985).
[46]Ann.Ill.Stat. ch. 29, ¶ 61; Ann.Minn.Stat. § 337.02; McKinney's N.Y.Gen.Oblig.Law § 5-322.1; Ann.Wash.Rev.Code § 4.24.115. The Minnesota statute is attacked by Kleinberger, *No Risk Allocation Need Apply: The Twisted Minnesota Law of Indemnification,* 17 Wm.Mitchell L.Rev. 775 (1987). The Washington statute is reviewed by Soha, *A*

In addition, a number of states enacted statutes with unusual characteristics. For example, New Mexico bars intermediate forms of indemnification unless the indemnification clause contains the exceptions specified in AIA Doc. A201, ¶ 3.18.3. It excludes from the contractor's obligation to indemnify the owner and the architect claims based on design activities of the architect or the latter's role in monitoring performance.[47]

Florida bars only clauses that indemnify against the indemnitiee's active negligence unless there is a specific monetary limit to the indemnification or separate consideration paid for indemnity clause.[48]

California, although barring any indemnification for design, exempts from the statute inspecting engineers who obtain indemnification from clients with designated financial capacity.[49]

Finally, Indiana does not apply its statute to road construction contracts,[50] while Colorado prohibits a public contract containing an indemnification provision under which a public entity will be exculpated from its own negligence.[51]

The patchwork of state anti-indemnity statutes creates a number of problems. First, contractors whose work crosses state lines must be certain that their liability insurance will cover the variety of indemnification clauses they may face. Second, additional legal difficulties can be generated if the contract was valid in state A where it was made but a dispute culminated in a lawsuit in state B. Usually the courts in state B will enforce the clause if it is valid where it was made. However, Maryland refused to enforce an indemnification clause that was not harmonious with its own anti-indemnification statute. It looked on enforcement of such a clause as violative of Maryland public policy.[52]

E. Common Law Regulation: Specificity Requirements

Long before the enactment of the anti-indemnification legislation described in the preceding subsection, the common law looked with hostility at such clauses. Although such clauses could be enforced, they were enforced only if the language clearly and unequivocally stated that indemnification would be made even if the loss had been caused in whole or in part by the indemnitee's negligence.[53] Some even required that negligence be mentioned specifically.[54] Regulation of such clauses reflected the belief that indemnification often exculpated the indemnitee and could be a disincentive for good safety practices as well as the belief that such clauses were forced on weaker parties. In operation, the effectiveness of such an approach depended on the skill of the person drafting the indemnification clause.

Study of Juristic Reasoning: The Historical Development and Interpretation of Construction Industry Indemnification Clauses in Washington, 10 U. Puget Sound L.Rev. 51, 77–80 (1986). Judicial treatment of the New York statute has not been consistent. For example, while *Kilfeather v. Astoria 31st St. Assocs.*, 156 A.D.2d 428, 548 N.Y.S.2d 545 (1989), gave a narrow reading of the statute and allowed a CM to receive indemnification from a contractor, that case was distinguished by *Chiarelli v. 128 Eighth Ave. Assocs.*, 152 Misc.2d 155, 574 N.Y.S.2d 921 (Sup.Ct.1991) which gave a stricter reading of the statute and denied a CM indemnification because of a belief that indemnification for one's own negligence would defeat the intent of the statute. Even if the statute bars enforcement of the indemnification clause, Illinois still enforces a promise to procure public liability insurance. *Bosio v. Branigar Org., Inc*, supra note 41.

[47]Ann.N.M.Stat. § 56-7-1, interpreted in *Sierra v. Garcia*, 746 P.2d 1105 (N.M.1987).

[48]West Ann.Fla.Stat. § 725.06, applied in *Cuhaci & Peterson Architects, Inc. v. Huber Constr. Co.*, 516 So.2d 1096 (Fla.Dist.Ct.App.1987).

[49]West Ann.Cal.Civ.Code §§ 2782, 2782.2, 2782.5. The latter was interpreted to allow a liability limitation that could have been bargained for but was not. See *Markborough v. Superior Court*, 227 Cal.App.3d 705, 277 Cal.Rptr. 919 (1991), discussed in Section 15.03(D).

[50]West Ann.Ind.Code § 26-2-5-1.

[51]West Ann.Colo.Rev.Stat. § 13-50.5-102(8) (does not apply to contract clauses under which one party indemnifies another for the negligent acts of the indemnitor or its subcontractors).

---

[52]*Bethlehem Steel Corp. v. G. C. Zarnas & Co.*, 304 Md. 183, 498 A.2d 605 (1985).

[53]For a few of the many cases, see *Becker v. Black & Veatch Consulting Eng'rs*, 509 F.2d 42 (8th Cir.1974); *Warburton v. Phoenix Steel Corp.*, 321 A.2d 345 (Del.Super.1974), affirmed 334 A.2d 225 (Del.1975). Pennsylvania, in *Ruzzi v. Butler Petroleum Co.*, 527 Pa. 1, 588 A.2d 1 (1991), followed the majority rule of requiring that a clause be clear and unequivocal if it is to cover negligence of the indemnitee. This was criticized as inflexible and mechanical, depriving the court of the opportunity of hearing testimony outside the writing itself which can indicate the intention of the parties. Pennsylvania Supreme Court Review, 1991, 65 Temp.L.Rev. 679 (1992).

[54]*Burns & Roe, Inc. v. Central Maine Power Co.*, 659 F.Supp. 141 (D.Me.1987); *Ethyl Corp. v. Daniel Constr. Co.*, 725 S.W.2d 705 (Tex.1987).

Although insurance is discussed in Subsection (H), it is important to note here that many cases look at the contract in which the indemnification clause is contained to see whether the indemnitor is required to purchase and maintain liability insurance.[55] This may indicate an intention that indemnification is to cover claims caused at least in part by the negligence of the indemnitee.

In recent years, there has been a modest tendency to look on indemnification clauses neutrally as representing a rational attempt to distribute losses efficiently by facilitating insurance coverage. Cases espousing a more favorable attitude toward such clauses seek to determine the intention of the parties as they would when interpreting any other clause.[56]

As noted in Section 20.02, finding the intention of the contracting parties is often difficult if the parties have contracted on the basis of a standard contract published by the American Institute of Architects (AIA) or the Engineers Joint Contracts Documents Committee (EJCDC). Indemnification is often a key contract clause (especially for insurers), but very frequently, such provisions are not understood by contracting parties, who may have selected an AIA or EJCDC document without careful analysis of its terms. The drafting by these organizations (often influenced or dominated by liability insurers) makes the process of interpretation a difficult one. It is in these contracts that a more mechanical and predictable rule may be preferable to one that focuses upon the intention of the contracting parties.

Yet the modest tendency toward interpreting indemnification clauses "as written" has by no means become the majority rule. This is demonstrated by *Ethyl Corp v. The Daniel Construction Co.* Earlier Texas decisions had required that the language be clear and unequivocal as to the obligation of the indemnitor to indemnify the indemnitee against the consequences of its own negligence. The court cited Texas decisions that had indicated a trend toward strict construction of such clauses, decisions that came close to adopting the requirement that negligence be expressly mentioned.

However, the court was concerned because

> [t]he scriveners of indemnity agreements have devised novel ways of writing provisions which failed to expressly state the true intent of those provisions. The intent of the scriveners is to indemnify the indemnitee for its negligence, yet be just ambiguous enough to conceal that intent from the indemnitor. The result has been a plethora of lawsuits to construe those ambiguous contracts. We hold the better policy is to cut through the ambiguity of those provisions and adopt the express negligence doctrine.
>
> . . . .
>
> Under the doctrine of express negligence, the intent of the parties must be specifically stated within the four corners of the contract.[57]

The often inelegantly drafted anti-indemnification statutes and the ambivalent attitude of judges toward indemnification make this process one of the most difficult in construction law.

Complexity is confounded by some jurisdictions that attempt to distinguish between active and passive negligence, concluding that the requisite degree of specificity applies to active but not to passive negligence.[58] Other courts do not make the active-passive distinction for this purpose.[59]

The plethora of cases that have interpreted indemnity clauses and have come to variant results caused the Illinois Supreme Court to state in despair:

> We have examined the authorities cited by the parties and many of those collected at 27 A.L.R.3d 663, and

---

[55]Pennsylvania Supreme Court Review, 1991, 65 Temp.L.Rev. 679 at 696, note 114 (citing cases) (1992).

[56]*New England Tel. & Tel. Co. v. Central Vermont Pub. Serv. Corp.,* supra note 33 (emphasizing two substantial corporations of equal bargaining power who knew what they were doing when they made the contract); *C. J. M. Constr., Inc. v. Chandler Plumbing & Heating,* 708 P.2d 60 (Alaska 1985) (two dissenting); *Morton Thiokol, Inc. v. Metal Bldg. Alteration Co.,* 193 Cal.App.3d 1025, 238 Cal.Rptr. 722 (1987) (focuses on intention of the parties rather than applying prior California rule that general indemnification clauses do not cover active negligence); *Phillippe v. Rhoads,* 233 Pa. Super. 503, 336 A.2d 374 (1975) (invalidation requires either legislation or showing of a monopoly).

---

[57]Supra note 54 at 707–708.

[58]*Morgan v. Stubblefield,* 6 Cal.3d 606, 493 P.2d 465, 100 Cal.Rptr. 1 (1972). But see *Morton Thiokol, Inc. v. Metal Bldg. Alteration Co.,* supra note 56; *Wrobel v. Trapani,* supra note 9; *Thompson-Starrett v. Otis Elevator Co.,* 271 N.Y. 36, 2 N.E.2d 35 (1936). The standard was relaxed considerably in *Levine v. Shell Oil Co.,* 28 N.Y.2d 205, 269 N.E.2d 799, 321 N.Y.S.2d 81 (1971).

[59]*Becker v. Black & Veatch Consulting Eng'rs,* supra note 53.

conclude that the contractual provisions involved are so varied that each must stand on its own language and little is to be gained by an attempt to analyze, distinguish or reconcile the decisions. The only guidance afforded is found in the accepted rule of interpretation which requires that the agreement be given a fair and reasonable interpretation based upon a consideration of all of its language and provisions.[60]

Despite this neutral approach, the court held that the language did not cover negligence of the indemnitee, as such intention was not expressed clearly and explicitly.

Yet the parties who are to plan their insurance coverage on the scope of risk and insurers who must cover particular risks must be able to know what such clauses will cover. If courts ignore existing precedents and focus too heavily on the surrounding facts and circumstances, the uncertainty of application will make indemnification an inefficient insurance facilitator and cause over-insurance and higher rates. This is demonstrated by *American Oil Co. v. Hart,*[61] a case that involved an indemnity claim by a large oil company against a small independent contractor who had been hired to repair a pole at a service station. The independent contractor's employee was injured while seeking to perform the repair work and successfully brought a claim against the oil company based on the latter's negligent maintenance of the pole. When the oil company sought indemnification from the independent contractor, the oil company pointed to the indemnity clause that was all-inclusive and seemed to cover this accident. In addition, the oil company pointed to an earlier case precedent that had upheld an identical clause when a similar accident had occurred. But the court, noting that the language must be "interpreted" in its setting, concluded that the indemnitor could not have intended to cover this loss when he received $40 a visit for making these repairs. Pointing to the need for a clear and unequivocal coverage of losses caused by the indemnitee's negligence, the court refused to apply the clause.

The indemnitor was a small businessperson, perhaps of doubtful financial responsibility. It is likely that the indemnity claim would not have been made had he not been insured. If so, it is simply a question of whether the oil company or its insurer or the indemnitor's insurer should pay for this loss. The court's conclusion might have been proper had the language not made it clear to the indemnitor that he would be taking this risk and that it should insure against it. But if the language served this function, the court's conclusion was incorrect.

It may seem unfair for the repairman to pay a large award when he was receiving but $40 a visit. But the charge for each visit should have included that portion of the insurance that could reasonably be chargeable to the repair work. The efficiency of the indemnity clause as an insurance facilitator is destroyed by disregarding judicial precedent and interpreting identical language in similar cases differently simply because it appears to be unfair when insurance is disregarded. The court's decision reflects the older, more traditional view of indemnity rather than one that emphasizes risk distribution and insurance costs.

F. Interpretation Issues

Losses and Indemnity Coverage. Clauses can cover certain losses but not others. For example, indemnity clauses can be drawn broadly enough to cover any loss, even those relating to property damage suffered by the indemnitee. Indemnity, however, is generally designed to transfer losses relating to claims third parties make against the indemnitee.[62]

Those who draft indemnification clauses should anticipate the types of losses that can occur. Typically, such clauses cover claims for harm to person or damage to property made by third parties. But clauses drafted with these claims in mind will not cover claims for purely economic losses unrelated

---

[60]*Tatar v. Maxon Constr. Co.,* 54 Ill.2d 64, 294 N.E.2d 272, 273–74 (1973).

[61]356 F.2d 657 (5th Cir.1966).

[62]*Varco-Pruden, Inc. v. Hampshire Constr. Co.,* 50 Cal.App.3d 654, 123 Cal.Rptr. 606 (1975); *Collins v. Montgomery Ward & Co.,* 21 Ill.App.3d 1037, 315 N.E.2d 670 (1974). In *Pac. Gas & Elec. Co. v. G. W. Thomas Drayage & Rigging Co.,* 69 Cal.2d 33, 442 P.2d 641, 69 Cal.Rptr. 561 (1968), the indemnitee sought recovery for damage to its own property. The court held that evidence submitted by the indemnitor that tended to show that the parties intended to cover only claims made by third parties should have been admitted into evidence.

to personal harm or damage to property.[63] Sometimes assertions of indemnity appear to be a grasping at straws. For example, an indemnity claim was made for damage resulting to window frames faultily built by the indemnitee prime contractor and then installed by the indemnitor subcontractor.[64] The court saw no specific language covering this loss and refused to include it within the general language.

Similarly, an indemnification clause in a prime contract was held not to cover a loss incurred by the owner to a third party based on the prime contractor having trespassed on the third party's land while doing the work.[65] The owner did not comply with its contract requirement to obtain an easement. The court held that the loss did not arise out of the prime contractor's performance even though the trespass was caused by its performance. A claim made based on an injury that occurred after the work had been completed, however, falls within the ambit of the indemnity clause. Injuries, whether they occur during or after performance, are the type of loss typically covered by insurance and part of the indemnification process.[66]

Work Relatedness of Injury.   The indemnitee who employs a work-related indemnity clause usually seeks protection against claims that are made incident to or arising out of the indemnitor's performance. Interpretation problems develop when a claimant is injured while at work but the principal cause of the injury is not her doing the work but the activity of the indemnitee who later seeks indemnification from, as a rule, the indemnitor employer of the claimant. The only connection that can be made between the accident and the activity of the indemnitor is that the accident would not have happened had the indemnitor not been on the job. These problems usually involve an accident to an employee of the subcontractor, a work-related clause, and a demand for indemnification from the indemnitee prime to the indemnitor subcontractor.[67]

Another interpretation issue arose in *General Accident Fire & Life Assurance Corp. v. Finegan & Burgess.*[68] A sign subcontract included an indemnity provision. After completion of the sign subcontractor's work, the owner's project engineer went to inspect the sign, accompanied by an employee of the sign subcontractor. After inspection, the engineer indicated he wished to check a particular switch that had been installed by another subcontractor. The employee of the sign subcontractor indicated the location of the switch and then left.

On his way to inspect the switch, the engineer fell from a walkway, which had no railing because of the prime contractor's negligence. The injured engineer sued both prime and sign subcontractor. The jury found that the prime contractor had been negligent but that the sign subcontractor had not. The prime contractor's insurer paid the claim and then brought an action against the sign subcontractor based on the indemnity provision. The purpose of inspection was to enable a tenant to move in earlier and was not directly related to the sign subcontractor's performance. The area where the injury had occurred was under the general contract of the prime contractor. For these reasons, the court held that the clause did not cover this accident. It seems that the engineer was no longer dealing with the sign subcontractor's work.

Amount Payable.   Usually the indemnitee seeks to transfer its entire loss to the indemnitor. This, as a rule, includes any money paid to the claimant, any costs of investigating the claim, any legal, investigative, or expert witness costs, and interest from the time the payment was made to the third party.[69] Most clauses deal with these issues. The most troublesome questions are those that relate to amounts paid under a settlement and costs to defend when it is determined that no liability existed.

As to the first, although one court seemed to require that the indemnitee establish that it would

[63]*Mobil Chemical Co. v. Blount Bros. Corp.*, 809 F.2d 1175 (5th Cir.1987); *Fairbanks North Star Borough v. Roen Design Assoc. Inc.* 727 P.2d 758 (Alaska 1986).
[64]*Mesker Bros. Iron Co. v. Des Lauriers Column Mould Co.*, 8 Ill.App.3d 113, 289 N.E.2d 223 (1972).
[65]*Serafine v. Metro. Sanitary Dist.*, 133 Ill.App.2d 93, 272 N.E.2d 716 (1971).
[66]*Becker v. Black & Veatch Consulting Eng'rs*, supra note 53.

[67]Compare *Hanley v. James McHugh Constr. Co.*, 444 F.2d 1006 (7th Cir.1971) (loss not caused by subcontractor's performance), with *White v. Morris Handler Co.*, 7 Ill.App.3d 199, 287 N.E.2d 203 (1972) (loss arose from performance of the work).
[68]351 F.2d 168 (6th Cir.1965).
[69]*Larive v. United States*, 449 F.2d 150 (8th Cir.1971).

have been liable,[70] it is better to require indemnity if a settlement were made in good faith.[71] It should not be necessary for the indemnitee to have to either litigate or settle and then establish legal responsibility.

Invoking the insurance contract, which also promises to defend and indemnify, some courts have held that the indemnitor is bound to a settlement made by the indemnitee only if the indemnitor was given notice and an opportunity to defend. But even here the indemnitee must establish that the settlement was reasonable and prudent under all circumstances. It need not establish that it would have lost the case but need only establish that the settlement was reasonable.[72] In that regard, it should be noted that the AIA indemnity clause in A201 (to be discussed in subsection (G)) requires the indemnitor—the contractor—to pay the costs of defense, leaving it to the indemnitees—the owner and the architect—to provide the defense.

Cost of investigating and defending the claim, the latter a particularly formidable expense, can cause difficulty. Usually indemnification clauses specifically state these costs are recoverable, and this result is likely to follow even in the absence of a specific provision dealing with costs of defense.[73] If the claimant does not prevail, the indemnitor will still have to pay the defense costs, at least where the indemnity clause is work related and not based

on liability.[74] If indemnity is based on liability, however, courts still tend to look at such clauses as a needless exercise of bargaining strength or even incentive to carelessness. Such courts will narrowly interpret them. Those courts that see these clauses as insurance facilitators and legitimate risk shift mechanisms will give such clauses a neutral interpretation. Even the latter courts will not always find that a particular loss is covered. But they do not resolve every doubt against coverage.

G. The AIA Indemnity Clause

Before 1966, AIA Doc. A201 did not contain an indemnification clause. But the liability explosion of the 1960s persuaded the AIA and its insurance counsel that the sting from increased liability could be reduced if the contractor was required to indemnify the owner and the architect against certain losses. As noted in (D), this attempt generated a struggle with the AGC and liability insurers of contractors. The issue was resolved in 1967.[75] Both groups and the insurers approved a clause, then ¶ 4.18. The clause required the contractor to indemnify owner and architect against certain losses (harm to person or damage to property other than the work itself) that arose out of the work and was caused by the negligence of the contractor, despite the loss's having been caused in part by the indemnitee. However, indemnification was excluded under ¶ 4.18.3 (now ¶ 3.18.3) for losses arising out of specified design activities of the architect or

> the giving of or the failure to give directions or instructions by the Architect . . . provided such giving or failure to give is the primary cause of the injury or damage.

The clause was an all-or-nothing intermediate indemnification clause. Noncontractual indemnification was not preempted by the indemnification clause.[76] Unlike many indemnity clauses, this does not require the indemnitee to *defend* as well as in-

---

[70]*Ford Motor Co. v. W. F. Holt & Sons, Inc.*, 453 F.2d 116 (6th Cir.1971). As shall be seen in Section 32.05(G), this is one of the deficiencies of the AIA clause.

[71]*Miller v. Shugart*, 316 N.W.2d 729 (Minn.1982) (indemnitor must pay settlement made by indemnitee if reasonable and prudent). But see *Peter Culley & Assocs. v. Superior Court*, 10 Cal.App.4th 1484, 13 Cal.Rptr.2d 624 (1992), which held that if the indemnitee settles without a trial, it must show that the liability is covered by the contract and that liability existed and the extent thereof. The court also held that the settlement is presumptive evidence of liability and the amount of liability but may be overcome by proof from the indemnitor that there was no liability or the settlement amount was unreasonable. This decision places too great a burden on the indemnitee.

[72]*United States Auto Assn. v. Morris*, 154 Ariz. 113, 741 P.2d 246 (1987).

[73]*Moses-Ecco Co. v. Roscoe-Ajax Corp.*, 320 F.2d 685 (D.C.Cir.1963).

---

[74]*Titan Steel Corp. v. Walton*, 365 F.2d 542 (10th Cir.1966); *Bethlehem Steel Corp. v. K.L.O. Welding Erectors, Inc.*, 132 N.J.Super. 496, 334 A.2d 346 (1975); *Tri-M Erectors, Inc. v. Donald M. Drake Co.*, 27 Wash.App. 529, 618 P.2d 1341 (1980).

[75]Changes in 1970 and 1976 were marginal. Currently, the clause is ¶ 3.18.

[76]AIA Doc. A201, ¶ 4.18.2, now ¶ 3.18.2.

demnify. Instead, the indemnitor pays the *cost* of defense.

The drafts for the A201 published in 1987 had included language that would have created comparative indemnity, an approach that the AIA had taken in its standard subcontract, A401. This would have complied with the increasing number of state statutes that require comparative indemnification.[77] But insurers resisted this; the troubled insurance industry was not in an adventurous mood. Yet what emerged as ¶ 3.18.1 states that the indemnification applies "only to the extent caused in whole or in part" by the negligence of the contractor or those for whom it was responsible. This appears to change the clause from an all-or-nothing to a comparative clause. Yet the clause also states that indemnification applies whether the claim is caused in whole or in part by the indemnitee. The latter would not be necessary were the clause simply one under which the contractor indemnifies for that portion of the loss caused by its negligence. It is unlikely that the AIA intended that ¶ 3.18.1 be a comparative indemnification clause. The phrase "only to the extent caused" may have been included inadvertently. Owners can avoid this uncertainty by using the language of 1976 unless state law dictates otherwise.

A few other aspects of the indemnification provision in A201 merit comment. First, ¶ 3.,18.1 states that the indemnification is done to "the fullest extent permitted by the law." This can simply signal to users that local legislation must be checked. Alternatively, it can be interpreted to invite the court to scale down "or reform" any clauses that would be invalidated by state legislation.[78]

Indemnification applies only if the contractor is negligent. This could preclude indemnity if there were a settlement in which no one admits having been negligent. Indemnification, as noted in Section 32.05(F), should apply to settlements made in good faith.

Finally, the architect given indemnification under ¶ 3.18 is not a party to the construction contract. Her right to enforce indemnification depends on ¶ 1.1.2 or common law third-party beneficiary

law.[79] Paragraph 1.1.2 states that the architect shall be entitled to performance of those obligations "intended to facilitate performance of the Architect's duties." (In 1976, A201 stated that the architect can enforce obligations "intended for his benefit," better language from the architect's perspective.)[80]

H. Insurance

Owners frequently require that contractors procure insurance covering the risks specified in the indemnification clause.[81] Even without this requirement, a prudent contractor will be certain that its insurance will cover this risk. Generally, liability policies cover only liability imposed by law and not that imposed or assumed by contract. It is usually possible for the contractor to obtain a specific endorsement covering the liability assumed by these indemnification clauses. Contractors should be certain, as should owners, that the liability policy covers this risk.

Generally, agreements to procure insurance to back up an indemnity provision are not invalidated by anti-indemnity legislation. Sometimes this is done expressly by statute.[82] When courts have faced this issue without the benefit of a statute, they have recognized the difference between an indemnity clause with exculpatory features and an insurance policy.[83] The former is often thought to induce carelessness on the site, while the latter is looked upon as a proper method of distributing

---

[77]See Section 32.05(D).
[78]*Robertson v. Swindell-Dressler Co.,* 82 Mich.App. 382, 267 N.W.2d 131 (1978).

[79]Owners have been able to recover on indemnification clauses in subcontracts to which they are not a party. *Schroeder v. C. F. Braun & Co.,* 502 F.2d 235 (7th Cir.1974); *Titan Steel Corp. v. Walton,* supra note 74. In *Pylon, Inc. v. Olympic Ins. Co.,* 271 Cal.App.2d 643, 77 Cal.Rptr. 72 (1969), the court seemed to assume that the engineer could maintain an action against the prime contractor based on the indemnity clause.
[80]For a more detailed discussion, see J. SWEET, SWEET ON CONSTRUCTION INDUSTRY CONTRACTS §§ 20.5–20.9 (2d.ed. 1992).
[81]As noted in Section 32.05(F), an indemnification clause linked with a requirement that the indemnitor maintain liability insurance is helpful in concluding that a general indemnity clause covers a claim that is caused in part by the negligence of the indemnitee.
[82]West Ann.Minn.Stat. § 337.05(1) The statutes are collected in 14 Constr.Litig.Rep. 10 (1993).
[83]*Holmes v. Watson-Forsberg Co.,* 488 N.W.2d 473 (Minn.1992). The cases are collected in 14 Constr.Litig.Rep. 10 (1993).

risks. But paradoxically, a promise to procure insurance assists in enforcement of an intermediate indemnity clause with its exculpation feature.

PROBLEMS

O and C entered into a construction contract using AIA Documents A101 and A201 set forth in Appendixes B and C. C was to construct a twenty-unit apartment complex on land owned by O in accordance with plans and specifications drafted by A.

During construction, C decided to take steps to protect the site from theft and neighborhood children who frequently played there. It erected a six foot barbed-wire fence with a gate that was locked after the day's work was completed. It also placed a fierce dog on the site after working hours to deter vandals and others from entering the site.

The dog caused a number of problems. Neighbors complained that the dog howled during the night and made it difficult for them to sleep. Those who passed by the construction site at night were often frightened by the menacing growls of the dog. Finally, one day the dog escaped from its handler when it was brought to the site after work had been completed. It bit a subcontractor's employee who had remained on the site a little late to finish some work.

The neighbors, the bystanders who walked by the site, and the worker who had been bitten all made claims against the owner. The injured worker also made a claim against the architect contending that the architect knew the dog was frequently brought to the site while a few workers still remained and did not take steps to make certain that the dog could not bite or molest them. In fact, the worker stated that her union steward had complained of this to the architect.

O and A have demanded that C defend these claims in accordance with ¶ 3.18 or of A201. Are their demands justified? Would C have to indemnify them if they made any settlements with the neighbors, the bystanders, or the injured worker? If legal action was brought against O that terminated in O's favor, would O be able to recover its legal expenses from C?

CHAPTER THIRTY-THREE

Surety Bonds: Backstopping Contractors

SECTION 33.01 Mechanics and Terminology

The surety bond transaction is a peculiar arrangement and differs procedurally from most contracts. The typical surety arrangement is essentially triangular. The "surety" obligates itself to perform or to pay a specified amount of money if the "principal debtor" (usually called the "principal") does not perform. The person to whom this performance is promised is usually called the "obligee" (sometimes called the "creditor"). In the building contract context, the surety is usually a professional bonding company. The principal is the prime contractor or, in the case of subcontractor bonds, a subcontractor. The obligee is the owner or, in the case of a subcontractor bond, the prime contractor.

If required by the owner, the prime contractor (the principal) applies for a bond from a bonding company. Usually the cost of the bond is paid indirectly by the owner because the bidder adds the cost of the bond to its costs when computing the bid. The bond will be issued to the owner (the obligee). The bond runs to the owner in that performance by the surety has been promised to the owner even though the application for the bond has been made by the prime contractor.

Another problem in dealing with surety bonds results from the antiquated way in which bonds are written. The earliest bonds were called penal bonds. The surety made an absolute promise to render a certain performance or to pay a specified amount of money. This would be followed by a paragraph stating that the bond would be void if the principal promptly and properly performed all the obligations under the contract between principal and obligee.

This format can cause problems. The transaction for which the surety provides financial security can be simple or complex. As an illustration of the former, the bail bondsperson will pay if the accused fails to appear at the hearing, or a banker's blanket bond requires the surety to pay if the bank official embezzles funds. However, a prime contractor backed by a surety has a wide variety of obligations set forth in the construction contract. For example, the prime contractor promises to build the project properly, not to damage the land of adjacent landowners,[1] to perform the work in such a way as to avoid exposing workers and others to unreasonable risk of harm, and to indemnify the owner and design professional if certain claims are made.

Often bond language does not state which duties are covered by the bond. For example, AIA Doc. A311 (set forth in Appendix D) refers to the construction contract and simply states that ". . . if Contractor shall promptly and faithfully perform said Contract then this obligation shall be null and void; otherwise it shall remain in full force and ef-

---

[1] Generally, the surety is not responsible for tort claims against its principal debtor, often those made by adjacent landowners. See Barker, *Third-Party Tort Claimants and the Contract Bond Surety*, 5 Constr. Law. No. 1, Spring 1984, p. 7. But if the bond exceeds a statutory amount and has broad language, one commentator believes a tort claimant may be able to recover form the surety. Perry, *Third Party Tort Claimants Can Recover Against Sureties Under Construction Bonds: Is the Bond a Comprehensive General Liability Policy?* 5 Constr.Lawyer No. 4, Apr. 1985, p. 5.

fect." Which aspects of the contractor's performance are backed up by the surety?

Also, bond language often does not go specifically into problems of coverage, problems of notices, and other things that are normally part of any contract. Improvement in this regard has been made, and some bonds today do contain specific provisions that do a better job of informing the parties of their rights and duties under the bond. (See AIA Doc. A312, also contained in Appendix D.)

SECTION 33.02 Function of Surety: Insurer Compared

A surety's function is to assure one party that the entity with whom it is dealing will be backed up by someone who is financially responsible. Sureties are used in transactions where persons deal with individuals or organizations of doubtful financial capacity. They provide credit. *Sureties* must be distinguished from *insurers*, though each serves the function of providing financial security.[2]

An insured is concerned that unusual, unexpected events will occur that will cause it to suffer losses or expose it to liability. Although it can self-insure—that is, bear the risk itself—it usually chooses to indemnify itself against this risk by buying a promise from an insurance company in exchange for payment of a premium. The insurer distributes this risk among its policy holders.

In public liability insurance, the insured itself may be at fault and cause a loss to the insurer. But the insurer cannot recover its loss from its own insured. Although it may seek to recover its losses from third parties through subrogation (stepping into the position of the person it has paid—its insured—and thereby acquiring any claims of its insured against those who caused the loss), it cannot recover from those named as insureds in the policy.

Sureties, on the other hand, in addition to dealing with ordinary construction contract performance problems, which can be considered business losses (not the "accidents" central to insurance), seek to recover any losses they have suffered from the principal party on whom it has written a bond as well

as others.[3] In the late 1970s, contractors who were liable because of defective workmanship sought to recover from their comprehensive (increasingly called "commercial") general liability (CGL) insurers, preferring this to seeing the loss paid by their sureties.[4] Their sureties would seek to recover from them. Their CGL insurers could not.

SECTION 33.03 Judicial Treatment of Sureties

Judicial attitude toward sureties has changed. Before the development of professional sureties, the surety would be a private person (perhaps a relative of the principal) who sought to aid the principal to obtain a contract or to stave off a creditor by obligating himself to perform if the principal did not. Such a surety was frequently not paid and received no direct benefit for taking this risk. For these reasons, the surety was considered a "favorite" of the law.

One illustration of this favored position was the Statute of Frauds, which requires that certain types of promises must be evidenced by a written memorandum.[5] The original Statute of Frauds enacted in 1677 included promises to answer for the debts, defaults, or miscarriages of another. Without a writing, the surety was not held liable. This was to protect him from the enforcement of an impulsive oral promise often made without due deliberation.[6] Also, any minor change in the contract between the principal and the obligee would discharge the surety (relieve him of liability).[7] The personal, unpaid surety could not handle assurance needs in a commercial economy. The personal surety himself might not be financially responsible. For this rea-

---

[2]For regulatory purposes, they are often lumped together. See Section 33.10(I).

[3]Typically, the surety seeks indemnification from its principal and even from the individual shareholders of a corporate principal. It may lose its right to indemnification if it refuses to settle in good faith. *Republic Ins. Co. v. Prince George's County,* 92 Md.App. 528, 608 A.2d 1301 (Md.App.1992).
[4]See Section 24.08.
[5]See Section 5.10.
[6]A written memorandum was not required where the main purpose or leading object of the surety was to benefit himself.
[7]See Section 33.10(C) for additional applications of this doctrine (called the *stricti juris* rule).

son, the professional, paid surety has developed as an important institution in both economic life generally and building contracts.

The development of professional sureties casts doubt on the protective rules developed largely when sureties were uncompensated. Although some protective rules are still applied, the professional surety is not regarded with the tender solicitude accorded the personal, uncompensated surety. This changeover from a legally favored position to one of neutrality—and perhaps even to one of disfavor—creates uncertainty in the law. Older cases are sometimes cited as precedents to protect sureties, but these precedents may be of limited value because being a surety is now a business.

The ambivalence toward sureties is illustrated by *Winston Corp. v. Continental Casualty Co.*,[8] which involved a claim by an owner against the surety on a performance bond. The bond incorporated all the provisions of the construction contract.

The contractor ran into financial difficulty causing delays, all known to the surety. Five months after the scheduled completion date, the owner met with the contractor and invited the surety. The surety, however, refused to attend. (The performance bond part of AIA Doc. 312, ¶¶ 2 and 3.1, requires the surety to participate in conferences when the owner considers declaring the contractor "in default." See Appendix D.) At this meeting, the owner and the contractor entered into an agreement designed to accelerate construction. Under the agreement, the contractor assigned the construction contract to the owner, permitted the owner to take possession of the premises, and assigned the contractor's subcontracts to the owner. However, the contractor's continued participation was expected. The owner immediately telephoned the surety notifying it of the new arrangement and mailed a copy of the letter agreement to the surety. A year after the scheduled completion date, the project was completed.

The owner sued on performance and payment bonds. The surety's defense was that the agreement between owner and contractor modified the original construction contract and discharged the surety. Also, the surety contended that the owner's

failure to give the surety seven days' written notice before terminating the contractor released the surety. The trial court, noting that sureties were favorites of the law, sustained the surety. The appellate court reversed, concluding that the owner had an absolute right to terminate the contract if the contractor was in default and that taking over the contract was merely an exercise of this right.

As for the failure to give the notice before taking over construction, the court agreed that historically, any slight deviation from the contract terms would discharge the surety. But, noted the court, the application of this doctrine often caused harsh and unjust results, especially when compensated sureties were relieved of their obligations because of technical breaches of the construction contract incorporated as part of the bond. This induced most courts to deviate from the doctrine under which the surety was the favorite of the law and required the compensated surety to show that the change in the original agreement was material and prejudicial to the surety. The surety must show that the change increased the surety's risk or changed the risk to the surety's detriment.

In addition to no longer considering the surety's being a favorite, courts generally construe ambiguities against the surety for the same reasons that insurance policies are construed against the insurance company.[9]

SECTION 33.04 Surety Bonds in Construction Contracts

Surety bonds play a vital part in the Construction Process. The contracting industry is volatile. Bankruptcies are not uncommon, and a few unsuccessful projects can cause financial catastrophe. Estimation of costs is difficult and requires a great amount of skill. Fixed-price contracts place many risks on the contractor, such as price increases, labor difficulties, subsurface conditions, and changing governmental policy. Some construction companies are poorly managed and supervised. Often they are undercapitalized and rely heavily

---

[8]508 F.2d 1298 (6th Cir.1975). See also *Ramada Dev. Co. v. United States Fid. & Guar. Co.*, 626 F.2d 517 (6th Cir.1980). For changes discharging the surety, see Section 33.10(C).

[9]*United States v. Algernon Blair, Inc.*, 329 F.Supp. 1360 (D.S.C.1971); *School Dist. No. 65R of Lincoln County v. Universal Sur. Co., Lincoln*, 178 Neb. 746, 135 N.W.2d 232 (1965).

on the technological skill of a limited number of individuals. If these individuals become unavailable, it is likely that difficulties will arise. Credit may be difficult to obtain for many construction companies. Some contractors do not insure against the risks and calamities that can be covered by insurance. Finally, anti-inflationary government policies such as tight money policies almost always hit the building industry first.

In most construction projects, bonds are needed to protect the owner.[10] The owner in most projects would like to be able to look to a financially solvent surety in the event that the successful bidder does not enter into the construction contract (bid bond), the prime contractor does not perform its work properly (performance bond), or the prime contractor does not pay its subcontractors or suppliers (payment bond).[11] (Bond requirements can act as preliminary screen for contractor selection.)

If these events do not occur, the amount paid for a surety bond may seem wasted. Some institutional owners believe there is no need for a surety bond system if the prime contractor is chosen carefully and if a well-administered payment system is used that eliminates the risk of unpaid subcontractors and suppliers. Such owners may choose to be self-insurers and not obtain bonds. They realize that there may be losses, but they believe that the losses over a long period will be less than the cost of bond premiums.

Even where a bond is not required at the outset, it is best to include a provision in the prime contract that will require the prime contractor to obtain a bond before or during performance if the owner so requests. Usually the cost of a bond issued after the price of the project is agreed on is paid by the owner.

Frequently, public construction requires that performance and payment bonds be furnished. The latter are required to protect subcontractors and suppliers who have no lien rights on public work. The Miller Act requires that federal prime contractors obtain performance bonds and payment bonds based on the contract price, and similar requirements exist under state public contracting. In ad-

dition, often local housing development legislation requires that the developer furnish bonds to protect the local government if improvements promised by the developer are not made. Legal advice should be obtained to determine whether bonds are required for the project.

The invitation to bid usually states whether the contractor is required to obtain a surety bond, the type or types of bonds, and the amount of the bonds. In some cases, the owner wishes to approve the form of the bond and the surety that is used. Also, the owner may want to have the right to refuse any substitution of surety without its express written consent given in advance of substitution.

Should the owner specify which surety bond company must be used? The choice will depend on rates, bond provisions, and the reputation of particular bonding companies for efficient operation and for fairness in adjusting claims. If relative standardization exists on these matters, it is advisable to give the contractor the freedom to choose the bonding company. It may have an established relationship with a particular bonding company. In most cases, it is probably sufficient to permit the bonding company to be chosen by the contractor, as long as the bonding company selected is licensed to operate in the state where the project will be built.

Occasionally, a designated bonding company will not be satisfactory to the owner. This dissatisfaction may be based on suspicion of financial instability or a past record of arbitrariness in claims handling. One method of exercising some control over the selection of the bonding company is to provide in the contract that the contractor submit to the owner the bond of a proposed bonding company for the owner's review. If the owner, in the exercise of its best judgment, determines that it is inadvisable to use that bonding company, the owner can veto the proposed bonding company and designate the bonding company to be used.[12]

---

[12]In *Weisz Trucking Co. v. Emil R. Wohl Constr.* 13 Cal.App.3d 256, 91 Cal.Rptr. 489 (1970), the standard for approval was held to be objective. However, a good-faith standard can be inserted in the contract. See also *Westbay Steel, Inc. v. United States,* 970 F.2d 648 (9th Cir.1992) (contracting officer's failure to use reasonable care to approve a Miller Act surety cannot be basis of claim under Federal Tort Claims Act). That case also held that a subcontractor could not maintain a claim against a contracting officer who had approved a contract without requiring that a Miller Act bond be filed.

---

[10]The relationship between the surety's obligation and any obligation to arbitrate was discussed in Section 30.15.
[11]As to the role of the design professional in advising on this choice, see Section 12.07.

SECTION 33.05 Bid Bond

The function of a bid bond is to provide the owner with a financially responsible party who will pay all or a portion of the damages caused if the bidder to whom a contract is awarded refuses to enter into it.[13]

SECTION 33.06 Performance Bond

The performance bond provides a financially responsible party to stand behind some aspects of the contractor's performance. If a payment bond is furnished, the performance bond will not include payment of subcontractors, their suppliers, and suppliers of the prime contractor. Illustrations of a performance bond are reproduced in Appendix D.

Bonds usually place a designated dollar limit on the surety's liability. Typically, bond limits are 50% or 100% of the contract price. Some statutory bonds are required to be 50% of the contract price.

SECTION 33.07 Payment Bond

The payment bond is an undertaking by the surety to pay unpaid subcontractors and suppliers. Standard payment bonds are set forth in Appendix D. An understanding of the function of a payment bond requires a differentiation between private and public construction work. All states give unpaid subcontractors and suppliers liens if they improve private construction projects. This process was described in Section 28.07(D).

Although there are various ways to avoid liens, one method has been to require prime contractors to obtain payment bonds. A payment bond obligates a surety to pay subcontractors and suppliers if they are not paid by the prime contractor.[14] The owner seeks to avoid liens filed against its property. Although the owner generally has no contractual relationship with unpaid subcontractors or suppliers, the latter parties if unpaid can assert liens against the owner's property they have improved. The owner would prefer directing them to the bonding company for payment.

Also, subcontractors are more likely to make bids if they can be assured of a surety if they are unpaid. The competent subcontractor who deals with a prime contractor of uncertain financial responsibility should add a contingency to its bid for the possibility of collection costs and the risk of not collecting. The presence of a payment bond should eliminate the need for this cost factor. In addition, subcontractors and suppliers should be more willing to perform properly and deliver materials as quickly as possible when they have assurance they will be paid. Though they have a right to a mechanics' lien, the procedures for perfecting the lien and satisfying the unpaid obligation out of foreclosure proceeds are cumbersome and often ineffective. Payment bonds are preferable to mechanics' liens.

Generally, subcontractors and suppliers cannot impose liens on public work. In some states, they can file a stop notice, which informs the owner that a subcontractor has not been paid and requires the owner to hold up payments to the prime contractor. Though unpaid subcontractors usually have no lien rights, they can present claims against public bodies through a request for special legislation to Congress, state legislatures, or city councils. Requiring that the prime contractor supply a payment bond provides the mechanism of relieving these legislative bodies from troublesome and time-consuming claims. Some reasons for having bonds on private projects also apply to public projects.

Competent subcontractors and willing suppliers are essential to the construction industry. These important components should have a mechanism that allows them to collect for their work. Payment bonds do this. Without a reliable payment mechanism, a substantial number of subcontractors may go out of business. Elimination of competent subcontractors can have the unfortunate effect of reducing competition and the quality of construction work.

SECTION 33.08 Subcontractor Bonds

Sometimes the prime contractor requires that subcontractors also obtain payment and performance

---

[13]See Section 18.04(E). Bid bonds are rare in Europe because of frequent prequalification of bidders. This is also the reason for European bonds being for 5% to 10% of the contract price, compared with the 50% to 100% in the United States.

[14]Whether an unpaid subcontractor or supplier can recover on a payment bond if it has endorsed a joint check but did not receive payment is discussed in Section 22.02(J).

bonds. If a subcontractor does not perform as obligated, the prime contractor (or its surety looking toward reimbursement) wants to have a financially responsible person to stand behind the subcontractor. This is the justification for a subcontractor performance bond. The justification for the subcontractor payment bond is similar to the justifications given for prime contractor payment bonds. If a subcontractor does not pay its sub-subcontractors or suppliers, the prime contractor (or its surety) is likely to be responsible, as it usually obligates itself to erect the project free and clear of liens. In many cases, unpaid sub-subcontractors and suppliers of subcontractors have lien rights against projects. In a large construction project, a substantial number of bonding companies who have written bonds on the various contractors in the project may exist. This makes the litigation in such cases complicated. Often the principal participants in a litigation are bonding companies, each seeking to shift the responsibility to the other.

SECTION 33.09 Other Bonds

Bonds can be used for other purposes in the Construction Process. For example, in some states, owners can post a bond that can preclude a lien from being filed or dissolve a lien that has been filed. Sometimes warranty bonds are used to back up the owner's claim under a warranty given by the contractor.

SECTION 33.10 Some Legal Problems

A. Who Can Sue on the Bond?

The most troublesome legal issue has been the seemingly simple question of whether unpaid subcontractors and suppliers can sue on a surety bond. Unpaid subcontractors and suppliers are not obligees under the bond. That is, the bond is not written to them, and the surety does not specifically oblige itself to them. Owners on prime contractor bonds or prime contractors on subcontractor bonds have no difficulty instituting legal action, because they are obligees and the bonds are written to them.

Early legal problems were complicated by the use of a single bond called the Faithful Performance Bond. Because the legal standard applied to

determine the right of someone other than the obligee to sue on the bond was whether the owner as obligee intended to benefit unpaid subcontractors and suppliers, some courts denied unpaid subcontractors and suppliers the right to sue on the bond. These courts reasoned that the owner must have intended to benefit itself. The bonds covered aspects of the prime contractor's performance other than nonpayment of subcontractors and suppliers. Also, the interests of owner and unpaid subcontractors and suppliers could conflict if each had claims against the prime contractor and if the amount of the bond could not satisfy all claims. As a result, the practice changed to encompass two bonds. The payment bond was to cover default consisting of nonpayment of subcontractors and suppliers, while the performance bond covered all other aspects of nonperformance by the prime contractor. Yet even where two bonds are issued, some courts still deny unpaid subcontractors and suppliers the right to sue on the bond.[15]

This already complicated area was muddled further by courts that differentiated between bonds on public works and those for private projects. These courts permitted subcontractors and suppliers to sue on bonds executed for public projects. They noted that liens could not be asserted on public projects and the intention of the public agency requiring the bonds must have been to benefit the unpaid subcontractors and suppliers. But on private projects that were lienable, an unpaid subcontractor or supplier could not sue on the bond, because the intention must have been to benefit the owner.[16] Courts deciding this question often focus solely on a private owner's desire to avoid liens being filed against its property. This ignores the other functions of surety bonds, such as those mentioned in Section 33.07.

Although the cases are not unanimous (bond language will vary), a strong modern tendency allows subcontractors and suppliers to sue directly on payment bonds if the bond seems to state that the surety will pay unpaid subcontractors or sup-

---

[15]See cases at infra note 17.

[16]See original opinion in *Fidelity & Deposit Co. of Baltimore v. Rainer*, 220 Ala. 262, 125 So. 55 (1929), which was reversed on rehearing. The original barred a claim by the subcontractor. The final decision allowed it.

pliers.[17] This outcome is reflected in AIA bonds (in Appendix D), which clearly give unpaid subcontractors and suppliers a right to sue the surety. Courts interpreting unclear language in a bond should recognize that the owner wants to be able to tell an unpaid subcontractor or supplier that it will be paid by the bonding company and that if the bonding company wrongfully refuses to pay, the subcontractor or supplier will be able to institute legal action itself. A bond purchases this right. The party paying for the bond intends that unpaid subcontractors and suppliers have this right, and this should be controlling.

Where bonds are required for public work or where bonds are filed on private work under statutes that allow this as a substitute for lien rights, the language of the statute frequently determines who can sue on the bond.[18] Typically, the claimant is someone who has furnished labor or materials that have gone into the project. Yet cases have permitted a contractor's lender[19] as well as a pension fund[20] to institute action on the payment bond.

As construction has become more complicated with linked sets of contracts connecting a variety of parties, problems as to who can sue on surety bonds take new forms. For example, an Illinois case involved an attempt by the ultimate user of a project (a school district) to bring a claim on the performance bond given by the prime contractor. The construction contract clearly indicated that the ul-

timate user was to be an intended beneficiary of the construction contract, and that contract was incorporated into the performance bond. But the school district was denied any direct rights on the bond because the bond itself made clear that only the obligee—the public entity acting as owner— could maintain an action on the bond.[21] Similarly, a number of cases have involved unsuccessful attempts by separate contractors (multiple primes) to bring claims on bonds where the principal was a separate contractor and the bond was issued to the owner as obligee.[22]

Planning in the context of a traditional construction organization should include ensuring that unpaid subcontractors and suppliers can bring action on these bonds. This can be accomplished relatively easily by including language similar to that in AIA bonds. Where a nontraditional system is used, such as separate contracts (multiple primes) or the party most interested in performance is not a nominal owner but an ultimate user, planning must consider who can bring legal action on any bonds that are supplied by contractors and that choice expressed in the contract and the bond.

B. Validity of Bond

Suppose the contractor misrepresents its resources when it applies for the bond. This would be misrepresentation by the applicant, not by the obligee-owner. In such a case, the bond is generally valid, and the surety bond company must pursue any remedies it has against the contractor-applicant. Fraud on the part of the obligee-owner, however, gives the surety the power to avoid having to perform under the bond. If the obligee-owner participates in or knows about the fraudulent statements made by the applicant contractor or misleads the bonding company in some other way, the bond would not have been validly obtained and cannot be legally enforced.

The principal contract—that is, the contract between owner and contractor—may not be enforce-

---

[17]*Socony-Vacuum Oil Co. v. Continental Cas. Co.*, 219 F.2d 645 (2d Cir.1955) (a leading case). See also *Houdaille Indus. Inc. v. United Bonding Ins. Co.*, 453 F.2d 1048 (5th Cir.1972); *Jacobs Assoc. v. Argonaut Ins. Co.*, 282 Or. 551, 580 P.2d 529 (1978) (reversing earlier opinion denying right of direct action), noted in 58 Or.L.Rev. 252 (1979); *Norland Co. v. West End Realty Corp.*, 206 Va. 938, 147 S.E.2d 105 (1966). But see *State of Florida v. Wesley Constr. Co.*, 316 F.Supp. 490 (S.D.Fla.1970), affirmed 453 F.2d 1366 (5th Cir.1972); *Layrite Concrete Products of Kennewick, Inc. v. H. Halvorson, Inc.*, 68 Wash.2d 70, 411 P.2d 405 (1966) (no right to sue); *James D. Shea Co. v. Perini Corp.*, 2 Mass.App. 912, 321 N.E.2d 831 (1975); *Day & Night Mfg. Co. v. Fid & Cas. Co of New York*, 85 Nev. 227, 452 P.2d 906 (1969) (right to sue only if language *clear*).

[18]*Houdaille Indus. Inc. v. United Bonding Ins. Co.*, supra note 17.

[19]*First Nat'l Bank of South Carolina v. United States Fid. & Guar. Co.*, 373 F.Supp. 235 (D.S.C.1974).

[20]*Trustees, Fla. West Coast Trowel Trades Pension Fund v. Quality Concrete Co.*, 385 So.2d 1163 (Fla.Dist.Ct.App.1980).

[21]*Board of Educ., School Dist. No. 15 DuPage County v. Fred L. Ockerlund Jr. & Assoc., Inc.*, 165 Ill.App.3d 439, 519 N.E.2d 95 (1988).

[22]*M.G.M. Constr. Corp. v. New Jersey Educ. Facilities Auth.*, 220 N.J.Super. 483, 532 A.2d 764 (1987); *Moore Constr. Co., Inc. v. Clarksville Dep't of Elec.* 707 S.W.2d 1 (Tenn.App.1985) (affirmed March 24, 1986).

able for various reasons. For example, suppose the building contract involved construction of a hideout for leaders of organized crime. To permit the owner to sue on the surety bond in such a case would further an illegal activity and would involve the court in that activity. For that reason, the bond would not be enforced.

Legal infirmities to contracts of a less serious nature, however, may exist. Suppose the contractor is not licensed. *Cohen v. Mayflower Corp.*[23] held that the owner can recover on a bond written on an unlicensed contractor. The court held that the licensing law was designed to protect owners and the owner could have sued the contractor even though the unlicensed contractor could not have sued the owner. The court concluded that the surety would be held on the bond because the principal debtor, the contractor, could have been held.

The stronger the public policy making the contract illegal, the less likely the bond will be enforced. If the policy is designed to protect the owner, however, such as in the *Cohen* case, it is more likely that the bond will be enforced.

C. Surety Defenses

Suppose the owner does not make progress payments despite the issuance of progress payment certificates. Suppose the design professional unjustifiably interferes with the work of the contractor or does not approve shop drawings in sufficient time to permit proper performance. The contractor's performance may be rendered impossible because of a court order, the death of the contractor, or some natural catastrophe. The performance may be rendered impracticable due to the discovery of unforeseen subsurface conditions. Most defenses the contractor has against the owner would be available to the surety who has issued a performance bond.

Likewise, defenses that the prime contractor could assert against a subcontractor or supplier claimant can generally be asserted by the surety who has issued a payment bond. The surety's function is to provide financial responsibility for the acts of the prime contractor. Generally, the surety's obligation is coextensive with that of the principal

debtor, that is, the prime contractor, but only to the extent of the bond limit.[24]

Suppose the claimant is an unpaid subcontractor or supplier under a payment bond. In *Houdaille Industries, Inc. v. United Bonding Insurance Co.*,[25] the surety was denied a defense based on an asserted claim that the owner had breached its contract with the contractor who was the principal on the bond. The court noted that although normally the surety can assert such defenses against the owner, this defense could not be asserted against a claimant "so long as it is not participated in or authorized by the materialman."[26]

Suppose the owner and the contractor modify the construction contract or the owner directs changes in the work. The surety's commitment can be limited by putting a fixed limit on the bond. Any change in the basic agreement, however, traditionally released the surety. The advent of the paid surety has made inroads on this rule. Bonds frequently provide that the surety "waives notice of any alteration or extension of time made by the owner." Without such a provision, the surety could be released because the principal obligation has been changed.[27]

Suppose the end of the Construction Process concludes with claims made by the participants. Two issues that may arise are the method of resolving disputes and whether claims have been barred by the passage of time. The first issue was discussed in Section 30.15, which deals with arbitration.

The second issue—that of the passage of time barring a claim—has caused difficulty. For example, in *State v. Bi-States Construction Co.*,[28] the state of Iowa brought a claim against a contractor who had abandoned a project and its surety. The contractor, a Nebraska corporation, had been dissolved under the laws of Nebraska on March 26,

---

[23]196 Va. 1153, 86 S.E.2d 860 (1955).

[24]That the bond limits will not always limit the surety's obligation is discussed in Subsection (E).

[25]Supra note 17.

[26]453 F.2d at 1053 n.4. But see *Chicago Bridge & Iron Co. v. Reliance Ins. Co.*, 46 Ill.2d 522, 264 N.E.2d 134 (1970) (surety *was* given a defense when the claimant subcontractor submitted false lien waivers that allowed the prime to dissipate progress payments).

[27]But see *Winston Corp. v. Continental Casualty Co.*, supra note 8, discussed in Section 33.03.

[28]269 N.W.2d 455 (Iowa 1978).

1969. Iowa law required that any claim against a dissolved corporation be made within two years. The owner's claim was barred because it did not institute the claim until almost five years after the contractor had been dissolved. The court held that the claim against the bonding company was also barred. The court recognized that divided authority on the question exists, but Iowa "adheres to the rule a surety may assert as a defense the statute of limitation if available to the principal."[29]

The California Supreme Court, however, after judicial vacillation, concluded that a claim could be made against a surety even if a completion statute[30] had given a defense to the contractor.[31] The court also held that the surety could bring a claim for reimbursement against the contractor because its claim did not arise until it had paid the claim of the owner. As a result, the statutory protection given to the contractor was not available to the surety.[32]

Once disputes develop, both obligee and principal debtor should notify the surety and keep the surety informed as to the posture and process of the dispute. In this regard, ¶ 14.2.2 of AIA Doc. A201 requires the surety to be notified in writing if the owner has terminated the contractor's performance.

D. Surety Responsibility

Some background must be kept in mind in understanding the responsibility of the surety. First, the surety's responsibility cannot exceed that owed by the principal debtor (the prime contractor on a prime contract performance bond and the subcontractor on a subcontractor performance bond) to the obligee (the owner on a prime bond and the prime on a subcontractor bond). Second, subject to some exceptions to be discussed in Subsection (E), the surety's obligation is limited to the penal sum on the bond (the bond limit). Third, much will depend upon the language of the bond.

Starting with the prime contractor performance bond, suppose that without justification the prime

contractor abandons the project. Here the principal item of damages will be the excess cost of reprocurement. Generally, the responsibility of the prime contractor when there is a default (termination) is controlled by the language of the bond. For example, AIA Doc. A312, 4, allows the surety to arrange for the original contractor to return and complete the work, usually by pumping additional resources into the original contractor. Alternatively, the surety can complete the project itself (here the surety must be aware of contractor licensing laws) or get a substitute to do so. Finally, ¶ 4.4 appears to allow the surety to pay whatever damages the owner would be entitled to recover from the defaulting contractor. This will be the cost of correction or the diminished value of the project. The latter formulas will also be used if the prime contractor has completed the project but has not performed in accordance with its contractual commitment.[33]

More complicated problems result when there has been delayed completion, by the contractor or after its default by a successor contractor. To determine whether delay comes within the bond commitment, the language of the bond is crucial. One court, pointing to the bond language and emphasizing that the most important responsibility of the prime contractor is to build the project, did not allow the obligee to recover delay damages from the surety.[34] Its decision may have been traceable to a belief that delays are inherent in all construction or to those Miller Act bond cases in which unpaid subcontractors and suppliers were allowed to recover for work they have performed or supplies they had furnished. But other courts have allowed recovery against the surety for the contractor's delayed performance.[35] Also, the commentators express the view that there is a trend toward the surety being liable for delay damages incurred by the owner.[36]

---

[29]269 N.W.2d at 457.
[30]See Section 23.03(G).
[31]*Regents of the Univ. of California v. Hartford Acc. and Indem. Co.,* 21 Cal.3d 624, 581 P.2d 197, 147 Cal.Rptr. 486 (1978).
[32]The California legislature added sureties to the completion statute. West Ann.Cal.Proc. Code § 337.15.

---

[33]See Section 27.03(D).
[34]*American Home Assur. v. Larkin General Hosp.,* 593 So.2d 195 (Fla.1992). Similarly, see *Lite-Air Products, Inc. v. Fidelity & Deposit Co. of Maryland,* 437 F.Supp. 801 (E.D.Pa.1977).
[35]*MAI Steel Service, Inc. v. Blake Constr. Co.,* 981 F.2d 414 (9th Cir.1992); *New Amsterdam Cas. Co. v. Bettes,* 407 S.W.2d 307 (Tex.Ct.App.1966).
[36]Sobel, *Owner Delay Damages Chargeable to Performance Bond Surety,* 21 Cal.W.L.Rev. 128 (1984). See also Sheak & Korzun, *Liquidated Damages and the Surety: Are They Defensible?* 9 Constr.Lawyer No. 2, Apr. 1989, p. 19. The authors seem to assume that the surety will be liable for delay damages.

The owner should be able to recover its delay damages from the surety, based upon a liquidated damages clause or actual damages. Time, though not as important as the work itself, still merits bond protection. Such protection would be reasonably expected by the owner-obligee. The recovery against the surety should be the same as the amount that would be recovered from the prime contractor. Generally, however, the surety will not be responsible for punitive damages that might be awarded against the prime contractor.[37] But the surety might, as noted in Section 33.10(I), be liable for punitive damages for *its* bad-faith refusal to settle the claim made on the bond.

Special problems can occur when one separate prime contractor has been damaged by the acts of another separate prime contractor. As noted earlier, it is unlikely that the surety on a prime contract bond would be liable for a claim made by another separate contractor unless that contractor could sustain a claim that it was an intended beneficiary.[38] But if the claim by the separate contractor were against the *owner*, based upon the owner's responsibility for coordinating the work properly, and the owner sought to transfer this loss to the separate prime contractor who had caused the additional expense to the claimant, the surety for the latter should be liable.

As to payment bonds, much depends upon any statute under which the payment bond was compelled or upon the language of the bond itself. Clearly, an unpaid subcontractor or supplier can recover on the payment bond for the reasonable value of the work it has performed or materials it has supplied. Unless profit can be encompassed in the preceding formula, however, some courts have difficulty awarding profits against the surety—probably a reflection of the view that a payment bond is a substitute for lien rights.[39] Others have allowed a claim against the surety for increased costs incurred by the subcontractor.[40]

While much depends on the language of the bond, the surety will be responsible for payment delay of the prime[41] measured from the time of the demand on the surety.[42] If the delay in payment resulted from the surety's refusal to settle, the surety will be responsible for any increased costs resulting from that refusal.[43]

A prime contractor who is having financial difficulties may request a surety to advance funds to complete the project. Usually, at the time the bond is issued the surety obtains an indemnification agreement from the prime contractor in which the surety carefully disclaims any responsibility to advance funds. Claims of prime contractors have been based on an assertion that sureties have a good-faith obligation to investigate the request and provide the financing in order to minimize the prime's liability under the indemnification agreement, despite the latter's contract usually providing that advance of funds by the surety will be within the sole discretion of the surety. While two commentators contend that there should be no obligation to investigate in good faith, they also note that some court decisions have tended to go in that direction.[44]

Finally, the exposure of the surety increasingly depends upon legislation. For example, Florida Statutes § 627.756 permits the recovery of attorneys' fees by a claimant on a surety bond.

E. Bond Limits

As noted, surety bonds generally contain what is called a penal sum, which limits the obligation of the surety. Generally, this bond limit is effective. However, as noted in Subsection (I), the law has begun to inquire into the settlement practices of sureties, just as it has examined the settlement practices of insurance companies. As a result, some exceptions have emerged to the general rule that the penal sum limits the surety's obligation. An important trial court opinion held that the surety's liability would not be limited by the bond limit, mainly because of improper settlement tactics of

---

[37]Annot. 2 A.L.R.4th 1254 (1980).
[38]See Section 33.10(A), note 22.
[39]*MAI Steel Service, Inc. v. Blake Constr. Co.*, supra note 35; *Lite-Air Products, Inc. v. Fidelity & Deposit Maryland*, supra note 34.
[40]*MAI Steel Service, Inc. v. Blake Constr. Co.*, supra note 35; *Tremack Co. v. Homestead Paving Co.*, 582 So.2d 26 (Fla.Dist.Ct.App.1991) (acceleration damages).

[41]*Insurance Co. of No. Am. v. United States*, 951 F.2d 1244 (Fed.Cir.1991) (surety's obligation to pay arises when principal defaults on contract).
[42]*Republic Ins. Co. v. Prince George's County*, supra note 3.
[43]Ibid.
[44]Toomey and Fisher, *Is a Surety Obligated to Investigate Financing a Contractor Who Requests Financial Assistance?* 12 Constr.Law. No. 4, Nov. 1992, p. 11.

the surety.[45] Similarly, in another case where the surety was not cooperative in settlement discussions concerning an award of prejudgment interest, the bond limit did not place a ceiling on the liability of the surety.[46]

F. Bankruptcy of Contractor

If, during the course of performance by the contractor, the contractor is adjudicated a bankrupt, the trustee in bankruptcy (the person who takes over the affairs of the bankrupt contractor) can determine whether to continue the contract.[47] Usually it does not. If the contract is not continued, the bankrupt contractor is released from any further obligation to perform under the contract. The owner has a claim against the bankrupt contractor but is not likely to recover much. Ending the contractor's obligation, however, should not release the surety. This is the risk contemplated when the surety bond is purchased.

G. Asserting Claims: Time Requirements

The surety's obligation usually requires that certain notices be given by claimants. Likewise, statutes requiring public work to be bonded, such as the Federal Miller Act, often specify that notices must be given within certain periods of time to designated persons.

Frequently, bonds require that legal action on the bond be brought within a designated time, usually shorter than the period specified by law. Such shortened periods to begin legal action are enforceable if reasonable.[48]

Bonds often specify the court in which legal action must be brought, usually to courts—state or federal—in the state in which the project is located.[49]

Suppose a claimant does not comply with all of the many requirements specified by bonds or statutes. A surety that denies responsibility because of some failure to comply appears to be using a technicality to avoid an obligation it was paid to perform. Courts generally interpret these requirements liberally in favor of claimants[50] and often require the surety to show prejudice caused by noncompliance before being given a defense.[51]

Courts are frequently faced with assertions by claimants that the surety should be estopped to assert these provisions or be found to have waived them.[52]

H. Reimbursement of Surety

It is sometimes said that sureties do not expect to take a loss. Sureties do not see themselves as insurers. As a result, they assert defenses that the principal debtor could have asserted. They seek bond language protection. Finally, if they *do* pay, they seek reimbursement by requiring the principal debtor to indemnify them.[53]

Usually the surety has a valid claim against the defaulting prime contractor. This claim is often uncollectible, however, because contractors who default are rarely in a financial position to reimburse the surety.

If the surety has obtained a guarantor for the principal debtor, such as an officer, shareholder, or friend of the prime contractor, the guarantors will be looked to for reimbursement. When the prime contractor defaults and the surety takes over, the surety usually notifies the owner that it should be

---

[45]*Continental Realty Corp. v. Andrew J. Crevolin Co.*, 380 F.Supp. 246 (S.D.W.Va.1974). As the case was settled, there was no appeal. The opinion is reviewed and criticized in Wisner, *Liability in Excess of the Contract Bond Penalty*, 43 Ins.Couns.J. 105 (1976).

[46]*Insurance Co. of North America v. United States*, supra note 41.

[47]Discussed in Section 34.04(C).

[48]In *Rumsey Elec. Co. v. Univ. of Delaware*, 358 A.2d 712 (Del.1976), a one-year bond provision was held valid where the statutory period was three years. Sometimes statutes of limitation are excessively long. But *City of Weippe v. Yarno*, 94 Idaho 257, 486 P.2d 268 (1971), held that the period for bringing the action cannot be shortened by the parties, as it would be unconscionable. The court stated that stipulating a longer period would be enforced if reasonable.

[49]AIA Doc. A312, ¶¶ 9, 11 found in Appendix D.

[50]*United States v. Merle A. Patnode Co.*, 457 F.2d 116 (7th Cir.1972); *American Bridge Div. United States Steel Corp. v. Brinkley*, 255 N.C. 162, 120 S.E.2d 529 (1961).

[51]*Winston Corp. v. Continental Cas. Co.*, supra note 8.

[52]*Contee Sand & Gravel Co. v. Reliance Ins. Co.*, 209 Va. 672, 166 S.E.2d 290 (1969) (surety estopped to plead a one-year period of limitations in the bond because subcontractor was misled by surety statement that its prime contractor did not have a surety bond).

[53]*Republic Ins. Co. v. Prince George's County*, supra note 3 (indemnification claim reduced to the extent that the surety caused a loss by refusing to settle in good faith).

paid all payments that would have gone to the prime contractor.[54] In addition, the surety usually demands any retainage at the end of the job that the owner has withheld to secure the owner against claims against the contractor.[55] In seeking the retainage, the surety usually competes with other creditors of the prime contractor, the taxing authorities, and the trustee in bankruptcy if the prime has been declared bankrupt. As a rule, many more claims than can be satisfied exist, and the result is a complicated lawsuit. Typically, the owner pays the retainage into court and notifies all claimants, and the court determines how the fund is to be distributed.

The surety who takes over after default usually succeeds to any claims the prime contractor may have against the owner or third parties. It is common for the prime contractor to ascribe its difficulties to the owner, the design professional, subcontractors, or other third parties.[56]

Sureties make strong efforts to be reimbursed and often succeed in salvaging a substantial amount of their loss when they are called on to respond for their principal's default.

I. Regulation: Bad-Faith Claims

Surety companies are regulated by the states in which they operate. In addition, sureties who wish to write bonds for federal projects must qualify under regulations of the United States Treasury Department. The financial capability of a surety limits the size of the projects a surety can bond. Bond dollar limits place a ceiling on exposure. In larger projects, there may be co–sureties or the surety may be required to reinsure a portion with another surety. Surety rates are usually regulated and are based on a specified percentage of the limit of the surety bond.

Subsection (E) noted that the surety's obligation is usually limited to the amount of the bond. It also cited a case in which the surety's obligation extended beyond the bond limit, largely because of the settlement tactics of the surety.

The law has taken steps to deter *insurers* from conduct unfair to its policyholders. Most states allow claims based on tortious bad-faith conduct by the insurer to policyholders and in some states to those who have claims against policyholders. These techniques have enabled claimants to obtain awards that exceed the policy limits, usually through awards for emotional distress or punitive damages.

Building on increased judicial regulation of insurers, a few states have allowed similar claims against *sureties*.[57] Where these claims are allowed and where the sureties' conduct falls below that required by law, the sureties' obligations can exceed the stated limit of the bond. This is accomplished by finding that the surety has committed a tort that in some instances may justify the award of punitive damages as well as normal tort damages.[58] However, some courts, fearful of blurring tort and contract distinctions, have not allowed tort claims of bad faith against sureties.[59]

---

[54]In *Gerstner Elec., Inc. v. Am. Ins. Co.*, 520 F.2d 790 (8th Cir.1975), the court rejected a contention of the contractor that the request for payments by the surety to the owner was a wrongful interference in the contract between contractor and owner.

[55]Retainage is discussed in Section 22.03.

[56]*Westerhold v. Carroll*, 419 S.W.2d 73 (Mo.1967) (indemnitor who had been forced to reimburse bonding company when prime defaulted was allowed to recover from architect who had negligently overcertified work); *Travelers Indem. Co. v. Evans Pipe Co.*, 432 F.2d 211 (6th Cir.1970) (surety who had to reimburse owner for defective pipe installed by its principal allowed to recover from supplier who had sold the defective pipe).

[57]*Farmer's Union Central Exch., Inc. v. Reliance Ins. Co.*, 626 F.Supp. 583 (D.N.D.1985) (stalling, deceiving, and then refusing to pay based on failure to bring legal action); *Republic Ins. Co. v. Prince George's County*, supra note 3 (awarding prejudgment interest a matter that rests in the discretion of the trier of fact); *Tonkin v. Bob Eldridge Construction Co.*, 808 S.W.2d 849 (Mo.App.1991) (surety liable to owner under state statute for attorneys' fees and interest for a bad-faith failure to investigate the claim; *Szarkowski v. Reliance Ins. Co.*, 404 N.W.2d 502 (N.D.1987); *State Surety Co. v. Lamb Constr. Co.*, 625 P.2d 184 (Wyo.1981). For a case holding that the conduct did not constitute bad faith but only constituted litigating tactics, see *United States v. Seaboard Ins. Co.*, 817 F.2d 956 (2d Cir.1987) (surety denied liability knowing contractor in default, failed to properly investigate and told consultants not to put comments in writing).

[58]*Riva Ridge Apartments v. Roger J. Fisher Co.*, 745 P.2d 1034 (Colo.App.1987), cert. denied November 9, 1987.

[59]*United States v. Wausau Ins. Co.*, 755 F.Supp. 906 (E.D.Cal.1991) (federal court predicted California would not allow a tort action for a violation of good faith and fair dealing; *Republic Ins. Co. v. Bd. of County Comm'rs of Saint Mary's County*, 68 Md.App. 248, 511 A.2d 1136 (1986) (court fearful bad-faith claims will be "boilerplate" and destroy differentiation between tort and contract).

PROBLEMS

1. C was a contractor in Idaho. He also did jobs in other nearby states. No requirement that contractors be licensed in Idaho existed, however. C undertook a job for O in Utah, which did have a licensing law. C did not apply for a license. C applied for and received a bond from S. C defaulted, and now S refuses to perform its obligations under the bond. S has offered to refund the premium to O. But S claims that the contract between C and O was illegal because C was not licensed. Should this defense be allowed? What would be your conclusion if Idaho had a licensing law and C's license had been revoked for fraud and incompetence? Would your conclusion be influenced by whether O knew of the revocation? If so, in what way?

2. O engaged C to build a construction project. O required that C furnish a payment and performance bond issued by S, which it did. During the project, O terminated C's contract because C had committed serious contractual breaches, including defective workmanship, unsafe construction practices, and careless work that damaged adjacent property. O has examined the bond that has been furnished by S, a bond similar to A311 set forth in Appendix D. What would O be able to recover from S? Would your answer have been different had the bond been A312, also set forth in Appendix D?

CHAPTER THIRTY-FOUR

Termination of a Construction Contract: Sometimes Necessary But Always Costly

SECTION 34.01 Termination: A Drastic Step

Termination does not occur frequently in construction contracts. One reason for this is the difficulty of determining whether a legal right to terminate exists, a point discussed principally in Sections 34.03 and 34.04. Often each party can correctly claim the other has breached. It may be difficult to determine whether a party wishing to terminate is sufficiently free from fault and can find a serious deviation on the part of the other. Another reason is the often troublesome question of whether the right to terminate has been lost (to be discussed in Section 34.03(E)). A third reason, the serious consequences of terminating without proper cause, is treated in this section.

Two cases illustrate the danger of an improper termination. The first, *Paul Hardeman, Inc. v. Arkansas Power & Light Co.,*[1] involved a contract under which Hardeman was to construct electric transmission lines for the Power & Light Company for a contract price of $2.7 million. Hardeman's bid had been much lower than the other bids. This was known only to the owner because this private project did not use a public bid opening. Despite the likelihood that there had been a mistake, the Power & Light Company proceeded to award the contract without discussing the mistake possibilities.

Many difficulties developed mainly because of Hardeman's inexperience. Several months before the completion date, at a time when somewhat less than one-half of the work had been completed,

Hardeman discovered that its bid had been substantially lower than the other bids because of a calculation mistake. It immediately communicated this to the Power & Light Company and accused it of wrongfully awarding the contract while knowing of the likelihood of the mistake. It threatened to leave the project unless some equitable adjustment was made. A conference was held, but no resolution was accomplished. Finally, the Power & Light Company and its engineer terminated Hardeman's contract, claiming that there had been unexcused delays as well as defective workmanship and materials. Evidently, the real reason for the termination was that the Power & Light Company sought the best tactical position for the inevitable lawsuit. Before the termination, Hardeman had claimed to have spent $3 million on the work. After termination, a successor contractor completed performance on a cost-plus basis and received $8 million.

Hardeman brought legal action claiming that the Power & Light Company wrongfully accepted its bid and that the termination had been improper. It sought restitution based on its asserted expenditures of $3 million less payments made of $600,000. The Power & Light Company sought $5.8 million on its part as damages.

Although the court seemed sympathetic to Hardeman's first claim, it felt it could not grant relief, because Hardeman's estimator had been grossly negligent.[2] A convenient alternative solution, how-

---

[1]380 F.Supp. 298 (E.D.Ark.1974).

[2]For a further discussion of relief from mistaken bids, see Section 18.04(E).

743

ever, arose. The termination had not been made in good faith. According to the court, such a drastic step had been taken prematurely at best.

Termination proved useful to Hardeman. The court concluded that improper termination entitled Hardeman to recover the reasonable value of its services, which the court found to be $2 million, less the amount paid of $600,000. The defendant's counterclaim for the excess cost of correction was denied.

Perhaps the case is not typical. The court seemed sympathetic to Hardeman's bidding mistake claim. Yet the case demonstrates that a precipitous termination by the owner can be very costly when the contractor has made a losing contract.

The second case, tried before a jury, lasted eighty-one trial days and generated a 36,000-page transcript, which was bound into 138 volumes. The reported appellate court decision filled thirty-three double-columned pages in the regional reports.[3] The contract was for the installation of pollution control equipment for an energy plant using coal-fired generators. The state pollution control board had been exerting pressure on the utility to install an electrostatic precipitator and fix strict deadlines for its installation.

The contractor submitted the low bid of approximately $7 million. At a point at which approximately 80% of the by now contract price of $8 million had been paid, the utility and the contractor disagreed as to whether the contractor was on schedule. Claiming that the contractor was not and that it was the contractor's fault, the utility demanded an acceleration. Concluding that the contract would not be completed on time, the utility terminated the contractor because of its alleged failure to maintain the schedule. Evidence at the trial indicated that at the time of termination the project was 60% complete. The utility spent an additional $5 million to complete the work. At the time of termination, the utility took possession of construction equipment and used it for approximately four months, based on a power given to it in the termination clause of the construction contract.

The contractor's claim and the judgment it obtained in the trial court indicate the high degree of

risk that termination can entail. The judgment was for $17 million (all figures approximations), broken down as follows:

1. Retainage and interest, lost profits on the project, extras, and additional expenses—$2 million.
2. Loss of future business—$3 million.
3. Punitive damages—$12 million.

In addition, the trial court did not grant the utility's counterclaim, because it concluded that the termination by the utility had not been proper.

On appeal, the court reduced the award drastically, concluding that the contractor was not entitled to one small expense item, future profits, and punitive damages, leaving an award of $2 million.

The case demonstrates the difficult position in which the utility found itself when it decided to terminate the contractor's performance. It was difficult to determine who was responsible for the delay. There had been acrimonious disputes throughout the entire performance. Although the court ultimately concluded that punitive damages were not appropriate for this wrongful termination, certainly at the time the termination step was taken, award of open-ended punitive damages could not be excluded as a possibility. To be sure, not terminating carried risks as well. But undoubtedly, the economic dislocation and legal exposure are factors that make termination in construction disputes relatively rare.

It may be necessary to terminate a construction contract. Performance may be going so badly and relations may be so strained that continued performance would be a disaster. But because of the reasons mentioned, the drastic step of termination should not be taken precipitously.

SECTION 34.02 Termination by Agreement of the Parties

Just as parties have the power to make a contract, they can "unmake" it. Exercise of this power in legal parlance may be described as rescission, cancellation, mutual termination, or some other synonym. However described, the parties have agreed that each is to be relieved from any further performance obligations. In lay terms, they have "called the deal off."

The legal requirements for such an arrangement are generally the same as those for performing a

[3]*Indiana & Michigan Elec. Co. v. Terre Haute Indus.*, 507 N.E.2d 588 (Ind.App.1987).

contract: manifestations of mutual assent, consideration, a lawful purpose, and compliance with any formal requirements. By and large, the mutual assent requirement does not present unusual difficulty. Suppose, however, one party makes a proposal to cancel but the other party does not *expressly* accept. In most cases, silence is not acceptance, but silence coupled with other acts that lead the proposer to believe there has been an agreement can be sufficient.

If each party has obligations yet to perform, the consideration consists of each party relieving the other of performance obligations. If one party has *fully* performed, however, enforcement problems may arise. For example, suppose a contractor has been paid the full contract price but has not finished performance. If the parties agree to cancel remaining obligations of the contractor, the owner is not receiving anything for its promise. Courts generally relax consideration requirements somewhat when parties adjust or cancel existing contracts. Such an agreement can be enforced by calling it a waiver or completed gift. Although problems can arise, agreements under which each party agrees to relieve the other are generally enforced.

Because most construction contracts do not have to be expressed by a written memorandum, formal requirements are rarely an impediment to enforcement of contracts of mutual termination. Desirability of proof usually, however, means that such agreements will be expressed in writing.

SECTION 34.03 Contractual Power to Terminate

Construction contracts frequently contain provisions giving one or both parties the power to terminate the contract. These provisions are a backdrop for the material to be discussed in Section 34.04—the common law right to terminate a contract. Although contracts are not always clear on this point, often an interrelationship exists between specific termination provisions and the common law. Specific provisions can be considered illustrations or amplifications of common law doctrines, with the common law doctrines still applicable. Alternatively, contractual termination provisions can be said to have supplanted common law doctrines.

It is likely that common law termination rights have not been eliminated by express contract termination provisions,[4] as the intention to waive common law rights requires clear expression.[5]

A. Default Termination

Construction contracts drafted by an owner generally give explicit termination rights only to the owner. On the other hand, construction contracts published by professional associations provide that either owner or contractor can terminate for certain designated defaults by the other. Although great variations exist, it may be useful to begin discussion with AIA documents.

AIA Doc. A201, ¶ 14.2.1, permits the owner to terminate if the contractor "persistently or repeatedly refuses or fails to supply enough properly skilled workers or proper materials," if it fails to make payments to subcontractors, if it persistently disregards laws or "otherwise is guilty of substantial breach of a provision of the Contract Documents." This does not specifically allow partial termination, something that may be useful if the owner wishes to retain the contractor for part of the work and find a successor for the rest.

Partial termination may be useful from the contractor's perspective. In a decision by the General Services Administration Board of Appeals, the contractor had properly performed part (35%) of the work and improperly performed another part (35%). The Board held that the agency could terminate only for the improperly performed part, based on the fiction of divisability.[6]

Can the owner terminate if the contractor falls behind its schedule? Federal construction contracts permit termination if the contractor fails to make progress and that failure endangers performance of

---

[4]AIA Doc. A.201, ¶ 13.4.1, applied in *Monmouth Public Schools v. D. H. Rouse,* 153 Ill.App.3d 901, 506 N.E.2d 315 (1987).
[5]*Armour & Co. v. Nard,* 463 F.2d 8 (8th Cir.1972); *Glantz Contracting Co. v. General Elec.,* 379 So.2d 912 (Miss.1980); *Bender-Mill Co. v. Thornwood Farms, Inc.,* 211 Va. 585, 179 S.E.2d 636 (1971). See also *North Harris County Junior College Dist. v. Fleetwood Constr. Co.,* 604 S.W.2d 247 (Tex.Ct.App.1980) (dictum would permit termination for reason not specified in AIA contract).
[6]*Nestos Painting Co.,* GSBCA 6945, 86-2 BCA ¶ 18,993.

the contract.[7] Suppose the owner wishes to bring in a successor contractor when the original contractor is falling far behind schedule. Paragraph 14.2.1 does permit the owner to terminate for persistent failure to supply proper workers or materials. But delays, even unexcused ones, may be caused by other factors. Does ¶ 8.3, with its provision for time extensions, preclude termination? Does a liquidated damage clause, if present, indicate an intention that termination is not appropriate for failure to comply with schedule or completion requirements? Again, the nonexclusivity of the termination clause can support a conclusion that protracted delay—practically if it appears that completion will not be "on time"—will create a power to terminate under common law principles.[8] Any exercise of such a power should be preceded by warnings and, where appropriate, an opportunity to cure.

A termination may be wrongful if the owner has not used good faith in exercising its contractual power to terminate. In *Paul Hardeman, Inc. v. Arkansas Power & Light Co.*, the court admitted that contractual grounds to terminate existed but held that the termination had not been made in good faith.[9] As shall be seen in Section 34.04(A), good faith of the party in default can be a factor in determining whether the right to terminate exists at common law.

The law has been reluctant to inject the good-faith principle as a limitation on the exercise of a contractual power to terminate. This is due in part to the possibility that termination not made in good faith would be sufficiently wrongful to be considered a tort and sufficiently intentional to be considered as the basis for punitive damages. This is illustrated by the unwillingness of the court in *Indiana & Michigan Elec. Co. v. Terre Haute Industries*[10]

to impose punitive damages after it concluded that the owner did not have grounds for terminating the contract. The award of punitive damages had been based on a finding that termination had been malicious and oppressive. The appellate court found that at worst the conduct was "substandard business practice, and arrogance."[11] It noted that the law did not impose punitive damages simply because "the contracting party or his agents are disagreeable people."[12] Awarding punitive damages, according to the court, would

> let all disputes and quarrels over broken contracts and disappointed business ventures become the subject of acrimonious litigation over punitive damages.[13]

Section 19.02 (D) chronicled the increasing use of the good-faith concept in American contract law. Its use in most termination disputes should be limited to whether the termination was proper, not the basis for invocation of tort law.[14]

AIA Doc. A201, ¶ 14.1, gives the contractor termination rights. In ¶ 14.1.1, a work stoppage of thirty days caused by certain designated events gives the contractor the right to terminate. In 1987, ¶ 14.1.3 was added. It permits the contractor to terminate if there has been a sixty-day work stoppage for the owner's failure to perform "matters important to the progress of the Work," a recognition of the materiality concept to be discussed in Section 34.04(A). Termination is not effective until the expiration of a seven-day written notice (discussed in Section 34.03(F)). Finally, ¶ 14.1.2 specifies a modest remedy. But ¶ 13.4.1 allows the contractor common law damages for a wrongful termination.

The contractual power to terminate cannot be exercised if the owner has caused or is responsible for the occurrence of the event that is the basis for the termination.[15]

The sparse grounds for termination and the absence of a contractual right to terminate when the work in progress is destroyed show that AIA documents seek to continue performance and avoid

[7]48 C.F.R. § 49.402-1 (1992). See also EJCDC General Conditions, No. 1910-8 cited infra note 17 at ¶ 15.2.6
[8]Whether the architect's certificate needed under ¶ 14.2.1 would be required is discussed in Section 34.03(D).
[9]Supra note 1. See also *Darwin Constr. Co., Inc. v. United States*, 811 F.2d 593 (Fed.Cir.1987) (despite contractual grounds for termination, owner cannot terminate where an abuse of discretion is shown); *Ohio Cas. Ins. Co. v. United States*, 12 Cl.Ct. 590 (1987) (arbitrary refusal by federal agency to terminate breached equitable duty owed the surety).
[10]Supra note 3.

[11]507 N.E.2d at 617.
[12]Ibid.
[13]Ibid.
[14]Similarly, in a termination context, see *McClain v. Kimbrough Constr. Co.*, 806 S.W.2d 194 (Tenn.App.1990).
[15]*Dep't. of Transp. v. Arapaho Constr., Inc.*, 257 Ga. 269, 357 S.E.2d 593 (1987).

the economic disruption that is caused by a contract termination.

B. Termination or Suspension for Convenience

Pioneered by federal procurement regulations,[16] construction contracts increasingly give private owners the right to terminate for convenience, something also found in subcontracts, particularly those that are tied to prime contracts where the owner has this power.[17] Invoking such a clause requires the contractor to stop work, place no further orders, cancel those orders that had been placed, and perform other acts designed to terminate performance and protect the interests of the government. The contractor is reimbursed for work performed, unavoidable losses suffered, and expenditures incurred to preserve and protect government property. The contractor is also paid a designated profit for work performed.

In addition, federal procurement law has developed the *constructive* convenience termination.[18] Under it, a contract with a convenience termination clause converts a *wrongful* government *default* termination into a *convenience* termination.

On its face, the power given to the owner by such contracts is very broad. The owner may terminate if the project is no longer needed or if it has become outmoded or uneconomical. Yet this power is not completely unrestricted. *Torncello v. United States*[19] outlined the history of such clauses and noted the importance of the government's ability to change its procurement objectives. But the court did not permit the agency to terminate when the agency used another contractor to do the work for less. (The agency knew a cheaper source existed when it made the contract.) The court stated that a

termination for convenience can be used only when the circumstances of the bargain have changed.

This takes away much of the usefulness of the clause. The court should have simply concluded that the government abused its discretion because terminating for convenience in *this* case was in bad faith.[20] Those owners who use a termination for convenience clause may have, at least judging by federal procurement, a false impression that the owner's rights are unlimited.

Although the AIA did not include a termination for convenience clause in the A201 published in 1987, ¶ 14.3 provides that the owner can suspend without cause. Currently, ¶ 14.3.2 grants the contractor an adjustment "for increases in the cost of performance . . . including profit on the increased cost of performance, caused by any suspension, delay or interruption," a generous adjustment formula.

In addition, ¶ 14.1.1.4 permits the contractor to *terminate* if there are repeated suspensions, delays, or interruptions that aggregate more than 100% of the total number of days scheduled for completion or 120 days in any 365-day period, whichever is less. Unfortunately, the power to terminate granted by ¶ 14.1.1 specifically requires a work stoppage of thirty days, a condition difficult to harmonize with the extended suspension formula.

C. Events For Which Neither Party Is Responsible

Some contracts specifically allow termination when events occur that have a devastating effect on performance, such as those that would justify common law relief for impracticability or impossibility. AIA documents do not specifically provide that these events permit termination. The party whose performance has been adversely affected receives time extensions and occasionally an equitable adjustment.

D. Role of Design Professional

Despite the AIA's movement toward reducing the activities and responsibilities of the architect, the owner's power to terminate under ¶ 14.2.2 requires that the architect certify "that sufficient cause exists

---

[16]48 C.F.R. § 52.249.1, .2 (1992).

[17]EJCDC General Conditions No. 1910-8 ¶ 15.4 allows it. The AIA considered adding such a clause in 1987 but decided not to do so. See also *Arc Elec. Constr. Co. v. George A. Fuller Co.*, 24 N.Y.2d 99, 247 N.E.2d 111, 299 N.Y.S.2d 129 (1969).

[18]*Torncello v. United States,* note 19 infra. The constructive convenience termination was rejected in a case involving a private contract. *Rogerson Aircraft Corp. v. Fairchild Ind. Inc.,* 632 F.Supp. 1494 (C.D.Cal.1986). The concept was attacked in Note, 52 Geo.Wash.L.Rev. 892 (1984).

[19]681 F.2d 756 (Ct.Cl.1982), overruling *Colonial Metals Co. v. United States,* 204 Ct.Cl. 320, 494 F.2d 1355 (1974), which had allowed the United States to terminate when it found a cheaper source it should have known of when it made the contract.

[20]*Automated Services, Inc.,* DOT BCA 1753, 87-1 BCA ¶ 19,459.

to justify such action." No similar requirement exists for a termination by the contractor.

Termination is a drastic step. Why require the architect to certify that there are adequate grounds? Such a decision, although involving some issues for which the architect may be trained, requires legal expertise rather than design skill. The owner may want the architect's advice. That need not require giving power to the architect to decide a sensitive and liability-exposing issue. The only possible justification is that such a preliminary step can act as a brake on any hasty, ill-conceived decision by the owner to terminate. It would be better to have the owner's attorney perform this function, especially if the architect's conduct is itself an issue in the termination.

Would an architect's certificate be needed for common law termination? Although it is not likely that such a procedure would apply to a common law termination, the common law can add grounds but should not destroy any agreed-upon conditions precedent.

E. Waiver of Termination and Reinstatement of Completion Date

Suppose one party has the power to terminate but does not exercise it. Has the power to terminate been lost? Under what conditions can it be revived?

The cases that have dealt with this problem have usually involved a performing party—usually the contractor—who has not met the completion date. But for various reasons, the owner decides not to terminate. The contractor continues to perform, believing that the power to terminate will not be exercised. At some point during performance, the owner wishes to either terminate immediately or set a firm date for completion which, if not met, will be grounds for termination.

Clearly, a termination after the contractor has been led to believe there will be no termination would be improper.[21] The more difficult question relates to the requirements for reinstating a firm deadline that will allow termination if performance is not met by that time. The parties can agree to

extend the time for completion coupled with the clear understanding that failure to comply would entitle the owner to terminate. The most difficult question involves unilateral attempts by the owner to reinstate a firm completion date and revive the right to terminate.

In *DeVito v. United States*,[22] the contractor did not meet a revised completion date of November 29, 1960, because of performance problems. Having anticipated default, on November 25, 1960, the contracting officer requested permission from higher headquarters to terminate the contract. The request moved slowly through various offices of the agency, and for unexplained reasons, authority to terminate was not granted until January 16, 1961. This was communicated to the contractor on January 17, 1961. During this period, the contractor continued to perform, made commitments, and delivered some units that were accepted by the government. At the time of termination, the contractor had many assemblies in various stages of completion and claimed to be on the verge of reaching full production.

In a legal action brought on behalf of the contractor the court first noted that the government is habitually lenient in granting reasonable extensions, "for it is more interested in production than in litigation."[23] Moreover, the court said, "default terminations—as a species of forfeiture—are strictly construed."[24] The court noted that permitting a delinquent contractor to continue performance can preclude the government from terminating if its actions or nonactions have led the contractor to believe no termination would occur; the contractor relied on this. The court held that the government had waived its right to terminate by allowing the contractor to continue performance under these circumstances.

As to reinstatement, the court stated:

When a due date has passed and the contract has not been terminated for default within a reasonable time, the inference is created that time is no longer of the essence so long as the constructive election not to terminate continues and the contractor proceeds with

---

[21]*De Vito v. United States*, 413 F.2d 1147 (Ct.Cl.1969); *United States v. Zara Contracting Co.*, 146 F.2d 606 (2d Cir.1944).

[22]Supra note 21.
[23]413 F.2d at 1153. See also *Martin J. Simko Contr., Inc. v. United States*, 11 Ct.Cl. 257 (1986), vacated, 852 F.2d 540 (Fed.Cir.1988).
[24]413 F.2d at 1153.

performance. The proper way thereafter for time to again become of the essence is for the Government to issue a notice under the Default clause setting a reasonable but specific time for performance on pain of default termination. The election to waive performance remains in force until the time specified in the notice, and thereupon time is reinstated as being of the essence. The notice must set a new time for performance that is both reasonable and specific from the standpoint of the performance capabilities of the contractor at the time the notice is given.[25]

The court held that such a process would not be required if the contractor had renounced its contract or was incapable of performance.[26]

The *DeVito* case emphasized that there could be no fixed rules regarding the government's waiver of its right to terminate. *H.N. Bailey & Associates v. United States,*[27] a case decided by the Court of Claims two years after *DeVito*, illustrated this. The contractor had been awarded a contract set aside for small businesses and was relatively inexperienced in manufacturing the products sought by the procurement. The delivery date was July 27, 1966, and it seemed clear to the government that the contractor would not be able to perform. As a result, on August 10, 1966, the government notified the contractor that there had been a default and that any assistance given the contractor, or acceptance of delinquent goods, would be solely for the purpose of mitigating damages and not to be construed as an indication that the government was waiving its rights. The notice also gave the contractor ten days to advise the procuring agency of any reason why the contract should not be terminated.

On August 19, 1966, the contractor responded by stating that it was confident it could produce the goods by the end of the following month. Yet despite this, production was not made, and the government terminated for default on September 6, 1966.

The contractor claimed the conduct of the government was inconsistent with its right to terminate and constituted a waiver. The court did not agree, however, and concluded that the government did not encourage or induce the contractor to continue performance; it characterized the government's conduct as "displaying a benevolent attitude towards the defaulting contractor."[28]

In *John Kubinski & Sons, Inc. v. Dockside Development Corp.,*[29] promised performance by the owner essential to the contractor's performance was not accomplished by the date promised, September 1, 1968. A conference was held between the parties that appeared to extend the time for the owner's performance until approximately November 15, 1968. The contractor terminated the contract on December 18, 1968, because the owner's promised work had not been done by that time. The owner hired a replacement contractor to complete the work.

The owner contended that the contractor's failure to perform the work was a default and that the contractor had waived any owner default. The waiver claim was based on activities engaged in by the contractor after the expiration of the original contract performance date. These activities consisted of some minor work and the delivery of a bond to receive payment for work that had been performed. The court concluded that none of the activities were inconsistent with an intention of the contractor to insist on its contract rights.

In rejecting the contention that the contractor should have terminated at the time of the original default by the owner, the court stated:

> Whether a breach of a time provision of a contract justifies a repudiation of the agreement is a determination not to be made lightly by an injured party, since, should a court later hold his determination unwarranted, the repudiator will himself be guilty of a material breach. To hold that [the contractor] should have declared an immediate forfeiture after September 1 would defeat the public policy that encourages the extrajudicial resolution of disagreements.[30]

The court further noted that it was not clear that failure to meet the original date was a material breach. But in any event, failing to meet the time extensions along with the advent of adverse winter

---

[25]Id. at 1154.
[26]A similar rule was announced in *Gamn Constr. Co. v. Townsend,* 32 Ill.App.3d 848, 336 N.E.2d 592 (1975).
[27]449 F.2d 376 (Ct.Cl. 1971).

[28]Id. at 385.
[29]33 Ill.App.3d 1015, 339 N.E.2d 529 (1975).
[30]339 N.E.2d at 534.

conditions certainly made the final breach material and terminated the contractor's obligation to perform.

Sometimes two issues arise in the context of waiver. This is illustrated by *Consolidated Engineering Co. v. Southern Steel Co.*,[31] a dispute between a prime and subcontractor. The subcontractor was to install steel furnished by the prime contractor. It was unable to do so because of the prime's delay in furnishing the materials. Although the subcontractor contended that it had the right to terminate, it continued to work for three more months. Then it abandoned the contract.

The prime claimed damages, contending that the subcontractor wrongfully abandoned the contract. The subcontractor claimed damages for not being able to perform its contract. The court concluded that the prime contractor had indeed breached and that the subcontractor was justified in stopping performance. This gave the subcontractor a defense to the claim by the prime contractor. It also gave the subcontractor a claim for damages against the prime contractor.

The prime did not contend that the subcontractor remaining on the job waived its rights to terminate; it contended that remaining on the job waived the subcontractor's claim for damages. The court disagreed, noting that the subcontractor reserved its rights at the time it continued performance. It in no way indicated it was giving up its right to recover damages for the prime contractor's breach.

F. Notice of Termination: *New England Structures, Inc. v. Loranger*

Termination clauses often require that a notice of termination be sent by the terminating party to the party whose performance is being terminated and,

in many contracts, to the lender or surety. The notice usually states that termination will become effective a designated number of days after dispatch or receipt of the notice.

In addition to problems common to all notices and communications,[32] such notice requirements create other legal problems. What are the rights and duties of the parties during the notice period? Can the defaulting party cure any defaults specified in the notice during the notice period and thereby keep the contract in effect? Can the party giving the notice to terminate demand continued performance or even accelerated performance during the notice period? Answers to such questions often depend on the reason for requiring a notice to terminate.[33]

The notice period may be designed to allow the terminating party to "cool off." Construction performance problems often generate animosity, and before the important step of termination is effective, the terminating party may wish to rethink its position. The notice period can permit a defaulting party to cure defaults in order to keep the contract in effect for the benefit of both parties. Whether the notice period is designed to "cure" may depend on the facts that give rise to termination and on the notice period.

Finally, the notice period can be used to wind down and secure the job. This allows each party to cut losses and make new arrangements. Such a position does not allow cure. But if the owner terminates, it should have the option of ordering that work be done that can be completed by the effective date of termination.

Such questions should be, but rarely are, answered by the termination clause. The following case illustrates this and other problems.

---

[31]699 S.W.2d 188 (Tex.1985).

[32]See Section 17.05(C).
[33]Section 12.13(F) treated this problem in the context of the design professional-client relationship.

NEW ENGLAND STRUCTURES, INC. v. LORANGER

Supreme Judicial Court of Massachusetts, 1968. 354 Mass. 62, 234 N.E.2d 888.
[Ed note: Court footnote renumbered.]

CUTTER, Justice.

In one case the plaintiffs, doing business as Theodore Loranger & Sons (Loranger), the general con-

tractor on a school project, seek to recover from New England Structures, Inc., a subcontractor (New England), damages caused by an alleged breach of the subcontract. Loranger avers that the breach made it necessary for Loranger at greater expense to engage another subcontractor to complete work on a roof deck. In a cross action, New England seeks to recover for breach of the subcontract by Loranger alleged to have taken place when Loranger terminated New England's right to proceed. The actions were consolidated for trial. A jury returned a verdict for New England in the action brought by Loranger, and a verdict for New England in the sum of $16,860.25 in the action brought by New England against Loranger. The cases are before us on Loranger's exceptions to the judge's charge.

Loranger, under date of July 11, 1961, entered into a subcontract with New England by which New England undertook to install a gypsum roof deck in a school, then being built by Loranger. New England began work on November 24, 1961. On December 18, 1961, New England received a telegram from Loranger which read, "Because of your . . . repeated refusal . . . or inability to provide enough properly skilled workmen to maintain satisfactory progress, we . . . terminate your right to proceed with work at the . . . school as of December 26, 1961, in accordance with Article . . . 5 of our contract. We intend to complete the work . . . with other forces and charge its costs and any additional damages resulting from your repeated delays to your account." New England replied, "Failure on your [Loranger's] part to provide . . . approved drawings is the cause of the delay." The telegram also referred to various allegedly inappropriate changes in instructions.

The pertinent portions of art. 5 of the subcontract are set out in the margin.[34] Article 5 stated grounds on which Loranger might terminate New England's right to proceed with the subcontract.

There was conflicting evidence concerning (a) how New England had done certain work; (b) whether cer-

---

[34]The Subcontractor agrees to furnish sufficient labor, materials, tools and equipment to maintain its work in accordance with the progress of the general construction work by the General Contractor. Should the Subcontractor fail to keep up with . . . [such] progress . . . then he shall work overtime with no additional compensation, if directed to do so by the General Contractor. If the Subcontractor should be adjudged a bankrupt . . . or *if he should persistently . . . fail to supply enough properly skilled workmen* . . . or . . . disregard instructions of the General Contractor or fail to observe or perform the provisions of the Contract, then the General Contractor may, by *at least five . . . days prior written notice to the Subcontractor* without prejudice to any other rights or remedies, *terminate the Subcontractor's right to proceed with the work.* In such event, Contractor for any excess cost occasioned . . . thereby . . ." (emphasis supplied).
[Ed. note: Balance of footnotes omitted.]

tain metal cross pieces (called bulb tees) had been properly "staggered" and whether joints had been welded on both sides by certified welders, as called for by the specifications; (c) whether New England had supplied an adequate number of certified welders on certain days; (d) whether and to what extent Loranger had waived certain specifications; and (e) whether New England had complied with good trade practices. The architect testified that on December 14, 1961, he had made certain complaints to New England's president. The work was completed by another company at a cost in excess of New England's bid. There was also testimony (1) that Loranger's job foreman told one of New England's welders "to do no work at the job site during the five-day period following the date of Loranger's termination telegram," and (2) that, "if New England had been permitted to continue its work, it could have completed the entire subcontract . . . within five days following the date of the termination telegram."

The trial judge ruled, as matter of law, that Loranger, by its termination telegram, confined the justification for its notice of termination to New England's "repeated refusal . . . or inability to provide enough properly skilled workmen to maintain satisfactory progress." He then gave the following instructions: "If you should find that New England . . . did not furnish a sufficient number of men to perform the required work under the contract within a reasonable time . . . then you would be warranted in finding that Loranger was justified in terminating its contract; and it may recover in its suit against New England. [T]he termination . . . cannot, as . . . matter of law, be justified for any . . . reason not stated in the telegram of December 18 . . . including failure to stagger the joints of the bulb tees or failure to weld properly . . . or any other reason, unless you find that inherent in the reasons stated in the telegram, namely, failure to provide enough skilled workmen to maintain satisfactory progress, are these aspects. Nevertheless, these allegations by Loranger of deficiency of work on the part of New England Structures may be considered by you, if you find that Loranger was justified in terminating the contract for the reason enumerated in the telegram. You may consider it or them as an element of damages sustained by Loranger." . . . Counsel for Loranger claimed exceptions to the portion of the judge's charge quoted above in the body of this opinion.

1. Some authority supports the judge's ruling, in effect, that Loranger, having specified in its telegram one ground for termination of the subcontract, cannot rely in litigation upon other grounds, except to the extent that the other grounds may directly affect the first ground asserted. See *Railway Co. v. McCarthy,* . . . ("Where a party gives a reason for his conduct and decision touching . . . a controversy, he cannot, after

litigation has begun, change his ground and put his conduct upon . . . a different consideration. He is not permitted thus to mend his hold. He is estopped from doing it by a settled principle of law."

Our cases somewhat more definitely require reliance or change of position based upon the assertion of the particular reason or defense before treating a person, giving one reason for his action, as estopped later to give a different reason. See *Bates v. Cashman.* There it was said, "The defendant is not prevented from setting up this defense. Although he wrote respecting other reasons for declining to perform the contract, he expressly reserved different grounds for his refusal. While of course one cannot fail in good faith in presenting his reasons as to his conduct touching a controversy he is not prevented from relying upon one good defense among others urged simply because he has not always put it forward, when it does not appear that he has acted dishonestly or that the other party has been misled to his harm, or that he is estopped on any other ground."

We think Loranger is not barred from asserting grounds not mentioned in its telegram unless New England establishes that, in some manner, it relied to its detriment upon the circumstance that only one ground was so asserted. Even if some evidence tended to show such reliance, the jury did not have to believe this evidence. They should have received instructions that they might consider grounds for termination of the subcontract and defenses to New England's claim (that Loranger by the telegram had committed a breach of the subcontract), other than the ground raised in the telegram, unless they found as a fact that New England had relied to its detriment upon the fact that only one particular ground for termination was mentioned in the telegram.

2. As there must be a new trial, we consider whether art. 5 of the subcontract . . . afforded New England any right during the five-day notice period to attempt to cure its default, and, in doing so, to rely on the particular ground stated in the telegram. Some evidence summarized above may suggest that such an attempt was made. Article 5 required Loranger to give "at least five . . . days prior written notice to the Subcontractor" of termination.

If a longer notice period had been specified, one might perhaps infer that the notice period was designed to give New England an opportunity to cure its defaults. An English text writer (Hudson's Building and Engineering Contracts, 9th ed. p. 530) says, "Where a previous warning notice of specified duration is expressly required by the contract before . . .

termination [in case of dissatisfaction], the notice should be explicit as to the grounds of dissatisfaction, so that during the time mentioned in the notice the builder may have the opportunity of removing the cause of objection." . . . This view was taken of a three-day notice provision in *Valentine v. Patrick Warren Constr. Co.,* . . . without, however, very full consideration of the provision's purpose. In Corbin, Contracts, § 1266, p. 66, it is said of a reserved power to terminate a contract, "If a period of notice is required, the contract remains in force and must continue to be performed according to its terms during the specified period after receipt of the notice of termination."

Whether the short five-day notice period was intended to give New England an opportunity to cure any specified breach requires interpretation . . . of art. 5, a matter of law for the court. . . . It would have been natural for the parties to have provided expressly that a default might be cured within the five-day period if that had been the purpose.

Strong practical considerations support the view that as short a notice period as five days in connection with terminating a substantial building contract cannot be intended to afford opportunity to cure defaults major enough (even under art. 5) to justify termination of a contract. Such a short period suggests that its purpose is at most to give the defaulting party time to lay off employees, remove equipment from the premises, cancel orders, and for similar matters.

Although the intention of the notice provision of art. 5 is obscure, we interpret it as giving New England no period in which to cure continuing defaults, but merely as directing that New England be told when it must quit the premises and as giving it an opportunity to take steps during the five-day period to protect itself from injury. Nothing in art. 5 suggests that a termination pursuant to its provisions was not to be effective in any event at the conclusion of the five-day period, even if New England should change its conduct.

If Loranger in fact was not justified by New England's conduct in giving the termination notice, it may have subjected itself to liability for breach of the subcontract. The reason stated in the notice, however, for giving the notice cannot be advanced as the basis of any reliance by New England in action taken by it to cure defaults. After the receipt of the notice, as we interpret art. 5, New England had no further opportunity to cure defaults.

Exceptions sustained.

The termination clause should clearly indicate whether it permits *cure*, provides a *cooling-off* period, or sets into motion a *winding down* of the project. To illustrate a cure provision, standard construction contracts used in Canada allow the owner to terminate if the contractor fails to perform after the owner notifies the contractor in writing that it is in default based on a certificate from the architect. The notice must instruct the contractor to correct the default within five working days from receipt of the notice. If the default cannot be cured within the five working days, the owner cannot terminate if the contractor begins correction of the default within the specified time, provides the owner with an acceptable schedule for such correction, and completes the correction in accordance with that schedule. This is a sensible procedure. Termination is the last desperate step in a troubled contract.

G. Taking Over Materials and Equipment

AIA Doc. A201, ¶¶ 14.2.2.1 and 14.2.2.3, permit the owner to

> take possession of the site and of all materials, equipment, tools, construction equipment and machinery thereon owned by the Contractor . . . 3 [and] . . . finish the Work by whatever reasonable method the owner may deem expedient.

These provisions are commonly included in construction contracts. The following are reasons given for such provisions:

1. To provide incentive to the contractor to take away its property from the site so that the owner can efficiently bring in a successor.
2. To provide the owner with material and equipment by which it can expeditiously continue the work with a successor (Note the conflict with (1) above).
3. To give the owner property which can be sold and the proceeds used to pay for any claim it may have against the contractor.

The owner, however, must first give a seven-day termination notice. Will this encourage the contractor to take away its property before the owner does? If so, the first justification will be accomplished, but the second will be frustrated. Also, suppose the successor contractor does not want to

use the materials and equipment. Does the contractor have the right to take away the materials and equipment during the seven-day period? (It is unlikely that the contractor will stand by passively when the owner takes possession of its property.)[35] Can the owner take materials or equipment owned by a subcontractor? If not, the clause may be of little value.[36] In addition, who actually owns the equipment? Much equipment is leased or purchased under conditional sales contracts. Even if the property is owned by the contractor, any attempt to use the equipment as security may force a struggle with others who contend that they have security interests that take precedence over any right created by such a clause or any other lien rights. If the contractor goes bankrupt, the trustee in bankruptcy surely will enter the fray. Also, if the contractor's property is used, a subsequent dispute may arise as to its use value.

As if these problems were not enough, suppose the termination is wrongful. Taking the contractor's property constitutes the tort of conversion. The owner must pay the reasonable value of property converted, restore any gains made through the conversion of the property, and perhaps pay punitive damages.[37]

The AIA did not expressly create a security interest. There is no express power to sell the materials and equipment and retain any amounts obtained to set off against claims it has against the contractor. (AIA Doc. A201, ¶ 12.2.4, dealing with failure to correct and remove defective work *did* create such a security interest.) Some justify the clause as a basis to back up emergency measures under which it is essential that the materials and equipment be used to keep the project performance going. They would deal with legal subtleties later. The risks are greater if equipment, particularly construction machinery, rather than materials is taken. In any event, continued inclusion of such clauses

---

[35]For a case describing such an imbroglio, see *Leo Spear Constr. Co. v. Fid. Cas. Co.*, 446 F.2d 439 (2d Cir.1971).

[36]This may depend on the implementation of any flow-through clause, as discussed in Section 28.04.

[37]See *Indiana & Michigan Elec. Co. v. Terre Haute Indus., Inc.*, supra note 3 (punitive damages not awarded despite wrongful termination followed by takeover of contractor's equipment; independent tort required).

in construction contracts may indicate that they do have utility.[38]

H. Effect on Existing Claims for Delay

Termination of the contract is usually accompanied by claims by each party against the other. If the owner terminated, the contractor, as a rule, claims no only that the termination was wrongful but also that the owner had committed earlier breaches which caused it losses. The terminating owner also is likely to have claims for damages based on delay and improper workmanship. Sometimes the terminated party will contend that invoking the termination remedy was a waiver of any claims that existed prior to termination. This contention was rejected in *Armour & Co. v. Nard*.[39] After pointing to a provision stating that failure to exercise a right shall not be considered a waiver, the court noted that a contractual remedy for breach generally does not exclude other remedies unless the clause shows a clear intention of the party to do so. Rarely will a terminating party implicitly give up damage claims.

I. Disputed Terminations

Suppose the owner terminates the contract in accordance with the termination clause and orders the contractor to leave the site. The contractor contends the termination is not justified and demands arbitration under a provision similar to AIA Doc. A201, ¶ 4.5.1. This provides for arbitration for all claims, disputes, and other matters arising out of or relating to the contract documents or the breach thereof. Paragraph 4.5.3 requires a contractor to continue working while the dispute is being arbitrated. The following issues can arise:

1. Does a contractor's demand for arbitration give it the right to remain on the site pending the arbitrator's decision?
2. If the *contractor* wishes to terminate, may the owner demand arbitration and insist that the contractor continue work?

3. Is the disputed termination subject to arbitration?
4. If arbitration takes place, should the arbitrator order that work be resumed?

Requiring the contractor under ¶ 4.5.3 to continue working while the dispute is being arbitrated assumes performance disputes that do not involve termination.[40] Termination is a special dispute. The *specific* provisions of a termination clause should take precedence over the *general* arbitration clause. Paragraph 4.3.4 to which reference is made in ¶ 4.5.3 does not require continued performance during arbitration when agreed otherwise in writing. This can refer to the termination clause with its specific mechanism. Also, ¶ 4.5.3 can be for the owner's benefit. The owner can waive its right to continued performance during arbitration by ordering the contractor to leave the site. Finally, the owner's ownership and control of the site should empower the owner to remove the contractor. Whether the termination was proper can be resolved by any applicable arbitration clause.[41] The termination has not attacked the validity of the arbitration clause.[42]

Generally, arbitrators have remedial discretion. Any award to continue performance, at least in New York,[43] would be upheld. But termination is usually acrimonious and the last desperate step. Rarely will reinstatement be chosen as the remedy; this is another reason to force the contractor to leave the site even if termination is challenged. If termination were improper, a damage award would be adequate.

Suppose the *contractor* wishes to terminate and the owner seeks arbitration. The issue here is less

---

[38]In *Northway Decking & Sheet Metal Corp. v. Inland-Ryerson Constr. Products Co.*, 426 F.Supp. 417 (D.R.I.1977), the court appears to have enforced such a clause. It denied a terminated subcontractor an injunction permitting him to remove unique hanging scaffolding.
[39]463 F.2d 8 (8th Cir.1972).

[40]But *Keyway Contractors, Inc. v. Leek Corp., Inc.*, 189 Ga.App. 467, 376 S.E.2d 212 (1988), cert. denied Jan. 11, 1989, held that a subcontractor who left the site when a dispute arose for which arbitration was required had breached the subcontract.
[41]Many cases hold that arbitration survives termination. See *County of Middlesex v. Gevyn Constr. Corp.*, 450 F.2d 53 (1st Cir.1971); *State v. Lombard Co.*, 106 Ill.App.3d 307, 436 N.E.2d 566 (1982); *Willis-Knighton Medical Center v. Southern Builders, Inc.*, 392 So.2d 505 (La.App.1980). But see *G&N Constr. Co. v. Kirpatovsky*, 181 So.2d 664 (Fla.Dist.Ct.App.1966).
[42]*Riess v. Murchison*, 384 F.2d 727 (9th Cir.1976).
[43]*Grayson-Robinson Stores, Inc. v. Iris Constr. Corp.*, 8 N.Y.2d 133, 168 N.E.2d 377, 202 N.Y.S.2d 303 (1960). See Section 30.12.

clear. The provision for removing the contractor from the site can be considered for the benefit of the owner. This interpretation would allow the owner to order that the contractor continue working. It is more likely, however, that the conclusion will not depend on which party terminates.

J. Public Contracts and Constitutional Protection

The increasing use of constitutional doctrines has been noted earlier.[44] In the context of terminating a public construction contract, the contractor may contend that the contract cannot be terminated without extending it due process of law, usually taken to mean some basic procedural rights. In *Riblet Tramway Co., Inc. v. Stickney,*[45] the court held that the contract could be terminated despite the lack of a prior hearing. The court recognized that the contractor's interest in the contract could not be taken away without cause. But what process would be appropriate? Using the standard of fundamental fairness, the court held that the contractor could sue for breach. This was sufficient to satisfy the contractor's due process rights.

SECTION 34.04 Termination by Law

A. Material Breach

If the contract does not expressly create a power to terminate, termination is allowed in the event of a material breach. Even with a contract clause dealing with termination, many of the factors that determine materiality can be influential when such clauses are interpreted. Also, as indicated, the termination clause may not be the exclusive source for determining when a party has the power to terminate.

Rather than establishing fixed rules, such as the importance of the clause breached, the law examines all the facts and circumstances surrounding the breach to determine whether it would be fair to permit termination. For example, the Restatement (Second) of Contracts articulates factors in

§ 241 that are significant in determining whether a particular breach is material.[46] They are

(a) the extent to which the injured party will be deprived of the benefit which he reasonably expected;

(b) the extent to which the injured party can be adequately compensated for the part of that benefit of which he will be deprived;

(c) the extent to which the party failing to perform or to offer to perform will suffer forfeiture;

(d) the likelihood that the party failing to perform or to offer to perform will cure his failure, taking account of all the circumstances including any reasonable assurances;

(e) the extent to which the behavior of the party failing to perform or to offer to perform comports with standards of good faith and fair dealing.

Looking first at breaches by the contractor, subparagraph (a) seeks to determine the importance of the deviation and the likelihood of future nonperformance. Subparagraph (b) examines whether the owner can be easily compensated for nonperformance. This would depend on whether the defect was easily correctable or whether compensation for an uncorrected defect would be easy to measure. In construction contracts, the latter can be difficult. Subparagraph (c) looks for uncompensated losses that the contractor would suffer if termination occurred. If it had ordered materials that had not yet been used and were not usable in other projects, this would militate against termination. Subparagraph (d) examines the likelihood that the contractor will be able to cure defects, while subparagraph (e) examines the reason for the breach. An example of the latter—a breach by a subcontractor that the prime contractor could not have reasonably prevented—might militate against materiality.

An owner breach would probably be nonpayment of money, failure to furnish the site, or noncooperation. Subparagraph (a) would look at the extent of the breach and the likelihood of future performance, while subparagraph (b) would look at whether the breach was easily compensable,

---

[44]See Sections 18.04(E) and 22.02(E).
[45]523 A.2d 107 (N.H.1987).

[46]Section 242 dealing with delay expands the list to include the extent to which the delay will hinder the injured party in making substitute arrangements and the extent to which the agreement provides for performance without delay.

such as interest for a breach consisting of not making a progress payment.

Subparagraph (c) examines the harm termination would cause the owner, such as lost loan commitments, liability to prospective tenants, or other lost business opportunities. Subparagraph (d) seeks to determine whether the breaches are likely to be cured. Finally, subparagraph (e) examines the reasons for nonperformance. Financial reverses or a steep rise in interest rates makes it less likely that the contractor can terminate for failure to pay. If, on the other hand, there was "bad blood" between the design professional and the contractor and the latter seized on the breach by the owner to injure the design professional, termination would be less likely. Likewise, if it appears that the contractor sought an excuse to end the contract, the breach would less likely be material.

Oak Ridge Construction Co. v. Tolley[47] involved a dispute between Tolley, the owner of land, and Oak Ridge, the contractor who had agreed to build a house and dig a well. Tolley disputed and refused to pay the invoice for extra drilling and casing. Oak Ridge stated that Tolley's failure to pay the invoice was a breach entitling Oak Ridge to terminate under the contract's termination provisions. It also notified Tolley that it was stopping work and would arbitrate the dispute. Tolley refused to arbitrate at that time, and the work proceeded no further.

After concluding that Tolley had not committed a breach by anticipatory repudiation (see subsection (B)), the court found that the work stoppage by Oak Ridge was a breach. It applied the sections of the Restatement noted earlier to determine whether the breach was material. It stated:

> We must therefore consider whether that breach was material in light of the factors enumerated above. The Tolleys, as the injured party, were deprived of the expected benefits of their contract (*i.e.*, receiving a completed home) by Oak Ridge's work stoppage, and could be adequately compensated for that deprivation by an award of damages. Furthermore, Oak Ridge's letter of September 7 gave no indication that the company would cure its failure to perform, and, in fact, the record indicates that Oak Ridge never did so act. Of

course, a finding of material breach will result in forfeiture for Oak Ridge (*i.e.*, the Tolleys will be discharged from all liability on the contract), however, Oak Ridge will be entitled to restitution for any benefit conferred upon the Tolleys by part performance or reliance (*i.e.*, the cost of digging the well) in excess of the loss Oak Ridge caused by its own breach. *See* Restatement, *supra*, §§241 comment d, 374(1). We note that the record contains no evidence concerning whether Oak Ridge's conduct "comports with standards of good faith and fair dealing." Under these circumstances, we find that Oak Ridge's breach constituted a material failure of performance thereby discharging the Tolleys from all liability under the contract.[48]

Many factors are relevant. Perhaps the most relevant factors are the particular nature of the nonperformance, the likelihood of future breaches, and the possibility of forfeiture. Some courts may look at the good faith not simply of the party failing to perform but of the party who seeks to exercise the power to terminate.[49] For example, Tennessee held that in the absence of a clause permitting the prime contractor to take over the subcontractor's work (the court noting that such clauses usually require a notice and an opportunity to cure), the prime contractor could not terminate the contract without notifying the subcontractor of its intention to do so and giving the subcontractor an opportunity to cure.[50]

Nonpayment—perhaps the most frequently asserted justification for termination—was discussed in Section 22.02(L). As to other breaches, one court held that a prime contract terminated when the owner did not make an equitable adjustment when different underground conditions were discovered and the contractor was in financial trouble.[51] A prime contractor's failure to have the site ready for the flooring subcontractor was a material breach.[52] Hindering the subcontractor's operation was a material breach when coupled with the prime con-

[47]351 Pa.Super. 32, 504 A.2d 1343 (1985).

[48]504 A.2d at 1348.
[49]See Sections 5.11(A) and 34.01.
[50]*McClain v. Kimbrough Constr. Co.,* supra note 14. See also Restatement (Second) Contracts. § 242 (1981).
[51]*Metro. Sewerage Comm'n. v. R. W. Constr., Inc.,* 72 Wis.2d 365, 241 N.W.2d 371 (1976).
[52]*Great Lakes Constr. Co. v. Republic Creosoting Co.,* 139 F.2d 456 (8th Cir.1943).

tractor's having stopped payment on a check that the subcontractor was about to negotiate.[53] Another court held that the collapse of a fallout shelter was sufficient grounds for the owner to terminate the contractor's performance and recover a down payment.[54] These obviously incomplete illustrations are not designed to indicate that breaches of the type described will always be considered material. As emphasized in this subsection, the determination of whether a breach is material requires a careful evaluation of the facts and circumstances surrounding the breach as well as the effect of termination.

B. Future Breach: Prospective Inability and Breach by Anticipatory Repudiation

Subsection (A) dealt with breaches that have occurred. This subsection deals with breaches that may occur in the future. It may appear that one party may not be able to perform when the time for performance arrives, or one party may state that it will not perform when the time for performance arrives. The contractor may discharge some of its employees, or a number of employees may quit. The contractor may cancel orders for supplies, or its suppliers may indicate that they will not perform at the time for performance. In such cases, the owner may realize that the contractor will be unable to perform.

Such inability to perform, whether it is on the part of the owner or the contractor or subcontractors, is likely to be a breach. Each party owes the other party the duty to *appear* to be ready to perform when the time for performance arises. Even if no such promise is implied, under certain circumstances, events may permit one party to suspend its performance or terminate the contract unless the other party who appears unable to perform can give assurance or security that when the time comes for performance, it will perform.[55] Insolvency as a basis for insecurity may be affected in a construction dispute by AIA Doc. A201, ¶ 2.2.1, which gives the contractor the power to ask for ev-

idence of financial arrangements from the owner. Failure to exercise this power may preclude the contractor's suspending its obligation to perform in the event that the owner becomes insolvent. Failure to make payments as indicated in Section 22.02(L) gives the contractor a right under A201, ¶ 9.7.1, to suspend performance.

Prospective inability deals with probabilities. In the examples given, the question is whether the owner must wait to see whether actual performance or defective performance will occur or whether it can demand assurance and, in the absence of this assurance, legally terminate any obligation to use the contractor.

Contracts often deal with termination rights. In the absence of such provisions, it will take a strong showing on the part of the owner to terminate the contractor's performance on the grounds that the contractor may not be able to perform in the future. In many cases, a combination of present and prospective nonperformance exists. If present nonperformance, such as the installation of defective materials or poor workmanship exists, the likelihood that this will continue will be a strong factor in influencing the court to allow the owner to terminate the contractor's performance. Pure prospective inability without present breach is likely to be held to be insufficient grounds unless the probabilities are very strong or unless a contract provision giving the owner this right exists.

A breach by anticipatory repudiation occurs when one party indicates to the other that it cannot or will not perform.[56] Each party is entitled to reasonable assurance that the other will perform in accordance with the contract. If one of the parties indicates that it will not or cannot perform, the other party loses this assurance. Should the latter be required to wait and see whether the threat or the indication of inability to perform will come to fruition?

The rights of a party to terminate a contract because of the other party's repudiation have ex-

---

[53]*Citizens Nat'l. Bank of Orlando v. Vitt*, 367 F.2d 541 (5th Cir.1966).
[54]*Economy Swimming Pool Co. v. Freeling*, 236 Ark. 888, 370 S.W.2d 438 (1963).
[55]Restatement (Second) of Contracts §§ 251, 252 (1981).

[56]As to the existence of a repudiation, compare *Oak Ridge Constr. Co. v. Tolley*, supra note 47 (owner questioning invoice charges and saying they were in dispute, not a repudiation), with *Twenty-Four Collection, Inc. M. Weinbaum Constr., Inc.*, 427 So.2d 1110 (Fla.Dist.Ct.App.1983) (contractor's mailgram threatening to shut down project unless a certain clause was eliminated constituted anticipatory repudiation).

panded. A feeling of assurance is important, as this is one purpose for making a contract. Sometimes a party who indicates it will not perform is jockeying for position. The owner may be trying to pay less or obtain more than the called for performance. If either party does repudiate, the other party will be given the right to terminate the obligation. The party to whom the repudiation is made need not terminate the obligation immediately. It may state that it intends to hold the other party to the contract. If the repudiator relies on the statement that the contract will be continued, the nonrepudiating party will lose its right to terminate based on the earlier repudiation.

The right to continue one's performance despite repudiation by the other party is qualified by the rule against enhancing damages. One party cannot recover damages caused by the other party's breach when those damages could have been avoided by taking reasonable steps to cut down or eliminate the loss. For example, if the contractor repudiates the contract, the owner cannot recover for those damages caused by breach that could have been avoided by hiring a replacement contractor.

Suppose the contractor states unequivocally without justification that it will leave the job in three days. It might be reasonable for the owner to insist that it will hold the contractor to the contract for a short period, such as until the date of the walkout or even for a few days after the walkout. But when it is clear that the contractor will not return to the project, the owner should take reasonable steps to replace the contractor if the owner wishes to continue the project. A replacement should be obtained when the repudiating contractor has committed its workers and machinery to another project and it appears that it has neither the willingness nor the capacity to return to the job. Any damages that could have been avoided by hiring a replacement will not be assessable against the repudiating contractor.

Repudiation may accompany present or prospective breach. The greater the scope of any present or prospective breach, the greater the likelihood that the court will release the innocent party by reason of both the present breach and the repudiation. Even a small breach, coupled with a repudiation, may be enough to terminate the innocent party's obligation to perform further.

C. Bankruptcy

Bankruptcy can affect contracts. Two types of bankruptcy are important for the purposes of this chapter. The first is liquidation under Chapter 7 and is known as straight bankruptcy. It has two principal objectives. First, the bankrupt can wipe the slate clean, or mostly clean, by discharging most of the bankrupt's debts. Second, bankruptcy should provide a fair and efficient liquidation of the bankrupt's estate. The trustee for the bankrupt takes over the bankrupt's estate, collects any money owed the bankrupt, compels repayment of any preferential payments made to creditors,[57] turns over specific property in the hands of the bankrupt to those who have security interests in them, and, after paying expenses of administering the estate, distributes any amount remaining pro rata to unsecured creditors. For example, if the amount of unsecured debts is one million dollars and the amount remaining is $100,000, each unsecured creditor receives ten cents for each dollar of unsecured debt. Payments to unsecured creditors, if made at all, are usually only a small fraction of the debts.

The second type is a petition for reorganization under Chapter 11. Such a petition protects the petitioner from creditor claims while it seeks to reorganize in order to put its business on a more sound financial basis. The petitioner submits a plan that involves a reorganization of the interests of owners and creditors, a rescheduling of debts, and, one hopes, an infusion of cash and credit for judicial approval.

Two recent developments dealt with the effect of bankruptcy on existing contracts. Bankruptcy law gives the trustee an election to assume (continue performance) or reject (refuse further performance) existing contracts. The U.S. Congress precluded any contract clause from automatically barring this election.[58] The trustee can assume any contract if it has the capacity to continue performance or pay compensation if it does not.

It was common for contracts to permit automatic termination if a party filed for bankruptcy. In con-

---

[57]Broadly speaking, a preference favors one creditor over the others without any legitimate business reason. For details, see 11 U.S.C.A. § 547.
[58]Id. at 365(e).

struction contracts, the owner was usually given the power to terminate if the contractor filed for bankruptcy or had serious financial problems. Despite the change in the law, some contracts still contain such provisions, either because of ignorance as to their ineffectiveness or simply because of inertia. In any event, assumptions of construction contracts are rare. The trustee seldom has the resources to continue performance.

More important and more controversial were well-publicized Chapter 11 petitions by troubled airlines seeking to rid themselves of onerous labor contracts or asbestos manufacturers facing thousands of large tort claims by those who claimed that asbestos gave them cancer.

In *N.L.R.B. v. Bildisco and Bildisco*,[59] the contractor filed a Chapter 11 petition and immediately stopped paying pension, health plan, and wage increases agreed on in a recent collective-bargaining agreement with the union representing its workers. The U.S. Supreme Court held that this would not be an unfair labor practice if the contractor can show that the contract burdened its financial affairs and the equities balanced in favor of rejection.

Congress responded to complaints by trade unions by creating difficult procedural requirements before a collective-bargaining agreement could be set aside under Chapter 11.[60]

SECTION 34.05 Restitution Where a Contract is Terminated

When a contract is terminated, each party, as a rule, has conferred benefit on the other. The owner has made progress payments, and the contractor has performed the work. What effect does termination have on the right of either to recover the value of what it has conferred on the other?

If termination has been made for the convenience of the owner, the clause permitting termination usually provides a formula for compensating the contractor for the work performed. Terminations for default made by the owner usually involve claims by the latter that exceed the value of any unpaid work performed by the contractor. In the event they do not, the contractor, though in de-

fault, in most jurisdictions would be entitled to recover the net benefit conferred on the owner.[61] If termination was made by the contractor based on an owner default, the contractor can use a restitutionary recovery that permits it in most cases to recover the reasonable value of the services performed.[62] If termination has been accomplished by mutual consent, the agreement will usually deal with compensation for benefits conferred. If it does not, either party is entitled to recover the net benefit it has conferred on the other. For example, if the work performed has a value of $100,000 and the contractor has been paid $90,000, the latter should recover $10,000. Conversely, if the figures were reversed, the owner would be entitled to that amount.

SECTION 34.06 Keeping Subcontractors After Termination

An owner (or its financial backers) who exercises its power to terminate the prime contract may wish to employ a successor to continue the project. Alternatively, the surety on a performance bond may exercise its option to complete the project with a successor.

Default by the prime contractor is likely to result in failure to pay the subcontractors. This would give subcontractors the power to terminate their obligations under their subcontracts. The owner or surety may wish to take an assignment of *some* subcontracts (those of subcontractors who are performing well and are not owed an excessive amount of money) without having to renegotiate with them. Renegotiation always causes delay and may result in a higher contract price. Yet most subcontracts will contain nonassignment clauses that preclude assignment of any rights under the contract without the consent of the party whose performance is being assigned—in this case, the subcontractors.

To avoid this, it has become common for the prime contract to include an assignment conditioned on default by the prime to the owner of those subcontracts that the owner wishes to take

---

[59]465 U.S. 513 (1984).
[60]11 U.S.C.A. § 1113.

[61]See Section 22.06(D).
[62]See Section 27.02(E).

over, with a promise by the prime contractor to obtain consent to these assignments by the subcontractors.[63]

[63]AIA Doc. A201, ¶ 5.4. For a comment on this and other approaches, see J. SWEET, SWEET ON CONSTRUCTION INDUSTRY CONTRACTS, § 18.11. (2d.ed. 1992).

PROBLEMS

1. On February 1, 1993, O and C entered into a contract under which C agreed to construct a residence in accordance with designated plans and specifications in exchange for O's promise to pay $300,000. The house was to be built in Madison, Wisconsin. The site was to be furnished by the contractor no later than April 1, 1993, and the work was to be completed by November 1 of that year. Because of difficulty in obtaining title to the site, the contractor was not given access to the site until May 15 of that year. When O rejected a demand for additional compensation, C terminated its obligation to build the house. It contends that the delay will require that the work be done during the winter and this will substantially increase the cost of performance. Did C have legal ground to terminate? What further facts would be useful in making this decision?

Suppose C went ahead with the work in May when it was given site access. Materials and labor shortages delayed the work in the summer. By September, it appeared C would not be able to finish until the following February. The delay would necessitate a substantial period of work during the winter. On November 15, C terminated the contract, claiming that the delay in furnishing the site justified termination. Is this contention correct? What further facts would be important in answering this question?

2. The contract between a subcontractor and a prime contractor provided that if the owner terminates the prime contract "for any cause whatsoever at any time," the subcontract will also be terminated and the prime contractor's liability will be limited to payment for work done. The subcontractor partially performed but was not paid in accordance with the subcontract. Yet the subcontractor continued performing until it heard that the owner had terminated the prime contract. It then stopped performing, relying not on nonpayment but rather on the owner's termination having been caused by the prime contractor's default.

The owner's termination was based on its contractual power to terminate for its convenience. The owner terminated because it felt that the project would not be needed.

a. Can the subcontractor recover from the prime contractor for work it has performed?

b. Can the subcontractor recover its lost profits from the prime contractor?

c. Would the subcontractor have been able to terminate for the prime contractor's failure to make payments?

3. O hired C to erect a steel building. C hired S to pour the concrete. S poured the concrete, but prior to the seven days that the concrete should have been allowed to cure in order to tolerate any heavy weight, S drove a gravel truck onto the concrete floor. The owner was present and ordered the truck off the concrete, asserting that it would damage the floor.

C met with O and promised to take reasonable steps to repair any damage caused by the truck and requested permission to continue with the work. However, O demanded that 4,000 square feet of concrete be removed and replaced before it would allow C to continue work.

O met with S and obtained a one-year guarantee from S for all materials and workmanship on the floor. However, O continued to insist that the concrete be replaced. C contended that no evidence of damage to the floor existed, that it could, by doubling the size of its work crew, complete its work by the contract deadline without placing any weight on the floor, and that any necessary repairs could be undertaken after the building was completed.

O was not satisfied and ordered C to leave the site. A subsequent investigation revealed that the concrete was not damaged and that it exceeded requirements as to strength. O obtained another contractor, who completed the project.

O has brought a claim against C for the excess costs it incurred in hiring a replacement contractor. C defended against that claim, asserting that it was justified in refusing to continue and asserted a counterclaim against O for its lost profits on the contract. How would you resolve these claims?

A

Standard Form of Agreement Between Owner and Architect

1987 Edition

T H E   A M E R I C A N   I N S T I T U T E   O F   A R C H I T E C T S

AIA Document B141

Standard Form of Agreement Between Owner and Architect

1987 EDITION

THIS DOCUMENT HAS IMPORTANT LEGAL CONSEQUENCES; CONSULTATION WITH AN ATTORNEY IS ENCOURAGED WITH RESPECT TO ITS COMPLETION OR MODIFICATION.

AGREEMENT

made as of the   day of   in the year of
Nineteen Hundred and

BETWEEN the Owner:
(Name and address)

and the Architect:
(Name and address)

For the following Project:
(Include detailed description of Project, location, address and scope.)

The Owner and Architect agree as set forth below.

> ## TERMS AND CONDITIONS OF AGREEMENT BETWEEN OWNER AND ARCHITECT

ARTICLE 1
ARCHITECT'S RESPONSIBILITIES

1.1 ARCHITECT'S SERVICES

1.1.1 The Architect's services consist of those services performed by the Architect, Architect's employees and Architect's consultants as enumerated in Articles 2 and 3 of this Agreement and any other services included in Article 12.

1.1.2 The Architect's services shall be performed as expeditiously as is consistent with professional skill and care and the orderly progress of the Work. Upon request of the Owner, the Architect shall submit for the Owner's approval a schedule for the performance of the Architect's services which may be adjusted as the Project proceeds, and shall include allowances for periods of time required for the Owner's review and for approval of submissions by authorities having jurisdiction over the Project. Time limits established by this schedule approved by the Owner shall not, except for reasonable cause, be exceeded by the Architect or Owner.

1.1.3 The services covered by this Agreement are subject to the time limitations contained in Subparagraph 11.5.1.

ARTICLE 2
SCOPE OF ARCHITECT'S BASIC SERVICES

2.1 DEFINITION

2.1.1 The Architect's Basic Services consist of those described in Paragraphs 2.2 through 2.6 and any other services identified in Article 12 as part of Basic Services, and include normal structural, mechanical and electrical engineering services.

2.2 SCHEMATIC DESIGN PHASE

2.2.1 The Architect shall review the program furnished by the Owner to ascertain the requirements of the Project and shall arrive at a mutual understanding of such requirements with the Owner.

2.2.2 The Architect shall provide a preliminary evaluation of the Owner's program, schedule and construction budget requirements, each in terms of the other, subject to the limitations set forth in Subparagraph 5.2.1.

2.2.3 The Architect shall review with the Owner alternative approaches to design and construction of the Project.

2.2.4 Based on the mutually agreed-upon program, schedule and construction budget requirements, the Architect shall prepare, for approval by the Owner, Schematic Design Documents consisting of drawings and other documents illustrating the scale and relationship of Project components.

2.2.5 The Architect shall submit to the Owner a preliminary estimate of Construction Cost based on current area, volume or other unit costs.

2.3 DESIGN DEVELOPMENT PHASE

2.3.1 Based on the approved Schematic Design Documents and any adjustments authorized by the Owner in the program,

schedule or construction budget, the Architect shall prepare, for approval by the Owner, Design Development Documents consisting of drawings and other documents to fix and describe the size and character of the Project as to architectural, structural, mechanical and electrical systems, materials and such other elements as may be appropriate.

2.3.2 The Architect shall advise the Owner of any adjustments to the preliminary estimate of Construction Cost.

2.4 CONSTRUCTION DOCUMENTS PHASE

2.4.1 Based on the approved Design Development Documents and any further adjustments in the scope or quality of the Project or in the construction budget authorized by the Owner, the Architect shall prepare, for approval by the Owner, Construction Documents consisting of Drawings and Specifications setting forth in detail the requirements for the construction of the Project.

2.4.2 The Architect shall assist the Owner in the preparation of the necessary bidding information, bidding forms, the Conditions of the Contract, and the form of Agreement between the Owner and Contractor.

2.4.3 The Architect shall advise the Owner of any adjustments to previous preliminary estimates of Construction Cost indicated by changes in requirements or general market conditions.

2.4.4 The Architect shall assist the Owner in connection with the Owner's responsibility for filing documents required for the approval of governmental authorities having jurisdiction over the Project.

2.5 BIDDING OR NEGOTIATION PHASE

2.5.1 The Architect, following the Owner's approval of the Construction Documents and of the latest preliminary estimate of Construction Cost, shall assist the Owner in obtaining bids or negotiated proposals and assist in awarding and preparing contracts for construction.

2.6 CONSTRUCTION PHASE—ADMINISTRATION OF THE CONSTRUCTION CONTRACT

2.6.1 The Architect's responsibility to provide Basic Services for the Construction Phase under this Agreement commences with the award of the Contract for Construction and terminates at the earlier of the issuance to the Owner of the final Certificate for Payment or 60 days after the date of Substantial Completion of the Work, unless extended under the terms of Subparagraph 10.3.3.

2.6.2 The Architect shall provide administration of the Contract for Construction as set forth below and in the edition of AIA Document A201, General Conditions of the Contract for Construction, current as of the date of this Agreement, unless otherwise provided in this Agreement.

2.6.3 Duties, responsibilities and limitations of authority of the Architect shall not be restricted, modified or extended without written agreement of the Owner and Architect with consent of the Contractor, which consent shall not be unreasonably withheld.

2.6.4 The Architect shall be a representative of and shall advise and consult with the Owner (1) during construction until final payment to the Contractor is due, and (2) as an Additional Service at the Owner's direction from time to time during the correction period described in the Contract for Construction. The Architect shall have authority to act on behalf of the Owner only to the extent provided in this Agreement unless otherwise modified by written instrument.

2.6.5 The Architect shall visit the site at intervals appropriate to the stage of construction or as otherwise agreed by the Owner and Architect in writing to become generally familiar with the progress and quality of the Work completed and to determine in general if the Work is being performed in a manner indicating that the Work when completed will be in accordance with the Contract Documents. However, the Architect shall not be required to make exhaustive or continuous on-site inspections to check the quality or quantity of the Work. On the basis of on-site observations as an architect, the Architect shall keep the Owner informed of the progress and quality of the Work, and shall endeavor to guard the Owner against defects and deficiencies in the Work. *(More extensive site representation may be agreed to as an Additional Service, as described in Paragraph 3.2.)*

2.6.6 The Architect shall not have control over or charge of and shall not be responsible for construction means, methods, techniques, sequences or procedures, or for safety precautions and programs in connection with the Work, since these are solely the Contractor's responsibility under the Contract for Construction. The Architect shall not be responsible for the Contractor's schedules or failure to carry out the Work in accordance with the Contract Documents. The Architect shall not have control over or charge of acts or omissions of the Contractor, Subcontractors, or their agents or employees, or of any other persons performing portions of the Work.

2.6.7 The Architect shall at all times have access to the Work wherever it is in preparation or progress.

2.6.8 Except as may otherwise be provided in the Contract Documents or when direct communications have been specially authorized, the Owner and Contractor shall communicate through the Architect. Communications by and with the Architect's consultants shall be through the Architect.

2.6.9 Based on the Architect's observations and evaluations of the Contractor's Applications for Payment, the Architect shall review and certify the amounts due the Contractor.

2.6.10 The Architect's certification for payment shall constitute a representation to the Owner, based on the Architect's observations at the site as provided in Subparagraph 2.6.5 and on the data comprising the Contractor's Application for Payment, that the Work has progressed to the point indicated and that, to the best of the Architect's knowledge, information and belief, quality of the Work is in accordance with the Contract Documents. The foregoing representations are subject to an evaluation of the Work for conformance with the Contract Documents upon Substantial Completion, to results of subsequent tests and inspections, to minor deviations from the Contract Documents correctable prior to completion and to specific qualifications expressed by the Architect. The issuance of a Certificate for Payment shall further constitute a representation that the Contractor is entitled to payment in the amount certified. However, the issuance of a Certificate for Payment shall not be a representation that the Architect has (1) made exhaustive or continuous on-site inspections to check the quality or

quantity of the Work, (2) reviewed construction means, methods, techniques, sequences or procedures, (3) reviewed copies of requisitions received from Subcontractors and material suppliers and other data requested by the Owner to substantiate the Contractor's right to payment or (4) ascertained how or for what purpose the Contractor has used money previously paid on account of the Contract Sum.

2.6.11 The Architect shall have authority to reject Work which does not conform to the Contract Documents. Whenever the Architect considers it necessary or advisable for implementation of the intent of the Contract Documents, the Architect will have authority to require additional inspection or testing of the Work in accordance with the provisions of the Contract Documents, whether or not such Work is fabricated, installed or completed. However, neither this authority of the Architect nor a decision made in good faith either to exercise or not to exercise such authority shall give rise to a duty or responsibility of the Architect to the Contractor, Subcontractors, material and equipment suppliers, their agents or employees or other persons performing portions of the Work.

2.6.12 The Architect shall review and approve or take other appropriate action upon Contractor's submittals such as Shop Drawings, Product Data and Samples, but only for the limited purpose of checking for conformance with information given and the design concept expressed in the Contract Documents. The Architect's action shall be taken with such reasonable promptness as to cause no delay in the Work or in the construction of the Owner or of separate contractors, while allowing sufficient time in the Architect's professional judgment to permit adequate review. Review of such submittals is not conducted for the purpose of determining the accuracy and completeness of other details such as dimensions and quantities or for substantiating instructions for installation or performance of equipment or systems designed by the Contractor, all of which remain the responsibility of the Contractor to the extent required by the Contract Documents. The Architect's review shall not constitute approval of safety precautions or, unless otherwise specifically stated by the Architect, of construction means, methods, techniques, sequences or procedures. The Architect's approval of a specific item shall not indicate approval of an assembly of which the item is a component. When professional certification of performance characteristics of materials, systems or equipment is required by the Contract Documents, the Architect shall be entitled to rely upon such certification to establish that the materials, systems or equipment will meet the performance criteria required by the Contract Documents.

2.6.13 The Architect shall prepare Change Orders and Construction Change Directives, with supporting documentation and data if deemed necessary by the Architect as provided in Subparagraphs 3.1.1 and 3.3.3, for the Owner's approval and execution in accordance with the Contract Documents, and may authorize minor changes in the Work not involving an adjustment in the Contract Sum or an extension of the Contract Time which are not inconsistent with the intent of the Contract Documents.

2.6.14 The Architect shall conduct inspections to determine the date or dates of Substantial Completion and the date of final completion, shall receive and forward to the Owner for the Owner's review and records written warranties and related documents required by the Contract Documents and assembled by the Contractor, and shall issue a final Certificate for Payment upon compliance with the requirements of the Contract Documents.

AIA DOCUMENT B141 • OWNER-ARCHITECT AGREEMENT • FOURTEENTH EDITION • AIA® • ©1987
THE AMERICAN INSTITUTE OF ARCHITECTS, 1735 NEW YORK AVENUE, N.W., WASHINGTON, D.C. 20006

2.6.15 The Architect shall interpret and decide matters concerning performance of the Owner and Contractor under the requirements of the Contract Documents on written request of either the Owner or Contractor. The Architect's response to such requests shall be made with reasonable promptness and within any time limits agreed upon.

2.6.16 Interpretations and decisions of the Architect shall be consistent with the intent of and reasonably inferable from the Contract Documents and shall be in writing or in the form of drawings. When making such interpretations and initial decisions, the Architect shall endeavor to secure faithful performance by both Owner and Contractor, shall not show partiality to either, and shall not be liable for results of interpretations or decisions so rendered in good faith.

2.6.17 The Architect's decisions on matters relating to aesthetic effect shall be final if consistent with the intent expressed in the Contract Documents.

2.6.18 The Architect shall render written decisions within a reasonable time on all claims, disputes or other matters in question between the Owner and Contractor relating to the execution or progress of the Work as provided in the Contract Documents.

2.6.19 The Architect's decisions on claims, disputes or other matters, including those in question between the Owner and Contractor, except for those relating to aesthetic effect as provided in Subparagraph 2.6.17, shall be subject to arbitration as provided in this Agreement and in the Contract Documents.

ARTICLE 3
ADDITIONAL SERVICES

3.1 GENERAL

3.1.1 The services described in this Article 3 are not included in Basic Services unless so identified in Article 12, and they shall be paid for by the Owner as provided in this Agreement, in addition to the compensation for Basic Services. The services described under Paragraphs 3.2 and 3.4 shall only be provided if authorized or confirmed in writing by the Owner. If services described under Contingent Additional Services in Paragraph 3.3 are required due to circumstances beyond the Architect's control, the Architect shall notify the Owner prior to commencing such services. If the Owner deems that such services described under Paragraph 3.3 are not required, the Owner shall give prompt written notice to the Architect. If the Owner indicates in writing that all or part of such Contingent Additional Services are not required, the Architect shall have no obligation to provide those services.

3.2 PROJECT REPRESENTATION BEYOND BASIC SERVICES

3.2.1 If more extensive representation at the site than is described in Subparagraph 2.6.5 is required, the Architect shall provide one or more Project Representatives to assist in carrying out such additional on-site responsibilities.

3.2.2 Project Representatives shall be selected, employed and directed by the Architect, and the Architect shall be compensated therefor as agreed by the Owner and Architect. The duties, responsibilities and limitations of authority of Project Representatives shall be as described in the edition of AIA Document B352 current as of the date of this Agreement, unless otherwise agreed.

3.2.3 Through the observations by such Project Representatives, the Architect shall endeavor to provide further protection for the Owner against defects and deficiencies in the Work, but the furnishing of such project representation shall not modify the rights, responsibilities or obligations of the Architect as described elsewhere in this Agreement.

3.3 CONTINGENT ADDITIONAL SERVICES

3.3.1 Making revisions in Drawings, Specifications or other documents when such revisions are:

.1 inconsistent with approvals or instructions previously given by the Owner, including revisions made necessary by adjustments in the Owner's program or Project budget;

.2 required by the enactment or revision of codes, laws or regulations subsequent to the preparation of such documents; or

.3 due to changes required as a result of the Owner's failure to render decisions in a timely manner.

3.3.2 Providing services required because of significant changes in the Project including, but not limited to, size, quality, complexity, the Owner's schedule, or the method of bidding or negotiating and contracting for construction, except for services required under Subparagraph 5.2.5.

3.3.3 Preparing Drawings, Specifications and other documentation and supporting data, evaluating Contractor's proposals, and providing other services in connection with Change Orders and Construction Change Directives.

3.3.4 Providing services in connection with evaluating substitutions proposed by the Contractor and making subsequent revisions to Drawings, Specifications and other documentation resulting therefrom.

3.3.5 Providing consultation concerning replacement of Work damaged by fire or other cause during construction, and furnishing services required in connection with the replacement of such Work.

3.3.6 Providing services made necessary by the default of the Contractor, by major defects or deficiencies in the Work of the Contractor, or by failure of performance of either the Owner or Contractor under the Contract for Construction.

3.3.7 Providing services in evaluating an extensive number of claims submitted by the Contractor or others in connection with the Work.

3.3.8 Providing services in connection with a public hearing, arbitration proceeding or legal proceeding except where the Architect is party thereto.

3.3.9 Preparing documents for alternate, separate or sequential bids or providing services in connection with bidding, negotiation or construction prior to the completion of the Construction Documents Phase.

3.4 OPTIONAL ADDITIONAL SERVICES

3.4.1 Providing analyses of the Owner's needs and programming the requirements of the Project.

3.4.2 Providing financial feasibility or other special studies.

3.4.3 Providing planning surveys, site evaluations or comparative studies of prospective sites.

3.4.4 Providing special surveys, environmental studies and submissions required for approvals of governmental authorities or others having jurisdiction over the Project.

3.4.5 Providing services relative to future facilities, systems and equipment.

3.4.6 Providing services to investigate existing conditions or facilities or to make measured drawings thereof.

3.4.7 Providing services to verify the accuracy of drawings or other information furnished by the Owner.

3.4.8 Providing coordination of construction performed by separate contractors or by the Owner's own forces and coordination of services required in connection with construction performed and equipment supplied by the Owner.

3.4.9 Providing services in connection with the work of a construction manager or separate consultants retained by the Owner.

3.4.10 Providing detailed estimates of Construction Cost.

3.4.11 Providing detailed quantity surveys or inventories of material, equipment and labor.

3.4.12 Providing analyses of owning and operating costs.

3.4.13 Providing interior design and other similar services required for or in connection with the selection, procurement or installation of furniture, furnishings and related equipment.

3.4.14 Providing services for planning tenant or rental space.

3.4.15 Making investigations, inventories of materials or equipment, or valuations and detailed appraisals of existing facilities.

3.4.16 Preparing a set of reproducible record drawings showing significant changes in the Work made during construction based on marked-up prints, drawings and other data furnished by the Contractor to the Architect.

3.4.17 Providing assistance in the utilization of equipment or systems such as testing, adjusting and balancing, preparation of operation and maintenance manuals, training personnel for operation and maintenance, and consultation during operation.

3.4.18 Providing services after issuance to the Owner of the final Certificate for Payment, or in the absence of a final Certificate for Payment, more than 60 days after the date of Substantial Completion of the Work.

3.4.19 Providing services of consultants for other than architectural, structural, mechanical and electrical engineering portions of the Project provided as a part of Basic Services.

3.4.20 Providing any other services not otherwise included in this Agreement or not customarily furnished in accordance with generally accepted architectural practice.

ARTICLE 4
OWNER'S RESPONSIBILITIES

4.1 The Owner shall provide full information regarding requirements for the Project, including a program which shall set forth the Owner's objectives, schedule, constraints and criteria, including space requirements and relationships, flexibility, expandability, special equipment, systems and site requirements.

4.2 The Owner shall establish and update an overall budget for the Project, including the Construction Cost, the Owner's other costs and reasonable contingencies related to all of these costs.

4.3 If requested by the Architect, the Owner shall furnish evidence that financial arrangements have been made to fulfill the Owner's obligations under this Agreement.

4.4 The Owner shall designate a representative authorized to act on the Owner's behalf with respect to the Project. The Owner or such authorized representative shall render decisions in a timely manner pertaining to documents submitted by the Architect in order to avoid unreasonable delay in the orderly and sequential progress of the Architect's services.

4.5 The Owner shall furnish surveys describing physical characteristics, legal limitations and utility locations for the site of the Project, and a written legal description of the site. The surveys and legal information shall include, as applicable, grades and lines of streets, alleys, pavements and adjoining property and structures; adjacent drainage; rights-of-way, restrictions, easements, encroachments, zoning, deed restrictions, boundaries and contours of the site; locations, dimensions and necessary data pertaining to existing buildings, other improvements and trees; and information concerning available utility services and lines, both public and private, above and below grade, including inverts and depths. All the information on the survey shall be referenced to a project benchmark.

4.6 The Owner shall furnish the services of geotechnical engineers when such services are requested by the Architect. Such services may include but are not limited to test borings, test pits, determinations of soil bearing values, percolation tests, evaluations of hazardous materials, ground corrosion and resistivity tests, including necessary operations for anticipating subsoil conditions, with reports and appropriate professional recommendations.

4.6.1 The Owner shall furnish the services of other consultants when such services are reasonably required by the scope of the Project and are requested by the Architect.

4.7 The Owner shall furnish structural, mechanical, chemical, air and water pollution tests, tests for hazardous materials, and other laboratory and environmental tests, inspections and reports required by law or the Contract Documents.

4.8 The Owner shall furnish all legal, accounting and insurance counseling services as may be necessary at any time for the Project, including auditing services the Owner may require to verify the Contractor's Applications for Payment or to ascertain how or for what purposes the Contractor has used the money paid by or on behalf of the Owner.

4.9 The services, information, surveys and reports required by Paragraphs 4.5 through 4.8 shall be furnished at the Owner's expense, and the Architect shall be entitled to rely upon the accuracy and completeness thereof.

4.10 Prompt written notice shall be given by the Owner to the Architect if the Owner becomes aware of any fault or defect in the Project or nonconformance with the Contract Documents.

4.11 The proposed language of certificates or certifications requested of the Architect or Architect's consultants shall be submitted to the Architect for review and approval at least 14 days prior to execution. The Owner shall not request certifications that would require knowledge or services beyond the scope of this Agreement.

ARTICLE 5
CONSTRUCTION COST

5.1 DEFINITION

5.1.1 The Construction Cost shall be the total cost or estimated cost to the Owner of all elements of the Project designed or specified by the Architect.

5.1.2 The Construction Cost shall include the cost at current market rates of labor and materials furnished by the Owner and equipment designed, specified, selected or specially provided for by the Architect, plus a reasonable allowance for the Contractor's overhead and profit. In addition, a reasonable allowance for contingencies shall be included for market conditions at the time of bidding and for changes in the Work during construction.

5.1.3 Construction Cost does not include the compensation of the Architect and Architect's consultants, the costs of the land, rights-of-way, financing or other costs which are the responsibility of the Owner as provided in Article 4.

5.2 RESPONSIBILITY FOR CONSTRUCTION COST

5.2.1 Evaluations of the Owner's Project budget, preliminary estimates of Construction Cost and detailed estimates of Construction Cost, if any, prepared by the Architect, represent the Architect's best judgment as a design professional familiar with the construction industry. It is recognized, however, that neither the Architect nor the Owner has control over the cost of labor, materials or equipment, over the Contractor's methods of determining bid prices, or over competitive bidding, market or negotiating conditions. Accordingly, the Architect cannot and does not warrant or represent that bids or negotiated prices will not vary from the Owner's Project budget or from any estimate of Construction Cost or evaluation prepared or agreed to by the Architect.

5.2.2 No fixed limit of Construction Cost shall be established as a condition of this Agreement by the furnishing, proposal or establishment of a Project budget, unless such fixed limit has been agreed upon in writing and signed by the parties hereto. If such a fixed limit has been established, the Architect shall be permitted to include contingencies for design, bidding and price escalation, to determine what materials, equipment, component systems and types of construction are to be included in the Contract Documents, to make reasonable adjustments in the scope of the Project and to include in the Contract Documents alternate bids to adjust the Construction Cost to the fixed limit. Fixed limits, if any, shall be increased in the amount of an increase in the Contract Sum occurring after execution of the Contract for Construction.

5.2.3 If the Bidding or Negotiation Phase has not commenced within 90 days after the Architect submits the Construction Documents to the Owner, any Project budget or fixed limit of Construction Cost shall be adjusted to reflect changes in the general level of prices in the construction industry between the date of submission of the Construction Documents to the Owner and the date on which proposals are sought.

5.2.4 If a fixed limit of Construction Cost (adjusted as provided in Subparagraph 5.2.3) is exceeded by the lowest bona fide bid or negotiated proposal, the Owner shall:

.1 give written approval of an increase in such fixed limit;

.2 authorize rebidding or renegotiating of the Project within a reasonable time;

.3 if the Project is abandoned, terminate in accordance with Paragraph 8.3; or

.4 cooperate in revising the Project scope and quality as required to reduce the Construction Cost.

5.2.5 If the Owner chooses to proceed under Clause 5.2.4.4, the Architect, without additional charge, shall modify the Contract Documents as necessary to comply with the fixed limit, if established as a condition of this Agreement. The modification of Contract Documents shall be the limit of the Architect's responsibility arising out of the establishment of a fixed limit. The Architect shall be entitled to compensation in accordance with this Agreement for all services performed whether or not the Construction Phase is commenced.

ARTICLE 6
USE OF ARCHITECT'S DRAWINGS, SPECIFICATIONS AND OTHER DOCUMENTS

6.1 The Drawings, Specifications and other documents prepared by the Architect for this Project are instruments of the Architect's service for use solely with respect to this Project and, unless otherwise provided, the Architect shall be deemed the author of these documents and shall retain all common law, statutory and other reserved rights, including the copyright. The Owner shall be permitted to retain copies, including reproducible copies, of the Architect's Drawings, Specifications and other documents for information and reference in connection with the Owner's use and occupancy of the Project. The Architect's Drawings, Specifications or other documents shall not be used by the Owner or others on other projects, for additions to this Project or for completion of this Project by others, unless the Architect is adjudged to be in default under this Agreement, except by agreement in writing and with appropriate compensation to the Architect.

6.2 Submission or distribution of documents to meet official regulatory requirements or for similar purposes in connection with the Project is not to be construed as publication in derogation of the Architect's reserved rights.

ARTICLE 7
ARBITRATION

7.1 Claims, disputes or other matters in question between the parties to this Agreement arising out of or relating to this Agreement or breach thereof shall be subject to and decided by arbitration in accordance with the Construction Industry Arbitration Rules of the American Arbitration Association currently in effect unless the parties mutually agree otherwise.

7.2 Demand for arbitration shall be filed in writing with the other party to this Agreement and with the American Arbitration Association. A demand for arbitration shall be made within a reasonable time after the claim, dispute or other matter in question has arisen. In no event shall the demand for arbitration be made after the date when institution of legal or equitable proceedings based on such claim, dispute or other matter in question would be barred by the applicable statutes of limitations.

7.3 No arbitration arising out of or relating to this Agreement shall include, by consolidation, joinder or in any other manner, an additional person or entity not a party to this Agreement, except by written consent containing a specific reference to this Agreement signed by the Owner, Architect, and any other person or entity sought to be joined. Consent to arbitration involving an additional person or entity shall not constitute consent to arbitration of any claim, dispute or other matter in question not described in the written consent or with a person or entity not named or described therein. The foregoing agreement to arbitrate and other agreements to arbitrate with an additional person or entity duly consented to by the parties to this Agreement shall be specifically enforceable in accordance with applicable law in any court having jurisdiction thereof.

7.4 The award rendered by the arbitrator or arbitrators shall be final, and judgment may be entered upon it in accordance with applicable law in any court having jurisdiction thereof.

ARTICLE 8

TERMINATION, SUSPENSION OR ABANDONMENT

8.1 This Agreement may be terminated by either party upon not less than seven days' written notice should the other party fail substantially to perform in accordance with the terms of this Agreement through no fault of the party initiating the termination.

8.2 If the Project is suspended by the Owner for more than 30 consecutive days, the Architect shall be compensated for services performed prior to notice of such suspension. When the Project is resumed, the Architect's compensation shall be equitably adjusted to provide for expenses incurred in the interruption and resumption of the Architect's services.

8.3 This Agreement may be terminated by the Owner upon not less than seven days' written notice to the Architect in the event that the Project is permanently abandoned. If the Project is abandoned by the Owner for more than 90 consecutive days, the Architect may terminate this Agreement by giving written notice.

8.4 Failure of the Owner to make payments to the Architect in accordance with this Agreement shall be considered substantial nonperformance and cause for termination.

8.5 If the Owner fails to make payment when due the Architect for services and expenses, the Architect may, upon seven days' written notice to the Owner, suspend performance of services under this Agreement. Unless payment in full is received by the Architect within seven days of the date of the notice, the suspension shall take effect without further notice. In the event of a suspension of services, the Architect shall have no liability to the Owner for delay or damage caused the Owner because of such suspension of services.

8.6 In the event of termination not the fault of the Architect, the Architect shall be compensated for services performed prior to termination, together with Reimbursable Expenses then due and all Termination Expenses as defined in Paragraph 8.7.

8.7 Termination Expenses are in addition to compensation for Basic and Additional Services, and include expenses which are directly attributable to termination. Termination Expenses shall be computed as a percentage of the total compensation for Basic Services and Additional Services earned to the time of termination, as follows:

　　.1 Twenty percent of the total compensation for Basic and Additional Services earned to date if termination occurs before or during the predesign, site analysis, or Schematic Design Phases; or

　　.2 Ten percent of the total compensation for Basic and Additional Services earned to date if termination occurs during the Design Development Phase; or

　　.3 Five percent of the total compensation for Basic and Additional Services earned to date if termination occurs during any subsequent phase.

ARTICLE 9

MISCELLANEOUS PROVISIONS

9.1 Unless otherwise provided, this Agreement shall be governed by the law of the principal place of business of the Architect.

9.2 Terms in this Agreement shall have the same meaning as those in AIA Document A201, General Conditions of the Contract for Construction, current as of the date of this Agreement.

9.3 Causes of action between the parties to this Agreement pertaining to acts or failures to act shall be deemed to have accrued and the applicable statutes of limitations shall commence to run not later than either the date of Substantial Completion for acts or failures to act occurring prior to Substantial Completion, or the date of issuance of the final Certificate for Payment for acts or failures to act occurring after Substantial Completion.

9.4 The Owner and Architect waive all rights against each other and against the contractors, consultants, agents and employees of the other for damages, but only to the extent covered by property insurance during construction, except such rights as they may have to the proceeds of such insurance as set forth in the edition of AIA Document A201, General Conditions of the Contract for Construction, current as of the date of this Agreement. The Owner and Architect each shall require similar waivers from their contractors, consultants and agents.

9.5 The Owner and Architect, respectively, bind themselves, their partners, successors, assigns and legal representatives to the other party to this Agreement and to the partners, successors, assigns and legal representatives of such other party with respect to all covenants of this Agreement. Neither Owner nor Architect shall assign this Agreement without the written consent of the other.

9.6 This Agreement represents the entire and integrated agreement between the Owner and Architect and supersedes all prior negotiations, representations or agreements, either written or oral. This Agreement may be amended only by written instrument signed by both Owner and Architect.

9.7 Nothing contained in this Agreement shall create a contractual relationship with or a cause of action in favor of a third party against either the Owner or Architect.

9.8 Unless otherwise provided in this Agreement, the Architect and Architect's consultants shall have no responsibility for the discovery, presence, handling, removal or disposal of or exposure of persons to hazardous materials in any form at the Project site, including but not limited to asbestos, asbestos products, polychlorinated biphenyl (PCB) or other toxic substances.

9.9 The Architect shall have the right to include representations of the design of the Project, including photographs of the exterior and interior, among the Architect's promotional and professional materials. The Architect's materials shall not include the Owner's confidential or proprietary information if the Owner has previously advised the Architect in writing of

AIA DOCUMENT B141 • OWNER-ARCHITECT AGREEMENT • FOURTEENTH EDITION • AIA® • ©1987
THE AMERICAN INSTITUTE OF ARCHITECTS, 1735 NEW YORK AVENUE, N.W., WASHINGTON, D.C. 20006

the specific information considered by the Owner to be confidential or proprietary. The Owner shall provide professional credit for the Architect on the construction sign and in the promotional materials for the Project.

ARTICLE 10

PAYMENTS TO THE ARCHITECT

10.1 DIRECT PERSONNEL EXPENSE

10.1.1 Direct Personnel Expense is defined as the direct salaries of the Architect's personnel engaged on the Project and the portion of the cost of their mandatory and customary contributions and benefits related thereto, such as employment taxes and other statutory employee benefits, insurance, sick leave, holidays, vacations, pensions and similar contributions and benefits.

10.2 REIMBURSABLE EXPENSES

10.2.1 Reimbursable Expenses are in addition to compensation for Basic and Additional Services and include expenses incurred by the Architect and Architect's employees and consultants in the interest of the Project, as identified in the following Clauses.

10.2.1.1 Expense of transportation in connection with the Project; expenses in connection with authorized out-of-town travel; long-distance communications; and fees paid for securing approval of authorities having jurisdiction over the Project.

10.2.1.2 Expense of reproductions, postage and handling of Drawings, Specifications and other documents.

10.2.1.3 If authorized in advance by the Owner, expense of overtime work requiring higher than regular rates.

10.2.1.4 Expense of renderings, models and mock-ups requested by the Owner.

10.2.1.5 Expense of additional insurance coverage or limits, including professional liability insurance, requested by the Owner in excess of that normally carried by the Architect and Architect's consultants.

10.2.1.6 Expense of computer-aided design and drafting equipment time when used in connection with the Project.

10.3 PAYMENTS ON ACCOUNT OF BASIC SERVICES

10.3.1 An initial payment as set forth in Paragraph 11.1 is the minimum payment under this Agreement.

10.3.2 Subsequent payments for Basic Services shall be made monthly and, where applicable, shall be in proportion to services performed within each phase of service, on the basis set forth in Subparagraph 11.2.2.

10.3.3 If and to the extent that the time initially established in Subparagraph 11.5.1 of this Agreement is exceeded or extended through no fault of the Architect, compensation for any services rendered during the additional period of time shall be computed in the manner set forth in Subparagraph 11.3.2.

10.3.4 When compensation is based on a percentage of Construction Cost and any portions of the Project are deleted or otherwise not constructed, compensation for those portions of the Project shall be payable to the extent services are performed on those portions, in accordance with the schedule set forth in Subparagraph 11.2.2, based on (1) the lowest bona fide bid or negotiated proposal, or (2) if no such bid or proposal is received, the most recent preliminary estimate of Construction Cost or detailed estimate of Construction Cost for such portions of the Project.

10.4 PAYMENTS ON ACCOUNT OF ADDITIONAL SERVICES

10.4.1 Payments on account of the Architect's Additional Services and for Reimbursable Expenses shall be made monthly upon presentation of the Architect's statement of services rendered or expenses incurred.

10.5 PAYMENTS WITHHELD

10.5.1 No deductions shall be made from the Architect's compensation on account of penalty, liquidated damages or other sums withheld from payments to contractors, or on account of the cost of changes in the Work other than those for which the Architect has been found to be liable.

10.6 ARCHITECT'S ACCOUNTING RECORDS

10.6.1 Records of Reimbursable Expenses and expenses pertaining to Additional Services and services performed on the basis of a multiple of Direct Personnel Expense shall be available to the Owner or the Owner's authorized representative at mutually convenient times.

ARTICLE 11

BASIS OF COMPENSATION

The Owner shall compensate the Architect as follows:

11.1 AN INITIAL PAYMENT of   Dollars ($)
shall be made upon execution of this Agreement and credited to the Owner's account at final payment.

11.2 BASIC COMPENSATION

11.2.1 FOR BASIC SERVICES, as described in Article 2, and any other services included in Article 12 as part of Basic Services, Basic Compensation shall be computed as follows:

(Insert basis of compensation, including stipulated sums, multiples or percentages, and identify phases to which particular methods of compensation apply, if necessary.)

11.2.2 Where compensation is based on a stipulated sum or percentage of Construction Cost, progress payments for Basic Services in each phase shall total the following percentages of the total Basic Compensation payable:

(Insert additional phases as appropriate.)

Schematic Design Phase:	percent (%)
Design Development Phase:	percent (%)
Construction Documents Phase:	percent (%)
Bidding or Negotiation Phase:	percent (%)
Construction Phase:	percent (%)
Total Basic Compensation:	one hundred percent (100%)

11.3 COMPENSATION FOR ADDITIONAL SERVICES

11.3.1 FOR PROJECT REPRESENTATION BEYOND BASIC SERVICES, as described in Paragraph 3.2, compensation shall be computed as follows:

11.3.2 FOR ADDITIONAL SERVICES OF THE ARCHITECT, as described in Articles 3 and 12, other than (1) Additional Project Representation, as described in Paragraph 3.2, and (2) services included in Article 12 as part of Additional Services, but excluding services of consultants, compensation shall be computed as follows:

(Insert basis of compensation, including rates and/or multiples of Direct Personnel Expense for Principals and employees, and identify Principals and classify employees, if required. Identify specific services to which particular methods of compensation apply, if necessary.)

11.3.3 FOR ADDITIONAL SERVICES OF CONSULTANTS, including additional structural, mechanical and electrical engineering services and those provided under Subparagraph 3.4.19 or identified in Article 12 as part of Additional Services, a multiple of (   ) times the amounts billed to the Architect for such services.

(Identify specific types of consultants in Article 12, if required.)

11.4 REIMBURSABLE EXPENSES

11.4.1 FOR REIMBURSABLE EXPENSES, as described in Paragraph 10.2, and any other items included in Article 12 as Reimbursable Expenses, a multiple of (   ) times the expenses incurred by the Architect, the Architect's employees and consultants in the interest of the Project.

11.5 ADDITIONAL PROVISIONS

11.5.1 IF THE BASIC SERVICES covered by this Agreement have not been completed within (   ) months of the date hereof, through no fault of the Architect, extension of the Architect's services beyond that time shall be compensated as provided in Subparagraphs 10.3.3 and 11.3.2.

11.5.2 Payments are due and payable (   ) days from the date of the Architect's invoice. Amounts unpaid (   ) days after the invoice date shall bear interest at the rate entered below, or in the absence thereof at the legal rate prevailing from time to time at the principal place of business of the Architect.

(Insert rate of interest agreed upon.)

(Usury laws and requirements under the Federal Truth in Lending Act, similar state and local consumer credit laws and other regulations at the Owner's and Architect's principal places of business, the location of the Project and elsewhere may affect the validity of this provision. Specific legal advice should be obtained with respect to deletions or modifications, and also regarding requirements such as written disclosures or waivers.)

11.5.3 The rates and multiples set forth for Additional Services shall be annually adjusted in accordance with normal salary review practices of the Architect.

ARTICLE 12
OTHER CONDITIONS OR SERVICES

(Insert descriptions of other services, identify Additional Services included within Basic Compensation and modifications to the payment and compensation terms included in this Agreement.)

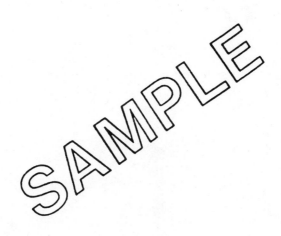

This Agreement entered into as of the day and year first written above.

OWNER   ARCHITECT

_____   _____
(Signature)   *(Signature)*

_____   _____
(Printed name and title)   *(Printed name and title)*

B

Standard Form of Agreement Between Owner and Contractor

1987 Edition

AIA Document A101

Standard Form of Agreement Between Owner and Contractor

where the basis of payment is a
STIPULATED SUM

1987 EDITION

THIS DOCUMENT HAS IMPORTANT LEGAL CONSEQUENCES; CONSULTATION WITH AN ATTORNEY IS ENCOURAGED WITH RESPECT TO ITS COMPLETION OR MODIFICATION.
The 1987 Edition of AIA Document A201, General Conditions of the Contract for Construction, is adopted in this document by reference. Do not use with other general conditions unless this document is modified.
This document has been approved and endorsed by The Associated General Contractors of America.

AGREEMENT

made as of the   day of   in the year of
Nineteen Hundred and

BETWEEN the Owner:
(Name and address)

and the Contractor:
(Name and address)

The Project is:
(Name and location)

The Architect is:
(Name and address)

The Owner and Contractor agree as set forth below.

ARTICLE 1
THE CONTRACT DOCUMENTS

The Contract Documents consist of this Agreement, Conditions of the Contract (General, Supplementary and other Conditions), Drawings, Specifications, Addenda issued prior to execution of this Agreement, other documents listed in this Agreement and Modifications issued after execution of this Agreement; these form the Contract, and are as fully a part of the Contract as if attached to this Agreement or repeated herein. The Contract represents the entire and integrated agreement between the parties hereto and supersedes prior negotiations, representations or agreements, either written or oral. An enumeration of the Contract Documents, other than Modifications, appears in Article 9.

ARTICLE 2
THE WORK OF THIS CONTRACT

The Contractor shall execute the entire Work described in the Contract Documents, except to the extent specifically indicated in the Contract Documents to be the responsibility of others, or as follows:

ARTICLE 3
DATE OF COMMENCEMENT AND SUBSTANTIAL COMPLETION

3.1 The date of commencement is the date from which the Contract Time of Paragraph 3.2 is measured, and shall be the date of this Agreement, as first written above, unless a different date is stated below or provision is made for the date to be fixed in a notice to proceed issued by the Owner.

(Insert the date of commencement, if it differs from the date of this Agreement or, if applicable, state that the date will be fixed in a notice to proceed.)

Unless the date of commencement is established by a notice to proceed issued by the Owner, the Contractor shall notify the Owner in writing not less than five days before commencing the Work to permit the timely filing of mortgages, mechanic's liens and other security interests.

3.2 The Contractor shall achieve Substantial Completion of the entire Work not later than

(Insert the calendar date or number of calendar days after the date of commencement. Also insert any requirements for earlier Substantial Completion of certain portions of the Work, if not stated elsewhere in the Contract Documents.)

, subject to adjustments of this Contract Time as provided in the Contract Documents.

(Insert provisions, if any, for liquidated damages relating to failure to complete on time.)

ARTICLE 4
CONTRACT SUM

4.1 The Owner shall pay the Contractor in current funds for the Contractor's performance of the Contract the Contract Sum of
Dollars

(**$**   ), subject to additions and deductions as provided in the Contract Documents.

4.2 The Contract Sum is based upon the following alternates, if any, which are described in the Contract Documents and are hereby accepted by the Owner:

(State the numbers or other identification of accepted alternates. If decisions on other alternates are to be made by the Owner subsequent to the execution of this Agreement, attach a schedule of such other alternates showing the amount for each and the date until which that amount is valid.)

4.3 Unit prices, if any, are as follows:

ARTICLE 5
PROGRESS PAYMENTS

5.1 Based upon Applications for Payment submitted to the Architect by the Contractor and Certificates for Payment issued by the Architect, the Owner shall make progress payments on account of the Contract Sum to the Contractor as provided below and elsewhere in the Contract Documents.

5.2 The period covered by each Application for Payment shall be one calendar month ending on the last day of the month, or as follows:

5.3 Provided an Application for Payment is received by the Architect not later than the
day of a month, the Owner shall make payment to the Contractor not later than the
day of the   month. If an Application for Payment is received by the
Architect after the application date fixed above, payment shall be made by the Owner not later than
days after the Architect receives the Application for Payment.

5.4 Each Application for Payment shall be based upon the Schedule of Values submitted by the Contractor in accordance with the Contract Documents. The Schedule of Values shall allocate the entire Contract Sum among the various portions of the Work and be prepared in such form and supported by such data to substantiate its accuracy as the Architect may require. This Schedule, unless objected to by the Architect, shall be used as a basis for reviewing the Contractor's Applications for Payment.

5.5 Applications for Payment shall indicate the percentage of completion of each portion of the Work as of the end of the period covered by the Application for Payment.

5.6 Subject to the provisions of the Contract Documents, the amount of each progress payment shall be computed as follows:

5.6.1 Take that portion of the Contract Sum properly allocable to completed Work as determined by multiplying the percentage completion of each portion of the Work by the share of the total Contract Sum allocated to that portion of the Work in the Schedule of Values, less retainage of   percent
(%). Pending final determination of cost to the Owner of changes in the Work, amounts not in dispute may be included as provided in Subparagraph 7.3.7 of the General Conditions even though the Contract Sum has not yet been adjusted by Change Order;

5.6.2 Add that portion of the Contract Sum properly allocable to materials and equipment delivered and suitably stored at the site for subsequent incorporation in the completed construction (or, if approved in advance by the Owner, suitably stored off the site at a location agreed upon in writing), less retainage of
percent (%);

5.6.3 Subtract the aggregate of previous payments made by the Owner; and

5.6.4 Subtract amounts, if any, for which the Architect has withheld or nullified a Certificate for Payment as provided in Paragraph 9.5 of the General Conditions.

5.7 The progress payment amount determined in accordance with Paragraph 5.6 shall be further modified under the following circumstances:

5.7.1 Add, upon Substantial Completion of the Work, a sum sufficient to increase the total payments to
percent (%) of the Contract
Sum, less such amounts as the Architect shall determine for incomplete Work and unsettled claims; and

5.7.2 Add, if final completion of the Work is thereafter materially delayed through no fault of the Contractor, any additional amounts payable in accordance with Subparagraph 9.10.3 of the General Conditions.

5.8 Reduction or limitation of retainage, if any, shall be as follows:

(If it is intended, prior to Substantial Completion of the entire Work, to reduce or limit the retainage resulting from the percentages inserted in Subparagraphs 5.6.1 and 5.6.2 above, and this is not explained elsewhere in the Contract Documents, insert here provisions for such reduction or limitation.)

ARTICLE 6
FINAL PAYMENT

Final payment, constituting the entire unpaid balance of the Contract Sum, shall be made by the Owner to the Contractor when (1) the Contract has been fully performed by the Contractor except for the Contractor's responsibility to correct nonconforming Work as provided in Subparagraph 12.2.2 of the General Conditions and to satisfy other requirements, if any, which necessarily survive final payment; and (2) a final Certificate for Payment has been issued by the Architect; such final payment shall be made by the Owner not more than 30 days after the issuance of the Architect's final Certificate for Payment, or as follows:

ARTICLE 7
MISCELLANEOUS PROVISIONS

7.1 Where reference is made in this Agreement to a provision of the General Conditions or another Contract Document, the reference refers to that provision as amended or supplemented by other provisions of the Contract Documents.

7.2 Payments due and unpaid under the Contract shall bear interest from the date payment is due at the rate stated below, or in the absence thereof, at the legal rate prevailing from time to time at the place where the Project is located.

(Insert rate of interest agreed upon, if any.)

(Usury laws and requirements under the Federal Truth in Lending Act, similar state and local consumer credit laws and other regulations at the Owner's and Contractor's principal places of business, the location of the Project and elsewhere may affect the validity of this provision. Legal advice should be obtained with respect to deletions or modifications, and also regarding requirements such as written disclosures or waivers.)

7.3 Other provisions:

ARTICLE 8
TERMINATION OR SUSPENSION

8.1 The Contract may be terminated by the Owner or the Contractor as provided in Article 14 of the General Conditions.

8.2 The Work may be suspended by the Owner as provided in Article 14 of the General Conditions.

ARTICLE 9
ENUMERATION OF CONTRACT DOCUMENTS

9.1 The Contract Documents, except for Modifications issued after execution of this Agreement, are enumerated as follows:

9.1.1 The Agreement is this executed Standard Form of Agreement Between Owner and Contractor, AIA Document A101, 1987 Edition.

9.1.2 The General Conditions are the General Conditions of the Contract for Construction, AIA Document A201, 1987 Edition.

9.1.3 The Supplementary and other Conditions of the Contract are those contained in the Project Manual dated
, and are as follows:

Document	Title	Pages

9.1.4 The Specifications are those contained in the Project Manual dated as in Subparagraph 9.1.3, and are as follows:
(Either list the Specifications herein or refer to an exhibit attached to this Agreement.)

Section	Title	Pages

9.1.5 The Drawings are as follows, and are dated   unless a different date is shown below:

(Either list the Drawings here or refer to an exhibit attached to this Agreement)

Number   **Title**   **Date**

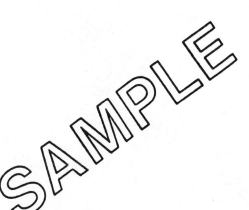

9.1.6 The Addenda, if any, are as follows:

Number   **Date**   **Pages**

Portions of Addenda relating to bidding requirements are not part of the Contract Documents unless the bidding requirements are also enumerated in this Article 9.

9.1.7 Other documents, if any, forming part of the Contract Documents are as follows:

(List here any additional documents which are intended to form part of the Contract Documents. The General Conditions provide that bidding requirements such as advertisement or invitation to bid, Instructions to Bidders, sample forms and the Contractor's bid are not part of the Contract Documents unless enumerated in this Agreement. They should be listed here only if intended to be part of the Contract Documents.)

This Agreement is entered into as of the day and year first written above and is executed in at least three original copies of which one is to be delivered to the Contractor, one to the Architect for use in the administration of the Contract, and the remainder to the Owner.

OWNER

_____
(Signature)

_____
(Printed name and title)

CONTRACTOR

_____
(Signature)

_____
(Printed name and title)

General Conditions of the Contract for Construction

1987 Edition

T H E   A M E R I C A N   I N S T I T U T E   O F   A R C H I T E C T S

AIA Document A201

General Conditions of the Contract for Construction

THIS DOCUMENT HAS IMPORTANT LEGAL CONSEQUENCES; CONSULTATION WITH AN ATTORNEY IS ENCOURAGED WITH RESPECT TO ITS MODIFICATION

1987 EDITION
TABLE OF ARTICLES

This document has been approved and endorsed by the Associated General Contractors of America.

 CAUTION: You should use an original AIA document which has this caution printed in red. An original assures that changes will not be obscured as may occur when documents are reproduced.

INDEX

AIA DOCUMENT A201 • GENERAL CONDITIONS OF THE CONTRACT FOR CONSTRUCTION • FOURTEENTH EDITION
AIA® • ©1987 THE AMERICAN INSTITUTE OF ARCHITECTS, 1735 NEW YORK AVENUE, N.W., WASHINGTON, D.C. 20006

AIA DOCUMENT A201 • GENERAL CONDITIONS OF THE CONTRACT FOR CONSTRUCTION • FOURTEENTH EDITION
AIA® • ©1987 THE AMERICAN INSTITUTE OF ARCHITECTS, 1735 NEW YORK AVENUE, N.W., WASHINGTON, D.C. 20006

GENERAL CONDITIONS OF THE CONTRACT FOR CONSTRUCTION

ARTICLE 1

GENERAL PROVISIONS

1.1 BASIC DEFINITIONS

1.1.1 THE CONTRACT DOCUMENTS

The Contract Documents consist of the Agreement between Owner and Contractor (hereinafter the Agreement), Conditions of the Contract (General, Supplementary and other Conditions), Drawings, Specifications, addenda issued prior to execution of the Contract, other documents listed in the Agreement and Modifications issued after execution of the Contract. A Modification is (1) a written amendment to the Contract signed by both parties, (2) a Change Order, (3) a Construction Change Directive or (4) a written order for a minor change in the Work issued by the Architect. Unless specifically enumerated in the Agreement, the Contract Documents do not include other documents such as bidding requirements (advertisement or invitation to bid, Instructions to Bidders, sample forms, the Contractor's bid or portions of addenda relating to bidding requirements).

1.1.2 THE CONTRACT

The Contract Documents form the Contract for Construction. The Contract represents the entire and integrated agreement between the parties hereto and supersedes prior negotiations, representations or agreements, either written or oral. The Contract may be amended or modified only by a Modification. The Contract Documents shall not be construed to create a contractual relationship of any kind (1) between the Architect and Contractor, (2) between the Owner and a Subcontractor or Subsubcontractor or (3) between any persons or entities other than the Owner and Contractor. The Architect shall, however, be entitled to performance and enforcement of obligations under the Contract intended to facilitate performance of the Architect's duties.

1.1.3 THE WORK

The term "Work" means the construction and services required by the Contract Documents, whether completed or partially completed, and includes all other labor, materials, equipment and services provided or to be provided by the Contractor to fulfill the Contractor's obligations. The Work may constitute the whole or a part of the Project.

1.1.4 THE PROJECT

The Project is the total construction of which the Work performed under the Contract Documents may be the whole or a part and which may include construction by the Owner or by separate contractors.

1.1.5 THE DRAWINGS

The Drawings are the graphic and pictorial portions of the Contract Documents, wherever located and whenever issued, showing the design, location and dimensions of the Work, generally including plans, elevations, sections, details, schedules and diagrams.

1.1.6 THE SPECIFICATIONS

The Specifications are that portion of the Contract Documents consisting of the written requirements for materials, equip-

ment, construction systems, standards and workmanship for the Work, and performance of related services.

1.1.7 THE PROJECT MANUAL

The Project Manual is the volume usually assembled for the Work which may include the bidding requirements, sample forms, Conditions of the Contract and Specifications.

1.2 EXECUTION, CORRELATION AND INTENT

1.2.1 The Contract Documents shall be signed by the Owner and Contractor as provided in the Agreement. If either the Owner or Contractor or both do not sign all the Contract Documents, the Architect shall identify such unsigned Documents upon request.

1.2.2 Execution of the Contract by the Contractor is a representation that the Contractor has visited the site, become familiar with local conditions under which the Work is to be performed and correlated personal observations with requirements of the Contract Documents.

1.2.3 The intent of the Contract Documents is to include all items necessary for the proper execution and completion of the Work by the Contractor. The Contract Documents are complementary, and what is required by one shall be as binding as if required by all; performance by the Contractor shall be required only to the extent consistent with the Contract Documents and reasonably inferable from them as being necessary to produce the intended results.

1.2.4 Organization of the Specifications into divisions, sections and articles, and arrangement of Drawings shall not control the Contractor in dividing the Work among Subcontractors or in establishing the extent of Work to be performed by any trade.

1.2.5 Unless otherwise stated in the Contract Documents, words which have well-known technical or construction industry meanings are used in the Contract Documents in accordance with such recognized meanings.

1.3 OWNERSHIP AND USE OF ARCHITECT'S DRAWINGS, SPECIFICATIONS AND OTHER DOCUMENTS

1.3.1 The Drawings, Specifications and other documents prepared by the Architect are instruments of the Architect's service through which the Work to be executed by the Contractor is described. The Contractor may retain one contract record set. Neither the Contractor nor any Subcontractor, Subsubcontractor or material or equipment supplier shall own or claim a copyright in the Drawings, Specifications and other documents prepared by the Architect, and unless otherwise indicated the Architect shall be deemed the author of them and will retain all common law, statutory and other reserved rights, in addition to the copyright. All copies of them, except the Contractor's record set, shall be returned or suitably accounted for to the Architect, on request, upon completion of the Work. The Drawings, Specifications and other documents prepared by the Architect, and copies thereof furnished to the Contractor, are for use solely with respect to this Project. They are not to be used by the Contractor or any Subcontractor, Subsubcontractor or material or equipment supplier on other projects or for additions to this Project outside the scope of the

AIA DOCUMENT A201 • GENERAL CONDITIONS OF THE CONTRACT FOR CONSTRUCTION • FOURTEENTH EDITION
AIA® • ©1987 THE AMERICAN INSTITUTE OF ARCHITECTS, 1735 NEW YORK AVENUE, N.W., WASHINGTON, D.C. 20006

Work without the specific written consent of the Owner and Architect. The Contractor, Subcontractors, Sub-subcontractors and material or equipment suppliers are granted a limited license to use and reproduce applicable portions of the Drawings, Specifications and other documents prepared by the Architect appropriate to and for use in the execution of their Work under the Contract Documents. All copies made under this license shall bear the statutory copyright notice, if any, shown on the Drawings, Specifications and other documents prepared by the Architect. Submittal or distribution to meet official regulatory requirements or for other purposes in connection with this Project is not to be construed as publication in derogation of the Architect's copyright or other reserved rights.

1.4   CAPITALIZATION

1.4.1 Terms capitalized in these General Conditions include those which are (1) specifically defined, (2) the titles of numbered articles and identified references to Paragraphs, Subparagraphs and Clauses in the document or (3) the titles of other documents published by the American Institute of Architects.

1.5   INTERPRETATION

1.5.1 In the interest of brevity the Contract Documents frequently omit modifying words such as "all" and "any" and articles such as "the" and "an," but the fact that a modifier or an article is absent from one statement and appears in another is not intended to affect the interpretation of either statement.

ARTICLE 2

OWNER

2.1   DEFINITION

2.1.1 The Owner is the person or entity identified as such in the Agreement and is referred to throughout the Contract Documents as if singular in number. The term "Owner" means the Owner or the Owner's authorized representative.

2.1.2 The Owner upon reasonable written request shall furnish to the Contractor in writing information which is necessary and relevant for the Contractor to evaluate, give notice of or enforce mechanic's lien rights. Such information shall include a correct statement of the record legal title to the property on which the Project is located, usually referred to as the site, and the Owner's interest therein at the time of execution of the Agreement and, within five days after any change, information of such change in title, recorded or unrecorded.

2.2   INFORMATION AND SERVICES REQUIRED OF THE OWNER

2.2.1 The Owner shall, at the request of the Contractor, prior to execution of the Agreement and promptly from time to time thereafter, furnish to the Contractor reasonable evidence that financial arrangements have been made to fulfill the Owner's obligations under the Contract. *[Note: Unless such reasonable evidence were furnished on request prior to the execution of the Agreement, the prospective contractor would not be required to execute the Agreement or to commence the Work.]*

2.2.2 The Owner shall furnish surveys describing physical characteristics, legal limitations and utility locations for the site of the Project, and a legal description of the site.

2.2.3 Except for permits and fees which are the responsibility of the Contractor under the Contract Documents, the Owner shall secure and pay for necessary approvals, easements, assess-ments and charges required for construction, use or occupancy of permanent structures or for permanent changes in existing facilities.

2.2.4 Information or services under the Owner's control shall be furnished by the Owner with reasonable promptness to avoid delay in orderly progress of the Work.

2.2.5 Unless otherwise provided in the Contract Documents, the Contractor will be furnished, free of charge, such copies of Drawings and Project Manuals as are reasonably necessary for execution of the Work.

2.2.6 The foregoing are in addition to other duties and responsibilities of the Owner enumerated herein and especially those in respect to Article 6 (Construction by Owner or by Separate Contractors), Article 9 (Payments and Completion) and Article 11 (Insurance and Bonds).

2.3   OWNER'S RIGHT TO STOP THE WORK

2.3.1 If the Contractor fails to correct Work which is not in accordance with the requirements of the Contract Documents as required by Paragraph 12.2 or persistently fails to carry out Work in accordance with the Contract Documents, the Owner, by written order signed personally or by an agent specifically so empowered by the Owner in writing, may order the Contractor to stop the Work, or any portion thereof, until the cause for such order has been eliminated; however, the right of the Owner to stop the Work shall not give rise to a duty on the part of the Owner to exercise this right for the benefit of the Contractor or any other person or entity, except to the extent required by Subparagraph 6.1.3.

2.4   OWNER'S RIGHT TO CARRY OUT THE WORK

2.4.1 If the Contractor defaults or neglects to carry out the Work in accordance with the Contract Documents and fails within a seven-day period after receipt of written notice from the Owner to commence and continue correction of such default or neglect with diligence and promptness, the Owner may after such seven-day period give the Contractor a second written notice to correct such deficiencies within a second seven-day period. If the Contractor within such second seven-day period after receipt of such second notice fails to commence and continue to correct any deficiencies, the Owner may, without prejudice to other remedies the Owner may have, correct such deficiencies. In such case an appropriate Change Order shall be issued deducting from payments then or thereafter due the Contractor the cost of correcting such deficiencies, including compensation for the Architect's additional services and expenses made necessary by such default, neglect or failure. Such action by the Owner and amounts charged to the Contractor are both subject to prior approval of the Architect. If payments then or thereafter due the Contractor are not sufficient to cover such amounts, the Contractor shall pay the difference to the Owner.

ARTICLE 3

CONTRACTOR

3.1   DEFINITION

3.1.1 The Contractor is the person or entity identified as such in the Agreement and is referred to throughout the Contract Documents as if singular in number. The term "Contractor" means the Contractor or the Contractor's authorized representative.

3.2 REVIEW OF CONTRACT DOCUMENTS AND FIELD CONDITIONS BY CONTRACTOR

3.2.1 The Contractor shall carefully study and compare the Contract Documents with each other and with information furnished by the Owner pursuant to Subparagraph 2.2.2 and shall at once report to the Architect errors, inconsistencies or omissions discovered. The Contractor shall not be liable to the Owner or Architect for damage resulting from errors, inconsistencies or omissions in the Contract Documents unless the Contractor recognized such error, inconsistency or omission and knowingly failed to report it to the Architect. If the Contractor performs any construction activity knowing it involves a recognized error, inconsistency or omission in the Contract Documents without such notice to the Architect, the Contractor shall assume appropriate responsibility for such performance and shall bear an appropriate amount of the attributable costs for correction.

3.2.2 The Contractor shall take field measurements and verify field conditions and shall carefully compare such field measurements and conditions and other information known to the Contractor with the Contract Documents before commencing activities. Errors, inconsistencies or omissions discovered shall be reported to the Architect at once.

3.2.3 The Contractor shall perform the Work in accordance with the Contract Documents and submittals approved pursuant to Paragraph 3.12.

3.3 SUPERVISION AND CONSTRUCTION PROCEDURES

3.3.1 The Contractor shall supervise and direct the Work, using the Contractor's best skill and attention. The Contractor shall be solely responsible for and have control over construction means, methods, techniques, sequences and procedures and for coordinating all portions of the Work under the Contract, unless Contract Documents give other specific instructions concerning these matters.

3.3.2 The Contractor shall be responsible to the Owner for acts and omissions of the Contractor's employees, Subcontractors and their agents and employees, and other persons performing portions of the Work under a contract with the Contractor.

3.3.3 The Contractor shall not be relieved of obligations to perform the Work in accordance with the Contract Documents either by activities or duties of the Architect in the Architect's administration of the Contract, or by tests, inspections or approvals required or performed by persons other than the Contractor.

3.3.4 The Contractor shall be responsible for inspection of portions of Work already performed under this Contract to determine that such portions are in proper condition to receive subsequent Work.

3.4 LABOR AND MATERIALS

3.4.1 Unless otherwise provided in the Contract Documents, the Contractor shall provide and pay for labor, materials, equipment, tools, construction equipment and machinery, water, heat, utilities, transportation, and other facilities and services necessary for proper execution and completion of the Work, whether temporary or permanent and whether or not incorporated or to be incorporated in the Work.

3.4.2 The Contractor shall enforce strict discipline and good order among the Contractor's employees and other persons carrying out the Contract. The Contractor shall not permit employment of unfit persons or persons not skilled in tasks assigned to them.

3.5 WARRANTY

3.5.1 The Contractor warrants to the Owner and Architect that materials and equipment furnished under the Contract will be of good quality and new unless otherwise required or permitted by the Contract Documents, that the Work will be free from defects not inherent in the quality required or permitted, and that the Work will conform with the requirements of the Contract Documents. Work not conforming to these requirements, including substitutions not properly approved and authorized, may be considered defective. The Contractor's warranty excludes remedy for damage or defect caused by abuse, modifications not executed by the Contractor, improper or insufficient maintenance, improper operation, or normal wear and tear under normal usage. If required by the Architect, the Contractor shall furnish satisfactory evidence as to the kind and quality of materials and equipment.

3.6 TAXES

3.6.1 The Contractor shall pay sales, consumer, use and similar taxes for the Work or portions thereof provided by the Contractor which are legally enacted when bids are received or negotiations concluded, whether or not yet effective or merely scheduled to go into effect.

3.7 PERMITS, FEES AND NOTICES

3.7.1 Unless otherwise provided in the Contract Documents, the Contractor shall secure and pay for the building permit and other permits and governmental fees, licenses and inspections necessary for proper execution and completion of the Work which are customarily secured after execution of the Contract and which are legally required when bids are received or negotiations concluded.

3.7.2 The Contractor shall comply with and give notices required by laws, ordinances, rules, regulations and lawful orders of public authorities bearing on performance of the Work.

3.7.3 It is not the Contractor's responsibility to ascertain that the Contract Documents are in accordance with applicable laws, statutes, ordinances, building codes, and rules and regulations. However, if the Contractor observes that portions of the Contract Documents are at variance therewith, the Contractor shall promptly notify the Architect and Owner in writing, and necessary changes shall be accomplished by appropriate Modification.

3.7.4 If the Contractor performs Work knowing it to be contrary to laws, statutes, ordinances, building codes, and rules and regulations without such notice to the Architect and Owner, the Contractor shall assume full responsibility for such Work and shall bear the attributable costs.

3.8 ALLOWANCES

3.8.1 The Contractor shall include in the Contract Sum all allowances stated in the Contract Documents. Items covered by allowances shall be supplied for such amounts and by such persons or entities as the Owner may direct, but the Contractor shall not be required to employ persons or entities against which the Contractor makes reasonable objection.

3.8.2 Unless otherwise provided in the Contract Documents:

    .1 materials and equipment under an allowance shall be selected promptly by the Owner to avoid delay in the Work;

    .2 allowances shall cover the cost to the Contractor of materials and equipment delivered at the site and all required taxes, less applicable trade discounts;

AIA DOCUMENT A201 • GENERAL CONDITIONS OF THE CONTRACT FOR CONSTRUCTION • FOURTEENTH EDITION
AIA® • ©1987 THE AMERICAN INSTITUTE OF ARCHITECTS, 1735 NEW YORK AVENUE, N.W., WASHINGTON, D.C. 20006

.3 Contractor's costs for unloading and handling at the site, labor, installation costs, overhead, profit and other expenses contemplated for stated allowance amounts shall be included in the Contract Sum and not in the allowances;

.4 whenever costs are more than or less than allowances, the Contract Sum shall be adjusted accordingly by Change Order. The amount of the Change Order shall reflect (1) the difference between actual costs and the allowances under Clause 3.8.2.2 and (2) changes in Contractor's costs under Clause 3.8.2.3.

3.9 SUPERINTENDENT

3.9.1 The Contractor shall employ a competent superintendent and necessary assistants who shall be in attendance at the Project site during performance of the Work. The superintendent shall represent the Contractor, and communications given to the superintendent shall be as binding as if given to the Contractor. Important communications shall be confirmed in writing. Other communications shall be similarly confirmed on written request in each case.

3.10 CONTRACTOR'S CONSTRUCTION SCHEDULES

3.10.1 The Contractor, promptly after being awarded the Contract, shall prepare and submit for the Owner's and Architect's information a Contractor's construction schedule for the Work. The schedule shall not exceed time limits current under the Contract Documents, shall be revised at appropriate intervals as required by the conditions of the Work and Project, shall be related to the entire Project to the extent required by the Contract Documents, and shall provide for expeditious and practicable execution of the Work.

3.10.2 The Contractor shall prepare and keep current, for the Architect's approval, a schedule of submittals which is coordinated with the Contractor's construction schedule and allows the Architect reasonable time to review submittals.

3.10.3 The Contractor shall conform to the most recent schedules.

3.11 DOCUMENTS AND SAMPLES AT THE SITE

3.11.1 The Contractor shall maintain at the site for the Owner one record copy of the Drawings, Specifications, addenda, Change Orders and other Modifications, in good order and marked currently to record changes and selections made during construction, and in addition approved Shop Drawings, Product Data, Samples and similar required submittals. These shall be available to the Architect and shall be delivered to the Architect for submittal to the Owner upon completion of the Work.

3.12 SHOP DRAWINGS, PRODUCT DATA AND SAMPLES

3.12.1 Shop Drawings are drawings, diagrams, schedules and other data specially prepared for the Work by the Contractor or a Subcontractor, Sub-subcontractor, manufacturer, supplier or distributor to illustrate some portion of the Work.

3.12.2 Product Data are illustrations, standard schedules, performance charts, instructions, brochures, diagrams and other information furnished by the Contractor to illustrate materials or equipment for some portion of the Work.

3.12.3 Samples are physical examples which illustrate materials, equipment or workmanship and establish standards by which the Work will be judged.

3.12.4 Shop Drawings, Product Data, Samples and similar submittals are not Contract Documents. The purpose of their submittal is to demonstrate for those portions of the Work for which submittals are required the way the Contractor proposes to conform to the information given and the design concept expressed in the Contract Documents. Review by the Architect is subject to the limitations of Subparagraph 4.2.7.

3.12.5 The Contractor shall review, approve and submit to the Architect Shop Drawings, Product Data, Samples and similar submittals required by the Contract Documents with reasonable promptness and in such sequence as to cause no delay in the Work or in the activities of the Owner or of separate contractors. Submittals made by the Contractor which are not required by the Contract Documents may be returned without action.

3.12.6 The Contractor shall perform no portion of the Work requiring submittal and review of Shop Drawings, Product Data, Samples or similar submittals until the respective submittal has been approved by the Architect. Such Work shall be in accordance with approved submittals.

3.12.7 By approving and submitting Shop Drawings, Product Data, Samples and similar submittals, the Contractor represents that the Contractor has determined and verified materials, field measurements and field construction criteria related thereto, or will do so, and has checked and coordinated the information contained within such submittals with the requirements of the Work and of the Contract Documents.

3.12.8 The Contractor shall not be relieved of responsibility for deviations from requirements of the Contract Documents by the Architect's approval of Shop Drawings, Product Data, Samples or similar submittals unless the Contractor has specifically informed the Architect in writing of such deviation at the time of submittal and the Architect has given written approval to the specific deviation. The Contractor shall not be relieved of responsibility for errors or omissions in Shop Drawings, Product Data, Samples or similar submittals by the Architect's approval thereof.

3.12.9 The Contractor shall direct specific attention, in writing or on resubmitted Shop Drawings, Product Data, Samples or similar submittals, to revisions other than those requested by the Architect on previous submittals.

3.12.10 Informational submittals upon which the Architect is not expected to take responsive action may be so identified in the Contract Documents.

3.12.11 When professional certification of performance criteria of materials, systems or equipment is required by the Contract Documents, the Architect shall be entitled to rely upon the accuracy and completeness of such calculations and certifications.

3.13 USE OF SITE

3.13.1 The Contractor shall confine operations at the site to areas permitted by law, ordinances, permits and the Contract Documents and shall not unreasonably encumber the site with materials or equipment.

3.14 CUTTING AND PATCHING

3.14.1 The Contractor shall be responsible for cutting, fitting or patching required to complete the Work or to make its parts fit together properly.

3.14.2 The Contractor shall not damage or endanger a portion of the Work or fully or partially completed construction of the Owner or separate contractors by cutting, patching or otherwise altering such construction, or by excavation. The Contractor shall not cut or otherwise alter such construction by the

Owner or a separate contractor except with written consent of the Owner and of such separate contractor; such consent shall not be unreasonably withheld. The Contractor shall not unreasonably withhold from the Owner or a separate contractor the Contractor's consent to cutting or otherwise altering the Work.

3.15 CLEANING UP

3.15.1 The Contractor shall keep the premises and surrounding area free from accumulation of waste materials or rubbish caused by operations under the Contract. At completion of the Work the Contractor shall remove from and about the Project waste materials, rubbish, the Contractor's tools, construction equipment, machinery and surplus materials.

3.15.2 If the Contractor fails to clean up as provided in the Contract Documents, the Owner may do so and the cost thereof shall be charged to the Contractor.

3.16 ACCESS TO WORK

3.16.1 The Contractor shall provide the Owner and Architect access to the Work in preparation and progress wherever located.

3.17 ROYALTIES AND PATENTS

3.17.1 The Contractor shall pay all royalties and license fees. The Contractor shall defend suits or claims for infringement of patent rights and shall hold the Owner and Architect harmless from loss on account thereof, but shall not be responsible for such defense or loss when a particular design, process or product of a particular manufacturer or manufacturers is required by the Contract Documents. However, if the Contractor has reason to believe that the required design, process or product is an infringement of a patent, the Contractor shall be responsible for such loss unless such information is promptly furnished to the Architect.

3.18 INDEMNIFICATION

3.18.1 To the fullest extent permitted by law, the Contractor shall indemnify and hold harmless the Owner, Architect, Architect's consultants, and agents and employees of any of them from and against claims, damages, losses and expenses, including but not limited to attorneys' fees, arising out of or resulting from performance of the Work, provided that such claim, damage, loss or expense is attributable to bodily injury, sickness, disease or death, or to injury to or destruction of tangible property (other than the Work itself) including loss of use resulting therefrom, but only to the extent caused in whole or in part by negligent acts or omissions of the Contractor, a Subcontractor, anyone directly or indirectly employed by them or anyone for whose acts they may be liable, regardless of whether or not such claim, damage, loss or expense is caused in part by a party indemnified hereunder. Such obligation shall not be construed to negate, abridge, or reduce other rights or obligations of indemnity which would otherwise exist as to a party or person described in this Paragraph 3.18.

3.18.2 In claims against any person or entity indemnified under this Paragraph 3.18 by an employee of the Contractor, a Subcontractor, anyone directly or indirectly employed by them or anyone for whose acts they may be liable, the indemnification obligation under this Paragraph 3.18 shall not be limited by a limitation on amount or type of damages, compensation or benefits payable by or for the Contractor or a Subcontractor under workers' or workmen's compensation acts, disability benefit acts or other employee benefit acts.

3.18.3 The obligations of the Contractor under this Paragraph 3.18 shall not extend to the liability of the Architect, the Archi-

tect's consultants, and agents and employees of any of them arising out of (1) the preparation or approval of maps, drawings, opinions, reports, surveys, Change Orders, designs or specifications, or (2) the giving of or the failure to give directions or instructions by the Architect, the Architect's consultants, and agents and employees of any of them provided such giving or failure to give is the primary cause of the injury or damage.

ARTICLE 4

ADMINISTRATION OF THE CONTRACT

4.1 ARCHITECT

4.1.1 The Architect is the person lawfully licensed to practice architecture or an entity lawfully practicing architecture identified as such in the Agreement and is referred to throughout the Contract Documents as if singular in number. The term "Architect" means the Architect or the Architect's authorized representative.

4.1.2 Duties, responsibilities and limitations of authority of the Architect as set forth in the Contract Documents shall not be restricted, modified or extended without written consent of the Owner, Contractor and Architect. Consent shall not be unreasonably withheld.

4.1.3 In case of termination of employment of the Architect, the Owner shall appoint an architect against whom the Contractor makes no reasonable objection and whose status under the Contract Documents shall be that of the former architect.

4.1.4 Disputes arising under Subparagraphs 4.1.2 and 4.1.3 shall be subject to arbitration.

4.2 ARCHITECT'S ADMINISTRATION OF THE CONTRACT

4.2.1 The Architect will provide administration of the Contract as described in the Contract Documents, and will be the Owner's representative (1) during construction, (2) until final payment is due and (3) with the Owner's concurrence, from time to time during the correction period described in Paragraph 12.2. The Architect will advise and consult with the Owner. The Architect will have authority to act on behalf of the Owner only to the extent provided in the Contract Documents, unless otherwise modified by written instrument in accordance with other provisions of the Contract.

4.2.2 The Architect will visit the site at intervals appropriate to the stage of construction to become generally familiar with the progress and quality of the completed Work and to determine in general if the Work is being performed in a manner indicating that the Work, when completed, will be in accordance with the Contract Documents. However, the Architect will not be required to make exhaustive or continuous on-site inspections to check quality or quantity of the Work. On the basis of on-site observations as an architect, the Architect will keep the Owner informed of progress of the Work, and will endeavor to guard the Owner against defects and deficiencies in the Work.

4.2.3 The Architect will not have control over or charge of and will not be responsible for construction means, methods, techniques, sequences or procedures, or for safety precautions and programs in connection with the Work, since these are solely the Contractor's responsibility as provided in Paragraph 3.3. The Architect will not be responsible for the Contractor's failure to carry out the Work in accordance with the Contract Documents. The Architect will not have control over or charge of and will not be responsible for acts or omissions of the Con-

AIA DOCUMENT A201 • GENERAL CONDITIONS OF THE CONTRACT FOR CONSTRUCTION • FOURTEENTH EDITION
AIA® • ©1987 THE AMERICAN INSTITUTE OF ARCHITECTS, 1735 NEW YORK AVENUE, N.W., WASHINGTON, D.C. 20006

tractor, Subcontractors, or their agents or employees, or of any other persons performing portions of the Work.

4.2.4 Communications Facilitating Contract Administration. Except as otherwise provided in the Contract Documents or when direct communications have been specially authorized, the Owner and Contractor shall endeavor to communicate through the Architect. Communications by and with the Architect's consultants shall be through the Architect. Communications by and with Subcontractors and material suppliers shall be through the Contractor. Communications by and with separate contractors shall be through the Owner.

4.2.5 Based on the Architect's observations and evaluations of the Contractor's Applications for Payment, the Architect will review and certify the amounts due the Contractor and will issue Certificates for Payment in such amounts.

4.2.6 The Architect will have authority to reject Work which does not conform to the Contract Documents. Whenever the Architect considers it necessary or advisable for implementation of the intent of the Contract Documents, the Architect will have authority to require additional inspection or testing of the Work in accordance with Subparagraphs 13.5.2 and 13.5.3, whether or not such Work is fabricated, installed or completed. However, neither this authority of the Architect nor a decision made in good faith either to exercise or not to exercise such authority shall give rise to a duty or responsibility of the Architect to the Contractor, Subcontractors, material and equipment suppliers, their agents or employees, or other persons performing portions of the Work.

4.2.7 The Architect will review and approve or take other appropriate action upon the Contractor's submittals such as Shop Drawings, Product Data and Samples, but only for the limited purpose of checking for conformance with information given and the design concept expressed in the Contract Documents. The Architect's action will be taken with such reasonable promptness as to cause no delay in the Work or in the activities of the Owner, Contractor or separate contractors, while allowing sufficient time in the Architect's professional judgment to permit adequate review. Review of such submittals is not conducted for the purpose of determining the accuracy and completeness of other details such as dimensions and quantities, or for substantiating instructions for installation or performance of equipment or systems, all of which remain the responsibility of the Contractor as required by the Contract Documents. The Architect's review of the Contractor's submittals shall not relieve the Contractor of the obligations under Paragraphs 3.3, 3.5 and 3.12. The Architect's review shall not constitute approval of safety precautions or, unless otherwise specifically stated by the Architect, of any construction means, methods, techniques, sequences or procedures. The Architect's approval of a specific item shall not indicate approval of an assembly of which the item is a component.

4.2.8 The Architect will prepare Change Orders and Construction Change Directives, and may authorize minor changes in the Work as provided in Paragraph 7.4.

4.2.9 The Architect will conduct inspections to determine the date or dates of Substantial Completion and the date of final completion, will receive and forward to the Owner for the Owner's review and records written warranties and related documents required by the Contract and assembled by the Contractor, and will issue a final Certificate for Payment upon compliance with the requirements of the Contract Documents.

4.2.10 If the Owner and Architect agree, the Architect will provide one or more project representatives to assist in carrying out the Architect's responsibilities at the site. The duties, responsibilities and limitations of authority of such project representatives shall be as set forth in an exhibit to be incorporated in the Contract Documents.

4.2.11 The Architect will interpret and decide matters concerning performance under and requirements of the Contract Documents on written request of either the Owner or Contractor. The Architect's response to such requests will be made with reasonable promptness and within any time limits agreed upon. If no agreement is made concerning the time within which interpretations required of the Architect shall be furnished in compliance with this Paragraph 4.2, then delay shall not be recognized on account of failure by the Architect to furnish such interpretations until 15 days after written request is made for them.

4.2.12 Interpretations and decisions of the Architect will be consistent with the intent of and reasonably inferable from the Contract Documents and will be in writing or in the form of drawings. When making such interpretations and decisions, the Architect will endeavor to secure faithful performance by both Owner and Contractor, will not show partiality to either and will not be liable for results of interpretations or decisions so rendered in good faith.

4.2.13 The Architect's decisions on matters relating to aesthetic effect will be final if consistent with the intent expressed in the Contract Documents.

4.3   CLAIMS AND DISPUTES

4.3.1 Definition. A Claim is a demand or assertion by one of the parties seeking, as a matter of right, adjustment or interpretation of Contract terms, payment of money, extension of time or other relief with respect to the terms of the Contract. The term "Claim" also includes other disputes and matters in question between the Owner and Contractor arising out of or relating to the Contract. Claims must be made by written notice. The responsibility to substantiate Claims shall rest with the party making the Claim.

4.3.2 Decision of Architect. Claims, including those alleging an error or omission by the Architect, shall be referred initially to the Architect for action as provided in Paragraph 4.4. A decision by the Architect, as provided in Subparagraph 4.4.4, shall be required as a condition precedent to arbitration or litigation of a Claim between the Contractor and Owner as to all such matters arising prior to the date final payment is due, regardless of (1) whether such matters relate to execution and progress of the Work or (2) the extent to which the Work has been completed. The decision by the Architect in response to a Claim shall not be a condition precedent to arbitration or litigation in the event (1) the position of Architect is vacant, (2) the Architect has not received evidence or has failed to render a decision within agreed time limits, (3) the Architect has failed to take action required under Subparagraph 4.4.4 within 30 days after the Claim is made, (4) 45 days have passed after the Claim has been referred to the Architect or (5) the Claim relates to a mechanic's lien.

4.3.3 Time Limits on Claims. Claims by either party must be made within 21 days after occurrence of the event giving rise to such Claim or within 21 days after the claimant first recognizes the condition giving rise to the Claim, whichever is later. Claims must be made by written notice. An additional Claim made after the initial Claim has been implemented by Change Order will not be considered unless submitted in a timely manner.

4.3.4 Continuing Contract Performance. Pending final resolution of a Claim including arbitration, unless otherwise agreed in writing the Contractor shall proceed diligently with performance of the Contract and the Owner shall continue to make payments in accordance with the Contract Documents.

4.3.5 Waiver of Claims: Final Payment. The making of final payment shall constitute a waiver of Claims by the Owner except those arising from:

.1 liens, Claims, security interests or encumbrances arising out of the Contract and unsettled;

.2 failure of the Work to comply with the requirements of the Contract Documents; or

.3 terms of special warranties required by the Contract Documents.

4.3.6 Claims for Concealed or Unknown Conditions. If conditions are encountered at the site which are (1) subsurface or otherwise concealed physical conditions which differ materially from those indicated in the Contract Documents or (2) unknown physical conditions of an unusual nature, which differ materially from those ordinarily found to exist and generally recognized as inherent in construction activities of the character provided for in the Contract Documents, then notice by the observing party shall be given to the other party promptly before conditions are disturbed and in no event later than 21 days after first observance of the conditions. The Architect will promptly investigate such conditions and, if they differ materially and cause an increase or decrease in the Contractor's cost of, or time required for, performance of any part of the Work, will recommend an equitable adjustment in the Contract Sum or Contract Time, or both. If the Architect determines that the conditions at the site are not materially different from those indicated in the Contract Documents and that no change in the terms of the Contract is justified, the Architect shall so notify the Owner and Contractor in writing, stating the reasons. Claims by either party in opposition to such determination must be made within 21 days after the Architect has given notice of the decision. If the Owner and Contractor cannot agree on an adjustment in the Contract Sum or Contract Time, the adjustment shall be referred to the Architect for initial determination, subject to further proceedings pursuant to Paragraph 4.4.

4.3.7 Claims for Additional Cost. If the Contractor wishes to make Claim for an increase in the Contract Sum, written notice as provided herein shall be given before proceeding to execute the Work. Prior notice is not required for Claims relating to an emergency endangering life or property arising under Paragraph 10.3. If the Contractor believes additional cost is involved for reasons including but not limited to (1) a written interpretation from the Architect, (2) an order by the Owner to stop the Work where the Contractor was not at fault, (3) a written order for a minor change in the Work issued by the Architect, (4) failure of payment by the Owner, (5) termination of the Contract by the Owner, (6) Owner's suspension or (7) other reasonable grounds, Claim shall be filed in accordance with the procedure established herein.

4.3.8 Claims for Additional Time

4.3.8.1 If the Contractor wishes to make Claim for an increase in the Contract Time, written notice as provided herein shall be given. The Contractor's Claim shall include an estimate of cost and of probable effect of delay on progress of the Work. In the case of a continuing delay only one Claim is necessary.

4.3.8.2 If adverse weather conditions are the basis for a Claim for additional time, such Claim shall be documented by data substantiating that weather conditions were abnormal for the period of time and could not have been reasonably anticipated, and that weather conditions had an adverse effect on the scheduled construction.

4.3.9 Injury or Damage to Person or Property. If either party to the Contract suffers injury or damage to person or property because of an act or omission of the other party, of any of the other party's employees or agents, or of others for whose acts such party is legally liable, written notice of such injury or damage, whether or not insured, shall be given to the other party within a reasonable time not exceeding 21 days after first observance. The notice shall provide sufficient detail to enable the other party to investigate the matter. If a Claim for additional cost or time related to this Claim is to be asserted, it shall be filed as provided in Subparagraphs 4.3.7 or 4.3.8.

4.4   RESOLUTION OF CLAIMS AND DISPUTES

4.4.1 The Architect will review Claims and take one or more of the following preliminary actions within ten days of receipt of a Claim: (1) request additional supporting data from the claimant, (2) submit a schedule to the parties indicating when the Architect expects to take action, (3) reject the Claim in whole or in part, stating reasons for rejection, (4) recommend approval of the Claim by the other party or (5) suggest a compromise. The Architect may also, but is not obligated to, notify the surety, if any, of the nature and amount of the Claim.

4.4.2 If a Claim has been resolved, the Architect will prepare or obtain appropriate documentation.

4.4.3 If a Claim has not been resolved, the party making the Claim shall, within ten days after the Architect's preliminary response, take one or more of the following actions: (1) submit additional supporting data requested by the Architect, (2) modify the initial Claim or (3) notify the Architect that the initial Claim stands.

4.4.4 If a Claim has not been resolved after consideration of the foregoing and of further evidence presented by the parties or requested by the Architect, the Architect will notify the parties in writing that the Architect's decision will be made within seven days, which decision shall be final and binding on the parties but subject to arbitration. Upon expiration of such time period, the Architect will render to the parties the Architect's written decision relative to the Claim, including any change in the Contract Sum or Contract Time or both. If there is a surety and there appears to be a possibility of a Contractor's default, the Architect may, but is not obligated to, notify the surety and request the surety's assistance in resolving the controversy.

4.5   ARBITRATION

4.5.1 Controversies and Claims Subject to Arbitration. Any controversy or Claim arising out of or related to the Contract, or the breach thereof, shall be settled by arbitration in accordance with the Construction Industry Arbitration Rules of the American Arbitration Association, and judgment upon the award rendered by the arbitrator or arbitrators may be entered in any court having jurisdiction thereof, except controversies or Claims relating to aesthetic effect and except those waived as provided for in Subparagraph 4.3.5. Such controversies or Claims upon which the Architect has given notice and rendered a decision as provided in Subparagraph 4.4.4 shall be subject to arbitration upon written demand of either party. Arbitration may be commenced when 45 days have passed after a Claim has been referred to the Architect as provided in Paragraph 4.3 and no decision has been rendered.

AIA DOCUMENT A201 • GENERAL CONDITIONS OF THE CONTRACT FOR CONSTRUCTION • FOURTEENTH EDITION
AIA® • ©1987 THE AMERICAN INSTITUTE OF ARCHITECTS, 1735 NEW YORK AVENUE, N.W., WASHINGTON, D.C. 20006

4.5.2 Rules and Notices for Arbitration. Claims between the Owner and Contractor not resolved under Paragraph 4.4 shall, if subject to arbitration under Subparagraph 4.5.1, be decided by arbitration in accordance with the Construction Industry Arbitration Rules of the American Arbitration Association currently in effect, unless the parties mutually agree otherwise. Notice of demand for arbitration shall be filed in writing with the other party to the Agreement between the Owner and Contractor and with the American Arbitration Association, and a copy shall be filed with the Architect.

4.5.3 Contract Performance During Arbitration. During arbitration proceedings, the Owner and Contractor shall comply with Subparagraph 4.3.4.

4.5.4 When Arbitration May Be Demanded. Demand for arbitration of any Claim may not be made until the earlier of (1) the date on which the Architect has rendered a final written decision on the Claim, (2) the tenth day after the parties have presented evidence to the Architect or have been given reasonable opportunity to do so, if the Architect has not rendered a final written decision by that date, or (3) any of the five events described in Subparagraph 4.3.2.

4.5.4.1 When a written decision of the Architect states that (1) the decision is final but subject to arbitration and (2) a demand for arbitration of a Claim covered by such decision must be made within 30 days after the date on which the party making the demand receives the final written decision, then failure to demand arbitration within said 30 days' period shall result in the Architect's decision becoming final and binding upon the Owner and Contractor. If the Architect renders a decision after arbitration proceedings have been initiated, such decision may be entered as evidence, but shall not supersede arbitration proceedings unless the decision is acceptable to all parties concerned.

4.5.4.2 A demand for arbitration shall be made within the time limits specified in Subparagraphs 4.5.1 and 4.5.4 and Clause 4.5.4.1 as applicable, and in other cases within a reasonable time after the Claim has arisen, and in no event shall it be made after the date when institution of legal or equitable proceedings based on such Claim would be barred by the applicable statute of limitations as determined pursuant to Paragraph 13.7.

4.5.5 Limitation on Consolidation or Joinder. No arbitration arising out of or relating to the Contract Documents shall include, by consolidation or joinder or in any other manner, the Architect, the Architect's employees or consultants, except by written consent containing specific reference to the Agreement and signed by the Architect, Owner, Contractor and any other person or entity sought to be joined. No arbitration shall include, by consolidation or joinder or in any other manner, parties other than the Owner, Contractor, a separate contractor as described in Article 6 and other persons substantially involved in a common question of fact or law whose presence is required if complete relief is to be accorded in arbitration. No person or entity other than the Owner, Contractor or a separate contractor as described in Article 6 shall be included as an original third party or additional third party to an arbitration whose interest or responsibility is insubstantial. Consent to arbitration involving an additional person or entity shall not constitute consent to arbitration of a dispute not described therein or with a person or entity not named or described therein. The foregoing agreement to arbitrate and other agreements to arbitrate with an additional person or entity duly consented to by parties to the Agreement shall be specifically enforceable under applicable law in any court having jurisdiction thereof.

4.5.6 Claims and Timely Assertion of Claims. A party who files a notice of demand for arbitration must assert in the demand all Claims then known to that party on which arbitration is permitted to be demanded. When a party fails to include a Claim through oversight, inadvertence or excusable neglect, or when a Claim has matured or been acquired subsequently, the arbitrator or arbitrators may permit amendment.

4.5.7 Judgment on Final Award. The award rendered by the arbitrator or arbitrators shall be final, and judgment may be entered upon it in accordance with applicable law in any court having jurisdiction thereof.

ARTICLE 5

SUBCONTRACTORS

5.1   DEFINITIONS

5.1.1 A Subcontractor is a person or entity who has a direct contract with the Contractor to perform a portion of the Work at the site. The term "Subcontractor" is referred to throughout the Contract Documents as if singular in number and means a Subcontractor or an authorized representative of the Subcontractor. The term "Subcontractor" does not include a separate contractor or subcontractors of a separate contractor.

5.1.2 A Sub-subcontractor is a person or entity who has a direct or indirect contract with a Subcontractor to perform a portion of the Work at the site. The term "Sub-subcontractor" is referred to throughout the Contract Documents as if singular in number and means a Sub-subcontractor or an authorized representative of the Sub-subcontractor.

5.2   AWARD OF SUBCONTRACTS AND OTHER CONTRACTS FOR PORTIONS OF THE WORK

5.2.1 Unless otherwise stated in the Contract Documents or the bidding requirements, the Contractor, as soon as practicable after award of the Contract, shall furnish in writing to the Owner through the Architect the names of persons or entities (including those who are to furnish materials or equipment fabricated to a special design) proposed for each principal portion of the Work. The Architect will promptly reply to the Contractor in writing stating whether or not the Owner or the Architect, after due investigation, has reasonable objection to any such proposed person or entity. Failure of the Owner or Architect to reply promptly shall constitute notice of no reasonable objection.

5.2.2 The Contractor shall not contract with a proposed person or entity to whom the Owner or Architect has made reasonable and timely objection. The Contractor shall not be required to contract with anyone to whom the Contractor has made reasonable objection.

5.2.3 If the Owner or Architect has reasonable objection to a person or entity proposed by the Contractor, the Contractor shall propose another to whom the Owner or Architect has no reasonable objection. The Contract Sum shall be increased or decreased by the difference in cost occasioned by such change and an appropriate Change Order shall be issued. However, no increase in the Contract Sum shall be allowed for such change unless the Contractor has acted promptly and responsively in submitting names as required.

5.2.4 The Contractor shall not change a Subcontractor, person or entity previously selected if the Owner or Architect makes reasonable objection to such change.

5.3 SUBCONTRACTUAL RELATIONS

5.3.1 By appropriate agreement, written where legally required for validity, the Contractor shall require each Subcontractor, to the extent of the Work to be performed by the Subcontractor, to be bound to the Contractor by terms of the Contract Documents, and to assume toward the Contractor all the obligations and responsibilities which the Contractor, by these Documents, assumes toward the Owner and Architect. Each subcontract agreement shall preserve and protect the rights of the Owner and Architect under the Contract Documents with respect to the Work to be performed by the Subcontractor so that subcontracting thereof will not prejudice such rights, and shall allow to the Subcontractor, unless specifically provided otherwise in the subcontract agreement, the benefit of all rights, remedies and redress against the Contractor that the Contractor, by the Contract Documents, has against the Owner. Where appropriate, the Contractor shall require each Subcontractor to enter into similar agreements with Sub-sub-contractors. The Contractor shall make available to each proposed Subcontractor, prior to the execution of the subcontract agreement, copies of the Contract Documents to which the Subcontractor will be bound, and, upon written request of the Subcontractor, identify to the Subcontractor terms and conditions of the proposed subcontract agreement which may be at variance with the Contract Documents. Subcontractors shall similarly make copies of applicable portions of such documents available to their respective proposed Sub-subcontractors.

5.4 CONTINGENT ASSIGNMENT OF SUBCONTRACTS

5.4.1 Each subcontract agreement for a portion of the Work is assigned by the Contractor to the Owner provided that:

.1 assignment is effective only after termination of the Contract by the Owner for cause pursuant to Paragraph 14.2 and only for those subcontract agreements which the Owner accepts by notifying the Subcontractor in writing; and

.2 assignment is subject to the prior rights of the surety, if any, obligated under bond relating to the Contract.

5.4.2 If the Work has been suspended for more than 30 days, the Subcontractor's compensation shall be equitably adjusted.

ARTICLE 6

CONSTRUCTION BY OWNER OR BY SEPARATE CONTRACTORS

6.1 OWNER'S RIGHT TO PERFORM CONSTRUCTION AND TO AWARD SEPARATE CONTRACTS

6.1.1 The Owner reserves the right to perform construction or operations related to the Project with the Owner's own forces, and to award separate contracts in connection with other portions of the Project or other construction or operations on the site under Conditions of the Contract identical or substantially similar to these including those portions related to insurance and waiver of subrogation. If the Contractor claims that delay or additional cost is involved because of such action by the Owner, the Contractor shall make such Claim as provided elsewhere in the Contract Documents.

6.1.2 When separate contracts are awarded for different portions of the Project or other construction or operations on the site, the term "Contractor" in the Contract Documents in each case shall mean the Contractor who executes each separate Owner-Contractor Agreement.

6.1.3 The Owner shall provide for coordination of the activities of the Owner's own forces and of each separate contractor with the Work of the Contractor, who shall cooperate with them. The Contractor shall participate with other separate contractors and the Owner in reviewing their construction schedules when directed to do so. The Contractor shall make any revisions to the construction schedule and Contract Sum deemed necessary after a joint review and mutual agreement. The construction schedules shall then constitute the schedules to be used by the Contractor, separate contractors and the Owner until subsequently revised.

6.1.4 Unless otherwise provided in the Contract Documents, when the Owner performs construction or operations related to the Project with the Owner's own forces, the Owner shall be deemed to be subject to the same obligations and to have the same rights which apply to the Contractor under the Conditions of the Contract, including, without excluding others, those stated in Article 3, this Article 6 and Articles 10, 11 and 12.

6.2 MUTUAL RESPONSIBILITY

6.2.1 The Contractor shall afford the Owner and separate contractors reasonable opportunity for introduction and storage of their materials and equipment and performance of their activities and shall connect and coordinate the Contractor's construction and operations with theirs as required by the Contract Documents.

6.2.2 If part of the Contractor's Work depends for proper execution or results upon construction or operations by the Owner or a separate contractor, the Contractor shall, prior to proceeding with that portion of the Work, promptly report to the Architect apparent discrepancies or defects in such other construction that would render it unsuitable for such proper execution and results. Failure of the Contractor so to report shall constitute an acknowledgment that the Owner's or separate contractors' completed or partially completed construction is fit and proper to receive the Contractor's Work, except as to defects not then reasonably discoverable.

6.2.3 Costs caused by delays or by improperly timed activities or defective construction shall be borne by the party responsible therefor.

6.2.4 The Contractor shall promptly remedy damage wrongfully caused by the Contractor to completed or partially completed construction or to property of the Owner or separate contractors as provided in Subparagraph 10.2.5.

6.2.5 Claims and other disputes and matters in question between the Contractor and a separate contractor shall be subject to the provisions of Paragraph 4.3 provided the separate contractor has reciprocal obligations.

6.2.6 The Owner and each separate contractor shall have the same responsibilities for cutting and patching as are described for the Contractor in Paragraph 3.14.

6.3 OWNER'S RIGHT TO CLEAN UP

6.3.1 If a dispute arises among the Contractor, separate contractors and the Owner as to the responsibility under their respective contracts for maintaining the premises and surrounding area free from waste materials and rubbish as described in Paragraph 3.15, the Owner may clean up and allocate the cost among those responsible as the Architect determines to be just.

AIA DOCUMENT A201 • GENERAL CONDITIONS OF THE CONTRACT FOR CONSTRUCTION • FOURTEENTH EDITION
AIA® • ©1987 THE AMERICAN INSTITUTE OF ARCHITECTS, 1735 NEW YORK AVENUE, N.W., WASHINGTON, D.C. 20006

ARTICLE 7

CHANGES IN THE WORK

7.1 CHANGES

7.1.1 Changes in the Work may be accomplished after execution of the Contract, and without invalidating the Contract, by Change Order, Construction Change Directive or order for a minor change in the Work, subject to the limitations stated in this Article 7 and elsewhere in the Contract Documents.

7.1.2 A Change Order shall be based upon agreement among the Owner, Contractor and Architect; a Construction Change Directive requires agreement by the Owner and Architect and may or may not be agreed to by the Contractor; an order for a minor change in the Work may be issued by the Architect alone.

7.1.3 Changes in the Work shall be performed under applicable provisions of the Contract Documents, and the Contractor shall proceed promptly, unless otherwise provided in the Change Order, Construction Change Directive or order for a minor change in the Work.

7.1.4 If unit prices are stated in the Contract Documents or subsequently agreed upon, and if quantities originally contemplated are so changed in a proposed Change Order or Construction Change Directive that application of such unit prices to quantities of Work proposed will cause substantial inequity to the Owner or Contractor, the applicable unit prices shall be equitably adjusted.

7.2 CHANGE ORDERS

7.2.1 A Change Order is a written instrument prepared by the Architect and signed by the Owner, Contractor and Architect, stating their agreement upon all of the following:

.1 a change in the Work;

.2 the amount of the adjustment in the Contract Sum, if any; and

.3 the extent of the adjustment in the Contract Time, if any.

7.2.2 Methods used in determining adjustments to the Contract Sum may include those listed in Subparagraph 7.3.3.

7.3 CONSTRUCTION CHANGE DIRECTIVES

7.3.1 A Construction Change Directive is a written order prepared by the Architect and signed by the Owner and Architect, directing a change in the Work and stating a proposed basis for adjustment, if any, in the Contract Sum or Contract Time, or both. The Owner may by Construction Change Directive, without invalidating the Contract, order changes in the Work within the general scope of the Contract consisting of additions, deletions or other revisions, the Contract Sum and Contract Time being adjusted accordingly.

7.3.2 A Construction Change Directive shall be used in the absence of total agreement on the terms of a Change Order.

7.3.3 If the Construction Change Directive provides for an adjustment to the Contract Sum, the adjustment shall be based on one of the following methods:

.1 mutual acceptance of a lump sum properly itemized and supported by sufficient substantiating data to permit evaluation;

.2 unit prices stated in the Contract Documents or subsequently agreed upon;

.3 cost to be determined in a manner agreed upon by the parties and a mutually acceptable fixed or percentage fee; or

.4 as provided in Subparagraph 7.3.6.

7.3.4 Upon receipt of a Construction Change Directive, the Contractor shall promptly proceed with the change in the Work involved and advise the Architect of the Contractor's agreement or disagreement with the method, if any, provided in the Construction Change Directive for determining the proposed adjustment in the Contract Sum or Contract Time.

7.3.5 A Construction Change Directive signed by the Contractor indicates the agreement of the Contractor therewith, including adjustment in Contract Sum and Contract Time or the method for determining them. Such agreement shall be effective immediately and shall be recorded as a Change Order.

7.3.6 If the Contractor does not respond promptly or disagrees with the method for adjustment in the Contract Sum, the method and the adjustment shall be determined by the Architect on the basis of reasonable expenditures and savings of those performing the Work attributable to the change, including, in case of an increase in the Contract Sum, a reasonable allowance for overhead and profit. In such case, and also under Clause 7.3.3.3, the Contractor shall keep and present, in such form as the Architect may prescribe, an itemized accounting together with appropriate supporting data. Unless otherwise provided in the Contract Documents, costs for the purposes of this Subparagraph 7.3.6 shall be limited to the following:

.1 costs of labor, including social security, old age and unemployment insurance, fringe benefits required by agreement or custom, and workers' or workmen's compensation insurance;

.2 costs of materials, supplies and equipment, including cost of transportation, whether incorporated or consumed;

.3 rental costs of machinery and equipment, exclusive of hand tools, whether rented from the Contractor or others;

.4 costs of premiums for all bonds and insurance, permit fees, and sales, use or similar taxes related to the Work; and

.5 additional costs of supervision and field office personnel directly attributable to the change.

7.3.7 Pending final determination of cost to the Owner, amounts not in dispute may be included in Applications for Payment. The amount of credit to be allowed by the Contractor to the Owner for a deletion or change which results in a net decrease in the Contract Sum shall be actual net cost as confirmed by the Architect. When both additions and credits covering related Work or substitutions are involved in a change, the allowance for overhead and profit shall be figured on the basis of net increase, if any, with respect to that change.

7.3.8 If the Owner and Contractor do not agree with the adjustment in Contract Time or the method for determining it, the adjustment or the method shall be referred to the Architect for determination.

7.3.9 When the Owner and Contractor agree with the determination made by the Architect concerning the adjustments in the Contract Sum and Contract Time, or otherwise reach agreement upon the adjustments, such agreement shall be effective immediately and shall be recorded by preparation and execution of an appropriate Change Order.

7.4 MINOR CHANGES IN THE WORK

7.4.1 The Architect will have authority to order minor changes in the Work not involving adjustment in the Contract Sum or extension of the Contract Time and not inconsistent with the intent of the Contract Documents. Such changes shall be effected by written order and shall be binding on the Owner and Contractor. The Contractor shall carry out such written orders promptly.

ARTICLE 8

TIME

8.1 DEFINITIONS

8.1.1 Unless otherwise provided, Contract Time is the period of time, including authorized adjustments, allotted in the Contract Documents for Substantial Completion of the Work.

8.1.2 The date of commencement of the Work is the date established in the Agreement. The date shall not be postponed by the failure to act of the Contractor or of persons or entities for whom the Contractor is responsible.

8.1.3 The date of Substantial Completion is the date certified by the Architect in accordance with Paragraph 9.8.

8.1.4 The term "day" as used in the Contract Documents shall mean calendar day unless otherwise specifically defined.

8.2 PROGRESS AND COMPLETION

8.2.1 Time limits stated in the Contract Documents are of the essence of the Contract. By executing the Agreement the Contractor confirms that the Contract Time is a reasonable period for performing the Work.

8.2.2 The Contractor shall not knowingly, except by agreement or instruction of the Owner in writing, prematurely commence operations on the site or elsewhere prior to the effective date of insurance required by Article 11 to be furnished by the Contractor. The date of commencement of the Work shall not be changed by the effective date of such insurance. Unless the date of commencement is established by a notice to proceed given by the Owner, the Contractor shall notify the Owner in writing not less than five days or other agreed period before commencing the Work to permit the timely filing of mortgages, mechanic's liens and other security interests.

8.2.3 The Contractor shall proceed expeditiously with adequate forces and shall achieve Substantial Completion within the Contract Time.

8.3 DELAYS AND EXTENSIONS OF TIME

8.3.1 If the Contractor is delayed at any time in progress of the Work by an act or neglect of the Owner or Architect, or of an employee of either, or of a separate contractor employed by the Owner, or by changes ordered in the Work, or by labor disputes, fire, unusual delay in deliveries, unavoidable casualties or other causes beyond the Contractor's control, or by delay authorized by the Owner pending arbitration, or by other causes which the Architect determines may justify delay, then the Contract Time shall be extended by Change Order for such reasonable time as the Architect may determine.

8.3.2 Claims relating to time shall be made in accordance with applicable provisions of Paragraph 4.3.

8.3.3 This Paragraph 8.3 does not preclude recovery of damages for delay by either party under other provisions of the Contract Documents.

ARTICLE 9

PAYMENTS AND COMPLETION

9.1 CONTRACT SUM

9.1.1 The Contract Sum is stated in the Agreement and, including authorized adjustments, is the total amount payable by the Owner to the Contractor for performance of the Work under the Contract Documents.

9.2 SCHEDULE OF VALUES

9.2.1 Before the first Application for Payment, the Contractor shall submit to the Architect a schedule of values allocated to various portions of the Work, prepared in such form and supported by such data to substantiate its accuracy as the Architect may require. This schedule, unless objected to by the Architect, shall be used as a basis for reviewing the Contractor's Applications for Payment.

9.3 APPLICATIONS FOR PAYMENT

9.3.1 At least ten days before the date established for each progress payment, the Contractor shall submit to the Architect an itemized Application for Payment for operations completed in accordance with the schedule of values. Such application shall be notarized, if required, and supported by such data substantiating the Contractor's right to payment as the Owner or Architect may require, such as copies of requisitions from Subcontractors and material suppliers, and reflecting retainage if provided for elsewhere in the Contract Documents.

9.3.1.1 Such applications may include requests for payment on account of changes in the Work which have been properly authorized by Construction Change Directives but not yet included in Change Orders.

9.3.1.2 Such applications may not include requests for payment of amounts the Contractor does not intend to pay to a Subcontractor or material supplier because of a dispute or other reason.

9.3.2 Unless otherwise provided in the Contract Documents, payments shall be made on account of materials and equipment delivered and suitably stored at the site for subsequent incorporation in the Work. If approved in advance by the Owner, payment may similarly be made for materials and equipment suitably stored off the site at a location agreed upon in writing. Payment for materials and equipment stored on or off the site shall be conditioned upon compliance by the Contractor with procedures satisfactory to the Owner to establish the Owner's title to such materials and equipment or otherwise protect the Owner's interest, and shall include applicable insurance, storage and transportation to the site for such materials and equipment stored off the site.

9.3.3 The Contractor warrants that title to all Work covered by an Application for Payment will pass to the Owner no later than the time of payment. The Contractor further warrants that upon submittal of an Application for Payment all Work for which Certificates for Payment have been previously issued and payments received from the Owner shall, to the best of the Contractor's knowledge, information and belief, be free and clear of liens, claims, security interests or encumbrances in favor of the Contractor, Subcontractors, material suppliers, or other persons or entities making a claim by reason of having provided labor, materials and equipment relating to the Work.

9.4 CERTIFICATES FOR PAYMENT

9.4.1 The Architect will, within seven days after receipt of the Contractor's Application for Payment, either issue to the

---

AIA DOCUMENT A201 • GENERAL CONDITIONS OF THE CONTRACT FOR CONSTRUCTION • FOURTEENTH EDITION
AIA® • ©1987 THE AMERICAN INSTITUTE OF ARCHITECTS, 1735 NEW YORK AVENUE, N.W., WASHINGTON, D.C. 20006

Owner a Certificate for Payment, with a copy to the Contractor, for such amount as the Architect determines is properly due, or notify the Contractor and Owner in writing of the Architect's reasons for withholding certification in whole or in part as provided in Subparagraph 9.5.1.

9.4.2 The issuance of a Certificate for Payment will constitute a representation by the Architect to the Owner, based on the Architect's observations at the site and the data comprising the Application for Payment, that the Work has progressed to the point indicated and that, to the best of the Architect's knowledge, information and belief, quality of the Work is in accordance with the Contract Documents. The foregoing representations are subject to an evaluation of the Work for conformance with the Contract Documents upon Substantial Completion, to results of subsequent tests and inspections, to minor deviations from the Contract Documents correctable prior to completion and to specific qualifications expressed by the Architect. The issuance of a Certificate for Payment will further constitute a representation that the Contractor is entitled to payment in the amount certified. However, the issuance of a Certificate for Payment will not be a representation that the Architect has (1) made exhaustive or continuous on-site inspections to check the quality or quantity of the Work, (2) reviewed construction means, methods, techniques, sequences or procedures, (3) reviewed copies of requisitions received from Subcontractors and material suppliers and other data requested by the Owner to substantiate the Contractor's right to payment or (4) made examination to ascertain how or for what purpose the Contractor has used money previously paid on account of the Contract Sum.

9.5 DECISIONS TO WITHHOLD CERTIFICATION

9.5.1 The Architect may decide not to certify payment and may withhold a Certificate for Payment in whole or in part, to the extent reasonably necessary to protect the Owner, if in the Architect's opinion the representations to the Owner required by Subparagraph 9.4.2 cannot be made. If the Architect is unable to certify payment in the amount of the Application, the Architect will notify the Contractor and Owner as provided in Subparagraph 9.4.1. If the Contractor and Architect cannot agree on a revised amount, the Architect will promptly issue a Certificate for Payment for the amount for which the Architect is able to make such representations to the Owner. The Architect may also decide not to certify payment or, because of subsequently discovered evidence or subsequent observations, may nullify the whole or a part of a Certificate for Payment previously issued, to such extent as may be necessary in the Architect's opinion to protect the Owner from loss because of:

.1 defective Work not remedied;

.2 third party claims filed or reasonable evidence indicating probable filing of such claims;

.3 failure of the Contractor to make payments properly to Subcontractors or for labor, materials or equipment;

.4 reasonable evidence that the Work cannot be completed for the unpaid balance of the Contract Sum;

.5 damage to the Owner or another contractor;

.6 reasonable evidence that the Work will not be completed within the Contract Time, and that the unpaid balance would not be adequate to cover actual or liquidated damages for the anticipated delay; or

.7 persistent failure to carry out the Work in accordance with the Contract Documents.

9.5.2 When the above reasons for withholding certification are removed, certification will be made for amounts previously withheld.

9.6 PROGRESS PAYMENTS

9.6.1 After the Architect has issued a Certificate for Payment, the Owner shall make payment in the manner and within the time provided in the Contract Documents, and shall so notify the Architect.

9.6.2 The Contractor shall promptly pay each Subcontractor, upon receipt of payment from the Owner, out of the amount paid to the Contractor on account of such Subcontractor's portion of the Work, the amount to which said Subcontractor is entitled, reflecting percentages actually retained from payments to the Contractor on account of such Subcontractor's portion of the Work. The Contractor shall, by appropriate agreement with each Subcontractor, require each Subcontractor to make payments to Sub-subcontractors in similar manner.

9.6.3 The Architect will, on request, furnish to a Subcontractor, if practicable, information regarding percentages of completion or amounts applied for by the Contractor and action taken thereon by the Architect and Owner on account of portions of the Work done by such Subcontractor.

9.6.4 Neither the Owner nor Architect shall have an obligation to pay or to see to the payment of money to a Subcontractor except as may otherwise be required by law.

9.6.5 Payment to material suppliers shall be treated in a manner similar to that provided in Subparagraphs 9.6.2, 9.6.3 and 9.6.4.

9.6.6 A Certificate for Payment, a progress payment, or partial or entire use or occupancy of the Project by the Owner shall not constitute acceptance of Work not in accordance with the Contract Documents.

9.7 FAILURE OF PAYMENT

9.7.1 If the Architect does not issue a Certificate for Payment, through no fault of the Contractor, within seven days after receipt of the Contractor's Application for Payment, or if the Owner does not pay the Contractor within seven days after the date established in the Contract Documents the amount certified by the Architect or awarded by arbitration, then the Contractor may, upon seven additional days' written notice to the Owner and Architect, stop the Work until payment of the amount owing has been received. The Contract Time shall be extended appropriately and the Contract Sum shall be increased by the amount of the Contractor's reasonable costs of shut-down, delay and start-up, which shall be accomplished as provided in Article 7.

9.8 SUBSTANTIAL COMPLETION

9.8.1 Substantial Completion is the stage in the progress of the Work when the Work or designated portion thereof is sufficiently complete in accordance with the Contract Documents so the Owner can occupy or utilize the Work for its intended use.

9.8.2 When the Contractor considers that the Work, or a portion thereof which the Owner agrees to accept separately, is substantially complete, the Contractor shall prepare and submit to the Architect a comprehensive list of items to be completed or corrected. The Contractor shall proceed promptly to complete and correct items on the list. Failure to include an item on such list does not alter the responsibility of the Contractor to complete all Work in accordance with the Contract Documents. Upon receipt of the Contractor's list, the Architect will make an inspection to determine whether the Work or designated portion thereof is substantially complete. If the Architect's inspection discloses any item, whether or not included on the Contractor's list, which is not in accordance with the requirements of the Contract Documents, the Contractor shall, before issuance of the Certificate of Substantial Completion, complete or correct such item upon notification by the Architect. The Contractor shall then submit a request for another inspection by the Architect to determine Substantial Completion. When the Work or designated portion thereof is substantially complete, the Architect will prepare a Certificate of Substantial Completion which shall establish the date of Substantial Completion, shall establish responsibilities of the Owner and Contractor for security, maintenance, heat, utilities, damage to the Work and insurance, and shall fix the time within which the Contractor shall finish all items on the list accompanying the Certificate. Warranties required by the Contract Documents shall commence on the date of Substantial Completion of the Work or designated portion thereof unless otherwise provided in the Certificate of Substantial Completion. The Certificate of Substantial Completion shall be submitted to the Owner and Contractor for their written acceptance of responsibilities assigned to them in such Certificate.

9.8.3 Upon Substantial Completion of the Work or designated portion thereof and upon application by the Contractor and certification by the Architect, the Owner shall make payment, reflecting adjustment in retainage, if any, for such Work or portion thereof as provided in the Contract Documents.

9.9   PARTIAL OCCUPANCY OR USE

9.9.1 The Owner may occupy or use any completed or partially completed portion of the Work at any stage when such portion is designated by separate agreement with the Contractor, provided such occupancy or use is consented to by the insurer as required under Subparagraph 11.3.11 and authorized by public authorities having jurisdiction over the Work. Such partial occupancy or use may commence whether or not the portion is substantially complete, provided the Owner and Contractor have accepted in writing the responsibilities assigned to each of them for payments, retainage if any, security, maintenance, heat, utilities, damage to the Work and insurance, and have agreed in writing concerning the period for correction of the Work and commencement of warranties required by the Contract Documents. When the Contractor considers a portion substantially complete, the Contractor shall prepare and submit a list to the Architect as provided under Subparagraph 9.8.2. Consent of the Contractor to partial occupancy or use shall not be unreasonably withheld. The stage of the progress of the Work shall be determined by written agreement between the Owner and Contractor or, if no agreement is reached, by decision of the Architect.

9.9.2 Immediately prior to such partial occupancy or use, the Owner, Contractor and Architect shall jointly inspect the area to be occupied or portion of the Work to be used in order to determine and record the condition of the Work.

9.9.3 Unless otherwise agreed upon, partial occupancy or use of a portion or portions of the Work shall not constitute acceptance of Work not complying with the requirements of the Contract Documents.

9.10   FINAL COMPLETION AND FINAL PAYMENT

9.10.1 Upon receipt of written notice that the Work is ready for final inspection and acceptance and upon receipt of a final Application for Payment, the Architect will promptly make

such inspection and, when the Architect finds the Work acceptable under the Contract Documents and the Contract fully performed, the Architect will promptly issue a final Certificate for Payment stating that to the best of the Architect's knowledge, information and belief, and on the basis of the Architect's observations and inspections, the Work has been completed in accordance with terms and conditions of the Contract Documents and that the entire balance found to be due the Contractor and noted in said final Certificate is due and payable. The Architect's final Certificate for Payment will constitute a further representation that conditions listed in Subparagraph 9.10.2 as precedent to the Contractor's being entitled to final payment have been fulfilled.

9.10.2 Neither final payment nor any remaining retained percentage shall become due until the Contractor submits to the Architect (1) an affidavit that payrolls, bills for materials and equipment, and other indebtedness connected with the Work for which the Owner or the Owner's property might be responsible or encumbered (less amounts withheld by Owner) have been paid or otherwise satisfied, (2) a certificate evidencing that insurance required by the Contract Documents to remain in force after final payment is currently in effect and will not be cancelled or allowed to expire until at least 30 days' prior written notice has been given to the Owner, (3) a written statement that the Contractor knows of no substantial reason that the insurance will not be renewable to cover the period required by the Contract Documents, (4) consent of surety, if any, to final payment and (5), if required by the Owner, other data establishing payment or satisfaction of obligations, such as receipts, releases and waivers of liens, claims, security interests or encumbrances arising out of the Contract, to the extent and in such form as may be designated by the Owner. If a Subcontractor refuses to furnish a release or waiver required by the Owner, the Contractor may furnish a bond satisfactory to the Owner to indemnify the Owner against such lien. If such lien remains unsatisfied after payments are made, the Contractor shall refund to the Owner all money that the Owner may be compelled to pay in discharging such lien, including all costs and reasonable attorneys' fees.

9.10.3 If, after Substantial Completion of the Work, final completion thereof is materially delayed through no fault of the Contractor or by issuance of Change Orders affecting final completion, and the Architect so confirms, the Owner shall, upon application by the Contractor and certification by the Architect, and without terminating the Contract, make payment of the balance due for that portion of the Work fully completed and accepted. If the remaining balance for Work not fully completed or corrected is less than retainage stipulated in the Contract Documents, and if bonds have been furnished, the written consent of surety to payment of the balance due for that portion of the Work fully completed and accepted shall be submitted by the Contractor to the Architect prior to certification of such payment. Such payment shall be made under terms and conditions governing final payment, except that it shall not constitute a waiver of claims. The making of final payment shall constitute a waiver of claims by the Owner as provided in Subparagraph 4.3.5.

9.10.4 Acceptance of final payment by the Contractor, a Subcontractor or material supplier shall constitute a waiver of claims by that payee except those previously made in writing and identified by that payee as unsettled at the time of final Application for Payment. Such waivers shall be in addition to the waiver described in Subparagraph 4.3.5.

AIA DOCUMENT A201 • GENERAL CONDITIONS OF THE CONTRACT FOR CONSTRUCTION • FOURTEENTH EDITION
AIA® • ©1987 THE AMERICAN INSTITUTE OF ARCHITECTS, 1735 NEW YORK AVENUE, N.W., WASHINGTON, D.C. 20006

ARTICLE 10

PROTECTION OF PERSONS AND PROPERTY

10.1 SAFETY PRECAUTIONS AND PROGRAMS

10.1.1 The Contractor shall be responsible for initiating, maintaining and supervising all safety precautions and programs in connection with the performance of the Contract.

10.1.2 In the event the Contractor encounters on the site material reasonably believed to be asbestos or polychlorinated biphenyl (PCB) which has not been rendered harmless, the Contractor shall immediately stop Work in the area affected and report the condition to the Owner and Architect in writing. The Work in the affected area shall not thereafter be resumed except by written agreement of the Owner and Contractor if in fact the material is asbestos or polychlorinated biphenyl (PCB) and has not been rendered harmless. The Work in the affected area shall be resumed in the absence of asbestos or polychlorinated biphenyl (PCB), or when it has been rendered harmless, by written agreement of the Owner and Contractor, or in accordance with final determination by the Architect on which arbitration has not been demanded, or by arbitration under Article 4.

10.1.3 The Contractor shall not be required pursuant to Article 7 to perform without consent any Work relating to asbestos or polychlorinated biphenyl (PCB).

10.1.4 To the fullest extent permitted by law, the Owner shall indemnify and hold harmless the Contractor, Architect, Architect's consultants and agents and employees of any of them from and against claims, damages, losses and expenses, including but not limited to attorneys' fees, arising out of or resulting from performance of the Work in the affected area if in fact the material is asbestos or polychlorinated biphenyl (PCB) and has not been rendered harmless, provided that such claim, damage, loss or expense is attributable to bodily injury, sickness, disease or death, or to injury to or destruction of tangible property (other than the Work itself) including loss of use resulting therefrom, but only to the extent caused in whole or in part by negligent acts or omissions of the Owner, anyone directly or indirectly employed by the Owner or anyone for whose acts the Owner may be liable, regardless of whether or not such claim, damage, loss or expense is caused in part by a party indemnified hereunder. Such obligation shall not be construed to negate, abridge, or reduce other rights or obligations of indemnity which would otherwise exist as to a party or person described in this Subparagraph 10.1.4.

10.2 SAFETY OF PERSONS AND PROPERTY

10.2.1 The Contractor shall take reasonable precautions for safety of, and shall provide reasonable protection to prevent damage, injury or loss to:

.1 employees on the Work and other persons who may be affected thereby;

.2 the Work and materials and equipment to be incorporated therein, whether in storage on or off the site, under care, custody or control of the Contractor or the Contractor's Subcontractors or Sub-subcontractors; and

.3 other property at the site or adjacent thereto, such as trees, shrubs, lawns, walks, pavements, roadways, structures and utilities not designated for removal, relocation or replacement in the course of construction.

10.2.2 The Contractor shall give notices and comply with applicable laws, ordinances, rules, regulations and lawful orders of public authorities bearing on safety of persons or property or their protection from damage, injury or loss.

10.2.3 The Contractor shall erect and maintain, as required by existing conditions and performance of the Contract, reasonable safeguards for safety and protection, including posting danger signs and other warnings against hazards, promulgating safety regulations and notifying owners and users of adjacent sites and utilities.

10.2.4 When use or storage of explosives or other hazardous materials or equipment or unusual methods are necessary for execution of the Work, the Contractor shall exercise utmost care and carry on such activities under supervision of properly qualified personnel.

10.2.5 The Contractor shall promptly remedy damage and loss (other than damage or loss insured under property insurance required by the Contract Documents) to property referred to in Clauses 10.2.1.2 and 10.2.1.3 caused in whole or in part by the Contractor, a Subcontractor, a Sub-subcontractor, or anyone directly or indirectly employed by any of them, or by anyone for whose acts they may be liable and for which the Contractor is responsible under Clauses 10.2.1.2 and 10.2.1.3, except damage or loss attributable to acts or omissions of the Owner or Architect or anyone directly or indirectly employed by either of them, or by anyone for whose acts either of them may be liable, and not attributable to the fault or negligence of the Contractor. The foregoing obligations of the Contractor are in addition to the Contractor's obligations under Paragraph 3.18.

10.2.6 The Contractor shall designate a responsible member of the Contractor's organization at the site whose duty shall be the prevention of accidents. This person shall be the Contractor's superintendent unless otherwise designated by the Contractor in writing to the Owner and Architect.

10.2.7 The Contractor shall not load or permit any part of the construction or site to be loaded so as to endanger its safety.

10.3 EMERGENCIES

10.3.1 In an emergency affecting safety of persons or property, the Contractor shall act, at the Contractor's discretion, to prevent threatened damage, injury or loss. Additional compensation or extension of time claimed by the Contractor on account of an emergency shall be determined as provided in Paragraph 4.3 and Article 7.

ARTICLE 11

INSURANCE AND BONDS

11.1 CONTRACTOR'S LIABILITY INSURANCE

11.1.1 The Contractor shall purchase from and maintain in a company or companies lawfully authorized to do business in the jurisdiction in which the Project is located such insurance as will protect the Contractor from claims set forth below which may arise out of or result from the Contractor's operations under the Contract and for which the Contractor may be legally liable, whether such operations be by the Contractor or by a Subcontractor or by anyone directly or indirectly employed by any of them, or by anyone for whose acts any of them may be liable:

.1 claims under workers' or workmen's compensation, disability benefit and other similar employee benefit acts which are applicable to the Work to be performed;

.2 claims for damages because of bodily injury, occupational sickness or disease, or death of the Contractor's employees;

.3 claims for damages because of bodily injury, sickness or disease, or death of any person other than the Contractor's employees;

.4 claims for damages insured by usual personal injury liability coverage which are sustained (1) by a person as a result of an offense directly or indirectly related to employment of such person by the Contractor, or (2) by another person;

.5 claims for damages, other than to the Work itself, because of injury to or destruction of tangible property, including loss of use resulting therefrom;

.6 claims for damages because of bodily injury, death of a person or property damage arising out of ownership, maintenance or use of a motor vehicle; and

.7 claims involving contractual liability insurance applicable to the Contractor's obligations under Paragraph 3.18.

11.1.2 The insurance required by Subparagraph 11.1.1 shall be written for not less than limits of liability specified in the Contract Documents or required by law, whichever coverage is greater. Coverages, whether written on an occurrence or claims-made basis, shall be maintained without interruption from date of commencement of the Work until date of final payment and termination of any coverage required to be maintained after final payment.

11.1.3 Certificates of Insurance acceptable to the Owner shall be filed with the Owner prior to commencement of the Work. These Certificates and the insurance policies required by this Paragraph 11.1 shall contain a provision that coverages afforded under the policies will not be cancelled or allowed to expire until at least 30 days' prior written notice has been given to the Owner. If any of the foregoing insurance coverages are required to remain in force after final payment and are reasonably available, an additional certificate evidencing continuation of such coverage shall be submitted with the final Application for Payment as required by Subparagraph 9.10.2. Information concerning reduction of coverage shall be furnished by the Contractor with reasonable promptness in accordance with the Contractor's information and belief.

11.2   OWNER'S LIABILITY INSURANCE

11.2.1 The Owner shall be responsible for purchasing and maintaining the Owner's usual liability insurance. Optionally, the Owner may purchase and maintain other insurance for self-protection against claims which may arise from operations under the Contract. The Contractor shall not be responsible for purchasing and maintaining this optional Owner's liability insurance unless specifically required by the Contract Documents.

11.3   PROPERTY INSURANCE

11.3.1 Unless otherwise provided, the Owner shall purchase and maintain, in a company or companies lawfully authorized to do business in the jurisdiction in which the Project is located, property insurance in the amount of the initial Contract Sum as well as subsequent modifications thereto for the entire Work at the site on a replacement cost basis without voluntary deductibles. Such property insurance shall be maintained, unless otherwise provided in the Contract Documents or otherwise agreed in writing by all persons and entities who are beneficiaries of such insurance, until final payment has been made as provided in Paragraph 9.10 or until no person or entity

other than the Owner has an insurable interest in the property required by this Paragraph 11.3 to be covered, whichever is earlier. This insurance shall include interests of the Owner, the Contractor, Subcontractors and Sub-subcontractors in the Work.

11.3.1.1 Property insurance shall be on an all-risk policy form and shall insure against the perils of fire and extended coverage and physical loss or damage including, without duplication of coverage, theft, vandalism, malicious mischief, collapse, falsework, temporary buildings and debris removal including demolition occasioned by enforcement of any applicable legal requirements, and shall cover reasonable compensation for Architect's services and expenses required as a result of such insured loss. Coverage for other perils shall not be required unless otherwise provided in the Contract Documents.

11.3.1.2 If the Owner does not intend to purchase such property insurance required by the Contract and with all of the coverages in the amount described above, the Owner shall so inform the Contractor in writing prior to commencement of the Work. The Contractor may then effect insurance which will protect the interests of the Contractor, Subcontractors and Sub-subcontractors in the Work, and by appropriate Change Order the cost thereof shall be charged to the Owner. If the Contractor is damaged by the failure or neglect of the Owner to purchase or maintain insurance as described above, without so notifying the Contractor, then the Owner shall bear all reasonable costs properly attributable thereto.

11.3.1.3 If the property insurance requires minimum deductibles and such deductibles are identified in the Contract Documents, the Contractor shall pay costs not covered because of such deductibles. If the Owner or insurer increases the required minimum deductibles above the amounts so identified or if the Owner elects to purchase this insurance with voluntary deductible amounts, the Owner shall be responsible for payment of the additional costs not covered because of such increased or voluntary deductibles. If deductibles are not identified in the Contract Documents, the Owner shall pay costs not covered because of deductibles.

11.3.1.4 Unless otherwise provided in the Contract Documents, this property insurance shall cover portions of the Work stored off the site after written approval of the Owner at the value established in the approval, and also portions of the Work in transit.

11.3.2 Boiler and Machinery Insurance. The Owner shall purchase and maintain boiler and machinery insurance required by the Contract Documents or by law, which shall specifically cover such insured objects during installation and until final acceptance by the Owner; this insurance shall include interests of the Owner, Contractor, Subcontractors and Sub-subcontractors in the Work, and the Owner and Contractor shall be named insureds.

11.3.3 Loss of Use Insurance. The Owner, at the Owner's option, may purchase and maintain such insurance as will insure the Owner against loss of use of the Owner's property due to fire or other hazards, however caused. The Owner waives all rights of action against the Contractor for loss of use of the Owner's property, including consequential losses due to fire or other hazards however caused.

11.3.4 If the Contractor requests in writing that insurance for risks other than those described herein or for other special hazards be included in the property insurance policy, the Owner shall, if possible, include such insurance, and the cost thereof shall be charged to the Contractor by appropriate Change Order.

AIA DOCUMENT A201 • GENERAL CONDITIONS OF THE CONTRACT FOR CONSTRUCTION • FOURTEENTH EDITION
AIA® • ©1987 THE AMERICAN INSTITUTE OF ARCHITECTS, 1735 NEW YORK AVENUE, N.W., WASHINGTON, D.C. 20006

11.3.5 If during the Project construction period the Owner insures properties, real or personal or both, adjoining or adjacent to the site by property insurance under policies separate from those insuring the Project, or if after final payment property insurance is to be provided on the completed Project through a policy or policies other than those insuring the Project during the construction period, the Owner shall waive all rights in accordance with the terms of Subparagraph 11.3.7 for damages caused by fire or other perils covered by this separate property insurance. All separate policies shall provide this waiver of subrogation by endorsement or otherwise.

11.3.6 Before an exposure to loss may occur, the Owner shall file with the Contractor a copy of each policy that includes insurance coverages required by this Paragraph 11.3. Each policy shall contain all generally applicable conditions, definitions, exclusions and endorsements related to this Project. Each policy shall contain a provision that the policy will not be cancelled or allowed to expire until at least 30 days' prior written notice has been given to the Contractor.

11.3.7 Waivers of Subrogation. The Owner and Contractor waive all rights against (1) each other and any of their subcontractors, sub-subcontractors, agents and employees, each of the other, and (2) the Architect, Architect's consultants, separate contractors described in Article 6, if any, and any of their subcontractors, sub-subcontractors, agents and employees, for damages caused by fire or other perils to the extent covered by property insurance obtained pursuant to this Paragraph 11.3 or other property insurance applicable to the Work, except such rights as they have to proceeds of such insurance held by the Owner as fiduciary. The Owner or Contractor, as appropriate, shall require of the Architect, Architect's consultants, separate contractors described in Article 6, if any, and the subcontractors, sub-subcontractors, agents and employees of any of them, by appropriate agreements, written where legally required for validity, similar waivers each in favor of other parties enumerated herein. The policies shall provide such waivers of subrogation by endorsement or otherwise. A waiver of subrogation shall be effective as to a person or entity even though that person or entity would otherwise have a duty of indemnification, contractual or otherwise, did not pay the insurance premium directly or indirectly, and whether or not the person or entity had an insurable interest in the property damaged.

11.3.8 A loss insured under Owner's property insurance shall be adjusted by the Owner as fiduciary and made payable to the Owner as fiduciary for the insureds, as their interests may appear, subject to requirements of any applicable mortgagee clause and of Subparagraph 11.3.10. The Contractor shall pay Subcontractors their just shares of insurance proceeds received by the Contractor, and by appropriate agreements, written where legally required for validity, shall require Subcontractors to make payments to their Sub-subcontractors in similar manner.

11.3.9 If required in writing by a party in interest, the Owner as fiduciary shall, upon occurrence of an insured loss, give bond for proper performance of the Owner's duties. The cost of required bonds shall be charged against proceeds received as fiduciary. The Owner shall deposit in a separate account proceeds so received, which the Owner shall distribute in accordance with such agreement as the parties in interest may reach, or in accordance with an arbitration award in which case the procedure shall be as provided in Paragraph 4.5. If after such loss no other special agreement is made, replacement of damaged property shall be covered by appropriate Change Order.

11.3.10 The Owner as fiduciary shall have power to adjust and settle a loss with insurers unless one of the parties in interest shall object in writing within five days after occurrence of loss to the Owner's exercise of this power; if such objection be made, arbitrators shall be chosen as provided in Paragraph 4.5. The Owner as fiduciary shall, in that case, make settlement with insurers in accordance with directions of such arbitrators. If distribution of insurance proceeds by arbitration is required, the arbitrators will direct such distribution.

11.3.11 Partial occupancy or use in accordance with Paragraph 9.9 shall not commence until the insurance company or companies providing property insurance have consented to such partial occupancy or use by endorsement or otherwise. The Owner and the Contractor shall take reasonable steps to obtain consent of the insurance company or companies and shall, without mutual written consent, take no action with respect to partial occupancy or use that would cause cancellation, lapse or reduction of insurance.

11.4 PERFORMANCE BOND AND PAYMENT BOND

11.4.1 The Owner shall have the right to require the Contractor to furnish bonds covering faithful performance of the Contract and payment of obligations arising thereunder as stipulated in bidding requirements or specifically required in the Contract Documents on the date of execution of the Contract.

11.4.2 Upon the request of any person or entity appearing to be a potential beneficiary of bonds covering payment of obligations arising under the Contract, the Contractor shall promptly furnish a copy of the bonds or shall permit a copy to be made.

ARTICLE 12

UNCOVERING AND CORRECTION OF WORK

12.1 UNCOVERING OF WORK

12.1.1 If a portion of the Work is covered contrary to the Architect's request or to requirements specifically expressed in the Contract Documents, it must, if required in writing by the Architect, be uncovered for the Architect's observation and be replaced at the Contractor's expense without change in the Contract Time.

12.1.2 If a portion of the Work has been covered which the Architect has not specifically requested to observe prior to its being covered, the Architect may request to see such Work and it shall be uncovered by the Contractor. If such Work is in accordance with the Contract Documents, costs of uncovering and replacement shall, by appropriate Change Order, be charged to the Owner. If such Work is not in accordance with the Contract Documents, the Contractor shall pay such costs unless the condition was caused by the Owner or a separate contractor in which event the Owner shall be responsible for payment of such costs.

12.2 CORRECTION OF WORK

12.2.1 The Contractor shall promptly correct Work rejected by the Architect or failing to conform to the requirements of the Contract Documents, whether observed before or after Substantial Completion and whether or not fabricated, installed or completed. The Contractor shall bear costs of correcting such rejected Work, including additional testing and inspections and compensation for the Architect's services and expenses made necessary thereby.

12.2.2 If, within one year after the date of Substantial Completion of the Work or designated portion thereof, or after the date

---

for commencement of warranties established under Subparagraph 9.9.1, or by terms of an applicable special warranty required by the Contract Documents, any of the Work is found to be not in accordance with the requirements of the Contract Documents, the Contractor shall correct it promptly after receipt of written notice from the Owner to do so unless the Owner has previously given the Contractor a written acceptance of such condition. This period of one year shall be extended with respect to portions of Work first performed after Substantial Completion by the period of time between Substantial Completion and the actual performance of the Work. This obligation under this Subparagraph 12.2.2 shall survive acceptance of the Work under the Contract and termination of the Contract. The Owner shall give such notice promptly after discovery of the condition.

12.2.3 The Contractor shall remove from the site portions of the Work which are not in accordance with the requirements of the Contract Documents and are neither corrected by the Contractor nor accepted by the Owner.

12.2.4 If the Contractor fails to correct nonconforming Work within a reasonable time, the Owner may correct it in accordance with Paragraph 2.4. If the Contractor does not proceed with correction of such nonconforming Work within a reasonable time fixed by written notice from the Architect, the Owner may remove it and store the salvable materials or equipment at the Contractor's expense. If the Contractor does not pay costs of such removal and storage within ten days after written notice, the Owner may upon ten additional days' written notice sell such materials and equipment at auction or at private sale and shall account for the proceeds thereof, after deducting costs and damages that should have been borne by the Contractor, including compensation for the Architect's services and expenses made necessary thereby. If such proceeds of sale do not cover costs which the Contractor should have borne, the Contract Sum shall be reduced by the deficiency. If payments then or thereafter due the Contractor are not sufficient to cover such amount, the Contractor shall pay the difference to the Owner.

12.2.5 The Contractor shall bear the cost of correcting destroyed or damaged construction, whether completed or partially completed, of the Owner or separate contractors caused by the Contractor's correction or removal of Work which is not in accordance with the requirements of the Contract Documents.

12.2.6 Nothing contained in this Paragraph 12.2 shall be construed to establish a period of limitation with respect to other obligations which the Contractor might have under the Contract Documents. Establishment of the time period of one year as described in Subparagraph 12.2.2 relates only to the specific obligation of the Contractor to correct the Work, and has no relationship to the time within which the obligation to comply with the Contract Documents may be sought to be enforced, nor to the time within which proceedings may be commenced to establish the Contractor's liability with respect to the Contractor's obligations other than specifically to correct the Work.

12.3 ACCEPTANCE OF NONCONFORMING WORK

12.3.1 If the Owner prefers to accept Work which is not in accordance with the requirements of the Contract Documents, the Owner may do so instead of requiring its removal and correction, in which case the Contract Sum will be reduced as appropriate and equitable. Such adjustment shall be effected whether or not final payment has been made.

ARTICLE 13

MISCELLANEOUS PROVISIONS

13.1 GOVERNING LAW

13.1.1 The Contract shall be governed by the law of the place where the Project is located.

13.2 SUCCESSORS AND ASSIGNS

13.2.1 The Owner and Contractor respectively bind themselves, their partners, successors, assigns and legal representatives to the other party hereto and to partners, successors, assigns and legal representatives of such other party in respect to covenants, agreements and obligations contained in the Contract Documents. Neither party to the Contract shall assign the Contract as a whole without written consent of the other. If either party attempts to make such an assignment without such consent, that party shall nevertheless remain legally responsible for all obligations under the Contract.

13.3 WRITTEN NOTICE

13.3.1 Written notice shall be deemed to have been duly served if delivered in person to the individual or a member of the firm or entity or to an officer of the corporation for which it was intended, or if delivered at or sent by registered or certified mail to the last business address known to the party giving notice.

13.4 RIGHTS AND REMEDIES

13.4.1 Duties and obligations imposed by the Contract Documents and rights and remedies available thereunder shall be in addition to and not a limitation of duties, obligations, rights and remedies otherwise imposed or available by law.

13.4.2 No action or failure to act by the Owner, Architect or Contractor shall constitute a waiver of a right or duty afforded them under the Contract, nor shall such action or failure to act constitute approval of or acquiescence in a breach thereunder, except as may be specifically agreed in writing.

13.5 TESTS AND INSPECTIONS

13.5.1 Tests, inspections and approvals of portions of the Work required by the Contract Documents or by laws, ordinances, rules, regulations or orders of public authorities having jurisdiction shall be made at an appropriate time. Unless otherwise provided, the Contractor shall make arrangements for such tests, inspections and approvals with an independent testing laboratory or entity acceptable to the Owner, or with the appropriate public authority, and shall bear all related costs of tests, inspections and approvals. The Contractor shall give the Architect timely notice of when and where tests and inspections are to be made so the Architect may observe such procedures. The Owner shall bear costs of tests, inspections or approvals which do not become requirements until after bids are received or negotiations concluded.

13.5.2 If the Architect, Owner or public authorities having jurisdiction determine that portions of the Work require additional testing, inspection or approval not included under Subparagraph 13.5.1, the Architect will, upon written authorization from the Owner, instruct the Contractor to make arrangements for such additional testing, inspection or approval by an entity acceptable to the Owner, and the Contractor shall give timely notice to the Architect of when and where tests and inspections are to be made so the Architect may observe such procedures.

The Owner shall bear such costs except as provided in Sub-paragraph 13.5.3.

13.5.3 If such procedures for testing, inspection or approval under Subparagraphs 13.5.1 and 13.5.2 reveal failure of the portions of the Work to comply with requirements established by the Contract Documents, the Contractor shall bear all costs made necessary by such failure including those of repeated procedures and compensation for the Architect's services and expenses.

13.5.4 Required certificates of testing, inspection or approval shall, unless otherwise required by the Contract Documents, be secured by the Contractor and promptly delivered to the Architect.

13.5.5 If the Architect is to observe tests, inspections or approvals required by the Contract Documents, the Architect will do so promptly and, where practicable, at the normal place of testing.

13.5.6 Tests or inspections conducted pursuant to the Contract Documents shall be made promptly to avoid unreasonable delay in the Work.

13.6   INTEREST

13.6.1 Payments due and unpaid under the Contract Documents shall bear interest from the date payment is due at such rate as the parties may agree upon in writing or, in the absence thereof, at the legal rate prevailing from time to time at the place where the Project is located.

13.7   COMMENCEMENT OF STATUTORY LIMITATION PERIOD

13.7.1 As between the Owner and Contractor:

.1 **Before Substantial Completion.** As to acts or failures to act occurring prior to the relevant date of Substantial Completion, any applicable statute of limitations shall commence to run and any alleged cause of action shall be deemed to have accrued in any and all events not later than such date of Substantial Completion;

.2 **Between Substantial Completion and Final Certificate for Payment.** As to acts or failures to act occurring subsequent to the relevant date of Substantial Completion and prior to issuance of the final Certificate for Payment, any applicable statute of limitations shall commence to run and any alleged cause of action shall be deemed to have accrued in any and all events not later than the date of issuance of the final Certificate for Payment; and

.3 **After Final Certificate for Payment.** As to acts or failures to act occurring after the relevant date of issuance of the final Certificate for Payment, any applicable statute of limitations shall commence to run and any alleged cause of action shall be deemed to have accrued in any and all events not later than the date of any act or failure to act by the Contractor pursuant to any warranty provided under Paragraph 3.5, the date of any correction of the Work or failure to correct the Work by the Contractor under Paragraph 12.2, or the date of actual commission of any other act or failure to perform any duty or obligation by the Contractor or Owner, whichever occurs last.

ARTICLE 14

TERMINATION OR SUSPENSION OF THE CONTRACT

14.1   TERMINATION BY THE CONTRACTOR

14.1.1 The Contractor may terminate the Contract if the Work is stopped for a period of 30 days through no act or fault of the Contractor or a Subcontractor, Sub-subcontractor or their agents or employees or any other persons performing portions of the Work under contract with the Contractor, for any of the following reasons:

.1 issuance of an order of a court or other public authority having jurisdiction;

.2 an act of government, such as a declaration of national emergency, making material unavailable;

.3 because the Architect has not issued a Certificate for Payment and has not notified the Contractor of the reason for withholding certification as provided in Subparagraph 9.4.1, or because the Owner has not made payment on a Certificate for Payment within the time stated in the Contract Documents;

.4 if repeated suspensions, delays or interruptions by the Owner as described in Paragraph 14.3 constitute in the aggregate more than 100 percent of the total number of days scheduled for completion, or 120 days in any 365-day period, whichever is less; or

.5 the Owner has failed to furnish to the Contractor promptly, upon the Contractor's request, reasonable evidence as required by Subparagraph 2.2.1.

14.1.2 If one of the above reasons exists, the Contractor may, upon seven additional days' written notice to the Owner and Architect, terminate the Contract and recover from the Owner payment for Work executed and for proven loss with respect to materials, equipment, tools, and construction equipment and machinery, including reasonable overhead, profit and damages.

14.1.3 If the Work is stopped for a period of 60 days through no act or fault of the Contractor or a Subcontractor or their agents or employees or any other persons performing portions of the Work under contract with the Contractor because the Owner has persistently failed to fulfill the Owner's obligations under the Contract Documents with respect to matters important to the progress of the Work, the Contractor may, upon seven additional days' written notice to the Owner and the Architect, terminate the Contract and recover from the Owner as provided in Subparagraph 14.1.2.

14.2   TERMINATION BY THE OWNER FOR CAUSE

14.2.1 The Owner may terminate the Contract if the Contractor:

.1 persistently or repeatedly refuses or fails to supply enough properly skilled workers or proper materials;

.2 fails to make payment to Subcontractors for materials or labor in accordance with the respective agreements between the Contractor and the Subcontractors;

.3 persistently disregards laws, ordinances, or rules, regulations or orders of a public authority having jurisdiction; or

.4 otherwise is guilty of substantial breach of a provision of the Contract Documents.

14.2.2 When any of the above reasons exist, the Owner, upon certification by the Architect that sufficient cause exists to jus-

any such action, may without prejudice to any other rights or remedies of the Owner and after giving the Contractor and the Contractor's surety, if any, seven days' written notice, terminate employment of the Contractor and may, subject to any prior rights of the surety:

    .1 take possession of the site and of all materials, equipment, tools, and construction equipment and machinery thereon owned by the Contractor;

    .2 accept assignment of subcontracts pursuant to Paragraph 5.4; and

    .3 finish the Work by whatever reasonable method the Owner may deem expedient.

14.2.3 When the Owner terminates the Contract for one of the reasons stated in Subparagraph 14.2.1, the Contractor shall not be entitled to receive further payment until the Work is finished.

14.2.4 If the unpaid balance of the Contract Sum exceeds costs of finishing the Work, including compensation for the Architect's services and expenses made necessary thereby, such excess shall be paid to the Contractor. If such costs exceed the unpaid balance, the Contractor shall pay the difference to the Owner. The amount to be paid to the Contractor or Owner, as the case may be, shall be certified by the Architect, upon application, and this obligation for payment shall survive termination of the Contract.

14.3 SUSPENSION BY THE OWNER FOR CONVENIENCE

14.3.1 The Owner may, without cause, order the Contractor in writing to suspend, delay or interrupt the Work in whole or in part for such period of time as the Owner may determine.

14.3.2 An adjustment shall be made for increases in the cost of performance of the Contract, including profit on the increased cost of performance, caused by suspension, delay or interruption. No adjustment shall be made to the extent:

    .1 that performance is, was or would have been so suspended, delayed or interrupted by another cause for which the Contractor is responsible; or

    .2 that an equitable adjustment is made or denied under another provision of this Contract.

14.3.3 Adjustments made in the cost of performance may have a mutually agreed fixed or percentage fee.

AIA DOCUMENT A201 • GENERAL CONDITIONS OF THE CONTRACT FOR CONSTRUCTION • FOURTEENTH EDITION
AIA® • ©1987 THE AMERICAN INSTITUTE OF ARCHITECTS, 1735 NEW YORK AVENUE, N.W., WASHINGTON, D.C. 20006

ARTICLE 4
CONTRACT SUM

4.1 The Owner shall pay the Contractor in current funds for the Contractor's performance of the Contract the Contract Sum of

Dollars

($), subject to additions and deductions as provided in the Con-
tract Documents.

4.2 The Contract Sum is based upon the following alternates, if any, which are described in the Contract Documents and are hereby accepted by the Owner:

(State the numbers or other identification of accepted alternates. If decisions on other alternates are to be made by the Owner subsequent to the execution of this Agreement, attach a schedule of such other alternates showing the amount for each and the date until which that amount is valid.)

4.3 Unit prices, if any, are as follows:

D

Performance and Payment Bonds

THE AMERICAN INSTITUTE OF ARCHITECTS

AIA Document A311

Performance Bond

KNOW ALL MEN BY THESE PRESENTS: that

(Here insert full name and address or legal title of Contractor)

as Principal, hereinafter called Contractor, and,

(Here insert full name and address or legal title of Surety)

as Surety, hereinafter called Surety, are held and firmly bound unto

(Here insert full name and address or legal title of Owner)

as Obligee, hereinafter called Owner, in the amount of

Dollars ($),

for the payment whereof Contractor and Surety bind themselves, their heirs, executors, administrators, successors and assigns, jointly and severally, firmly by these presents.

WHEREAS,

Contractor has by written agreement dated   19   , entered into a contract with Owner for
(Here insert full name, address and description of project)

in accordance with Drawings and Specifications prepared by

(Here insert full name and address or legal title of Architect)

which contract is by reference made a part hereof, and is hereinafter referred to as the Contract.

PERFORMANCE BOND

NOW, THEREFORE, THE CONDITION OF THIS OBLIGATION is such that, if Contractor shall promptly and faithfully perform said Contract, then this obligation shall be null and void; otherwise it shall remain in full force and effect.

The Surety hereby waives notice of any alteration or extension of time made by the Owner.

Whenever Contractor shall be, and declared by Owner to be in default under the Contract, the Owner having performed Owner's obligations thereunder, the Surety may promptly remedy the default, or shall promptly

1) Complete the Contract in accordance with its terms and conditions, or

2) Obtain a bid or bids for completing the Contract in accordance with its terms and conditions, and upon determination by Surety of the lowest responsible bidder, or, if the Owner elects, upon determination by the Owner and the Surety jointly of the lowest responsible bidder, arrange for a contract between such bidder and Owner, and make available as Work progresses (even though there should be a default or a succession of defaults under the contract or contracts of completion arranged under this paragraph) sufficient funds to pay the cost of completion less the balance of the contract price; but not exceeding, including other costs and damages for which the Surety may be liable hereunder, the amount set forth in the first paragraph hereof. The term "balance of the contract price," as used in this paragraph, shall mean the total amount payable by Owner to Contractor under the Contract and any amendments thereto, less the amount properly paid by Owner to Contractor.

Any suit under this bond must be instituted before the expiration of two (2) years from the date on which final payment under the Contract falls due.

No right of action shall accrue on this bond to or for the use of any person or corporation other than the Owner named herein or the heirs, executors, administrators or successors of the Owner.

Signed and sealed this _____ day of _____ 19____

_____
(Witness)

_____ (Principal) _____ (Seal)

_____
(Title)

_____
(Witness)

_____ (Surety) _____ (Seal)

_____
(Title)

THE AMERICAN INSTITUTE OF ARCHITECTS

AIA Document A311

Labor and Material Payment Bond

THIS BOND IS ISSUED SIMULTANEOUSLY WITH PERFORMANCE BOND IN FAVOR OF THE
OWNER CONDITIONED ON THE FULL AND FAITHFUL PERFORMANCE OF THE CONTRACT

KNOW ALL MEN BY THESE PRESENTS: that

(Here insert full name and address or legal title of Contractor)

as Principal, hereinafter called Principal, and,

(Here insert full name and address or legal title of Surety)

as Surety, hereinafter called Surety, are held and firmly bound unto

(Here insert full name and address or legal title of Owner)

as Obligee, hereinafter called Owner, for the use and benefit of claimants as hereinbelow defined, in the

amount of
(Here insert a sum equal to at least one-half of the contract price) Dollars ($),
for the payment whereof Principal and Surety bind themselves, their heirs, executors, administrators,
successors and assigns, jointly and severally, firmly by these presents.

WHEREAS,

Principal has by written agreement dated 19 , entered into a contract with Owner for
(Here insert full name, address and description of project)

in accordance with Drawings and Specifications prepared by
(Here insert full name and address or legal title of Architect)

which contract is by reference made a part hereof, and is hereinafter referred to as the Contract.

LABOR AND MATERIAL PAYMENT BOND

NOW, THEREFORE, THE CONDITION OF THIS OBLIGATION is such that, if Principal shall promptly make payment to all claimants as hereinafter defined, for all labor and material used or reasonably required for use in the performance of the Contract, then this obligation shall be void; otherwise it shall remain in full force and effect, subject, however, to the following conditions:

1. A claimant is defined as one having a direct contract with the Principal or with a Subcontractor of the Principal for labor, material, or both, used or reasonably required for use in the performance of the Contract, labor and material being construed to include that part of water, gas, power, light, heat, oil, gasoline, telephone service or rental of equipment directly applicable to the Contract.

2. The above named Principal and Surety hereby jointly and severally agree with the Owner that every claimant as herein defined, who has not been paid in full before the expiration of a period of ninety (90) days after the date on which the last of such claimant's work or labor was done or performed, or materials were furnished by such claimant, may sue on this bond for the use of such claimant, prosecute the suit to final judgment for such sum or sums as may be justly due claimant, and have execution thereon. The Owner shall not be liable for the payment of any costs or expenses of any such suit.

3. No suit or action shall be commenced hereunder by any claimant:

a) Unless claimant, other than one having a direct contract with the Principal, shall have given written notice to any two of the following: the Principal, the Owner, or the Surety above named, within ninety (90) days after such claimant did or performed the last of the work or labor, or furnished the last of the materials for which said claim is made, stating with substantial

accuracy the amount claimed and the name of the party to whom the materials were furnished, or for whom the work or labor was done or performed. Such notice shall be served by mailing the same by registered mail or certified mail, postage prepaid, in an envelope addressed to the Principal, Owner or Surety, at any place where an office is regularly maintained for the transaction of business, or served in any manner in which legal process may be served in the state in which the aforesaid project is located, save that such service need not be made by a public officer.

b) After the expiration of one (1) year following the date on which Principal ceased Work on said Contract, it being understood, however, that if any limitation embodied in this bond is prohibited by any law controlling the construction hereof such limitation shall be deemed to be amended so as to be equal to the minimum period of limitation permitted by such law.

c) Other than in a state court of competent jurisdiction in and for the county or other political subdivision of the state in which the Project, or any part thereof, is situated, or in the United States District Court for the district in which the Project, or any part thereof, is situated, and not elsewhere.

4. The amount of this bond shall be reduced by and to the extent of any payment or payments made in good faith hereunder, inclusive of the payment by Surety of mechanics' liens which may be filed of record against said improvement, whether or not claim for the amount of such lien be presented under and against this bond.

Signed and sealed this   day of   19

_____ (Witness)

(Principal)   (Seal)

(Title)

_____ (Witness)

(Surety)   (Seal)

(Title)

THE AMERICAN INSTITUTE OF ARCHITECTS

AIA Document A312

Performance Bond

Any singular reference to Contractor, Surety, Owner or other party shall be considered plural where applicable.

CONTRACTOR (Name and Address):   SURETY (Name and Principal Place of Business):

OWNER (Name and Address):

CONSTRUCTION CONTRACT
  Date:
  Amount:
  Description (Name and Location):

BOND
  Date (Not earlier than Construction Contract Date):
  Amount:
  Modifications to this Bond: □ None   □ See Page 3

CONTRACTOR AS PRINCIPAL   SURETY
Company: (Corporate Seal)   Company: (Corporate Seal)

Signature: _____   Signature: _____
Name and Title: Name and Title:

(Any additional signatures appear on page 3)

(FOR INFORMATION ONLY—Name, Address and Telephone)
AGENT or BROKER: OWNER'S REPRESENTATIVE (Architect, Engineer or other party):

1 The Contractor and the Surety, jointly and severally, bind themselves, their heirs, executors, administrators, successors and assigns to the Owner for the performance of the Construction Contract, which is incorporated herein by reference.

2 If the Contractor performs the Construction Contract, the Surety and the Contractor shall have no obligation under this Bond, except to participate in conferences as provided in Subparagraph 3.1.

3 If there is no Owner Default, the Surety's obligation under this Bond shall arise after:

3.1 The Owner has notified the Contractor and the Surety at its address described in Paragraph 10 below that the Owner is considering declaring a Contractor Default and has requested and attempted to arrange a conference with the Contractor and the Surety to be held not later than fifteen days after receipt of such notice to discuss methods of performing the Construction Contract. If the Owner, the Contractor and the Surety agree, the Contractor shall be allowed a reasonable time to perform the Construction Contract, but such an agreement shall not waive the Owner's right, if any, subsequently to declare a Contractor Default; and

3.2 The Owner has declared a Contractor Default and formally terminated the Contractor's right to complete the contract. Such Contractor Default shall not be declared earlier than twenty days after the Contractor and the Surety have received notice as provided in Subparagraph 3.1; and

3.3 The Owner has agreed to pay the Balance of the Contract Price to the Surety in accordance with the terms of the Construction Contract or to a contractor selected to perform the Construction Contract in accordance with the terms of the contract with the Owner.

4 When the Owner has satisfied the conditions of Paragraph 3, the Surety shall promptly and at the Surety's expense take one of the following actions:

4.1 Arrange for the Contractor, with consent of the Owner, to perform and complete the Construction Contract; or

4.2 Undertake to perform and complete the Construction Contract itself, through its agents or through independent contractors; or

4.3 Obtain bids or negotiated proposals from qualified contractors acceptable to the Owner for a contract for performance and completion of the Construction Contract, arrange for a contract to be prepared for execution by the Owner and the contractor selected with the Owner's concurrence, to be secured with performance and payment bonds executed by a qualified surety equivalent to the bonds issued on the Construction Contract, and pay to the Owner the amount of damages as described in Paragraph 6 in excess of the Balance of the Contract Price incurred by the Owner resulting from the Contractor's default; or

4.4 Waive its right to perform and complete, arrange for completion, or obtain a new contractor and with reasonable promptness under the circumstances:

.1 After investigation, determine the amount for which it may be liable to the Owner and, as soon as practicable after the amount is determined, tender payment therefor to the Owner; or

.2 Deny liability in whole or in part and notify the Owner citing reasons therefor.

5 If the Surety does not proceed as provided in Paragraph 4 with reasonable promptness, the Surety shall be deemed to be in default on this Bond fifteen days after receipt of an additional written notice from the Owner to the Surety demanding that the Surety perform its obligations under this Bond, and the Owner shall be entitled to enforce any remedy available to the Owner. If the Surety proceeds as provided in Subparagraph 4.4, and the Owner refuses the payment tendered or the Surety has denied liability, in whole or in part, without further notice the Owner shall be entitled to enforce any remedy available to the Owner.

6 After the Owner has terminated the Contractor's right to complete the Construction Contract, and if the Surety elects to act under Subparagraph 4.1, 4.2, or 4.3 above, then the responsibilities of the Surety to the Owner shall not be greater than those of the Contractor under the Construction Contract, and the responsibilities of the Owner to the Surety shall not be greater than those of the Owner under the Construction Contract. To the limit of the amount of this Bond, but subject to commitment by the Owner of the Balance of the Contract Price to mitigation of costs and damages on the Construction Contract, the Surety is obligated without duplication for:

6.1 The responsibilities of the Contractor for correction of defective work and completion of the Construction Contract;

6.2 Additional legal, design professional and delay costs resulting from the Contractor's Default, and resulting from the actions or failure to act of the Surety under Paragraph 4; and

6.3 Liquidated damages, or if no liquidated damages are specified in the Construction Contract, actual damages caused by delayed performance or non-performance of the Contractor.

7 The Surety shall not be liable to the Owner or others for obligations of the Contractor that are unrelated to the Construction Contract, and the Balance of the Contract Price shall not be reduced or set off on account of any such unrelated obligations. No right of action shall accrue on this Bond to any person or entity other than the Owner or its heirs, executors, administrators or successors.

8 The Surety hereby waives notice of any change, including changes of time, to the Construction Contract or to related subcontracts, purchase orders and other obligations.

9 Any proceeding, legal or equitable, under this Bond may be instituted in any court of competent jurisdiction in the location in which the work or part of the work is located and shall be instituted within two years after Contractor Default or within two years after the Contractor ceased working or within two years after the Surety refuses or fails to perform its obligations under this Bond, whichever occurs first. If the provisions of this Paragraph are void or prohibited by law, the minimum period of limitation avail-

AIA DOCUMENT A312 • PERFORMANCE BOND AND PAYMENT BOND • DECEMBER 1984 ED. • AIA ®
THE AMERICAN INSTITUTE OF ARCHITECTS, 1735 NEW YORK AVE., N.W., WASHINGTON, D.C. 20006
THIRD PRINTING • MARCH 1987

A312-1984   2

able to sureties as a defense in the jurisdiction of the suit shall be applicable.

10 Notice to the Surety, the Owner or the Contractor shall be mailed or delivered to the address shown on the signature page.

11 When this Bond has been furnished to comply with a statutory or other legal requirement in the location where the construction was to be performed, any provision in this Bond conflicting with said statutory or legal requirement shall be deemed deleted herefrom and provisions conforming to such statutory or other legal requirement shall be deemed incorporated herein. The intent is that this Bond shall be construed as a statutory bond and not as a common law bond.

12 DEFINITIONS

12.1 Balance of the Contract Price: The total amount payable by the Owner to the Contractor under the Construction Contract after all proper adjustments have been made, including allowance to the Con-

tractor of any amounts received or to be received by the Owner in settlement of insurance or other claims for damages to which the Contractor is entitled, reduced by all valid and proper payments made to or on behalf of the Contractor under the Construction Contract.

12.2 Construction Contract: The agreement between the Owner and the Contractor identified on the signature page, including all Contract Documents and changes thereto.

12.3 Contractor Default: Failure of the Contractor, which has neither been remedied nor waived, to perform or otherwise to comply with the terms of the Construction Contract.

12.4 Owner Default: Failure of the Owner, which has neither been remedied nor waived, to pay the Contractor as required by the Construction Contract or to perform and complete or comply with the other terms thereof.

MODIFICATIONS TO THIS BOND ARE AS FOLLOWS:

(Space is provided below for additional signatures of added parties, other than those appearing on the cover page.)

CONTRACTOR AS PRINCIPAL		SURETY	
Company:	(Corporate Seal)	Company:	(Corporate Seal)

Signature: _____

Name and Title:

Address:

Signature: _____

Name and Title:

Address:

AIA DOCUMENT A312 • PERFORMANCE BOND AND PAYMENT BOND • DECEMBER 1984 ED. • AIA ®
THE AMERICAN INSTITUTE OF ARCHITECTS, 1735 NEW YORK AVE., N.W., WASHINGTON, D.C. 20006
THIRD PRINTING • MARCH 1987

A312-1984   3

THE AMERICAN INSTITUTE OF ARCHITECTS

AIA Document A312

Payment Bond

Any singular reference to Contractor, Surety, Owner or other party shall be considered plural where applicable.

CONTRACTOR (Name and Address):

SURETY (Name and Principal Place of Business):

OWNER (Name and Address):

CONSTRUCTION CONTRACT
  Date:
  Amount:
  Description (Name and Location):

BOND
  Date (Not earlier than Construction Contract Date):
  Amount:
  Modifications to this Bond: ☐ None   ☐ See Page 6

CONTRACTOR AS PRINCIPAL   SURETY
Company: (Corporate Seal)   Company: (Corporate Seal)

Signature: _____   Signature: _____
Name and Title: Name and Title:

(Any additional signatures appear on page 6)

(FOR INFORMATION ONLY—Name, Address and Telephone)
AGENT or BROKER: OWNER'S REPRESENTATIVE (Architect, Engineer or
                                       other party):

AIA DOCUMENT A312 · PERFORMANCE BOND AND PAYMENT BOND · DECEMBER 1984 ED. · AIA ®
THE AMERICAN INSTITUTE OF ARCHITECTS, 1735 NEW YORK AVE., N.W., WASHINGTON, D.C. 20006
THIRD PRINTING · MARCH 1987

A312-1984   4

1 The Contractor and the Surety, jointly and severally, bind themselves, their heirs, executors, administrators, successors and assigns to the Owner to pay for labor, materials and equipment furnished for use in the performance of the Construction Contract, which is incorporated herein by reference.

2 With respect to the Owner, this obligation shall be null and void if the Contractor:

2.1 Promptly makes payment, directly or indirectly, for all sums due Claimants, and

2.2 Defends, indemnifies and holds harmless the Owner from claims, demands, liens or suits by any person or entity whose claim, demand, lien or suit is for the payment for labor, materials or equipment furnished for use in the performance of the Construction Contract, provided the Owner has promptly notified the Contractor and the Surety (at the address described in Paragraph 12) of any claims, demands, liens or suits and tendered defense of such claims, demands, liens or suits to the Contractor and the Surety, and provided there is no Owner Default.

3 With respect to Claimants, this obligation shall be null and void if the Contractor promptly makes payment, directly or indirectly, for all sums due.

4 The Surety shall have no obligation to Claimants under this Bond until:

4.1 Claimants who are employed by or have a direct contract with the Contractor have given notice to the Surety (at the address described in Paragraph 12) and sent a copy, or notice thereof, to the Owner, stating that a claim is being made under this Bond and, with substantial accuracy, the amount of the claim.

4.2 Claimants who do not have a direct contract with the Contractor:

.1 Have furnished written notice to the Contractor and sent a copy, or notice thereof, to the Owner, within 90 days after having last performed labor or last furnished materials or equipment included in the claim stating, with substantial accuracy, the amount of the claim and the name of the party to whom the materials were furnished or supplied or for whom the labor was done or performed; and

.2 Have either received a rejection in whole or in part from the Contractor, or not received within 30 days of furnishing the above notice any communication from the Contractor by which the Contractor has indicated the claim will be paid directly or indirectly; and

.3 Not having been paid within the above 30 days, have sent a written notice to the Surety (at the address described in Paragraph 12) and sent a copy, or notice thereof, to the Owner, stating that a claim is being made under this Bond and enclosing a copy of the previous written notice furnished to the Contractor.

5 If a notice required by Paragraph 4 is given by the Owner to the Contractor or to the Surety, that is sufficient compliance.

6 When the Claimant has satisfied the conditions of Paragraph 4, the Surety shall promptly and at the Surety's expense take the following actions:

6.1 Send an answer to the Claimant, with a copy to the Owner, within 45 days after receipt of the claim, stating the amounts that are undisputed and the basis for challenging any amounts that are disputed.

6.2 Pay or arrange for payment of any undisputed amounts.

7 The Surety's total obligation shall not exceed the amount of this Bond, and the amount of this Bond shall be credited for any payments made in good faith by the Surety.

8 Amounts owed by the Owner to the Contractor under the Construction Contract shall be used for the performance of the Construction Contract and to satisfy claims, if any, under any Construction Performance Bond. By the Contractor furnishing and the Owner accepting this Bond, they agree that all funds earned by the Contractor in the performance of the Construction Contract are dedicated to satisfy obligations of the Contractor and the Surety under this Bond, subject to the Owner's priority to use the funds for the completion of the work.

9 The Surety shall not be liable to the Owner, Claimants or others for obligations of the Contractor that are unrelated to the Construction Contract. The Owner shall not be liable for payment of any costs or expenses of any Claimant under this Bond, and shall have under this Bond no obligations to make payments to, give notices on behalf of, or otherwise have obligations to Claimants under this Bond.

10 The Surety hereby waives notice of any change, including changes of time, to the Construction Contract or to related subcontracts, purchase orders and other obligations.

11 No suit or action shall be commenced by a Claimant under this Bond other than in a court of competent jurisdiction in the location in which the work or part of the work is located or after the expiration of one year from the date (1) on which the Claimant gave the notice required by Subparagraph 4.1 or Clause 4.2.3, or (2) on which the last labor or service was performed by anyone or the last materials or equipment were furnished by anyone under the Construction Contract, whichever of (1) or (2) first occurs. If the provisions of this Paragraph are void or prohibited by law, the minimum period of limitation available to sureties as a defense in the jurisdiction of the suit shall be applicable.

12 Notice to the Surety, the Owner or the Contractor shall be mailed or delivered to the address shown on the signature page. Actual receipt of notice by Surety, the Owner or the Contractor, however accomplished, shall be sufficient compliance as of the date received at the address shown on the signature page.

13 When this Bond has been furnished to comply with a statutory or other legal requirement in the location where the construction was to be performed, any provision in this Bond conflicting with said statutory or legal requirement shall be deemed deleted herefrom and provisions conforming to such statutory or other legal requirement shall be deemed incorporated herein. The intent is that this

AIA DOCUMENT A312 • PERFORMANCE BOND AND PAYMENT BOND • DECEMBER 1984 ED. • AIA®
THE AMERICAN INSTITUTE OF ARCHITECTS, 1735 NEW YORK AVE., N.W., WASHINGTON, D.C. 20006
THIRD PRINTING • MARCH 1987

A312-1984   5

Bond shall be construed as a statutory bond and not as a common law bond.

14 Upon request by any person or entity appearing to be a potential beneficiary of this Bond, the Contractor shall promptly furnish a copy of this Bond or shall permit a copy to be made.

15 DEFINITIONS

15.1 Claimant: An individual or entity having a direct contract with the Contractor or with a subcontractor of the Contractor to furnish labor, materials or equipment for use in the performance of the Contract. The intent of this Bond shall be to include without limitation in the terms "labor, materials or equipment" that part of water, gas, power, light, heat, oil, gasoline, telephone service or rental equipment used in the Construction Contract, architectural and engineering services required for performance of the work of the Contractor and the Contractor's subcontractors, and all other items for which a mechanic's lien may be asserted in the jurisdiction where the labor, materials or equipment were furnished.

15.2 Construction Contract: The agreement between the Owner and the Contractor identified on the signature page, including all Contract Documents and changes thereto.

15.3 Owner Default: Failure of the Owner, which has neither been remedied nor waived, to pay the Contractor as required by the Construction Contract or to perform and complete or comply with the other terms thereof.

MODIFICATIONS TO THIS BOND ARE AS FOLLOWS:

(Space is provided below for additional signatures of added parties, other than those appearing on the cover page.)

CONTRACTOR AS PRINCIPAL
Company: (Corporate Seal)

Signature: _____
Name and Title:
Address:

SURETY
Company: (Corporate Seal)

Signature: _____
Name and Title:
Address:

AIA DOCUMENT A312 • PERFORMANCE BOND AND PAYMENT BOND • DECEMBER 1984 ED. • AIA ®
THE AMERICAN INSTITUTE OF ARCHITECTS, 1735 NEW YORK AVE., N.W., WASHINGTON, D.C. 20006
THIRD PRINTING • MARCH 1987

A312-1984 6

E

Standard Form of Agreement Between Contractor and Subcontractor

1987 Edition

T H E   A M E R I C A N   I N S T I T U T E   O F   A R C H I T E C T S

AIA Document A401

Standard Form of Agreement
Between Contractor and Subcontractor

1987 EDITION

THIS DOCUMENT HAS IMPORTANT LEGAL CONSEQUENCES; CONSULTATION WITH AN ATTORNEY IS ENCOURAGED WITH RESPECT TO ITS COMPLETION OR MODIFICATION.

This document has been approved and endorsed by the American Subcontractors Association and the Associated Specialty Contractors, Inc.

AGREEMENT

made as of the
Nineteen Hundred and

day of

in the year of

BETWEEN the Contractor:
(Name and Address)

and the Subcontractor:
(Name and Address)

The Contractor has made a contract for construction dated
The Owner:
(Name and Address)

with

For the following Project:
(Name and Location)

which Contract is hereinafter referred to as the Prime Contract and which provides for the furnishing of labor, materials, equipment and services in connection with the construction of the Project. A copy of the Prime Contract, consisting of the Agreement Between Owner and Contractor (from which compensation amounts may be deleted) and the other Contract Documents enumerated therein has been made available to the Subcontractor.

The Architect for the Project is:
(Name and Address)

The Contractor and the Subcontractor agree as set forth below.

TERMS AND CONDITIONS OF AGREEMENT
BETWEEN CONTRACTOR AND SUBCONTRACTOR

ARTICLE 1
THE SUBCONTRACT DOCUMENTS

1.1. The Subcontract Documents consist of (1) this Agreement; (2) the Prime Contract, consisting of the Agreement between the Owner and Contractor and the other Contract Documents enumerated therein, including Conditions of the Contract (General, Supplementary and other Conditions), Drawings, Specifications, Addenda issued prior to execution of the Agreement between the Owner and Contractor and Modifications issued subsequent to the execution of the Agreement between the Owner and Contractor, whether before or after the execution of this Agreement, and other Contract Documents, if any, listed in the Owner-Contractor Agreement; (3) other documents listed in Article 16 of this Agreement; and (4) Modifications to this Subcontract issued after execution of this Agreement. These form the Subcontract, and are as fully a part of the Subcontract as if attached to this Agreement or repeated herein. The Subcontract represents the entire and integrated agreement between the parties hereto and supersedes prior negotiations, representations or agreements, either written or oral. An enumeration of the Subcontract Documents, other than Modifications issued subsequent to the execution of this Agreement, appears in Article 16.

1.2 The Subcontractor shall be furnished copies of the Subcontract Documents upon request, but the Contractor may charge the Subcontractor for the cost of reproduction.

ARTICLE 2
MUTUAL RIGHTS AND RESPONSIBILITIES

2.1 The Contractor and Subcontractor shall be mutually bound by the terms of this Agreement and, to the extent that provisions of the Prime Contract apply to the Work of the Subcontractor, the Contractor shall assume toward the Subcontractor all obligations and responsibilities that the Owner, under the Prime Contract, assumes toward the Contractor, and the Subcontractor shall assume toward the Contractor all obligations and responsibilities which the Contractor, under the Prime Contract, assumes toward the Owner and the Architect. The Contractor shall have the benefit of all rights, remedies and redress against the Subcontractor which the Owner, under the Prime Contract, has against the Contractor, and the Subcontractor shall have the benefit of all rights, remedies and redress against the Contractor which the Contractor, under the Prime Contract, has against the Owner, insofar as applicable to this Subcontract. Where a provision of the Prime Contract is inconsistent with a provision of this Agreement, this Agreement shall govern.

2.2 The Contractor may require the Subcontractor to enter into agreements with Sub-subcontractors performing portions of the Work of this Subcontract by which the Subcontractor and the Sub-subcontractor are mutually bound, to the extent of the Work to be performed by the Sub-subcontractor, assuming toward each other all obligations and responsibilities which the Contractor and Subcontractor assume toward each other and having the benefit of all rights, remedies and redress each against the other which the Contractor and Subcontractor have by virtue of the provisions of this Agreement.

ARTICLE 3
CONTRACTOR

3.1 SERVICES PROVIDED BY THE CONTRACTOR

3.1.1 The Contractor shall cooperate with the Subcontractor in scheduling and performing the Contractor's Work to avoid conflicts or interference in the Subcontractor's Work and shall expedite written responses to submittals made by the Subcontractor in accordance with Paragraph 4.1 and Article 5. As soon as practicable after execution of this Agreement, the Contractor shall provide the Subcontractor copies of the Contractor's construction schedule and schedule of submittals, together with such additional scheduling details as will enable the Subcontractor to plan and perform the Subcontractor's Work properly. The Subcontractor shall be notified promptly of subsequent changes in the construction and submittal schedules and additional scheduling details.

3.1.2 The Contractor shall provide suitable areas for storage of the Subcontractor's materials and equipment during the course of the Work. Additional costs to the Subcontractor resulting from relocation of such facilities at the direction of the Contractor, except as previously agreed upon, shall be reimbursed by the Contractor.

3.1.3 Except as provided in Article 14, the Contractor's equipment will be available to the Subcontractor only at the Contractor's discretion and on mutually satisfactory terms.

3.2 COMMUNICATIONS

3.2.1 The Contractor shall promptly make available to the Subcontractor information which affects this Subcontract and which becomes available to the Contractor subsequent to execution of this Subcontract.

3.2.2 The Contractor shall not give instructions or orders directly to employees or workmen of the Subcontractor, except to persons designated as authorized representatives of the Subcontractor.

3.2.3 The Contractor shall permit the Subcontractor to request directly from the Architect information regarding the percentages of completion and the amount certified on account of Work done by the Subcontractor.

3.2.4 If hazardous substances of a type of which an employer is required by law to notify its employees are being used on the site by the Contractor, a subcontractor or anyone directly or indirectly employed by them (other than the Subcontractor), the Contractor shall, prior to harmful exposure of the Subcontractor's employees to such substance, give written notice of the chemical composition thereof to the Subcontractor in sufficient detail and time to permit the Subcontractor's compliance with such laws.

3.3 CLAIMS BY THE CONTRACTOR

3.3.1 Liquidated damages for delay, if provided for in Paragraph 9.3 of this Agreement, shall be assessed against the Subcontractor only to the extent caused by the Subcontractor, the Subcontractor's employees and agents, Sub-subcontractors, suppliers or any person or entity for whose acts the Subcontractor may be liable, and in no case for delays or causes arising outside the scope of this Subcontract.

3.3.2 Except as may be indicated in this Agreement, the Contractor agrees that no claim for payment for services rendered or materials and equipment furnished by the Contractor to the Subcontractor shall be valid without prior notice to the Subcontractor and unless written notice thereof is given by the Contractor to the Subcontractor not later than the tenth day of the calendar month following that in which the claim originated.

3.4 CONTRACTOR'S REMEDIES

3.4.1 If the Subcontractor defaults or neglects to carry out the Work in accordance with this Agreement and fails within three working days after receipt of written notice from the Contractor to commence and continue correction of such default or neglect with diligence and promptness, the Contractor may, after three days following receipt by the Subcontractor of an additional written notice, and without prejudice to any other remedy the Contractor may have, make good such deficiencies and may deduct the cost thereof from the payments then or thereafter due the Subcontractor, provided, however, that if such action is based upon faulty workmanship or materials and equipment, the Architect shall first have determined that the workmanship or materials and equipment are not in accordance with requirements of the Prime Contract.

ARTICLE 4
SUBCONTRACTOR

4.1 EXECUTION AND PROGRESS OF THE WORK

4.1.1 The Subcontractor shall cooperate with the Contractor in scheduling and performing the Subcontractor's Work to avoid conflict, delay in or interference with the Work of the Contractor, other subcontractors or Owner's own forces.

4.1.2 The Subcontractor shall promptly submit Shop Drawings, Product Data, Samples and similar submittals required by the Subcontract Documents with reasonable promptness and in such sequence as to cause no delay in the Work or in the activities of the Contractor or other subcontractors.

4.1.3 The Subcontractor shall submit to the Contractor a schedule of values allocated to the various parts of the Work of this Subcontract, aggregating the Subcontract Sum, made out in such detail as the Contractor and Subcontractor may agree upon or as required by the Owner, and supported by such evidence as the Contractor may direct. In applying for payment, the Subcontractor shall submit statements based upon this schedule.

4.1.4 The Subcontractor shall furnish to the Contractor periodic progress reports on the Work of this Subcontract as mutually agreed, including information on the status of materials and equipment which may be in the course of preparation or manufacture.

4.1.5 The Subcontractor agrees that the Architect will have the authority to reject Work which does not conform to the Prime Contract. The Architect's decisions on matters relating to aesthetic effect shall be final if consistent with the intent expressed in the Prime Contract.

4.1.6 The Subcontractor shall pay for materials, equipment and labor used in connection with the performance of this Subcontract through the period covered by previous payments received from the Contractor, and shall furnish satisfactory

evidence, when requested by the Contractor, to verify compliance with the above requirements.

4.1.7 The Subcontractor shall take necessary precautions to protect properly the Work of other subcontractors from damage caused by operations under this Subcontract.

4.1.8 The Subcontractor shall cooperate with the Contractor, other subcontractors and the Owner's own forces whose Work might interfere with the Subcontractor's Work. The Subcontractor shall participate in the preparation of coordinated drawings in areas of congestion, if required by the Prime Contract, specifically noting and advising the Contractor of potential conflicts between the Work of the Subcontractor and that of the Contractor, other subcontractors or the Owner's own forces.

4.2 LAWS, PERMITS, FEES AND NOTICES

4.2.1 The Subcontractor shall give notices and comply with laws, ordinances, rules, regulations and orders of public authorities bearing on performance of the Work of this Subcontract. The Subcontractor shall secure and pay for permits and governmental fees, licenses and inspections necessary for proper execution and completion of the Subcontractor's Work, the furnishing of which is required of the Contractor by the Prime Contract.

4.2.2 The Subcontractor shall comply with Federal, state and local tax laws, social security acts, unemployment compensation acts and workers' or workmen's compensation acts insofar as applicable to the performance of this Subcontract.

4.3 SAFETY PRECAUTIONS AND PROCEDURES

4.3.1 The Subcontractor shall take reasonable safety precautions with respect to performance of this Subcontract, shall comply with safety measures initiated by the Contractor and with applicable laws, ordinances, rules, regulations and orders of public authorities for the safety of persons or property in accordance with the requirements of the Prime Contract. The Subcontractor shall report to the Contractor within three days an injury to an employee or agent of the Subcontractor which occurred at the site.

4.3.2 If hazardous substances of a type of which an employer is required by law to notify its employees are being used on the site by the Subcontractor, the Subcontractor's Sub-subcontractors or anyone directly or indirectly employed by them, the Subcontractor shall, prior to harmful exposure of any employees on the site to such substance, give written notice of the chemical composition thereof to the Contractor in sufficient detail and time to permit compliance with such laws by the Contractor, other subcontractors and other employers on the site.

4.3.3 In the event the Subcontractor encounters on the site material reasonably believed to be asbestos or polychlorinated biphenyl (PCB) which has not been rendered harmless, the Subcontractor shall immediately stop Work in the area affected and report the condition to the Contractor in writing. The Work in the affected area shall resume in the absence of asbestos or polychlorinated biphenyl (PCB), or when it has been rendered harmless, by written agreement of the Contractor and Subcontractor, or in accordance with final determination by the Architect on which arbitration has not been demanded, or by arbitration as provided in this Agreement. The Subcontractor shall not be required pursuant to Article 5 to perform without consent any Work relating to asbestos or polychlorinated biphenyl (PCB).

4.3.4 To the fullest extent permitted by law, the Contractor shall indemnify and hold harmless the Subcontractor, the Subcontractor's Sub-subcontractors, and agents and employees of any of them from and against claims, damages, losses and expenses, including but not limited to attorneys' fees, arising out of or resulting from performance of the Work in the affected area if in fact the material is asbestos or polychlorinated biphenyl (PCB) and has not been rendered harmless, provided that such claim, damage, loss or expense is attributable to bodily injury, sickness, disease or death, or to injury to or destruction of tangible property (other than the Work itself) including loss of use resulting therefrom, but only to the extent caused in whole or in part by negligent acts or omissions of the Contractor, Architect, Owner, anyone directly or indirectly employed by any of them, or anyone for whose acts any of them may be liable, regardless of whether or not such claim, damage, loss or expense is caused in part by a party indemnified hereunder. Such obligation shall not be construed to negate, abridge, or reduce other rights or obligations of indemnity which would otherwise exist as to a party or person described in this Subparagraph 4.3.4.

4.4 CLEANING UP

4.4.1 The Subcontractor shall keep the premises and surrounding area free from accumulation of waste materials or rubbish caused by operations performed under this Subcontract. The Subcontractor shall not be held responsible for unclean conditions caused by other contractors or subcontractors.

4.5 WARRANTY

4.5.1 The Subcontractor warrants to the Owner, Architect and Contractor that materials and equipment furnished under this Subcontract will be of good quality and new unless otherwise required or permitted by the Subcontract Documents, that the Work of this Subcontract will be free from defects not inherent in the quality required or permitted, and that the Work will conform with the requirements of the Subcontract Documents. Work not conforming to these requirements, including substitutions not properly approved and authorized, may be considered defective. The Subcontractor's warranty excludes remedy for damage or defect caused by abuse, modifications not executed by the Subcontractor, improper or insufficient maintenance, improper operation, or normal wear and tear under normal usage. This warranty shall be in addition to and not in limitation of any other warranty or remedy required by law or by the Subcontract Documents.

4.6 INDEMNIFICATION

4.6.1 To the fullest extent permitted by law, the Subcontractor shall indemnify and hold harmless the Owner, Contractor, Architect, Architect's consultants, and agents and employees of any of them from and against claims, damages, losses and expenses, including but not limited to attorney's fees, arising out of or resulting from performance of the Subcontractor's Work under this Subcontract, provided that such claim, damage, loss or expense is attributable to bodily injury, sickness, disease or death, or to injury to or destruction of tangible property (other than the Work itself) including loss of use resulting therefrom, but only to the extent caused in whole or in part by negligent acts or omissions of the Subcontractor, the Subcontractor's Sub-subcontractors, anyone directly or indirectly employed by them or anyone for whose acts they may be liable, regardless of whether or not such claim, damage, loss or expense is caused in part by a party indemnified hereunder.

Such obligation shall not be construed to negate, abridge, or otherwise reduce other rights or obligations of indemnity which would otherwise exist as to a party or person described in this Paragraph 4.6.

4.6.2 In claims against any person or entity idemnified under this Paragraph 4.6 by an employee of the Subcontractor, the Subcontractor's Sub-subcontractors, anyone directly or indirectly employed by them or anyone for whose acts they may be liable, the indemnification obligation under this Paragraph 4.6 shall not be limited by a limitation on amount or type of damages, compensation or benefits payable by or for the Subcontractor or the Subcontractor's Sub-subcontractors under workers' or workmen's compensation acts, disability benefit acts or other employee benefit acts.

4.6.3 The obligations of the Subcontractor under this Paragraph 4.6 shall not extend to the liability of the Architect, the Architect's consultants, and agents and employees of any of them arising out of (1) the preparation or approval of maps, drawings, opinions, reports, surveys, Change Orders, designs or specifications, or (2) the giving of or the failure to give directions or instructions by the Architect, the Architect's consultants, and agents and employees of any of them, provided such giving or failure to give is the primary cause of the injury or damage.

4.7 REMEDIES FOR NONPAYMENT

4.7.1 If the Contractor does not pay the Subcontractor through no fault of the Subcontractor, within seven days from the time payment should be made as provided in this Agreement, the Subcontractor may, without prejudice to other available remedies, upon seven additional days' written notice to the Contractor, stop the Work of this Subcontract until payment of the amount owing has been received. The Subcontract Sum shall, by appropriate adjustment, be increased by the amount of the Subcontractor's reasonable costs of shutdown, delay and start-up.

ARTICLE 5
CHANGES IN THE WORK

5.1 The Owner may make changes in the Work by issuing Modifications to the Prime Contract. Upon receipt of such a Modification issued subsequent to the execution of the Subcontract Agreement, the Contractor shall promptly notify the Subcontractor of the Modification. Unless otherwise directed by the Contractor, the Subcontractor shall not thereafter order materials or perform Work which would be inconsistent with the changes made by the Modifications to the Prime Contract.

5.2 The Subcontractor may be ordered in writing by the Contractor, without invalidating this Subcontract, to make changes in the Work within the general scope of this Subcontract consisting of additions, deletions or other revisions, including those required by Modifications to the Prime Contract issued subsequent to the execution of this Agreement, the Subcontract Sum and the Subcontract Time being adjusted accordingly. The Subcontractor, prior to the commencement of such changed or revised Work, shall submit promptly to the Contractor written copies of a claim for adjustment to the Subcontract Sum and Subcontract Time for such revised Work in a manner consistent with requirements of the Subcontract Documents.

5.3 The Subcontractor shall make claims promptly to the Contractor for additional cost, extensions of time and damages for delays or other causes in accordance with the Subcontract

Documents. A claim which will affect or become part of a claim which the Contractor is required to make under the Prime Contract within a specified time period or in a specified manner shall be made in sufficient time to permit the Contractor to satisfy the requirements of the Prime Contract. Such claims shall be received by the Contractor not less than two working days preceding the time by which the Contractor's claim must be made. Failure of the Subcontractor to make such a timely claim shall bind the Subcontractor to the same consequences as those to which the Contractor is bound.

ARTICLE 6
ARBITRATION

6.1 Any controversy or claim between the Contractor and the Subcontractor arising out of or related to this Subcontract, or the breach thereof, shall be settled by arbitration, which shall be conducted in the same manner and under the same procedure as provided in the Prime Contract with respect to claims between the Owner and the Contractor, except that a decision by the Architect shall not be a condition precedent to arbitration. If the Prime Contract does not provide for arbitration or fails to specify the manner and procedure for arbitration, it shall be conducted in accordance with the Construction Industry Arbitration Rules of the American Arbitration Association currently in effect unless the parties mutually agree otherwise.

6.2 Except by written consent of the person or entity sought to be joined, no arbitration arising out of or relating to the Subcontract shall include, by consolidation or joinder or in any other manner, any person or entity not a party to the Agreement under which such arbitration arises, unless it is shown at the time the demand for arbitration is filed that (1) such person or entity is substantially involved in a common question of fact or law, (2) the presence of such person or entity is required if complete relief is to be accorded in the arbitration, (3) the interest or responsibility of such person or entity in the matter is not insubstantial, and (4) such person or entity is not the Architect, the Architect's employee, the Architect's consultant, or an employee or agent of any of them. This agreement to arbitrate and any other written agreement to arbitrate with an additional person or persons referred to herein shall be specifically enforceable under applicable law in any court having jurisdiction thereof.

6.3 The Contractor shall give the Subcontractor prompt written notice of any demand received or made by the Contractor for arbitration if the dispute involves or relates to the Work, materials, equipment, rights or responsibilities of the Subcontractor. The Contractor shall consent to inclusion of the Subcontractor in the arbitration proceeding whether by joinder, consolidation or otherwise, if the Subcontractor requests in writing to be included within ten days after receipt of the Contractor's notice.

6.4 The award rendered by the arbitrator or arbitrators shall be final, and judgment may be entered upon it in accordance with applicable law in any court having jurisdiction thereof.

6.5 This Article 6 shall not be deemed a limitation of rights or remedies which the Subcontractor may have under Federal

law, under state mechanics' lien laws, or under applicable labor or material payment bonds unless such rights or remedies are expressly waived by the Subcontractor.

ARTICLE 7
TERMINATION, SUSPENSION OR ASSIGNMENT OF THE SUBCONTRACT

7.1   TERMINATION BY THE SUBCONTRACTOR

7.1.1 The Subcontractor may terminate the Subcontract for the same reasons and under the same circumstances and procedures with respect to the Contractor as the Contractor may terminate with respect to the Owner under the Prime Contract, or for nonpayment of amounts due under this Subcontract for 60 days or longer. In the event of such termination by the Subcontractor for any reason which is not the fault of the Subcontractor, Sub-subcontractors or their agents or employees or other persons performing portions of the Work under contract with the Subcontractor, the Subcontractor shall be entitled to recover from the Contractor payment for Work executed and for proven loss with respect to materials, equipment, tools, and construction equipment and machinery, including reasonable overhead, profit and damages.

7.2   TERMINATION BY THE CONTRACTOR

7.2.1 If the Subcontractor persistently or repeatedly fails or neglects to carry out the Work in accordance with the Subcontract Documents or otherwise to perform in accordance with this Agreement and fails within seven days after receipt of written notice to commence and continue correction of such default or neglect with diligence and promptness, the Contractor may, after seven days following receipt by the Subcontractor of an additional written notice and without prejudice to any other remedy the Contractor may have, terminate the Subcontract and finish the Subcontractor's Work by whatever method the Contractor may deem expedient. If the unpaid balance of the Subcontract Sum exceeds the expense of finishing the Subcontractor's Work, such excess shall be paid to the Subcontractor, but if such expense exceeds such unpaid balance, the Subcontractor shall pay the difference to the Contractor.

7.3   ASSIGNMENT OF THE SUBCONTRACT

7.3.1 In the event of termination of the Prime Contract by the Owner, the Contractor may assign this Subcontract to the Owner, with the Owner's agreement, subject to the provisions of the Prime Contract and to the prior rights of the surety, if any, obligated under bonds relating to the Prime Contract. If the Work of the Prime Contract has been suspended for more than 30 days, the Subcontractor's compensation shall be equitably adjusted.

7.3.2 The Subcontractor shall not assign the Work of this Subcontract without the written consent of the Contractor, nor subcontract the whole of this Subcontract without the written consent of the Contractor, nor further subcontract portions of this Subcontract without written notification to the Contractor when such notification is requested by the Contractor.

AIA DOCUMENT A401 • CONTRACTOR-SUBCONTRACTOR AGREEMENT • TWELFTH EDITION • AIA® • ©1987
THE AMERICAN INSTITUTE OF ARCHITECTS, 1735 NEW YORK AVENUE, N.W., WASHINGTON, D.C. 20006

ARTICLE 8
THE WORK OF THIS SUBCONTRACT

8.1   The Subcontractor shall execute the following portion of the Work described in the Subcontract Documents, including all labor, materials, equipment, services and other items required to complete such portion of the Work, except to the extent specifically indicated in the Subcontract Documents to be the responsibility of others:

(Insert a precise description of the Work of this Subcontract, referring where appropriate to numbers of Drawings, sections of Specifications and pages of Addenda, Modifications and accepted Alternates.)

ARTICLE 9
DATE OF COMMENCEMENT AND SUBSTANTIAL COMPLETION

9.1   The Subcontractor's date of commencement is the date from which the Contract Time of Paragraph 9.3 is measured; it shall be the date of this Agreement, as first written above, unless a different date is stated below or provision is made for the date to be fixed in a notice to proceed issued by the Contractor.

(Insert the date of commencement, if it differs from the date of this Agreement or, if applicable, state that the date will be fixed in a notice to proceed.)

9.2   Unless the date of commencement is established by a notice to proceed issued by the Contractor, or the Contractor has commenced visible Work at the site under the Prime Contract, the Subcontractor shall notify the Contractor in writing not less than five days before commencing the Subcontractor's Work to permit the timely filing of mortgages, mechanic's liens and other security interests.

9.3   The Work of this Subcontract shall be substantially completed not later than

(Insert the calendar date or number of calendar days after the Subcontractor's date of commencement. Also insert any requirements for earlier Substantial Completion of certain portions of the Subcontractor's Work, if not stated elsewhere in the Subcontract Documents.)

, subject to adjustments of this Subcontract Time as provided in the Subcontract Documents.

(Insert provisions, if any, for liquidated damages relating to failure to complete on time.)

9.4   Time is of the essence of this Subcontract.

9.5   No extension of time will be valid without the Contractor's written consent after claim made by the Subcontractor in accordance with Paragraph 5.2.

ARTICLE 10
SUBCONTRACT SUM

10.1   The Contractor shall pay the Subcontractor in current funds for performance of the Subcontract the Subcontract Sum of
Dollars (**$**   ),
subject to additions and deductions as provided in the Subcontract Documents.

10.2   The Subcontract Sum is based upon the following alternates, if any, which are described in the Subcontract Documents and
have been accepted by the Owner and the Contractor:
(Insert the numbers or other identification of accepted alternates.)

10.3   Unit prices, if any, are as follows:

ARTICLE 11
PROGRESS PAYMENTS

11.1   Based upon applications for payment submitted to the Contractor by the Subcontractor, corresponding to Applications for
Payment submitted by the Contractor to the Architect, and Certificates for Payment issued by the Architect, the Contractor shall make
progress payments on account of the Subcontract Sum to the Subcontractor as provided below and elsewhere in the Subcontract
Documents.

11.2   The period covered by each application for payment shall be one calendar month ending on the last day of the month, or as
follows:

11.3   Provided an application for payment is received by the Contractor not later than the
day of a month, the Contractor shall include the Subcontractor's Work covered by that application in the next Application for Payment which the Contractor is entitled to submit to the Architect. The Contractor shall pay the Subcontractor each progress payment

within three working days after the Contractor receives payment from the Owner. If the Architect does not issue a Certificate for Payment or the Contractor does not receive payment for any cause which is not the fault of the Subcontractor, the Contractor shall pay the Subcontractor, on demand, a progress payment computed as provided in Paragraphs 11.7 and 11.8.

11.4   If an application for payment is received by the Contractor after the application date fixed above, the Subcontractor's Work covered by it shall be included by the Contractor in the next Application for Payment submitted to the Architect.

11.5   Each application for payment shall be based upon the most recent schedule of values submitted by the Subcontractor in accordance with the Subcontract Documents. The schedule of values shall allocate the entire Subcontract Sum among the various portions of the Subcontractor's Work and be prepared in such form and supported by such data to substantiate its accuracy as the Contractor may require. This schedule, unless objected to by the Contractor, shall be used as a basis for reviewing the Subcontractor's applications for payment.

11.6   Applications for payment submitted by the Subcontractor shall indicate the percentage of completion of each portion of the Subcontractor's Work as of the end of the period covered by the application for payment.

11.7   Subject to the provisions of the Subcontract Documents, the amount of each progress payment shall be computed as follows:

11.7.1 Take that portion of the Subcontract Sum properly allocable to completed Work as determined by multiplying the percentage completion of each portion of the Subcontractor's Work by the share of the total Subcontract Sum allocated to that portion of the Subcontractor's Work in the schedule of values, less that percentage actually retained, if any, from payments to the Contractor on account of the Work of the Subcontractor. Pending final determination of cost to the Contractor of changes in the Work which have been properly authorized by Construction Change Directive, amounts not in dispute may be included to the same extent provided in the Prime Contract, even though the Subcontract Sum has not yet been adjusted;

11.7.2 Add that portion of the Subcontract Sum properly allocable to materials and equipment delivered and suitably stored at the site by the Subcontractor for subsequent incorporation in the Subcontractor's Work or, if approved in advance by the Owner, suitably stored off the site at a location agreed upon in writing, less the same percentage retainage required by the Prime Contract to be applied to such materials and equipment in the Contractor's Application for Payment;

11.7.3 Subtract the aggregate of previous payments made by the Contractor; and

11.7.4 Subtract amounts, if any, calculated under Subparagraph 11.7.1 or 11.7.2 which are related to Work of the Subcontractor for which the Architect has withheld or nullified, in whole or in part, a Certificate of Payment for a cause which is the fault of the Subcontractor.

11.8   SUBSTANTIAL COMPLETION

11.8.1 When the Subcontractor's Work or a designated portion thereof is substantially complete and in accordance with the requirements of the Prime Contract, the Contractor shall, upon application by the Subcontractor, make prompt application for payment for such Work. Within 30 days following issuance by the Architect of the Certificate for Payment covering such substantially completed Work, the Contractor shall, to the full extent allowed in the Prime Contract, make payment to the Subcontractor, deducting any portion of the funds for the Subcontractor's Work withheld in accordance with the Certificate to cover costs of items to be completed or corrected by the Subcontractor. Such payment to the Subcontractor shall be the entire unpaid balance of the Subcontract Sum if a full release of retainage is allowed under the Prime Contract for the Subcontractor's Work prior to the completion of the entire Project. If the Prime Contract does not allow for a full release of retainage, then such payment shall be an amount which, when added to previous payments to the Subcontractor, will reduce the retainage on the Subcontractor's substantially completed Work to the same percentage of retainage as that on the Contractor's Work covered by the Certificate.

ARTICLE 12

FINAL PAYMENT

12.1   Final payment, constituting the entire unpaid balance of the Subcontract Sum, shall be made by the Contractor to the Subcontractor when the Subcontractor's Work is fully performed in accordance with the requirements of the Contract Documents, the Architect has issued a Certificate for Payment covering the Subcontractor's completed Work and the Contractor has received payment from the Owner. If, for any cause which is not the fault of the Subcontractor, a Certificate for Payment is not issued or the Contractor does not receive timely payment or does not pay the Subcontractor within three working days after receipt of payment from the Owner, final payment to the Subcontractor shall be made upon demand.
(Insert provisions for earlier final payment to the Subcontractor, if applicable.)

12.2   Before issuance of the final payment, the Subcontractor, if required, shall submit evidence satisfactory to the Contractor that all payrolls, bills for materials and equipment, and all known indebtedness connected with the Subcontractor's Work have been satisfied.

ARTICLE 13
INSURANCE AND BONDS

13.1   The Subcontractor shall purchase and maintain insurance of the following types of coverage and limits of liability:

13.2   Coverages, whether written on an occurrence or claims-made basis, shall be maintained without interruption from date of commencement of the Subcontractor's Work until date of final payment and termination of any coverage required to be maintained after final payment.

13.3   Certificates of insurance acceptable to the Contractor shall be filed with the Contractor prior to commencement of the Subcontractor's Work. These certificates and the insurance policies required by this Article 13 shall contain a provision that coverages afforded under the policies will not be cancelled or allowed to expire until at least 30 days' prior written notice has been given to the Contractor. If any of the foregoing insurance coverages are required to remain in force after final payment and are reasonably available, an additional certificate evidencing continuation of such coverage shall be submitted with the final application for payment as required in Article 12. If any information concerning reduction of coverage is not furnished by the insurer, it shall be furnished by the Subcontractor with reasonable promptness according to the Subcontractor's information and belief.

13.4   The Contractor shall furnish to the Subcontractor satisfactory evidence of insurance required of the Contractor under the Prime Contract.

13.5   Waivers of Subrogation. The Contractor and Subcontractor waive all rights against (1) each other and any of their Subcontractors, Sub-subcontractors, agents and employees, each of the other, and (2) the Owner, the Architect, the Architect's consultants, separate contractors, and any of their subcontractors, sub-subcontractors, agents and employees for damages caused by fire or other perils to the extent covered by property insurance provided under the Prime Contract or other property insurance applicable to the Work, except such rights as they may have to proceeds of such insurance held by the Owner as fiduciary. The Subcontractor shall require of the Subcontractor's Sub-subcontractors, agents and employees, by appropriate agreements, written where legally required for validity, similar waivers in favor of other parties enumerated herein. The policies shall provide such waivers of subrogation by endorsement or otherwise. A waiver of subrogation shall be effective as to a person or entity even though that person or entity would otherwise have a duty of indemnification, contractual or otherwise, did not pay the insurance premium directly or indirectly, and whether or not the person or entity had an insurable interest in the property damaged.

13.6   The Contractor shall promptly, upon request of the Subcontractor, furnish a copy or permit a copy to be made of any bond covering payment of obligations arising under the Subcontract.

13.7   Performance Bond and Payment Bond:
(If the Subcontractor is to furnish bonds, insert the specific requirements here)

ARTICLE 14
TEMPORARY FACILITIES AND WORKING CONDITIONS

14.1   The Contractor shall furnish and make available to the Subcontractor the following temporary facilities, equipment and services; these shall be furnished at no cost to the Subcontractor unless otherwise indicated below:

14.2   Specific working conditions:
(Insert any applicable arrangements concerning working conditions and labor matters for the Project.)

ARTICLE 15

MISCELLANEOUS PROVISIONS

15.1 Where reference is made in this Agreement to a provision of the General Conditions or another Subcontract Document, the reference refers to that provision as amended or supplemented by other provisions of the Subcontract Documents.

15.2 Payments due and unpaid under this Subcontract shall bear interest from the date payment is due at such rate as the parties may agree upon in writing or, in the absence thereof, at the legal rate prevailing from time to time at the place where the Project is located.

(Insert rate of interest agreed upon, if any.)

(Usury laws and requirements under the Federal Truth in Lending Act, similar state and local consumer credit laws and other regulations at the Owner's, Contractor's and Subcontractor's principal places of business, the location of the Project and elsewhere may affect the validity of this provision. Legal advice should be obtained with respect to deletions or modifications, and also regarding requirements such as written disclosures or waivers.)

ARTICLE 16

ENUMERATION OF SUBCONTRACT DOCUMENTS

16.1 The Subcontract Documents, except for Modifications issued after execution of this Agreement, are enumerated as follows:

16.1.1 This executed Standard Form of Agreement Between Contractor and Subcontractor, AIA Document A401, 1987 Edition;

16.1.2 The Prime Contract, consisting of the Agreement between the Owner and Contractor dated as first entered above and the other Contract Documents enumerated in the Owner-Contractor Agreement; Conditions of the Contract (General, Supplementary and other Conditions), Drawings, Specifications, Addenda and other documents enumerated therein;

16.1.3 The following Modifications to the Prime Contract, if any, issued subsequent to the execution of the Owner-Contractor Agreement but prior to the execution of this Agreement:

Modification   **Date**

16.1.4 Other Documents, if any, forming part of the Subcontract Documents are as follows:
(List any additional documents which are intended to form part of the Subcontract Documents. Requests for proposal and the Subcontractor's bid or proposal should be listed here only if intended to be part of the Subcontract Documents.)

This Agreement entered into as of the day and year first written above.

CONTRACTOR   SUBCONTRACTOR

_____   _____
(Signature)   *(Signature)*

_____   _____
(Printed name and title)   *(Printed name and title)*

AIA DOCUMENT A401 • CONTRACTOR-SUBCONTRACTOR AGREEMENT • TWELFTH EDITION • AIA® • ©1987
THE AMERICAN INSTITUTE OF ARCHITECTS, 1735 NEW YORK AVENUE, N.W., WASHINGTON, D.C. 20006   **A401-1987   10**

Construction Industry Arbitration Rules

American Arbitration Association

as Amended and Effective on January 1, 1993

National Construction Dispute Resolution Committee

Representatives of the fourteen organizations listed below constitute the National Construction Dispute Resolution Committee (NCDRC). This committee is the sponsor of the arbitration procedure specially designed for the construction industry by the American Arbitration Association (AAA).

AMERICAN CONSULTING
ENGINEERS COUNCIL

AMERICAN INSTITUTE OF ARCHITECTS

AMERICAN SOCIETY
OF CIVIL ENGINEERS

AMERICAN SOCIETY
OF INTERIOR DESIGNERS

AMERICAN SOCIETY
OF LANDSCAPE ARCHITECTS

AMERICAN SUBCONTRACTORS
ASSOCIATION

ASSOCIATED BUILDERS
AND CONTRACTORS, INC.

ASSOCIATED GENERAL CONTRACTORS

ASSOCIATED SPECIALTY
CONTRACTORS, INC.

BUSINESS ROUNDTABLE

CONSTRUCTION SPECIFICATIONS
INSTITUTE

NATIONAL ASSOCIATION
OF HOME BUILDERS

NATIONAL SOCIETY
OF PROFESSIONAL ENGINEERS

NATIONAL UTILITY
CONTRACTORS ASSOCIATION

Construction Industry Arbitration Rules

1. Agreement of Parties
The parties shall be deemed to have made these rules a part of their arbitration agreement whenever they have provided for arbitration by the American Arbitration Association (hereinafter AAA) or under its Construction Industry Arbitra-tion Rules. These rules and any amendment of them shall apply in the form obtaining at the time the demand for arbitration or submission agreement is received by the AAA. The parties, by written agreement, may vary the procedures set forth in these rules.

2. Name of Tribunal
Any tribunal constituted by the parties for the settlement of their dispute under these rules shall be called the Construction Industry Arbitration Tribunal.

3. Administrator and Delegation of Duties
When parties agree to arbitrate under these rules, or when they provide for arbitration by the AAA and an arbitration is initiated under these rules, they thereby authorize the AAA to administer the arbitration. The authority and duties of the AAA are prescribed in the agreement of the parties and in these rules, and may be carried out through such of the AAA's representatives as it may direct.

4. National Panel of Arbitrators
In cooperation with the National Construction Industry Dispute Resolution Committee, the AAA shall establish and maintain a National Panel of Construction Industry Arbitrators and shall appoint arbitrators as provided in these rules.

5. Regional Offices
The AAA may, in its discretion, assign the administration of an arbitration to any of its regional offices.

6. Initiation under an Arbitration Provision in a Contract
Arbitration under an arbitration provision in a contract shall be initiated in the following manner:

(a) The initiating party (hereinafter claimant) shall, within the time period, if any, specified in the contract(s), give written notice to the other party (hereinafter respondent) of its intention to arbitrate (demand), which notice shall contain a statement setting forth the nature of the dispute, the amount involved, if any, the remedy sought, and the hearing locale requested, and

(b) shall file at any regional office of the AAA three copies of the notice and three copies of the arbitration provisions of the contract, together with the appropriate filing fee as provided in the schedule on page 23.

The AAA shall give notice of such filing to the respondent or respondents. A respondent may file an answering statement in duplicate with the AAA within ten days after notice from the AAA, in which event the respondent shall at the same time send a copy of the answering statement to the claimant. If a counterclaim is asserted, it shall contain a statement setting forth the nature of the counterclaim, the amount involved, if any, and the remedy sought. If a counterclaim is made, the appropriate fee provided in the schedule on page 23 shall be forwarded to the AAA with the answering statement. If no answering statement if filed within the stated time, it will be treated as a denial of the claim. Failure to file an answering statement shall not operate to delay the arbitration.

7. Initiation under a Submission

Parties to any existing dispute may commence an arbitration under these rules by filing at any regional office of the AAA three copies of a written submission to arbitrate under these rules, signed by the parties. It shall contain a statement of the matter in dispute, the amount involved, if any, the remedy sought, and the hearing locale requested, together with the appropriate filing fee as provided in the schedule on page 23.

8. Changes of Claim

After filing of a claim, if either party desires to make any new or different claim or counterclaim, it shall be made in writing and filed with the AAA, and a copy shall be mailed to the other party, who shall have a period of ten days from the date of such mailing within which to file an answer with the AAA. After the arbitrator is appointed, however, no new or different claim may be submitted except with the arbitrator's consent.

9. Applicable Procedures

Unless the AAA in its discretion determines otherwise, the Expedited Procedures shall be applied in any case where no disclosed claim or counterclaim exceeds $50,000, exclusive of interest and arbitration costs. Parties may also agree to using the Expedited Procedures in cases involving claims in excess of $50,000. The Expedited Procedures shall be applied as described in Sections 53 through 57 of these rules, in addition to any other portion of

these rules that is not in conflict with the Expedited Procedures.

All other cases shall be administered in accordance with Sections 1 through 52 of these rules.

10. Administrative Conference, Preliminary Hearing, and Mediation Conference

At the request of any party or at the discretion of the AAA, an administrative conference with the AAA and the parties and/or their representatives will be scheduled in appropriate cases to expedite the arbitration proceedings. There is no administrative fee for this service.

In large or complex cases, at the request of any party or at the discretion of the arbitrator or the AAA, a preliminary hearing with the parties and/ or their representatives and the arbitrator may be scheduled by the arbitrator to specify the issues to be resolved, to stipulate to uncontested facts, and to consider any other matters that will expedite the arbitration proceedings. Consistent with the expedited nature of arbitration, the arbitrator may, at the preliminary hearing, establish (i) the extent of and schedule for the production of relevant documents and other information, (ii) the identification of any witnesses to be called, and (iii) a schedule for further hearings to resolve the dispute. There is no administrative fee for the first preliminary hearing.

With the consent of the parties, the AAA at any stage of the proceeding may arrange a mediation conference under the Construction Industry Mediation Rules, in order to facilitate settlement. The mediator shall not be an arbitrator appointed to the case. Where the parties to a pending arbitration agree to mediate under the AAA's rules, no additional administrative fee is required to initiate the mediation.

11. Fixing of Locale

The parties may mutually agree on the locale where the arbitration is to be held. If any party requests that the hearing be held in a specific locale and the other party files no objection thereto within ten days after notice of the request has been sent to it by the AAA, the locale shall be the one requested. If a party objects to the locale requested by the other party, the AAA shall have the power to determine the locale and its decision shall be final and binding.

12. Qualifications of an Arbitrator

Any neutral arbitrator appointed pursuant to Section 13, 14, 15, or 54, or selected by mutual choice of the parties or their appointees, shall be subject to disqualification for the reasons specified in Section 19. If the parties specifically so agree in writing, the arbitrator shall not be subject to disqualification for those reasons.

Unless the parties agree otherwise, an arbitrator selected unilaterally by one party is a party-appointed arbitrator and is not subject to disqualification pursuant to Section 19.

The term "arbitrator" in these rules refers to the arbitration panel, whether composed of one or more arbitrators and whether the arbitrators are neutral or party appointed.

13. Appointment from Panel

If the parties have not appointed an arbitrator and have not provided any other method of appointment, the arbitrator shall be appointed in the following manner: immediately after the filing of the demand or submission, the AAA shall send simultaneously to each party to the dispute an identical list of names of persons chosen from the panel.

Each party to the dispute shall have ten days from the transmittal date in which to strike any names objected to, number the remaining names in order of preference, and return the list to the AAA. If a party does not return the list within the time specified, all persons named therein shall be deemed acceptable. From among the persons who have been approved on both lists, and in accordance with the designated order of mutual preference, the AAA shall invite the acceptance of an arbitrator to serve. If the parties fail to agree on any of the persons named, or if acceptable arbitrators are unable to act, or if for any other reason the appointment cannot be made from the submitted lists, the AAA shall have the power to make the appointment from among other members of the panel without the submission of additional lists.

14. Direct Appointment by a Party

If the agreement of the parties names an arbitrator or specifies a method of appointing an arbitrator, that designation or method shall be followed. The notice of appointment, with the name and address of the arbitrator, shall be filed with the AAA by the appointing party. Upon the request of any appointing party, the AAA shall submit a list of members of the panel from which the party may, if it so desires, make the appointment.

If the agreement specifies a period of time within which an arbitrator shall be appointed and any party fails to make the appointment within that period, the AAA shall make the appointment.

If no period of time is specified in the agreement, the AAA shall notify the party to make the appointment. If within ten days thereafter an arbitrator has not been appointed by a party, the AAA shall make the appointment.

15. Appointment of Neutral Arbitrator by Party-Appointed Arbitrators or Parties

If the parties have selected party-appointed arbitrators, or if such arbitrators have been appointed as provided in Section 14, and the parties have authorized them to appoint a neutral arbitrator within a specified time and no appointment is made within that time or any agreed extension, the AAA may appoint a neutral arbitrator, who shall act as chairperson.

If no period of time is specified for appointment of the neutral arbitrator and the party-appointed arbitrators or the parties do not make the appointment within ten days from the date of the appointment of the last party-appointed arbitrator, the AAA may appoint the neutral arbitrator, who shall act as chairperson.

If the parties have agreed that their party-appointed arbitrators shall appoint the neutral arbitrator from the panel, the AAA shall furnish to the party-appointed arbitrators, in the manner provided in Section 13, a list selected from the panel, and the appointment of the neutral arbitrator shall be made as provided in that section.

16. Nationality of Arbitrator in International Arbitration

Where the parties are nationals or residents of different countries, any neutral arbitrator shall, upon the request of either party, be appointed from among the nationals of a country other than that of any of the parties. The request must be made prior to the time set for the appointment of the arbitrator as agreed by the parties or set by these rules.

17. Number of Arbitrators

If the arbitration agreement does not specify the number of arbitrators, the dispute shall be heard and determined by one arbitrator, unless the AAA, in its discretion, directs that a greater number of arbitrators be appointed.

18. Notice to Arbitrator of Appointment

Notice of the appointment of the neutral arbitrator, whether appointed mutually by the parties or by the AAA, shall be sent to the arbitrator by the AAA, together with a copy of these rules, and the signed acceptance of the arbitrator shall be filed with the AAA prior to the opening of the first hearing.

19. Disclosure and Challenge Procedure

Any person appointed as neutral arbitrator shall disclose to the AAA any circumstance likely to affect impartiality, including any bias or any financial or personal interest in the result of the arbitration or any past or present relationship with the parties or their representatives. Upon receipt of such information from the arbitrator or another source, the AAA shall communicate the information to the parties and, if it deems it appropriate to do so, to the arbitrator and others. Upon objection of a party to the continued service of a neutral arbitrator, the AAA shall determine whether the arbitrator should be disqualified and shall inform the parties of its decision, which shall be conclusive.

20. Vacancies

If for any reason an arbitrator is unable to perform the duties of the office, the AAA may, on proof satisfactory to it, declare the office vacant. Vacancies shall be filled in accordance with the applicable provisions of these rules.

In the event of a vacancy in a panel of neutral arbitrators after the hearings have commenced, the remaining arbitrator or arbitrators may continue with the hearing and determination of the controversy, unless the parties agree otherwise.

21. Date, Time, and Place of Hearing

The arbitrator shall set the date, time, and place for each hearing. The AAA shall send a notice of hearing to the parties at least ten days in advance of the hearing date, unless otherwise agreed by the parties.

22. Representation

Any party may be represented by counsel or other authorized representative. A party intending to be so represented shall notify the other party and the AAA of the name and address of the representative at least three days prior to the date set for the hearing at which that person is first to appear. When such a representative initiates an arbitration or responds for a party, notice is deemed to have been given.

23. Stenographic Record

Any party desiring a stenographic record shall make arrangements directly with a stenographer and shall notify the other parties of these arrangements in advance of the hearing. The requesting party or parties shall pay the cost of the record. If the transcript is agreed by the parties to be, or determined by the arbitrator to be, the official record of the proceeding, it must be made available to the arbitrator and to the other parties for inspection, at a date, time, and place determined by the arbitrator.

24. Interpreters

Any party wishing an interpreter shall make all arrangements directly with the interpreter and shall assume the costs of the service.

25. Attendance at Hearings

The arbitrator shall maintain the privacy of the hearings unless the law provides to the contrary. Any person having a direct interest in the arbitration is entitled to attend hearings. The arbitrator shall otherwise have the power to require the exclusion of any witness, other than a party or other essential person, during the testimony of any other witness. It shall be discretionary with the arbitrator to determine the propriety of the attendance of any other person.

26. Postponements

The arbitrator for good cause shown may postpone any hearing upon the request of a party or upon the arbitrator's own initiative, and shall also grant such postponement when all of the parties agree.

27. Oaths

Before proceeding with the first hearing, each arbitrator may take an oath of office and, if required by law, shall do so. The arbitrator may require witnesses to testify under oath administered by any duly qualified person and, if it is required by law or requested by any party, shall do so.

28. Majority Decision

All decisions of the arbitrators must be by a majority. The award must also be made by a majority unless the concurrence of all is expressly required by the arbitration agreement or by law.

29. Order of Proceedings and Communication with Arbitrator

A hearing shall be opened by the filing of the oath of the arbitrator, where required; by the recording of the date, time, and place of the hearing, and the presence of the arbitrator, the parties, and their representatives, if any; and by the receipt by the arbitrator of the statement of the claim and the answering statement, if any.

The arbitrator may, at the beginning of the hearing, ask for statements clarifying the issues involved. In some cases, part or all of the above will have been accomplished at the preliminary hearing conducted by the arbitrator pursuant to Section 10.

The complaining party shall then present evidence to support its claim. The defending party shall then present evidence supporting its defense. Witnesses for each party shall submit to questions or other examination. The arbitrator has the discretion to vary this procedure but shall afford a full and equal opportunity to all parties for the presentation of any material and relevant evidence.

Exhibits, when offered by either party, may be received in evidence by the arbitrator.

The names and addresses of all witnesses and a description of the exhibits in the order received shall be made a part of the record.

There shall be no direct communication between the parties and a neutral arbitrator other than at oral hearing, unless the parties and the arbitrator agree otherwise. Any other oral or written communication from the parties to the neutral arbitrator shall be directed to the AAA for transmittal to the arbitrator.

30. Arbitration in the Absence of a Party or Representative

Unless the law provides to the contrary, the arbitration may proceed in the absence of any party or representative who, after due notice, fails to be present or fails to obtain a postponement. An award shall not be made solely on the default of a party. The arbitrator shall require the party who is present to submit such evidence as the arbitrator may require for the making of an award.

31. Evidence

The parties may offer such evidence as is relevant and material to the dispute and shall produce such evidence as the arbitrator may deem necessary to an understanding and determination of the dispute. An arbitrator or other person authorized by law to subpoena witnesses or documents may do so upon the request of any party or independently.

The arbitrator shall be the judge of the relevance and materiality of the evidence offered, and conformity to legal rules of evidence shall not be necessary. All evidence shall be taken in the presence of all of the arbitrators and all of the parties, except where any of the parties is absent in default or has waived the right to be present.

32. Evidence by Affidavit and Posthearing Filing of Documents or Other Evidence

The arbitrator may receive and consider the evidence of witnesses by affidavit, but shall give it only such weight as the arbitrator deems it entitled to after consideration of any objection made to its admission.

If the parties agree or the arbitrator directs that documents or other evidence be submitted to the arbitrator after the hearing, the documents or other evidence shall be filed with the AAA for transmission to the arbitrator. All parties shall be afforded an opportunity to examine such documents or other evidence.

33. Inspection or Investigation

An arbitrator finding it necessary to make an inspection or investigation in connection with the arbitration shall direct the AAA to so advise the parties. The arbitrator shall set the date and time and the AAA shall notify the parties. Any party who so desires may be present at such an inspection or investigation. In the event that one or all parties are not present at the inspection or investigation, the arbitrator shall make a verbal or written report to the parties and afford them an opportunity to comment.

34. Interim Measures of Protection

(a) At the request of any party, the tribunal may take whatever interim measures it deems necessary with respect to the dispute, including measures for the conservation of property.

(b) Such interim measures may be taken in the form of an interim award and the tribunal may require security for the costs of such measures.

35. Closing of Hearing

The arbitrator shall specifically inquire of all parties whether they have any further proofs to offer or witnesses to be heard. Upon receiving negative replies or if satisfied that the record is complete, the arbitrator shall declare the hearing closed.

If briefs are to be filed, the hearing shall be declared closed as of the final date set by the arbitrator for the receipt of briefs. If documents are to be filed as provided in Section 32 and the date set for their receipt is later than that set for the receipt of briefs, the later date shall be the date of closing the hearing. The time limit within which the arbitrator is required to make the award shall commence to run, in the absence of other agreements by the parties, upon the closing of the hearing.

36. Reopening of Hearing

The hearing may be reopened on the arbitrator's initiative, or upon application of a party, at any time before the award is made. If reopening the hearing would prevent the making of the award within the specific time agreed on by the parties in the contract(s) out of which the controversy has arisen, the matter may not be reopened unless the parties agree on an extension of time. When no specific date is fixed in the contract, the arbitrator may reopen the hearing and shall have thirty days from the closing of the reopened hearing within which to make an award.

37. Waiver of Oral Hearing

The parties may provide, by written agreement, for the waiver of oral hearings in any case. If the parties are unable to agree as to the procedure, the AAA shall specify a fair and equitable procedure.

38. Waiver of Rules

Any party who proceeds with the arbitration after knowledge that any provision or requirement of these rules has not been complied with and who fails to state an objection in writing shall be deemed to have waived the right to object.

39. Extensions of Time

The parties may modify any period of time by mutual agreement. The AAA or the arbitrator may for good cause extend any period of time established by these rules, except the time for making the award. The AAA shall notify the parties of any extension.

40. Serving of Notice

Each party shall be deemed to have consented that any papers, notices, or process necessary or proper for the initiation or continuation of an arbitration under these rules; for any court action in connection therewith; or for the entry of judgment on any award made under these rules may be served on a party by mail addressed to the party or its representative at the last known address or by personal service, in or outside the state where the arbitration is to be held, provided that reasonable opportunity to be heard with regard thereto has been granted to the party.

The AAA and the parties may also use facsimile transmission, telex, telegram, or other written forms of electronic communication to give the notices required by these rules.

41. Time of Award

The award shall be made promptly by the arbitrator and, unless otherwise agreed by the parties or specified by law, no later than thirty days from the date of closing the hearing, or, if oral hearings have been waived, from the date of the AAA's transmittal of the final statements and proofs to the arbitrator.

42. Form of Award

The award shall be in writing and shall be signed by a majority of the arbitrators. It shall be executed in the manner required by law.

43. Scope of Award

The arbitrator may grant any remedy or relief that the arbitrator deems just and equitable and within the scope of the agreement of the parties, including, but not limited to, specific performance of a contract. The arbitrator shall, in the award, assess arbitration fees, expenses, and compensation as provided in Sections 48, 49, and 50 in favor of any party and, in the event that any administrative fees or expenses are due the AAA, in favor of the AAA.

44. Award upon Settlement

If the parties settle their dispute during the course of the arbitration, the arbitrator may set forth the terms of the agreed settlement in an award. Such an award is referred to as a consent award.

45. Delivery of Award to Parties

Parties shall accept as legal delivery of the award the placing of the award or a true copy thereof in the mail addressed to a party or its representative at the last known address, personal service of the award, or the filing of the award in any other manner that is permitted by law.

46. Release of Documents for Judicial Proceedings

The AAA shall, upon the written request of a party, furnish to the party, at its expense, certified copies of any papers in the AAA's possession that may be required in judicial proceedings relating to the arbitration.

47. Applications to Court and Exclusion of Liability

(a) No judicial proceeding by a party relating to the subject matter of the arbitration shall be deemed a waiver of the party's right to arbitrate.

(b) Neither the AAA nor any arbitrator in a proceeding under these rules is a necessary party in judicial proceedings relating to the arbitration.

(c) Parties to these rules shall be deemed to have consented that judgment upon the arbitration award may be entered in any federal or state court having jurisdiction thereof.

(d) Neither the AAA nor any arbitrator shall be liable to any party for any act or omission in connection with any arbitration conducted under these rules.

48. Administrative Fees

As a not-for-profit organization, the AAA shall prescribe filing and other administrative fees to compensate it for the cost of providing administrative services. The fees in effect when the demand for arbitration or submission agreement is received shall be applicable.

The filing fee shall be advanced by the initiating party or parties, subject to final apportionment by the arbitrator in the award.

The AAA may, in the event of extreme hardship on the part of any party, defer or reduce the administrative fees.

49. Expenses

The expenses of witnesses for either side shall be paid by the party producing such witnesses. All other expenses of the arbitration, including required travel and other expenses of the arbitrator, AAA representatives, and any witness and the cost of any proof produced at the direct request of the arbitrator, shall be borne equally by the parties, unless they agree otherwise or unless the arbitrator in the award assesses such expenses or any part thereof against any specified party or parties.

50. Neutral Arbitrator's Compensation

Unless the parties agree otherwise, members of the National Panel of Construction Industry Arbitrators appointed as neutrals will serve without compensation for the first day of service.

Thereafter, compensation shall be based on the amount of service involved and the number of hearings. An appropriate daily rate and other arrangements will be discussed by the administrator with the parties and the arbitrator. If the parties fail to agree to the terms of compensation, an appropriate rate shall be established by the AAA and communicated in writing to the parties.

Any arrangement for the compensation of a neutral arbitrator shall be made through the AAA and not directly between the parties and the arbitrator.

51. Deposits

The AAA may require the parties to deposit in advance of any hearings such sums of money as it deems necessary to cover the expense of the arbitration, including the arbitrator's fee, if any, and shall render an accounting to the parties and return any unexpended balance at the conclusion of the case.

52. Interpretation and Application of Rules

The arbitrator shall interpret and apply these rules insofar as they relate to the arbitrator's powers and duties. When there is more than one arbitrator and a difference arises among them concerning the meaning or application of these rules, it shall be decided by a majority vote. If that is not possible, either an arbitrator or a party may refer the question to the AAA for final decision. All other rules shall be interpreted and applied by the AAA.

Expedited Procedures

53. Notice by Telephone

The parties shall accept all notices from the AAA by telephone. Such notices by the AAA shall subsequently be confirmed in writing to the parties. Should there be a failure to confirm in writing any notice hereunder, the proceeding shall nonetheless be valid if notice has, in fact, been given by telephone.

54. Appointment and Qualifications of Arbitrator

The AAA shall submit simultaneously to each party an identical list of five proposed arbitrators drawn from the National Panel of Construction Industry Arbitrators, from which one arbitrator shall be appointed.

Each party may strike two names from the list on a peremptory basis. The list is returnable to the AAA within seven days from the date of the AAA's mailing to the parties.

If for any reason the appointment of an arbitrator cannot be made from the list, the AAA may make the appointment from among other members of the panel without the submission of additional lists.

The parties will be given notice by telephone by the AAA of the appointment of the arbitrator, who shall be subject to disqualification for the reasons specified in Section 19. Within seven days, the parties shall notify the AAA, by telephone, of any objection to the arbitrator appointed. Any objection by a party to the arbitrator shall be confirmed in writing to the AAA with a copy to the other party or parties.

55. Date, Time, and Place of Hearing

The arbitrator shall set the date, time, and place of the hearing. The AAA will notify the parties by telephone, at least seven days in advance of the hearing date. A formal notice of hearing will also be sent by the AAA to the parties.

56. The Hearing

Generally, the hearing shall be completed within one day, unless the dispute is resolved by submission of documents under Section 37. The arbitrator, for good cause shown, may schedule an additional hearing to be held within seven days.

57. Time of Award

Unless otherwise agreed by the parties, the award shall be rendered not later than fourteen days from the date of the closing of the hearing.

Administrative Fees

The AAA's administrative charges are based on filing and service fees. Arbitrator compensation, if any, is not included. Unless the parties agree otherwise, arbitrator compensation and administrative fees are subject to allocation by the arbitrator in the award.

Filing Fees

A nonrefundable filing fee is payable in full by a filing party when a claim, counterclaim, or additional claim is filed, as provided below.

Amount of Claim	Filing Fee
Up to $25,000	$300
Above $25,000 to $50,000	$500
Above $50,000 to $250,000	$1,000
Above $250,000 to $500,000	$2,000
Above $500,000 to $5,000,000	$3,000
Above $5,000,000	$4,000

When no amount can be stated at the time of filing, the filing fee is $1,000, subject to adjustment when the claim or counterclaim is disclosed.

When a claim or counterclaim is not for a monetary amount, an appropriate filing fee will be determined by the AAA.

Hearing Fees

For each day of hearing held before a single arbitrator, an administrative fee of $100 is payable by each party.

For each day of hearing held before a multiarbitrator panel, an administrative fee of $150 is payable by each party.

Postponement Fees

A fee of $100 is payable by a party causing a postponement of any hearing scheduled before a single arbitrator.

A fee of $150 is payable by a party causing a postponement of any hearing scheduled before a multiarbitrator panel.

Processing Fees

No processing fee is payable until 180 days after a case is initiated.

On single-arbitrator cases, a processing fee of $150 per party is payable 180 days after the case is initiated, and every 90 days thereafter, until the case is withdrawn or settled or the hearings are closed by the arbitrator.

On multiarbitrator cases, a processing fee of $200 per party is payable 180 days after the case is initiated, and every 90 days thereafter, until the case is withdrawn or settled or the hearings are closed by the arbitrators.

Suspension for Nonpayment

If arbitrator compensation or administrative charges have not been paid in full, the AAA may so inform the parties in order that one of them may make the required payment. If such payments are not made, the arbitrator may order the suspension or termination of the proceedings. If no arbitrator has yet been appointed, the AAA may suspend the proceedings in such a situation.

Hearing Room Rental

Rooms for hearings are available on a rental basis. Check with our local office for availability and rates.

Index

**ALTERNATIVE DISPUTE
RESOLUTION**
See Arbitration; Arbitration-
Mediation; Claims
Master; Expert Determination Process;
Mediation;
Mini-trial; Referees; Special Master;
Summary Jury Trial
**AMERICAN ARBITRATION
ASSOCIATION**
Arbitrator selection, 671
Award, 679
Complex Guidelines, 686
Expedited arbitration, 687
Expenses, 678
Immunity, 683
Mediation Rules, 687
Research, 685
**AMERICAN INSTITUTE OF
ARCHITECTS**
Document making, 5, 425, 428–429
**AMERICANS WITH DISABILITIES
ACT**
Access to, 55
Described, 268–269
Employment, 55
Liability, 256
ANTI-TRUST LAWS
Bid depositories, registries, 626–627
Competitive bidding restraints as
affected by, 181
Fee schedules, as affected
by, 181
Specifications, 412
"Turn bidding," 380

APPEALS
See Litigation
ARBITRATION
See also Arbitration-Mediation;
Arbitrator
Aesthetic effect, 667
Adhesion contracts, 663–665
Agreements to arbitrate, enforcement
of, 662–663, 665–667
AIA B 141, 664
Architect's errors, 667
Attorneys' fees, 678
Award, 679
Award finality, 679–681, 685
Clauses, types, 683
Common law, 662, 665
Completion, claims after, 667
Complex AAA, 686–687
Compromise study, AAA, 685
Conditions precedent, 669
Consolidation, 672–674
Costs of, 678, 686
Daily fine, 679
Delay, 667
Default, 675
Discovery rights, 670–671, 683–684
Documents, 676
Dollar limits, 686
EJCDC, 670, 686
Evidence, rules of, 676
Ex parte communications, 676–677
Expedited, AAA, 687
Federal Arbitration Act, 663, 680, 686
Federal procurement, 689–690
Filing fees, 663, 683
Finality of award, 679–681, 685
Fraudulently obtained agreement for,
665–666
Hearing, 674, 675
Activities preceding, 670–671
Waiver of, 675
Incorporated by reference, 665
Implied terms, 667
Insurers, 681–683
Interim relief, 678
International, 691–692
Interpretation, 666
Joinder of parties, 672–674
Judicial attitude toward, 663–665
Judicial non-binding, 15
Judicial review of, 679–681, 685
Jurisdiction, 666–678
Licensing laws, avoidance of, 165, 677
Litigation compared, 683–686
Local procurement, 343, 690–691
Mutuality, 666
Non-binding, 15
Non-parties, 667–668
Oaths, 675
Place of, 11, 672, 684

Postponement, 675
Precedent, need not follow, 677
Professional liability
Insurance, effect on, 681–683
Recesses, 675
Remedies, 678–679
Re-opening, 677
Site, visit to, 676
Specific performance remedy, 678–679
Speed, 684
State Procurement, 343, 690–691
Statute of Limitations, 669, 677
Subcontract, incorporation by
Reference to prime contract, 633–
634, 665
Subpoena power of arbitrator, 675–
676
Substantive rules, 677–678, 685
Surety bond, effect on, 681–683
Termination of construction
Contract as affecting, 754–755
Threshold issues, who decides, 669
Time of, 675
Time to claim, 668
Tort claims, 667
Transcript, 676
Unconscionability, 663–665, 666
Uniform Arbitration Act, 679–680
Validity, 665–666
"Vouching in," 674
Waiver of, 669–670
Wisconsin, 672
Witnesses, 676
ARBITRATION-MEDIATION
Described, 687
ARBITRATOR
See also Arbitration
Disclosure, duty of, 671–672
Immunity of, 683
Selection, 671
Skills, 684–685
Subpoena power, 675–676
ARCHITECTS
See Design Professionals
**ARCHITECTURAL AND
ENGINEERING**
**PERFORMANCE INFORMATION
CENTER**
Project data, 317
**ARCHITECTURAL WORKS
COPYRIGHT**
ACT OF 1990
Berne Convention, 333
Completed project, 333
ARIZONA STATUTES
§ 32–1101, pp. 140, 169
§ 32–1121, p. 169
ARIZONA SESSION LAWS
1981, ch. 223, § 1, p. 140
1986, ch. 318, § 21, p. 140